CRC
HANDBOOK
OF
MAMMALIAN METABOLISM
OF
PLANT COMPOUNDS

Author

Ronald R. Scheline
Professor of Biochemical Pharmacology
Department of Pharmacology and Toxicology
School of Medicine
University of Bergen
Bergen, Norway

CRC Press
Taylor & Francis Group
Boca Raton London New York

CRC Press is an imprint of the
Taylor & Francis Group, an **informa** business

PREFACE

Because of the widespread and increasing use of foreign compounds (xenobiotics) in many fields, the understanding of their metabolic fate has assumed growing importance. This subject, known as xenobiochemistry, has accordingly shown rapid growth during the past four or five decades. These developments have also resulted in a significant expansion of the relevant scientific literature including review journals and books. However, much of this information is topic-oriented rather than compound-oriented and the investigator requiring a convenient source of information on the metabolic properties of a particular xenobiotic compound often finds that no summary of these data is available. I have personally felt this need many times in the course of my studies of xenobiochemistry and this deficiency was a major reason for my decision to summarize current knowledge on the metabolism of a major group of xenobiotics. Also instrumental in this decision was the demonstrated value of Professor R. T. Williams' classic book *Detoxication Mechanisms* which, although published more than 30 years ago, remains a useful source of earlier information on the metabolism of specific xenobiotics.

The term *plant compounds* as used in the present volume can be largely equated with the phrase *secondary plant compounds*, i.e., compounds which are apparently not essential to the functioning of the living plant cell. Accordingly, the text will not deal with the metabolism of compounds entering into the glycolytic and tricarboxylic acid cycles, and basic nucleotides, the amino acids of proteins, and the common lipids. Also, attention has been given to compounds from higher plants rather than from lower classes including algae and fungi. As a general rule I have deliberately chosen to exclude from the text ancillary material including botanical sources, pharmacological and toxicological properties, and other uses of the various compounds. The latter subjects are more readily available in the literature and their inclusion would unduly increase the length of the book.

I wish to give special thanks to the staff of the University of Bergen library for their assistance in providing me with many of the required literature references. I would also like to acknowledge the help of several technical aids which were of immense help to me in the preparation of the manuscript. These were the Macintosh programs EndNote by Niles and Associates and ChemIntosh by SoftShell Company. The former program enabled me to manage the several thousand references in my collection and to prepare the required lists of references. The latter program was used to make the drawings of the chemical structures including the illustrations of the metabolic pathways.

Ronald R. Scheline
Bergen, Norway

AUTHOR

Ronald R. Scheline, Ph.D., is Professor of Biochemical Pharmacology in the School of Medicine at the University of Bergen in Bergen, Norway.

Dr. Scheline received his B.S. degree in pharmacy from the University of California at San Francisco in 1955. He obtained his Ph.D. degree in pharmaceutical chemistry from the University of California at San Francisco in 1958. After doing postdoctoral work at St. Mary's Medical School, University of London and at the University of Bergen, he was appointed an Assistant Professor of Pharmacology at the latter institution in 1965. He became Associate Professor in 1966 and Professor in 1974.

Dr. Scheline is a member of the Norwegian Society of Pharmacology and Toxicology, the American Chemical Society, and a charter member of the International Society for the Study of Xenobiotics (ISSX).

Dr. Scheline has published more than 70 research papers in the field of metabolism of foreign compounds. His major research interests include the influence of the normal gastrointestinal microflora on the metabolism of xenobiotics and the metabolism of plant compounds in mammals.

TABLE OF CONTENTS

METABOLISM OF HYDROCARBONS

ALIPHATIC HYDROCARBONS

Many aliphatic hydrocarbons are found in plants, the most common being long-chain n-alkanes with the general formula $H_3C-(CH_2)_n-CH_3$. The lowest natural member of straight-chain alkanes is n-pentane, although its higher homologue n-heptane is more common. The series of plant n-alkanes extends to C_{35} and it is noteworthy that the odd-numbered members are much more abundant than are the even-numbered. This fact is apparently related to their pathway of synthesis which involves decarboxylation of fatty acids with even numbers of carbon atoms. Branched hydrocarbons are also found, usually in small quantities. Branching commonly occurs near the end of the chain (methyl group in the ω–1 or ω–2 positions). Many branched-chain hydrocarbons are acyclic terpenoids. These are based on the isoprene molecule and are therefore $(C_5)_n$-compounds. Although there is no reason to believe that the pathways and mechanisms involved in the mammalian metabolism of these terpenoid compounds differ significantly from those involved in the metabolism of other compounds containing similar chemical groups, terpenoids are grouped separately in this book. A further type of aliphatic hydrocarbon from plants includes compounds containing unsaturated groups. Olefinic groups are most common, however a few hydrocarbons containing acetylenic groups are also known.

Considerable interest has been devoted to the mammalian metabolism of several n-alkanes, with the result that the reaction products and the mechanisms by which they are formed are well understood in some cases. The report by McCarthy[1] appears to have been influential in directing attention to the possibility of n-alkane metabolism. This study showed that both **n-hexadecane** $(H_3C-(CH_2)_{14}-CH_3)$ and **n-octadecane** $(H_3C-(CH_2)_{16}-CH_3)$ were converted by ω-oxidation in rats and goats to fatty acids of the same chain lengths. These products could then undergo further reactions including alterations in chain length or incorporation into lipid fractions, especially liver phospholipids. These metabolic pathways were confirmed in rats given n-octadecane.[2] Most of the radioactivity was found in the liver, especially as fatty acids of lecithin, 2 h after administration of the 1-^{14}C-labeled compound (200 mg/kg, p.o. or i.v.). Respiratory $^{14}CO_2$ was detected during this period. The localization of the enzymes metabolizing n-hexadecane and the mechanisms involved have also been investigated. Mitchell and Hübscher[3] reported that it was metabolized via cetyl alcohol (n-hexadecanol) to palmitic acid (n-hexadecanoic acid) by the mucosa of guinea pig small intestine. Activity was located mainly in the microsomal fraction. These same reaction products were shown by Kusunose et al.[4] to be formed by the microsomal fraction from mouse liver. Similar fractions from mouse lung or kidney showed little or no activity, respectively. This investigation indicated that the ω-hydroxylation of n-hexadecane is catalyzed by a monooxygenase system requiring NADPH and O_2. The oxidation of n-hexadecane was not carried out by the intestinal microflora of rats[5] or guinea pigs.[3]

The metabolism of **n-heptadecane** $(H_3C-(CH_2)_{15}-CH_3)$ in rats was studied by Tulliez and Bories[6] using 8-^{14}C-labeled material administered in the food. Following a small dose (1 mg), 65% of the radioactivity was lost as respiratory $^{14}CO_2$. Little radioactivity was excreted in the urine and feces (~1% each) in 6 d. Most of the material remaining in the body was oxidized to heptadecanoic acid which entered into the normal pathways of fatty acid metabolism including dehydrogenation to heptadecenoic acid and incorporation into lipids and phospholipids. Tulliez and Bories,[7] in similar feeding experiments with long-chain aliphatic hydrocarbons, reported that no urinary excretion and little (5 to 10%) fecal excretion occurred with compounds below C_{24}. Retention of n-alkanes in the body was found with the compounds $<C_{20}$, however retention decreased rapidly to zero with higher homologues. Branched-chain compounds showed lower retention.

Several studies dealing with the *in vitro* or *in vivo* metabolism of lower alkanes, including **n-pentane,** n-hexane, n-heptane, and n-octane, have been carried out. Hydroxylated derivatives of *n*-pentane were formed from rat and rabbit liver microsomes.[8] The major metabolite was 2-pentanol (~80 to 90%) and the remainder was 3-pentanol. The primary alcohol was not detected in these experiments. Daugherty et al.[9] administered [1,5-^{14}C]pentane (~0.75 mg) by inhalation to rats and recovered ~50% of the radioactivity as expired $^{14}CO_2$. An additional 8% was found in the urine, however the nature of this material was not determined.

An investigation of the hydroxylation of *n*-**heptane** (Figure 1 including Structure (1)) by rat liver microsomes indicated that this reaction was carried out by a common microsomal hydroxylating system involving NADPH-cytochrome c reductase and cytochrome P-450.[10] This microsomal system was further studied by Frommer et al.[11] who found that all four isomeric alcohols were formed from *n*-heptane. In these *in vitro* experiments, ω–1 hydroxylation was the main reaction and the relative amounts of 2-, 1-, 3-, and 4-heptanol formed were ~27:5:4:1. An extensive study of the *in vivo* metabolism of *n*-heptane was reported by Bahima et al.[12] They exposed rats to an atmosphere (2,000 ppm) of *n*-heptane for 6-h periods and investigated the metabolites subsequently excreted in the urine. They identified 14 metabolites and found that these were largely (~97%) eliminated within the first 18 h after exposure. The metabolites were excreted as glucuronides and, mainly, as sulfates with no free compounds being detected. This study confirmed the *in vitro* results noted above which showed that all four isomeric heptanols were formed and that 2-heptanol was the major hydroxylated product. Additionally, ω–2 hydroxylation giving 3-heptanol was a prominent reaction and both of these alcohols underwent further oxidation to give several ketones, diols, hydroxyketones, and diones. The probable pathways of metabolism of *n*-heptane are shown in Figure 1. Bahima et al.[12] believed that the large detected amounts of a terminal metabolite shown in Figure 1, the cyclic metabolite γ-valerolactone, arose from 4-hydroxypentanoic acid which, in turn, was formed via β-oxidation of 6-hydroxy-3-ketoheptanoic acid. However the putative acidic precursors (shown in brackets) were not actually detected in the samples.

Metabolic data are also available on the nonplant hydrocarbon *n*-hexane. This literature is especially abundant because of the role of metabolism in the neurotoxicity of this hydrocarbon. Much of this work falls outside the scope of this book and only the highlights of the metabolism of *n*-hexane will be covered.

The metabolism of *n*-hexane has been studied in several mammalian species. While the results obtained show both qualitative and quantitative differences, it is clear that ω–1 oxidation, either at one or both ends of the molecule, is the major metabolic event. This was first shown by DiVincenzo et al.[13] who reported that, following a dose of 250 mg/kg (i.p.), the serum of guinea pigs contained both 5-hydroxy-2-hexanone and 2,5-hexanedione. Couri et al.[14] reported that 2-hexanol was also a metabolite of *n*-hexane in guinea pigs. Pathways involving ω–1 oxidation are also prominent in humans, with 2-hexanol and 2,5-hexanedione being major urinary metabolites.[15] However, both ω-oxidation and ω–2 oxidation of hexane have been reported, with 1-hexanol[16] and 3-hexanol[17] shown to be urinary metabolites of *n*-hexane in rats. Perbellini et al.[18] made a comparative study of *n*-hexane metabolism in rats, rabbits, and monkeys in which they also summarized its pathways of metabolism in animals. In general, the picture is similar to that seen in Figure 1 with its higher homologue *n*-heptane. Thus, all isomeric alcohols are formed and, from 2-hexanol, oxidation and further ω–1 oxidation take place to give 2-hexanone, 2,5-hexanediol, 5-hydroxy-2-hexanone, and 2,5-hexanedione. Additionally, Fedtke and Bolt[19,20] showed that 4,5-dihydroxy-2-hexanone is a metabolite of *n*-hexane in rats and man. Couri and Milks[21] reviewed the metabolism and toxicity of *n*-hexane. The metabolism of the *n*-hexane metabolite 2-hexanone is discussed in the chapter entitled Metabolism of Aldehydes, Ketones, and Quinones.

The metabolism of *n*-**octane** is of interest because of some differences when compared with that found with the lower homologues. Olson et al.[22] administered *n*-octane to rats (1400 mg/kg, intragastrically) and investigated the urinary metabolites excreted during the following 48-h period. The urine samples were treated with a glucuronidase + sulfatase preparation and four

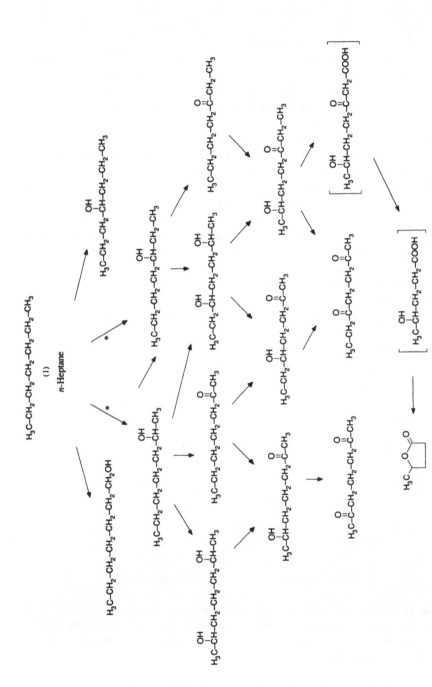

FIGURE 1.   Metabolic pathways of *n*-heptane in rats. Compounds in parentheses were not detected. (*) Major routes of hydroxylation.

metabolites were identified. In addition to the expected 2-octanol and 3-octanol, 5-ketohexanoic acid and 6-ketoheptanoic acid were identified. All four metabolites were excreted by both male and female rats, but in different relative amounts. In males, the relative amounts were 6.9:12.1:15.5:1.0, respectively, whereas female rats gave values of 2.0:1.7:0.7:1.0. In contrast to that noted above with the lower homologues, no ketones, diols, hydroxyketones or diones were detected. The keto acids identified have six and seven carbon atoms and it seems likely that the hexanoic acid derivative, at least, may arise in a manner similar to that suggested for the formation of 4-hydroxypentanoic acid from *n*-heptane (see Figure 1). Dahl[23] studied the elimination of radioactivity by rats following the inhalation of *n*-[1-^{14}C]octane. At a level of ~2 ppm, loss of the inhaled radioactivity was about equal via the urine and as $^{14}CO_2$. Although most of the radioactivity present in the tissues at the end of the inhalation period was lost fairly rapidly, ~5% of the dose remained in the body after 70 h.

Studies of the rat liver monooxygenase system forming isomeric alcohols from *n*-heptane[11] and *n*-hexane[24] indicated that at least three activities were involved. Similar results were reported by Nilsen et al.[25] who studied the preferential formation of hexanol isomers from *n*-hexane using different purified forms of cytochrome P-450 from rabbit liver. Lu et al.[26] reported on the properties of a solubilized form of the liver microsomal system which hydroxylated a series of *n*-alkanes from C_6 to C_{16}. Further information on the microsomal hydroxylation of alkanes of low chain length was presented by Ichihara et al.[27] Preparations from livers of mice, rats, rabbits, and cattle oxidized hexane, octane, and decane and the data indicated that the last compound was metabolized via decanol to decanoic acid and decamethyleneglycol.

The above results indicate that numerous normal alkanes ranging from the lower homologues to the C_{16}- and C_{18}-compounds undergo oxidation, initially to alcohols. It seems likely that this general reaction should also be involved in the metabolism of the higher homologues of plant alkanes.

Relatively little is known of the metabolism of plant olefins and acetylenes. One of the simplest of these is **isoprene** (2-methyl-1,3-butadiene) ($H_2C=C(CH_3)-CH=CH_2$). Its metabolism was studied by Del Monte et al.[28] using mouse liver microsomes and by Longo et al.[29] using similar preparations from rats, hamsters, and rabbits. In all cases the microsomes formed both the 1,2-epoxide and the 3,4-epoxide derivatives. These two metabolites were then converted to the corresponding diols. The major pathway was via the 1,2-epoxide in all species. It was noted that the diepoxide was formed in small amounts from the 3,4-epoxide, however this reaction did not occur via the 1,2-epoxide.

Neubauer[30] reported that rabbits given ***n*-1-octene** (caprylene, 1-octylene) ($H_3C-(CH_2)_5-CH=CH_2$) excreted material conjugated with glucuronic acid. This finding indicates that oxidation of the alkene took place. Maynert et al.[31] studied the metabolism of two octene isomers, *n*-1-octene and *n*-4-octene, using rat liver microsomes and NADPH in the presence of O_2. These compounds were oxidized to epoxides which were then hydrolyzed by epoxide hydrolase to *n*-octane-1,2-diol and *n*-octane-4,5-diol, respectively. A similar metabolic sequence was found by Watabe and Yamada[32] with the synthetic homologue, *n*-1-hexadecene ($H_3C-(CH_2)_{13}-CH=CH_2$). White et al.[33] reported that, in addition to this more common route, *n*-1-octene may also be metabolized by rat liver microsomal systems to oct-1-en-3-ol and further to oct-1-en-3-one. However, the epoxide-diol pathway showed a rate more than 40 times higher than the 3-oxo route. Oct-1-en-3-one showed high chemical reactivity and readily formed a thioether in the presence of *N*-acetylcysteine.

MONOTERPENOID HYDROCARBONS

ACYCLIC TERPENE HYDROCARBONS

This is a fairly limited group of C_{10} olefinic compounds, the best known being β-**myrcene** (2). Ishida et al.[34,35] administered it (670 mg/kg, intragastrically) to rabbits and identified three neutral and two acidic metabolites in the urine following conjugate hydrolysis using β-glucuronidase + sulfatase. Two neutral metabolites were the glycols (3) (main neutral metabolite) and (4) which no doubt arose via epoxidation and then hydration at the 3,10- and 1,2-bonds, respectively. These glycols, which did not show optical rotation, underwent further oxidation and the hydroxyacids (5) and (6) were also identified.

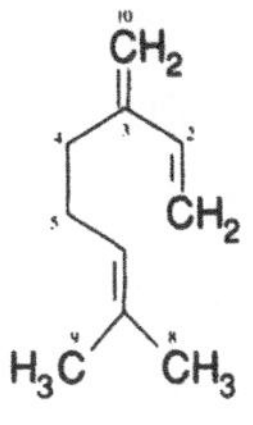

(2)

β-Myrcene

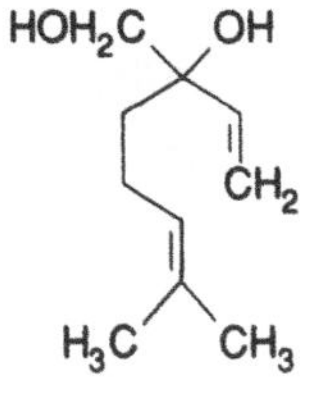

(3)

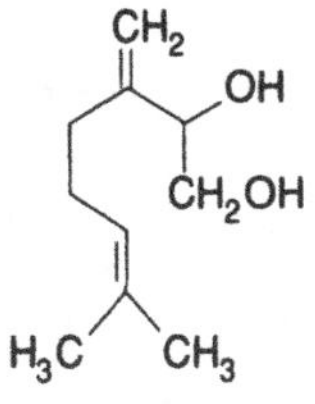

(4)

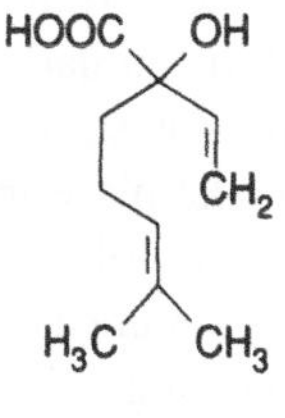

(5)

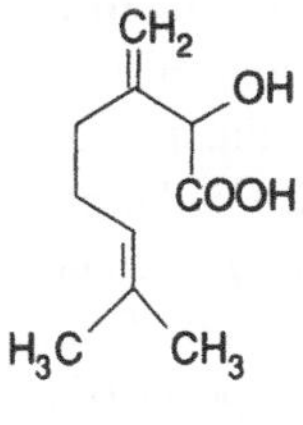

(6)

The above findings were confirmed by Madyastha and Srivatsan[36] who gave β-myrcene intragastrically (800 mg/kg/d) to rats for 20 d. Compounds (3) and (5) were the most prominent urinary metabolites. The glycols (3) (major) and (4) (trace) were formed when β-myrcene was incubated with induced hepatic microsomes. They tentatively identified the third neutral urinary metabolite as the cyclic compound (7). Ishida et al.[35] also isolated a cyclic neutral compound, however their spectroscopic evidence suggested that it had structure (8) (uroterpenol). It is noteworthy that allylic hydroxylation, which might be expected to give 4-, 5-, 8-, or 9-hydroxylated derivatives of myrcene, was not detected in these experiments.

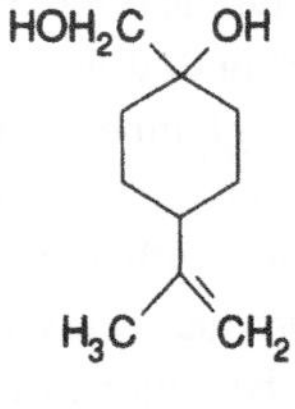

(7)

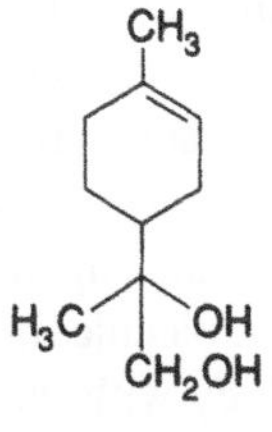

(8)

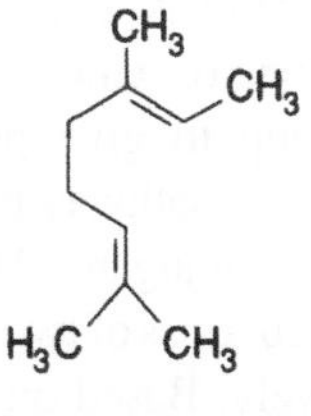

(9)

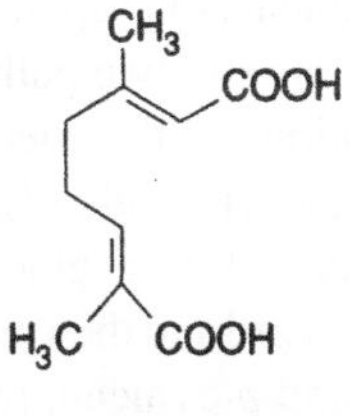

(10)

A closely related and partially reduced derivative of β-myrcene, 2,6-dimethyl-2,6-octadiene (dihydromyrcene) (9), was studied by Kuhn et al.[37] When rabbits were given 25 g of this compound (2×5 g daily, p.o.), 1.5 g of the optically inactive "Hildebrandt acid" (10) was isolated from the urine as the only detected metabolite. Whether or not the epoxide-diol pathway is also involved with the metabolism of dihydromyrcene is not known, however it is evident that allylic hydroxylation of the terminal methyl groups is implicated in the formation of metabolite (10). It also seems likely that much of the 90 to 95% of the dose unaccounted for may consist of intermediate products, especially alcohols which would be excreted mainly as glucuronide conjugates rather than in the free form, as is the case with the dicarboxylic acid metabolite.

MONOCYCLIC TERPENE HYDROCARBONS

Only three compounds from this group have been studied. These are *p*-menthene (11), α-phellandrene (*p*-mentha-1,5-diene) (12), and limonene (*p*-mentha-1,8-diene) (13). Several of the reports on these compounds are old, resulting in the fragmentary knowledge of these conpounds. However, the metabolism of limonene has been extensively investigated in a number of mammalian species.

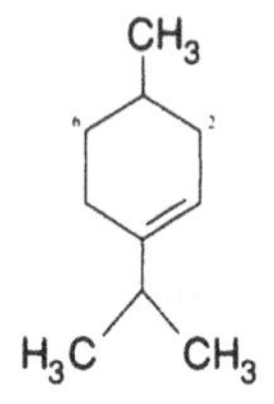

(11)	(12)	(13)
p-Menthene	α-Phellandrene	Limonene

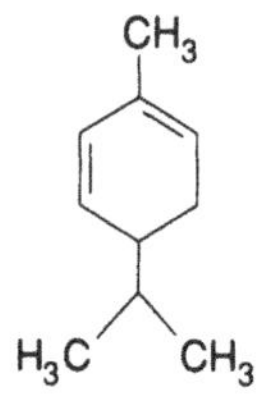

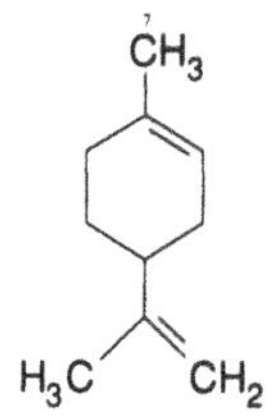

Neubauer[30] administered *p*-menthene (11) (0.8 g/kg, p.o.) to rabbits and reported the urinary excretion of material conjugated with glucuronic acid. In similar experiments, Hämäläinen[38] found no evidence for reduction of the double bond but showed that hydroxylation and glucuronidation occurred. Oxidation was reported to take place in the position *ortho* to the methyl group, giving either *p*-menthen-2-ol or *p*-menthen-6-ol.

α-**Phellandrene** (12) differs from *p*-menthene in the number and position of the double bonds, a difference which appears to have considerable bearing on its metabolic fate. Fromm and Hildebrandt[39] reported that rabbits given α-phellandrene excreted in the urine a glucuronide material which, on acid hydrolysis, gave the aromatic compound *p*-cymene (14) and a phenolic compound. Later studies employed sheep and Harvey[40] found that acid-hydrolyzed urines contained *p*-cymene, carvotanacetone (15), phellandric acid (16), and possibly the corresponding aldehyde, phellandral. Harvey felt that the phenolic metabolite reported by Fromm and Hildebrandt[39] may have been due to an impurity in the sample of α-phellandrene used and, in fact, the two samples employed by Harvey were only 80 and 95% pure. Wright[41] isolated the glycine conjugate of phellandric acid, termed phellanduric acid (17), from the urine of sheep dosed with α-phellandrene. The urine also contained material conjugated with glucuronic acid which, in agreement with the earlier study by Harvey, was converted to *p*-cymene and carvotanacetone upon acid hydrolysis. These results indicate that α-phellandrene is metabolized along at least two pathways, one of which involves both reduction of a ring double bond and oxidation of the methyl group to give phellandric acid (16) and its glycine conjugate, phellanduric acid (17). The other pathway involves ring hydroxylation to give a polyol which is excreted as a glucuronide conjugate. This conjugate, upon treatment with hot acid, is hydrolyzed and then dehydrated. Loss of two or three molecules of water produces carvotanacetone and *p*-cymene, respectively. Based on an analogy with the known behavior of other *p*-menthane triols, Wright postulated that the polyol is *p*-menthane-2,3,6-triol (18).

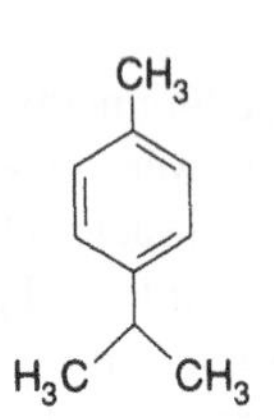

| (14) | (15) | (16) |

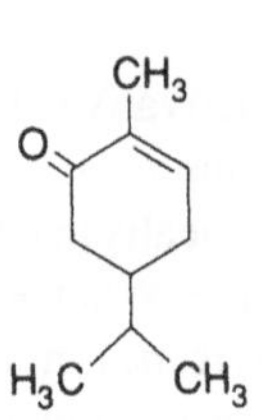

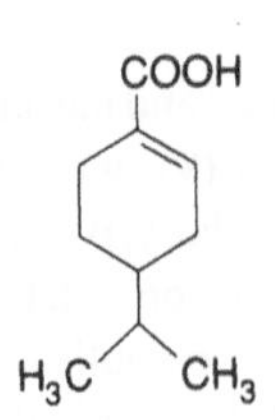

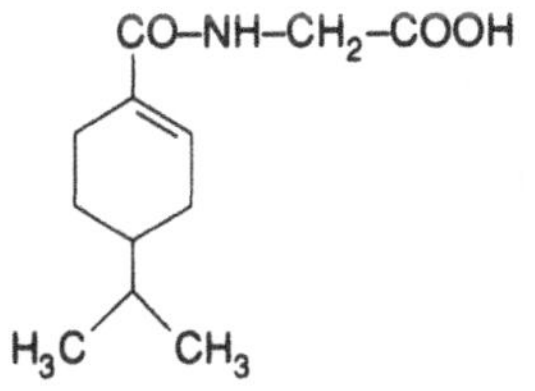

CH₃ structures with labels (17) and (18)

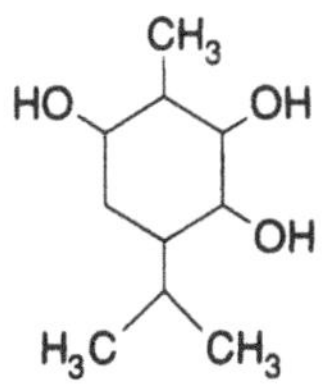

The metabolism of **limonene** (13) was the subject of an early investigation by Hildebrandt[42] who reported that rabbits oxidized the methyl group at C-7 to a carboxy group and hydroxylated the molecule at some other position to form a metabolite which was excreted as a glucuronic acid conjugate. No further reports on limonene metabolism appeared for a period of more than 60 years. Then Wade et al.[43] discovered that human urine contained, in conjugated form, a monoterpenoid compound named uroterpenol. It was shown to be p-menth-1-ene-8,9-diol (see Figure 2 and Structure (8)) and the evidence obtained indicated that it originated from dietary limonene. While some uroterpenol was excreted in the free form, most appeared to be conjugated with glucuronic acid, an assumption which was subsequently confirmed.[44] Smith et al.[45] reported large increases of uroterpenol excretion over normal values following ingestion by man of limonene or limonene-containing products.

Subsequently, several extensive investigations of the metabolism of d-limonene were carried out. Some of these employed the 9-^{14}C-labeled compound and studied its absorption, distribution, and excretion in rats,[46] its metabolism in rabbits,[47] and its metabolism in six species including man.[48] The metabolic fate of d-limonene in rats was also investigated by Regan and Bjeldanes.[49] These exemplary studies, employing a range of modern analytical techniques, clearly showed that multiple pathways of metabolism are involved. Oxidation at C-7 to form the carboxylic acid reported by Hildebrandt was regularly seen and the formation of metabolites with hydroxy groups at one or more of five different aliphatic and alicyclic sites was shown to occur. It seems reasonable to believe that similar multiple pathways of metabolism for the other terpene hydrocarbons also exist and that the earlier studies, employing only a few animal species and using less advanced methodology, have but scratched the surface of this subject.

The studies of Igimi et al.[46] and Kodama et al.[47,48] showed that urinary excretion of metabolites was far more important than fecal excretion and that no unchanged compound was detected in the urine in the species studied (rat, guinea pig, hamster, rabbit, dog, and man). However, Römmelt et al.[50] reported that small amounts of limonene administered intravenously were excreted in the expired air of pigs and humans. Following a dose of ~0.6 mg/kg, this value was 1.4% after 2 h with the bulk of the excretion occurring within 20 to 30 min. Igimi et al.[46] and Kodama et al.[47,48] showed that most of the radioactivity appeared in the urine within 24 h and 75 to 95% was recovered in 2 to 3 d. Less than 10% of the dose was excreted in the feces. These experiments employed rather high doses (800 mg/kg, p.o.) in all species except dogs (400 mg/ kg) and man (27 mg/kg). Igimi et al.[46] found that loss of radioactivity as [14]CO₂ was insignificant and less than 2% of the dose was recovered in this form in rats. Biliary excretion, in the rat at least, accounted for a fairly large proportion of the radioactivity (25% of the dose) which consisted mainly of the glucuronide of uroterpenol (8).

The general patterns of d-limonene metabolism found with the ^{14}C-labeled compound are similar in all of the six species studied, however the quantitative relationships vary considerably among them. These quantitative data on metabolite excretion[48] are summarized in Table 1. These values show that ring hydroxylation reactions at C-2 to give metabolite (29) or at C-6 to give metabolites (23) and (24) are usually minor reactions. However, Regan and Bjeldanes[49] reported that ~16% of the unconjugated metabolites of d-limonene in rats corresponded to metabolite (23). Furthermore, this metabolite consisted of a mixture of the 6-α-hydroxy and 6-

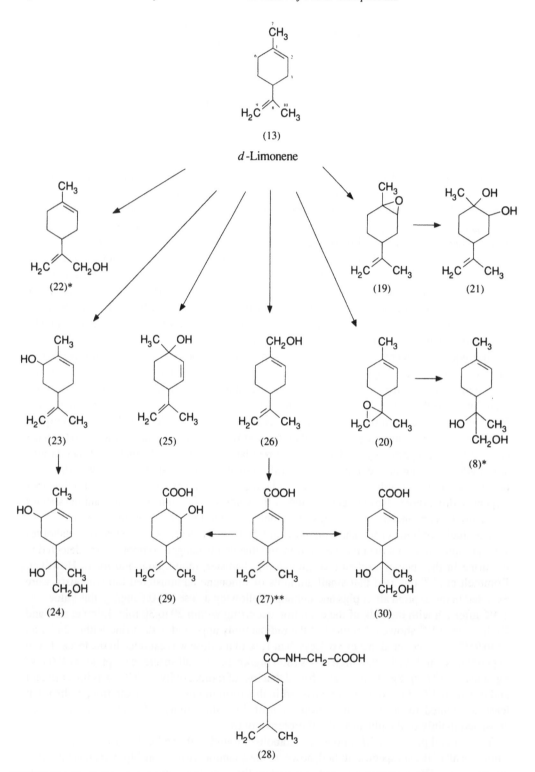

FIGURE 2. Metabolic pathways of *d*-limonene. (*) Also present in form of 10-*O*-glucuronide. (**) Also present in form of 7-*O*-glucuronide.

TABLE 1

Species Differences in [^{14}C]-d-Limonene Metabolism[48]

Metabolite	Structure (see Figure 2)	Rat (3[a])	Guinea pig (3[a])	Hamster (4[a])	Rabbit (3[a])	Dog (2[a])	Man (2[a])
p-Mentha-1,8-dien-10-ol	(22)	0[b]	0.3	0.3	3.6	2.1	0.1
10-O-glucuronide of metabolite (22)		1.7	3.1	2.3	3.4	0.3	1.0
p-Mentha-1,8-dien-6-ol	(23)	0.9	0.1	0.1	0.4	0.2	0
p-Menth-1-ene-6,8,9-triol	(24)	tr[c]	tr	tr	0.1	0.2	3.4
p-Menth-1-ene-8,9-diol (uroterpenol)	(8)	3.5	1.3	11.3	9.6	28.3	1.3
10-O-glucuronide of metabolite (8)		3.2	27.7	7.9	6.8	2.4	27.4
Perillic acid	(27)	5.1	5.3	7.6	5.1	10.4	2.4
Perillylglycine	(28)	8.1	20.5	1.6	2.2	1.4	0.4
7-O-glucuronide of metabolite (27)		1.8	4.2	22.4	3.5	0.2	2.9
2-Hydroxy-p-menth-8-en-7-oic acid	(29)	6.7	1.1	2.5	2.7	4.6	1.1
Perillic acid-8,9-diol	(30)	26.6	0.7	9.5	14.5	3.7	2.2
Total		56.3	64.3	65.5	51.9	54.4	42.2
% of urinary radioactivity identified		70	78	68	61	71	61

[a] Number of animals.
[b] Mean % of dose in 48-h urine.
[c] (tr) Trace.

β-hydroxy diastereoisomers. The reason for the appreciably higher excretion of 6-hydroxylated metabolites in this study is not known, however it may be related to the fact that the urine samples were collected from rats dosed daily for 10 d with *d*-limonene (400 mg/kg, p.o.), a regimen which may have led to induction of the hepatic monooxygenase enzyme systems. A further point of general interest found in Table 1 is that formation of one or more of the 8,9-diols occurred to a much greater extent in all species than did ring oxidation. Pathways leading to or via perillic acid (27) accounted for as much as 85% (rat) and as little as ~20% (man) of the quantitated urinary metabolites. Regan and Bjeldanes[49] characterized only one 8,9-diol, uroterpenol (8), and reported that ~5% of the unconjugated material was excreted in this form in rats. In addition to confirming the formation of metabolites (8) and (23), the latter study also showed that perillic acid (27) was excreted. A previously unreported metabolite, perillyl alcohol (26) which is an intermediate in the formation of perillic acid, was detected in small amounts. Two additional novel metabolites, a set of diastereoisomers of *p*-mentha-2,8-dien-1-ol (25), were the most abundant compounds identified. Regan and Bjeldanes proposed that the diastereoisomeric pairs of compounds (23) and (25) which are formed arise from the corresponding pair of the postulated 1,2-epoxide intermediate (19), however they subsequently showed that this was not the case.[51] They found that limonene-1,2-glycol (21) was the only neutral metabolite of limonene-1,2-epoxide (19) when the latter was given i.p. to rats or incubated with rat liver homogenates. They concluded that the 1,2-epoxide is not a major intermediate in the metabolism of limonene in rats. It seems reasonable to assume that the 6α- and 6β-hydroxy metabolites (23) are formed via allylic hydroxylation of *d*-limonene. Watabe et al.[52,53] studied the metabolism of limonene to epoxides and glycols by rat liver microsomes. They found that both the 1,2-epoxide (19) and 8,9-epoxide (20) were formed as well as smaller amounts of the 8,9-glycol. Little or no 1,2-glycol was detected in these incubates. They calculated that the rate of hydrolysis of the 1,2-epoxide was only ~1% of that seen with the 8,9-isomer.

Other clues to the pathways involved in *d*-limonene metabolism were obtained by Kodama et al.[48] who administered some of the metabolites to rats or dogs. The results obtained suggest that formation of the 8,9-diol (metabolite (8)) leads to its conjugation with glucuronic acid rather than to further oxidation at C-7 or in the ring. On the other hand, initial oxidation at the latter two sites may be followed by formation of the respective 8,9-diols. A general summary of these results is illustrated in Figure 2 which shows the metabolites of *d*-limonene together with their possible routes of formation.

BICYCLIC TERPENE HYDROCARBONS

Several early investigations dealt with the metabolism of some bicyclic terpene hydrocarbons, however the analytical methods then available were inadequate to allow many firm conclusions to be drawn. While recent reports on several of these compounds have furnished much useful information, including evidence showing that the bicyclic structure can be metabolized to monocyclic derivatives, considerable work nonetheless remains to be done in this complex area.

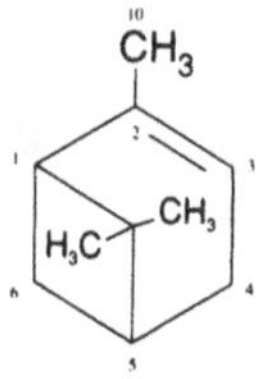

(31)

α-Pinene

α-**Pinene** (2-pinene) (31) was reported by Fromm and Hildebrandt[39] to undergo hydroxylation and conjugation with glucuronic acid in rabbits. Hämäläinen[38] also found evidence for the hydroxylation of α-pinene in rabbits. Harvey[40] was able to recover only 3 to 5% of the dose of α-pinene in the 24-h urine of sheep. This material was found in the neutral fraction and no acidic or phenolic products were detected. Wright[41] repeated this experiment giving sheep two or three daily doses (10 g) of α-pinene but had similar difficulties. Little metabolic material was recovered, however α-pinene appeared to be metabolized to a glucuronide conjugate which was converted to *p*-cymene (14) upon heating with acid. Hämäläinen[38] had previously noted the dehydration of an α-pinene metabolite to *p*-cymene and believed that the precursor might be a monocyclic triol. Newer data on this interesting reaction of conversion of a bicyclic terpene to a monocyclic derivative are equivocal in the case of α-pinene and, for the reason noted below, are included in the following summary of β-pinene metabolism. Römmelt et al.[50] reported that small amounts of α-pinene administered intravenously were excreted in the expired air of pigs and humans. Following a dose of ~0.6 mg/kg, this value was 5% after 2 h with the bulk of the excretion occurring within 10 to 15 min.

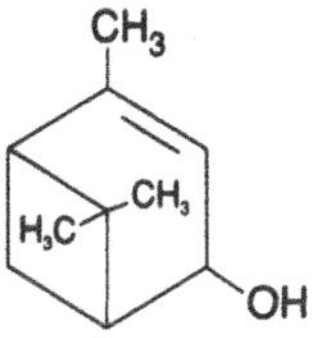

(32) 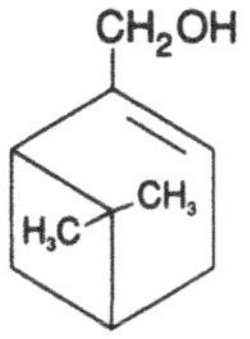　　　　　　　(33) 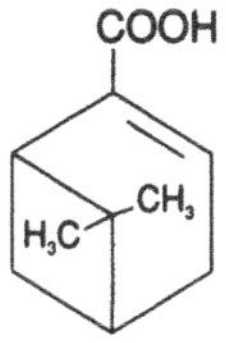　　　　　　　(34)

The hydroxylation of α-pinene was confirmed by Ishida et al.[54] who demonstrated that it underwent allylic hydroxylation when fed to rabbits. In this study, the (−)-isomer (firpene), (+)-isomer (australene), and (±)-isomer of α-pinene were each given orally in doses of 440 or 560 mg/kg. Following hydrolysis with β-glucuronidase + sulfatase, the neutral fractions of urine contained in each case verbenol (2-pinen-4-ol) (32) as the major metabolite. This monohydroxylated derivative resulted from an endocyclic allylic hydroxylation, however trace amounts of the corresponding exocyclic oxidation product, myrtenol (2-pinen-10-ol) (33), were also identified. It was also found that verbenol formation was appreciably higher from the (−)- than from the (+)-isomer of α-pinene. Ishida et al.[34,35] gave a more complete account of these investigations and reported that the acidic fraction from urine contained small amounts of myrtenic acid (34). Southwell et al.[55] fed (+)-α-pinene to the brushtail possum (*Trichosurus vulpecula*) and isolated myrtenic acid from the urine. They detected *trans*-verbenol in the feces.

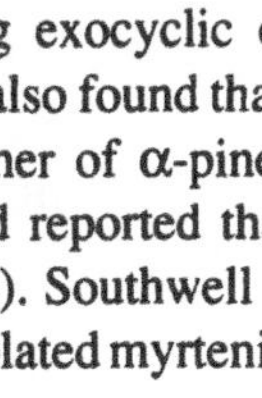

(35)　　　　　　　(36) 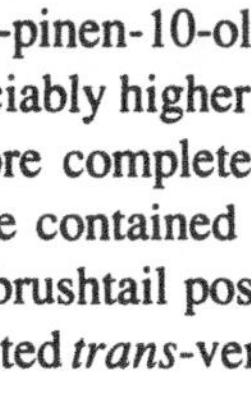　　　　　　　(37)

β-Pinene

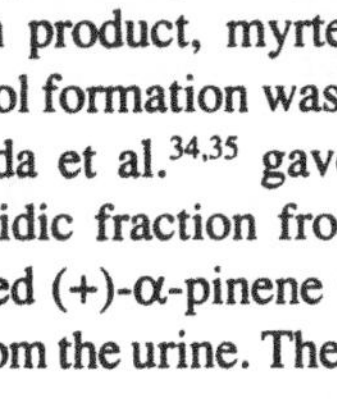

β-Pinene (nopinene, 2(10)-pinene) (35) was included in the study of Hämäläinen[38] who believed that it underwent hydroxylation at the 3- or 4-position in rabbits. In an experiment identical to those described above with α-pinene, Ishida et al.[54] showed that the alcohol formed is *trans*-pinocarveol (2(10)-pinen-3-ol) (36) which is the product of endocyclic allylic hydroxylation. Among the three other alcohols identified as urinary metabolites of (−)-β-pinene, the most abundant was *trans*-10-pinanol (37). Ishida et al.[34,35] identified myrtenic acid (34) in the acidic fraction of the urine in these experiments and believed that it arose via compound (37). Myrtenic acid was detected as a urinary metabolite of (−)-β-pinene in the brushtail possum.[55] As noted above, this acidic metabolite was also found after dosing with α-pinene. In the latter case, its formation is straightforward via allylic hydroxylation and subsequent oxidation of the primary alcohol to the carboxylic acid. Such a direct route is not available with β-pinene. Because both α- and β-pinene are commonly found together in natural products, it is possible that the samples of the latter used in the metabolic experiments may have contained sufficient amounts of the former to give some myrtenic acid. Indeed, Southwell et al.[55] reported that their sample of β-pinene contained 11% of the α-isomer. The articles by Ishida et al. did not state the purities of the terpenoids used. Certainly, the question of the presence of closely-related impurities must be given careful consideration when interpreting the results of metabolic studies with this group of compounds. In experiments identical to those noted above with α-pinene, Römmelt et al.[50] found that ~3% of the β-pinene administered to pigs and humans i.v. was excreted in the expired air, mainly within 10 to 15 min.

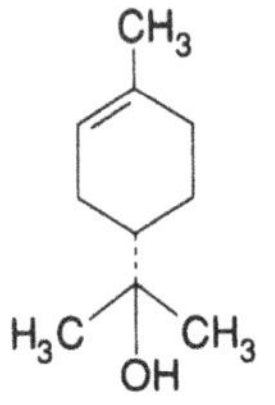

(38) 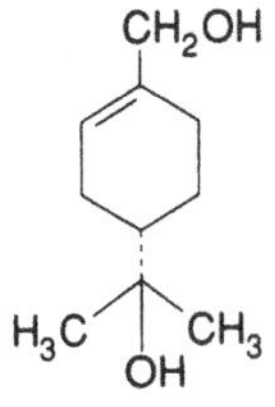   (39)

The investigation of β-pinene metabolism reported by Ishida et al.[34,35,54] described two further neutral metabolites, the monocyclic derivatives α-terpineol (38) and 1-*p*-menthen-7,8-diol (39). The conversion of β-pinene to monocyclic derivatives was first suggested by Hämäläinen[38] who reported that treatment of the urinary glucuronide material from rabbits with hot acid led to the formation of *p*-cymene.

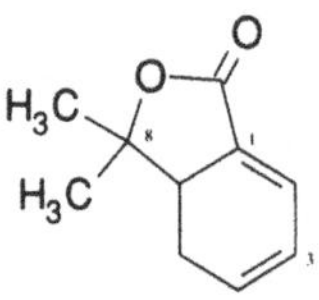

(40) 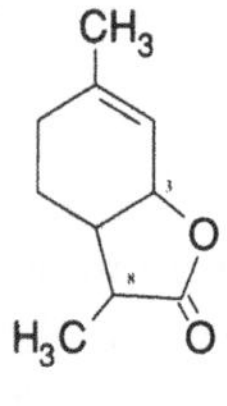   (41) 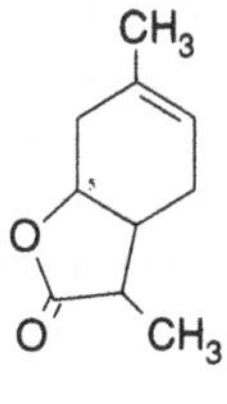   (42) 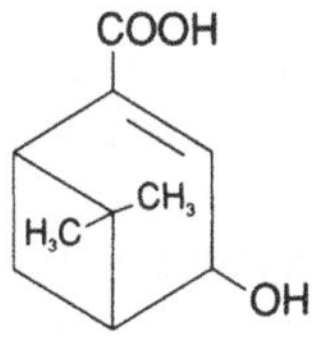   (43)

Further information on the subject of the formation of monocyclic derivatives from bicyclic terpenes was obtained by Southwell[56] using the koala (*Phascolarctus cinereus*). The urine from these animals fed on eucalyptus leaves contained three monoterpene lactones. One of these was shown to be *o*-mentha-1,3-dien-1→8-olide (40) and the others were probably *p*-menth-1-en-8→3-olide (41) and *p*-menth-1-en-8→5-olide (42). The essential oil of eucalyptus leaves contains numerous terpenoid and related compounds, however it was noted that the major constituents are α- and β-pinene, 1,8-cineole, and *p*-cymene. The structures of the three lactones show more resemblance to the pinenes than to other compounds and it was therefore suggested that they are metabolites of the pinenes. Whether the lactones arise from one or both of these bicyclic terpenes was not determined. However, the excretion of large amounts of ester glucuronides in koala urine was interpreted to indicate that the lactones identified may arise via cyclization of carboxylic acids which are formed by hydrolysis of urinary glucuronide conjugates. Southwell et al.[55] suggested that the lactone (40) may arise from 4-hydroxymyrtenic acid (43) and the other lactones possibly from isomeric hydroxyacids. This interpretation is in agreement with the findings of Lander and Mechoulam[57] who studied the ring opening of pinene derivatives. However, Ishida et al.[35] did not detect (43) as a metabolite of β-pinene in rabbits.

Hämäläinen[38] also administered the fully saturated bornane (camphane) (44) to rabbits and found the glucuronide of borneol (2-hydroxybornane) in the urine. This was confirmed by Robertson and Hussain[58] and Robertson and Solomon[59] following the administration of 0.5- to

1.0-g doses of bornane to rabbits. However, most of the hydroxylated material was shown to be 3-hydroxybornane (epiborneol). While bornane can therefore be hydroxylated at either C-2 or C-3, Robertson and Solomon reported that hydroxylation at both sites to give the 2,3-diol did not occur. Compounds which would be formed from the diol via dehydrogenation (i.e., ketols) were also shown not to be excreted in the urine.

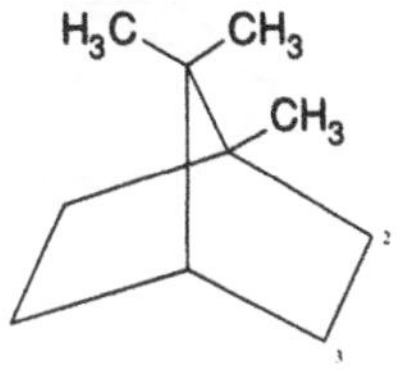

(44)
Bornane

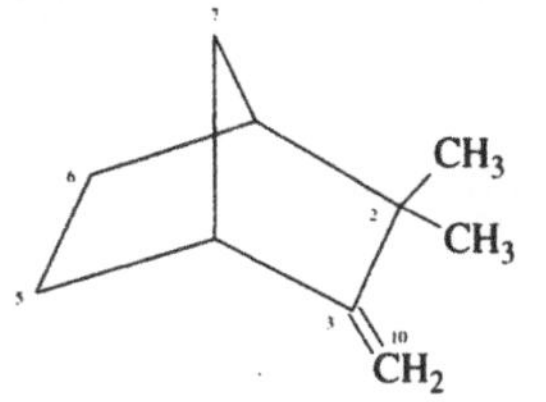

(45)
Camphene

Camphene (2,2-dimethyl-3-methylenenorbornane) (45) was investigated by Fromm and Hildebrandt[39] and by Hildebrandt[42] who reported that it was excreted in rabbits as the glucuronide conjugate of a hydroxy derivative (camphenol). Subsequently, Fromm et al.[60] reported that the monoglucuronide of camphene glycol was also formed by rabbits. Ishida et al.[61] confirmed and expanded these early findings. They gave *dl*-camphene to rabbits (800 mg/kg, intragastrically) and hydrolyzed possible conjugates with β-glucuronidase + sulfatase. The fraction containing neutral metabolites was obtained and several compounds were isolated or detected. The main metabolic pathway found was that of epoxidation and hydration at the 3,10-bond to give the epimers (46) and (47). Two camphenols were found and they were shown to be 5-*exo*-hydroxycamphene (48) and 7-hydroxycamphene (49), the latter compound probably having the hydroxy group *anti*- to the *gem*-dimethyl group. Additionally, Ishida et al. reported two monohydroxylated metabolites which could be formed following a rearrangement reaction of camphene to tricycline (50). The products found were the 3- and 10-hydroxy derivatives. The camphene employed in this study was described as having a purity of >90%. The possibility that the metabolites based on (50) might be due to any impurity present in the material given was not discussed. The excretion of small amounts of camphene in the expired air of pigs and humans was reported by Römmelt et al.[50] Following an i.v. dose of ~0.6 mg/kg, this value was 3.6% after 2 h, with the bulk of the excretion occurring within 10 to 15 min.

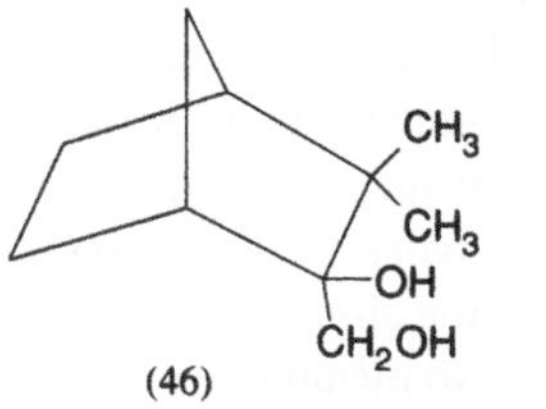

(46)

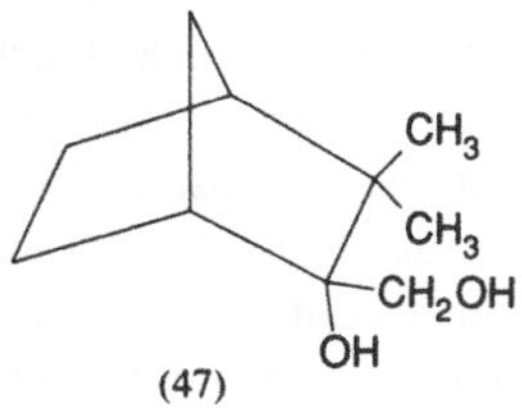

(47)

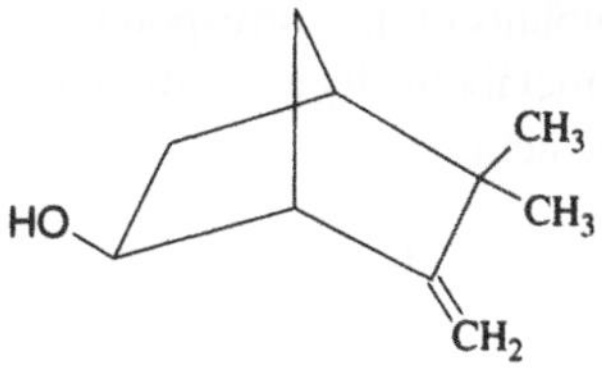

(48)

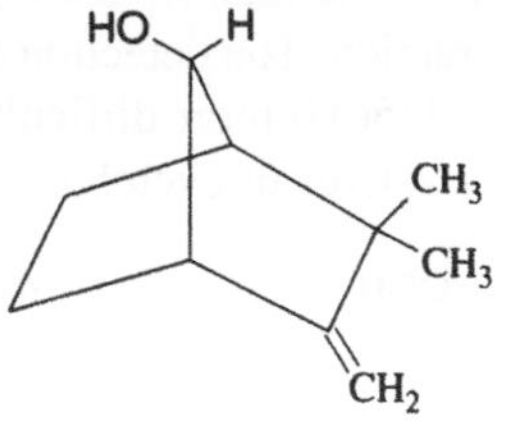

(49)

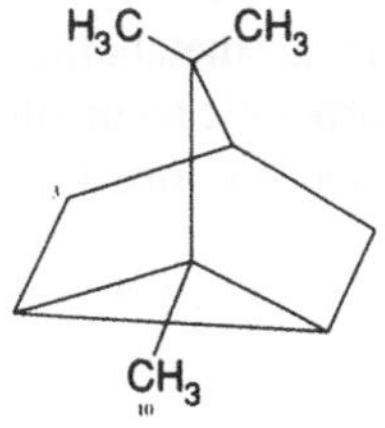

(50)

4(10)-Thujene (sabinene) (51) was reported to be hydroxylated and excreted as a glucuronide in rabbits.[38,39,42] Hämäläinen[38] believed that oxidation occurred at one of the two sites adjacent to the isopropyl group.

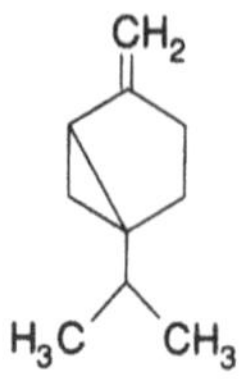

| (51) | (52) | (53) |
| 4(10)-Thujene | 3-Carene | |

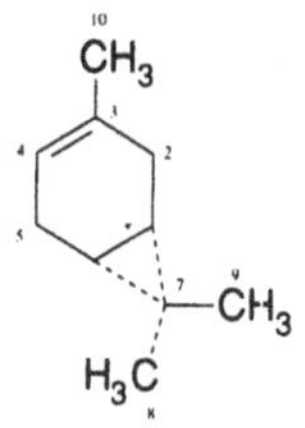

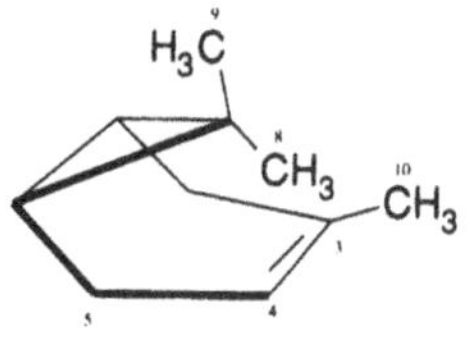

The metabolism of **3-carene** (Δ^3-carene) (52) in rabbits was studied by Ishida et al.[34,35,54] They administered (+)-3-carene (670 mg/kg) intragastrically and isolated the metabolites from the neutral and acidic fractions of urine following sample treatment with β-glucuronidase + sulfatase for conjugate hydrolysis. The initial report described three hydroxylated derivatives and the subsequent communications also identified four acidic metabolites. Most of these were formed by allylic hydroxylation at C-10 and/or hydroxylation at the *gem*-dimethyl group. Metabolite (54) furnishes an example of the latter reaction, which was shown to be stereospecific. The reason for the preferential oxidation at C-9 is no doubt a result of the more exposed site of this methyl group, a fact easily visualized when 3-carene is depicted in its more stable boat form (53).[62]

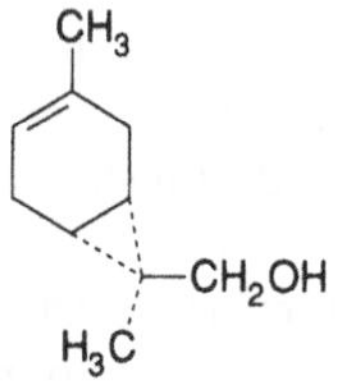

| (54) | (55) | (56) |

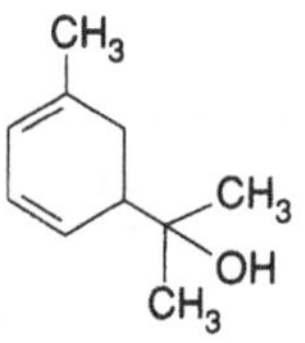

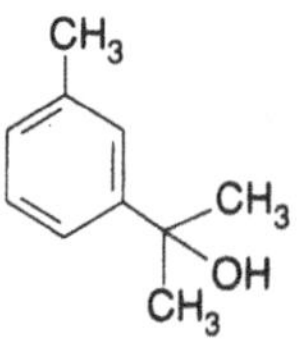

Further oxidation of (54) formed the carboxylic acid derivative (57). Although 10-hydroxy-3-carene was not detected, it undoubtedly served as an intermediate in the formation of chaminic acid (58). The latter compound was the most prominent acidic metabolite isolated and its ease of formation may be responsible for the lack of the intermediate compound. However, 10-hydroxylation is seen in the hydroxyacid (59). Finally, the dicarboxylic acid (60) was found to be a prominent metabolite. It is noteworthy that the allylic hydroxylation which occurred with the aforementioned metabolites was exocyclic. No metabolite was found which arose directly from endocyclic allylic hydroxylation (at C-2 or C-5). However, Ishida et al. believed that allylic hydroxylation did occur at C-5 but that this undetected intermediate underwent rearrangement to give the monocyclic derivative *m*-mentha-4,6-dien-8-ol (55). The latter compound was the major metabolite in the neutral fraction. The detection of minor amounts of the corresponding aromatic metabolite *m*-cymen-8-ol (56) is more difficult to explain metabolically as aromatization is normally a reaction of derivatives of cyclohexanecarboxylic acid.

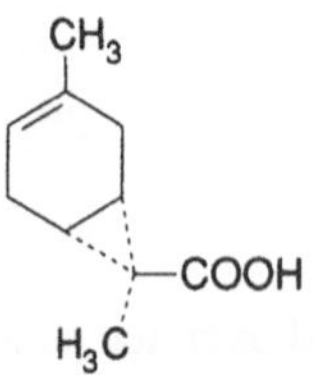

| (57) | (58) | (59) | (60) |

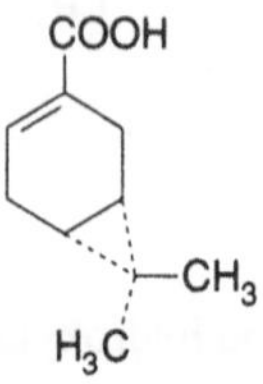

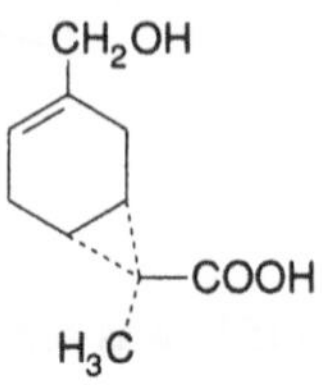

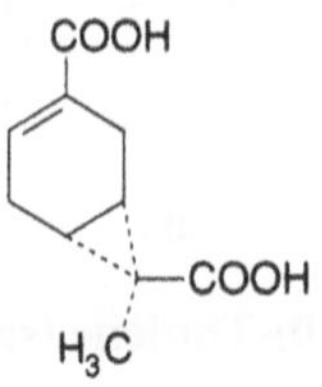

AROMATIC HYDROCARBONS

Few aromatic hydrocarbons are found in plants. The best known examples are *p*-cymene and styrene, however a few compounds containing one or usually several unsaturated bonds (e.g., polyacetylenic compounds) have also been described.

p-**Cymene** (14) shows close structural similarity to the monocyclic terpene hydrocarbons. *p*-Menthane is the fully reduced derivative and two compounds covered above, *p*-menthene (11) and α-phellandrene (12), show intermediate degrees of unsaturation, having one or two ring double bonds, respectively. Studies of *p*-cymene metabolism in dogs were carried out over a century ago and indicated that the major pathway was oxidation of the methyl group to give cumic acid (67) followed by its conjugation with glycine to give cuminuric acid[63] (68) (see Figure 3). Other investigations in sheep also showed that cumic acid[40] and its glycine conjugate[41] were urinary metabolites of *p*-cymene. Southwell et al.[55] reported that *p*-cymene was metabolized to cumic acid in both the brushtail possum and the koala. The earlier studies did not suggest that aromatic hydroxylation occurred with *p*-cymene and this possibility was specifically checked by Bakke and Scheline.[64] Following acid or enzymatic hydrolysis of the 48-h urines of rats given *p*-cymene (100 mg/kg, p.o.), neither of the possible monohydric phenols (thymol and carvacrol) was detected. This shows that only aliphatic oxidation occurred and small amounts of the intermediate *p*-isopropylbenzyl alcohol (66) were also detected. Earlier studies indicated that oxidation of the isopropyl group did not occur, however the presence of two unidentified metabolites in the samples obtained by Bakke and Scheline suggested that the metabolic picture might be more complex than previously assumed.

The assumption that *p*-cymene undergoes more extensive metabolism than that revealed by the earlier studies was clearly shown in experiments using a rabbit[35] and rats and guinea pigs.[65] The doses used were 670 mg/kg (rabbit, p.o.) and 100 mg/kg (rats and guinea pigs, p.o. and inhalation). In both studies the urine was collected for 2 or 3 d and the neutral and acid metabolites investigated after hydrolyzing the samples with β-glucuronidase + sulfatase. In these investigations, 20 metabolites were identified, however not all were found in the urine of all three species. In fact, only seven metabolites were reported from rabbits, however two additional urinary metabolites from rabbits were subsequently reported.[66] The structures of these metabolites are shown in Figure 3 together with their probable routes of formation and information on the qualitative differences in the three species. Urinary excretion values from rats and guinea pigs are summarized in Table 2. These illustrate some striking quantitative differences including the prominence of the pathway via metabolites (73) and (74) leading to (79) in rats compared with the formation of large amounts of the glycine conjugate (68) in guinea pigs. A noteworthy finding in this study was the identification of carvacrol (61) and a hydroxylated carvacrol derivative (62) which were formed by guinea pigs. This was the first evidence showing that *p*-cymene can undergo aromatic hydroxylation in animals; however, the excretion values clearly show that this pathway is of minor importance. Ishida et al.[66] investigated the stereochemistry of metabolites (63), (64), (74), and (79) excreted by rabbits. The enantiomeric ratios (*R:S*) of these were 0:100, 65:35, 91:9, and 30:70, respectively.

The metabolism of **styrene** (81) has been studied in great detail; however, interest in this subject arises not because of the occasional presence of styrene in plants and plant products but because of its importance as an industrial chemical. In the latter regard the occupational health hazards posed by styrene have stimulated extensive investigations of its biological disposition. Much of this work clearly falls outside the scope of the present volume and no attempt will be made to summarize the voluminous literature on this subject. Instead, a concise summary of the pathways of metabolism of styrene will be presented, with the literature references cited also serving as entry points to related studies.

The multiple pathways of metabolism of styrene are summarized in Figures 4 (metabolites lacking sulfur) and 5 (metabolites containing sulfur). Many of the metabolites shown in Figure 4 may be excreted as glucuronides and, in some cases, sulfates, however these conjugates are

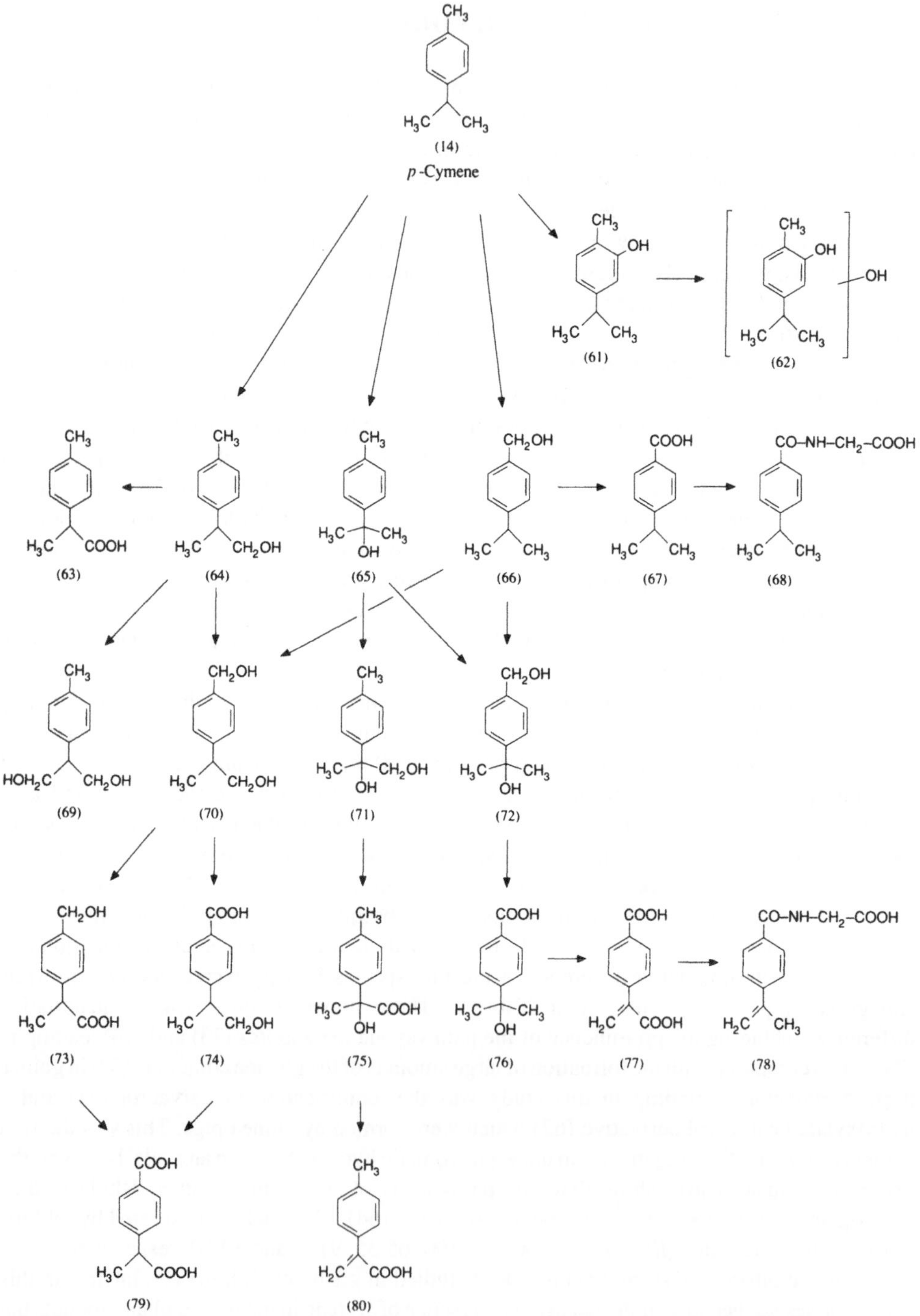

FIGURE 3.   Metabolic pathways of *p*-cymene. Rats formed all metabolites except (61) and (62). Guinea pigs formed all metabolites except (77). Rabbits formed metabolites (63), (64), (65), (67), (74), (75), (76), (79), and (80).

not included in the scheme. In an early study, Spencer et al.[67] found that rats and rabbits degraded styrene to benzoic acid (88) which was predominantly excreted in the urine as its glycine conjugate (hippuric acid) (90). They reported that this pathway accounted for 50 to 90% of the dose (270 to 810 mg/kg) given orally or by inhalation. El Masri et al.[68] carried out an extensive investigation of styrene metabolism in rabbits. The major urinary metabolite was again shown to be hippuric acid (90) which accounted for ~40% of the dose (520 mg/kg, p.o.). Mandelic acid

TABLE 2
Urinary Metabolites of *p*-Cymene[65]

Metabolite	Structure (see Figure 3)	Rat	Guinea pig
Carvacrol	(61)	0[a]	tr[b]
Hydroxycarvacrol	(62)	0	1
2-*p*-Tolylpropionic acid	(63)	1	4
p-Cymen-9-ol	(64)	8	8
p-Cymen-8-ol	(65)	9	14
p-Isopropylbenzyl alcohol	(66)	1	6
Cumic acid	(67)	19	tr
Cuminuric acid	(68)	2	31
2-*p*-Tolylpropan-1,3-diol	(69)	tr	tr
2-*p*-(Hydroxymethyl)phenylpropan-1-ol	(70)	tr	tr
2-*p*-Tolylpropan-1,2-diol	(71)	tr	7
2-*p*-(Hydroxymethyl)phenylpropan-2-ol	(72)	tr	tr
2-*p*-(Hydroxymethyl)phenylpropionic acid	(73)	4	tr
2-*p*-Carboxyphenylpropan-1-ol	(74)	11	tr
2-*p*-Carboxyphenylpropan-2-ol	(76)	9	tr
p-Isopropenylbenzoic acid	(77)	tr	0
p-Isopropenylbenzoyl glycine	(78)	tr	tr
2-*p*-Carboxyphenylpropionic acid	(79)	16	tr
Total		80	71

[a] Mean % of dose (100 mg/kg, intragastrically) in 48-h urine.
[b] (tr) Trace.

(85) was also detected and James and White,[69] in similar experiments using a dose of ~145 mg/kg, reported that an average of 32% of the dose was excreted in this form. While these two metabolites accounted for the bulk of the administered styrene, El Masri et al. found that ~2% could be recovered unchanged in the respiratory air and a further 5% and 6% excreted in the urine as a mercapturic acid and as the monoglucuronide of phenylethane-1,2-diol (phenylglycol) (83), respectively. James and White confirmed the excretion value of 5% for the mercapturic acid derivative and reported it to be *N*-acetyl-*S*-(2-hydroxy-2-phenylethyl)cysteine (107).

The above comments indicate that the metabolic fate of most of the styrene administered to rabbits can be accounted for. The same is true in humans and Bardodej and Bardodejova,[70] in experiments involving inhalation exposure to 22 ppm for 8 h, found that 85% of the retained dose was excreted in the urine as mandelic acid (85) and 10% as phenylglyoxylic acid (86). Wigaeus et al.,[71] in studies in which humans were exposed to low levels of styrene vapor, reported that 58% of the absorbed material was excreted in the urine as compounds (85) and (86). The latter metabolite is also formed from styrene in rats.[72] About 11% of the dose (455 mg/kg, i.p.) was excreted in the urine in 10 h as metabolite (86). Mandelic acid and hippuric acid accounted for ~8% each. These three metabolites were also reported by Braun et al.[73] to be excreted in the urine of rats given styrene. James and White[69] reported that 8.5% of the styrene (210 mg/kg) fed to rats was excreted as the mercapturic acid, metabolite (107). A subsequent investigation of styrene metabolism in rats by Bakke and Scheline[64] was directed towards the identification of hydroxylated metabolites. After an oral dose of 100 mg/kg, the 48-h urine was found to contain 0.1% of the dose as *p*-vinylphenol (94). Interestingly, the previously identified diol (83) formed by rabbits was not detected in rat urine. Instead, 1-phenylethanol (95) and a trace of 2-phenylethanol (96) were present. These two phenylethanols were also detected in the urine of humans exposed to styrene.[74] They calculated that the formation of phenyl glycol (83) was favored over the phenylethanols by a factor of nine. The absence of the diol as a urinary metabolite of styrene in rats was unexpected as Leibman and Ortiz[75] showed that it was formed

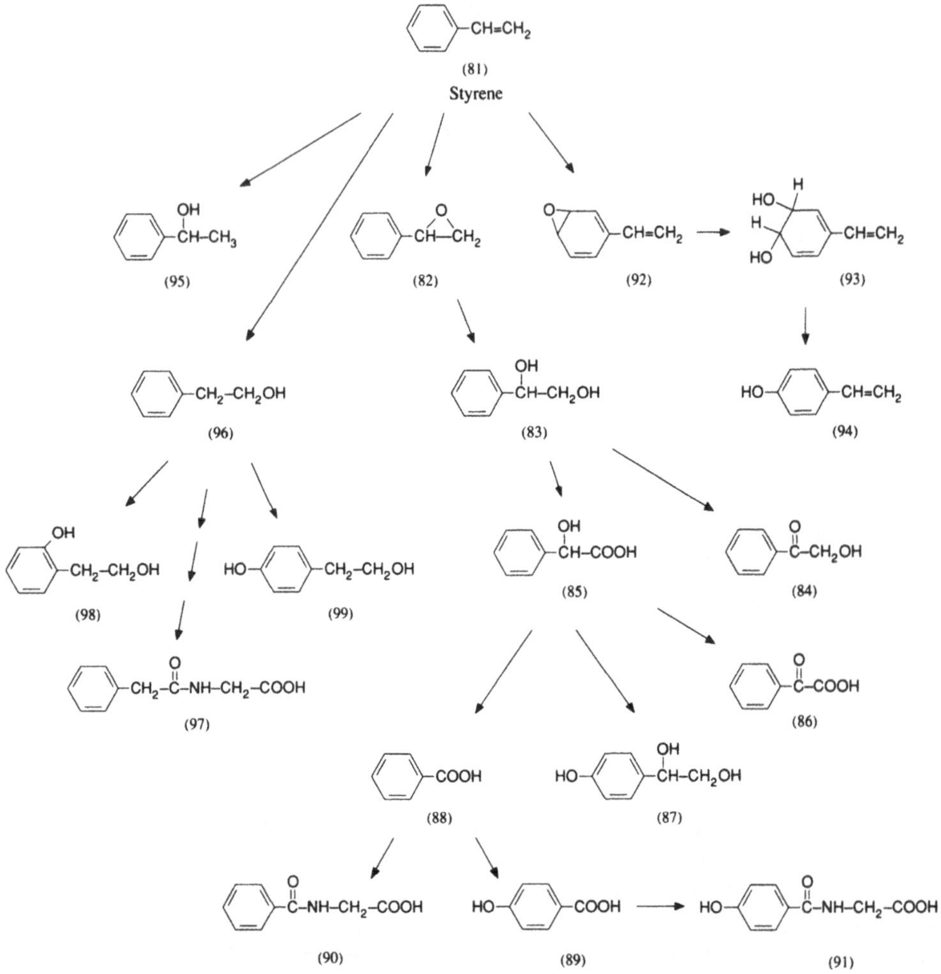

FIGURE 4.   Metabolism of styrene to metabolites lacking sulfur.

from styrene by both rat and rabbit liver microsomal preparations. However, Pantarotto et al.[76] identified it as a urinary metabolite of styrene given by i.p. injection to rats. They also found the expected mandelic (85), benzoic (88), and hippuric (90) acids and confirmed the formation of *p*-vinylphenol (94). Interestingly, they reported that additional ring-hydroxylated metabolites were excreted, with small amounts of *p*-hydroxybenzoic (89), *p*-hydroxymandelic (87), and *p*-hydroxyhippuric (91) acids being detected in the urine. Pantarotto et al. suggested that *p*-vinylphenol was formed via 1-vinylbenzene-3,4-oxide (92) and Watabe et al.,[77] who were able to synthesize this highly unstable arene oxide, showed that it isomerized quantitatively in aqueous medium to *p*-vinylphenol. Nakatsu et al.[78] detected very small amounts of the intermediate dihydrodiol, 3,4-dihydroxy-3,4-dihydro-1-vinylbenzene (93), in the urine of rats given styrene. Pfäffli et al.[79] showed that the formation of *p*-vinylphenol is also a minor reaction in humans. They detected the phenol in the urine of workers exposed to styrene, however the amount excreted was only ~0.3% of the mandelic acid found.

The detection of 2-phenylethanol (96) as a trace metabolite of styrene noted above is of interest in connection with the report by Delbressine et al.[80] that phenaceturic acid (97) was excreted by styrene-treated rats. They found that 1.4% of an i.p. dose of 150 mg/kg was excreted in the urine as phenaceturic acid. They proposed that this metabolic pathway proceeds from styrene to 2-phenylethanol and then to the corresponding aldehyde and carboxylic acid, the latter undergoing conjugation with glycine. They considered the alternative pathway, from styrene oxide (82) via a possible rearrangement to phenylacetaldehyde, to be less likely because no

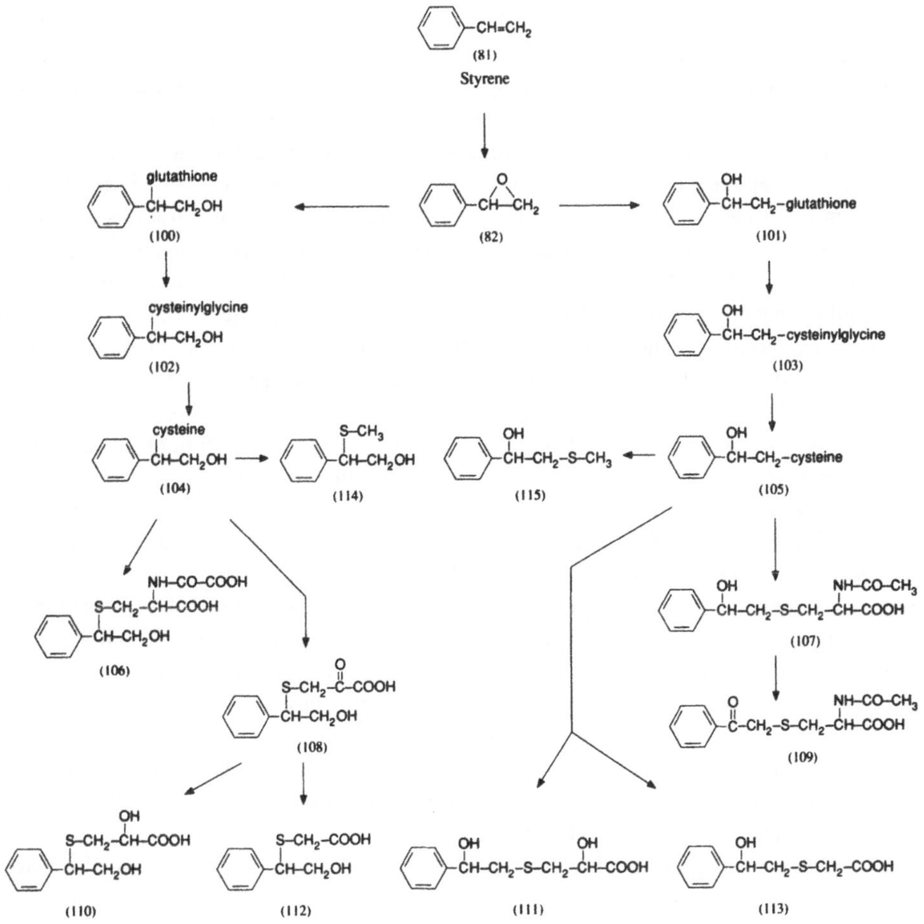

FIGURE 5.   Metabolism of styrene via styrene oxide to metabolites containing sulfur.

evidence was found for the formation of phenaceturic acid following the administration of styrene oxide itself. However, this point remains conjectural and it is possible that a pathway via styrene oxide is nonetheless involved. Indeed, Nakatsu et al.[78] detected the *o*- and *p*-hydroxy derivatives of 2-phenylethanol (compounds (98) and (99), respectively) in the urine of rats given styrene oxide i.p. These minor metabolites are depicted in Figure 4 as simply arising from 2-phenylethanol itself, however their actual routes of formation are not known. This study also demonstrated the excretion, after oral but not i.v. administration, of fairly large amounts of ω-hydroxyacetophenone (84). It was believed to be the oxidation product of the glycol (83).

Further support for the scheme shown in Figure 4 comes from the finding that the glycol (83), when itself fed to rabbits, yielded mandelic acid (85) and hippuric acid (90).[68] Similarly, Ohtsuji and Ikeda[72] showed that the glycol was metabolized to mandelic, hippuric, and phenylglyoxylic (86) acids in rats. Interestingly, administration of the latter compound to rats did not result in its conversion to hippuric acid or mandelic acid.

It is clear that epoxidation to styrene oxide (phenyloxirane) (82) is the key step in the metabolism of styrene which then leads to further metabolic change along several routes. Not unexpectedly, the epoxide itself has not been detected as a urinary metabolite of styrene in the studies cited above; however, Leibman and Ortiz[81] demonstrated its formation from styrene using rabbit liver microsomes. Watabe et al.[82] described similar findings using rat liver microsomes. The activities of formation and hydration of styrene oxide in various tissues from several animal species was reported by Cantoni et al.[83] Because of the pivotal role played by styrene oxide in the metabolism of styrene, it has been employed in numerous metabolic studies. Accordingly, the pathways of metabolism outlined in Figures 4 and 5 encompass studies in which either styrene or its oxide were employed. El Masri et al.[68] and James and White[69] administered styrene oxide to rabbits and found that the metabolites and their quantities were generally similar to those excreted when styrene was given. Ohtsuji and Ikeda[72] reported that rats injected with the epoxide excreted phenylglyoxylic acid (86), mandelic acid (85) and hippuric acid (90) in the urine. The *in vitro* metabolism of styrene oxide by preparations of epoxide hydrolase using various tissues from rats, guinea pigs, and rabbits was studied by James et al.[84] and Ryan et al.[85] The formation of DNA-adducts with styrene oxide was discussed by Liu et al.[86]

The finding that mercapturic acid derivatives are common metabolites of both styrene and its epoxide also demonstrates the important role the latter compound plays in the metabolism of styrene. Boyland and Williams[87] found that styrene oxide is a good substrate for the glutathione *S*-transferase which, by forming the glutathione conjugate, initiates the conversion of the epoxide to the mercapturic acid. Fjellstedt et al.[88] described the preparation of this enzyme in purified form and James et al.[84] and Ryan et al.[85] investigated its properties in preparations from various tissues from rats, guinea pigs, and rabbits using [14C]styrene oxide as a substrate. The properties of several forms of glutathione transferase from the cytosol of human liver have been described.[89,90] Ryan et al.[85] and Ryan and Bend[91] showed that the isolated perfused rat liver converted styrene oxide to *S*-(1-phenyl-2-hydroxyethyl)glutathione (100). Subsequently, Pachecka et al.[92] and Watabe et al.[93] showed that styrene oxide was converted by rat liver cytosol to both of the isomeric glutathione conjugates (100) and (101) (*S*-(2-phenyl-2-hydroxyethyl)glutathione. Both groups found that the conjugates (100) and (101) were formed in a ratio of 3:2. More recently, Hiratsuka et al.[94] studied the regioselective conjugation of glutathione by rat liver cytosol with the (*R*)- and (*S*)-enantiomers of styrene oxide. They also carried out kinetic studies of the conjugation reaction using six major isoenzymes of glutathione transferase isolated from rat liver cytosol.

The respective mercapturic acid derivatives (106) and (107) derived from the two glutathione conjugates noted above were previously shown to be urinary metabolites of styrene in rats.[95] The ratio of compounds (106) and (107) in the urine was ~2:1 and much smaller amounts of a third mercapturic acid, the keto derivative (109) of (107), were also excreted. The total excretion of mercapturic acids amounted to ~11% of the dose (250 mg/kg). Both Delbressine et al.[96] and Watabe et al.[97] confirmed the predominance of the formation of metabolite (106) over (107) in rats given styrene or styrene oxide. Steele et al.[98] found that, although rats metabolized most of the glutathione conjugates of styrene oxide to the corresponding mercapturic acids, small amounts of the intermediate cysteinylglycine ((102) + (103)) and cysteine ((104) + (105)) conjugates were present in the urine.

An investigation of additional sulfur-containing metabolites formed from styrene oxide was carried out by Nakatsu et al.[78] As expected, they found that mercapturic acids were formed by rats, however guinea pigs excreted mainly the corresponding acetic acid derivatives (112) and (113). Additionally, the lactic acid derivatives (110) and (111) and a pyruvic acid derivative (108) were detected in guinea pig or mouse urine. Furthermore, they showed that both rats and guinea pigs excreted two methylthio derivatives [(114) and (115)] in addition to the major neutral metabolite, phenyl glycol (83).

Another topic of interest regarding styrene is its stereometabolism. The initial metabolic product, styrene oxide, is the first member of a chiral sequence which includes phenyl glycol (83) and mandelic acid (85). Watabe et al.[99] synthesized the *R*- and *S*-enantiomers of both styrene oxide and phenyl glycol. Using rat liver microsomes, they found that styrene was metabolized to the *S*-epoxide at a slightly higher rate than to the *R*-form. The subsequent hydrolysis to phenyl glycol showed a 4:1 ratio in reaction rates of the *S*-enantiomer relative to the *R*-enantiomer. Watabe et al.[97] expanded these studies by administering styrene and both racemic and optically active styrene oxide i.p. to rats. Their results showed that the *S*-isomer was metabolized mainly via hydrolysis to phenyl glycol and then to mandelic and phenylglyoxylic acids, whereas the R-isomer was mainly converted to mercapturic acids. However, Delbressine et al.[96] found nearly equal amounts of urinary thioethers following similar experiments with R-(+)-styrene oxide or S-(–)-styrene oxide. They concluded that the microsomal oxidation of styrene occurs preferentially via the R-(+)-isomer. The same conclusion was reached by Korn et al.[100] who measured the enantiomers of mandelic acid in the urine of humans exposed to styrene. They found a ratio of R-isomer to S-isomer of ~1.5:1. Korn et al.[74,101] extended these studies to include the enantiomers of phenyl glycol and showed that the R-isomer (which will be formed from R-(+)-styrene oxide) predominated by a factor of nearly three over the S-isomer in the urine. Drummond et al.[102] found a similar preponderance of the R-enantiomer of mandelic acid in the urine of rats given oral doses of styrene. However, in humans exposed to styrene vapor, the R:S-ratio was only slightly greater than unity.

As noted above, a few aliphatic plant hydrocarbons are polyacetylenic compounds. While the metabolism of these naturally occurring compounds has not been studied, that of the simplest member of this type, phenylacetylene (116), was investigated and the results obtained may furnish some indication of the likely metabolism of the other derivatives. El Masri et al.[68] found that a single dose (400 mg/kg, p.o.) of phenylacetylene was slowly metabolized by rabbits and that 30 to 40% was eliminated unchanged in the respiratory air over 3 d. The urine did not contain metabolites conjugated with glucuronic acid but 5 to 15% of the dose was excreted daily during the first 3 d as phenaceturic acid (97) (see Figure 4). It was proposed that the latter metabolite arises via asymmetrical hydration of phenylacetylene to give compound (117) which is the enol form of phenylacetaldehyde (118). The latter compound will readily be oxidized to phenylacetic acid and conjugated with glycine to form the urinary metabolite (97). Hydration in the reverse manner would give phenylmethylketone which could undergo further metabolism to phenylmethylcarbinol, benzoic acid, and mandelic acid (85). Of these possible metabolites only traces of the latter were detected and it can therefore be concluded that attachment of the oxygen function occurs nearly entirely at the ω-carbon.

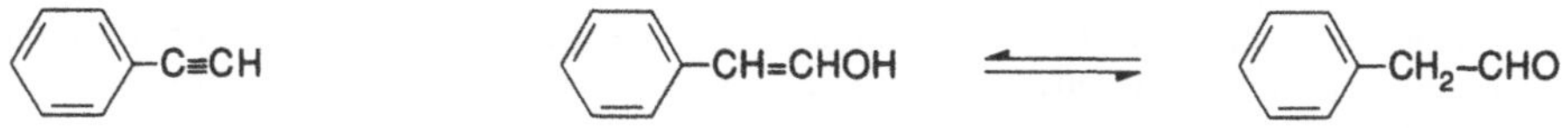

(116)   (117)   (118)

REFERENCES

1. **McCarthy, R. D.,** Mammalian metabolism of straight-chain saturated hydrocarbons, *Biochim. Biophys. Acta,* 84, 74, 1964.
2. **Popović, M.,** The metabolism of paraffins in rats, *FEBS Lett.,* 12, 49, 1970.
3. **Mitchell, M. P. and Hübscher, G.,** Oxidation of *n*-hexadecane by subcellular preparations of guinea pig small intestine, *Eur. J. Biochem.,* 7, 90, 1968.
4. **Kusunose, M., Ichihara, K., and Kusunose, E.,** Oxidation of *n*-hexadecane by mouse liver microsomal fraction, *Biochim. Biophys. Acta,* 176, 679, 1969.
5. **Albro, P. W. and Fishbein, L.,** Absorption of aliphatic hydrocarbons by rats, *Biochim. Biophys. Acta,* 219, 437, 1970.
6. **Tulliez, J. E. and Bories, G. F.,** Metabolism of an *n*-paraffin, heptadecane, in rats, *Lipids,* 13, 110, 1978.
7. **Tulliez, J. and Bories, G.,** Métabolisme des hydrocarbures paraffiniques et naphténiques chez les animaux supérieurs. I. Rétention des paraffines (normal, cyclo et ramifiées) chez le rat, *Ann. Nutr. Alim.,* 29, 201, 1975.
8. **Frommer, U., Ullrich, V., and Staudinger, H.,** Hydroxylation of aliphatic compounds by liver microsomes. I. The distribution pattern of isomeric alcohols, *Hoppe-Seyler's Z. Physiol. Chem.,* 351, 903, 1970.

9.   **Daugherty, M. S., Ludden, T. M., and Burk, R. F.,** Metabolism of ethane and pentane to carbon dioxide by the rat, *Drug Metab. Dispos.,* 16, 666, 1988.

10. **Das, M. L., Orrenius, S., and Ernster, L.,** On the fatty acid and hydrocarbon hydroxylation in rat liver microsomes, *Eur. J. Biochem.,* 4, 519, 1968.

11. **Frommer, U., Ullrich, V., Staudinger, H. J., and Orrenius, S.,** The monooxidation of *n*-heptane by rat liver microsomes, *Biochim. Biophys. Acta,* 280, 487, 1972.

12. **Bahima, J., Cert, A., and Menéndez-Gallego, M.,** Identification of volatile metabolites of inhaled *n*-heptane in rat urine, *Toxicol. Appl. Pharmacol.,* 76, 473, 1984.

13. **DiVincenzo, G. D., Kaplan, C. J., and Dedinas, J.,** Characterization of the metabolites of methyl *n*-butyl ketone, methyl iso-butyl ketone, and methyl ethyl ketone in guinea pig serum and their clearance, *Toxicol. Appl. Pharmacol.,* 36, 511, 1976.

14. **Couri, D., Abdel-Rahman, M. S., and Hetland, L. B.,** Biotransformation of *n*-hexane and methyl *n*-butyl ketone in guinea pigs and mice, *Am. Ind. Hyg. Assoc. J.,* 39, 295, 1978.

15. **Perbellini, L., Brugnone, F., and Pavan, I.,** Identification of the metabolites of *n*-hexane, cyclohexane, and their isomers in men's urine, *Toxicol. Appl. Pharmacol.,* 53, 220, 1980.

16. **Dolara, P., Franconi, F., Basosi, D., and Moneti, G.,** Urinary excretion of some *n*-hexane metabolites, *Pharmacol. Res. Commun.,* 10, 503, 1978.

17. **Perbellini, L., Brugnone, F., Pastorello, G. and Grigolini, L.,** Urinary excretion of n-hexane metabolites in rats and humans, *Int. Arch. Occup. Environ. Health,* 42, 349, 1979.

18. **Perbellini, L., Amatini, M. C., Brugnone, F., and Frontali, N.,** Urinary excretion of *n*-hexane metabolites. A comparative study in rat, rabbit and monkey, *Arch. Toxicol.,* 50, 203, 1982.

19. **Fedtke, N. and Bolt, H. M.,** 4,5-Dihydroxy-2-hexanone: a new metabolite of *n*-hexane and of 2,5-hexanedione in rat urine, *Biomed. Environ. Mass Spectrom.,* 14, 563, 1987.

20. **Fedtke, N. and Bolt, H. M.,** The relevance of 4,5-dihydroxy-2-hexanone in the excretion kinetics of *n*-hexane metabolites in rat and man, *Arch. Toxicol.,* 61, 131, 1987.

21. **Couri, D. and Milks, M.,** Toxicity and metabolism of the neurotoxic hexacarbons *n*-hexane, 2-hexanone, and 2,5-hexanedione, *Annu. Rev. Pharmacol. Toxicol.,* 22, 145, 1982.

22. **Olson, C. T., Yu, K. O., Hobson, D. W., and Serve, M. P.,** The metabolism of *n*-octane in Fischer 344 rats, *Toxicol. Lett.,* 31, 147, 1986.

23. **Dahl, A. R.,** The fate of inhaled octane and the nephrotoxicant, isooctane, in rats, *Toxicol. Appl. Pharmacol.,* 100, 334, 1989.

24. **Frommer, U., Ullrich, V., and Orrenius, S.,** Influence of inducers and inhibitors on the hydroxylation pattern of *n*-hexane in rat liver microsomes, *FEBS Lett.,* 41, 14, 1974.

25. **Nilsen, O. G., Toftgård, R., Eng, L., and Gustafsson, J.-Å.,** Regio-selectivity of purified forms of rabbit liver microsomal cytochrome P-450 in the metabolism of benzo(a)pyrene, *n*-hexcane and 7-ethoxyresorufin, *Acta Pharmacol. Toxicol.,* 48, 369, 1981.

26. **Lu, A. Y. H., Strobel, H. W., and Coon, M. J.,** Properties of a solubilized form of the cytochrome P-450-containing mixed-function oxidase of liver microsomes, *Mol. Pharmacol.,* 6, 213, 1970.

27. **Ichihara, K., Kusunose, E., and Kusunose, M.,** Microsomal hydroxylation of decane, *Biochim. Biophys. Acta,* 176, 713, 1969.

28. **Del Monte, M., Citti, L., and Gervasi, P. G.,** Isoprene metabolism by liver microsomal mono-oxygenases, *Xenobiotica,* 15, 591, 1985.

29. **Longo, V., Citti, L., and Gervasi, P. G.,** Hepatic microsomal metabolism of isoprene in various rodents, *Toxicol Lett.,* 29, 33, 1985.

30. **Neubauer, O.,** Ueber Glykuronsäurepaarung bei Stoffen der Fettreihe, *Arch. Exp. Pathol. Pharmak.,* 46, 133, 1901.

31. **Maynert, E. W., Foreman, R. L., and Watabe, T.,** Epoxides as obligatory intermediates in the metabolism of olefins to glycols, *J. Biol. Chem.,* 245, 5234, 1970.

32. **Watabe, T. and Yamada, N.,** The biotransformation of 1-hexadecene to carcinogenic 1,2-epoxyhexadecane by hepatic microsomes, *Biochem. Pharmacol.,* 24, 1051, 1975.

33. **White, I. N. H., Green, M. L., Bailey, E., and Farmer, P. B.,** Metabolic activation of olefins. Conversion of 1-octene to a putative reactive intermediate 1-octen-3-one: an alternative pathway to epoxidation, *Biochem. Pharmacol.,* 35, 1569, 1986.

34. **Ishida, T., Asakawa, Y., Takemoto, T., and Aratani, T.,** Biotransformation of terpenoids in mammals, *Res. Bull. Hiroshima Inst. Technol.,* 14, 9, 1980.

35. **Ishida, T., Asakawa, Y., Takemoto, T., and Aratani, T.,** Terpenoids biotransformation in mammals. III. Biotransformation of α-pinene, β-pinene, pinane, 3-carene, carane, myrcene, and *p*-cymene in rabbits, *J. Pharm. Sci.,* 70, 406, 1981.

36. **Madyastha, K. M. and Srivatsan, V.,** Metabolism of β-myrcene *in vivo* and *in vitro*: its effects on rat-liver microsomal enzymes, *Xenobiotica,* 17, 539, 1987.

37. **Kuhn, R., Köhler, F., and Köhler, L.,** Über Methyl-oxydationen im Tierkörper, *Hoppe-Seyler's Z. Physiol. Chem.,* 242, 171, 1936.

38. **Hämäläinen, J.,** Über das Verhalten der alicyklischen Verbindungen bei der Glykuronsäurepaarung im Organismus, *Skand. Arch. Physiol.*, 27, 141, 1912.
39. **Fromm, E. and Hildebrandt, H.,** Ueber das Schicksal cyclischer Terpene und Campher im thierischen Organismus, *Hoppe-Seyler's Z. Physiol. Chem.*, 33, 579, 1901.
40. **Harvey, J. M.,** The detoxication of terpenes by sheep, *Pap. Dep. Chem. Univ. Qd*, 1, 1942.
41. **Wright, S. E.,** Detoxication mechanisms in the sheep, *Pap. Dep. Chem. Univ. Qd*, 1, 1945.
42. **Hildebrandt, H.,** Ueber das Schicksal einiger cyklischer Terpene und Kampfer im Thierkörper, *Hoppe-Seyler's Z. Physiol. Chem.*, 36, 452, 1902.
43. **Wade, A. P., Wilkinson, G. S., Dean, F. M., and Price, A. V.,** The isolation, characterization and structure of uroterpenol, a monoterpene from human urine, *Biochem. J.*, 101, 727, 1966.
44. **Dean, F. M., Price, A. W., Wade, A. P., and Wilkinson, G. S.,** Uroterpenol β-D-glucuronide, *J. Chem. Soc. (C)*, 1893, 1967.
45. **Smith, O. W., Wade, A. P., and Dean, F. M.,** Uroterpenol, a pettenkofer chromogen of dietary origin and a common constituent of human urine, *J. Endocrinol.*, 45, 17, 1969.
46. **Igimi, H., Nishimura, M., Kodama, R., and Ide, H.,** Studies on the metabolism of d-limonene (p-mentha-1,8-diene). I. The absorption, distribution and excretion of d-limonene in rats, *Xenobiotica*, 4, 77, 1974.
47. **Kodama, R., Noda, K., and Ide, H.,** Studies on the metabolism of d-limonene (p-mentha-1,8-diene). II. The metabolic fate of d-limonene in rabbits, *Xenobiotica*, 4, 85, 1974.
48. **Kodama, R., Yano, T., Furukawa, K., Noda, K., and Ide, H.,** Studies on the metabolism of d-limonene (p-mentha-1,8-diene). IV. Isolation and characterization of new metabolites and species differences in metabolism, *Xenobiotica*, 6, 377, 1976.
49. **Regan, J. W. and Bjeldanes, L. F.,** Metabolism of (+)-limonene in rats, *J. Agric. Food Chem.*, 24, 377, 1976.
50. **Römmelt, H., Zuber, A., Dirnagl, K., and Drexel, H.,** Zur Resorption von Terpenen aus Badezusätzen, *Münch. Med. Wschr.*, 116, 537, 1974.
51. **Regan, J. W., Morris, M. M., Nao, B., and Bjeldanes, L. F.,** Metabolism of limonene-1,2-epoxide in the rat, *Xenobiotica*, 10, 859, 1980.
52. **Watabe, T., Hiratsuka, A., Isobe, M., and Ozawa, N.,** Metabolism of d-limonene by hepatic microsomes to non-mutagenic epoxides toward *Salmonella typhimurium*, *Biochem. Pharmacol.*, 29, 1068, 1980.
53. **Watabe, T., Hiratsuka, A., Ozawa, N., and Isobe, M.,** A comparative study on the metabolism of d-limonene and 4-vinylcyclohex-1-ene by hepatic microsomes, *Xenobiotica*, 11, 333, 1981.
54. **Ishida, T., Asakawa, Y., Okano, M., and Aratani, T.,** Biotransformation of terpenoids in mammal. I. Biotransformation of 3-carene and related compounds in rabbits, *Tetrahedron Lett.*, 2437, 1977.
55. **Southwell, I. A., Flynn, T. M., and Degabriele, R.,** Metabolism of α- and β-pinene, p-cymene and 1,8-cineole in the brushtail possum, *Trichosurus vulpecula*, *Xenobiotica*, 10, 17, 1980.
56. **Southwell, I. A.,** Essential oil metabolism in the koala. III. Novel urinary monoterpenoid lactones, *Tetrahedron Lett.*, 1885, 1975.
57. **Lander, N. and Mechoulam, R.,** Formation of *ortho*-menthenes by acid-catalysed ring opening of pin-2-ene derivatives, *J. Chem. Soc. Perkin I*, 485, 1976.
58. **Robertson, J. S. and Hussain, M.,** Metabolism of camphors and related compounds, *Biochem. J.*, 113, 57, 1969.
59. **Robertson, J. S. and Solomon, E.,** Metabolism of camphanediols, *Biochem. J.*, 121, 503, 1971.
60. **Fromm, E., Hildebrandt, H., and Clemens, P.,** Ueber da Schicksal cyklischer Terpene und Campher im thierischen Organismus. Ueber das Verhalten des Camphens im Tierkörper, *Hoppe-Seyler's Z. Physiol. Chem.*, 37, 189, 1903.
61. **Ishida, T., Asakawa, Y., Takemoto, T., and Aratani, T.,** Terpenoid biotransformation in mammals. II. Biotransformation of dl-camphene in rabbits, *J. Pharmacol. Sci.*, 68, 928, 1979.
62. **Acharya, S. P.,** Conformations of 3-carene and 2-carene. Their conformational preference and the reactivity of the double bond, *Tetrahedron Lett.*, 4117, 1966.
63. **Williams, R. T.,** *Detoxication Mechanisms*, Chapman and Hall, London, 1959, 204.
64. **Bakke, O. M. and Scheline, R. R.,** Hydroxylation of aromatic hydrocarbons in the rat, *Toxicol. Appl. Pharmacol.*, 16, 691, 1970.
65. **Walde, A., Ve, B., Scheline, R. R., and Monge, P.,** p-Cymene metabolism in rats and guinea-pigs, *Xenobiotica*, 13, 503, 1983.
66. **Ishida, T., Matsumoto, T., and Asakawa, Y.,** Enantioselective metabolism of p-cymene in rabbits, Third European Symposium on Foreign Compound Metabolism. Metabolic and Kinetic Perspectives in Safety Assessment: Some Current Issues, London, 1989.
67. **Spencer, H. C., Irish, D. D., Adams, E. M., and Rowe, V. K.,** The response of laboratory animals to monomeric styrene, *J. Ind. Hyg. Toxicol.*, 24, 295, 1942.
68. **El Masri, A. M., Smith, J. N., and Williams, R. T.,** Studies in detoxication 73. The metabolism of alkylbenzenes: phenylacetylene and phenylethylene (styrene), *Biochem. J.*, 68, 199, 1958.
69. **James, S. P. and White, D. A.,** The metabolism of phenethyl bromide, styrene and styrene oxide in the rabbit and rat, *Biochem. J.*, 104, 914, 1967.

70. **Bardodej, Z. and Bardodejova, E.,** Biotransformation of ethyl benzene, styrene, and alpha-methylstyrene in man, *Am. Ind. Hyg. Assoc.,* 31, 206, 1970.
71. **Wigaeus, E., Löf, A., Bjurström, R., and Nordqvist, M. B.,** Exposure to styrene. Uptake, distribution, metabolism and elimination in man, *Scand. J. Work Environ. Health,* 9, 479, 1983.
72. **Ohtsuji, H. and Ikeda, M.,** The metabolism of styrene in the rat and the stimulatory effect of phenobarbital, *Toxicol. Appl. Pharmacol.,* 18, 321, 1971.
73. **Braun, W. H., Madrid, E. O., and Karbowski, R. J.,** Parallel gas chromatography-mass spectrometry and gas proportional counting, *Anal. Chem.,* 48, 2284, 1976.
74. **Korn, M., Wodarz, R., Drysch, K., Schoknecht, W., and Schmahl, F. W.,** Stereometabolism of styrene in man: gas chromatographic determination of phenylethyleneglycol enantiomers and phenylethanol isomers in the urine of occupationally-exposed persons, *Arch. Toxicol.,* 58, 110, 1985.
75. **Leibman, K. C. and Ortiz, E.,** Oxidation of styrene in liver microsomes, *Biochem. Pharmacol.,* 18, 552, 1969.
76. **Pantarotto, C., Fanelli, R., Bidoli, F., Morazzoni, P., Salmona, M., and Szczawinska, K.,** Arene oxides in styrene metabolism, a new perspective in styrene toxicity?, *Scand. J. Work Environ. Health,* 4, 67, 1978.
77. **Watabe, T., Hiratsuka, A., Aizawa, T., Sawahata, T., Ozawa, N., Isobe, M., and Takabatake, E.,** Studies on metabolism and toxicity of styrene. IV. 1-Vinylbenzene 3,4-oxide, a potent mutagen formed as a possible intermediate in the metabolism in vivo of styrene to 4-vinylphenol, *Mutat. Res.,* 93, 45, 1982.
78. **Nakatsu, K., Hugenroth, S., Sheng, L.-S., Horning, E. C., and Horning, M. G.,** Metabolism of styrene oxide in the rat and guinea pig, *Drug Metab. Dispos.,* 11, 463, 1983.
79. **Pfäffli, P., Hesso, A., Vainio, H., and Hyvönen, M.,** 4-Vinylphenol excretion suggestive of arene oxide formation in workers occupationally exposed to styrene, *Toxicol. Appl. Pharmacol.,* 60, 85, 1981.
80. **Delbressine, L. P. C., Ketalaars, H. C. J., Seutter-Berlage, F., and Smeets, F. L. M.,** Phenaceturic acid, a new urinary metabolite of styrene in the rat, *Xenobiotica,* 10, 337, 1980.
81. **Leibman, K. C. and Ortiz, E.,** Epoxide intermediates in microsomal oxidation of olefins to glycols, *J. Pharmacol. Exp. Ther.,* 173, 242, 1970.
82. **Watabe, T., Isobe, M., Yoshikawa, K., and Takabatake, E.,** Studies on metabolism and toxicity of styrene. I. Biotransformation of styrene to styrene glycol *via* styrene oxide by rat liver microsomes, *J. Pharmacobio. Dyn.,* 1, 98, 1978.
83. **Cantoni, L., Salmona, M., Facchinetti, T., Pantarotto, C., and Belvedere, G.,** Hepatic and extrahepatic formation and hydration of styrene oxide in vitro in animals of different species and sex, *Toxicol. Lett.,* 2, 179, 1978.
84. **James, M. O., Fouts, J. R., and Bend, J. R.,** Hepatic and extrahepatic metabolism, *in vitro,* of an epoxide (8-^{14}C-styrene oxide) in the rabbit, *Biochem. Pharmacol.,* 25, 187, 1976.
85. **Ryan, A. J., James, M. O., Ben-Zvi, Z., Law, F. C. P., and Bend, J. R.,** Hepatic and extrahepatic metabolism of ^{14}C-styrene oxide, *Environ. Health Perspect.,* 17, 135, 1976.
86. **Liu, S.-F., Rappaport, S. M., Rasmussen, J., and Bodell, W. J.,** Detection of styrene oxide–DNA adducts by ^{32}P-postlabelling, *Carcinogenesis,* 9, 1401, 1988.
87. **Boyland, E. and Williams, K.,** An enzyme catalyzing conjugation of epoxides with glutathione, *Biochem. J.,* 94, 190, 1965.
88. **Fjellstedt, T. A., Allen, R. H., Duncan, B. K., and Jakoby, W. B.,** Enzymatic conjugation of epoxides with glutathione, *J. Biol. Chem.,* 248, 3702, 1973.
89. **Pacifici, G. M., Warholm, M., Guthenberg, C., Mannervik, B., and Rane, A.,** Detoxification of styrene oxide by human liver glutathione transferase, *Hum. Toxicol.,* 6, 483, 1987.
90. **Pacifici, G. M., Guthenberg, C., Warholm, M., Mannervik, B., and Rane, A.,** Conjugation of styrene oxide by the basic and acidic forms of glutathione transferase in the human fetal liver, *Dev. Pharmacol. Ther.,* 11, 243, 1988.
91. **Ryan, A. J. and Bend, J. R.,** The metabolism of styrene oxide in the isolated perfused rat liver. Identification and quantitation of major metabolites, *Drug Metab. Dispos.,* 5, 363, 1977.
92. **Pachecka, J., Gariboldi, P., Cantoni, L., Belvedere, G., Mussini, E., and Salmona, M.,** Isolation and structure determination of enzymatically formed styrene oxide glutathione conjugates, *Chem. Biol. Interact.,* 27, 313, 1979.
93. **Watabe, T., Hiratsuka, A., Ozawa, N., and Isobe, M.,** Glutathione S-conjugates of phenyloxirane, *Biochem. Pharmacol.,* 30, 390, 1981.
94. **Hiratsuka, A., Yokoi, A., Iwata, H., Watabe, T., Satoh, K., Hatayama, I., and Sato, K.,** Glutathione conjugation of styrene 7,8-oxide enantiomers by major glutathione transferase isoenzymes isolated from rat livers, *Biochem. Pharmacol.,* 38, 4405, 1989.
95. **Seutter-Berlage, F., Delbressine, L. P. C., Smeets, F. L. M., and Ketalaars, H. C. J.,** Identification of three sulphur-containing urinary metabolites of styrene in the rat, *Xenobiotica,* 8, 413, 1978.
96. **Delbressine, L. P. C., van Bladeren, P. J., Smeets, F. L. M., and Seutter-Berlage, F.,** Stereoselective oxidation of styrene to styrene oxide in rats as measured by mercapturic acid excretion, *Xenobiotica,* 11, 589, 1981.

97. **Watabe, T., Ozawa, N., and Yoshikawa, K.,** Studies on metabolism and toxicity of styrene. V. The metabolism of styrene, racemic, (*R*)-(+)-, and (*S*)-(–)-phenyloxiranes in the rat, *J. Pharmacobiol. Dyn.*, 5, 129, 1982.

98. **Steele, J. W., Yagen, B., Hernandez, O., Cox, R. H., Smith, B. R., and Bend, J. R.,** The metabolism and excretion of styrene oxide-glutathione conjugates in the rat and by isolated perfused liver, lung and kidney preparations, *J. Pharmacol. Exp. Ther.*, 219, 35, 1981.

99. **Watabe, T., Ozawa, N., and Yoshikawa, K.,** Stereochemistry in the oxidative metabolism of styrene by hepatic microsomes, *Biochem. Pharmacol.*, 30, 1695, 1981.

100. **Korn, M., Wodarz, R., Schoknecht, W., Weichardt, H., and Bayer, E.,** Styrene metabolism in man: gas chromatographic separation of mandelic acid enantiomers in the urine of exposed persons, *Arch. Toxicol.*, 55, 59, 1984.

101. **Korn, M., Wodarz, R., Drysch, K., and Schmahl, F. W.,** Stereometabolism of styrene in man. Urinary excretion of chiral styrene metabolites, *Arch. Toxicol.*, 60, 86, 1987.

102. **Drummond, L., Caldwell, J., and Wilson, H. K.,** The metabolism of ethylbenzene and styrene to mandelic acid: stereochemical considerations, *Xenobiotica*, 19, 199, 1989.

METABOLISM OF ALCOHOLS

Two main metabolic pathways are available for alcohols: oxidation and direct conjugation. The oxidation of alcohols is carried out by enzymes designated oxidoreductases. In contrast to many oxidative reactions which are catalyzed by NADPH-dependent monooxygenases located in the endoplasmatic reticulum, alcohol oxidation is to a large extent carried out by soluble liver enzymes. The best known of these is liver alcohol dehydrogenase which has been studied mainly in connection with its role in ethanol metabolism, however it is also responsible for the oxidation of many xenobiotic alcohols of various types. The general reaction shown by primary alcohols is:

$$R\text{--}CH_2OH + NAD^+ \rightleftharpoons R\text{--}CHO + NADH + H^+$$

It is noteworthy that this reaction is reversible, allowing both aldehydes and ketones to be reduced. Since the reaction is pH-dependent, *in vitro* systems show a shift in the equilibrium to the carbinol form as the pH is lowered to neutrality. However, the formation of the aldehydes is usually favored *in vivo* because these products can be further oxidized to acids.

Among the wide variety of alcohol substrates oxidized by alcohol dehydrogenase, primary alcohols show maximum activity with the activity peak occurring with 1-butanol. Kassam et al.[1] found that the prominent isoenzyme of liver alcohol dehydrogenase in Caucasian subjects gave an activity peak which was about equal for 1-butanol and 1-octanol. Oxidation proceeds via the aldehyde to the corresponding carboxylic acid and the latter may be oxidized completely to carbon dioxide and water or, if this is prevented, be conjugated with glucuronic acid. Extensive oxidation can be expected to occur with the normal alcohols which undergo chain shortening via β-oxidation, the fragments entering into the metabolic pool of 2-carbon compounds. Substitution in the α-position (2-alkyl alcohols) may greatly influence this oxidative pathway and whereas 2-methyl alcohols undergo extensive oxidation, this is not the case with 2-ethyl derivatives. This was demonstrated experimentally by Kamil et al.[2,3] and these effects were discussed by Williams.[4]

A fundamental difference exists between primary and secondary alcohols in that oxidation of the latter leads to the formation of ketones rather than aldehydes and carboxylic acids. Further oxidation of the carbonyl group is therefore prevented, however the aliphatic moiety of this ketone may undergo oxidation. This is noted below with 2-hexanol. The latter reaction will increase the polarity of the molecule and also provide a site for conjugation. The dehydrogenation of secondary alcohols proceeds much more slowly than with primary alcohols and many substrates including secondary aromatic alcohols are not metabolized by this route.

Another important group of substrates of alcohol dehydrogenase includes benzyl alcohol and its derivatives. As noted later in this chapter, several plant alcohols of this type including saligenin, vanillyl alcohol, and benzyl alcohol itself are extensively oxidized in animals.

Many soluble oxidoreductases in addition to alcohol dehydrogenase are present in mammals; however, they often have narrow substrate specificities and are certainly more important in the metabolism of normal rather than xenobiotic compounds. It is also noteworthy that oxidoreductases localized in the microsomal rather than the soluble fraction of liver cells have been reported. These seem to be important in the oxidation of some cyclic and secondary aromatic alcohols. Additionally, ethanol undergoes metabolism by a microsomal ethanol oxidizing system. This subject was recently reviewed by Lieber.[5] Details of the mammalian oxidation of alcohols by the various enzyme systems and especially of the distribution and properties of alcohol dehydrogenase are available in the reviews of McMahon[6,7] and Bosron and Li.[8] Liver alcohol dehydrogenase exhibits a high degree of polymorphism[9] and the structure and function of the isoenzymes have received considerable attention in recent years. Jörnvall et al.[10] reviewed these newer developments. Other reports have dealt with the metabolism of some aliphatic alcohols[11] and of a variety of aliphatic and aromatic alcohols[1] by isoenzymes of human liver alcohol dehydrogenase.

The second major metabolic pathway for alcohols is direct conjugation of the hydroxy group with glucuronic acid. It is noteworthy that the formation of sulfate esters, as seen with phenolic hydroxy groups, does not occur with aliphatic alcohols. This lack of sulfate conjugation was demonstrated in rabbits by Kamil et al.[2] with several pentanols. This investigation also determined the extent of glucuronide formation in rabbits of a large number of alcohols, including many commonly found in plants. The data on the latter compounds are summarized in Table 1 which divides the compounds into normal alcohols, other primary alcohols, and secondary alcohols. These findings are in general agreement with the early qualitative results of Neubauer[12] who administered many of these alcohols to rabbits and dogs. It is evident that conjugation is a general reaction of these compounds, however this pathway is of minor importance with the lower homologues. Kamil et al.[13] subsequently reported the isolation of small amounts of both methyl and ethyl glucuronides from the urine of rabbits given the corresponding alcohols. Conjugation of the normal alcohols is low (10% or less of the dose) but branching increases it considerably. This is seen with the values given by 2-heptanol. Several other closely related 2- and 3-ols not listed in Table 1 gave conjugation values of between 50 and 70%. 2-Ethyl-1-hexanol is notable for its extensive conjugation; however, the urine samples contained a reducing glucuronide which was shown by Kamil et al.[3] to be that of 2-ethylhexanoic acid rather than that of the administered alcohol.

In addition to the main pathways noted above, alcohols may also be metabolized by ancillary routes. Prominent among these is the oxidation at a carbon atom of aliphatic, alicyclic, or aromatic groups. Also, several alcohols have been shown to undergo conjugation with fatty acids in the body. The esters formed may be retained in the tissues for prolonged periods. The literature on this subject was reviewed by Chang et al.,[14] however the possible significance of this metabolic pathway for plant alcohols is not known. Finally, some aromatic alcohols (benzyl alcohols) undergo reduction to the corresponding toluene derivatives by the gastrointestinal microflora.[15]

In addition to these metabolic pathways, alcohols may also be lost unchanged from the body in the expired air. This occurs with volatile alcohols and, more importantly, with volatile ketones formed from several secondary alcohols. Kamil et al.[2] found small amounts of the ketones derived from 2-butanol and 2-heptanol in the expired air of rabbits given the alcohols. This point was studied in detail by Haggard et al.[16] with seven of the eight isomers of pentanol. When rats were given 1-pentanol (1 g/kg, i.p.) ~0.9% of the dose was recovered unchanged in the expired air. The corresponding value with 2-pentanol was 5.4%, however an additional 38% of the dose was lost by this route as methyl *n*-propyl ketone.

ALIPHATIC ALCOHOLS

A multitude of aliphatic alcohols have been isolated from plants, however these commonly occur in esterified forms and the amounts of free alcohols present may be quite small. Chemically these compounds span a wide range and include alcohols from C_1 to C_{34}. Both odd- and even-numbered examples are found and fully saturated derivatives are most common. Primary and secondary alcohols are encountered and, among the latter, the hydroxy group is usually located at C-2.

The books of Williams[17] and Browning[18] cover the metabolism of numerous alcohols including many of plant origin and some of the following data were taken from these sources. It is felt that no useful purpose will be served by reviewing the voluminous literature on the metabolism of methanol and ethanol. These compounds have other areas of association by virtue of their toxic properties and the dangers involved in their social abuse. Also, reviews of the metabolism of these alcohols are available. Cornish[19] and Tephly et al.[20] summarized the metabolism of methanol. Useful reviews of the metabolism of ethanol include those by Lieber,[21] Crow,[22] and Crabb et al.[23] The book edited by Majchrowicz and Noble[24] includes several chapters on various aspects of ethanol metabolism. Further information, especially regarding the

TABLE 1

Conjugation of Some Plant Alcohols with Glucuronic Acid in Rabbits[2]

Compound	Synonyms	Structure	Dose (mg/kg)	Conjugation (% of dose[a])
Normal alcohols				
Methanol		H_3C-OH	800	0[b]
Ethanol		H_3C-CH_2OH	765	0.5
1-Propanol		$H_3C-CH_2-CH_2OH$	800	0.9
1-Butanol		$H_3C-(CH_2)_2-CH_2OH$	400	1.8
1-Pentanol	*n*-Amyl alcohol	$H_3C-(CH_2)_3-CH_2OH$	735	7
1-Hexanol		$H_3C-(CH_2)_4-CH_2OH$	850	10
1-Heptanol		$H_3C-(CH_2)_5-CH_2OH$	965	5
1-Octanol	Caprylic alcohol	$H_3C-(CH_2)_6-CH_2OH$	1080	10
1-Nonanol		$H_3C-(CH_2)_7-CH_2OH$	1200	4
1-Decanol		$H_3C-(CH_2)_8-CH_2OH$	1320	4[c]
1-Octadecanol	Stearyl alcohol	$H_3C-(CH_2)_{16}-CH_2OH$	2250	5—10[c]
Other primary alcohols				
2-Methyl-1-propanol	Isobutyl alcohol	$(H_3C)_2-CH-CH_2OH$	615	4
2-Methyl-1-butanol		$H_3C-CH_2-CH(CH_3)-CH_2OH$	735	10
3-Methyl-1-butanol	Isoamyl alcohol	$(H_3C)_2-CH-CH_2-CH_2OH$	735	9
2-Ethyl-1-hexanol		$H_3C-(CH_2)_3-CH(-CH_2CH_3)-CH_2OH$	1080	87[b,d]
Secondary alcohols				
2-Butanol	*sec.*-Butyl alcohol	$H_3C-CH_2-CH(OH)-CH_3$	615	14
2-Pentanol	*sec.*-Amyl alcohol	$H_3C-(CH_2)_2-CH(OH)-CH_3$	735	45
2-Heptanol		$H_3C-(CH_2)_4-CH(OH)-CH_3$	965	55
2-Octanol		$H_3C-(CH_2)_5-CH(OH)-CH_3$	1080	16

[a] Usually average values from three animals.
[b] See text.
[c] Incomplete absorption, some alcohol recovered in feces.
[d] Ester glucuronide of corresponding carboxylic acid.

roles of alcohol dehydrogenase, the microsomal ethanol-oxidizing system, and catalase, is found in the articles by Mezey,[25] Teschke et al.,[26] and Lieber.[5]

Among the lower normal alcohols from plants, both **1-propanol** and **1-butanol** are readily oxidized, little being conjugated with glucuronic acid (Table 1). Both alcohols are oxidized by liver alcohol dehydrogenase and, in fact, maximum activity among primary alcohols was reported with 1-butanol.[6] Gaillard et al.[27] recorded that very small amounts (~0.1% or less) of these alcohols were excreted unchanged in the urine of rats given large doses (2000 mg/kg, p.o.). DiVincenzo and Hamilton[28] studied the metabolism of n-[1-^{14}C]butanol in rats. They used oral doses of 4.5, 45, or 450 mg/kg, however the results obtained were similar over this 100-fold range. The major route of loss of radioactivity was as expired CO_2 which accounted for ~75 to 85% of the dose in 24 h. About 0.5% or less was found as unchanged alcohol in the expired air. Excretion of radioactivity in the urine was low (~3 to 5%) and this material consisted of urea (25%) and conjugates of 1-butanol (75%). The loss of radioactivity in the feces was low (~1%) and 12 to 16% was recovered in the carcass after 24 h.

1-Pentanol (n-amyl alcohol) was noted above to be excreted unchanged in small amounts in the expired air in rats. It is conjugated to a small extent with glucuronic acid in rabbits (Table 1) and undergoes rapid metabolism in rats following its injection.[16] Following a dose of 1000 mg/kg, the alcohol disappeared from the blood within 3.5 h. Gaillard and Derache,[29] using oral doses of 2000 mg/kg, found very low blood levels of 1-pentanol in rats. Furthermore, no unchanged alcohol was excreted in the urine at this dose level.[27] Hinson and Neal[30] reported a kinetic study of the oxidation of **1-octanol** by horse liver alcohol dehydrogenase.

Of the higher normal alcohols, **1-hexadecanol** (cetyl alcohol, $H_3C-(CH_2)_{14}-CH_2OH$) was reported not to form a glucuronide conjugate in rabbits.[12] It is oxidized to the corresponding fatty acid, palmitic acid, in rats.[31] However, the data summarized in Table 1 indicate that glucuronide formation occurs to a relatively small extent in rabbits with both lower and higher homologues of cetyl alcohol and the earlier negative report should perhaps be only tentatively accepted. In fact, McIsaac and Williams[32] reported that 6% of a large dose (~2000 mg/kg, p.o.) was excreted conjugated with glucuronic acid by rats. A further 20% of the dose was excreted unchanged in the feces within 48 h. Baxter et al.[33] administered [1-^{14}C]cetyl alcohol intragastrically to rats using a low dose (0.67 mg/kg) and a high dose (1000 mg/kg). After 24 h about two-thirds of the radioactivity at both dose levels was recovered in the feces and intestinal contents. The remainder was absorbed and ~75% of this material was found in the lymph. Very little radioactivity was excreted as expired CO_2 (1 to 1.5%) and none was found in the urine. Much of the cetyl alcohol was oxidized (presumably to palmitic acid) during absorption. Stetten and Schoenheimer[31] found that 1-octadecanol (stearyl alcohol) was metabolized to the corresponding fatty acid. It was noted that some of the C_{16}-alcohol was converted to the C_{18}-acid and some of the C_{18}-alcohol to the C_{16}-acid. The extensive excretion of 1-octadecanol and its metabolites in the lymph of rats was reported by Sieber et al.[34] Thoracic duct lymph was collected for 24 h following an intraduodenal dose of 1.35 mg/kg of the ^{14}C-labeled alcohol. More than 55% of the absorbed radioactivity was recovered in the lymph and most (50 to 70%) of this was recovered as triglycerides. In addition, small amounts of phospholipids, cholesterol esters, and unchanged 1-octadecanol were also found.

Two common primary plant alcohols with branched chains are **isobutyl alcohol** (2-methyl-1-propanol) and **isoamyl alcohol** (3-methyl-1-butanol). Both are characterized by their rapid oxidation, however small amounts are conjugated with glucuronic acid (Table 1). Their rapid metabolism was indicated by the findings of Gaillard and Derache[29] who measured low blood levels of these alcohols following large doses (2000 mg/kg, p.o.) to rats. Little or no unchanged alcohol was excreted by rats, even at this high dose level. Gaillard and Derache[27] recorded values of ~0.3 and 0% for isobutyl alcohol and isoamyl alcohol, respectively. Williams[35] summarized the main points in the metabolism of isobutyl alcohol. These involve oxidation to the corresponding aldehyde and then the carboxylic acid derivative, the latter being decarboxylated to acetone and CO_2. The investigation of Haggard et al.[16] showed that rats given isoamyl alcohol

(1 g/kg, i.p.) excreted little (1%) unchanged compound in the expired air and even less in the urine. However, the blood levels declined to an undetectable amount within 5 h. The rapid metabolism of isoamyl alcohol was also reported by Greenberg[36] who detected it in the blood of rats 2 h after dosing (15 mg/kg, i.p.) but not thereafter.

As noted above, the secondary alcohols are distinguished by higher degrees of conjugation with glucuronic acid (see Table 1) and by oxidation to ketones rather than to aldehydes and then carboxylic acids. The ketones formed are fairly volatile and may be lost in the expired air, as previously noted with **2-butanol**, **2-pentanol**, and **2-heptanol**. The investigation of Haggard et al.[16] also showed that 2-pentanol, and especially its ketone metabolite, were removed from the blood at a slower rate than were the primary pentanols.

Experiments with 2-butanol and its nonplant homologue 2-hexanol have revealed a new metabolic route for these secondary alcohols. DiVincenzo et al.[37] determined the metabolites present in the serum of guinea pigs given 2-hexanol (450 mg/kg, i.p.). In addition to the expected 2-hexanone, several metabolites formed by ω–1 oxidation of the alkyl group were detected. These included 2,5-hexanediol, 2,5-hexanedione and 5-hydroxy-2-hexanone. Dietz et al.[38] calculated that ~96% of an oral dose (1776 mg/kg) of 2-butanol was oxidized in rats to butan-2-one. Two metabolites formed by ω–1 oxidation, 3-hydroxy-2-butanone and 2,3-butanediol, were identified in the blood. They measured the blood concentrations of the four compounds and found that the administered alcohol disappeared first, followed by 2-hexanone and its 3-hydroxy derivative. The most persistent metabolite was the diol which was still detectable 30 h after dosing.

Sugar alcohols are polyhydric alcohols in which the aldehyde group of sugars is replaced by an alcohol group. They are of fairly common occurrence in plants and show many of the properties of the corresponding sugars. This close association includes their metabolism and it must be stressed that sugar-sugar alcohol interconversions occur in the pathways of mammalian carbohydrate metabolism. Thus, reversible pyridine nucleotide-linked dehydrogenations are involved in various sugar isomerizations which involve the polyols as intermediates. This is seen with D-sorbitol, an intermediate in the glucopyranose-fructopyranose isomerization which is carried out by an NADP-specific D-hexitol dehydrogenase present in many mammalian tissues. It is therefore clear that the metabolism of polyhydric alcohols is to a large extent closely associated with normal carbohydrate metabolism and that much of the subject falls outside the scope of this book. Indeed, much of the interest in the metabolism of the polyols has centered on their caloric values and conversion to glycogen, factors of importance in the use of polyols as sweetening agents in dietary and diabetic foods. Carr and Krantz[39] reviewed the subject of glycogen formation by polyhydric alcohols. The metabolism of several polyols was summarized briefly by Williams[40] and more extensively by Touster and Shaw[41] and by Bässler.[42]

Galactitol (dulcitol) (1) was shown earlier to be partly converted to glycogen in rats but more recent work[43] indicated that this occurs to a very limited extent. Using compounds labeled with ^{14}C in the 1-position, the rate of $^{14}CO_2$ formation from galactitol by liver preparations was found to be only 3% of that observed with sorbitol. When the polyol was administered to rats or humans by injection nearly all was excreted unchanged in the urine and little or no $^{14}CO_2$ was detected in the expired air.

```
   CH2OH           CH2OH           CH2OH
    |               |               |
   CHOH            HOCH            CHOH
    |               |               |
  HOCH            HOCH            HOCH
    |               |               |
  HOCH             CHOH            CHOH
    |               |               |
   CHOH            CHOH            CHOH
    |               |               |
   CH2OH           CH2OH           CH2OH

    (1)             (2)             (3)

  Galactitol       Mannitol       D-Sorbitol
```

The metabolism of **D-mannitol** (2) was conveniently summarized by Olmsted[44] who noted that it was only slowly absorbed from the intestine and that 80% or more of this absorbed material was excreted unchanged in the urine in humans. Extensive urinary excretion of unchanged compound is also seen in other species (rats, rabbits, dogs, and monkeys), however it is converted to liver glycogen to a limited extent. Nasrallah and Iber[45] studied the absorption and metabolism in man of mannitol uniformly labeled with ^{14}C. Following oral doses of 28 to 100 g, ~18 and 32% of the radioactivity were found in the urine and feces, respectively, in 48 h. The urinary radioactivity consisted entirely of unchanged compound. Some mannitol was metabolized in these experiments and as much as 18% of the dose was recovered in 12 h as respiratory CO_2.

D-Sorbitol (glucitol) (3) was noted above to be converted in mammalian tissues to both glucose and fructose and it is therefore understandable that this polyol enters more extensively into the pathways of carbohydrate metabolism than is the case with galactitol and mannitol. The results of Ertel et al.[46] indicated that an oral dose (100 mg) of sorbitol was completely absorbed by rats. Contrariwise, Schell-Dompert and Siebert[47] reported that it was incompletely absorbed in rats and man with the result that it underwent fermentation by intestinal bacteria. The bacterial products undergo absorption and utilization by the host. They also showed that humans exhale considerable quantities of hydrogen following oral sorbitol administration. Some sorbitol may be excreted unchanged in the urine. Strack et al.[48] reported a value of 12 to 18% of the dose (0.25 to 0.75 g/kg/h) in rabbits. They also found that blood fructose levels were increased whereas little effect on blood glucose was noted. The results of Maeda,[49] also obtained using rabbits, suggested that D-sorbitol is initially metabolized slowly to fructose and then rapidly to glycogen, glucose, and CO_2. The delayed nature of the conversion of D-sorbitol to glucose and glycogen was also underlined by Olmsted.[44] Adcock and Gray[50] showed that a small amount (<3%) of the oral dose (35 g) of uniformly labeled [^{14}C]sorbitol was excreted unchanged in the urine. The finding that at least 75% was metabolized to $^{14}CO_2$ confirmed an earlier report which stated that it was abundantly recovered as expired CO_2.[51] The oxidation of sorbitol to a similar extent in rats given the polyol (p.o. or i.p.) was reported by Wick et al.[52] The conversion of sorbitol to mannitol (2) in humans was reported by Soummer-Dauphant,[53] who suggested that the latter was formed via fructose produced from the sorbitol not broken down by glycolysis. A study of the blood and liver levels of sorbitol metabolites including C-3- and C-6-phosphorylated intermediates in rats was carried out by Heinz and Wittneben.[54]

Of the naturally occurring heptitols, only **D-volemitol** (sedoheptitol, D-β-mannoheptitol) (4) appears to have been studied. Hiatt et al.[55] reported that it, unlike mannitol, was not capable of serving as a glycogen precursor in rat liver.

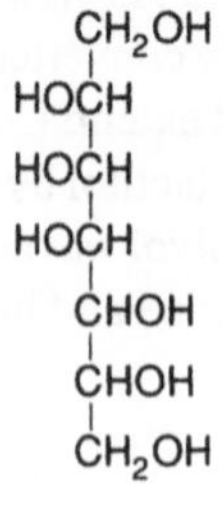

CH$_2$OH
|
HOCH
|
HOCH
|
HOCH
|
CHOH
|
CHOH
|
CH$_2$OH

(4)

D-Volemitol

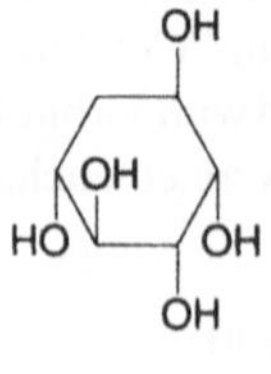

(5)

Quercitol

A closely related group of compounds is that of the carbocyclic polyols or cyclitols. Williams[56] referred to 19th century experiments which reported that **D-quercitol** (1,2,3,4,5-cyclohexanepentol, acorn sugar) (5) was not metabolized in the body but was largely excreted unchanged. The C_6-cyclitols are the inositols, of which myo-inositol (*meso*-inositol) is the best known. It is widely distributed in plants and animals, having physiologic functions, and a discussion of its metabolism therefore falls outside the scope of this book.

MONOTERPENOID ALCOHOLS

ACYCLIC TERPENE ALCOHOLS

The first metabolic studies in this area were carried out at the beginning of the present century when Hildebrandt[57] and Neubauer[12] administered **linalool** (3,7–dimethyl-1,6–octadien–3-ol) (6) and geraniol (*trans*-3,7–dimethyl–2,6-octadien–1-ol) (11), respectively, to rabbits. In the former study it was noted that treatment with acid resulted in the isomerization of linalool to geraniol. This conversion did not occur metabolically and linalool was reported to be excreted as its glucuronide conjugate. Rahman[58] studied the absorption, distribution, and excretion in rats of linalool labeled with ^{14}C in positions 1 and 2. Some of these results were described by Parke et al.[59] and the excretion of linalool as a glucuronide conjugate was confirmed. They noted that the urinary, fecal, and biliary metabolites were mainly polar, ether-insoluble, conjugates. Rahman also investigated the metabolism of linalool by the intestinal microflora. The rat cecal microflora reduced linalool to 1,2-dihydrolinalool (7) and to tetrahydrolinalool (8) and the latter metabolite was also formed by the fecal microflora in man. In contrast, mouse cecal and sheep rumen microflora formed only metabolite (7) and little reduction was observed with guinea pig cecal microflora. Additionally, α-terpineol (30) was identified in the incubates using sheep and bovine rumen microflora or human fecal microflora. In addition to linalool itself, both dihydro- and tetrahydrolinalool were identified as urinary metabolites of linalool (500 mg/kg, intragastrically) in rats. They were mainly excreted in conjugated form and, together with ~10% of the dose recovered as ^{14}C-urea, accounted for nearly 60% of the radioactivity. Fecal radioactivity (~15%) consisted of linalool and its reduced metabolites and a further 25% of the dose was recovered as respiratory [14]CO_2. Fecal excretion of radioactivity was delayed and this was probably due to biliary excretion and enterohepatic circulation of the metabolites. More than 25% of the dose (60 mg/kg, i.p.) was excreted in the bile in 6 to 11 h and much of this was due to polar conjugates of linalool. Chadha and Madyastha[60] gave large oral doses (800 mg/kg) of linalool to rats for 20 d and isolated two metabolites which arose from allylic hydroxylation at C-8. These were the diol (9) and the corresponding carboxylic acid (10). Rat lung microsomes are able to carry out the hydroxylation of linalool at C-8.[61]

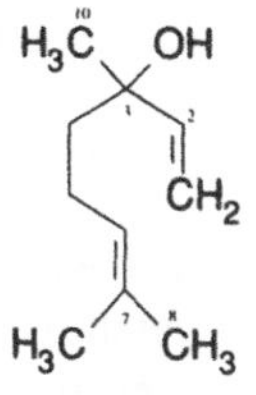

(6)
Linalool

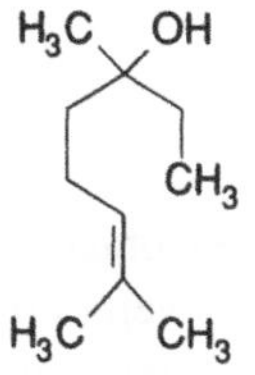

(7)

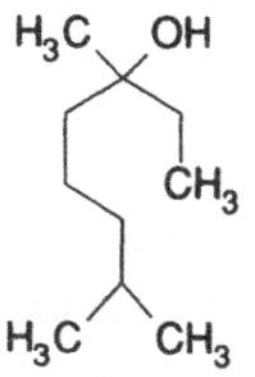

(8)

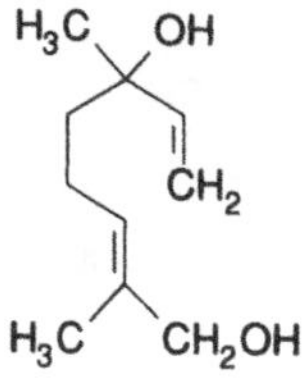

(9)

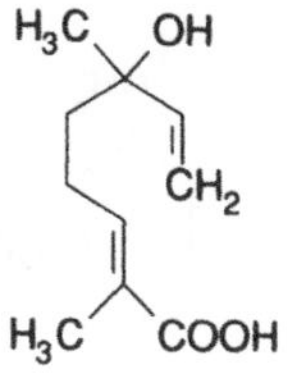

(10)

Neubauer[12] did not detect a urinary glucuronide of **geraniol** (11) in rabbits and believed that it might undergo cyclization. However, Hildebrandt[62] subsequently showed that geraniol was extensively oxidized in rabbits to a dicarboxylic acid derivative. **Nerol** (12), the *cis* isomer of geraniol, was excreted conjugated with glucuronic acid. Hildebrandt believed that the carboxy groups were formed by oxidation of the alcohol group at C-1 and the methyl group at C-3. However, several subsequent studies with geraniol showed that the allylic oxidation instead occurs at a C-7 methyl group. This produces metabolite (13), known as Hildebrandt acid.[63--65] These three investigations also showed that the optically active metabolite (14), termed reduced Hildebrandt acid, was also excreted. Approximately 25 to 30% of the oral dose (17 to 55 g given in daily 2- to 10-g portions) was isolated as these two dicarboxylic acids in a ratio of ~2:1 between metabolites (14) and (13). Following i.p. dosage this ratio was reversed and the total recovery was reduced to ~15%. Licht and Jamroz[66] and Licht and Coscia[67] showed that the initial step in the metabolism of geraniol and nerol, allylic hydroxylation at C-8 to give the alcohol intermediates (e.g., (15)), was carried out with a monooxygenase system from rabbit liver microsomes. Hydroxylation of geraniol and nerol at C-8 was also shown to occur with rat lung microsomes.[61] Formation of the two epoxides and the diepoxide of geraniol by monooxygenase systems from rat and rabbit liver was reported by Licht et al.[68]

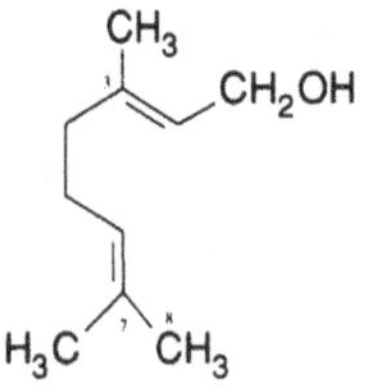

(11) Geraniol

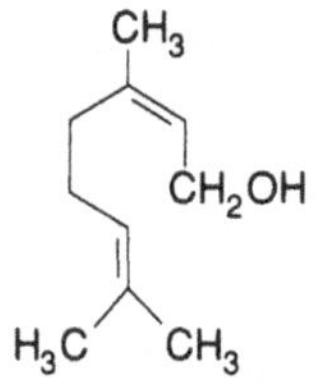

(12) Nerol

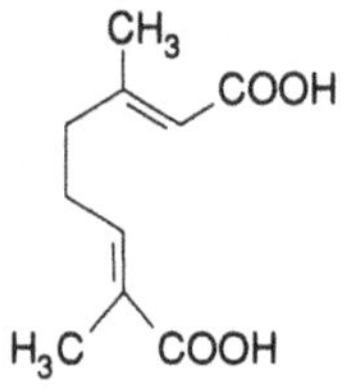

(13)

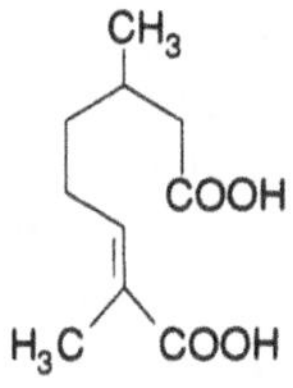

(14)

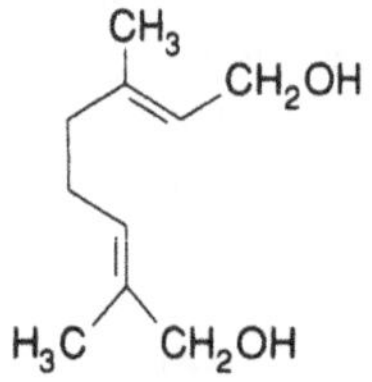

(15)

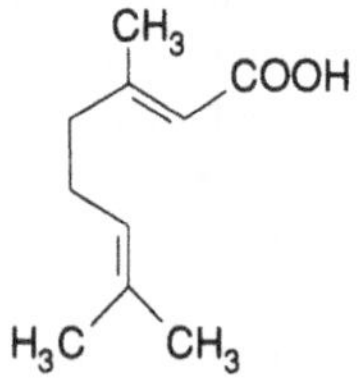

(16)

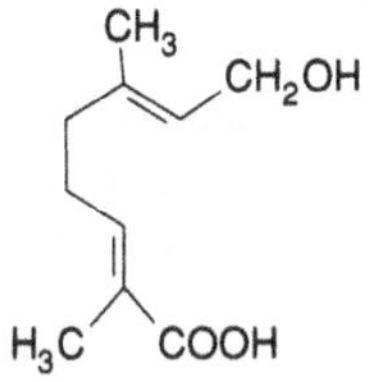

(17)

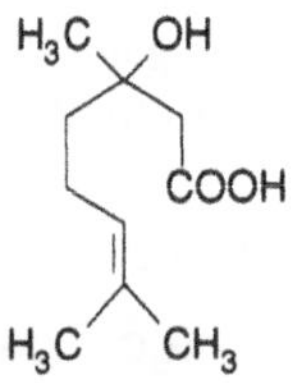

(18)

The metabolism of geraniol by rats given oral doses of 800 mg/kg for 20 d was reported by Chadha and Madyastha.[60] They detected several urinary metabolites of geraniol which were excreted conjugated or free. The most prominent metabolites were identified and, among the acidic compounds, geranic acid (16), 8-carboxygeraniol (17), 3-hydroxycitronellic acid (18), and Hildebrandt acid (13) were reported. The neutral metabolites identified were geraniol itself and 8-hydroxygeraniol (15).

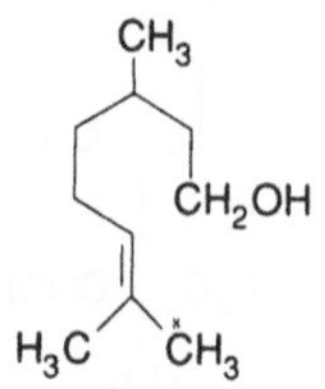

(19)

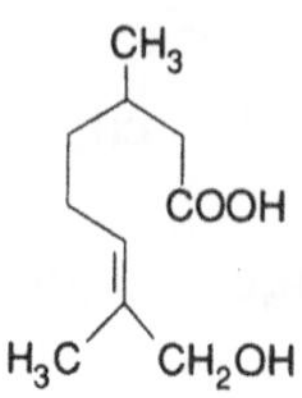

(20)

Citronellol

Fischer and Bielig[64] studied the metabolism in rabbits of **citronellol** (3,7-dimethyl-6-octen-1-ol) (19). This alcohol differs from geraniol or nerol by the lack of a double bond at C-2,3. A small amount of the total dose of 30 g was recovered in the urine unchanged, but most of the identified material consisted of compounds formed by double ω-oxidation, i.e., oxidation of the alcohol group and allylic oxidation of the methyl group at C-7. Both the intermediate alcohol, 7-hydroxymethyl-3-methyl-6-octenoic acid (20), and reduced Hildebrandt acid (14) were isolated to the extent of ~8 and 10%, respectively, of the dose. Chadha and Madyastha[61] reported that hydroxylation of citronellol at C-8 was carried out by rat lung microsomes.

The metabolism of the closely related sesquiterpenoid farnesol and the diterpenoid phytol is covered in the chapter entitled Metabolism of Higher Terpenoids.

MONOCYCLIC TERPENE ALCOHOLS

The initial studies on the metabolism of monocyclic terpene alcohols dealt primarily with their conjugation with glucuronic acid. This subject was first approached over 75 years ago by Hämäläinen as noted below, however the most detailed information was obtained with the menthols, especially *l*-menthol.

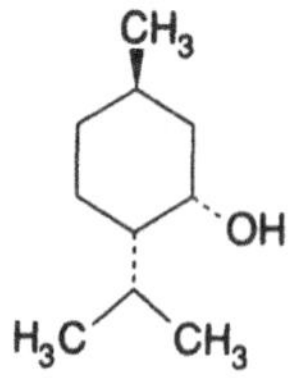

(21)

d-Neomenthol

The menthols (3-*p*-menthanols) contain three asymmetric carbon atoms and eight isomers are therefore possible. These are the *d* and *l* forms of menthol, isomenthol, neomenthol, and neoisomenthol, however only ***d*-neomenthol** (21) and ***l*-menthol** (22) (see Figure 1) occur naturally. Quick[69] administered *l*-menthol to rabbits and later to dogs and humans[70] and determined the extent of its conjugation. In the first case, oral doses of ~0.25 g/kg resulted in 48% being excreted in the urine as menthol glucuronide. The values decreased only slightly to 41 to 46% (0.5 g/kg) and 42 to 48% (1.0 g/kg) and then more rapidly to slightly under 20% as lethal doses of 2 to 2.5 g/kg were approached. Excretion of the conjugate was rapid and ~90% of the conjugated material was recovered in 6 h following a dose of 1 g/kg. Interestingly, dogs fed menthol (5 g) excreted only ~5% of the dose as the glucuronide. In humans fed 1 g, which corresponds to a dose of only ~15 mg/kg, the value was much higher (79%). Bell et al.,[71] using a dose of 0.5 g, reported values of ~41% in 4 h and 50% in 24 h for the urinary excretion of *l*-menthol glucuronide in humans. No unchanged menthol was found in the urine. Williams[72] studied the conjugation of menthols in detail and, in the case of *l*-menthol, found that rabbits fed 1 g/kg excreted an average of 48% of the dose as menthol glucuronide, thus confirming the finding of Quick. None of these experiments provided any information on the fate of the unconjugated menthol.

The fundamental point of interest in the studies of Williams on the glucuronic acid conjugation of the menthols was that of the effects of optical and geometrical isomerism on this metabolic pathway. When identical experiments to that noted above with *l*-menthol were carried out with *dl*-menthol, 59% of the dose was excreted conjugated with glucuronic acid.[72] From these values it can be calculated that the corresponding value for *d*-menthol must be 70%. Similarly, administration of *dl*-isomenthol and *d*-isomenthol gave values of 55% and 65%, respectively, which indicate that the *l*-form of this geometrical isomer must undergo conjugation only to the extent of 45% of the dose. In both cases the *d*-isomer was more extensively conjugated and this fact was utilized to effect the resolution of *dl*-menthol[73] and *dl*-isomenthol.[74] A subsequent study using the naturally occurring *d*-neomenthol (21) showed that rabbits fed 1 g/

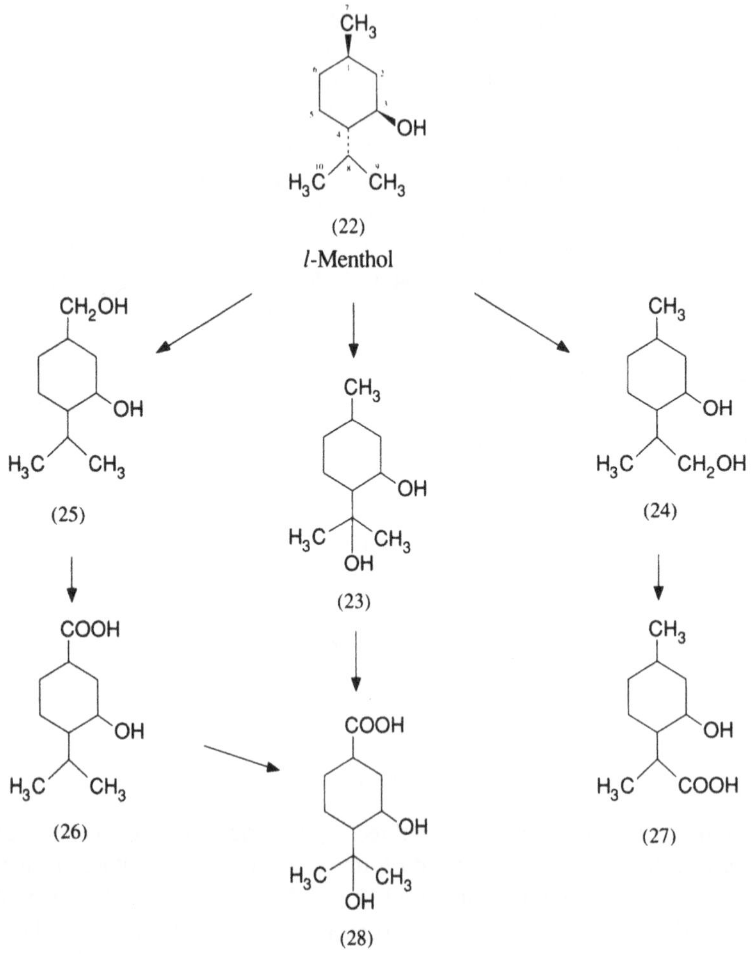

FIGURE 1.   Metabolic pathways of *l*-menthol in rats.

kg excreted 67 to 68% of the dose combined with glucuronic acid.[75] Thus, the extent of conjugation of the *d*-forms of menthol, isomenthol, or neomenthol in rabbits is nearly identical. Wright[76] reported that sheep are similar to rabbits in their ability to conjugate *l*-menthol with glucuronic acid.

However, the question of the fate of the remainder of the administered *l*-menthol was not approached and this point remains a matter of speculation. Williams[77] suggested that ring fission and degradation may occur, however it seems reasonable to assume that oxidation of the methyl and/or isopropyl groups may be involved in the metabolism of the unaccounted portion of the dose. This possibility was studied using gas chromatographic-mass spectrometric techniques.[78] Rats were given *dl*-menthol (400 mg/kg, p.o.) and the 24-h urines hydrolyzed using a β-glucuronidase + sulfatase preparation followed by separation into neutral and acidic fractions. In the neutral fraction a mixture of hydroxylated menthol derivatives was detected which appeared to account for nearly twice as much material as that due to menthol itself. Most of this consisted of monohydroxylated derivatives of menthol, of which three were recognized; however, evidence was also obtained for the presence of small amounts of a di- and a trihydroxylated menthol derivative. The positions of the hydroxy groups in these metabolites were not determined but the finding that the two major metabolites in the acidic fraction corresponded to monocarboxylated derivatives of menthol clearly indicates that both the methyl group and the isopropyl group underwent oxidation. The third monohydroxylated derivative was no doubt formed from oxidation of the isopropyl group to give a tertiary carbinol which is, of course, unable to be oxidized further. Two recent studies of the metabolism of *l*-menthol in rats confirmed the formation of these diols. Madyastha and Srivatsan[79] identified, in addition to unchanged compound, *p*-menthane-3,8-diol (23) and *p*-menthane-3,9-diol (24) as urinary metabolites of *l*-menthol. Yamaguchi and Caldwell[80] similarly found these two metabolites and *p*-menthane-3,7-diol (25). The latter study also identified the two monocarboxylated derivatives detected earlier, 3-hydroxy-*p*-menthane-7-carboxylic acid (26) and 3-hydroxy-*p*-menthane-9-carboxylic acid (27). A third acidic metabolite, identified in both studies, was 3,8-dihydroxy-*p*-menthane-7-carboxylic acid (28). The oxidized metabolites were excreted in the urine as such and as glucuronide conjugates. The metabolic pathways of *l*-menthol in the rat are shown in Figure 1.

Quantitative assessment of the excretion of *l*-menthol and its metabolites in rats was obtained by using 3-^{3}H-labeled compound (500 mg/kg, p.o.).[80] The excretion of radioactivity in 48 h was 38% (urine) and 34% (feces). The high value for fecal excretion appears to be due to pronounced enterohepatic circulation of *l*-menthol and bile duct-cannulated rats excreted two-thirds of the dose in the bile in 24 h. The main biliary metabolite was *l*-menthol glucuronide. Rat liver microsomes readily converted *l*-menthol to *p*-menthane-3,8-diol (23).[79]

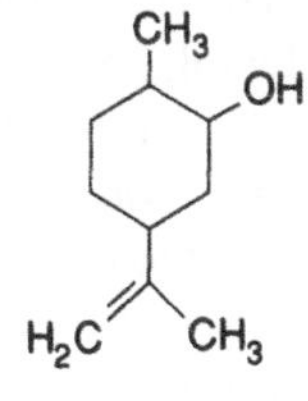

(29)
Dihydrocarveol

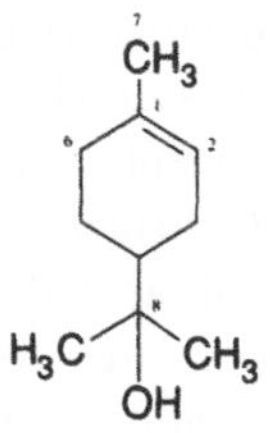

(30)
α-Terpineol

Of the other monocyclic terpene alcohols **dihydrocarveol** (*p*-8(9)-menthen-2-ol) (29) was one of the many terpenoids studied by Hämäläinen[81] with regard to their ability to form conjugates with glucuronic acid when given to rabbits. Not unexpectedly, this reaction was found to occur with dihydrocarveol. In a similar study, Hämäläinen[82] reported that the isomeric menthenol α-**terpineol** (*p*-1-menthen-8-ol) (30) formed a glucuronide conjugate in rabbits. Its metabolism in sheep was studied by Wright[76] who found that terpin (31) could be isolated from the acid-hydrolyzed urines. α-Terpineol labeled with ^{3}H was given to guinea pigs orally, intravenously, and by inhalation.[83] They noted rapid absorption and elimination of the radioactivity and claimed that α-terpineol was excreted mainly in the urine in unchanged form. Horning et al.[84] reported that the major neutral urinary metabolite of α-terpineol in man is *p*-menthan-1,2,8-triol (32). This metabolite was also excreted in rats and it was suggested that formation of the 1,2-diol moiety occurred via an epoxide intermediate which, however, was not detected in any of the urine extracts. Madyastha and Srivatsan[85] confirmed the formation of the triol (32) in rats given α-terpineol and showed that it was excreted partly free and partly as acid-hydrolyzable conjugates. However, the major neutral metabolite in both of these fractions was α-terpineol itself. Several acidic metabolites were also detected and the major pathway involved allylic oxidation at C-7. The probable intermediate in this sequence, *p*-menthen-7,8-diol, was not detected; however, the resultant acid, oleuropeic acid (33), and its reduced derivative, dihydrooleuropeic acid (34), were identified. Additional minor neutral and acidic metabolites were detected, however it is not known if these may arise from endocyclic allylic hydroxylation as is found with other menthene derivatives (e.g., *d*-limonene). The 6-hydroxy compound which could be formed from α-terpineol in this manner is known as *trans*-sobrerol. Ventura et al.[86,87] carried out extensive metabolism studies with the latter compound in rats, dogs, and humans. These detailed investigations using modern chromatographic and spectroscopic methods revealed that *trans*-sobrerol is subjected to extensive metabolism including endocyclic and exocyclic allylic hydroxylation and oxidation of the alicyclic hydroxy group to a ketone group. The latter reaction forms a hydroxylated derivative of carvotanacetone and its metabolism in rats was investigated by Ventura et al.[88] These findings suggest that the metabolic fate of many terpenoids is far more complex than most studies have revealed.

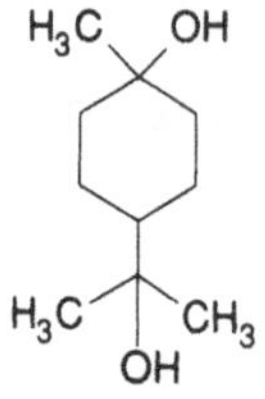

(31)   (32) 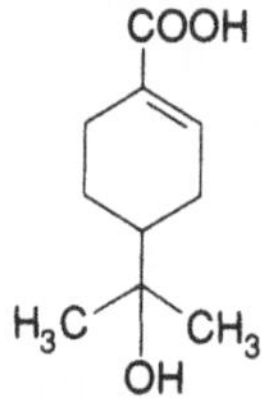   (33) 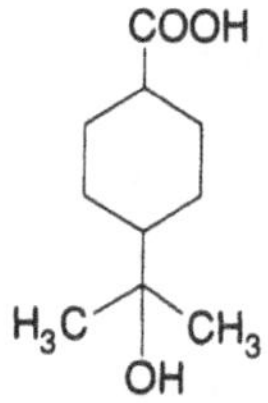   (34)

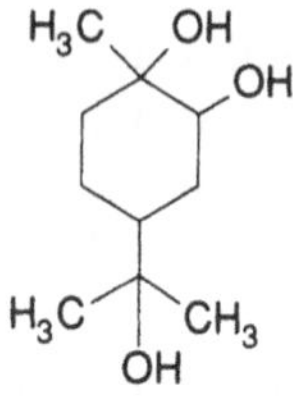

Terpin

Terpin (*p*-menthan-1,8-diol) (31), which usually occurs in the hydrated form, was noted above to be a metabolite of α-terpineol. When the diol itself was given to animals it was excreted in the urine conjugated with glucuronic acid, as shown by Hämäläinen[81] in rabbits and Wright[76] in sheep. The anhydride of terpin is **1,8-cineole** (eucalyptol; 1,8-epoxy-*p*-menthane) (35) which was also included in the experiments of Hämäläinen[89] and Wright.[76] In the former case, the compound was believed to undergo hydroxylation at the 2- or 3-position followed by conjugation of this alcohol with glucuronic acid. However, Wright found little metabolic material in the urine of sheep, even after giving large doses, and believed that 1,8-cineole may be largely oxidized in this species. More recently, several investigations have dealt with its metabolism and it is clear that numerous metabolites are formed.

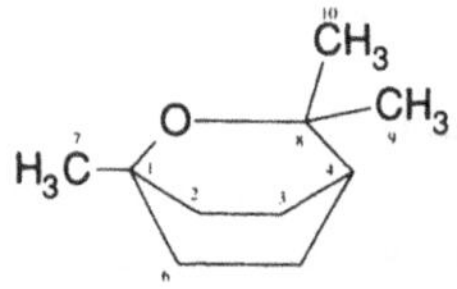

(35)   (36)

1,8-Cineole

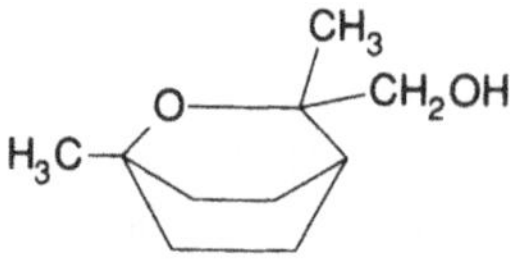

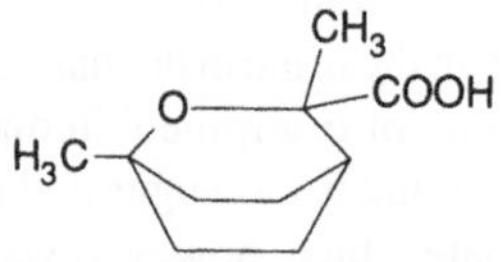

(37) 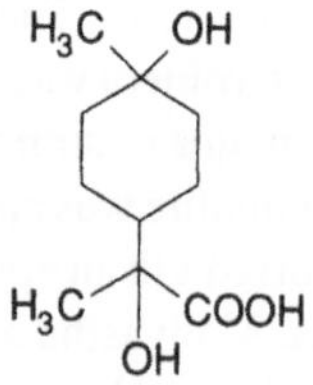   (38)

The metabolism of 1,8-cineole in brushtail possum (*Trichosurus vulpecula*) was studied by Flynn and Southwell[90] and Southwell et al.[91] About 29 g of the terpene was given in the food over a period of 8 d and the urine was found to contain 9-hydroxycineole (36) and cineol-9-oic acid (37). Frigerio et al.[92] studied the metabolism of 1,8-cineole (500 mg/kg, i.p.) in rats. They found four urinary metabolites including compounds (36) and (37). The remaining two were dihydroxy cineoles and were believed to be hydroxy derivatives of (36). The same pattern of metabolites was found in the urine samples collected daily for 3 d and all metabolites were also excreted as glucuronides. Frigerio et al. also studied the metabolism of 1,8-cineole by rat liver microsomes and detected metabolite (36) in the incubates. They also detected small amounts of a second metabolite which corresponded to demethylated cineole (lacking a methyl group at C-

8). Madyastha and Chadha[93] gave large oral doses of 1,8-cineole (800 mg/kg for 20 d) to rats and detected four neutral and four acidic metabolites in the urine. The two major neutral metabolites were shown to be 2-hydroxycineole and 3-hydroxycineole. The main acidic metabolite was shown to be compound (38). Formation of this metabolite requires the hydrolysis of the cyclic ether linkage, however the possible effect of the hydrolysis procedure employed in preparing the urinary extracts (refluxing at pH 3 to 4 for 6 h) was not determined. The remaining compounds were minor metabolites and were not identified. Miyazawa et al.[94] recently reported on the structures of the 2- and 3-hydroxycineoles which were excreted in the urine of rabbits given 1,8-cineole (200 mg/kg, p.o.). They showed that each of these hydroxylated derivatives exists as a pair of isomers with the hydroxy groups having either an *endo-* or *exo-*orientation.

Ventura and Selva[95] investigated the metabolism of a synthetic terpenoid closely related to 1,8-cineole. Epomediol is 2,6-dihydroxy-1,8-cineole and it was oxidized by rats to the corresponding 2-keto derivative and to the hydroxymethyl derivative corresponding to (36). Formation of the hydroxymethyl metabolite confirms the susceptibility of the *gem-*methyl groups of cineoles to oxidation. Additionally, epomediol itself was excreted unchanged and as a glucuronide conjugate in the urine and bile.

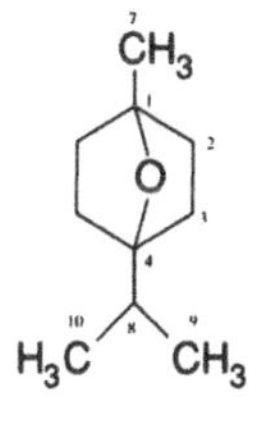

(39)   (40)   (41)

1,4-Cineole

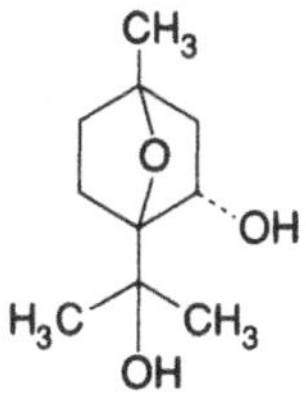

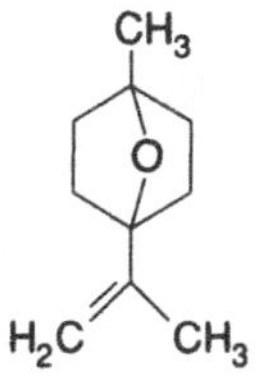

The metabolism of **1,4-cineole** (39) (2 g/animal, p.o.) in rabbits was studied by Asakawa et al.[96] The urine was hydrolyzed with a β-glucuronidase + sulfatase preparation and five metabolites were isolated and identified. These were 9-hydroxy-1,4-cineole and its oxidized derivative 1,4-cineole-9-carboxylic acid, the two diols 8,9-dihydroxy-1,4-cineole and 3,8-dihydroxy-1,4-cineole (40), and the allylic alcohol derivative 1,4-cineole-8-en-9-ol (41). 9-Hydroxy-1,4-cineole was the most abundant metabolite and no evidence was obtained for oxidation of the methyl group at C-7.

BICYCLIC TERPENE ALCOHOLS

It is noted in the preceding section on monocyclic derivatives that their conjugation with glucuronic acid is a major metabolic pathway. This is equally true of the bicyclic terpene alcohols. A further similarity between these two groups relates to the fact that a large number of both types were included in the extensive study carried out by Hämäläinen.[81] Interestingly, many of the bicyclic terpene alcohols studied by Hämäläinen have never been reinvestigated and this early report on their conjugation with glucuronic acid remains the sole source of information on their metabolic fate. These alcohols include α-**santenol** (1,7-dimethyl–2-norbornanol) (42), β-**santenol** (santene hydrate; 2,3-dimethyl–2-norbornanol) (43), **camphenilol** (44), **fenchyl alcohol** (fenchol) (45), **isofenchyl alcohol** (46), and **thujyl alcohol** (47). In all cases the alcohols, when given to rabbits, were found to be excreted as their glucuronides. However, in the case of thujyl alcohol (47) another metabolite, the glucuronide of *p*-menthane-2,4-diol (48), was also reported. An analogous reaction was reported by Hämäläinen to occur with the corresponding ketone, thujone (see the chapter entitled Metabolism of Aldehydes, Ketones, and Quinones). The terpenoid hydrocarbons β-pinene and 3-carene provide further examples of the metabolic conversion of bicyclic terpenoids to monocyclic derivatives (see the chapter entitled Metabolism of Hydrocarbons).

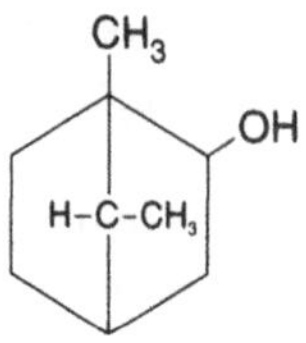

(42)

α-Santenol

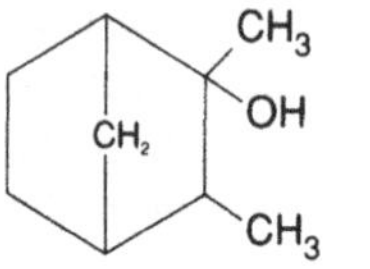

(43)

β-Santenol

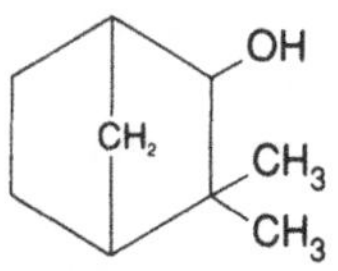

(44)

Camphenilol

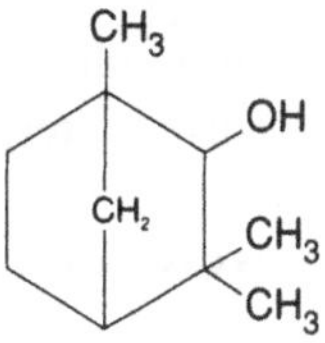

(45)

Fenchyl alcohol

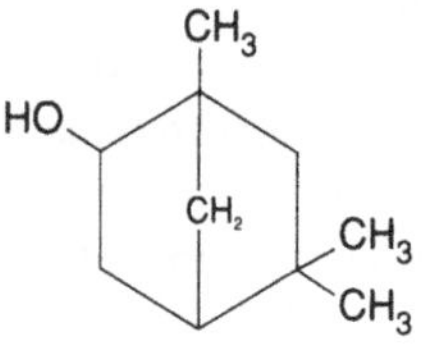

(46)

Isofenchyl alcohol

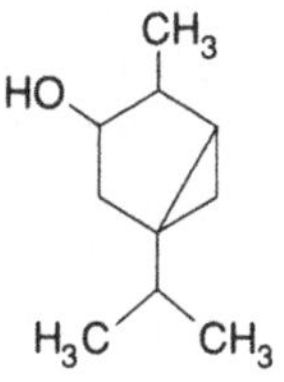

(47)

Thujyl alcohol

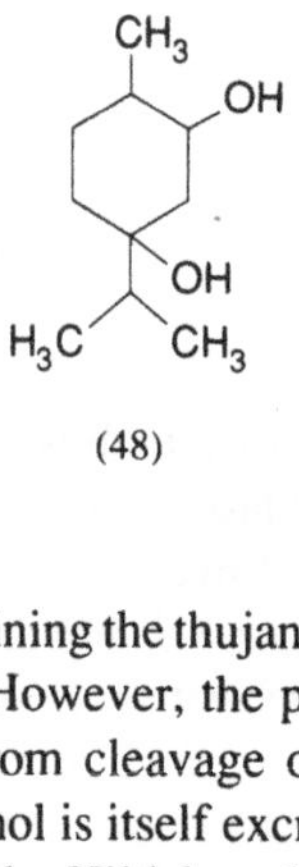

(48)

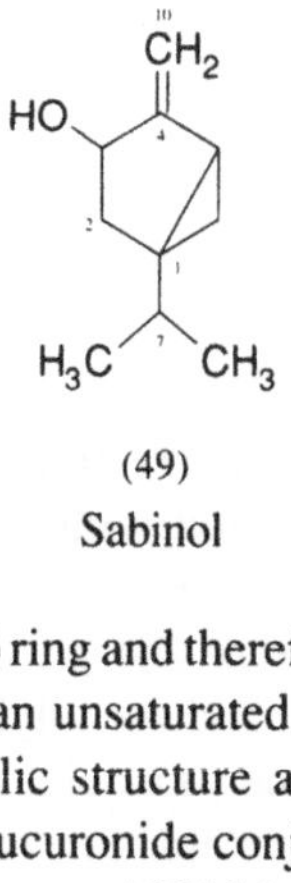

(49)

Sabinol

Another alcohol containing the thujane (sabinane) ring and therefore closely related to thujyl alcohol is **sabinol** (49). However, the presence of an unsaturated bond at C-4,10 appears to protect the compound from cleavage of the bicyclic structure as noted above with thujyl alcohol.[81,97] Instead, sabinol is itself excreted as a glucuronide conjugate, a finding which had been reported previously by Hildebrandt[57] and Fromm and Hildebrandt.[98]

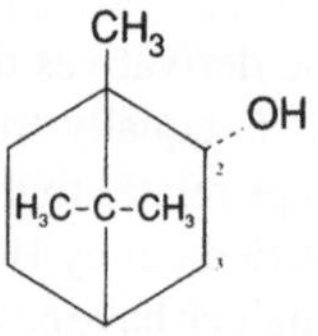

(50)

d-Borneol

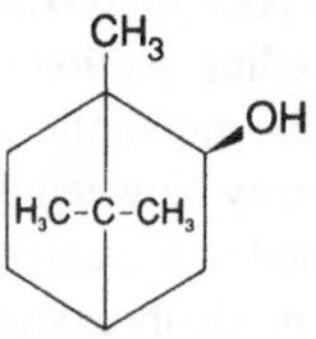

(51)

l-Borneol

The bicyclic terpene alcohol which has most often been the subject of metabolic studies is borneol (*endo*-2-hydroxycamphane). Again, the reports are older and are concerned with the ability of the alcohol to undergo glucuronide conjugation. Both **d-borneol** (Borneo camphor) (50) and **l-borneol** (51) were investigated and Magnus-Levy[99] reported that rabbits excreted equal amounts of bornyl glucuronide after administration of 1 g of *d*- or *l*-borneol. Also, dogs given large, repeated doses excreted ~75% of both forms as the glucuronide. Likewise, Hämäläinen[100] found that 22% of both *d*- and *l*-borneol was excreted in the urine conjugated with glucuronic acid when rabbits were given daily doses of 1.5 to 2 g. Nonetheless, Pryde and Williams[101] repeated these experiments in dogs using a mixture of *d*- and *l*-borneol and found evidence for the preferential conjugation of the *d*-form. They isolated pure β-*l*-bornyl-*d*-glucuronide from the urine of dogs fed pure *l*-borneol. Further quantitative data on the extent of excretion of bornyl glucuronide following the administration of *d*-borneol to dogs and man or to man was obtained by Quick[70] and by Pryde and Williams,[102] respectively. Dogs given an oral dose of 5 g excreted only ~50% of the dose as the conjugate, a value somewhat lower than that found earlier by Magnus-Levy.[99] Quick[70] obtained in man values of 69% (3.5 g, urine collected 6 h) and 81% (2.0 g, urine collected 10 h). The subsequent study of Pryde and Williams[102] confirmed the rapid and extensive urinary excretion of bornyl glucuronide in man, reporting a value of ~80% of the dose in 12 h following administration of 2 g amounts to 24 subjects.

A compound showing close structural similarity to borneol is camphane-2,3-diol which contains an additional hydroxy group at C-3. Four racemic diols are possible, the hydroxy groups at C-2 and C-3 being *endo-endo*, *exo-exo*, *endo-exo*, or *exo-endo*, i.e., two *cis* and two *trans* forms. These diols were studied in connection with their possible formation from other camphane derivatives and are therefore mentioned in the summaries of the metabolism of camphane (see the chapter entitled Metabolism of Hydrocarbons) and camphorquinone (see the chapter entitled Metabolism of Aldehydes, Ketones, and Quinones). However, Robertson and Solomon[103] administered the racemic camphane-2,3-diols to rabbits and found that they were excreted in the urine partly free and partly combined with glucuronic acid. In addition, dehydrogenation of the diols to ketols occurred.

AROMATIC ALCOHOLS

The simplest member of this rather limited class of plant alcohols is **benzyl alcohol** (52). Not unexpectedly, its metabolism has been shown to be straightforward, mainly involving oxidation to benzoic acid followed by conjugation of the latter with glycine to form hippuric acid. Snapper et al.[104] reported that this pathway accounted for 80 to 90% of the dose (1.5 g) in man and Diack and Lewis[105] recorded an average of 67% in rabbits fed a dose of 400 mg/kg of benzyl alcohol. Bray et al.[106] also fed benzyl alcohol to rabbits and found that 76% of a dose of ~500 mg/kg was excreted as ether-soluble acids (mainly hippuric but also a small amount of benzoic). The remainder of the dose was excreted as benzoyl glucuronide, a finding which suggests that glucuronidation assumes quantitative importance when the amounts of benzoic acid being formed exceed those capable of being conjugated with glycine. This was substantiated in experiments in which glycine was fed concomitantly, in which case the formation of hippuric acid increased to 91 to 98% of the dose (~250 to 800 mg/kg) and that of the glucuronide fell correspondingly. The metabolism of benzyl alcohol to hippuric acid is deficient in preterm neonates.[107] Teuchy et al.[108] reported that rats given benzyl alcohol (44 mg/rat, i.p.) excreted an average of 58% of the dose as hippuric acid. It may be reasonably assumed that the remainder is conjugated with glucuronic acid, however other pathways may be operative. One of these is mercapturic acid formation.[109] Rats given benzyl alcohol (108 mg/kg, i.p.) excreted ~0.8% of the dose as benzylmercapturic acid (*S*-benzyl-*N*-acetylcysteine).[110] This reaction sequence was also demonstrated by Chidgey et al.[111] These studies revealed that the initial step, conjugation with glutathione, took place with the sulfate ester of benzyl alcohol and not with the alcohol itself. Benzyl sulfate, which is formed from benzyl alcohol by the action of aryl sulfotransferase,[112] is a substrate for glutathione *S*-transferase.[113] Sloane[114] reported that guinea pig liver microsomes were able to convert benzyl alcohol to phenol, however the possible *in vivo* significance of this reaction is not known.

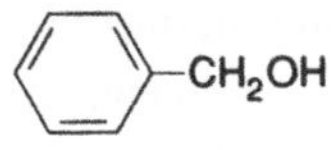

(52)	(53)
Benzyl alcohol	2-Phenylethanol

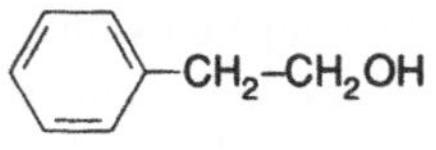

The metabolism of **2-phenylethanol** (phenethyl alcohol) (53) in rabbits was studied by Smith et al.[115] and by El Masry et al.[116] In the former investigation, an average of 7% of the dose (460 mg/kg, p.o.) was excreted combined with glucuronic acid while no increase in the ethereal sulfate output was detected. The other study indicated that the alcohol (244 mg/kg) was oxidized to phenylacetic acid which was conjugated with glycine and excreted in large amounts as phenaceturic acid. In addition, they reported that ~3% of the dose apparently was excreted as hippuric acid.

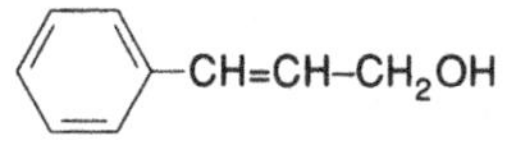

(54)

Cinnamyl alcohol

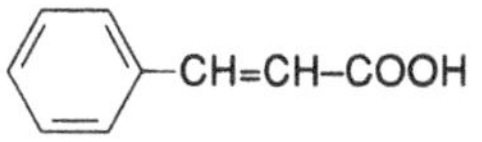

(55)

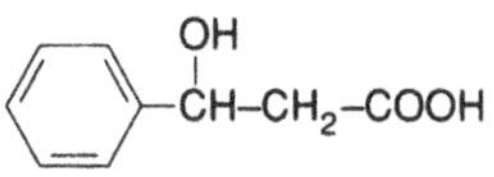

(56)

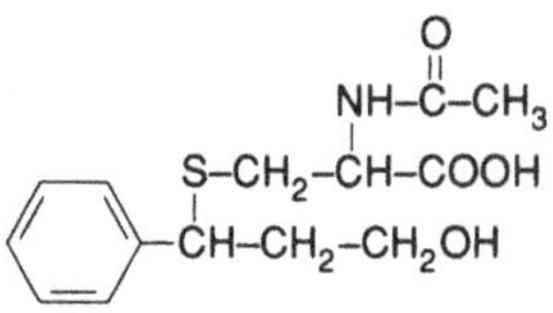

(57)

Another unsubstituted aromatic alcohol is **cinnamyl alcohol** (styryl alcohol) (54) which, when given to rabbits orally in repeated daily doses of 3 to 5 g, was shown by Fischer and Bielig[64] to result in the urinary excretion of small amounts of unchanged compound, its reduced product dihydrocinnamyl alcohol (3-phenylpropanol), and perhaps some 1-phenylpropane-1,3-diol. The latter metabolite may be an intermediate in the oxidation of the C_6–C_3-alcohol to C_6–C_1-compounds and 30 to 35% of the dose was recovered as benzoic acid. However, the major urinary metabolite was cinnamic acid (55) (65 to 70% of the dose). They also detected trace amounts of the corresponding reduced acid, dihydrocinnamic acid.

Peele and Oswald[117] administered cinnamyl alcohol (274 mg/kg, i.p.) to rats and found that 3-hydroxy-3-phenylpropionic acid (56), the oxidation product of the diol noted above, was excreted in the urine. The major acidic metabolite was hippuric acid and cinnamyl alcohol itself was the sole neutral metabolite detected. Teuchy et al.[108] showed earlier that cinnamyl alcohol is a precursor of benzoic acid in rats. They reported that an average of 10% of the dose (550 mg/rat, i.p.) was excreted in the urine as hippuric acid. In contrast to the results noted above using rabbits, Peele and Oswald did not detect cinnamic acid (55), the direct oxidation product of cinnamyl alcohol, as a urinary metabolite in rats.

Delbressine et al.[118] reported that conjugation with glutathione is an additional metabolic pathway of cinnamyl alcohol. They found that rats given the alcohol i.p. excreted ~9% of the dose (125 mg/kg) as the mercapturic acid derivative (57) (*N*-acetyl-*S*-(1-phenyl-3-hydroxypropyl)cysteine. Their results indicated that this pathway proceeded via cinnamaldehyde. The latter compound, but not cinnamyl alcohol itself, reacted spontaneously or enzymatically with glutathione.

The metabolism of **saligenin** (salicyl alcohol) (58a) has received little attention, however Williams[119] noted that it was oxidized to the expected salicylic acid. The rate of oxidation in rabbits may be somewhat slower than that seen with benzyl alcohol as some unchanged compound was excreted following a 1 g dose. Using about half this dose, Williams[120] found that 6 to 10% was excreted conjugated with sulfate. Fötsch and Pfeifer[121] noted that rat serum or homogenates of liver, kidney, lung, or intestinal wall oxidized saligenin to salicylic acid. Further oxidation to gentisic acid was observed when liver homogenates were used.

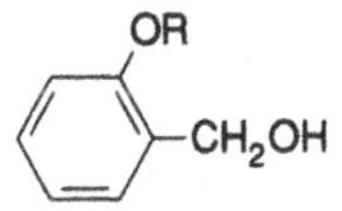

(58)

(a) Saligenin, R = H

(b) Salicin, R = β–D–glucose

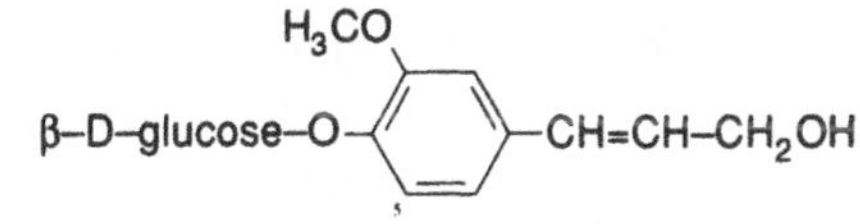

(59)

Coniferin

Salicin (58b) is the phenolic glucoside of saligenin. Drasar and Hill[122] reported that it was hydrolyzed by a number of intestinal bacteria including enterobacteria (*Klebsiella*), streptococci, staphylococci, and nonsporing anaerobes (bifidobacteria). Fötsch et al. studied its metabolism in rats following oral administration[123] or s.c. injection.[121] Following repeated doses (286 mg/kg, p.o., daily), the rats excreted 15% of the administered salicin unchanged in the urine. The most prominent urinary metabolite was salicylic acid which was excreted both free (30% of the dose) and conjugated (5%). The glycine conjugate of salicylic acid, salicyluric acid, accounted for only 0.1%. Gentisic acid (2,5-dihydroxybenzoic acid) (~2%, mainly free) and 2,3-dihydroxybenzoic acid (0.1%) were additional minor metabolites. The final metabolite detected was the aglycon saligenin which furnished a further 0.1% (free) and 5% (conjugated). In contrast, the same dose of injected salicin was mainly excreted unchanged in the urine and only 0.05% was recovered as salicylic acid. Incubation of salicin with cecum or colon contents resulted in its hydrolysis to saligenin.[121] This reaction was also observed when kidney homogenates were used, however no hydrolysis occurred with liver or lung homogenates or serum.

Gastrodin is closely related to salicin, being the corresponding phenolic glycoside of 4-hydroxybenzyl alcohol. Lu et al.[124] noted that the ^{3}H-labeled compound was rapidly absorbed after oral dosage and that 66% of the radioactivity was excreted in the urine within 24 h. The main metabolite detected was 4-hydroxybenzyl alcohol.

Coniferin (59), a closely related glycoside, is 4-β-D-glucoside of coniferyl alcohol. Drasar and Hill[122] reported that enterobacteria (*Proteus, Escherichia,* and *Klebsiella*) are able to hydrolyze it. This will similarly bring about its absorption and metabolism, however the metabolic fate of coniferyl alcohol has not been studied.

Syringin is 5-methoxyconiferin and the ^{3}H-labeled compound was given i.p. to rats.[125] The urinary excretion of radioactivity was 35% (2 h), 55% (4 h), and 90% (48 h) after a dose of 5 mg/kg and only ~3% was found in the feces.

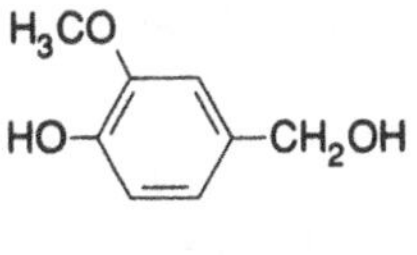

(60)

Vanillyl alcohol

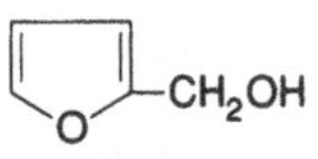

(61)

Furfuryl alcohol

Vanillyl alcohol (60) was administered to rats orally at doses of 100 or 300 mg/kg and found to give essentially the same pattern of urinary metabolites as that shown by the corresponding aldehyde, vanillin.[126] The compounds identified were vanillyl alcohol itself, its oxidation products vanillin and vanillic acid, the latter being excreted to a small extent as its glycine conjugate. In addition, the reactions of *O*-demethylation, decarboxylation, and alcohol reduction shown by Scheline[15] to be carried out with vanillyl alcohol and its metabolites by the rat intestinal microorganisms *in vitro* were also found to occur *in vivo*. These reactions accounted for the presence in the urine, in conjugated form, of catechol, 4-methylcatechol, guaiacol and 4-methylguaiacol. These metabolic pathways are illustrated in Figure 2 of the chapter entitled Metabolism of Aldehydes, Ketones, and Quinones. The *O*-methylation of vanillyl alcohol to 3,4-dimethoxybenzyl alcohol was reported by Friedhoff et al.[127] to be carried out by a guaiacol-*O*-methyltransferase system found in the 100,000 g supernatant fraction from rat liver which utilized *S*-adenosylmethionine. This pathway is probably not quantitatively significant *in vivo*.

The metabolism of **furfuryl alcohol** (61) appears to be straightforward, involving oxidation to furoic acid. Paul et al.[128] found that the major urinary metabolite of the alcohol (~50 mg/kg, p.o.) in rats was furoylglycine.

REFERENCES

1. **Kassam, J. P., Tang, B. K., Kadar, D., and Kalow, W.,** *In vitro* studies of human liver alcohol dehydrogenase variants using a variety of substrates, *Drug Metab. Dispos.*, 17, 567, 1989.
2. **Kamil, I. A., Smith, J. N., and Williams, R. T.,** Studies in detoxication. XLVI. The metabolism of aliphatic alcohols. The glucuronic acid conjugation of acyclic aliphatic alcohols, *Biochem. J.*, 53, 129, 1953.
3. **Kamil, I. A., Smith, J. N., and Williams, R. T.,** Studies in detoxication. XLVII. The formation of ester glucuronides of aliphatic acids during the metabolism of 2-ethylbutanol and 2-ethylhexanol, *Biochem. J.*, 53, 137, 1953.
4. **Williams, R. T.,** *Detoxication Mechanisms*, Chapman and Hall, London, 1959, 47.
5. **Lieber, C.,** The microsomal ethanol oxidizing-system: its role in ethanol and xenobiotic metabolism, *Biochem. Soc. Trans.*, 16, 232, 1988.
6. **McMahon, R. E.,** Enzymatic oxidation and reduction of alcohols, aldehydes and ketones, in *Concepts in Biochemical Pharmacology*, Part 2, Brodie, B. B. and Gillette, J. R., Eds., Springer Verlag, Berlin, 1971, 500.
7. **McMahon, R. E.,** Alcohols, aldehydes, and ketones, in *Metabolic Basis of Detoxication*, Jakoby, W. B., Bend, J. R., and Caldwell, J., Eds., Academic Press, New York, 1982, 91.
8. **Bosron, W. F. and Li, T.-K.,** Alcohol dehydrogenase, in *Enzymatic Basis of Detoxication*, Vol. 1, Jakoby, W. B., Ed., Academic Press, New York, 1980, 231.
9. **Bosron, W. F. and Li, T.-K.,** Genetic polymorphism of human liver alcohol and aldehyde dehydrogenases, and their relationship to alcohol metabolism and alcoholism, *Hepatology*, 6, 502, 1986.
10. **Jörnvall, H., Höög, J.-O., von Bahr-Lindström, H., Johansson, J., Kaiser, R., and Persson, B.,** Alcohol dehydrogenases and aldehyde dehdrogenases, *Biochem. Soc. Trans.*, 16, 223, 1988.
11. **Ehrig, T., Bohren, K. M., Wermuth, B., and von Wartburg, J.-P.,** Degradation of aliphatic alcohols by human liver alcohol dehydrogenase: effect of ethanol and pharmacokinetic implications, *Alcoholism Clin. Exp. Res.*, 12, 789, 1988.
12. **Neubauer, O.,** Ueber Glykuronsäurepaarung bei Stoffen der Fettreihe, *Arch. Exp. Pathol. Pharmak.*, 46, 133, 1901.
13. **Kamil, I. A., Smith, J. N., and Williams, R. T.,** Studies in detoxication. 50. The isolation of methyl and ethyl glucuronides from the urine of rabbits receiving methanol and ethanol, *Biochem. J.*, 54, 390, 1953.
14. **Chang, M. J. W., Leighty, E. G., Haggerty, G. C., and Fentiman, A. F.,** Acylation of xenobiotic alcohols: a metabolic pathway for drug retention, in *Xenobiotic Conjugation Chemistry*, Paulson, G. D., Caldwell, J., Hutson, D. H., and Menn, J. J., Eds., American Chemical Society, Washington, D.C., 1986, 214.
15. **Scheline, R. R.,** The metabolism of some aromatic aldehydes and alcohols by the rat intestinal microflora, *Xenobiotica*, 2, 227, 1972.
16. **Haggard, H. W., Miller, D. P., and Greenberg, L. A.,** The amyl alcohols and their ketones: their metabolic fates and comparative toxicities, *J. Ind. Hyg. Toxicol.*, 27, 1, 1945.
17. **Williams, R. T.,** *Detoxication Mechanisms*, Chapman and Hall, London, 1959.
18. **Browning, E.,** *Toxicity and Metabolism of Industrial Solvents*, Elsevier, Amsterdam, 1965.
19. **Cornish, H. H.,** Solvents and vapors, in *Toxicology, The Basic Science of Poisons*, Casarett, L. J. and Doull, J., Eds., Macmillan, New York, 1975, 503.
20. **Tephly, T. R., Makar, A. B., McMartin, K. E., Hayreh, S. S., and Martin-Amat, G.,** Methanol. Its metabolism and toxicity, in *Biochemistry and Pharmacology of Ethanol*, Vol. 1, Majchrowicz, E. and Noble, E. P., Eds., Plenum Press, New York, 1979, 145.
21. **Lieber, C. S.,** Alcohol, in *Chemical and Biological Aspects of Drug Dependence*, Mulé, S. J. and Brill, H., Eds., CRC Press, Boca Raton, FL, 1974, 135.

22. **Crow, K. E.,** Ethanol metabolism by the liver, *Rev. Drug Metab. Drug Interact.*, 5, 113, 1985.

23. **Crabb, D. W., Bosron, W. F., and Li, T.-K.,** Ethanol metabolism, *Pharmacol. Ther.*, 34, 59, 1987.

24. **Majchrowicz, E. and Noble, E. P.,** *Biochemistry and Pharmacology of Ethanol*, Vol. 1, Plenum Press, New York, 1979.

25. **Mezey, E.,** Ethanol metabolism and ethanol-drug interactions, *Biochem. Pharmacol.*, 25, 869, 1976.

26. **Teschke, R., Hasumura, Y., and Lieber, C. S.,** Hepatic ethanol metabolism: respective roles of alcohol dehydrogenase, the microsomal ethanol-oxidizing system, and catalase, *Arch. Biochem. Biophys.*, 175, 635, 1976.

27. **Gaillard, D. and Derache, R.,** Vitesse de la métabolisation de différents alcools chez le Rat, *C. R. Séanc. Soc. Biol.*, 158, 1605, 1964.

28. **DiVincenzo, G. D. and Hamilton, M. L.,** Fate of *n*-butanol in rats after oral administration and its uptake by dogs after inhalation or skin application, *Toxicol. Appl. Pharmacol.*, 48, 317, 1979.

29. **Gaillard, D. and Derache, R.,** Métabolisation de différents alcools, présents dans les boissons alcooliques, chez le Rat, *Trav. Soc. Pharm. Montpellier*, 25, 51, 1965.

30. **Hinson, J. A. and Neal, R. A.,** An examination of octanol and octanal metabolism to octanoic acid by horse liver alcohol dehydrogenase, *Biochim. Biophys. Acta*, 384, 1, 1975.

31. **Stetten, D. and Schoenheimer, R.,** The biological relations of the higher aliphatic alcohols to fatty acids, *J. Biol. Chem.*, 133, 347, 1940.

32. **McIsaac, W. M. and Williams, R. T.,** The metabolism of spermaceti, *West Afr. J. Biol. Chem.*, 2, 42, 1958.

33. **Baxter, J. H., Steinberg, D., Mize, C. E., and Avigan, J.,** Absorption and metabolism of uniformly ^{14}C-labeled phytol and phytanic acid by the intestine of the rat studied with thoracic duct cannulation, *Biochim. Biophys. Acta*, 137, 277, 1967.

34. **Sieber, S. M., Cohn, V. H., and Wynn, W. T.,** The entry of foreign compounds into the thoracic duct lymph of the rat, *Xenobiotica*, 4, 265, 1974.

35. **Williams, R. T.,** *Detoxication Mechanisms*, Chapman and Hall, London, 1959, 59.

36. **Greenberg, L. A.,** The appearance of some congeners of alcoholic beverages and their metabolites in blood, *Q. J. Stud. Alcohol*, Suppl. 5, 20, 1970.

37. **DiVincenzo, G. D., Kaplan, C. J., and Dedinas, J.,** Characterization of the metabolites of methyl *n*-butyl ketone, methyl iso-butyl ketone, and methyl ethyl ketone in guinea pig serum and their clearance, *Toxicol. Appl. Pharmacol.*, 36, 511, 1976.

38. **Dietz, F. K., Rodriguez-Giaxola, M., Traiger, G. J., Stella, V. J., and Himmelstein, K. J.,** Pharmacokinetics of 2-butanol and its metabolites, *J. Pharmacokinet. Biopharm.*, 9, 553, 1981.

39. **Carr, C. J. and Krantz, J. C.,** Metabolism of the sugar alcohols and their derivatives, in *Advances in Carbohydrate Chemistry*, Vol. 1, Pigman, W. W. and Wolfrom, M. L., Eds., Academic Press, New York, 1945, 175.

40. **Williams, R. T.,** *Detoxication Mechanisms*, Chapman and Hall, London, 1959, 79.

41. **Touster, O. and Shaw, D. R. D.,** Biochemistry of the acyclic polyols, *Physiol. Rev.*, 42, 181, 1962.

42. **Bässler, K. H.,** Absorption, metabolism, and tolerance of polyol sugar substitutes, *Pharmacol. Ther. Dent.*, 3, 85, 1978.

43. **Weinstein, A. N. and Segal, S.,** The metabolic fate of galactitol-1-^{14}C in mammalian tissues, *Biochim. Biophys. Acta*, 156, 9, 1968.

44. **Olmsted, W. H.,** The metabolism of mannitol and sorbitol. Their use as sugar substitutes in diabetic therapy, *Diabetes*, 2, 132, 1953.

45. **Nasrallah, S. M. and Iber, F. L.,** Mannitol absorption and metabolism in man, *Am. J. Med. Sci.*, 258, 80, 1969.

46. **Ertel, N. H., Akgun, S., Kemp, F. W., and Mittler, J. C.,** The metabolic fate of exogenous sorbitol in the rat, *J. Nutr.*, 113, 566, 1983.

47. **Schell-Dompert, E. and Siebert, G.,** Metabolism of sorbitol in the intact organism, *Hoppe-Seyler's Z. Physiol. Chem.*, 361, 1069, 1980.

48. **Strack, E., Kukfahl, E., Mueller, F., and Beyreiss, K.,** The conversion of sorbitol in the animal body during enteral and intravascular infusions, *Z. Ges. Exp. Med.*, 139, 23, 1965 (Chem. Abstr. 64, 1120f, 1966).

49. **Maeda, K.,** Sorbitol as an infusion medium. II. Metabolism of sorbitol, *Sapporo Igaku Zasshi*, 29, 144, 1966 (Chem. Abstr. 69, 50666s, 1968).

50. **Adcock, L. H. and Gray, C. H.,** The metabolism of sorbitol in the human subject, *Biochem. J.*, 65, 554, 1957.

51. **Stetten, M. R. and Stetten, D.,** Metabolism of sorbitol and glucose compared in normal and alloxan-diabetic rats, *J. Biol. Chem.*, 193, 157, 1951.

52. **Wick, A. N., Almen, M. C., and Joseph, L.,** The metabolism of sorbitol, *J. Am. Pharm. Assoc. Sci. Ed.*, 40, 542, 1951.

53. **Soummer-Dauphant, A. M., Chambon, P., and Chabert, J. M.,** Conversion of sorbitol to mannitol when used parenterally, *Clin. Chim. Acta*, 67, 325, 1976.

54. **Heinz, F. and Wittneben, H.-E.,** Metabolitengehalte in der Rattenleber nach Sorbitolapplikation, *Hoppe-Seyler's Z. Physiol. Chem.*, 351, 1215, 1970.

55. **Hiatt, E. P., Carr, C. J., Evans, W. E., and Krantz, J. C.,** Sugar alcohols XIII. Primulatol and glycogen stroage in the rat, *Proc. Soc. Exp. Biol. Med.,* 38, 356, 1938.
56. **Williams, R. T.,** *Detoxication Mechanisms,* Chapman and Hall, London, 1959, 118.
57. **Hildebrandt, H.,** Ueber Synthesen im Thierkörper. II. Verbindungen der Kamphergruppe, *Arch. Exp. Pathol. Pharmak.,* 45, 110, 1901.
58. **Rahman, K. M. Q.,** Studies on the Metabolism of Linalool, a Naturally Occurring Food Flavour, Ph. D. Thesis, University of Surrey, England, 1974.
59. **Parke, D. V., Rahman, K. M. Q., and Walker, R.,** The absorption, distribution and excretion of linalool in the rat, *Biochem. Soc. Trans.,* 2, 612, 1974.
60. **Chadha, A. and Madyastha, K. M.,** Metabolism of geraniol and linalool in the rat and effects on liver and lung microsomal enzymes, *Xenobiotica,* 14, 365, 1984.
61. **Chadha, A. and Madyastha, K. M.,** ω-Hydroxylation of acyclic monoterpene alcohols by rat lung microsomes, *Biochem. Biophys. Res. Commun.,* 108, 1271, 1982.
62. **Hildebrandt, H.,** Über das biologische Verhalten von Nerol, Geraniol, Cyclogeraniol, *Beitr. Chem. Physiol. Pathol.,* 4, 251, 1904.
63. **Kuhn, R., Köhler, F., and Köhler, L.,** Über Methyl-oxydationen im Tierkörper, *Hoppe-Seyler's Z. Physiol. Chem.,* 242, 171, 1936.
64. **Fischer, F. G. and Bielig, H.-J.,** Über die Hydrierung ungesättigter Stoffe im Tierkörper. Biochemische Hydrierungen. VII, *Hoppe-Seyler's Z. Physiol. Chem.,* 266, 73, 1940.
65. **Asano, M. and Yamakawa, T.,** The fate of branched chain fatty acids in animal body. I. A contribution to the problem of "Hildebrandt acid", *J. Biochem. (Tokyo),* 37, 321, 1950.
66. **Licht, H. J. and Jamroz, G.,** Involvement of a cytochrome P-450-dependent monoxygenase in isoprenoid alcohol metabolism in mammals, *Fed. Proc. Fed. Am. Soc. Exp. Biol.,* 36, 832, 1977.
67. **Licht, H. J. and Coscia, C. J.,** Cytochrome P-450LM$_2$ mediated hydroxylation of monoterpene alcohols, *Biochemistry,* 17, 5638, 1978.
68. **Licht, H. J., Madyastha, K. M., Coscia, C. J., and Krueger, R. J.,** Comparison of plant and hepatic cytochrome P-450-dependent monoterpene monoxygenases, in *Microsomes, Drug Oxidations, and Chemical Carcinogenesis,* Vol. 1, Coon, M. J., Conney, A. H., Estabrook, R. W., Gelboin, H. V., Gillette, J. R., and O'Brien, P. J., Eds., Academic Press, New York, 1980, 211.
69. **Quick, A. J.,** The synthesis of menthol glycuronic acid in the rabbit, *J. Biol. Chem.,* 61, 679, 1924.
70. **Quick, A. J.,** Quantitative studies of β-oxidation. IV. The metabolism of conjugated glycuronic acids, *J. Biol. Chem.,* 80, 535, 1928.
71. **Bell, G. D., Henry, D. A., Langman, M. J. S., and Roddie, M. E.,** Glucuronidation of l-menthol in normal individuals and patients with liver disease, *Br. J. Clin. Pharmacol.,* 12, 272, 1981.
72. **Williams, R. T.,** Studies in detoxication. II. (a) The conjugation of isomeric 3-menthanols with glucuronic acid and the symmetric conjugation of *dl*-menthol and *dl-iso*menthol in the rabbit. (b) *d-iso*Menthylglucuronide, a new conjugated glucuronic acid, *Biochem. J.,* 32, 1849, 1938.
73. **Williams, R. T.,** Studies in detoxication. III. The use of the glucuronic acid detoxication mechanism of the rabbit for the resolution of *dl*-menthol, *Biochem. J.,* 33, 1519, 1939.
74. **Williams, R. T.,** Studies in detoxication. IV. (a) The resolution of *dl-iso*menthol through its conjugation with glucuronic acid in the rabbit, *Biochem. J.,* 34, 48, 1940.
75. **Williams, R. T.,** Studies in detoxication. VII. The biological reduction of *l*-menthone to *d*-neomenthol and of *d*-isomenthone to *d*-isomenthol in the rabbit. The conjugation of *d*-neomenthol with glucuronic acid, *Biochem. J.,* 34, 690, 1940.
76. **Wright, S. E.,** Detoxication mechanisms in the sheep, *Pap. Dep. Chem. Univ. Qd,* 1, 1945.
77. **Williams, R. T.,** *Detoxication Mechanisms,* Chapman and Hall, London, 1959, 529.
78. **Scheline, R. R.,** Unpublished data, 1977.
79. **Madyastha, K. M. and Srivatsan, V.,** Studies on the metabolism of *l*-menthol in rats, *Drug Metab. Dispos.,* 16, 765, 1988.
80. **Yamaguchi, T. and Caldwell, J.,** Disposition and metabolism of *l*-menthol in the rat, Third European Symposium on Foreign Compound Metabolism. Metabolic and Kinetic Perspectives in Safety Assessment: Some current issues, London, England, 1989.
81. **Hämäläinen, J.,** Über das Verhalten der alicyklischen Verbindungen bei der Glykuronsäurepaarung im Organismus, *Skand. Arch. Physiol.,* 27, 141, 1912.
82. **Hämäläinen, J.,** Zur Konstitution der Terpineol-35°-glucuronsäure, *Biochem. Z.,* 50, 220, 1913.
83. **Zhou, H.-L., Zheng, Q.-J., Mo, Q.-Z., and Bian, R.-L.,** Kinetic aspects of absorption, distribution, and excretion of [³H]α-terpineol in guinea pigs, *Zhongguo Xaolixue Yu Dulixue Zashi,* 1, 293, 1988 (Chem. Abstr. 108, 142778g, 1988).
84. **Horning, M. G., Butler, C. M., Stafford, M., Stillwell, R. N., Hill, R. M., Zion, T. E., Harvey, D. J., and Stillwell, W. G.,** Metabolism of drugs by the epoxide-diol pathway, in *Advances in Mass Spectrometry in Biochemistry and Medicine,* Vol. 1, Frigerio, A. and Castagnoli, N., Eds., Spectrum Publications, New York, 1976, 91.

47

85. **Madyastha, K. M. and Srivatsan, V.,** Biotransformations of α-terpineol in the rat: Its effects on the liver microsomal cytochrome P-450 system, *Bull. Environ. Contam. Toxicol.*, 41, 17, 1988.

86. **Ventura, P., Schiavi, M., and Serafini, S.,** The metabolism of *trans*-sobrerol in the rat, *Xenobiotica*, 13, 139, 1983.

87. **Ventura, P., Schiavi, M., Serafini, S., and Selva, A.,** Further studies of *trans*-sobrerol metabolism: rat, dog and human urine, *Xenobiotica*, 15, 317, 1985.

88. **Ventura, P., Pellegata, R., Schiavi, M., and Serafini, S.,** Biotransformation of *trans*-sobrerol. III. Metabolites of 8-hydroxycarvotanacetone in the rat, *Xenobiotica*, 16, 317, 1986.

89. **Hämäläinen, J.,** Über das Schicksal des Cineols (Eukalyptols) im Organismus, *Skand. Arch. Physiol.*, 24, 1, 1911.

90. **Flynn, T. M. and Southwell, I. A.,** 1,3-Dimethyl-2-oxabicyclo[2,2,2]-octane-3-methanol and 1,3-dimethyl-2-oxabicyclo[2,2,2]-octane-3-carbxoylic acid, urinary metabolites of 1,8-cineole, *Aust. J. Chem.*, 32, 2093, 1979.

91. **Southwell, I. A., Flynn, T. M., and Degabriele, R.,** Metabolism of α- and β-pinene, *p*-cymene and 1,8-cineole in the brushtail possum, *Trichosurus vulpecula*, *Xenobiotica*, 10, 17, 1980.

92. **Frigerio, A., Paladino, R., Testoni, G., Cobelli, L., Pastorello, D., and Tolentino, D.,** 1,8-Cineole (eucalyptol) metabolites in rat, in *Chromatography and Mass Spectrometry in Nutrition Science and Food Safety*, Frigerio, A. and Milon, H., Eds., Elsevier, Amsterdam, 1984, 79.

93. **Madyastha, K. M. and Chadha, A.,** Metabolism of 1,8-cineole in rat: its effects on liver and lung microsomal cytochrome P-450 systems, *Bull. Environ. Contam. Toxicol.*, 37, 759, 1986.

94. **Miyazawa, M., Kameoka, H., Morinaga, K., Negoro, K., and Mura, N.,** Hydroxycineole: four new metabolites of 1,8-cineole in rabbits, *J. Agric. Food Chem.*, 37, 222, 1989.

95. **Ventura, P. and Selva, A.,** Biotransformation of epomediol. I. Isolation and identification of main metabolites in the rat, *Biomed. Mass Spectrom.*, 9, 18, 1982.

96. **Asakawa, Y., Toyota, M., and Ishida, T.,** Biotransformation of 1,4-cineole, a monoterpene ether, *Xenobiotica*, 18, 1129, 1988.

97. **Hämäläinen, J.,** Zum forensisch-chemischen Nachweis von Sadebaumölvergiftung, *Biochem. Z.*, 41, 241, 1912.

98. **Fromm, E. and Hildebrandt, H.,** Ueber das Schicksal cyclischer Terpene und Campher im thierischen Organismus, *Hoppe-Seyler's Z. Physiol. Chem.*, 33, 579, 1901.

99. **Magnus-Levy, A.,** Über Paarung der Glukuronsäure mit optischen Antipoden, *Biochem. Z.*, 2, 319, 1907.

100. **Hämäläinen, J.,** Über isomere Borneolglykuronsäuren, *Skand. Arch. Physiol.*, 23, 86, 1909.

101. **Pryde, J. and Williams, R. T.,** The biochemistry and physiology of glucuronic acid. IV. (a) The occurrence of conjugated glucuronic acids in the animal body. (b) Observations on the conjugation of *d*- and *l*-borneol, *Biochem. J.*, 28, 131, 1934.

102. **Pryde, J. and Williams, R. T.,** The biochemistry and physiology of glucuronic acid. VII. A note on the conjugation of borneol in man, *Biochem. J.*, 30, 799, 1936.

103. **Robertson, J. S. and Solomon, E.,** Metabolism of camphanediols, *Biochem. J.*, 121, 503, 1971.

104. **Snapper, J., Grünbaum, A., and Sturkop, S.,** Über die Spaltung und die Oxydation von Benzylalkohol und Benzylestern im menschlichen Organismus, *Biochem. Z.*, 155, 163, 1925.

105. **Diack, S. L. and Lewis, H. B.,** Studies in the synthesis of hippuric acid in the animal organism. VII. A comparison of the rate of elimination of hippuric acid after ingestion of sodium benzoate, benzyl alcohol, and benzyl esters of succinic acid, *J. Biol. Chem.*, 77, 89, 1928.

106. **Bray, H. G., Thorpe, W. V., and White, K.,** Kinetic studies of the metabolism of foreign organic compounds. I. The formation of benzoic acid from benzamide, toluene, benzyl alcohol and benzaldehyde and its conjugation with glycine and glucuronic acid in the rabbit, *Biochem. J.*, 48, 88, 1951.

107. **LeBel, M., Ferron, L., Masson, M., Pichette, J., and Carrier, C.,** Benzyl alcohol metabolism and elimination in neonates, *Dev. Pharmacol. Ther.*, 11, 347, 1988.

108. **Teuchy, H., Quatacker, J., Wolf, G., and Van Sumere, C. F.,** Quantitative investigation of the hippuric acid formation in the rat after administration of some possible aromatic and hydroaromatic precursors, *Arch. Int. Physiol. Biochim.*, 79, 573, 1971.

109. **Clapp, J. J. and Young, L.,** Formation of mercapturic acids in rats after the administration of aralkyl esters, *Biochem. J.*, 118, 765, 1970.

110. **Van Doorn, R., Leijdekkers, C.-M., Bos, R. P., Brouns, R. M. E., and Henderson, P. T.,** Alcohol and sulphate intermediates in the metabolism of toluene and xylenes to mercapturic acids, *J. Appl. Toxicol.*, 1, 236, 1981.

111. **Chidgey, M. A. J., Kennedy, J. F., and Caldwell, J.,** Studies on benzyl acetate. II. Use of specific metabolic inhibitors to define the pathway leading to the formation of benzylmercapturic acid in the rat, *Food Chem. Toxicol.*, 24, 1267, 1986.

112. **Binder, T. P. and Duffel, M. W.,** Sulfation of benzylic alcohols catalyzed by aryl sulfotransferase. IV, *Mol. Pharmacol.*, 33, 477, 1988.

113. **Gillham, B.,** The reaction of aralkyl sulphate esters with glutathione catalysed by rat liver preparations, *Biochem. J.*, 121, 667, 1971.
114. **Sloane, N. H.,** Hydroxymethylation of the benzene ring. I. Microsomal formation of phenol via prior hydroxymethylation of benzene, *Biochim. Biophys. Acta*, 107, 599, 1965.
115. **Smith, J. N., Smithies, R. H., and Williams, R. T.,** Studies in detoxication. LVI. The metabolism of alkylbenzenes. Stereochemical aspects of the biological hydroxylation of ethylbenzene to methylphenylcarbinol, *Biochem. J.*, 56, 320, 1954.
116. **El Masry, A. M., Smith, J. N., and Williams, R. T.,** Studies in detoxication. LXIX. The metabolism of alkylbenzenes: *n*-propylbenzene and *n*-butylbenzene with further observations on ethylbenzene, *Biochem. J.*, 64, 50, 1956.
117. **Peele, J. D. and Oswald, E. O.,** Metabolism of naturally occurring propenylbenzene derivatives. III. Allylbenzene, propenyl benzene, and related metabolic products, *Biochim. Biophys. Acta*, 497, 598, 1977.
118. **Delbressine, L. P. C., Klippert, P. J. M., Reuvers, J. T. A., and Seutter-Berlage, F.,** Isolation and identification of mercapturic acids of cinnamic aldehyde and cinnamyl alcohol from urine of female rats, *Arch. Toxicol.*, 49, 57, 1981.
119. **Williams, R. T.,** *Detoxication Mechanisms*, Chapman and Hall, London, 1959, 320.
120. **Williams, R. T.,** Studies in detoxication. I. The influence of (a) dose and (b) *o-, m-* and *p*-substitution on the sulphate detoxication of phenol in the rabbit, *Biochem. J.*, 32, 878, 1938.
121. **Fötsch, G. and Pfeifer, S.,** Die Biotransformation der Phenolglycoside Leiocarposid und Salicin - Beispiele für Besonderheiten von Absorption und Metabolismus glycosidischer Verbindungen, *Pharmazie*, 44, 710, 1989.
122. **Drasar, B. S. and Hill, M. J.,** *Human Intestinal Flora*, Academic Press, London, 1974, 63.
123. **Fötsch, G., Pfeifer, S., Bartoszek, M., Franke, P., and Hiller, K.,** Biotransformation der Phenolglycoside Leiocarposid und Salicin, *Pharmazie*, 44, 555, 1989.
124. **Lu, G.-W., Zou, Y.-J., and Mo, Q.-Z.,** Absorption, distribution, metabolism and excretion of ^{3}H-gastrodin in rats, *Yaoxue Xuebao*, 20, 167, 1985 (Chem. Abstr. 103, 134370z, 1985).
125. **Bezdetko, G. N., German, A. V., Shevchenko, V. P., Mitrokhin, Y. I., Myasoedov, N. F., Dardymov, I. V., Todorov, I. N., and Barenboim, G. M.,** Study of the pharmacokinetics and action mechanism of eleuthero-coccus glycosides. I. Incorporation of tritium into Eleutheroside B, kinetics of its accumulation and excretion from the body of an animal, *Khim.-Farm. Zh.*, 15, 28, 1981 (Chem. Abstr. 94, 95710j, 1981).
126. **Strand, L. P. and Scheline, R. R.,** The metabolism of vanillin and isovanillin in the rat, *Xenobiotica*, 5, 49, 1975.
127. **Friedhoff, A. J., Schweitzer, J. W., Miller, J., and van Winkle, E.,** Guaiacol-*O*-methyltransferase: a mammalian enzyme capable of forming di-*O*-methyl catecholamine derivatives, *Experientia*, 28, 517, 1972.
128. **Paul, H. E., Austin, F. L., Paul, M. F., and Ells, V. R.,** Metabolism of nitrofurans. I. Ultraviolet absorption studies of urinary end-products after oral administration, *J. Biol. Chem.*, 180, 345, 1949.

METABOLISM OF PHENOLS AND ETHERS

PHENOLS

The phenolic group is a common feature of many classes of plant substances. Accordingly, phenolic compounds furnish a large and heterogeneous group, examples of which are found in most of the chapters in this book. The latter fact indicates that other structural features are usually considered to be more important or typical for the compound in question. We find that compounds in which the phenolic group is the sole or most prominent functional group are relatively rare in plants. It is with this rather restricted group of compounds that the present section is concerned.

Mono-, di-, and trihydric phenols of the simpler types generally seem to be found in relatively few plants. The most widely distributed of these compounds is probably hydroquinone; however, even the simplest member, phenol, has been occasionally reported to occur in small amounts in a few plant sources. In view of the rather limited amount of metabolic data available on most of the other members of this group, it seems justified to begin this section with a summary of the metabolic fate of phenol itself.

A general understanding of the main metabolic pathways of **phenol** has been available for some time, the earliest investigations dating from the last century. In brief, phenol may undergo conjugation reactions directly by virtue of the presence of a free hydroxy group or it may undergo oxidation which gives rise to hydroxylated metabolites, those in turn being subject to conjugation and, to a smaller extent, the formation of highly reactive metabolites. The former pathways largely involve the formation of glucuronide or ethereal sulfate conjugates, however phenyl phosphate has also been identified as a urinary metabolite of phenol in cats[1] and sheep,[2] the latter species also forming some hydroquinone phosphate. The oxidative pathway furnishes mainly hydroquinone, the p-hydroxylated product, although some o-hydroxylation to catechol may sometimes be detected. Not unexpectedly, the conjugative pathways are quantitatively far more important than are the oxidative pathways.

Differences in experimental procedures make it difficult to compare the results of earlier studies with those carried out more recently which have employed ^{14}C-labeled phenol. The former investigations employed rabbits and measured primarily the urinary excretion of glucuronide and sulfate conjugates. These are largely conjugates of phenol itself, but the contribution of small amounts of oxidized metabolites to these values was usually not ascertained. In any case, Williams[3] showed that the extent of phenol conjugation with sulfate in rabbits is dose dependent, oral doses of 25 or 50 mg/kg giving values of 41 and 37% of the dose, respectively, whereas doses of 100 to 250 mg/kg gave values of 18 to 19%. The latter values are in close agreement with those of Bray et al.[4] who reported 15 to 16% using a dose of 250 mg/kg in similar experiments. Conjugation of phenol with glucuronic acid at these higher dose levels (125 to 250 mg/kg, p.o.) accounted for ~70% of the dose.[5,6] Porteous and Williams[5] also found that ~14% of the dose was oxidized to other phenols, mainly hydroquinone.

The availability of ^{14}C-labeled phenol has resulted in a much more comprehensive understanding of the pathways of phenol metabolism, especially in regard to the oxidized metabolites. The initial experiments of this type were carried out by Parke and Williams[7] who found that orally administered phenol (50 to 60 mg/kg) was about equally excreted in the urine of rabbits as the glucuronide and sulfate conjugates (~45% of dose, each). The value for the latter conjugate is similar to that noted above at this dose level. In addition, ~10% was recovered as hydroquinone, 0.5 to 1% as catechol, and a trace as hydroxyquinol (1,2,4-trihydroxybenzene). No evidence was found for the formation of phenylmercapturic acid or any of the isomers of muconic acid (HOOC–CH=CH–CH=CH–COOH) which might have arisen by ring cleavage. The lack of formation of muconic acid was confirmed by Inoue et al.[8] in experiments in which rabbits were given phenol (50 mg/kg, i.p.). Their results also showed that negligible amounts of catechol were formed and that hydroxyquinol formation proceeds via quinol and not catechol. The preference for *p*-hydroxylation over *o*-hydroxylation was also observed in *in vitro* experiments using [^{14}C]phenol and rat liver microsomes in the presence of an NADPH-generating system.[9] The ratio of hydroquinone to catechol formed was 20:1, however no significant formation of the triol was found.

Another development facilitated by the use of [^{14}C]phenol has been the study of species differences in its metabolism. Table 1 summarizes some of the major findings from these investigations in a large number of species. Similar information from several reports on the metabolism of phenol in rats is shown in Table 2. Additional information using numerous rodents and marsupials native to Australia is found in the publication of Baudinette et al.;[10] however, only their data on the common laboratory mouse and the desert hopping mouse (*Notomys alexis*) are included in Table 1. Wheldrake et al.[11] investigated in more detail the metabolism of phenol in *N. alexis* and Ramli and Wheldrake[12] described *in vitro* studies of phenol in this species which showed that the conjugation ratio (glucuronide:sulfate) was similar to that found in whole animals.

These results show that free phenol was not excreted following oral dosage. The influence of dose on the glucuronide:sulfate ratio is clearly seen in the data shown in Table 2. Extremes in these ratios are seen with cats and the closely related genet, civet, hyena, and lion which essentially produced only ethereal sulfates (Table 1). Pigs represent the other extreme with nearly all the excreted material being the glucuronide. Oehme and Davis[13] also reported that pigs excreted phenol conjugated primarily with glucuronic acid. They noted that goats, a species not included in Table 1, excreted mainly phenyl sulfate. The excretion values for phenyl phosphate ($C_6H_5OPO(OH)_2$) are appreciable for sheep and experiments in cats similar to those summarized in Table 1 but using i.p. dosage indicated that as much as 12% of the dose was excreted in this form.[1]

Another significant finding concerning the metabolism of phenol was reported by Powell et al.[14] who showed that conjugation of orally administered phenol in rats occurred largely in the gastrointestinal tract prior to its uptake into the portal circulation and transport to the liver. Shirkey et al.[15] reported that cell preparations from both rat small intestine mucosa and liver conjugated phenol with glucuronic acid and sulfate. In the former preparations glucuronide formation was the major reaction, however hepatocytes produced nearly equal amounts of the two metabolites as well as some hydroquinone glucuronide. Phenol metabolism can also be carried out by the lungs, however this site probably has little influence following the usual routes of administration which will be expected to result mainly in metabolism by the intestinal wall and the liver. Hogg et al.[16] found that rat whole lung preparations extensively metabolized phenol to the glucuronide and sulfate conjugates (in about equal amounts) and that they also produced some hydroquinone sulfate. It was noted that formation of the glucuronide conjugate by lung tissue may be reduced or absent in other species (rabbit, man) which instead show mainly sulfate conjugation. The rapid and extensive conjugation of phenol by the lungs was also reported by Cassidy and Houston.[17] The biliary excretion of phenol is not extensive;[18] however, Gbodi and Oehme[19] reported that the material found in rat bile is mainly the glucuronide and that enterohepatic circulation took place.

A novel but minor metabolite of phenol in rats is the *o*-methoxy derivative guaiacol.[20] This metabolite, found consistently but only in trace amounts following oral doses of 100 mg/kg, is formed via catechol which can then be *O*-methylated.

Considerable interest has been shown recently in the metabolism of phenol to highly reactive oxidation products. Hepatic microsomes, in the presence of an NADPH-generating system, metabolized phenol to intermediates which bound covalently to microsomal macromolecules.[9,21] *N*-acetylcysteine adducts of hydroquinone and catechol were formed when the amino acid was added to the incubates, a finding which indicates that the dihydric phenols were precursors of the reactive metabolites responsible for covalent binding.[9] Similarly, addition of glutathione

TABLE 1
Urinary Metabolites of [^{14}C]Phenol in Various Species

Species	Dose mg/kg (route)	% of dose excreted (time)	% of excreted radioactivity found as					Ref.
			Phenyl glucuronide	Phenyl sulfate	Hydroquinone glucuronide	Hydroquinone sulfatez	Other	
Rodents								
Mouse	25 (p.o.)	66[a] (24 h)	35	46	15	5		138
	25 (i.p.)	(16 h)	31	26	27	3		10
Desert hopping mouse	25 (i.p.)	(16 h)	34	36	23	3		10
Rat (see Table 2)								
Egyptian jerboa	25 (p.o.)	47 (24 h)	26	61	1	12		138
Gerbil	25 (p.o.)	55 (24 h)	35	42	1	19		138
Hamster	25 (p.o.)	73,78 (24 h)	44,41	27,24	28,27	1,tr[b]		138
Lemming	25 (p.o.)	40 (24 h)	39	35	15	10		138
Guinea pig	25 (p.o.)	64 (24 h)	82,73	13,22	5	tr		138
Other								
Indian fruitbat	25 (p.o.)	50,58 (24 h)	91,89	9,11	—[c]	—		138
Rabbit	25 (p.o.)	48 (24 h)	46	45	—	9		138
European hedgehog	20 (p.o.)	34,43 (24 h)	20,10	63,86	—	17,4		138
Sheep	25 (p.o.)	85,86 (8 h)	47,49	36, 29	1.5, 2.8	3.8, 4.1	Phenyl phosphate 8, 13; Quinol phosphate 3	2
Pig	25 (p.o.)	83,84 (8 h)	89, 87	0, 2.4	10, 5	0, 1.4		2
	25 (p.o.)	76,95 (24 h)	96,99	4,1	—	—		138
Elephant	10 (p.o.)	49 (24 h)	25	73	—	1		139
Carnivores								
Ferret	25 (p.o.)	51 (24 h)	40	28	—	30		138
Cat	25 (p.o.)	49 (24 h)	0.2	88	—	9	Phenyl phosphate, 2.5	1,138
	20 (i.v.)	(6 h)	—	80	—	20		36,37
	40 (i.v.)	(6 h)	—	78	—	16	Phenol, 2; hydroquinone, 4	36,37
Dog	25 (p.o.)	53,6S2 (24 h)	24,12	33,68	—	43,20		138
Forest genet	10 (p.o.)	37,58 (24 h)	—	100,98	—			140
African civet	10 (p.o.)	60 (24 h)	—	97				140
Hyena	10 (p.o.)	15,47 (24 h)	0	93,86		6,13		139

TABLE 1 (continued)
Urinary Metabolites of [14C]Phenol in Various Species

Species	Dose mg/kg (route)	% of dose excreted (time)	Phenyl glucuronide	Phenyl sulfate	Hydroquinone glucuronide	Hydroquinone sulfate	Other	Ref.
				% of excreted radioactivity found as				
Lion	10 (p.o.)	77 (24 h)	—	99				140
Primates								
Bushbaby[d]	10 (i.m.)	60 (24 h)	23	77				141
Tree shrew[d]	10 (i.m.)	61 (24 h)	48	52				141
Capuchin[e]	10 (i.m.)	30,48 (24 h)	59	41				141
	25 (p.o.)	73 (24 h)	65	14	21	—		138
Tamarin[e]	10 (i.m.)	93 (24 h)	89	11				141
Squirrel monkey[e]	25 (p.o.)	31 (24 h)	68	7	25	—		138
Cynomolgus monkey[f]	10 (i.m.)	98 (24 h)	34	66	—	—		141
Rhesus monkey[f]	50 (p.o.)	37,49 (24 h)	40,30	60,70	—	—		138
	10 (i.m.)	78,82 (24 h)	22	78				141
Man	0.01 (p.o.)	90 (24 h)	16	77	tr	1		138

a   Values are either group averages, individual values for two animals, or single values (for the tamarin, bushbaby, and tree shrew).
b   (tr) Trace.
c   (—) Not detected.
d   Prosimian.
e   New World monkey.
f   Old World monkey.

TABLE 2
Urinary Metabolites of [^{14}C]Phenol in Rats

Dose (mg/kg)	Route	Sex	Collection period (h)	% of dose excreted	% of excreted radioactivity found as				Ref.
					Phenyl glucuronide	Phenyl sulfate	Hydroquinone glucuronide	Hydroquinone sulfate	
1.2	i.v.	M	2	92[a]	28	72			142
1.2	p.o.	M	24	80	15	85			27
2.5	i.v.	M	2	96[a]	34	66			142
6.3	i.v.	M	2	91[a]	42	58			142
12.5	i.v.	M	2	93[a]	50	50			142
12.5	i.v.	M	3	91	49	58			143
12.5	i.v.	F	3	87	45	51			143
25	i.v.	M	2	91[a]	60	40			142
25	p.o.		8	90	42	55	2	1	2
25	p.o.	F	24	95	42	54	2	1	138

[a] Includes values for biliary excretion which, at higher doses, may include appreciable excretion of phenyl glucuronide.

resulted in the formation of 2-(S-glutathionyl)hydroquinone.[21] Several additional investigations have shown that the formation of reactive oxidation products from phenol was peroxidase-mediated.[22-26] Several metabolites including 2,2'-biphenol (1), 4,4'-biphenol (2) and the corresponding quinone of the latter (p-diphenoquinone) (3) were identified in these studies. p-Diphenoquinone was shown to form conjugates with glutathione[24] and several of the intermediates are able to bind with macromolecules.[23,24]

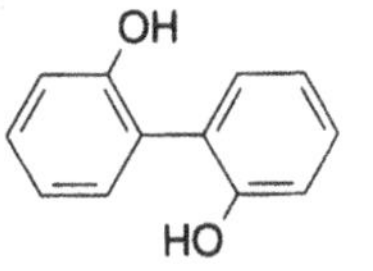

(1) 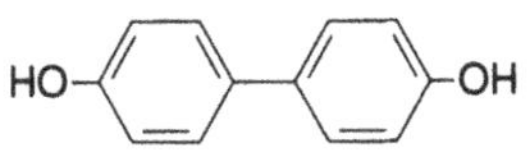   (2) 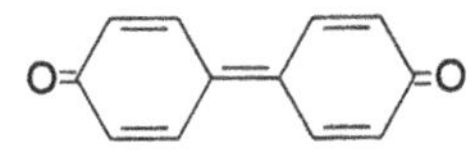   (3)

Edwards et al.[27] administered the glucoside of phenol ([14C]phenyl β-D-glucopyranoside) orally to rats and found that 88% (urine) and 3% (feces) of the dose (2.9 mg/kg) was excreted within 24 h. It is notable that 78% of the material excreted in the urine was unchanged phenyl glucoside. The remainder was phenyl sulfate (22%) and phenyl glucuronide (trace). These results indicate that this glucoside possesses properties which permit its extensive absorption from the intestine. Excretion of a smaller part of the dose as typical conjugation products was probably due to the absorption of the aglycon following its liberation by β-glycosidases in the lower regions of the intestinal tract.

As noted above, the simple dihydric phenols are not common plant compounds. Catechol (1,2-dihydroxybenzene) and especially hydroquinone (quinol; 1,4-dihydroxybenzene) occur naturally, however the resorcinol (1,3-dihydroxybenzene) structure appears to be found mainly in various 5-alkyl derivatives. Merker et al.[28] reported that resorcinol itself was rapidly excreted in the urine, mainly as a glucuronide conjugate, when administered to rats. They found that s.c. administration of 14C-labeled material (10 mg/kg) resulted in recoveries of radioactivity of 63, 88, and 98% in the urine at 1, 3, and 24 h, respectively. Kim and Matthews[29] also studied the metabolism of [14C]resorcinol (112 mg/kg, p.o.) in rats. They similarly reported rapid excretion, with >90% and <3% of the dose excreted in the urine and feces, respectively, in 24 h. About 70% of the urinary radioactivity was due to resorcinol glucuronide, however some sulfate conjugate and, probably, a diconjugate containing glucuronide and sulfate groups were also detected. Biliary excretion of radioactive material was not extensive and accounted for only 5 to 8% of a smaller dose (11.2 mg/kg, i.v.).

The simplest of the 5-alkyl resorcinols is **orcinol** (5-methylresorcinol) which is generally found in lower plants (e.g., lichens) but may also occur as glycosides in some heathers. According to early reports summarized by Williams,[30] orcinol undergoes the expected metabolic reactions of conjugate formation with one of the phenolic groups. Martin et al.[31] found that when orcinol (300 mg/d) was administered to sheep by continuous ruminal infusion, ~90 to 100% of the dose was recovered as urinary orcinol. However, it was not stated if this was as free or conjugated material. Corresponding urinary recoveries of hydroquinone and catechol were 55 to 88% and 22 to 44%, respectively.

Not unexpectedly, formation of monoglucuronide and monosulfate conjugates are the main routes of metabolism with the simple dihydric phenols and, again, Williams[32] conveniently summarized the earlier literature on this subject. The conjugation of catechol and hydroquinone is probably similar to that noted above with phenol, however the data obtained from various animal species are rather limited. In rabbits, **catechol** (100 to 200 mg/kg, p.o.) was excreted in the urine as glucuronide (70 to 75% of dose) and sulfate (18%) conjugates.[6,33] In the former study free catechol (2% of dose) was isolated from the urine and traces of the ethereal sulfate of hydroxyquinol (1,2,4-trihydroxybenzene) were detected. However, Inoue et al.[8] reported that the excretion of urinary hydroxyquinol in rabbits given catechol (50 mg/kg, i.p.) was not significantly greater than that recorded in control animals. The small amounts of catechol normally present in human urine are excreted mainly in conjugated form.[34] Urine samples not treated with β-glucuronidase and sulfatase showed free catechol levels of only 1 to 5% of the total amounts.

Garton and Williams,[35] in experiments similar to those noted above with catechol, reported values of 43 and 30% for glucuronide and sulfate conjugation, respectively, of **hydroquinone**. Bray et al.[6] reported corresponding values of 78 and 18%. Miller et al.[36,37] found that the radioactivity in the urine from cats given [^{14}C]hydroquinone (20 mg/kg, i.v.) consisted mainly of its sulfate conjugate (87%) together with a glucuronide conjugate (3%) and free hydroquinone (10%). DiVincenzo et al.[38] also employed [^{14}C]hydroquinone in a study of its disposition in rats. Intragastric administration of 5, 30, and 200 mg/kg resulted in rapid urinary excretion of the radioactivity (~87% of the dose in 24 h and an average of 97% or more in 96 h). They showed that, at the highest dose, 1.1 to 1.3% of the urinary radioactivity was due to unchanged hydroquinone and that 56 and 42% consisted of the glucuronide and sulfate conjugates, respectively. No formation of hydroxyquinol (1,2,4-trihydroxybenzene) was found. Not unexpectedly, prior dosing with unlabeled hydroquinone (200 mg/kg) for 4 d caused the glucuronide/sulfate ratio to shift. In this case 72% was excreted as the glucuronide and 23% as the sulfate. In contrast to this negative finding on the formation of hydroxyquinol in rats, Inoue et al.[8] reported that this metabolite was formed in rabbits given quinol (50 mg/kg, i.p.).

In addition to the main reactions of conjugation noted above, polyhydric phenols may undergo other routes of metabolism including oxidation to quinone derivatives and, in the case of o-dihydric compounds (catechols), O-methylation. The peroxide-dependent metabolism of phenol to reactive intermediates is noted above. This phenomenon was also reported with hydroquinone[26,39] and 1,4-benzoquinone was shown to be the reactive metabolite formed from hydroquinone by macrophages.[40] The reaction also occurs with catechol which results in the formation of o-benzoquinone and glutathione conjugates of the quinone as well as in covalent binding to protein.[41,42] Sawahata and Neal[9] also reported that both hydroquinone and catechol were precursors of reactive metabolites capable of covalent binding. Tunek et al.[21] found that hydroquinone, when incubated with liver microsomes in a medium containing glutathione and an NADPH-generating system, formed a conjugate which was identified as 2-(S-glutathionyl)hydroquinone. Oxidation to quinones may proceed further to polymerization products which give the urine a dark appearance.[36]

Catechol is known to be an excellent substrate of the O-methyltransferase system in rat liver.[43] The reactivity and kinetic parameters of catechol with regard to its O-methylation by catechol-O-methyltransferase was studied by Schüsler-Van Hees and Beijersbergen Van Henegouwen.[44] Bakke[20] studied this reaction in vivo in rats. Following the oral administration of catechol (100 mg/kg, p.o.), the 48-h urines contained, in conjugated form, ~7% of the dose as guaiacol. Di-O-methylation, which would form veratrol, was not detected. As expected, resorcinol and hydroquinone did not form O-methylated metabolites.

Arbutin is the monoglucoside of hydroquinone. Both Frohne[45] and Temple et al.[46] reported that it was hydrolyzed after oral administration to rats and humans. The resulting hydroquinone then underwent conjugation with glucuronic acid or sulfate and the conjugates were excreted in the urine. Because of the hydrophilic nature of its glucose moiety, it is possible that arbutin following oral administration may be poorly absorbed from the gastrointestinal tract and therefore come into contact with the β-glucosidases produced by the gut microflora. This should result in its hydrolysis, after which the aglycon may be absorbed and metabolized to the conjugates. Drasar and Hill[47] reported that arbutin was hydrolyzed by enterobacteria (*Escherichia* sp.). Nonetheless, because arbutin is soluble in both water and polar organic solvents, it is possible that these properties may allow it to be absorbed in an unchanged form from the intestine. The results of Edwards et al.[27] noted above with phenyl glucoside support this belief. Jahodář et al.[48] confirmed this behavior with arbutin which they showed was excreted in an unchanged form in the urine of female rats. They found excretion values of 82% in 16 h and 100% in 30 h following the intragastric administration of 100 mg arbutin/kg. They also reported the lack of increase in urinary sulfate output following this dosage of the glycoside. Furthermore, they did not detect free hydroquinone in the urine of rats given 20, 100, or 400 mg arbutin/kg. The conclusions of Jahodář et al.[48] were based on thin-layer and liquid chromatography and also characterization by melting point and UV- and IR-spectra of the urinary material isolated by chromatography. Their study was thorough, however they did not exclude the presence of hydroquinone glucuronide in the excreted material by comparing their samples with an authentic sample of the glucuronide. A noteworthy finding was that enzymatic hydrolysis of the glucoside arbutin occurred following incubation with β-glucuronidase. They therefore stressed that hydroquinone liberation from a sample using this enzyme cannot be regarded as unequivocal evidence for the presence of the glucuronide conjugate of the aglycon.

Both **guaiacol**, the monomethyl ether of catechol, and **hydroquinone monomethyl ether** occur naturally. Their metabolism to glucuronide and sulfate conjugates appears to be very similar to that shown by phenol and Williams[3] found that similar amounts (~20%) of the three phenols given to rabbits in equivalent doses (250 mg/kg of phenol) were excreted as ethereal sulfates. Bray et al.,[6] in similar experiments with hydroquinone monomethyl ether using a dose of 380 mg/kg, reported respective values of 69 and 13% for glucuronide and sulfate excretion. Grischkanski[49] reported that some unchanged guaiacol was excreted in the respiratory air of rabbits given large oral doses (1 g/kg). Another possible metabolic pathway of these phenolic ethers is O-demethylation and Wong and Sourkes[50] found that the urine of rats given guaiacol (50 mg/kg, i.p.) contained the glucuronide and/or sulfate conjugates of catechol. Bray et al.[51] estimated that ~3% of a dose of hydroquinone monomethyl ether (~250 mg/kg) was excreted in the urine of rabbits as demethylated product (hydroquinone). Similar experiments with **hydroquinone dimethyl ether** also showed O-demethylation. About a third of the dose was excreted as the monomethyl ether and some hydroquinone was also formed.

The metabolism of the trihydric phenols pyrogallol and phloroglucinol closely follows the general patterns outlined above, involving formation of glucuronides and ethereal sulfates.[30] **Pyrogallol** (1,2,3-trihydroxybenzene), being a catecholic compound, would be expected to undergo O-methylation and this reaction has been shown to take place in both *in vitro* and *in vivo* experiments. Some confusion as to the nature of the product or products formed has occurred. Archer et al.[52] believed that catechol-O-methyltransferase catalyzed the stepwise methylation of pyrogallol, first to 1-O-methylpyrogallol and then to 1,2-di-O-methylpyrogallol. However, similar experiments carried out by Masri et al.[53] indicated the selective O-methylation of the middle hydroxy group, forming 2-O-methylpyrogallol. It was noted that pyrogallol is a relatively poor substrate for catechol-O-methyltransferase and Masri et al.[54] had earlier reported that only a trace of the 2-methoxy metabolite could be detected in the urine of rats given pyrogallol (50 mg, p.o.), although it is not clear if this compound was also present in the large amount of conjugated material that was excreted. A subsequent *in vivo* study in rats[55] clearly indicated that a moderate amount of pyrogallol was converted to 2-O-methylpyrogallol, which was then excreted entirely in conjugated form. This finding was confirmed by Bakke[20] who showed that ~6% of the dose (100 mg/kg, p.o.) underwent this metabolic pathway. Another reaction with pyrogallol is its dehydroxylation to resorcinol (1,3-dihydroxybenzene).[55] However, this dehydroxylation reaction, which is of minor quantitative importance, is carried out not by the tissues but by the intestinal bacteria.

Phloroglucinol (1,3,5-trihydroxybenzene), which lacks the catechol structure, is not a substrate of catechol-O-methyltransferase and no methylated derivatives were detected in the urine of rats given oral doses of 100 mg/kg of the phenol.[20] Monge et al.[56] recovered 90% of the dose (~8 mg/kg, orally) in the urine of rats in 24 h. No metabolites were excreted after this time. The only compound detected in the urine samples following hydrolysis with β-glucuronidase + sulfatase was phloroglucinol itself. It was also found that incubation of phloroglucinol with rat cecal microorganisms resulted in the disappearance of the phenol. However, the nature of the

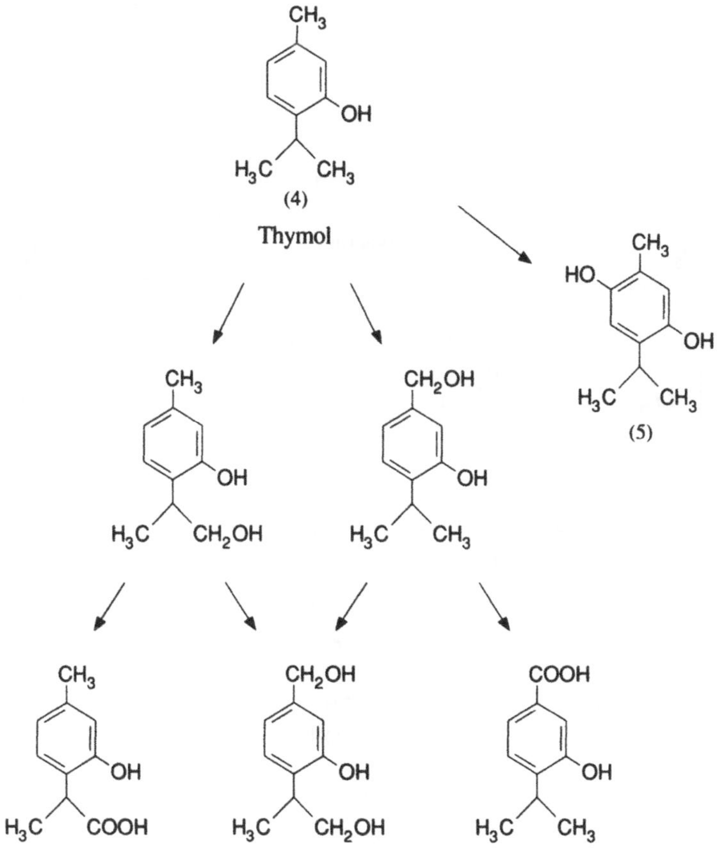

FIGURE 1. Metabolic pathways of thymol in rats.

products formed was not ascertained. Takaji et al.[57] gave phloroglucinol intravenously to humans and found that the urine contained unchanged compound (1 to 2%), phloroglucinol glucuronide (37 to 50%), and other conjugates (12 to 14%). The distribution and excretion of [³H]phloroglucinol (50 mg/kg, i.v.) in rats was studied by Fujie and Ito.[58] It is notable that more than a fifth of the injected dose was excreted in the feces, a finding no doubt related to the often demonstrated ability of the rat to excrete phenolic compounds in the bile as their glucuronide conjugates.

Of the O-methylated derivatives of pyrogallol known to occur in plants, the 1,3-dimethyl ether (**2,6-dimethoxyphenol**) was shown by Miller et al.[36,37] to be metabolized in cats mainly by hydroxylation and ethereal sulfate formation to 2,6-dimethoxyquinol disulfate. Over 90% of the urinary radioactivity was due to this metabolite when the ¹⁴C-labeled parent phenol was administered (20 or 40 mg/kg, i.v.). Small amounts of the unchanged compound and a glucuronide conjugate were detected and, at the higher dosage, a few percent of the urinary radioactivity was due to free 2,6-dimethoxyquinol. The urine samples obtained using the higher dosage were characteristically dark in color.

Thymol (4) and carvacrol (6) are two isomeric methyl isopropyl phenols. Williams[59] reviewed the earlier literature which showed that **thymol** was excreted as glucuronide and ethereal sulfate conjugates in rabbits, dogs, and man. Robbins[60] reported that dogs and man did not excrete unchanged thymol in the feces but that approximately a third of the dose (1 g) was excreted in the urine as free thymol. Williams also stated that a small amount of oxidation, giving thymoquinol (5) (see Figure 1), also occurred. More recently, Takada et al.[61] confirmed several of these findings. Following thymol administration to rabbits, they measured large increases in glucuronic acid excretion and moderate increases in ethereal sulfate excretion in the 24-h urines. They also isolated thymol glucuronide from rabbit urine (following its conversion to the methyl

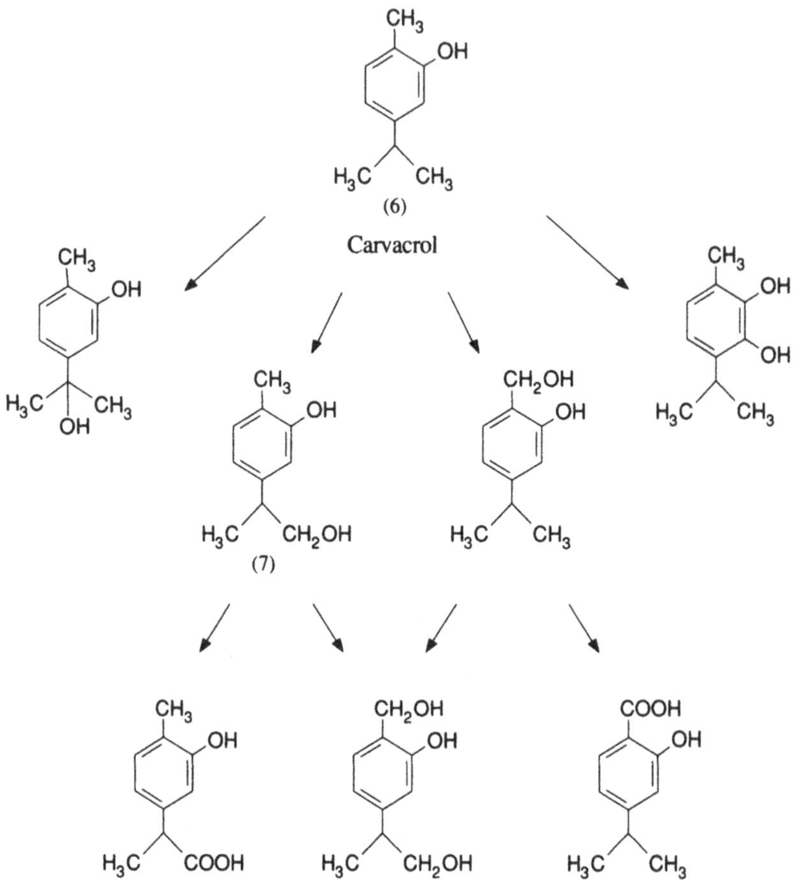

FIGURE 2.   Metabolic pathways of carvacrol in rats.

ester of the triacetyl derivative). Their results following the ingestion of thymol (0.6 g) by humans showed that thymol glucuronide, thymol sulfate, thymoquinol sulfate, and small amounts of unchanged thymol were excreted in the urine. The amounts of thymoquinol sulfate formed appeared to be very small.

Austgulen et al.[62] reported that thymol was oxidized by rats to several phenolic compounds containing hydroxy or carboxy groups and that metabolite excretion was largely complete within 24 h. This investigation used gas chromatographic-mass spectrometric techniques for metabolite identification. Following an intragastric dose of 150 mg/kg and hydrolysis of the urine with a β-glucuronidase + sulfatase preparation, thymol itself was found to be the most prominent metabolite. Thymoquinol was found in relatively small amounts and was the sole metabolite resulting from ring hydroxylation. Extensive hydroxylation of both the methyl group and the isopropyl group occurred and further oxidation at these sites to give carboxylic acid derivatives was also noted. Oxidation of the isopropyl group to give the tertiary alcohol derivative did not occur. The structures of these metabolites and their probable pathways of formation are shown in Figure 1.

The investigation by Austgulen et al.[62] also covered the metabolism of **carvacrol** (6). Previously, the only information on this subject was the report by Schröder and Vollmer[63] which noted that carvacrol was rapidly excreted in the urine by rats and rabbits and that very little appeared in the feces or expired air. The newer experiments, carried out in the same manner as those with thymol, confirmed this rapid rate of excretion, and showed many results which were similar to those summarized above for thymol. These include the detection of a single ring-hydroxylated metabolite and of analogous alcohol and carboxylic acid metabolites arising from oxidation of the methyl and isopropyl groups. Additionally, the latter group was also metabolized to a tertiary alcohol derivative. Another difference was the finding that the major metabolite from carvacrol was the alcohol derivative (7) (see Figure 2) obtained from the oxidation of one of the terminal methyl groups of the isopropyl moiety. The structures and probable pathways of formation of these metabolites of carvacrol are shown in Figure 2.

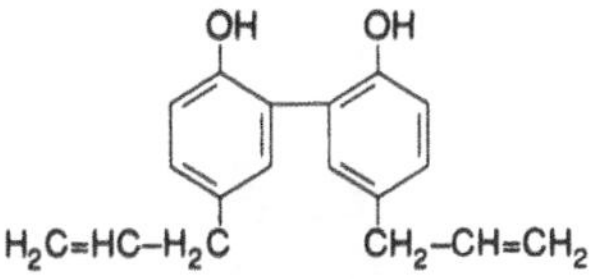

OH OH

$H_2C=HC-H_2C$ $CH_2-CH=CH_2$

(8)

Magnolol

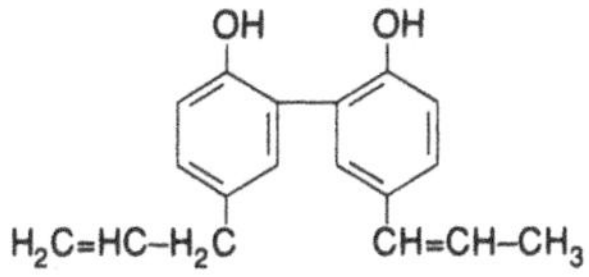

OH OH

$H_2C=HC-H_2C$ $CH=CH-CH_3$

(9)

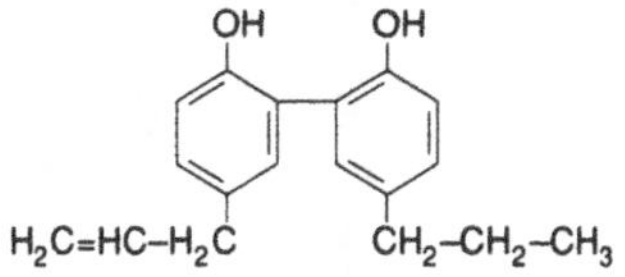

OH OH

$H_2C=HC-H_2C$ $CH_2-CH_2-CH_3$

(10)

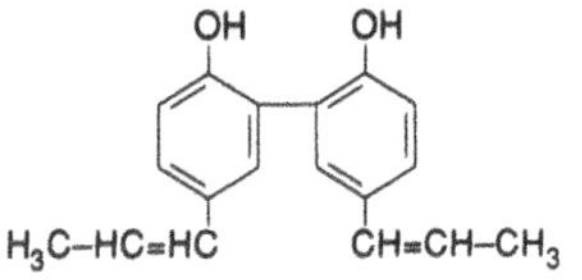

OH OH

$H_3C-HC=HC$ $CH=CH-CH_3$

(11)

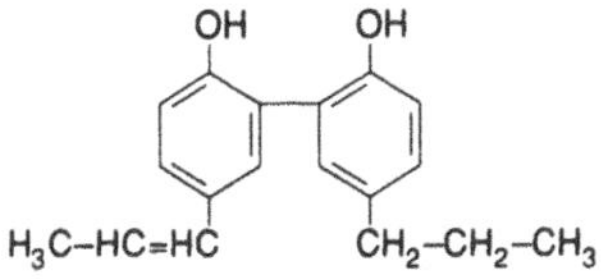

OH OH

$H_3C-HC=HC$ $CH_2-CH_2-CH_3$

(12)

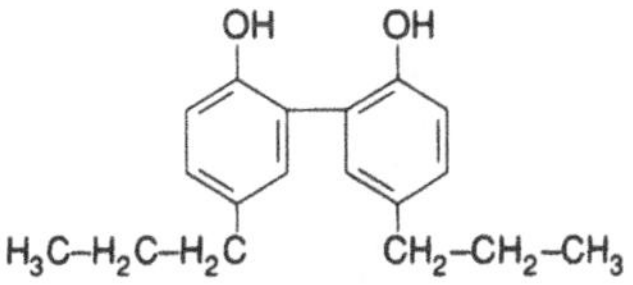

OH OH

$H_3C-H_2C-H_2C$ $CH_2-CH_2-CH_3$

(13)

The metabolism of the biphenyl derivative **magnolol** (8) has been extensively studied in rats. Hattori et al.[64] employed [ring-^{14}C]magnolol and found that most of the radioactivity was excreted in the feces after either oral or i.p. dosage (~1.7 mg/kg). In the former case, 72 and 7% of the radioactivity were recovered in the feces and urine, respectively, in 6 d. The corresponding values after i.p. administration were 67 and 12%. Most of this excretion occurred within the first 12 h. Biliary excretion of metabolites explained these similarities and 47% in 12 h (oral) or 80% in 10 h (i.p.) of the administered radioactivity was excreted in the bile. Nearly half of the biliary material was due to glucuronide conjugates, among which magnolol-2-O-glucuronide was by far the most prominent. Measurements of radioactivity at various times showed peak levels in blood and liver at 0.25 and 8 h. These results indicate that enterohepatic circulation of the biliary material took place, a conclusion which was confirmed in related experiments which used whole-body autoradiography[65] and pairs of intercannulated rats.[66] Most of the metabolites of magnolol excreted in the feces were in the free form and Hattori et al.[67] showed that these are derivatives in which the side chains were isomerized or reduced, e.g., compounds (9), (10), (11), (12), and (13). The identification of isomagnolol (11) shows that both allyl groups underwent isomerization. The most abundant of these metabolites was the fully reduced tetrahydromagnolol (13). Incubation of magnolol with rat fecal bacteria resulted in its isomerization and small amount of side-chain reduction. However, the major fecal metabolite (13) was not detected. It is therefore possible that these metabolites arise from both bacterial and tissue metabolism, however the absence of the fully reduced metabolite in the incubation experiments may be due to experimental factors.

The stilbenes furnish a relatively limited group of phenolic substances. The majority of these C_6-C_2-C_6 compounds have substituents (hydroxy and/or methoxy) in the 3- and 5-positions. None of these has been studied metabolically. In addition, a few naturally occurring stilbenes contain a single substituent in the 4-position and the metabolism of one of these, **4-hydroxy-stilbene** (14) (see Figure 3), is partially understood by virtue of the fact that it is the primary hydroxylation product of stilbene, an aromatic hydrocarbon which has been the subject of several metabolic investigations.[68-71] Several interesting findings emerge from these studies, perhaps the most notable being the discovery of several tri- and tetrahydroxylated metabolites. Furthermore, considerable reduction of the double bond of the stilbene metabolites to give bibenzyl derivatives was noted in rats. This reaction is carried out by the intestinal bacteria

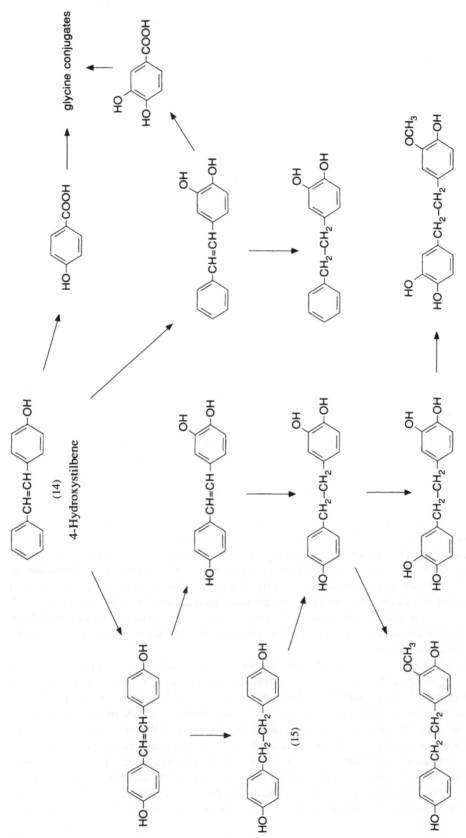

FIGURE 3.   Metabolic pathways of 4-hydroxystilbene in rats.

following the biliary excretion of glucuronide conjugates of the hydroxylated stilbenes.[72] Breault[73] reported that rabbits injected with stilbene excreted 4,4′-dihydroxybibenzyl (15) in the urine. While many of these results using stilbene fall outside the scope of this book, the findings also suggest the probable metabolic pathways of 4-hydroxystilbene in rats (Figure 3).

It is noteworthy that rats metabolize 4-hydroxystilbene more extensively than do rabbits and also that they reduce a larger portion of the dose to bibenzyl derivatives. The only polyhydroxylated metabolites found in rabbit urine were dihydroxylated (3,4- or 4,4′-) derivatives; however, cleavage to C_6–C_1-derivatives was more pronounced in rabbits than in rats.[71] On the other hand, Breault,[73] did not find that stilbene or its metabolites were metabolized to C_6–C_1-derivatives in rabbits. Scheline,[74] in experiments similar to those carried out earlier with stilbene, administered 4-hydroxystilbene to rats and confirmed the pattern of phenolic metabolites found in earlier studies. The most abundant of these metabolites were the trihydroxylated derivatives. Small amounts of *O*-methylated derivatives of the tri- and tetrahydroxy compounds were detected in the study with 4-hydroxystilbene.

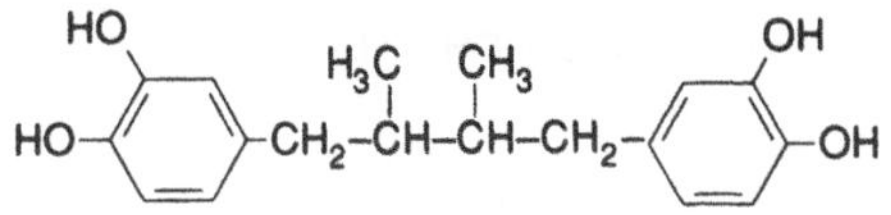

(16)

Nordihydroguaiaretic acid

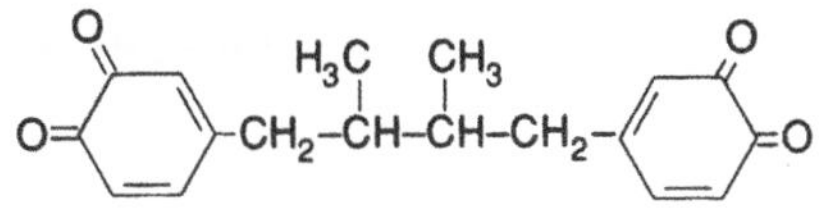

(17)

Nordihydroguaiaretic acid (16) is a plant phenol possessing antioxidant properties. Its metabolic fate is poorly understood, however Grice et al.[75] reported that the corresponding *o*-quinone derivative (17) was present in the kidneys of rats fed for 4 weeks or more on a diet containing 2% of the phenol. Urine samples from these animals contained traces of the *o*-quinone but no free phenol. Interestingly, when rats were given single 250 mg doses of nordihydroguaiaretic acid, the ileum and cecum contained ~1 and 0.6%, respectively, of the dose as the *o*-quinone after 7.5 h. This finding is somewhat surprising as the reducing environment of the lower intestine would be expected to favor the phenolic rather than the quinoid form, thus making it unlikely that the lower intestine is the site of the *o*-quinone formation as suggested by Grice et al.[75] Ve and Scheline[76] found that rats given oral doses of nordihydroguaiaretic acid (100 mg) excreted in the urine small amounts of glucuronide and/or sulfate conjugates of the unchanged compound and its monomethyl ether.

ETHERS

Dealkylation of ethers is observed with both methyl and ethyl derivatives, however the former type is most commonly encountered among plant xenobiotics. While *O*-demethylation occurs with aliphatic ethers, it is mainly seen with aromatic ethers. Interestingly, this reaction may be carried out both in the tissues by the microsomal monooxygenases and in the intestine by the microflora. It appears that bacterial *O*-demethylation may occur with a more restricted range of compounds, however it has been reported with many simple phenols, alcohols, aldehydes, ketones, acids, and flavonoids containing methoxy groups. Also, dealkylation by the microflora takes place more readily with dimethoxy-, trimethoxy-, and hydroxymethoxy-derivatives than with compounds containing a single methoxy group. On the other hand, methoxylated alkaloids may be more resistant to bacterial *O*-demethylation. Smith and Griffiths[77] found that papaverine, quinine, and reserpine were not demethylated when incubated with rat cecal microorganisms.

As with the phenolic group discussed in the preceding section, ethers are commonly found among many different types of plant compounds. Accordingly, their metabolism is dealt with in most sections of this book. The present section will therefore be restricted to a review of the metabolism of a limited number of aromatic ethers.

The metabolism of **anisole** (methyl phenyl ether), the simplest aromatic ether, was studied in rabbits by Bray et al.[78] They found that it, after intragastric administration of 400 to 500 mg/kg, was hydroxylated mainly to p-methoxyphenol but also to a small amount of o-methoxyphenol. These phenols were excreted in the urine partly unconjugated (2%) but mainly as the glucuronide (48%) and sulfate (29%) conjugates.

The remaining compounds covered in this section belong to a group of closely related ethers based on the alkenebenzene structure. These compounds can be divided into two types, derivatives of allylbenzene (18) or propenylbenzene (19). The latter group often has the prefix iso added to their common names. These compounds usually contain one or more substituents (hydroxy, methoxy, or methylenedioxy) in the 3-, 4-, and 5-positions, although 2-substitution may also occur.

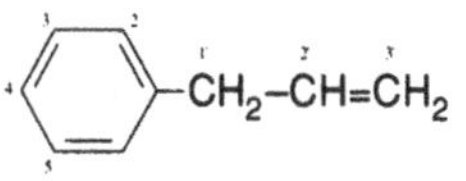

$$\text{(18)} \qquad\qquad\qquad \text{(19)}$$

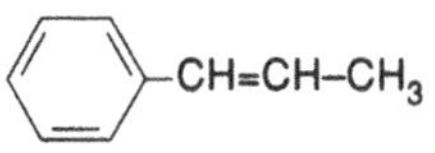

The fate of **estragole** (4-allylanisole) (20) (see Figure 4) in rats was studied by Solheim and Scheline.[79] Numerous urinary and biliary metabolites were identified and this allowed the major routes of metabolism to be proposed. Several subsequent studies have confirmed and expanded these initial findings with the result that the metabolism of estragole in rodents and humans is now reasonably well understood. The metabolites of estragole and their interrelationships are shown in Figure 4. The initial study indicated that metabolites containing alcoholic or phenolic hydroxy groups were excreted largely as glucuronide and/or sulfate conjugates whereas acidic metabolites were excreted mainly free. Zangouras et al.[80] found that mice and rats excreted the hydroxylated metabolite (21) entirely as the glucuronide conjugate.

A major metabolic reaction with estragole is O-demethylation which forms the corresponding p-hydroxy derivative chavicol (22). Solheim and Scheline[79] found that rats excreted 39% of the dose (100 mg/kg) as this phenolic compound following oral administration of estragole. The extent of O-demethylation was assessed by Anthony et al.[81] who, using [$methoxy$-^{14}C]estragole, measured the amounts of $^{14}CO_2$ exhaled 48 h after dosing. They employed five dose levels from 0.05 to 1000 mg/kg in mice (i.p. dosage) and rats (oral dosage). At the lowest levels ~50 and 55 to 60% of the dose were lost as $^{14}CO_2$ in mice and rats, respectively. The values decreased by ~15 percentage points in both species when the largest doses were given.

The pathway via the 1'-hydroxy derivative (21) is also important and gives rise to several compounds including 4-methoxyhippuric acid (23). Sangster et al.[82] showed that the latter compound is a prominent urinary metabolite in humans given a small dose of estragole. They noted that the qualitative aspects of the metabolic pathways of estragole are similar in humans and rodents. Also, the proportions of the small dose (100 mg) in humans which were excreted in the urine and as respiratory $^{14}CO_2$ resembled those found using high doses (>100 mg/kg) in rodents. The extent of formation of 1'-hydroxyestragole (21) in rodents was also shown to be related to the dose level employed.[80,81] Using the range of doses noted above, urinary excretion of this metabolite increased from ~1 to 8 to 14% with increasing doses. A previous study[83] reported that mice converted ~20% of the estragole (275 mg/kg, i.p.) to urinary 1'-hydroxyestragole. It is noteworthy that the excretion of metabolite (21) by humans given a small dose (100 mg) of estragole accounted for only 0.2 to 0.4% of the dose.[82]

Estragole epoxide (24) was detected as a biliary metabolite of estragole in rats[79] and both it and its demethylated derivative (25) were found in the urine of rats given estragole.[84] The resultant diols (26) and (27) were also detected as urinary metabolites in the latter investigation.

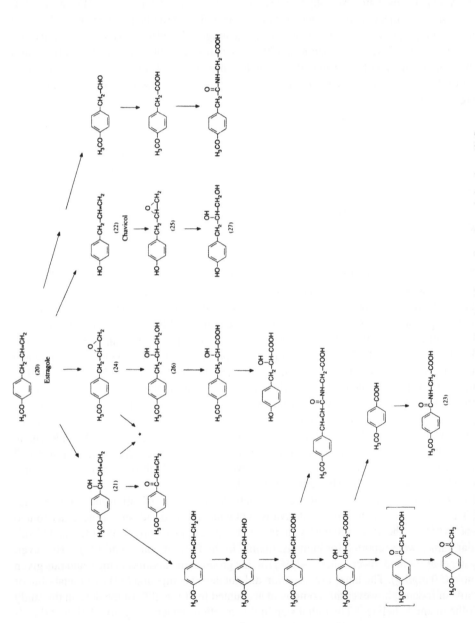

FIGURE 4. Metabolic pathways of estragole. Postulated intermediate shown in brackets. (*) See text.

Furthermore, all four of the aforementioned metabolites were formed in incubates of estragole and rat liver microsomes. The microsomal metabolism of estragole was also studied by Swanson et al.[85] who found that mouse and rat liver microsomes formed both the epoxide (24) and the 1'-hydroxy derivative (21). Additionally, they demonstrated that each of these metabolites, when themselves used as substrates, was metabolized to 1'-hydroxyestragole-2',3'-oxide (not shown in Figure 4). Another reaction not shown in Figure 4 is the formation of adducts with purine bases in DNA which is mediated by 1'-hydroxyestragole (21).[86] When compound (21) was administered to mice by injection, two major and two minor nucleoside adducts were isolated from the liver. The major products were shown to be N^2-(estragol-1'-yl)deoxyguanosine (28) and N^2-(*trans*-isoestragol-3'-yl)deoxyguanosine (29). Wiseman et al.[87] presented further information on the chemical nature of these and other adducts and showed that compound (28) consisted of a pair of diastereoisomers. These adducts were also found in mouse liver after administering estragole itself.[88,89]

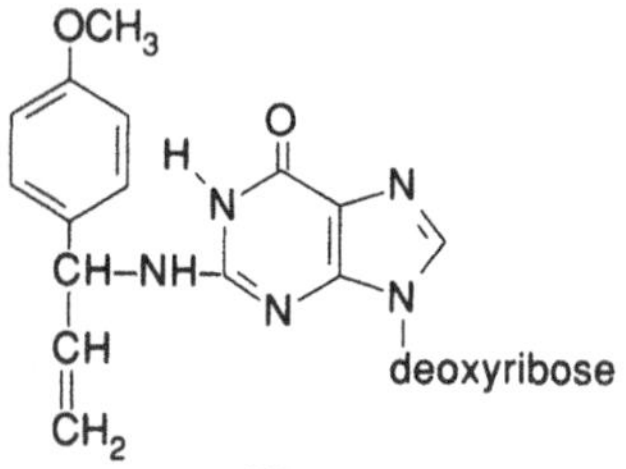

(28) 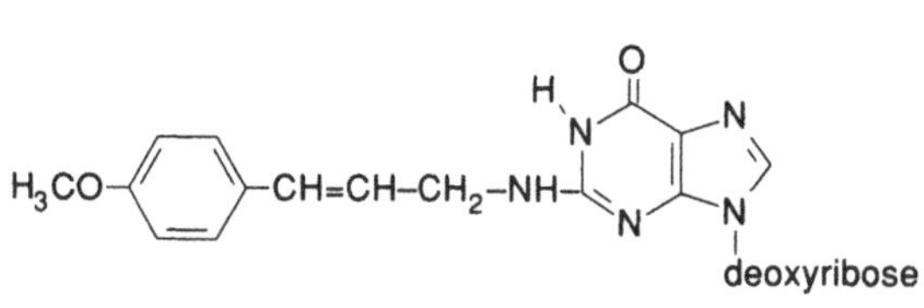   (29)

Both the *cis* and *trans* forms of **anethole** (4-propenylanisole) (30) (see Figure 5) occur naturally, however *trans*-anethole is the most abundant isomer and the one which has been studied metabolically. Figure 5 shows the pathways in its metabolism. The importance of the cinnamoyl pathway via compound (31) was first demonstrated by Le Bourhis[90] who reported that 35 to 50% and 4 to 5% of the anethole given to rats, rabbits, and humans were excreted as conjugates of anisic acid (32) and of *p*-hydroxybenzoic acid (33), respectively. The conjugated material included glucuronides, sulfates, and glycine derivatives. Strolin-Benedetti and Le-Bourhis,[91] using [^{14}C]anethole labeled in either the benzene ring or the side chain, recorded rapid elimination of the radioactivity following oral dosage (100 mg/kg) to rats. Excretion of radioactivity, which was largely complete within 24 h, occurred predominately in the urine, however it also occurred in the expired air with the compound labeled in the side chain. Subsequently, detailed investigations of the metabolism of anethole were carried out in rats,[79] mice and rats,[92] and humans.[82] These studies, which show a high degree of concordance, provided the results upon which Figure 5 is based.

The two prominent metabolic pathways of anethole involve ω–oxidation and O-demethylation. The first route, which furnishes derivatives of cinnamic acid and benzoic acid, accounted for ~60 to 65% of the dose (100 mg/kg) in rats in the form of metabolites (34), (32), and (23).[79] Similar values were reported for humans, mainly in the form of compound (23).[82] However, Sangster et al.[92] found lower values (~20%) for this group of metabolites in mice and rats given anethole (50 mg/kg). The latter investigation did not detect compound (34) as a metabolite of anethole in rodents, however this compound accounted for 15 to 20% of the dose in the study by Solheim and Scheline.[79] The initial step in this reaction sequence, hydroxylation at the 3'-position to give compound (35), was shown by Swanson et al.[85] to be carried out by hepatic microsomes from mice and rats. Randerath et al.[88] and Phillips et al.[89] reported that the DNA adducts formed in mouse liver with many alkenebenzene derivatives were produced in low levels from anethole.

The second main metabolic route of anethole is O-demethylation and the values reported by Solheim and Scheline[79] and Sangster et al.[92] agree closely. The results, which were obtained by

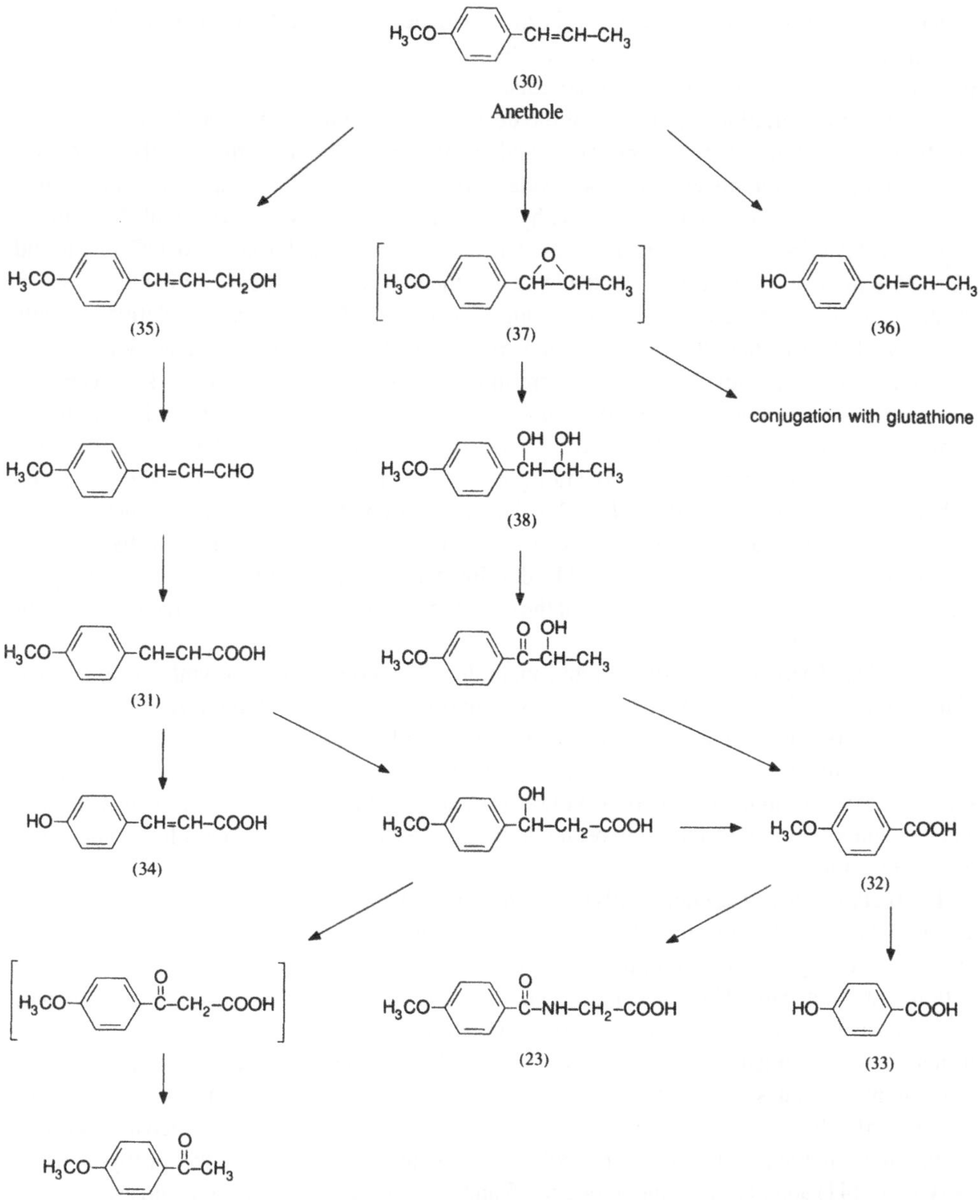

FIGURE 5. Metabolic pathways of anethole. Postulated intermediates shown in brackets.

different methods, indicated that this route was responsible for 40 to 50% of the dose. Solheim and Scheline, by quantitating the individual urinary metabolites, found that compound (36) accounted for ~30% and compound (34) ~15 to 20% of the dose. Sangster et al. employed [*methoxy*-^{14}C]anethole and measured the amount of $^{14}CO_2$ lost in the expired air. They also found that an additional 0.5% of the dose was excreted in the urine as radioactive urea. The demethylation of anethole was shown by Axelrod[93] to be carried out by the liver microsomal *O*-demethylating system.

An interesting point of difference between the results of Solheim and Scheline[79] and Sangster et al.[92] concerns the formation of sulfur-containing metabolites. Their possible formation was considered in the former study, however they were not detected with the methods used. The latter study was able to show the presence of small amounts of several derivatives (both methylthio compounds and probably mercapturic acids) which show that the putative epoxide (37) underwent conjugation with glutathione. Further evidence for the formation of anethole epoxide is seen in the excretion of small amounts of the resultant diol (38). Sangster et al.[92] also found that this diol was produced in two diastereoisomeric forms.

The effect of variations in the size of the dose on the amounts of metabolites formed from anethole was studied in mice and rats[94] and in humans.[95] In the former study [*methoxy-*[14]C]anethole was administered at dose levels of 0.05, 5, 50, and 1500 mg/kg and the results showed that the relative extent of *O*-demethylation (as measured by the amount of [14]CO_2 in the expired air) declined with increasing doses. These values decreased from 56 to 32% in rats and 72 to 35% in mice when going from the lowest to the highest dose. A concomitant increase in the amounts of urinary radioactivity was found. This resulted from increasing relative amounts of nearly all of the metabolites excreted in the urine. The total recoveries of radioactivity (~85%) was similar at all dose levels, however excretion was prolonged when 1500 mg/kg was given. The number of metabolites observed did not change over the range of doses used, a finding in contrast to that of Solheim and Scheline[79] who, using doses of 100 and 400 mg/kg, detected several of the minor metabolites only after giving the higher dose. The three oral doses given to humans[95] ranged from ~0.01 to 4 mg/kg. The disposition of radioactive *trans*-anethole was not influenced by changes in the size of the dose and the results were similar to those found previously in rodents given high doses. The dominant urinary metabolite was 4-methoxyhippuric acid (23) which accounted for ~60% of the dose and >90% of the urinary radioactivity. About 15% of the dose was lost as [14]CO_2.

The reduced derivative of estragole and anethole is 4-*n*-propylanisole, a synthetic rather than plant compound. Its metabolic fate was studied in mice and rats[96] and in humans.[82]

An alkenebenzene derivative closely related to estragole is **chavicol** (4-allylphenol) (22) (see Figure 4). While it has not itself been studied metabolically, the results obtained with estragole suggest that it may undergo oxidation of the allyl moiety via the epoxide-diol pathway. Chavicol and its metabolites would also be expected to be excreted to a large extent as glucuronide and/ or sulfate conjugates.

The metabolism of **eugenol** (4-allyl-2-methoxyphenol) (39) (see Figure 6) has received less attention than that of many other alkenebenzene derivatives, however it is clear that the presence of a free hydroxy group is an important determinant in its metabolism. Caldwell et al.[97] and Sutton et al.[98] noted that 1′-hydroxylation, which results in the formation of compound (40), was less extensive with eugenol than with some other allylbenzene derivatives because of this fact. Instead, it was excreted extensively and directly as glucuronide and sulfate conjugates in mice, rats, and man. Values of 50 to 60% of the dose (0.5 to 1000 mg/kg, p.o.) of [*ring-*[14]C]eugenol were recorded for these conjugates in rats. Conjugation with sulfate predominated at low doses of eugenol. In rats, reduction of the allylic group also occurred and the propylbenzene derivatives (41) and (42) accounted for 5 to 15 and 1 to 3%, respectively, of the dose (0.5 to 50 mg/kg). Incubation of eugenol with rat cecal microorganisms under anaerobic conditions resulted in both *O*-demethylation and reduction of the double bond. The effectiveness of eugenol in increasing the urinary excretion of ether glucuronides in rats was noted by Yuasa.[99]

The allylic hydroxylation of eugenol at the 1′-position was indicated by the results of Oswald et al.[100] who found that the nitrogen-containing metabolites formed from eugenol were similar to those identified when safrole or myristicin were administered to rats or guinea pigs (see below). A subsequent report by Oswald et al.[101] dealt with the identification of the highly labile tertiary aminopropiophenone metabolites of eugenol. These were found to be the *N,N*-dimethylamino (43a), pyrrolidinyl (43b), and piperidyl (43c) derivatives analogous to those formed from safrole (see Figure 9). The relative resistance of eugenol to 1′-hydroxylation was reported by Randerath et al.[88] and Phillips et al.[89] They found that eugenol did not form DNA adducts when given to mice. On the other hand, it is known that eugenol undergoes metabolism by the epoxide-diol pathway which is general for allylbenzenes. This was shown by Padieu and Maume[102] who found that rat liver cell lines converted it to its epoxide (44) and then to the

67

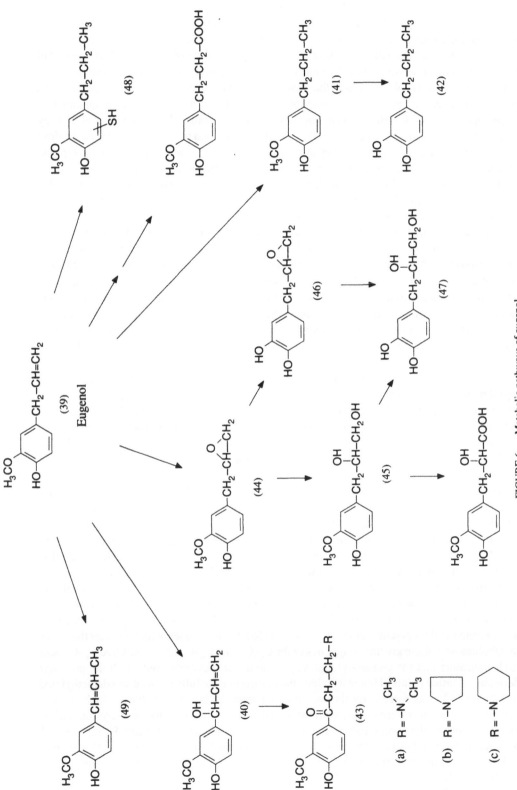

FIGURE 6.   Metabolic pathways of eugenol.

corresponding diol (45). Additionally, Delaforge et al.[84] reported that eugenol epoxide (44), its demethylated derivative 4-allylcatechol epoxide (46) and the diols (45) and (47) formed from hydrolysis of these epoxides were detected as urinary metabolites of rats given eugenol. Also, these four metabolites were formed from eugenol by rat liver microsomes. However, Swanson et al.[85] reported that the activity of mouse and rat hepatic microsomes for producing eugenol epoxide was much lower than that found with several related substrates (e.g., estragole and safrole). Thompson et al.[103,104] found that several peroxidase systems metabolized eugenol to a quinone methide intermediate having cytotoxic properties. Information on the quantitative importance of the O-demethylation of eugenol in rats was obtained by Weinberg et al.[105] They administered [*methoxy*-^{14}C]eugenol (450 mg/kg, i.p.) and found that only 0.2 to 1% of the dose was excreted as respiratory $^{14}CO_2$. They also found that the tissue radioactivity declined rapidly, although traces could be detected after 4 d. The bulk of the tissue radioactivity was due to unchanged compound, but some unidentified, water-soluble, radioactive material was also usually present. Both ether-soluble and water-soluble metabolites were detected in the excreta, however the nature of these was not determined.

The most detailed study of eugenol metabolism was reported by Fischer et al.[106] They administered eugenol (150 mg) orally to human subjects and found that that it was rapidly absorbed and metabolized. The urinary excretion of metabolites was nearly complete in 24 h and 95% of the dose was recovered. Unchanged eugenol accounted for <0.1% of the dose, however ~55% was excreted as glucuronide and sulfate conjugates of eugenol, mainly during the first few hours. All of the compounds shown in Figure 6 except the O-demethylated derivatives [(46), (47), and (42)] were identified in this study and the novel sulfur-containing compound (48) was the most prominent of these (~11%). Isomerization to isoeugenol (49) occurred and both the *trans* and *cis* forms were found in a ratio of ~7:1.

The metabolism of **eugenol methyl ether** (4-allylveratrole) (50) (see Figure 7) and **isoeugenol methyl ether** (4-propenylveratrole) (58) (see Figure 8) was studied in rats and shown to result in the excretion of a large number of urinary metabolites.[107] Additionally, Delaforge et al.[84] detected several epoxides and diols formed from eugenol methyl ether. These are compounds (51) and (52) and the corresponding demethylated compounds (44), (45), (46), and (47). All but the last of these were detected as urinary metabolites and all were formed by preparations of rat liver microsomes.

The proposed routes of metabolism of eugenol methyl ether (50) and isoeugenol methyl ether (58) are shown in Figure 7 and Figure 8, respectively. Although common metabolic pathways are observed with the mono- and dimethoxy derivatives of the allyl and propenyl compounds, O-demethylation is much less pronounced with the dimethoxy compounds. A total of only 11 to 13% due to the O-demethylated compounds (39) and (53) was found with eugenol methyl ether compared with the 39 to 46% noted with estragole. With the corresponding propenyl derivatives these values were 1% (metabolites (59) and (60)) and 31 to 32%, respectively. In summary, the major metabolic pathways of eugenol methyl ether give either the α-hydroxy acid (54), presumably via the epoxide (51) and diol (52), or the C_6–C_1- and C_6–C_3-acids which are also excreted as their glycine conjugates (55) and (56). With isoeugenol methyl ether the major metabolites were these glycine conjugates of the C_6–C_1- and C_6–C_3-acids and a free C_6–C_3-acid (61), indicating that the cinnamoyl pathway is the major metabolic route with the propenyl derivative. Further points of interest include the finding that the fully reduced and demethylated metabolite (42) was formed by the intestinal bacteria. Biliary metabolites, which in both numbers and amounts were greater with the allyl derivative than with the propenyl derivative, included most of the compounds found in the urine. Randerath et al.[88] and Phillips et al.[89] reported that eugenol methyl ether readily formed adducts with liver DNA in mice. This finding indirectly confirms the formation of the 1'-hydroxylated derivative (57) which further asserts its effect, probably on guanine residues of DNA, via 1'-esters.

Safrole (62) (see Figure 9) is structurally closely related to eugenol methyl ether (50),

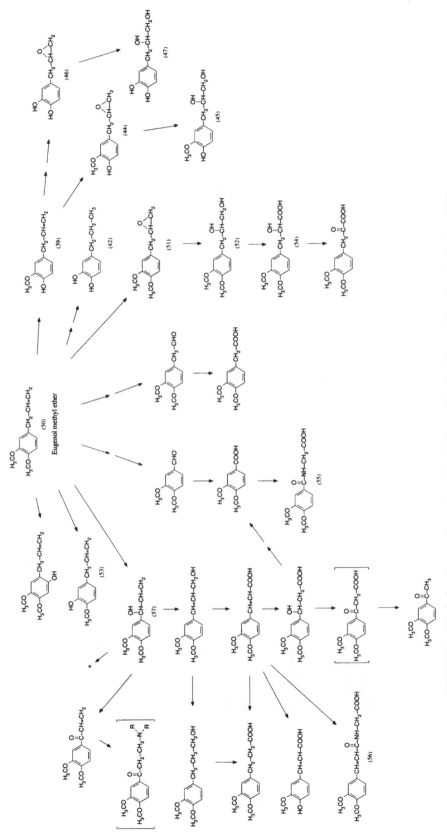

FIGURE 7. Metabolic pathways of eugenol methyl ether in rats. Postulated intermediates shown in brackets. (*) See text.

69

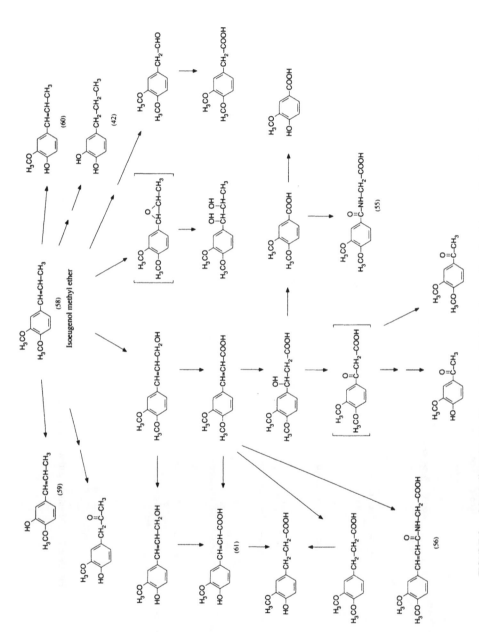

FIGURE 8.   Metabolic pathways of isoeugenol methyl ether in rats. Postulated intermediates shown in brackets.

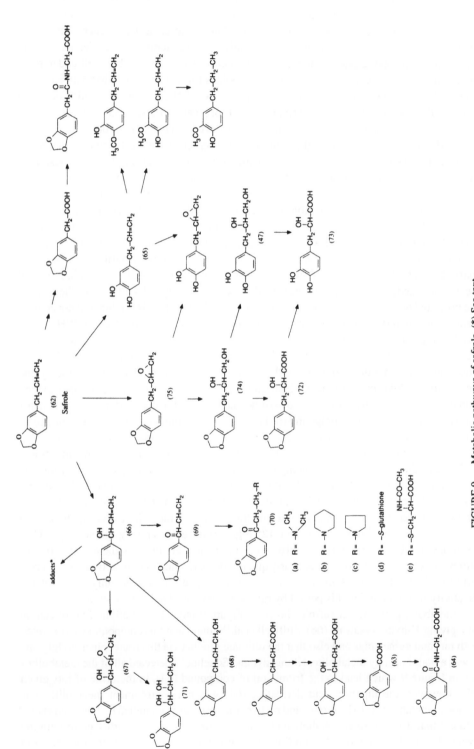

FIGURE 9.   Metabolic pathways of safrole. (*) See text.

differing only by the presence of a methylenedioxy group rather than two methoxy groups in the 3,4-position. Although hepatocarcinogenicity has been shown with other alkenebenzene derivatives (e.g., estragole and eugenol methyl ether), the demonstration of this effect with safrole has received the greatest amount of attention. Furthermore, the dependence of this effect upon metabolic activation has resulted in a relatively large number of studies devoted to elucidating the metabolic fate of safrole. Its metabolism is now reasonably well understood, although the quantitative assessments of the metabolites formed are somewhat deficient.

Metabolism studies with safrole have a long history. Heffter[108] administered it orally to rabbits and dogs and was able to detect small amounts of piperonylic acid (63) in the urine. This was the only metabolite found, however, and it was believed that much of the dose was excreted unchanged in the expired air. Several recent investigations clearly indicate, on the other hand, that safrole undergoes extensive metabolism which shows a close similarity to that described above with the related allylbenzene derivatives, estragole and eugenol methyl ether.

Cleavage of the methylenedioxy moiety appears to be a prominent metabolic reaction with safrole and this leads to the formation of catechol (3,4-dihydroxyphenyl) derivatives. This subject was studied in mice by Casida et al.[109] and Kamienski and Casida[110] who administered safrole (0.8 mg/kg, p.o.) labeled with ^{14}C in the methylenedioxy moiety. About 61% of the radioactivity was recovered as $^{14}CO_2$ in the expired air in 48 h (mostly during the first 12 h). Urinary radioactivity amounted to 23% of the dose, mainly as water-soluble metabolites, and the feces contained 3%. Five metabolites were detected chromatographically in the ether extract of the urine and one of these, piperonylic acid (63), was identified. This fraction did not contain piperonylglycine (64). *In vitro* experiments using the mouse liver microsome-NADPH system showed that safrole was demethylenated, yielding [^{14}C]formate and a catechol derivative which, however, was not identified. Anders et al.[111] studied the mechanism of the demethylenation reaction and found that the initial step in the reaction sequence is oxidation of the methylene carbon to form a hydroxy derivative which ultimately yields formate and, to a minor extent, carbon monoxide. It seems reasonable to assume that the catechol detected in the experiments with safrole was the expected compound (65) as it, mainly in conjugated form, was found to be a major urinary metabolite of safrole in rats and guinea pigs.[112] Similarly, Strolin Benedetti et al.[113] reported that this conjugate was the main urinary metabolite of [1′-^{14}C]safrole in rats and man. These findings of extensive demethylenation of safrole were subsequently confirmed by Klungsøyr and Scheline.[114] They found that rats excreted in the urine 72% of an oral dose (162 mg/kg) as 4-allylcatechol (65), largely within 2 d. The proposed metabolic pathways of safrole are shown in Figure 9 which is mainly based on the findings of these three investigations. The i.p. doses employed by Stillwell et al. were 125 mg/kg (rats) and 50 mg/kg (guinea pigs). Oral doses of 0.8 or 750 mg/kg (rats) and ~0.16 or 1.7 mg (man) were given in the investigation of Strolin Benedetti et al. who found that nearly quantitative excretion of radioactivity in the urine occurred in 24 h when small amounts of safrole were given. However, only ~25% of the radioactivity was excreted in this period by rats which had received the larger dose.

The metabolic pathways of safrole shown in Figure 9 indicate that allylic hydroxylation occurs, giving 1′-hydroxysafrole (66). Stillwell et al.[112] found that this compound was excreted by both rats and guinea pigs and also that the ultimate product via this route, piperonylglycine (64), was a major urinary metabolite. Klungsøyr and Scheline[114] detected all of the metabolites shown in Figure 9 which lead to the formation of compound (64), in the urine of rats given safrole; however, the hippuric acid derivative (64) was not a prominent metabolite. The formation of metabolite (66) by mouse and rat liver microsomes was studied by Swanson et al.[85] who found that it was further metabolized to the epoxide (67). A noteworthy point with the pathway of 1′-hydroxylation is the finding that the isomeric allyl alcohol (68) arises from compound (66) by a chemical rearrangement reaction. This was reported by Borchert et al.[115] who directed their attention to the allylic hydroxylation of safrole to form metabolite (66) in mice, rats, hamsters, and guinea pigs. The urinary excretion of compound (68) following the administration of 1′-hydroxysafrole itself was reported by Peele and Oswald.[116] Borchert et al.

found that as much as a third of the dose of safrole (300 mg/kg, i.p.) was excreted as the glucuronide conjugate of (66) in mice whereas the corresponding values in the other species were only 1 to 3%. Klungsøyr and Scheline[114] reported a value of 4% for the urinary excretion of metabolite (66) in rats given about half of this dose. Stillwell et al.[112] noted that most of the excreted 1'-hydroxysafrole in rats and guinea pigs was in the form of conjugated material and Levi et al.[117] reported on the urinary excretion of several glucuronides of safrole metabolites by rats. Strolin Benedetti et al.[113] found that metabolite (66) and its isomer, 3'-hydroxyisosafrole (68), were excreted in conjugated form in the urine of rats but not humans.

The carcinogenic properties of safrole are closely linked to the formation of 1'-hydroxysafrole which, after further metabolism to a 1'-ester, may react covalently with hepatic protein, DNA, and RNA.[118] The latter report showed that alkaline digestion of hepatic protein from rats or mice given 1'-hydroxysafrole liberated 3'-methylmercaptoisosafrole. Phillips et al.[119] and Wiseman et al.[87] showed that several adducts with purine bases in DNA were formed, mainly with guanine residues. They characterized several of these adducts and found that the major ones were N^2-(safrol-1'-yl)deoxyguanosine, which was present as a diastereoisomeric pair, and N^2-(*trans*-isosafrol-3'-yl)deoxyguanosine. The structures of these major adducts correspond to those formed with estragole noted above (compounds (28) and (29), respectively). While the above investigations employed 1'-hydroxysafrole, Randerath et al.[88] and Phillips et al.[89] showed that the adducts were formed *in vivo* in mice injected with safrole itself.

Another consequence of the formation of 1'-hydroxysafrole is its oxidation to 1'-oxosafrole (69). While Stillwell et al.[112] and Borchert et al.[120] reported that this compound was not detected in the urines of safrole-treated rats, Borchert et al.[115] believed that it is formed *in vivo*. Subsequently, Peele and Oswald[116] detected the keto compound in the urine of animals given 1'-hydroxysafrole. 1'-Oxosafrole is implicated in the formation of tertiary aminopropiophenone derivatives. These are condensation products of 1'-oxosafrole with several endogenous amines. Their excretion in the urine was first described by Oswald et al.[121] and they were subsequently characterized.[122] The tertiary aminopropiophenone derivative excreted by the guinea pig was shown to be the *N,N*-dimethyl derivative (70a) whereas in rats the major component was the piperidyl derivative (70b) with some *N,N*-dimethyl derivative and trace amounts of the pyrrolidinyl derivative (70c) also being present. Peele and Oswald[116] reported that administration of synthetic 1'-hydroxysafrole to rats resulted in the urinary excretion of the tertiary aminopropiophenone derivatives (70a) and (70b) and 3,4-methylenedioxyhippuric acid (64). The basic metabolites were not formed when 3'-hydroxyisosafrole (68) was given. A chemical study of the formation of the tertiary aminopropiophenone metabolites of safrole was reported by McKinney et al.[123] Another pathway of metabolism of 1'-oxosafrole is conjugation with glutathione.[124] They showed that 1'-hydroxysafrole was oxidized in both rats and mice to 1'-oxosafrole which was then metabolized in part to the glutathione and *N*-acetylcysteine conjugates (70d) and (70e), respectively.

Another important metabolic pathway of safrole shown in Figure 9 is that which leads to the formation of several epoxides and their resultant diols. This pathway also leads to the triol (71), an interesting polyhydroxylated metabolite which is excreted in small amounts by rats but not guinea pigs.[112] Final products of this metabolic route are the α-hydroxy acids (72) and (73). The formation of relatively large amounts of α-hydroxy acids was noted above with related allylbenzenes and it is therefore not surprising that Stillwell et al.[112] reported that compound (72) was a major urinary metabolite of safrole in rats and guinea pigs. The finding of the excretion of the 2',3'-diols (74) and (47) and the α-hydroxy acids following administration of safrole epoxide (75) further supports the existence of the epoxide-diol pathway. Klungsøyr and Scheline[114] found that compounds (74) and (72) were prominent urinary metabolites of safrole in rats. Watabe and Akamatsu[125] demonstrated the conversion of safrole oxide (75) to the diol (74) by hepatic microsomal epoxide hydrolase and several subsequent studies have dealt with the metabolism of safrole via the epoxide-diol pathway.[84,85,102,126-128]

The earliest report on the metabolism of **isosafrole** (76) (see Figure 10) was by Heffter[108] who

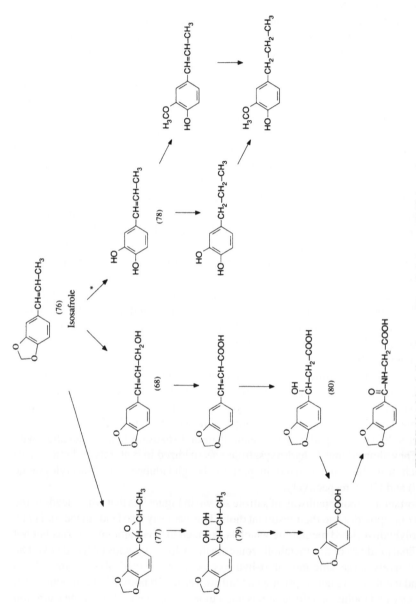

FIGURE 10. Metabolic pathways of isosafrole in rats. (*) Major pathway.

was unable to detect any piperonylic acid in the urine of isosafrole-treated rabbits. He believed that most of the dose was lost unchanged in the expired air, however an indication that this was probably not the case was reported by Fishbein et al.[129] who detected several metabolites of isosafrole in rat bile and urine. Klungsøyr and Scheline[130] identified ten urinary metabolites of isosafrole following intragastric doses (162 mg/kg) to rats. They recovered 89% of the dose within 72 h, however most of the excretion occurred within the first 24 h. The urine samples were treated with a β-glucuronidase + sulfatase preparation prior to metabolite identification. The metabolites found and their likely pathways of formation are shown in Figure 10. Isosafrole epoxide (77), which was not detected in this study, is included in Figure 10 because it was previously shown by Janiaud et al.[127] to be formed from isosafrole by cultures of rat liver cells.

The major metabolic route of isosafrole in the rat was found by Klungsøyr and Scheline[130] to be demethylenation to compound (78). The four demethylenated metabolites shown in Figure 10 accounted for 92% of the identified material. Metabolite (78) comprised ~96% of the demethylenated material (78.5% of the dose). Identification of the diol (79) which accounted for ~2% of the dose confirmed the report of Padieu and Maume[102] that isosafrole is a substrate for the epoxide-diol pathway in rats. The cinnamoyl pathway via compound (68) accounted for only a few percent of the dose of isosafrole. Peele and Oswald[116] identified most of the metabolites depicted in Figure 10 as arising from compound (68) following the administration of the latter compound itself. Additionally, they detected 3,4-methylenedioxyacetophenone which is probably an artifact formed from decarboxylation of the keto acid derived from (80). Acetophenone derivatives are believed to be formed in a similar manner following administration of the methyl ethers of eugenol and isoeugenol (see Figures 7 and 8).

Oswald et al.[121] reported that isosafrole behaved similarly to safrole with respect to the formation of ketone derivatives and nitrogen-containing metabolites. However, allylic hydroxylation of isosafrole produces the 3-phenylallyl alcohol derivative (68) rather than the 1-phenylallyl alcohol derivative (66) (see Figure 9) which is probably the precursor of the postulated ketone intermediate. Therefore, the significance of this finding and the actual nature of the metabolites detected remains unclear. In fact, Peele and Oswald[116] found that the 3'-hydroxy derivative (68), when itself given to rats, did not form nitrogen-containing metabolites.

The reduced derivative of safrole and isosafrole is dihydrosafrole, a synthetic rather than plant compound. Its metabolic fate was studied in rats[130] and in dogs and monkeys.[131]

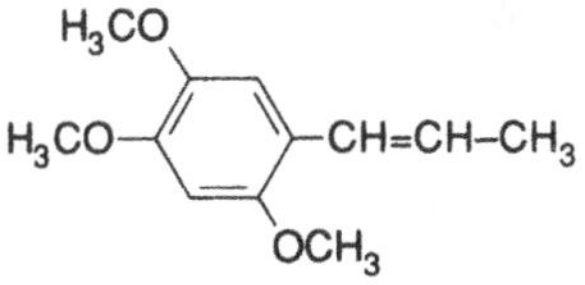

H₃CO

H₃CO—[ring]—CH=CH–CH₃

OCH₃

(81)

Asarone (*trans*)

β-Asarone(*cis*)

Asarone (*trans* isomer of (81)) and **β-asarone** (*cis* isomer of (81)) are structural isomers of isoelemicin (87). The only report dealing with their metabolic fate is that of Oswald et al.[121] who studied the urinary excretion of nitrogen-containing metabolites of alkenebenzene derivatives. With the asarones the effect of *cis* and *trans* double bonds on the formation of these metabolites could be studied and it was found that the *trans* isomer underwent this reaction (producing ninhydrin-positive material) more extensively than did the *cis* isomer. Asarone gave rise to three such metabolites, however their identities have not been ascertained.

The metabolism of **elemicin** (82) (see Figure 11) and **isoelemicin** (87) (see Figure 12) in rats was studied by Solheim and Scheline.[132] Urinary and biliary metabolites were identified

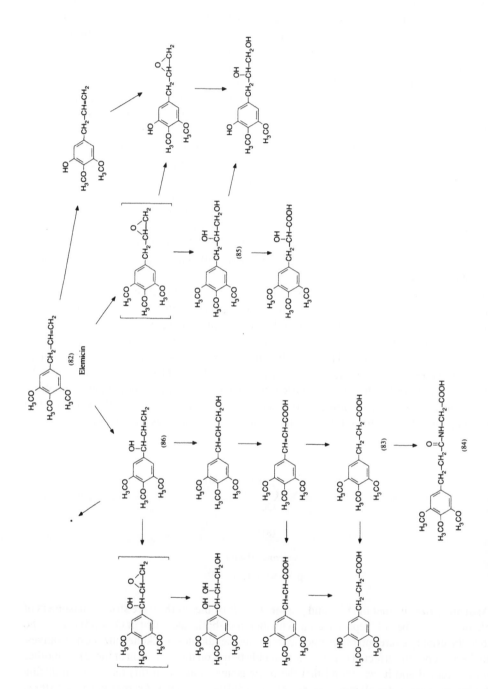

FIGURE 11.   Metabolic pathways of elemicin in rats. Postulated intermediates shown in brackets. (*) See text.

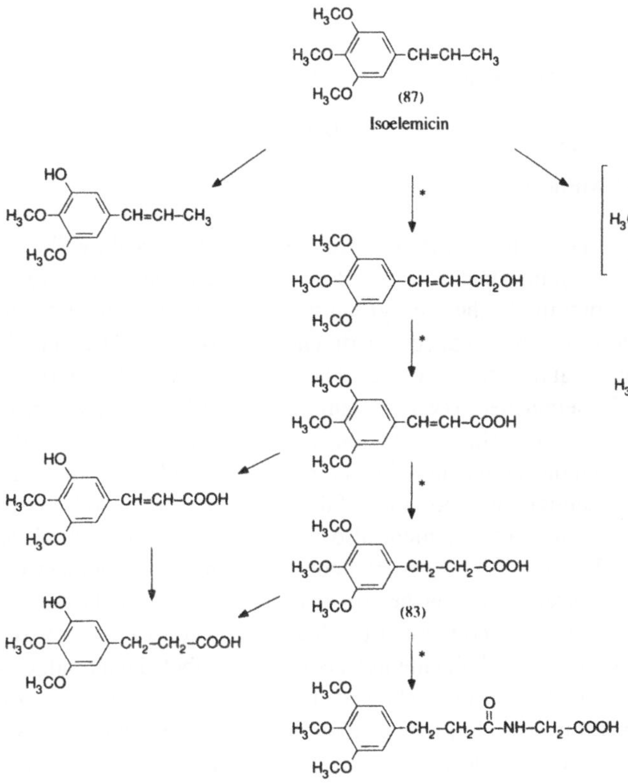

FIGURE 12. Metabolic pathways of isoelemicin in rats. Postulated intermediate shown in brackets. (*) Major pathway.

following oral doses of 400 mg/kg. The same general pathways of metabolism were found with these trimethoxy derivatives as with the mono- and dimethoxy derivatives noted above; however, the extent of O-demethylation was less with the trimethoxy than with the dimethoxy compounds. Furthermore, the total amounts of identified metabolites were much smaller with the trimethoxy derivatives. The probable pathways of metabolism of elemicin are shown in Figure 11 which illustrates both the cinnamoyl pathway leading to the C_6–C_3-acid (83) and its glycine conjugate (84) and the epoxide-diol pathway leading to the diol (85). These three compounds were the most prominent urinary metabolites. A pathway via compound (86) was shown by Randerath et al.[88] and Phillips et al.[89] to give significant levels of adducts with DNA in mouse liver.

Solheim and Scheline[132] found that isoelemicin (87) was metabolized mainly via the cinnamoyl pathway as shown in Figure 12. Compound (83) was the most prominent urinary metabolite. It is noteworthy that β-oxidation to C_6–C_1-derivatives, as found with the mono- and dimethoxy derivatives discussed above, was not observed with elemicin or isoelemicin. Other metabolites which may be formed from these compounds include sulfur- and nitrogen-containing compounds. Solheim and Scheline obtained evidence for the formation of small amounts of the former and reported reductions in the levels of liver glutathione after dosing. Oswald et al.[133] reported that elemicin was converted to a small extent to tertiary aminopropiophenone derivatives. These metabolites are the 3-N,N-dimethylamino-, 3-piperidyl-, and 3-pyrrolidinyl-1-(3′,4′,5′-trimethoxyphenyl)-1-propanones which are analogous to the safrole metabolites (70a), (70b), and (70c), respectively (see Figure 9).

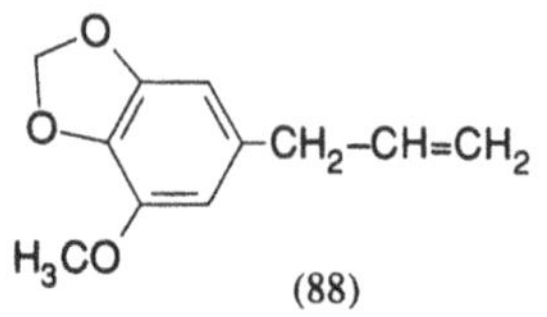

(88)

Myristicin

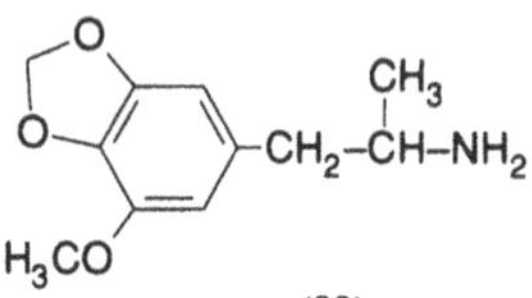

(89)

Several possible routes of metabolism of **myristicin** (88) were discussed by Shulgin[134] in relationship to the possible psychotropic activity of this allylbenzene derivative. It was suggested that addition of ammonia to the allyl group may occur to give the amphetamine derivative (89). This hypothesis was investigated experimentally by Braun and Kalbhen[135] using the isolated perfused rat liver or rat liver homogenates. Their results showed that myristicin was converted to 3-methoxy-4,5–methylenedioxyamphetamine (89) by both preparations. Oxygenation of the liver homogenates during incubation markedly increased the yield of the metabolite and this was felt to indicate that an oxidation reaction precedes introduction of the amino group. However, the quantitative importance of this pathway remains to be assessed.

Peele[136] reported on the metabolism of myristicin, also labeled with ^{14}C in the methylenedioxy or methoxy groups, in rats. With the ^{14}C-labeled compounds, excretion of radioactivity was nearly complete in 48 h and occurred largely by the urinary route. Only ~2% of the radioactivity was lost as respiratory $^{14}CO_2$ when the compound labeled in the methoxy group was given, however this value increased to 25% with the methylenedioxy-^{14}C-labeled material. Casida et al.[109] and Kamienski and Casida[110] similarly showed that myristicin underwent extensive demethylenation *in vivo*. When mice were given a small dose (~1 mg/kg, p.o.) of myristicin labeled with ^{14}C in the methylenedioxy moiety, 73% of the radioactivity was lost in 48 h as $^{14}CO_2$. Most of this excretion occurred during the first 12 h and the 48-h recoveries of urinary and fecal radioactivity were 15 and 3%, respectively. *In vitro* experiments using the mouse liver microsome-NADPH system indicated demethylenation to give formate and the corresponding catechol, however the two metabolites which were detected were not identified.

Peele[136] also identified numerous urinary metabolites of myristicin in rats. The results obtained from the 24-h urine samples hydrolyzed with β-glucuronidase showed similarities to that illustrated in Figure 9 for safrole. Thus, 5-hydroxyeugenol, the catechol metabolite arising from demethylenation of myristicin, was identified. This metabolite is analogous to metabolite (65) produced from safrole. Additional metabolites were myristicin glycol (analogous to metabolite (74)), 1′-hydroxymyristicin (analogous to metabolite (66)), and 3′-hydroxyisomyristicin (analogous to metabolite (68)). Additional evidence for the 1′-hydroxylation of myristicin was presented by Randerath et al.[88] and Phillips et al.[89] who found that significant levels of DNA-binding occurred in the livers of mice given myristicin. Acidic metabolites reported by Peele[136] included the methoxymethylenedioxy derivatives of benzoic and phenyllactic acids analogous to the safrole metabolites (63) and (72), respectively. Several additional metabolites were identified including the similarly substituted phenylacetic acid derivative and myristicin acetophenone. These pathways are known to occur with eugenol methyl ether (see Figure 7) and the acetophenone derivative is probably an artifact produced from decarboxylation of the corresponding β-keto acid. Another metabolite of myristicin was tentatively identified as dihydromyristicin which is formed by the reduction of the allylic double bond.

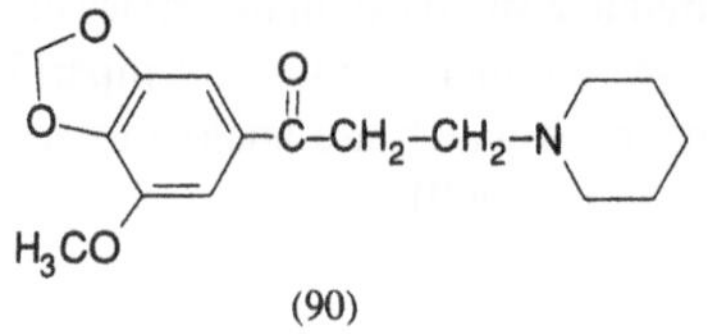

(90)

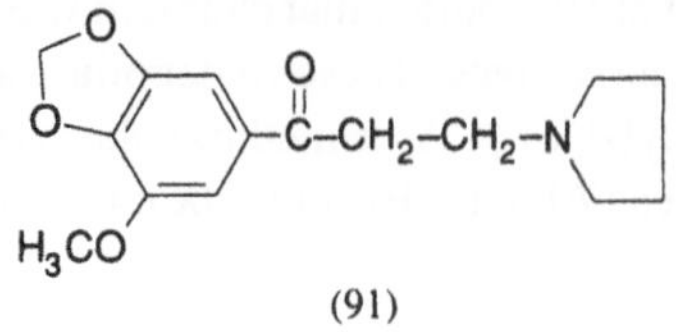

(91)

In addition to these neutral and acidic metabolites of myristicin, Peele[136] also observed the presence of Mannich base derivatives. The urinary excretion of these nitrogen-containing metabolites of myristicin and several related alkenebenzene derivatives was first reported by Oswald et al.[121] and further investigated by Oswald et al.[100] who showed that these compounds are, in the rat, mainly 3-piperidyl-1-(3'-methoxy-4',5'-methylenedioxyphenyl)-1-propanone (90) and, in the guinea pig, mainly the corresponding 3-pyrrolidinyl derivative (91). Trace amounts of the other of these tertiary aminopropiophenone derivatives were also excreted by these two species, however a third type, the *N,N*-dimethylamino derivative analogous to that formed from safrole (Figure 9, compound (70a)), was not detected.

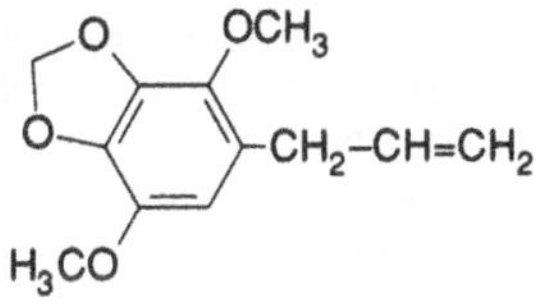

(92)   (93)

Parsley apiole   Dill apiole

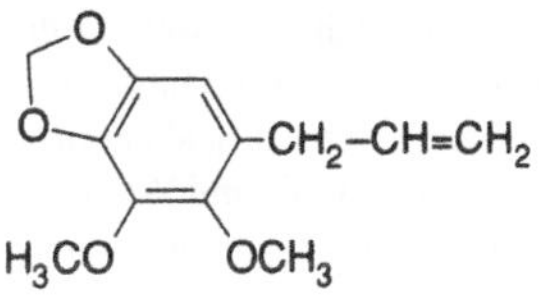

Apiol (92) is 2-methoxymyristicin and is also known as parsley apiole. Heffter[108] reported that it was extensively oxidized in the body and that an unidentified metabolite was excreted in the urine of both rabbits and dogs. Dill apiole (93) is a positional isomer of apiole. It seems reasonable to assume that both isomers will be metabolized along the pathways described above for related allylbenzenes, however the presence of the 2-methoxy group may possibly reduce the extent of allylic hydroxylation and thus the production of the key intermediate, 1'-hydroxyapiol. Randerath et al.[88] and Phillips et al.[89] found that both parsley apiole and dill apiole formed adducts with nucleotides in DNA in mouse liver. The amounts formed were relatively small, however their presence indicates that 1'-hydroxylation occurred.

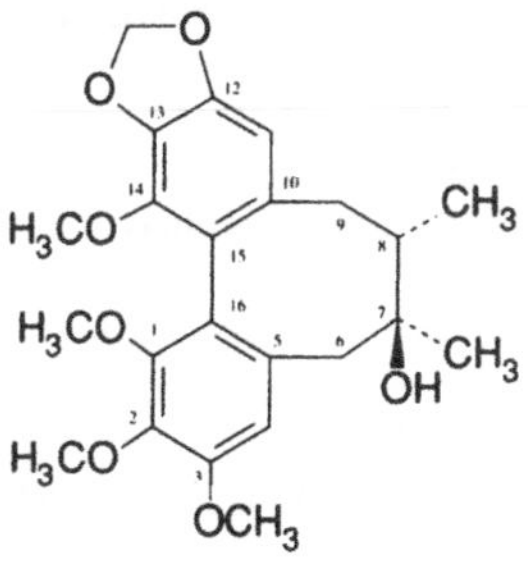

(94)

Gomisin A

The metabolism of **gomisin A** (94), a complex aromatic ether, was studied by Ikeya et al.[137] using the 9000-g supernatant fraction from rat liver homogenates. A major metabolite was the demethylenated derivative having hydroxy groups at both C-12 and C-13. Four additional oxidized metabolites were identified. These were two monohydroxylated derivatives (at C-8 or C-9) and two mono-*O*-demethylated derivatives (at C-2 or C-14).

REFERENCES

1. **Capel, I. D., Millburn, P., and Williams, R. T.,** Monophenyl phosphate, a new conjugate of phenol in the cat, *Biochem. Soc. Trans.*, 2, 305, 1974.
2. **Kao, J., Bridges, J. W., and Faulkner, J. K.,** Metabolism of [¹⁴C]phenol by sheep, pig and rat, *Xenobiotica*, 9, 141, 1979.
3. **Williams, R. T.,** Studies in detoxication. I. The influence of (a) dose and (b) *o-*, *m-* and *p-*substitution on the sulphate detoxication of phenol in the rabbit, *Biochem. J.*, 32, 878, 1938.
4. **Bray, H. G., Humphris, B. G., Thorpe, W. V., White, K., and Wood, P. B.,** Kinetic studies of the metabolism of foreign organic compounds. IV. The conjugation of phenols with sulphuric acid, *Biochem. J.*, 52, 419, 1952.
5. **Porteous, J. W. and Williams, R. T.,** Studies in detoxication. IXX. The metabolism of benzene. I. (a) The determination of phenol in urine with 2:6-dichloroquinonechloroimide. (b) The excretion of phenol, glucuronic acid and ethereal sulphate by rabbits receiving benzene and phenol. (c) Observations on the determination of catechol, quinol and muconic acid in urine, *Biochem. J.*, 44, 46, 1949.
6. **Bray, H. G., Thorpe, W. V., and White, K.,** Kinetic studies of the metabolism of foreign organic compounds. V. A mathematical model expressing the metabolic fate of phenols, benzoic acids and their precursors, *Biochem. J.*, 52, 423, 1952.
7. **Parke, D. V. and Williams, R. T.,** Studies in detoxication. LIV. The metabolism of benzene. (a) The formation of phenylglucuronide and phenylsulphuric acid from [¹⁴C]benzene. (b) The metabolism of [¹⁴C]phenol, *Biochem. J.*, 55, 337, 1953.
8. **Inoue, O., Seiji, K., and Ikeda, M.,** Pathways for formation of catechol and 1,2,4-benzenetriol in rabbits, *Bull. Environ. Contam. Toxicol.*, 43, 220, 1989.
9. **Sawahata, T. and Neal, R. A.,** Biotransformation of phenol to hydroquinone and catechol by rat liver microsomes, *Mol. Pharmacol.*, 23, 453, 1983.
10. **Baudinette, R. V., Wheldrake, J. F., Hewitt, S., and Hawke, D.,** The metabolism of [¹⁴C] phenol by native Australian rodents and marsupials, *Aust. J. Zool.*, 28, 511, 1980.
11. **Wheldrake, J. F., Baudinette, R. V., and Hewitt, S.,** The metabolism of phenol in a desert rodent *Notomys alexis*, *Comp. Biochem. Physiol.*, 61C, 103, 1978.
12. **Ramli, J. B. and Wheldrake, J. F.,** Phenol conjugation in the desert hopping mouse, *Notomys alexis*, *Comp. Biochem. Physiol.*, 69C, 379, 1981.
13. **Oehme, F. W. and Davis, L. E.,** Comparative study of biotransformation and excretion of phenol, *Pharmacologist*, 11, 241, 1969.
14. **Powell, G. M., Miller, J. J., Olavesen, A. H., and Curtis, C. G.,** Liver as major organ of phenol detoxication?, *Nature (London)*, 252, 234, 1974.
15. **Shirkey, R. J., Kao, J., Fry, J. R., and Bridges, J. W.,** A comparison of xenobiotic metabolism in cells isolated from rat liver and small intestinal mucosa, *Biochem. Pharmacol.*, 28, 1461, 1979.
16. **Hogg, S. I., Curtis, C. G., Upshall, D. G., and Powell, G. M.,** Conjugation of phenol by rat lung, *Biochem. Pharmacol.*, 30, 1551, 1981.
17. **Cassidy, M. K. and Houston, J. B.,** Phenol conjugation by lung *in vivo*, *Biochem. Pharmacol.*, 29, 471, 1980.
18. **Smith, R. L.,** Excretion of drugs in bile, in *Concepts in Biochemical Pharmacology*, Part 1, Brodie, B. B. and Gillette, J. R., Eds., Springer Verlag, Berlin, 1971, 354.
19. **Gbodi, T. A. and Oehme, F. W.,** The fate of phenol, *o-*phenylphenol, and disophenol in rats, *Toxicol. Appl. Pharmacol.*, 45, 223, 1978.
20. **Bakke, O. M.,** *O*-Methylation of simple phenols in the rat, *Acta Pharmacol. Toxicol.*, 28, 28, 1970.
21. **Tunek, A., Platt, K. L., Przybylski, M., and Oesch, F.,** Multi-step metabolic activation of benzene. Effect of superoxide dismutase on covalent binding to microsomal macromolecules, and identification of glutathione conjugates using high pressure liquid chromatography and field desorption mass spectrometry, *Chem. Biol. Interact.*, 33, 1, 1980.
22. **Sawahata, T. and Neal, R. A.,** Horseradish peroxidase-mediated oxidation of phenol, *Biochem. Biophys. Res. Commun.*, 109, 988, 1982.
23. **Subrahmanyam, V. V. and O'Brian, P. J.,** Phenol oxidation product(s), formed by a peroxidase reaction, that bind to DNA, *Xenobiotica*, 15, 873, 1985.
24. **Eastmond, D. A., Smith, M. T., Ruzo, L. O., and Ross, D.,** Metabolic activation of phenol by human myeloperoxidase and horseradish peroxidase, *Mol. Pharmacol.*, 30, 674, 1986.
25. **Eastmond, D. A., French, R. C., Ross, D., and Smith, M. T.,** Metabolic activation of 1-naphthol and phenol by a simple superoxide-generating system and human leukocytes, *Chem. Biol. Interact.*, 63, 47, 1987.
26. **Schlosser, M. J., Shurina, R. D., and Kalf, G. F.,** Metabolism of phenol and hydroquinone to reactive products by macrophage peroxidase or purified prostaglandin H synthase, *Environ. Health Perspect.*, 82, 229, 1989.
27. **Edwards, V. T., Jones, B. C., and Hutson, D. H.,** A comparison of the metabolic fate of phenol, phenyl glucoside and phenyl 6-*O*-malonyl-glucoside in the rat, *Xenobiotica*, 16, 801, 1986.
28. **Merker, P. C., Yeung, D., Doughty, D., and Nacht, S.,** Pharmacokinetics of resorcinol in the rat, *Res. Commun. Chem. Pathol. Pharmacol.*, 38, 367, 1982.

29. **Kim, Y. C. and Matthews, H. B.,** Comparative metabolism and excretion of resorcinol in male and female F344 rats, *Fundam. Appl. Toxicol.*, 9, 409, 1987.
30. **Williams, R. T.,** *Detoxication Mechanisms*, Chapman and Hall, London, 1959, 306.
31. **Martin, A. K., Milne, J. A., and Moberley, P.,** Urinary quinol and orcinol outputs as indices of voluntary intake of heather (*Calluna vulgaris* L. (Hull)) by sheep, *Proc. Nutr. Soc.*, 34, 70, 1975.
32. **Williams, R. T.,** *Detoxication Mechanisms*, Chapman and Hall, London, 1959, 303.
33. **Garton, G. A. and Williams, R. T.,** Studies in detoxication. XVII. The fate of catechol in the rabbit and the characterization of catechol glucuronide, *Biochem. J.*, 43, 206, 1948.
34. **Carmella, S. G., LaVoie, E. J., and Hecht, S. S.,** Quantitative analysis of catechol and 4-methylcatechol in human urine, *Food Chem. Toxicol.*, 20, 587, 1982.
35. **Garton, G. A. and Williams, R. T.,** Studies in detoxication. XXI. The fates of quinol and resorcinol in the rabbit in relation to the metabolism of benzene, *Biochem. J.*, 44, 234, 1949.
36. **Miller, J. J., Powell, G. M., Olavesen, A. H., and Curtis, C. G.,** The metabolism and toxicity of phenols in cats, *Biochem. Soc. Trans.*, 1, 1163, 1973.
37. **Miller, J. J., Powell, G. M., Olavesen, A. H., and Curtis, C. G.,** The toxicity of dimethoxyphenol and related compounds in the cat, *Toxicol. Appl. Pharmacol.*, 38, 47, 1976.
38. **DiVincenzo, G. D., Hamilton, M. L., Reynolds, R. C., and Ziegler, D. A.,** Metabolic fate and disposition of [^{14}C]hydroquinone given orally to Sprague-Dawley rats, *Toxicology*, 33, 9, 1984.
39. **Schlosser, M. J. and Kalf, G. F.,** Metabolic activation of hydroquinone by macrophage peroxidase, *Chem. Biol. Interact.*, 72, 191, 1989.
40. **Thomas, D. J., Sadler, A., Subrahmanyam, V. V., Siegel, D., Reasor, M. J., Wierda, D., and Ross, D.,** Bone marrow stromal cell bioactivation and detoxification of the benzene metabolite hydroquinone: comparison of macrophages and fibroblastoid cells, *Mol. Pharmacol.*, 37, 255, 1990.
41. **Sadler, A., Subrahmanyam, V. V., and Ross, D.,** Oxidation of catechol by horseradish peroxidase and human leukocyte peroxidase: reactions of *o*-benzoquinone and *o*-benzosemiquinone, *Toxicol. Appl. Pharmacol.*, 93, 62, 1988.
42. **Bhat, R. V., Subrahmanyam, V. V., Sadler, A., and Ross, D.,** Bioactivation of catechol in rat and human bone marrow cells, *Toxicol. Appl. Pharmacol.*, 94, 297, 1988.
43. **Axelrod, J. and Tomchick, R.,** Enzymatic *O*-methylation of epinephrine and other catechols, *J. Biol. Chem.*, 233, 702, 1958.
44. **Schüsler-Van Hees, M. T. I. W. and Beijersbergen Van Henegouwen, G. M. J.,** Enzymatic *O*-methylation of catechols and catecholamines, *Pharm. Weekbl. Sci. Ed.*, 4, 176, 1982.
45. **Frohne, D.,** Untersuchungen zur Frage der harndesinfizierenden Wirkungen von Bärentraubenblatt-Extrakten, *Planta Med.*, 18, 1, 1970.
46. **Temple, A., Gal, F., and Reboul, C.,** Glucosides phénoliques de certaines éricacées Etude de l'élimination de l'arbutin et de l'hydroquinone, *Trav. Soc. Pharm. Montpellier*, 31, 5, 1971.
47. **Drasar, B. S. and Hill, M. J.,** *Human Intestinal Flora*, Academic Press, London, 1974, 63.
48. **Jahodář, L., Leifertová, I., and Lisá, M.,** Elimination of arbutin from the organism, *Folia Pharm. (Prague)*, 8, 7, 1985.
49. **Grischkanski, A.,** Pharmakologische Untersuchungen einiger Guajacol- und Thymolverbindungen, *Naunyn-Schmiedebergs Arch. Exp. Pathol. Pharmakol.*, 65, 283, 1941.
50. **Wong, K. P. and Sourkes, T. L.,** Metabolism of vanillin and related substances in the rat, *Can. J. Biochem.*, 44, 635, 1966.
51. **Bray, H. G., Craddock, V. M., and Thorpe, W. V.,** Metabolism of ethers in the rabbit. II. Nuclear-substituted anisoles, *Biochem. J.*, 60, 225, 1955.
52. **Archer, S., Arnold, A., Kullnig, R. K., and Wylie, D. W.,** The enzymic methylation of pyrogallol, *Arch. Biochem. Biophys.*, 87, 153, 1960.
53. **Masri, M. S., Robbins, D. J., Emerson, O. H., and DeEds, F.,** Selective *para*- or *meta*-*O*-methylation with catechol *O*-methyl transferase from rat liver, *Nature (London)*, 202, 878, 1964.
54. **Masri, M. S., Booth, A. N., and DeEds, F.,** O-Methylation *in vitro* of dihydroxy- and trihydroxy-phenolic compounds by liver slices, *Biochim. Biophys. Acta*, 65, 495, 1962.
55. **Scheline, R. R.,** The decarboxylation of some phenolic acids by the rat, *Acta Pharmacol. Toxicol.*, 24, 275, 1966.
56. **Monge, P., Solheim, E., and Scheline, R. R.,** Dihydrochalcone metabolism in the rat: phloretin, *Xenobiotica*, 14, 917, 1984.
57. **Takaji, K., Suzuki, T., Saitoh, Y., and Nishihara, K.,** Excretion of phloroglucinol and its conjugates after intravenous administration in man, *Yakuzaigaku*, 31, 213, 1971 (Chem. Abstr. 80, 78285t, 1974).
58. **Fujie, K. and Ito, H.,** Distribution and excretion of 1,3,5-trihydroxybenzene, *Arzneim. Forsch.*, 22, 777, 1972.
59. **Williams, R. T.,** *Detoxication Mechanisms*, Chapman and Hall, London, 1959, 300.
60. **Robbins, B. H.,** Quantitative studies on the absorption and excretion of certain resorcinols and cresols in dogs and man, *J. Pharmacol. Exp. Ther.*, 52, 54, 1934.
61. **Takada, M., Agata, I., Sakamoto, M., Yagi, N., and Hayashi, N.,** On the metabolic detoxication of thymol in rabbit and man, *J. Toxicol. Sci.*, 4, 341, 1979.

62. **Austgulen, L.-T., Solheim, E., and Scheline, R. R.,** Metabolism in rats of *p*-cymene derivatives: carvacrol and thymol, *Pharmacol. Toxicol.*, 61, 98, 1987.

63. **Schröder, V. and Vollmer, H.,** Über die Ausscheidung von Thymol, Carvacrol, Eugenol und guajacol und die Verteilung dieser Substanzen im Organismus, *Naunyn-Schmiedebergs Arch. Exp. Pathol. Pharmakol.*, 168, 331, 1932.

64. **Hattori, M., Endo, Y., Takebe, S., Kobashi, K., Fukusaku, N., and Namba, T.,** Metabolism of magnolol from magnoliae cortex. II. Absorption, metabolism and excretion of [ring-¹⁴C]magnolol in rats, *Chem. Pharm. Bull.*, 34, 158, 1986.

65. **Ma, Y.-H., Kakiuchi, N., Hattori, M., Terasawa, K., Hirate, J., Kato, T., Horikoshi, I., Fukasaku, N., and Namba, T.,** Metabolism of magnolol from magnoliae cortex. III. Distribution of administered [ring-¹⁴C] magnolol in rats by whole-body autoradiography, *J. Med. Pharm. Soc. Wakan-Yaku*, 3, 129, 1986.

66. **Ma, Y.-H., Ye, J.-N., Fukasaku, N., Hattori, M. and Namba, T.,** Metabolism of magnolol from magnoliae cortex. IV. Enterohepatic circulation and gastrointestinal excretion of (ring-¹⁴C)magnolol in rats, *Shoyakugaku Zasshi*, 42, 130, 1988.

67. **Hattori, M., Sakamoto, T., Endo, Y., Kakiuchi, N., Kobashi, K., Mizuno, T., and Namba, T.,** Metabolism of magnolol from magnoliae cortex. I. Application of liquid chromatography-mass spectrometry to the analysis of metabolites of magnolol in rats, *Chem. Pharm. Bull.*, 32, 5010, 1984.

68. **Sinsheimer, J. E. and Smith, R. V.,** 4,4′-Dihydroxybibenzyl, a reduction metabolite of *trans*-stilbene, *J. Pharm. Sci.*, 57, 713, 1968.

69. **Sinsheimer, J. E. and Smith, R. V.,** Metabolic hydroxylations of *trans*-stilbene, *Biochem. J.*, 111, 35, 1969.

70. **Scheline, R. R.,** Polyhydroxylated metabolites of *trans*-stilbene in the rat, *Experientia*, 30, 880, 1974.

71. **Tay, L. K. and Sinsheimer, J. E.,** Metabolism of *trans*-stilbene in rabbits and rats, *Drug Metab. Dispos.*, 4, 154, 1976.

72. **Tay, L. K. and Sinsheimer, J. E.,** Intestinal bacterial reduction of 4,4′-dihydroxystilbene to 4,4′-dihydroxybibenzyl, *J. Pharm. Sci.*, 64, 471, 1975.

73. **Breault, G. O.,** Stilbene metabolism, *Diss. Abstr. Int. B*, 31, 4603, 1971.

74. **Scheline, R. R.,** Unpublished data, 1977.

75. **Grice, H. C., Becking, G., and Goodman, T.,** Toxic properties of nordihydroguaiaretic acid, *Food Cosmet. Toxicol.*, 6, 155, 1968.

76. **Ve, B. and Scheline, R. R.,** Unpublished data, 1976.

77. **Smith, G. E. and Griffiths, L. A.,** Metabolism of *N*-acylated and *O*-dealkylated drugs by the intestinal microflora during anaerobic incubation *in vitro*, *Xenobiotica*, 4, 477, 1974.

78. **Bray, H. G., James, S. P., Thorpe, W. V., and Wasdell, M. R.,** The metabolism of ethers in the rabbit. I. Anisole and diphenyl ether, *Biochem. J.*, 54, 547, 1953.

79. **Solheim, E. and Scheline, R. R.,** Metabolism of alkenebenzene derivatives in the rat. I. *p*-Methoxyallylbenzene (estragole) and *p*-methoxypropenylbenzene (anethole), *Xenobiotica*, 3, 493, 1973.

80. **Zangouras, A., Caldwell, J., Hutt, A. J., and Smith, R. L.,** Dose dependent conversion of estragole in the rat and mouse to the carcinogenic metabolite, 1′-hydroxyestragole, *Biochem. Pharmacol.*, 30, 1383, 1981.

81. **Anthony, A., Caldwell, J., Hutt, A. J. and Smith, R. L.,** Metabolism of estragole in rat and mouse and influence of dose size on excretion of the proximate carcinogen 1′-hydroxyestragole, *Food Chem. Toxicol.*, 25, 799, 1987.

82. **Sangster, S. A., Caldwell, J., Hutt, A. J., Anthony, A., and Smith, R. L.,** The metabolic disposition of [methoxy-¹⁴C]-labelled *trans*-anethole, estragole and *p*-propylanisole in human volunteers, *Xenobiotica*, 17, 1223, 1987.

83. **Drinkwater, N. R., Miller, E. C., Miller, J. A., and Pitot, H. C.,** Hepatocarcinogenicity of estragole (1-allyl-4-methoxybenzene) and 1′-hydroxyextragole in the mouse and mutagenicity of 1′-acetoxyextragole in bacteria, *J. Natl. Cancer Inst.*, 57, 1323, 1976.

84. **Delaforge, M., Janiaud, P., Levi, P., and Morizot, J. P.,** Biotransformation of allylbenzene analogues *in vivo* and *in vitro* through the epoxide-diol pathway, *Xenobiotica*, 10, 737, 1980.

85. **Swanson, A. B., Miller, E. C., and Miller, J. A.,** The side-chain epoxidation and hydroxylation of the hepatocarcinogens safrole and estragole and some related compounds by rat and mouse liver microsomes, *Biochim. Biophys. Acta*, 673, 504, 1981.

86. **Phillips, D. H., Miller, J. A., Miller, E. C., and Adams, B.,** Structures of the DNA adducts formed in mouse liver after administration of the proximate hepatocarcinogen 1′-hydroxyestragole, *Cancer Res.*, 41, 176, 1981.

87. **Wiseman, R. W., Fennell, T. R., Miller, J. A., and Miller, E. C.,** Further characterization of the DNA adducts formed by electrophilic esters of the hepatocarcinogens 1′-hydroxysafrole and 1′-hydroxyestragole *in vitro* and in mouse liver *in vivo*, including new adducts at C-8 and N-7 of guanine residues, *Cancer Res.*, 45, 3096, 1985.

88. **Randerath, K., Haglund, R. E., Phillips, D. H., and Reddy, M. V.,** ³²P-Post-labelling analysis of DNA adducts formed in the livers of animals treated with safrole, estragole and other naturally-occurring alkenylbenzenes. I. Adult female CD-1 mice, *Carcinogenesis*, 5, 1613, 1984.

89. **Phillips, D. H., Reddy, M. V., and Randerath, K.,** ³²P-Post-labelling analysis of DNA adducts formed in the livers of animals treated with safrole, estragole and other naturally-occurring alkenylbenzenes. II. Newborn male B6C3F₁ mice, *Carcinogenesis*, 5, 1623, 1984.

90. **Le Bourhis, B.,** Identification de quelques métabolites du trans-anéthole chez l'Homme, le Lapin et le Rat, *Ann. Pharm. Fr.*, 28, 355, 1970.

91. **Strolin-Benedetti, M. and LeBourhis, B.,** Répartition dans l'organisme et élimination du trans-anéthole-^{14}C, *C. R. Acad. Sci. Paris*, 274D, 2378, 1972.

92. **Sangster, S. A., Caldwell, J., Smith, R. L., and Farmer, P. B.,** Metabolism of anethole. I. Pathways of metabolism in the rat and mouse, *Food Chem. Toxicol.*, 22, 695, 1984.

93. **Axelrod, J.,** The enzymic cleavage of aromatic ethers, *Biochem. J.*, 63, 634, 1956.

94. **Sangster, S. A., Caldwell, J., and Smith, R. L.,** Metabolism of anethole. II. Influence of dose size on the route of metabolism of *trans*-anethole in the rat and mouse, *Food Chem. Toxicol.*, 22, 707, 1984.

95. **Caldwell, J. and Sutton, J. D.,** Influence of dose size on the disposition of *trans*-[*methoxy*-^{14}C]anethole in human volunteers, *Food Chem. Toxicol.*, 26, 87, 1988.

96. **Sangster, S. A., Caldwell, J., Hutt, A. J., and Smith, R. L.,** The metabolism of *p*-propylanisole in the rat and mouse and its variation with dose, *Food Chem. Toxicol.*, 21, 263, 1983.

97. **Caldwell, J., Anthony, A., and Sutton, J. D.,** Comparative metabolism of two allylbenzenes, estragole (ES) and eugenol (EUG) in rat, mouse and man, North American Symposium on Risk Assessment and the Biological Fate of Xenobiotics, Key Biscayne, FL, 1985.

98. **Sutton, J. D., Sangster, S. A., and Caldwell, J.,** Dose-dependent variation in the disposition of eugenol in the rat, *Biochem. Pharmacol.*, 34, 465, 1985.

99. **Yuasa, A.,** Experimental studies on glucuronidation. III. UDP-glucuronyltranserase activity and glucuronide excretion enhanced by oral administration of eugenol, *Jpn. J. Vet. Sci.*, 36, 427, 1974.

100. **Oswald, E. O., Fishbein, L., Corbett, B. J., and Walker, M. P.,** Urinary excretion of tertiary amino methoxy methylenedioxy propiophenones as metabolites of myristicin in the rat and guinea pig, *Biochim. Biophys. Acta*, 244, 322, 1971.

101. **Oswald, E. O., Fishbein, L., Corbett, B. J., and Walker, M. P.,** Chemical lability of the tertiary aminopropiophenones of eugenol as characterized by combined gas-liquid chromatography and chemical ionization mass spectrometry, *J. Chromatogr.*, 73, 59, 1972.

102. **Padieu, P. and Maume, B. F.,** Evaluation by mass fragmentography of metabolic pathways of endogenous and exogenous compounds in eukaryote cell cultures, in *Quantitative Mass Spectrometry in Life Sciences*, de Leenheer, A. P. and Roncucci, R. R., Eds., Elsevier, Amsterdam, 1977, 49.

103. **Thompson, D., Norbeck, K., Olsson, L.-I., Constantin-Teodosiu, D., Van der Zee, J., and Moldéus, P.,** Peroxidase-catalyzed oxidation of eugenol: formation of a cytotoxic metabolite(s), *J. Biol. Chem.*, 264, 1016, 1989.

104. **Thompson, D., Constantin-Teodosiu, D., Norbeck, K., Svensson, B., and Moldéus, P.,** Metabolic activation of eugenol by myeloperoxidase and polymorphonuclear leucocytes, *Chem. Res. Toxicol.*, 2, 186, 1989.

105. **Weinberg, J. E., Rabinowitz, J. L., Zanger, M., and Gennaro, A. R.,** ^{14}C-Eugenol. I. Synthesis, polymerization, and use, *J. Dent. Res.*, 51, 1055, 1972.

106. **Fischer, I. U., von Unruh, G. E., and Dengler, H. J.,** The metabolism of eugenol in man, *Xenobiotica*, 20, 209, 1990.

107. **Solheim, E. and Scheline, R. R.,** Metabolism of alkenebenzene derivatives in the rat. II. Eugenol and isoeugenol methyl ethers, *Xenobiotica*, 6, 137, 1976.

108. **Heffter, A.,** Zur Pharmakologie der Safrolgruppe, *Arch. Exp. Pathol. Pharmak.*, 35, 342, 1895.

109. **Casida, J. E., Engel, J. L., Essac, E. G., Kamienski, F. X., and Kuwatsuka, S.,** Methylene-C^{14}-dioxyphenyl compounds: metabolism in relation to their synergistic action, *Science*, 153, 1130, 1966.

110. **Kamienski, F. X. and Casida, J. E.,** Importance of demethylenation in the metabolism *in vivo* and *in vitro* of methylenedioxyphenyl synergists and related compounds in mammals, *Biochem. Pharmacol.*, 19, 91, 1970.

111. **Anders, M. W., Sunram, J. M., and Wilkinson, C. F.,** Mechanism of the metabolism of 1,3-benzodioxoles to carbon monoxide, *Biochem. Pharmacol.*, 33, 577, 1984.

112. **Stillwell, W. G., Carman, M. J., Bell, L., and Horning, M. G.,** The metabolism of safrole and 2',3'-epoxysafrole in the rat and guinea pig, *Drug. Metab. Dispos.*, 2, 489, 1974.

113. **Strolin Benedetti, M., Malnoë, A., and Broillet, A. L.,** Absorption, metabolism and excretion of safrole in the rat and man, *Toxicology*, 7, 69, 1977.

114. **Klungsøyr, J. and Scheline, R. R.,** Metabolism of safrole in the rat, *Acta Pharmacol. Toxicol.*, 52, 211, 1983.

115. **Borchert, P., Wislocki, P. G., Miller, J. A., and Miller, E. C.,** The metabolism of the naturally occurring hepatocarcinogen safrole to 1'-hydroxysafrole and the electrophilic reactivity of 1'-acetoxysafrole, *Cancer Res.*, 33, 575, 1973.

116. **Peele, J. D. and Oswald, E. O.,** Metabolism of the proximate carcinogen 1'-hydroxysafrole and the isomer 3'-hydroxyisosafrole, *Bull. Environ. Contam. Toxicol.*, 19, 396, 1978.

117. **Levi, P., Janiaud, P., Delaforge, M., Morizot, J. P., Maume, B. F., and Padieu, P.,** Présence de métabolites glucuroconjugués du safrol dans l'urine de rats traités, *C. R. Séanc. Soc. Biol.*, 171, 1034, 1977.

118. **Wislocki, P. G., Borchert, P., Miller, J. A., and Miller, E. C.,** The metabolic activation of the carcinogen 1'-hydroxysafrole *in vivo* and *in vitro* and the electrophilic reactivites of possible ultimate carcinogens, *Cancer Res.*, 36, 1686, 1976.

119. **Phillips, D. H., Miller, J. A., Miller, E. C., and Adams, B.,** N^2 atom of guanine and N^6 atom of adenine residues as sites for covalent binding of metabolically activated 1'-hydroxysafrole to mouse liver DNA *in vivo, Cancer Res.*, 41, 2664, 1981.

120. **Borchert, P., Miller, J. A., Miller, E. C., and Shires, T. K.,** 1'-Hydroxysafrole, a proximate carcinogenic metabolite of safrole in the rat and mouse, *Cancer Res.*, 33, 590, 1973.

121. **Oswald, E. O., Fishbein, L., and Corbett, B. J.,** Metabolism of naturally occurring propenylbenzene derivatives. I. Chromatographic separation of ninhydrin-positive materials of rat bile, *J. Chromatogr.*, 45, 437, 1969.

122. **Oswald, E. O., Fishbein, L., Corbett, B. J., and Walker, M. P.,** Identification of tertiary aminomethylenedioxypropiophenones as urinary metabolites of safrole in the rat and guinea pig, *Biochim. Biophys. Acta*, 230, 237, 1971.

123. **McKinney, J. D., Oswald, E., Fishbein, L., and Walker, M.,** On the mechanism of formation of Mannich bases as safrole metabolites, *Bull. Environ. Contam. Toxicol.*, 7, 305, 1972.

124. **Fennell, T. R., Miller, J. A., and Miller, E. C.,** Characterization of the biliary and urinary glutathione and *N*-acetylcysteine metabolites of the hepatic carcinogen 1'-hydroxysafrole and its 1'-oxo metabolite in rats and mice, *Cancer Res.*, 44, 3231, 1984.

125. **Watabe, T. and Akamatsu, K.,** Photometric assay of hepatic epoxide hydrolase activity with safrole oxide (SAFO) as substrate, *Biochem. Pharmacol.*, 23, 2839, 1974.

126. **Delaforge, M., Janiaud, P., Chessebeuf, M., Padieu, P., and Maume, B. F.,** Possible occurrence of the epoxide-diol metabolic pathway for hepatocarcinogenic safrole in cultured rat liver cells, as compared with whole animal: a metabolic study by mass spectrometry, in *Advances in Mass Spectrometry in Biochemistry and Medicine*, Vol. 2, Frigerio, A., Ed., Spectrum Publications, New York, 1976, 65.

127. **Janiaud, P., Delaforge, M., Levi, P., Maume, B. F., and Padieu, P.,** Etude comparative en culture cellulaire de foie de Rat du métabolisme de différents analogues et métabolites d'un hépatocancérogéne naturel: le safrol, *C. R. Séanc. Soc. Biol.*, 170, 1035, 1976.

128. **Janiaud, P., Delaforge, M., Levi, P., Maume, B. F., and Padieu, P.,** Métabolisme d'un hépatocancérogéne naturel, le safrol. Étude, chez le rat et dans des cultures de cellules hépatiques de rat, de l'action d'effecteurs sur plusieurs voies métaboliques et des formes de transport, *Coll. Int. CNRS*, 431, 1977.

129. **Fishbein, L., Fawkes, J., Falk, H. L., and Thompson, S.,** Thin-layer chromatography of rat bile and urine following intravenous administration of safrole, isosafrole, and dihydrosafrole, *J. Chromatogr.*, 29, 267, 1967.

130. **Klungsøyr, J. and Scheline, R. R.,** Metabolism of isosafrole and dihydrosafrole in the rat, *Biomed. Mass Spectrom.*, 9, 323, 1982.

131. **Petridou-Fischer, J., Whaley, S. L., and Dahl, A. R.,** In vivo metabolism of nasally instilled dihydrosafrole [1-(3,4-methylenedioxyphenyl)-propane] in dogs and monkeys, *Chem. Biol. Interact.*, 64, 1, 1987.

132. **Solheim, E. and Scheline, R. R.,** Metabolism of alkenebenzene derivatives in the rat. III. Elemicin and isoelemicin, *Xenobiotica*, 10, 371, 1980.

133. **Oswald, E. O., Fishbein, L., Corbett, B. J., and Walker, M. P.,** Metabolism of naturally occurring propenylbenzene derivatives. II. Separation and identification of tertiary aminopropiophenones by combined gas-liquid chromatography and chemical ionization mass spectrometry, *J. Chromatogr.*, 73, 43, 1972.

134. **Shulgin, A.,** Possible implication of myristicin as a psychotic substance, *Nature (London)*, 210, 380, 1966.

135. **Braun, U. and Kalbhen, D. A.,** Evidence for the biogenic formation of amphetamine derivatives from components of nutmeg, *Pharmacology*, 9, 312, 1973.

136. **Peele, J. D.,** Investigations on the metabolism of myristicin and the mechanism of formation of basic metabolites of allylbenzene in the rat, *Diss. Abstr. Int.*, 37B, 1234, 1976.

137. **Ikeya, Y., Taguchi, H., Mitsuhashi, H., Sasaki, H., Matsuzaki, T., Aburada, M., and Hosoya, E.,** Studies on the metabolism of gomisin A (TJN-101). I. Oxidative products of gomisin A formed by rat liver S9 mix, *Chem. Pharm. Bull.*, 36, 2061, 1988.

138. **Capel, I. D., French, M. R., Millburn, P., Smith, R. L., and Williams, R. T.,** The fate of [^{14}C]phenol in various species, *Xenobiotica*, 2, 25, 1972.

139. **Caldwell, J., French, M. R., Idle, J. R., Renwick, A. G., Bassir, O., and Williams, R. T.,** Conjugation of foreign compounds in the elephant and hyaena, *FEBS Lett.*, 60, 391, 1975.

140. **French, M. R., Bababunmi, E. A., Golding, R. R., Bassir, O., Caldwell, J., Smith, R. L., and Williams, R. T.,** The conjugation of phenol, benzoic acid, 1-naphthylyacetic acid and sulphadimethoxine in the lion, civet and genet, *FEBS Lett.*, 46, 134, 1974.

141. **Mehta, R., Hirom, P. C., and Millburn, P.,** The influence of dose on the pattern of conjugation of phenol and 1-naphthol in non-human primates, *Xenobiotica*, 8, 445, 1978.

142. **Weitering, J. G., Krijgsheld, K. R., and Mulder, G. J.,** The availability of inorganic sulphate as a rate limiting factor in the sulphate conjugation or xenobiotics in the rat? Sulphation and glucuronidation of phenol, *Biochem. Pharmacol.*, 28, 757, 1979.

143. **Meerman, J. H. N., Nijland, C., and Mulder, G. J.,** Sex differences in sulfation and glucuronidation of phenol, 4-nitrophenol and *N*-hydroxy-2-acetylaminofluorene in the rat *in vivo, Biochem. Pharmacol.*, 36, 2605, 1987.

METABOLISM OF ALDEHYDES, KETONES, AND QUINONES

ALDEHYDES

Aldehydes may undergo either metabolic reduction to alcohols or oxidation to carboxylic acids. In the former case alcohol dehydrogenase or an aldehyde reductase may be responsible. Oxidation, which is generally the major metabolic route for this group of compounds, may be carried out by various enzymes including aldehyde dehydrogenase, aldehyde oxidase, or xanthine oxidase. McMahon[1] reviewed the reductive and oxidative metabolism of aldehydes and more detailed coverage of the enzyme systems involved was given by Bosron and Li[2] on alcohol dehydrogenase, by von Wartburg and Wermuth[3] on alcohol reductase, by Weiner[4] and Jörnvall et al.[5] on aldehyde oxidizing enzymes, and by Rajagopalan[6] on xanthine oxidase and aldehyde oxidase. The genetic polymorphism of human aldehyde dehydrogenase was discussed by Bosron and Li.[7]

ALIPHATIC ALDEHYDES

The existence of a profusion of aliphatic plant aldehydes is well documented. These include straight-chain compounds from the simplest type, methanal (formaldehyde), to the C_{18}-aldehyde, n-octadecanal, as well as numerous branched-chain examples. As a rule, they are found in low concentrations, often in fruits to which they impart characteristic aromas. Among the simpler homologues, acetaldehyde (ethanal) is fairly frequently encountered but aldehydes containing from 6 to 12 carbon atoms are generally those which one commonly associates with this group of plant compounds. Decanal is the most widespread of the higher homologues, being found in citrus fruits and numerous aromatic oils.

Most of the data on the metabolic fate of aliphatic aldehydes deals with the simpler members of the group, **formaldehyde** and **acetaldehyde**. This results from the long-standing interest in the toxicity of the former compound, a metabolite of methanol, and the formation of the latter in ethanol metabolism. Furthermore, these compounds enter into the normal pathways of one- and two-carbon metabolism. Formaldehyde metabolism was summarized by Hathway[8] and by Weiner[4] whereas Williams[9] reviewed the metabolism of both it and acetaldehyde. In both cases their metabolism is straightforward and involves conversion to the corresponding acid followed by oxidation to CO_2.

Williams[10] stated that little is known about the metabolism of the higher aliphatic aldehydes but assumed that these also undergo oxidation to the corresponding acids which then enter into the normal pathways for fatty acid metabolism. The general metabolic pathway for these compounds is chain-shortening due to β-oxidation. The ready oxidation of these aldehydes to CO_2 and water ensures that little or no urinary excretion of intermediate products occurs. This general understanding has subsequently been neither confirmed nor contradicted and no new metabolic results were presented by Opdyke[11,12] in reviews of biological data on several naturally occurring saturated aliphatic aldehydes containing from 6 to 14 carbon atoms. Boyland[13] administered **n-heptanal** (oenanthal, heptaldehyde) orally to rats and rabbits and was unable to detect any acidic metabolites including the ω-oxidation product, pimelic acid, in the urine. It was concluded that the aldehyde undergoes complete oxidation in the body. In an early study Neubauer[14] administered heptaldehyde and also isobutyraldehyde and isovaleraldehyde to rabbits, however no useful conclusions could be drawn from the results. Hinson and Neal[15,16] carried out kinetic studies of the oxidation of **n-butanal** (butyraldehyde), **n-hexanal**, and **n-octanal** (octylaldehyde, capryl aldehyde) by horse liver alcohol dehydrogenase and showed that they were oxidized to the corresponding carboxylic acids. This general picture of aliphatic aldehyde metabolism was confirmed by Kutzman et al.[17] who employed [*carbonyl*-^{11}C]octanal. They administered a small dose (15 mg/kg) to rats by inhalation over a 2-min period. The radioactivity was eliminated nearly entirely as respiratory [11]CO_2 and only traces of radioactivity were detected in the urine. After 40 min ~60% of the radioactivity was lost by the expiratory route. It was also found that many organs showed increased levels of radioactivity towards the later part of this time period. This indicates that the process of β-oxidation also results in the incorporation of the radiolabel into many biological molecules.

A few aliphatic plant aldehydes contain double bonds. A common example of this group is ***trans*-hex-2-en-1-al** (leaf aldehyde). It seems likely that this compound may be partly metabolized to a mercapturic acid derivative. This is suggested by the finding that its administration to rats gave rise to a moderate fall in the liver glutathione content in rats.[18] Also, the aldehyde is an excellent substrate for the glutathione *S*-transferase activity in rat liver which forms glutathione conjugates with a number of αβ-unsaturated carbonyl compounds.[19,20]

MONOTERPENOID ALDEHYDES
Acyclic Terpene Aldehydes

Metabolic data on this group is limited to two compounds, citral and citronellal. Natural citral, 3,7–dimethyl–2,6–octadienal, consists of a mixture of geometric isomers, geranial (citral a) (1) and neral (citral b) (2). Citronellal (3,7–dimethyl–6–octenal) (3) differs from citral only by the lack of a double bond at C-2,3.

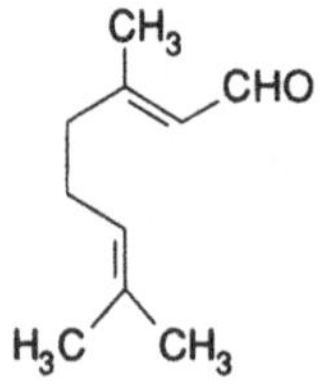

(1)	(2)	(3)
Geranial	Neral	Citronellal

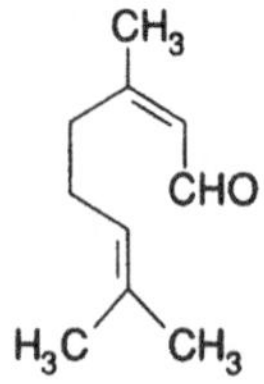

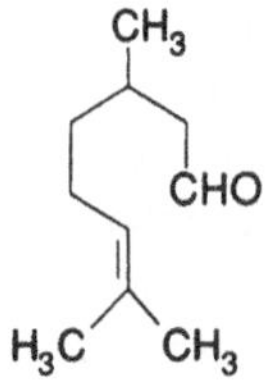

The metabolism of **citral** was first studied by Hildebrandt[21] in rabbits. Two urinary metabolites were detected, one of which was geranic acid (4) formed by the oxidation of the aldehyde group. The other compound was a dicarboxylic acid which was then believed to be a geranic acid derivative with a carboxy rather than a methyl group at C-3. This problem was reinvestigated by Kuhn et al.[22] who administered large, repeated doses (2 × 5 g daily, p.o.) to rabbits. They found that this dicarboxylic acid, now known as Hildebrandt acid, had structure (5). Ishida et al.[23] recently reported that, following administration of citral containing a 1:1 mixture of geranial and neral to rabbits, this dicarboxylic acid consisted entirely of the *trans*-isomer (at C-2,3). Kuhn and Livada[24] found that ~10% of the citral (2 to 4 g/d, i.p.) given to dogs was excreted in the urine as metabolite (5). A second dicarboxylic acid acid metabolite was detected in which the double bond at C-2,3 was reduced. The latter compound (6), reduced Hildebrandt acid, was shown to be optically active (dextrorotatory) and it therefore resulted from an asymmetric reduction of the double bond. Neubauer[14] detected a glucuronide conjugate in the urine of rabbits given citral (1 g/kg, p.o.).

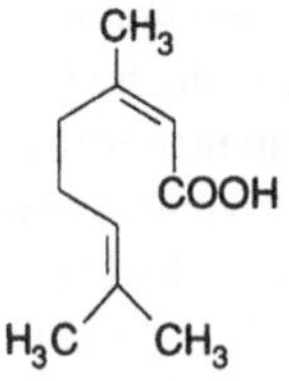

(4)	(5)	(6)

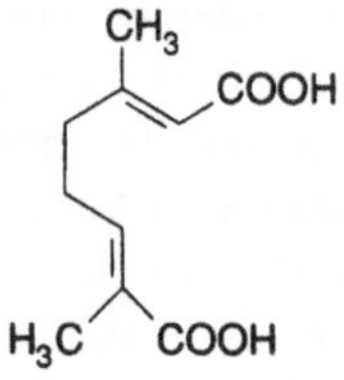

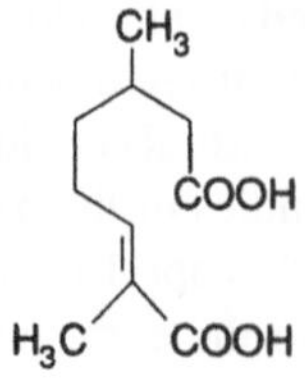

Phillips et al.[25] studied the disposition of [14C]citral following oral dosage to rats and mice. At a low dose (5 mg/kg), rats excreted more than 95% of the radioactivity within 24 h. Most of this was in the urine (60%), with a further 20% found in the expired CO_2 and 17% in the feces. The results seen after very large doses (770 or 960 mg/kg) were similar, however excretion of radioactivity was slower and the final values for urinary excretion were higher. The identities of the metabolites were not determined, however at the lowest dose level <0.5% of the urinary material was unchanged compound. The results in mice given 100 mg/kg were similar. These findings of rapid and extensive metabolism of citral were confirmed by Diliberto et al.[26,27] They gave [^{14}C]citral to rats and detected ~12 metabolites in the urine and bile, ~6 of which were common to both. Hildebrandt acid and its glucuronide conjugate were minor metabolites in urine and bile. About 10 to 15% of oral doses was lost both as expired $^{14}CO_2$ and in the feces, however ~25% of the dose (5 mg/kg, i.v.) was excreted in the bile within 4 h. They found that the disposition of citral was similar over a dose range of 5 to 500 mg/kg. Likewise, Ishida et al.[23] detected chromatographically approximately ten neutral urinary metabolites of citral in rabbits. These compounds were not identified but it is evident that the metabolism of citral is more complex than hitherto revealed.

Parke and Rahman[28] found that the administration of citral for 3 d to rats resulted in moderate (~25%) increases in the activities of several liver microsomal enzyme systems. This phenomenon was also indicated by the findings of Diliberto et al.[26,27] who noted that repeated administration of citral to rats resulted in increases in the extent of biliary excretion of citral metabolites.

The metabolism of **citronellal** (3) was first investigated by Hildebrandt.[21] In view of the close structural similarity between citral and citronellal, it was considered likely that their pathways of metabolism would be similar. However, none of the expected carboxylic acid derivatives was detected in the urine of rabbits given citronellal. This anomalous finding was subsequently clarified by Kuhn and Löw[29] who reported that citronellal was cyclized to p-menthane-3,8-diol (7). This reaction takes place readily under acidic conditions and they believed that metabolite (7) was formed in the stomach. Following its absorption it is conjugated with glucuronic acid at position C-3. About 25% of a total dose of 50 g of citronellal fed to rabbits was isolated from the urine as this glucuronide. That this is not the sole metabolic pathway of citronellal, however, was shown by Asano and Yamakawa[30] who, unlike previous workers, were able to isolate the reduced dicarboxylic acid (6) from the urine of rabbits given the aldehyde by s.c. injection. The latter finding was recently confirmed by Ishida et al.[23] who also found that both the *trans*- (7) and *cis*-forms (8) of p-menthane-3,8-diol were excreted in the urine of rabbits given citronellal orally. They also identified a third neutral metabolite, (–)-isopregol (9), which is the dehydrated derivative of metabolite (7).

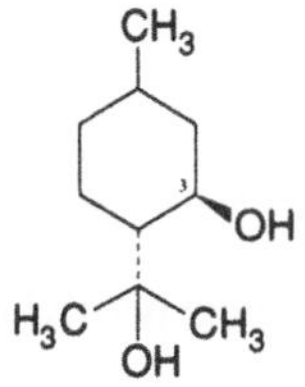

(7) 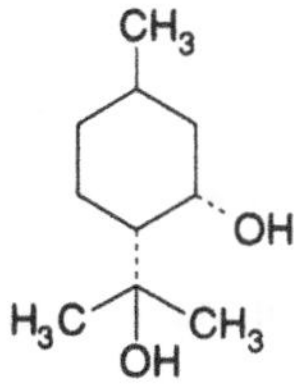   (8) 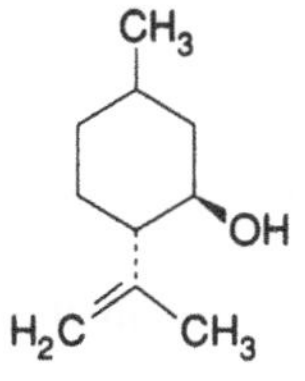   (9)

Monocyclic Terpene Aldehydes

Perillaldehyde (perilla aldehyde, perillyl aldehyde) (10) (see Figure 1) shows close structural similarity to the terpene hydrocarbon limonene, the difference being the aldehyde moiety at C-7. Its metabolism in rabbits was studied by Ishida et al.[23,31] Following oral doses (0.7 to 0.8 g/kg) the urine was collected for 3 d and then hydrolyzed with a β-glucuronidase + sulfatase preparation. They identified both oxidized and reduced metabolites (carboxylic acids and alcohols, respectively) and, interestingly, several aromatic metabolites (cumic acid derivatives). The structures of perillaldehyde and the metabolites reported are shown in Figure 1. The main acidic metabolite was perillic acid (11). In contrast, only traces of cumic acid (12) were

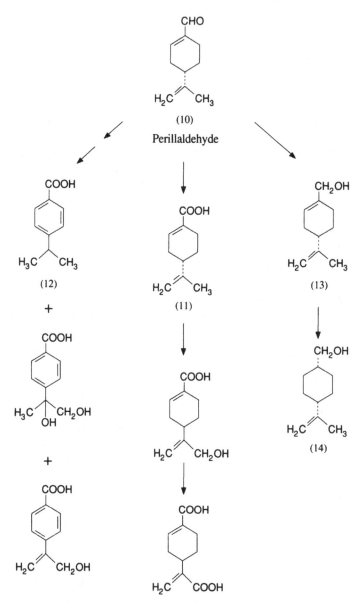

FIGURE 1.   Metabolic pathways of perillaldehyde in rabbits.

formed. Likewise, only small amounts of the reduction products perillyl alcohol (13) and *cis*-shisool (14) were found.

Bicyclic Terpene Aldehydes

Ishida et al.[23,31] gave oral doses (0.7 to 0.8 g/kg) of **myrtenal** (15) to rabbits. The urine was collected for 3 d and then hydrolyzed with a β-glucuronidase + sulfatase preparation. They isolated and identified three acidic and two neutral metabolites of which myrtenic acid (16) was the major product. They also isolated the monocyclic derivative perillic acid (11) (see Figure 1) and a hydroxylated acid which they tentatively identified as 4-hydroxymyrtenic acid. The neutral metabolites were the alcohols myrtenol (17) and *cis*-myrtanol (*cis*-10-pinanol) (18). The fact that the corresponding aldehyde lacking the ring double bond (myrtanal) was not detected, suggests that the reductive pathway was (17) → (18). The structural similarity between myrtenal and α-pinene and the formation of common metabolites from these two terpenes (see the chapter entitled Metabolism of Hydrocarbons) suggest that myrtenal is a transient intermediate in the metabolism of α-pinene.

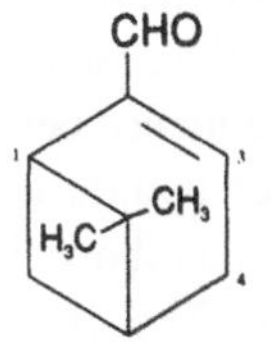

(15) (16) 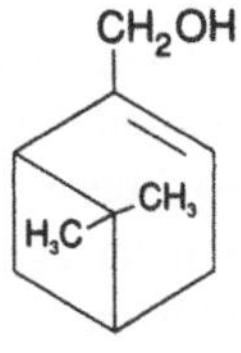 (17) 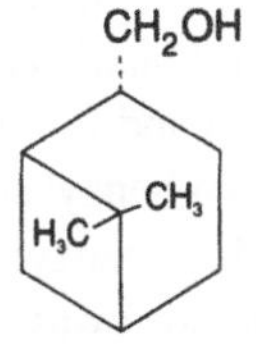 (18)

Myrtenal

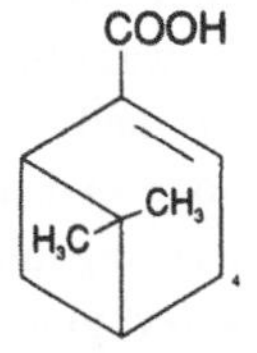

AROMATIC ALDEHYDES

The major metabolic pathway of **benzaldehyde** (19) is straightforward and involves oxidation to benzoic acid (20) which may be excreted as such or as hippuric acid (21) following conjugation with glycine. This was shown by Friedmann and Türk[32] who gave two dogs a total of 10 g each of benzaldehyde by i.p. injection over a period of 5 d. They isolated from the urine ~2 and 13% of the dose as benzoic acid and 72 and 39% as hippuric acid. Bray et al.[33] carried out a kinetic study of the formation of benzoic acid in rabbits from several precursors including benzaldehyde. They found that the combined urinary benzoic acid and hippuric acid fraction accounted for ~90% of the administered dose of benzaldehyde (~250 to 750 mg/kg, p.o.). No excretion of benzoyl glucuronide occurred in three animals whereas 1% and 5% of the dose were found with two others. The extent of formation of the latter metabolite was noted to be dependent on the amount of benzoic acid present in the body. With benzaldehyde, the rate of formation of the acid under the experimental conditions was not sufficient to appreciably exceed the capacity for its conjugation with glycine. A subsequent investigation of benzaldehyde metabolism in rabbits confirmed the urinary excretion of small amounts of benzoic acid and large amounts of hippuric acid.[34] Following oral doses of 350 or 750 mg/kg, the values were ~1.5 and 67 to 70%, respectively, in both groups. However, these animals also excreted ~9% (low dose) and 11% (high dose) as benzoyl glucuronide. Teuchy et al.[35] found that ~30% of the dose of benzaldehyde (44 mg/animal, i.p.) was excreted as urinary hippuric acid by rats. This suggests that conjugation of the metabolically formed benzoic acid with glycine may be less extensive in rats at high doses than in rabbits, however this study did not measure the extent of urinary benzoic acid excretion by rats. The fate of very small doses of benzaldehyde administered by inhalation was studied in rats by Kutzman et al.[36] They employed ^{11}C-labeled material and gave an average of 2.5 mg of benzaldehyde over a 2-min period. Neither radioactive CO_2 nor organic metabolites were detected in the expired air and low levels of activity in the intestine indicated that biliary excretion was not an important excretory route. They found that inhaled benzaldehyde had a biological half-life of only ~10 min. Excretion of metabolites appeared to be entirely via the kidneys and more than 90% of the urinary material was hippuric acid.

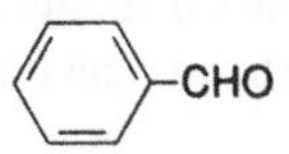

(19)

Benzaldehyde

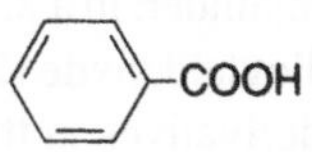

(20)

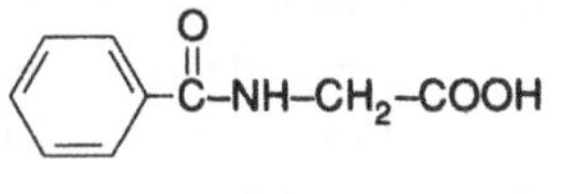

(21)

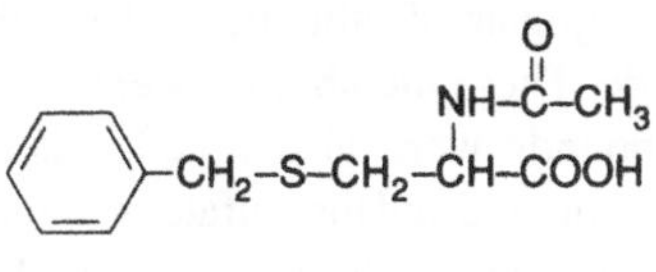

(22)

Seutter-Berlage et al.[37] administered benzaldehyde (210 mg/kg, i.p.) to rats and found no evidence for the excretion of mercapturic acid derivatives in the urine. However, later reports indicated that benzylmercapturic acid (22) was a minor urinary metabolite of benzaldehyde in both rats[38] and rabbits.[34] The value recorded in rabbits was <0.01% of the dose (350 or 750 mg/ kg, p.o.). Laham and Potvin[38] proposed that it was formed via the sequence benzaldehyde → benzyl alcohol → benzyl sulfate → S-benzylglutathione → benzylmercapturic acid. Reduction of benzaldehyde to benzyl alcohol in the experiments using rabbits[34] was shown by the finding that ~3% of the dose was excreted in the urine as benzyl glucuronide. The site of reduction is not known, however Scheline[39] noted that this reaction occurred *in vitro* when benzaldehyde was incubated with a mixed culture of rat cecal microorganisms. Whether or not benzaldehyde reduction will have significance *in vivo* depends either on the extent of its absorption in the upper regions of the intestine or on its diffusion from the systemic circulation into the lower intestine. It seems reasonable to assume that the latter mechanism may operate in the case of benzaldehyde.

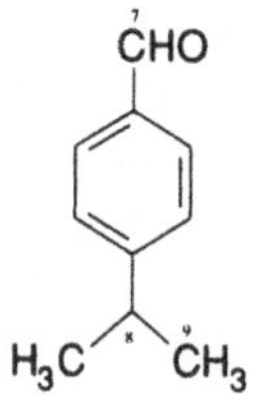

(23)	(24)
p-Cuminaldehyde	Salicylaldehyde

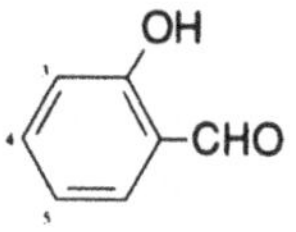

The metabolism of **p-cuminaldehyde** (23) was studied in rabbits.[23] The urinary metabolites of the aldehyde (0.7 to 0.8 g/kg, p.o.) identified after hydrolysis of the urine with β-glucuronidase + sulfatase indicated that reduction to the corresponding alcohol, *p*-cumyl alcohol, took place. However, the major routes of metabolism resulted in the formation of oxidized compounds which were shown to be identical to several of the terminal products in the metabolism of *p*-cymene (see the chapter entitled Metabolism of Hydrocarbons). The three acidic metabolites isolated were derivatives of cumic acid (*p*-isopropylbenzoic acid) which additionally contained a carboxy group at C-9 or a hydroxy group at C-8 or C-9.

Information on the metabolism of **salicylaldehyde** (2-hydroxybenzaldehyde) (24), obtained both from the earlier literature and from unpublished data, was summarized by Williams.[40] Salicylic acid was found to be the major metabolite in rabbits, cats, and dogs. Excretion of the unchanged aldehyde, both free and conjugated with glucuronic acid and sulfate, has been detected in rabbits. Williams[41] recorded a value of ~6% of the dose (325 mg/kg, p.o.) for the latter metabolite. Quantitative data was also obtained by Bray et al.[42] who fed salicylaldehyde (0.4 g/ kg) to rabbits and found that 75% was excreted as the corresponding acid. Only ~3% was due to the ethereal sulfate and the remainder, in a 2:1 ratio, consisted of ester and ether glucuronides. Seutter-Berlage et al.[37] gave salicylaldehyde (250 mg/kg) to rats and found no evidence for the excretion of mercapturic acid derivatives in the urine.

Sato et al.[43] studied the metabolism of salicylaldehyde using rat liver preparations, mainly in order to determine the ability of the phenolic group to undergo sulfate conjugation. Two metabolites were detected when the supernatant fraction of liver homogenates was used. One was the sulfate conjugate of salicylic acid and the other probably the corresponding conjugate of salicylaldehyde. These metabolites were also detected when liver slices were used and two additional compounds were also noted. These were tentatively identified as the glycine conjugate of salicylic acid and the sulfate conjugate of 2,5-dihydroxybenzoic acid. Interestingly, the sulfate conjugate of salicylic acid found in these experiments was not formed when salicylic acid itself was used as the substrate. It was therefore assumed that the conjugation reaction precedes oxidation of the aldehyde group. However, it is not clear if the latter step occurred enzymatically or merely as a result of chemical oxidation. Salicylaldehyde is metabolized *in vitro* both to salicylic acid and to salicyl alcohol when incubated with rat cecal microorganisms.[44]

Hartles and Williams[45] studied the metabolism of **4-hydroxybenzaldehyde** in rabbits. Using oral doses of ~300 mg/kg it was found that slightly more than 40% was excreted as 4-conjugates, the glucuronide:ethereal sulfate ratio being ~3.5:1. Nearly identical experiments were carried out by Bray et al.[42] who reported a value of 25% of the dose (400 mg/kg) for 4-conjugates, the glucuronide to sulfate ratio being ~2:1. A small amount (4%) of ester glucuronide was excreted and the main metabolite was 4-hydroxybenzoic acid (67%). The conjugates reported by Hartles and Williams were mainly of the hydroxy acid rather than of the hydroxy aldehyde; however, ~2 to 3% of the dose was found to be excreted as the 4-glucuronide of the latter compound. These *in vivo* experiments showed that the extent of glucuronide excretion was higher with the aldehyde than with the corresponding acid and it therefore seems likely that some conjugation occurs prior to aldehyde oxidation. This view was also held by Quick[46] who also found evidence for the formation in dogs of a diglucuronide of 4-hydroxybenzoic acid following administration of the aldehyde. Seutter-Berlage et al.[37] administered 4-hydroxybenzaldehyde (250 mg/kg, i.p.) to rats and found no evidence for the excretion of mercapturic acid derivatives in the urine.

Sato et al.,[47] in experiments similar to those described above with salicylaldehyde, studied the metabolism of 4-hydroxybenzaldehyde *in vitro* by rat liver preparations. The ethereal sulfates of the substrate and its carboxylic acid derivative were formed both by supernatant and slices. Smaller amounts of two other metabolites were also formed by liver slices, these being possibly the sulfate conjugate of 3,4-dihydroxybenzaldehyde and a hippuric acid derivative. Metabolism of 4-hydroxybenzaldehyde by rat cecal microorganisms leads to both the corresponding acid and alcohol.[44] Interestingly, reduction can proceed further in this case by converting the alcohol moiety to a methyl group, forming *p*-cresol.

The metabolism of **protocatechualdehyde** (3,4-dihydroxybenzaldehyde) follows a pattern similar to that noted above with other phenolic aldehydes. Dodgson and Williams[48] reported that ~40% of the dose (250 mg/kg, p.o.) was excreted in the urine by rabbits as two glucuronide conjugates, apparently of the aldehyde and of protocatechuic acid, and as an ethereal sulfate. The glucuronide:ethereal sulfate ratio was nearly 2:1. Large amounts of free protocatechuic acid were also excreted in the urine and this metabolite probably accounted for much of the remainder of the dose. Wong and Sourkes[49] studied the urinary metabolites of protocatechualdehyde (50 mg/kg, i.p.) in rats and found that it was *O*-methylated to vanillin and ultimately converted to catechol. These pathways are discussed in the following summary of the metabolism of vanillin. The metabolism of protocatechualdehyde by rat cecal microorganisms is similar to that found with 4-hydroxybenzaldehyde.[44] Some oxidation to protocatechuic acid takes place but the main reaction is complete reduction of the aldehyde group, giving 4-methylcatechol.

The aromatic aldehyde which has received the largest amount of attention from a metabolic viewpoint is **vanillin** (4-hydroxy–3-methoxybenzaldehyde) (25) (see Figure 2). A few very early studies, the results of which were summarized by Sammons and Williams,[50] delineated the major pathways in its metabolism including oxidation to vanillic acid (26), conjugation of some of this metabolite with glucuronic acid, an increase in the urinary output of ethereal sulfates, and possibly minor excretion of unchanged compound. With this background, Sammons and Williams[50] set out to obtain a quantitative assessment of the metabolism of vanillin (1 g/kg, p.o.) in rabbits. They accounted for ~83% of the dose, 69% being excreted as vanillic acid, either free (44%) or conjugated (25%). The latter material consisted mainly of the ether glucuronide of vanillic acid. About 8% of the dose was recovered as ethereal sulfates of vanillin and/or vanillic acid. The remaining 14% was associated with the 4-glucuronide of vanillin. No demethylation of vanillin was detected in this study.

The most abundant data on the metabolic pathways of vanillin (Figure 2) and the amounts of metabolites formed have been obtained in studies using rats. Using a dose of 100 mg/kg (i.p.),

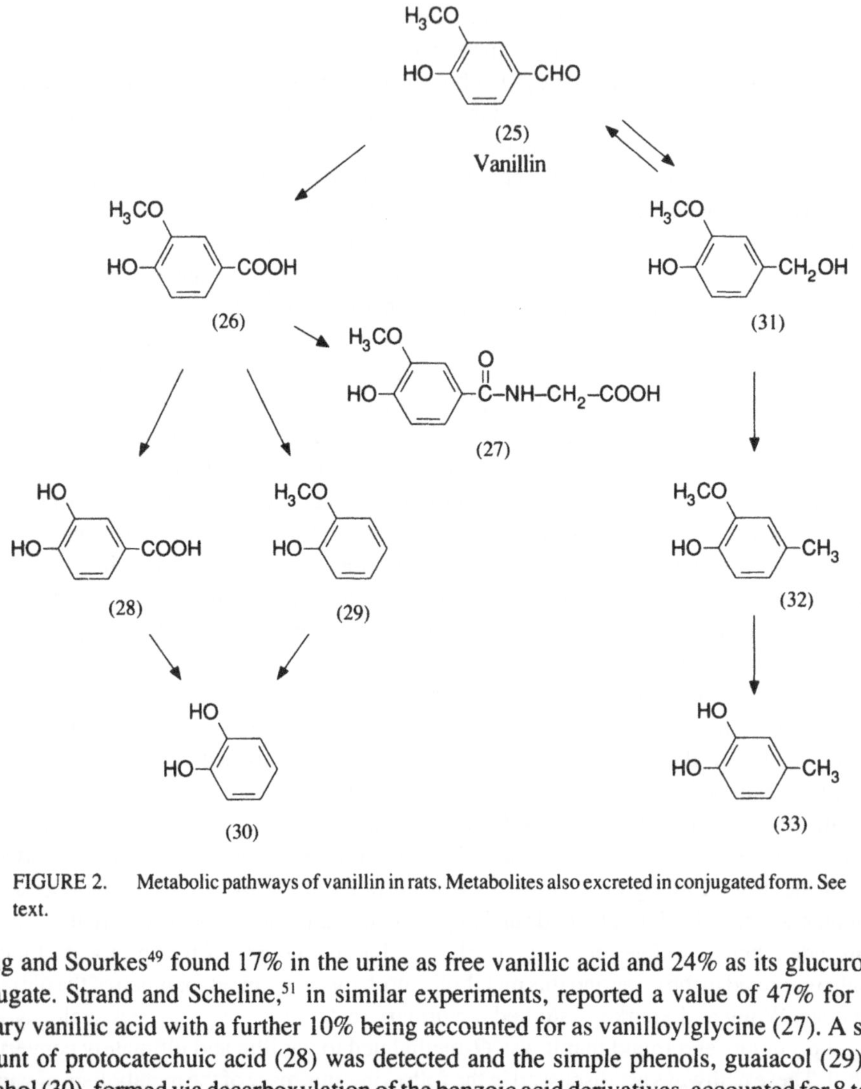

FIGURE 2. Metabolic pathways of vanillin in rats. Metabolites also excreted in conjugated form. See text.

Wong and Sourkes[49] found 17% in the urine as free vanillic acid and 24% as its glucuronide conjugate. Strand and Scheline,[51] in similar experiments, reported a value of 47% for total urinary vanillic acid with a further 10% being accounted for as vanilloylglycine (27). A small amount of protocatechuic acid (28) was detected and the simple phenols, guaiacol (29) and catechol (30), formed via decarboxylation of the benzoic acid derivatives, accounted for 8 to 9% of the dose. The oxidative pathway therefore made up 65 to 70% of the dose in these experiments. A further 7% was excreted as vanillin and 19% was recovered as the reduction product vanillyl alcohol (31). As previously demonstrated in *in vitro* experiments with rat cecal microorganisms,[44] further reduction of vanillin or vanillyl alcohol took place and 2 to 3% of the dose was found in the urine as 4-methylguaiacol (32) or 4-methylcatechol (33). With the exception of the glycine conjugate and some of the vanillic acid, the metabolites were excreted in the urine as glucuronide and/or sulfate conjugates. The reductive pathway to give vanillyl alcohol was reported earlier by Wong and Sourkes[49] who gave a value of 10% for this conversion. They also showed that catechol (4% of the dose) was excreted in the urine. The study of Strand and Scheline[51] also ascertained the biliary metabolites of vanillin in the rat, the effect of prevention of biliary excretion on the metabolic pattern, the effects of suppression of the intestinal microorganisms, and the effects of inhibition of the bacterial β-glucuronidase in the intestine. It was found that much of the extensive metabolism of vanillin in rats is dependent on the biliary excretion of the glucuronides of vanillin and its primary metabolites, vanillic acid and vanillyl alcohol. These conjugates are substrates for bacterial enzymes which produce the decarboxyl-

H₃CO

HO—⟨ ⟩—CH₂–NH₂

(34)

Little study of the metabolism of vanillin in man has been made although Grebennik et al.[52] reported that 2% was excreted in the urine unchanged and 73% as vanillic acid. A previously unreported metabolite of vanillin, 3-methoxy-4-hydroxybenzylamine (34), was found to be excreted in the urine of human subjects who ingested vanilla extract.[53] The quantitative aspects of this conversion were not ascertained but based on the likely content of 50 to 100 mg of vanillin in the 60 ml of vanilla extract consumed and a normal urinary creatinine excretion of 2 g/d. The data presented suggest a value of ~1% of the administered vanillin for this metabolite. This appears to be the only study in which the formation of a basic metabolite of vanillin has been investigated and this finding should be given further attention. Investigation of the metabolism of the vanillin homologue ethyl vanillin (3-ethoxy-4-hydroxybenzaldehyde) gave indication of another unexpected metabolite, 3-ethoxy-4-hydroxymandelic acid.[54] However, this metabolite was detected only when large amounts of the major metabolite, 3-ethoxy-4-hydroxybenzoic acid, were present and it was believed that its formation may be due to the presence of a homologous aldehyde as an impurity in the product ingested. In any case, no report has appeared of the formation of a mandelic acid derivative of vanillin itself.

Other studies of vanillin metabolism have considered its conversion *in vitro*. Dirscherl and Brisse[55] studied the formation of vanillic acid from several precursors including vanillin and found that 81% of this substrate was converted by rat liver homogenates and 12 to 70% by human liver homogenates under the experimental conditions employed. Friedhoff et al.[56] reported that an enzyme found in the supernatant fraction of rat liver preparations and referred to as guaiacol-*O*-methyltransferase showed weak activity in transferring the methyl group of *S*-adeno-sylmethionine to vanillin to form 3,4-dimethoxybenzaldehyde.

The metabolism of isovanillin (3-hydroxy-4-methoxybenzaldehyde), which unlike its iso-mer vanillin does not occur naturally, was studied in rats by Strand and Scheline.[51] A major aim of this investigation was to assess the significance of the presence or absence of a free hydroxy group in the 4-position on the patterns of metabolism of the two aldehydes. As noted above with vanillin, complete reduction of the aldehyde group to a toluene derivative is carried out by the intestinal microflora and this reaction is characteristic of the 4-hydroxy aldehydes.[44] Isovanillin lacks this group and was therefore not fully reduced in the *in vitro* experiments. This difference in metabolism was also seen in the *in vivo* experiments as a methylguaiacol derivative was not detected in the urine from rats given isovanillin, in contrast to that seen with vanillin. In most other respects, however, the metabolism of the two aldehydes showed a similar picture. Following oral doses of isovanillin (100 mg/kg) the urinary metabolites and their percentages were: isovanillin, 19; isovanillyl alcohol, 10; isovanillic acid, 22; vanillic acid, 11; isovanil-loylglycine, 19; catechol, 7; 4-methylcatechol, 1; and a small amount of protocatechuic acid. An interesting feature of these results is the finding that considerable demethylation to catechol derivatives occurred. This reaction liberates the 4-hydroxy group and thereby furnishes substrates for the bacterial decarboxylation and reduction reactions that form catechol and 4-methylcatechol, respectively, as well as for the *O*-methylation reaction which forms vanillic acid.

Anisaldehyde (4-methoxybenzaldehyde) was found by Sammons and Williams[57] to form an ester glucuronide when fed to rabbits (2 g/animal). This aldehyde is also metabolized to the corresponding acid and alcohol when incubated *in vitro* with rat cecal microorganisms.[44] No information is available on the urinary excretion of that portion of the anisic acid which is not conjugated with glucuronic acid; however, it is probable that some excretion of the free acid occurs as noted below with veratraldehyde. Seutter-Berlage et al.[37] administered anisaldehyde

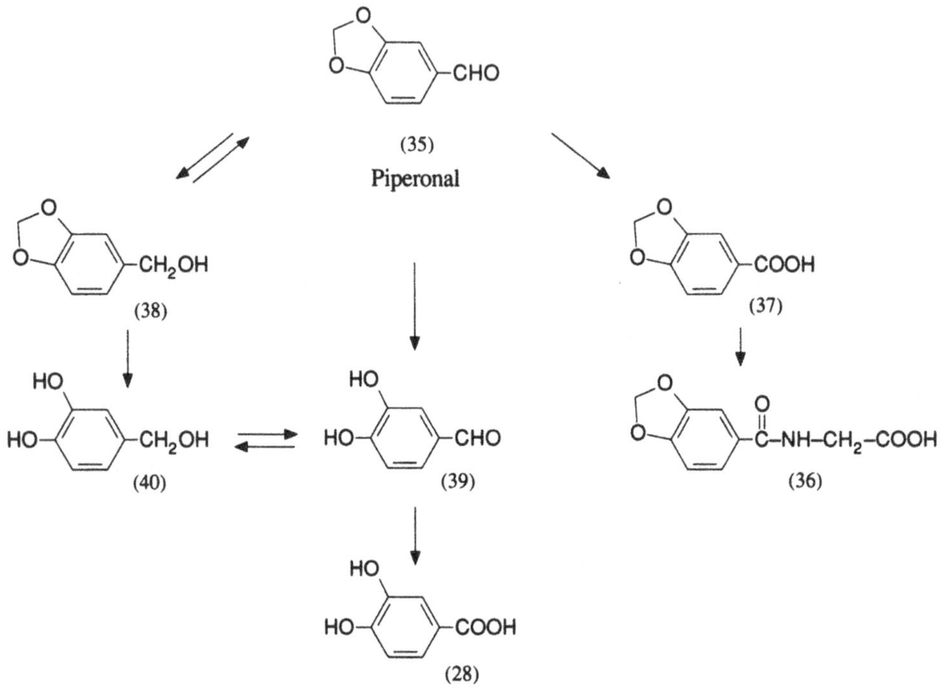

FIGURE 3.   Metabolic pathways of piperonal in rats.

(300 mg/kg, i.p.) to rats and found no evidence for the excretion of mercapturic acid derivatives in the urine. However, when the same dose of *o*-**anisaldehyde** (2-methoxybenzaldehyde) was given, 2% of the dose was excreted in the urine within 24 h as *o*-methoxybenzylmercapturic acid.

Veratraldehyde (3,4-dimethoxybenzaldehyde) was shown by Sammons and Williams[57] to be excreted by rabbits as free veratric acid to an extent of ~28% of the dose in experiments similar to those noted above with anisaldehyde. The amount of veratroylglucuronide excreted accounted for a further 38%. The only other metabolite detected in these experiments with veratraldehyde was catechol, which was excreted to a minor extent and in conjugated form. This reaction sequence, involving both *O*-demethylation and decarboxylation, is no doubt dependent on the intestinal microflora, as shown *in vitro* with rat cecal microorganisms.[44] Additionally, *O*-demethylation can be carried out in the liver. Müller-Enoch et al.[58] found that the perfused rat liver oxidized veratraldehyde and then *O*-demethylated the veratric acid to both vanillic acid and isovanillic acid. It was shown that 4-*O*-demethylation occurred more extensively as the ratio of the vanillic and isovanillic acids formed was 15:1.

Heffter[59] administered **piperonal** (3,4-methylenedioxybenzaldehyde) (35) (see Figure 3) orally to a rabbit and isolated from the urine ~28% of the dose (4 g) as piperonylic acid. Williams[60] reported that some of the latter metabolite was excreted as the glucuronide conjugate. Kamienski and Casida[61] studied the metabolic fate in mice of piperonal labeled with ^{14}C in the methylenedioxy moiety. About 1% of the oral dose (0.75 mg/kg) was recovered as $^{14}CO_2$ in 48 h indicating that this compound was fairly resistant to demethylenation. Most of the radioactivity (89%) was recovered in the urine during this period and the major metabolite was piperonylglycine (36). Some free piperonylic acid (37) was also detected in addition to two unidentified metabolites. These unidentified metabolites were also found when piperonyl alcohol (38) but not piperonylic acid was administered. *In vitro* experiments using mouse liver preparations showed some reduction of piperonal to piperonyl alcohol but mainly oxidation to the acid and its glycine derivative. A microsome-NADPH system showed low demethylenation activity, converting some piperonal to 3,4-dihydroxybenzoic acid. Klungsøyr and Scheline[62] gave piperonal (150 mg/kg) orally to rats in order to determine the identities and quantities of the metabolites excreted in the urine. They found that excretion was rapid with 93% of the dose recovered in 24 h. In agreement with the earlier results, piperonylglycine was found to be the major metabolite (71%) and piperonylic acid furnished a further 20%. Four minor metabolites were also identified. These were piperonyl alcohol (0.9%) and three compounds formed via scission of the methylenedioxy moiety: protocatechualdehyde (39) (0.1%), protocatechuyl alcohol (40) (0.1%), and protocatechuic acid (28) (0.5%). The latter values confirm the resistance of piperonal to demethylation. The urine samples were hydrolyzed with a β-glucuronidase + sulfatase preparation before analysis and the extent to which the metabolites may have been conjugated in these forms was not determined. The metabolic pathways of piperonal in the rat are shown in Figure 3.

The metabolism of **cinnamaldehyde** (cinnamic aldehyde) (41) follows the expected pathway of oxidation to the corresponding acid, although the presence of the three-carbon side chain allows for shortening by β-oxidation. This was first reported by Friedmann and Mai[63] who administered the aldehyde (150 mg/kg, i.p.) to rabbits and found benzoic acid, hippuric acid, and the glycine conjugate of cinnamic acid in the urine. Teuchy et al.[35] recovered nearly 30% of the dose (30 to 40 mg/animal, i.p.) as urinary hippuric acid when cinnamaldehyde was given to rats.

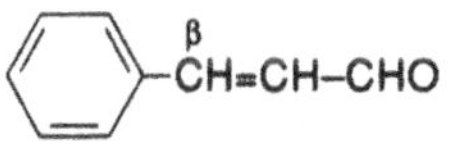

(41)
Cinnamaldehyde

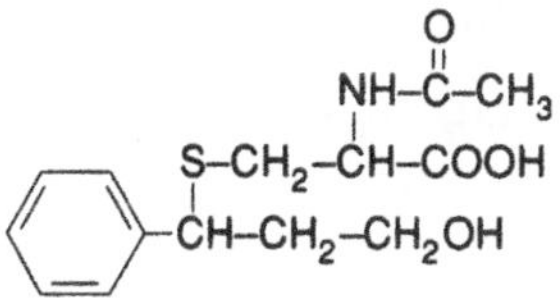

(42)

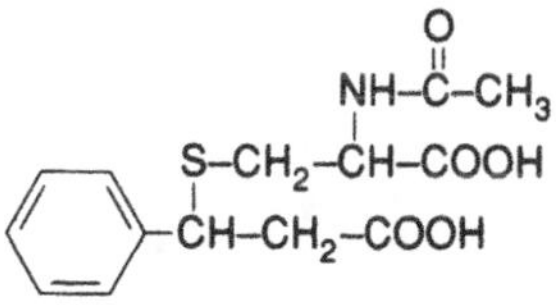

(43)

Boyland and Chasseaud[19,20] reported that the glutathione S-transferase activity present in the livers of numerous mammalian species effected the conjugation of cinnamaldehyde with glutathione. In addition, the administration of cinnamaldehyde to rats caused an appreciable reduction in the liver glutathione level in rats.[18] The mercapturic acid derivatives which result from this conjugation were isolated and characterized by Delbressine et al.[64] They gave rats ten daily doses of the aldehyde (250 mg/kg, i.p.) during a 2-week period. The two mercapturic acids excreted had structures (42) and (43), with the former accounting for ~80% of the total (15% of the dose). Both of these metabolites have the thio-linkage attached to the β-carbon atom, however the terminal carbon shows reduction in one case and oxidation in the other. It was shown that the terminal oxidation did not occur prior to glutathione addition.

The metabolism of *o*-**methoxycinnamaldehyde** (44) (see Figure 4) was studied in rats[65] and found to differ considerably from that summarized above for cinnamaldehyde. No evidence was obtained for conjugation of this aromatic aldehyde with glutathione and the pathway involving β-oxidation to give C_6–C_1-acids was strongly curtailed. Interestingly, fairly large amounts of the β-hydroxy acid (45) which would be expected to be the intermediate in this sequence were present in the urine. As shown in Figure 4 the major routes of metabolism were oxidation to the corresponding cinnamic and phenylpropionic acid derivatives, both of which were extensively conjugated with glycine. This study indicated that *o*-methoxycinnamaldehyde was rapidly metabolized and excreted. In 24 h 91% of the oral dose (210 mg/kg) was recovered in the urine.

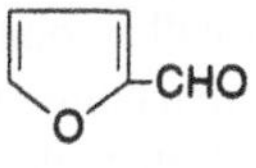

(46)
Furfural

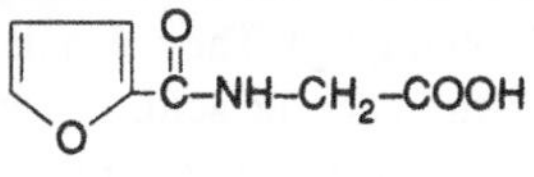

(47)

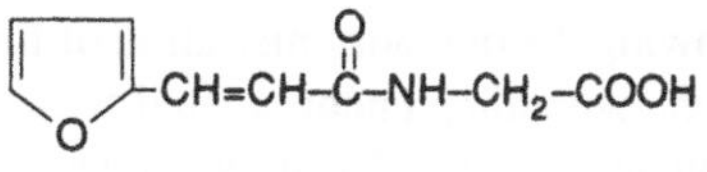

(48)

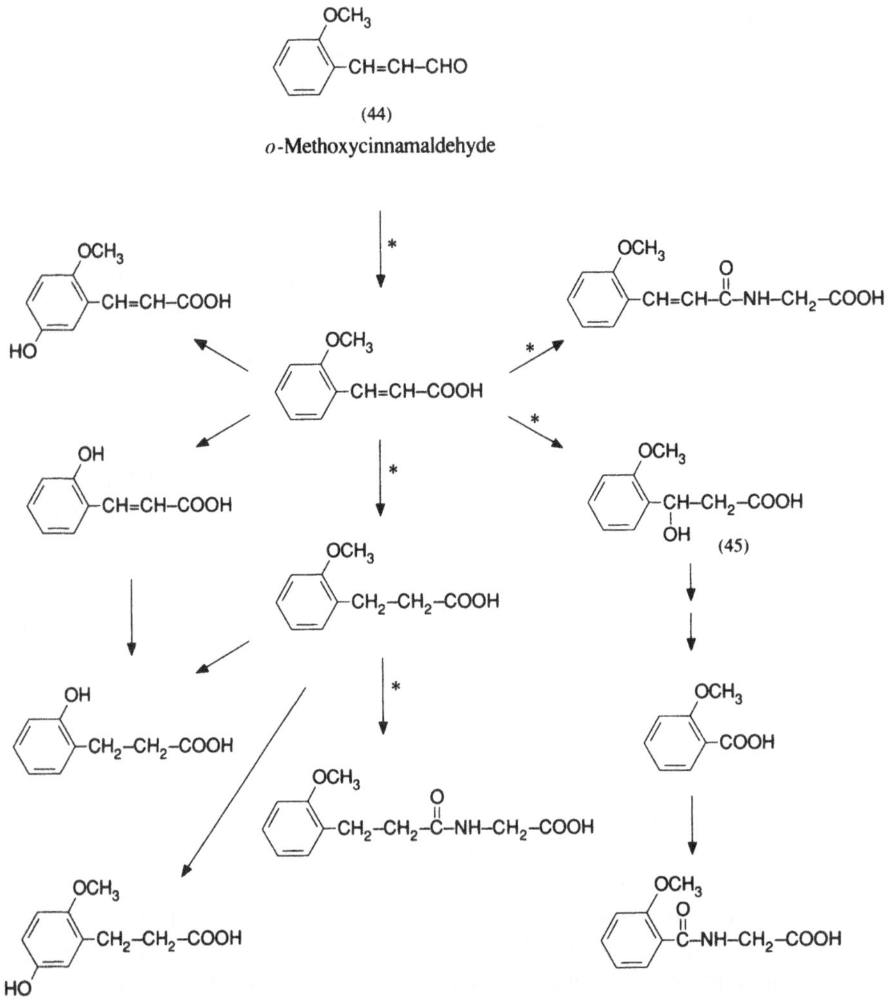
OCH3
CH=CH-CHO
(44)
o-Methoxycinnamaldehyde
OCH3
CH=CH-COOH
HO
OCH3
CH=CH-COOH
OCH3
CH=CH-C-NH-CH2-COOH
OH
CH=CH-COOH
OCH3
CH-CH2-COOH
OH
(45)
OCH3
CH2-CH2-COOH
OH
CH2-CH2-COOH
OCH3
COOH
OCH3
CH2-CH2-C-NH-CH2-COOH
OCH3
CH2-CH2-COOH
HO
OCH3
C-NH-CH2-COOH

FIGURE 4. Metabolic pathways of *o*-methoxycinnamaldehyde in rats. (*) Major pathway.

Furfural (46) contains a heterocyclic rather than a carbocyclic ring; however, its metabolism follows the expected pathway of oxidation to the corresponding acid (furan-2-carboxylic acid, α-furoic acid) which is also conjugated with glycine to give furoylglycine (47). These metabolites were detected in the urine of furfural-treated rabbits and dogs by Jaffé and Cohn,[66] and Paul et al.[67] identified the glycine conjugate as the major urinary metabolite of furfural (50–60 mg/kg, p.o.) in rats. In the former study a third metabolite, furfuracryluric acid (2-furanacryloylglycine) (48), was also isolated. The formation of an acrylic acid derivative, in essence an aldol-like condensation of the aldehyde with acetate, is an unusual reaction which has not been observed with other aldehydes. The results of Paul et al.[67] did not indicate whether or not compound (48) was also formed in rats, however Laham and Potvin[68] recently reported that Sprague-Dawley rats excreted small amounts of *trans*-2-furanacryloylglycine in the urine following the oral administration of furfural (50 mg/kg). They also detected small amounts of the corresponding unconjugated acid, *trans*-2-furanacrylic acid. These minor metabolites each accounted for ~1 to 2% of the dose and free furan-2-carboxylic acid furnished a further 3%. The major metabolite was furoylglycine (47) which accounted for about a third of the dose. The possible formation of additional conjugates, e.g., furan-2-carboxylic acid glucuronide, was not investigated. Similarly, Flek and Šedivec[69] obtained evidence for the formation of 2-furanacryloylglycine in man. They studied the fate of inhaled furfural and found that the glycine conjugate (48) was present in the urine in amounts varying from 0.5 to 5% of that of furoylglycine, the major metabolite. Elimination of unchanged furfural in the expired air was slight (<1%) and virtually no free furan-2-carboxylic acid was detected in the urine, probably because the low dose of furfural employed (1.9 mg/kg) did not exceed the capacity of the subsequent conjugation of the acid formed with glycine. A biological half-life of furfural of only 2 to 2.5 h was noted in these experiments. The results also indicated that no other conjugates of furan-2-carboxylic acid were formed.

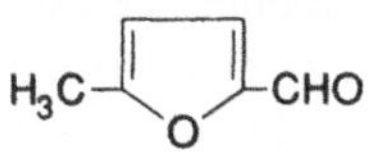

(49)   (50)

5-Methyl-2-furaldehyde

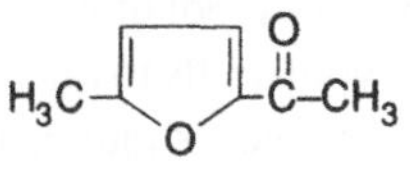

The metabolism of the closely related **5-methyl-2-furaldehyde** (49) was described by Jodynis-Liebert.[70] The aldehyde (80 to 160 mg/kg, p.o.) was given to rats and the major urinary metabolite was found to be 5-methylfuroylglycine. A second metabolite, which accounted for ~7% of the dose, was identified as 5-methyl-2-furylmethyl ketone (50). The pathway of formation of the latter metabolite is not known, however it was also excreted in the urine in similar amounts when 5-methyl-2-furancarboxylic acid was administered.

Gossypol (51) is a toxic polyphenolic aldehyde which is found in cottonseed. Its presence in meal made from cottonseed limits the usefulness of this product in livestock feeding, except in ruminant animals which are resistant to the toxic effects. Since cottonseed meal has been used as a feedstuff for various species of animals, it is not surprising that metabolic data on gossypol is available from both mammalian and nonmammalian species. The following account of gossypol metabolism will not include the results obtained with fish and chickens, however these subjects are included in the review of Abou-Donia[71] on the physiological effects and metabolism of gossypol. More recently, interest in gossypol has focused on its potential use as an antifertility agent for males. This subject was reviewed by Qian and Wang[72] who also covered the chemistry, biological effects, metabolism, and toxicology of gossypol.

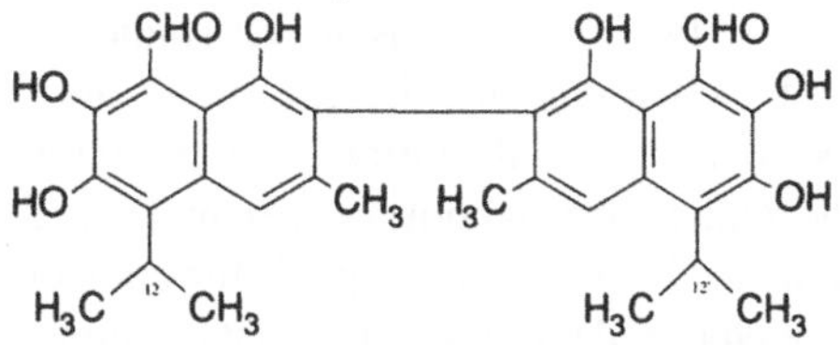

(51)

Gossypol

Information on the metabolic disposition of gossypol has been obtained in several mammalian species including the mouse, rat, dog, pig, and monkey. Bressani et al.[73] fed dogs on a diet containing cottonseed meal and studied the excretion of free and bound gossypol. They found that the bulk of the gossypol was eliminated in the feces and that the amount of free gossypol excreted was several times higher than that present in the food. They believed that this indicated a capacity for intestinal hydrolysis of the bound forms of gossypol present in the cottonseed meal. The biological disposition in rats of gossypol labeled with ^{14}C in the aldehyde groups was investigated by Abou-Donia et al.[74] When given orally at a dose level of 25 mg/kg, relatively slow excretion of radioactivity was observed in respiratory CO_2, urine, and feces. Respiratory $^{14}CO_2$ accounted for ~12% of the radioactivity when the animals were given a basal diet and nearly 21% when this diet was supplemented with ferrous sulfate. Urinary excretion of radioactivity accounted for a mere 3% of the dose whereas 77% was recovered in about a week in the feces. Similar results were obtained by Skutches and Smith[75] who employed two labeled forms of biosynthetic [^{14}C]gossypol. Material labeled in the formyl groups resulted in the excretion of $^{14}CO_2$ in the expired air whereas negligible amounts of $^{14}CO_2$ were expired when the rats were given material orally in which the ^{14}C was located only in the naphthalene rings or in the 12- and 12'-positions. These results show that the carbonyl groups are partly converted to CO_2 but that degradation of the naphthalene rings does not occur. In view of the patterns of tissue radioactivity obtained and the finding that ~90% of the radioactivity was recovered in 3 d in the feces, it was concluded that gossypol or gossypol metabolites are excreted in the bile. Tang et al.[76] administered [^{14}C]gossypol orally to mice (40 mg/kg), rats (15 mg/kg), a dog (2 mg/kg), and a monkey (2 mg/kg). The half-lives of radioactivity in the blood of these animals were 31 h, 16.5 h, 45 h, and 11 h, respectively. In the rodents the highest concentrations of radioactivity 48 h after dosing were in the gastrointestinal tract. They confirmed that elimination was mainly by the fecal route and that little of the radiolabel was found in the urine. The monkey showed the highest rate of excretion of radioactivity via the gastrointestinal tract. Wang et al.[77] recorded a value of 72% for the fecal excretion of gossypol 72 h after dosing rats orally (10 mg/kg).

The most extensive studies of gossypol metabolism have been carried out in swine and the above findings pointing to the importance of the biliary excretion of metabolites in rats have their parallels in swine.[78-80] The nature of the biliary metabolites of gossypol in swine was studied by Abou-Donia and Dieckert[81] who found that this material consisted mainly of glucuronide (33%) and sulfate (22%) conjugates. Only ~7% was due to nonconjugated material and, of the remainder, ~16 and 22% were due to further conjugated material hydrolyzed by hot-acid treatment and to the remaining water-soluble metabolites, respectively. The radioactive material excreted in the urine consisted of the same components that were found in the bile, however in different proportions. About 43% was due to the water-soluble metabolites and 27% to nonconjugated material. Abou-Donia and Dieckert[82] subsequently reported the fate in swine of a single oral dose (6.7 mg/kg) of gossypol labeled with ^{14}C in the aldehyde groups. Fecal excretion of radioactivity was extensive and amounted to 7% in 1 d, 36% in 2 d, 56% in 3 d, 74% in 4 d, and 95% in 20 d. The loss as expired $^{14}CO_2$ was 2.1% during this period and the urinary excretion of radioactivity amounted to only 0.7%. The levels of radioactivity were highest in muscle and liver and especially those in the liver showed a slow decline. The nature of the nonconjugated metabolites in the liver 2 d after dosing was studied and evidence was obtained for the presence of four radioactive compounds. The major metabolite was shown to be unchanged gossypol and good evidence for the formation of the oxidized product, gossypolone (52a), was also obtained. The remaining metabolites were tentatively identified as gossypolonic (52b) and demethylated gossic acid (53). Apogossypol is the compound formed by loss of the two aldehyde groups from gossypol and it was assumed that this nonradioactive metabolite was also formed. However, the loss of only 2.1% of the dose as $^{14}CO_2$ indicates that this pathway is of minor importance in pigs.

(52)   (53)

(a) R = CHO

(b) R = COOH

The studies summarized above utilized racemic gossypol. The existence of two optical isomers probably results from the restricted rotation of the two naphthalene moieties around the interconnecting C–C bond. The (+)- and (–)-enantiomers were recently isolated and several features of their metabolic disposition in rats, dogs, and humans have been studied. Chen et al.[83] gave (+)- and (–)-gossypol (~25 mg/kg, orally) to rats and found similar excretion patterns for the two isomers. The values recorded in 5 d were 70 and 80% (feces) and 2.3 and 2.8% (urine) for the (+)- and (–)-isomers, respectively. However, the blood levels of the (–)-isomer decreased far more rapidly after an i.v. injection than was the case with the (+)-isomer. Wu et al.[84] studied the pharmacokinetics of (±)-, (+),-and (–)-gossypol in dogs and humans. These results pointed to definite differences in the disposition of the (+)- and (–)-enantiomers, with the latter showing a higher rate of elimination in both species. Their findings suggested that the (–)-isomer, which is known to be both the most effective and the most toxic of the two forms, is more likely to undergo metabolism.

KETONES

Ketones differ from the aldehydes discussed above in that they are not susceptible to further oxidation. Instead, the carbonyl group in ketones is commonly subjected to reduction. Investigations in this area, especially those employing *in vitro* methods, have increased greatly in number in recent years and several reviews are available. Felsted and Bachur[85] listed a large number of mammalian ketone reductases and summarized their properties. They also discussed the species and tissue distribution of the reductases in general as well as their subcellular location. Further subjects covered were enzyme purification and the properties and physiological role of the ketone reductases. The subject of ketone reduction was briefly reviewed by Felsted and Bachur[86] and by McMahon.[1] The stereospecific reduction of ketones in rats and rabbits, both *in vitro* and *in vivo*, was investigated by Prelusky et al.[87]

ALIPHATIC KETONES

Williams[88] noted more than 30 years ago that the metabolism of very few aliphatic ketones had been investigated in detail. This statement is equally true today. The aliphatic ketones occurring naturally are very often methyl ketones. These have the general formula CH_3–CO–R and R varies from methyl in acetone to *n*-tetradecyl (C_{14}) in hexadecan-2-one. In addition to these 2-ones, some 3-ones and other ketones with branched and/or unsaturated chains are also found.

The lower aliphatic ketones have relatively low boiling points. This volatility influences their biological disposition and elimination may occur in the expired air. Schwarz[89] reported that 50 to 60% of the acetone (b.p. 56°) given to dogs at a dose level of 0.5 to 0.6 g/kg was eliminated in the expired air. Corresponding values for 2-pentanone (b.p. 102°, 0.36 g/kg) and 3-pentanone (b.p. 101°, 0.33 g/kg) were ~24 and 9%, respectively. Haggard et al.[90] found that 2-pentanol (1 g/kg) was extensively metabolized in rats to 2-pentanone and that over 35% of the dose was recovered as the ketone in the expired air. If, however, the dose levels were reduced, the relative amounts of expired ketone were also lowered. Schwarz[89] also used a smaller dose (3.5 mg/kg) and found that 18% of the acetone was lost by the respiratory route. In rats given 6 to 7 mg/kg, a maximum of 7% of the acetone was lost by this route.[91]

The general and major metabolic pathway with aliphatic ketones is reduction to the corresponding secondary alcohol followed by conjugation of the latter with glucuronic acid. This was described by Neubauer[14] for numerous ketones including **acetone, 2-butanone, 2-pentanone, 3-pentanone, 2-hexanone,** and **2-octanone.** The compounds were given orally to rabbits (~1 to 4 g/kg) and in some cases to dogs and the results indicated that at least part of the dose was reduced and conjugated. A similar finding was made by Kamil et al.[92] with the closely related **2-heptanone.** When given to rabbits orally, this ketone (0.95 g/kg) was reduced to the

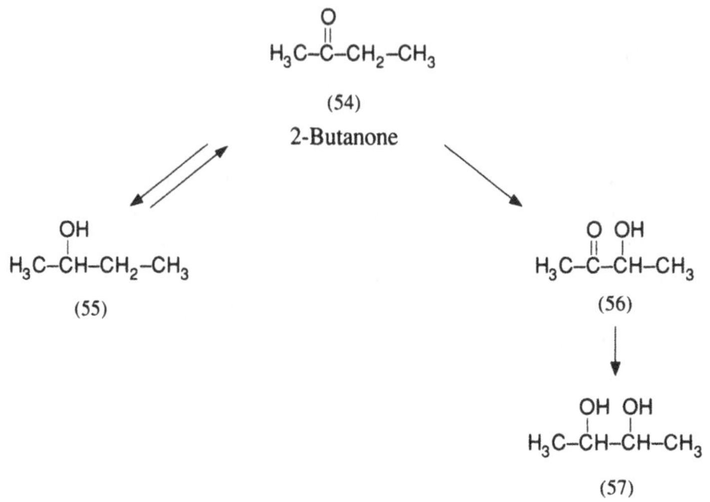

FIGURE 5.   Metabolic pathways of 2-butanone.

alcohol which was excreted as the glucuronide conjugate to the extent of slightly over 40% of the dose. Albro et al.[93] gave small oral doses (0.25 mg/kg) of [2-^{14}C]-2-heptanone to rats and found that 64% of the radioactivity was found in the expired CO_2 in 48 h. Only 15% of the dose was excreted in the urine, mainly within the first 24 h, and a mere 1.5% was found in the feces. At least 10% of the radioactivity was retained in the body after 48 h and incorporation into liver proteins, urea, cholesterol, and DNA was demonstrated.

As noted above, acetone is partly reduced and excreted as a glucuronide, however this pathway is less important with this compound than with higher ketones. Neubauer[14] found the glucuronide after administering acetone to rabbits but not to dogs. When relatively small doses of acetone are administered, oxidation may be considerable. Price and Rittenberg[91] found that at least half of the radioactivity was lost as [14]CO_2 when methyl-labeled acetone (1 to 7 mg/kg) was given orally to rats. However, this process takes place relatively slowly. This study also suggested that acetone may be metabolized to one- and two-carbon fragments which then enter normal pathways of metabolism. Sakami and Lafaye[94] obtained evidence indicating that acetone may also be converted to a three-carbon intermediate of glycolysis.

DiVincenzo et al.[95] gave 2-butanone (methyl ethyl ketone) (54) (see Figure 5) to guinea pigs (450 mg/kg, i.p.) and identified the metabolites present in serum. They found that both reduction and oxidation occurred. In the former case 2-butanol (55) was formed and, in the latter, ω–1 oxidation gave both 3-hydroxy-2-butanone (56) and butane-2,3-diol (57). Similar results were reported by Dietz et al.[96] who administered large oral doses (1690 mg/kg) to rats. They found that the diol was the most prominent and persistent metabolite detected in the blood. Humans exposed to 2-butanone similarly metabolize it to 2-butanol, 3-hydroxy-2-butanone, and butane-2,3-diol which are excreted in the urine in conjugated form.[97] These pathways are illustrated in Figure 5.

The metabolism of 2-hexanone (methyl *n*-butyl ketone) (58) (see Figure 6) has been closely studied because of the neurotoxicity shown by this ketone. When guinea pigs were given a dose of 450 mg/kg (i.p.), the serum was found to contain some unchanged compound and the expected reduction product, 2-hexanol (59).[95] However ω–1 oxidation was also shown to occur and both 5-hydroxy-2-hexanone (60) and 2,5-hexanedione (61) were detected. In fact, the latter compound was the principal metabolite of 2-hexanone in guinea pig serum. DiVincenzo et al.[98] employed 1-^{14}C-labeled 2-hexanone and oral doses of 20 or 200 mg/kg in rats. The results obtained after 48 h were similar at both dose levels, except for the amount of unchanged ketone in the expired air which was lower at the lower level (2 vs. 6%). Otherwise, 38 to 42% of the radioactivity was recovered as respiratory CO_2, 35 to 40% was excreted in the urine, 0.8 to 1.4%

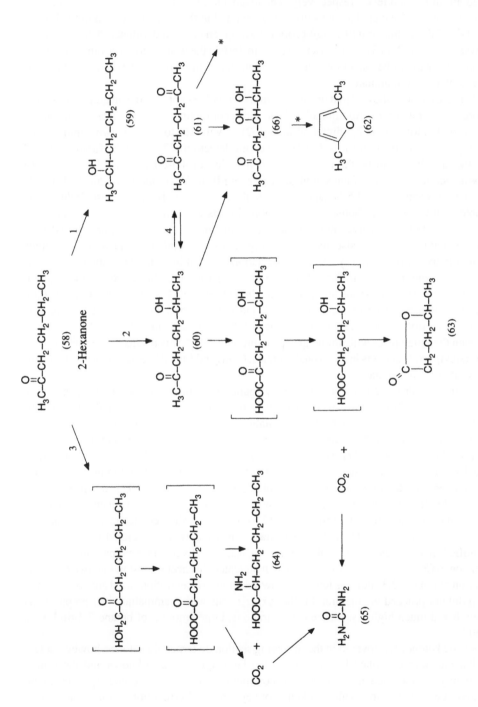

FIGURE 6. Metabolic pathways of 2-hexanone in rats. Postulated intermediates shown in brackets. (*) See text.

was found in the feces, and 14 to 18% remained in the carcass. The last value declined slowly to 8% after 6 d. Similar experiments were carried out in rats by Abdel-Rahman et al.[99] using ^{3}H-labeled ketone (267 mg/kg. i.p.). The excretion values after 72 h were 13, 40, and 8% in the expired air, urine, and feces, respectively. The disposition of 2-hexanone was also studied in dogs and humans.[100] No unchanged ketone was excreted in the urine or exhaled in the breath. When 1-^{14}C-labeled material (0.1 mg/kg, orally) was given to two individuals, 29 to 50% of the dose was lost as $^{14}CO_2$ in 3 to 5 d and a further 25 to 28% in the urine in 8 d. The other metabolic product noted was 2,5-hexanedione. The absorption and elimination of 2-hexanone in dogs was similar to that seen in humans.

DiVincenzo et al.[98] showed that the three serum metabolites noted above in guinea pigs were also present in rat serum. Additionally, these three compounds were excreted in the urine together with radioactive 2,5-dimethylfuran (62), γ-valerolactone (63), two amino acids including norleucine (64), and urea (65). However, the detection of 2,5-dimethylfuran was noted by Fedtke and Bolt[101] to be the result of the conversion of 4,5-dihydroxy-2-hexanone (66) following acidic treatment of the samples. Fedtke and Bolt[101,102] found that 4,5-dihydroxy-2-hexanone was formed from 2,5-hexanedione in both rats and man. Based on these findings, the probable pathways of metabolism of 2-hexanone in rats are outlined in Figure 6. Thus, 2-hexanone metabolism involves not only ketone reduction (Pathway 1, Figure 6) and ω–1 oxidation (Pathway 2), but also the the participation of several other metabolic reactions including α-oxidation (Pathway 3), decarboxylation, transamination, enolization, lactonization, and cyclization (as an artifact). It is also of interest to note that the reversible reaction between 5-hydroxy-2-hexanone and 2,5-hexanedione (Pathway 4) will lead to randomization of the label (the presence of ^{14}C at both C-1 and C-6) when 1-^{14}C-labeled 2-hexanone is administered. Furthermore, the neurotoxicology of 2-hexanone is mediated via 2,5-hexanedione which then may react with amino groups in proteins.[103,104] This results in the formation of 2,5-dimethylpyrrole adducts in the axonal proteins. Couri and Milks[105] reviewed the metabolism and toxicity of 2-hexanone.

Acetoin (3-hydroxy-2-butanone) is widely found as a minor plant constituent, however interest in its metabolism has mainly been due to its formation as a by-product in the metabolism of pyruvate. Consequently, much of the available information lies outside the scope of the present volume. Järnefelt[106] included a thorough review of the earlier literature in his study of the mammalian synthesis and breakdown of acetoin. Additionally, Williams[107] summarized its metabolism. In brief, the earlier results indicated that acetoin was reduced to butane-2,3-diol which may be excreted in the urine to an extent of 5 to 25% of the dose. This occurred at high dose levels but little free acetoin was excreted and oxidation to diacetyl did not appear to occur. Newer studies by Gabriel et al.[108] employing [2,3-^{14}C]acetoin showed that 15% of a small i.p. dose to rats was lost as $^{14}CO_2$ in 12 h and that a further 22% was excreted in the urine as unidentified metabolites. Additional experiments using rat liver mince preparations showed conversion to CO_2 and provided evidence which indicated that acetoin is split into two C_2-units. Formation of butane-2,3-diol was low at the low substrate concentrations used and oxidation of acetoin did not proceed via the diol. Further experiments with mammalian liver preparations showed that at much higher substrate concentrations, large amounts of butane-2,3-diol were formed.[109]

Alicyclic ketones are covered in the following section on terpenoid ketones, however a few miscellaneous alicyclic plant ketones not of the latter type are also known and these may conveniently be included here. These are commonly cyclopentanone and cyclohexanone derivatives containing various alkyl, alkenyl, or acyl groups. Furthermore, a ring double bond is common, as seen with jasmone (3-methyl-2-(2-pentenyl-2-cyclopenten-1-one). The metabolism of the substituted derivatives has not been studied, however some information is available on the parent compounds, cyclopentanone and cyclohexanone, and this may offer useful insights into the possible routes of metabolism of the substituted derivatives. James and Waring[110] administered oral doses (~190 mg/kg) of the two ketones to rats and rabbits. In rabbits, most of the material was found in the urine conjugated with glucuronic acid. This accounted for ~40 to 60% of the dose with the C_5-compound and 50 to 85% with the C_6-compound. The latter range agreed well with the value of 66% reported by Elliott et al.[111] in similar experiments employing a dose of 250 mg/kg. This material consists of cyclopentyl and cyclohexyl glucuronide, respectively, and their excretion indicates that the ketones undergo appreciable reduction to the corresponding alcohols. No cyclopentane-1,2-diol was excreted, however a few percent of the dose was accounted for as sulfur-containing metabolites. These were identified as 2-hydroxycyclopentylmercapturic acid (67a), which was excreted only in trace amounts as both the *cis* and *trans* isomers, and a metabolite accounting for ~2.5% of the dose which is probably the corresponding ethereal sulfate derivative (67b). Similar results showing a trace of the 2-hydroxycycloalkylmercapturic acid and ~2 to 3% of the sulfate metabolite were seen with cyclohexanone following its administration to rabbits or after administration of both ketones to rats. Boyland and Chasseaud[18] found that the administration of cyclohexanone and its unsaturated derivative, cyclohex-2-en-1-one, to rats led to a moderate fall in the liver glutathione levels, a change which generally points to mercapturic acid formation.

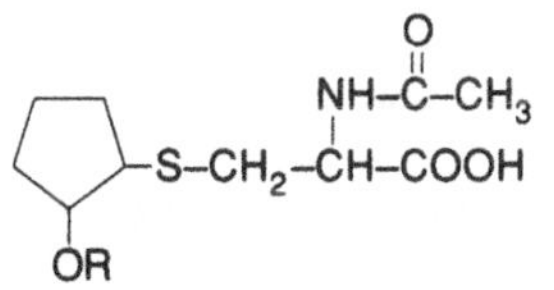

$$\text{S-CH}_2\text{-CH-COOH}$$

(67)

(a) R = H

(b) R = SO$_3$H

MONOTERPENOID KETONES
Monocyclic Terpene Ketones

Information on this group of compounds is limited to five closely related ketones based on the *p*-menthane (1-isopropyl-4-methylcyclohexane) (68) structure and to the ionones. Many of the investigations on the metabolism of ketone derivatives of menthane are of older vintage. Williams[112] reviewed the earlier literature on the metabolism of monocyclic terpene ketones. Because of the nature of the methodology then employed, these older investigations gave little indication that these compounds may be converted to numerous metabolites along multiple pathways. However, some indication of their metabolic complexity was revealed in a study of piperitone metabolism in rats.[113] Using gas chromatography-mass spectrometry, at least six metabolites of piperitone (75) were detected in the neutral fraction from the β-glucuronidase-hydrolyzed urines of rats given this compound (100 mg/kg, p.o.). These metabolites included dehydrogenated and oxygenated derivatives.

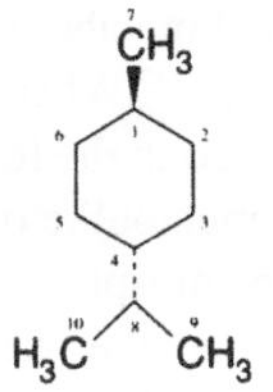

(68) 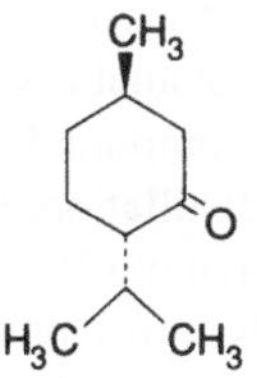   (69)
l-Menthone

The simplest of these ketones is ***l*-menthone** (*p*-menthan-3-one) (69). The earliest work on this compound indicated that it was excreted by rabbits as a glucuronide conjugate.[14] This metabolite could be formed either following the reduction of the keto group or oxidation at some other site in the molecule. The latter possibility was suggested by Hildebrandt,[114] and Hämäläinen[115] believed that hydroxylation of the tertiary carbon at C-4 occurred to give 4-hydroxymenthone. This suggestion was based on the finding that the metabolite underwent loss of water when warmed in 5% H_2SO_4, giving pulegone (79) or its isomer, *p*-menth-4(5)-en-3-one. This subject was placed on a firmer footing by Williams[116] who showed that at least part (10 to 15%) of the *l*-menthone underwent reduction of the keto group in rabbits to give *d*-neomenthol (70). Interestingly, the other possible reduction product, *l*-menthol (71), was not detected although the experimental methods employed would have easily shown the presence of its glucuronide had it been excreted. These results were best explained by assuming that *l*-menthone undergoes an asymmetric reduction. Metabolites other than *d*-neomenthol glucuronide were undoubtedly formed in these experiments as quantitative measurements of the urinary glucuronic acid output showed that ~30 to 40% of the *l*-menthone fed was converted to glucuronides of hydroxy derivatives.

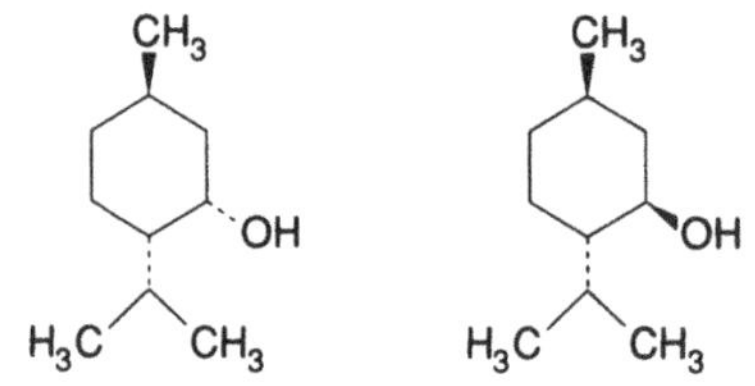

(70)   (71)

The metabolism of the isomeric **d-isomenthone** (72) was also investigated by Williams.[116] The possible reduction products in this case are *d*-isomenthol (73) and *d*-neoisomenthol (74) and, of these, only the former was definitely identified in the menthanol glucuronide fraction obtained from rabbit urine. However, this compound seemed to be the major if not the only metabolite present.

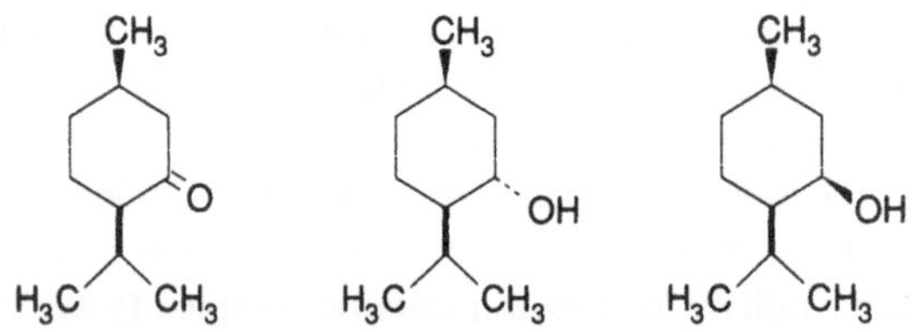

(72)   (73)   (74)
d-Isomenthone

The metabolic fate of **piperitone** (*p*-menth-1-en-3-one) (75) is very poorly understood. It is noted above that rats excreted at least six unidentified metabolites of piperitone in the urine, however only a single report is available on their nature.[117] When the ketone (9.3 g) was given orally to sheep, no unchanged compound was recovered in the feces. The urine was subjected to acid hydrolysis and steam distillation. No acidic metabolite or neutral hydrocarbons were detected in the distillate but thymol (76) and perhaps diosphenol (77) were identified. Some unchanged piperitone was also present as well as another ketone, carvotanacetone (78). Formation of the latter compound involves an unexpected migration of oxygen and this point especially would benefit by reinvestigation to see if it is a metabolite or an artifact produced by the dehydration of an alcohol under acidic conditions.

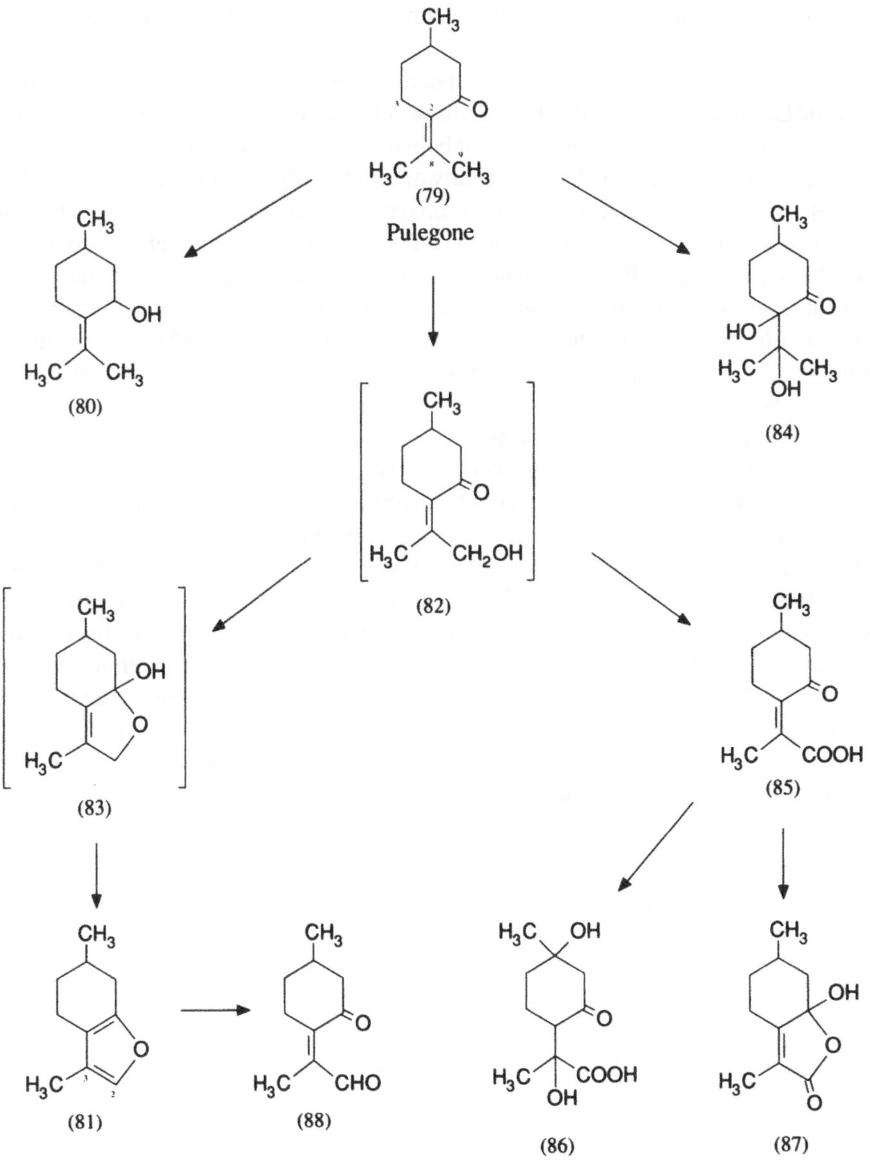

FIGURE 7. Metabolic pathways of pulegone. Postulated intermediates shown in brackets.

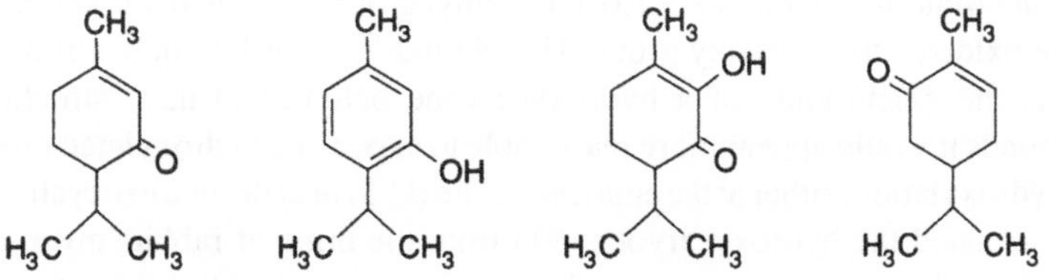

(75) Piperitone (76) (77) (78)

Pulegone (*R*-(+)-*p*-menth-4(8)-en-3-one) (79) (see Figure 7) is isomeric with piperitone, having instead an exocyclic double bond. The finding of Hildebrandt[114] that it was excreted conjugated with glucuronic acid indicated that pulegone was probably reduced to the corresponding alcohol. This was confirmed by Teppati[118] who reported that the metabolites formed were the glucuronides of pulegol (80) and, apparently, the fully reduced compound, menthol. Nelson and Gordon[119] cited unpublished results which indicated that *R*-(+)-pulegone underwent extensive oxidative and some reductive metabolism in mice. Reduction of the ketone group to give pulegol was noted and oxidation at nearly all of the carbon atoms was observed. Gordon et al.[120] reported that mouse liver microsomes converted pulegone to several metabolites, a major one of which was shown to be menthofuran (81). Their results suggested that this metabolite was formed by a cytochrome P-450-mediated oxidation of the ketone to the allylic alcohol 9-hydroxypulegone (82) which could then cyclize to give the hemiketal (83) which, in turn, formed menthofuran by dehydration. The formation of metabolites (80) and (81) was confirmed by Moorthy et al.[121] who administered large doses of (+)-pulegone orally to rats. In addition to these urinary metabolites, they also identified the diol (84) which was probably formed via epoxidation of the double bond at C-2,8. Moorthy et al. also identified the expected carboxylic acid derivative (85) formed from the allylic alcohol. Two additional urinary metabolites, probably derived from metabolite (85), were tentatively identified as compounds (86) and (87). The most prominent metabolites were pulegol (80), menthofuran (81), and the lactone derivative (87). Additional unidentified neutral and acidic metabolites were detected, however none of these appeared to be formed via endocyclic allylic hydroxylation at C-3. McClanahan et al.[122] studied the metabolic activation of (+)-pulegone to reactive intermediates which may bind covalently to cellular protein. Their results indicated that the ultimate reactive intermediate was an unsaturated γ-ketoaldehyde (88) formed from menthofuran (81) by the cytochrome P-450 monooxygenase system. Similar studies by Madyastha and Moorthy[123] confirmed the covalent binding of reactive metabolites formed from pulegone by rat liver microsomal cytochrome P-450. They proposed that the compounds responsible for this binding were the γ-ketoaldehyde (88) or menthofuran-2,3-epoxide, a possible intermediate in the conversion of menthofuran (81) to compound (88).

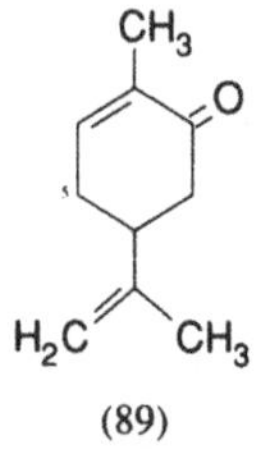

(89) (90) 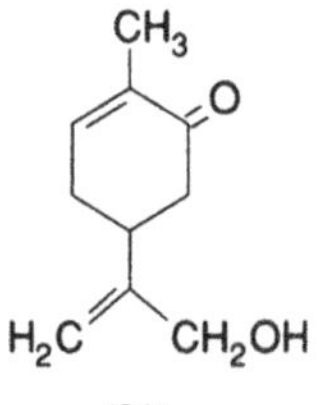 (91)

Carvone

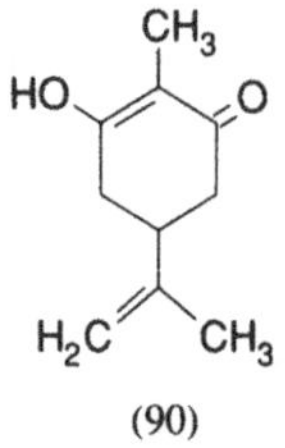

Carvone (*p*-mentha-6,8-dien-2-one) (89) was reported by Hildebrandt[114] to undergo conjugation with glucuronic acid in rabbits. Also, the methyl group adjacent to the ketone moiety was reported to be oxidized to a carboxy group. Hildebrandt[124] identified in the urine of carvone-treated rabbits the glucuronide of a hydroxycarvone believed to have structure (90). On metabolic grounds it would appear more reasonable to expect the hydroxylated product to arise from allylic hydroxylation, either at the endocyclic site (C-5) or at the two exocyclic sites. In fact, Ishida et al.[23] isolated the hydroxycarvone (91) from the urine of rabbits given carvone. An analogous hydroxylation of the isopropenyl group is seen with *d*-limonene (see the chapter entitled Metbolism of Hydrocarbons), however no evidence was obtained for the involvement of the epoxide-diol pathway in the metabolism of carvone. The latter pathway is commonly found in terpenoids which contain the isopropenyl group (e.g. *d*-limonene and nootkatone (see the chapter entitled Metabolism of Higher Terpenoids)). Fischer and Bielig[125] stated that unsaturated ketones which escape complete oxidation are excreted chiefly as optically active carbinols in which one double bond has been reduced. This general statement has been assumed to apply to carvone, which was one of the numerous test compounds given to rabbits. However, due to the high toxicity shown by carvone at the dose employed (2×2 g/kg, p.o.) no metabolic information was obtained and the applicability of this conclusion to carvone can only be assumed.

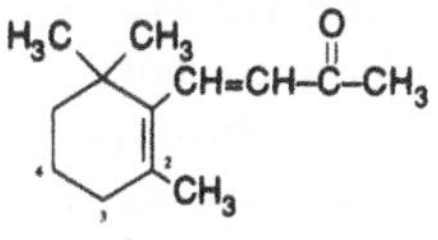

(92)
β-Ionone

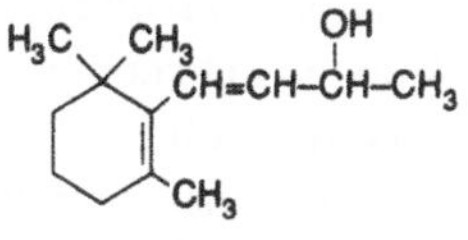

(93)

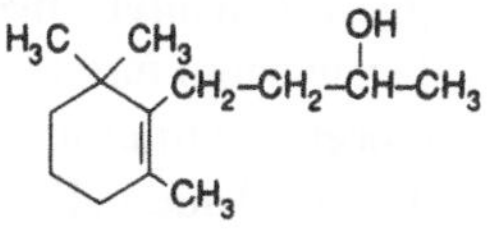

(94)

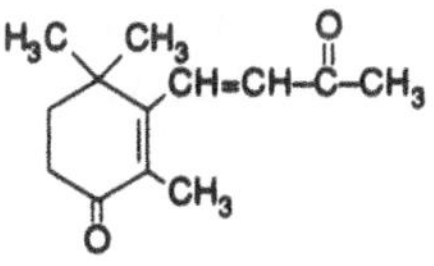

(95)

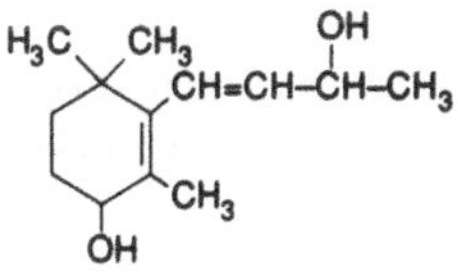

(96)

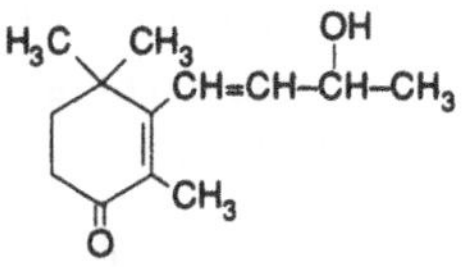

(97)

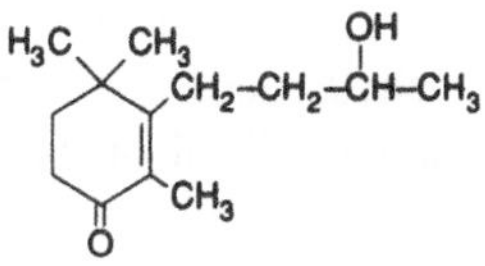

(98)

Several investigations have dealt with the metabolism of β–**ionone** (92) in rabbits. Three major metabolic routes are suggested by its structure (reduction of the keto group, double bond reduction, and oxidation of the alicyclic ring) and the available data demonstrate that all of these pathways are utilized. Additionally, some unchanged compound was detected in the urine,[126,127] although the latter investigation indicated that the amount excreted was quite small. Ketone reduction of β-ionone was first reported by Bielig and Hayashida[126] who detected the presence of the corresponding carbinol, β-ionol (93), in urine together with dihydro-β-ionol (94). Furthermore, four oxygenated derivatives of β-ionone, two derived from (93) and two from (94), were also detected. These compounds were believed to be derivatives in which the methyl group at C-2 was oxidized to a hydroxymethyl group and it was speculated that further oxidation to the corresponding aldehydes and carboxylic acids was likely. This study showed that the carbinol metabolites are optically active and that the ring double bond is not reduced. This subject was reinvestigated by Prelog and Meier[128] who detected three oxygenated metabolites in the urine of rabbits fed β-ionone. However, these were shown to be substituted at the 3-position and identified as 3-oxo-β-ionone (95), 3-hydroxy-β-ionol (96), and, probably, 3-oxy-β-ionol (97). This work was done with acid-hydrolyzed urine and the above metabolites were found in the neutral fraction which showed approximately a 2:1 ratio between the ketonic and nonketonic components. In similar experiments Ide and Toki[127] found that all of the urinary metabolites detected were compounds oxygenated at the 3-position. They confirmed the excretion of metabolites (95), (96), and (97), the latter being a major metabolite in the free fraction. Another major metabolite in this fraction was found to be dihydro-3-oxo-β-ionol (98). The glucuronides of metabolites (97) and (98) were detected in the urine.

The data summarized above indicate that β-ionone can undergo fairly extensive metabolism and the results of Ide and Toki[127] suggest that the major urinary metabolites are relatively highly metabolized. Nonetheless, our understanding of this subject is somewhat less than satisfactory, largely because of experimental inadequacies in the published work. All of the studies reported were done using rabbits which received oral doses of 20 to 50 g of β-ionone over periods ranging from 7 to 18 d. Parke and Rahman[28] showed that treatment of rats for 3 d with β-ionone gave 50 to 75% increases in the activities of several hepatic enzyme systems including that which carries out aromatic hydroxylation. Results following a more conventional dosage approach would therefore be of interest as would a closer investigation of the conjugated metabolites of β-ionone. The free metabolites isolated by Ide and Toki accounted for only a relatively small fraction of the administered dose and it seems logical to assume that the bulk is excreted as glucuronide conjugates. An approach utilizing enzyme hydrolysis of the urine samples followed by analysis using gas chromatography-mass spectrometry is warranted. The evidence concerning β-ionone oxidation suggests that allylic hydroxylation is involved, giving rise to 3-hydroxy derivatives which may then undergo further oxidation to keto compounds. As noted above, 3-hydroxy-β-ionol (96) has been identified as a urinary metabolite of β-ionone. 3-Hydroxy-β-ionone has not been detected although Ide and Toki[127] reported that this metabolite was produced from β-ionone by the rabbit liver microsomal enzyme system. Allylic hydroxylation may also afford metabolites containing the hydroxymethyl group at C-2. In view of the suggestions by Bielig and Hayashida[126] noted above concerning the formation of several metabolites of this type, the possibility of this alternative pathway must be taken into consideration. An additional point is that the carbinols formed may be oxidized to ketones. Takenoshita and Toki[129] purified a dehydrogenase from the soluble fraction of rabbit liver which dehydrogenated several alicyclic and acyclic alcohols including β-ionol (93) and 3-hydroxy-β-ionol (96).

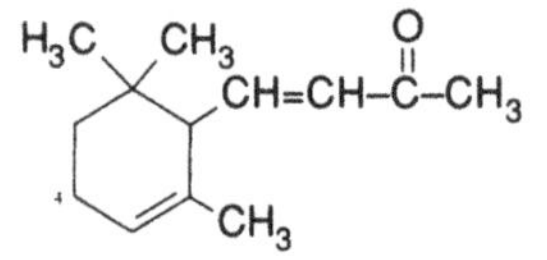

(99)

α-Ionone

The metabolism of α-**ionone** (99) in rabbits was studied by Prelog and Würsch.[130] They obtained evidence indicating that allylic hydroxylation also occurs with this compound, resulting in the 4-hydroxy derivative.

Bicyclic Terpene Ketones

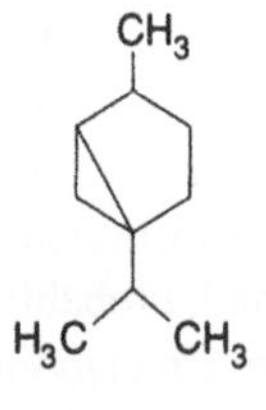

(100) 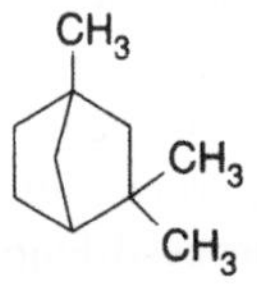   (101) 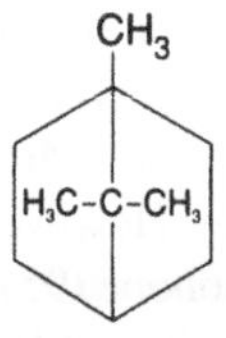   (102)

The ketones in this group are derivatives of three types of bicyclic terpenoid structures based on thujane (sabinane) (100), fenchane (101), and bornane (camphane) (102). Metabolic data on the first two types are limited to single representatives, thujone and fenchone, respectively.

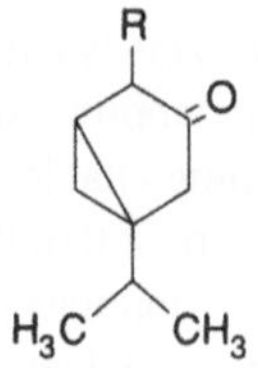

(103) 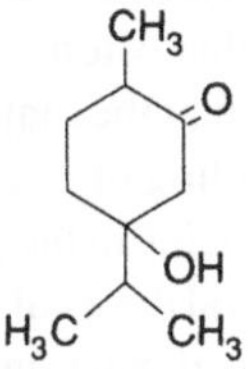   (104) 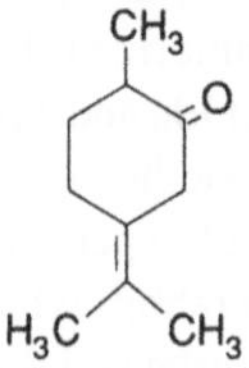   (105)

(a) α-Thujone, R = —CH₃

(b) β-Thujone, R = ····CH₃

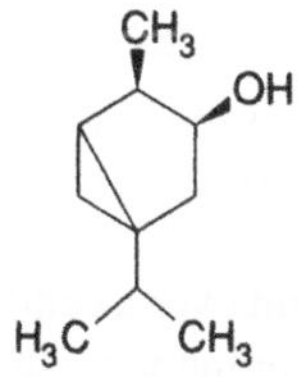

(106) 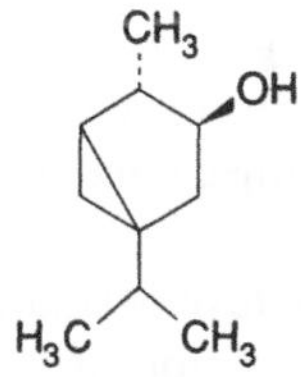   (107)

Hildebrandt[21] gave **thujone** (3-thujanone) (103) to rabbits and found that an optically active glucuronide was excreted in the urine. It was suggested that thujone was hydroxylated prior to conjugation. This was corroborated by Hämäläinen[115] in similar experiments which indicated that the glucuronide conjugate of 4-p-menthanol-2-one (104) was a urinary metabolite of thujone. The identity of the metabolite was determined by the use of chemical reactions including dehydration to carvenone (105) when treated with 5% H_2SO_4. A similar reaction involving degradation of the thujane skeleton to a monocyclic structure was also reported by Hämäläinen with the closely related thujyl alcohol (see the chapter entitled Metabolism of Alcohols). Conversion of bicyclic terpenoids to monocyclic derivatives is an interesting reaction about which little is known. However, analogous reactions were recently demonstrated with the terpenoid hydrocarbons β-pinene and 3-carene (see the chapter entitled Metabolism of Hydrocarbons). Ishida et al.[23] administered thujone (consisting of a 9:2 mixture of α-thujone (103a) and β-thujone (103b)) orally to rabbits. The neutral urinary fraction contained the corresponding alcohols, β-hydroxy-α-thujane (neoisothujanol) (106) and β-hydroxy-β-thujane (thujanol) (107). Thus, the hydroxy groups in both metabolites have a β-orientation.

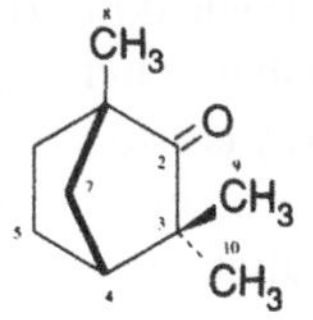

(108)   (109)

(+)-Fenchone

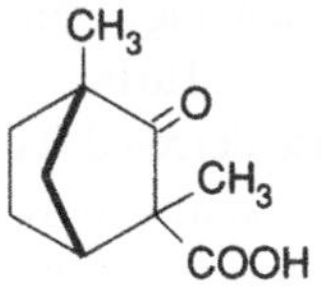

Rimini[131] reported that (+)-**fenchone** (108) was probably oxidized in dogs to its 4-hydroxy derivative. Hämäläinen[115] also recorded the conversion of fenchone to a hydroxylated derivative in rabbits. Reinarz and Zanke[132] administered fenchone to dogs and obtained evidence which indicated that besides 4-hydroxyfenchone, 5-hydroxyfenchone and Π-apofenchone-3-carboxylic acid (109), all as their glucuronides, were excreted in the urine. More recently, Miyazawa and Kameoka identified 8-hydroxyfenchone,[133] 9-hydroxyfenchone,[134] and 10-hydroxyfenchone[135] as major urinary metabolites of (+)-fenchone in rabbits.

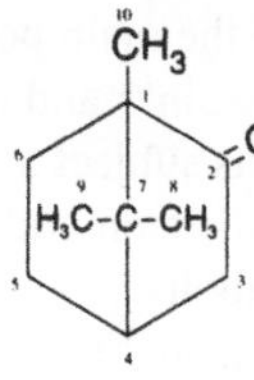

(110)   (111)

Camphor   (+)-Camphor

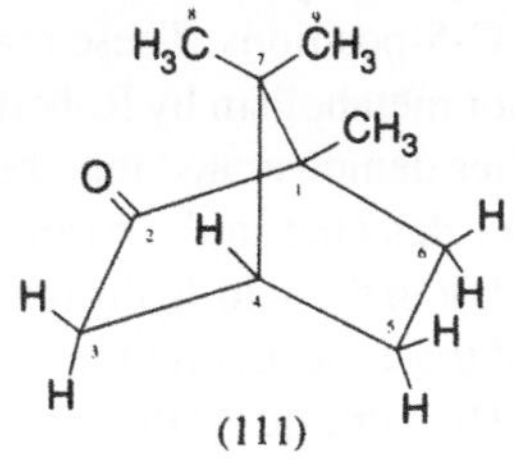

Ordinary camphor, (+)-**camphor**, is also known under several other names including Japan camphor, Formosa camphor, 2-bornanone, and 2-camphanone. Its structure is commonly shown as illustrated in (110), however structure (111) gives a more informative representation of (+)-camphor, especially with regard to the orientation of the *endo* bonds (downward in the illustration) and *exo* bonds (towards the convex side of the ring). Substituents 8 and 9 are often designated Π, these being termed either *cis* or *trans*, respectively, because of their orientation to the keto group.

Studies on the metabolic fate of (+)-camphor began over a century ago when Wiedemann[136] obtained a glycosidic material from the urine of camphor-treated dogs. Schmiedeberg and Mayer[137] soon showed that this material consisted of conjugates which liberated hydroxycamphor upon hydrolysis. This subject received no further attention for a period of 50 years until new investigations were initiated by Japanese workers. Asahina and Ishidate[138] found that the hydroxycamphor (termed camphorol) fraction obtained from the camphoglucuronic acid mixture excreted by camphor-treated dogs consisted of both the 3- and 5-hydroxy derivatives. Later reports by Asahina and Ishidate[139-141] indicated that oxidation at the Π-position also occurred, giving both the *cis* and *trans* isomers of Π-hydroxycamphor. Small amounts of another metabolite, *trans*-Π-apocamphor-7-carboxylic acid (*trans*-isoketopinic acid) (112), were also found in the urine. Greater amounts of the latter metabolite were excreted when a mixture of the Π-hydroxycamphors was administered. However, Shimamoto[142] found no evidence for the formation of *cis*-Π-hydroxycamphor in similar experiments. The relative values obtained for urinary camphor metabolites from dogs given daily doses (5 g) of camphor were: 3-hydroxycamphor, 15%; 5-hydroxycamphor, 55%; *trans*-Π-hydroxycamphor, 20%; and losses during work-up, 10%. Shimamoto[143] carried out similar experiments in rabbits given camphor (0.4 g/kg). The relative amount of hydroxycamphor excreted was only ~10% of that found previously in dogs and changes in the proportions of these metabolites were noted. This was especially true of the Π-hydroxy derivative which was no longer detected. The values for the 3- and 5-hydroxy compounds were 25 and 60%, respectively. Kawabata[144] carried out metabolic experiments in which dogs were given chronic doses (i.p. or p.o.) of camphor. The Π-hydroxy derivative was reported to be a major metabolite and small amounts of the carboxylic acid derivative (112) were also detected. This pathway is also seen in horses. Gallicano et al.[145] administered (+)-camphor to mares topically (6 g) or intratracheally (1 g) and, following hydrolysis of the urine with a β-glucuronidase + sulfatase preparation, detected at least two metabolites. One was shown to be a hydroxycamphor, however the major metabolite was compound (112).

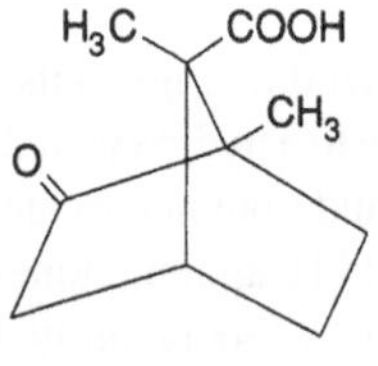

(112) 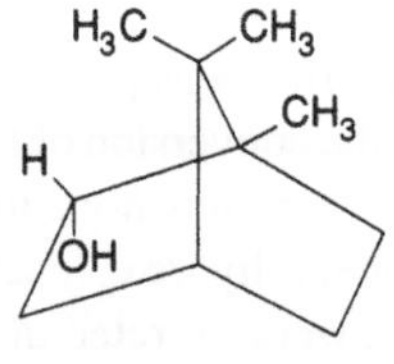   (113) 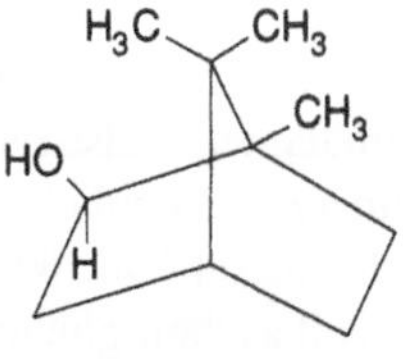   (114)

As noted above, major pathways in the metabolism of (+)-camphor involve oxidation at the C-3- and especially the C-5-positions. These reactions were the main points of interest in the investigations of camphor metabolism by Robertson and Hussain[146] and Leibman and Ortiz.[147] Additionally, these studies demonstrated that the keto group is subject to metabolic reduction, a reaction which was not detected in the earlier studies. Robertson and Hussain found three metabolites of (+)-camphor in the acid-hydrolyzed urine of rabbits given the compound (290 to 560 mg/kg, p.o.). One of these, accounting for slightly more than 20% of the total, was shown to be (+)-borneol (113). The same qualitative result was obtained in this species by Leibman and Ortiz. The latter workers also studied this reduction *in vitro* using rat and rabbit liver preparations. With the former species, the 9000-g supernatant fraction was without effect whereas this fraction from rabbit liver brought about appreciable reduction. The major reduction product was (+)-borneol (2-*endo*-hydroxybornane) (113), but ~1% of the total consisted of isoborneol (2-*exo*-hydroxybornane) (114). Further experiments showed that reduction was much more readily carried out by the cytosol than by the microsomal fraction.

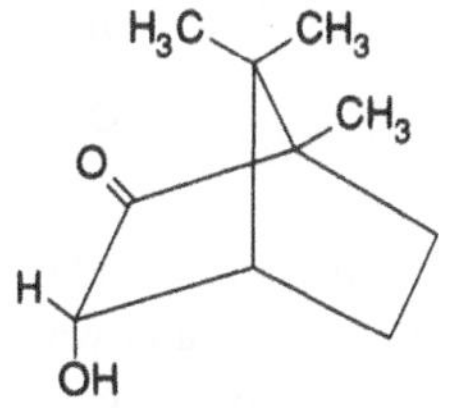

(115) 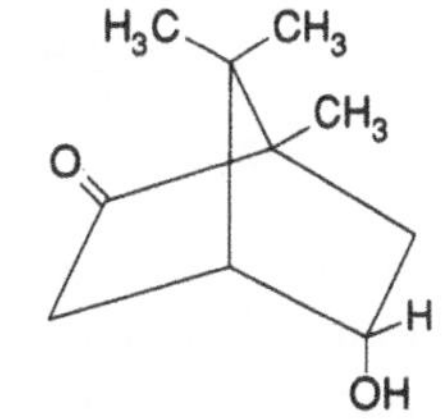   (116) 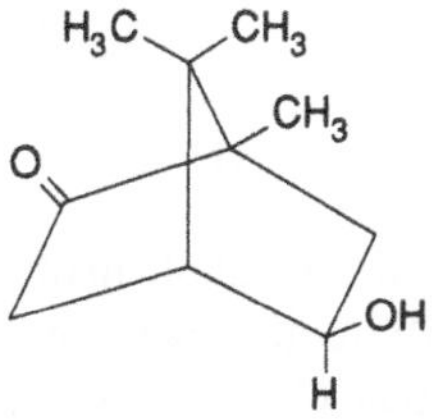   (117) 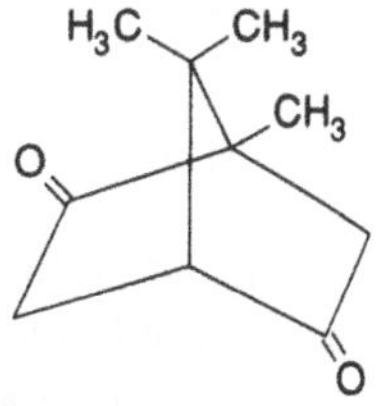   (118)

The two other urinary metabolites of (+)-camphor in rabbits described by Robertson and Hussain[146] were found to be conjugates of (+)-3-*endo*-hydroxycamphor (115) and (+)-5-*endo*-hydroxycamphor (116). These accounted for ~20 and 60%, respectively, of the identified material and the data thus harmonize with earlier results. Similarly, Leibman and Ortiz[147] found that camphor was excreted in the urine of dogs and rabbits as conjugates of both 3- and 5-hydroxycamphor. The former metabolite was most likely the *endo* isomer, however in the latter case both the *endo* (116) and *exo* (117) isomers were detected. Both of these 5-hydroxy derivatives were formed *in vitro* when (+)-camphor was incubated with rat liver microsomes or rabbit liver 9000-g supernatant fraction. This makes it unlikely that the *in vivo* results may have been influenced by isomerization of the hydroxycamphors during hydrolysis of the urinary glycosides. An important related point is that a small amount of the corresponding 5-oxo derivative (118) was detected in the urine of camphor-treated rabbits as well as in the incubates in the liver microsome experiments. Furthermore, reduction of the 5-keto group to a hydroxy group took place readily in the liver cytosol and the possibility arose that interconversion of the diketo derivative might be responsible for the proportions of the two 5-hydroxy isomers found. However, the data indicated that this was not the case and that both the 5-*endo*- and 5-*exo*-hydroxycamphors are primary hydroxylation products of (+)-camphor. Sligar et al.[148] reviewed the individual steps involved in a P-450-dependent system showing camphor 5-hydroxylase activity and a theoretical study of this reaction was recently reported.[149]

As noted above in the discussion of the metabolism of (+)-camphor, hydroxylated derivatives may be converted to ketocamphors (oxocamphors). This is true of the 5-hydroxy derivative which forms camphane-2,5-dione (5-oxocamphor) (118) and of Π-hydroxycamphor and its keto derivative Π-oxocamphor (119). The latter compound is, strictly speaking, a ketoaldehyde rather than a diketone and this fact has bearing on its metabolic fate. Nonetheless, its metabolism can conveniently be summarized at this point. In addition to the two compounds noted above, camphorquinone (3-oxocamphor) (120) has also been studied metabolically. When the latter compound, in the (+)-form, was given to rabbits (dose range of 140 to 440 mg/kg, p.o.), Robertson and Hussain[146] found that it was excreted in the urine as conjugates of 3-hydroxy-camphor (121) and 2-hydroxy-*epi*-camphor (122), both in the *endo* configurations. These results thereby confirmed those of Reinarz and Zanke[150] who reported that the glucuronides of metabolites (121) and (122) were excreted in the urine of dogs treated with camphorquinone. The data of Robertson and Hussain also showed that nearly 40% of the dose could be accounted for as urinary glucuronides. Subsequently, Robertson and Solomon[151] reported that the urine extract obtained in the earlier study contained small amounts of the *cis* and *trans* isomers of the diol (camphane-2,3-diol). These *in vivo* results showing mainly reduction of a single keto group at either C-2 or C-3 differ from those obtained using rat liver microsomes or cytosol.[147] In the latter investigation, camphorquinone was reduced equally well in both liver preparations but only at C-3 as 3-hydroxycamphor (121), presumably the *endo* isomer, was the sole reported metabolite.

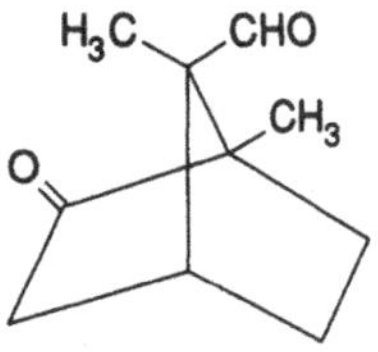

(119) 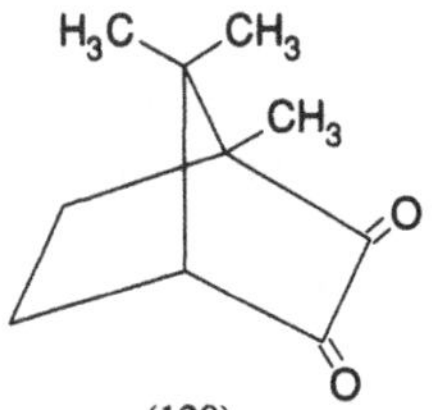   (120) 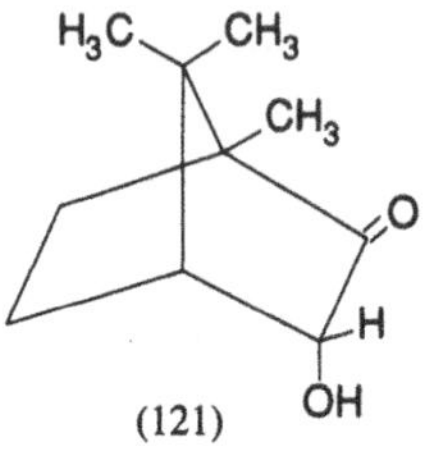   (121) 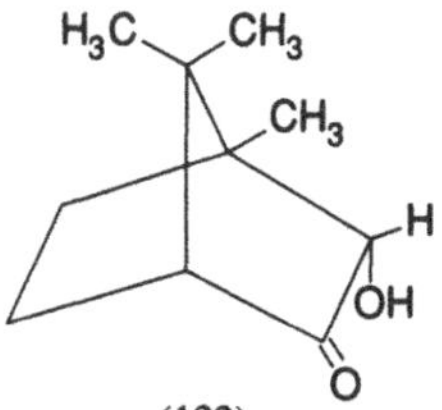   (122)

The initial study on the metabolism of 5-oxocamphor (118) suggested that this compound is metabolized rather differently than is its 3-oxo isomer.[152] Whereas the latter compound was reduced *in vivo* at either C-2 or C-3 to hydroxyketones, 5-oxocamphor was reported to be oxidized at C-4 to 4-hydroxy-5-oxocamphor and at C-8 or C-9 to Π-hydroxy-5-oxocamphor. In addition, some reduction of the 5-keto group giving 5-*endo*-hydroxycamphor (116) was noted and a fourth metabolite, probably β-cyclocamphanone-Π-carboxylic acid, was reported. For-mation of the latter compound, a tricyclic terpene, seems implausible as it involves the creation of a carbon-carbon bond between C-3 and C-5 concomitant with loss of the oxygen atom at C-5. This problem was therefore reinvestigated by Ishidate et al.[153] who administered a total of 330 g of 5-oxocamphor to dogs in 5 g daily portions. Neither the 4-hydroxy diketone nor the cyclocamphanone derivative was found in the urine. Instead, the hydrolysis product of the glucuronide excreted consisted nearly entirely of 5-hydroxycamphor. Robertson and Hussain[146] fed 5-oxocamphor (30 to 35 mg/kg) to rabbits and found a single urinary metabolite in the acid-hydrolyzed, 24-h urines. This compound was shown to be 5-*endo*-hydroxycamphor (116) but, curiously, no increase in the urinary output of glucuronides was registered in these experiments. Leibman and Ortiz[147] investigated the interconversions between the 5-oxo- and 5-hydroxycamphors *in vitro* using rat liver fractions. The main finding was that the cytosol or 9000-g supernatant fractions, but not the microsomal fraction, extensively reduced the 5-oxo compound to 5-*endo*-hydroxycamphor.

The third oxocamphor to have been studied metabolically is Π-oxocamphor (Π-apocamphor-7-aldehyde) (119). This compound also represents an intermediate stage in the metabolism of Π-hydroxycamphor to Π-apocamphor-7-carboxylic acid (112), a reaction discussed above. However, the present summary deals only with data obtained in experiments in which Π-oxocamphor itself was administered. Kawabata[144] administered chronically 120 g of Π-oxocamphor to dogs and recovered the corresponding Π-hydroxy compound and carboxylic acid (isoketopinic acid; *trans*-Π-apocamphor-7-carboxylic acid) (112) in the urine. Approximately 20 and <1%, respectively, of the total dose were obtained as these metabolites. Tamura and Imanari[154,155] identified the urinary metabolites of Π-oxocamphor in humans following its p.o. or s.c. administration. After oral dosage, ~70% was recovered as the glucuronide of the Π-hydroxy metabolite with a further 5 and 25% excreted as the carboxylic acid and its ester glucuronide, respectively. Following injection (20 mg), however, only a trace of the alcohol glucuronide was excreted and the acid and its conjugate accounted for 30 and 70%, respectively, of the dose.

Early studies on the metabolism of (–)-**camphor** in dogs[156,157] and in rabbits[158] indicated that it was converted to a hydroxycamphor, thought to be the 3-hydroxy derivative, which was excreted in the urine both free and conjugated with glucuronic acid. This isomer of camphor was also included in the investigations of Robertson and Hussain[146] and Leibman and Ortiz.[147] In the former study, rabbits were given oral doses (85 to 280 mg/kg) of (–)-camphor. The acid hydrolyzed urines contained four metabolites, three of which were identified as borneol, 3-*endo*–hydroxycamphor and 5-*endo*–hydroxycamphor. The relative proportions of these metabolites were 1.0:2.4:5.0, respectively. These results demonstrate that (–)-camphor is metabolized in rabbits qualitatively in the same way as noted above with (+)-camphor and that both reduction of the keto group and ring oxidation at C-3 and C-5 occur. Leibman and Ortiz[147] studied the *in vitro* metabolism of (–)-camphor using the 9000-g supernatant fraction of rabbit liver. Reduction of the keto group was noted but both the *endo* isomer (borneol) and the *exo* isomer (isoborneol) were identified, the amounts formed being ~1:3. Both isomers were formed when a similar preparation from rat liver was used, however the relative amounts of these products were now ~6:1. These *in vitro* studies using the rat liver preparation also showed that (–)-camphor was oxidized in a similar fashion to that seen with (+)- camphor. Thus, oxidation to 3-hydroxycamphor and, to a somewhat greater extent, 5-hydroxycamphor was observed and the latter metabolite was present in both the *endo* and *exo* configurations.

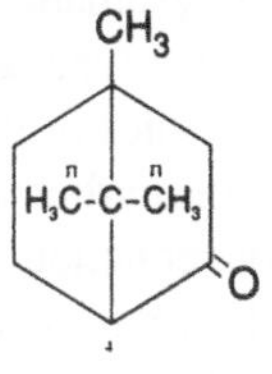

(123)	(124)	(125)
epi-Camphor	(+)-*epi*-Camphor	

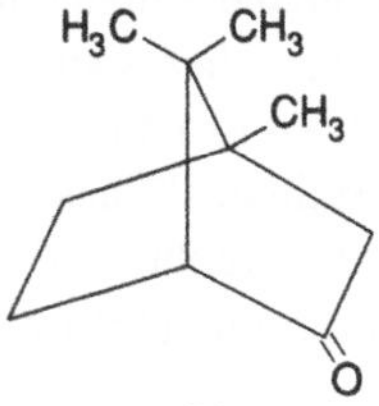

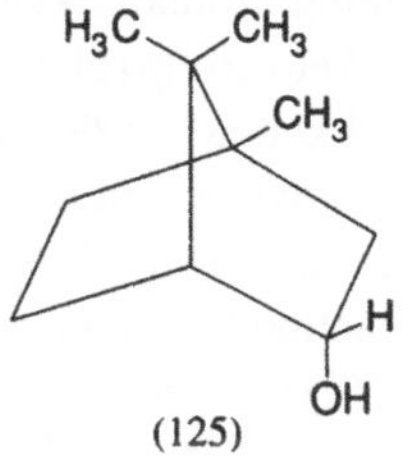

epi-Camphor (β-camphor) (123) differs from ordinary camphor in that the keto group is located at C-3 rather than C-2. Reinarz and Zanke[159] investigated the urinary metabolites of this compound in a dog which had received a total dose of 108 g divided equally over a 13-d period. The main metabolic reaction found was oxidation *ortho* to the keto group to give 4-hydroxy-*epi*-camphor which was excreted conjugated with glucuronic acid. Small amounts of the *cis* and *trans* isomers of Π-hydroxy-*epi*-camphor were also formed. More recently, Robertson and Hussain[146] gave oral doses (200 to 400 mg/kg) of (+)-*epi*-camphor (124) to rabbits. Three metabolites were detected in the hydrolyzed urines but only one of these, (+)-epi-borneol (125), was identified. It is tempting to speculate that the other metabolites may be hydroxylated ketones similar to or identical with those described by Reinartz and Zanke, however this possibility was not clarified.

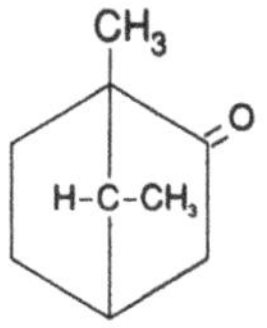

(126)	(127)	(128)
Santenone	Norcamphor	

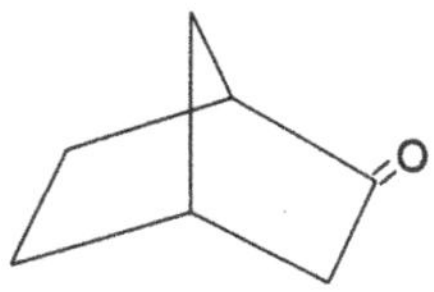

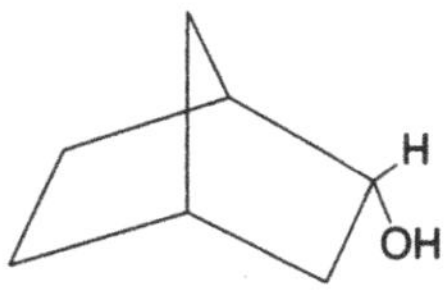

Santenone (1,7-dimethyl-2-norbornanone) (126), which was earlier known as Π-norcamphor, was one of the many alicyclic compounds studied by Hämäläinen[115] in regard to their ability to undergo glucuronide formation in rabbits. The glucuronide obtained from santenone was of a ketoalcohol, santenonol. The position of the hydroxy group was not determined but the results show that the keto group did not undergo reduction in this case. The latter reaction is, however, prominent in the metabolism of (±)-**norcamphor** (2-norbornanone) (127). Robertson and Hussain[146] gave oral doses of ~200 to 400 mg/kg of this compound to rabbits and found that 30% of the dose could be accounted for as urinary glucuronides. Hydrolysis of the urines showed the presence of a single aglycon which was isolated and identified as (+)-*endo*-norborneol (128). The reduction of norcamphor to norborneol in rabbits was also reported by Krieger.[160] However, about equal quantities of both the *endo* and the *exo* isomers of norborneol were found to be excreted in the urine, chiefly in conjugated form.

AROMATIC KETONES

Aromatic ketones do not furnish a large group of plant compounds, nonetheless it is convenient to divide these into several types according to their chemical features. True aromatic ketones are derivatives of benzophenone (diphenyl ketone) and several compounds, usually with hydroxy and/or methoxy groups in one or both rings, occur naturally. The simplest of these is *p*-hydroxybenzophenone. Robinson[161] found that rabbits excreted 72% of the dose (400 mg/kg) in the 48-h urine as *p*-hydroxybenzophenone glucuronide. Sulfate conjugation was not observed and the ketone group was not reduced.

The remaining aromatic ketones are mixed ketones in which the aromatic ring is either adjacent to the keto group (phenones) or at another position. Of the former type, **acetophenone** (methyl phenyl ketone) (129) is the simplest example. The earliest studies of its metabolism were carried out over a century ago and Williams[162] has summarized the earlier work. In brief, acetophenone was shown to be metabolized in rabbits and dogs both by reduction and oxidation. The former reaction gave methyl phenyl carbinol (130) whereas both benzoic acid and a small amount (1 to 2%) of mandelic acid (131) were found as oxidized metabolites. Smith et al.[163,164] gave acetophenone (450 mg/kg, p.o.) to rabbits and found that nearly half of the dose was excreted in the urine as the glucuronide of (–)-methylphenylcarbinol. Gal et al.[165] found that acetophenone was reduced by rabbit liver cytosol to this enantiomer with >95% stereospecificity. A small amount (3%) of ethereal sulfate was also detected. An earlier report by Culp and McMahon[166] using an aromatic aldehyde-ketone reductase from rabbit kidney cortex showed 76% production of the *S*-(–)- and 24% production of the *R*-(+)- forms of methylphenylcarbinol. El Masry et al.,[167] in experiments similar to those of Smith et al. but with a dose of 240 mg/kg, reported that 19% of the dose was found as urinary hippuric acid. The initial oxidation product in the metabolic sequence leading to hippuric acid is ω-hydroxyacetophenone (132) and Kiese and Lenk[168] showed that rabbits excreted small amounts (0.5 to 1%) of it in the urine in conjugated form following an i.p. dose of ~350 mg/kg of the ketone. Kiese and Lenk[169] found that microsomal preparations from rabbit liver readily produced the ω-hydroxylated product of acetophenone. Additional urinary metabolites of acetophenone in rabbits include the 3- and 4-hydroxylated derivatives, however these phenols accounted together for only ~0.5% of the dose.[168]

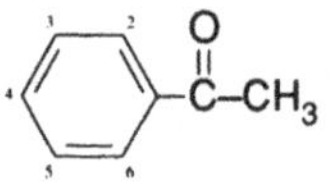

(129)

Acetophenone

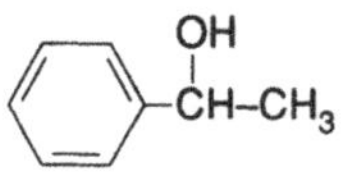

(130)

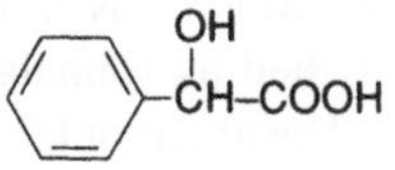

(131)

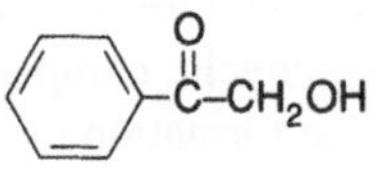

(132)

A number of derivatives of acetophenone have been found in plants and, as noted above with the true aromatic ketones, these generally contain hydroxy and/or methoxy groups. Several reports are available which describe the metabolism of **peonol** (133) (paeonol, 2-hydroxy-4-methoxyacetophenone) (see Figure 8) in various mammalian species. Yokoyama et al.[170] employed ^{14}C-labeled material and found that rats excreted 88% in the urine and 5% in the feces within 24 h after an oral dose (16 mg/kg). The metabolites detected in the 7-h urine samples were 2,5-dihydroxy-4-methoxyacetophenone (134) and 2,4-dihydroxyacetophenone (resacetophenone) (135). Similar experiments were carried out by Mimura and Baba[171] in mice, rats, guinea pigs, and rabbits. Following oral doses of 20 mg/kg, excretion of the radioactivity was rapid and amounted to 83 to 98% of the dose within 2 d. Most of this was found in the urine. Both of the

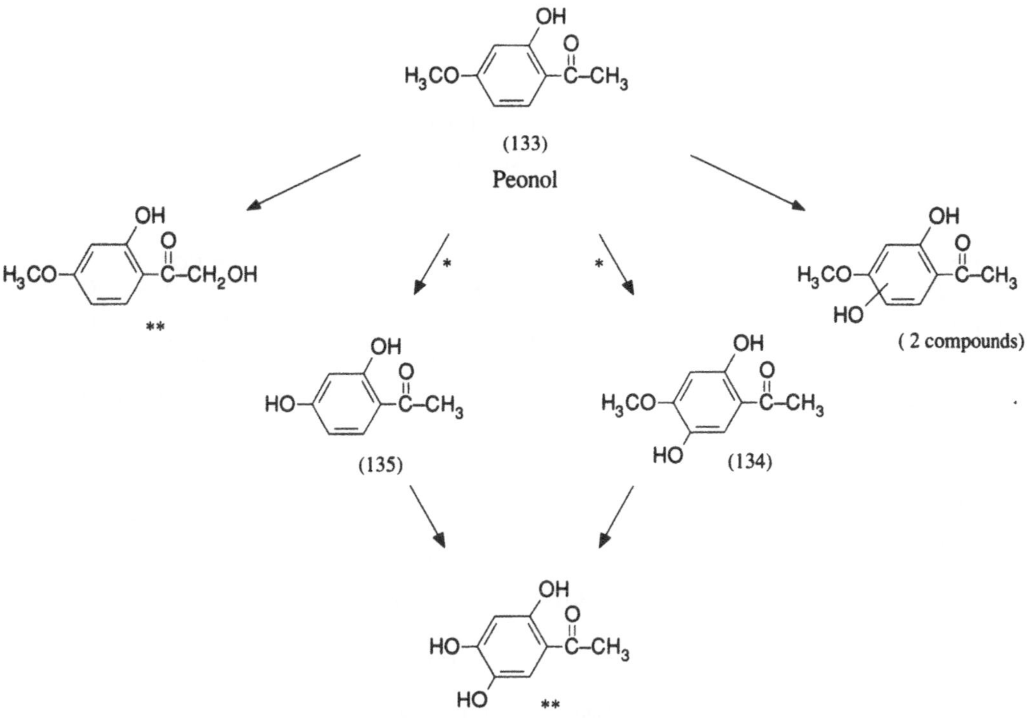

FIGURE 8.   Metabolic pathways of peonol in rats. (*) Major pathway. (**) Tentative structures.

metabolites noted above, as well as peonol itself, were detected as urinary metabolites in all four species. However, different relative amounts of the compounds were excreted by the various species. 2,5-Dihydroxy-4-methoxyacetophenone was the major metabolite in mice and rats, whereas guinea pigs excreted mainly 2,4-dihydroxyacetophenone. Rabbits excreted large, nearly equal amounts of both of these metabolites. All of the urinary metabolites excreted by rats were in the form of conjugates (as β-glucuronides, sulfates, or enzyme-resistant material), but the other species excreted half or more of the dose in the free form. Gjertsen et al.[172] administered peonol (166 mg/kg) intragastrically to rats and accounted for 61% of the dose as urinary metabolites in 2 d. These were excreted mainly as glucuronide and/or sulfate conjugates and, at this dose level, nearly equal amounts of peonol, 2,4-dihydroxyacetophenone and 2,5-dihydroxy-4-methoxyacetophenone were present. Additional findings showed that, besides these major metabolites, four other compounds were usually formed in small or trace amounts. One of these (3% of the dose) was a trihydroxylated derivative of acetophenone (probably the 2,4,5-isomer). The three trace metabolites were all hydroxylated derivatives, i.e., isomeric with 2,5-dihydroxy-4-methoxyacetophenone. In two cases the substitution was certainly in the ring, however the final compound showed mass spectral characteristics which suggested that it was an ω-hydroxylated derivative. However, no acidic metabolites which might arise from further oxidation of the latter compound were detected. The structures of these compounds and their probable routes of formation are shown in Figure 8.

The metabolism of peonol in man was studied using material labeled with both ^{2}H and ^{13}C.[173] Following an oral dose (100 mg), the recovery in the urine was 79% of the dose in 24 h and 83% in 72 h. The major urinary metabolite was 2,5-dihydroxy-4-methoxyacetophenone. It accounted for 53% and was excreted mainly as about equal amounts of a glucuronide conjugate and enzyme-resistant conjugate. Resacetophenone (135) (18%) and peonol (~9%) were also formed and excreted mainly as glucuronide conjugates. The excretion of nonconjugated metabolites was small (<1%).

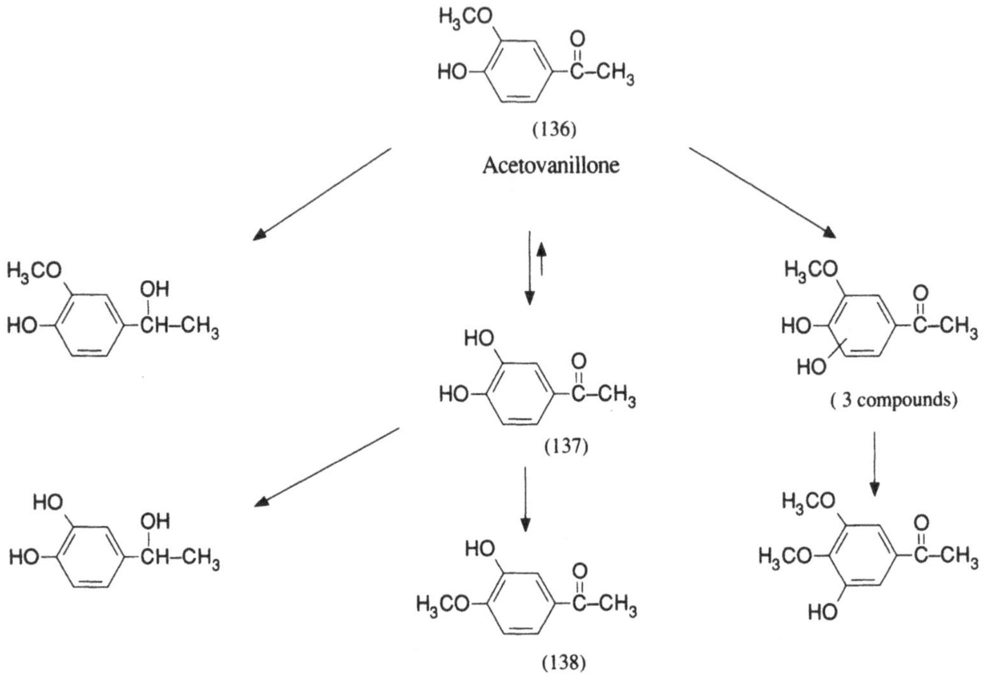

H$_3$CO
HO—⬡—C(=O)—CH$_3$

(136)

Acetovanillone

H$_3$CO
HO—⬡—CH(OH)—CH$_3$

HO
HO—⬡—C(=O)—CH$_3$

(137)

H$_3$CO
HO—⬡—C(=O)—CH$_3$
HO

(3 compounds)

HO
HO—⬡—CH(OH)—CH$_3$

HO
H$_3$CO—⬡—C(=O)—CH$_3$

(138)

H$_3$CO
H$_3$CO—⬡—C(=O)—CH$_3$
HO

FIGURE 9. Metabolic pathways of acetovanillone in rats.

Acetovanillone (4-hydroxy-3-methoxyacetophenone) (136) (see Figure 9) is mainly excreted in the urine as the corresponding glucuronide conjugate. Daly et al.[174] recorded a value of ~80% for this metabolite in 20 h following a dose of 120 mg/kg in rats. No urinary 3,4-dihydroxyacetophenone (137) was detected, however the presence in the urine of a small amount (0.5%) of acetoisovanillone (3-hydroxy-4-methoxyacetophenone) (138) indicates that some demethylation (followed by *p-O*-methylation) must have occurred. *In vitro* experiments indicated that acetovanillone was *O*-demethylated by microsomal preparations from both rat and guinea pig liver. However, acetovanillone can also be demethylated by the intestinal bacteria[175] and this possibility, which would be quite likely in the event that acetovanillone glucuronide is excreted in the bile, must be kept in mind. Recently, Gjertsen et al.[172] confirmed several of the earlier findings and also showed that acetovanillone underwent oxidation to ring hydroxylated products and reduction to carbinol derivatives. Following an intragastric dose of 166 mg/kg in rats, 97% of the dose was accounted for in the urine, nearly entirely within the first day. Metabolites were mainly excreted as glucuronides and/or sulfates and 80% of the dose was shown to be acetovanillone. Both its demethylated product, 3,4-dihydroxyacetophenone (6%), and acetoisovanillone (traces) were routinely found in the samples; however, most of the other metabolites, some present in only trace amounts, were not always detected. The structures of the metabolites of acetovanillone and their probable metabolic interrelationships are shown in Figure 9.

Acetoveratrone (3,4-dimethoxyacetophenone) differs from peonol and acetovanillone by virtue of its less polar nature and by the lack of a hydroxy group which would allow direct conjugation. It is therefore reasonable to assume that it may be subjected to extensive metabolism. In fact, Gjertsen[176] was able to detect 20 metabolites in the urine of rats given acetoveratrone (270 mg/kg, intragastrically). The urine samples were routinely treated with a β-glucuronidase + sulfatase preparation before analysis. Many of the metabolites identified were present only in trace amounts, usually during the first and, in some cases, second day after dosing and they were sometimes not produced by all of the test animals. Table 1 lists the names of the compounds identified, however the positions of the ring substituents were not ascertained for

TABLE 1
Urinary Metabolites of Acetoveratrone in Rats[176]

Metabolite excreted	Days
Acetoveratrone	1
Acetovanillone (4-hydroxy-3-methoxyacetophenone)	1—6
Acetoisovanillone (3-hydroxy-4-methoxyacetophenone)	1—4
3,4-Dihydroxyacetophenone	1—4
1-(4-hydroxy-3-methoxyphenyl)ethanol	1—3
1-(3,4-dihydroxyphenyl)ethanol	1—2
Hydroxydimethoxyacetophenone (two compounds)	1—2
Dihydroxymethoxyacetophenone (five compounds)	1—2
Dihydroxydimethoxyacetophenone	1—2
ω-Hydroxyacetoveratrone[a]	1
ω-Hydroxyacetovanillone[a]	1
ω-Hydroxyacetoisovanillone[a]	1
Hydroxy-ω-hydroxyacetoveratrone[a]	1
3,4-Dimethoxymandelic acid	1—2
3,4-Dimethoxyphenylglyoxylic acid	1

[a]   Structure (132) shows position of the ω-hydroxyl group.

the eight metabolites having dimethoxyhydroxy-, dihydroxymethoxy-, or dihydroxydimethoxy-patterns of substitution. In a similar experiment using a lower dose of acetoveratrone (180 mg/kg), Gjertsen[176] found that an average of 64% of the dose was excreted in the urine, mainly within the first 2 d. The major metabolite was acetovanillone (45%), however acetoisovanillone (8%), 3,4-dihydroxyacetophenone (5%), and one of the dihydroxymethoxyacetophenones (5%) were also found.

In addition, several reports are available which describe the metabolism of acetophenone derivatives closely related to those occurring naturally. Because of this close structural similarity, it seems worthwhile to briefly review the metabolism of these compounds. Williams[177] summarized the older literature which dealt with the metabolism of 2,4-dihydroxyacetophenone (resacetophenone), 2,3,4-trihydroxyacetophenone (gallacetophenone), and 2,4,6-trihydroxyacetophenone (phloroacetophenone) in rats, rabbits, and dogs. These compounds are apparently not metabolized by reduction of the keto group or by oxidation of the methyl group. Instead, the availability of the phenolic hydroxy groups results in direct conjugation giving glucuronides and ethereal sulfates. Dodgson[178] showed that, in the case of resacetophenone, glucuronic acid conjugation occurs at the 4-position. Haley and Bassin[179] administered phloroacetophenone (100 mg/kg, s.c.) to rats and found that the excretion of metabolites in the urine was complete within 24 h. Nearly 80% of the dose was accounted for and the free unchanged phenol and glucuronide fractions each amounted to slightly over a fifth and the ethereal sulfate fraction a third of the dose. These general results were largely confirmed by Gjertsen[176] who administered resacetophenone and phloroacetophenone to rats using intragastric doses of 152 and 168 mg/kg, respectively. By using sensitive gas chromatographic-mass spectrometric methods, trace amounts of two hydroxylated and one methoxylated derivatives of resacetophenone were detected as urinary metabolites. However, urinary excretion of resacetophenone itself was rapid and nearly all of the 94% of the dose which was recovered was excreted as this compound (free and/or conjugated) within 24 h. Similar experiments with phloroacetophenone showed an excretion of 85% within 24 h and the unchanged compound, again as free and/or conjugated material, was the sole urinary metabolite.

The metabolism of 2,6-dihydroxyacetophenone, 2-hydroxy-6–methoxyacetophenone, and 2,6-dimethoxyacetophenone was studied in rats by Bobik et al.[180] Both unlabeled material and material labeled with ^{14}C in the keto group were given by i.p. injection at doses of ~60 to 80 mg/

kg. About 80% of the radioactivity was excreted in the urine in 24 h and this was extracted nearly quantitatively into ether following acid hydrolysis. In the case of the 2,6-dihydroxy derivative, this material consisted solely of the unchanged compound. With 2-hydroxy-6-methoxyaceto-phenone, most of the hydrolyzed material was unchanged compound but 4% was shown to be two hydroxylated metabolites. These were identified as 2,3-dihydroxy-6-methoxyacetophe-none and 2,5-dihydroxy-6-methoxyacetophenone. When 2,6-dimethoxyacetophenone was given, no unchanged compound was excreted and the 2-hydroxy-6-methoxy derivative was the major (66%) metabolite. Aromatic hydroxylation at the 3-position was also detected in this case and 2,6-dimethoxy-3-hydroxyacetophenone (26%) and 2,3-dihydroxy-6-methoxyacetophe-none (1%) were found. The remainder (7%) of the radioactivity was present as unidentified polar material. No aromatic carboxylic acids were detected as metabolites in any of the experiments.

Another type of phenone includes compounds in which the methyl group is replaced by various other substituents. Typical of this group are the chalcones and dihydrochalcones. However, these compounds show both biosynthetic and metabolic similarities to the flavonoids and their metabolic fate is therefore summarized in the chapter entitled Metabolism of Oxygen Heterocyclic Compounds.

The final type of aromatic mixed ketone includes a relatively small number of compounds in which the aromatic ring is not situated adjacent to the keto group. The metabolic data available on this group of ketones is limited to three compounds: *p*-hydroxybenzylacetone, its 3-methoxy derivative zingerone, and curcumin.

p-**Hydroxybenzylacetone** (4-(4-hydroxyphenyl)butan-2-one, raspberry ketone, oxyphenalon) (139) (see Figure 10) was administered to rats, guinea pigs, and rabbits.[181] The dose (164 mg/kg, intragastrically) was rapidly excreted and nearly 90% was found in the urine within 24 h. The most prominent urinary metabolites were *p*-hydroxybenzylacetone and the corresponding carbinol, both mainly conjugated with glucuronic acid and/or sulfate. In addition to the reductive pathway, oxidative metabolism including ring hydroxylation and side-chain oxidation was observed. The latter pathway led to 1,2- and 2,3-diol derivatives which were probably intermediates in the formation of the C_6–C_3- and C_6–C_2-metabolites detected. The structures of the metabolites identified and their proposed metabolic relationships in rats are shown in Figure 10. Most of the metabolites found in rat urine were also present in the samples from guinea pigs and rabbits. The most notable differences were the more extensive excretion of conjugated *p*-hydroxybenzylacetone in guinea pigs and the greater reduction of it to the corresponding carbinol in rabbits.

The results described above for *p*-hydroxybenzylacetone show many similarities to those of Monge et al.[182] who studied the metabolism of **zingerone** (4-(4-hydroxy-3–methoxyphenyl)butan-2-one) (140) (see Figure 11) in rats. Zingerone and its metabolites were also largely excreted in the urine within 24 h as glucuronide and/or sulfate conjugates. From 52 to 56% of the dose (100 mg/kg, p.o. or i.p.) was excreted as conjugates of zingerone itself and reduction to the corresponding carbinol accounted for a further 11 to 13%. Additionally, side-chain oxidation took place at all three available sites, however oxidation at the 3-position, which probably resulted in the C_6–C_2-metabolites, predominated. The probable metabolic pathways of zinger-one in the rat are shown in Figure 11. The formation of several *O*-demethylated metabolites was shown to be due to the extensive biliary excretion of zingerone and some of its metabolites followed by their *O*-demethylation by intestinal microorganisms.

Figures 10 and 11 show that the hydroxy- and hydroxymethoxy-derivatives of benzylacetone are metabolized via the same main routes. Several studies with related compounds have shown similar results. The parent compound, benzylacetone, was metabolized to small amounts of both benzoic and phenylacetic acids by rabbits.[167] Wright and Holder[183] found that piperonylacetone (4-(3,4-methylenedioxyphenyl)butan-2-one) was converted to 3,4-methylenedioxyphenylacetic acid by rabbits. However, the major metabolite was the reduction product 4-(3,4-methylenedioxyphenyl)butan-2-ol which accounted for 81.5% of the dose (100 mg/kg, orally).

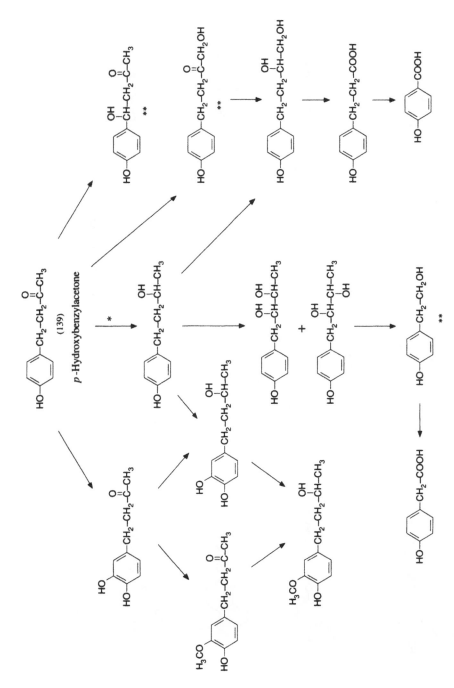

FIGURE 10. Metabolic pathways of *p*-hydroxybenzylacetone in rats. (*) Major pathway. Pathways similar in guinea pigs (except for metabolites marked with ** which were absent) and in rabbits.

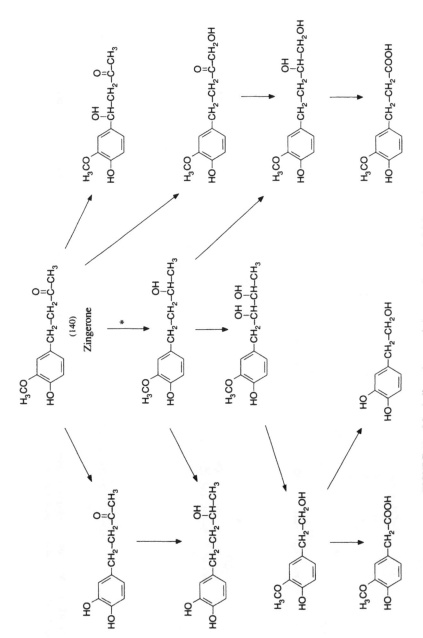

FIGURE 11. Metabolic pathways of zingerone in rats. (*) Major pathway.

In addition, a study of the metabolism by rabbit liver preparations of phenylacetone and 1-phenylbutan-2-one[184] showed reduction to the corresponding carbinols and formation of small amounts of the 1,2-glycols and ketols (2-one-1-ols).

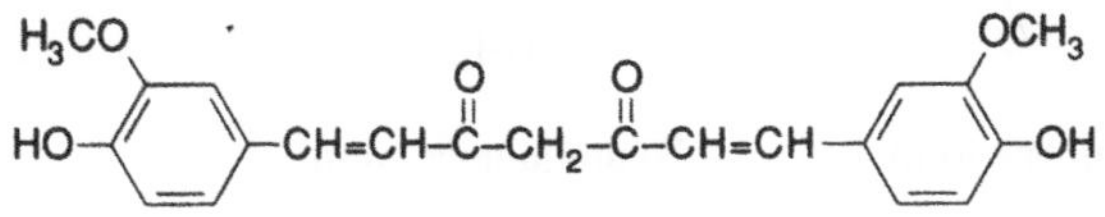

(141)

Curcumin

Metabolic studies of **curcumin** (turmeric yellow, diferuloylmethane) (141) have shown that it undergoes appreciable excretion in the bile and feces. Holder et al.[185] gave [³H]curcumin orally to rats and found that 89 and 6% of the dose (~2 mg/kg) was excreted in the feces and urine, respectively, within 72 h. The corresponding values following i.p. administration were 73 and 11%. Ravindranath and Chandrasekhara[186] also employed [³H]curcumin but gave rats intragastric doses of ~60, 460, or 2300 mg/kg. At the lowest dose 91% of the radioactivity was found in the feces and 2% in the urine, with most of the excretion taking place within 3 d. Comparison of the measurements for fecal radioactivity with those for fecal curcumin suggested that only about a third of the material present was due to the unchanged compound. This finding is best explained by the biliary excretion of curcumin metabolites. Holder et al.[185] found that low doses (~2 mg/kg, i.v. or i.p.) resulted in >80% of the radioactivity being excreted in the bile within 6 to 8 h. Most of this material was found to be conjugated with glucuronic acid and the aglycons were shown to be the reduced metabolites tetrahydrocurcumin (142) and hexahydrocurcumin (143). In addition, small amounts of dihydroferulic acid (4-hydroxy-3-methoxyphenylpropionic acid) and probably ferulic acid were excreted in the bile. These findings indicate that the reductive metabolism occurred in the tissues and that it was not a result of the action of the intestinal microflora. Plummer[187] obtained evidence that hexahydrocurcumin is also formed in humans. Three subjects each received a dose of 700 mg of deuterium-labeled curcumin over a 2-h period. Compound (143) was detected in the urine samples collected 24 h after dosing, however no deuterated material was found in the fractions containing free or conjugated acidic metabolites or in the free neutral fraction. Ravindranath and Chandrasekhara[188] reported that curcumin may also undergo metabolism during absorption from the intestine. They found that everted sacs of rat intestine formed a colorless metabolite which was less polar than curcumin.

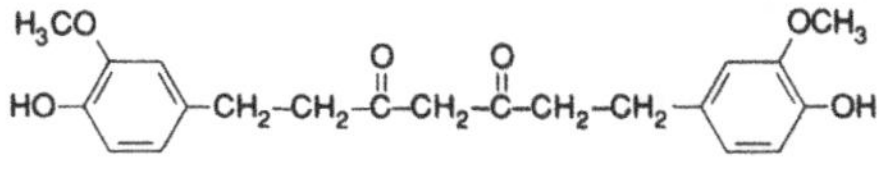

(142)

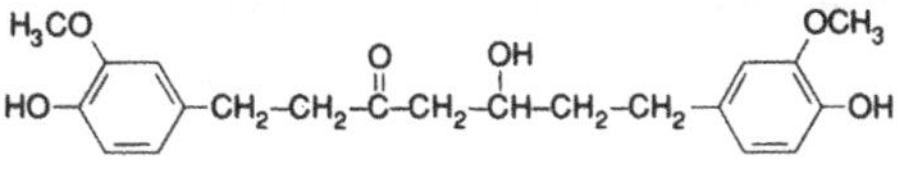

(143)

In addition to the studies noted above which used radioactive curcumin, several other reports are available which deal with the fate of curcumin and which used colorimetric or spectrophotometric methods for its determination. Wahlström and Blennow[189] gave large intragastric doses (1000 mg/kg) to rats and found that an average of 75% was excreted in the feces within 72 h whereas negligible amounts were present in the urine. They also noted the pronounced excretion of curcumin in the bile. Even larger oral doses (~2300 mg/kg) were given to rats by Ravindranath and Chandrasekhara.[190] They did not detect unchanged compound in the urine, however they noted a considerable increase in the excretion of urinary glucuronides and sulfates. Colorimetric determination accounted for ~40% of the dose as unchanged curcumin in the feces.

QUINONES

The information available on the metabolism of plant quinones is meager. No data appear to

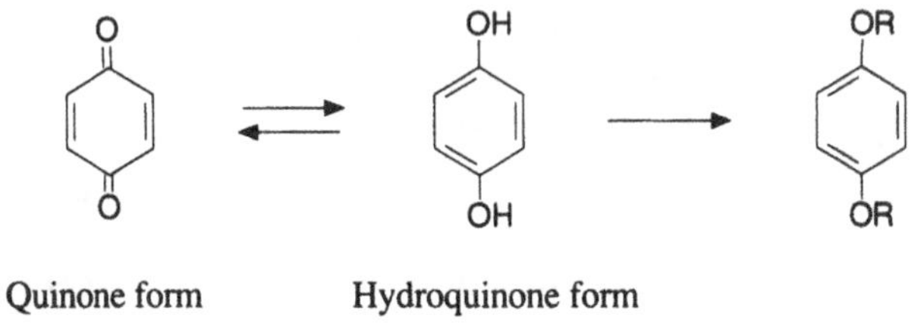

Quinone form   Hydroquinone form

FIGURE 12.   General pathway of quinone metabolism. (R) Conjugating group.

be available on the fate of the benzoquinone derivatives, although the expected general route of metabolism is a reversible reduction to the corresponding dihydric phenol followed by conjugation and excretion (Figure 12), as shown by Glock et al.[191] and Bray and Garrett[192] with a few synthetic derivatives of *p*-benzoquinone.

NAPHTHOQUINONES

The most well-known naphthoquinone derivatives are the phylloquinones (vitamins K1) and menaquinones (vitamins K2), however the metabolism of these physiologically important compounds will not be summarized here.

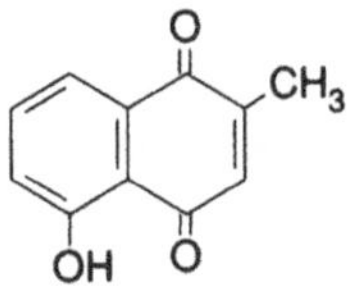

(144)

Plumbagin

The metabolism of **plumbagin** (144) in rats was studied by Chandrasekaran and Nagarajan.[193] After an i.m. dose (16 mg/kg), the blood and urine were examined for metabolites. The colorimetric analysis used for determining plumbagin did not detect it in any of the blood samples taken during the 24-h period following dosing. About 40% of the dose was found in the urine samples collected from 2 to 48 h after dosing and the excreted material was reported to be unchanged compound. It was also found that plumbagin was present in the small intestine after 6 h and that considerable excretion in the feces occurred.

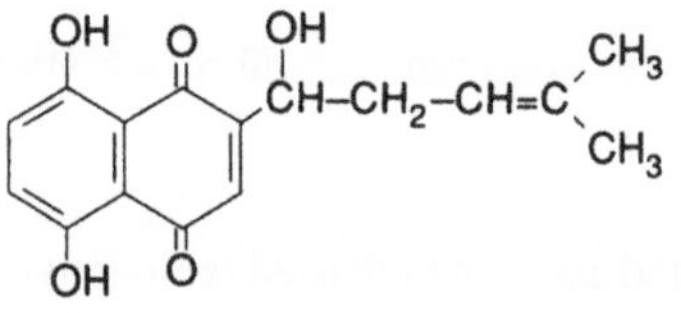

(145)

Shikonin

The metabolism of **shikonin** (145) in mice was studied using ³H-labeled material.[194] High concentrations of radioactivity were detected in the liver and bile following i.v. administration and excretion of metabolites in the urine and feces was equal (each ~40% of the dose). These results suggest that appreciable biliary excretion of shikonin occurs in mice. Unchanged compound comprised only a small part of the urinary and fecal radioactivity, however the identities of the metabolites were not ascertained.

ANTHRAQUINONES

This group furnishes the largest number of quinones, many of which have been historically important because of their use as mordant dyes (alizarin is the principle coloring matter of madder) and as cathartics (cascara, rhubarb, senna). The parent compound, anthraquinone (146), has been detected in plant material although it is most likely an artifact. The naturally occurring anthraquinones are, with few exceptions, hydroxylated derivatives which may be present as methyl ethers or, especially, as glycosides. In addition, C-glycosides are found, e.g., barbaloin from aloe. Anthraquinones undergo reduction to anthrones (147), the latter existing also in the tautomeric anthranol forms (148).

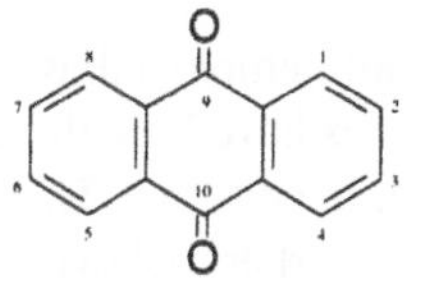

(146)
Anthraquinone

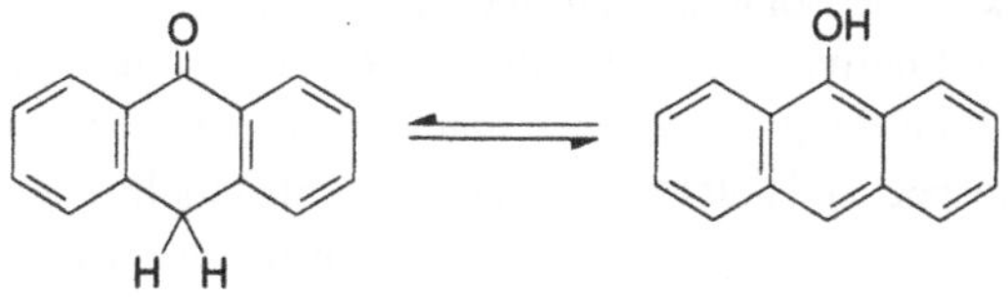

(147)

(148)

The metabolism of **1-hydroxyanthraquinone** and **2-hydroxyanthraquinone** given orally to rats (~330 mg/kg) was studied by Fujita et al.[195] With the former compound, ~5% of the dose was found in the 48-h urine as unchanged compound and its conjugates. A further 2.5% of the dose was identified as alizarin (149) which was also excreted partly free and partly as conjugates. Fecal excretion of 1-hydroxyanthraquinone (13%) and alizarin (0.8%) was recorded during this period. Similar experiments with the 2-hydroxy isomer showed a urinary excretion of the unchanged compound and its conjugates of ~6% and a corresponding value of 0.7% for alizarin. Fecal excretion was 44 and 0.2%, respectively. These results indicate that hydroxylation is more extensive with 1-hydroxyanthraquinone than with the 2-hydroxy isomer. Only about a fifth of the dose of the former compound was accounted for whereas half of the latter was determined. However, it is not known if the low recoveries were due to metabolite excretion after 48 h or to formation of unidentified metabolites.

In experiments similar to those described above with the two monohydroxy anthraquinones, Fujita et al.[196] showed that the corresponding methyl ethers (~165 mg/kg) were demethylated. Thus the urinary metabolites of the methoxy compounds were qualitatively similar to those detected when the hydroxyanthraquinones were given. In both cases the urinary excretion of the mono- and dihydroxy metabolites accounted for ~10% of the dose, however the proportion of alizarin (149) formed was higher in these experiments than in those with the hydroxyanthraquinones. The unchanged compounds were not detected in the urine but were present in the feces to the extent of ~13% (1-methoxyanthraquinone) and 27% (2-methoxyanthraquinone) of the dose. In the latter case, 2-hydroxyanthraquinone (6%) was also detected. Total metabolite recoveries were ~24% with 1-methoxyanthraquinone and 44% with 2-methoxyanthraquinone. The results indicated that demethylation occurred more readily with the latter compound.

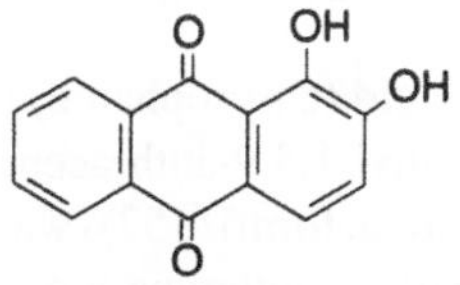

(149)
Alizarin

Mähner and Dulce[197] studied the metabolism of **alizarin** (149) in rats (100 mg/kg, p.o.). They found alizarin monoglucuronide in the urine, the excretion of this metabolite being maximal 6 to 8 h after administration of the quinone. However, the fate of most of the compound is not known as only ~10% of the dose was recovered as the monoglucuronide in the 24-h urines. Similar experiments by Fujita et al.[198] with alizarin (~165 mg/kg) indicated that unchanged compound was the major urinary metabolite although unknown metabolites were detected by paper chromatography. **Alizarin-1-methyl ether** was shown in this study to be demethylated to alizarin which was then excreted in the urine. No data are available on the metabolism of the other monomethyl ether of alizarin, however Fujita et al.[198] found that the synthetic dimethyl ether (~165 mg/kg) was demethylated and excreted in the urine as alizarin-1-methyl ether (1.5%), alizarin (trace), and unchanged compound (trace). No 2-methyl ether was detected, however, and this preference for demethylation at the 2-position is similar to that described above with the isomeric methoxyanthraquinones.

Several derivatives of 1,8-dihydroxyanthraquinone are commonly employed as cathartic agents. It has long been known that the naturally occurring glycosides have little direct effect in this respect and that they must undergo metabolic change in order to be active. The chain of events in this process was a subject of considerable discussion, some workers believing that the glycosides were absorbed from the small intestine and re-excreted in the bowel where they were then subject to metabolism to the active entities. Contrariwise, others held that the glycosides were very poorly absorbed in the upper intestine and that metabolic activation occurred as a result of bacterial metabolism of the unabsorbed glycosides in the lower intestine. The reviews by Fairbairn[199] and by Breimer and Baars[200] should be consulted for an assessment of the earlier work in this field together with the relevant literature references. It is now well-recognized that the second explanation, involving passage of the unabsorbed glycosides along the intestine, is correct. This conclusion was also reached by Dobbs et al.[201] who showed that senna glucosides were virtually nonabsorbed in the stomach and small intestine of pigs. The necessity of bacterial metabolism of the glycosides to liberate active compounds was also shown by Hardcastle and Wilkins[202] who found no increase in colon motility produced *in situ* by senna glycosides. However increased peristalsis occurred when the senna preparation was previously incubated with feces or *E. coli*. This effect was similar to that produced by the rhein anthrone formed by chemical hydrolysis and reduction of senna. Tamura et al.[203] reported that a cell-free extract of human feces, which showed a variety of glycosidase activities, hydrolyzed **frangulin A** (emodin-L-rhamnoside) to the aglycon. Metabolism of the anthraquinone glycosides by intestinal bacteria has been most thoroughly investigated with the senna glycosides sennoside A and B (see below).

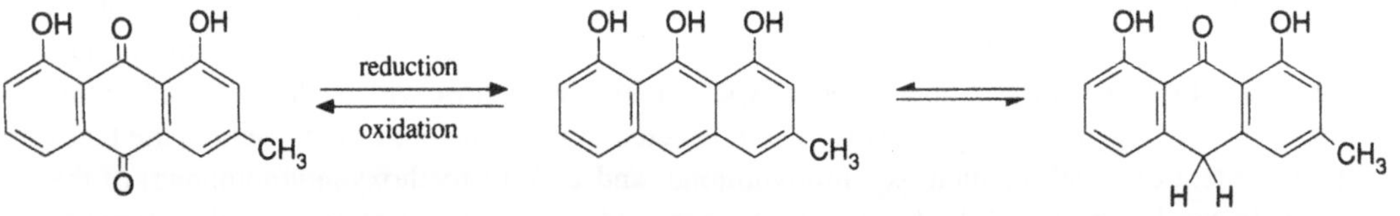

(150)	(151)	(152)
Chrysophanic acid	Chrysarobin	Chrysarobin
	(anthranol form)	(anthrone form)

The metabolism of **chrysophanic acid** (chrysophanol) (1,8-dihydroxy-3-methylanthraquinone) (150) and **chrysarobin** (3-methyl-1,8,9-anthracenetriol (anthranol form) (151); 1,8-dihydroxy-3-methyl-9-anthrone (anthrone form) (152)) was investigated by Ippen[204] in a larger study concerning the interconversion of the anthraquinone and anthranol/anthrone forms.[204-206] This interconversion was shown to occur and it was found that the quinone is the form which predominates in the body and which is excreted in the urine. Reduction of the quinone took place in the large intestine so that both forms were excreted in the feces. Hsu et al.[207] gave chrysophanic acid orally to mice and found that only 6% of the dose was excreted in the 24-h urine as total anthraquinone derivatives. Vyth and Kamp[208] administered various anthraquinone-containing laxatives to humans and obtained evidence which suggested that chrysophanic acid was oxidized to give the 3-carboxy derivative rhein (153). Similar results were described by

Fairbairn[209] who reported that mice given oral doses of chrysophanic acid excreted the 3-carboxy derivative in the urine. These experiments also indicated low values for the excretion of anthraquinone material in the urine (2 to 3% in 24 h) and feces (34 to 45% in 24 h).

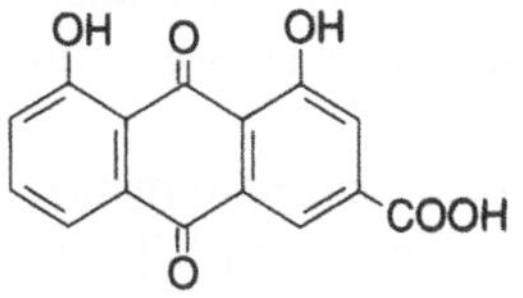

(153)

Rhein

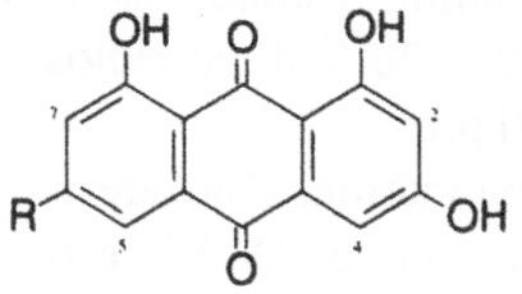

(154)

(a) Emodin, R = CH₃

(b) Emodic acid, R = COOH

(c) ω-Hydroxyemodin, R = CH₂OH

(d) R = CHO

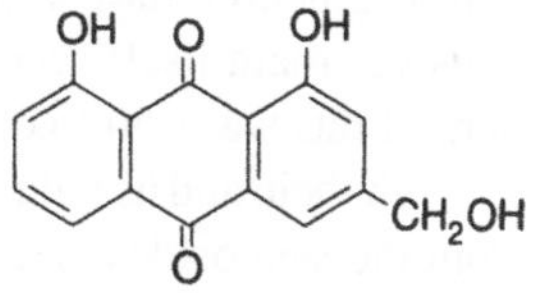

(155)

Aloe-emodin

Chen et al.[210] studied the absorption, distribution, and excretion of anthraquinone derivatives of Chinese rhubarb, including **rhein** (1,8-dihydroxyanthraquinone-3-carboxylic acid) (153) and **emodin** (1,3,8-trihydroxy-6-methylanthraquinone) (154a). The compounds were given to animals and humans by oral, i.m., or i.v. administration. The anthraquinone derivatives were readily absorbed (rhein more easily than emodin) and excreted, excretion occurring in the bile, urine, and feces. Excretion was complete within 2 d, however only slightly less than half of the administered dose was accounted for, this being present nearly equally in the urine and feces. Even lower excretion values were reported by Hsu et al.[207] who administered rhein and **aloe-emodin** (1,8-dihydroxy-3-(hydroxymethyl)anthraquinone) (155) orally to mice. They found that 27 and 14%, respectively, were excreted in the 24-h urine as total anthraquinone derivatives. Fairbairn[209] reported similar findings when these two anthraquinones were given orally to mice. The recovery of anthraquinone material in the urine 24 h after giving rhein was 25 to 33% with an additional 4 to 10% present in the feces. For aloe-emodin, the corresponding values were 3 to 9% and 16 to 31%, respectively, for an 8-h period. The investigation of Vyth and Kamp[208] noted above with chrysophanic acid likewise indicated that aloe-emodin was oxidized to rhein, which was then found as a urinary metabolite. Lemmens and Dreessen[211] recovered only 17 to 20% of the dose as anthracene derivatives in the urine and feces of rats given rhein orally. Urinary excretion of metabolites was larger than fecal excretion by a factor of ten and unchanged rhein, rhein monoglucuronide, and rhein monosulfate were isolated from the urine. These three urinary metabolites were also reported by Lemli and Lemmons[212] in similar experiments, however the total recoveries were ~10% of the dose. They did not detect reduced metabolites of rhein in the urine. Sund and Elvegård[213] studied the metabolism and transport of rhein using rat jejunum and colon preparations. They found that both tissues formed glucuronide (mainly) and sulfate conjugates of rhein.

The common problem of low recoveries of urinary and fecal metabolites of anthraquinone derivatives was dealt with by De Witte and Lemli[214,215] who employed [¹⁴C]rhein. This material (25 mg/kg) was injected intracecally in rats and 90% of the radioactivity was recovered in the urine (37%) and feces (53%) in 5 d. The main urinary metabolite was rhein monoglucuronide. About half as much was found in the form of the corresponding monosulfate and small amounts of free rhein and another compound which liberated rhein upon treatment with glucuronidase were excreted. The results indicated that rhein was metabolized by the cecal microflora. Some of these bacterial metabolites were reduction products (anthrone forms) and others no longer acted as 1,8-dihydroxyanthranoids. Rhein labeled with ¹⁴C was also employed by Lang[216] in an investigation of its pharmacokinetics in rats. Relatively small doses were used (~0.1 to 0.8 mg/

kg). About 20% of the radioactivity was excreted in the bile in 8 h after oral or i.v. dosage or administration in the colon. The biliary material contained a small amount of free rhein but consisted mainly of two metabolites, both of which could be hydrolyzed to rhein by treatment with β-glucuronidase + sulfatase. A similar picture was found with the urinary metabolites, however rhein itself accounted for 10 to 20% of the radioactivity. The urinary recoveries in normal rats was ~50 (p.o.) and 70% (i.p.).

Both rhein and its reduced form rhein anthrone (160) are of interest because of their formation from the sennosides (see below). The study by De Witte and Lemli[214,215] included [14C]rhein anthrone and showed that more than 97% of the radioactivity was recovered in the urine and feces within 5 d. However, a mere 3% of the dose (20 mg/kg) was found in the urine, presumably due to the pronounced laxative effect of this compound which promotes rapid passage through the intestine.

The metabolism of emodin (154a) has been the subject of several recent investigations. Furthermore, the use of [14C]-labeled emodin[217] has made possible a much better assessment of the rates and routes of metabolite excretion. The latter study used generally labeled emodin which was given orally (~50 mg/kg) to rats. Radioactivity in the blood increased rapidly and reached a maximum after about 2 h. The level decreased to ~30% of maximum after 24 h, however low levels of radioactivity were still detected 5 d after dosing. Interestingly, the levels in mesentery and fat showed sharp increases after 72 h and ~5% of the radioactivity remained in the tissues 5 d after dosing. No loss of radioactivity occurred via expired CO_2. The major excretory route was found to be the feces and 68% of the radioactivity was detected there in 5 d. The corresponding value for urine was only 22%, of which about four-fifths was excreted during the first 24 h. The metabolites in the urine were mainly free anthraquinones, with conjugates (glucuronides and/or sulfates) making a very minor contribution to this material. Free anthraquinones were also a major component (27% of the dose) in the feces, however an additional 40% of the dose was found to be nonanthraquinone degradation products or nonextractable material. This finding may explain the low recovery values (10 to 50%) generally found in earlier studies of anthraquinone metabolism. A notable finding by Bachmann and Schlatter[217] was the excretion of nearly 50% of the administered radioactivity in the bile in 30 h. Furthermore, emodin was subjected to enterohepatic circulation in the rat. Only emodin and emodic acid (154b) were identified as metabolites in this study. Both were found in the urine, bile, and feces, however emodin was always the major component.

The metabolism of emodin was studied by Sun et al.[218] in mice given an oral dose (91 mg/kg). Most of the material was excreted within 24 h, both in the urine (30% of the dose) and feces (21%). About 9% of the dose was found in the urine as emodin, mainly in the free form, however a further 12 and 2% were excreted as ω-hydroxyemodin (154c) and emodic acid (154b), respectively. They also reported the presence of small amounts (1% or less) of chrysophanic acid (150) and physcion (3-*O*-methylemodin) (156). Subsequently, Sun and Chen[219] identified several additional metabolites in similar experiments. These were the 3-formyl analogue (154d) of emodin and the 3-hydroxymethyl and 3-carboxy analogues of chrysophanic acid. Several of these metabolites require dehydroxylation or *O*-methylation for their formation. If their possible presence as impurities in the administered material can be ruled out, then the detection of the chrysophanic acid derivatives suggests that dehydroxylation of emodin by the intestinal flora occurred. The presence of physcion is more difficult to explain as *O*-methylation is generally limited to catecholic phenols.

The *in vitro* metabolism of emodin by hepatic microsomes was studied by Masuda et al.[220] and Murakami et al.[221] In the former investigation rat microsomes converted emodin to at least ten anthraquinoid metabolites, among which the 2-, 4-, 5-, and 7-hydroxy derivatives were identified. The latter study employed hepatic microsomes from six mammalian species and found that ω-hydroxyemodin (154c) was formed in all cases. Microsomes from guinea pigs and rats were the most active, followed by those from mice and rabbits, and with those from cattle and pigs having the lowest activities. They confirmed that the microsomes formed ring-hydroxylated derivatives of emodin, however direct oxidation of the ring was less prominent in nearly all cases than was ω-hydroxylation. 2-Hydroxy-, 4-hydroxy-, and 7-hydroxyemodin were identified in this study; however, none of the latter type of phase I products has yet been found as a metabolite in animals given emodin. Sun and Chen[219] reported that the 9000-g supernatant fraction from mouse or rat liver oxidized emodin to compounds (154b), (154c), and (154d).

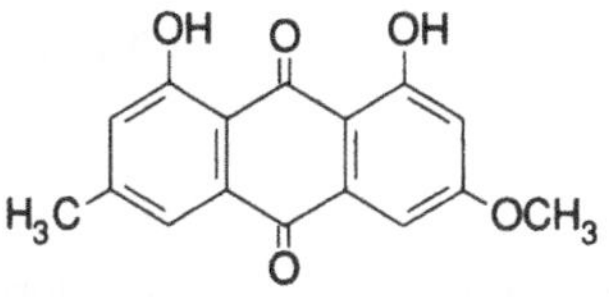

(156)

Physcion

Sun and Chen[222] reported on the metabolism of **physcion** (156) in mice and rats and in liver homogenates. Following oral dosage, six metabolites were identified in the urine. These included the O-demethylated product emodin and three derivatives of emodin which resulted from ω-oxidation. These were ω-hydroxyemodin (154c), emodic acid (154b), and the corresponding intermediate compound containing a 6-formyl moiety (154d). This oxidative pathway also occurred with physcion itself as 3-methoxy-6-carboxy-1,8-dihydroxyanthraquinone was identified. The final metabolite was chrysophanic acid (150). The comments made above on the detection of compound (150) following emodin dosage may also apply in this case. All of the metabolites except 3-methoxy-6-carboxy-1,8-dihydroxyanthraquinone were detected when physcion was incubated with liver homogenates. In view of the well-known lack of dehydroxylating ability of tissue enzymes, the detection of the dehydroxylated metabolite (150) in the latter experiments lends further support to the belief that its presence may be due to an impurity in the starting material.

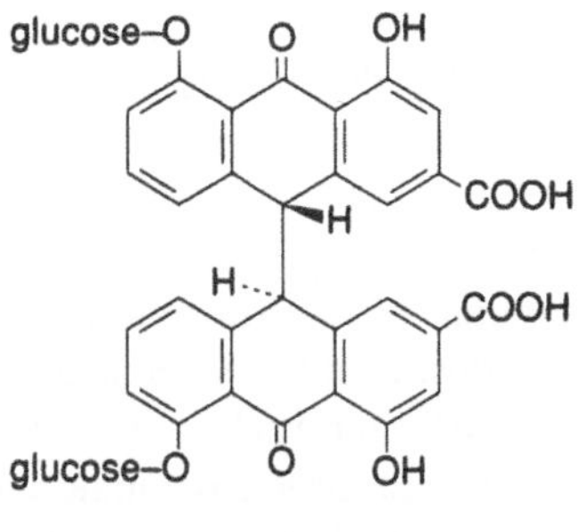

(157)	(158)
Sennoside A	Sennoside B

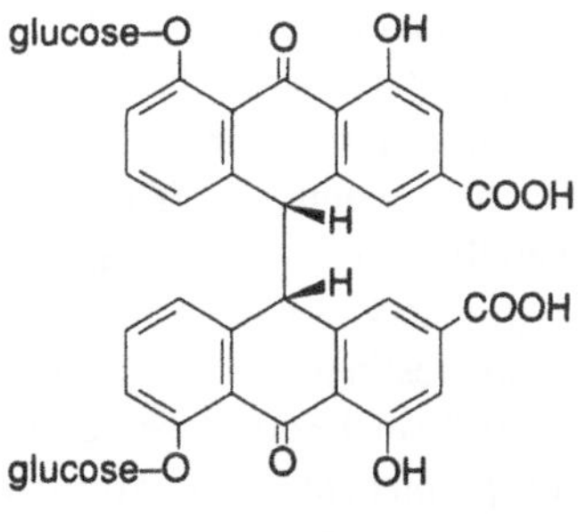

Sennoside A (157) and **sennoside B** (158) are dimeric anthraquinone glycosides which are present in several widely used cathartic drugs. They are torsional isomers which, due to hindrance in rotation around a single bond, differ only in their stereochemistry at C-10–C-10′. Sennoside A has a *trans* configuration and sennoside B a *meso* configuration. The sennosides have been more thoroughly studied than any other group in the field of anthraquinone metabolism and special emphasis has been given to the transformations effected by mixed populations and pure strains of intestinal bacteria.

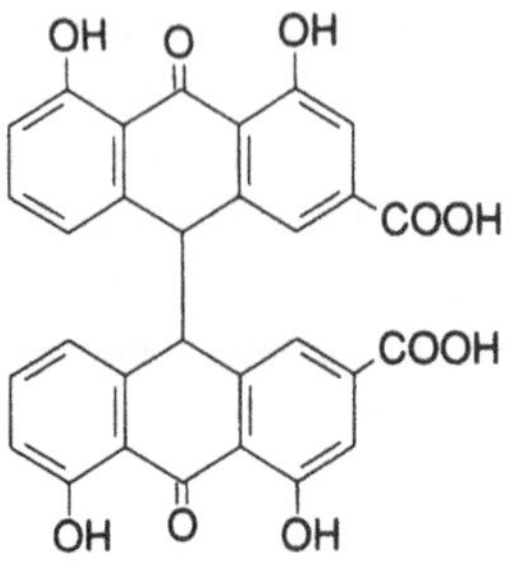

(159)

Sennidin

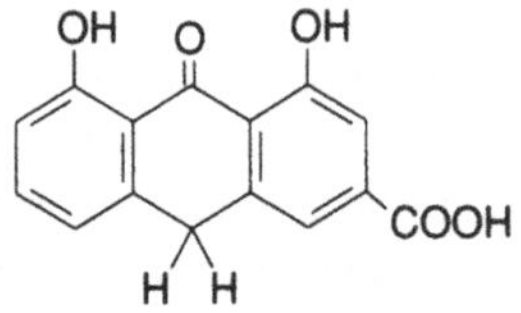

(160)

Rhein-9-anthrone

The report by Lemmens[223] clearly outlined the major features in the disposition of sennoside A and sennoside B. No metabolism took place in the rat stomach or small intestine; however, in the cecum the glycosides were hydrolyzed by the bacterial flora to sennidin (159) and further metabolized to rhein-9-anthrone (160) and rhein (153). All three metabolites as well as the glycoside were found in the feces of rats dosed with sennoside A. These experiments also illustrated the low recovery of anthracene derivatives in the urine and feces of rats given sennoside A. Using an oral dose of 25 mg/kg only ~5% was found in the excreta (24 h). These values increased to ~12%, nearly equally divided between urine and feces, when the dose was increased to 85 mg/kg. The metabolites found in the urine following administration of sennoside A were sennidin, rhein, and rhein monoglucuronide. Similar experiments by Dreessen et al.[224] using mice and rats (conventional and germ-free) confirmed and expanded these findings. Not unexpectedly, administration of sennoside A or B to germ-free animals failed to produce a laxative effect. Also, their incubation with cecal contents from germ-free rats did not give rise to metabolites. When cecal microorganisms from normal rats were used, the expected hydrolysis to sennidins and further reduction to rhein-9-anthrone were seen. Interestingly, both sennidin A and sennidin B were formed from each of the sennosides. This isomerism reaction was explained by the formation of a free radical (161). This mechanism was first proposed by Lemli and Lemmens[212] and involves the formation of the anthrone free radical from the dianthrone aglycons (sennidins). This reaction, because it is reversible, then results in the formation of both the A and B isomers of sennidin. Additionally, hydrogenation of the free radical by the microflora forms rhein-9-anthrone. The study by Dreessen et al.[224] also indicated that rhein, which was generally detected in small amounts, was not primarily formed by bacterial metabolism. It is probably formed from rhein-9-anthrone by oxidation during manipulation of the test solutions. Interestingly, while nearly all of the dose could be accounted for in the *in vitro* incubations, only ~10% of the oral dose was recovered in the form of known metabolites. Hietala et al.[225] studied the problem of low recoveries and compared chromatographic and colorimetric detection methods. The latter, which was less specific, gave higher recoveries of material present in the gastrointestinal contents and feces when sennoside B was administered to rats. Uchino et al.[226] recorded a value of ~6% for the excretion of rhein (free and conjugated) in the urine of humans 24 h after the oral administration of a product containing sennoside A and B. A detailed study by Sasaki et al.[227] of the laxative effect of sennoside A in mice led to the conclusion that rhein-9-anthrone is the metabolite responsible for activity.

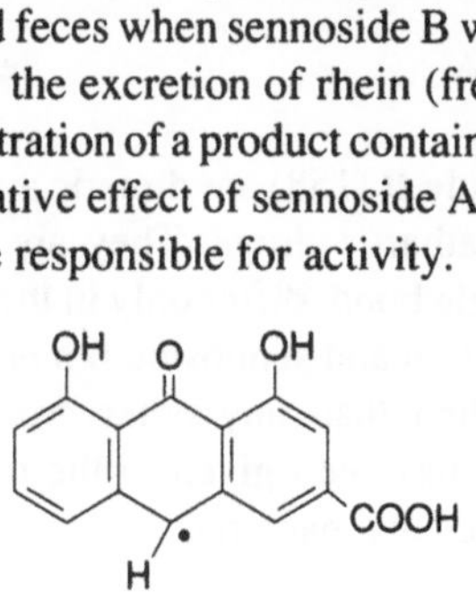

(161)

A study of the disposition of sennoside A and B, their aglycons (sennidin A and B) and rhein in rats and their metabolism by the intestinal microflora of mice was carried out by Moreau et al.[228] who employed high performance liquid chromatography for the determination of the anthraquinone derivatives. Their results largely confirmed the earlier findings summarized above, however they noted that the sennosides, and especially the sennidins, lacked stability upon incubation. They also found that rats excreted greater amounts of sennidins in the bile than in the urine following the oral administration of sennoside A or B (~15 mg/kg).

In addition to the above investigations which have clarified many aspects of the disposition of the sennosides in animals and the role of the intestinal microflora in their metabolism, other studies have expanded on the latter subject using isolated strains of intestinal bacteria. The results obtained have both given a clearer picture of the actual microorganisms involved in these reactions and a more detailed understanding of the metabolic steps which take place. Kobashi et al.[229] studied the ability of 20 species of anaerobic bacteria isolated from human feces to metabolize sennoside A. Several of these species showed activity and it was proposed that the initial metabolic step was reduction of the sennoside at the C-10–C-10'-bond to give the 8-glucoside of rhein-9-anthrone. The latter compound was subsequently hydrolyzed by a β-glucosidase and the resulting rhein-9-anthrone then oxidized to sennidin. Dreessen and Lemli[230] carried out similar experiments with sennoside A and B using various strains of nine species of human fecal bacteria. In the latter study *Bacteroides fragilis*, *Streptococcus faecium*, and *Streptococcus faecalis* showed activity in converting the glycosides to sennidins and rhein-9-anthrone. The results of Dreessen and Lemli indicated that the metabolic sequence was more likely to be along the conventional pathway from sennoside to sennidin which, via anthrone free radicals, was then reduced to rhein-9-anthrone. However, they allowed for the possibility of a facultative pathway involving anthrone glucoside free radicals to give rhein-9-anthrone-8-glucoside. This subject was re-investigated by Hattori et al.,[231] mainly using mixed cultures from rat or human feces. Their findings supported the sennoside → sennidin → rhein-9-anthrone pathway and they were also able to detect the intermediate sennidin monoglucosides which are the initial products in this sequence. Sennidin A-8-monoglucoside was formed from sennoside A and both the 8- and 8'-monoglucosides of sennidin B were identified from incubates of sennoside B. This investigation also studied the interconversion of sennidin A and B by rat feces preparations as well as the metabolism of sennidin A to rhein-9-anthrone by individual species of human intestinal bacteria.

Several more recent investigations studied the mechanisms by which sennosides or sennidins are reduced to anthrone derivatives. Using *Peptostreptococcus intermedius*, a common bacterium found in the human intestine, Akao et al.[232] showed that the reduction rates were stimulated by the addition of NADH, NAD, glucose, and removal of O_2 from the environment. The results indicated that reduction occurred nonenzymatically by the reduced form of the flavin cofactor. This nonenzymatic reduction was studied in more detail by Akao et al.[233] and it was shown that the sennidins were reduced much more effectively than were the sennosides. Hattori et al.[234] reviewed these latest findings and concluded that the pathway from sennoside → sennidin → rhein-9-anthrone may be predominant in the intestine, whereas the alternative pathway which initially forms sennidin-8-monoglucoside is of lesser importance.

Compernolle et al.[235] identified two anthraquinone 2,2'-dimers in human plasma following ingestion of a preparation containing cascara. These compounds were shown to contain either two aloe-emodin (155) moieties or one each of aloe-emodin and chrysophanic acid (150). These dimers were not themselves present in the preparation ingested, however they could be formed upon oxidative hydrolysis. It was therefore concluded that the precursors of the metabolites identified in plasma were 2,2'-bianthraquinone glycosides.

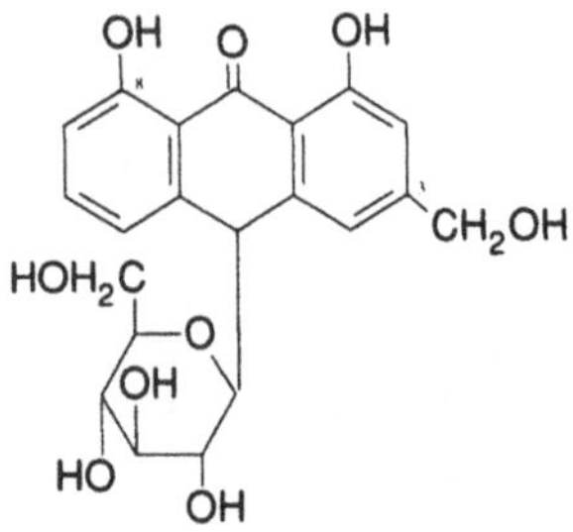

(162)
Barbaloin

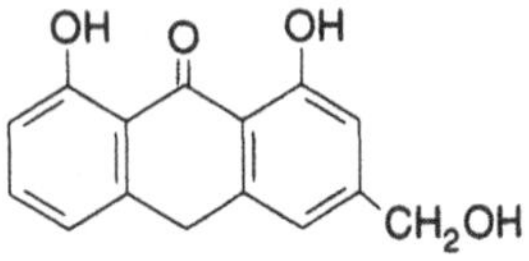

(163)

Most of the anthraquinone glycosides are *O*-glycosides, however some *C*-glycosides are also known. Very little metabolic data are available on these derivatives which include **barbaloin** (aloin) (162) as a well-known example. Fairbairn[209] reported on experiments in which mice received oral doses of barbaloin or **cascaroside A** (barbaloin 8-*O*-glucoside). In the latter case barbaloin was detected in the feces, however barbaloin itself appeared unchanged. This indicates that, in mice at least, the intestinal flora showed little or no activity for splitting *C*-glycosides. Contrariwise, newer studies employing rat cecal microorganisms[236] or human intestinal bacteria[237] showed that cleavage of the *C*-glucosyl bond occurred. Aloe-emodin anthrone (163) was formed when barbaloin was incubated anaerobically with human fecal microorganisms. However, similar experiments using the fecal microflora from mice and rats showed reduced (rats) or no (mice) ability to form metabolite (163).[237] The study using rat cecal microorganisms showed that cascaroside A (or its stereoisomer cascaroside B) was first hydrolyzed to barbaloin and then reduced to aloe-emodin anthrone (163). These reactions did not take place when cecal incubates from germ-free rats were used. Only the initial hydrolysis step was observed when cascarosides A and B were incubated with cecal contents from guinea pigs or with *Streptococcus faecium* or *S. faecalis* from human feces. Similar results were obtained in experiments with cascarosides C and D (identical to cascarosides A and B except for a methyl group at C-3).

REFERENCES

1. **McMahon, R. E.,** Alcohols, aldehydes, and ketones, in *Metabolic Basis of Detoxication*, Jakoby, W. B., Bend, J. R., and Caldwell, J., Eds., Academic Press, New York, 1982, 91.
2. **Bosron, W. F. and Li, T.-K.,** Alcohol dehydrogenase, in *Enzymatic Basis of Detoxication*, Vol. 1, Jakoby, W. B., Ed., Academic Press, New York, 1980, 231.
3. **von Wartburg, J.-P. and Wermuth, B.,** Aldehyde reductase, in *Enzymatic Basis of Detoxication*, Vol. 1, Jakoby, W. B., Ed., Academic Press, New York, 1980, 249.
4. **Weiner, H.,** Aldehyde oxidizing enzymes, in *Enzymatic Basis of Detoxication*, Vol. 1, Jakoby, W. B., Ed., Academic Press, New York, 1980, 261.
5. **Jörnvall, H., Höög, J.-O., von Bahr-Lindström, H., Johansson, J., Kaiser, R., and Persson, B.,** Alcohol dehydrogenases and aldehyde dehydrogenases, *Biochem. Soc. Trans.*, 16, 223, 1988.
6. **Rajagopalan, K. V.,** Xanthine oxidase and aldehyde oxidase, in *Enzymatic Basis of Detoxication*, Vol. 1, Jakoby, W. B., Ed., Academic Press, New York, 1980, 295.
7. **Bosron, W. F. and Li, T.-K.,** Genetic polymorphism of human liver alcohol and aldehyde dehydrogenases, and their relationship to alcohol metabolism and alcoholism, *Hepatology*, 6, 502, 1986.
8. **Hathway, D. E.,** Biotransformations Part I: Drugs, food additives and contaminants, carcinogens, and toxins, in *Foreign Compound Metabolism in Mammals*, Vol. 3, Hathway, D. E., Ed., The Chemical Society, London, 1975, 201.
9. **Williams, R. T.,** *Detoxication Mechanisms*, Chapman and Hall, London, 1959, 89.

131

10. **Williams, R. T.,** *Detoxication Mechanisms,* Chapman and Hall, London, 1959, 92.

11. **Opdyke, D. L. J.,** Monographs on fragrance raw materials, *Food Cosmet. Toxicol.,* 11, 95, 1973.

12. **Opdyke, D. L. J.,** Monographs on fragrance raw materials, *Food Cosmet. Toxicol.,* 11, 477, 1973.

13. **Boyland, E.,** Experiments on the chemotherapy of cancer. IV. Further experiments with aldehydes and their derivatives, *Biochem. J.,* 34, 1196, 1940.

14. **Neubauer, O.,** Ueber Glykuronsäurepaarung bei Stoffen der Fettreihe, *Arch. Exp. Pathol. Pharmak.,* 46, 133, 1901.

15. **Hinson, J. A. and Neal, R. A.,** An examination of the oxidation of aldehydes by horse liver alcohol dehydrogenase, *J. Biol. Chem.,* 247, 7106, 1972.

16. **Hinson, J. A. and Neal, R. A.,** An examination of octanol and octanal metabolism to octanoic acid by horse liver alcohol dehydrogenase, *Biochim. Biophys. Acta,* 384, 1, 1975.

17. **Kutzman, R. S., Meyer, G.-J., Wolf, A. P., and Akasaka, T.,** The biodistribution and metabolic fate of inhaled ^{11}C-octanal in the rat after acute exposure, *Drug Metab. Dispos.,* 9, 331, 1981.

18. **Boyland, E. and Chasseaud, L. F.,** The effect of some carbonyl compounds on rat liver glutathione levels, *Biochem. Pharmacol.,* 19, 1526, 1970.

19. **Boyland, E. and Chasseaud, L. F.,** Enzyme-catalysed conjugations of glutathione with unsaturated compounds, *Biochem. J.,* 104, 95, 1967.

20. **Boyland, E. and Chasseaud, L. F.,** Enzymes catalysing conjugation of glutathione with αβ-unsaturated carbonyl compounds, *Biochem. J.,* 109, 651, 1968.

21. **Hildebrandt, H.,** Ueber Synthesen im Thierkörper. II. Verbindungen der Kamphergruppe, *Arch. Exp. Pathol. Pharmak.,* 45, 110, 1901.

22. **Kuhn, R., Köhler, F., and Köhler, L.,** Über Methyl-oxydationen im Tierkörper, *Hoppe-Seyler's Z. Physiol. Chem.,* 242, 171, 1936.

23. **Ishida, T., Toyota, M., and Asakawa, Y.,** Terpenoid biotransformation in mammals V. Metabolism of (+)-citronellal, (±)-7-hydroxycitronellal, citral, (–)-perillaldehyde, (–)-myrtenal, cuminaldehyde, thujone, and (±)-carvone in rabbits, *Xenobiotica,* 19, 843, 1989.

24. **Kuhn, R. and Livada, K.,** Über den Einfluss von Seitenketten auf die Oxydationsvorgänge im Tierkörper. Modellversuche zum biologischen Abbau der Carotinfarbstoffe, *Hoppe-Seyler's Z. Physiol. Chem.,* 220, 235, 1933.

25. **Phillips, J. C., Kingsnorth, J., Gangolli, S. D., and Gaunt, I. F.,** Studies on the absorption, distribution and excretion of citral in the rat and mouse, *Food Cosmet. Toxicol.,* 14, 537, 1976.

26. **Diliberto, J. J., Usha, G., Burka, L. T., and Birnbaum, L. S.,** Biotransformation of citral in rats, *Toxicologist,* 8, 208, 1988.

27. **Diliberto, J. J., Usha, G., and Birnbaum, L. S.,** Disposition of citral in male Fischer rats, *Drug Metab. Dispos.,* 16, 721, 1988.

28. **Parke, D. V. and Rahman, H.,** The effects of some terpenoids and other dietary anutrients on hepatic drug-metabolizing enzymes, *Biochem. J.,* 113, 12, 1969.

29. **Kuhn, R. and Löw, I.,** Über die Ausscheidung von Menthoglykol-glucuronsäure durch Kaninchen nach Fütterung von Citronellal, *Hoppe-Seyler's Z. Physiol. Chem.,* 254, 139, 1938.

30. **Asano, M. and Yamakawa, T.,** The fate of branched chain fatty acids in animal body. I. A contribution to the problem of "Hildebrandt acid", *J. Biochem. (Tokyo),* 37, 321, 1950.

31. **Ishida, T., Asakawa, Y., and Takemoto, T.,** Metabolism of myrtenal, perilla aldehyde and dehydroabietic acid in rabbits, *Res. Bull. Hiroshima Inst. Technol.,* 16, 79, 1982.

32. **Friedmann, E. and Türk, W. T.,** Zur Kenntnis des Abbaues der Karbonsäuren im Tierkörper. Verhalten des Benzaldehyds im Tierkörper, *Biochem. Z.,* 55, 425, 1913.

33. **Bray, H. G., Thorpe, W. V., and White, K.,** Kinetic studies of the metabolism of foreign organic compounds. I. The formation of benzoic acid from benzamide, toluene, benzyl alcohol and benzaldehyde and its conjugation with glycine and glucuronic acid in the rabbit, *Biochem. J.,* 48, 88, 1951.

34. **Laham, S., Potvin, M., and Robinet, M.,** Metabolism of benzaldehyde in New Zealand white rabbits, *Chemosphere,* 17, 517, 1988.

35. **Teuchy, H., Quatacker, J., Wolf, G., and Van Sumere, C. F.,** Quantitative investigation of the hippuric acid formation in the rat after administration of some possible aromatic and hydroaromatic precursors, *Arch. Int. Physiol. Biochim.,* 79, 573, 1971.

36. **Kutzman, R. S., Meyer, G.-J., and Wolf, A. P.,** Biodistribution and excretion of [11C]benzaldehyde by the rat after two-minute inhalation exposures, *Xenobiotica,* 10, 281, 1980.

37. **Seutter-Berlage, F., Rietveld, E. C., Plate, R., and Klippert, P. J. M.,** Mercapturic acids as metabolites of aromatic aldehydes and alcohols, in *Biological Reactive Intermediates-II: Chemical Mechanisms and Biological Effects,* Vol. 136, Snyder, R., Parke, D. V., Kocsis, J. J., Jollow, D. J., Gibson, G. G., and Witmer, C. M., Eds., Plenum Press, New York, 1981, 359.

38. **Laham, S. and Potvin, M.,** Biological conversion of benzaldehyde to benzylmercapturic acid in the Sprague-Dawley rat, *Drug Chem. Toxicol.,* 10, 209, 1987.

39. **Scheline, R. R.,** Unpublished data, 1972.
40. **Williams, R. T.,** *Detoxication Mechanisms,* Chapman and Hall, London, 1959, 333.
41. **Williams, R. T.,** Studies in detoxication. I. The influence of (a) dose and (b) *o-, m-* and *p-*substitution on the sulphate detoxication of phenol in the rabbit, *Biochem. J.,* 32, 878, 1938.
42. **Bray, H. G., Thorpe, W. V., and White, K.,** Kinetic studies of the metabolism of foreign organic compounds. V. A mathematical model expressing the metabolic fate of phenols, benzoic acids and their precursors, *Biochem. J.,* 52, 423, 1952.
43. **Sato, T., Suzuki, T., Fukuyama, T., and Yoshikawa, H.,** Studies on conjugation of S^{35}-sulfate with phenolic compounds. IV. Metabolism of *o-*cresol, *m-*cresol, salicylaldehyde, salicylic acid, toluene, benzoic acid and related substances in rat liver, *J. Biochem. (Tokyo),* 43, 421, 1956.
44. **Scheline, R. R.,** The metabolism of some aromatic aldehydes and alcohols by the rat intestinal microflora, *Xenobiotica,* 2, 227, 1972.
45. **Hartles, R. L. and Williams, R. T.,** Studies in detoxication. XVIII. A study of the relation between conjugation and deamination of *p-*hydroxybenzylamine and related compounds in the rabbit, *Biochem. J.,* 43, 296, 1948.
46. **Quick, A. J.,** The relationship between chemical structure and physiological response. II. The conjugation of hydroxy- and methoxybenzoic acids, *J. Biol. Chem.,* 97, 403, 1932.
47. **Sato, T., Suzuki, T., Fukuyama, T., and Yoshikawa, H.,** Studies on conjugation of S^{35}-sulfate with phenolic compounds. III. Metabolism of *p-*cresol, *p-*hydroxybenzaldehyde and *p-*hydroxybenzoic acid in rat liver, *J. Biochem. (Tokyo),* 43, 413, 1956.
48. **Dodgson, K. S. and Williams, R. T.,** Studies in detoxication. XXVII. The orientation of conjugation in the metabolites of 4-chlorocatechol and 4-chlororesorcinol, with some observation on the fate of (+)-adrenaline, protocatechuic acid and protocatechuic aldehyde in the rabbit, *Biochem. J.,* 45, 381, 1949.
49. **Wong, K. P. and Sourkes, T. L.,** Metabolism of vanillin and related substances in the rat, *Can. J. Biochem.,* 44, 635, 1966.
50. **Sammons, H. G. and Williams, R. T.,** Studies in detoxication. XII. The metabolism of vanillin and vanillic acid in the rabbit. The identification of glucurovanillin and the structure of glucurovanillic acid, *Biochem. J.,* 35, 1175, 1941.
51. **Strand, L. P. and Scheline, R. R.,** The metabolism of vanillin and isovanillin in the rat, *Xenobiotica,* 5, 49, 1975.
52. **Grebennik, L. I., Gnevkovskaya, T. V., and Smirnov, G. A.,** Metabolism of vanillin ingested with phthivazide, *Vop. Med. Khim. Akad. Med. Nauk SSSR,* 9, 127, 1963 (Chem. Abstr. 59, 13238b, 1963).
53. **Perry, T. L., Hansen, S., Hestrin, M., and MacIntyre, L.,** Exogenous urinary amines of plant origin, *Clin. Chim. Acta,* 11, 24, 1965.
54. **Mamer, O. A., Montgomery, J. A., Deckelbaum, R. J., and Granot, E.,** Identification of urinary 3-ethoxy-4-hydroxybenzoic and 3-ethoxy-4-hydroxymandelic acids after dietary intake of ethyl vanillin, *Biomed. Mass Spectrom.,* 12, 163, 1985.
55. **Dirscherl, W. and Brisse, B.,** Exogene Vorstufen der Vanillinsäure: Vanillin, Protocatechusäure, Kaffeesäure und Feralusäure (Homogenatversuche mit Ratten- und Menschenleber), *Hoppe-Seyler's Z. Physiol. Chem.,* 346, 55, 1966.
56. **Friedhoff, A. J., Schweitzer, J. W., Miller, J., and van Winkle, E.,** Guaiacol-*O*-methyltransferase: a mammalian enzyme capable of forming di-*O*-methyl catecholamine derivatives, *Experientia,* 28, 517, 1972.
57. **Sammons, H. G. and Williams, R. T.,** Studies in detoxication. XIV. The metabolism of veratraldehyde and veratric acid in the rabbit, *Biochem. J.,* 40, 223, 1946.
58. **Müller-Enoch, D., Thomas, H., and Holzmann, P.,** Metabolism of 3,4-dimethoxyphenyl compounds in rat liver, *Hoppe-Seyler's Z. Physiol. Chem.,* 355, 1232, 1974.
59. **Heffter, A.,** Zur Pharmakologie der Safrolgruppe, *Arch. Exp. Pathol. Pharmak.,* 35, 342, 1895.
60. **Williams, R. T.,** *Detoxication Mechanisms,* Chapman and Hall, London, 1959, 334.
61. **Kamienski, F. X. and Casida, J. E.,** Importance of demethylenation in the metabolism *in vivo* and *in vitro* of methylenedioxyphenyl synergists and related compounds in mammals, *Biochem. Pharmacol.,* 19, 91, 1970.
62. **Klungsøyr, J. and Scheline, R. R.,** Metabolism of piperonal and piperonyl alcohol in the rat with special reference to the scission of the methylenedioxy group, *Acta Pharm. Suec.,* 21, 67, 1984.
63. **Friedmann, E. and Mai, H.,** Verhalten der Cinnamalessigsäure und des Zimtaldehyds im Tierkörper, *Biochem. Z.,* 242, 282, 1931.
64. **Delbressine, L. P. C., Klippert, P. J. M., Reuvers, J. T. A., and Seutter-Berlage, F.,** Isolation and identification of mercapturic acids of cinnamic aldehyde and cinnamyl alcohol from urine of female rats, *Arch. Toxicol.,* 49, 57, 1981.
65. **Samuelsen, O. B., Brenna, J., Solheim, E., and Scheline, R. R.,** Metabolism of the cinnamon constituent *o-*methoxycinnamaldehye in the rat, *Xenobiotica,* 16, 845, 1986.
66. **Jaffé, M. and Cohn, R.,** Ueber das Verhalten des Furfurols im thierischen Organismus. I, *Ber. Dtsch. Chem. Ges.,* 20, 2311, 1887.
67. **Paul, H. E., Austin, F. L., Paul, M. F., and Ells, V. R.,** Metabolism of nitrofurans. I. Ultraviolet absorption studies of urinary end-products after oral administration, *J. Biol. Chem.,* 180, 345, 1949.

68. **Laham, S. and Potvin, M.,** Metabolism of furfural in the Sprague-Dawley rat, *Toxicol. Environ. Chem.,* 24, 35, 1989.
69. **Flek, J. and Š edivec, V.,** The absorption, metabolism and excretion of furfural in man, *Int. Arch. Occup. Environ. Health,* 41, 159, 1978.
70. **Jodynis-Liebert, J.,** Metabolism of 5-methyl-2-furaldehyde in rat. III. Identification and determination of 5-methyl-2-furylmethylketone, *Xenobiotica,* 18, 887, 1988.
71. **Abou-Donia, M. B.,** Physiological effects and metabolism of gossypol, in *Pesticide Reviews,* Vol. 61, Günther, F. A., Ed., Springer Verlag, New York, 1976, 125.
72. **Qian, S.-Z. and Wang, Z.-G.,** Gossypol: a potential antifertility agent for males, *Annu. Rev. Pharmacol. Toxicol.,* 24, 329, 1984.
73. **Bressani, R., Elias, L. G., and Braham, J. E.,** All-vegetable protein mixtures for human feeding. XV. Studies in dogs on the absorption of gossypol from cottonseed flour-containing vegetable protein mixtures, *J. Nutr.,* 83, 209, 1964.
74. **Abou-Donia, M. B., Lyman, C. M., and Dieckert, J. W.,** Metabolic fate of gossypol: the metabolism of ^{14}C-gossypol in rats, *Lipids,* 5, 938, 1970.
75. **Skutches, C. L. and Smith, F. H.,** Metabolism of gossypol, biosynthesized from methyl-^{14}C- and carboxyl-^{14}C- labeled sodium acetate, in rat, *J. Am. Oil Chem. Soc.,* 51, 413, 1974.
76. **Tang, X.-C., Zhu, M.-K., and Shi, Q.-X.,** Comparative studies on the absorption, distribution and excretion of ^{14}C-gossypol in four species of animals, *Acta Pharm. Sinica,* 15, 212, 1980.
77. **Wang, N.-G., Li, G.-X., Chen, Q.-Q., and Lei, H.-P.,** In vivo metabolism of gossypol, *Chin. Med. J.,* 59, 596, 1979 (Chem. Abstr. 92, 174852u, 1980).
78. **Smith, F. H. and Clawson, A. J.,** Effect of diet on accumulation of gossypol in the organs of swine, *J. Nutr.,* 87, 317, 1965.
79. **Albrecht, J. E., Clawson, A. J., and Smith, F. H.,** Rate of depletion and route of elimination of intravenously injected gossypol in swine, *J. Anim. Sci.,* 35, 941, 1972.
80. **Skutches, C. L., Herman, D. L., and Smith, F. H.,** Effect of intravenous gossypol injection on iron utilization in swine, *J. Nutr.,* 103, 851, 1973.
81. **Abou-Donia, M. B. and Dieckert, J. W.,** Urinary and biliary excretion of ^{14}C-gossypol in swine, *J. Nutr.,* 104, 754, 1974.
82. **Abou-Donia, M. B. and Dieckert, J. W.,** Metabolic fate of gossypol: the metabolism of [^{14}C]gossypol in swine, *Toxicol. Appl. Pharmacol.,* 31, 32, 1975.
83. **Chen, Q.-Q., Chen, H., and Lei, H.-P.,** Comparative study on the metabolism of optical gossypol in rats, *J. Ethnopharmacol.,* 20, 31, 1987.
84. **Wu, D.-F., Yu, Y.-W., Tang, Z.-M., and Wang, M.-Z.,** Pharmacokinetics of (±)-, (+)-, and (−)-gossypol in humans and dogs, *Clin. Pharmacol. Ther.,* 39, 613, 1986.
85. **Felsted, R. L. and Bachur, N. R.,** Mammalian carbonyl reductases, *Drug Metab. Rev.,* 11, 1, 1980.
86. **Felsted, R. L. and Bachur, N. R.,** Ketone reductases, in *Enzymatic Basis of Detoxication,* Vol. 1, Jakoby, W. B., Ed., Academic Press, New York, 1980, 281.
87. **Prelusky, D. B., Coutts, R. T., and Pasutto, F. M.,** Stereospecific metabolic reduction of ketones, *J. Pharm. Sci.,* 71, 1390, 1982.
88. **Williams, R. T.,** *Detoxication Mechanisms,* Chapman and Hall, London, 1959, 95.
89. **Schwarz, L.,** Ueber die Oxydation des Acetons und homologer Ketone der Fettsäurereihe, *Arch. Exp. Pathol. Pharmak.,* 40, 168, 1898.
90. **Haggard, H. W., Miller, D. P., and Greenberg, L. A.,** The amyl alcohols and their ketones: their metabolic fates and comparative toxicities, *J. Ind. Hyg. Toxicol.,* 27, 1, 1945.
91. **Price, T. D. and Rittenberg, D.,** The metabolism of acetone. I. Gross aspects of catabolism and excretion, *J. Biol. Chem.,* 185, 449, 1950.
92. **Kamil, I. A., Smith, J. N., and Williams, R. T.,** Studies in detoxication. XXXXVI. The metabolism of aliphatic alcohols. The glucuronic acid conjugation of acyclic aliphatic alcohols, *Biochem. J.,* 53, 129, 1953.
93. **Albro, P. W., Corbett, J. T., and Schroeder, J. L.,** Metabolism of methyl *n*-amyl ketone (2-heptanone) and its binding to DNA of rat liver in vivo and in vitro, *Chem. Biol. Interact.,* 51, 295, 1984.
94. **Sakami, W. and Lafaye, J. M.,** The metabolism of acetone in the intact rat, *J. Biol. Chem.,* 193, 199, 1951.
95. **DiVincenzo, G. D., Kaplan, C. J., and Dedinas, J.,** Characterization of the metabolites of methyl *n*-butyl ketone, methyl iso-butyl ketone, and methyl ethyl ketone in guinea pig serum and their clearance, *Toxicol. Appl. Pharmacol.,* 36, 511, 1976.
96. **Dietz, F. K., Rodriguez-Giaxola, M., Traiger, G. J., Stella, V. J., and Himmelstein, K. J.,** Pharmacokinetics of 2-butanol and its metabolites, *J. Pharmacokinet. Biopharm.,* 9, 553, 1981.
97. **Kež ić , S. and Monster, A. C.,** Determination of methyl ethyl ketone and its metabolites in urine using capillary gas chromatography, *J. Chromatogr.,* 428, 275, 1988.
98. **DiVincenzo, G. D., Hamilton, M. L., Kaplan, C. J., and Dedinas, J.,** Metabolic fate and disposition of ^{14}C-labeled methyl *n*-butyl ketone in the rat, *Toxicol. Appl. Pharmacol.,* 41, 547, 1977.

99. **Abdel-Rahman, M. S., Hetland, L. B., and Couri, D.,** Toxicity and metabolism of methyl n-butyl ketone, *Am. Ind. Hyg. Assoc. J.*, 37, 95, 1976.

100. **DiVincenzo, G. D., Hamilton, M. L., Kaplan, C. J., Krasavage, W. J., and O'Donoghue, J. L.,** Studies on the respiratory uptake and excretion and the skin absorption of methyl *n*-butyl ketone in humans and dogs, *Toxicol. Appl. Pharmacol.*, 44, 593, 1978.

101. **Fedtke, N. and Bolt, H. M.,** The relevance of 4,5-dihydroxy-2-hexanone in the excretion kinetics of *n*-hexane metabolites in rat and man, *Arch. Toxicol.*, 61, 131, 1987.

102. **Fedtke, N. and Bolt, H. M.,** 4,5-Dihydroxy-2-hexanone: a new metabolite of n-hexane and of 2,5-hexanedione in rat urine, *Biomed. Environ. Mass Spectrom.*, 14, 563, 1987.

103. **DeCaprio, A. P., Olajos, E. J., and Weber, P.,** Covalent binding of a neurotoxic *n*-hexane metabolite: conversion of primary amines to substituted pyrrole adducts by 2,5-hexanedione, *Toxicol. Appl. Pharmacol.*, 65, 440, 1982.

104. **DeCaprio, A. P. and O'Neill, E. A.,** Alterations in rat axonal cytoskeletal proteins induced by *in vitro* and *in vivo* 2,5-hexanedione exposure, *Toxicol. Appl. Pharmacol.*, 78, 235, 1985.

105. **Couri, D. and Milks, M.,** Toxicity and metabolism of the neurotoxic hexacarbons *n*-hexane, 2-hexanone, and 2,5-hexanedione, *Annu. Rev. Pharmacol. Toxicol.*, 22, 145, 1982.

106. **Järnefelt, J.,** Studies on the enzymatic synthesis and breakdown of acetoin in the animal organism, *Ann. Acad. Sci. Fenn. Ser. A*, 57, 1, 1955.

107. **Williams, R. T.,** *Detoxication Mechanisms*, Chapman and Hall, London, 1959, 62.

108. **Gabriel, M. A., Ilbawi, M., and al-Khalidi, U. A. S.,** The oxidation of acetoin to CO_2 in intact animals and in liver mince preparations, *Comp. Biochem. Physiol.*, 41B, 493, 1972.

109. **Gabriel, M. A., Jabara, H., and al-Khalidi, U. A. S.,** Metabolism of acetoin in mammalian liver slices and extracts. Interconversion with butane-2,3-diol and biacetyl, *Biochem. J.*, 124, 793, 1971.

110. **James, S. P. and Waring, R. H.,** The metabolism of alicyclic ketones in the rabbit and rat, *Xenobiotica*, 1, 573, 1971.

111. **Elliott, T. H., Parke, D. V., and Williams, R. T.,** Studies in detoxication. LXXIX. The metabolism of *cyclo*[¹⁴C]hexane and its derivatives, *Biochem. J.*, 72, 193, 1959.

112. **Williams, R. T.,** *Detoxication Mechanisms*, Chapman and Hall, London, 1959, 524.

113. **Meyer, T. and Scheline, R. R.,** Unpublished data, 1973.

114. **Hildebrandt, H.,** Ueber das Schicksal einiger cyklischer Terpene und Kampfer im Thierkörper, *Hoppe-Seyler's Z. Physiol. Chem.*, 36, 452, 1902.

115. **Hämäläinen, J.,** Über das Verhalten der alicyklischen Verbindungen bei der Glykuronsäurepaarung im Organismus, *Skand. Arch. Physiol.*, 27, 141, 1912.

116. **Williams, R. T.,** Studies in detoxication. VII. The biological reduction of *l*-menthone to *d*-neomenthol and of *d*-isomenthone to *d*-isomenthol in the rabbit. The conjugation of *d*-neomenthol with glucuronic acid, *Biochem. J.*, 34, 690, 1940.

117. **Harvey, J. M.,** The detoxication of terpenes by sheep, *Pap. Dep. Chem. Univ. Qd*, 1, 1942.

118. **Teppati, R.,** Sulla trasformazione del pulegone nell'organismo, *Arch. Int. Pharmacodyn. Ther.*, 57, 440, 1937.

119. **Nelson, S. D. and Gordon, W. P.,** Mammalian drug metabolism, *J. Nat. Prod.*, 46, 71, 1983.

120. **Gordon, W. P., Huitric, A. C., Seth, C. L., McClanahan, R. H., and Nelson, S. D.,** The metabolism of the abortifacient terpene, (*R*)-(+)-pulegone, to a proximate toxin, menthofuran, *Drug Metab. Dispos.*, 15, 589, 1987.

121. **Moorthy, B., Madyastha, P., and Madyastha, K. M.,** Metabolism of a monoterpene ketone, R-(+)-pulegone—a hepatotoxin in rat, *Xenobiotica*, 19, 217, 1989.

122. **McClanahan, R. H., Thomassen, D., Slattery, J. T., and Nelson, S. D.,** Metabolic activation of (*R*)-(+)-pulegone to a reactive enonal that covalently binds to mouse liver proteines, *Chem. Res. Toxicol.*, 2, 349, 1989.

123. **Madyastha, K. M. and Moorthy, B.,** Pulegone mediated hepatoxicity: evidence for covalent binding of *R*(+)-[¹⁴C)pulegone to microsomal proteins in vitro, *Chem. Biol. Interact.*, 72, 325, 1989.

124. **Hildebrandt, H.,** Ueber das Verhalten von Carvon und Santalol im Thierkörper, *Hoppe-Seyler's Z. Physiol. Chem.*, 36, 441, 1902.

125. **Fischer, F. G. and Bielig, H.-J.,** Über die Hydrierung ungesättigter Stoffe im Tierkörper. Biochemische Hydrierungen. VII, *Hoppe-Seyler's Z. Physiol. Chem.*, 266, 73, 1940.

126. **Bielig, H.-J. and Hayashida, A.,** Über das Verhalten des β-Jonons im Tierkörper (Biochemische Hydrirungen VIII), *Hoppe-Seyler's Z. Physiol. Chem.*, 266, 99, 1940.

127. **Ide, H. and Toki, S.,** Metabolism of β-ionone. Isolation, characterization and identification of the metabolites in the urine of rabbits, *Biochem. J.*, 119, 281, 1970.

128. **Prelog, V. and Meier, H. L.,** Untersuchungen über Organextrakte und Harn. XVIII. Mitteilung. Über die biochemische Oxydation von β-jonon im Tierkörper, *Helv. Chim. Acta*, 33, 1276, 1950.

129. **Takenoshita, R. and Toki, S.,** Rabbit liver 3-hydroxyhexobarbital dehydrogenase. Purification and properties, *J. Biol. Chem.*, 249, 5428, 1974.

130. **Prelog, V. and Würsch, J.,** Untersuchungen über Organextrakte und Harn. XXI. Mitteilung. Über die biochemische Oxydation von α-jonon im Tierkörper, *Helv. Chim. Acta*, 34, 859, 1951.

131. **Rimini, E.,** Ossidazione biologica del carone e del fencone, *Gazz. Chim. Ital.*, 39, 186, 1909 (Chem. Abstr. 5, 689, 1911).

132. **Reinarz, F. and Zanke, W.,** Der Abbau des Fenchons im tierischen Organismus, *Ber. Dtsch. Chem. Ges.*, 69, 2259, 1936.

133. **Miyazawa, M. and Kameoka, H.,** Biotransformation and bioavailability of (+)-fenchone to hydroxyfenchone in rabbits, *Chem. Express*, 3, 503, 1988.

134. **Miyazawa, M. and Kameoka, H.,** 9-Hydroxyfenchone: a new metabolite of (+)-fenchone in rabbits, *Chem. Express*, 3, 231, 1988.

135. **Miyazawa, M. and Kameoka, H.,** 10-Hydroxyfenchone: a new metabolite of (+)-fenchone in rabbits, *Chem. Express*, 2, 547, 1987.

136. **Wiedemann, C.,** Beiträge zur Pharmakologie des Camphers, *Arch. Exp. Pathol. Pharmak.*, 6, 216, 1877.

137. **Schmiedeberg, O. and Meyer, H.,** Ueber Stoffwechselprodukte nach Campherfütterung, *Hoppe-Seyler's Z. Physiol. Chem.*, 3, 422, 1879.

138. **Asahina, Y. and Ishidate, M.,** Über campherol, *Ber. Dtsch. Chem. Ges.*, 61, 533, 1928.

139. **Asahina, Y. and Ishidate, M.,** Über das Vorkommen der π-Oxy-Derivative im Campherol, *Ber. Dtsch. Chem. Ges.*, 66, 1673, 1933.

140. **Asahina, Y. and Ishidate, M.,** Über 5- und π-Oxy-campher und ihre Derivate, *Ber. Dtsch. Chem. Ges.*, 67, 71, 1934.

141. **Asahina, Y. and Ishidate, M.,** Über zwei neue Umwandlungsprodukte des camphers im Tierkörper, *Ber. Dtsch. Chem. Ges.*, 68, 947, 1935.

142. **Shimamoto, T.,** On the isolation of π-oxycamphor from campherol, *Sci. Pap. Inst. Phys. Chem. Res. (Tokyo)*, 25, 52, 1934.

143. **Shimamoto, T.,** On the constituents of "rabbit's" campherol, *Sci. Pap. Inst. Phys. Chem. Res. (Tokyo)*, 25, 59, 1934.

144. **Kawabata, H.,** Biological change of π-oxocamphor, *J. Pharm. Soc. Jpn.*, 63, 455, 1943 (Chem. Abstr. 44, 6967f, 1950).

145. **Gallicano, K. D., Park, H. C., and Young, L. M.,** A sensitive liquid chromatographic procedure for the analysis of camphor in equine urine and plasma, *J. Anal. Toxicol.*, 9, 24, 1985.

146. **Robertson, J. S. and Hussain, M.,** Metabolism of camphors and related compounds, *Biochem. J.*, 113, 57, 1969.

147. **Leibman, K. C. and Ortiz, E.,** Mammalian metabolism of terpenoids. I. Reduction and hydroxylation of camphor and related compounds, *Drug Metab. Dispos.*, 1, 543, 1973.

148. **Sligar, S. G., Gelb, M. H., and Heimbrook, D. C.,** Bio-organic chemistry and cytochrome P-450-dependent catalysis, *Xenobiotica*, 14, 63, 1984.

149. **Collins, J. R. and Loew, G. H.,** Theoretical study of the product specificity in the hydroxylation of camphor, norcamphor, 5,5-difluorcamphor, and pericyclocamphanone by cytochrome P-450$_{cam}$, *J. Biol. Chem.*, 263, 3164, 1988.

150. **Reinarz, F. and Zanke, W.,** Über die Abbauprodukte des Camphers und Campherchinons im tierischen Organismus, *Ber. Dtsch. Chem. Ges.*, 67, 548, 1934.

151. **Robertson, J. S. and Solomon, E.,** Metabolism of camphanediols, *Biochem. J.*, 121, 503, 1971.

152. **Reinarz, F., Zanke, W., and Faust, K.,** Über die Abbauprodukte des Cyclo-camphanons und des Diketo-camphans im tierischen Organismus, *Ber. Dtsch. Chem. Ges.*, 67, 1536, 1934.

153. **Ishidate, M., Kawahata, H., and Nakazawa, K.,** Über den Abbau des *p*-Diketo-camphans im tierischen Organismus und die Synthese der β-Cycloisoketopinsäure, *Ber. Dtsch. Chem. Ges.*, 74, 1707, 1941.

154. **Tamura, Z. and Imanari, T.,** Metabolism of *trans*-π-oxocamphor, *Chem. Pharm. Bull.*, 12, 370, 1964.

155. **Tamura, Z. and Imanari, T.,** Metabolism of *trans*-π-oxocamphor. II. Metabolism under oral administration, *J. Pharm. Soc. Jpn.*, 90, 506, 1970.

156. **Magnus-Levy, A.,** Über Paarung der Glukuronsäure mit optischen Antipoden, *Biochem. Z.*, 2, 319, 1907.

157. **Mayer, P.,** Über asymmetrische Glucuronsäurepaarung, *Biochem. Z.*, 9, 439, 1908.

158. **Hämäläinen, J.,** Zur Spaltbarkeit der Borneol- und Camphoglykuronsäuren durch Enzyme, *Skand. Arch. Physiol.*, 23, 297, 1909.

159. **Reinarz, F. and Zanke, W.,** Über die Abbau-produkte des Epicamphers im tierischen Organismus, *Ber. Dtsch. Chem. Ges.*, 67, 589, 1934.

160. **Krieger, H.,** Metabolism of norcamphors in rabbits, *Acta Chim. Fenn.*, 35B, 174, 1962 (Chem. Abstr. 58, 3792d, 1963).

161. **Robinson, D.,** Studies in detoxication. LXXIV. The metabolism of benzhydrol, benzophenone and *p*-hydroxybenzophenone, *Biochem. J.*, 68, 584, 1958.

162. **Williams, R. T.,** *Detoxication Mechanisms*, Chapman and Hall, London, 1959, 336.

163. **Smith, J. N., Smithies, R. H., and Williams, R. T.,** Studies in detoxication. LVI. The metabolism of alkylbenzenes. Stereochemical aspects of the biological hydroxylation of ethylbenzene to methylphenylcarbinol, *Biochem. J.*, 56, 320, 1954.

164. **Smith, J. N., Smithies, R. H., and Williams, R. T.,** Studies in detoxication. LIX. The metabolism of alkylbenzenes. The biological reduction of ketones derived from alkylbenzenes, *Biochem. J.,* 57, 74, 1954.

165. **Gal, J., Harper, T., Friedman, T. C., and Thompson, J. A.,** Stereoselective reduction of achiral ketones to chiral alcohols by rabbit liver cytosol, *Pharmacologist,* 22, 242, 1980.

166. **Culp, H. W. and McMahon, R. E.,** Reductase for aromatic aldehydes and ketones. The partial purification and properties of a reduced triphosphopyridine nucleotide-dependent reductase from rabbit kidney cortex, *J. Biol Chem.,* 243, 848, 1968.

167. **El Masry, A. M., Smith, J. N., and Williams, R. T.,** Studies in detoxication. LXIX. The metabolism of alkylbenzenes: *n*-propylbenzene and *n*-butylbenzene with further observations on ethylbenzene, *Biochem. J.,* 64, 50, 1956.

168. **Kiese, M. and Lenk, W.,** Hydroxyacetophenones: urinary metabolites of ethylbenzene and acetophenone in the rabbit, *Xenobiotica,* 4, 337, 1974.

169. **Kiese, M. and Lenk, W.,** ω- and (ω-1)-Hydroxylation of 4-chloropropionanilide in liver microsomes of rabbits treated with phenobarbital or 3-methylcholanthrene, *Biochem. Pharmacol.,* 22, 2575, 1973.

170. **Yokoyama, T., Hayase, Y., Aikawa, H., Odaka, Y., Miyamoto, S., and Mori, Y.,** Study on the absorption, distribution, metabolism and excretion of paeonol, a main component of Paeonia moutan Sins, in rats, *Oyo Yakuri,* 13, 457, 1977 (Chem. Abstr. 88, 83346h, 1978).

171. **Mimura, K. and Baba, S.,** Studies on the biotransformation of paeonol by isotope tracer techniques. II. Species differences in metabolism, *Chem. Pharm. Bull.,* 28, 1704, 1980.

172. **Gjertsen, F. B., Solheim, E., and Scheline, R. R.,** Metabolism of aromatic plant ketones in rats: acetovanillone and paeonol, *Xenobiotica,* 18, 225, 1988.

173. **Mimura, K. and Baba, S.,** Determination of paeonol metabolites in man by the use of stable isotopes, *Chem. Pharm. Bull.,* 29, 2043, 1981.

174. **Daly, J. W., Axelrod, J., and Witkop, B.,** Dynamic aspects of enzymatic *O*-methylation and -demethylation of catechols *in vitro* and *in vivo, J. Biol. Chem.,* 235, 1155, 1960.

175. **Scheline, R. R.,** Unpublished data, 1970.

176. **Gjertsen, F. B.,** Acetofenonderivatmetabolisme hos rotte: metoksy-, hydroksy- og metylendioksy-substituenter, Thesis, University of Bergen, Bergen, Norway, 1985.

177. **Williams, R. T.,** *Detoxication Mechanisms,* Chapman and Hall, London, 1959, 338.

178. **Dodgson, K. S.,** The orientation of glucuronic acid conjugation in resacetophenone, *Biochem. J.,* 47, 11, 1950.

179. **Haley, T. J. and Bassin, M.,** Metabolism and urinary excretion of several flavonoid compounds, *Proc. Soc. Exp. Biol. Med.,* 81, 298, 1952.

180. **Bobik, A., Holder, G. M., Ryan, A. J., and Wiebe, L. I.,** Inhibitors of hepatic mixed function oxidases. I. The metabolism of 2,6-dihydroxy-, 2,hydroxy-6-methoxy- and 2,6-dimethoxyacetophenones, *Xenobiotica,* 5, 65, 1975.

181. **Sporstøl, S. and Scheline, R. R.,** The metabolism of 4-(4-hydroxyphenyl)butan-2-one (raspberry ketone) in rats, guinea pigs and rabbits, *Xenobiotica,* 12, 249, 1982.

182. **Monge, P., Scheline, R., and Solheim, E.,** The metabolism of zingerone, a pungent principle of ginger, *Xenobiotica,* 6, 411, 1976.

183. **Wright, D. J. and Holder, G. M.,** The metabolism of some food additives related to piperonal in the rabbit, *Xenobiotica,* 10, 265, 1980.

184. **Kammerer, R. C., Cho, A. K., and Jonsson, J.,** *In vitro* metabolism of phenylacetone, phenyl-2-butanone, and 3-methyl-1-phenyl-2-butanone by rabbit liver preparations, *Drug Metab. Dispos.,* 6, 396, 1978.

185. **Holder, G. M., Plummer, J. L., and Ryan, A. J.,** The metabolism and excretion of curcumin (1,7-bis-(4-hydroxy-3-methoxyphenyl)-1,6-heptadiene-3,5-dione) in the rat, *Xenobiotica,* 8, 761, 1978.

186. **Ravindranath, V. and Chandrasekhara, N.,** Metabolism of curcumin — studies with [³H]curcumin, *Toxicology,* 22, 337, 1982.

187. **Plummer, J. L.,** The Metabolism of Curcumin and Some Related Substances, Ph. D. Thesis, University of Sydney, Sydney, Australia, 1977.

188. **Ravindranath, V. and Chandrasekhara, N.,** In vitro studies on the intestinal absorption of curcumin in rats, *Toxicology,* 20, 251, 1981.

189. **Wahlström, B. and Blennow, G.,** A study on the fate of curcumin in the rat, *Acta Pharmacol. Toxicol.,* 43, 86, 1978.

190. **Ravindranath, V. and Chandrasekhara, N.,** Absorption and tissue distribution of curcumin in rats, *Toxicology,* 16, 259, 1980.

191. **Glock, G. E., Thorp, R. H., Ungar, J., and Wien, R.,** The antibacterial action of 4:6-dimethoxytoluquinone and its fate in the animal body, *Biochem. J.,* 39, 308, 1945.

192. **Bray, H. G. and Garrett, A. J.,** The metabolism of 1:4-quinones and their reactivity with sulphydryl groups, *Biochem. J.,* 80, 6, 1961.

193. **Chandrasekaran, B. and Nagarajan, B.,** Metabolism of echitamine and plumbagin in rats, *J. Biosci.,* 3, 395, 1981.

194. **Wang, W. J., Yi, M. G., and Zhu, X. Y.,** A study on absorption, distribution and excretion of [³H]shikonin in mice, *Yaosue Xuebao*, 23, 246, 1988 (Chem. Abstr. 109, 47792u, 1988).

195. **Fujita, M., Furuya, T., and Matsuo, M.,** Studies on the metabolism of naturally occurring anthraquinones. I. The metabolism of 1-hydroxyanthraquinone and 2-hydroxyanthraquinone, *Chem. Pharm. Bull.*, 9, 962, 1961.

196. **Fujita, M., Furuya, T., and Matsuo, M.,** Studies on the metabolism of naturally occurring anthraquinones. II. The metabolism of 1-methoxyanthraquinone and 2-methoxyanthraquinone, *Chem. Pharm. Bull.*, 9, 967, 1961.

197. **Mähner, B. and Dulce, H.-J.,** Ausscheidungsprodukte von Hydroxyanthrachinonen im Harn bei Ratten, *Z. Klin. Chem. Klin. Biochem.*, 6, 99, 1968.

198. **Fujita, M., Furuya, T., and Matsuo, M.,** Studies on the metabolism of naturally occurring anthraquinones III. The metabolism of alizarin dimethyl ether, *Chem. Pharm. Bull.*, 10, 909, 1962.

199. **Fairbairn, J. W.,** Chemical structure, mode of action and therapeutical activity of anthraquinone glycosides, *Pharm. Weekbl.*, 100, 1493, 1965.

200. **Breimer, D. D. and Baars, A. J.,** Pharmacokinetics and metabolism of anthraquinone laxatives, *Pharmacology*, 14 (Suppl. 1), 30, 1976.

201. **Dobbs, H. E., Lane, A. C., and MacFarlane, I. R.,** Trasporto e meccanismo d'azione dei sennosidi, *Il Farmaco Ed. Sci.*, 30, 147, 1975.

202. **Hardcastle, J. D. and Wilkins, J. L.,** The action of sennosides and related compounds on human colon and rectum, *Gut*, 11, 1038, 1970.

203. **Tamura, G., Gold, C., Ferro-Luzzi, A., and Ames, B. A.,** Fecalase: a model for activation of dietary glycosides to mutagens by intestinal flora, *Proc. Natl. Acad. Sci. USA*, 77, 4961, 1980.

204. **Ippen, H.,** Toxicity and metabolism of cignolin, *Dermatologica*, 119, 211, 1959.

205. **Ippen, H. and Montag, T.,** Stoffwechselbezichungen zwischen 1,8-dioxyanthrachinon und 1,8-dioxyanthranol-9, *Arzneim. Forsch.*, 8, 778, 1958.

206. **Ippen, H.,** Metabolism of 1,8-dihydroxyanthraquinone, *Planta Med.*, 7, 423, 1959.

207. **Hsu, G.-Y., Sun, C.-C., Chen, C.-H., and Wu, W. T.,** Chinese rhubarb. X. Metabolic transformations of anthraquinone derivatives, *Acta Biochim. Biophys. Sin.*, 6, 110, 1966 (Chem. Abstr. 65, 14287f, 1966).

208. **Vyth, A. and Kamp, P. E.,** Detection of anthraquinone laxatives in the urine, *Pharm. Weekbl.*, 1, 456, 1979.

209. **Fairbairn, J. W.,** In discussion following article of Lemli, J. and Lemmens, L., *Pharmacology*, 20 (Suppl. 1), 50, 1980.

210. **Chen, C.-H., Kao, S.-M., Du, H.-F., and Yo, W.-H.,** Studies on Chinese rhubarb. IV. Absorption, distribution and excretion of anthraquinone derivatives, *Acta Pharm. Sin.*, 10, 525, 1963.

211. **Lemmens, L. and Dreessen, M.,** The laxative action of anthracene derivative. III. Absorption, metabolism and excretion of rhein in the rat, *Pharm. Weekbl. Sci. Ed.*, 1, 134, 1979.

212. **Lemli, J. and Lemmens, L.,** Metabolism of sennosides and rhein in the rat, *Pharmacology*, 20 (Suppl. 1), 50, 1980.

213. **Sund, R. B. and Elvegård, S.-O.,** Anthraquinone laxatives: metabolism and transport of danthron and rhein in the rat small and large intestine in vitro, *Pharmacology*, 36 (Suppl. 1), 144, 1988.

214. **De Witte, P. and Lemli, J.,** Metabolism of ¹⁴C-rhein and ¹⁴C-rhein anthrone in rats, *Pharmacology*, 36 Suppl. 1, 152, 1988.

215. **De Witte, P. and Lemli, J.,** Excretion and distribution of [¹⁴C]rhein and [¹⁴C]rhein anthrone in rat, *J. Pharm. Pharmacol.*, 40, 652, 1988.

216. **Lang, W.,** Pharmacokinetics of ¹⁴C-labelled rhein in rats, *Pharmacology*, 36 (Suppl. 1), 158, 1988.

217. **Bachmann, M. and Schlatter, C.,** Metabolism of [¹⁴C]emodin in the rat, *Xenobiotica*, 11, 217, 1981.

218. **Sun, Y., Li, Q., and Chen, Q.-H.,** Excretion of emodin and its metabolites in mice, *Nanjing Yaoxueyuan Xuebao*, 17, 132, 1986 (Chem. Abstr. 105, 164389w, 1986).

219. **Sun, Y. and Chen, Q.,** Isolation and identification of emodin metabolites in mice and rats both in vitro and in vivo, *Shengwu Huaxue Yu Shengwu Wuli Xuebao*, 19, 447, 1987 (Chem. Abstr. 109, 31489q, 1988).

220. **Masuda, T., Haraikawa, K., Morooka, N., Nakano, S., and Ueno, Y.,** 2-Hydroxyemodin, an active metabolite of emodin in the hepatic microsomes of rats, *Mutat. Res.*, 149, 327, 1985.

221. **Murakami, H., Kobayashi, J., Masuda, T., Morooka, N., and Ueno, Y.,** ω-Hydroxyemodin, a major hepatic metabolite of emodin in various animals and its mutagenic activity, *Mutat. Res.*, 180, 147, 1987.

222. **Sun, Y. and Chen, Q.-H.,** Isolation and identification of metabolites of physcion in rats and mice in vivo and in vitro, *Yaoxue Xuebao*, 21, 748, 1986 (Chem. Abstr. 106, 43394f, 1987).

223. **Lemmens, L.,** The laxative action of anthracene derivatives. II. Absorption, metabolism and excretion of sennoside A and B in the rat, *Pharm. Weekbl. Sci. Ed.*, 1, 2, 1979.

224. **Dreessen, M., Eyssen, H., and Lemli, J.,** The metabolism of sennosides A and B by the intestinal microflora: in vitro and in vivo studies on the rat and the mouse, *J. Pharm. Pharmacol.*, 33, 679, 1981.

225. **Hietala, P., Lainonen, H., and Marvola, M.,** New aspects on the metabolism of the sennosides, *Pharmacology*, 36 (Suppl. 1), 138, 1988.

226. **Uchino, K., Yamamura, Y., Saitoh, Y., and Nakagawa, F.,** Determination of rhein and its conjugates in urine by high-performance liquid chromatography, *J. Chromatogr.*, 380, 462, 1986.

227. **Sasaki, K., Yamauchi, K., and Kuwano, S.,** Metabolic activation of sennoside A in mice, *Planta Med.,* 37, 370, 1979.

228. **Moreau, J. P., Moreau, S., and Skinner, S.,** Comparative physiological dispositon of some anthraquinone glycosides and aglycones, *Biopharm. Drug Dispos.,* 6, 325, 1985.

229. **Kobashi, K., Nishimura, T., Kusaka, M., Hattori, M., and Namba, T.,** Metabolism of sennosides by human intestinal bacteria, *Planta Med.,* 40, 225, 1980.

230. **Dreessen, M. and Lemli, J.,** Qualitative and quantitative interactions between the sennosides and some human intestinal bacteria, *Pharm. Acta Helv.,* 57, 350, 1982.

231. **Hattori, M., Kim, G., Motoike, S., Kobashi, K., and Namba, T.,** Metabolism of sennosides by intestinal flora, *Chem. Pharm. Bull.,* 30, 1338, 1982.

232. **Akao, T., Akao, T., Mibu, K., Hattori, M., Namba, T., and Kobashi, K.,** Enzymatic reduction of sennidin and sennoside in *Peptostreptococcus intermedius, J. Pharmacobiol. Dyn.,* 8, 800, 1985.

233. **Akao, T., Mibu, K., Erabi, T., Hattori, M., Namba, T., and Kobashi, K.,** Non-enzymic reduction of sennidins and sennosides by reduced flavin, *Chem. Pharm. Bull.,* 35, 1998, 1987.

234. **Hattori, M., Namba, T., Akao, T., and Kobashi, K.,** Metabolism of sennosides by human intestinal bacteria, *Pharmacology,* 36 Suppl. 1, 172, 1988.

235. **Compernolle, F., Toppet, S., Cuveele, J., and Lemli, J.,** 2,2′-Dimers of aloe-emodin and chrysophanol: bianthraquinones isolated from human plasma and aromatic cascara fluid extract USP XXI, *Tetrahedron,* 43, 3055, 1987.

236. **Dreessen, M. and Lemli, J.,** Studies in the field of drugs containing anthraquinone derivatives. XXXVI. The metabolism of cascarosides by intestinal bacteria, *Pharm. Acta Helv.,* 63, 287, 1988.

237. **Hattori, M., Kanda, T., Shu, Y.-Z., Akao, T., Kobashi, K., and Namba, T.,** Metabolism of barbaloin by intestinal bacteria, *Chem. Pharm. Bull.,* 36, 4462, 1988.

METABOLISM OF ACIDS, LACTONES, AND ESTERS

ACIDS

Compounds containing a carboxylic acid group may be metabolized along several routes. However, the most common pathways involve conjugation reactions with amino acids or glucuronic acid. Also, chain shortening due to β-oxidation occurs when the carboxylic acid group is attached to a suitable aliphatic moiety. The various metabolic reactions of compounds containing the carboxylic acid group were reviewed by Caldwell.[1] Additional reactions of the carboxy group have been discovered which reveal that numerous lipophilic conjugates may be formed. This subject was reviewed by Caldwell[2,3] and Quistad and Hutson[4] and the findings demonstrate that the acids may undergo chain extension or incorporation into triglycerides, cholesterol esters, and fatty acid derivatives. In addition to these reactions which occur in the tissues, the metabolism (e.g., decarboxylation or reduction) of some carboxylic acids may be carried out by the intestinal microflora.[5]

ALIPHATIC ACIDS

The aliphatic acids occurring in plants encompass a large number of compounds which include saturated and unsaturated volatile acids and nonvolatile mono-, di-, and tricarboxylic acids. All of the members of the homologous series of saturated, normal fatty acids from C_1 (formic acid) to C_{16} (palmitic acid) occur naturally as do many even-numbered members from C_{18} (stearic acid) upwards. The essential feature in the metabolism of these compounds is their oxidative degradation, a universal biochemical capacity among living organisms. This process involves β-oxidation which results in the formation of two-carbon fragments from the fatty acids by way of acylated coenzyme A derivatives. Additionally, some of these aliphatic acids may also undergo chain lengthening. These metabolic reactions fall outside the scope of this book, however, and will not be discussed further. The same applies to the metabolism of the di- and tricarboxylic acids which enter into the ubiquitous citric acid cycle. It is noteworthy, however, that some of the acids (e.g., citric and malic) produced by these mitochondrial reactions sometimes accumulate and are concentrated in the vacuolar sap of plants.

Among the unsaturated plant fatty acids, both sorbic acid, a C_6-acid, and a few acyclic terpenoid C_{10}-acids have been studied metabolically. **Sorbic acid** (1) is known to enter into conventional fatty acid metabolism and Deuel et al.[6] reported that its metabolism is identical with that of butanoic and hexanoic acid. Sorbic acid is thus converted extensively to CO_2 and water. Fingerhut et al.,[7] using 1-^{14}C-labeled compound, found that 85% of the radioactivity was excreted by rats as respiratory $^{14}CO_2$ following oral administration of 60 to 1200 mg/kg of the acid. It was absorbed nearly quantitatively from the intestine but no unchanged compound was excreted in the urine. Sorbic acid is not utilized in the formation of glycogen; however, the detection of radioactivity in the lipids indicates that the acetyl coenzyme A formed is used for fatty acid synthesis. A minor route of metabolism is allylic hydroxylation and further oxidation of the terminal methyl group. Kuhn et al.[8] reported that small amounts (<1% dose) of *trans-trans*-muconic acid (2) were excreted in the urine of rabbits fed large amounts of sorbic acid.

H₃C–CH=CH–CH=CH–COOH   HOOC–CH=CH–CH=CH–COOH

(1)   (2)

Sorbic acid

Oxidation similar to that noted above with sorbic acid was reported by Kuhn et al.[9] to take place with **geranic acid** (3). Following repeated large oral doses to rabbits, two oxidation products were isolated from the urine. These were 2,6-dimethyl-2,6-octadienedioic acid

(Hildebrandt acid) (4) and optically active 2,6-dimethyl-2-octenedioic acid (reduced Hildebrandt acid) (5) which were excreted in a ratio of 3:2.

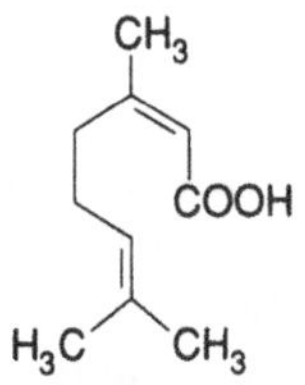

(3)   (4) 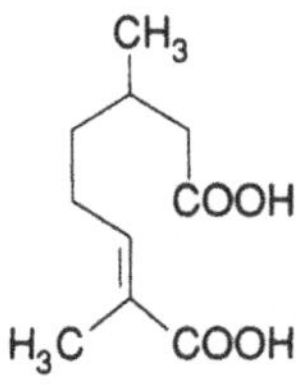   (5)

Geranic acid

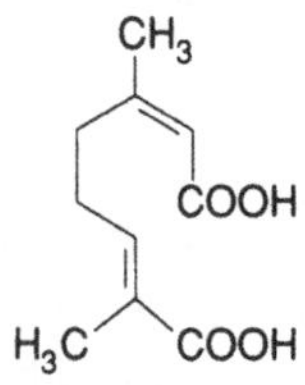

d-Citronellic acid and **l-rhodinic acid** are enantiomorphs of compound (6) which differ structurally from geranic acid only by the lack of the double bond at the 2,3-position. Asano and Yamakawa[10] found that the metabolism of these acids in rabbits was analogous to that reported for geranic acid, the corresponding enantiomorphic reduced Hildebrandt acids (5) being excreted in the urine.

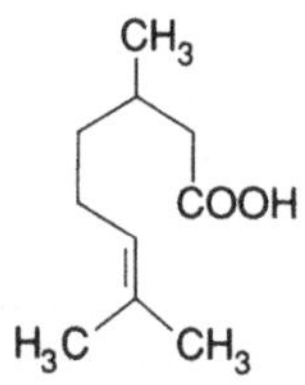

(6)

d-Citronellic acid

l-Rhodinic acid

Oxalic acid (HOOC–COOH) is the simplest member of the series of unsubstituted aliphatic dicarboxylic acids and, except for succinic and fumaric acids which enter into the citric acid cycle, is the derivative which has received the most attention from a metabolic point of view. Oxalic acid is toxic by virtue of the formation of insoluble calcium oxalate in the body but is otherwise essentially metabolically inert. Both Weinhouse and Friedmann[11] and Curtin and King[12] reported that 1% or less was lost as $^{14}CO_2$ when rats were injected with a few mg of [^{14}C]oxalic acid; however, even this small amount of CO_2 may have arisen from impurities (e.g., formic acid) rather than from oxalic acid itself. No evidence was found for its conversion to other metabolites, however only ~20 to 40% of the dose (2 to 7 mg) was recovered unchanged in the urine.[11] This was believed to be due to the precipitation of calcium oxalate in the tissues. Similarly, Jones et al.[13] reported that oxalic acid given intraperitoneally to rats was only incompletely excreted in the urine. Hodgkinson[14] stated that more than half of the normally ingested oxalic acid in rats undergoes destruction in the intestine. The incomplete recovery of oxalate has been reported in several other animal species. Oral doses of tracer amounts of [^{14}C]oxalic acid to mice resulted in the excretion of only 8% of the ^{14}C in the urine during 3 d.[15] About 45% of the doses (20 and 50 mg/kg) were recovered unchanged in the urine when sodium oxalate was given orally in capsules to pigs.[16] In two pigs given sodium oxalate in food at a level of 50 mg/kg/d, the mean excretion values were only 6 and 11%. The bulk of the oxalate in diets fed horses could not be accounted for in the urine and feces.[17] It was suggested that this was due to utilization by the intestinal bacterial flora. This conclusion was also reached by Shirley and Schmidt-Nielsen[18] in a study of oxalate metabolism in white rats, hamsters, pack rats (*Neotoma albigula*), and sand rats (*Psammomys obesus*). The last three species are more closely related taxonomically and differed from white rats in their ability to degrade almost all of the ingested oxalate. Only about half of the oxalate was degraded in white rats. Additionally, it was believed that dietary factors contribute to these differences. Both pack rats and sand rats normally consume large amounts of oxalate in their diets. In contrast to these results, Elder and Wyngaarden[19] reported that [^{14}C]oxalic acid was excreted unchanged almost quantitatively (89 to 99%) in man.

Several studies have shown that rumen microorganisms from sheep and cattle are able to metabolize oxalate.[20-22] Mixed populations of rumen bacteria from sheep fed increasing quantities of oxalate showed increasing abilities to degrade oxalate.[23] This trend and methane production by the cultures showed a positive correlation. Recently, pure cultures of oxalate-degrading anaerobes were isolated from rumen or cecal contents from various animal species or from human feces.[24]

The metabolism of oxalic acid in animals and man was reviewed by Hodgkinson.[14] Additional information, especially on the gastrointestinal degradation of oxalate, was summarized by Allison.[24]

The higher dicarboxylic acids including the C_6-, C_8-, and C_{10}-homologues (**adipic acid, suberic acid,** and **sebacic acid,** respectively) are largely excreted unchanged when fed to dogs and humans.[25] Similar results showing the extensive excretion of unchanged C_9-compound (**azelaic acid**) were reported by Weitzel[26] who gave large, daily doses of several of these higher dicarboxylic acids to humans.

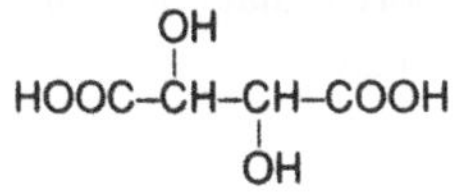

$$HOOC-\overset{\overset{\displaystyle OH}{|}}{C}H-\overset{\overset{\displaystyle }{|}}{C}H-COOH$$
$$\underset{OH}{}$$

(7)

(+)-Tartaric acid

L(+)-Tartaric acid (7) is the naturally occurring form of tartaric acid. Most metabolic studies have used this form, however data are also available for D(−)-tartaric acid and for the synthetic DL-form. The existence of species differences in the fate of L(+)-tartaric acid has been known for many years. Underhill et al.[27] found that 90 to 100% of an oral dose (50 mg/kg) was excreted unchanged in the urine by rabbits. A similar picture was seen with dogs given oral doses of 100 to 600 mg/kg; however, lower recoveries were recorded when higher doses were employed with both species. Rats also excreted most of the compound unchanged in the urine following oral dosage. On the other hand, similar experiments using guinea pigs showed low amounts of L(+)-tartaric acid in the urine. Urinary excretion was extensive following injection of the compound in rabbits, guinea pigs, and dogs. These results indicated that L(+)-tartaric acid was not metabolized in the tissues of these animals but that it underwent alteration in the guinea pig intestine. Similar results were obtained in man by Underhill et al.[28] who recovered ~20% of the orally administered L(+)-tartaric acid unchanged in the urine. The remainder was not detected in the feces and this loss was ascribed to bacterial metabolism in the large intestine. These conclusions were confirmed by Finkle[29] who found an average of 17% excreted unchanged in the urine after oral doses of 200 to 400 mg. Similar findings were made using a 4 g dose, however nearly quantitative excretion of unchanged compound was recorded following the injection (i.m.) of the acid (1 to 2 g). Gry and Larsen[30] gave large (1000 mg/kg) intragastric doses of L(+)-tartaric acid to rats and guinea pigs and confirmed the high recovery (~73%) in the urine of rats and the low recovery (~4%) in the urine of guinea pigs. They also used pigs and found an intermediate value (26%) following a dose of 500 mg/kg. The results obtained using D(−)-tartaric acid in parallel experiments gave similar values in guinea pigs and pigs, however a lower value (52%) was recorded in rats. Additionally, the tartaric acids were metabolized upon incubation with cecal microorganisms from rats or guinea pigs.

The metabolism of tartaric acid has also been studied using ^{14}C-labeled material. Chasseaud et al.[31] gave [1,4-^{14}C]-L-(+)-tartaric acid orally or intravenously to rats. Following doses of 400 mg/kg, excretion of radioactivity was nearly complete within 12 h in the urine and 24 h in the expired air. The excretion values after 48 h were ~70% (urine), 14% (feces), and 16% (expired air) after oral dosage. The corresponding values after injection were 82%, 1%, and 8%. The values for urinary excretion correspond well with earlier results, however these findings suggest that the conversion of tartaric acid to CO_2 occurs both in the intestine due to bacterial action and in the tissues. Similar experiments were carried out by Chadwick et al.[32] who, however, employed ^{14}C-labeled DL-tartaric acid at a lower dose level (18.8 mg/kg). They also recorded a value of 70% for the urinary excretion of tartrate. They confirmed that some metabolism to CO_2 occurred in the tissues, however the intestine was found to be the main site of this reaction. Incubation experiments showed that several genera of aerobic and anaerobic intestinal bacteria converted tartaric acid to CO_2. It was also found that L-tartaric acid disappeared more rapidly than did D-tartaric acid when incubated with fecal microorganisms. Chadwick et al. also administered tartaric acid to man. They gave 2.5, 5, or 10 g oral doses of sodium L-tartrate together with the labeled compound and confirmed the low excretion level of tartrate in the urine. In all cases only ~12% of the material was excreted unchanged in the urine. The bulk of the radioactivity was found in the expired CO_2 and the rate of excretion reached a peak after about 4 h. In contrast, i.v. dosage in man resulted in a much lower value for expired $^{14}CO_2$, however this was greatest during the first hour. Malan et al.[33] carried out similar experiments with ^{14}C-labeled DL-tartaric acid in man and reported that ~30% of the radioactivity was excreted in the urine and 70% (as $^{14}CO_2$) in the expired air.

ALICYCLIC ACIDS

(–)-Quinic acid (1,3,4,5-tetrahydroxycyclohexanecarboxylic acid) (8) was first shown by Lautemann[34] to undergo metabolic aromatization. This transformation involves conversion to benzoic acid (18) which was excreted in the urine as its glycine conjugate (hippuric acid (19)). The reaction sequence, again shown in man, was confirmed by Quick.[35] Vasiliu et al.[36,37] reported that 50 to 60% of the quinic acid (10 g) given to sheep was excreted as hippuric acid but that little aromatization occurred in dogs. Values of 5 to 10% were found when dogs were maintained on a meat diet, however none was detected when a diet giving an acidic urine was fed. Bernhard[38] also found that (–)-quinic acid was not converted to hippuric acid in dogs but instead excreted unchanged. Profound variations in aromatization ability among various animal species were also reported by Beer et al.[39] who confirmed the high value (as much as 70%) in man but found very low values following oral dosage of (–)-quinic acid to rats and rabbits or after s.c. administration to guinea pigs and cats. Interestingly, Bernhard et al.[40] found that as much as 50 to 80% of the dose (0.7 to 1 g/kg) of quinic acid was aromatized in guinea pigs when oral administration was employed. The significance of the route of administration used was clearly demonstrated by Cotran et al.[41] They found that quinic acid was aromatized and excreted in the urine as hippuric acid in man and guinea pigs after oral administration. This did not occur when the guinea pigs were given the compound intraperitoneally. Furthermore, inhibition of the intestinal bacteria with neomycin suppressed quinic acid aromatization and it was concluded that this reaction is achieved by the intestinal bacteria. This effect of neomycin has been confirmed in rats[42] and in rhesus monkeys.[43]

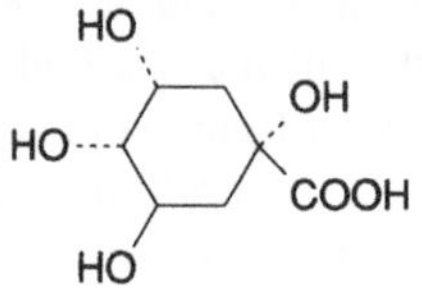

(8)

(–)-Quinic acid

It therefore seems evident that the aromatization of quinic acid is dependent on the metabolic activity of the intestinal microorganisms and that the species differences in conversion may be related to differences in this activity of the microfloras. However, the results summarized above do not indicate which of the several steps in the aromatization sequence are dependent on bacterial metabolism. This subject has been clarified using shikimic acid and is treated below.

In addition to the species differences noted above, the investigation by Adamson et al.[43] employed a large number of animal species and indicated that extensive aromatization (20 to 60%) was confined to man and the Old World monkeys (baboon and rhesus and green monkeys). Low conversion levels (0 to 5%) were found in all of the other species studied which included New World monkeys (spider and squirrel monkeys, capuchin, bushbaby, slow lorris, and tree shrew), carnivores (ferret, cat, and dog), rodents (lemming, mouse, rat, hamster, and guinea pig), and the fruit bat, hedgehog, and rabbit. Quinic acid doses of ~300 mg/kg (p.o.) were typical for most of these experiments. Martin[44] confirmed the earlier reports of Vasiliu showing that sheep readily aromatize quinic acid. About a quarter of the quinic acid infused into the rumen was excreted in the urine as benzoic acid.

Fairly large variations in the extent of aromatization of (–)-quinic acid have been found in various experiments using the same animal species. This is evident from that noted above in guinea pigs and has also been recorded in rats. The data of Beer et al.[39] indicated that only ~1% of the dose (2 g divided among six rats) was aromatized. The values of Adamson et al.[43] following a dose of 600 mg/kg averaged ~5%, whereas Teuchy et al.,[45] using a 50 mg dose, failed to find any conversion. Indahl and Scheline[46] recorded the 48-h urinary hippuric acid excretion in 32 rats given 100 mg doses of quinic acid and found that this metabolite accounted for 12% of the dose. Interestingly, this value was not changed following the feeding of a purified diet containing 1% quinic acid for 24 or 48 d in the attempt to promote a metabolic adaptation of the microflora to this compound.

Aromatization of (–)-quinic acid in rats leads not only to the urinary excretion of hippuric acid but of catechol as well.[47] However, this pathway is of minor importance and Indahl and Scheline[46] found that only 1% of the dose (100 mg, p.o.) was excreted by rats as catechol. Martin[48] reported a similar result in sheep given a large (37 g) daily infusion of quinic acid into the rumen. The increase in urinary catechol amounted to ~0.5% of the dose. Catechol was also identified as a metabolite when quinic acid was incubated anaerobically with mixed cultures of rat fecal or cecal bacteria.[49,50] The *in vitro* formation of benzoic acid from quinic acid by rat cecal bacteria was also reported.[46]

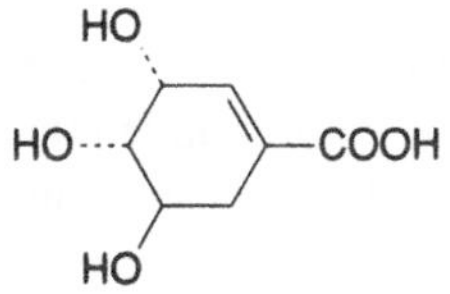

(9)

Shikimic acid

Shikimic acid (3,4,5-trihydroxy-1-cyclohexene-1-carboxylic acid) (9) is structurally closely related to quinic acid (8) and its metabolic fate in animals is similar to that described above for the latter compound. Asatoor[42] reported that the oral administration of shikimic acid to rats led to the urinary excretion of larger quantities of hippuric acid than those formed from quinic acid. Aromatization of shikimic acid was likewise suppressed when the animals were treated with neomycin. The alternative pathway leading to the formation of catechol is also seen with shikimic acid.[47,50]

A comprehensive investigation of the metabolism of shikimic acid in rats was carried out by Brewster et al.[51,52] and Brewster.[53] Using oral or i.p. dosage to normal animals or oral dosage to animals treated with antibiotics to suppress the intestinal microflora, they found that 93 to 100%

of the radioactivity was excreted within 24 h following administration of generally labeled [^{14}C]shikimic acid (100 mg/kg). However, the amounts of radioactivity excreted by the various routes (urine, feces, and expired air) differed greatly in the three groups. Following oral dosage to normal rats the amounts recovered were 40 to 57% (urine), 45 to 63% (feces), and 4 to 7% (as $^{14}CO_2$). After injection nearly all of the radioactivity was recovered in the urine in the form of unchanged shikimate. Conversely, antibiotic treatment resulted in the fecal excretion of most of the oral dose (mainly as unchanged shikimate).

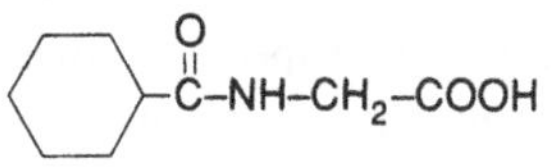

$$\text{C}_6\text{H}_{11}-\overset{\overset{\displaystyle O}{\|}}{\text{C}}-\text{NH}-\text{CH}_2-\text{COOH}$$

(10)

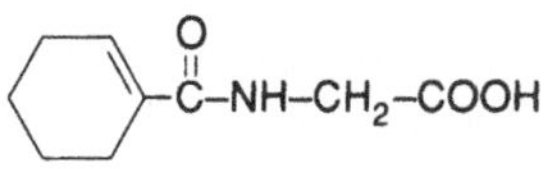

$$\text{C}_6\text{H}_9-\overset{\overset{\displaystyle O}{\|}}{\text{C}}-\text{NH}-\text{CH}_2-\text{COOH}$$

(11)

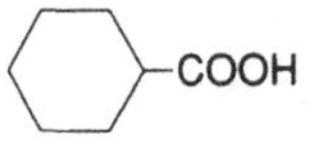

cyclohexane–COOH (12)

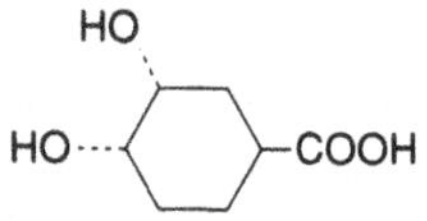

HO / HO··· cyclohexane–COOH (13)

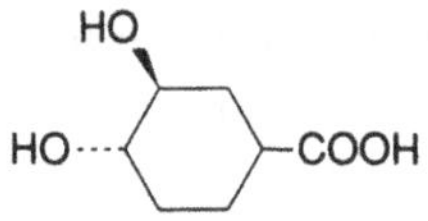

HO / HO··· cyclohexane–COOH (14)

Investigation of the above urinary fraction which contained 40 to 57% of the radioactivity indicated a fairly complex metabolic picture. However, hippuric acid was the major metabolite (21 to 30% of the dose). Additionally, small amounts of two partially aromatized glycine conjugates were also identified. These were hexahydrohippuric acid (10) and 3,4,5,6-tetrahydrohippuric acid (11). Hexahydrohippuric acid was reported by Balba and Evans[54] to be a urinary metabolite of shikimic acid in herbivores (sheep, cattle, horse, and elephant); however, they did not detect it in similar experiments in rats, pigs, or humans. Other urinary metabolites identified by Brewster et al.[52] and Brewster[53] were unchanged shikimate (8 to 12%), catechol (~1%), small amounts of the β-glucuronides of benzoic acid and cyclohexanecarboxylic acid (12), and 2 to 3% of a pair of novel metabolites shown to be *cis*- and *trans*-3,4-dihydroxycyclohexanecarboxylic acid ((13) and (14), respectively). The last two compounds were found to be the most prominent metabolites in the feces.

Significantly, it was shown that the intestinal bacteria effected the reduction of shikimic acid to cyclohexanecarboxylic acid (12). This reaction was demonstrated *in vitro* using mixed bacterial cultures from cecal contents or feces from the mouse, rat, guinea pig, rabbit, and man but not the ferret.[52,55] Balba and Evans[54] also reported this conversion using rat cecal contents or sheep rumen fluid. Brewster et al.[52] did not detect metabolites (13) and (14) in these incubates, however their findings indicated that they were products of bacterial metabolism. Although their structures suggest that they might be intermediates in the formation of cyclohexanecarboxylic acid, Brewster et al. believed that they were formed via a different pathway than that which leads to cyclohexanecarboxylic acid. It is possible that aromatization of metabolites (13) and (14) is involved in the formation of catechol.

Additionally, Brewster et al.[56,57] showed that rats and also the perfused rat liver were able to extensively aromatize cyclohexanecarboxylic acid to hippuric acid. The glycine conjugates (10) and (11) were also formed. The amounts of these two metabolites excreted in the urine varied from about a half of the hippuric acid excreted following a large dose (200 mg/kg) of cyclohexanecarboxylic acid to a sixth of that excreted after a small dose (0.5 mg/kg). Additional urinary metabolites were the ester glucuronides of benzoic acid and cyclohexanecarboxylic acid. Brewster et al. proposed therefore that the intestinal bacteria are responsible for the reduction of shikimic acid to cyclohexanecarboxylic acid which, following absorption, is partially or fully aromatized and conjugated with glycine in the tissues.

The aromatization of cyclohexanecarboxylic acid by liver enzymes by way of coenzyme A derivatives has been reported.[58,59] The reaction sequence from cyclohexanecarboxylic acid to hippuric acid was studied in detail by Svardal and Scheline[60] using subcellular fractions from guinea pig liver. They found that activity was localized exclusively in the mitochondria. Svardal and Scheline[61] reported that the reaction sequence showed twice the activity in whole homogenates of guinea pig liver than in those from rabbit liver. The activity in preparations from rats was only a fifth of that present in guinea pigs and only very weak activity was found in mouse liver. Further experiments[62] on the submitochondrial localization of the cyclohexanecarboxyl-CoA to hippuric acid conversion showed that it is present in the mitochondrial matrix. Nonetheless, several of the studies noted above have identified aromatic metabolites when quinic acid or shikimic acid was incubated with intestinal microorganisms. These findings suggest that alternative sites and sequences for the aromatization of these compounds may exist, although it seems clear that aromatization is predominantly a tissue reaction.

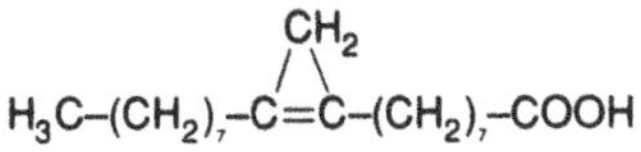

$$H_3C-(CH_2)_7-C=C-(CH_2)_7-COOH$$

with CH_2 bridge across the C=C

(15)

Sterculic acid

$$HOOC-(CH_2)_n-CH-CH-CH_2-COOH$$

with CH_2 bridge

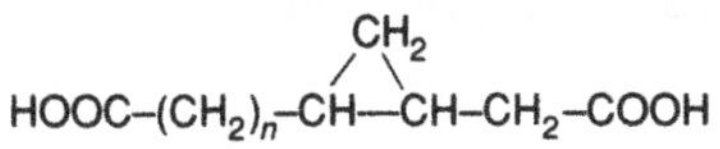

(16)

The biological disposition of **sterculic acid** (15) was studied in rats[63] and several of its urinary metabolites were identified.[64] Both of these investigations used sterculic acid labeled with ^{14}C in the 9,10-methylene bridge. After giving an intragastric dose of ~60 mg/kg dissolved in corn oil, less than 1% of the radioactivity was found in the expired CO_2 in 26 h. Increasing levels of radioactivity were incorporated into body fat and a value of 8% was recorded after 26 h. The main route of excretion was the urine (48% in 16 h) and a further 11% of the dose was found in the feces at that time. Interestingly, rats fed on a diet with an oil containing cyclopropene fatty acids for at least 60 d excreted the urinary metabolites at a higher rate than did rats fed the normal diet which included corn oil. The nature of the metabolites was the same in both cases. The main urinary metabolites ((16), $n = 1$-5) were found to be dicarboxylic acids of shorter chain length. All of these contained a cyclopropane ring which resulted from reduction of the double bond at C-9,10 and in most cases these products were formed in both *cis* and *trans* forms, i.e., different orientation of the groups around the cyclopropane ring. The most prominent of these metabolites was *cis*-3,4-methyleneadipic acid ((16), $n = 1$) and lesser amounts of its *trans* isomer were also formed. Thus, ω-oxidation of the alkyl group and β-oxidation of both side chains occurred in addition to reduction of the double bond. The corresponding derivative of suberic acid ((16), $n = 3$) was also excreted in fairly large amounts. Interestingly, metabolites with even numbers of methylene groups in the side chain were also identified. These derivatives of pimelic acid ((16), $n = 2$) and azelaic acid ((16), $n = 4$) were believed to be formed via initial ω-oxidation of the alkyl group followed by α-oxidation which would give an intermediate having $n = 6$.

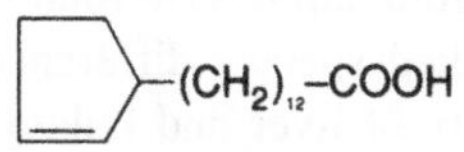

$$-(CH_2)_{12}-COOH$$

(17)

Chaulmoogric acid

Bernhard and Müller[65] fed small amounts of **chaulmoogric acid** (13-(2-cyclopenten-1-yl)tridecanoic acid) (17) to dogs and reported that it was well absorbed. However, no urinary metabolites were found.

AROMATIC ACIDS

Aromatic plant acids include aryl and aralkyl carboxylic acids, sometimes lacking ring substituents but generally containing hydroxy and/or methoxy groups. It is, in fact, these phenolic acids and their derivatives which we usually associate with the term aromatic plant acids. The general features of the metabolism of these compounds were noted at the beginning of this chapter and indicated that both tissue reactions and bacterial reactions may be involved. In the former case metabolic conjugation is the most important feature, the pathways involved utilizing a carbohydrate (usually glucuronic acid) or an amino acid (e.g., glycine, glutamine, or taurine). Numerous other metabolic possibilities exist when other functional groups are present. Prominent among these are O-methylation of catechols and β-oxidation of some aralkyl acids. In the case of bacterial metabolism, decarboxylation reactions are sometimes observed and ring substituents may also be metabolized (e.g., dehydroxylation or O-demethylation). Examples of these various metabolic routes are illustrated below in the summaries of the metabolic fate of particular aromatic plant acids.

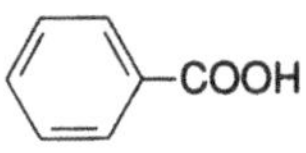

(18)   (19)

Benzoic acid   Hippuric acid

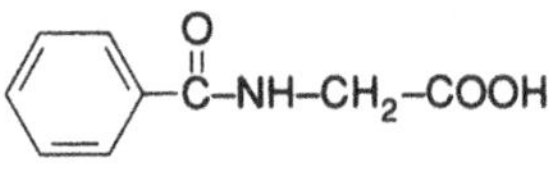

Conjugation reactions are primarily involved in the metabolism of **benzoic acid** (18). Conjugation with glycine to form hippuric acid (19) is paramount, however conjugation with glucuronic acid to form an ester glucuronide also occurs. In both cases a major point of interest has been the correlation of the extent of these conjugative pathways with the animal species used. Data on the conjugation of benzoic acid in numerous mammalian species are summarized in Table 1. These findings were obtained mainly from newer studies which usually employed ^{14}C-labeled benzoic acid. Williams[66] reviewed the earlier literature on benzoic acid conjugation.

The data in Table 1 show that hippuric acid formation is the dominant pathway in most mammalian species. It was found to be lacking only in the Indian fruit bat which forms mainly benzoyl glucuronide, however these animals also excrete ~10% of the dose (100 mg/kg, i.p.) as benzoyl-L-(+)-glutamic acid.[67] Ferrets are also strong producers of benzoyl glucuronide, especially at higher dose levels. Idle et al.[68] found that ferrets do not have the ability to conjugate benzoic acid with taurine. The conjugation of benzoic acid with carnitine was studied by Quistad et al.[69] who found that rats excreted in the urine 0.04% of an oral dose (2 mg/kg) of [*carboxy*-^{14}C]benzoic acid as benzoylcarnitine. Kao et al.[70] studied the metabolism of benzoic acid in two omnivores (rat and hamster) and two carnivores (ferret and dog) using hepatocytes and kidney tubule fragments. Hippuric acid formation occurred in both preparations from the omnivores. However, it was observed with the kidney preparations but not the hepatocytes from the carnivores. Small amounts of glucuronidation were found using hepatocytes from all species. The results indicated that the marked species differences in the patterns of benzoic acid conjugation are related to the ability of liver and kidney cells to carry out the conjugation reactions. The ability of isolated rat intestinal mucosal cells to conjugate benzoic acid with glycine is much lower than that found with isolated liver cells.[71]

Jones[72] investigated the metabolism of benzoic acid in rats over a wide range of dose levels (10 μg to 1000 mg/kg). No difference in the extent of metabolism was found and the animals excreted 80 to 100% of the dose in the urine in 24 h, solely as hippuric acid. Rats maintained on a protein-deficient diet show reduced excretion of hippuric acid.[73] Following an i.p. dose of 200 mg/kg, the normal output was reduced from 99 to 74% in 24 h. This reduction was compensated for by an increase in the excretion of benzoyl glucuronide to 25% of the dose compared with trace amounts in the normal group.

TABLE 1
Conjugation of Benzoic Acid in Various Species

Species	Dose (mg/kg, p.o. unless otherwise indicated)	% of dose excreted in 24-h urine	Benzoic acid	Hippuric acid	Benzoyl glucuronide	Ref.
				% excreted in 24 h as		
Rodents						
Mouse	56	55	tr[a]	95	5	274
	100, i.p.	100	3	70	24	275
Rat	50	100	1	99	tr	274
	50 mg/rat			47		45
	100, i.p.	97	9	75	13	275
	185			88		40
Gerbil	29	75	2	98	0	274
Hamster	52	99	1	97	1	274
Lemming	56	98	tr	100	0	274
Guinea pig	49	79	tr	98	3	274
Other						
Indian fruit bat	50	49,54	12,30	tr	88,70	274
	50 mg/bat, i.p.	98,83	10,19	<0.01	89,81	276
	100, i.p.	69—96	0—3[b]	0	84—95	67
Rabbit	49	60	0	100	0	274
	200	86	tr	98	2	274
	92			79[c]		91
	500	94	1	84	15	277
European hedgehog	50	67,78	5,7	76,86	19,7	274
Pig	50	48,51	15,7	85,93	tr	274
	500—800	93	7	61	32	278
Sheep	200[d]			75		279
Horse	5	~100	1	95,98	1—3	280
Elephant	100	64	9	90		281

TABLE 1 (continued)
Conjugation of Benzoic Acid in Various Species

Species	Dose (mg/kg, p.o. unless otherwise indicated)	% of dose excreted in 24-h urine	% excreted in 24 h as Benzoic acid	Hippuric acid	Benzoyl glucuronide	Ref.
Carnivores						
Ferret	50	69	9	70	22	274
	198	78	9	47	44	274
	200	67	22	30	49	274
	100	86	1	40	59	87
	100, i.p.	86	2	34	51	68
Cat	51	29,86	tr	100	0	274
Dog	51	94	0	82	18	274
Forest genet	75	54,79	28,23	67,75	0	282
African civet	75	35,44	17,0	77,95	0	282
Lion	75	92	15	84	0	282
Primates						
Capuchin	50	57	0	100	tr	274
Squirrel monkey	50	46,49	14,18	81,83	5, tr	274
Marmoset	40	69	39	33	27	116
	100	58	19	42	38	116
Rhesus monkey	20	47	0	100	0	274
Man	1	100	0	100	0	274
	20	[e]		62		283
	42	[e]		50—85	5	284
	43	[e]				277
	50	[e]		40		283

[a] (tr) Trace.
[b] 5 to 15% also excreted as benzoyl-L-(+)-glutamic acid.
[c] Percent of total dose.
[d] 24-h Infusion into abomasum.
[e] 4-h Excretion.

Only a small percentage of the benzoic acid absorbed from the intestinal tract appears in the lymph.[74] They recorded a value of 1.4% of the dose (~0.6 mg/kg, intraduodenally) after collecting the thoracic duct lymph for 24 h. Hippuric acid was the most abundant metabolite in the lymph, however fairly large amounts of unchanged compound and small amounts of benzoyl glucuronide were also detected. The bile is similarly a minor excretory route for benzoic acid in rats. Hirom et al.[75] found that only ~1% of the dose (50 mg/kg, i.p.) was excreted in the bile in 3 h, whereas 90% was recovered in the urine in this period.

Decarboxylation is a possible metabolic pathway of benzoic acid, however Bernhard et al.[40] reported that virtually no radioactivity was detected in the expired air of rats given the acid labeled with ^{14}C in the carboxy group. Jones[72] reported that rats converted 3% of a small oral dose (10 µg/kg) of [*carboxy*-^{14}C]benzoic acid to radioactive CO_2 within 24 h. Hydroxylation is another possible reaction. Bray et al.[76] reported that benzoic acid did not form an ethereal sulfate, a finding which would preclude its metabolism by hydroxylation to a phenolic acid. However, Sato et al.[77] found that rat liver slices were capable of metabolizing small amounts of benzoic acid to the sulfate conjugate of 4-hydroxybenzoic acid. This subject was reinvestigated by Acheson and Gibbard[78] who found that rats given [*carboxy*-^{14}C]benzoic acid (500 mg/kg, i.p.) excreted ~0.25% of the dose as 2-, 3-, and 4-hydroxybenzoic acids. At about one-tenth of this dose level, only ~0.04% of the radioactivity was due to hydroxylated metabolites and the 2-hydroxy isomer was not present. This latter pattern was also observed in guinea pigs given a dose of ~200 mg/kg and the amount excreted was similarly ~0.04% of the dose. Thus, hydroxylation is a very minor metabolic reaction with benzoic acid.

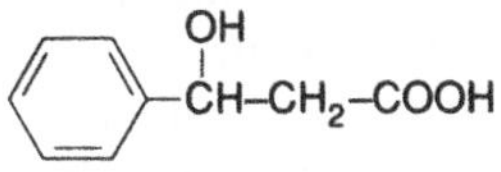

(20) 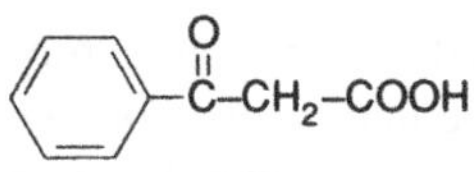   (21)

An additional reaction of benzoic acid in the horse is the addition of a two-carbon fragment.[79] This unusual reaction, which accounted for 2% of the dose (~5 mg/kg, intragastrically), resulted in the excretion in the urine of 3-hydroxy-3-phenylpropionic acid (β-phenylhydracrylic acid) (20) and smaller amounts of 3-keto-3-phenylpropionic acid (21). It is noteworthy that compound (20) has been shown to be metabolized to hippuric acid in rats[80] and dogs.[81]

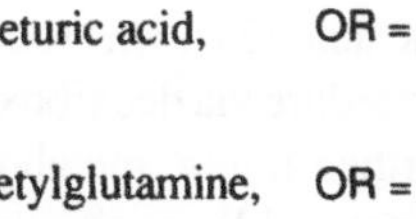

(22)

(a) Phenylacetic acid,   OR = OH

(b) Phenaceturic acid,   OR = NH–CH$_2$–COOH

(c) Phenacetylglutamine,   OR = NH–CH–(CH$_2$)$_2$–C–NH$_2$ with COOH and O substituents

(d) Phenacetyltaurine,   OR = NH–CH$_2$–CH$_2$–SO$_3$H

Studies on the metabolism of **phenylacetic acid** (22a) show similarity to those dealing with benzoic acid, partly because interest has been focused on conjugation reactions and partly because investigations dealing with phenylacetic acid span a very long period of time. However, little advantage is to be gained by summarizing the earlier findings which are well covered in the review by Williams.[82] Instead, several more recent reports including the extensive comparative study of James et al.[83] form the basis of the present summary. These studies have considerably clarified the uncertainty formerly present about the conjugation of phenylacetic acid. Nonetheless, the early reports by Thierfelder and Sherwin[84,85] must be noted as they showed that phenylacetic acid was conjugated in man with glutamine rather than glycine as is seen in most other mammals. The species variations in phenylacetic acid conjugation are clearly shown in Table 2 which summarizes the major findings of James et al.[83] as well as the results of other investigations as noted. Several points of interest emerge from these data. The prominence of the pathway involving conjugation with glutamine which gives rise to phenacetylglutamine (22c) appears suddenly in the evolutionary scale with the New World monkeys, being completely absent in the two prosimian species studied. However, conjugation with glutamine is not completely absent in lower mammals and has been reported in rats[72] and ferrets.[68,86] Idle et al.[87] first reported appreciable urinary material from ferrets which was later shown to be mainly phenacetylglutamine. Hirom et al.[86] found that ferrets excreted ~5 to 15% of the dose as this conjugated material over a wide range of doses (0.01 to 400 mg/kg). Most of this was conjugated with glutamine, however several other amino acids were also involved.

The switch to conjugation with glutamine noted above is concomitant with the reduction in the extent of glycine conjugation. The latter reaction, which predominates in lower mammals, forms phenaceturic acid (22b) and shows low to intermediate values in the New World monkeys, values of 1% or less in the Old World monkeys and is essentially absent in man. Also noteworthy is the detection of a new conjugate, phenylacetyltaurine (22d), in all of the mammalian species except bats. The extent of this conjugative pathway is considerable in several species, especially the ferret and some monkeys. The results in Table 2 show that conjugation of phenylacetic acid with glucuronic acid does not occur in most species. However Bray et al.[76] reported that 5% of the dose was excreted in this form following large (0.75 g/kg) oral doses to rabbits.

Phenylketonurics, who produce larger than normal amounts of phenylacetic acid, were found by James and Smith[88] to have a normal glutamine conjugation mechanism which was able to deal with these increased levels.

As with benzoic acid, limited aromatic hydroxylation occurs with phenylacetic acid. Idle et al.[68] reported that ferrets excreted ~2% of the dose (100 mg/kg, i.p.) as 4-hydroxyphenylacetic acid. Decarboxylation is also a minor metabolic pathway. Jones[72] found that ~1% of a small oral dose (10 μg/kg) of [1-^{14}C]phenylacetic acid was excreted as respiratory $^{14}CO_2$ within 24 h by rats.

$$\langle\!\!\!\!\bigcirc\!\!\!\!\rangle\text{--CH=CH--COOH}$$

(23)

Cinnamic acid

Investigations using several animal species make it clear that the principal metabolic product of **cinnamic acid** (23) is hippuric acid (19). This was reported by Dakin[89] who administered the acid (0.25 to 0.5 g/kg, s.c.) to cats and dogs. In addition, a small quantity of an intermediate oxidation product, 3-keto-3-phenylpropionic acid (21), was found as was some acetophenone which no doubt arose during the isolation procedure via decarboxylation of the keto acid. Dakin reported that cinnamoylglycine was a further minor metabolite, however no unchanged cinnamic acid was excreted in the urine. Raper and Wayne[90] administered cinnamic acid (115 mg/kg, s.c.) to dogs and found that 74% of the dose was excreted as hippuric acid. Extensive conversion is also seen in rabbits and El Masry et al.[91] reported a value of 74% following an oral dose of 300 mg/kg. They did not detect the cinnamoylglycine reported by Dakin in cat and dog urine in the urine of rabbits given cinnamic acid. This negative finding was confirmed by Fahelbum and James[92] who also reported that 60% of the dose (~150 mg/kg) of cinnamic acid was excreted in the urine of rabbits as hippuric acid. Trace amounts of *p*-hydroxyhippuric acid were also detected. The urinary excretion of hippuric acid by rats receiving cinnamic acid was

TABLE 2
Conjugation of Phenylacetic Acid in Various Species

Species	Dose (mg/kg)	% of dose excreted in 24-h urine	% excreted in 24 h as					Ref.
			Phenylacetic acid	Glycine conjugate	Glutamine conjugate	Taurine conjugate	Glucuronic acid conjugate	
Rodents								
Mouse	80, i.p.	54	32	56	—[a]	7	6	83
Rat	80, i.p.	95	—	99	—[a]	1	—	83
Hamster	0.01—1,000, p.o.	70—95	52	~65—85	~6—9	1	—	72
Guinea pig	80, i.p.	63	5	94	—	1	—	83
Other								
Indian fruit bat	~15/bat, i.p.	27—81	6	76	—	—	12	285
Vampire bat	80, i.p.	75	—	100	—	—	—	83
Rabbit	80, i.p.	85	2	97	—	1	—	83
Elephant	100, p.o.	31 (6 h)	—	100 (6 h)	—	—	—	281
Carnivores								
Ferret	80, i.p.	95	3	43	b	32	22	83
	100, i.p.	68	3	63	b	21	—	87
	100, i.p.	68	2	43	b	14	—	68
	100, i.p.	68	3	63	b	21	—	86
	100, p.o.	89	4	64	b	11	6	86
Cat	80, i.p.	75	1	98	—	1	—	83
Dog	80, i.p.	81	—	94	—	4	2	83
Hyena	25, p.o.	13	13	87	—	—	—	83
Prosimians								
Bushbaby	80, i.p.	69	—	87	—	13	—	83
Slow loris	80, i.p.	82	20	69	—	10	—	83
New World Monkeys								
Squirrel monkey	50, i.m.	67	4	2	75	18	—	83
Capuchin[c]	80, i.p.	28	4	10	64	20	—	83
Capuchin[d]	80, i.p.	34	4	23	29	44	—	83
Marmoset	80, i.m.	71	5	0.8	79	0.4	—	83

TABLE 2 (continued)
Conjugation of Phenylacetic Acid in Various Species

Species	Dose (mg/kg)	% of dose excreted in 24-h urine	% excreted in 24 h as					Ref.
			Phenylacetic acid	Glycine conjugate	Glutamine conjugate	Taurine conjugate	Glucuronic acid conjugate	
Old World Monkeys								
Rhesus monkey	80, i.p.	74	55	1	32	23,1[e]	—	83
Cynomolgus monkey	80, i.p.	36,79[e]	42,5[e]	1	56,90[e]	2,4[e]	—	83
Green monkey	25, i.m.	87	12	0.5	79	4	—	83
Red bellied monkey	8, i.m.	45	8	1	87	3	—	83
Mona monkey	8, i.m.	15	45	1	32	21	—	83
Mangabey	8, i.m.	79	55	0.5	31	7	—	83
Drill	8, i.m.	52	65	0.4	28	7	—	83
Baboon	2, i.m.	100	5	0.1	85	10	—	83
Man	1, p.o.	98	—	<0.05	93	7	—	83
Man	~1, p.o.	96—100	—	—	90—98			88

[a] (—) Not detected.
[b] See text.
[c] *Cebus albifrons.*
[d] *Cebus nigrivittatus.*
[e] Separate values shown due to large individual differences.

reported to be 44% of the dose (50 mg, i.p.) in 24 h[45] and 67% of the dose (50 mg, p.o.).[92] Similar experiments by Teuchy and Van Sumere[93] with [3-^{14}C]cinnamic acid resulted in ~48% of the radioactivity being recovered in the urine in 24 h and most of this (39% of the dose) was due to hippuric acid. Smaller amounts of benzoic acid and cinnamic acid were also found but no phenolic acid metabolites were detected. Not unexpectedly, no radioactivity was detected in the expired air. Bhatia et al.[94] administered [1-^{14}C]cinnamic acid to rats and recovered ~47% of the radioactivity in the urine in 4 d following an oral dose of 250 mg/kg. Interestingly, several phenolic metabolites were identified in this study. These included the 4-hydroxy-, 3,4-dihydroxy-, and the isomeric monomethyl ethers of the 3,4-dihydroxy derivatives of cinnamic acid, i.e., *p*-coumaric, caffeic, ferulic, and isoferulic acids. Several other urinary metabolites of cinnamic acid were detected and, of these, benzoic acid, 4-hydroxybenzoic acid, and hippuric acid were identified. No additional glycine conjugates were found. The possibility that cinnamic acid or its metabolites may also be conjugated with glucuronic acid has not received much attention, however Quick[95] reported that a larger proportion of the metabolically formed benzoic acid was conjugated with glucuronic acid than with glycine in the dog. More information on this point was obtained by Fahelbum and James in the study noted above using rats and rabbits. They found that 3 and 10%, respectively, of the dose were excreted in the urine of these two species as material conjugated with glucuronic acid.

The metabolism of cinnamic acid in man was studied by Snapper et al.[96] who employed oral doses (6 g). They reported that 50 to 75% of the dose was rapidly (4 h) excreted as hippuric acid. Additionally, a small amount (~3 to 6%) was excreted as cinnamoyl glucuronide. Snapper and Saltzman,[97] in similar experiments, confirmed the excretion of a few percent of the dose as cinnamoyl glucuronide. However, no benzoyl glucuronide was excreted, in contrast to that noted above in the dog. Experiments in man were repeated by Hoskins et al.[98] using more modern analytical methods. Using a dose of sodium cinnamate equivalent to 435 mg of cinnamic acid, they found that 87% of the dose was excreted in the urine as hippuric acid. They concluded that the glucuronide conjugate noted above represented 1% or less of the dose. Also, some free benzoic acid was excreted but only traces of free cinnamic acid were detected.

Cinnamic acid, in contrast to that found with the closely related cinnamaldehyde and cinnamyl alcohol, was not excreted by rats as *N*-acetylcysteine (mercapturic acid) derivatives.[99] While side-chain reduction of numerous phenolic cinnamic acid derivatives by intestinal bacteria is known to occur,[100] this reaction with cinnamic acid itself has received little attention. However, Hansen and Crawford[101] reported that reduction to 3-phenylpropionic acid occurred when cinnamic acid was incubated aerobically or anaerobically with a suspension of monkey (*Cercopithecus aethiops*) feces. The metabolism of cinnamic acid and related compounds was reviewed by Hoskins.[102]

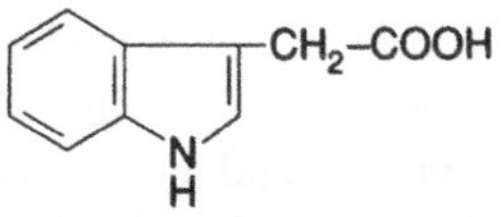

(24)

Indole-3-acetic acid

Another unsubstituted aromatic acid is **indole-3-acetic acid** (heteroauxin) (24). Erspamer[103] found that rats excreted it partly unchanged and partly as the glycine conjugate, indoleaceturic acid. A comprehensive study of the metabolism of indole-3-acetic acid in 17 mammalian species was carried out by Bridges et al.[104] This investigation, which employed most of the species used in the similar study with phenylacetic acid summarized above,[83] showed appreciable urinary excretion of unchanged compound in most species. The glutamine conjugate was formed in man and monkeys but not in prosimians, carnivores, rodents, or rabbits. In these latter species and also in New World monkeys, the glycine conjugate was found whereas this pathway was absent in man and Old World monkeys. Conjugation with taurine, which was noted above as a common metabolic route with phenylacetic acid, was prominent in ferrets and also in some of the monkeys. Interestingly, conjugation with glucuronic acid was observed only in man, in which case it amounted to 20 to 30% of the dose.

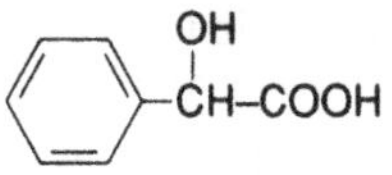

(25)

Mandelic acid

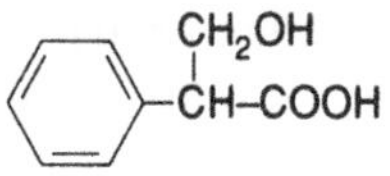

(26)

Tropic acid

Two simple aromatic hydroxy acids from plants are **mandelic acid** (25) and its higher homologue **tropic acid** (26), although the latter compound is more correctly a hydrolysis product obtained from some tropane alkaloids (e.g., atropine) rather than a true plant constituent. An important factor governing the metabolic fate of these acids is their acidity (pKa values of 3.4 and 4.1, respectively) which results in their extensive ionization at physiologic pH values. Accordingly, the most notable feature of their fate in the body is their extensive excretion in the unchanged state. Many of the investigations in this area are of an early date and a summary of this data was given by Williams.[105,106] While some of this information is conflicting, it was reported that mandelic acid may be dehydrogenated to the corresponding α-keto acid, phenylglyoxylic acid (benzoylformic acid), in dogs and man. This metabolite was reported by Ohtsuji and Ikeda[107] to be a urinary metabolite of mandelic acid in rats. This was confirmed by Drummond et al.[108] who administered (*R*)-, (*S*)-, and racemic mandelic acid orally to rats. They found that the urinary excretion of phenylglyoxylic acid was greatest when (*S*)-mandelic acid was given. Nearly 80% of the dose (100 mg/kg) was found as phenylglyoxylic acid in the latter case, compared with only 46% following administration of racemic mandelic acid. When phenylglyoxylic acid was administered, little (1%) was excreted as mandelic acid. These experiments also demonstrated that the chiral inversion of (*S*)-mandelic acid to the (*R*)-enantiomer occurred. Most of the mandelic acid excreted in the urine after dosing with the (*S*)-isomer was in the (*R*)-form. Contrariwise, administration of the latter resulted in its excretion unchanged. An additional reaction of mandelic acid in the rat was reported to be formation of hippuric acid.[107] However, this reaction of chain-shortening to a $C_6–C_1$-derivative was not detected in rabbits by El Masry et al.[91] who did not find increased urinary hippuric acid levels following the administration of racemic mandelic acid.

Gosselin et al.[109] included [^{14}C]tropic acid in their metabolic study of atropine in mice and rats. This preparation was labeled in the α-position and when injected (1 mg/kg, i.p.) it was excreted unchanged to the extent of 95 to 98% in 2 h, no radioactivity being found in the expired air. Ve and Scheline,[110] checking specifically for the ability of rats given tropic acid (100 or 400 mg/kg, p.o.) to oxidize the alcohol moiety, found only unchanged compound in the urine and no evidence for the dicarboxylic acid product, phenylmalonic acid.

The major part of this review of the metabolism of aromatic acids will be devoted to phenolic acids and their various ether derivatives, starting with the substituted benzoic acids ($C_6–C_1$-phenolic acids) and continuing with the higher homologues, of which the cinnamic acids ($C_6–C_3$-phenolic acids) are the most prominent. It is evident that due to the presence of both carboxy and phenolic hydroxy (or methoxy) groups, numerous pathways of metabolism are available to the phenolic acids. Conjugation of the carboxy group with an amino acid or with glucuronic acid is possible and the latter reaction, as well as ethereal sulfate formation, may occur with the hydroxy group. Ring hydroxylation and both *O*-methylation and *O*-demethylation reactions occur, the latter often carried out by intestinal bacteria. These microorganisms are also responsible for the dehydroxylation and decarboxylation of certain phenolic acids as well as double bond reduction in cinnamic acids. The reverse of the latter reaction, dehydrogenation of

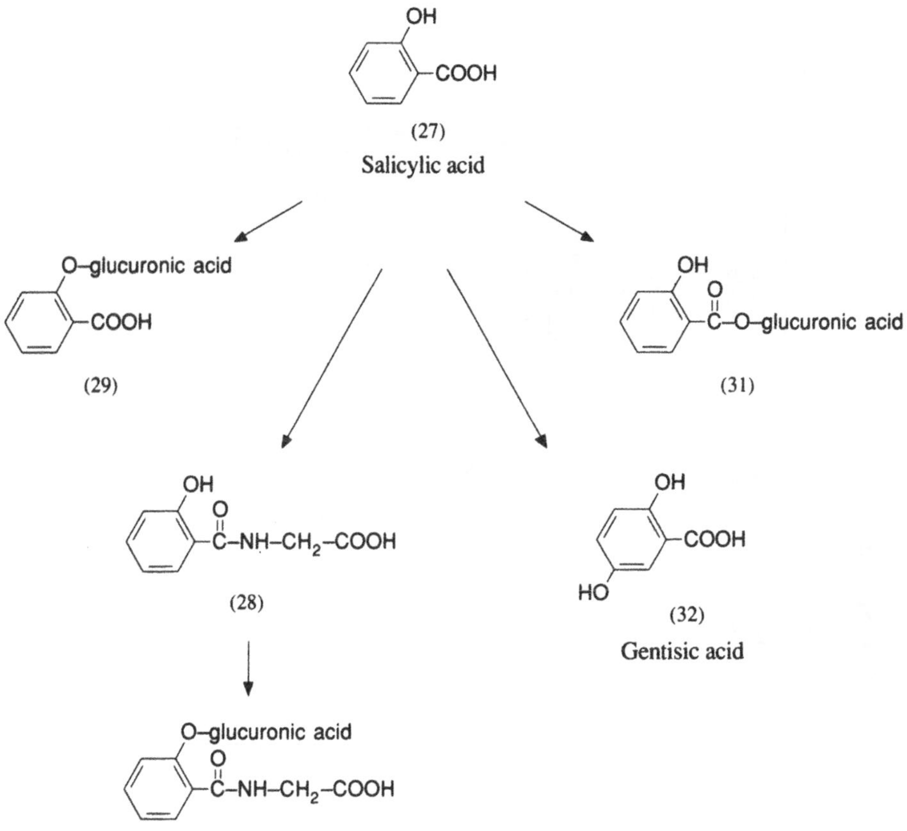

FIGURE 1. Main metabolic pathways of salicylic acid. See text for description of additional minor or occasional metabolites.

C_6–C_3-acids, is a tissue reaction. Our knowledge of this network of interrelated metabolic pathways has expanded considerably during the past few decades and the following summaries of the metabolism of individual phenolic acids will give special consideration to this newer data.

The simplest plant phenolic acids are **salicylic acid** (2-hydroxybenzoic acid) (27) (see Figure 1) and 4-hydroxybenzoic acid. In view of the important medical uses of the former compound, it is hardly surprising that an abundant literature on its metabolic fate is available. A detailed discussion of this subject falls outside the scope of this book and the present summary aims primarily to illustrate the known metabolic pathways of salicylate in various animal species. As the pKa value of salicylic acid is ~3, it is understandable that appreciable urinary excretion of unchanged compound occurs. This is seen in Table 3 which brings together quantitative data on the excretion of salicylic acid and its metabolites in several animal species. Additional data on the metabolism of salicylate in man may be garnered from studies which employed aspirin, its acetylated derivative. Hutt et al.[111] gave a single oral dose of [*carboxy*-^{14}C]aspirin equivalent to 0.77 g of salicylate to four subjects. The mean recovery of radioactivity in the urine was 96% in 24 h. The major metabolite was salicyluric acid (28) (64% of the dose) followed by free salicylic acid (14%), phenolic glucuronides of salicylic acid (29) and salcyluric acid (30) (6% each), ester glucuronide of salicylic acid (31) (4%), and gentisic acid (32) (1%). The results from the four individuals were generally similar; however, Caldwell et al.[112] found large differences in values for the metabolites in a group of 85 individuals of both sexes. Using a single oral dose of aspirin equivalent to 0.69 g of salicylate, they recovered a mean of 66% of the dose in the urine within 12 h. The mean value for salicyluric acid was 39% and the range was 6 to 72%.

TABLE 3
Metabolism of Salicyclic Acid in Various Species

Species	Dose	Urinary excretion period (h)	% of dose excreted as					Ref.
			Salicylic acid	Salicyluric acid	Salicyl ether glucuronide	Salicyl ester glucuronide	Gentisic acid	
Mouse	150—250 mg/kg, i.p.	24					8	153
Rat	~ 330 mg/kg, p.o.	24	41—63	<1	1—3	0	18—34 + <1 as glucuronide	286
	5 mg/kg, i.p.	24	29	25	33[a]		8 + 3 as glucuronide	120
	50—200 mg/kg, i.p.	24	25—35	4—7	32—44[a]		10—17 + 5—15 as glucuronide	120
Rabbit	0.1—1.5 g/kg, p.o.	24	85	~5			<4	132
	0.25—0.5 g/kg, p.o.	24	38	0	5—14	3—4	4—5	122
Dog	38 mg/kg, i.v.	24	50	29	33[a]	0	small amount	119
	1 g, i.v.	30—36		10	25		4—5	123
Goat	38 mg/kg, i.v.	24	46	41	12		0	119
Pig	38 mg/kg, i.v.	24	46	31	23		small amount	119
Horse	4—20 g, i.v.	3—24	~100	0			<0.5	118
	38 mg/kg, i.v.	24	65	25	10[a]			119
	35 mg/kg, p.o.	24	94	0.5	2	0.2	small amount in male	280
Rhesus monkey	9 mg/kg, i.v.	16	20	72			2—3	287
Man	2—3 g daily	24—36		55	25[a]		4—8	129
	1 g, i.v.		10—85	0—50	12—30	0—10	not >1	123
	1 g, p.o.		10[b]	69	21[a]			117
	~1.5 g, p.o.	24					3	153

[a]   Total glucuronide.
[b]   See text.

Conjugation with glucuronic acid (22%, range = 0.1 to 90%) and excretion of free salicylic acid (7%, range = 1 to 18%) were also highly variable. Montgomery et al.,[113] who also used a similar dose of aspirin as the source of salicylate, reported that age or sex have little influence on the urinary recovery of salicylate and its metabolites in man. However, Emudianughe et al.[114] reported different results when black Nigerian subjects were given an oral dose (1 g) of salicylic acid. Their test group included 78 females and 44 males aged 18 to 32 years and the results showed significant differences between the two sexes in the excretion of salicyluric acid and salicylic acid glucuronide. The total salicylate excretion in 12 h for all subjects was 52% (48% in females and 61% in males). The mean values for salicyluric acid excretion during this period were 19% for females and only 6.5% for males. In contrast, the corresponding values for glucuronic acid conjugates were 24 and 44%. Similar findings were presented in a subsequent report on a smaller group of individuals.[115] These results, when compared with previous data from other sources which employed Caucasian subjects, suggest that the latter have a higher capacity for glycine conjugation than that found in black Nigerians. Additional data on salicylate metabolism following the administration of aspirin was presented by Hall and James[116] for rats and the marmoset (*Callithrix jacchus*). The main urinary metabolite in both species was unchanged salicylic acid.

When considering the values for the excretion of free salicylate it is important to note that these are closely related to the urinary pH. In the investigation of Hollister and Levy[117] the mean value of 10% shown in Table 3 includes values which range from 2 to 26%. It was noted that the individuals giving the lowest values had a consistently low urine pH (~5) whereas the opposite was true in the individual giving the highest value. In the latter case the urinary pH values generally ranged between 6 and 7. The exceptionally high excretion of unchanged compound found by Schubert[118] in the horse may be explained by the high urinary pH (range 6.9 to 8.1) of these animals. However, the data of Davis and Westfall[119] indicate that urinary pH is not the sole factor influencing the extent of free salicylate excretion as the value found in horses (urine pH 7.6) was ~50% greater than that in goats (urine pH 8.2). Salicylic acid excretion in the lymph of rats has also been studied.[74] After a small (0.7 mg/kg) intraduodenal dose of ^{14}C-labeled compound, the thoracic duct lymph collected for 24 h contained 4% of the radioactivity. Most of this material was unchanged salicylic acid, however small amounts of salicyluric acid and gentisic acid were also detected.

The main metabolic transformations of salicylic acid are illustrated in Figure 1. Not shown is the formation of an ethereal sulfate of salicylic acid which was reported by Haberland et al.[120] to be excreted by rats. Hall and James[116] believed that this compound was a minor urinary metabolite in rats and marmosets given aspirin. However, it was reported not to be formed in rabbits[121,122] and in dogs and man.[123] Sato et al.,[77] using rat liver supernatant, found no formation of the ethereal sulfate of salicylic acid. When liver slices were employed, three sulfate conjugates were detected but these were of gentisic acid and of other metabolites rather than of salicylic acid. Another possible metabolic pathway which is not encountered is decarboxylation. Schayer[124] and Alpen et al.,[125] using salicylate labeled with ^{14}C in the carboxy group, did not find any $^{14}CO_2$ in the expired air of rats. Scheline[126] showed that rat cecal bacteria which are capable of decarboxylating numerous phenolic benzoic acids did not carry out this reaction with salicylic acid. Similar negative results were reported by Martin[48] when salicylic acid was infused into the rumen of sheep. Also, sheep rumen bacteria did not dehydroxylate it to a nonphenolic aromatic acid.[127]

A reported metabolite of salicylic acid about which some uncertainty exists is the uraminsalicylic acid first isolated from dog urine by Baldoni.[128] This compound was also reported by Kapp and Coburn[129] to be excreted by man and their data indicated that it was a compound consisting of glycine, salicylic acid, and gentisic acid. According to Haberland et al.,[120] uraminsalicylic acid is a molecular complex of gentisic acid and salicyluric acid. However, Alpen et al.[123] did not detect it in the urine of dogs or man given salicylic acid. Davis and Westfall[119] and Davis et al.[130] detected an unidentified metabolite of salicylic acid which they believed was uraminsalicylic acid. It was found only in the urine of adult cats and newborn dogs and pigs and was converted by acid hydrolysis to salicylic and gentisic acids and an amino acid. These investigators believed that the metabolite is formed in situations in which competing pathways are deficient. Their data showed that plasma salicylate levels declined much more slowly in cats than in the other species studied (dogs, goats, pigs, and horses). A related point of interest is the finding of Wilson et al.[131] that the glycine conjugate of gentisic acid (gentisuric acid) was found in most urines from patients treated with acetylsalicylic acid (aspirin). Other experiments indicated that gentisuric acid could be formed along two pathways. It was found both following incubation of gentisic acid and glycine with rat and beef liver mitochondrial preparations and after incubating salicyluric acid with rat liver microsomes.

The data shown in Table 3 indicate that hydroxylation of salicylic acid generally occurs to a small extent. This reaction takes place in the position *para* to the hydroxy group, forming gentisic acid (32), however the *o*-hydroxylated product, 2,3-dihydroxybenzoic acid, was reported by Bray et al.[132] to be formed in trace amounts in rabbits. Grootveld and Halliwell[133] found that it is also a minor metabolite of acetylsalicylic acid in man following an oral dose (1.2 g). Its concentrations in plasma and urine were much lower than those of gentisic acid. Formation of the 2,3-dihydroxy compound was also shown to be carried out by isolated rat kidney mitochondria.[134] Dumazert and El Ouachi[135] found that both gentisic acid and 2,3,5-trihydroxybenzoic acid were formed from salicylic acid in rabbits and man. The trihydroxy derivative was conjugated with glucuronic acid in rabbits and with glycine and glucuronic acid in man. Mitoma et al.[136] reported the *in vitro* formation of gentisic acid from salicylic acid by the rabbit liver microsomal system requiring NADPH and O_2. Table 3 also shows that formation of the glucuronic acid conjugate of salicylic acid may take place at the hydroxy or the carboxy group, giving the ether glucuronide (29) or ester glucuronide (31), respectively. While many of the results indicated the total excretion of salicyl glucuronide, those which differentiated between the two forms reveal that the ether glucuronide predominates. Both glucuronides were isolated from urine by Robinson and Williams.[137]

Although falling outside the scope of this book, the pharmacokinetics of salicylate metabolism and elimination in animals and man has received a great deal of attention. Much useful information in this area was reviewed by Levy and Leonards,[138] Davison,[139] and Hucker et al.[140]

4-Hydroxybenzoic acid contains the same functional groups as does salicylic acid and it is therefore not surprising that it too is metabolized to glycine and glucuronic acid conjugates and to a hydroxylated derivative. In addition, the presence of the phenolic hydroxy group in the 4- rather than the 2-position allows for some ethereal sulfate formation and also for metabolism by the bacteria of the gastrointestinal microflora. The quantitative data available are more limited with the 4-isomer and were obtained mainly using rabbits. The major findings from these studies are shown in Table 4. The isolation of both the ether and ester glucuronides from the urine of rabbits given 4-hydroxybenzoic acid was reported by Tsukamoto and Terada.[141]

The metabolism of 4-hydroxybenzoic acid in rats appears to be qualitatively similar to that seen in rabbits. Booth et al.[142] detected unchanged compound and relatively large amounts of its ethereal sulfate in rat urine. Similar experiments in rats by Derache and Gourdon[143] showed the formation of these two compounds as well as the glycine conjugate, the ether and ester glucuronides, and an unidentified metabolite. Kiwada et al.[144] gave rats small (2 mg/kg) doses intraduodenally or intravenously using [*carboxy*-^{14}C]-4-hydroxybenzoic acid. Excretion of radioactivity was rapid and the urine and bile contained 95 to 96% and 4 to 5%, respectively, of the dose in 5 h following both routes of administration. In contrast with earlier results which showed appreciable excretion of unchanged compound after giving fairly large doses, only 3% of the dose was recovered unchanged in the urine in this study after intraduodenal dosage. The major urinary metabolite was 4-hydroxyhippuric acid which accounted for 52% of the dose. The

TABLE 4

Metabolites of 4-Hydroxybenzoic Acid in Rabbits

Dose, p.o. (mg/kg)	Urinary excretion period (h)	% of dose excreted as						Ref.
		4-Hydroxy-benzoic acid	Glycine conjugate	Ether glucuronide	Ester glucuronide	Ethereal sulfate	3,4-Dihydroxy-benzoic acid	
365	48					7		121
100—1,500	24	30—60	10—30	0—19	2—16	4—8		288
250—350	24			18	[a]	9		289
100—1,500	24						<4	132
400	24	52	23	8	7	5		290
800	24	59	13	4	8	5		290

[a] Little, if any, present.

ester glucuronide (24%) and ethereal sulfate (9%) of 4-hydroxybenzoic acid were also excreted. All of these compounds were found in the bile in relative amounts which reflected their abundance in the urine.

In cats, oral doses (13, 26, and 260 mg/kg) of 4-hydroxybenzoic acid resulted in its metabolism to a single urinary metabolite, 4-hydroxyhippuric acid.[145] Nearly the entire dose was recovered in the urine, mainly during the first 24 h. Quick[146] isolated this glycine conjugate from the urine of dogs fed 4-hydroxybenzoic acid (~300 to 500 mg/kg) but, interestingly, the major metabolite isolated was reported to be the diglucuronide conjugate. Quick found that the phenolic acid was excreted partly unchanged and partly conjugated with glycine in man. Tompsett[147] also obtained evidence for this latter reaction in man. An *in vitro* study of the metabolism of 4-hydroxybenzoic acid using rat liver preparations was carried out by Sato et al.[148] The ethereal sulfate was formed by the supernatant fraction whereas liver slices produced this metabolite, the ethereal sulfate of 3,4-dihydroxybenzoic acid and a further product thought to be the ethereal sulfate of 4-hydroxyhippuric acid.

As noted above, the presence of the phenolic hydroxy group *para* to the carboxy group in 4-hydroxybenzoic acid furnishes a substrate which may undergo degradation by the gastrointestinal bacteria. Mixed bacterial cultures from the rat cecum extensively decarboxylate this acid to phenol[126] and the same is true with human fecal bacteria.[149] In a survey which included many common intestinal bacteria, Soleim and Scheline[150] found that only *Aerobacter aerogenes* was capable of carrying out this reaction. Martin[44,48] reported that 4-hydroxybenzoic acid, when infused into the rumen of sheep, was extensively decarboxylated and excreted in the urine as phenol. More than half of the amount infused (10.9 g/d) was recovered in the urine as phenol and a further 10% was found as its oxidation product, hydroquinone. Similar experiments did not lead to the formation of urinary nonphenolic aromatic acids.[127]

Gentisic acid (2,5-dihydroxybenzoic acid) (32) (see Figure 1) was mentioned above as a metabolite of salicylic acid, however it itself has been the subject of a few studies, the first dating from the last century. Clarke and Mosher[151] summarized the early work on gentisic acid metabolism and concluded that the conflicting and confusing results largely stemmed from questionable methodology. Stated briefly, some of the earliest investigations indicated that part of the gentisic acid was excreted in the urine conjugated with sulfate whereas subsequent studies pointed to the excretion of gentisate largely in the unchanged form. Thus, Consden and Stanier[152] reported that ~90% of the dose was excreted unchanged in the urine by man. Roseman and Dorfman[153] recovered 61 to 77% of the dose (37 mg sodium gentisate/kg, p.o.) in the 24-h urine in man and found no increase in the amount of gentisate following acid hydrolysis of the samples. In the study of Clarke and Mosher, ~60% of the oral dose of either gentisic acid or its sodium salt was recovered unchanged in the 24-h urine in man and less than 1% was found in the feces. However, acid hydrolysis of the urine samples liberated a small additional amount (~10%) of gentisic acid and it was suggested that some conjugation, mainly forming the 5-*O*-sulfate derivative, took place. An interesting finding was the occasional collection of dark urines, sometimes when voided but usually as a result of exposure to air, which were thought to be due to the presence of the oxidized or quinone form (33) of gentisic acid. The evidence for the urinary excretion of conjugates of gentisic acid by man was more substantial in the study reported by Batterman and Sommer.[154] Following repeated administration of gentisate (3 × 300 mg daily, p.o.), they found that the average daily excretion of free gentisate accounted for 26% of the dose while a further 36% was excreted in a form liberated by acid hydrolysis. Not unexpectedly, alkalinization of the urine increased by 50% the amount of free gentisate excreted. In view of the finding of Wilson et al.[131] noted above on the conversion in man of salicylic acid to gentisic acid followed by conjugation with glycine to form gentisuric acid, it seems likely that the increased values reported after sample hydrolysis may be explained, in part at least, by the presence of this conjugate.

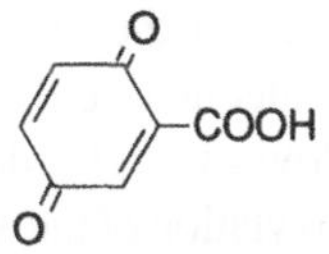

(33)

The results summarized above do not give a clear picture of the metabolism of gentisic acid, however a few additional studies, mainly using laboratory animals, provide some supplementary information. Haberland et al.[120] administered [*carboxy*-^{14}C]gentisic acid (100 mg/kg, i.p.) to rats and recovered more than 90% of the radioactivity in the 24-h urine. This material consisted of about two-thirds unchanged compound and one-third gentisic acid glucuronide. The most detailed study of gentisic acid metabolism was carried out in dogs by Astill et al.[155] They found that single oral doses of 190 or 310 mg/kg were excreted nearly quantitatively in the urine within 3 d. Most excretion occurred within 24 h and amounted to 84 and 71% of the lower and higher dose, respectively. The compounds excreted were shown to be free gentisic acid, gentisic acid 5-*O*-sulfate, and gentisic acid 5-*O*-glucuronide, i.e., conjugation apparently only in the 5-position. Mention was made of unpublished results which indicated that these two conjugates were also formed in the rat. This indicates that the glucuronide previously reported by Haberland et al.[120] was the 5-*O*-glucuronide. The values obtained for the excretion of unchanged gentisic acid, its *O*-sulfate, and *O*-glucuronide were 63, 28, and 7%, respectively (low dose) and 61, 20, and 15%, respectively (high dose). The urine did not contain any salicylic acid or hydroquinone, metabolites which could arise via dehydroxylation or decarboxylation, respectively; however, some *O*-methylated product, 5-methoxysalicylic acid, was detected when the higher dose level was employed. This metabolite was previously reported by Sakamoto et al.[156] to be formed from gentisic acid in man.

Gentisic acid was not among the phenolic acids which were decarboxylated by rat intestinal bacteria.[126] Similarly, sheep rumen bacteria neither decarboxylated it to a simple phenol[48] nor dehydroxylated it to a nonphenolic aromatic acid.[127]

Although many of the metabolic pathways seen with the monohydric phenolic acids are also encountered with **protocatechuic acid** (3,4-dihydroxybenzoic acid), the presence of the catechol moiety in the 3,4-position affords two further possibilities, *O*-methylation and dehydroxylation. An early study using rabbits given protocatechuic acid (250 mg/kg, p.o.) indicated that 18 and 13% of the dose were excreted in the urine as glucuronic acid and ethereal sulfate conjugates, respectively.[157] The remainder was thought to be excreted unchanged, however DeEds et al.[158,159] reported that *O*-methylation of the 3-hydroxy group occurred, vanillic acid being excreted when rats or rabbits were given protocatechuic acid. Conjugation of the metabolically formed vanillic acid with glycine also occurs in rats.[160] Concurrent investigations by Scheline[161] and Wong and Sourkes[162] confirmed that protocatechuic acid is metabolized to vanillic acid in rats and the former study showed that some of this material was in the form of acid-labile conjugated material. Additionally, both studies showed that some decarboxylation to catechol also occurred. The most detailed study of the metabolism of protocatechuic acid was carried out by Dacre and Williams.[163,164] Using material labeled with ^{14}C in the carboxy group, they found that ~72% of the dose (100 mg/kg) was excreted in the urine (64%) and feces (8%) in 7 d by rats. Most of the urinary radioactivity was excreted within 24 h and quantitative analysis of this material indicated that it consisted mainly of free protocatechuic acid (21% of the dose) or its conjugates (15%, about half of which was due to the glycine conjugate). The second major metabolic pathway was *O*-methylation, giving free (4%) and conjugated (17%) vanillic acid. Dehydroxylation, a reaction of the intestinal microorganisms, occurred only to a limited extent, giving 3-hydroxy-, 4-hydroxy-, and 3-methoxybenzoic acids to the extent of ~2, 1, and 2%, respectively. Interestingly, when these experiments were repeated in rats treated orally with neomycin to inhibit the activity of the intestinal bacteria, complete recovery of the radioactivity was obtained in the urine (87%) and feces (12%). The main differences were an increase in total protocatechuic acid excretion from 36% in the normal group to 55% in the treated group and the very low values for dehydroxylated products. These results are best explained by the bacterial decarboxylation of the acid to catechol and $^{14}CO_2$ in the normal animals. Significantly, the latter metabolite was detected when the acid was incubated *in vitro* with rat gut contents. This decarboxylation reaction has been demonstrated in several investigations, first by Booth and Williams[49] using rat cecal or fecal extracts and subsequently by Scheline[126,161] using mixed cultures of of rat cecal microorganisms. Protocatechuic acid was also extensively decarboxylated to catechol when infused into the rumen of sheep.[48] Soleim and Scheline[150] studied a number of common intestinal bacteria for their ability to decarboxylate *p*-hydroxylated aromatic acids and found that *Aerobacter aerogenes* extensively decarboxylated protocatechuic acid. Martin[127] found no evidence for the dehydroxylation of protocatechuic acid to nonphenolic aromatic acids when it was infused into the rumen of sheep.

The above summary of protocatechuic acid metabolism shows that in addition to excretion of the acid as such or as conjugates, the main pathways (in rats, at least) involve decarboxylation by the intestinal bacteria or *O*-methylation in the tissues. The latter reaction, effected by the enzyme catechol *O*-methyltransferase, has been extensively studied due to involvement of *O*-methylation in catecholamine metabolism and some of these findings deal with the methylation of protocatechuic acid. Early examples of such studies include those of Pellerin and D'Iorio[165] who used rat liver or kidney homogenates fortified with [*methyl*-^{14}C]-L-methionine and of Axelrod and Tomchick[166] who used a rat liver preparation of the transferase and the methyl donor, *S*-adenosylmethionine. The latter compound was also employed by Dirscherl and Brisse[167] who showed the conversion of protocatechuic acid to vanillic acid by rat and human liver homogenates. Masri et al.,[168] using rat and rabbit liver slices, reported that protocatechuic acid was converted to vanillic acid and its glycine conjugate. Interestingly, isovanillic acid, the 4-*O*-methyl derivative of protocatechuic acid, was sometimes detected in small amounts. The likelihood that protocatechuic acid may undergo both 3- and 4-*O*-methylation in man had been previously suggested by Hill et al.[169] and several more recent investigations using catechol-*O*-methyltransferase preparations clearly demonstrated that both 3- and 4-*O*-methylation take place.[170,171] The ratio of 3- to 4-*O*-methylated products was found to be ~5.5:1, however the very small amounts of isovanillic acid actually excreted by animals suggests that the products ultimately excreted are also dependent upon *O*-demethylation of the methylated metabolites. Marzullo and Friedhoff[172] described two forms of catechol *O*-methyltransferase which differ in the ratio of methylated isomers produced from protocatechuic acid.

In contrast to that seen with most of the hydroxybenzoic acids, interest in the metabolism of **gallic acid** (3,4,5-trihydroxybenzoic acid) is of fairly recent date, mainly dealing with the reactions of *O*-methylation and decarboxylation. Furthermore, the investigations have used only rats and rabbits, as was the case with the initial study carried out by Booth et al.[173] When gallic acid was given to rats either in the diet (0.5% level) or in single 100 mg doses (p.o.), the major urinary excretory products were unchanged compound and its 4-*O*-methyl ether. Also excreted, mainly as an acid-labile conjugate, was 2-*O*-methylpyrogallol. Similar results were obtained when rats were given gallic acid by i.p. injection except that some pyrogallol (1,2,3-trihydroxybenzene) was also detected. Rabbits fed a diet containing gallic acid also formed the 4-*O*-methylated acid, pyrogallol, and 2-*O*-methylpyrogallol. The data indicated that the benzoic acid derivatives were largely excreted in the free forms although an acid-labile conjugate of 4-*O*-methylgallic acid was excreted by rabbits. Blumenberg and Dohrmann[174] noted a marked increase in glucuronic acid excretion in rabbits fed gallic acid and identified its glucuronide in urinary extracts. Contrariwise, Watanabe and Oshima[175] who carried out similar experiments in rabbits, claimed that little conjugated material was excreted. They found that unchanged gallic acid was the main excretory product, together with some 4-*O*-methylgallic acid and pyrogallol. Scheline[161] reported that rats given gallic acid (100 mg/kg, p.o.) excreted it and its 4-*O*-methyl ether both free and as acid-labile conjugates whereas the decarboxylated metabolites, pyrogallol and 2-O-methylpyrogallol, were excreted in conjugated form.

Both the O-methylation and decarboxylation reactions seen with gallic acid have been studied in further detail. Masri et al.[168] found that rat or rabbit liver slices selectively methylated gallic acid in the 4-position, in agreement with the *in vivo* results. The same results were obtained when a catechol O-methyltransferase preparation from rat liver was used.[176] The decarboxylation reaction was investigated in rats by Scheline[161] who showed that it is carried out by the intestinal microorganisms, a finding which substantiates the proposal presented earlier by Tompsett.[177] Decarboxylation occurred readily *in vitro* when gallic acid was incubated anaerobically with rat intestinal contents or feces. Interestingly, resorcinol which arises from the 2-dehydroxylation of pyrogallol, was also formed in these incubates and this compound was sometimes also detected in small amounts in the urine samples from rats given gallic acid. The structural requirement for decarboxylation, which was subsequently shown to be the presence of a free *p*-hydroxy group,[126] indicates that the 2-O-methylpyrogallol excreted in the urine of animals given gallic acid is formed via the sequence gallic acid → pyrogallol → 2-O-methylpyrogallol rather than by the decarboxylation of 4-O-methylgallic acid. The decarboxylation of gallic acid to pyrogallol and metabolism of the latter compound to resorcinol was shown by Martin[48] to occur when the acid was infused into the rumen of sheep. The phenols formed were excreted in the urine. Similar experiments showed that gallic acid was not dehydroxylated to nonphenolic aromatic acids in the sheep rumen.[127] Krumholz and Bryant[178] described a new species of anaerobic bacterium isolated from the rumen of a steer which metabolized gallic acid to pyrogallol and resorcinol.

The remaining benzoic acid derivatives to be covered in this section are methyl ethers of some of the phenolic acids covered above and also the methylenedioxy derivative, piperonylic acid. The acid in this group having the simplest structure is **anisic acid** (4-methoxybenzoic acid), which was found by Quick[146] to be excreted by man conjugated with glycine or glucuronic acid. Following an oral dose of 3.7 g, the recovery of metabolites was nearly complete and about half was excreted in each form. The results obtained by Bray et al.[179] in rabbits given anisic acid (400 mg/kg, p.o.) were similar, most of the dose being excreted as the ester glucuronide of anisic acid (57%), 38% as the glycine conjugate (4-methoxyhippuric acid), and only ~1% as unchanged compound. The metabolism of anisic acid in rats was studied by Cramer and Michael[180] who found that ~85% of the dose (100 mg, i.p.) could be accounted for in the 24-h urines. Nearly 80% of the dose was due to anisic acid (6%) or its glycine (16%) and glucuronic acid (58%) conjugates; however, a further 6% was found to consist of 4-hydroxybenzoic acid (1.4%) and its glycine and glucuronic acid conjugates (0.6 and 4.2%, respectively). Axelrod[181] reported that rabbit liver microsomes can O-demethylate anisic acid.

The metabolism of **vanillic acid** (4-hydroxy-3-methoxybenzoic acid) shows similarity to that noted above with its demethylated derivative protocatechuic acid. Sammons and Williams[182] fed vanillic acid (1 g/kg) to rabbits and accounted for 83% of the dose in the urine, 56% as free vanillic acid and 16% and 11% as glucuronide and sulfate conjugates, respectively. They also reported that a further 5% or so may undergo demethylation, probably to protocatechuic acid. When considering the metabolism of vanillic acid, it should be recalled that this compound is an important metabolite of vanillin, the metabolism of which is covered in the chapter entitled Metabolism of Aldehydes, Ketones, and Quinones. The metabolic pathways of vanillin in the rat are illustrated in Figure 2 of above-mentioned chapter which indicates that some conjugation of vanillic acid with glycine also occurs. The pathway leading to protocatechuic acid is also shown and, interestingly, the two acids may undergo decarboxylation to give guaiacol (*o*-methoxyphenol) and catechol. Wong and Sourkes[162] reported that vanillic acid is a source of urinary catechol in rats and it is now clear that this reaction is carried out by the intestinal bacteria. Scheline[126] incubated vanillic acid anaerobically with a mixed culture of rat cecal microorganisms and found that it underwent both decarboxylation and O-demethylation, giving guaiacol, protocatechuic acid, and catechol. Martin[48] reported that vanillic acid was converted to guaiacol and catechol by sheep rumen microorganisms. Infusion of 7.5g/d of the acid into the rumen resulted in the excretion in the urine of 4% of the dose as guaiacol and 60% as catechol. Krumholz and Bryant[183] described a new species of anaerobic bacterium isolated from the rumen of a steer which demethylated vanillic acid.

The metabolism of **veratric acid** (3,4-dimethoxybenzoic acid) has not been studied in detail, however Sammons and Williams[184] found that rabbits given oral doses (1 g/kg) excreted ~41% in the urine as unchanged compound and a further 28% as veratroylglucuronide. Vera-troylglycine, vanillic acid derivatives, and protocatechuic acid were not detected but a small amount of catechol was reported. Formation of the latter metabolite indicates that O-demethylation and subsequent decarboxylation took place. Scheline[126] found that veratric acid undergoes O-demethylation at both the 3- and 4-positions when incubated anaerobically with rat cecal bacteria and, in view of that noted above with vanillic acid similarly treated, the initial formation of vanillic acid is sufficient to explain the subsequent degradation to catechol. However, a further site of O-demethylation is the liver and Thomas and Müller-Enoch[185] reported that perfused rat liver O-demethylated veratric acid in both the 3- and 4-positions, with ~15 times as much of the latter metabolite (vanillic acid) being formed.

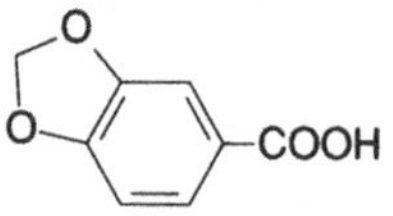

(34)

Piperonylic acid

Piperonylic acid (3,4-methylenedioxybenzoic acid) (34) is structurally similar to veratric acid and its metabolism not unexpectedly follows similar pathways. Heffter[186] reported that about half of a 5 g dose was excreted in the urine by man as an unidentified acid. This was believed by Williams[187] to be the glycine conjugate which was reported to be excreted by rabbits together with the ester glucuronide of piperonylic acid. Acheson and Atkins[188] found that the glycine conjugate was a urinary metabolite of piperonylic acid in rats, some unchanged compound also being detected. Klungsøyr and Scheline[189] confirmed these results in rats. Following an intragastric dose of 166 mg/kg, 77 and 7%, respectively, were found excreted as the conjugate and the free acid in 24 h. Kamienski and Casida[190] reported that the glycine conjugate was also the major urinary metabolite of piperonylic acid in mice. The latter investigation also looked into the question of cleavage of the methylenedioxy moiety, using material labeled with ^{14}C at this site. Only ~1% of the radioactivity was recovered in the expired air in 48 h and this indicates that this pathway is of minor importance. Furthermore, *in vitro* experiments with mouse liver microsomes failed to show any demethylenation of piperonylic acid to the corresponding catechol and CO_2. The study of Klungsøyr and Scheline likewise did not find any conversion of piperonylic acid to catechol derivatives in rats. It seems reasonable to assume that the extensive ionization of this acid at physiological pH values will reduce its ability to effectively reach the oxidative enzymes in the lipoidal microsomal system.

The two relevant methyl ethers of gallic acid are **syringic acid** (3,5-dimethoxy-4-hy-droxybenzoic acid) (35) and 3,4,5-trimethoxybenzoic acid. The metabolism of both of these acids given orally at doses of 800 mg/kg to rats was studied by Griffiths.[191] With syringic acid, large amounts of unchanged compound were excreted in the urine together with two minor metabolites identified as 3-O-methylgallic acid (36) and its 4-O-methylated product, 3,4-dimethoxy-5-hydroxybenzoic acid (37). Scheline[126] showed that the O-demethylation of syr-ingic acid to compound (36) can be carried out by the intestinal microflora of rats. Krumholz and Bryant[183] described a new species of anaerobic bacterium isolated from the rumen of a steer which demethylated syringic acid.

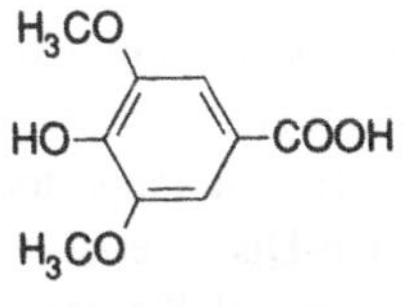

(35) (36) 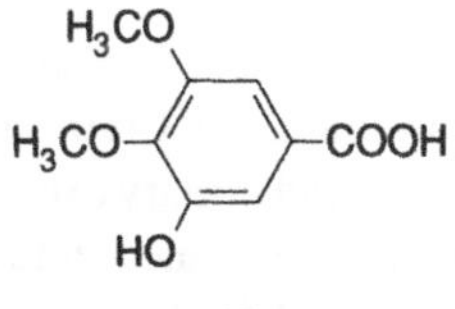 (37)

Syringic acid

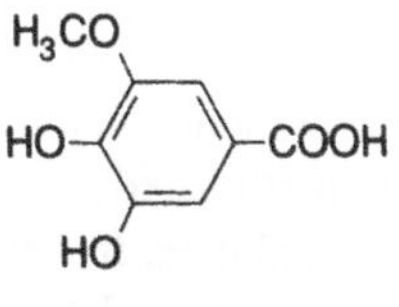

Some uncertainty exists about the metabolic fate of **3,4,5-trimethoxybenzoic acid**. Numerof et al.[192] reported that essentially the entire dose (1.75 mg/kg, p.o.) was rapidly excreted in the urine as unchanged material by mice. However, Griffiths[191] reported that the major urinary metabolite of this acid (800 mg/kg, p.o.) in rats was the demethylated compound (37). A small amount of the other monodemethylated product, syringic acid (35), was also detected. The differences noted in these two studies were possibly due to the large differences in the doses employed. The *O*-demethylation observed with many phenolic methoxybenzoic acids following anaerobic incubation with rat intestinal microorganisms was not seen with 3,4,5-trimethoxybenzoic acid,[126] however its demethylation by a new species of anaerobic bacterium isolated from the rumen of a steer was described by Krumholz and Bryant.[183]

Phenolic derivatives of phenylacetic acid are not common plant constituents and only a few examples are known. These include the 2- and 4-hydroxy derivatives and homogentisic acid (2,5-dihydroxyphenylacetic acid), however very little metabolic data is available on these compounds. **4-Hydroxyphenylacetic acid** appears to be mainly excreted unchanged and Ewins and Laidlaw[193] found that man excreted 50% of a 0.5 g oral dose in this form in the urine within 36 h. They also referred to 19th century studies which reported corresponding values of 40 to 50% in dogs receiving 2 g and ~80% in humans receiving 7.5 g. 4-Hydroxyphenylacetic acid may undergo hydroxylation to a small extent. Wiseman-Distler et al.[194] administered the acid (21 mg/kg) to rats and found that after i.p. injection, ~4% of the dose was excreted in the 24-h urine as 4-hydroxy-3-methoxyphenylacetic acid. The probable intermediate, 3,4-dihydroxyphenylacetic acid, was not detected and neither of these compounds was excreted when the 4-hydroxy acid was given orally. The only other noteworthy point about this compound seems to be its ability to undergo decarboxylation. This reaction results in the formation of *p*-cresol and has been shown to be carried out *in vitro* by fecal bacteria from rats[161] and humans.[149] Martin[44,48] reported the same reaction when the acid was infused into the rumen of sheep. However, it was not subject to dehydroxylation to nonphenolic aromatic acids under similar conditions.[127] The metabolic pathways of **homogentisic acid** have not been elucidated except, of course, for its role in the oxidative metabolism of phenylalanine and tyrosine which involves transamination to *p*-hydroxyphenylpyruvic acid and conversion of the latter to homogentisic acid which then undergoes ring scission and further oxidation. Homogentisic acid is not susceptible to decarboxylation or dehydroxylation when incubated with mixed cultures of rat cecal microorganisms.[161]

The remaining phenolic phenylacetic acid derivative is **homoprotocatechuic acid** (3,4-dihydroxyphenylacetic acid). Although this compound appears to be primarily associated with lower rather than higher plants, it is the only member of this group which has received significant attention from a metabolic point of view. It is felt that a short summary of the main findings is therefore desirable. Homoprotocatechuic acid metabolism is interesting from a historical point of view since it was one of the compounds included in the first reports on the biological *O*-methylation of catecholic phenols in animals.[158,195] This reaction has been a major point of interest in subsequent studies of the metabolic fate of homoprotocatechuic acid and has been demonstrated both *in vitro* and *in vivo*. Masri et al.[168] found that rat or rabbit liver slices methylated it mainly in the 3-position, giving rise to homovanillic acid, but also noted that a small amount of 4-*O*-methylation (forming homoisovanillic acid) occurred. Axelrod and Tomchick[166] reported the 3-*O*-methylation of homoprotocatechuic acid by a catechol-*O*-methyltransferase system from rat liver which requires *S*-adenosylmethionine. The ability of this system to O-methylate at the 3- or 4-positions was extensively studied by Creveling et al.[170,171] who found a 7:1 ratio between 3-O-methylated and 4-O-methylated products. The *in vivo* formation and urinary excretion of O-methylated phenolic acids from homoprotocatechuic acid has been investigated in rats, rabbits and man. In rats, Wiseman-Distler et al.[194] reported values of ~16 to 19% following doses of ~10 to 50 mg/kg, i.p. and Dacre et al.[196] recorded a mean value of 19% with an oral dose of 100 mg/kg. At the latter dose level, the urinary excretion of homovanillic acid in rabbits accounted for only ~6% of the dose[196,197] whereas, in man, 40% of a much smaller dose (2.5 mg, i.v.) was recovered in this form in the 24-h urine.[198] The amount of unchanged homoprotocatechuic acid excreted in these studies ranged from 18 to 44%[194] and 55%[196] in rats, 63% in rabbits, and 25 to 30% in man.

The second major point of interest in the metabolism of homoprotocatechuic acid concerns its dehydroxylation by the intestinal microflora. In their initial report, Booth et al.[195] reported that rabbits given the catechol acid orally excreted in the urine a p-dehydroxylated product, 3-hydroxyphenylacetic acid. The dehydroxylation reaction in rabbits was quantitated by Scheline et al.[197] who found that 14% of an oral dose (100 mg/kg) was excreted in the 44-h urine as 3-hydroxyphenylacetic acid. The other dehydroxylation product, 4-hydroxyphenylacetic acid, was also detected, but only to an extent of 10% of this. The p-dehydroxylation of homoprotocatechuic acid *in vitro* by rat intestinal bacteria was demonstrated by Booth and Williams[49] and by Scheline.[100,199] The two last studies also showed that a second bacterial reaction, decarboxylation, takes place with homoprotocatechuic acid. The product, 4-methylcatechol, was formed in the *in vitro* incubation experiments and was also excreted as an acid-labile conjugate in the urine of rats treated with oral doses (100 or 400 mg/kg) of the acid.[199]

The second major group of phenolic acids from plants includes the C_6–C_3-derivatives, mainly hydroxycinnamic acids. The most common representatives of this type are p-coumaric, caffeic, ferulic, and sinapic acids. In some cases (e.g., melilotic and dihydrocaffeic acids), forms in which the side-chain double bond is reduced are also known. Some information is available on most of these cinnamic acid derivatives, however caffeic acid is the only compound studied extensively. Of the monohydroxy derivatives, both **o-coumaric acid** (38) and its reduced derivative, **melilotic acid** (39) (see Figure 2), occur naturally. As they are metabolically interconvertible, essentially the same picture is seen with both compounds. This was shown to be the case by Furuya[200] and by Booth et al.[201] in rats and rabbits. The pathways of metabolism elucidated in these studies and in the investigation of Mead et al.[202] who administered the two acids to rabbits are illustrated in Figure 2.

Figure 2 shows that o-coumaric acid and melilotic acid are metabolized via several routes, however the details concerning some of the pathways are unclear. The pathways proposed in Figure 2 should therefore not be considered to exclude other possible routes or intermediates which are not shown. The formation of 2-hydroxyphenylacetic acid (40) illustrates this lack of certainty about the actual mechanism involved. It is also known that metabolite (40) is formed from coumarin (41), however in that case an accompanying metabolite of coumarin, 3-hydroxycoumarin should also have been detected. The latter metabolite was not found but this need not preclude its formation as a transient intermediate. Booth et al.[201] suggested that the phenylacetic acid derivative (40) is formed from 2-hydroxyphenyllactic acid, a compound isomeric with 2-hydroxyphenylhydracrylic acid (42). Both of the last two compounds can be formed by hydration of the double bond of o-coumaric acid and, with the lactic acid derivative, oxidation would give an α-keto carboxylic acid derivative which could easily loose CO_2 to form metabolite (40). Some evidence for this mechanism was obtained by Booth et al. who detected only metabolite (40) and unchanged compound in the urine of rats fed 2-hydroxyphenyllactic acid. Furthermore, Flatow[203] reported that guinea pigs converted the corresponding α-keto acid, 2-hydroxyphenylpyruvic acid, to metabolite (40).

Although most of the data on the metabolism of o-coumaric acid and melilotic acid is of a qualitative nature, Furuya[200] also obtained some quantitative values in rabbits given o-coumaric

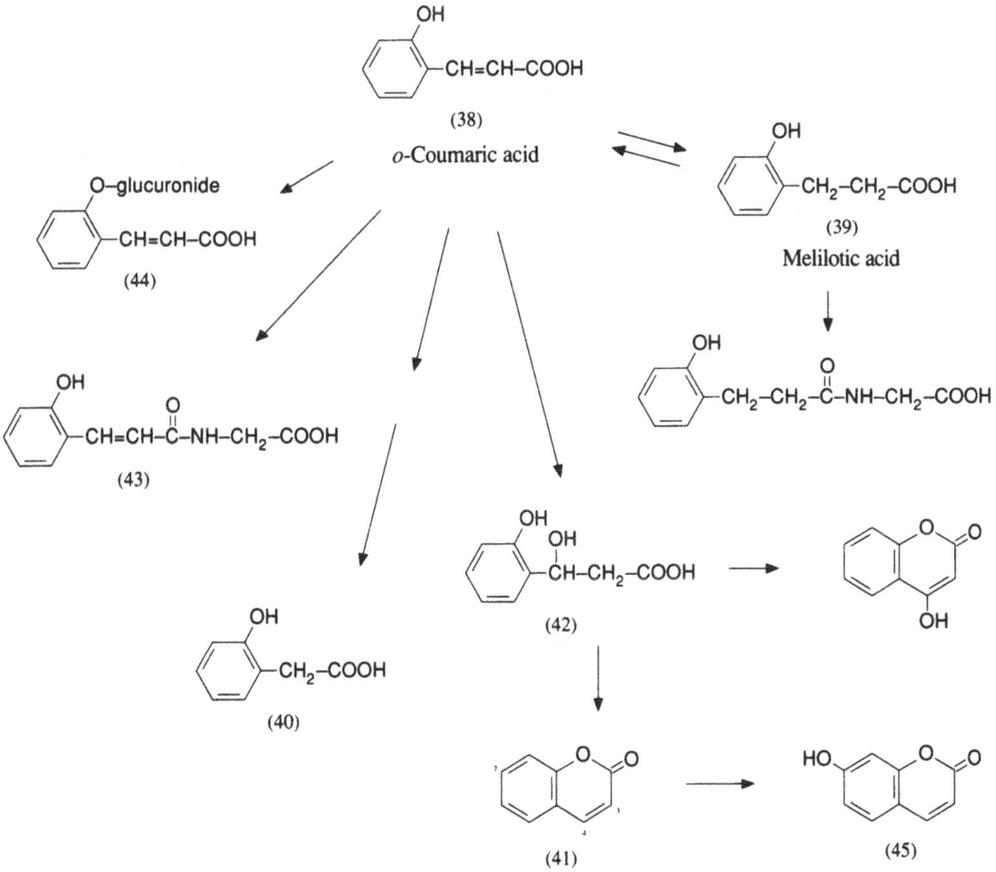

FIGURE 2.   Metabolic pathways of *o*-coumaric acid and melilotic acid in rats and rabbits.

acid (200 mg/kg, p.o.). Nearly half of the dose was accounted for in the 48-h urine and 20% was due to unchanged compound. The conjugates (43) and (44) accounted for 11 and 3.5%, respectively, and a further 8% was due to free melilotic acid (39). The other metabolites were excreted in fairly small amounts, however 4 to 5% was due to free and conjugated 7-hydroxycoumarin (45). In view of the demonstrated formation of the β-hydroxy derivative (42), it might be expected that shortening of the side chain by β-oxidation would produce some salicylic acid. However, all three of the *in vivo* studies cited above failed to detect this compound as a urinary metabolite of the two C_6–C_3-acids. It is noteworthy that the reduction of *o*-coumaric acid to melilotic acid can be carried out by the intestinal microorganisms.[161] Also, Martin[127] found that sheep rumen microorganisms dehydroxylated *o*-coumaric acid. More than 60% of the infused dose (~7 g/d) was recovered in the urine as benzoic acid and small amounts of cinnamic acid and 3-phenylpropionic acid were also detected.

The metabolism of **p-coumaric acid** (4-hydroxycinnamic acid) is straightforward and, in rats, the urinary metabolites are unchanged compound, its glycine conjugate, the reduced derivative 4-hydroxyphenylpropionic acid (phloretic acid) and the β-oxidation product 4-hydroxybenzoic acid.[204,205] In the former study the ethereal sulfate of 4-hydroxybenzoic acid was also detected. A possible metabolic reaction which does not seem to occur is ring hydroxylation. Bajaj et al.,[206] using an enzyme system from mouse liver that hydroxylated *m*-coumaric acid in the *p*-position (forming caffeic acid), did not produce this dihydroxylated acid when *p*-coumaric was used as the substrate. Details of the β-oxidation of *p*-coumaric acid to 4-hydroxybenzoic acid by rat liver preparations were studied by Ranganathan and Ramasarma.[207,208] The enzyme

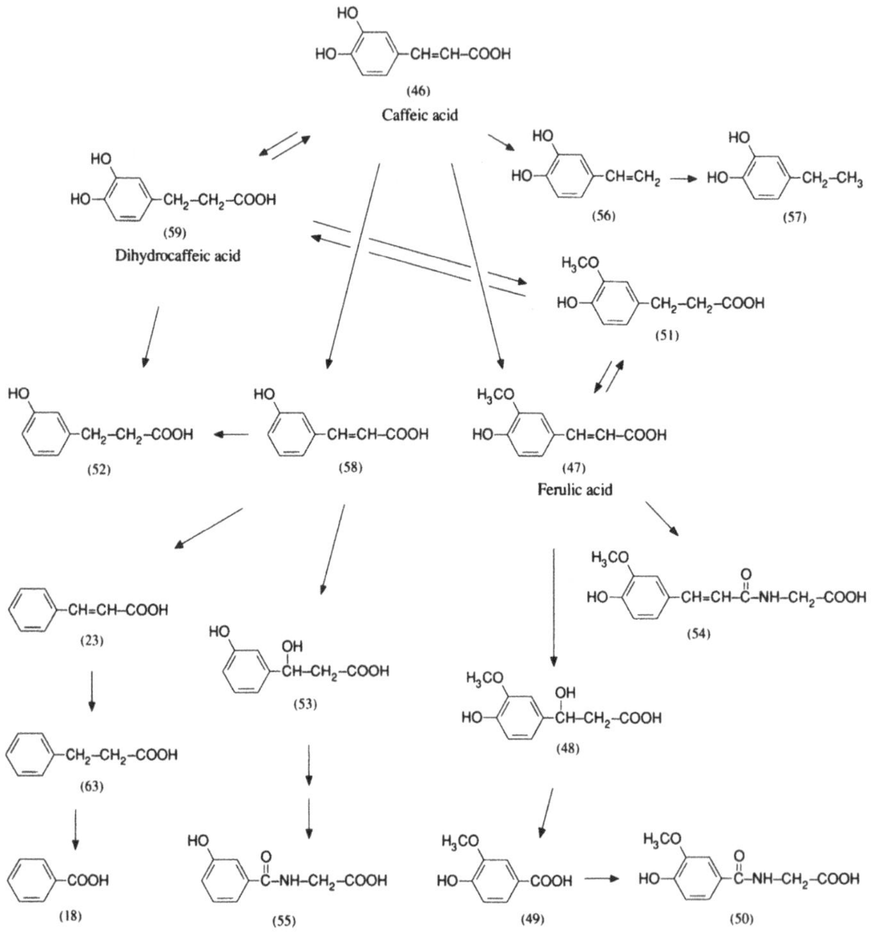

FIGURE 3. Metabolic pathways of caffeic acid.

carrying out this conversion is localized in the mitochondria and requires ATP. Mitochondrial enzymes were also reported to catalyze the interconversion of *p*-coumaric acid and phloretic acid.[208] It must be added, however, that the reduction of the double bond can be carried out by rat intestinal microorganisms, as shown by Scheline[100] and by Griffiths and Smith.[205] The former investigation also showed that the microflora is able to decarboxylate *p*-coumaric acid but not phloretic acid. This reaction is carried out by a strain of *Bacillus* isolated from rat intestine[209] and by *Aerobacter aerogenes*.[150] The significance or this decarboxylation reaction *in vivo* has not been assessed. Interestingly, the bacterial metabolism of both *p*-coumaric and phloretic acids in the sheep rumen leads to dehydroxylated rather than decarboxylated products.[44,127]

Caffeic acid (46) (see Figure 3) is structurally a relatively simple compound, nonetheless it contains several metabolically active sites, a property which leads to the formation of a large number of metabolites. The reported metabolites of caffeic acid are shown in Figure 3 which also illustrates their probable routes of formation. Of course, all of these metabolites have not been observed in a single experiment and the number of metabolites detected, as well as their amounts, will depend on many factors including the dose, route of administration, and animal species. A subject of special interest in the study of caffeic acid metabolism has been the role of the intestinal bacteria which are now known to be solely responsible for several of the reactions shown in Figure 3.

Several caffeic acid metabolites shown in Figure 3 are 3-*O*-methyl derivatives and this reaction was first reported by DeEds et al.[158] who identified ferulic acid (47) in the urine of rabbits fed caffeic acid. A subsequent detailed investigation of caffeic acid metabolism in rats and humans by Booth et al.[210] revealed the formation of all of the hydroxymethoxy derivatives except 4-hydroxy-3-methoxyphenylhydracrylic acid (48), which was among the numerous metabolites of caffeic acid in man identified by Shaw and Trevarthen.[211] It is noteworthy that none of the metabolites shown in Figure 3 possess the 3-hydroxy-4-methoxyphenyl moiety. Neither the study of Booth et al. nor that of Shaw and Trevarthen, which are the most extensive studies available on the patterns of caffeic acid metabolism, reported any of these derivatives. However, Hill et al.[169] suggested that the O-methylation of caffeic acid may occur at either hydroxy group since several members of the isomeric catechol ethers corresponding to compounds (47), (48), (49), (50), and (51) were detected in human urine and were believed to arise from dietary chlorogenic acid, an ester of caffeic acid. This suggestion that both sets of O-methyl esters are formed is further supported by evidence obtained *in vitro* in studies using catechol-O-methyltransferase from liver. Pellerin and D'Iorio[165] and Discherl and Brisse[167] found that rat liver homogenates O-methylated caffeic acid to ferulic acid, however studies by Masri et al.[168,176] and Creveling et al.[171] showed that rat or rabbit liver slices or preparations of liver catechol-O-methyltransferase formed both isomers (ferulic and isoferulic acids). The latter study recorded a *meta:para* ratio of O-methylation of only 2.8:1 and the usual lack of 4-methyl ether metabolites as urinary metabolites of caffeic acid seems therefore to be best explained by their greater susceptibility towards subsequent O-demethylation rather than due to their lack of formation.

Quantitative values for the excretion of caffeic acid and its metabolites have not been reported, however Booth et al.[210] found that 3-hydroxyphenylpropionic acid (52) was the major metabolite in rats, relatively little chain-shortening to C_6–C_1-metabolites occurring. Similar findings in rats were reported by Scheline.[100] On the other hand, Booth et al. found that humans produced larger amounts of a greater number of metabolites including the C_6–C_1-derivatives. The intermediate hydracrylic acids (48) and (53) have accordingly only been detected in human urine.[169,211] In addition to the glycine conjugates (50), (54), and (55) shown in Figure 3, some of the other metabolites are excreted conjugated with glucuronic acid. Scheline[161] found that caffeic acid, ferulic acid, and to a minor extent, 3-hydroxyphenylpropionic acid (52) were partly excreted in rat urine as glucuronides and that both decarboxylated metabolites, 4-vinylcatechol (56) and 4-ethylcatechol (57) were detected only in conjugated form.

As noted above, the metabolism of caffeic acid by intestinal bacteria is an important feature which leads to the formation of several of the metabolites shown in Figure 3. All of the 3-hydroxy derivatives are a result of this phenomenon and the role of the bacteria in caffeic acid dehydroxylation was first reported by Shaw et al.[212] This reaction was subsequently shown to take place *in vitro* using incubates of mixed bacterial cultures from rat or rabbit cecal contents and sheep rumen fluid.[213] Scheline[100] confirmed this reaction using the rat cecal microflora. The ultimate hydroxylated product in these experiments is 3-hydroxyphenylpropionic acid (52), a finding which shows that reduction of the double bond is also a result of bacterial metabolism. Masri et al.[168] reported that this reduction can also be carried out *in vitro* by rat or rabbit liver slices. This indicates that alternative sites of reduction are available; however, that occurring in the intestine is probably quantitatively the most important. The dehydroxylation reaction takes place entirely by virtue of bacterial metabolism and can be prevented when the gut flora is suppressed by treatment with neomycin.[214] Accordingly, 3-hydroxyphenylpropionic acid is not a urinary metabolite of caffeic acid in germ-free rats.[215,216] These studies identified phenolic metabolites of caffeic acid and they did not determine if complete dehydroxylation to nonphenolic aromatic compounds also occurred. It is therefore of interest to note that Martin[127] found that caffeic acid, when infused into the rumen of sheep, underwent didehydroxylation. Most of the dose (~7 g/d) was excreted in the urine in the form of benzoic acid, however small amounts of urinary cinnamic acid and 3-phenylpropionic acid were also excreted.

These findings have stimulated interest in elucidating the nature of the microorganisms responsible for caffeic acid dehydroxylation. The first report on this subject was that of Perez-

Silva et al.[217] who isolated a strain of *Pseudomonas* from rat feces which could convert caffeic acid to *m*-coumaric acid (58) and its reduction product (52). Perez Silva and Rodriguez Sánchez[218] subsequently described five species of *Pseudomonas*, *P. fluorescens*, *P. viburni*, *P. insolita*, *P. myxogenes*, and *P. chlororaphis*, which converted caffeic acid to 3-hydroxyphenylpropionic acid. Peppercorn and Goldman[219] carried out a more extensive study of this problem and tested the ability of 12 microorganisms isolated from human feces to metabolize caffeic acid. Interestingly, no single organism was able to degrade the acid to 3-hydroxyphenylpropionic acid, however the reduction of dihydrocaffeic acid (59) was demonstrated with *Clostridium perfringens* and a *Peptostreptococcus* sp. and this metabolite was then dehydroxylated by a mixed culture of *Escherichia coli* and *Streptococcus faecalis* var. *liguifaciens*. Peppercorn and Goldman[216] reported that germ-free rats selectively infected with a mixture consisting of two *Lactobacillus* strains, a *Bacteroides* sp., and a *Streptococcus* group N gained the ability to convert caffeic acid to 3-hydroxyphenylpropionic acid.

A third reaction of caffeic acid known to take place by bacterial metabolism is that of decarboxylation. This reaction, leading to metabolites (56) and (57), was first shown to occur *in vivo* in rats fed caffeic acid by Scheline.[100] These metabolites were also formed when caffeic acid was incubated anaerobically with rat cecal microorganisms. Several studies on the microbiology of the decarboxylation reaction have been made and it was shown to be carried out by a *Bacillus* sp. isolated from rat intestine,[209] by *Streptococcus fecium* isolated from human feces[219] and by *Aerobacter aerogenes*.[150]

The metabolism of **dihydrocaffeic acid** (hydrocaffeic acid) (59) (see Figure 3) has been investigated in conjunction with several of the studies with caffeic acid noted above. According to the metabolic scheme shown in Figure 3, the reduced compound, which may be dehydrogenated to caffeic acid, should give rise to the same metabolites as those formed from the latter compound. While this is probably true in a qualitative sense, the quantitative picture is undoubtedly different since the rate of the dehydrogenation reaction will influence the availability of compounds (46), (47), and (58) and thus many of the remaining reactions as well. In fact, Booth et al.[210] found that dihydrocaffeic acid fed to rats produced a pattern of urinary metabolites identical to that seen with ferulic acid (47), rather than with caffeic acid. Other points of interest with dihydrocaffeic acid are that it is a substrate of catechol-*O*-methyltransferase[171] and that it undergoes dehydroxylation but not decarboxylation as a result of bacterial metabolism. The latter results were obtained using both mixed cultures of intestinal bacteria[100] and several pure cultures.[150,219]

Methoxylated cinnamic acid derivatives from plants include a few compounds derived from mono-, di-, and trihydric phenolic acids. The sole example of the first type is **4-methoxycinnamic acid** (60) (see Figure 4) which Woo[220] found to be oxidized to 4-methoxybenzoic acid (61) in rabbits. Solheim and Scheline,[221] who found that this cinnamic acid derivative was an intermediate in the metabolism of the alkenebenzenes estragole and anethole in rats, studied its metabolism following doses of 100 and 400 mg/kg (p.o. and i.p.). Excretion of unchanged compound was detected only at the higher dosage. The urinary metabolites excreted, including the postulated β-keto acid (62), are shown in Figure 4 and indicate that, in rats, 4-methoxycinnamic acid is metabolized either by *O*-demethylation or by β-oxidation which leads mainly to benzoic acid derivatives.

Both of the monomethyl ethers of caffeic acid, **ferulic acid** (47) (see Figure 3) and isoferulic acid (3-hydroxy-4-methoxycinnamic acid) occur naturally, but it is the former compound which is by far the most common and which has received attention from a metabolic point of view. As pointed out above in the review of caffeic acid metabolism, ferulic acid is formed from it by 3-*O*-methylation. In accordance with the pathways illustrated in Figure 3, several of which are reversible, the metabolic fates of the two compounds should be qualitatively identical. The experimental data available indicate that this is largely true, especially with regard to the terminal metabolic products. Thus Shaw et al.[211] found that the same 3-hydroxyphenyl and 3-

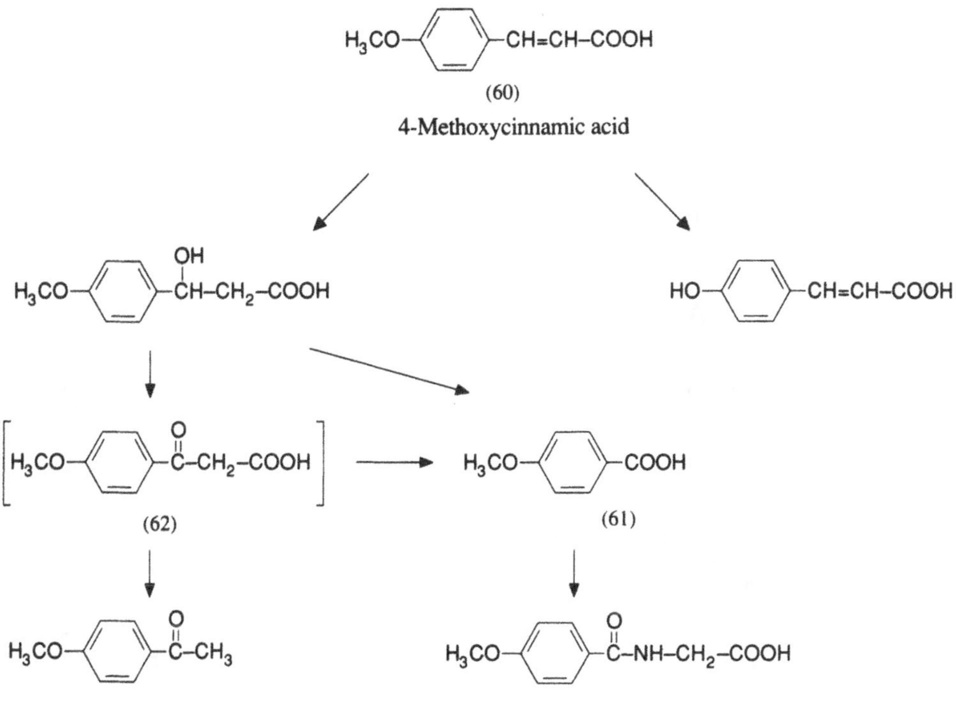

FIGURE 4. Metabolic pathways of 4-methoxycinnamic acid in rats. Postulated intermediate shown in brackets.

methoxy-4-hydroxyphenyl acids were excreted in human urine following the ingestion of caffeic or ferulic acids. These metabolites are shown in Figure 3 and include the 3-hydroxyphenyl derivatives (52), (53), and (55) and the hydroxymethoxy derivatives (47), (48), (49), (50), (51), and (54). Similarly, Booth et al.[210] fed ferulic acid to rats and found that it was metabolized to the dehydroxylated compound (52) and the hydroxymethoxy derivatives (49), (50), (51), and (54). Teuchy and Van Sumere[93] reported that 3-hydroxyphenylpropionic acid (52) was the main urinary metabolite of ferulic acid (150 to 190 mg/kg, i.p.) in rats and that vanillic acid (49) was also excreted. The conversion of ferulic acid to vanillic acid by rat liver homogenates was reported by Dirscherl and Brisse.[167]

The dehydroxylation reaction leading to 3-hydroxyphenylpropionic acid (52) is carried out by intestinal microorganisms.[100] Indeed, the entire sequence of transformations was seen when ferulic acid was incubated anaerobically with a mixed culture of rat cecal microorganisms. The intermediate compounds dihydroferulic acid (51) and dihydrocaffeic acid (59) were also detected. The finding of Teuchy and Van Sumere noted above that 3-hydroxyphenylpropionic acid (52) is the major urinary metabolite of ferulic acid given by i.p. dosage is explained by the fact that the conjugate, feruloyl glucuronide, is readily excreted in the bile of rats.[100] Plummer[222] found that more than 40% of an i.v. dose (~45 mg/kg) of ferulic acid was excreted in the bile of rats within 6 h. Most of this material was conjugated with glucuronic acid. Likewise, Westendorf and Czok[223] reported extensive biliary excretion of ferulic acid in rats (~30% of the dose) compared with the small value (3%) seen with m- or p-coumaric acid or caffeic acid. Thus, administration of ferulic acid to rats either orally or by injection will lead to a qualitatively similar pattern of metabolism since the biliary ferulic acid glucuronide is readily hydrolyzed and further metabolized by the bacteria. Krumholz and Bryant[183] described a new species of anaerobic bacterium isolated from the rumen of a steer which demethylated ferulic acid. Martin[127] found that ferulic acid, when infused into the rumen of sheep, underwent complete dehydroxylation. Most of the dose (~8 to 12 g/d) was excreted in the urine in the form of benzoic acid (18), however small amounts of urinary cinnamic acid (23) and 3-phenylpropionic acid (63) were also excreted.

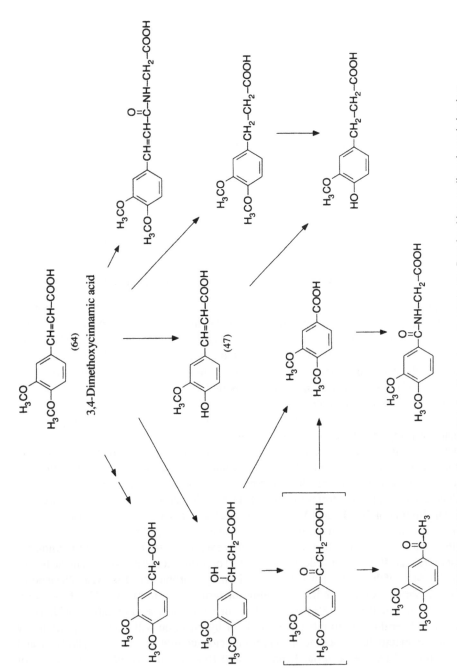

FIGURE 5.   Metabolic pathways of 3,4-dimethoxycinnamic acid in rats. Postulated intermediate shown in brackets.

Another bacterial metabolite of ferulic acid is 4-vinylguaiacol.[100] This decarboxylation product is formed in the same way that 4-vinylcatechol (56) arises from the degradation of caffeic acid (see Figure 3). Metabolite (56) was shown to be a urinary metabolite of caffeic acid in rats[100] and it therefore seems reasonable to assume that 4-vinylguaiacol, in conjugated form, is a urinary metabolite of ferulic acid. This point has not been clarified however.

Little is known of the metabolism of **isoferulic acid** (3-hydroxy-4-methoxycinnamic acid), however the results of Scheline[100] obtained using anaerobic incubates of rat cecal microorganisms suggest that its fate will probably be fairly similar to that seen with its isomer, ferulic acid. These experiments showed that double bond reduction as well as limited degradation to 3-hydroxyphenylpropionic acid (52) occurred. Direct decarboxylation of isoferulic acid, which lacks a free *p*-hydroxy group, does not occur.

The metabolic pathways of **3,4-dimethoxycinnamic acid** (64) (see Figure 5) would be expected to be very similar to those shown above (Figure 4) for the corresponding 4-methoxy derivative. Based upon data obtained by Solheim and Scheline[224] in rats given the dimethoxy compound (100 and 400 mg/kg, p.o. and i.p.), this appears to be the case as both of the previously noted pathways of *O*-demethylation and β-oxidation leading to benzoic acid derivatives were found. However, a rather more extensive pattern of metabolism was seen with 3,4-dimethoxycinnamic acid, some of which was also excreted unchanged, and these pathways are therefore shown in Figure 5. Smith and Griffiths[225] reported that the rat cecal microflora can extensively *O*-demethylate 3,4-dimethoxycinnamic acid when incubated under anaerobic conditions.

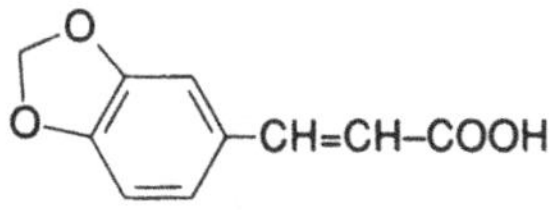

(65)

The metabolism of **3,4-methylenedioxycinnamic acid** (65) in rats was studied by Acheson and Atkins[188] and Klungsøyr and Scheline.[189] In the former study, it was administered in the food at doses of 320 to 460 mg/kg. The only metabolites detected in the 48-h urine samples were 3,4-methylenedioxybenzoylglycine (major) and 3,4-methylenedioxycinnamoylglycine (minor). In the latter investigation which used an intragastric dose of 192 mg/kg, 83% of the dose was accounted for in the urine within 24 h. Again, 3,4-methylenedioxybenzoylglycine was the major metabolite (67% of the dose), however the glycine conjugate of the administered compound was not detected. Small amounts (~3% each) of the free C_6–C_3- and C_6–C_1-acids were found as well as intermediate products formed via β-oxidation. The latter were the 3-hydroxy and 3-keto derivatives of 3-(3,4-methylenedioxyphenyl)propionic acid. It is noteworthy that all of the metabolites identified contained an intact methylenedioxy moiety. The demethylenated compound (caffeic acid) which would be formed from 3,4-methylenedioxycinnamic acid was not a urinary metabolite.

The metabolic fate of **sinapic acid** (66) (see Figure 6) in animals has been investigated in rats and rabbits. Griffiths[191] fed the compound (800 mg/kg) to rats and found that, in addition to unchanged compound, dihydrosinapic acid (67) and the two *O*-demethylated acids (68) and (69) were excreted in the urine. Interestingly, the maximal excretion of the demethylated acids occurred on the second and third days and the excretion of dihydrosinapic acid was also delayed. Some of the demethylated metabolites were excreted as acid-labile conjugates. Subsequently, Griffiths[226] reported that the fully *O*-demethylated compound, 3,5-dihydroxyphenylpropionic acid (70) is also a urinary metabolite of sinapic acid in rats and also that it is the major metabolite of the acid (200 mg/animal) in rabbits. The latter investigation also showed that suppression of the intestinal microflora with an antibiotic resulted in the complete abolition of the excretion of *O*-demethylated metabolites in the urine. Also, metabolite (70) was formed when sinapic acid was incubated with rat intestinal bacteria. A detailed study of this bacterial degradation of

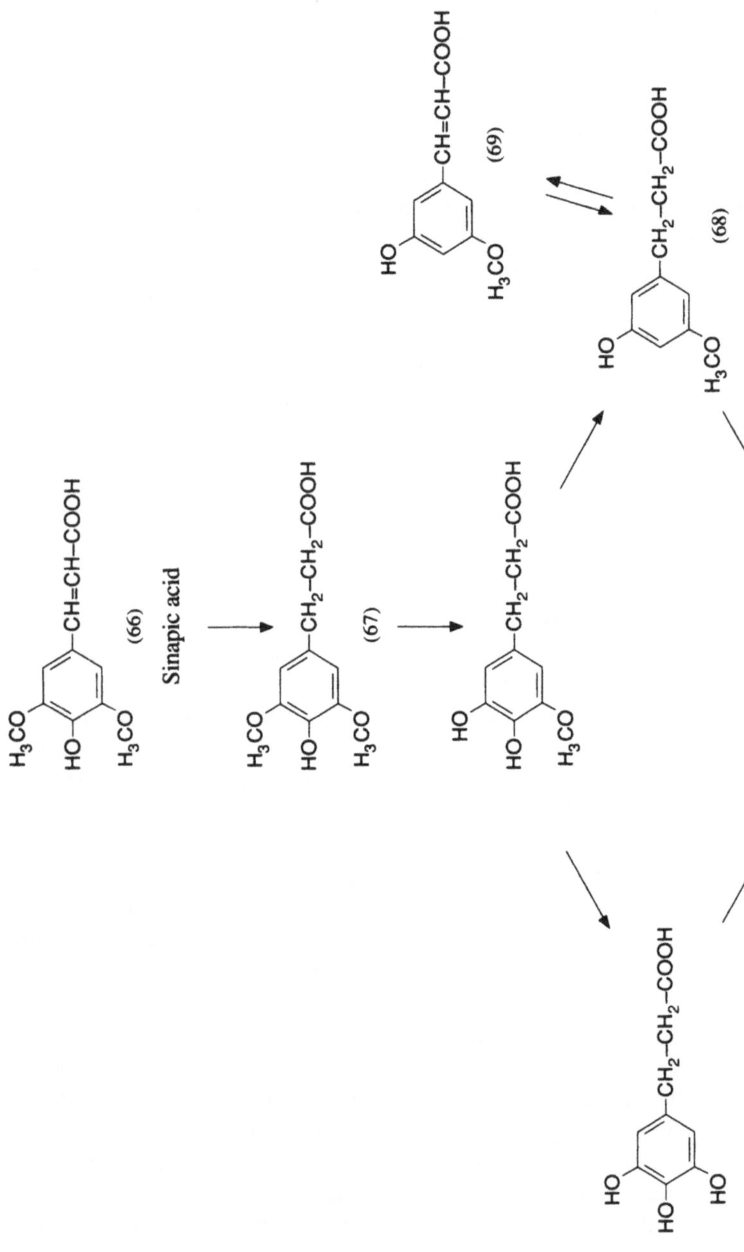

FIGURE 6.   Metabolic pathways of sinapic acid in rats.

sinapic acid was made by Meyer and Scheline[227] who found that it was reduced to dihydrosinapic acid (67) and then O-demethylated and dehydroxylated, forming finally 3,5-dihydroxyphenyl-propionic acid (70). Additionally, Meyer and Scheline[228] reported that the latter metabolite was the only compound besides unchanged compound detected in the urine of rats following an oral dose (100 mg/kg) of sinapic acid. Both acids were excreted partly free and partly conjugated, sinapic acid only during the first 24 h but metabolite (70) for as long as 2 or 3 d. These results make it quite clear that the metabolic activities of the intestinal microflora play a key role in the metabolism of sinapic acid in rats and rabbits. The probable routes of metabolism are shown in Figure 6, however the available data do not rule out alternative minor pathways. For example, the detection of the cinnamic derivative (69) as a urinary metabolite[191] can best be explained by the dehydrogenation in the tissues of some of the 3-hydroxy-5-methoxyphenylpropionic acid (68) formed bacterially. However, its formation via O-demethylation and dehydroxylation without double bond reduction is also a possibility. It is noteworthy that benzoic acid derivatives have not been reported as sinapic acid metabolites. Attention is also drawn to the fact that the articles of Griffiths[191] and Meyer and Scheline[227,228] also deal with the metabolism, both in rats and by rat cecal microorganisms, of several closely related derivatives of sinapic acid including **3,4,5-trimethoxycinnamic acid**, a rarely encountered plant acid. The metabolism of the latter compound is similar to that of sinapic acid in rats insofar as the dihydric acid (70) is the ultimate metabolite in both cases. However, major urinary metabolites of the trimethoxy compound are unchanged compound and the mono-O-demethylated compound isomeric with sinapic acid, i.e., 3-hydroxy-4,5-dimethoxycinnamic acid. These metabolites were excreted both free and as conjugates. The phenylpropionic acid analogues of both of these compounds were also detected. The studies of Meyer and Scheline devoted special attention to the pathways of metabolism of 3,4,5-trimethoxycinnamic acid and indicated that tissue O-demethylation is limited to the initial m-O-demethylation. The alternative route, p-O-demethylation, was not seen and further demethylation and degradation to the final metabolites are reactions carried out by the intestinal bacteria.

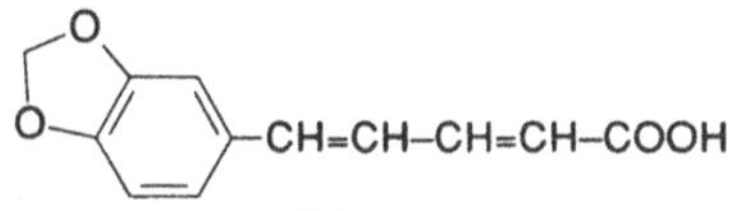

(71)

Piperic acid

Piperic acid (71), which occurs as derivatives in pepper, was fed to rats at a dose level of 500 to 700 mg/kg and the urinary metabolites studied.[188] Not unexpectedly, evidence for the β-oxidation of this unsaturated acid was obtained, the urine containing 3,4-methylenedioxycin-namoylglycine, 3,4-methylenedioxybenzoic acid (piperonylic acid), and greater amounts of the glycine conjugate of the latter acid. It was estimated that, within 72 h, ~0.3% of the dose of piperic acid was excreted as piperonylic acid and ~40% as piperonylglycine.

LACTONES

Lactones are derived from hydroxy acids by the intramolecular loss of a molecule of water and may therefore be regarded as internal esters of hydroxy acids. The naturally occurring members of this group of O-heterocyclic compounds include those with ring systems which may be five-membered (γ-lactones derived from 4-hydroxy acids), six-membered (δ-lactones derived from 5-hydroxy acids), or many-membered (macrocyclic lactones). No metabolic information is available on the last group, the most well-known examples being the C_{15}- and C_{16}-compounds exaltolide and ambrettolide which are noted for their musk-like odors. Diverse types of plant compounds contain the more common γ- and δ-lactone groups but, in many cases, it seems more reasonable to summarize their metabolism in other sections of this book. This is true with coumarin and its derivatives, lactones of *o*-hydroxycinnamic acid, and with several iridoid glucosides which are covered in the chapter entitled Metabolism of Oxygen Heterocyclic Compounds. Other lactones are more closely associated with their terpenoid properties and have therefore been included in the chapter entitled Metabolism of Higher Terpenoids. Examples include the sesquiterpenoids hymenoxon, santonin, and qinghaosu and the cardiac glycosides.

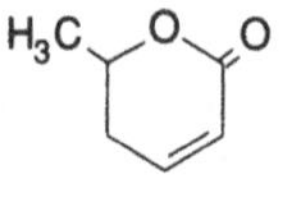

(72)

Parasorbic acid

Parasorbic acid (72) is the lactone of 5-hydroxy-2-hexenoic acid. It is readily transformed, especially under alkaline conditions, to sorbic acid (2,4-hexadienoic acid). The metabolism of the latter compound is summarized above in this chapter. However, parasorbic acid itself, when administered to rats, is known to cause an extensive reduction of liver glutathione levels.[229] Furthermore, rat liver and kidney preparations contain glutathione *S*-transferase activity which catalyzes the reaction of glutathione with parasorbic acid.[230,231] These findings suggest that parasorbic acid may be metabolized to a mercapturic acid derivative.

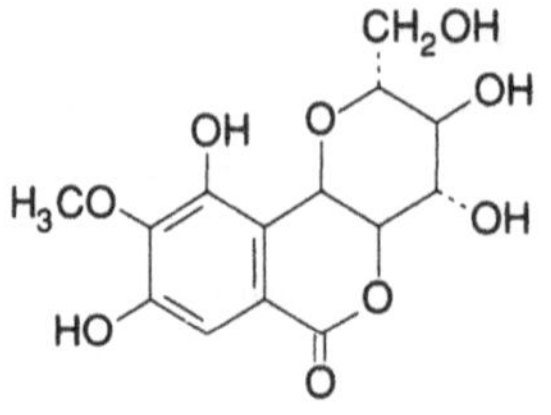

(73)   (74)

Bergenin

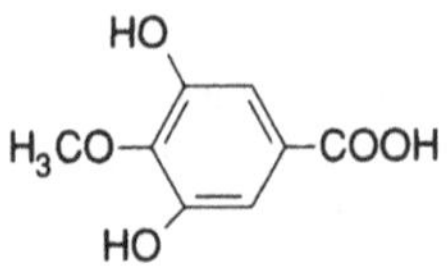

Bergenin (73) is a glucosylated derivative of gallic acid in which the sugar is connected to the ring at C-2 by a C–C bond. Hattori et al.[232] found that the glucose moiety was removed and 4-*O*-methylgallic acid (74) formed when bergenin was incubated anaerobically with human fecal microorganisms.

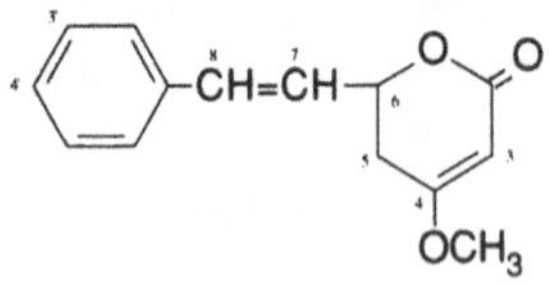

(75)   (76)

Kawain   Methysticin

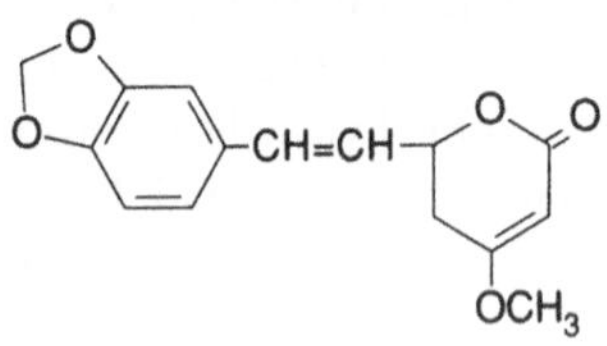

Some of the kava pyrones are 5,6-dihydro-α-pyrones and therefore, strictly speaking, lactone derivatives. Rasmussen et al.[233] studied the metabolism of three of these, kawain (75), methysticin (76), and dihydrokawain (79) (see Figure 7), in rats. They were given doses intragastrically (400 mg/kg) or by i.p. injection (100 mg/kg). When **methysticin** was administered, little appeared to be metabolized and only small amounts of two urinary metabolites were detected. These were identified as 3′,4′-dihydroxykawain which arose via demethylenation of the methylenedioxy moiety and as 3′,4′-dihydroxydihydrokawain which in addition shows reduction at C-7,8. No evidence was obtained for metabolites formed by scission of the lactone

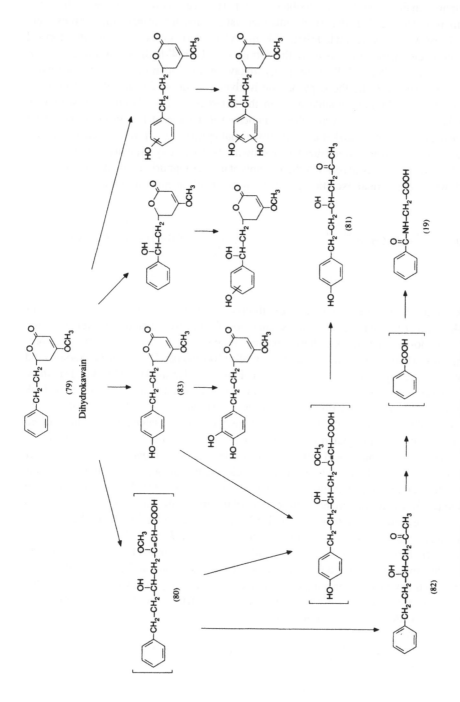

FIGURE 7.   Metabolic pathways of dihydrokawain in rats. Postulated intermediates shown in brackets.

ring. Unchanged methysticin was detected in the feces following oral dosage and it was concluded that it was probably poorly absorbed from the intestine. This, in turn, no doubt resulted from the very low water solubility of methysticin which was measured. A similar conclusion was made with **kawain** (75) which was excreted unchanged in the feces and gave large variations in the amounts of metabolites present in the urine. However its metabolism was usually more extensive than that seen with methysticin and ten urinary metabolites were detected. These were shown to include both 4′-hydroxylated derivatives and compounds formed by scission of the lactone ring. Among the former type both 4′-hydroxykawain, the major hydroxylated metabolite, and 4′-hydroxydihydrokawain were identified. A closely related metabolite was shown to be the α-pyrone derivative, 4′-hydroxy-5,6-dehydrokawain (77), which contains an additional double bond in the heterocyclic ring. Unlike that seen with methysticin, the results with kawain clearly indicated that some of the material underwent scission of the lactone ring. Two ketone derivatives analogous to those formed from dihydrokawain (see below) were detected and these were shown to be 4-hydroxy-6-phenyl-5-hexen-2-one (78) and its 4′-hydroxy derivative. Also, the amounts of urinary hippuric acid and 4-hydroxybenzoic acid, which are normal excretion products, were greatly increased in the animals given kawain.

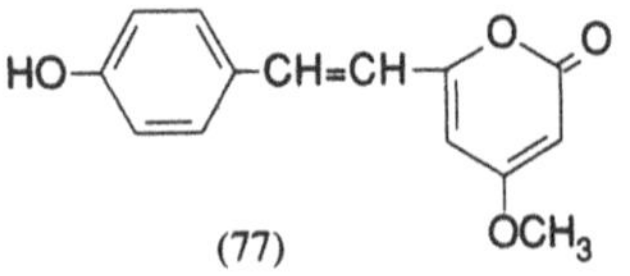

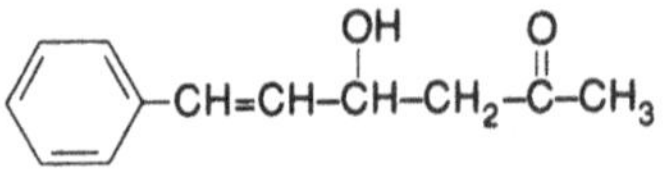

The kava dihydropyrone which was most thoroughly studied by Rasmussen et al.[233] was **dihydrokawain** (79) (see Figure 7). Only small amounts of urinary metabolites were detected following i.p. dosage (100 mg/kg), however nine compounds were detected, some in appreciable amounts, when dihydrokawain was given orally (400 mg/kg). These nine metabolites and their probable pathways of formation are shown in Figure 7. Also shown (in brackets) are three postulated intermediate products. Additional evidence that the latter metabolic reactions took place was obtained by administering compound (80), the product formed from dihydrokawain upon opening of the lactone ring. The urine from these animals contained, besides unchanged compound and its *p*-hydroxylated derivative, some metabolite (81) and large amounts of metabolites (82) and hippuric acid (19). It seems evident that the ketone (82) must be further metabolized by sequential β-oxidation to benzoic acid, the precursor to metabolite (19). It was also noted that compound (80) is unstable and easily converted to (82). About half of the dose was accounted for when assessments of the extents of the urinary excretion (48 h) of the various metabolites were made. The ketones (81) and (82) each accounted for 2 to 3% of the dose and a value of nearly 12% was found for hippuric acid (19). Figure 7 shows that six hydroxylated derivatives were formed from dihydrokawain. Most of these accounted for only a few percent of the dose, however metabolite (83) was the major urinary metabolite and gave a value of 17%. No evidence was obtained for the biliary excretion of any of the dihydrokawain metabolites and neither dihydrokawain nor its *p*-hydroxylated derivative gave rise to detectable metabolites when incubated with rat cecal microorganisms.

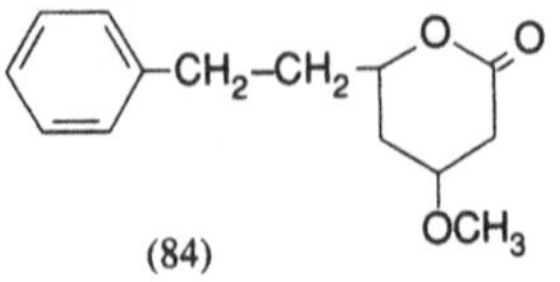

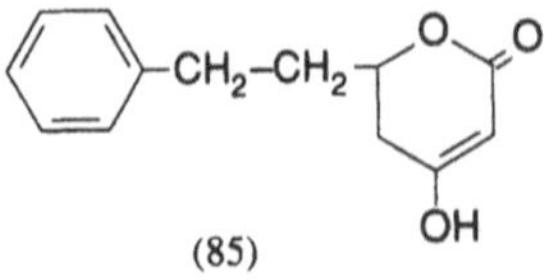

Duffield et al.[234] studied the nature of the urinary metabolites formed in humans given kava preparations. These preparations contained numerous kava lactones including kawain, methysticin, and their 7,8-dihydro derivatives. They identified the unchanged kava constituents and numerous metabolites. None of the latter were formed via ring-scission reactions. They identified compounds (84) and (85) which were probably formed from kawain or dihydrokawain by reduction of the double bond at C-3,4 or O-demethylation at C-4, respectively. Several additional metabolites were detected which showed both of these changes.

It should be noted that other closely related kava pyrones (e.g., yangonin) have a double bond in the 5,6-position. These compounds are therefore α-pyrones (i.e., coumalin derivatives) and their metabolism is covered in the chapter entitled Metabolism of Oxygen Heterocyclic Compounds.

The metabolism of **ellagic acid** (86) has been studied in both rats and mice, however our knowledge on this subject is limited. Doyle and Griffiths[235] administered large oral (50 or 100 mg) or i.p. (50 mg) doses to rats and found that the two routes of dosage resulted in very different patterns of metabolism. After oral dosage, ellagic acid itself was not detected in either the urine or feces. However, two metabolites were formed from microbial action of the intestinal microflora. One of these was shown to be 3,8-dihydroxy-6H-dibenzo[b,d]pyran-6-one (87). This indicates that the parent molecule was subjected to ring scission, decarboxylation, and dehydroxylation, all well-known reactions of the microflora. Following a 50 mg oral dose of ellagic acid nearly 10% was recovered as metabolite (87), with 4.4 and 5.5% excreted in the urine and feces, respectively, during a 6-d period. This metabolite was also found in the bile in the form of two glucuronide conjugates. The identity of the other microbial metabolite was not determined, however it was shown by comparison with an authentic sample not to be the dehydroxylation product in which the 3- and 8-hydroxy groups of ellagic acid are missing. The unidentified metabolite had acidic properties and it is therefore possible that it is an intermediate product formed from scission of the pyrone ring. Poor absorption of ellagic acid following oral dosage was also reported by Smart et al.[236] using mice. They found little or no ellagic acid in blood, lung, or liver after intragastric administration (90 mg/kg) or after feeding 1% of the compound in the diet for a week. Teel and Martin[237] confirmed the poor absorption of ellagic acid in mice. Using a small intragastric dose (0.2 mg/kg) of ^{3}H-labeled material, they reported that ~28% of the dose was absorbed and, of this, most was rapidly excreted in the urine. After 24 h ~19% of the dose was found in the feces and more than 50% remained in the gastrointestinal tract. Free ellagic acid and its sulfate, glucuronide, and glutathione conjugates were detected in the urine, bile, and blood. In addition, four unidentified metabolites were detected in the urine, however none of these appeared to be identical to those reported by Doyle and Griffiths.[235]

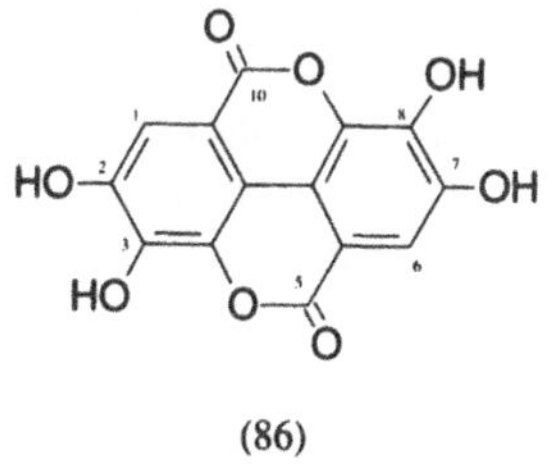

(86)　　　　　　　　　　　　(87)

Ellagic acid

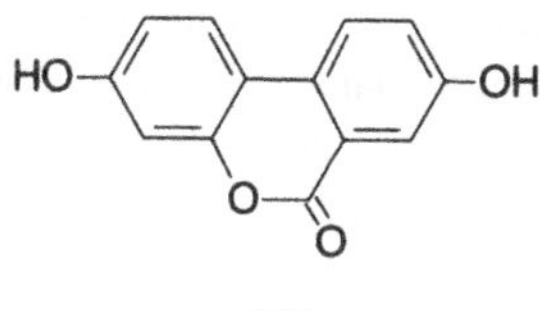

As noted above, injection of ellagic acid leads to a different pattern of metabolism than that seen with oral dosage. Doyle and Griffiths[235] detected only traces of metabolite (87) in the urine after i.p. dosage. Instead they found a new metabolite which was also excreted as two acid-labile conjugates. The identity of this metabolite is not known but it shows similarity to the dehydroxylated reference compound noted above, 2,7-dihydroxy-1-benzopyrano[5,4,3-c,d,e][1]benzopyran-5,10-dione. Mass spectral evidence showed that the metabolite was not the O-methyl ether of ellagic acid. Doyle and Griffiths also showed that ellagic acid, which is structurally a dimer of gallic acid, was not metabolized to this simple phenolic acid or its metabolites. Smart et al.[236] and Teel[238] also administered ellagic acid by injection. In the former study, both free and conjugated ellagic acid were found in the urine after i.v. dosage (18 mg/kg) and excretion was complete within 2 h. About 46% of the dose was excreted by this route and a further 26%, mainly free, was detected in the feces within 7 h. Teel administered ^{3}H-labeled material (0.4 mg/kg, i.p.) and found that bile contained unchanged ellagic acid and an unidentified metabolite whereas urine contained ellagic acid and three different unidentified compounds. Teel also showed that the bile contained additional radioactive material in the form of glucuronide and glutathione conjugates. In urine, sulfate esters, glucuronides, and glutathione conjugates were reported, however none of these were identified.

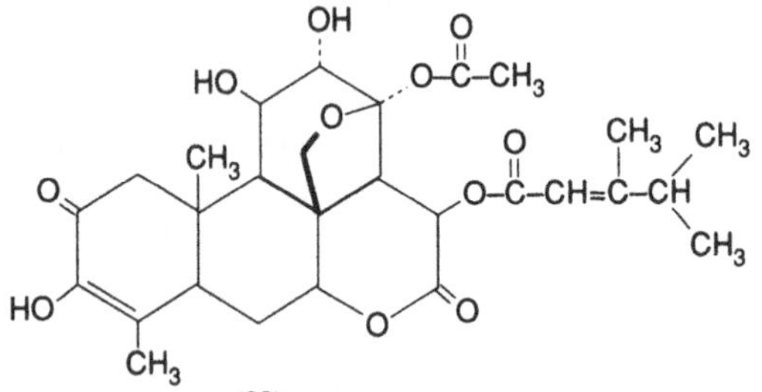

(88)

Bruceantin

Bruceantin (88) is a complex lactone derivative which contains ester, ketone, and hydroxy groups. Suling et al.[239] studied its disposition in mice and found that <2% of the dose (1.5 mg/kg, i.v.) was excreted unchanged in the urine and feces within 24 h. Appreciable amounts of bruceantin were found in the bile. Bruceantin was metabolized by the 9000-g supernatant fraction from mouse liver homogenates but not by similar preparations from lung or kidney. The *in vitro* metabolism of bruceantin was dependent on NADPH and a more polar metabolite was detected but not identified.

ESTERS

Plants contain a large number of carboxylic esters of diverse types. These include the triglycerides, esters of glycerol with three fatty acid molecules which serve as food storage materials, and many simple esters of long-chain fatty acids which provide waxy protective coatings. A discussion of the metabolism of these types falls outside the scope of this book. Other typical plant esters include those consisting of numerous lower alcohols combined with various aliphatic or aromatic acids. These are often volatile compounds and their aromaticity imparts pleasing or distinctive characteristics to many fruits. A large percentage of the nearly one thousand flavoring agents listed in Fenaroli's Handbook of Flavor Ingredients[240] are esters and many of these occur naturally.

The hydrolysis of esters to their constituent parts, an acid and an alcohol, was reviewed by Heymann.[241,242] A useful earlier survey of the esterases in human tissues was presented by La Du and Snady.[243] Ester hydrolysis is widespread in mammalian tissues, however the most important sites are the liver, gastrointestinal tract, and blood. Both interspecies and intraspecies differences in esterase activities are well known. An example of this is the lack of activity in human plasma for hydrolyzing atropine compared with that found in some strains of rabbits. Ester hydrolysis in the gastrointestinal tract may occur in the mucosa or, in some cases, in the lumen due to the metabolism by the microflora. The latter situation is noted below in the summary of the metabolism of chlorogenic acid. Additional examples of ester metabolism by the gut flora are found in other chapters, e.g., the metabolism of some higher terpenoids including steroids and of the pyrrolizidine alkaloid lasiocarpine.

The subject of ester metabolism is to a large extent identical with that of the fate of the particular acid and alcohol components. This is conveniently illustrated by the study of Gallaher and Loomis[244] who investigated the metabolism of **ethyl acetate** in rats. Blood levels of both ethyl acetate and ethanol were determined after i.p. injections of the ester. Hydrolysis of the latter was very rapid (half-life = 5 to 10 m) and its disappearance was accompanied by a rapid increase in the blood ethanol levels which then fell relatively slowly. Morris[245] recently showed that large amounts of inspired ethyl acetate were hydrolyzed in the upper respiratory tracts of rats and Syrian hamsters. Both because of the rapid hydrolysis of many esters to their component parts, and because only a few of the myriad of plant esters have been the subject of specific metabolic investigations, the discussion which follows deals with a relatively limited number of compounds. Also, some esters are included in other chapters according to their major structural features. Such examples, in addition to those noted above, include some diterpenoid and tetraterpenoid esters, the alkaloid cocaine, and the glucosinolate sinalbin.

The natural pyrethrins are alicyclic esters which include the well-known pyrethrins I (89) and II (90) and the closely related homologues cinerin I and II (91a,b) and jasmolin I and II (91c,d). Allethrin I and II are synthetic analogues in which the pentadienyl side chain at C-3' in the pyrethrins is replaced by an allyl group. Studies on the metabolism of the pyrethrins have been hampered by their complex chemistry, particularly their stereochemistry, however, the detailed investigation by Elliott et al.[246] provided an excellent foundation for the understanding of pyrethrin metabolism. Additional summaries of this subject were given by Yamamoto et al.[247] and by Casida.[248] Recently, Class et al.[249] investigated the *in vitro* metabolism of all six of the natural pyrethrins using mouse or rat liver microsomal preparations. Hutson[250] reviewed the metabolic fate of both natural pyrethrins and synthetic pyrethroids in mammals. Also, Class et al.[249] studied the metabolism of allethrin I *in vitro* using microsomal preparations and *in vivo* in rats.

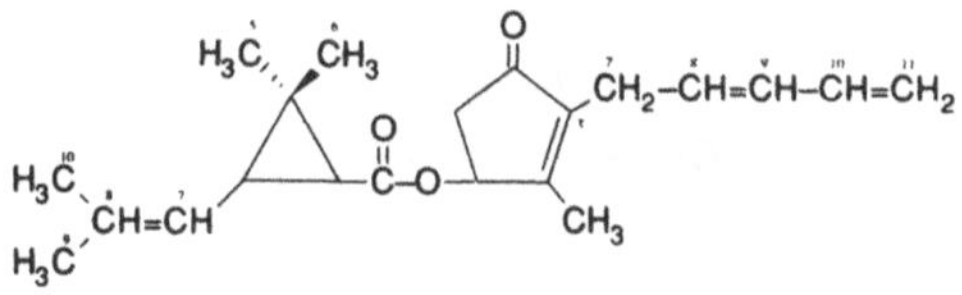

(89)
Pyrethrin I

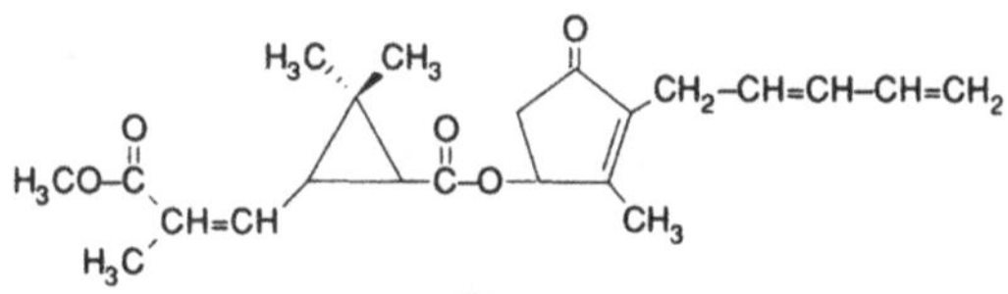

(90)
Pyrethrin II

Pyrethrin I (89) and **pyrethrin II** (90) contain several potential sites of metabolism. These include the ester linkage, the isobutenyl group of the acid moiety, the pentadienyl side chain of the alcohol moiety, and, in pyrethrin II, the methyl ester group. Results in rats indicate that hydrolysis of the central ester linkage occurs to a very limited extent. This was made evident by the finding that the same radioactive urinary metabolites were excreted following administration of pyrethrin I labeled in the acid or alcohol moieties.[246] This is in contrast to that seen with allethrin and several other related synthetic derivatives. Allylic hydroxylation of pyrethrin I at C-10 to give the corresponding alcohol derivative is well documented and this intermediate is further oxidized to the aldehyde and carboxylic acid derivatives. An additional reaction of the isobutenyl group is epoxidation at C-7,8. Likewise, several metabolites were identified which resulted from epoxidation of the pentadienyl side chain[249] Hydrolysis of these epoxides resulted in the formation of diols and those substituted at C-8',11' and C-8',9' (or C-10',11') were identified. Also, 7'-hydroxy derivatives formed via allylic hydroxylation of the pentadienyl moiety were detected. The C-10 carboxylic acid derivative was not detected in the urine or feces of rats and appears to be a transient product of metabolism. The main products found in the excreta are more highly metabolized compounds including diol derivatives of this acid which showed prolonged excretion profiles.[246] Similar pathways were found in *in vitro* experiments with pyrethrin II and these metabolites contained either an intact or a hydrolyzed methyl ester group at C-10. However, it is noteworthy that none of the metabolites of pyrethrin II found in the *in vivo* study contained the original methyl ester group. The hydrolysis of this group was also shown by the excretion of $^{14}CO_2$ in the expired air of rats given the ^{14}C-methoxycarbonyl-labeled compound. Unchanged pyrethrins were not excreted in the urine of rats, however they were detected in the feces.

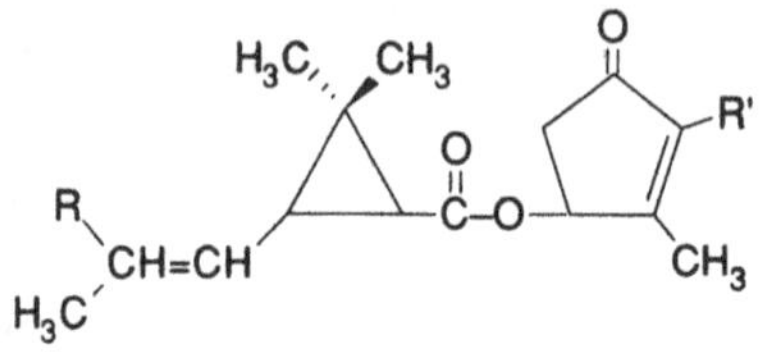

(91)

(a) Cinerin I, $R = CH_3$, $R' = CH_2-CH=CH-CH_3$

(b) Cinerin II, $R = COOCH_3$, $R' = CH_2-CH=CH-CH_3$

(c) Jasmolin I, $R = CH_3$, $R' = CH_2-CH=CH-CH_2-CH_3$

(d) Jasmolin II, $R = COOCH_3$, $R' = CH_2-CH=CH-CH_2-CH_3$

Casida[248] proposed that the pathways of metabolism of **cinerin I** (91a) and **cinerin II** (91b) and of **jasmolin I** (91c) and **jasmolin II** (91d) are the same as those seen with the corresponding pyrethrins as noted above. This belief was confirmed by Class et al.[249] using mouse or rat liver microsomal preparations. However, cinerin II, unlike that found with the other natural pyrethrins, was also hydroxylated at C-5 (or C-6).

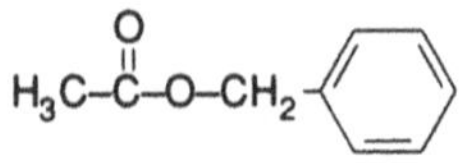

(92)   (93)

Benzyl acetate

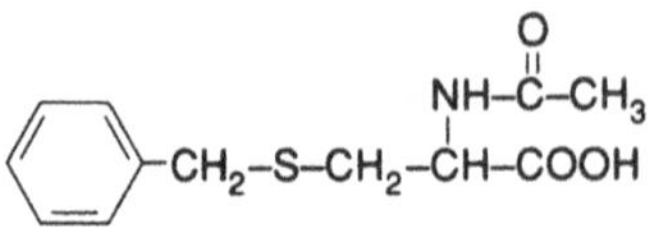

Simple aromatic esters are easily hydrolyzed in the body and, as noted above, their metabolic fate is largely that of their acid and alcohol components. Excellent examples of this are seen with **benzyl acetate** (92), **benzyl benzoate,** and **benzyl cinnamate,** naturally occurring esters which are rapidly and completely metabolized via benzyl alcohol to benzoic acid in man.[251] Clapp and Young[252] administered a large dose (1 g/kg, s.c.) of benzyl acetate to rats and isolated hippuric acid from the 40-h urines. Another metabolite isolated, which accounted for only 0.2% of the dose, was benzylmercapturic acid (93). These results were confirmed by Abdo et al.[253] who administered [*ring*-^{14}C]benzyl acetate to mice and rats. They employed small i.v. doses (5 or 10 mg/kg) or a 100-fold range of oral doses and, in most cases, recovered ~90% of the radioactivity in the urine in 24 h. Nearly all of this material consisted of hippuric acid. About 3 to 5% of the urinary radioactivity in rats was due to a mercapturic acid. The latter metabolite was found in lower amounts in some of the mice. No unchanged benzyl acetate was detected in any of the urine samples. Small amounts of benzyl alcohol and one or more unidentified metabolites were excreted in some cases by both mice and rats. Chidgey and Caldwell[254] and Chidgey et al.[255,256] carried out a comprehensive investigation of the absorption, metabolism, and excretion in rats of [*methylene*-^{14}C]benzyl acetate. In the initial study metabolites in both the plasma and urine were determined following intragastric doses of 5, 250, or 500 mg/kg. Also, doses were given neat or dissolved in 1 ml corn oil or propylene glycol. Loss of radioactivity from the plasma was slower when the ester was dissolved in vehicle. The metabolites present in plasma were dependent on the dose level, vehicle, and time after dosing. With the higher dose levels, benzoic acid was the main plasma metabolite, but hippuric acid was the major metabolite when 5 mg/

kg was administered. No unchanged ester was detected in the plasma, but small amounts of benzyl alcohol and benzylmercapturic acid (93) were occasionally found. Two unidentified metabolites, of moderate and high polarity, were also detected. Excretion of radioactivity in the urine was rapid and accounted for 70 to 80% of the dose in 24 h at all dose levels. The most prominent metabolite in these samples was hippuric acid which accounted for ~60 to 70% of the dose. Excretion of benzoyl glucuronide increased from 2.5% at the lowest dose to 10 to 12% at the highest, probably due to the limited availability of glycine required for hippuric acid formation. The other urinary metabolites were benzoic acid (~2 to 4%) and benzylmercapturic acid (~1 to 2%). Formation of the latter metabolite from benzyl acetate was studied using specific metabolic inhibitors.[255] Experiments using inhibitors of alcohol dehydrogenase or of sulfotransferase activity suggested that formation of benzylmercapturic acid involves the formation of the sulfate ester of benzyl alcohol as an obligatory intermediate. Additional information on the role of this intermediate (benzyl sulfate) is given in the summary of benzyl alcohol metabolism (see the chapter entitled Metabolism of Alcohols). Chidgey et al.[255] also studied the percutaneous absorption and disposition of benzyl acetate in rats. This method of application is relevant because the ester is mainly used in commerce as a fragrance. Doses of 100, 250, or 500 mg/kg were applied neat or as 50% solutions in ethanol to the shaved backs of rats. From 28 to 56% of the radioactivity was recovered in the urine in 24 h and hippuric acid was the major metabolite in all cases. It accounted for ~95% of the urinary radioactivity, however benzoic acid, benzoyl glucuronide, and benzylmercapturic acid each furnished an additional 1 to 2%. McMahon et al.[257] studied the influence of age on the metabolism of [*ring*-^{14}C]benzyl acetate in rats and mice. This factor had no effect on the formation of the major metabolite, hippuric acid, however changes in the minor routes of metabolism were found. The urinary excretion of benzylmercapturic acid was significantly higher in older animals (~2% of the dose) than in younger animals (~1%). This pathway also increased when the dose was increased from 5 mg/kg to 500 mg/kg.

Fahelbum and James[92] demonstrated the metabolic similarity of cinnamic acid and **methyl cinnamate** in rats and rabbits. Ester hydrolysis occurred very rapidly and no unchanged compound was detected in the peripheral blood. Only traces were found in the portal blood, however cinnamic acid and methanol were readily detected. The urinary metabolites of methyl cinnamate, being identical with those of cinnamic acid, are summarized above in the discussion of the metabolism of the latter compound.

Longland et al.[258] and Grundschober[259] reported the hydrolysis of many aliphatic and aromatic esters, both synthetic and naturally occurring. Artificial gastric and pancreatic juices were used as were preparations from liver and small intestine. While large variations in the extent and rate of hydrolysis were noted, high activities were generally present in the tissue preparations. Studies on the metabolism of propyl anthranilate in rats[260] and piperonyl acetate and piperonyl isobutyrate in rabbits[261] furnish useful information on ester hydrolysis. Although these flavoring agents are synthetic rather than naturally occurring, they show close structural similarity to many common plant esters.

Methyl salicylate (wintergreen oil) was the subject of an early study by Baas[262] who recovered ~25% of the dose (5 to 10 g) in the urine of dogs as hydrolysis products, a value quite similar to those reported by Hanzlik and Wetzel[263] in similar experiments using 1.4 to 2.9 g (200 mg/kg) doses. Small amounts (0.2 to 0.5%) of the unchanged ester were found in the urine following oral dosage, however nearly 15% was reported as urinary methyl salicylate when it was given intramuscularly. Urinary metabolites were detected for as long as 6 d. However, experiments by Davison et al.[264] clearly demonstrated that orally administered methyl salicylate was rapidly and completely hydrolyzed in rats (550 mg/kg) and dogs (300 mg/kg) and also, although to a somewhat lesser extent, in humans (~8 mg/kg). Experiments in rats, rabbits, dogs, and monkeys showed that the site of hydrolysis is mainly the liver. The further metabolism of methyl salicylate should then closely follow that shown for salicylic acid in Figure 1. Williams[265] stated that glucuronide conjugates accounted for 12 to 55% of the dose (300 mg/kg) of the ester in rabbits. The ether glucuronide (21) was isolated from this material.[137] In addition, an interesting exception to that found with salicylic acid was reported in that as much as 10% of the dose was excreted as the ethereal sulfate. As this reaction is not observed with the acid, it appears that sulfate conjugation must precede hydrolysis.

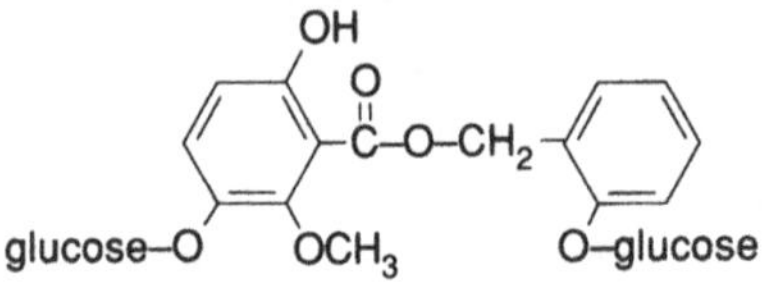

(94)

Leiocarposide

The metabolism of the phenolic ester glycoside **leiocarposide** (94) was studied in rats by Fötsch et al. following oral administration[266] or s.c. injection.[267] It was poorly absorbed in the former case and most of the dose (~600 mg/kg/d) was excreted unchanged in the feces. However, a small amount of the material was found in the urine as hydrolysis products. These included leiocarpic acid (3,6-dihydroxy-2-methoxybenzoic acid) (2% of the dose) and its conjugates (2%), salicylic acid (0.5%), salicyluric acid (0.5%), and conjugates of salicylic acid (0.1%). Most of the injected leiocarposide (~600 mg/kg) was excreted unchanged in the urine and only 0.2 and 0.1% were recovered as leiocarpic acid and salicylic acid, respectively. Incubation of leiocarposide with rat cecum or colon contents resulted in hydrolysis of both the ester and the glycoside bonds to give leiocarpic acid and saligenin.[267] These hydrolysis reactions were also observed when kidney homogenates were used. No activity was seen with rat serum or homogenates of liver or lung.

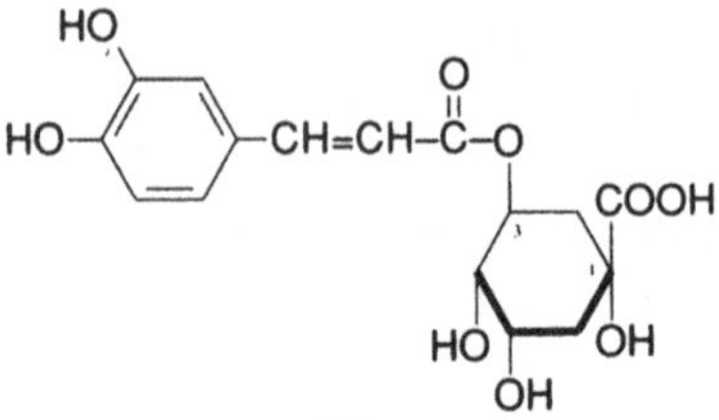

(95)

Chlorogenic acid

Chlorogenic acid (3-caffeoylquinic acid) (95) is a commonly occurring plant ester which was first studied metabolically by Booth et al.[210] The fate of the caffeic acid moiety was the point of interest in this study and the results showed a similar pattern of urinary metabolites derived from caffeic acid (see Figure 3) following the administration of either caffeic acid or chlorogenic acid. Thus, rats given orally 100 mg of the latter compound excreted 3-hydroxyphenylpropionic acid (52) as the major urinary metabolite. Following ingestion of 1 g chlorogenic acid, humans excreted mainly caffeic acid (46), dihydroferulic acid (51), the glucuronide of *m*-coumaric acid (58), and *m*-hydroxyhippuric acid (55). These results indicate that a larger number of urinary metabolites are detected following caffeic acid administration, however this might be expected in view of the slower appearance of caffeic acid in the organism when the ester is given. End products of metabolism would be more likely to predominate in the latter situation. It seems reasonable to assume that the most influential site of metabolism of chlorogenic acid is the intestine rather than the tissues as Scheline[5] noted that its incubation with mixed cultures of rat cecal bacteria resulted in the formation of 3-hydroxyphenylpropionic acid (52). Martin[127] found that infusion of chlorogenic acid into the rumen of sheep led to the urinary excretion of several aromatic acids which lacked ring substituents. The most abundant metabolite was benzoic acid which accounted for 44% of the dose (~6g/d); however, 3-phenylpropionic acid and cinnamic acid furnished a further 4%. The indication of the key role of intestinal metabolism is supported by the findings of Czok et al.[268] who detected chlorogenic acid in the serum and bile of rats following i.v. dosage but not when it was administered into the stomach or intestine. They also found that the amount of ester in the stomach decreased with time and that this occurred concomitantly with the appearance of the hydrolysis products, caffeic acid and quinic acid. However, tissue esterases are capable of hydrolyzing chlorogenic acid following its i.v. administration as Michaud et al.[269] detected both of the monomethyl ethers of caffeic acid in the bile of rats given the ester by this route. The same results were obtained when the closely related diester, 1,3-dicaffeoylquinic acid, was given.

Tannins are a complex group of plant polyphenols which are usually classified either as hydrolyzable tannins or as condensed tannins. The former type is characterized by a central glucose unit attached by ester linkages to several gallic acid groups or groups built up of gallic acid. This product is known under several names including tannic acid, gallotannin, and gallotannic acid and, upon hydrolysis, yields eight or nine gallic acid molecules for each glucose molecule. This indicates that tannic acid contains mainly residues of digallic acid (96) esterified with glucose. The empirical formula of commercial tannic acid is usually given as $C_{76}H_{52}O_{46}$. Condensed tannins are quite different compounds which are built up with much more stabile linkages. They probably arise by the oxidative polymerization of flavonoid compounds of the catechin (3-hydroxyflavan) and 3,4-dihydroxyflavan types.

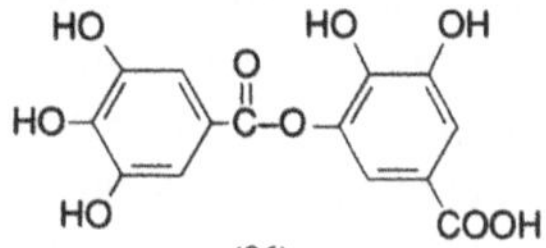

(96)
Digallic acid

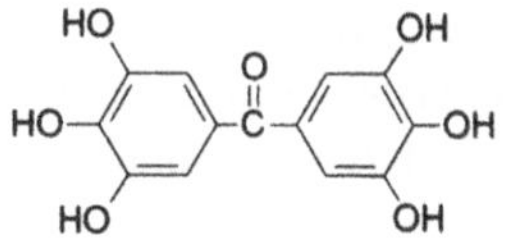

(97)

(98)

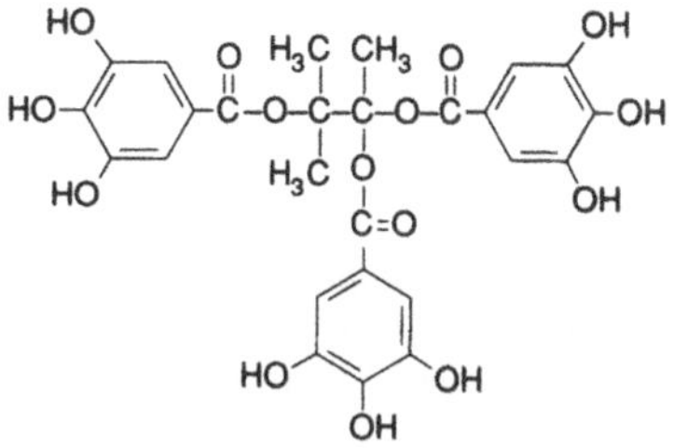

(99)

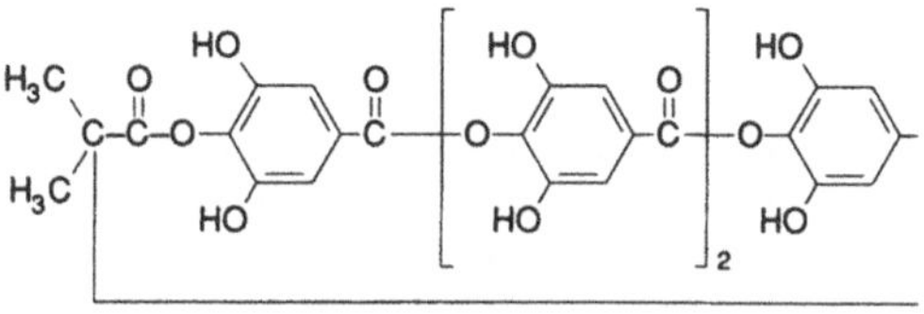

(100)

Nearly all of the metabolic studies on tannins have been carried out with the former type, tannic acid, and Williams[270] summarized the older literature which indicated that it was excreted as gallic acid (3,4,5-trihydroxybenzoic acid) by rabbits, dogs, and man. No unchanged tannic acid was detected in the urine but pyrogallol, the decarboxylation product of gallic acid was sometimes found. These results have been confirmed and extended in more recent investigations. Booth et al.[173] administered **tannic acid** (100 mg, p.o.) to rats and found that the urinary metabolites were the same as those excreted when gallic acid was given. As summarized above in this chapter, these are mainly unchanged compound and its 4-*O*-methyl ether. Similarly, Watanabe and Oshima[175] administered gallotannin from tea to rabbits and detected only gallic acid metabolites in the urine. These included pyrogallol in addition to the two metabolites mentioned above. Blumenberg et al.[271] administered repeated daily doses (60 mg/kg, p.o.) of tannic acid, gallic acid, and **digallic acid** (96) to rats and found that identical patterns of urinary metabolites were obtained chromatographically in all cases. Small amounts of gallic acid and increased amounts of glucuronides of it and a metabolite were detected. Tannic acid was not itself found in the urine. Gupta and Dani[272] incubated tannic acid with rat liver microsomes and identified four metabolites. These novel compounds were identified as bis-(3,4,5-trihydroxyphenyl)methanone (97) and the more complex metabolites (98), (99), and (100).

Milić and Stojanović[273] studied the metabolic fate of lucerne tannins in mice. This product is a complex mixture consisting of ~85% tannic substances. These consist of gallotannins, condensed tannins, and free gallic acid in a ratio of ~16:3:1. Needless to say, the administration of this mixture of both hydrolyzable and condensed tannins, which was fed in the diet at 0.5 or 4% levels, makes the results more difficult to interpret. This difficulty is compounded by the fact that only the feces were analyzed for metabolites. Information on the excretion of metabolites in the urine is likely to have given a clearer indication of the fate of the condensed tannins. The identified compounds in this fraction were (–)-gallocatechin gallate, (–)-epigallocatechin gallate, (+)-gallocatechin, and (–)-epicatechin and their presence unchanged in the feces was interpreted by Milić and Stojanović to indicate that they passed unaltered through the gastrointestinal tract. However, it seems more reasonable to assume that these fecal metabolites represent merely the unmetabolized portion of the material fed. As described in the chapter entitled Metabolism of Oxygen Heterocyclic Compounds, catechins are extensively degraded in the intestine by the microflora and give rise to C_6–C_3-acids containing catechol and *m*-hydroxyphenyl groups. Interestingly, several such compounds were detected in the feces of mice fed the lucerne tannins. Another point or interest is the finding that *O*-methylated derivatives of protocatechuic and gallic acids were among the most prominent fecal metabolites. This is probably explained by the biliary excretion of various polyhydroxylated intermediates following their *O*-methylation in the tissues. In conclusion, it seems reasonable to assume that the fecal metabolites reported for lucerne tannins arise from a series of diverse but interrelated steps encompassing incomplete absorption, metabolism by intestinal microorganisms, metabolism by tissue enzymes, and biliary excretion.

REFERENCES

1. **Caldwell, J.,** Conjugation of xenobiotic carboxylic acids, in *Metabolic Basis of Detoxication. Metabolism of Functional Groups*, Jakoby, W. B., Bend, J. R., and Caldwell, J., Eds., Academic Press, New York, 1982, 271.
2. **Caldwell, J.,** Xenobiotic acyl-coenzymes A: critical intermediates in the biochemical pharmacology and toxicology of carboxylic acids, *Biochem. Soc. Trans.*, 12, 9, 1984.
3. **Caldwell, J.,** Novel xenobiotic–lipid conjugates, *Biochem. Soc. Trans.*, 13, 852, 1985.
4. **Quistad, G. B. and Hutson, D. H.,** Lipophilic xenobiotic conjugates, in *Xenobiotic Conjugation Chemistry*, Paulson, G. D., Caldwell, J., Hutson, D. H., and Menn, J. J., Eds., American Chemical Society, Washington, DC, 1986, 204.
5. **Scheline, R. R.,** Metabolism of foreign compounds by gastrointestinal microorganisms, *Pharmacol. Rev.*, 25, 451, 1973.
6. **Deuel, H. J., Calbert, C. E., Anisfeld, L., McKeehan, H., and Blunden, H. D.,** Sorbic acid as a fungistatic agent for foods. II. Metabolism of α,β-unsaturated fatty acids with emphasis on sorbic acid, *Food Res.*, 19, 13, 1954.

187

7. **Fingerhut, M., Schmidt, B., and Lang, K.,** Über den Stoffwechsel der 1-^{14}C-Sorbinsäure, *Biochem. Z.*, 336, 118, 1962.

8. **Kuhn, R., Köhler, F., and Köhler, L.,** Über die biologische Oxydation hochungsättiger Fettsäuren. Ein neuer Weg zur Darstellung von Polyen-dicarbonsäuren, *Hoppe-Seyler's Z. Physiol. Chem.*, 247, 197, 1937.

9. **Kuhn, R., Köhler, F., and Köhler, L.,** Über Methyl-oxydationen im Tierkörper, *Hoppe-Seyler's Z. Physiol. Chem.*, 242, 171, 1936.

10. **Asano, M. and Yamakawa, T.,** The fate of branched chain fatty acids in animal body. I. A contribution to the problem of "Hildebrandt acid", *J. Biochem. (Tokyo)*, 37, 321, 1950.

11. **Weinhouse, S. and Friedmann, B.,** Metabolism of labeled 2-carbon acids in the intact rat, *J. Biol. Chem.*, 191, 707, 1951.

12. **Curtin, C. O. and King, C. G.,** The metabolism of ascorbic acid-1-C^{14} and oxalic acid-C^{14} in the rat, *J. Biol. Chem.*, 216, 539, 1955.

13. **Jones, A. R., Gadiel, P., and Stevenson, D.,** The fate of oxalic acid in the Wistar rat, *Xenobiotica*, 11, 385, 1981.

14. **Hodgkinson, A.,** Oxalate metabolism in animals and man, in *Oxalic Acid in Biology and Medicine*, Academic Press, London, 1977, chap. 6.

15. **Salminen, S. and Salminen, E.,** Urinary excretion of orally administered oxalic acid in MYRJ 45-treated NMRI mice, *Toxicol. Lett.*, 37, 91, 1987.

16. **Wilson, G. D. A. and Harvey, D. G.,** Studies on experimental oxaluria in pigs, *Br. Vet. J.*, 133, 418, 1977.

17. **McKenzie, R. A., Blaney, B. J., and Gartner, R. J. W.,** The effect of dietary oxalate on calcium, phosphorus and magnesium balances in horses, *J. Agric. Sci. Cambridge*, 97, 69, 1981.

18. **Shirley, E. K. and Schmidt-Nielsen, K.,** Oxalate metabolism in the pack rat, sand rat, hamster, and white rat, *J. Nutr.*, 91, 496, 1967.

19. **Elder, T. D. and Wyngaarden, J. B.,** The biosynthesis and turnover of oxalate in normal and hyperoxaluric subjects, *J. Clin. Invest.*, 39, 1337, 1960.

20. **Morris, M. P. and García-Rivera, J.,** The destruction of oxalates by the rumen contents of cows, *J. Dairy Sci.*, 38, 1169, 1955.

21. **Watts, P. S.,** Decomposition of oxalic acid *in vitro* by rumen contents, *Aust. J. Agric. Res.*, 8, 266, 1957.

22. **Allison, M. J., Littledike, E. T., and James, L. F.,** Changes in ruminal oxalate degradation rates associated with adaptation to oxalate ingestion, *J. Animal Sci.*, 45, 1173, 1977.

23. **Allison, M. J., Cook, H. M., and Dawson, K. A.,** Selection of oxalate-degrading rumen bacteria in continuous cultures, *J. Anim. Sci.*, 53, 810, 1981.

24. **Allison, M. J.,** Anaerobic oxalate-degrading bacteria of the gastrointestinal tract, in *Plant Toxicology Proceedings of the Australia – U.S.A. Poisonous Plants Symposium Brisbane, Australia, May 14-18, 1984*, Seawright, A. A., Hegarty, M. P., James, L. F., and Keeler, R. F., Eds., Queensland Poisonous Plants Committee, Yeerongpilly, Australia, 1985, 120.

25. **Bernhard, K. and Andreae, M.,** Stoffwechselversuche mit Dicarbonsäuren, *Hoppe-Seyler's Z. Physiol. Chem.*, 245, 103, 1937.

26. **Weitzel, G.,** Die Bernsteinsäureausscheidung bei Stoffwechselbelastung mit höheren *n*-Dicarbonsäuren, *Hoppe-Seyler's Z. Physiol. Chem.*, 282, 185, 1947.

27. **Underhill, F. P., Leonard, C. S., Gross, E. G., and Jaleski, T. C.,** Studies on the metabolism of tartrates. II. The behavior of tartrate in the organism of the rabbit, dog, rat and guinea pig, *J. Pharmacol. Exp. Ther.*, 43, 359, 1931.

28. **Underhill, F. P., Peterman, F. I., Jaleski, T. C., and Leonard, C. S.,** Studies on the metabolism of tartrates. III. The behavior of tartrates in the human body, *J. Pharmacol. Exp. Ther.*, 43, 381, 1931.

29. **Finkle, P.,** The fate of tartaric acid in the human body, *J. Biol. Chem.*, 100, 349, 1933.

30. **Gry, J. and Larsen, J. C.,** Metabolism of L(+)- and D(–)-tartaric acid in different animal species, *Arch. Toxicol.*, Suppl. 1, 351, 1978.

31. **Chasseaud, L. F., Down, W. H., and Kirkpatrick, D.,** Absorption and biotransformation of L(+)-tartaric acid in rats, *Experientia*, 33, 998, 1977.

32. **Chadwick, V. S., Vince, A., Killingley, M., and Wrong, O. M.,** The metabolism of tartrate in man and the rat, *Clin. Sci. Mol. Med.*, 54, 273, 1978.

33. **Malan, J., Müller, F. O., and Hundt, H. K. L.,** Kinetika van DL-Wynsteensuur en Menslike Vrywilligers, *S. Afr. J. Sci.*, 75, 319, 1979.

34. **Lautemann, E.,** Ueber die Reduction der Chinasäure zu Benzoësäure und die Verwandlung derselben in Hippursäure im theirischen Organismus, *Justus Liebigs Ann. Chem.*, 125, 9, 1863.

35. **Quick, A. J.,** The conjugation of benzoic acid in man, *J. Biol. Chem.*, 92, 65, 1931.

36. **Vasiliu, H., Timoşencu, A., Zaimov, C., and Coteleu, V.,** The nonnitric mother substances of hippuric acid. The behavior of quinic acid and of other substances in the animal organism, *Bul. Fac. Sti Agric. Chişinău (Comun. Lab. Chim. Agr.)*, 2, 56, 1938 (Chem. Abstr. 32, 8514, 1938).

37. **Vasiliu, H., Timoşencu, A., Zaimov, C., and Coteleu, V.,** The nitrogen-free parent substances of hippuric acid. The behavior of quinic acid and other substances in the animal organism, *Bul. Fac. Sti. Agric. Chişinău (Comun. Lab. Chim. Agr.)*, 3, 77, 1940 (Chem. Abstr. 34, 4424, 1940).

38. **Bernhard, K.,** Stoffwechselversuche zur Dehydrierung des Cyclohexanringes, *Hoppe-Seyler's Z. Physiol. Chem.*, 248, 256, 1937.
39. **Beer, C. T., Dickens, F., and Pearson, J.,** The aromatization of hydrogenated derivatives of benzoic acid in animal tissues, *Biochem. J.*, 48, 222, 1951.
40. **Bernhard, K., Vuilleumier, J. P., and Brubacher, G.,** Zur Frage der Entstehung der Benzoesäure im Tierkörper, *Helv. Chim. Acta*, 38, 1438, 1955.
41. **Cotran, R., Kendrick, M. I., and Kass, E. H.,** Role of intestinal bacteria in aromatization of quinic acid in man and guinea pig, *Proc. Soc. Exp. Biol. Med.*, 104, 424, 1960.
42. **Asatoor, A. M.,** Aromatisation of quinic acid and shikimic acid by bacteria and the production of urinary hippurate, *Biochim. Biophys. Acta*, 100, 290, 1965.
43. **Adamson, R. H., Bridges, J. W., Evans, M. E., and Williams, R. T.,** Species differences in the aromatization of quinic acid *in vivo* and the role of gut bacteria, *Biochem. J.*, 116, 437, 1970.
44. **Martin, A. K.,** Metabolism of aromatic compounds in the rumen, *Proc. Nutr. Soc.*, 34, 69, 1975.
45. **Teuchy, H., Quatacker, J., Wolf, G., and Van Sumere, C. F.,** Quantitative investigation of the hippuric acid formation in the rat after administration of some possible aromatic and hydroaromatic precursors, *Arch. Int. Physiol. Biochim.*, 79, 573, 1971.
46. **Indahl, S. R. and Scheline, R. R.,** Quinic acid aromatization in the rat. Urinary hippuric acid and catechol excretion following the singular or repeated administration of quinic acid, *Xenobiotica*, 3, 549, 1973.
47. **Booth, A. N., Robbins, D. J., Masri, M. S., and DeEds, F.,** Excretion of catechol after ingestion of quinic and shikimic acids, *Nature (London)*, 187, 691, 1960.
48. **Martin, A. K.,** The origin of urinary aromatic compounds excreted by ruminants. III. The metabolism of phenolic compounds to simple phenols, *Br. J. Nutr.*, 48, 497, 1982.
49. **Booth, A. N. and Williams, R. T.,** Dehydroxylation of catechol acids by intestinal contents, *Biochem. J.*, 88, 66, 1963.
50. **Scheline, R. R.,** The metabolism of drugs and other organic compounds by the intestinal microflora, *Acta Pharmacol. Toxicol.*, 26, 332, 1968.
51. **Brewster, D., Jones, R. S., and Parke, D. V.,** Aromatization of shikimic acid in the rat and the role of gastrointestinal micro-organisms, *Biochem. Soc. Trans.*, 4, 518, 1976.
52. **Brewster, D., Jones, R. S., and Parke, D. V.,** The metabolism of shikimate in the rat, *Biochem. J.*, 170, 257, 1978.
53. **Brewster, D.,** Mass spectrometry in the elucidation of shikimate biotransformation products in the rat, *Biomed. Mass Spectrom.*, 6, 447, 1979.
54. **Balba, M. T. and Evans, W. C.,** The origin of hexahydrohippurate (cyclohexanoylglycine) in the urines of herbivores, *Biochem. Soc. Trans.*, 5, 300, 1977.
55. **Brewster, D., Jones, R. S., and Parke, D. V.,** Shikimic acid aromatization in animals, *Xenobiotica*, 7, 109, 1977.
56. **Brewster, D., Jones, R. S., and Parke, D. V.,** The metabolism of cyclohexanecarboxylate in the rat, *Biochem. J.*, 164, 595, 1977.
57. **Brewster, D., Jones, R. S., and Parke, D. V.,** The metablism of cyclohexanecarboxylic acid in the isolated perfused rat liver, *Xenobiotica*, 7, 601, 1977.
58. **Mitoma, C., Posner, H. S., and Leonard, F.,** Aromatization of hexahydrobenzoic acid by mammalian liver mitochondria, *Biochim. Biophys. Acta*, 27, 156, 1958.
59. **Babior, B. M. and Bloch, K.,** Aromatization of cyclohexanecarboxylic acid, *J. Biol. Chem.*, 241, 3643, 1966.
60. **Svardal, A. M. and Scheline, R. R.,** Intracellular localization and some properties of the system in guinea pig liver responsible for the aromatization of cyclohexanecarboxylic acid to hippuric acid, *Mol. Cell. Biochem.*, 65, 107, 1984.
61. **Svardal, A. M. and Scheline, R. R.,** The aromatization of cyclohexanecarboxylic acid to hippuric acid: substrate specificity and species differences, *Mol. Cell. Biochem.*, 67, 171, 1985.
62. **Svardal, A. M. and Scheline, R. R.,** The aromatization of cyclohexanecarboxyl-CoA to hippuric acid by guinea pig liver mitochondria: submitochondrial localization, *Mol. Cell. Biochem.*, 67, 181, 1985.
63. **Nixon, J. E., Yoss, J. K., Eisele, T. A., Pawlowski, N. E., and Sinnhuber, R. O.,** Metabolism and tissue ditribution of label from [9,10-methylene-^{14}C]sterculic acid in the rat, *Lipids*, 12, 629, 1977.
64. **Eisele, T. A., Yoss, J. K., Nixon, J. E., Pawlowski, N. E., Libbey, L. M., and O., S. R.,** Rat urinary metabolites of [9,10-*methylene-14*^{c}]sterculic acid, *Biochim. Biophys. Acta*, 488, 76, 1977.
65. **Bernhard, K. and Müller, L.,** Stoffwechselversuche mit ω-cyclopentenyl- und ω-cyclopenyl-substituierten Fettsäuren, *Hoppe-Seyler's Z. Physiol. Chem.*, 256, 85, 1938.
66. **Williams, R. T.,** *Detoxication Mechanisms*, Chapman and Hall, London, 1959, 349.
67. **Idle, J. R., Millburn, P., and Williams, R. T.,** Benzoylglutamic acid, a metabolite of benzoic acid in Indian fruit bats, *FEBS Lett.*, 59, 234, 1975.
68. **Idle, J. R., Millburn, P., and Williams, R. T.,** Taurine conjugates as metabolites of arylacetic acids in the ferret, *Xenobiotica*, 8, 253, 1978.

69. **Quistad, G. B., Staiger, L. E., and Schooley, D. A.,** The role of carnitine in the conjugation of acidic xenobiotics, *Drug Metab. Dispos.*, 14, 521, 1986.

70. **Kao, J., Jones, C. A., Fry, J. R., and Bridges, J. W.,** Species differences in the metabolism of benzoic acid by isolated hepatocytes and kidney tubule fragments, *Life Sci.*, 23, 1221, 1978.

71. **Shirkey, R. J., Kao, J., Fry, J. R., and Bridges, J. W.,** A comparison of xenobiotic metabolism in cells isolated from rat liver and small intestinal mucosa, *Biochem. Pharmacol.*, 28, 1461, 1979.

72. **Jones, A. R.,** Some observations on the urinary excretion of glycine conjugates by laboratory animals, *Xenobiotica*, 12, 387, 1982.

73. **Thabrew, M. I., Bababunmi, E. A., and French, M. R.,** The metabolic fate of [^{14}C]benzoic acid in protein-energy deficient rats, *Toxicol. Lett.*, 5, 363, 1980.

74. **Sieber, S. M., Cohn, V. H., and Wynn, W. T.,** The entry of foreign compounds into the thoracic duct lymph of the rat, *Xenobiotica*, 4, 265, 1974.

75. **Hirom, P. C., Millburn, P., and Smith, R. L.,** Bile and urine as complementary pathways for the excretion of foreign organic compounds, *Xenobiotica*, 6, 55, 1976.

76. **Bray, H. G., Neale, F. C., and Thorpe, W. V.,** The fate of certain organic acids and amides in the rabbit. I. Benzoic and phenylacetic acids and their amides, *Biochem. J.*, 40, 134, 1946.

77. **Sato, T., Suzuki, T., Fukuyama, T., and Yoshikawa, H.,** Studies on conjugation of S^{35}-sulfate with phenolic compounds. IV. Metabolism of *o*-cresol, *m*-cresol, salicylaldehyde, salicylic acid, toluene, benzoic acid and related substances in rat liver, *J. Biochem. (Tokyo)*, 43, 421, 1956.

78. **Acheson, R. M. and Gibbard, S.,** The hydroxylation of benzoic acid by rats and guinea-pigs, *Biochim. Biophys. Acta*, 59, 320, 1962.

79. **Marsh, M. V., Caldwell, J., Hutt, A. J., Smith, R. L., Horner, M. W., Houghton, E., and Moss, M. S.,** 3-Hydroxy- and 3-keto-3-phenylpropionic acids: novel metabolites of benzoic acid in horse urine, *Biochem. Pharmacol.*, 31, 3225, 1982.

80. **Kazakoff, C. W. and Mamer, O. A.,** Biological conversion of β-phenylhydracrylic acid to hippuric acid, *Biomed. Mass Spectrom.*, 5, 612, 1978.

81. **Quick, A. J.,** Quantitative studies of β-oxidation. II. The metabolism of phenylvaleric acid, phenyl-α,β-pentenic acid, phenyl-β,γ-pentenic acid, mandelic acid, phenyl-β-hydroxypropionic acid, and acetophenone in dogs, *J. Biol. Chem.*, 80, 515, 1928.

82. **Williams, R. T.,** *Detoxication Mechanisms*, Chapman and Hall, London, 1959, 374.

83. **James, M. O., Smith, R. L., Williams, R. T., and Reidenberg, M.,** The conjugation of phenylacetic acid in man, sub-human primates and some non-primate species, *Proc. R. Soc. Lond. (Biol.)*, 182, 25, 1972.

84. **Thierfelder, H. and Sherwin, C. P.,** Phenylacetyl-glutamin, ein Stoffwechsel-Produkt des menschlichen Körpers nach Eingabe von Phenyl-essigsäure, *Ber. Dtsch. Chem. Ges.*, 47, 2630, 1914.

85. **Thierfelder, H. and Sherwin, C. P.,** Phenylacetylglutamin und seine Bildung im menschlichen Körper nach Eingabe von Phenylessigsaüre, *Hoppe-Seyler's Z. Physiol. Chem.*, 94, 1, 1915.

86. **Hirom, P. C., Idle, J. C., Millburn, P., and Williams, R. T.,** Glutamine conjugation of phenylacetic acid in the ferret, *Biochem. Soc. Trans.*, 5, 1033, 1977.

87. **Idle, J. R., Millburn, P., and Williams, R. T.,** Taurine conjugation of arylacetic acids in the ferret, *Biochem. Soc. Trans.*, 4, 139, 1976.

88. **James, M. O. and Smith, R. L.,** The conjugation of phenylacetic acid in phenylketonurics, *Eur. J. Clin. Pharmacol.*, 5, 243, 1973.

89. **Dakin, H. D.,** The mode of oxidation in the animal organism of phenyl derivatives of fatty acids. IV. Further studies on the fate of phenylpropionic acid and some of its derivatives, *J. Biol. Chem.*, 6, 203, 1909.

90. **Raper, H. S. and Wayne, E. J.,** A quantitative study of the oxidation of phenyl-fatty acids in the animal organism, *Biochem. J.*, 22, 188, 1928.

91. **El Masry, A. M., Smith, J. N., and Williams, R. T.,** Studies in detoxication. LXIX. The metabolism of alkylbenzenes: *n*-propylbenzene and *n*-butylbenzene with further observations on ethylbenzene, *Biochem. J.*, 64, 50, 1956.

92. **Fahelbum, I. M. S. and James, S. P.,** The absorption and metabolism of methyl cinnamate, *Toxicology*, 7, 123, 1977.

93. **Teuchy, H. and Van Sumere, C. F.,** The metabolism of [1-^{14}C]phenylalanine, [3-^{14}C]cinnamic acid and [2-^{14}C]ferulic acid in the rat, *Arch. Int. Physiol. Biochim.*, 79, 589, 1971.

94. **Bhatia, I. S., Bajaj, K. L., and Chakravarti, P.,** Metabolism of cinnamic acid in albino rats, *Indian J. Exp. Biol.*, 15, 118, 1977.

95. **Quick, A. J.,** Quantitative studies of β-oxidation. I. The conjugation of benzoic acid and phenylacetic acid formed as the end-products from the oxidation of phenyl-substituted aliphatic acids, *J. Biol. Chem.*, 77, 581, 1928.

96. **Snapper, I., Yü, T. F., and Chiang, Y. T.,** Cinnamic acid metabolism in man, *Proc. Soc. Exp. Biol. Med.*, 44, 30, 1940.

97. **Snapper, I. and Saltzman, A.,** Hippuric acid, cinnamoylglucuronic acid and benzoylglucuronic acid in the urine of normal individuals and in patients with hepatic dysfunction after ingestion of sodium cinnamate, *Arch. Biochem.,* 24, 1, 1949.

98. **Hoskins, J. A., Holliday, S. B., and Greenway, A. M.,** The metabolism of cinnamic acid by healthy and phenylketonuric adults: a kinetic study, *Biomed. Mass Spectrom.,* 11, 296, 1984.

99. **Delbressine, L. P. C., Klippert, P. J. M., Reuvers, J. T. A., and Seutter-Berlage, F.,** Isolation and identification of mercapturic acids of cinnamic aldehyde and cinnamyl alcohol from urine of female rats, *Arch. Toxicol.,* 49, 57, 1981.

100. **Scheline, R. R.,** Metabolism of phenolic acids by the rat intestinal microflora, *Acta Pharmacol. Toxicol.,* 26, 189, 1968.

101. **Hansen, I. L. and Crawford, M. A.,** Bacterial degradation of the aromatic amino acid side chain, *Biochem. Pharmacol.,* 17, 338, 1968.

102. **Hoskins, J. A.,** The occurrence, metabolism and toxicity of cinnamic acid and related compounds, *J. Appl. Toxicol.,* 4, 283, 1984.

103. **Erspamer, V.,** Observations on the fate of indolalkylamines in the organism, *J. Physiol. (London),* 127, 118, 1955.

104. **Bridges, J. W., Evans, M. E., Idle, J. R., Millburn, P., Osiyemi, F. O., Smith, R. L., and Williams, R. T.,** The conjugation of indolylacetic acid in man, monkeys and other species, *Xenobiotica,* 4, 645, 1974.

105. **Williams, R. T.,** *Detoxication Mechanisms,* Chapman and Hall, London, 1959, 380.

106. **Williams, R. T.,** *Detoxication Mechanisms,* Chapman and Hall, London, 1959, 383.

107. **Ohtsuji, H. and Ikeda, M.,** The metabolism of styrene in the rat and the stimulatory effect of phenobarbital, *Toxicol. Appl. Pharmacol.,* 18, 321, 1971.

108. **Drummond, L., Caldwell, J., and Wilson, H. K.,** The stereoselectivity of 1,2-phenylethanediol and mandelic acid metabolism and disposition in the rat, *Xenobiotica,* 20, 159, 1990.

109. **Gosselin, R. E., Gabourel, J. D., Kalser, S. C., and Wills, J. H.,** The metabolism of C^{14}-labeled atropine and tropic acid in mice, *J. Pharmacol. Exp. Ther.,* 115, 217, 1955.

110. **Ve, B. and Scheline, R. R.,** Unpublished data, 1977.

111. **Hutt, A. J., Caldwell, J., and Smith, R. L.,** The metabolism of [*carboxyl*-^{14}C]aspirin in man, *Xenobiotica,* 12, 601, 1982.

112. **Caldwell, J., O'Gorman, J., and Smith, R. L.,** Inter-individual differences in the glycine conjugation of salicylic acid, *Br. J. Clin. Pharmacol.,* 9, 114, 1980.

113. **Montgomery, P. R., Berger, L. G., Mitenko, P. A., and Sitar, D. S.,** Salicylate metabolism: effects of age and sex in adults, *Clin. Pharmacol. Ther.,* 39, 571, 1986.

114. **Emudianughe, T. S., Oduleye, S. O., Ebadan, E. E., and Eneji, S. D.,** Sex differences in salicylic acid metabolism in Nigerian subjects, *Xenobiotica,* 16, 177, 1986.

115. **Emudianughe, T. S.,** Possible genetic influence on conjugate formation in salicylic acid metabolism, *Arch. Int. Pharmacodyn. Ther.,* 292, 7, 1988.

116. **Hall, B. E. and James, S. P.,** Some pathways of xenobiotic metabolism in the adult and neonatal marmoset (*Callithrix jacchus*), *Xenobiotica,* 10, 421, 1980.

117. **Hollister, L. and Levy, G.,** Some aspects of salicylate distribution and metabolism in man, *J. Pharm. Sci.,* 54, 1126, 1965.

118. **Schubert, B.,** Identification and metabolism of some doping substances in horses, *Acta Vet. Scand.,* Suppl. 21, 1967.

119. **Davis, L. E. and Westfall, B. A.,** Species differences in biotransformation and excretion of salicylate, *Am. J. Vet. Res.,* 33, 1253, 1972.

120. **Haberland, G. L., Medenwald, H., and Köster, L.,** Stoffwechseluntersuchungen an Pharmazeutika. II. Stoffwechselendprodukte der Salicyl-, Acetylsalicyl-, Gentisin- und Salicylursäure im Rattenurin, *Hoppe-Seyler's Z. Physiol. Chem.,* 306, 235, 1957.

121. **Williams, R. T.,** Studies in detoxication. I. The influence of (a) dose and (b) *o-, m-* and *p*-substitution on the sulphate detoxication of phenol in the rabbit, *Biochem. J.,* 32, 878, 1938.

122. **Bray, H. G., Ryman, B. E., and Thorpe, W. V.,** The fate of certain organic acids and amides in the rabbit. V. *o-* and *m*-Hydroxybenzoic acids and amides, *Biochem. J.,* 43, 561, 1948.

123. **Alpen, E. L., Mandel, H. G., Rodwell, V. W., and Smith, P. K.,** The metabolism of C^{14} carboxyl salicylic acid in the dog and in man, *J. Pharmacol. Exp. Ther.,* 102, 150, 1951.

124. **Schayer, R. W.,** The metabolism of radioactive salicylic acid, *Arch. Biochem.,* 28, 371, 1950.

125. **Alpen, E. L., Mandel, H. G. and Smith, K. P.,** Studies on the metabolism of C^{14} carboxyl salicylate, *J. Pharmacol. Exp. Ther.,* 101, 1, 1951.

126. **Scheline, R. R.,** Decarboxylation and demethylation of some phenolic benzoic acid derivatives by rat caecal contents, *J. Pharm. Pharmacol.,* 18, 664, 1966.

127. **Martin, A. K.,** The origin of urinary aromatic compounds excreted by ruminants. II. The metabolism of phenolic cinnamic acids to benzoic acid, *Br. J. Nutr.,* 47, 155, 1982.

128. **Baldoni, A.,** Sul comportamento del salicilato sodico nell' organismo, *Archo. Farmac. Sper.*, 8, 174, 1909.
129. **Kapp, E. M. and Coburn, A. F.,** Urinary metabolites of sodium salicylate, *J. Biol. Chem.*, 145, 549, 1942.
130. **Davis, L. E., Westfall, B. A., and Short, C. R.,** Biotransformation and pharmacokinetics of salicylate in newborn animal, *Am. J. Vet. Res.*, 34, 1105, 1973.
131. **Wilson, J. T., Howell, R. L., Holladay, M. W., Brilis, G. M., Chrastil, J., Watson, J. T., and Taber, D. F.,** Gentisuric acid: metabolic formation in animals and identification as a metabolite of aspirin in man, *Clin. Pharmacol. Ther.*, 23, 635, 1978.
132. **Bray, H. G., Thorpe, W. V., and White, K.,** The fate of certain organic acids and amides in the rabbit. X. The application of paper chromatography to metabolic studies of hydroxybenzoic acids and amides, *Biochem. J.*, 46, 271, 1950.
133. **Grootveld, M. and Halliwell, B.,** 2,3-Dihydroxybenzoic acid is a product of human aspirin metabolism, *Biochem. Pharmacol.*, 37, 271, 1988.
134. **Kyle, M. E. and Kocsis, J. J.,** Metabolism of salicylate by isolated kidney and liver mitochondria, *Chem. Biol. Interact.*, 59, 325, 1986.
135. **Dumazert, C. and El Ouachi, M.,** Sur le métabolisme des salicylates, *Ann. Pharm. Fr.*, 12, 723, 1954.
136. **Mitoma, C., Posner, H. S., Reitz, H. C., and Udenfriend, S.,** Enzymatic hydroxylation of aromatic compounds, *Arch. Biochem. Biophys.*, 61, 431, 1956.
137. **Robinson, D. and Williams, R. T.,** Glucuronides of salicylic acid, *Biochem. J.*, 62, 23, 1956.
138. **Levy, G. and Leonards, J. R.,** Absorption, metabolism, and excretion of salicylates, in *The Salicylates*, Smith, M. J. H. and Smith, P. K., Eds., Interscience Publishers, New York, 1966, 5.
139. **Davison, C.,** Salicylate metabolism in man, *Ann. N.Y. Acad. Sci.*, 179, 249, 1971.
140. **Hucker, H. B., Kwan, K. C., and Duggan, D. E.,** Pharmacokinetics and metabolism of non-steroidal anti-inflammatory agents, in *Progress in Drug Metabolism*, Vol. 5, Bridges, J. W. and Chasseaud, L. F., Eds., John Wiley & Sons, London, 1980, 165.
141. **Tsukamoto, H. and Terada, S.,** Metabolism of drugs. XXVII. Metabolic fate of *p*-hydroxybenzoic acid and its derivatives in rabbit, *Chem. Pharm. Bull.*, 10, 91, 1962.
142. **Booth, A. N., Jones, F. T., and DeEds, F.,** Metabolic and glucosuria studies on naringin and phloridzin, *J. Biol. Chem.*, 233, 280, 1958.
143. **Derache, R. and Gourdon, J.,** Métabolisme d'un conservateur alimentaire: l'acide parahydroxybenzoïque et ses esters, *Food Cosmet. Toxicol.*, 1, 189, 1963.
144. **Kiwada, H., Awazu, S., and Hanano, M.,** The study on the biological fate of paraben at the dose of practical usage in rat. I. The metabolism and excretion of ethyl *p*-hydroxybenzoate (ethyl paraben) and *p*-hydroxybenzoic acid, *J. Pharmacobiol. Dyn.*, 2, 356, 1979.
145. **Phillips, J. C., Hardy, K., Richards, R., Cottrell, R. C., and Gangoli, S. D.,** Studies of the metabolic fate of *p*-hydroxybenzoic acid in male and female cats, *Toxicology*, 7, 257, 1977.
146. **Quick, A. J.,** The relationship between chemical structure and physiological response. II. The conjugation of hydroxy- and methoxybenzoic acids, *J. Biol. Chem.*, 97, 403, 1932.
147. **Tompsett, S. L.,** Polyhydroxy (catecholic) phenolic acids—studies of their metabolism in man, *J. Pharm. Pharmacol.*, 13, 115, 1961.
148. **Sato, T., Suzuki, T., Fukuyama, T., and Yoshikawa, H.,** Studies on conjugation of S^{35}-sulfate with phenolic compounds. III. Metabolism of *p*-cresol, *p*-hydroxybenzaldehyde and *p*-hydroxybenzoic acid in rat liver, *J. Biochem. (Tokyo)*, 43, 413, 1956.
149. **Curtius, H. C., Mettler, M., and Ettlinger, L.,** Study of the intestinal tyrosine metabolism using stable isotopes and gas chromatography-mass spectrometry, *J. Chromatogr.*, 126, 569, 1976.
150. **Soleim, H. A. and Scheline, R. R.,** Metabolism of xenobiotics by strains of intestinal bacteria, *Acta Pharmacol. Toxicol.*, 31, 471, 1972.
151. **Clarke, N. E. and Mosher, R. E.,** Phenolic compounds in the treatment of rheumatic fever. II. The metabolism of gentisic acid and the ethanolamide of gentisic acid, *Circulation*, 7, 337, 1953.
152. **Consden, R. and Stanier, W. M.,** Metabolism of gentisic acid, *Biochem. J.*, 48, 14, 1951.
153. **Roseman, S. and Dorfman, A.,** The determination and metabolism of gentisic acid, *J. Biol. Chem.*, 192, 105, 1951.
154. **Batterman, R. C. and Sommer, E. M.,** Fate of gentisic acid in man as influenced by alkalinization and acidification, *Proc. Soc. Exp. Biol. Med.*, 82, 376, 1953.
155. **Astill, B. D., Fassett, D. W., and Roudabush, R. L.,** The metabolism of phenolic antioxidants. IV. The metabolites of gentisic acid in the dog, *Biochem. J.*, 90, 194, 1964.
156. **Sakamoto, Y., Inamori, K., and Nasu, H.,** Methylation of gentisic acid. Formation of 5-methoxysalicylic acid, *J. Biochem. (Tokyo)*, 46, 1667, 1959.
157. **Dodgson, K. S. and Williams, R. T.,** Studies in detoxication. XXVII. The orientation of conjugation in the metabolites of 4-chlorocatechol and 4-chlororesorcinol, with some observation on the fate of (+)-adrenaline, protocatechuic acid and protocatechuic aldehyde in the rabbit, *Biochem. J.*, 45, 381, 1949.

158. **DeEds, F., Booth, A. N., and Jones, F. T.,** Methylation of phenolic hydroxyl groups by rabbit, *Fed. Proc. Fed. Am. Soc. Exp. Biol.*, 14, 332, 1955.

159. **DeEds, F., Booth, A. N., and Jones, F. T.,** Methylation and dehydroxylation of phenolic compounds by rats and rabbits, *J. Biol. Chem.*, 225, 615, 1957.

160. **Masri, M. S., Booth, A. N., and DeEds, F.,** The metabolism and acid degradation of quercetin, *Arch. Biochem. Biophys.*, 85, 284, 1959.

161. **Scheline, R. R.,** The decarboxylation of some phenolic acids by the rat, *Acta Pharmacol. Toxicol.*, 24, 275, 1966.

162. **Wong, K. P. and Sourkes, T. L.,** Metabolism of vanillin and related substances in the rat, *Can. J. Biochem.*, 44, 635, 1966.

163. **Dacre, J. C. and Williams, R. T.,** Dehydroxylation of [¹⁴C]protocatechuic acid in the rat, *Biochem. J.*, 84, 81, 1962.

164. **Dacre, J. C. and Williams, R. T.,** The role of the tissues and gut micro-organisms in the metabolism of [¹⁴C]protocatechuic acid in the rat. Aromatic dehydroxylation, *J. Pharm. Pharmacol.*, 20, 610, 1968.

165. **Pellerin, J. and D'Iorio, A.,** Methylation of the 3-OH position of catechol acids by rat liver and kidney preparations, *Can. J. Biochem. Physiol.*, 36, 491, 1958.

166. **Axelrod, J. and Tomchick, R.,** Enzymatic O-methylation of epinephrine and other catechols, *J. Biol. Chem.*, 233, 702, 1958.

167. **Dirscherl, W. and Brisse, B.,** Exogene Vorstufen der Vanillinsäure: Vanillin, Protocatechusäure, Kaffeesäure und Feralusäure (Homogenatversuche mit Ratten- und Menschenleber), *Hoppe-Seyler's Z. Physiol. Chem.*, 346, 55, 1966.

168. **Masri, M. S., Booth, A. N., and DeEds, F.,** O-Methylation *in vitro* of dihydroxy- and trihydroxy-phenolic compounds by liver slices, *Biochim. Biophys. Acta*, 65, 495, 1962.

169. **Hill, G. A., Ratcliffe, J., and Smith, P.,** Urinary catechol ethers, *Chem. Ind.*, 399, 1959.

170. **Creveling, C. R., Dalgard, N., Shimizu, H., and Daly, J. W.,** Catechol O-methyltransferase. III. *m*- and *p*-O-methylation of catecholamines and their metabolites, *Mol. Pharmacol.*, 6, 691, 1970.

171. **Creveling, C. R., Morris, N., Shimizu, H., Ong, H. H., and Daly, J.,** Catechol O-methyltransferase. IV. Factors affecting *m*- and *p*-methylation of substituted catechols, *Mol. Pharmacol.*, 8, 398, 1972.

172. **Marzullo, G. and Friedhoff, A. J.,** Catechol-O-methyltransferase from rat liver: two forms having different *meta:para* methylation ratios, *Life Sci.*, 17, 933, 1975.

173. **Booth, A. N., Masri, M. S., Robbins, D. J., Emerson, O. H., Jones, F. T., and DeEds, F.,** The metabolic fate of gallic acid and related compounds, *J. Biol. Chem.*, 234, 3014, 1959.

174. **Blumenberg, F.-W. and Dohrmann, R.,** Über den Ausscheidungsmechanismus oral zugeführter Gallussäure, *Arzneim. Forsch.*, 10, 109, 1960.

175. **Watanabe, A. and Oshima, Y.,** Metabolism of gallic acid and tea catechin by rabbit, *Agric. Biol. Chem.*, 29, 90, 1965.

176. **Masri, M. S., Robbins, D. J., Emerson, O. H., and DeEds, F.,** Selective *para*- or *meta*-O-methylation with catechol O-methyl transferase from rat liver, *Nature (London)*, 202, 878, 1964.

177. **Tompsett, S. L.,** The determination and excretion of polyhydroxy (catecholic) phenolic acids in urine, *J. Pharm. Pharmacol.*, 10, 157, 1958.

178. **Krumholz, L. R. and Bryant, M. P.,** *Eubacterium oxidoreducens* sp. nov. requiring H₂ or formate to degrade gallate, pyrogallol, phloroglucinol and quercetin, *Arch. Microbiol.*, 144, 8, 1986.

179. **Bray, H. G., Humphris, B. G., Thorpe, W. V., White, K., and Wood, P. B.,** Kinetic studies on the metabolism of foreign organic compounds. VI. Reactions of some nuclear-substituted benzoic acids, benzamides and toluenes in the rabbit, *Biochem. J.*, 59, 162, 1955.

180. **Cramer, M. B. and Michael, W. R.,** Metabolism of p-anisic acid by the rat, *Life Sci.*, 10(Suppl. 2), 1255, 1971.

181. **Axelrod, J.,** The enzymic cleavage of aromatic ethers, *Biochem. J.*, 63, 634, 1956.

182. **Sammons, H. G. and Williams, R. T.,** Studies in detoxication. XII The metabolism of vanillin and vanillic acid in the rabbit. The identification of glucurovanillin and the structure of glucurovanillic acid, *Biochem. J.*, 35, 1175, 1941.

183. **Krumholz, L. R. and Bryant, M. P.,** *Syntrophococcus sucromutans* sp. nov. gen. nov. uses carbohydrates as electron donors and formate, methoxymonobenzenoids or *Methanobrevibacter* as electron acceptor systems, *Arch. Microbiol.*, 143, 313, 1986.

184. **Sammons, H. G. and Williams, R. T.,** Studies in detoxication. XIV. The metabolism of veratraldehyde and veratric acid in the rabbit, *Biochem. J.*, 40, 223, 1946.

185. **Thomas, H. and Müller-Enoch, D.,** Stoffwechsel von 3,4-Dimethoxybenzosäure in der Rattenleber, *Naturwissenschaften*, 61, 222, 1974.

186. **Heffter, A.,** Zur Pharmakologie der Safrolgruppe, *Arch. Exp. Path. Pharmak.*, 35, 342, 1895.

187. **Williams, R. T.,** *Detoxication Mechanisms*, Chapman and Hall, London, 1959, 371.

188. **Acheson, R. M. and Atkins, G. L.,** The metabolites of piperic acid and some related compounds in the rat, *Biochem. J.*, 79, 268, 1961.

189. **Klungsøyr, J. and Scheline, R. R.,** Metabolism in rats of several carboxylic acid derivatives containing the 3,4-methylenedioxyphenyl group, *Acta Pharmacol. Toxicol.,* 49, 305, 1981.

190. **Kamienski, F. X. and Casida, J. E.,** Importance of demethylenation in the metabolism *in vivo* and *in vitro* of methylenedioxyphenyl synergists and related compounds in mammals, *Biochem. Pharmacol.,* 19, 91, 1970.

191. **Griffiths, L. A.,** Metabolism of sinapic acid and related compounds in the rat, *Biochem. J.,* 113, 603, 1969.

192. **Numerof, P., Gordon, M., and Kelly, J. M.,** The metabolism of reserpine. I. Studies in the mouse with C-14 labeled reserpine, *J. Pharmacol. Exp. Ther.,* 115, 427, 1955.

193. **Ewins, A. J. and Laidlaw, P. P.,** The fate of parahydroxyphenylethylamine in the organism, *J. Physiol. (London),* 41, 78, 1910.

194. **Wiseman-Distler, M. H., Sourkes, T. L., and Carabin, S.,** Precursors of 3,4-dihydroxyphenylacetic acid and 4-hydroxy-3-methoxyphenylacetic acid in the rat, *Clin. Chim. Acta,* 12, 335, 1965.

195. **Booth, A. N., Murray, C. W., DeEds, F., and Jones, F. T.,** Metabolic fate of rutin and quercetin, *Fed. Proc. Fed. Am. Soc. Exp. Biol.,* 14, 321, 1955.

196. **Dacre, J. C., Scheline, R. R., and Williams, R. T.,** The role of the tissues and gut flora in the metabolism of [¹⁴C]homoprotocatechuic acid in the rat and rabbit, *J. Pharm. Pharmacol.,* 20, 619, 1968.

197. **Scheline, R. R., Williams, R. T., and Wit, J. G.,** Biological dehydroxylation, *Nature (London),* 188, 849, 1960.

198. **Alton, H. and Goodall, M.,** Metabolism of 3,4-dihydroxyphenylacetic acid (DOPAC) in the human, *Biochem. Pharmacol.,* 18, 1373, 1969.

199. **Scheline, R. R.,** 4-Methylcatechol, a metabolite of homoprotocatechuic acid, *Experientia,* 23, 493, 1967.

200. **Furuya, T.,** Studies on the metabolism of naturally occurring coumarins. VI. Urinary metabolites of *o*-coumaric acid and melilotic acid, *Chem. Pharm. Bull.,* 6, 706, 1958.

201. **Booth, A. N., Masri, M. S., Robbins, D. J., Emerson, O. H., Jones, F. T., and DeEds, F.,** Urinary metabolites of coumarin and *o*-coumaric acid, *J. Biol. Chem.,* 234, 946, 1959.

202. **Mead, J. A. R., Smith, J. N., and Williams, R. T.,** Studies in detoxication. LXXII. The metabolism of coumarin and of *o*-coumaric acid, *Biochem. J.,* 68, 67, 1958.

203. **Flatow, L.,** Über den Abbau von Aminosäuren im Organismus, *Hoppe-Seyler's Z. Physiol. Chem.,* 64, 367, 1910.

204. **Booth, A. N., Masri, M. S., Robbins, D. J., Emerson, O. H., Jones, F. T., and DeEds, F.,** Urinary phenolic acid metabolites of tyrosine, *J. Biol. Chem.,* 235, 2649, 1960.

205. **Griffiths, L. A. and Smith, G. E.,** Metabolism of apigenin and related compounds in the rat, *Biochem. J.,* 128, 901, 1972.

206. **Bajaj, K. L., Singh, J., and Chakravarti, P.,** Enzymic hydroxylation of *m*-coumaric acid by mice liver hydroxylase, *Indian J. Exp. Biol.,* 15, 381, 1977.

207. **Ranganathan, S. and Ramasarma, T.,** Enzymic formation of *p*-hydroxybenzoate from *p*-hydroxycinnamate, *Biochem. J.,* 122, 487, 1971.

208. **Ranganathan, S. and Ramasarma, T.,** The metabolism of phenolic acids in the rat, *Biochem. J.,* 140, 517, 1974.

209. **Indahl, S. R. and Scheline, R. R.,** Decarboxylation of 4-hydroxycinnamic acids by *Bacillus* strains isolated from rat intestine, *Appl. Microbiol.,* 16, 667, 1968.

210. **Booth, A. N., Emerson, O. H., Jones, F. T., and DeEds, F.,** Urinary metabolites of caffeic and chlorogenic acids, *J. Biol. Chem.,* 229, 51, 1957.

211. **Shaw, K. N. F. and Trevarthen, J.,** Exogenous sources of urinary phenol and indole acids, *Nature (London),* 182, 797, 1958.

212. **Shaw, K. N. F., Gutenstein, M. and Jepson, J. B.,** Intestinal flora and diet in relation to *m*-hydroxyphenyl acids of human urine, in *Proc. 5th Int. Cong. Biochem., Moscow 1961,* Vol. 9, Sissakian, N. M., Ed., Pergamon Press, Oxford, 1963, 427.

213. **Booth, A. N. and Williams, R. T.,** Dehydroxylation of caffeic acid by rat and rabbit cæcal contents and sheep rumen liquor, *Nature (London),* 198, 684, 1963.

214. **Dayman, J. and Jepson, J. B.,** The metabolism of caffeic acid in humans: the dehydroxylating action of intestinal bacteria, *Biochem. J.,* 113, 11, 1969.

215. **Scheline, R. R. and Midtvedt, T.,** Absence of dehydroxylation of caffeic acid in germ-free rats, *Experientia,* 26, 1068, 1970.

216. **Peppercorn, M. A. and Goldman, P.,** Caffeic acid metabolism by gnotobiotic rats and their intestinal bacteria, *Proc. Natl. Acad. Sci. U.S.A.,* 69, 1413, 1972.

217. **Perez-Silva, G., Rodriquez, D., and Perez-Silva, J.,** Dehydroxylation of caffeic acid by a bacterium isolated from rat faeces, *Nature (London),* 212, 303, 1966.

218. **Perez-Silva, M. G. and Rodriquez Sánchez, D.,** Deshidroxilación del ácido cafeico por varias especies de pseudomonas, *Boln R. Soc. Esp. Hist. Nat.,* 65, 401, 1967.

219. **Peppercorn, M. A. and Goldman, P.,** Caffeic acid metabolism by bacteria of the human gastrointestinal tract, *J. Bacteriol.,* 108, 996, 1971.

220. **Woo, W. S.,** *p*-Methoxycinnamate and its metabolite in rabbit serum, *J. Pharm. Sci.,* 57, 27, 1968.

221. **Solheim, E. and Scheline, R. R.,** Metabolism of alkenebenzene derivatives in the rat. I. *p*-Methoxyallylbenzene (estragole) and *p*-methoxypropenylbenzene (anethole), *Xenobiotica,* 3, 493, 1973.

222. **Plummer, J. L.,** The Metabolism of Curcumin and Some Related Substances, Ph. D. Thesis, University of Sydney, Sydney, Australia, 1977.

223. **Westendorf, J. and Czok, G.,** Studies on the pharmacokinetics of ¹⁴C-cinnamic acid derivatives in rats, *Z. Ernährungswiss.,* 17, 26, 1978 (Chem. Abstr. 88, 182436u, 1978).

224. **Solheim, E. and Scheline, R. R.,** Metabolism of alkenebenzene derivatives in the rat. II. Eugenol and isoeugenol methyl ethers, *Xenobiotica,* 6, 137, 1976.

225. **Smith, G. E. and Griffiths, L. A.,** Metabolism of *N*-acylated and *O*-dealkylated drugs by the intestinal microflora during anaerobic incubation *in vitro, Xenobiotica,* 4, 477, 1974.

226. **Griffiths, L. A.,** 3,5-Dihydroxyphenylpropionic acid, a further metabolite of sinapic acid, *Experientia,* 26, 723, 1970.

227. **Meyer, T. and Scheline, R. R.,** 3,4,5-Trimethoxycinnamic acid and related compounds. I. Metabolism by the rat intestinal microflora, *Xenobiotica,* 2, 383, 1972.

228. **Meyer, T. and Scheline, R. R.,** 3,4,5-Trimethoxycinnamic acid and related compounds. II. Metabolism in the rat, *Xenobiotica,* 2, 391, 1972.

229. **Boyland, E. and Chasseaud, L. F.,** The effect of some carbonyl compounds on rat liver glutathione levels, *Biochem. Pharmacol.,* 19, 1526, 1970.

230. **Boyland, E. and Chasseaud, L. F.,** Enzyme-catalysed conjugations of glutathione with unsaturated compounds, *Biochem. J.,* 104, 95, 1967.

231. **Boyland, E. and Chasseaud, L. F.,** Enzymes catalysing conjugation of glutathione with αβ-unsaturated carbonyl compounds, *Biochem. J.,* 109, 651, 1968.

232. **Hattori, M., Shu, Y.-Z., Tomimori, T., Kobashi, K., and Namba, T.,** A bacterial cleavage of the *C*-glucosyl bond of mangiferin and bergenin, *Phytochemistry,* 28, 1289, 1989.

233. **Rasmussen, A. K., Scheline, R. R., Solheim, E., and Hänsel, R.,** Metabolism of some kava pyrones in the rat, *Xenobiotica,* 9, 1, 1979.

234. **Duffield, A. M., Jamieson, D. D., Lidgard, R. O., Duffield, P. H., and Bourne, D. J.,** Identification of some human urinary metabolites of the intoxication beverage kava, *J. Chromatogr.,* 475, 273, 1989.

235. **Doyle, B. and Griffiths, L. A.,** The metabolism of ellagic acid in the rat, *Xenobiotica,* 10, 247, 1980.

236. **Smart, R. C., Huang, M.-T., Chang, R. L., Sayer, J. M., Jerina, D. M., and Conney, A. H.,** Disposition of the naturally occurring antimutagenic plant phenol, ellagic acid, and its synthetic derivative, 3-*O*-decylellagic acid and 3,3′-di-*O*-methylellagic acid in mice, *Carcinogenesis,* 7, 1663, 1986.

237. **Teel, R. W. and Martin, R. M.,** Disposition of the plant phenol ellagic acid in the mouse following oral administration by gavage, *Xenobiotica,* 18, 397, 1988.

238. **Teel, R. W.,** Distribution and metabolism of ellagic acid in the mouse following intraperitoneal administration, *Cancer Lett.,* 34, 165, 1987.

239. **Suling, W. J., Wooley, C. W., and Shannon, W. M.,** Disposition and metabolism of bruceantin in the mouse, *Cancer Chemother. Pharmacol.,* 3, 171, 1979.

240. **Fenaroli, G.,** *Fenaroli´s Handbook of Flavor Ingredients,* Vol. 2, CRC Press, Boca Raton. FL, 1975.

241. **Heymann, E.,** Carboxylesterases and amidases, in *Enzymatic Basis of Detoxication,* Vol. II, Jakoby, W. B., Ed., Academic Press, New York, 1980, 291.

242. **Heymann, E.,** Hydrolysis of carboxylic esters and amides, in *Metabolic Basis of Detoxication Metabolism of Functional Groups,* Jakoby, W. B., Bend, J. R., and Caldwell, J., Eds., Academic Press, New York, 1982, 229.

243. **La Du, B. N. and Snady, H.,** Esterases of human tissues, in *Concepts in Biochemical Pharmacology,* Part 2, Brodie, B. B. and Gillette, J. R., Eds., Springer Verlag, Berlin, 1971, 477.

244. **Gallaher, E. J. and Loomis, T. A.,** Metabolism of ethyl acetate in the rat: hydrolysis to ethyl alcohol *in vitro* and *in vivo, Toxicol. Appl. Pharmacol.,* 34, 309, 1975.

245. **Morris, J. B.,** First-pass metabolism of inspired ethyl acetate in the upper respiratory tracts of the F344 rat and Syrian hamster, *Toxicol. Appl. Pharmacol.,* 102, 331, 1990.

246. **Elliott, N., Janes, N. F., Kimmel, E. C., and Casida, J. E.,** Metabolic fate of pyrethrin I, pyrethrin II, and allethrin administered orally to rats, *J. Agric. Food Chem.,* 20, 300, 1972.

247. **Yamamoto, I., Elliott, M., and Casida, J. E.,** The metabolic fate of pyrethrin I, pyrethrin II, and allethrin, *Bull. WHO,* 44, 347, 1971.

248. **Casida, J. E.,** Biochemistry of the pyrethrins, in *Pyrethrum. The Natural Insecticide,* Casida, J. E., Ed., Academic Press, New York, 1973, 101.

249. **Class, T. J., Ando, T., and Casida, J. E.,** Pyrethroid metabolism: microsomal oxidase metabolism of (*S*)-bioallethrin and the six natural pyrethrins, *J. Agric. Food Chem.,* 38, 529, 1990.

250. **Hutson, D. H.,** The metabolic fate of synthetic pyrethroid insecticides in mammals, in *Progress in Drug Metabolism,* Vol. 3, Bridges, J. W. and Chasseaud, L. F., Eds., John Wiley & Sons, Chichester, 1979, 215.

251. **Snapper, J., Grünbaum, A., and Sturkop, S.,** Über die Spaltung und die Oxydation von Benzylalkohol und Benzylestern im menschlichen Organismus, *Biochem. Z.,* 155, 163, 1925.

252. **Clapp, J. J. and Young, L.,** Formation of mercapturic acids in rats after the administration of aralkyl esters, *Biochem. J.,* 118, 765, 1970.

253. **Abdo, K. M., Huff, J. E., Haseman, J. K., Boorman, G. A., Eustis, S. L., Mathews, H. B., Burka, L. T., Prejean, J. D., and Thompson, R. B.,** Benzyl acetate carcinogenicity, metabolism, and disposition in Fischer 344 rats and B6C3F mice, *Toxicology,* 37, 159, 1985.

254. **Chidgey, M. A. J. and Caldwell, J.,** Studies on benzyl acetate. I. Effect of dose size and vehicle on the plasma pharmacokinetics and metabolism of [*methylene*-¹⁴C]benzyl acetate in the rat, *Food Chem. Toxicol.,* 24, 1257, 1986.

255. **Chidgey, M. A. J., Kennedy, J. F., and Caldwell, J.,** Studies on benzyl acetate. II. Use of specific metabolic inhibitors to define the pathway leading to the formation of benzylmercapturic acid in the rat, *Food Chem. Toxicol.,* 24, 1267, 1986.

256. **Chidgey, M. A. J., Kennedy, J. F., and Caldwell, J.,** Studies on benzyl acetate. III. The percutaneous absorption and disposition of [*methylene*-¹⁴C]benzyl acetate in the rat, *Food Chem. Toxicol.,* 25, 521, 1987.

257. **McMahon, T. F., Diliberto, J. J., and Birnbaum, L. S.,** Age-related changes in the disposition of benzyl acetate A model compound for glycine conjugation, *Drug Metab. Dispos.,* 17, 506, 1989.

258. **Longland, R. C., Shilling, W. H., and Gangolli, S. D.,** The hydrolysis of flavouring esters by artificial gastrointestinal juices and rat tissue preparations, *Toxicology,* 8, 197, 1977.

259. **Grundschober, F.,** Toxicological assessment of flavouring esters, *Toxicology,* 8, 387, 1977.

260. **Fahelbum, I. M. S. and James, S. P.,** Absorption, distribution and metabolism of propyl anthranilate, *Toxicology,* 12, 75, 1979.

261. **Wright, D. J. and Holder, G. M.,** The metabolism of some food additives related to piperonal in the rabbit, *Xenobiotica,* 10, 265, 1980.

262. **Baas, H. K. L.,** Beiträge zur Spaltung der Säure-Ester im Darm, *Hoppe-Seyler's Z. Physiol. Chem.,* 14, 416, 1890.

263. **Hanzlik, P. J. and Wetzel, N. C.,** The salicylates. XII. The excretion of salicyl after the administration of methyl salicylate to animals, *J. Pharmacol. Exp. Ther.,* 14, 43, 1920.

264. **Davison, C., Zimmerman, E. F., and Smith, P. K.,** On the metabolism and toxicity of methyl salicylate, *J. Pharmacol. Exp. Ther.,* 132, 207, 1961.

265. **Williams, R. T.,** *Detoxication Mechanisms,* Chapman and Hall, London, 1959, 362.

266. **Fötsch, G., Pfeifer, S., Bartoszek, M., Franke, P., and Hiller, K.,** Biotransformation der Phenolglycoside Leiocarposid und Salicin, *Pharmazie,* 44, 555, 1989.

267. **Fötsch, G. and Pfeifer, S.,** Die Biotransformation der Phenolglycoside Leiocarposid und Salicin - Beispiele für Besonderheiten von Absorption und Metabolismus glycosidischer Verbindungen, *Pharmazie,* 44, 710, 1989.

268. **Czok, G., Walter, W., Knoche, K., and Degener, H.,** Über die Resorbierbarkeit von Chlorogensäure durch die Ratte, *Z. Ernährungswiss.,* 13, 108, 1974.

269. **Michaud, J., Lesca, M. F., and Roudge, A. M.,** Hepatic metabolism of caffeic derivatives, *Bull. Soc. Pharm. Bordeaux,* 110, 65, 1971 (Chem. Abstr. 76, 80973b, 1972).

270. **Williams, R. T.,** *Detoxication Mechanisms,* Chapman and Hall, London, 1959, 306.

271. **Blumenberg, F.-W., Enneker, C., and Kessler, J.-J.,** Zur Frage der hepatotoxischen Wirkung oral verabreichten Tannins und seiner Galloyl-Bausteine, *Arzneim. Forsch.,* 10, 223, 1960.

272. **Gupta, M. M. and Dani, H. M.,** Characterization of tannic acid metabolites formed in vitro by rat liver microsomes and assay of their carcinogenicity by the microsomal degranulation technique, *Chem. Biol. Interact.,* 63, 39, 1987.

273. **Milić, B. L. and Stojanović, S.,** Lucerne tannins. III. Metabolic fate of lucerne tannins in mice, *J. Sci. Food Agric.,* 23, 1163, 1972.

274. **Bridges, J. W., French, M. R., Smith, R. L., and Williams, R. T.,** The fate of benzoic acid in various species, *Biochem. J.,* 118, 47, 1970.

275. **Kato, Y.,** Studies on anti-inflammatory agents. XX. Metabolism of ³⁵S-2-amino-3-ethoxycarbonyl-4,5,6,7-tetrahydrothieno[2,3-c]-pyridine (³⁵S-Nor-Y-3642) and ¹⁴C-benzoic acid, *J. Pharm. Soc. Jpn.,* 92, 1152, 1972.

276. **Bababunmi, E. A., Smith, R. L., and Williams, R. T.,** The absence of hippuric acid synthesis in the Indian fruit bat, *Life Sci.,* 12(Suppl. 2), 317, 1973.

277. **Bray, H. G., Thorpe, W. V., and White, K.,** Kinetic studies of the metabolism of foreign organic compounds. I. The formation of benzoic acid from benzamide, toluene, benzyl alcohol and benzaldehyde and its conjugation with glycine and glucuronic acid in the rabbit, *Biochem. J.,* 48, 88, 1951.

278. **Csonka, F. A.,** On the administration of various proteins with benzoic acid to a pig, *J. Biol. Chem.,* 60, 545, 1924.

279. **Martin, A. K.,** Metabolism of benzoic acid by sheep, *J. Sci. Food Agric.,* 17, 496, 1966.

280. **Marsh, M. V., Caldwell, J., Smith, R. L., Horner, M. W., Houghton, E., and Moss, M. S.,** Metabolic conjugation of some carboxylic acids in the horse, *Xenobiotica,* 11, 655, 1981.

281. **Caldwell, J., French, M. R., Idle, J. R., Renwick, A. G., Bassir, O., and Williams, R. T.,** Conjugation of foreign compounds in the elephant and hyaena, *FEBS Lett.,* 60, 391, 1975.

282. **French, M. R., Bababunmi, E. A., Golding, R. R., Bassir, O., Caldwell, J., Smith, R. L., and Williams, R. T.,** The conjugation of phenol, benzoic acid, 1-naphthylacetic acid and sulphadimethoxine in the lion, civet and genet, *FEBS Lett.*, 46, 134, 1974.

283. **Tremblay, C., Melançon, S. B., and Dallaire, L.,** Limites physiologiques de détoxication de l'acide benzoïque par la glycine-N-acylase chez les individus normaux (l'enfant et l'adulte), *Union Med. Can.*, 103, 703, 1974.

284. **Van Sumere, C. F., Teuchy, H., Verbeke, H. P. R., and Bekaert, J.,** Quantitative investigation on the hippuric acid formation in healthy and diseased individuals, *Clin. Chim. Acta*, 26, 85, 1969.

285. **Ette, S. I., French, M. R., Smith, R. L., and Williams, R. T.,** Glycine conjugation in the Indian fruit bat, *FEBS Lett.*, 49, 134, 1974.

286. **Quilley, E. and Smith, M. J. H.,** The application of paper partition chromatography to the study of the metabolism of salicylate in the rat, *J. Pharm. Pharmacol.*, 4, 624, 1952.

287. **Wan, S. H. and Riegelman, S.,** Renal contribution to overall metabolism of drugs. II. Biotransformation of salicylic acid to salicyluric acid, *J. Pharm. Sci.*, 61, 1284, 1972.

288. **Bray, H. G., Ryman, B. E., and Thorpe, W. V.,** The fate of certain organic acids and amides in the rabbit. II. *p*-Hydroxybenzoic acid and its amide, *Biochem. J.*, 41, 212, 1947.

289. **Hartles, R. L. and Williams, R. T.,** Studies in detoxication. XVIII. A study of the relation between conjugation and deamination of *p*-hydroxybenzylamine and related compounds in the rabbit, *Biochem. J.*, 43, 296, 1948.

290. **Tsukamoto, H. and Terada, S.,** Metabolism of drugs. XLVII. Metabolic fate of *p*-hydroxybenzoic acid and its derivatives in rabbit, *Chem. Pharm. Bull.*, 12, 765, 1964.

METABOLISM OF HIGHER TERPENOIDS

Terpenoid compounds include a multitude of diverse plant constituents which are related by virtue of a common biosynthetic origin. Thus, their basic skeletons are derived from mevalonic acid and consist of C_5-units, i.e., the isoprene molecule ($CH_2=C(CH_3)-CH=CH_2$). They are further classified according to the number of such units present, the simplest C_{10}-derivatives containing two. These C_{10}-compounds are known as monoterpenoids and in this book the choice was made to include them in other chapters rather than with their higher relatives. This decision was dictated by their close metabolic relationship to other classes of plant compounds, especially the alcohols, aldehydes, and ketones, and, in contrast, the lack of similarity to the metabolism seen with many of the terpenoids included in this chapter. These latter compounds include the sesquiterpenoids (C_{15}), diterpenoids (C_{20}), triterpenoids (C_{30}), and tetraterpenoids (C_{40}).

SESQUITERPENOIDS

The sesquiterpenoids are C_{15} compounds which include both acyclic and cyclic derivatives. An important member of the former group is the alcohol **farnesol** (1). Its metabolism in mammals is unclear, however Fischer and Bielig[1] isolated an acid which they believed to be farnesenic acid, the acid analogue of farnesol, from the urine of a rabbit given the alcohol (6.7 g, i.p., during a 4-d period).

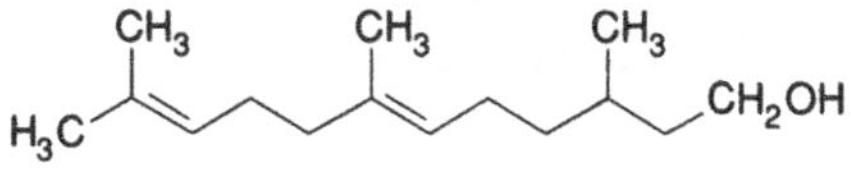

(1)

Farnesol

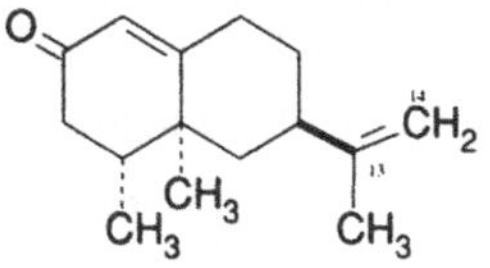

(2)

(+)-Nootkatone

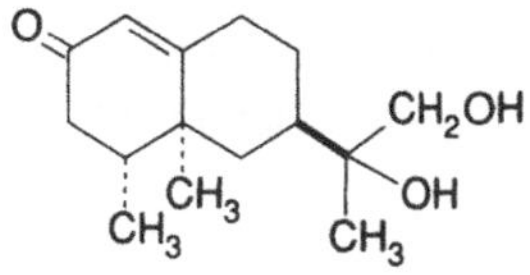

(3)

The metabolism of **(+)-nootkatone** (2) in rabbits was reported by Asakawa et al.[2] They used an oral dose of ~1000 mg/kg and investigated the neutral urinary metabolites following hydrolysis with a β-glucuronidase + sulfatase preparation. A compound representing ~35% of this material as separated by gas chromatography was shown to be (+)-nootkatone-13,14-diol (3). The metabolism of an isopropenyl group to a glycol via an epoxide is well-documented in several other compounds (e.g., camphene, limonene). Asakawa et al.,[2] in parallel experiments, also studied the metabolism of **(−)-elemol** (4), a tertiary alcohol showing some structural likeness to nootkatone. About 70 to 80% of the urinary metabolites was shown to be unchanged compound. Additionally, ~10% of the neutral metabolites was identified as (−)-15-hydroxy-elemol (5), a product of allylic hydroxylation. No evidence was obtained for oxidation of the vinyl group at C-7,8 which would result in formation of an epoxide and the 7,8-glycol.

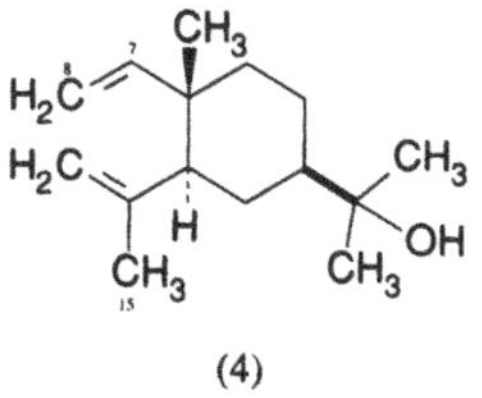

(4)

(−)-Elemol

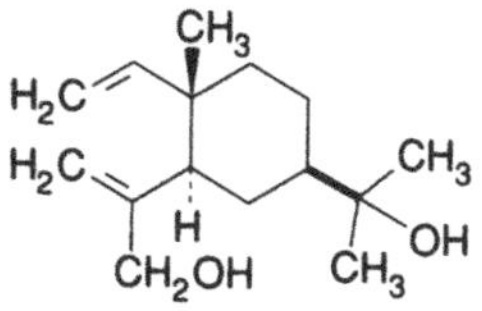

(5)

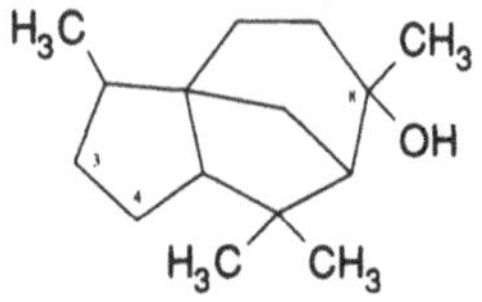

(6)   (7)

Cedrol

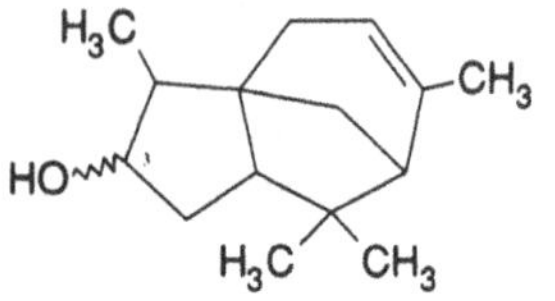

The hydroxylation of **cedrol** (6) in rabbits was studied by Bang and Ourisson.[3] Following an oral dose of ~330 mg/kg, the urine was collected for 4 d and hydrolyzed enzymatically with a β-glucuronidase preparation. This resulted in the recovery of cedrol (5% of the dose), a mixture of alcohols (12%), and a mixture of diols (35%). Both of the alcohols had structure (7) and differed only in the orientation of the hydroxy group at C-3. The diols were shown to be 3-hydroxycedrol, the mixture likewise consisting of two stereoisomers differing in configuration at C-3. Interestingly, 3-hydroxycedrol was also excreted in the urine of dogs given cedrol, however only the 3α-alcohol (—OH at C-3) was present.[4] The latter investigation employed an oral dose (2 g) of cedrol and also treated the urine samples with a β-glucuronidase preparation prior to metabolite analysis. Metabolites were present only in the initial 24-h urine sample and three oxidized metabolites in addition to 3α-hydroxycedrol were characterized. These were shown to be 4-hydroxycedrol, 12-hydroxycedrol, and an acidic diol formed from cedrol by hydroxylation at C-4 and oxidation of the methyl group at C-8 to a carboxy group.

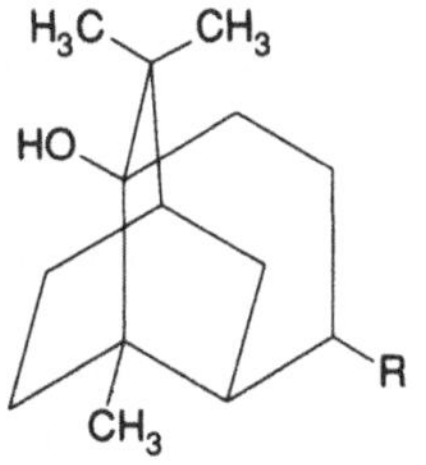

(8)

(a) Patchouli alcohol,   R = –CH₃

(b) R = –CH₂OH

(c) R = –COOH

Bang et al.[5] administered **patchouli alcohol** (8a) to rabbits and dogs and found that it was oxidized to the primary alcohol metabolite (8b). The urine of rabbits fed patchouli alcohol (330 mg/kg) contained, after hydrolysis with β-glucuronidase, both the diol (8b) and the corresponding carboxylic acid (8c).

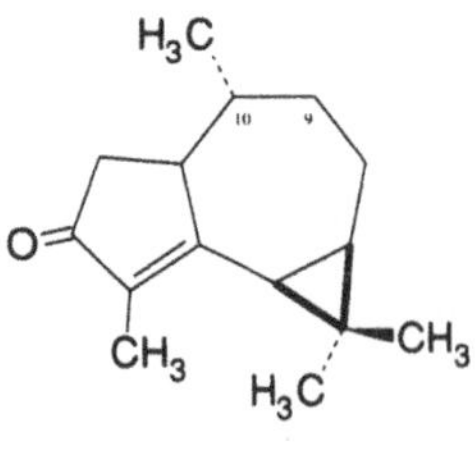

(9)   (10) 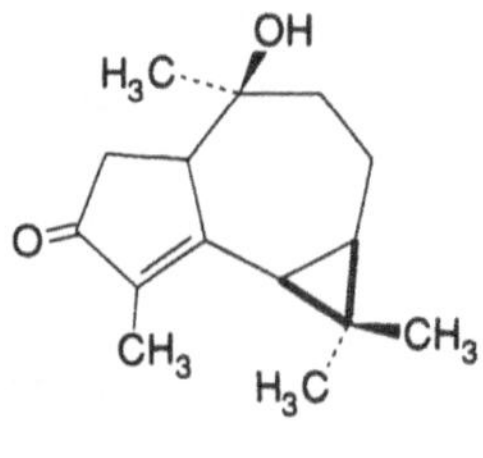   (11)

(–)-Cyclocolorenone

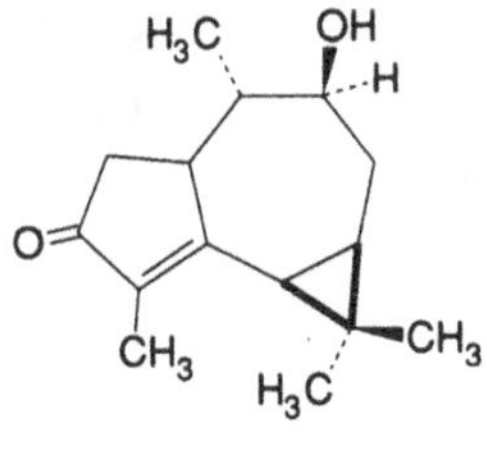

Asakawa et al.[2] studied the metabolism of (–)-**cyclocolorenone** (9) given orally to male rabbits. The material administered was the essential oil of *Solidago altissima* which contained ~20 compounds, including cyclocolorenone as the main component (~24% of the total material detected by gas-liquid chromatography (g.l.c.)). Two hydroxylated metabolites, following hydrolysis of the urinary material with β-glucuronidase + sulfatase, were isolated and shown to be the 9β-hydroxy- and 10β-hydroxy-derivatives of cyclocolorenone (compounds (10) and (11), respectively). It was postulated that these two alcohols may arise via an intermediate 9,10-epoxy derivative.

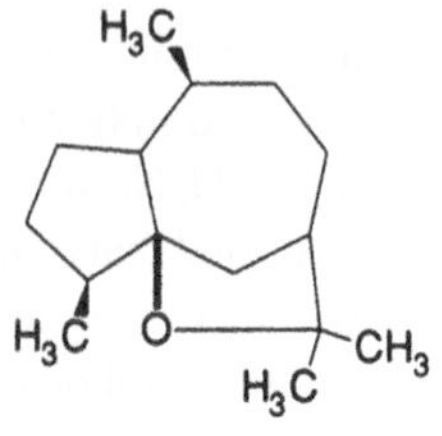

(12)   (13) 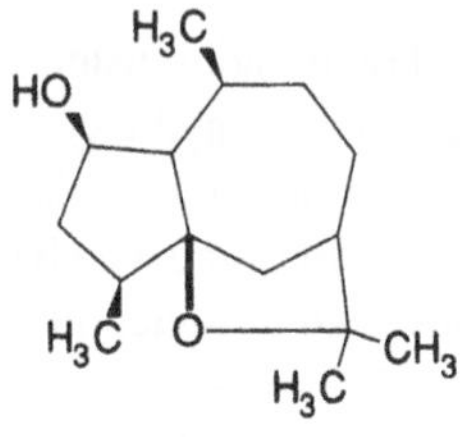   (14)

Guaioxide

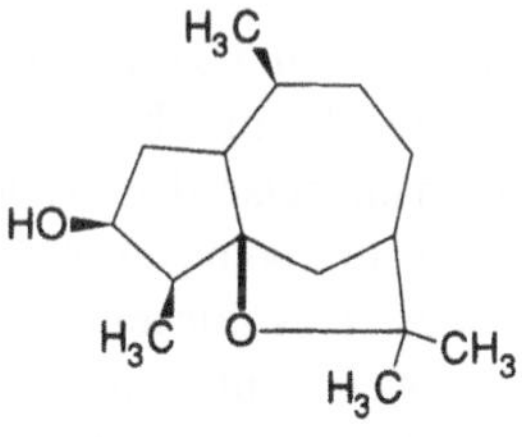

Guaioxide (12) was metabolized by rabbits to two alcohol derivatives.[6] These metabolites had structures (13) and (14).

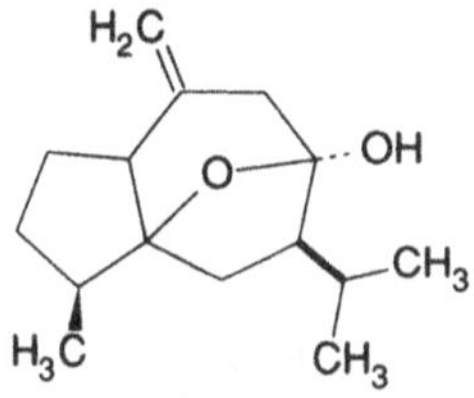

(15)

Curcumol

Curcumol (15) is another ditertiary ether and Su et al.[7] administered the ^{3}H-labeled compound to normal rats and tumor-bearing mice. Although no metabolic products were described, they reported that curcumol was rapidly and completely absorbed from the gastro-intestinal tract in rats. Its tissue distribution was similar in both animals. The radioactivity was mainly excreted by rats in the urine and bile. The radioactivity found after 24 h in the urine was 45% (oral dose) and 51% (i.v.) of the dose. The corresponding values for biliary radioactivity were 36 and 56%. Fecal excretion of radioactivity accounted for only 7% (oral) and 14% (i.v.) after 72 h. This indicates that most of the biliary radioactivity was reabsorbed from the gastrointestinal tract of normal animals.

The hydrocarbon α-**santalene** (16) was found by Zundel[8] to undergo allylic hydroxylation in rabbits. The metabolites formed were α-santalol (17), its isomer (18), and the diol (19).

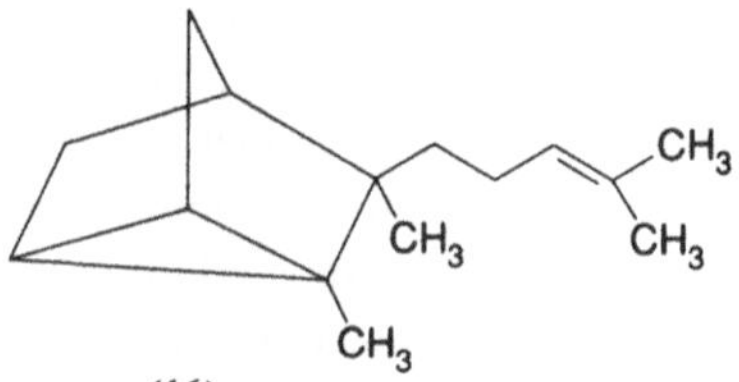

(16)   (17)

α-Santalene   α-Santalol

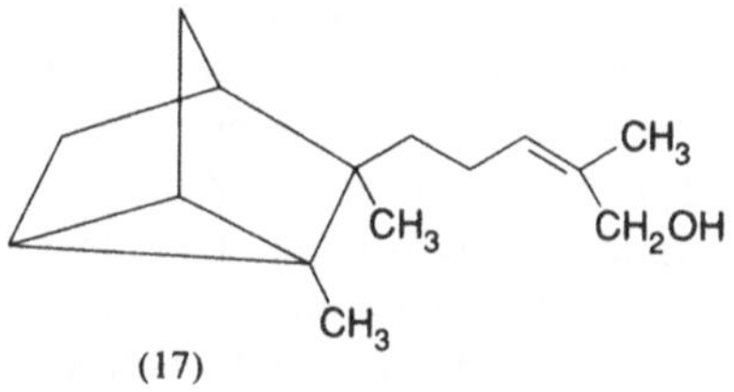

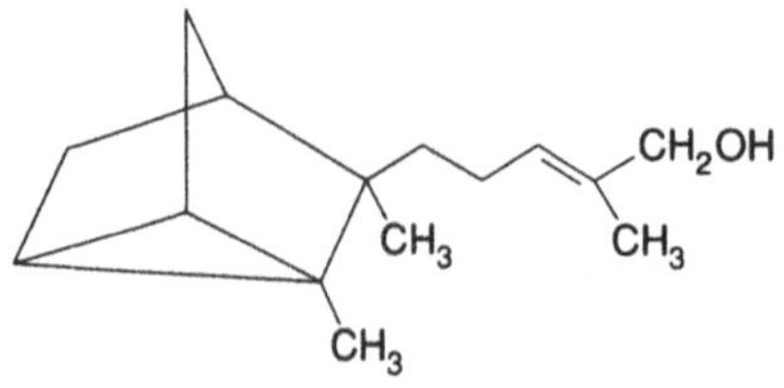

(18)

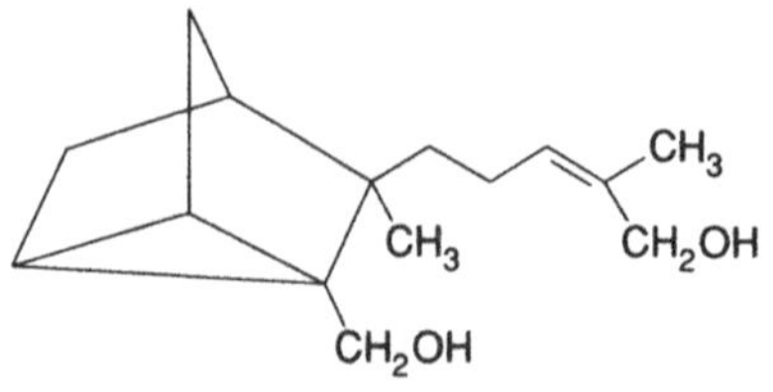

(19)

Hildebrandt[9] administered the tricyclic alcohol α-**santalol** (17) orally to rabbits and found evidence for a C_{10}-hydroxy acid in the urine. This metabolite corresponds to a compound having one methyl group and the alcoholic side chain replaced by –COOH and –CH$_2$OH. No subsequent results have appeared which might confirm this point, however Zundel[8] reported that α-santalol was converted to metabolite (20) in dogs. This illustrates that oxidative chain shortening is possible. These newer experiments showed a different metabolic picture in rabbits which converted α-santalol to the diols (19) and (21) and the triol (22).

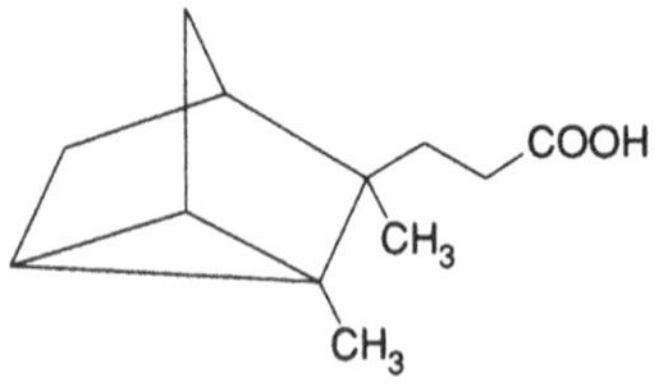

(20)

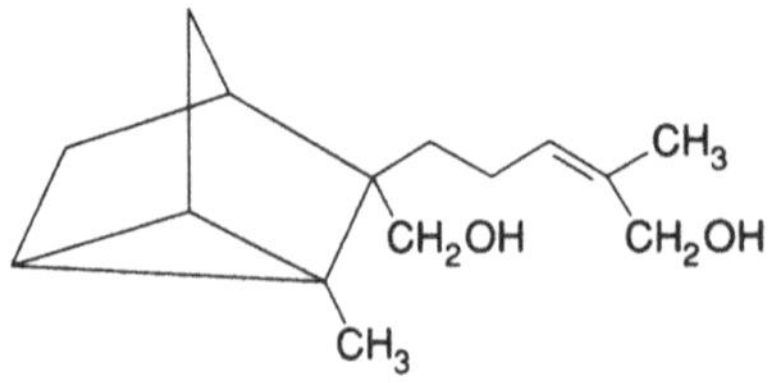

(21)

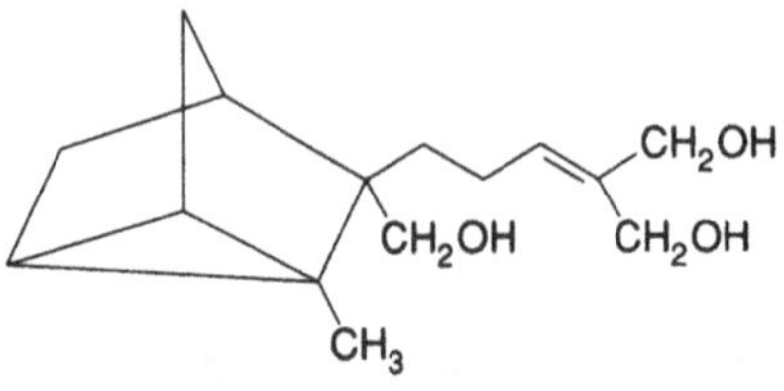

(22)

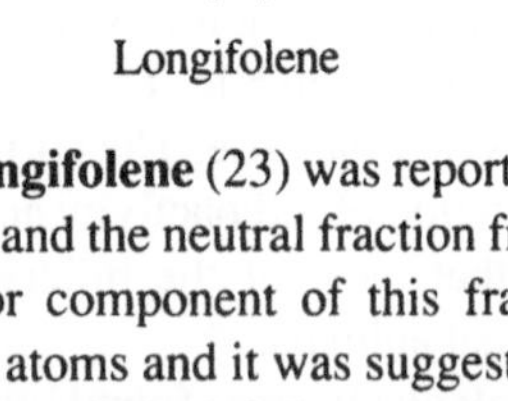

(23)

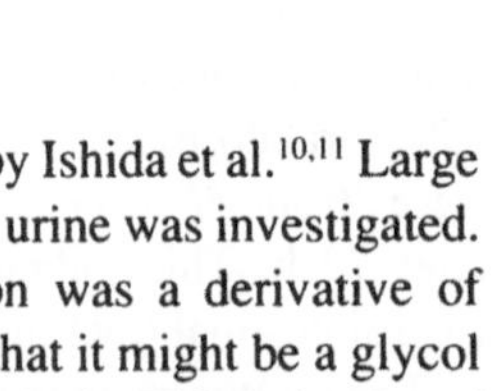

(24)

Longifolene

The metabolism of the hydrocarbon **longifolene** (23) was reported by Ishida et al.[10,11] Large doses were administered orally to rabbits and the neutral fraction from urine was investigated. The initial study showed that the major component of this fraction was a derivative of longifolene which contained two oxygen atoms and it was suggested that it might be a glycol formed via epoxidation of the double bond at C-7,13 followed by hydrolysis. This behavior of an *exo*-methylene group has been reported for other related compounds including camphene (see the chapter entitled Metabolism of Hydrocarbons). However, additional spectroscopic studies in the latter report indicated that the metabolite was hydroxyisolongifolaldehyde (24). Ishida et al. postulated that the epoxide formed, due to its low stability, isomerized to a stable *endo*-aldehyde. Hydroxylation of the *gem*-dimethyl group is well-known among cyclic compounds, however this example shows that it also occurs on a seven-membered ring. Asakawa et al.[2] presented further information on these experiments and showed that (24) was (2*S*, 7*S*)-(+)- 14-hydroxyisolongifoldaldehyde. Following oral doses of ~1000 mg/kg to male rabbits, this compound represented about a third of the neutral urinary metabolites separated by g.l.c. This fraction, which represented ~10% of the dose, showed many additional chromatographic peaks.

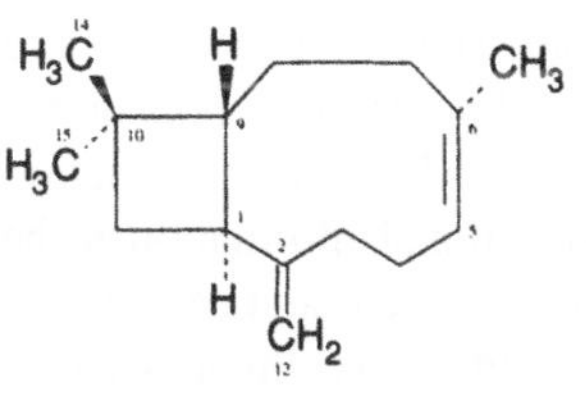

(25)

(−)-Caryophyllene

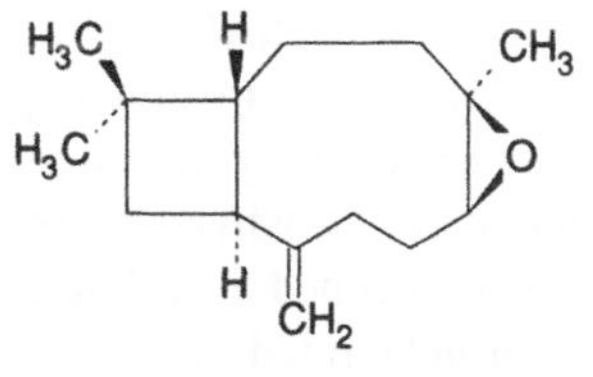

(26)

(−)-Caryophyllene-5,6-oxide

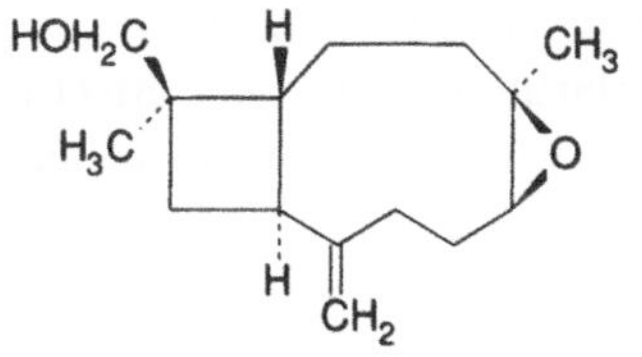

(27)

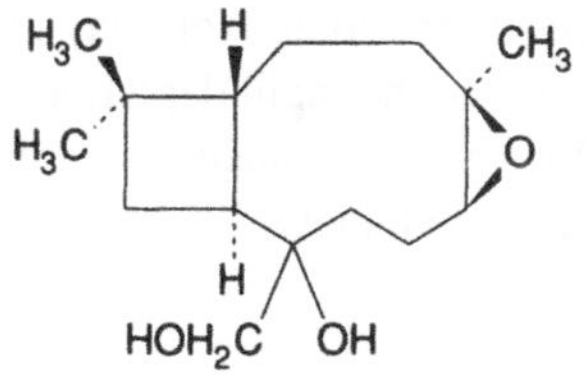

(28)

The neutral urinary metabolites of **(−)-caryophyllene** (25) and **(−)-caryophyllene-5,6-oxide** (26) given orally to male rabbits in doses of ~1000 mg/kg were studied by Asakawa et al.[2,12] Stereoselective oxidation of the *gem*-dimethyl group in both compounds gave rise to metabolite (27) which was found to have the same chirality (S) at C-10 as was the case with longifoline noted above. A second metabolite (28) of caryophyllene involved epoxidation at both the 5,6- and 2,12-positions. In the latter case, the actual epoxide formed from the *exo*-methylene group was not detected and the metabolite identified was its hydrolysis product, a 2,12-glycol.

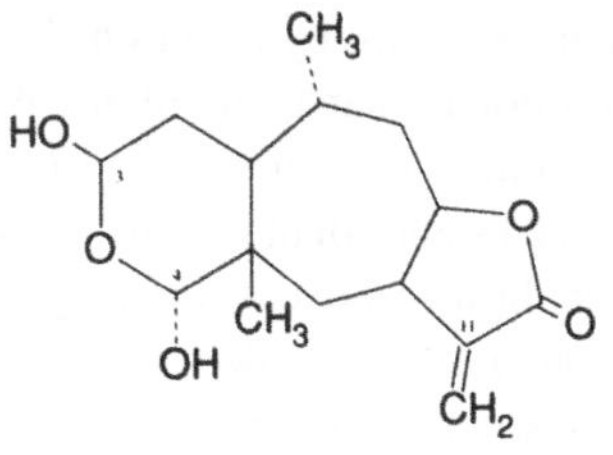

(29)

Hymenoxon

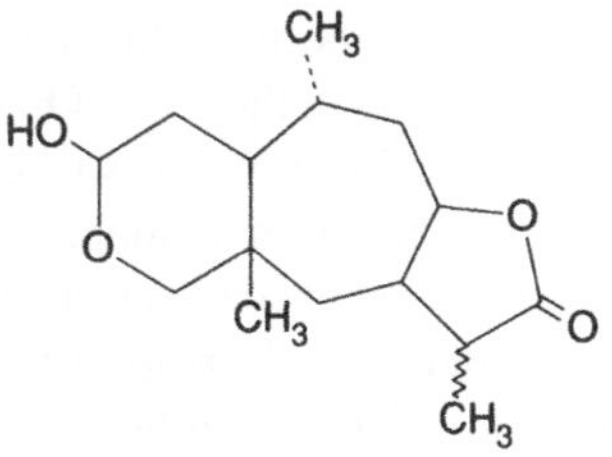

(30)

The metabolism of **hymenoxon** (29), a sesquiterpene lactone, was studied in rabbits.[13] It was slowly absorbed from the gastrointestinal tract and then rapidly metabolized to glucuronides which were excreted in the urine and bile. The urine and bile contained five and six conjugates, respectively, and four of these appeared to be common to both excretory routes. However, only two aglycons of undetermined structure were derived from these glucuronides. Little unchanged compound was excreted in the urine following the i.v. administration of hymenoxon (~17 mg/ kg) to rabbits. The metabolism of hymenoxon in sheep was studied by Terry et al.[14] It was given both orally (50 to 100 mg/kg) and by rumen fistula (45 to 90 mg/kg), following which urine, bile, and feces samples were collected. The samples were extracted with ethyl acetate, in some cases after treatment with β-glucuronidase or sulfatase, and analyzed by a gas chromatographic method. No unchanged hymenoxon was detected in the samples from urine, bile, or feces following intragastric or intraruminal dosage. Chromatograms of the urine samples showed the presence of four metabolites. The two major metabolites, which were largely excreted in the form of glucuronide conjugates, were found to have structure (30). These compounds are epimers having an 11-α-methyl or 11-β-methyl group. Thus, hymenoxon was dehydroxylated at C-4 and reduced at the *exo*-methylene group at C-11. Interestingly, both of these reactions are typical of those carried out by intestinal microorganisms. However, this pathway appeared to be a minor metabolic route of hymenoxon in sheep. Additional information on the metabolic fate of hymenoxon was obtained by Sylvia et al.[15,16] who reported that it bound deoxyguanosine in a cell-free system and formed adducts with guanine residues in cellular DNA. Characterization of one of the adducts indicated that two guanine residues reacted with each hymenoxon molecule. They believed that this reaction occurred by the formation of Schiff base products via a reactive dialdehyde form of hymenoxon. The latter compound would be formed by loss of water from the bishemiacetal structure to give aldehyde groups at C-3 and C-4. Another pathway in the metabolism of hymenoxon may involve glutathione. Merrill et al.[17] showed that an i.p. dose in mice caused the rapid depletion of hepatic glutathione. The same effect was observed with **helenalin**, a closely related sesquiterpene lactone which also contains a highly reactive α-methylene-γ-lactone moiety.

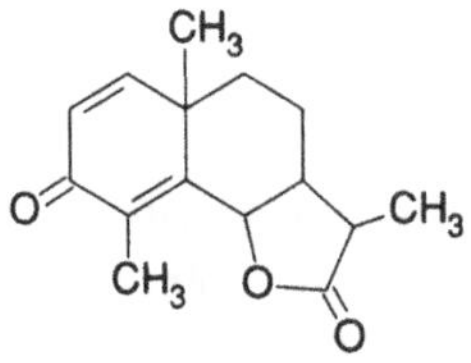

(31)

Santonin

Morishima[18] reported that the sesquiterpene lactone **santonin** (31) was excreted partly unchanged in the urine and feces after oral or s.c. administration. In rabbits, a combined value of 3.5% of the dose was reported following oral dosage, whereas, by injection, 22% was found in the urine and much less in the feces. Oral administration to man resulted in 19% of the dose being excreted in the 24-h urine. Kaya[19] carried out similar experiments in dogs and found, following either route of administration, some unchanged santonin in the bile, urine, and feces. About 70% of the dose was excreted in the urine and very little in the feces after injection whereas values of 4 to 6% and 7 to 9%, respectively, were found after oral dosage. A metabolite, α-hydroxysantonin, was detected in the serum, bile, intestinal juice, urine, and feces. Santonin contains an a,β-unsaturated group in common with many compounds of this type which are known to be conjugated with glutathione. However, Boyland and Chasseaud[20] found that this glutathione S-transferase activity from rat liver did not utilize santonin as a substrate.

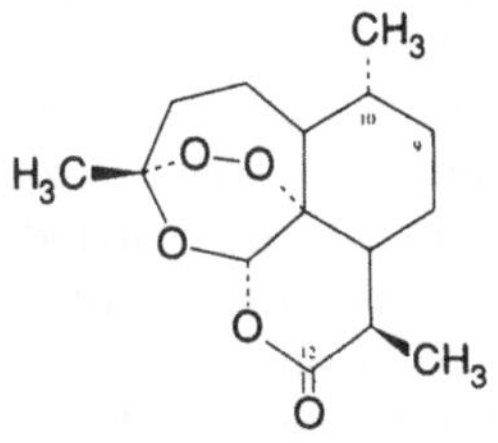

(32)   (33)

Qinghaosu

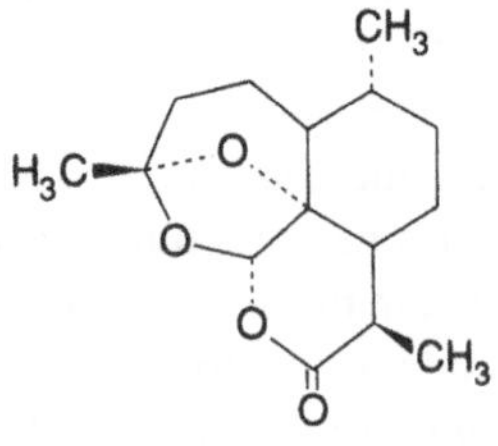

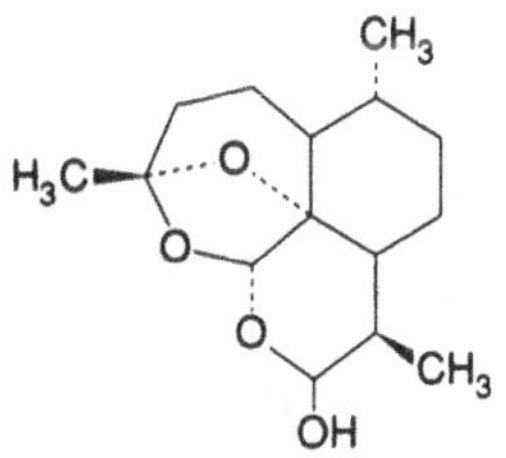

(34) 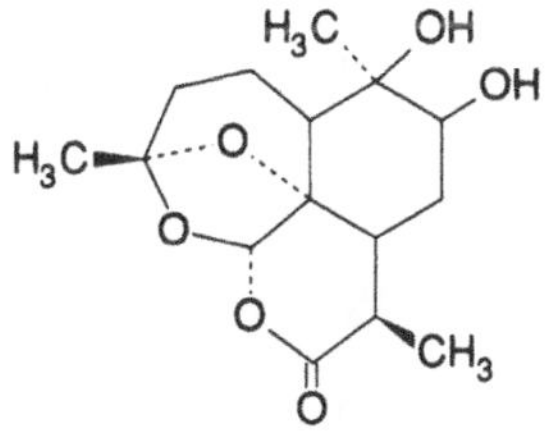   (35) 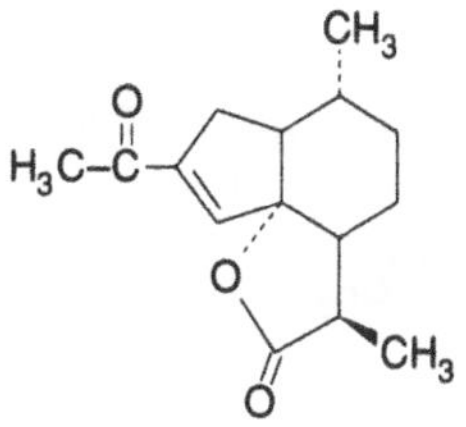   (36)

The sesquiterpene lactone **qinghaosu** (artemisinine, artemisine) (32) was shown by Niu et al.[21] to undergo rapid absorption in rats. Following an oral dose (150 mg/kg), only 8% of the administered material was recovered in the gastrointestinal tract after 4 h. They showed that it was not destroyed in the stomach or ileum, however metabolism in the liver was rapid. They also found that little unchanged compound was excreted in the urine in 48 h. Following an i.v. injection, a value of 2.6% of the dose (150 mg/kg) was recorded. The corresponding values after an i. m. or oral dose (150 or 300 mg/kg, respectively) were <0.1%. Similarly, only 0.4% of the oral dose was found in the feces in 48 h. Four urinary metabolites of qinghaosu, all deoxy analogues, have been reported. Following oral administration of the lactone to humans, Zhu et al.[22] identified compounds (33) (deoxyartemisinine), (34) (dihydrodeoxyartemisinine), and (36). Subsequently the 9,10-dihydroxy derivative (35) was identified in similar experiments.[23]

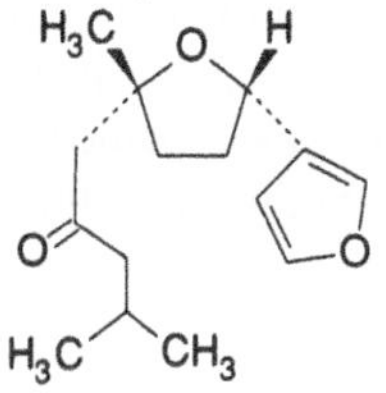

(37)   (38) 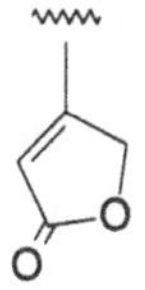   (39)
(–)-Ngaione

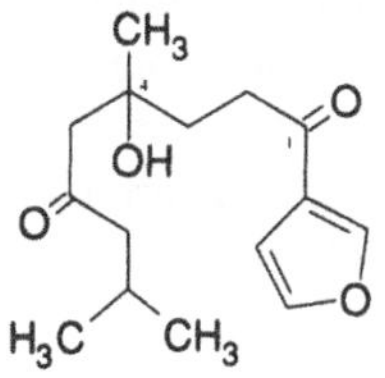

The *in vivo* metabolism of the furanosesquiterpene ketone (–)-**ngaione** (37) was studied in rats.[24] The pattern of urinary metabolites was similar following oral (355 mg/kg) or i.p. (200 mg/kg) dosage and no unchanged compound was found. Five furanoid metabolites were isolated and one of these was identified as 4-hydroxymyoporone (38). A second metabolite was tentatively identified as a γ-lactone derivative closely related to (38) with structure (39) attached at C-1 and no hydroxy group at C-4. The structures of the other metabolites were only partially elucidated.

DITERPENOIDS

The diterpenoids comprise a large and chemically diverse group of C_{20}-compounds which, as with the lower terpenes, includes hydrocarbons, alcohols, ethers and acids and their derivatives. Although the mammalian metabolism of relatively few diterpenes has been studied, the available information will be grouped according to the ring skeleton of the compound. Only a few of the many basic types are represented and include linear derivatives (40), abietane derivatives (41), kaurane derivatives (42), cassane derivatives (43), taxane derivatives (44), and other derivatives with a less common skeleton.

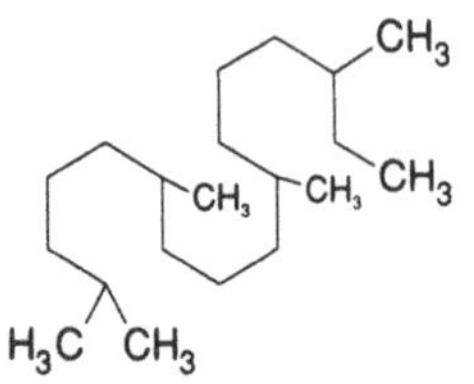

(40) 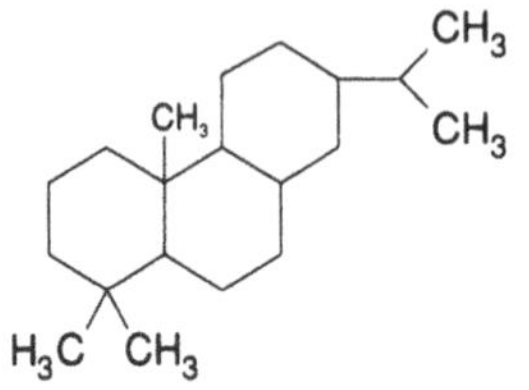   (41) 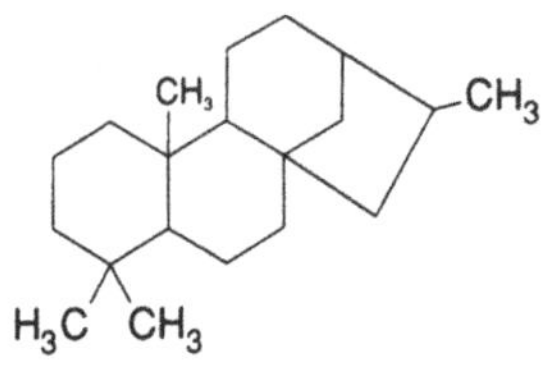   (42)

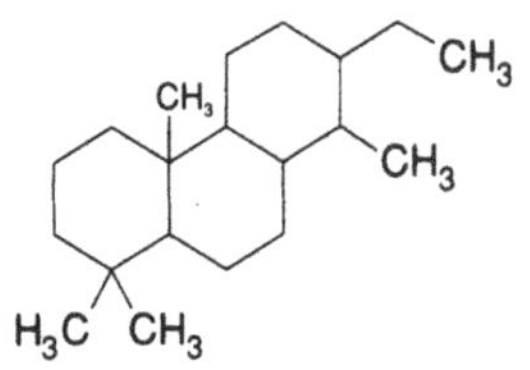

(43) 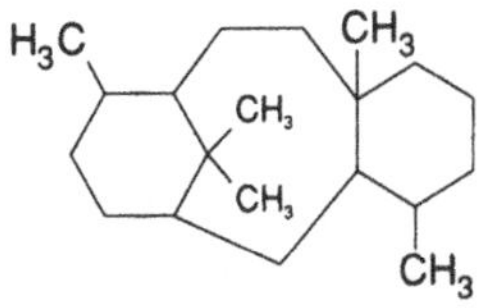   (44)

LINEAR DERIVATIVES

The metabolic disposition of **pristane** (2,6,10,14-tetramethylpentadecane) (45) was studied in rats using ^{3}H-labeled material.[25] When pristane (~0.44 mg/kg) was fed in the diet, 66% of the radioactivity was excreted in the feces in 7 d. More than 80% of this material was in the form of unchanged compound. About 15% of the dose was excreted in the urine during this period, mainly as metabolites. An additional 8% was recovered in the tissues of the animals. The nature of the tissue metabolites was studied following oral doses of ~75 mg/kg. Metabolism by ω- and ω–1-oxidation to give pristan-1-ol and pristan-2-ol, respectively, was demonstrated. Additionally, pristanic acid (46) and its β-oxidation product 4,8,12-trimethyltridenanoic acid (47) were also identified. The pristane metabolites were incorporated into neutral lipids and phospholipids in the liver.

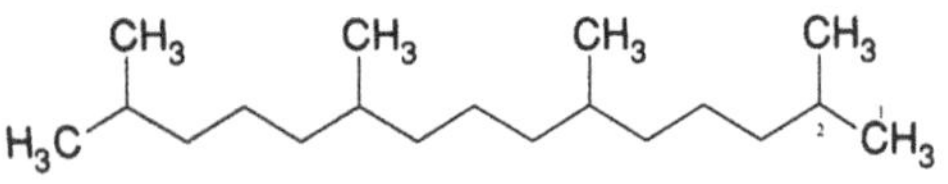

(45)   (46)

Pristane

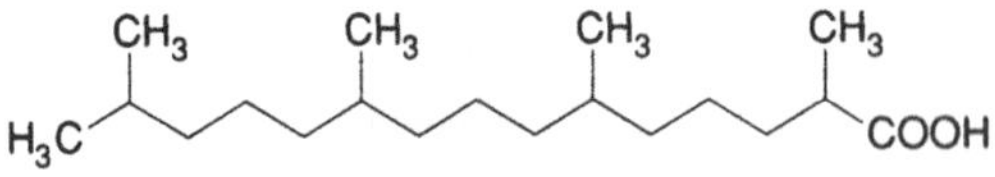

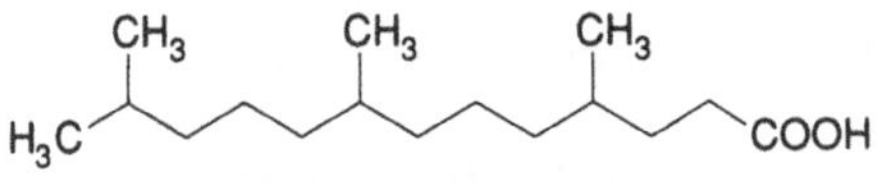

(47)

Phytane (2,6,10,14-tetramethylhexadecane), while not a plant compound, is a higher homologue of pristane. Its metabolism in rats was studied by Albro and Thomas[26] who found that it was metabolized to a tertiary alcohol (probably the 2-ol) and a variety of short-chain carboxylic acids.

An important acyclic diterpene is **phytol** (48) which is undoubtedly the most widely occurring of all diterpenes, being found as an esterified component of chlorophyll. It is also the most extensively studied compound in this group. Investigations in several animal species indicate that phytol is metabolized along multiple routes which give rise to numerous metabolites. These include the oxidation product, phytanic acid (49), and many lipids and triglycerides formed via the esterification of phytol or phytanic acid.

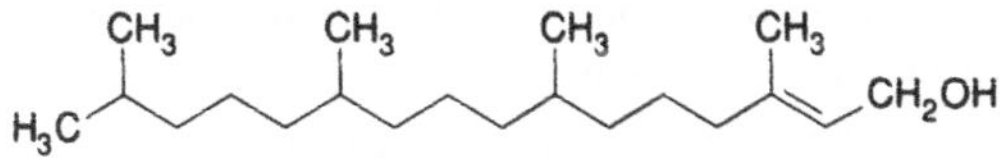

(48)

Phytol

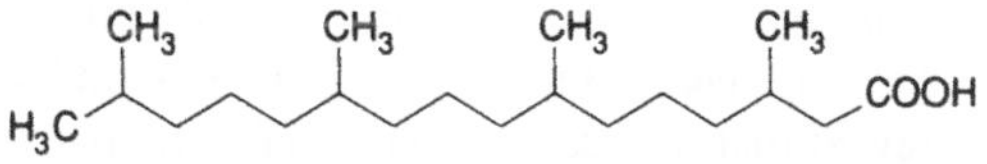

(49)

Mize et al.[27] found that tracer doses (0.2 mg) of uniformly-labeled [^{14}C]phytol given orally to rats were well absorbed (30 to 66%). About 30% of the absorbed material was converted to $^{14}CO_2$ in 18 h and conversion of the alcohol to phytanic acid was demonstrated. Feeding the animals a diet containing 5% phytol for 2 weeks or more, which resulted in an accumulation of both phytanic acid and the corresponding unsaturated compound (phytenic acid) in the blood and tissues, did not lead to a reduction in the fraction of the dose which was converted to $^{14}CO_2$. Accumulation of phytanic acid in animals fed large dietary doses (2 to 5%) of phytol was also reported for the mouse, rabbit, and chinchilla.[28] Phytol was converted to phytanic acid in the rumen and tissues of sheep[29] and the rumen of cows.[30] The investigation of Mize et al. also showed that the radioactivity was found in all the major classes of lipids in the liver, however most was found in the triglycerides and phospholipids with relatively little present as cholesterol esters. These findings agree in most respects with the report of Klenk and Kremer;[31] however, the latter investigation indicated that the phytanic acid formed was incorporated mainly into cholesterol esters and triglycerides. The formation of phytenic acid noted above suggests that it is an intermediate in the conversion of phytol to phytanic acid. Baxter and Milne[32] found that five isomers of phytenic acid were formed in rats fed phytol. These were found in the intestinal lymph and were present largely in the form of complex lipids. The isomers of phytenic acid were shown to be the *trans*-Δ^2 (50a), *cis*-Δ^2 (50b), *trans*-Δ^3 (50c), *cis*-Δ^3 (50d), and 3-methylene (50e) compounds. The lymph from the thoracic duct of rats given phytol also contained appreciable amounts of free and esterified phytol[33] in addition to these oxidized metabolites.

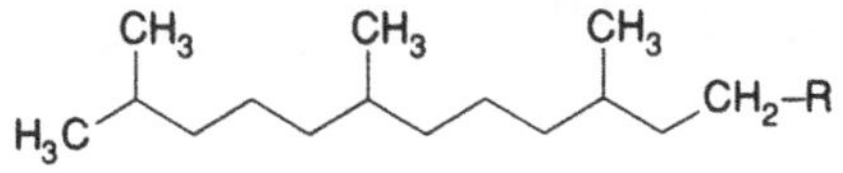

(50)

(a) 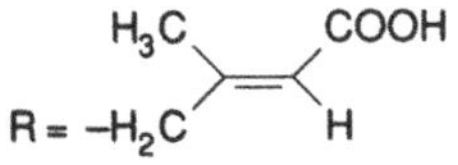   R =

(b) 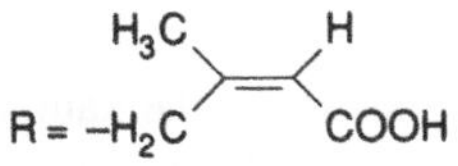   R =

(c) 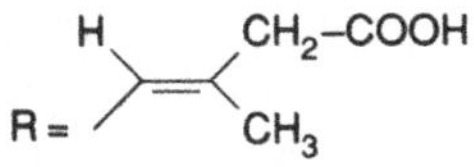   R =

(d) 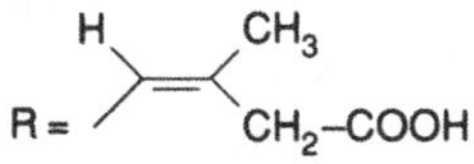   R =

(e) 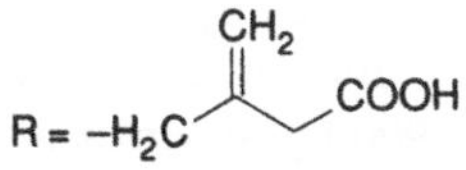   R =

The absorption and metabolism of phytol in man was reported by Steinberg et al.[34] Small oral doses of uniformly-labeled [^{14}C]phytol were converted in part to $^{14}CO_2$ and phytanic acid was detected in the plasma. This study also included patients with Refsum's disease, who have an impaired ability to metabolize phytanic acid, and it was noted that these individuals had a lower capacity to oxidize phytol to CO_2 and a prolonged accumulation of phytanic acid in the plasma. Several investigations of the further degradation of phytanic acid have been carried out which reveal that it undergoes an initial α-oxidation to give α-hydroxyphytanic acid which is then decarboxylated to the C_{19}-acid (pristanic acid) and further metabolized by classical β-oxidation.[35-37] Vamecq and Draye[38] recently found that 2-oxophytanic acid, a previously unreported metabolite of α-hydroxyphytanic acid, was formed by rat kidney.

ABIETANE DERIVATIVES

The metabolism of **(–)-abietic acid** (51) and **(+)-dehydroabietic acid** (52) in rabbits was studied by Asakawa et al.[2] The compounds (2 g) were given orally to rabbits (2 to 3 kg) and the urine collected for 3 d. The acidic fractions were investigated following hydrolysis of the samples with β-glucuronidase + sulfatase. Ishida et al.,[39] in a similar experiment with dehydroabietic acid, had previously reported that excretion of metabolites in the neutral fraction was negligible. Following the administration of abietic acid a single metabolite was isolated and shown to be an acidic derivative in which the methyl group at C-16 was oxidized to a carboxy group. The postulated alcohol intermediate was not detected, however two alcohols were identified as metabolites of dehydroabietic acid. These were shown to be the 15-hydroxy- (major metabolite) and 16-hydroxy (minor metabolite) derivatives. In this case the acidic derivative (carboxy at C-16) was not found. Also detected were a compound containing an isopropenyl group attached to C-13 and trace amounts of the 7-keto derivative of dehydroabietic acid. The former compound no doubt arose via dehydration of the 15-hydroxy derivative. Xu et al.[40] carried out a pharmacokinetic study of [^{3}H]abietic acid in mice and found high levels of radioactivity in the stomach, intestine, liver, and skin. Only 32% of the radioactivity was excreted in the urine and feces in 72 h.

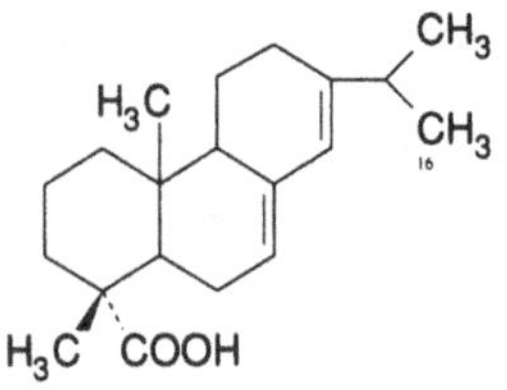

(51)

(–)-Abietic acid

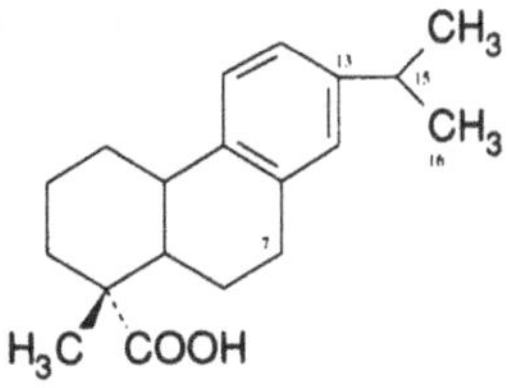

(52)

(+)-Dehydroabietic acid

KAURANE DERIVATIVES

Steviol (53), a diterpene which possesses the kaurane skeleton, is the aglycon of several sweet-tasting glycosides including stevioside and rebaudioside A. Wingard et al.[41] found that both of these glycosides were hydrolyzed to steviol when incubated anaerobically with rat cecal microorganisms. Steviol was readily absorbed from the lower intestine in rats following its intracecal administration. Using 17-^{14}C-labeled material (0.7 mg/animal) given orally, only ~1.5% of the radioactivity was excreted in the urine whereas 96% was recovered in the feces. These values indicate extensive biliary excretion of the radioactivity and intracecal administration of steviol to rats with bile-duct cannulas showed that the entire dose appeared in the bile. Virtually no radioactivity appeared in the respiratory CO_2, a finding which demonstrates the metabolic stability of the exocyclic methylene group at C-17. Compadre et al.[42] investigated the metabolism of steviol using the 9000-g supernatant fraction from rat liver. Only slight metabolism was noted and unchanged compound was the major gas chromatographic peak in the samples of incubates; however, nine metabolites were detected. The most abundant of these, representing about two-thirds of the metabolized material, was identified as 15α-hydroxysteviol. Two other hydroxylated derivatives, 7β-hydroxysteviol and 17-hydroxyisosteviol (54), were also identified. However, Compadre et al. believed that the identification of compound (45), which is known to be formed via the acid arrangement of steviol-16,17α-epoxide, suggests that a 16,17-epoxide derivative of steviol was formed during incubation. The epoxide itself was not detected as a metabolite. The remaining six metabolites, which were present in very small amounts, were not identified. However, three of these showed mass spectra similar to those observed with 15α- and 7β-hydroxysteviol and may therefore be hydroxylated derivatives of steviol.

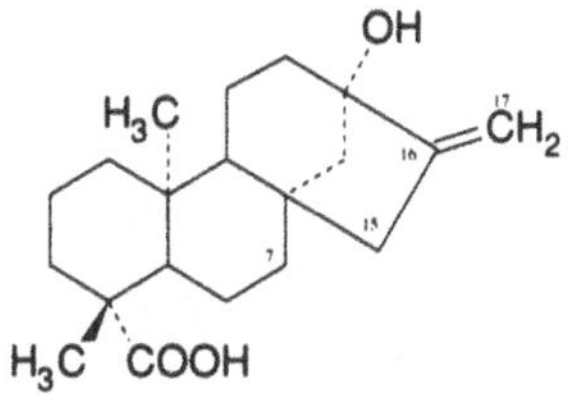

(53)	(54)	(55)
Steviol		Grandiflorenic acid

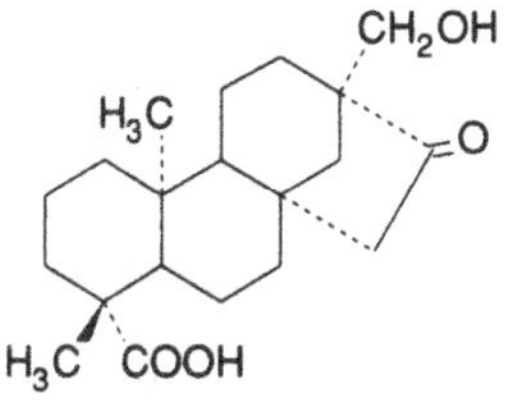

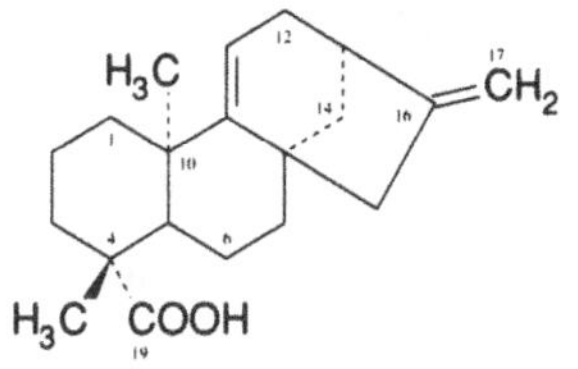

Grandiflorenic acid (55) is another tetracyclic diterpene based on the kaurane skeleton. Neidlein and Stumpf[43] studied several aspects of its metabolism in rats, also using material labeled with ^{14}C in the 17-position. When male rats were given grandiflorenic acid (50 mg/kg, i.p.), only 5% of the dose was excreted in the urine whereas ~73% was lost in the feces. The fecal material contained 11 metabolites, however unchanged compound accounted for over 40% of the fecal radioactivity. Several of the other metabolites were characterized. One was a hydroxylated derivative, probably formed by oxidation of a methyl group at either C-4 or C-10. This metabolite made up 6% of the fecal radioactivity. A further 9% consisted of the 12α-hydroxy derivative formed by the allylic hydroxylation of grandiflorenic acid. Further oxidation of this 12α-hydroxy metabolite occurred and 12-ketograndiflorenic acid was also identified. Another metabolite (6% of the fecal radioactivity) was hydroxylated at both the 6β and 14β positions and reduced at the exocyclic double bond at C-16. The latter reaction was probably carried out by intestinal microorganisms as it was shown that grandiflorenic acid itself was converted to its 16,17-dihydro derivative when incubated with rat intestinal contents. Incubation of grandiflorenic acid with rat liver microsomes resulted in the formation of six metabolites, of which 12α-hydroxygrandiflorenic acid was the most prominent. Grandiflorenic acid was conjugated with glucuronic acid when incubated with rat liver preparations containing UDP-glucuronic acid. These findings suggest that grandiflorenic acid and its metabolites are extensively excreted in the bile of rats as glucuronide conjugates. This assumption was confirmed in a subsequent study of the absorption, metabolism and excretion of grandiflorenic acid in rats.[44] The biliary metabolites were subjected to an extensive enterohepatic circulation.

TAXANE DERIVATIVES

Taxol is a complex taxane derivative which contains multiple ester linkages. Its pharmacokinetics in humans was studied by Longnecker et al.[45] and Wiernik et al.[46] who found that average urinary excretion values for unchanged compound were ~4 to 6% of the i.v. dose. Taxol was extensively bound to plasma proteins, however its rapid elimination from the plasma suggests that metabolism, biliary excretion, or tissue binding may be important factors in its biological disposition. No metabolites were identified in these studies.

CASSANE DERIVATIVES

Both **cassaine** (56) and **cassaidine** (57a) contain an ester group and it is therefore not surprising that these diterpene alkaloids are metabolized by hydrolysis. Zelck[47] reported that cassaine was rapidly hydrolyzed by liver microsomes. Cronlund[48] found that cassaidic acid (57b) was a major urinary metabolite of cassaidine following i.v. administration of the latter to guinea pigs.

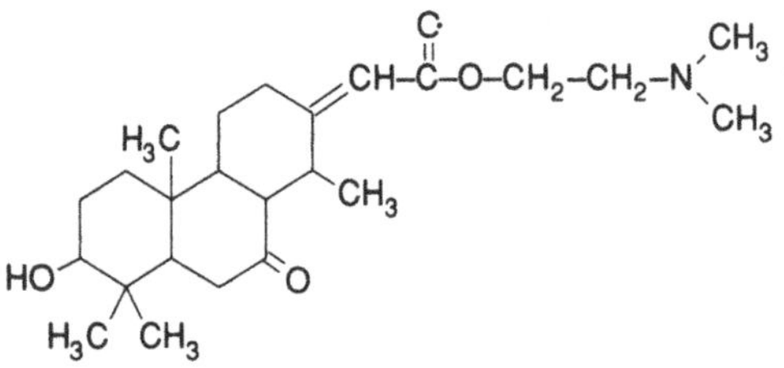

(56)

Cassaine

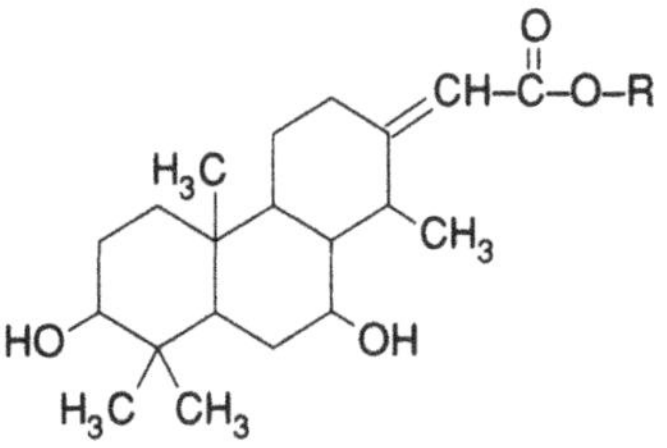

(57)

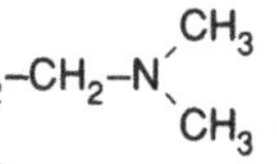

(a) Cassaidine, R = –CH$_2$–CH$_2$–N⟨CH$_3$,CH$_3$

(b) Cassaidic acid, R = H

OTHER DERIVATIVES

Cryptotanshinone (58) was administered orally to rats by Xie and Shen[49] who found that only ~0.3% of the dose was recovered in the urine as unchanged compound following a single dose. They isolated the unchanged compound and six metabolites from urine and bile. The identities of four of the metabolites were determined. These were shown to be tanshinone IIA (59), two hydroxylated derivatives of tanshinone IIA, and probably a glutamic acid conjugate of a compound formed from tanshinone IIA by oxidation of the methyl group attached to the furan ring to a carboxy group.

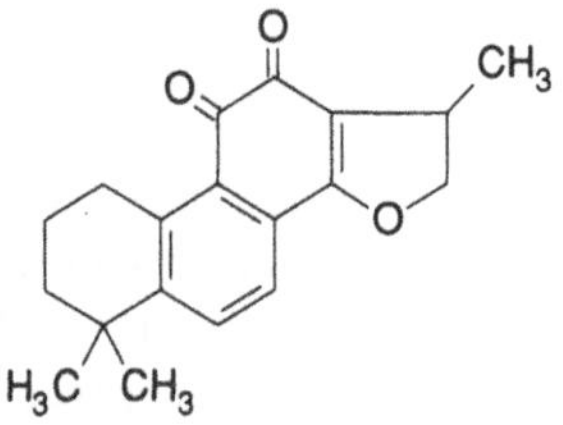

(58)

Cryptotanshinone

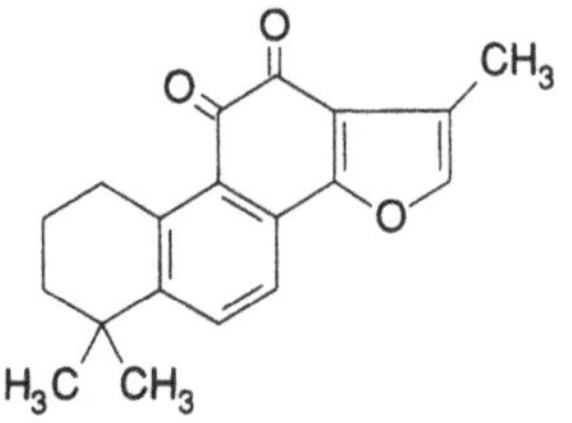

(59)

TRITERPENOIDS

Triterpenoids are C$_{30}$-compounds or derivatives thereof and include hydrocarbons, alcohols, ketones, and esters. Triterpenoid alcohols occur both free and as glycosides, many of the latter being called saponins. The classification of triterpenoids is sometimes arbitrary, being based on differences in use or properties rather than chemical differences. Only a limited number of reports on the metabolism of some types of triterpenoids are available; however, it seems advantageous to divide the group into four basic types of true triterpenes according to their ring skeletons (ursane type (60), oleanane type (61), dammarane type (62), and cycloartane type (63)), into an extensive group of cardiac steroids, and, finally, into some miscellaneous steroids.

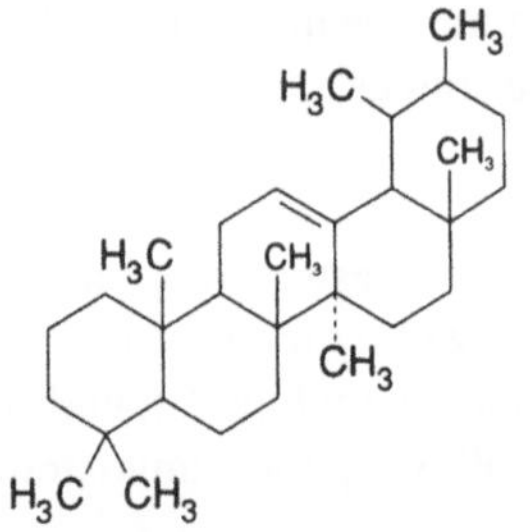

(60)

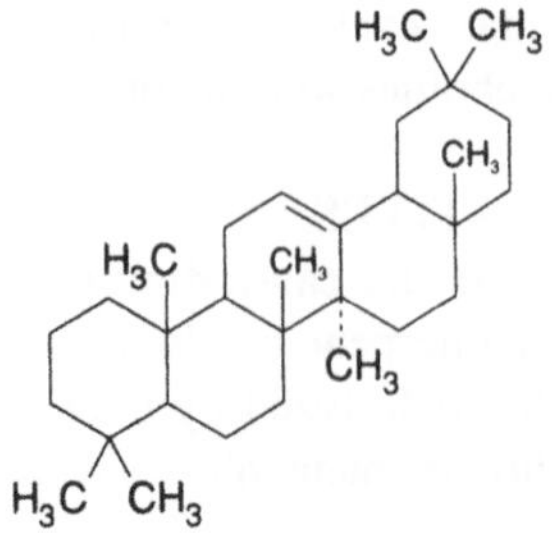

(61)

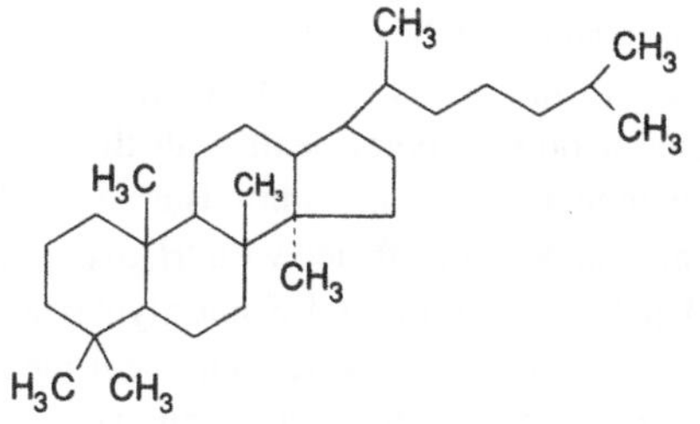

(62) 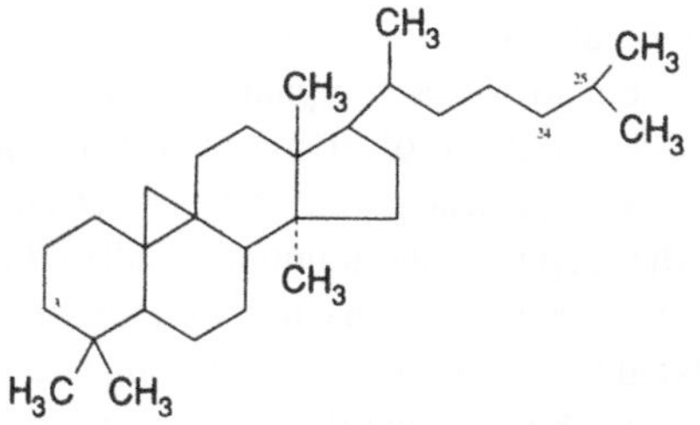   (63)

The hydrocarbon squalene is the only common acyclic terpenoid. Although it is found in small amounts in some plant oils, interest in its metabolism stems primarily from its role as an intermediate in the biosynthesis of cholesterol. A discussion of its metabolism therefore falls outside the scope of this book.

URSANE DERIVATIVES

The metabolism of the ursolic acid derivatives **asiatic acid** (64a), its glycoside **asiaticoside** (64b), and **madecassic acid** (64c) in rats was studied by Chasseaud et al.[50] Similar studies in dogs and humans were reported by Hathway.[51] The most striking finding in all species was the extensive fecal excretion of unchanged compound when asiatic acid or madecassic acid was administered orally. Nearly 90% of the dose (~10 to 20 mg/kg) was recovered in the feces of rats in 4 or 6 d when tritium-labeled asiatic acid or madecassic acid, respectively, was used. Average values of ~90% in 4 d were also obtained using dogs and, in man, ~85% of a single oral dose of the acids was recovered in the feces after 4 d. However, the fact that the fecal material was due to the unchanged compounds is not attributable to lack of absorption but, in considerable degree, to biliary excretion of absorbed material. Absorption of either asiatic or madecassic acid in rats accounted for about half of the dose since the 36-h bile contained 45% of the administered radioactivity, mainly as the glucuronides but partly as the sulfate conjugates of the acids. Steric considerations indicated that the 2α- or 3β-hydroxy groups are the most likely sites of conjugation. The presence of unconjugated metabolites in the feces indicates that conjugate hydrolysis was carried out by the intestinal microflora. Likewise, the ester bond in asiaticoside (64b) was hydrolyzed by the microflora and the metabolism of both the acid and its ester are therefore similar.

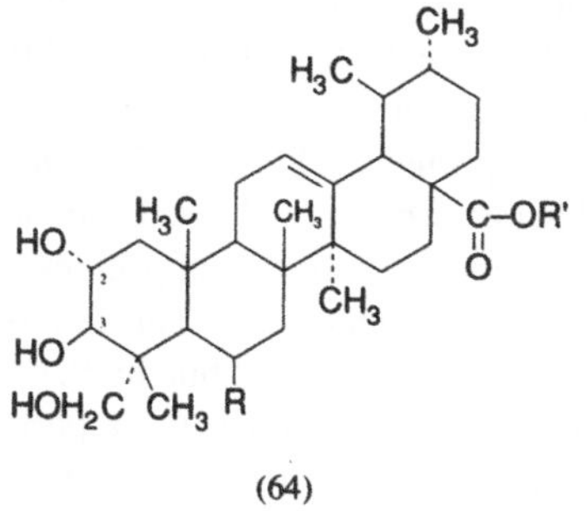

(64)

(a) Asiatic acid,   R = R' = H

(b) Asiaticoside,   R = H   R' = –glucose–glucose–rhamnose

(c) Madecassic acid,   R = OH   R' = H

OLEANANE DERIVATIVES

Glycyrrhetic acid (18β-glycyrrhetinic acid) (65) is the aglycon of the 3-O-diglucuronide **glycyrrhizic acid** (glycyrrhizin). Intravenous administration of the glycoside to rats did not result in the presence of detectable amounts of glycyrrhetic acid in the blood, however this did occur after oral dosage.[52] This investigation also indicated that the decline in the blood levels of glycyrrhetic acid was accompanied by an increase in the levels of a metabolite, probably a glucuronide conjugate of glycyrrhetic acid. Several other reports dealt with the biological disposition of glycyrrhetic acid and its glycoside in man. Carlat et al.[53] administered 3 to 4 g oral doses of the aglycon or the ammonium salt of the glycoside, both labeled with ^{3}H, to humans and found that ~98% of the radioactivity was excreted in the feces within 1 d. Some glycyrrhetic acid was isolated from the small amount of material (<1% of the dose) excreted in the urine in both experiments. This indicates that the sugar residues are removed in the body. Similar experiments were carried out by Terasawa et al.[54] who gave oral doses (133 mg) of glycyrrhizic acid to humans. They found that glycyrrhetic acid was detectable in the urine for as long as 130 h, however only ~2% of the dose (as unchanged compound or its metabolite (65)) was excreted by this route. Helbing[55] gave a single oral dose (4 g) or six daily doses of 2 g of glycyrrhetic acid to two patients and, using a chromatographic method of assay, recovered 53 to 61% of the dose in the feces. This study also showed prolonged excretion lasting several days. The lower recoveries in these experiments were not believed to be due to bacterial degradation of glycyrrhetic acid in the gut since its concentration in a fecal suspension was not significantly changed following incubation for 4 d. However, newer data summarized below indicate that metabolism of glycyrrhetic acid may be carried out by the gut flora.

(65)

Glycyrrhetic acid

The prolonged excretion of glycyrrhetic acid suggests that it may be subjected to enterohepatic circulation, however Carlat et al. concluded that this did not occur in humans since <0.5% of the oral dose was found in the blood after 4 h or excreted in the bile within 4 h. Nonetheless, the results of Terasawa et al. included one subject who showed peaks in the urinary excretion profile of glycyrrhetic acid at 20, 70, and 110 h. In addition, the enterohepatic circulation of glycyrrhetic acid in rats has been established. Parke et al.[56] found that ~30 to 70% of an oral dose (25 mg/kg) and essentially all of the aglycon given intraperitoneally were excreted in the bile of rats. Ichikawa et al.[57] reported a value of 80% of the dose (100 mg/kg) for the biliary excretion of glycyrrhizic acid in rats. They also showed that the biliary material was reabsorbed from the intestine. Rapid and extensive excretion of β-glycyrrhetic acid in the bile was also found in mice given the compound intravenously.[58] This led to the excretion of the material in the feces. Three metabolites were detected in rat bile by Parke et al. and none was identical with unchanged compound. Subsequently, Iveson et al.[59] reported these to be the 30-glucuronide, the 3-*O*-sulfate and, probably, the 3-*O*-glucuronide conjugates of glycyrrhetic acid. In intact rats, fecal excretion of radioactivity was very extensive whereas only ~1% of the dose (20 to 60 mg/kg, p.o.) was excreted in the urine. Iveson et al. reported that this fecal material consisted nearly entirely of unchanged glycyrrhetic acid following oral dosage whereas some conjugated material was also present after injection. This finding is probably related to the much faster biliary excretion of conjugates in the latter case with the result that some of this material could more easily escape hydrolysis by the intestinal bacteria.

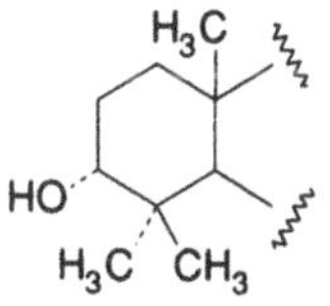

(66) 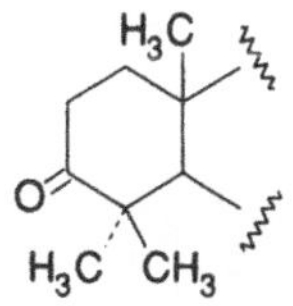   (67)

The role of the intestinal microflora in the metabolism of glycyrrhizic acid and its aglycon was studied by Hattori et al.[60] They found that the glycoside was hydrolyzed to glycyrrhetic acid when incubated anaerobically with a mixed culture of human fecal bacteria. They also showed that the aglycon underwent epimerization to give the 3α-epimer. This reaction was reversible and small amounts of the intermediate compound, 3-dehydroglycyrrhetic acid were detected. Structures (66) and (67), respectively, illustrate these changes to glycyrrhetic acid (65). Hattori et al.[61] investigated the ability of selected strains of human intestinal bacteria to carry out these reactions. No activity was found among numerous stock strains of fecal bacteria, however *Peptostreptococcus intermedius* showed a moderate ability to hydrolyze glycyrrhizic acid to glycyrrhetic acid. When bacteria isolated directly from human feces were tested, a *Ruminococcus* sp. was found which had the ability to hydrolyze glycyrrhizic acid and to reduce compound (67) to glycyrrhetic acid. Additionally, a strain of *Clostridium innocuum* was found which reduced the keto derivative (67) to the 3-epi-derivative (66). Additional biochemical studies on the properties of the novel dehydrogenase from the *Ruminococcus* sp. which catalyzed the 3-β-hydroxy $\leftrightarrow$ 3-keto interconversion were reported by Akao et al.[62] A similar investigation of the enzyme from *Clostridium innocuum* which was responsible for the corresponding reaction with the 3-α-hydroxy derivative (66) was carried out by Akao et al.[63] Akao et al.[64] found that a *Eubacterium* sp. isolated from human feces could hydrolyze glycyrrhizic acid to its aglycon. Biochemical studies indicated that this β-glucuronidase differed from the common type which numerous intestinal bacteria produce and which hydrolyzes many phenolic monoglucuronides.

While most of the saponins investigated are dammarane-type triterpenes (see below), saikosaponins have the oleanane skeleton. Shimizu et al.[65] investigated the transformation of **saikosaponin a** (68a) and **saikosaponin d** (68b) in rat gastric juice and by the mouse intestinal flora. They also studied the excretion of saikosaponin derivatives in the feces of rats following the oral administration of saikosaponin a (5 or 20 mg/kg). Both of these compounds are unstable under acidic conditions and it was shown that incubation in rat gastric juice (pH 1.5) resulted in scission of the heterocyclic ring. With saikosaponin a two isomeric products were identified. These were saikosaponin b$_1$ (69a) and saikosaponin g (70). However, with saikosaponin d only saikosaponin b$_2$ (69b), the epimer of saikosaponin b1, was formed. No loss of the sugar residues was found under these conditions, however saikosaponin a was readily converted to its aglycon, saikogenin F, when it was incubated anaerobically with mouse intestinal microorganisms. Small amounts of the intermediate compound prosaikogenin F, the monofucoside of of saikogenin F, were also detected. Similar patterns of conversion to prosaikogenins and then to saikogenins were found when the other saikosaponins (b$_1$, g, d, and b$_2$) were incubated with the mouse intestinal flora. When saikosaponin a was administered orally to rats, identification of the metabolites in the feces indicated that metabolism occurred by both of the above routes. However, bacterial metabolism predominated and more than 50% of the dose was recovered in 24 h as unchanged compound, prosaikogenin F, and saikogenin F. About 1% of the dose was recovered as various diene derivatives. These included saikosaponin b$_1$ and its hydrolysis products prosaikogenin A and saikogenin A, and saikosaponin g and its hydrolysis products prosaikogenin H and saikogenin H.

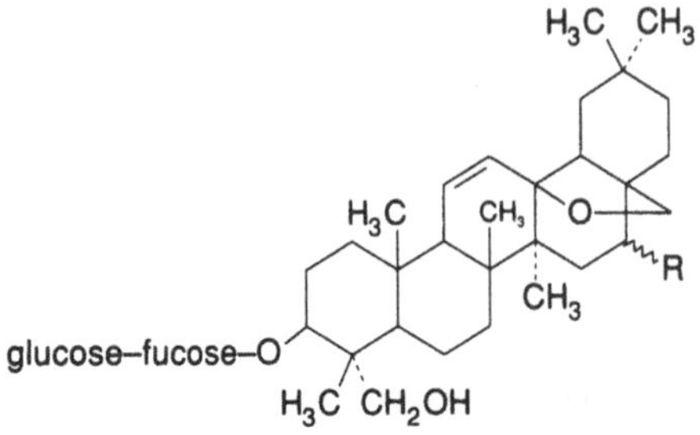

(68)

(a) Saikosaponin a, R = ─OH

(b) Saikosaponin d, R = ····OH

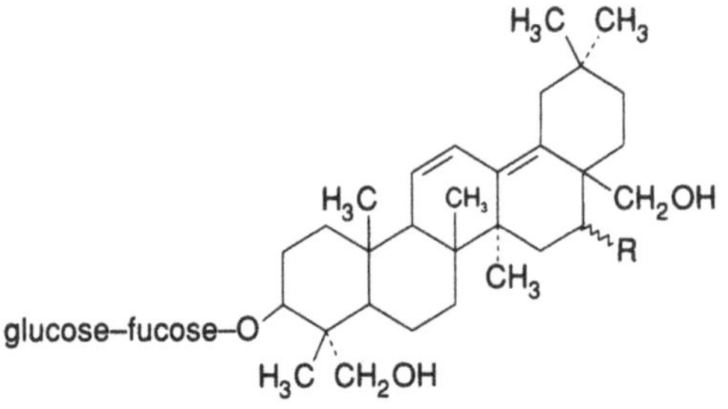

(69)

(a) R = ─OH

(b) R = ····OH

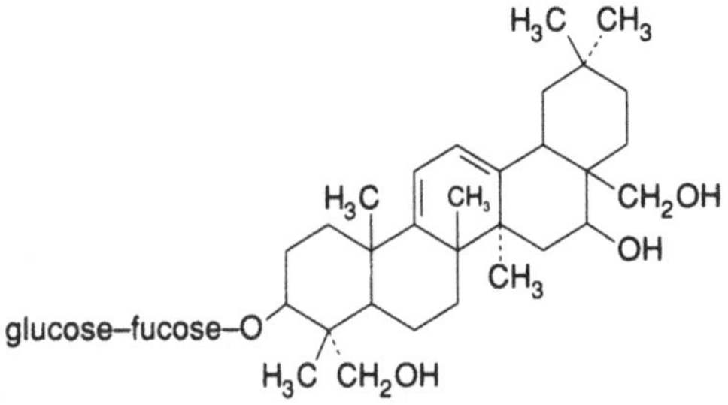

(70)

The metabolism of **reduced lantadene A** in rats and sheep was investigated by Pass et al.[66] who found that different patterns of metabolites were formed in the liver of animals which showed different sensitivities to the toxic effects of this triterpene. Hepatotoxicity occurs in sheep and female rats, which formed a similar major metabolite, whereas a different major metabolite was formed by male rats and nonsusceptible female rats. The identities of these metabolites were not determined, however the major metabolite formed by sheep and female rats was not a glucuronide conjugate.

Aralosides (saparal saponins) are oleanane derivatives. Following intragastric doses (50 or 1000 mg/kg) to rats, only 3% was excreted unchanged in the feces.[67] Urine was the main route of excretion and 20 (6 h), and 80% (6 to 24 h) of the dose was recovered, apparently as araloside metabolites.

DAMMARANE DERIVATIVES

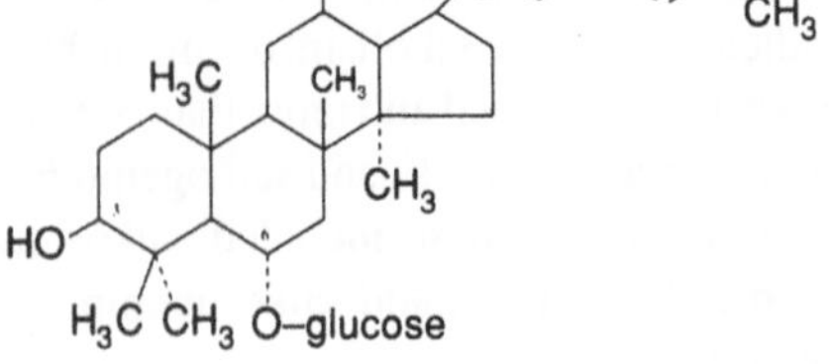

(71)

Ginsenoside R_{g1}

Saponins are triterpenoid glycosides which are powerful surface active and hemolytic agents. Their aglycons are known as sapogenins. Many saponins, and notably the ginsenosides, are based upon the dammarane skeleton. Several reports on their metabolism are available, however some of these findings are difficult to interpret because of methodological uncertainties. This is especially the case when radioactive saponins are employed because the radioactivity measured may be due to metabolites or decomposition products as well as to the administered compound.

One of the main saponins from ginseng is **ginsenoside R$_{g1}$** (71) and most of the metabolic studies have dealt with it. It is also designated as Panax saponin A and ginsenoside A$_2$ in some reports. Han and Chang[68] reported that only negligible amounts of unchanged compound were excreted in the urine when rabbits were given oral doses (29 mg/kg) of [³H]ginsenoside R$_{g1}$. Chen et al.[69] found no unchanged ginsenoside R$_{g1}$ in the urine of rabbits given oral doses (~165 mg/kg). However, the small amounts excreted in the urine do not reflect a corresponding lack of absorption. Han et al.[70] reported gastrointestinal absorption values in mice of 10 to 50%, although only 3 to 4% of the radioactivity following an oral dose (70 or 140 mg/kg) of ³H-labeled compound was excreted in the urine in 26 h. Their results indicated that most of this material was unchanged ginsenoside. Nearly 90% of the dose remained in the gastrointestinal tract after 26 h and much of this material was due to degradation products, probably prosapogenin derivatives lacking a glucose moiety. Similarly, Odani et al.[71] showed that rats given an oral dose (100 mg/ kg) of ginsenoside R$_{g1}$ rapidly absorbed a small portion (~2 to 20%) of the dose and excreted a small amount (0.4%) unchanged in the urine within 24 h. During this time 41% of the dose was recovered in the feces as unchanged compound. Following i.v. dosage (5 mg/kg), rapid biliary excretion of the unchanged compound occurred. This route accounted for 57% of the dose in 4 h and was ~2.5 times the amount excreted in the urine during this period. They also demonstrated that no detectable amounts of the ginsenoside were found in the brain and that no detectable metabolism occurred in the liver. Their results demonstrated that ginsenoside R$_{g1}$ underwent metabolism or degradation in the gastrointestinal tract and a subsequent investigation[72] determined the nature of these products. They found that ginsenoside R$_{g1}$, both in the rat stomach or under mild acidic conditions, was transformed into six products. Four of the products were shown to be epimeric pairs of compounds with structures (72) (ginsenoside R$_h$) and (73), i.e., pairs having different configurations of the hydroxy group at C-20. The final pair was unstable but believed to be of a Δ²⁰-prosapogenin type formed via elimination of the hydroxy group at C-20. The findings of Odani et al. were supported by those of Strömbom et al.[73] who administered ginsenoside R$_{g1}$ orally (500 mg/kg) or intravenously (250 mg/kg) to mice. They found that only 2 to 4% of the dose was excreted in the urine in 24 h and that both routes of administration gave similar results. This indicates that biliary excretion of the material took place. This is also supported by their finding that the ginsenoside was rapidly absorbed (~30% within 1 h) after oral administration. Strömbom et al. noted that the bulk of the material was excreted as metabolites, mainly in the feces but also in the urine. Five metabolites were detected in intestinal contents and feces, and two of these were tentatively identified as compound (72) and its 25-hydroxy derivative (73). They also found that the same compounds were produced by mild acidic hydrolysis of ginsenoside R$_{g1}$. The urine contained fewer and lower amounts of these metabolites.

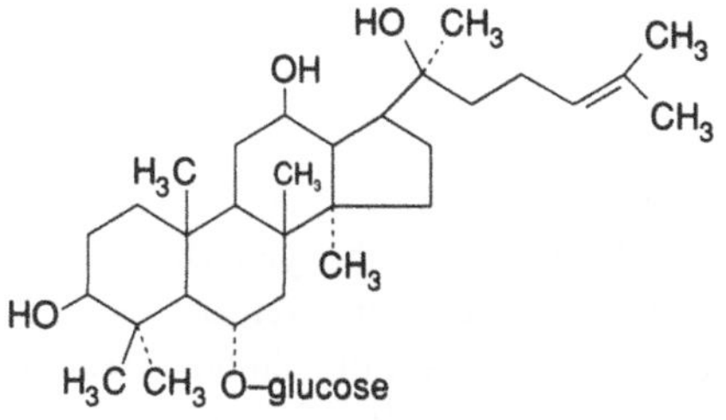

(72) 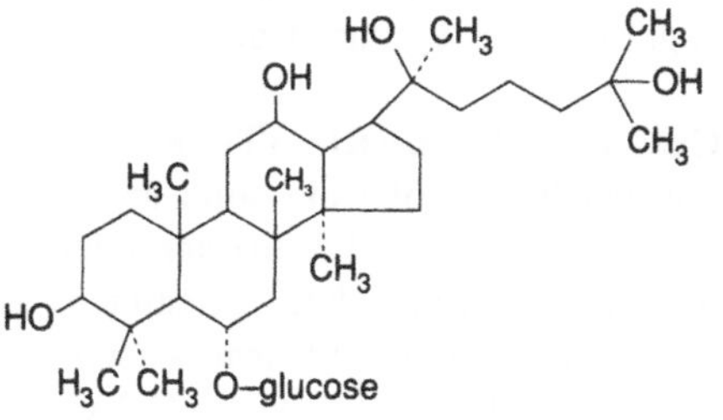   (73)

Ginsenoside R$_e$ differs from ginsenoside R$_{g1}$ only by the presence of an –*O*-glucose-rhamnose moiety rather than the –*O*-glucose moiety at C-6. Han et al.,[70] in experiments identical to those noted above with ginsenoside R$_{g1}$, found that ~7% of the radioactivity was excreted in the urine of mice following an oral dose of 40 mg/kg.

Ginsenoside R$_{g2}$ differs from ginsenoside R$_{g1}$ by having a free hydroxy group at C-20 and an –*O*-glucose-rhamnose moiety at C-6. Chen[74] reported that its incubation with rat gastric juice converted it to its epimer (the 20(*R*)-form) and to the corresponding pair of epimers which also contained a hydroxy group at C-25. A further eight products were formed when the ginsenoside and the three aforementioned compounds were incubated with intestinal fluids.

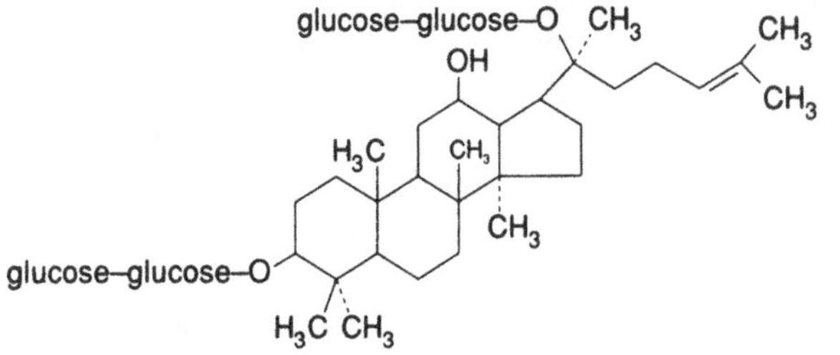

(74)

Ginsenoside R$_{b1}$

The ginsenosides noted above are examples of the panaxatriol-type, whereas **ginsenoside R$_{b1}$** (74), which contains two ring hydroxy groups, is a panaxadiol derivative. Han et al.[70] gave oral doses (40 to 180 mg/kg) of ^{3}H-labeled ginsenoside R$_{b1}$ to mice and reported that only ~1.5% of the radioactivity was excreted in the urine in 26 h. Most of this material appeared to be unchanged compound. Most of the dose remained in the gastrointestinal tract which contained unchanged glycoside and larger amounts of unidentified degradation products. Odani et al.[75] carried out experiments with ginsenoside R$_{b1}$ similar to those described above with ginsenoside R$_{g1}$ and found that it had a different pattern of metabolic disposition. Little absorption from the gastrointestinal tract of rats occurred and a mere 0.05% of the oral dose (100 mg/kg) was excreted unchanged in the urine within 48 h. Also, only ~11% was found in the feces in 24 h. Following i.v. administration (5 mg/kg), <1% was excreted in the bile and the compound was only slowly excreted from the body. This was believed to correlate with a high degree of binding to plasma proteins. Similar pharmacokinetic behavior was reported by Chen et al.[69] for **ginsenoside C**, a panaxadiol-type compound which differs from ginsenoside R$_{b1}$ only by the replacement of the terminal glucose unit at C-20 with an arabinose moiety. The experiments with ginsenoside R$_{b1}$ indicated that unabsorbed compound after oral dosage was rapidly transformed in the intestine. Odani et al.[72] described the nature of these decomposition products. Unlike that noted above with ginsenoside R$_{g1}$, the product formed from ginsenoside R$_{b1}$ in the rat stomach was not similar to any of the hydrolysis products formed in acidic solution. Three compounds were detected in the large intestine. These included the compound formed in the stomach, another unidentified compound and ginsenoside R$_d$ which differs from ginsenoside R$_{b1}$ only in the absence of the terminal glucose moiety at C-20. These three metabolites were also formed when ginsenoside R$_{b1}$ was incubated with cecal contents.

CYCLOARTANE DERIVATIVES

The metabolism of **γ-oryzanol** in rabbits was studied by Fujiwara et al.[76] This mixed compound was described as an ester of ferulic acid (4-hydroxy-3-methoxycinnamic acid) and several triterpenes including cycloartanol, cycloartenol, 24-methylenecycloartanol, and cyclobranol. These compounds share a common structure (63) and have the additional features of –OH at C-3 (cycloartanol), –OH at C-3 and $\Delta^{24,25}$ (cycloartenol), –OH at C-3 and –CH=CH$_2$ at

C-24 (24-methylenecycloartanol), and –OH at C-3, –CH$_3$ at C-24, and $\Delta^{24.25}$ (cyclobranol). The ester linkage was at C-3 and the ferulic acid moiety was labeled with ^{14}C. The radioactivity and metabolites excreted in the urine in 48 h were studied following oral doses (40 mg/kg) of γ-oryzanol. Only ~6% of the radioactivity was excreted in the urine during this period and no unchanged compound was detected. No metabolites derived from the terpenoid structure were described and the compounds identified were ferulic acid and its metabolites. These included vanillic acid, a product of β-oxidation, the glycine conjugate of vanillic acid, and acetovanillone (4-hydroxy-3-methoxyacetophenone) which no doubt arose via decomposition of the β-keto derivative of the C$_6$–C$_3$-acid. Fujiwara et al.[77] subsequently reported on the metabolism of γ-oryzanol in rats. Using material labeled with ^{14}C at C-3 in the ferulic acid moiety, they found that 10% of the dose (50 mg/kg, p.o.) was excreted in the urine, largely within 48 h. During this period 85% of the radioactivity was recovered in the feces. The nature of the fecal metabolites was not determined. No unchanged compound was found in the urine, however free and conjugated ferulic acid and several previously identified metabolites of ferulic acid (see the chapter entitled Metabolism of Acids, Lactones, and Esters) were identified. These included dihydroferulic acid, *m*-hydroxycinnamic acid, *m*-hydroxyphenylpropionic acid, and *m*-hydroxyhippuric acid. Hippuric acid was also identified as a urinary metabolite of γ-oryzanol. The excretion of *m*-hydroxyphenyl derivatives in the urine demonstrates that the acid moiety was subjected to metabolism by the intestinal microflora. It is not clear if this results from hydrolysis of γ-oryzanol in the intestine prior to absorption or from esterase activity in the tissues following absorption. In the latter case, biliary excretion of the ferulic acid formed would explain the formation of the microbial metabolites.

CARDIAC STEROIDS

The cardiac glycosides or cardenolides, formerly in the form of digitalis and more recently as pure glycosides, have held a central role in therapeutics for the past 200 years. This prominence has resulted in a vast literature, to which studies of the absorption, distribution, metabolism, and excretion make a large contribution because of their influence on the therapeutic properties. Useful results in these areas have largely appeared during the past few decades following the development of sensitive and selective detection methods. Because of the abundance of the literature, a detailed discussion of the biological disposition of these compounds is clearly beyond the scope of this book. The following summary, while giving a comprehensive account of the metabolic changes known to occur with the cardiac glycosides, will therefore not attempt to be exhaustive.

The review by Doherty[78] and the book edited by Storstein[79] are convenient sources of information on the biological disposition, particularly in relation to clinical implications, of cardiac glycosides. These publications deal primarily with digitoxin and digoxin, the two most commonly used compounds. Other useful reviews include those by Temple et al.,[80] who covered the chemistry of the cardiac glycosides, the metabolism of the most important compounds, their biliary excretion, and drug interactions, and by Wirth[81] who summarized the metabolism in man of many of the glycosides.

Both the nomenclature and the chemistry of the cardiac glycosides are complex and it is therefore useful to summarize these points before reviewing the pathways of metabolism. The central structural feature of these compounds is a steroid nucleus (see structure (75b)) which is similar to that found in the saponins but which contains the distinguishing feature of a *cis*-configuration of the substituents attached at C-13,14 and the presence of a 14β-hydroxy group. Further characteristic features are the unsaturated lactone group at C-17 and the presence of unusual sugar residues attached at the C-3-hydroxy group. Structural variations of this basic pattern involve the addition of hydroxy groups to the ring system, the replacement of the C-19-methyl with an aldehyde group, the presence of a six-membered lactone ring at C-17, a *trans* configuration of the substituents at C-5,10, and, of course, variations in the types and numbers

TABLE 1
Chemical Components of some Cardiac Glycosides

Type	Plant source	Aglycone	Sugar moieties at C-3

Digitalis — *Digitalis purpurea* — Digitoxigenin —— digitoxose – digitoxose – digitoxose – glucose
↑——————————————— Digitoxin ——————————↑
↑————— Purpurea glycoside A (Desacetyldigilanide A) —————↑

Digitalis purpurea — Gitoxigenin —— digitoxose – digitoxose – digitoxose
↑——————————— Gitoxin —————————↑

Digitalis lanata — Digitoxigenin —— digitoxose – digitoxose – digitoxose-3-acetyl – glucose
↑——————————————— Digitoxin ——————↑
↑————————————— Lanatoside A (Digilanide A) —————————↑

Digitalis lanata — Digoxigenin —— digitoxose – digitoxose – digitoxose-3-acetyl – glucose
↑——————————— Digoxin ——————↑
↑————————————— Lanatoside C (Digilanide C, Cedilanid) ————↑

Thevetin — *Thevetia neriifolia* — Digitoxigenin —— thevetose – glucose – glucose
↑————— Neriifolin —————↑
↑————— Thevebioside —————↑
↑————— Cerberoside (Thevetin B) ———↑

Thevetia neriifolia — Cannogenin —— thevetose – glucose – glucose
↑————— Peruvoside ———↑
↑————— Thevetin A ——————↑

Strophanthus kombé — Strophanthidin ——— cymarose – glucose – glucose
↑– Cymarin (K-Strophanthin-α) –↑
↑—— Strophanthin (K-Strophanthin-β) ——↑
↑——— K-Strophanthoside (K-Strophanthin-γ) ———↑

Strophanthus — *Strophanthus gratus* — Ouabagenin ——— rhamnose – glucose
↑— Ouabain (G-Strophanthin) —↑
↑——— G-Strophanthoside ————↑

Convallaria — *Convallaria majalis* — Strophanthidin —— rhamnose
↑——— Convallatoxin ———↑

Convallaria majalis — Strophanthidol —— rhamnose
↑——— Convallatoxol ———↑

Squill — *Urginea maritima* — Scillarenin —— rhamnose – glucose
↑—— Proscillaridin A ———↑
↑————Scillaren A————————↑

of sugar groups attached at C-3. Table 1 summarizes the nomenclature and chemical relation-ships of most of the cardiac glycosides subsequently discussed in this section and also lists their botanical sources. The latter are informative because cardiac glycoside terminology is mainly based on botanical origins rather than chemical structures. From Table 1 it can be seen that these compounds consist of an aglycon, also termed genin, and from one to four sugar residues. These residues can vary with respect to their stability towards hydrolysis, and the natural glycosides (also called native or genuine glycosides) which occur in the plant easily undergo hydrolysis of the terminal glucose unit and, when present, of an acetyl group to form the glycosides commonly employed in therapeutics (e.g., digitoxin, digoxin). Further hydrolysis of the rare sugar moieties, commonly digitoxose, requires stronger conditions chemically (e.g., acid hydrolysis) and gives rise to the aglycons.

Digitoxin (75a) is one of the two most widely used cardiac glycosides and it is therefore understandable that its metabolism has been extensively studied. Nevertheless, not all of the results are easily reconciled and newer findings have shown that some aspects of the traditional interpretation of digitoxin metabolism were incorrect. This is especially true of the pathway which results in removal of the sugar groups of the glycoside. As noted below, this process is dependent on the cytochrome P-450-system for the stepwise removal of the sugar groups to give the bisdigitoxoside and then the monodigitoxoside. Some studies have described the loss of the final sugar group to give the aglycon, digitoxigenin (75b). Another metabolic pathway of digitoxin is hydroxylation of the steroid nucleus. This takes place at several sites, but that at C-12 has received the greatest amount of interest as this reaction forms the cardioactive derivative digoxin or its cleavage products. A third pathway is epimerization of the hydroxy group at C-3, which occurs with digitoxigenin. This reaction changes the orientation of the hydroxy from β to α via the 3-keto intermediate. An additional reaction is the reduction of the double bond in the lactone ring attached to C-17. Finally, conjugation of the metabolites with glucuronate or sulfate also takes place. This occurs with 3-epidigitoxigenin, however more recent studies have indicated that glucuronidation of digitoxigenin monodigitoxoside is an important metabolic step.

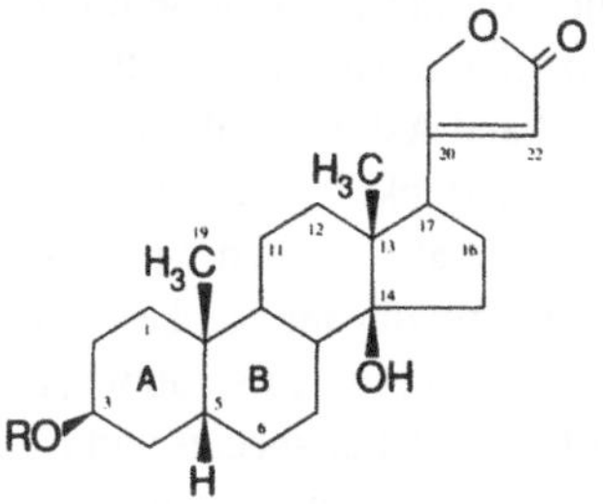

(75)

(a) Digitoxin, R =

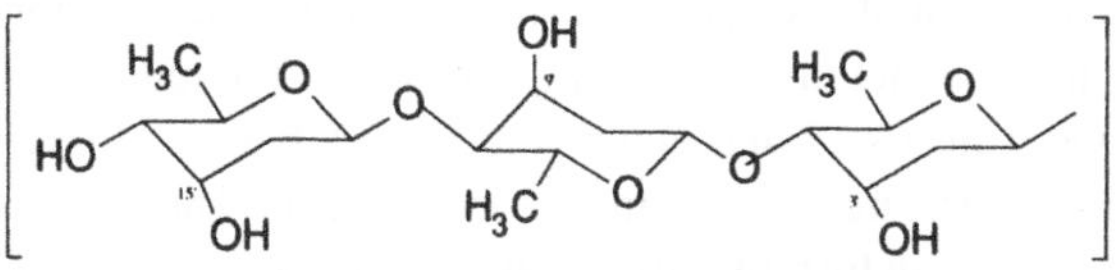

(b) Digitoxigenin, R = H

The stepwise removal of the sugar residues of digitoxin is a major metabolic pathway. It was shown to occur in rat liver slices[82] and in similar preparations from guinea pig, rabbit, cat, dog, and man[83] with the result that the aglycon digitoxigenin was formed. Sequential loss of the sugar groups was also reported by Kolenda et al.[84] when digitoxin was perfused through the isolated guinea pig liver. The aglycon is often not detected *in vivo*[85] and, in a review of cardiac glycoside metabolism, Repke[86] concluded that this finding is due to slow cleavage of the glycosidic linkages compared with the rate of subsequent metabolism of the released aglycon (e.g., 3-epimerization and conjugation). Other investigations of digitoxin metabolism likewise failed to detect the aglycon as a urinary metabolite. This result was obtained by Rietbrock and Vöhringer[87] in the rat and in man by Wirth et al.[88] On the other hand, Okita[89] reported digitoxigenin in the urine of patients receiving [³H]digitoxin and it was also detected in most of the tissues analyzed 18 h following administration of digitoxin to rats.[90] The removal of the sugar groups has been traditionally referred to as a hydrolytic process, however numerous newer studies clearly demonstrate that this is not the case. Schmoldt et al.[91] reported that the cytochrome P-450-system from rat liver microsomes cleaved the glycosidic bonds in digitoxin to give the bis- and monodigitoxosides of digitoxigenin. Further investigation of this pathway by Schmoldt and Rohloff[92] revealed that an axial hydroxy group in the terminal digitoxose unit was oxidized to the corresponding keto derivative which then underwent loss of the sugar moiety. The positions involved are those at C-15′, C-9′, and C-3′. Significantly, the 3′-dehydro derivative of the monodigitoxoside of digitoxigenin was found to have a high degree of stability. This fact serves to explain why the genin was not detected in many studies. Similar results were obtained using rat hepatocytes.[93] Schmoldt et al.[94] reported that this same mechanism operates in human liver microsomes. Kershaw et al.,[95] using rat liver slices, confirmed the main pathway of digitoxin metabolism to be stepwise loss of the sugar groups to give the monodigitoxoside which was then subject to glucuronidation (see below). They also proposed a mechanism of cleavage which involved oxidation of the carbon atom containing the relevant axial hydroxy group (e.g., C-15' or C-9'). Loss of water from the carbon atom now containing two hydroxy groups would result in the dehydro (i.e., keto) compound. Other reports have confirmed the key role of the cytochrome P-450-system, however the bisdigitoxoside of digitoxigenin was sometimes found as the major cleavage product. This was the case with a study using rat liver homogenates, liver fractions, and hepatocytes[96] and rat and mouse liver homogenates.[97] Interestingly, the latter study showed a species difference in that rabbit liver homogenates produced digitoxigenin as the major product.

The hydroxylation of digitoxin is of interest because this metabolic pathway may produce metabolites with cardioactive properties. An example of this is the 12β-hydroxylation reaction which forms digoxin. Brown et al.[98] and Ashley et al.[99] identified digoxin in the urine of both rats and humans given digitoxin and Repke[85] showed that the partially cleaved glycosides containing the 12β-hydroxy group were also formed. Species differences in the hydroxylation of digitoxin at C-12 were noted by Herrmann and Repke[83] who found it to occur in liver slices from rats and humans but not from guinea pigs, cats, rabbits, and dogs. Hydroxylation at some unidentified position occurred with these latter species and Herrmann and Repke[100] suggested that this might be at C-6. This subject was investigated further using the aglycon digitoxigenin. The formation of hydroxylated metabolites was suggested in experiments using rat liver homogenates[101] or rat liver microsomes,[102] and Talcott and Stohs[103] identified several monohydroxylated metabolites using the latter system. In addition to 12β-hydroxydigitoxigenin, slightly larger amounts of the 5β- and smaller amounts of the 1β- and 16β-hydroxylated derivatives were also found. Interestingly, only the 5-hydroxylase activity was increased when the animals were pretreated with inducing agents. The formation of 5β-hydroxydigitoxigenin from digitoxigenin by rabbit liver homogenates was reported by Bulger and Stohs;[104] however, Bulger et al.[105] subsequently found that most of the hydroxylation under these conditions took place at C-6, the product being 6β-hydroxy-3-epidigitoxigenin. This finding thus confirmed the earlier suggestion of Herrmann and Repke noted above. A novel hydroxylated product formed by guinea pigs was shown by Carvalhas et al.[106] to be 17α-hydroxydigitoxin. It was produced by guinea pig liver slices and microsomes and also found in the plasma following digitoxin administration. Other investigations using *in vitro* systems have noted the formation of metabolites hydroxylated in the 12-position. Schmoldt et al.[91] concluded that the P-450-system in rat liver microsomes was involved in this reaction. Hydroxylation of digitoxin to various derivatives of digoxin was shown to occur using isolated rat hepatocytes[93] and rat liver slices,[95] however this pathway was less prominent than that resulting in cleavage of the sugar residues.

The subject of epimerization (inversion from β– to α-configuration) of digitoxigenin shown to occur by Repke and co-workers was reviewed by Repke[86] who noted that the reaction occurs mainly in the liver. The reaction sequence involves the formation of the 3-dehydro (i.e., 3-keto) derivative and subsequent hydrogenation to the epigenin. Repke and Samuels[107] carried out an enzymatic study of the 3α,β-hydroxysteroid dehydrogenases in rat liver and noted that this sequence is a process of detoxification since both the keto and epi derivatives possess very weak biological activities. Enzyme activity was present both in the supernatant fraction and the microsomal fraction. Subsequent investigations have shown the formation of the 3-keto and 3-epi derivatives from digitoxigenin using rat liver homogenates[101] or microsomes.[102] 3-Epidigitoxigenin is also the major metabolite of digitoxigenin when homogenates of guinea pig liver are used.[108] As noted above, rat liver microsomes did not cleave digitoxin past the monodigitoxoside and the keto- and epigenins are therefore not formed *in vitro* from digitoxin.[91]

Hydrogenation of the double bond at C-20,22 in the lactone ring is a well known reaction with digoxin (see below). This pathway has received less attention in the case of digitoxin. Bodem and von Unruh[109] found that detectable but very low amounts of dihydrodigitoxin were present in the serum of patients given maintenance doses of digitoxin. However, in patients with renal insufficiency, much higher levels were measured. These sometimes exceeded those of digitoxin itself. Strobach[110] studied the formation of dihydrodigitoxin in a large number of patients and found that ~6% of the dose was excreted in the urine of those with normal renal function. A value of 3% was seen in patients with renal insufficiency. Strobach concluded that this metabolic pathway is not important in the metabolic inactivation of digitoxin. Flasch and Heinz[111] found no evidence for the reduction of digitoxin to dihydrodigitoxin in cats. However, they showed that when the dihydro compound itself was administered, some hydrolytic opening of the lactone ring to give dihydrodigitoxinic acid was detected.

Many studies of the metabolism of digitoxin have detected very polar metabolites, and several newer investigations have given special attention to the specific glucuronide conjugates formed. Repke[85] noted that the cleavage of the sugar moieties leads finally to a conjugate rather than the aglycon and Herrmann and Repke[112] found that incubation of digitoxigenin with liver slices from rat, guinea pig, rabbit, dog, or man resulted in the formation of a total of six highly polar metabolites. The sulfate conjugates of the genin and the epigenin were formed in all species except the dog which formed a conjugate which was probably a diglucuronide of a hydroxylated digitoxigenin derivative. The glucuronide of 3-epidigitoxigenin was formed by liver slices of rabbit, dog, and man. However, quantitative data noted below indicate that complete loss of the sugar groups is not a major pathway in the metabolism of digitoxin. It is therefore of interest to note that Vöhringer and Rietbrock[113] found the glucuronide of digitoxigenin monodigitoxoside as a urinary metabolite of digitoxin in man. This conjugate is a quantitatively important metabolite in other species as well. Richards et al.,[114] using rat and rabbit liver homogenates, showed that digitoxigenin monodigitoxoside is a far better substrate for glucuronidation than are the bisdigitoxoside, the genin, or digitoxin itself. Castle[115]reported the formation of the glucuronide of digitoxigenin monodigitoxoside by rat liver microsomes and Schmoldt et al.,[116] using this system, showed the preference for the monodigitoxoside compared with the other compounds noted above and also with 3-epidigitoxigenin. The same preference was seen when microsomes from guinea pig and cat liver were used, although the conversion rates were lower. Rabbit liver microsomes also readily formed the glucuronide of digitoxigenin monodig-itoxoside, however in this species 3-epidigitoxigenin was the best substrate. Further studies on this UDP-glucuronosyltransferase system in rat liver were carried out by Schmoldt and Promies[117] who suggested that a specific form of the enzyme was involved. This subject was investigated further in more detail by Watkins and Klaassen[118] and von Meyerinck et al.[119] who showed that a form of the transferase enzyme is present in rat liver which is highly specific for digitoxigenin monodigitoxoside.

The above summary of digitoxin metabolism included relatively little information about the quantitative importance in man of the various pathways or of the routes of excretion. These points will be considered now. It is important to note that digitoxin is completely absorbed from the intestine following oral dosage, a fact clearly related to its nonpolar properties due to the relatively low degree of hydroxylation of the steroid ring. Most of the absorbed digitoxin is excreted in the urine, in part as unchanged compound but also as metabolites. The excretion of unchanged digitoxin in the urine varies with different dosage regimens. In patients receiving the drug daily, Jelliffe et al.[120] found that ~30% of the daily losses were attributable to urinary digitoxin. Marcus[121] reported a value of only 8% under similar conditions. This corresponds well with the 6 to 10% value reported in an early study by Okita et al.,[122] who found that the urinary excretion of radioactivity following administration of [14C]digitoxin accounted for 60 to 80% of the dose. Vöhringer and Rietbrock[123] reported that ~63% of the daily dose was eliminated in the urine and that about half of this material was due to digitoxin. A higher value was found by Wirth et al.[88] who reported that 79% of the urinary radioactivity following the i.v. administration of [3H]digitoxin to cardiac patients receiving the drug daily was unchanged compound. In a study which employed a single oral dose of [3H]digitoxin, Vöhringer and Rietbrock[113] measured a prolonged urinary excretion of radioactivity. Only 22% of the dose was excreted in 8 d and

~60% of this material was unchanged compound. Storstein[124] found that the proportion of digitoxin excreted unchanged in the urine following steady-state dosage was higher than that excreted after a single dose. Major urinary metabolites of digitoxin include polar conjugated derivatives, among which the glucuronide of digitoxigenin monodigitoxoside is prominent.[113,123] Hydroxylation of digitoxin at C-12 is not a major metabolic pathway in man. Jelliffe et al.[120] stated that ~8% of the daily loss of digitoxin from the body was in the form of digoxin. Vöhringer and Rietbrock[123] found that this reaction accounted for only 6 to 7% of the material excreted daily. Storstein[124] reported that only ~1% of the total cardioactive and conjugated material excreted in the urine was represented by digoxin following maintenance doses of digitoxin. However, values of ~25% were noted following a single dose. In any case, extensive conversion of digitoxin to digoxin seems unlikely in view of the relatively short biological half-life of digoxin and the well-known persistence of digitoxin in the body. Of course, these findings in man need not apply to other species and, in fact, Vöhringer and Rietbrock[113] found that metabolite excretion in rats occurred largely by the fecal route and that over twice as much digoxin than digitoxin was excreted in the urine.

An important feature in the biological disposition of digitoxin is its biliary excretion. This allows for both its enterohepatic circulation and the fecal excretion of metabolites in spite of complete initial absorption. The enterohepatic circulation of digitoxin in man seems to have been originally proposed by Okita et al.[125] and this phenomenon has subsequently been shown to occur in other species as well. Katzung and Meyers[126] found that the half-life of digitoxin was markedly shortened in animals with biliary fistulas. The bile contained both digitoxin and digoxin in the nonpolar fraction as well as two polar metabolites which released digitoxigenin on hydrolysis.[127] Russell and Klaassen[128] found 46% of the dose in the bile (12 h) in dogs. The corresponding value in rabbits was 53%. Marzo and Ghirardi[129] found that nearly 70% of the administered radioactivity was excreted in the bile in 5 h when guinea pigs were given [³H]digitoxin intravenously. Unchanged compound accounted for only a small part of this material, most of which was due to more polar metabolites. Ingwersen[130] showed that the bile in guinea pigs contained mainly digitoxigenin monodigitoxoside and water-soluble conjugates. Studies on the biliary excretion of digitoxin in rats have given various results. Cox and Wright[131] found that ~10% of the dose was excreted in 5 h in the bile of rats, whereas Russell and Klaassen[128] reported a value of 71% (12 h). Castle and Lage[132] found that a major portion of the biliary material consisted of water-soluble metabolites. Furthermore, this material was susceptible to hydrolysis by β-glucuronidase.[133] The hydrolyzed material contained small amounts of unchanged digitoxin, digitoxigenin bisdigitoxoside, and digitoxigenin, however the most abundant compound was digitoxigenin monodigitoxoside. The reabsorption of digitoxigenin monodigitoxoside and its glucuronide conjugate from the intestine of rats was studied by Volp and Lage.[134] Züllich et al.[135] found values between 31 and 39% for the biliary excretion of digitoxin in rats and they also determined the nature of the metabolites. They confirmed that the most abundant metabolite was a conjugate of digitoxigenin monodigitoxoside. Moderate amounts of the bisdigitoxosides of both digitoxigenin and digoxigenin were present as well as a few percent of both of these aglycons. The biliary excretion of digitoxin and its metabolites in rats decreases with increasing age.[136,137] This is mainly due to the decrease in the excretion of the metabolites formed by loss of the sugar groups and by conjugation. The biliary excretion of digitoxin in man does not appear to be extensive. Beermann et al.[138] stated that less than 10% of the dose of [³H]digitoxin was excreted in the bile in man in 24 h. The same value for this period was reported by Sattler et al.[139] from patients who received a single i.v. dose of ³H-labeled material. A further 8% of the dose was excreted in the bile during the next 2 d. Storstein[140] reported a 24-h value of only 1.5% in man; however, this was a measurement of digitoxin and cardioactive metabolites and it seems likely that considerable amounts of inactive products were also present.

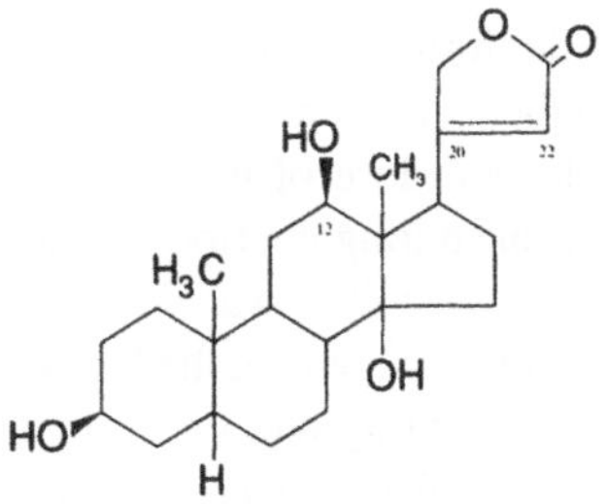

(76)

Digoxigenin

Digoxin is the 12β-hydroxy derivative of digitoxin. Its aglycon is digoxigenin (76). As noted above, digoxin is formed metabolically from digitoxin. However, interest in the biological disposition of digoxin is considerable because of its widespread use as a cardioactive glycoside. Although the chemical difference between digoxin and digitoxin is relatively small, it is nonetheless sufficient to give rise to considerable differences in their fate in the body. As summarized by Doherty et al.,[141] the absorption following oral dosage of digoxin in man is only 80 to 90% complete. However, other reports have given lower values. Beerman et al.[142] recorded figures of 40 to 60% and Magnusson et al.[143] found that only ~45% of the radioactivity was excreted in the urine in 3 d following a single oral dose of [³H]digoxin. Its persistence in the body is also much shorter than that of digitoxin, the half-life being ~1.5 d compared with 5 to 7 d for the latter compound. Metabolism is usually much less extensive with digoxin which is largely excreted unchanged in the urine[144] than with digitoxin which is appreciably excreted in the form of metabolites. The study of Magnusson et al.[143] revealed that over 90% of the urinary radioactivity consisted of unchanged digoxin. However, exceptions to this situation are encountered as some subjects produce relatively large amounts of dihydrodigoxin (see below). The high proportion of urinary excretion of unchanged compound is also seen in rats[145] and dogs[146] but not in guinea pigs which excrete mainly metabolites which are either less or more polar than digoxin.

The breakdown of digoxin in the upper gastrointestinal tract was noted by Beerman et al.[142] About 6 to 8% of the dose lost one or more of the digitoxose groups, probably due to the action of gastric acid. Gault et al.[147] found that minimal hydrolysis occurred in *in vitro* experiments at pH 3, however at pH values of ~1 the formation of the genin and its bis- and monodigitoxosides was rapid and extensive. They also showed that this breakdown could take place in the human stomach. Further studies on the influence of gastric acidity on digoxin metabolism were reported by Gault et al.[148,149] and Magnusson et al.,[150] however the latter group concluded that the acidic degradation *in vivo* is usually less than could be anticipated from *in vitro* data and that the clinical significance of this hydrolysis appears to be limited.

An area of digoxin metabolism which has received considerable attention is the reduction of the unsaturated lactone group to give dihydrodigoxin. It is well known that hydrogenation at C-20,22 greatly reduces the cardioactivity of digoxin and related compounds and Butler et al.[151] showed that the reduced metabolite appears to be eliminated from the body at a faster rate than digoxin. Abel et al.[152] reported that this reaction took place in dogs. However, most of the subsequent information comes from studies in man. Watson et al.[153] found that dihydrodigoxin was the major dihydro derivative of digoxin. Clark and Kalman[154] reported that the urinary excretion of dihydrodigoxin by 50 patients given maintenance doses of digoxin averaged 13% of the material found in the methylene chloride fraction (contains also digoxin, mono-, and bisdigitoxosides and genins). Greenwood et al.[155] found that an average of 16% of the maintenance dose of digoxin was excreted in the urine as the dihydro derivative. Similar results were reported by Peters et al.[156] who recorded an average value of 12%. Interestingly, they found that the values obtained from the 100 patients investigated ranged from 2 to 52%. A considerable advance in the understanding of the formation of reduced digoxin metabolites was made by Lindenbaum et al.[157] who studied the formation of the dihydro compounds in a large group of normal individuals given a single oral dose of digoxin. Significantly, little or no reduced metabolites were excreted in the urine during the first 8 h and maximal excretion occurred on the second day. Excretion of these metabolites was decreased if the glycoside was given by injection and their formation varied inversely with the bioavailability of the preparation. This relationship between digoxin bioavailability and the formation of reduced metabolites was confirmed by Magnusson et al.[158] Lindenbaum et al. suggested that the reduced metabolites are formed as a result of metabolism by the intestinal microflora. This hypothesis was confirmed in another study[159] which showed that digoxin reduction occurred upon incubation with fecal cultures from individuals known to carry out the reaction *in vivo*. Also, these individuals produced little or no reduction products when they were treated with erythromycin or tetracycline. Dobkin et al.[160,161] screened a large number of aerobic and anaerobic bacteria, including the most common species present in the human intestine, and found that strains of *Eubacterium lentum* were the only organisms capable of reducing digoxin. The configuration of the asymmetric center at C-20 was investigated in dihydrodigoxigenin[162] and in dihydrodigoxin[163] and it was shown that the metabolite formed in man is the 20*R* epimer. It is not surprising that this is the epimer formed by cultures of *Eubacterium lentum* and by fecal incubates from a subject known to convert digoxin to dihydrodigoxin.[164] Alam et al.[165] showed that variations in the ability to produce reduced metabolites exist among different ethnic groups and that these appear to be determined by environmental factors present early in life. The literature dealing with the dehydrogenation of digoxin was reviewed by Lindenbaum.[166]

Although digoxin undergoes considerable excretion as such, several metabolic products in addition to the reduced derivatives noted above have been identified. Ashley et al.[99] noted that a digoxin metabolite in the urine of rats and humans was identical to one of the metabolites excreted following digitoxin administration. Wright[167] subsequently identified it as the bisdigitoxoside of digoxin. This picture of extensive urinary excretion of unchanged compound and some excretion of the bisdigitoxoside derivative was also described by Okita[89] who reported that small or trace amounts of several other metabolites including the monodigitoxoside and the aglycon were also excreted by man. Gault et al.[168] showed that metabolic products were mainly excreted during the first day but that, even then, values of 90 to 95% for unchanged digoxin were typical. In a subsequent report,[169] small amounts of all the products formed by cleavage of the sugar groups as well as the 3-keto derivative and the 3-epigenin were detected in the plasma and urine. However, the most abundant group of plasma metabolites consisted of polar compounds. The metabolism of the cleavage products of digoxin was studied in detail in rats by Abshagen and Rietbrock.[170] The general resistance of digoxin to loss of the sugar groups found in the *in vivo* experiments noted above was also seen when *in vitro* techniques were used. Wong and Spratt[171] were able to demonstrate the formation of only trace amounts of the genin when digoxin was incubated with rat liver slices or homogenates. Subsequently, Lage and Spratt[172] were able to demonstrate cleavage of the sugar residues using rat liver slices but not homogenates. The *in vitro* hepatic metabolism of digoxin using preparations from mice, rats, guinea pigs, rabbits, cats, and dogs was investigated by Lage and Spratt.[173] Significantly, Schmoldt and Ahsendorf[174] showed that cleavage of the sugar groups of digoxin and its bis- and monodigitoxosides is carried out by the cytochrome P-450-system in the same fashion as that described above with digitoxin. They found that digoxigenin bisdigitoxoside was a relatively poor substrate for the enzyme. This may explain why the latter compound is prominent among digoxin metabolites.

The highly polar metabolites of digoxin noted above have been detected in other investigations. Dwenger and Haberland[175] obtained results from three subjects which indicated that 9 to 34% of the material excreted during the first 24 h was in this form. Benthe[176] estimated that ~15% of the material excreted in the urine after a single dose of digoxin was in the form of water-soluble compounds. The glucuronide and sulfate conjugates of the bisdigitoxoside were present in this material. Kolenda et al.[84] found that the isolated perfused guinea pig liver converted digoxin to conjugates containing glucuronic and/or sulfuric acid.

When digoxigenin monodigitoxoside is given to man it is rapidly metabolized and excreted in the urine mainly as conjugates of itself and of 3-epidigoxin.[177] *In vivo* experiments in rats using the aglycon, digoxigenin, also showed rapid metabolism with 3-epimerization a major metabolic pathway.[178] Gault et al.[179] likewise showed that the genin underwent rapid epimerization and glucuronide conjugation in man. Digoxigenin was not hydroxylated when incubated with rat liver homogenates.[180] Likewise, experiments using rat liver microsomes failed to show any hydroxylation of digoxigenin, in contrast to that observed with digitoxigenin.[103]

Investigations of the biliary excretion of digoxin and its metabolites indicate that this route may be of importance in several animal species. However, this does not appear to be the case in man. Doherty et al.[181] recovered only 5 and 8% of the radioactivity from a single i.v. dose of [³H]digoxin in the bile after 1 and 7 d, respectively. They found that digoxin itself was the major component of the biliary material and they concluded that the enterohepatic circulation of digoxin was of minor importance. Klotz and Antonin[182] found that patients during steady-state administration of digoxin excreted only 2 to 10% of the dose in the bile. They similarly concluded that enterohepatic circulation played a minor role in the biological disposition of digoxin in man. However, the data of Caldwell and Cline[183] indicated that it may be more extensive. They found that ~30% of an i.v. dose of digoxin was excreted in the bile in 24 h and that most of this material was presumably reabsorbable, biologically active drug. In other species it is apparent that a considerable portion of the administered digoxin is excreted in the bile. This was shown by Cox and Wright[131] who found that rats excreted 40% of the dose in the bile in 5 h. This material consisted of unchanged compound and a metabolite, which as noted above was later identified as the bisdigitoxoside, in about a 1:3 ratio. Other studies using rats gave excretion values of 32% in 2 h and 59% in 12 h,[128] 45% in 12 h,[145] and 61% in 11 h.[184] The last investigation showed clearly that enterohepatic circulation is an important feature in the biological disposition of the compound in rats. These results also showed that the bulk of the material excreted in the bile was unchanged compound and, in contrast to the findings of Cox and Wright noted above, the ratio between digoxin and its bisdigitoxoside derivative was ~4:1. About 10% of the total was due to water-soluble metabolites. Kitani et al.[185] reported that the biliary excretion of digoxin and its metabolites decreased with age in male rats. This was shown to be the result of a marked reduction in the amount of digoxigenin bisdigitoxoside excreted in the older animals. This phenomenon was not observed in female rats. The investigation of Russell and Klaassen[128] also included dogs. They found that ~14% of the dose was excreted in 12 h, a value close to that of the 15% reported by Harrison et al.[186] who collected bile for ~4.5 d. Only ~10% of this material was due to unchanged compound. According to Marcus et al.,[146,187] much of the material excreted in dog bile is due to water-soluble metabolites. The extent of biliary excretion of digoxin is large in both rabbits and guinea pigs. Russell and Klaassen[128] reported values of 39% (2 h) and 60% (12 h) of the administered radioactivity following the i.v. dosage of [³H]digoxin in rabbits. Using the isolated perfused guinea pig liver Kolenda et al.[188] found that nearly 70% of the added digoxin was excreted in the bile in 4 h and that this material was mainly in the form of polar metabolites. Haass et al.[189] also demonstrated the extensive and rapid biliary excretion of digoxin in the guinea pig. Using [³H]digoxin, Marzo and Ghirardi[129] found that ~72% of the radioactivity was excreted in the bile in 5 h. However, only 5 to 8% of this material was due to unchanged compound, the bulk being due to more polar metabolites.

The above summaries have dealt with the metabolism of the two most commonly used cardiac glycosides, digitoxin and digoxin. As shown in Table 1, the natural glycosides occurring in the plant contain an added sugar residue, a terminal glucose moiety. Some of these compounds obtained from *Digitalis lanata* and termed lanatosides are employed therapeutically. This is true of **lanatoside A** and **lanatoside C** and some metabolic information is available on these compounds. Deslanatoside C (deacetyllanatoside C) is closely related to lanatoside C, lacking only the acetyl group on the digitoxose moiety. It is useful to note that the presence of an extra sugar residue furnishes compounds of greater polarity than that found with the desglucose glycosides. This difference in physicochemical properties has a bearing on the biological disposition of these natural glycosides. The increase in polarity might be expected to result in reduced absorption following oral administration. However, the data of Beermann[190] indicated that the total recovery of radioactivity in the urine of man following administration of [³H]lanatoside C was only slightly less than that found after giving digoxin. Dengler et al.[191] also reported that the excretion of urinary radioactivity (4 to 6 d) following the administration of lanatoside C was nearly as great as that shown with digoxin (~80%). In fact, similarity between these two compounds in this respect may be due to the fact that the terminal glucose unit of the natural glycosides appears to be removed not by the action of tissue enzymes but by those of the intestinal bacteria.[86] Lauterbach and Repke[192] reported that species of *Clostridium* are mainly involved. Hawksworth et al.[193] found β-glucosidase activity in all of the six major groups of intestinal bacteria tested but noted that the enterococci were most active in converting lanatoside C to acetyldigoxin and, in some cases, to digoxin. These two metabolites were rapidly formed when lanatoside C was incubated with human feces.[190] Aldous and Thomas[194] studied the absorption and metabolism of lanatoside C in humans after oral administration and found that its conversion to digitoxin and perhaps further breakdown products took place in the intestine, partly by acid hydrolysis and partly by bacterial action. Little or no unchanged compound was detected in most of the plasma samples. Interestingly, Beermann[190] found that, although the total uptakes of digoxin and lanatoside C were similar, the sites of absorption were different. Lanatoside C was absorbed to a far greater extent in the distal parts of the intestine. Another metabolite of lanatoside C is dihydrodigoxin.[195] Although very small amounts of this reduced metabolite were excreted in the urine following i.v. dosage, amounts larger than those found for digoxin were excreted in some patients given lanatoside C orally.

Another consequence of the presence of an additional sugar residue in lanatosides A and C compared with digitoxin and digoxin is their increased biliary excretion. Cox and Wright[131] found that 70 to 75% of the dose was recovered in the bile in 5 h when rats were given the lanatosides intravenously. Corresponding values for digitoxin and digoxin were 10 and 40%, respectively. Furthermore, the latter compounds were partly excreted as metabolites whereas the lanatosides were excreted in the bile as such. The results of Beermann[190] in man indicate that the biliary excretion of lanatoside C is less extensive than seen in the rat and also that the nature of the biliary metabolites is dependent on the route of administration. Unchanged compound was most abundant following i.v. dosage whereas various hydrolysis products were mostly detected after oral dosage. In regard to urinary metabolites of lanatoside C, Ashley et al.[99] found that these were identical in both rats and man. Brown and Wright[196] showed the similarity between the excretory products of lanatoside C and digoxin in rats and Beermann[190] found that most of the urinary material following the administration of lanatoside C to man is digoxin.

The biological disposition of **deslanatoside C**, the deacetylated derivative of lanatoside C, was studied by Marzo et al.[197] It was given intravenously to guinea pigs, rabbits, and dogs and found to be excreted in the urine in 24 h to the extent of 43 to 50% of the dose in all species. About 75 to 85% of this material was unchanged compound. Unpublished findings were cited of similar extensive (>60%) and rapid urinary excretion of deslanatoside C in humans.

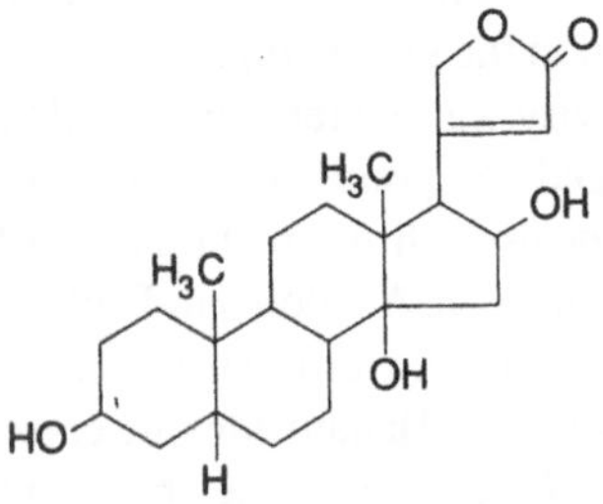

(77)

Gitoxin

Gitoxin is 16β-hydroxydigitoxin and thus a positional isomer of digoxin. Its genin is gitoxigenin (77) and, as indicated in Table 1, the glycoside contains three digitoxose units attached at C-3. Studies by Lesne[198] and Hupin et al.[199] indicated that it was well absorbed following oral administration in man. Several subsequent investigations have dealt with its biological disposition in guinea pigs. Using ^{3}H-labeled material, Kadima and Lesne[200] found that only 5% of the radioactivity was excreted by the kidneys following i.v. dosage. During the same period (96 h), 78% was recovered in the feces. These results indicate that gitoxin undergoes extensive biliary excretion and cannulation experiments showed that ~83% of the dose was excreted in the bile in 6 h. Gastrointestinal absorption of gitoxin was also shown to be high in the guinea pig.[201] About 95% of the dose was absorbed, however the material subsequently excreted in the bile consisted of polar metabolites which were poorly reabsorbed from the intestine. Pellegrin and Lesne[202] reported that the isolated rabbit liver was able to metabolize gitoxin to very polar metabolites which were readily excreted in the bile. Kadima and Lesne[200] also studied the nature of the gitoxin metabolites formed by the guinea pig. Stepwise hydrolysis of the sugar residues occurred and they identified the aglycon gitoxigenin as well as its mono- and bisdigitoxosides. The detection of diginatin (12β-hydroxygitoxin) and some of its derivatives in plasma, urine, bile, and intestinal contents indicated that hydroxylation took place. Most of the polar material which was excreted in the urine was susceptible to normal hydrolytic procedures and therefore probably consisted of glucuronides and/or sulfates. On the other hand, more than 80% of the polar metabolites excreted in the bile was stable to such treatment. Kadima et al.[203] carried out a further study of these polar biliary metabolites of gitoxin and found mass spectrometric evidence for the presence of dihydrodiginatin, i.e., a metabolite formed by 12β-hydroxylation and reduction of the double bond in the lactone ring. Their findings indicated that this metabolite may exist in ring and open-chain tautomeric forms including a form resulting from interaction of the 16β-hydroxy group with the carbonyl function in the lactone moiety. The metabolism of gitoxigenin, the aglycon of gitoxin, was studied using liver preparations from rats and several other animal species.[100,107] They showed that both epimerization at C-3 and hydroxylation of the steroid ring occurred.

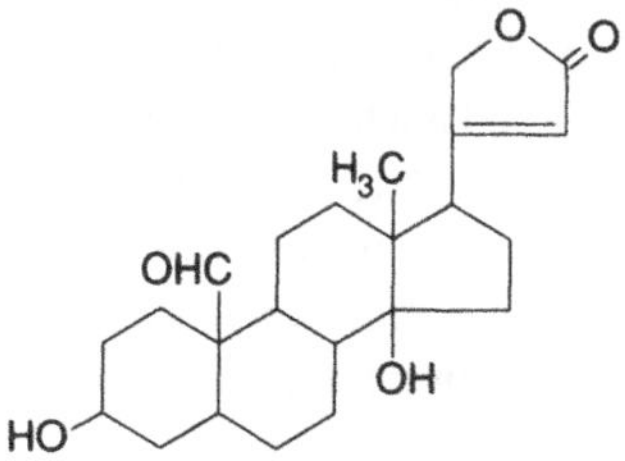

(78)

Cannogenin

Several cardiac glycosides including **cerberoside** (thevetin B) and **thevetin A** have been isolated from the yellow oleander, a tropical bush. Their chemical relationships are shown in Table 1 and it is seen that loss of the two terminal glucose units gives the glycosides **neriifolin** and **peruvoside**, respectively. Digitoxigenin (75b) is the aglycon of neriifolin and cannogenin (78) of peruvoside. Raudonat and Engler[204] gave cerberoside orally to rats and identified neriifolin (0.9% of dose) in the urine together with another metabolite which, however, was not the aglycon digitoxigenin. No unchanged compound was detected. In the feces, ~2.6% of the dose was recovered as neriifolin. Engler et al.[205] noted that neriifolin was found in both the urine and feces of rats following oral or s.c. administration of cerberoside and they proposed the existence of an enterohepatic circulation of these glycosides. Of interest in this regard is the finding of Lauterbach and Repke[192] that cerberoside was converted to neriifolin when incubated with rat feces. The biological disposition of neriifolin itself was studied in rats by Zhao et al.[206] They found rapid and complete absorption of the ^{3}H-labeled material following oral administration. The plasma half-life was 5 d and the radioactivity was most persistent in the liver, bile, and intestine. Most of the radioactivity was excreted in the bile and feces, with only 6% found in the urine. A value of 8.5% for urinary excretion was noted when neriifolin was given by i.v. injection. Peruvoside metabolism in man was studied by Fröhlich et al.[207] who found that similar amounts of radioactivity were excreted in the urine following oral or i.v. administration. About 33 to 37% of the dose was recovered in 48 h and none of this was due to unchanged drug. Two unidentified metabolites were detected in the urine and feces. It is noteworthy that considerable material was excreted in the bile, however this appears to be lost by the fecal route rather than undergoing enterohepatic circulation.

Another main group of cardioactive glycosides includes the strophanthus glycosides; however, the therapeutic use of these compounds has been much more limited than that of the digitalis glycosides. The only member of the group which has had much importance is ouabain. The nomenclature of the strophanthus glycosides and their chemical relationships are shown in Table 1. Engler et al.[205] gave the natural glycoside **K-strophanthoside** to rats both orally and subcutaneously and found that cymarin, the glycoside formed by loss of both glucose residues, was excreted in the urine and feces. Two unidentified metabolites were also detected, however these were not identical with unchanged K-strophanthoside, strophanthin, or the aglycon strophanthidin (79).[204] The material found in the bile was K-strophanthoside, not cymarin, and the participation of an enterohepatic circulation was concluded to be of importance in determining the metabolic fate of this glycoside in rats. This belief is strengthened by the finding that cymarin formation was not seen when several tissue preparations were used but was extensive when K-strophanthoside was incubated with rat feces. The latter finding was confirmed by Lauterbach and Repke.[192] The biological disposition of [^{3}H]K-strophanthoside in guinea pigs was studied following rectal administration,[208] i.v. administration,[209] and i.v. and intraduodenal administration.[129] In the first case the results indicated that the compound was rapidly absorbed, mainly as unchanged K-strophanthoside. About 55% of the dose was absorbed in 15 h. The urinary excretion at this time was 17 to 19%, this material consisting of unchanged compound and cymarin in a ratio of 9:1. When given intravenously to bile duct cannulated guinea pigs, the glycoside was excreted unchanged in the urine and bile in approximately equal amounts. These values were ~20% in 5 h and 30–40% in 24 h. Only ~2% of the dose was excreted in the urine and bile in 5 h when K-strophanthoside was given intraduodenally. Only 16% of the dose was absorbed when K-strophanthoside was given orally to man.[210] About 70% of this absorbed material was excreted in the urine, mainly as conjugated derivatives.

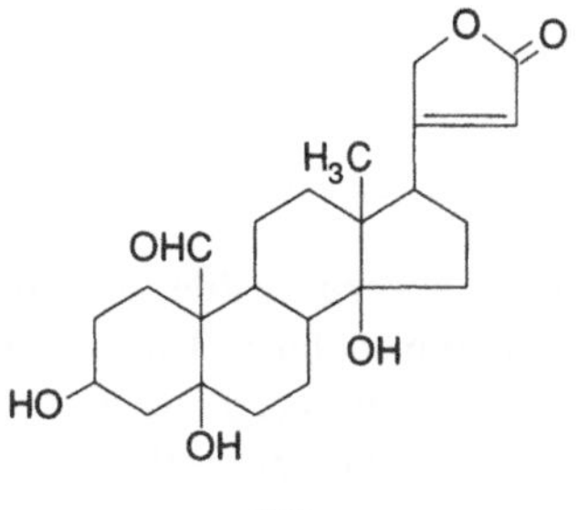

(79)

Strophanthidin

The metabolism of **cymarin** was reported by Lauterbach[211] who found that the aldehyde group was reduced by rat liver slices or homogenates to give cymarol, the corresponding alcohol, and that the aglycons of these glycosides, strophanthidin and strophanthidol, were also formed. Moerman[212] found that cymarin was partly metabolized *in vivo* in rats and that unchanged compound, cymarol and strophanthidin, were excreted in the bile and urine. The biological disposition of cymarin in man was studied by Strobach et al.[210] who found that ~47% of an oral dose was absorbed. Excretion of unchanged compound and metabolites (mainly conjugates) in the urine was slow and accounted for 9 and 21% of the dose in 1 and 7 d, respectively. The corresponding values after i.v. dosage were 36 and 46%. Reduction of the aldehyde group can also take place with the aglycon strophanthidin.[211] The same occurred with a closely related glycoside, **helveticoside**. This compound differs from cymarin only by the presence of a digitoxose moiety in place of the cymarose unit. Additional examples of a strophanthidin → strophanthidol reduction using rat liver were reported with the D- and L-arabinosides of strophanthidin.[213]

Cymarol was noted above to be a metabolite of cymarin. These glycosides differ only by the group attached at C-10, cymarol having –CH$_2$OH and cymarin having –CHO. Gundert-Remy et al.[214] reported on the biological disposition of [^{3}H]cymarol in man. Following i.v. administration, ~30% of the radioactivity was excreted in the urine whereas 65 to 70% was found in the feces. These values point to a high degree of biliary excretion of the material, a finding also noted in rats.[215] More than 85% of the urinary material in man was excreted within 24 h and ~90% of this was more polar than cymarol itself.

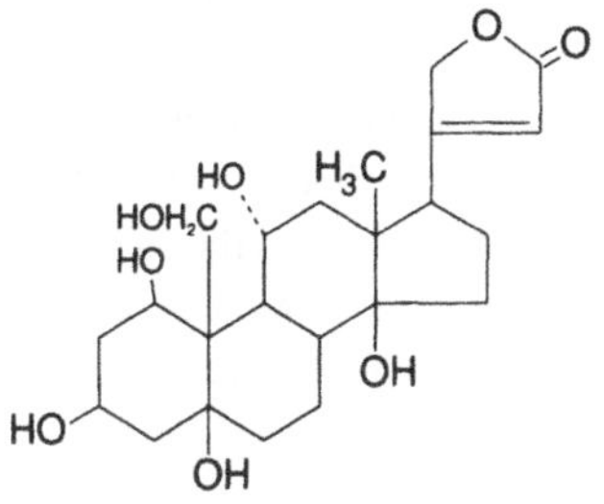

(80)

Ouabagenin

Ouabain, the rhamnoside of ouabagenin (80), is a rapidly acting cardiac glycoside which has high polarity due to the presence of five free hydroxy groups attached to the steroid nucleus. An early study by Farah[216] indicated that it was extensively excreted in the bile in rats and that little metabolic alteration occurred. Cox et al.[217] reinvestigated this subject and confirmed the earlier results. Following i.v. administration, ~90% of the dose was excreted in the bile in 5 h whereas only 4% was recovered in the urine. Similarly, Kupferberg and Schanker[218] recovered 85% of the unchanged ouabain in the bile of rats only 90 min after its i.v. dosage. Values of 42% in 2 h and 55% in 12 h, again as unchanged compound, were recorded by Russell and Klaassen.[128,219] The extent of biliary excretion of ouabain injected intravenously in rats decreased with increasing age.[220] Sato et al.[221] extended this work and found that the gradual decrease occurred with both male and female rats during the test period (4 to 28 months). However, biliary excretion of ouabain during the initial 10-min period was about twice as high in females as in males in all age groups.

The biliary excretion of ouabain in several other species has been studied and it appears that appreciable species variations occur, both with regard to the extent of excretion and the material excreted. The investigations of Russell and Klaassen showed that the 12-h values for rabbits and dogs were only 4.4 and 1.3%, respectively. In the latter case this material appeared to be unchanged compound but about one-third of that excreted in rabbit bile was in a more water-soluble form. Seldon et al.[222] recovered only ~5% of the radioactivity in the bile collected for 4 d when dogs were given [^{3}H]ouabain intravenously. Biliary excretion of ouabain in the guinea pig is also very low. This was shown by Kolenda et al.[188] using the isolated, perfused guinea pig liver and it was also shown that the glycoside was not degraded by the liver under these conditions.[84] Marzo et al.[209] and Marzo and Ghirardi[129] recovered less than 4% of the dose in the

5-h bile after i.v. administration of ouabain to guinea pigs. The corresponding urinary excretion was ~25%. The material excreted by both routes was shown to be unchanged compound. Ouabain given intravenously was also excreted preferentially in the urine of sheep.[223] Marks et al.[224] found that the same is true in man and that the urinary material is primarily unchanged glycoside. Seldon et al.[222] recovered only ~5% of the radioactivity in the bile of man following i.v. administration of [³H]ouabain, whereas 46 to 48% was excreted in the urine in 5 d. Interestingly, the results indicated that the fecal excretion of ouabain may arise from intestinal excretion at sites in addition to the biliary tract. Strobach et al.[210] found that ~33% of an i.v. dose was excreted in the urine of man and, of this, ~80% was unchanged compound. The remainder appeared to be conjugated metabolites. The gastrointestinal absorption of ouabain is very poor. Marchetti et al.[225] recorded values of less than 10% of the dose in both guinea pigs and man. Strobach et al.[210] found that only 1.4% of the oral dose was absorbed from the gastrointestinal tract in man and that about a third of the absorbed material was excreted in the urine.

The convallaria (lily of the valley) glycosides are closely related to the foregoing strophan-thus glycosides in that they are derivatives of the same genin, strophanthidin (79) or its corresponding C-19 alcohol, strophanthidol. As shown in Table 1, **convallatoxin** is the rhamnoside of strophanthidin whereas **convallatoxol** is the rhamnoside of strophanthidol. The latter glycoside is fairly polar due to the presence of three hydroxy groups and Greenberger et al.[226] found that its intestinal absorption was lower than that of digitoxin and other nonpolar glycosides. Similar properties also account for the fact that ~80% of the absorbed convallatoxin was excreted in the bile of rats.[227] Lauterbach[211] also showed that the aldehyde group of convallatoxin can undergo metabolic reduction to the alcohol. Convallatoxol was formed when incubates using rat liver slices or homogenates were used, however no reduction was seen with guinea pig liver and only traces of the alcohol were formed by cat liver.

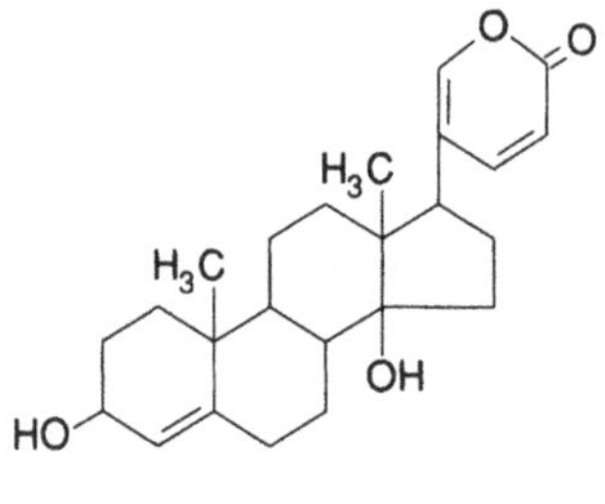

(81)

Scillarenin

The squill glycosides, termed scilladienolides, differ chemically from the compounds mentioned above by the presence at C-17 of a six-membered lactone ring containing two double bonds. The chemical components of the natural glycoside **scillaren A** are listed in Table 1 which indicates that the aglycon scillarenin (81) is attached via the hydroxy group at C-3 to rhamnose, giving the glycoside **proscillaridin A** and then to the terminal glucose moiety. Simon and Wright[228] reported that scillaren A was extensively excreted in the bile following i.v. administration to rats. They found that 80 to 90% of the dose was excreted unchanged by this route in 5 h. No glycosides or metabolites were detected in the 12-h urine. However, Lauterbach and Repke[82] found that scillaren A was extensively hydrolyzed to proscillaridin A when incubated with a rat liver preparation. The latter desglucose glycoside was not detected when incubations with rat feces were carried out, however this appears to be due to rapid further metabolism rather than lack of ability to remove the terminal glucose moiety. A few reports on the biological disposition of proscillaridin A have appeared and Greenberger et al.[226] found that it, like other nonpolar glycosides (e.g., digitoxin and digoxin), was more rapidly absorbed from guinea pig intestine than were several more polar glycosides. Davis et al.[229] suggested that it is eliminated rapidly. However, the metabolic fate of proscillaridin A is unclear. It is perhaps noteworthy that clinical trials with this glycoside have shown a relatively poor bioavailability after oral administration.[230] Andersson et al.[231] calculated that the absorption of active proscillaridin A amounted to only ~7% of the dose during the first 4 h in man. The absorbed material was then excreted extensively in the bile in conjugated form.[232] Accordingly, attention has been turned to a semisynthetic derivative, methylproscillaridin, the 4'-methyl ether of proscillaridin. Investigations have been carried out on its biological disposition in rats and dogs[233] and in man.[234-236]

A few studies on the biological disposition of cardiac glycosides have included some of the less common derivatives. Angarskaya et al.[237] found that the desglucose derivative of **cheirotoxin** (strophanthidin-lyxose-glucose) was excreted unchanged in the bile following i.v. administration to rats. **Adonitoxin**, the rhamnoside of adonitoxigenin which is a structural isomer of strophanthidin, was however, excreted partly unchanged and partly as an oxidation product in similar experiments. The investigations of Herrmann and Repke[100] and Repke and Samuels[107] dealt with the epimerization and hydroxylation of cardiac glycoside aglycons, mainly digitoxigenin. In addition, several other aglycons (structural differences from digitoxigenin (75b) shown in parentheses) were studied including **xysmalogenin** ($\Delta^{5,6}$), **sarmentogenin** (hydroxy group at C-11), and **diginatigenin** (hydroxy groups at C-12 and C-16). The studies of Lauterbach and Repke[82,192] on the cleavage of the sugar residues of cardiac glycosides by tissue enzymes and by the intestinal bacteria also included **evomonoside** (digitoxigenin α-rhamnoside) and **somalin** (digitoxigenin cymaroside).

In contrast to the common digitalis glycosides which have a *cis* A/B junction, some cardenolide glycosides show a *trans* configuration (i.e., 5α-H) at this point. An example of this type is **uzarigenin** which differs structurally from digitoxigenin only in this respect. Hermann and Repke[100] and Repke and Samuels[107] found that it underwent epimerization at the 3-hydroxy group (3-β-OH $\rightarrow$ 3-α-OH) when incubated with various fractions from liver. **Gomphoside** (82) is another example of a cardenolide derivative having a *trans* A/B junction. It is also of interest because the sugar moiety is doubly linked to the genin. Structure (82) shows that the 6-deoxy-β-D-hexosulose unit is attached via both the 2α- and 3β- hydroxy groups of the genin. Mutlib et al.[238] investigated the metabolism of [^{3}H]gomphoside in rats and found that the bile was the major excretory route. Following an i.p. dose (2 mg/kg), 68% of the radioactivity was recovered in the bile in 8 h. A subsequent report[239] noted a value of 90% in 24 h in similar experiments. In noncannulated rats 83% of the dose was excreted in the feces and 5% in the urine in 48 h. About 90% of the material in the bile was shown to consist of a single metabolite, gomphoside 3'-β-D-glucuronide. The remainder of the biliary radioactivity consisted of a group of relatively nonpolar metabolites which was similar to the compounds formed when gomphoside was incubated with rat liver microsomes. The changes in all of these compounds were confined to the A-ring. The major metabolite in this group was the genin of gomphoside, gomphogenin (2α-hydroxyuzarigenin) (83). A small amount of the intermediate product (84) was also found. The epimer of the genin, 3-epigomphogenin (85), was isolated and it was shown to have a 2a,3α-dihydroxy structure. Two additional metabolites were identified and shown to be 19-oxo compounds (i.e., an aldehyde group at C-10). These were the epimeric pair calactin (86a) and calotropin (86b) which differ only in the orientation of the hydroxy group at C-3' (β- and α-, respectively).

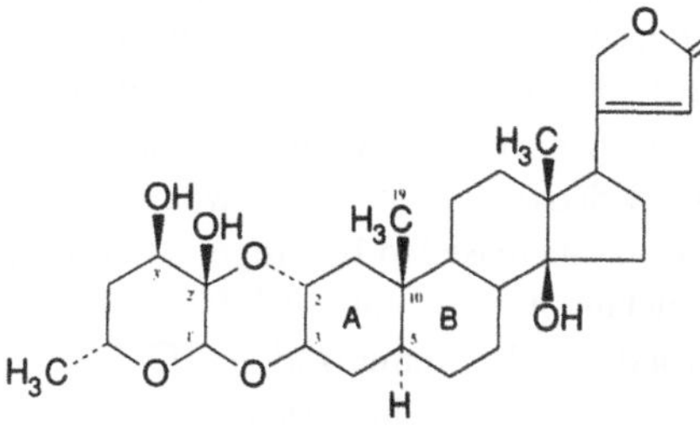

(82)

Gomphoside

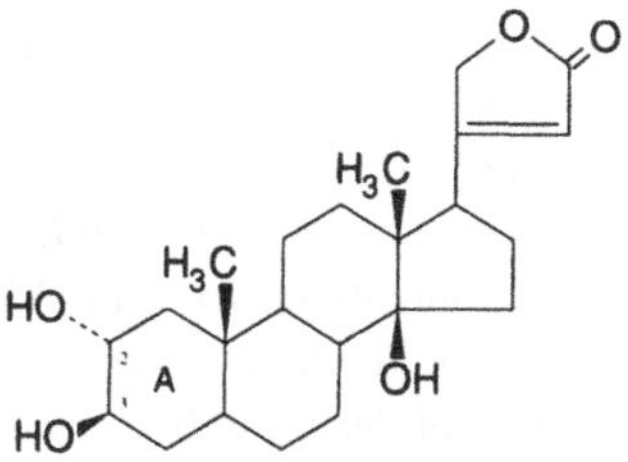

(83)

Gomphogenin

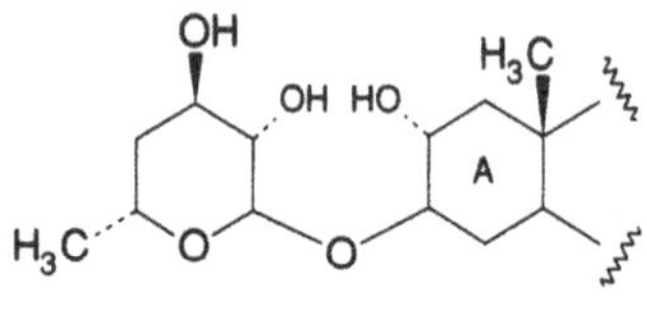

(84) 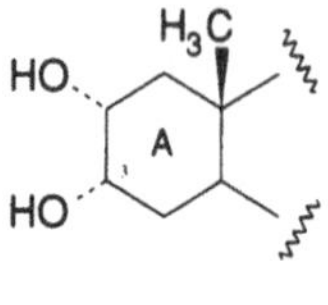   (85) 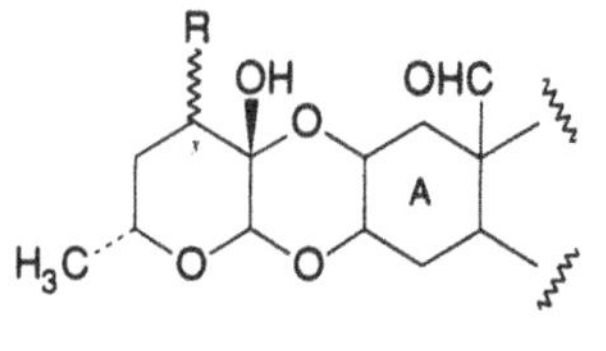   (86)

(a) Calactin, R = —OH

(b) R = ····OH

Mutlib et al.[240] studied the metabolism of **gomphogenin** (83) and **calactin** (86a) which, as noted above, are metabolic products of gomphoside. The metabolism of gomphogenin *in vitro*, using microsomes or 9000-g homogenates from rat liver, was found to be more extensive than that noted previously with the glycoside.[238] The metabolic changes found with gomphogenin were confined to the A-ring and included dehydrogenation at both the C-2 and C-3 positions. The metabolites formed were 2α-hydroxyuzarigenone (87) and 2-oxo-uzarigenin (88). The former compound was converted to 3-epigomphogenin (85) which was shown in the previous study to be a metabolite of gomphoside. The main product in these experiments was the 2-oxo derivative (88) which, however, did not serve as an intermediate in the formation of 2-epigomphogenin. The latter compound was not detected and, instead, (88) was hydroxylated to 4α-hydroxy-2-oxo-uzarigenin (89). Similar experiments with calactin (86a) showed that it was mainly metabolized by oxidation of the aldehyde group to the corresponding carboxy derivative (10-carboxy-19-norgomphoside) (90). A minor additional metabolite (calotropagenin) (91) was also formed by loss of the sugar moiety.

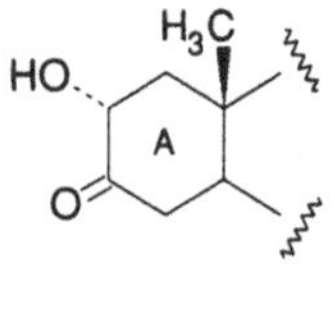

(87) 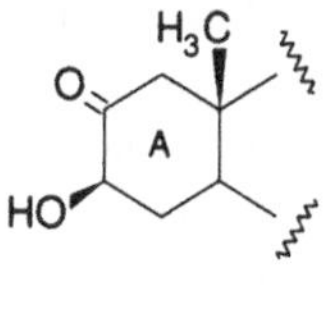   (88) 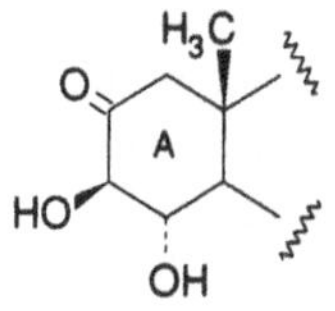   (89)

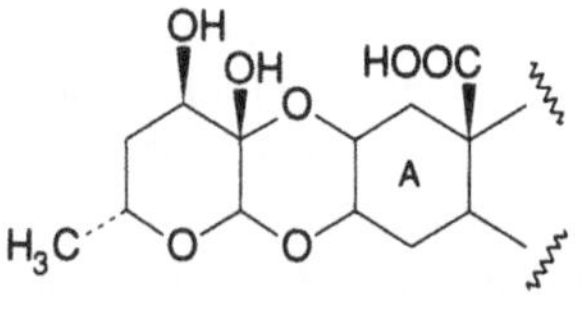

(90) 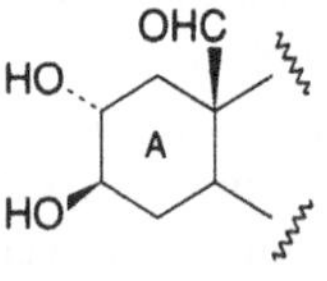   (91)

MISCELLANEOUS STEROIDS

The metabolism of ³H-labeled **ecdysone** (92) in mice was studied by Lafont et al.[241] and Girault et al.[242] Following an i.p. injection of either tracer amounts or 6.7 mg/kg of the compound, excretion of radioactivity was rapid. Most of the dose was excreted within 24 h, mainly in the feces, and the results indicated a high level of biliary excretion. About 45% of the excreted material was unchanged compound, however the 14-deoxy derivative (93) was also present in the intestinal contents and excreta. When the dose was increased to 16.7 mg/kg, two additional metabolites were identified. These were further metabolites of compound (93) which resulted from reduction of the 7-ene-6-one structure (94) and, finally, epimerization at C-3 (95). It was suggested that dehydroxylation at C-14 was carried out by intestinal bacteria.

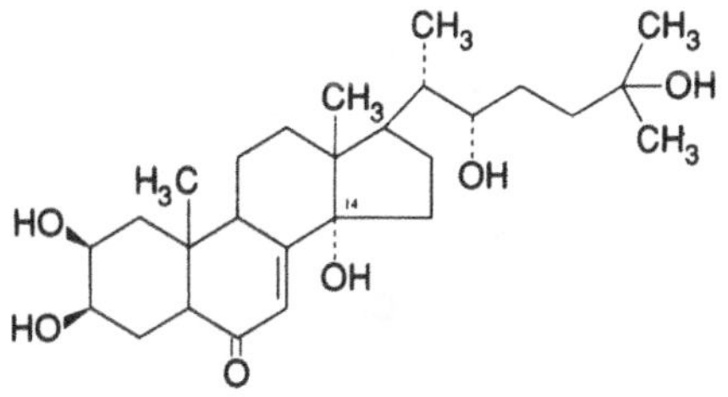

(92)

Ecdysone

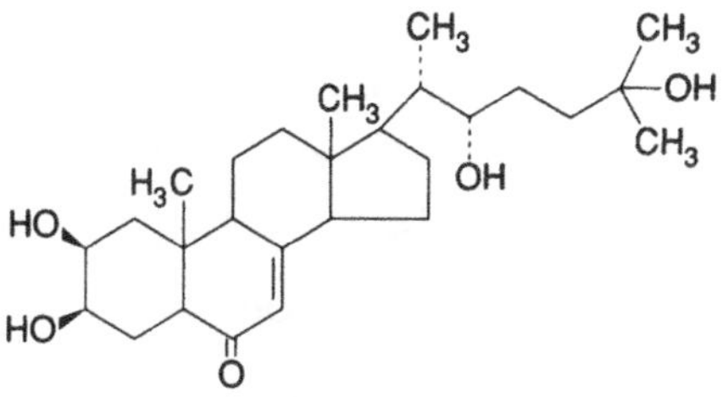

(93)

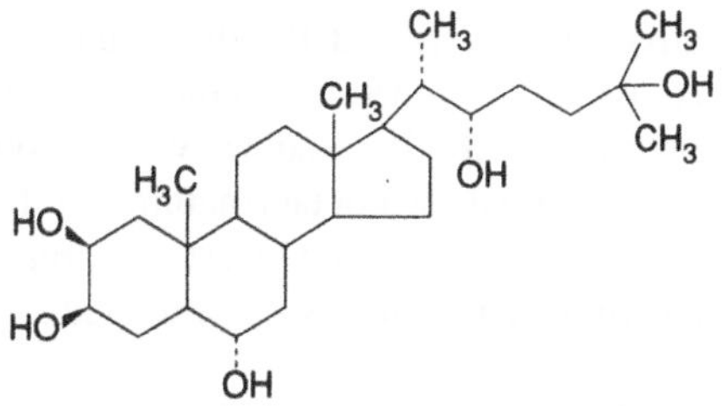

(94)

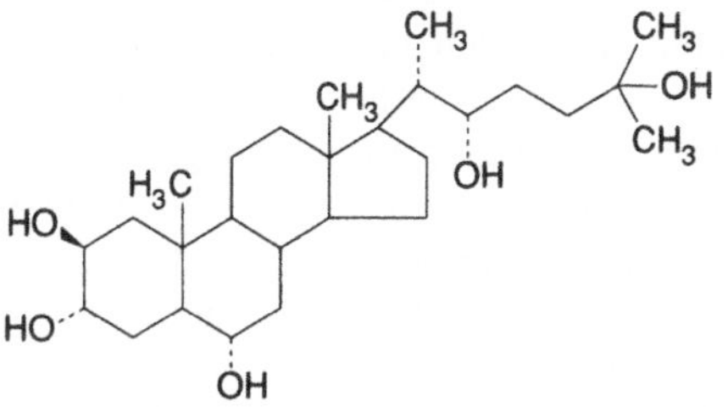

(95)

Isojuripidine (96) was labeled with ^{3}H and administered orally to rats and dogs by Valzelli and Goldaniga.[243] A single dose (5 mg/kg) in dogs produced plasma levels which declined very slowly. Retention of the radioactivity was also seen in rats given doses of 1 or 10 mg/kg and very little excretion (1% in 32 h) occurred in the urine. However, fecal excretion accounted for 55% of the radioactivity during this period. Metabolites in the plasma and liver were investigated and it was found that N-acetylisojuripidine was the major metabolite. Four additional metabolites were identified which showed alterations at the C-3- and/or C-6-positions. Structures (97), (98), (99), and (100) illustrate these changes. The prolonged retention of the material in the body may be related to the fact that these metabolites are less polar than isojuripidine.

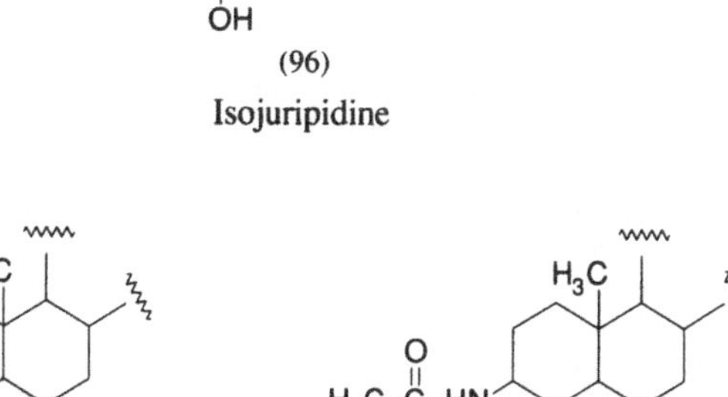

(96)

Isojuripidine

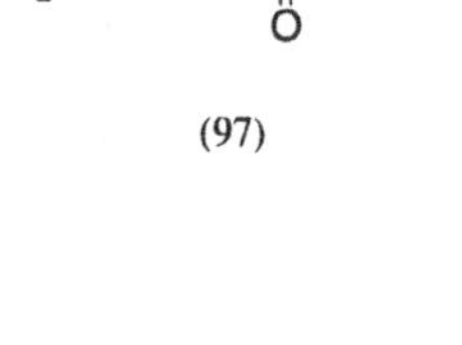

(97)

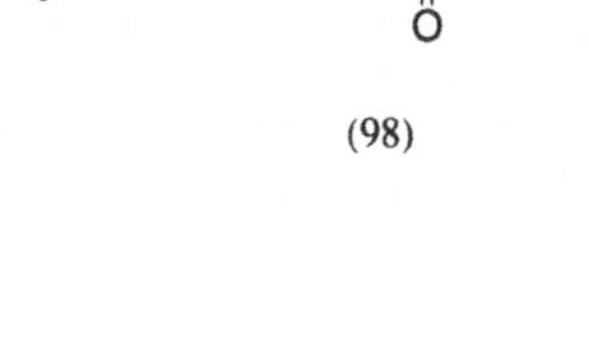

(98)

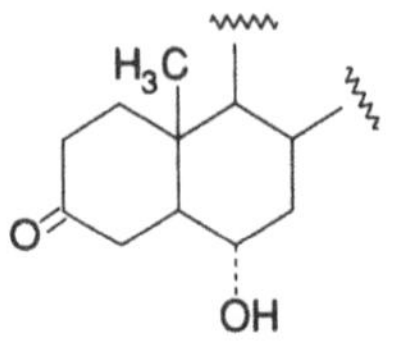

(99) 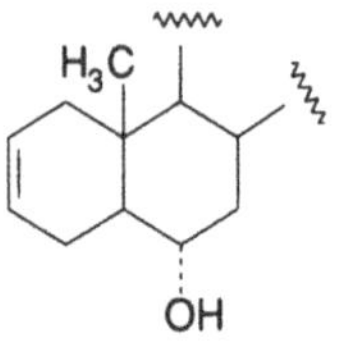   (100)

TETRATERPENOIDS

The most familiar tetraterpenoids are the carotenoids which include yellow to red, lipid-soluble plant pigments. Some carotenoid derivatives are esters and Booth[244] studied the metabolism in rats of the ester of citraurin. This compound, obtained from orange peel, was hydrolyzed during passage through the gastrointestinal tract. Similar findings were reported with **physalien**, the dipalmitate of zeaxanthol, and **taraxien**, a diester of taraxanthin.[245] A third of the esters was recovered in the feces, mainly as free carotenol formed by ester hydrolysis, whereas the remainder was not accounted for. Some of the taraxien was also excreted as the monoester.

β-Carotene is both the most widespread and widely studied carotenoid. The latter fact is related to its important role as a source of vitamin A. This topic, which largely falls outside the scope of this book, covers not only the metabolism of β-carotene but also a number of other carotenoid vitamin A precursors. Examples of the latter type include **cryptoxanthin** and the closely related crustaxanthin. Their metabolism in rats was studied by John et al.[246] and Boonjawat and Olson,[247] respectively. Further information on the carotenoids and their metabolism is available in the reviews of Simpson and Chichester[248] and Davies[249] and in the monograph by Goodwin.[250]

REFERENCES

1. **Fischer, F. G. and Bielig, H.-J.,** Über die Hydrierung ungesättigter Stoffe im Tierkörper. Biochemische Hydrierungen VII, *Hoppe-Seyler's Z. Physiol. Chem.*, 266, 73, 1940.
2. **Asakawa, Y., Ishida, T., Toyota, M., and Takemoto, T.,** Terpenoid biotransformation in mammals. IV. Biotransformation of (+)-longifoline, (–)-caryophylline, (–)-caryophylline oxide, (–)-cyclocolorenone, (+)-nootkatone, (+)-elemol, (–)-abietic acid and (+)-dehydroabietic acid in rabbits, *Xenobiotica*, 16, 753, 1986.
3. **Bang, L. and Ourisson, G.,** Hydroxylation of cedrol by rabbits, *Tetrahedron Lett.*, 1881, 1975.
4. **Trifilieff, E., Bang, L., and Ourisson, G.,** Hydroxyation du cedrol par le chien, *Tetrahedron Lett.*, 4307, 1975.
5. **Bang, L., Ourisson, G., and Teisseire, P.,** Hydroxylation of patchoulol by rabbits. Hemi-synthesis of nor-patchoulol, the odour carrier of patchouli oil, *Tetrahedron Lett.*, 2211, 1975.
6. **Ping, C. K., Bang, L., Ourisson, G., Mercier-Rohmer, M., Trifilieff, E., and Zundel, J.-L.,** Hydroxylation of di-tertiary sesquiterpene ethers, *J. Chem. Res.*, (S), 315, 1980.
7. **Su, C.-Y., Liu, C.-Y., Xu, H.-X., and Zhu, X.-Y.,** The metabolism of ³H-curcumol in normal rats and tumour bearing mice, *Acta Pharm. Sin.*, 15, 257, 1980.
8. **Zundel, J. L.,** Thèse de Doctorat d'Etat, Université Louis Pasteur, Strasbourg, 1976, Cited by Santhanakrishnan, T. S., *Tetrahedron*, 40, 3597, 1984.
9. **Hildebrandt, H.,** Ueber das Verhalten von Carvon und Santalol im Thierkörper, *Hoppe-Seyler's Z. Physiol. Chem.*, 36, 441, 1902.
10. **Ishida, T., Asakawa, Y., Takemoto, T., and Aratani, T.,** Biotransformation of terpenoids in mammals, *Res. Bull. Hiroshima Inst. Technol.*, 14, 9, 1980.
11. **Ishida, T., Asakawa, Y., and Takemoto, T.,** Hydroxyisolongifolaldehyde: a new metabolite of (+)-longifoline in rabbits, *J. Pharm. Sci.*, 71, 965, 1982.

12. **Asakawa, Y., Taira, Z., Takemoto, T., Ishida, T., Kido, M., and Ichikawa, Y.,** X-Ray crystal structure analysis of 14-hydroxycaryophylline oxide, a new metabolite of (–)caryophylline, in rabbits, *J. Pharm. Sci.*, 70, 710, 1981.

13. **Hill, D. W., Bailey, E. M., and Camp, B. J.,** Tissue distribution and disposition of hymenoxon, *J. Agric. Food Chem.*, 28, 1269, 1980.

14. **Terry, M. K., Williams, H. G., Kim, H. L., Post, L. O., and Bailey, E. M.,** Ovine urinary metabolites of hymenoxon, a toxic sesquiterpene lactone isolated from *Hymenoxys odorata* DC, *J. Agric. Food Chem.*, 31, 1208, 1983.

15. **Sylvia, V. L., Joe, C. O., Stipanovic, R. D., Kim, H. L., and Busbee, D. L.,** Alkylation of deoxyguanosine by the sesquiterpene lactone hymenoxon, *Toxicol. Lett.*, 29, 69, 1985.

16. **Sylvia, V. L., Kim, H. L., Norman, J. O., and Busbee, D. L.,** The sesquiterpene lactone hymenoxon acts as a bifunctional alkylating agent, *Cell Biol. Toxicol.*, 3, 39, 1987.

17. **Merrill, J. C., Kim, H. L., Safe, S., Murray, C. A., and Hayes, M. A.,** Role of glutathione in the toxicity of the sesquiterpene lactones hymenoxon and helenalin, *J. Toxicol. Environ. Health*, 23, 159, 1988.

18. **Morishima, E.,** The fate of santonin in rabbit and man and the influence of dehydrocholic acid on it, *Nippon Yakurigaku Zasshi*, 57, 353, 1961 (Chem. Abstr. 58, 839c, 1963).

19. **Kaya, K.,** The fate of santonin in the dog. II. Absorption and excretion of santonin and appearance of α-hydroxysantonin, *Nippon Yakurigaku Zasshi*, 56, 368, 1960 (Chem. Abstr. 55, 21372h, 1961).

20. **Boyland, E. and Chasseaud, L. F.,** Enzyme-catalysed conjugations of glutathione with unsaturated compounds, *Biochem. J.*, 104, 95, 1967.

21. **Niu, X., Ho, L., Ren, Z., and Song, Z.,** Metabolic fate of qinghaosu in rats; a new TLC densitometric method for its determination in biological material, *Eur. J. Drug Metab. Pharmacokinet.*, 10, 55, 1985.

22. **Zhu, D., Huang, B., Chen, Z., and Yin, M.,** Studies on the physiological disposition of qing-hao-sou. I. Biotransformation of qing-hao-sou after oral medication in human, *Acta Pharm. Sin.*, 15, 509, 1980.

23. **Zhu, D., Huang, B., Chen, Z., Yin, M., Yang, Y., Dai, M., Wang, B., and Huang, Z.,** Isolation and identification of the metabolite of artemisinine in human, *Acta Pharmacol. Sin.*, 4, 194, 1983.

24. **Ng, J. C., Venzke, B. N., and Seawright, A. A.,** Furanoid urinary metabolites of (–)-ngaione in the rat, in *Plant Toxicology Proceedings of the Australia – U.S.A. Poisonous Plants Symposium Brisbane, Australia, May 14-18, 1984*, Seawright, A. A., Hegarty, M. P., James, L. F., and Keeler, R. F., Eds., Queensland Poisonous Plants Committee, Yeerongpilly, Australia, 1985, 512.

25. **Le Bon, A. M., Cravedi, J. P., and Tulliez, J. E.,** Disposition and metabolism of pristane in rat, *Lipids*, 23, 424, 1988.

26. **Albro, P. W. and Thomas, R. O.,** Metabolism of phytane in rats, *Biochim. Biophys. Acta*, 372, 1, 1974.

27. **Mize, C. E., Avigan, J., Baxter, J. H., Fales, H. M., and Steinberg, D.,** Metabolism of phytol-U-^{14}C and phytanic acid-U-^{14}C in the rat, *J. Lipid Res.*, 7, 692, 1966.

28. **Steinberg, D., Avigan, J., Mize, C. E., Baxter, J. H., Cammermeyer, J., Fales, H. M., and Highet, P. F.,** Effects of dietary phytol and phytanic acid in animals, *J. Lipid Res.*, 7, 684, 1966.

29. **Hidiroglou, M. and Jenkins, K. J.,** Fate of ^{14}C-phytol administered orally to sheep, *Can. J. Physiol. Pharmacol.*, 50, 458, 1972.

30. **Patton, S. and Benson, A. A.,** Phytol metabolism in the bovine, *Biochim. Biophys. Acta*, 125, 22, 1966.

31. **Klenk, E. and Kremer, G. J.,** Untersuchungen zum Stoffwechsels des Phytols, Dihydrophytols und der Phytansäure, *Hoppe-Seyler's Z. Physiol. Chem.*, 343, 39, 1965.

32. **Baxter, J. H. and Milne, G. W. A.,** Phytenic acid: identification of five isomers in chemical and biological products of phytol, *Biochim. Biophys. Acta*, 176, 265, 1969.

33. **Baxter, J. H., Steinberg, D., Mize, C. E., and Avigan, J.,** Absorption and metabolism of uniformly ^{14}C-labeled phytol and phytanic acid by the intestine of the rat studied with thoracic duct cannulation, *Biochim. Biophys. Acta*, 137, 277, 1967.

34. **Steinberg, D., Mize, C. E., Avigan, J., Fales, H. M., Eldjarn, L., Try, K., Stokke, O., and Refsum, S.,** Studies on the metabolic error in Refsum's disease, *J. Clin. Invest.*, 46, 313, 1967.

35. **Mize, C. E., Avigan, J., Steinberg, D., Pittman, R. C., Fales, H. M., and Milne, G. W. A.,** A major pathway for the mammalian oxidative degradation of phytanic acid, *Biochim. Biophys. Acta*, 176, 720, 1969.

36. **Tsai, S.-C., Avigan, J., and Steinberg, D.,** Studies on the α oxidation of phytanic acid by rat liver mitochondria, *J. Biol. Chem.*, 244, 2682, 1969.

37. **Hutton, D. and Steinberg, D.,** Identification of propionate as a degradation product of phytanic acid oxidation in rat and human tissues, *J. Biol. Chem.*, 248, 6871, 1973.

38. **Vamecq, J. and Draye, J.-P.,** The enzymatic and mass spectrometric identification of 2-oxophytanic acid, a product of the peroxisomal oxidation of L-2-hydroxyphytanic acid, *Biomed. Environ. Mass Spectrom.*, 15, 345, 1988.

39. **Ishida, T., Asakawa, Y., and Takemoto, T.,** Metabolism of myrtenal, perilla aldehyde and dehydroabietic acid in rabbits, *Res. Bull. Hiroshima Inst. Technol.*, 16, 79, 1982.

40. **Xu, Q., Wang, Z., Yuan, L., and Yi, M. G.,** Pharmacokinetic study on ^{3}H-abietic acid, *Yiyao Gongye*, 17, 499, 1986 (Chem. Abstr. 106, 60723e, 1987).

41. **Wingard, R. E., Brown, J. P., Enderlin, F. E., Dale, J. A., Hale, R. L., and Seitz, C. T.,** Intestinal degradation and absorption of the glycosidic sweeteners stevioside and rebaudoside A, *Experientia*, 36, 519, 1980.

42. **Compadre, C. M., Hussain, R. A., Nanayakkara, N. P. D., Pezzuto, J. M., and Kinghorn, A. D.,** Mass spectral analysis of some derivatives and *in vitro* metabolites of steviol, the aglycone of the natural sweeteners, stevioside, rebaudioside A, and rubusoside, *Biomed. Environ. Mass Spectrom.*, 15, 211, 1988.

43. **Neidlein, R. and Stumpf, U.,** Biotransformation und Pharmakokinetik von Grandiflorensäure [Kauradien-9(11),16-säure-18]. II. Mitteilung: Untersuchungen zur Biotransformation an Ratten, *Arzneim. Forsch.*, 27, 1162, 1977.

44. **Neidlein, R. and Stumpf, U.,** Biotransformation und Pharmakokinetik von Grandiflorensäure [Kauradien-9(11),16-säure-18]. III. Mitteilung: Untersuchungen zur Pharmakokinetik an Ratten, *Arzneim. Forsch.*, 27, 1384, 1977.

45. **Longnecker, S. M., Donehower, R. C., Cates, A. E., Chen, T.-L., Brundrett, R. B., Grochow, L. B., Ettinger, D. S., and Colvin, M.,** High-performance liquid chromatographic assay for taxol in human plasma and urine and pharmacokinetics in a phase I trial, *Cancer Treat. Rep.*, 71, 53, 1987.

46. **Wiernik, P. H., Schwartz, E. L., Strauman, J. J., Dutcher, J. P., Lipton, R. B., and Paietta, E.,** Phase I clinical and pharmacokinetic study of taxol, *Cancer Res.*, 47, 2486, 1987.

47. **Zelck, U.,** Biochemistry of the Erythrophleum alkaloids. Enzymic splitting of cassaine, the principle alkaloid in *Erythrophleum guineense*, in *Biochemie und Physiologie der Alkaloide*, Mothes, K., Ed., Akademi Verlag, Berlin, East Germany, 1972, 141.

48. **Cronlund, A.,** Metabolism of cassaidine in the guinea pig, *Acta Pharm. Suec.*, 13, 43, 1976.

49. **Xie, M.-Z. and Shen, Z.-F.,** Absorption, distribution, excretion and metabolism of cryptotanshinone, *Acta Pharm. Sin.*, 18, 90, 1983.

50. **Chasseaud, L. F., Fry, B. J., Hawkins, D. R., Lewis, J. D., Sword, I. P., Taylor, T., and Hathway, D. E.,** The metabolism of asiatic acid, madecassic acid and asiaticoside in the rat, *Arzneim. Forsch.*, 21, 1379, 1971.

51. **Hathway, D. E.,** The confirmation in man of the metabolic data found for drugs in animals, in *International Aspects of Drug Evaluation and Usage*, Jouhar, A. J. J. and Grayson, M. F., Eds., Churchill Livingstone, Edinburgh, 1973, 19.

52. **Sakiya, Y., Akada, Y., Kawano, S., and Miyauchi, Y.,** Rapid estimation of glycyrrhizin and glycyrrhetinic acid in plasma by high-speed liquid chromatography, *Chem. Pharm. Bull.*, 27, 1125, 1979.

53. **Carlat, L. E., Margraf, H. W., Weathers, H. H., and Weichselbaum, T. E.,** Human metabolism of orally ingested glycyrrhetinic acid and monoammonium glycyrrhizinate, *Proc. Soc. Exp. Biol. Med.*, 102, 245, 1959.

54. **Terasawa, K., Bandoh, M., Tosa, H., and Hirate, J.,** Disposition of glycyrrhetic acid and its glycosides in healthy subjects and patients with pseudoaldosteronism, *J. Pharmacobio. Dyn.*, 9, 95, 1986.

55. **Helbing, A. R.,** A new method for the determination of 18-β-glycyrrhetinic acid in biological material, *Clin. Chim. Acta*, 8, 756, 1963.

56. **Parke, D. V., Pollock, S., and Williams, R. T.,** The fate of tritium-labelled β-glycyrrhetic acid in the rat, *J. Pharm. Pharmacol.*, 15, 500, 1963.

57. **Ichikawa, T., Ishida, S., Sakiya, Y., Sawada, Y., and Hanano, M.,** Biliary excretion and enterohepatic cycling of glycyrrhizin in rats, *J. Pharm. Sci.*, 75, 672, 1986.

58. **Miyake, T., Asano, K., Saito, M., Yoshida, M., and Shimura, K.,** Studies on the absorption, excretion, and distribution of β-glycyrrhetinic acid in the mouse, *Kaku Igaku*, 13, 451, 1976 (Chem. Abstr. 85, 153696m, 1976).

59. **Iveson, P., Lindup, W. E., Parke, D. V., and Williams, R. T.,** The metabolism of carbenoxolone in the rat, *Xenobiotica*, 1, 79, 1971.

60. **Hattori, M., Sakamoto, T., Kobashi, K., and Namba, T.,** Metabolism of glycyrrhizin by human intestinal flora, *Planta Med.*, 48, 38, 1983.

61. **Hattori, M., Sakamoto, T., Yamagishi, T., Sakamoto, K., Konishi, K., Kobashi, K., and Namba, T.,** Metabolism of glycyrrhizin by human intestinal flora. II. Isolation and characterization of human intestinal bacteria capable of metabolizing glycyrrhizin and related compounds, *Chem. Pharm. Bull.*, 33, 210, 1985.

62. **Akao, T., Akao, T., Hattori, M., Namba, T., and Kobashi, K.,** 3β-Hydroxysteroid dehydrogenase of *Ruminococcus* sp. from human intestinal bacteria, *J. Biochem. (Tokyo)*, 99, 1425, 1986.

63. **Akao, T., Akao, T., Hattori, M., Namba, T., and Kobashi, K.,** Purification and properties of 3α-hydroxyglycyrrhetinate dehydrogenase of *Clostridium innocuum* from human intestine, *J. Biochem. (Tokyo)*, 103, 504, 1988.

64. **Akao, T., Akao, T., and Kobashi, K.,** Glycyrrhizin β-D-glucuronidase of *Eubacterium* sp. from human intestinal flora, *Chem. Pharm. Bull.*, 35, 705, 1987.

65. **Shimizu, K., Amagaya, S., and Ogihara, Y.,** Structural transformation of saikosaponins by gastric juice and intestinal flora, *J. Pharmacobio. Dyn.*, 8, 718, 1985.

66. **Pass, M. A., Goosem, M. W., and Pollitt, S.,** A relationship between hepatic metabolism of reduced lantadene A and its toxicity in rats and sheep, *Comp. Biochem. Physiol.*, 82C, 457, 1985.

67. **Iskenderov, G. B. and Orudzheva, K. F.,** Pharmacokinetics of triterpene saponins of saparal in rats, *Farmatsiya (Moscow)*, 38, 44, 1989 (Chem. Abstr. 112, 48235p, 1990).

68. **Han, B. H. and Chang, I. M.,** Metabolism of dammarane triterpene glycosides of Korean ginseng. I. Absorption, organ distribution and excretion of ^{3}H-panax saponin A, *Korean J. Ginseng Sci.*, 2, 17, 1977.

69. **Chen, S. E., Sawchuk, R. J., and Staba, E. J.,** American ginseng. III. Pharmacokinetics of ginsenosides in the rabbit, *Eur. J. Drug Metab. Pharmacokinet.*, 5, 161, 1980.

70. **Han, B. H., Park, M. H., Kim, D. H., and Hong, S. K.,** Studies on the metabolic fates of ginsenosides, *Korean Biochem. J.*, 19, 213, 1986.

71. **Odani, T., Tanizawa, H., and Takino, Y.,** Studies on the absorption, distribution, excretion and metabolism of ginseng saponins. II. The absorption, distribution and excretion of ginsenoside Rg_1 in the rat, *Chem. Pharm. Bull.*, 31, 292, 1983.

72. **Odani, T., Tanizawa, H., and Takino, Y.,** Studies on the absorption, distribution, excretion and metabolism of ginseng saponins. IV. Decomposition of ginsenoside-Rg_1 and -Rb_1 in the digestive tract of rats, *Chem. Pharm. Bull.*, 31, 3691, 1983.

73. **Strömbom, J., Sandberg, F., and Dencker, L.,** Studies on absorption and distribution of ginsenoside R_{g-1} by whole-body autoradiography and chromatography, *Acta Pharm Suec.*, 22, 113, 1985.

74. **Chen, Y.,** Study on the metabolites of 20(S)-ginsenoside-R_{g2}, *Shenyang Yaoxueyuan Xuebao*, 4, 202, 1987 (Chem. Abstr. 108, 142752u, 1988).

75. **Odani, T., Tanizawa, H., and Takino, Y.,** Studies on the absorption, distribution, excretion and metabolism of ginseng saponins. III. The absorption, distribution and excretion of ginsenoside Rb_1 in the rat, *Chem. Pharm. Bull.*, 31, 1059, 1983.

76. **Fujiwara, S., Sakurai, S., Noumi, K., Sugimoto, I., and Awata, N.,** Metabolism of γ-oryzanol in rabbit, *J. Pharm. Soc. Jpn.*, 100, 1011, 1980.

77. **Fujiwara, S., Sakurai, S., Sugimoto, I., and Awata, N.,** Absorption and metabolism of γ-oryzanol in rats, *Chem. Pharm. Bull.*, 31, 645, 1983.

78. **Doherty, J. E.,** Digitalis glycosides. Pharmacokinetics and their clinical implications, *Ann. Intern. Med.*, 79, 229, 1973.

79. **Storstein, O.,** *Symposium on Digitalis*, Gyldendal Norsk Forlag, Oslo, Norway, 1973.

80. **Temple, D. J., Harron, D. W. G., and Collier, P. S.,** Utilisation of digitalis glycosides: the relevance of their biotransformation, *Int. J. Pharmaceutics*, 2, 127, 1979.

81. **Wirth, K. E.,** Relevant metabolism of cardiac glycosides, in *Cardiac Glycosides 1785-1985 Biochemistry-Pharmacology-Clinical Relevance*, Erdmann, E., Greef, K., and Skou, J. C., Eds., Steinkopff Verlag, Darmstadt, W. Germany, 1986, 257.

82. **Lauterbach, F. and Repke, K.,** Die fermentative Abspaltung von D-Digitoxose, D-Cymarose und L-Thevetose aus Herzglykosiden durch Leberschnitte, *Naunyn Schmiedebergs Arch. Exp. Pathol. Pharmakol.*, 239, 196, 1960.

83. **Herrmann, I. and Repke, K.,** Speciesunterschiede in der Biotransformation von Digotoxin, *Naunyn Schmiedebergs Arch. Exp. Pathol. Pharmakol.*, 247, 35, 1964.

84. **Kolenda, K.-D., Lüllmann, H., and Peters, T.,** Metabolism of cardiac glycosides studied in the isolated perfused guinea-pig liver, *Br. J. Pharmacol.*, 41, 661, 1971.

85. **Repke, K.,** Über Spaltung und Hydroxylierung von Digitoxin bei der Ratte, *Naunyn Schmiedebergs Arch. Exp. Pathol. Pharmakol.*, 237, 34, 1959.

86. **Repke, K.,** Metabolism of cardiac glycosides, in *New Aspects of Cardiac Glycosides*, Vol. 3, Wilbrandt, W. and Lindgren, P., Eds., Pergamon Press, Oxford, 1963, 47.

87. **Rietbrock, N. and Vöhringer, H.-F.,** Metabolism and excretion of ^{3}H-digitoxin in the rat, *Biochem. Pharmacol.*, 23, 2567, 1974.

88. **Wirth, K. E., Frölich, J. C., Hollifield, J. W., Falkner, F. C., Sweetman, B. S., and Oates, J. A.,** Metabolism of digitoxin in man and its modification by spironolactone, *Eur. J. Clin. Pharmacol.*, 9, 345, 1976.

89. **Okita, G. T.,** Metabolism of radioactive cardiac glycosides, *Pharmacologist*, 6, 45, 1964.

90. **Castle, M. C. and Lage, G. L.,** Metabolism and distribution of digitoxin in the rat, *Arch. Int. Pharmacodyn. Ther.*, 203, 323, 1973.

91. **Schmoldt, A., Benthe, H. F., and Haberland, G.,** Digitoxin metabolism by rat liver microsomes, *Biochem. Pharmacol.*, 24, 1639, 1975.

92. **Schmoldt, A. and Rohloff, C.,** Dehydro-digitoxosides for digitoxigenin: formation and importance for the digitoxin metabolism in the rat, *Naunyn Schmiedeberg's Arch. Pharmacol.*, 305, 167, 1978.

93. **van Bezooijen, C. F. A., Soekawa, Y., Ohta, M., Nokubo, M., and Kitani, K.,** Metabolism of digitoxin by isolated rat hepatocytes, *Biochem. Pharmacol.*, 29, 3023, 1980.

94. **Schmoldt, A., von Meyerinck, L., Drohn, W., and Blömer, I.,** Enzymatic basis for digitoxin metabolism and possible drug interactions in man, in *Cardiac Glycosides 1785-1985 Biochemistry-Pharmacology-Clinical Relevance*, Erdmann, E., Greef, K., and Skou, J. C., Eds., Steinkopff Verlag, Darmstadt, W. Germany, 1986, 273.

95. **Kershaw, W. C., Campbell, P., and Lage, G. L.,** *In vitro* digitoxin metabolism. Rate-limiting step and alteration following spironolactone pretreatment, *Drug Metab. Dispos.*, 13, 635, 1985.

96. **Castle, M. C.,** Digitoxin metabolism by rat liver homogenates, subcellular fractions and isolated hepatocytes: stimulation by spironolactone and pregnenolone-16α-carbonitrile, *J. Pharmacol. Exp. Ther.*, 211, 120, 1979.

97. **Volp, R. F. and Lage, G. L.,** Digitoxin metabolism by rat, mouse, and rabbit: NADPH requirement and spironolactone effect, *Res. Commun. Chem. Pathol. Pharmacol.*, 38, 501, 1982.

98. **Brown, B. T., Wright, S. E., and Okita, G. T.,** C₁₂-Hydroxylation of digitoxin, *Nature (London)*, 180, 607, 1957.

99. **Ashley, J. J., Brown, B. T., Okita, G. T., and Wright, S. E.,** The metabolites of cardiac glycosides in human urine, *J. Biol. Chem.*, 232, 315, 1958.

100. **Herrmann, I. and Repke, K.,** Epimerisierung und Hydroxylierung von digitaloiden Steroidlactonen durch Fermente der Leber, *Naunyn Schmiedebergs Arch. Exp. Pathol. Pharmakol.*, 248, 351, 1964.

101. **Stohs, S. J., Reinke, L. A., and El-Olemy, M. M.,** Metabolism *in vitro* of digitoxigenin by rat liver homogenates, *Biochem. Pharmacol.*, 20, 437, 1971.

102. **Spratt, J. L.,** Digitoxigenin metabolism by rat liver microsomes and its induction by phenobarbital, *Biochem. Pharmacol.*, 22, 1669, 1973.

103. **Talcott, R. E. and Stohs, S. J.,** Effects of PCN and phenobarbital on ³H-digitoxigenin and ³H-digoxigenin metabolism by rat liver microsomes, *Res. Commun. Chem. Pathol. Pharmacol.*, 5, 663, 1973.

104. **Bulger, W. H. and Stohs, S. J.,** 5β-Hydroxydigitoxigenin—a metabolite of digitoxigenin by rabbit liver homogenates, *Biochem. Pharmacol.*, 22, 1745, 1973.

105. **Bulger, W. H., Stohs, S. J., and Wheeler, D. M. S.,** 6β-Hydroxy-3-epidigitoxigenin—the major metabolite of digitoxigenin by rabbit liver homogenates, *Biochem. Pharmacol.*, 23, 921, 1974.

106. **Carvalhas, M. L., Figueira, M. A., Araujo, M. E., and Maya, M. R.,** A new metabolic pathway of digitoxin found in the guinea pig: 17α-hydroxylation of the steroid nucleus, *Drug Metab. Dispos.*, 11, 85, 1983.

107. **Repke, K. and Samuels, L. T.,** Enzymatic basis for epimerization of cardiotonic steroids at carbon 3 in rat liver, *Biochemistry*, 3, 689, 1964.

108. **Bulger, W. H.,** The *in vitro* metabolism of digitoxigenin, *Diss. Abstr. Int.*, 35B, 4075, 1975.

109. **Bodem, G. and von Unruh, E.,** Dihydrodigitoxin, a metabolite of digitoxin in humans, in *Cardiac Glycosides*, Bodem, G. and Dengler, H. J., Eds., Springer Verlag, Berlin, 1978, 74.

110. **Strobach, H.,** Renal excretion and plasma levels of dihydrodigitoxin in volunteers and renally healthy and renally insufficient patients being treated with dogitoxin, in *Cardiac Glycosides 1785-1985 Biochemistry-Pharmacology-Clinical Relevance*, Erdmann, E., Greef, K., and Skou, J. C., Eds., Steinkopff Verlag, Darmstadt, W. Germany, 1986, 269.

111. **Flasch, H. and Heinz, N.,** Pharmacokinetics of dihydrodigitoxin in the cat. A comparison with digitoxin, *Naunyn Schmiedeberg's Arch. Pharmacol.*, 310, 147, 1979.

112. **Herrmann, I. and Repke, K.,** Konjugation von Cardenolidgeninen mit Schwefelsäure oder Glucuronsäure, *Naunyn Schmiedebergs Arch. Exp. Pathol. Pharmakol.*, 248, 370, 1964.

113. **Vöhringer, H. F. and Rietbrock, N.,** Metabolism and excretion of digitoxin in man, *Clin. Pharmacol. Ther.*, 16, 796, 1974.

114. **Richards, L. G., Castle, M. C., and Lage, G. L.,** Glucuronidation of digitoxin and its derivatives by rat and rabbit liver homogenates, *Drug Metab. Dispos.*, 5, 469, 1977.

115. **Castle, M. C.,** Glucuronidation of digitalis glycosides by rat liver microsomes: stimulation by spironolactone and pregnenolone-16α-carbonitrile, *Biochem. Pharmacol.*, 29, 1497, 1980.

116. **Schmoldt, A., von der Eldern-Dellbrügge, U., and Benthe, H. F.,** On the glucuronidation of digitalis compounds in different species, *Arch. Int. Pharmacodyn. Ther.*, 255, 180, 1982.

117. **Schmoldt, A. and Promies, J.,** On the substrate specificity of the digitoxigenin monodigitoxoside conjugating UDP-gluycuronyltransferase in rat liver, *Biochem. Pharmacol.*, 31, 2285, 1982.

118. **Watkins, J. B. and Klaassen, K. D.,** Development of UDP-glucuronosyltransferase activity toward digitoxigenin-monodigitoxoside in neonatal rats, *Drug Metab. Dispos.*, 13, 186, 1985.

119. **von Meyerinck, L., Coffman, B. L., Green, M. D., Kirkpatrick, R. B., Schmoldt, A., and Tephly, T. R.,** Separation, purification, and characterization of digitoxigenin-monodigitoxoside UDP-glucuronosyltransferase activity, *Drug Metab. Dispos.*, 13, 700, 1985.

120. **Jelliffe, R. W., Buell, J., Kalaba, R., Sridhar, R., and Rockwell, R.,** A mathematical study of the metabolic conversion of digitoxin to digoxin in man, *Math. Biosci.*, 6, 387, 1970.

121. **Marcus, F. I.,** Digitalis pharmacokinetics and metabolism, *Am. J. Med.*, 58, 452, 1975.

122. **Okita, G. T., Kelsey, F. E., Talso, P. J., Smith, L. B., and Geiling, E. M. K.,** Studies on the renal excretion of radioactive digitoxin in human subjects with cardiac failure, *Circulation*, 7, 161, 1953.

123. **Vöhringer, H. F. and Rietbrock, N.,** Pharmacokinetics and metabolism of digitoxin in the human, in *Cardiac Glycosides*, Bodem, G. and Dengler, H. J., Eds., Springer Verlag, Berlin, 1978, 64.

124. **Storstein, L.,** Studies on digitalis. VIII. Digitoxin metabolism on a maintenance regimen and after a single dose, *Clin. Pharmacol. Ther.*, 21, 125, 1977.

125. **Okita, G. T., Talso, P. J., Curry, J. H., Smith, F. D., and Geiling, E. M. K.,** Metabolic fate of radioactive digitoxin in human subjects, *J. Pharmacol. Exp. Ther.*, 115, 371, 1955.

237

126. **Katzung, B. G. and Meyers, F. H.,** Excretion of radioactive digitoxin by the dog, *J. Pharmacol. Exp. Ther.*, 149, 257, 1965.

127. **Katzung, B. G. and Meyers, F. H.,** Biotransformation of digitoxin in the dog, *J. Pharmacol. Exp. Ther.*, 154, 575, 1966.

128. **Russell, J. Q. and Klaassen, C. D.,** Biliary excretion of cardiac glycosides, *J. Pharmacol. Exp. Ther.*, 186, 455, 1973.

129. **Marzo, A. and Ghirardi, P.,** Biliary and urinary excretion of five cardiac glycosides and its correlation with their physical and chemical properties, *Naunyn Schmiedeberg's Arch. Pharmacol.*, 298, 51, 1977.

130. **Ingwersen, F.,** The pharmacokinetics of ³H-digitoxin and its metabolites in the guinea pig, *Naunyn Schmiedeberg's Arch. Pharmacol.*, 282, R38, 1974.

131. **Cox, E. and Wright, S. E.,** The hepatic excretion of digitalis glycosides and their genins in the rat, *J. Pharmacol. Exp. Ther.*, 126, 117, 1959.

132. **Castle, M. C. and Lage, G. L.,** H³-Digitoxin and its metabolites following spironolactone pretreatment of rats, *Res. Commun. Chem. Pathol. Pharmacol.*, 6, 601, 1973.

133. **Castle, M. C. and Lage, G. L.,** Cleavage by β-glucuronidase of the water-soluble metabolites of digitoxin excreted in the bile of control and spironolactone-pretreated rats, *Toxicol. Appl. Pharmacol.*, 27, 641, 1974.

134. **Volp, R. F. and Lage, G. L.,** The fate of a major biliary metabolite of digitoxin in the rat intestine, *Drug Metab. Dispos.*, 6, 418, 1978.

135. **Züllich, G., Damm, K. H., Braun, W., and Lisboa, B. P.,** Studies on biliary excreted metabolites of [G-³H]digitoxin in rats, *Arch. Int. Pharmacodyn. Ther.*, 215, 160, 1975.

136. **Kitani, K., Kanai, S., Sato, Y., and Nokubo, M.,** Biliary excretion of digitoxin and its metabolites in young and old male Wistar rats, *Exp. Gerontol.*, 17, 407, 1982.

137. **Kitani, K., Sato, Y., and van Bezooijen, K.,** The effect of age on the biliary excretion of digitoxin and its metabolites in female BN/Bi rats, *Arch. Gerontol. Geriatr.*, 1, 43, 1982.

138. **Beermann, B., Hellström, K., and Rosén, A.,** Fate of orally administered ³H-digitoxin in man with special reference to the absorption, *Circulation*, 43, 852, 1971.

139. **Sattler, R. W., Schmiedeberg, W., and Seiler, K.-U.,** Über die Beeinflussbarkeit des enterohepatischen Kreislaufs von Digitoxin beim Menschen, *Arzneim. Forsch.*, 27(II), 1615, 1977.

140. **Storstein, L.,** Studies on digitalis. III. Biliary excretion and enterohepatic circulation of digitoxin and its cardioactive metabolites, *Clin. Pharmacol. Ther.*, 17, 313, 1975.

141. **Doherty, J. E., Hall, W. H., Murphy, M. L., and Beard, O. W.,** New information regarding digitalis metabolism, *Chest*, 59, 433, 1971.

142. **Beermann, B., Hellström, K., and Rosén, A.,** The absorption of orally administered [12α-³H]digoxin in man, *Clin. Sci.*, 43, 507, 1972.

143. **Magnusson, J. O., Bergdahl, B., Bogentoft, C., Jonsson, U. E., and Tekenbergs, L.,** Excretion of digoxin and its metabolites in urine after a single oral dose in healthy subjects, *Biopharm. Drug Dispos.*, 3, 211, 1982.

144. **Marcus, F. I., Kapadia, G. J., and Kapadia, G. G.,** The metabolism of digoxin in normal subjects, *J. Pharmacol. Exp. Ther.*, 145, 203, 1964.

145. **von Bergmann, K., Abshagen, U., and Rietbrock, N.,** Quantitative analysis of digoxin, 4‴-acetyldigoxin and 4‴-methyldigoxin and their metabolites in bile and urine of rats, *Naunyn Schmiedeberg's Arch. Pharmacol.*, 273, 154, 1972.

146. **Marcus, F. I., Pavlovich, J., Burkhalter, L., and Cuccia, C.,** The metabolic fate of tritiated digoxin in the dog: a comparison of digitalis administration with and without a "loading dose", *J. Pharmacol. Exp. Ther.*, 156, 548, 1967.

147. **Gault, M. H., Charles, J. D., Sugden, D. L., and Kepkay, D. C.,** Hydrolysis of digoxin by acid, *J. Pharm. Pharmacol.*, 29, 27, 1977.

148. **Gault, H., Kalra, J., Ahmed, M., Kepkay, D. C., and Barrowman, J.,** Influence of gastric pH on digoxin biotransformation. I. Intragastric hydrolysis, *Clin. Pharmacol. Ther.*, 27, 16, 1980.

149. **Gault, H., Kalra, J., Ahmed, M., Kepkay, D. C., Longerich, L., and Barrowman, J.,** Influence of gastric pH on digoxin biotransformation. II. Extractable urinary metabolites, *Clin. Pharmacol. Ther.*, 29, 181, 1981.

150. **Magnusson, J. O., Bergdahl, B., and Gustafsson, S.,** Urinary excretion of digoxin and its metabolites in hyperacidic patients and in patients during coronary care, *Arzneim. Forsch.*, 34(I), 87, 1984.

151. **Butler, V. P., Tse-Eng, D., Lindenbaum, J., Kalman, S. M., Preibisz, J. J., Rund, D. G., and Wissel, P. S.,** The development and application of a radioimmunoassay for dihydrodigoxin, a digoxin metabolite, *J. Pharmacol. Exp. Ther.*, 221, 123, 1982.

152. **Abel, R. M., Luchi, R. J., Peskin, G. W., Conn, H. L., and Miller, L. D.,** Metabolism of digoxin: role of the liver in tritiated digoxin degradation, *J. Pharmacol. Exp. Ther.*, 150, 463, 1965.

153. **Watson, E., R., Clark, D. R., and Kalman, S. M.,** Identification by gas chromatography-mass spectrometry of dihydrodigoxin—a metabolite of digoxin in man, *J. Pharmacol. Exp. Ther.*, 184, 424, 1973.

154. **Clark, D. R. and Kalman, S. M.,** Dihydrodigoxin: a common metabolite of digoxin in man, *Drug Metab. Dispos.*, 2, 148, 1974.

155. **Greenwood, H., Snedden, W., Hayward, R. P., and Landon, J.,** The measurement of urinary digoxin and dihydrodigoxin by radioimmunoassay and by mass spectrometry, *Clin. Chim. Acta*, 62, 213, 1975.

156. **Peters, U., Falk, L. C., and Kalman, S. M.,** Digoxin metabolism in patients, *Arch. Int. Med.*, 138, 1074, 1978.

157. **Lindenbaum, J., Tse-Eng, D., Butler, V. P., and Rund, D. G.,** Urinary excretion of reduced metabolites of digoxin, *Am. J. Med.*, 71, 67, 1981.

158. **Magnusson, J. O., Bergdahl, B., Bogentoft, C., and Jonsson, U. E.,** Metabolism of digoxin and absorption site, *Br. J. Clin. Pharmacol.*, 14, 284, 1982.

159. **Lindenbaum, J., Rund, D. G., Butler, V. P., Tse-Eng, D., and Saha, J. R.,** Inactivation of digoxin by the gut flora: reversal by antibiotic therapy, *N. Engl. J. Med.*, 305, 789, 1981.

160. **Dobkin, J. F., Saha, J. R., Butler, V. P., Neu, H. C., and Lindenbaum, J.,** Inactivation of digoxin by *Eubacterium lentum*, an anaerobe of the human gut flora, *Trans. Assoc. Am. Physicians*, 95, 22, 1982.

161. **Dobkin, J. F., Saha, J. R., Bulter, V. P., Neu, H. C., and Lindenbaum, J.,** Digoxin-inactivating bacteria: identification in human gut flora, *Science*, 220, 325, 1983.

162. **Bockbrader, H. N. and Reuning, R. H.,** Spectral analysis of the configuration and solution conformation of dihydrodigoxigenin epimers, *J. Pharm. Sci.*, 72, 271, 1983.

163. **Reuning, R. H., Shepard, T. A., Morrison, B. E., and Bockbrader, H. N.,** Formation of [20R]-dihydrodigoxin from digoxin in humans, *Drug Metab. Dispos.*, 13, 51, 1985.

164. **Robertson, L. W., Chandrasekaran, A., Reuning, R. H., Hui, J., and Rawal, B. D.,** Reduction of digoxin to 20R-dihydrodigoxin by cultures of *Eubacterium lentum.*, *Appl. Environ. Microbiol.*, 51, 1300, 1986.

165. **Alam, A. N., Saha, J. R., Dobkin, J. F., and Lindenbaum, J.,** Interethnic variation in the metabolic inactivation of digoxin by the gut flora, *Gastroenterology*, 95, 117, 1988.

166. **Lindenbaum, J.,** The dehydrogenation of digoxin, in *Cardiac Glycosides 1785-1985 Biochemistry-Pharmacology-Clinical Relevance*, Erdmann, E., Greef, K., and Skou, J. C., Eds., Steinkopff Verlag, Darmstadt, W. Germany, 1986, 263.

167. **Wright, S. E.,** The identification of digoxin metabolite B (digitoxin metabolite C) with digoxigenin di-digitoxoside, *J. Pharm. Pharmacol.*, 14, 613, 1962.

168. **Gault, M. H., Ahmed, M., Symes, A. L., and Vance, J.,** Extraction of digoxin and its metabolites from urine and their separation by Sephadex LH-20 column chromatography, *Clin. Biochem.*, 9, 46, 1976.

169. **Gault, M. H., Longerich, L. L., Loo, J. C. K., Ko, P. T. H., Fine, A., Vasdev, S. C., and Dawe, M. A.,** Digoxin biotransformation, *Clin. Pharmacol. Ther.*, 35, 74, 1984.

170. **Abshagen, U. and Rietbrock, N.,** Metabolism of digoxigenin, digoxigeninmonodigitoxoside and digoxigeninbisdigitoxoside in rats, *Naunyn Schmiedeberg's Arch. Pharmacol.*, 276, 157, 1973.

171. **Wong, K. C. and Spratt, J. L.,** Conversion of digoxin to digoxigenin by liver tissue *in vitro*, *Biochem. Pharmacol.*, 13, 489, 1964.

172. **Lage, G. L. and Spratt, J. L.,** H³-Digoxin metabolism by adult male rat tissues *in vitro*, *J. Pharmacol. Exp. Ther.*, 149, 248, 1965.

173. **Lage, G. L. and Spratt, J. L.,** Species and sex variation in the hepatic metabolism of H³-digoxin *in vitro*, *J. Pharmacol. Exp. Ther.*, 159, 182, 1968.

174. **Schmoldt, A. and Ahsendorf, B.,** Cleavage of digoxigenin digitoxosides by rat liver microsomes, *Eur. J. Drug Metab. Pharmacokinet.*, 5, 225, 1980.

175. **Dwenger, A. and Haberland, G.,** Metabolism of ³H-digoxin and some acetyl-digoxins: time-dependent formation of hydrophilic metabolites after oral application in man, *Naunyn Schmiedebergs Arch. Pharmakol.*, 270, 102, 1971.

176. **Benthe, H. F.,** Occurence and chemical nature of polar water-soluble digoxin metabolites, in *Cardiac Glycosides*, Bodem, G. and Dengler, H. J., Eds., Springer Verlag, Berlin, 1978, 52.

177. **Kuhlmann, J., Abshagen, U., and Rietbrock, N.,** Pharmacokinetics and metabolism of digoxigenin-monodigitoxoside in man, *Eur. J. Clin. Pharmacol.*, 7, 87, 1974.

178. **Thomas, R. E. and Wright, S. E.,** 3-Epimerisation of digoxigenin in the rat, *J. Pharm. Pharmacol.*, 17, 459, 1965.

179. **Gault, H., Kalra, J., Longerich, L., and Dawe, M.,** Digoxigenin biotransformation, *Clin. Pharmacol. Ther.*, 31, 695, 1982.

180. **Talcott, R. E., Stohs, S. J., and El-Olemy, M. M.,** Metabolites and some characteristics of the metabolism of ³H-digoxigenin by rat liver homogenates, *Biochem. Pharmacol.*, 21, 2001, 1972.

181. **Doherty, J. E., Flanigan, W. J., Murphy, M. L., Bulloch, R. T., Dalrymple, G. L., Beard, O. W., and Perkins, W. H.,** Tritiated digoxin. XIV. Enterohepatic circulation, absorption, and excretion studies in human volunteers, *Circulation*, 42, 867, 1970.

182. **Klotz, U. and Antonin, K. H.,** Biliary excretion studies with digoxin in man, *Int. J. Clin. Pharmacol.*, 15, 332, 1977.

183. **Caldwell, J. H. and Cline, C. T.,** Biliary excretion of digoxin in man, *Clin. Pharmacol. Ther.*, 19, 410, 1976.

184. **Abshagen, U., von Bergmann, K., and Rietbrock, N.,** Evaluation of the enterohepatic circulation after ³H-digoxin administration in the rat, *Naunyn Schmiedeberg's Arch. Pharmacol.*, 275, 1, 1972.

185. **Kitani, K., Ohta, M., Kanai, S., and Sato, Y.,** Sex difference in the biliary excretion of digoxin and its metabolites in aging Wistar rats, *Arch. Gerontol. Geriatr.*, 4, 1, 1985.

186. **Harrison, C. E., Brandenburg, R. O., Ongley, P. A., Orvis, A. L., and Owen, C. A.,** The distribution and excretion of tritiated substances in experimental animals following the administration of digoxin-³H, *J. Lab. Clin. Med.*, 67, 764, 1966.

187. **Marcus, F. I., Petterson, A., Salel, A., Scully, J., and Kapadia, G. G.,** The metabolism of tritiated digoxin in renal insufficiency in dogs and man, *J. Pharmacol. Exp. Ther.*, 152, 372, 1966.

188. **Kolenda, K.-D., Lüllmann, H., Peters, T., and Seiler, K.-U.,** Plasma concentration, uptake by liver, and biliary excretion of tritiated cardiac glycosides in the isolated perfused guinea-pig liver, *Br. J. Pharmacol.*, 41, 648, 1971.

189. **Haass, A., Lüllmann, H., and Peters, T.,** Absorption rates of some cardiac glycosides and portal blood flow, *Eur. J. Pharmacol.*, 19, 366, 1972.

190. **Beermann, B.,** On the fate of orally administered ³H-lanatoside C in man, *Eur. J. Clin. Pharmacol.*, 5, 11, 1972.

191. **Dengler, H. J., Bodem, G., and Wirth, K.,** Pharmakokinetische Untersuchungen mit H³-Digoxin und H³-Lanatosid beim Menschen, *Arzneim. Forsch.*, 23, 64, 1973.

192. **Lauterbach, F. and Repke, K.,** Über die Veteilung herzglykosidspaltender Fermente im tierischen Organismus, *Naunyn Schmiedebergs Arch. Exp. Pathol. Pharmakol.*, 240, 45, 1960.

193. **Hawksworth, G., Drasar, B. S., and Hill, M. J.,** Intestinal bacteria and the hydrolysis of glycosidic bonds, *J. Med. Microbiol.*, 4, 451, 1971.

194. **Aldous, S. and Thomas, R.,** Absorption and metabolism of lanatoside C. II. Fate after oral administration, *Clin. Pharmacol. Ther.*, 21, 647, 1977.

195. **Bodem, G., Grube, E., Ochs, H. R., and Gerloff, J.,** Studies on the enteral absorption and metabolism of lanatoside C, *Verh. Dtsch. Ges. Inn. Med.*, 84, 757, 1978 (Chem. Abstr. 90, 132544y, 1979).

196. **Brown, B. T. and Wright, S. E.,** The cardioactive metabolites of digitalis glycosides, *J. Biol. Chem.*, 220, 431, 1956.

197. **Marzo, A., Ghirardi, P., Brusoni, B., and Marchetti, G.,** Pharmacokinetics of deslanatoside C-³H administered parenterally to the guinea-pig, rabbit and dog, *Arch. Int. Pharmacodyn. Ther.*, 233, 156, 1978.

198. **Lesne, M.,** Pharmacological reevaluation of gitoxin in man, *Int. J. Clin. Pharmacol.*, 16, 456, 1978.

199. **Hupin, C., de Suray, J. M., Versluys, J., Lorent, M., Dodion, L, and Lesne, M.,** Bioavailability study of gitoxin in a solid dosage form, *Int. J. Clin. Pharmacol. Biopharm.*, 17, 197, 1979.

200. **Kadima, L. N. and Lesne, M.,** Studies on the pharmacokinetics and the metabolism of gitoxin in the guinea-pig. I. Disposition kinetics following an i.v. bolus of ³H-gitoxin, *Arch. Int. Pharmacodyn. Ther.*, 245, 20, 1980.

201. **Kadima, L. N. and Lesne, M.,** Studies on the pharmacokinetics and the metabolism of gitoxin in the guinea-pig. II. The enteral absorption and the enterohepatic recirculation, *Arch. Int. Pharmacodyn. Ther.*, 258, 4, 1982.

202. **Pellegrin, P. and Lesne, M.,** Metabolism of gitoxin by the rabbit isolated liver, *Arch. Int. Pharmacodyn. Ther.*, 260, 293, 1982.

203. **Kadima, L. N., Lhoest, G., and Lesne, M.,** Isolation and mass spectrometric identification of gitoxin metabolites excreted in bile, *Eur. J. Drug Metab. Pharmacokinet.*, 7, 111, 1982.

204. **Raudonat, H. W. and Engler, R.,** Ausscheidungsprodukte des k-Strophanthin-γ und des Thevetins nach oraler Verabfolgung bei Ratten, *Naunyn Schmiedebergs Arch. Exp. Pathol. Pharmakol.*, 232, 295, 1957.

205. **Engler, R., Holtz, P., and Raudonat, H. W.,** Über die Spaltung herzwirksamer Glykoside im Tierkörper, *Naunyn Schmiedebergs Arch. Exp. Pathol. Pharmakol.*, 233, 393, 1958.

206. **Zhao, K., Zhu, X., Yi, M., Liu, Z., and Song, Z.,** Pharmacokinetics of ³H-neriifolin in rats, *Yaoxue Xuebao*, 21, 572, 1986 (Chem. Abstr. 105, 183361x, 1986).

207. **Fröhlich, J. C., Falkner, F. C., Watson, J. T., and Scheler, F.,** Metabolism of peruvoside in man, *Eur. J. Clin. Pharmacol.*, 5, 65, 1972.

208. **Marzo, A., Ghirardi, P., Croce, G., and Marchetti, G.,** Absorption, distribution and excretion of K-strophanthoside(³H) administered rectally to guinea-pigs, *Naunyn Schmiedeberg's Arch. Pharmacol.*, 279, 19, 1973.

209. **Marzo, A., Ghirardi, P., and Marchetti, G.,** The absorption, distribution and excretion of K-strophanthoside-³H in guinea-pigs after parenteral administration, *J. Pharmacol. Exp. Ther.*, 189, 185, 1974.

210. **Strobach, H., Wirth, K. E., and Rojsathapron, K.,** Absorption, metabolism and elimination of strophanthus glycosides in man, *Naunyn Schmiedeberg's Arch. Pharmacol.*, 334, 496, 1986.

211. **Lauterbach, F.,** Die Reduktion der Aldehydgruppe des Strophanthidins und seiner Glykoside im tierischen Stoffwechsel, *Naunyn Schmiedebergs Arch. Exp. Pathol. Pharmakol.*, 247, 71, 1964.

212. **Moerman, E.,** Distribution, excretion and metabolism of cymarin in the rat, *Arch. Int. Pharmacodyn. Ther.*, 156, 489, 1965.

213. **Topchii, L. Y., Angarskaya, M. A., and Makarevich, I. F.,** Biotransformations of strophanthidin D- and L-arabinosides in rats, *Farmakol. Toksikol.*, 40, 703, 1977 (Chem. Abstr. 88, 83317z, 1978).

214. **Gundert-Remy, U., Weber, E., and Rabl, W.,** ³H-Cymarol in man. Excretion pathways and serum protein binding, *Arzneim. Forsch.*, 28(II), 1174, 1978.

215. **Gundert-Remy, U.,** Influence of the interruption of enterohepatic circulation on the excretion of ^{3}H-cymarol in rats, *Naunyn Schmiedeberg's Arch. Pharmacol.*, 287, R42, 1975.

216. **Farah, A.,** On the elimination of G-strophanthin by the rat, *J. Pharmacol. Exp. Ther.*, 86, 248, 1946.

217. **Cox, E., Roxburgh, G., and Wright, S. E.,** The metabolism of ouabain in the rat, *J. Pharm. Pharmacol.*, 11, 535, 1959.

218. **Kupferberg, H. J. and Schanker, L. S.,** Biliary secretion of ouabain-^{3}H and its uptake by liver slices in the rat, *Am. J. Physiol.*, 214, 1048, 1968.

219. **Russell, J. Q. and Klaassen, C. D.,** Species variation in the biliary excretion of ouabain, *J. Pharmacol. Exp. Ther.*, 183, 513, 1972.

220. **Kitani, K., Kanai, S., Miura, R., Morita, Y., and Kasahara, M.,** The effect of ageing on the biliary excretion of ouabain in the rat, *Exp. Gerontol.*, 13, 9, 1978.

221. **Sato, Y., Kanai, S. and Kitani, K.,** Biliary excretion of ouabain in aging male and female F-344 rats, *Arch. Gerontol. Geriatr.*, 6, 141, 1987.

222. **Seldon, R., Margolies, M. N., and Smith, T. W.,** Renal and gastrointestinal excretion of ouabain in dog and man, *J. Pharmacol. Exp. Ther.*, 188, 615, 1974.

223. **Dutta, S., Marks, B. H., and Smith, C. R.,** Distribution and excretion of ouabain-H^3 and dihydro-ouabain-H^3 in rats and sheep, *J. Pharmacol. Exp. Ther.*, 142, 223, 1963.

224. **Marks, B. H., Dutta, S., Gauther, J., and Elliott, D.,** Distribution in plasma, uptake by the heart and excretion of ouabain-H^3 in human subjects, *J. Pharmacol. Exp. Ther.*, 145, 351, 1964.

225. **Marchetti, G. V., Marzo, A., de Ponti, C., Scalvini, A., Merlo, L., and Noseda, V.,** Blood levels and tissue distribution of ^{3}H-ouabain administered per os. An experimental and clinical study, *Arzneim. Forsch.*, 21, 1399, 1971.

226. **Greenberger, N. J., MacDermott, R. P., Martin, J. F., and Dutta, S.,** Intestinal absorption of six tritium-labeled digitalis glycosides in rats and guinea pigs, *J. Pharmacol. Exp. Ther.*, 167, 265, 1969.

227. **Lauterbach, F.,** Enterale Resorption, biliäre Ausscheidung und entero-hepatischer Kreislauf von Herzglykosiden bei der Ratte, *Naunyn Schmiedebergs Arch. Exp. Pathol. Pharmakol.*, 247, 391, 1964.

228. **Simon, M. and Wright, S. E.,** The excretion of scillaren A by rats, *J. Pharm. Pharmacol.*, 12, 767, 1960.

229. **Davis, S. H., Van Dyke, K., and Robinson, R. L.,** The uptake and release of proscillaridin-^{3}H by several tissues of the guinea pig after oral administration, *Arch. Int. Pharmacodyn. Ther.*, 177, 231, 1969.

230. **Andersson, K.-E., Bertler, Å., and Redfors, A.,** On the pharmacokinetics of proscillaridin A in man, *Eur. J. Clin. Pharmacol.*, 8, 421, 1975.

231. **Andersson, K.-E., Bergdahl, B., Dencker, H., and Wettrell, G.,** Proscillaridin activity in portal and peripheral venous blood after oral administration to man, *Eur. J. Clin. Pharmacol.*, 11, 277, 1977.

232. **Andersson, K.-E., Bergdahl, B., and Wettrell, G.,** Biliary excretion and enterohepatic recycling of proscillaridin A after oral administration to man, *Eur. J. Clin. Pharmacol.*, 11, 273, 1977.

233. **Weymann, J., Schenk, G., and Kesselring, K.,** Untersuchungen zur Disposition von Meproscillarin an den Spezies Ratte und Hund, *Arzneim. Forsch.*, 28(I), 520, 1978.

234. **Rietbrock, N. and Staud, R.,** Metabolism and excretion of methylproscillaridin by man, *Eur. J. Clin. Pharmacol.*, 8, 427, 1975.

235. **Staud, R., Rietbrock, N., and Fassbender, H. P.,** Excretion of methylproscillaridin in patients with a biliary fistula, *Eur. J. Clin. Pharmacol.*, 9, 99, 1975.

236. **Rietbrock, N.,** Metabolismus von Meproscillarin beim Menschen, *Arzneim. Forsch.*, 28(I), 540, 1978.

237. **Angarskaya, M. A., Sokolova, V. E., Lyubartseva, L. A., and Lutokhin, S. I.,** Metabolism of some strophanthin-like glycosides in the liver, *Farmakol. Toksikol.*, 30, 438, 1967 (Chem. Abstr. 67, 80862d, 1967).

238. **Mutlib, A. E., Cheung, H. T. A., and Watson, T. R.,** *In vivo* and *in vitro* metabolism of gomphoside, a cardiotonic steroid with doubly-linked sugar, *J. Steroid Biochem.*, 28, 65, 1987.

239. **Mutlib, A. E. and Watson, T. R.,** The pharmacokinetics and tissue distribution of gomphoside in Wistar rats, *Eur. J. Drug Metab. Pharmacokinet.*, 14, 117, 1989.

240. **Mutlib, A. E., Cheung, H. T. A., and Watson, T. R.,** *In vitro* metabolism of the cardiotonic steroids gomphogenin and calactin, *J. Steroid Biochem.*, 29, 135, 1988.

241. **Lafont, R., Girault, J.-P., and Kerb, U.,** Excretion and metabolism of injected ecdysone in the white mouse, *Biochem. Pharmacol.*, 37, 1174, 1988.

242. **Girault, J.-P., Lafont, R., and Kerb, U.,** Ecdysone catabolism in the white mouse, *Drug Metab. Dispos.*, 16, 716, 1988.

243. **Valzelli, G. and Goldaniga, G.,** Metabolic studies of synthetic isojuripidine in the dog and rat, *Biochem. Pharmacol.*, 22, 911, 1973.

244. **Booth, V. H.,** Vitamin A activity of orange peel pigment, *Biokhimiya*, 12, 21, 1947.

245. **Booth, V. H.,** The splitting of carotenoid esters in the alimentary tract of the rat, *Biochim. Biophys. Acta*, 84, 188, 1964.

246. **John, J., Kishore, G. S., Subbarayan, C., and Cama, H. R.,** Metabolism and biological potency of cryptoxanthin in rat, *Indian J. Biochem.*, 7, 222, 1970.

247. **Boonjawat, J. and Olson, J. A.,** The metabolism of radioactive crustaxanthin (3,3′,4,4′-tetrahydroxy-β-carotene), *Comp. Biochem. Physiol.*, 50B, 363, 1975.
248. **Simpson, K. L. and Chichester, C. O.,** Metabolism and nutritional significance of carotenoids, *Annu. Rev. Nutr.*, 1, 351, 1981.
249. **Davies, B. H.,** Carotenoid metabolism in animals: a biochemist's view, *Pure Appl. Chem.*, 57, 679, 1985.
250. **Goodwin, T. W.,** *The Biochemistry of the Carotenoids, Vol. 2 Animals,* Chapman and Hall, London, 1984.

METABOLISM OF OXYGEN HETEROCYCLIC COMPOUNDS

FURANS

The furan structure is present in a number of plant compounds and the metabolism of some more complex types (e.g., psoralens and tanshinones) is noted below. Many of the metabolic changes found with 8-methoxypsoralen involve the furan moiety. Other furan derivatives include furan aldehydes (see the chapter entitled Metabolism of Aldehydes, Ketones, and Quinones) and the furanosesquiterpenoid ketone ngaione (see the chapter entitled Metabolism of Higher Terpenoids). The role of metabolic activation of furans in the formation of toxic metabolites was reviewed by Burka and Boyd[1] and Strolin Benedetti and Dostert.[2]

2-Methylfuran (sylvan) (1) was metabolized by rat liver microsomes to 4-oxo-2-pentenal (acetylacrolein) (2).[3] This reactive aldehyde intermediate was shown to bind irreversibly to microsomal macromolecules. Further studies[4,5] on this metabolic activation indicated that it was mediated by the cytochrome P-450-system.

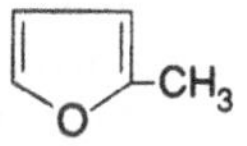

(1)

2-Methylfuran

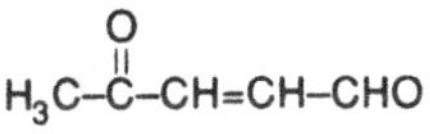

(2)

PYRONES

As with the furan structure noted above, the pyrone structure is found in several classes of natural compounds. These include the chromones, coumarins, flavones, isoflavones, and xanthones. The metabolism of these compounds is discussed subsequently in this chapter and the present section will deal only with the simpler derivatives of α-pyrone (3) and γ-pyrone (4).

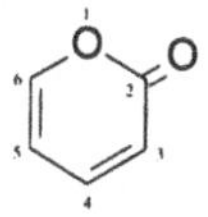

(3)

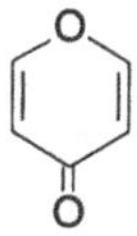

(4)

α-PYRONES

Metabolic data are available on only a few of the simpler derivatives of α-pyrone. Hirata et al.[6] administered ^{14}C-labeled glycoside **aloenin** (5a) intragastrically (~50 mg/kg) to rats. They investigated both the urine and feces for radioactive metabolites, however their methodology employed extraction with methanol of the feces and of the urine following its evaporation to dryness. No provision was made for conjugate hydrolysis, although it seems reasonable to assume that the two urinary metabolites identified, the aglycon (5b) and 2,5-dimethyl-7-hydroxychromone (6), would have been excreted largely as glucuronide and/or sulfate conjugates. Indeed, their data on the urinary excretion of radioactivity (as measured by methanolic extraction) showed that only 17 to 22% of the dose was found in 24 h and that very little was excreted thereafter. The radioactivity in the feces (11 to 13% of the dose in 24 h) was due to the aglycon (5b) and glucose. The sequence leading from the pyrone (5b) to the chromone was not studied, however it seems likely that this would be initiated by scission of the heterocyclic ring. It is perhaps of interest to note that the putative β-methoxy acid thus produced shows close similarity to the postulated intermediate in the metabolism of dihydrokawain (see the chapter entitled Metabolism of Acids, Lactones, and Esters). Furthermore, the resultant products show structural similarities even though the formation of the chromone (6) requires a ring-closure step not seen in the reaction sequence with dihydrokawain.

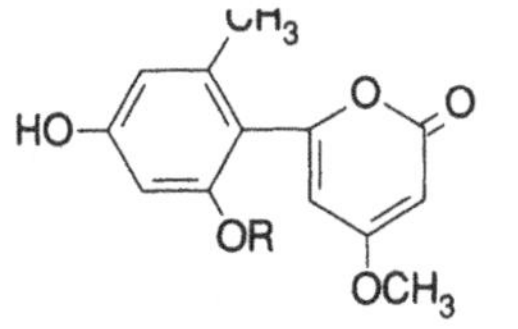

(5)

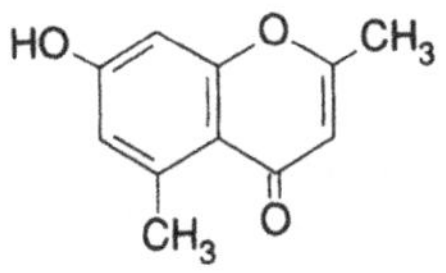

(6)

(a) Aloenin, R = β-D-glucose

(b) R = H

The metabolism of two α-pyrones from kava was studied by Rasmussen et al.[7] They administered **yangonin** (7) and **7,8-dihydroyangonin** (8) to rats intragastrically (400 mg/kg) or by i.p. injection (100 mg/kg). The biological disposition of these compounds should be compared with that of three closely related lactones (5,6-dihydro-α-pyrone derivatives), kawain, dihydrokawain, and methysticin (see the chapter entitled Metabolism of Acids, Lactones, and Esters). It was noted that 7,8-dihydroyangonin (8), as with dihydrokawain, gave larger amounts of urinary metabolites than did the compounds which contain a double bond at C-7,8. It is likely that this is due to the extremely low solubility of the latter type of kava pyrones. In fact, the aqueous solubility of yangonin was found to be only 0.4 µg/ml (at 22° C). Three urinary metabolites of yangonin were detected and two of these (probably geometrical isomers) were formed by *O*-demethylation. However, this reaction was not extensive and only 2% of an oral dose (100 mg/kg) of [4′-*methoxy*-^{14}C]yangonin was excreted as respiratory $^{14}CO_2$. The third metabolite, formed in very small amounts, corresponded to a derivative of the above in which an additional hydroxy group was present and reduction at C-7,8 had occurred. No evidence was obtained for metabolites of yangonin formed via scission of the heterocyclic ring. The same was true with 7,8-dihydroyangonin which, like yangonin, was converted by *O*-demethylation to a corresponding 4′-hydroxy derivative. Also, two minor metabolites formed by hydroxylation of the *O*-demethylated metabolite were detected. Duffield et al.[8] studied the nature of the urinary metabolites formed in humans given kava preparations. These preparations contained both yangonin and the 4′-desmethoxy derivative of yangonin. The 4′-hydroxy metabolite noted above in rats was also identified in human urine; however, it was not determined if it was formed by *O*-demethylation of yangonin or *p*-hydroxylation of the desmethoxy compound, or both. No indication was obtained that the pyrones were metabolized in man via ring-scission reactions.

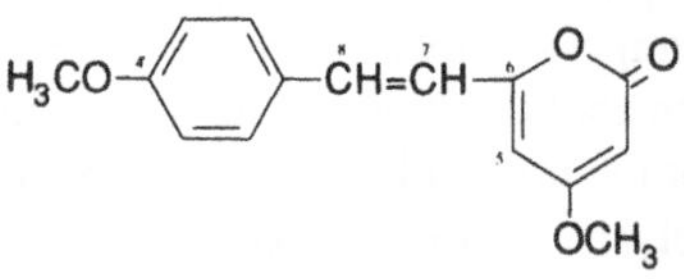

(7)

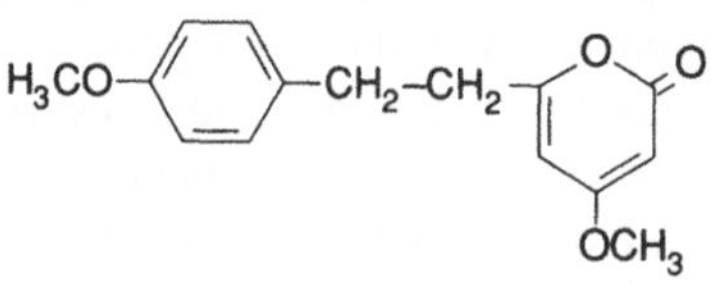

(8)

Yangonin

7,8-Dihydroyangonin

γ-PYRONES

The metabolism of **maltol** (9) and its synthetic homologue ethyl maltol was studied by Rennhard.[9] They found that maltol (10 mg/kg) was excreted in the urine as glucuronide and sulfate conjugates following i.v. administration to dogs. Conjugate excretion was rapid, occurring mainly within 6 h, and ~60% of the dose was accounted for after 48 h. It is not known if maltol is metabolized along other pathways and the fate of the remaining 40% of the dose is unknown. However, fecal excretion of maltol is probably very low as virtually none of the ethyl homologue or its conjugates was detected in the feces following a large (200 mg/kg) oral dose.

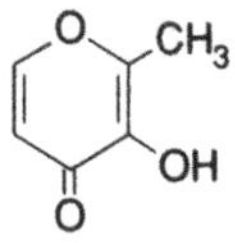

(9)

Maltol

CHROMAN, CHROMENE, CHROMANONE, AND CHROMONE DERIVATIVES

CHROMAN DERIVATIVES

The derivatives of chroman (10) which have been studied metabolically include the dye hematoxylin and some tocopherol derivatives, however most of the data available deals with the cannabinoids.

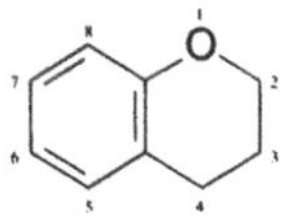

(10)

Gautrelet and Gravellat[10] reported that **hematoxylin** was excreted partly unchanged and partly in a reduced form. However, nothing is known of the actual metabolic change occurring with this phenolic derivative of chroman.

The metabolism of tocopherols is interesting, both due to the hydrolytic scission of the chroman ring at the 1,2-bond and the extensive degradation of the isoprenoid side chain to acidic metabolites including a hydroxy acid which readily forms a γ-lactone. Simon et al.[11,12] showed that α-**tocopherol** (11) (see Figure 1), administered as the succinate ester, was metabolized and excreted in the urine by rabbits and man mainly as conjugates of tocopheronic acid (2-(3-hydroxy-3-methyl-5-carboxypentyl)-3,5,6-trimethylbenzoquinone) (13) and its γ-lactone (tocopheronolactone) (14). The conjugates are most likely glucuronides, probably not of the benzoquinones themselves but of their reduced hydroquinone derivatives. The latter, upon liberation, would be expected to be fairly easily oxidized by air. The initial step in the metabolism of α-tocopherol is considered to be hydrolysis of the chroman ring to give a α-tocopherylquinone (12),[12] a metabolite which is known to be formed in the animal body.[13] Furthermore, this latter metabolite, when itself given orally to rats, is converted to the lactone (14) in higher amounts than that observed with α-tocopherol.[14] The metabolic pathways illustrated in Figure 1 show that tocopherylquinone (12) undergoes extensive side-chain oxidation to form tocopheronic acid (13) which is excreted in the urine together with its γ-lactone (14). Further details of the metabolism of α-tocopherol are available in the review by Draper and Csallany.[15] In addition, Watanabe et al.[16] found that the acidic metabolites (15) and (16) were also present in the urine of rabbits given α-tocopherol acetate. These compounds probably arise by dehydration of (13) followed by reduction and β-oxidation.

The term cannabinoid is used to describe a group of C_{21} compounds characteristically present in *Cannabis sativa*, their carboxylic acids, analogues, homologues, and transformation products.[17] The biologically most important cannabinoids are the tetrahydrocannabinols which are derivatives of chroman. However, some representatives belong strictly to other chemical classes. Examples include cannabinol (22), a substituted chromene, and cannabidiol (20) a

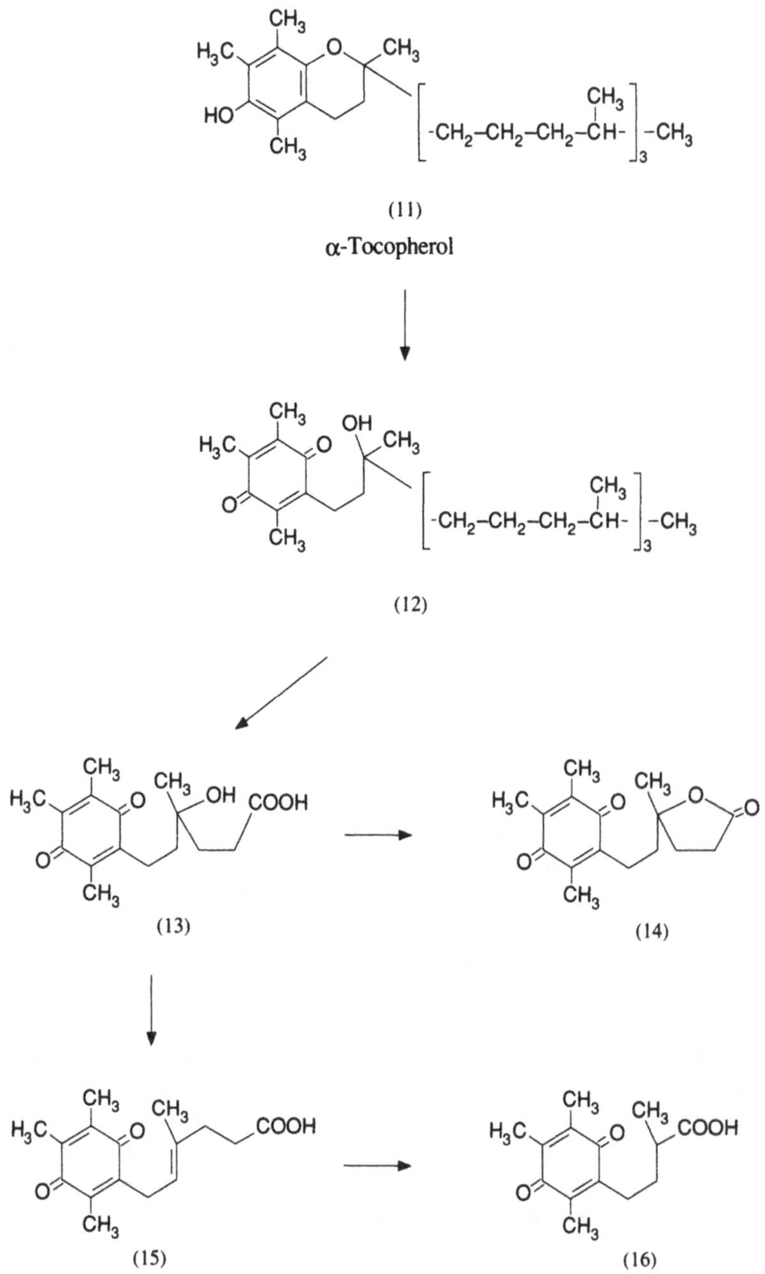

(11)

α-Tocopherol

(12)

(13)   (14)

(15)   (16)

FIGURE 1.   Metabolic pathways of α-tocopherol.

phenol which lacks the heterocyclic ring. Nonetheless, it is both logical and convenient to include all of the cannabinoids in the present section.

A prerequisite in any discussion of cannabinoid metabolism is an explanation of their chemical numbering. Several numbering systems have been proposed, however two are in common use. The dibenzopyran system, which is based on the formal chemical rules used for numbering pyran-type compounds, is preferred in North America and has been adopted by Chemical Abstracts. This system is employed with the tetrahydrocannabinols and, with it, the major psychoactive cannabinoid is designated (−)-Δ⁹-*trans*-tetrahydrocannabinol or merely Δ⁹-tetrahydrocannabinol (Δ⁹-THC) (17). However, as noted above, some cannabinoids are not pyran derivatives and this system is then no longer applicable. Where applicable the dibenzo-pyran system is used in this book. The second system has a biogenic basis and considers the cannabinoids to be substituted monoterpenoids. With this system, which has been widely used in Europe, the compound noted above is designated Δ^1-tetrahydrocannabinol (Δ^1-THC) (18).

(17)	(18)
Δ^9-Tetrahydrocannabinol	Δ^1-Tetrahydrocannabinol
	(monoterpenoid numbering)

Δ^9-Tetrahydrocannabinol, which is an oily, water-insoluble liquid, shows instability to air, light, high temperatures, and acidic conditions. Acids cause it to isomerize to Δ^8-tetrahydrocannabinol (19), a psychoactive compound which is usually a minor component of cannabis products. Both the Δ^9- and Δ^8-isomers undergo oxidation in air to cannabinol which is not psychoactive. In addition to Δ^9-tetrahydrocannabinol, the other main cannabinoids found in *Cannabis sativa* are cannabinol (22) and cannabidiol (20). However, the structures of more than 60 naturally occurring cannabinoids have been determined.[18] Notable among these are homologues in which the pentyl side chain is replaced with methyl, *n*-propyl, or *n*-butyl groups. Also, the cannabinoids in fresh material consist mainly of two acidic derivatives of Δ^9-tetrahydrocannabinol. These are designated Δ^9-tetrahydrocannabinolic acids A and B and contain a carboxy group at position 2 or 4, respectively. Additional information on cannabinoid chemistry is found in the articles by Harvey,[19] Mechoulam,[17,20] and Mechoulam et al.[21]

Our knowledge on cannabinoids has reached a high degree of maturity during the past two decades. This is especially true in regard to their chemistry, including the subjects of structural identification, synthetic and biosynthetic routes, and analytical methods. Likewise, a wealth of information is now available on the biological disposition of the cannabinoids including numerous reviews of their metabolism. For this reason it is felt that detailed coverage of the subject in this book is unwarranted. Instead, the general features and pathways of metabolism of the cannabinoids will be summarized. A limited number of primary articles will be cited; however, additional details and references are available in the reviews of Agurell et al.,[22] Harvey,[19] Harvey and Paton,[23] and, for the earlier literature, Lemberger and Rubin.[24]

Δ^9-**Tetrahydrocannabinol** (17) is a highly lipophilic compound and little or no unchanged material is excreted from the body. Instead, it undergoes an extensive array of reactions which lead to the formation of more than 80 known metabolites. Our present knowledge is based on results obtained from a multitude of investigations, using both *in vivo* and *in vitro* methods in the mouse, rat, guinea pig, rabbit, dog, monkey, and man. The important features of these metabolic pathways are summarized schematically in Figure 2, however not all of the pathways have been demonstrated in all species and in many cases a certain route will make a minor metabolic contribution.

The presence of a phenolic hydroxy group at C-1 suggests that Δ^9-tetrahydrocannabinol should be directly susceptible to phase II metabolism. However, several early studies failed to detect such a glucuronide or sulfate conjugate. Subsequent studies showed that the *O*-glucuronide is formed in rabbits[25] and man,[26] however this reaction is not a major route. Instead, the major site of metabolism is at C-11 which, via allylic hydroxylation, results in the formation of 11-hydroxy-Δ^9-tetrahydrocannabinol. This was the first metabolite to be characterized and

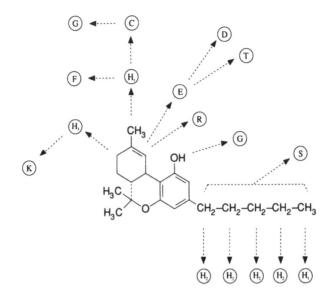

Ⓒ	= −COOH		Ⓗ₃	= ⟩CHOH
Ⓓ	= (structure with OH, OH)		Ⓗ₄	= HO····CH (α-isomer) HO−CH (β-isomer)
Ⓔ	= (structure with O)		Ⓚ	= O=C⟨
Ⓕ	= fatty acid conjugates		Ⓡ	= −CH₂−CH₂−
Ⓖ	= O-glucuronide		Ⓢ	= −(CH₂)ₙ−COOH
Ⓗ₁	= −CH₂OH		Ⓣ	= glutathione conjugate

FIGURE 2.   Metabolic pathways of Δ⁹-tetrahydrocannabinol.

is formed by all the mammalian species studied. As shown in Figure 2, allylic hydroxylation may also occur at an endocyclic site (C-8). This results in the formation of both the α–hydroxy and β–hydroxy epimers. On the other hand, allylic hydroxylation at the other ring site (C-10α) has not been reported. The other main region of hydroxylation is the pentyl side chain and all monohydroxy derivatives have been identified. However, the 5′-hydroxy derivative itself is rarely detected.[27] Additional hydroxylated metabolites are formed by combinations of these initial steps. At least 20 di- and trihydroxy derivatives have been identified and the relative amounts of these generally reflect the extent of formation of the monohydroxy compounds.

Further oxidation of the hydroxylated metabolites of Δ⁹-tetrahydrocannabinol leads to ketone or carboxylic acid derivatives. Of the first type, several 8-oxo compounds have been identified, mainly those with hydroxy groups at C-11 or in the side chain. Formation of carboxylic acids may occur at C-11 or at side-chain sites. The former reaction, giving Δ⁹-tetrahydrocannabinol-11-oic acid, is a major metabolic route which has been demonstrated to take place in all of the species studied. The side-chain acids identified are all degradation products in which 1–4 methylene units are lacking. This is a common finding with most cannabinoids except cannabidiol. The propionic acid derivative (*n* = 2, Figure 2) figures prominently among these acidic metabolites. Harvey and Leuschner[27] showed that these acids with shortened side chains containing odd numbers of carbon atoms are formed via β-oxidation following initial hydroxylation at C-5'. However, Harvey[28] reported on studies in mice which indicated that initial ω-2 hydroxylation was significant in the formation of acidic metabolites containing three carbon atoms in the side chain. This pathway was also a major step in the formation of metabolites with side chains containing two carbon atoms. Dicarboxylic acids are commonly found as urinary metabolites of Δ^9-tetrahydrocannabinol in rabbits[29] and man.[30]

The presence of an endocyclic double bond in Δ^9-tetrahydrocannabinol allows for its epoxidation.[31,32] While both this metabolite, the 9α,10α-epoxide, and the corresponding 11-hydroxy compound have been identified, this route appears to be of minor quantitative importance. In fact, the products shown in Figure 2 as arising from the epoxide have only been detected when the epoxide itself, and not Δ^9-tetrahydrocannabinol, was dosed or incubated with liver microsomes. Furthermore, Yamamoto et al.[33] showed that in certain systems (e.g., human liver microsomes) the epoxide was resistant to hydrolysis by epoxide hydrolase and instead metabolized by the monooxygenase system involving cytochrome P-450.

Leighty[34] reported that rats showed prolonged tissue retention of a nonpolar metabolite of Δ^9-tetrahydrocannabinol. When 7-hydroxy-Δ^9-tetrahydrocannabinol was incubated with a rat liver microsomal enzyme system,[35] the nonpolar material formed was shown to consist of several fatty acid conjugates of the 7-hydroxy metabolite. The most abundant esters contained palmitic (C_{16}) and stearic (C_{18}) acids; however, a small amount of material was esterified with unsaturated C_{18}-acids (presumably oleic and linoleic). These compounds, which may also be formed from dihydroxylated derivatives of Δ^9-tetrahydrocannabinol,[36] appear to be synthesized by the same microsomal system that is involved in the esterification of long-chain fatty acids with cholesterol.[37] Hydrolysis of these conjugates can be carried out by cholesterol esterase.[38]

Studies in the mouse indicated that the metabolism of the biologically inactive isomer (+)-Δ^9-tetrahydrocannabinol, which does not occur naturally, is similar to that shown by the (–)-isomer.[39] These findings were confirmed by Harvey[40] who found that, although the patterns of metabolism were qualitatively similar, the (+)-isomer showed less conversion to the carboxylic acid derivative (at C-11) and less hydroxylation at C-8.

Comparisons of Δ^9-tetrahydrocannabinol metabolism in different animal species must take into account the experimental conditions employed. Experiments using *in vitro* methods will show the early metabolites including many hydroxylated compounds, whereas *in vivo* studies show mainly terminal metabolites including acidic compounds and conjugates. Furthermore, different organs including lung[41] and brain[42] show metabolic profiles which may differ considerably from those normally encountered using liver preparations. Marriage and Harvey[43] showed that cytochrome P-450 isoenzymes from mouse liver gave different metabolic profiles with Δ^9-tetrahydrocannabinol. This factor may explain the organ and species differences in metabolism commonly demonstrated. Most experiments in humans have been carried out in males; however, Wall et al.[44,45] investigated the metabolism of Δ^9-tetrahydrocannabinol in both male and female subjects. They did not find any sex differences in the metabolic patterns or excretion of metabolites. Differences in the routes of excretion of metabolites are a common feature of cannabinoid metabolism. In some species including rats and man, metabolites are preferentially lost via the fecal route. This finding is, in turn, associated with a high degree of biliary excretion of metabolites in these species. Conversely, the urine is the main excretory route in other species including the rabbit.

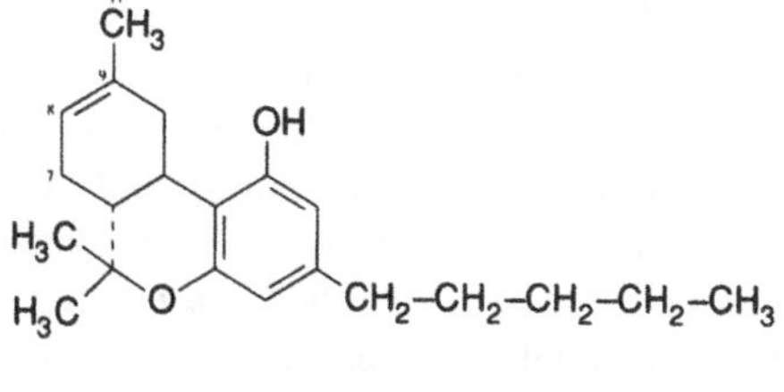

(19)

Δ^8-Tetrahydrocannabinol

Δ^8-**Tetrahydrocannabinol** (19) is designated Δ^6-tetrahydrocannabinol under the monoterpenoid numbering system. Although the data available on its metabolism are considerably less extensive than that for the Δ^9-isomer, it can nevertheless be concluded that the general features of metabolism are similar with both isomers. Thus, allylic hydroxylation at C-11 and C-7 and side-chain hydroxylation at all sites are initial reactions which lead to the formation of more than 20 mono-, di-, and trihydroxy derivatives. The structures of these metabolites were given in the review by Harvey and Paton[23] which also listed the other types of metabolites including 7-keto derivatives, carboxylic acids, metabolites formed by the epoxide-diol pathway, and conjugates.

Although qualitative differences in metabolism have been recorded with Δ^8- and Δ^9-tetrahydrocannabinol, the most notable quantitative differences include a lesser degree of endocyclic allylic hydroxylation (at C-7) and the more extensive use of the epoxide-diol pathway with the Δ^8-isomer. The latter finding may be related to the presence of a less hindered double bond in Δ^8-tetrahydrocannabinol and it is also known that the epoxides formed (both the $8\alpha,9\alpha$-, and $8\beta,9\beta$-isomers) are more readily hydrolyzed by epoxide hydrolase to the *trans*-8,9-diol than is the case with the corresponding epoxide from Δ^9-tetrahydrocannabinol.[33] The diol formed has the $8\beta,9\alpha$ configuration.[46]

In experiments identical to those described above using Δ^9-tetrahydrocannabinol and its 7-hydroxy derivative, Leighty et al.[35] showed that the Δ^8-isomer also formed fatty acid conjugates. These are esters derived from the 7-hydroxy metabolite and palmitic (C_{16}) and stearic (C_{18}) acids and lesser amounts of unsaturated C_{18}-acids. The microsomal enzyme system involved appears to be the same as that which forms esters of cholesterol and long-chain fatty acids.[37] An interesting conjugation reaction reported with Δ^8-tetrahydrocannabinol is the formation of a *C*-glucuronide. This C-2-derivative was found to be formed both *in vitro* using a UDP-glucuronosyltransferase preparation from rabbit liver[47] and *in vivo* in mouse liver.[48]

The metabolism of the (+)-isomer of Δ^8-tetrahydrocannabinol, which does not occur naturally, was studied in the mouse.[49] Although several qualitative and quantitative differences were noted, the metabolic patterns of the (+)- and (–)-isomers were generally similar.

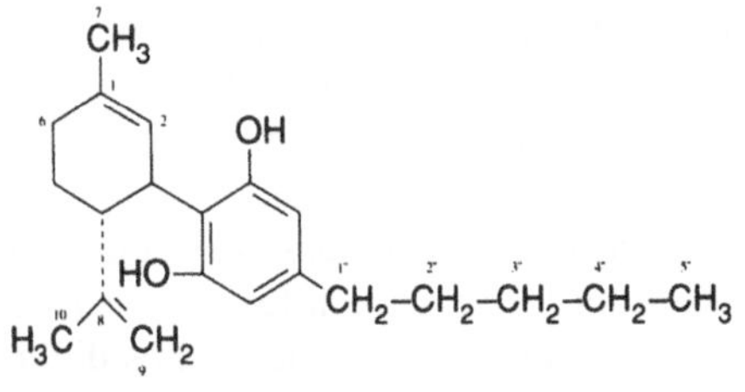

(20)

Cannabidiol

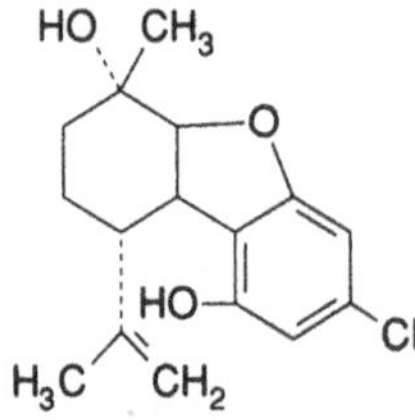

(21)

Cannabidiol (CBD) (20) is a major cannabis constituent which lacks the benzopyran structure. It is therefore commonly numbered according to the monoterpenoid system as shown. The pathways of metabolism of cannabidiol are qualitatively similar to those described above for the tetrahydrocannabinols, however qualitative and species differences have been found. The general routes of metabolism are summarized in Figure 3. The formation of hydroxylated products leads to both mono- and dihydroxy derivatives and Harvey and Paton[23] listed 17 of these. As with the tetrahydrocannabinols, exocyclic allylic hydroxylation at C-7 is a major reaction. The corresponding pathway at C-10 is less common, however a 10-hydroxy derivative was recently identified as a urinary metabolite of cannabidiol in man.[50] Endocyclic allylic hydroxylation at C-6 is a common metabolic pathway and several ketone derivatives formed via further oxidation of this group have also been identified. Further ring-hydroxylated metabolites have not been found. Additional metabolites include acidic compounds formed by oxidation of the pentyl side chain. Harvey and Paton[23] listed 14 of these derivatives with $n = 1–4$ (see Figure

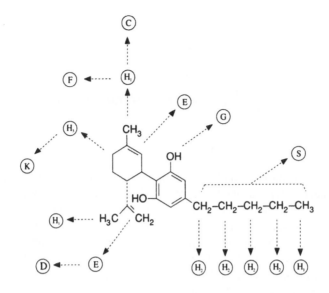

Ⓒ	= –COOH		Ⓗ₂	=	＼CHOH ／
Ⓓ	= diol		Ⓗ₃	= HO⋯⋯CH (α-isomer)	HO–CH (β-isomer)
Ⓔ	= epoxide		Ⓚ	= O=C ／＼	
Ⓕ	= fatty acid conjugates		Ⓡ	= –CH₂–CH₂–	
Ⓖ	= O-glucuronide (and glucoside)		Ⓢ	= –(CH₂)ₙ–COOH	
Ⓗ₁	= –CH₂OH				

FIGURE 3.   Metabolic pathways of cannabidiol.

3) including many with additional sites of oxidation. The mouse is especially efficient in degrading the side chain. More than 30 metabolites were identified in human urine following the administration of cannabidiol.[50]

Although cannabidiol contains endocyclic and exocyclic double bonds, the only identified metabolites formed via the epoxide-diol pathway are two 8,9-diol derivatives recently shown to be urinary metabolites of cannabidiol in man.[50] However, Yamamoto et al.[51,52] reported that cannabielsoin (21) was a metabolite of cannabidiol and that it was probably formed via cannabidiol-1,2-epoxide. This reaction sequence was carried out by hepatic microsomes from mice, rats, rabbits, and, to a greater extent, guinea pigs. Also, cannabielsoin was isolated from the liver of guinea pigs given cannabidiol by injection. There is no evidence for the cyclization of cannabidiol to Δ⁹-tetrahydrocannabinol.[53] Several glucuronide conjugates have been identified and, among these, the glucuronide of cannabidiol itself was prominent. Harvey et al.[54] concluded that glucuronidation was a major metabolic route for cannabidiol. The conjugation of cannabidiol with fatty acids has also been demonstrated.[36] Recently, Samara et al.[55] reported that three metabolites of cannabidiol, the 4″-hydroxy-, 5″-hydroxy-, and 6-oxo-derivatives, were conjugated with glucose in dogs. These were major urinary metabolites, however they were not detected in similar experiments in rats and man. Phenolic rather than aliphatic hydroxy groups appeared to be involved in this conjugation reaction.

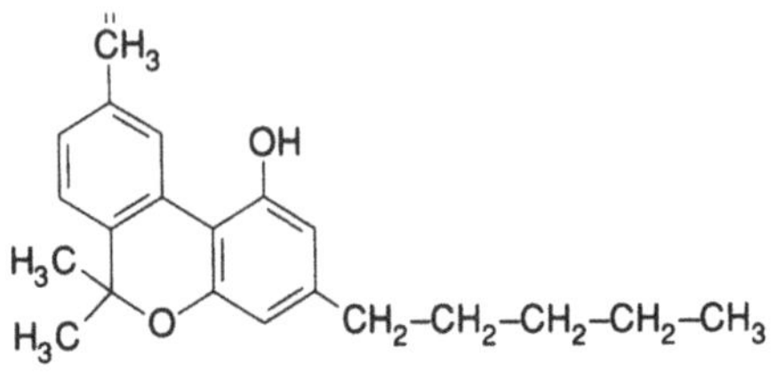

(22)

Cannabinol

The structure of **cannabinol** (22) shows that both carbocyclic rings are aromatic in nature. This fact influences its metabolic fate insofar as allylic hydroxylation is not possible. Nonetheless, the methyl group at C-11 is benzylic and thus provides a site for oxidation. The widespread formation of C-11-hydroxylated metabolites and further derivatives typical for cannabinoids is therefore also seen with cannabinol. The patterns of cannabinol metabolism generally follow those seen with other cannabinoids, however they are less diverse.[23] These reactions are summarized in Figure 4. Hydroxylation at C-11 is an important route, however metabolites showing monohydroxylation at all the sites in the side chain have also been identified. Additionally, dihydroxylated metabolites are known which combine hydroxylation at C-11 with the side-chain sites (except C-5′). Further oxidation at C-11, via the aldehyde,[56] to the 11-oic acid furnishes a major metabolite which may be accompanied by further oxidation products (hydroxy or ketone groups in the pentyl side chain). Oxidation of the side chain of cannabinol takes place to give lower homologues of carboxylic acid derivatives with $n = 0$ or 2 (see Figure 4). The formation of phenolic glucuronides occurs with cannabinol itself as well as with 7-hydroxycannabinol and cannabinol-7-oic acid. It is likely that the extensive fecal excretion of cannabinol metabolites by rats after i.v. dosage[56] may be explained by the biliary excretion of these conjugates. Wall et al.[57] similarly reported that fecal excretion of metabolites was much more extensive than urinary elimination in humans given cannabinol by i.v. injection. The conjugation of cannabinol with fatty acids has been demonstrated in an *in vitro* system[36] and several of these derivatives were identified as fecal metabolites in rats.[58] These compounds were shown to be oleic and palmitic acid esters of the 11-hydroxy, 4′-hydroxy, and 5′-hydroxy derivatives of cannabinol. Harvey et al.[59] made a comparative study of the routes of metabolism of cannabinol in mice, rats, and guinea pigs.

It was noted above that cannabis contains minor amounts of homologues in which the pentyl side chain is replaced with methyl, *n*-propyl, or *n*-butyl groups. The first two types are sometimes designated tetrahydrocannaborcol and tetrahydrocannabivarol derivatives, respectively. The *n*-propyl derivatives are especially abundant in some samples. The *in vivo* metabolism in mice of these homologues of both Δ8- and Δ9-tetrahydrocannabinol was investigated by Brown and Harvey.[60-62] The metabolism was generally similar to that found with the pentyl homologues, however the derivatives with the shortest side chains showed a greater degree of oxidation at C-11 to give the 11-ol and 11-oic acid and further hydroxylation products. On the other hand, hydroxylation at C-8 was reduced in the lower homologues. The relative importance of oxidation at C-11 was also observed with the *n*-butyl derivative, however increases were seen in the amounts of hydroxylated products at C-8 and in the side chain. Interestingly, similar studies using the synthetic homologues containing a hexyl side chain[63] or a heptyl side chain[64] showed a trend towards the formation of more hydroxylated metabolites at the expense of carboxylic acids.

The only report on the metabolism of **cannabichromene** (23) is that of Harvey et al.[65] This cannabinoid was present in the cannabis tincture they administered to mice and they found that it was metabolized to a glucuronide conjugate in the liver.

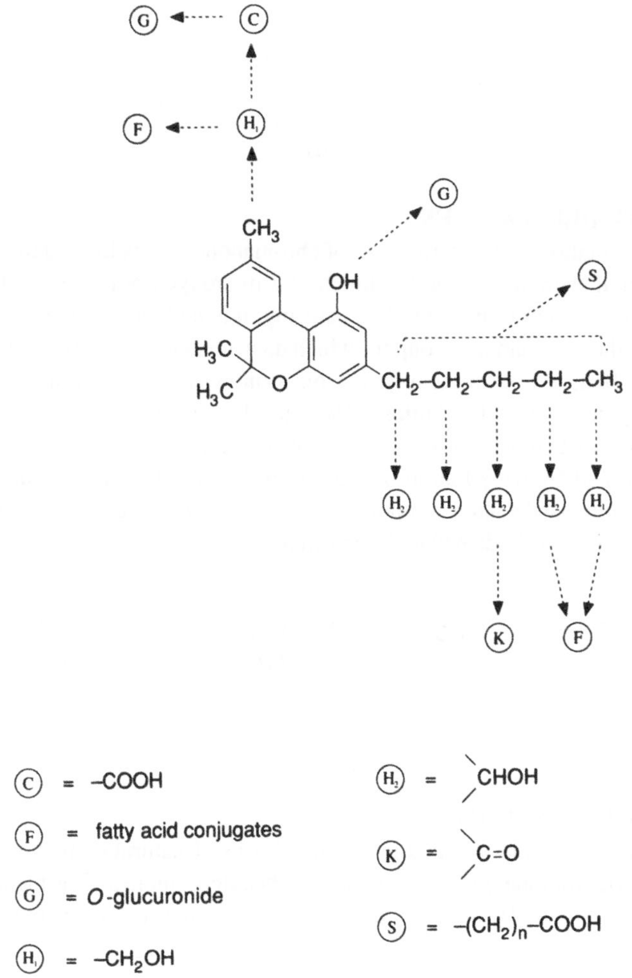

FIGURE 4.   Metabolic pathways of cannabinol.

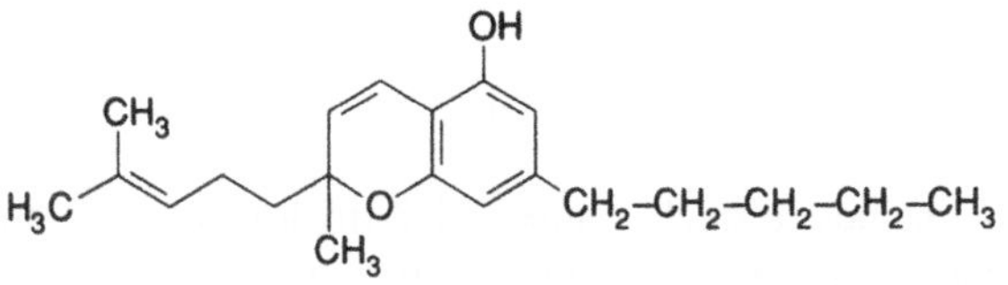

(23)
Cannabichromene

CHROMENE DERIVATIVES

Nearly all naturally occurring derivatives of chromene (36) are 2,2-dimethylchromenes, but no metabolic information is available on the simpler types. As noted in the previous section, cannabinol (31) is a substituted chromene but its metabolism is more conveniently summarized together with the other cannabinoids. An example in which the chromene group is incorporated into a more complex structure is seen with calophyllolide (70). However, this compound is also a coumarin derivative and its metabolism is therefore covered below.

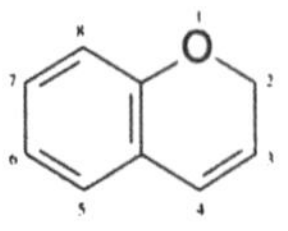

(24)

CHROMANONE DERIVATIVES

Metabolic information on the derivatives of chromanone (25) is limited to a short report by Krishnaswamy et al.[66] on the fate of 2,2-dimethyl-7-hydroxychromanone (26). Although this compound does not appear to be naturally occurring, it is included here because it is the sole representative of the chromanone group for which data are available. After oral administration (200 mg/kg) to rabbits, ~23% of the original compound was recovered in the 48-h urine, free (19.5%) or conjugated (3.5%). In addition, ~1% of the dose was excreted as a highly fluorescent metabolite which was believed to be 2,2-dimethyl-7-hydroxychromene (27). This compound was probably formed by dehydration of the corresponding 4-ol intermediate and it seems reasonable to assume that the latter compound, and not the chromene, was the actual metabolite. Dehydration of similar 4-hydroxylated derivatives has been reported with flavanone (see below).

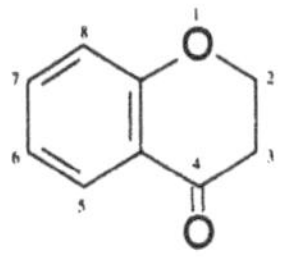

(25) 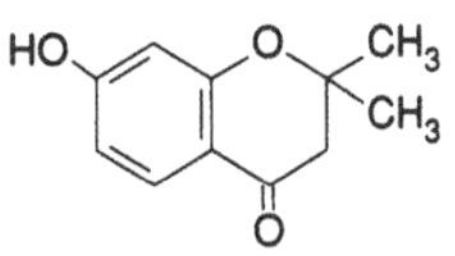   (26) 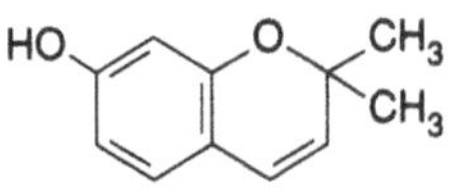   (27)

CHROMONE DERIVATIVES

The chromone (28) group is found in several classes of natural compounds including the flavones (2-phenylchromones) and isoflavones (3-phenylchromones). The metabolism of these groups is discussed below. However, other chromones are not widely encountered in nature.

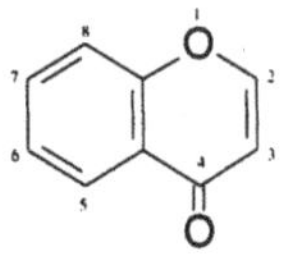

(28)

The present section includes a single compound, 3-(hydroxymethyl)-8-methoxychromone (29a). This derivative is not naturally occurring but contains both methoxy and hydroxymethyl substituents which are typical of plant chromones. A discussion of its metabolism may therefore have relevance to other compounds in this group. Crew et al.[67,68] administered ^{14}C-labeled 3-(hydroxymethyl)-8-methoxychromone (29a) orally to rats and dogs. In the former species ~75 to 80% of the radioactivity was excreted within 24 h, with 40 to 50% appearing in the urine. Two labeled compounds were used, either with the ^{14}C in the 4-position or in both the 2-position and the hydroxymethyl group at C-3. Many of the urinary metabolites identified are those to be expected from a compound containing a methoxy and a hydroxymethyl group. Thus, demethylated compound (29b) was detected as was the oxidation product (30a). In addition, some evidence was also obtained for the presence of the demethylated acid (30b). These conventional metabolites were excreted both free and as their glucuronide and/or sulfate conjugates. Also excreted in the urine were small quantities of the decarboxylated metabolite 8-methoxychromone (31a) and relatively large amounts of 8-hydroxychromone (31b), which was the major urinary metabolite in rats. Interestingly, the two acetophenone derivatives (32a) and (32b) were also excreted in the urine.

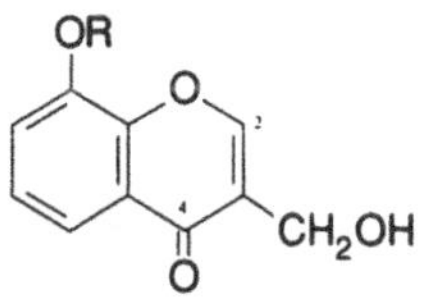

OR ... O ... CH₂OH ... O

(29)

(a) R = CH₃

(b) R = H

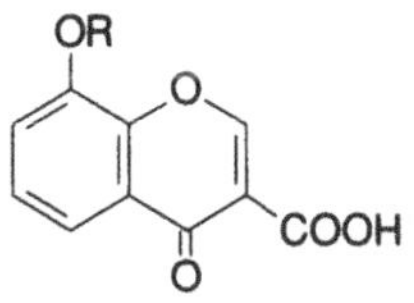

OR ... O ... COOH ... O

(30)

(a) R = CH₃

(b) R = H

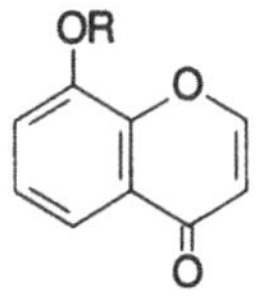

OR ... O ... O

(31)

(a) R = CH₃

(b) R = H

RO ... OH ... O ... ‖ ... –C–CH₃

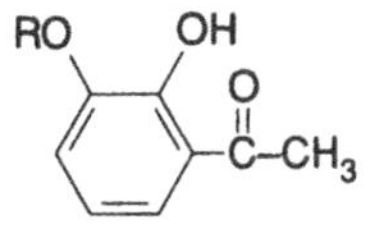

(32)

(a) R = CH₃

(b) R = H

DiCarlo et al.[69] found that plasma from rats given 3-(hydroxymethyl)-8-methoxychromone orally contained unchanged compound and metabolites (30a), (31a), and (32a) in unconjugated form. The plasma levels of these compounds were determined over a period of 48 h. The formation of metabolites (32a) and (32b) demonstrates that scission of the heterocyclic ring at the 1,2-position must have taken place. Also pertinent in this regard is the finding that administration of compound (29a) in which the ^{14}C was located in the 4-position did not lead to the formation of respiratory $^{14}CO_2$. However, ~30 to 40% of the administered radioactivity was recovered in the CO_2 when the other labeled compound was given. The finding that 8-methoxy- and 8-hydroxychromone were excreted in the urine indicates that much of the CO_2 must derive from the hydroxymethyl group. Crew et al.[68] noted that the 3-carboxy compounds were fairly readily decarboxylated. It is therefore possible that an undetermined extent of the decarboxylation may occur chemically rather than enzymatically. In fact, the presence of acetophenone derivatives as urinary metabolites also suggests that other purely chemical transformations may be involved in the overall metabolism of 3-hydroxymethyl-6-methoxychromone. Although the findings of Crew et al.[67,68] do not reveal the mechanism involved in scission of the γ-pyrone ring or the point in the sequence of metabolism at which this occurs, it is possible that a β-keto acid intermediate may be produced. If so, this highly unstable compound would be expected to be decarboxylated to an acetophenone derivative as has been found with several closely related aromatic compounds metabolized via β-keto acids.[70,71]

COUMARINS

COUMARIN

Coumarin (33) (see Figure 2) was formerly used in foods and drugs for its flavoring properties, but this practice has been discontinued because of the liver toxicity produced by the compound in some animal species. The toxicology of coumarin, with special reference to its metabolism and species differences, was reviewed by Cohen.[72] Coumarin is not a typical representative of this class of compounds insofar as it lacks an oxygen atom at the 7-position, a characteristic of nearly all of the naturally occurring coumarins.

The metabolism of coumarin, while being neither overly complex nor unexpected, is nonetheless extensive and leads to a large number of metabolites. These fall into two main

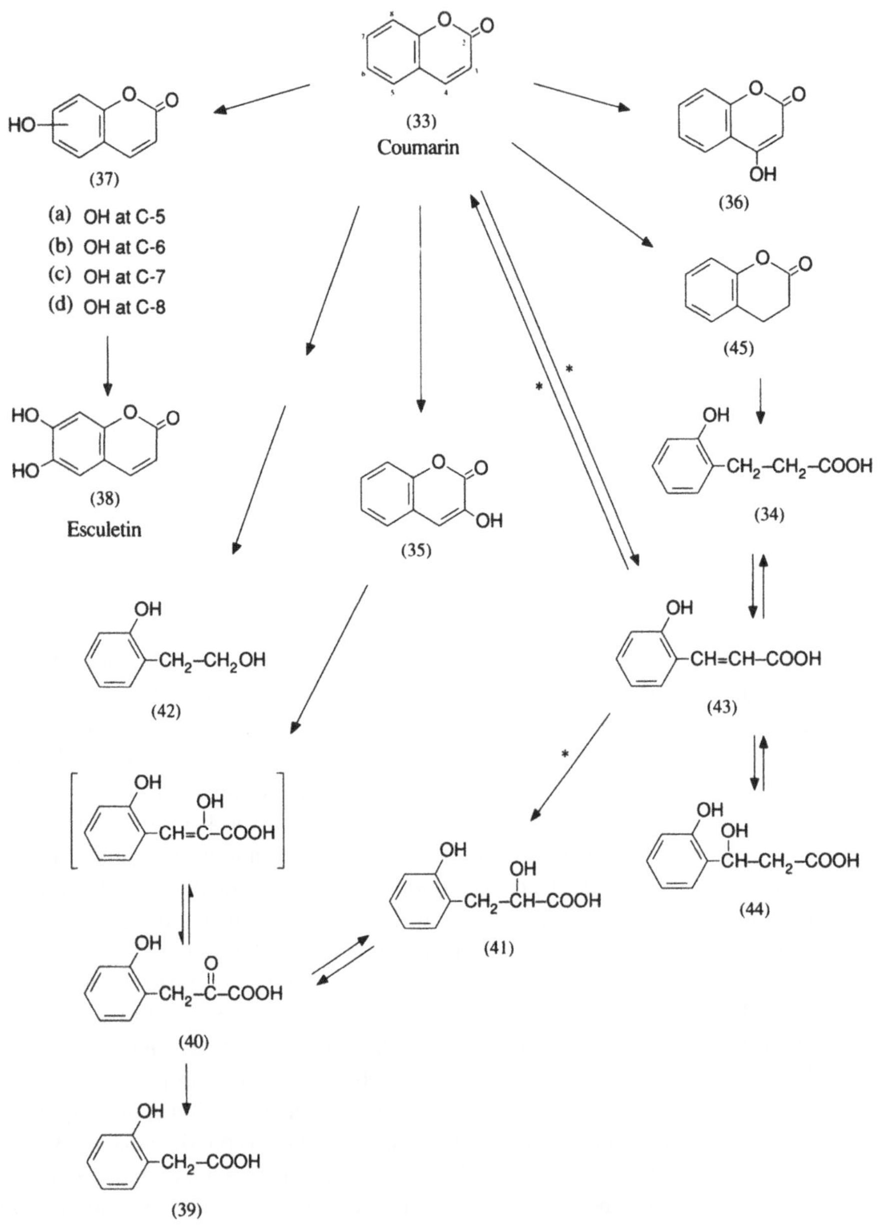

FIGURE 5.   Metabolic pathways of coumarin. Postulated intermediate shown in brackets. (*) See text.

groups: (1) hydroxylated products and (2) products in which the α-pyrone ring has been cleaved. However, formation of these scission products is sometimes dependent upon initial ring hydroxylation. The metabolic pathways shown in Figure 5 illustrate both the hitherto detected metabolites of coumarin and the probable pathways by which the metabolites are formed. In addition, most of these metabolites may undergo conjugation reactions. These furnish glucuronide, sulfate, or glycine conjugates and note of these is made in the following text. It is important to note that certain metabolites have not been consistently detected, even in experiments using the same animal species. A good example of the variability in metabolite formation is that seen with melilotic acid (34) in rats.[73-77] Furthermore, considerable species variations in metabolism have been noted and these will be discussed below.

The proposed scheme of coumarin metabolism shown in Figure 5 is based upon the identification of urinary metabolites in *in vivo* studies reported by Furuya,[78] Mead et al.,[76] Booth et al.,[73] Kaighen and Williams,[75] Feuer et al.,[79] Scheline,[77] Pekker and Schäfer,[80] Shilling et al.,[81] Van Sumere and Teuchy,[82] Feuer,[74] and Norman and Wood.[83] However, several studies using microsomal preparations have duplicated many of these findings.[84-88] Figure 5 shows that hydroxylation occurs at all possible positions in the coumarin molecule (compounds (35), (36), and (37a)—(37d)). This was shown in the study of Kaighen and Williams[75] using rabbits. However, several of the hydroxycoumarins are formed in small amounts and have not been detected in all of the studies. This is especially true of the 5- and 6-hydroxy derivatives which are never found in very large amounts in the urine of coumarin-treated animals. Also, polyhydroxylated compounds are seldom encountered although Pekker and Schäfer[80] detected 6,7-dihydroxycoumarin (esculetin) (38) as a minor coumarin metabolite in rabbits. The 4- and 8-hydroxycoumarins are somewhat more abundant and may account for a few percent of the dose. The major hydroxylated metabolites are the 3- and 7- derivatives. As noted below, the relative proportions of these can differ greatly.

As shown in Figure 5, 3-hydroxylation gives rise not only to the hydroxylated compound itself but also to several phenolic acids resulting from ring scission. The most characteristic of these is *o*-hydroxyphenylacetic acid (39) which is often a prominent coumarin metabolite. It is generally assumed that this C_6-C_2-acid is formed via 3-hydroxycoumarin (35). Kaighen and Williams[75] showed that when the latter compound was fed to rabbits it was mainly excreted as its glucuronide conjugate, however it was also metabolized to *o*-hydroxyphenylpyruvic (40), *o*-hydroxyphenyllactic (41), and *o*-hydroxyphenylacetic (39) acids. They found that this pathway was more prominent in rats in which case the main urinary metabolite of (35) was (39). Gibbs et al.[88] studied the metabolism of coumarin by rat liver microsomes and found that the 3- and 7-hydroxy derivatives were formed in addition to *o*-hydroxyphenyllactic acid (41) and *o*-hydroxyphenylacetic acid (39). Walters et al.[89] similarly found that rat liver microsomes converted coumarin to several hydroxycoumarins and to *o*-hydroxyphenylacetic acid. However, when 3-hydroxycoumarin was similarly incubated, no metabolism to any of the aforementioned metabolites was found.[88] A similar negative result was reported by Norman and Wood[83] using a 10,000-g supernatant fraction from rat liver. They concluded that the ring-opened products may therefore be formed via an intermediate other than 3-hydroxycoumarin. On the other hand, Kerékjártó[87] found that 3-hydroxycoumarin was converted to *o*-hydroxyphenylpyruvic (40) when incubated with liver microsomes. This subject clearly requires further study. A novel finding of Norman and Wood[83] was the identification of *o*-hydroxyphenylethanol (42) as a metabolite of coumarin in rats. Following an i.p. injection of coumarin, the urine (24 h) contained ~15% as much metabolite (42) as of *o*-hydroxyphenylacetic acid, the major urinary metabolite. Dosing with (42) resulted in the excretion of large amounts of the acidic derivative (39), however no metabolic reduction of the latter to (42) was found. The *in vitro* formation of *o*-hydroxyphenylethanol was also studied and it was found that this required incubation with the 10,000-g fraction of rat liver. No product was formed when only microsomes or only the soluble fraction was used. Also, no *o*-hydroxyphenylethanol was formed from 3-hydroxycoumarin when the 10,000-g fraction was used.

The mechanisms involved in the formation of *o*-coumaric acid (43) and *o*-hydroxyphenylhydracrylic acid (44) are also unclear. In addition, it must be noted that not all investigations have detected compound (43) as a coumarin metabolite. Mead et al.[76] and Kaighen and Williams[75] reported that it was absent from the urine of coumarin-treated rabbits, although Booth et al.,[73] Furuya,[78] and Scheline[77] reported that it was excreted by both rabbits and rats. It is perhaps significant that in those experiments in which *o*-coumaric acid was detected, its reduction product melilotic acid (34) was also reported. If metabolites (34) and (43) do appear concomitantly, the formation of the latter may be solely via the intermediate 3,4-dihydrocoumarin (45) and no direct formation from coumarin itself need be proposed. Feuer[74] presented results which showed that whereas only traces of melilotic acid were detected in the urine of rats given coumarin, nearly half of the dose was excreted as this metabolite when 3,4-dihydrocoumarin was given. A further point of interest regarding the pathway via metabolite (45) is that this compound appears to be formed by the intestinal bacteria and not by the tissue enzymes.[77] Thus, the occasional absence of (34) and (43) as urinary metabolites of coumarin may be explained by variations in the metabolic activities of the intestinal microflora of the animals involved. Some further support for this proposal is available in the results of Fink and Kerékjártó[86] who found that liver microsomes formed all of the possible hydroxy derivatives from coumarin as well as the acidic metabolites (41) and (39). However, compounds (34), (43), and (44) were not detected.

Another point of interest regarding the pathways shown in Figure 5 is that the transformation between metabolites (41) and (43) does not appear to be reversible. Whereas administration of (34) or (43) to animals resulted in the urinary excretion of the phenylacetic acid derivative (39), only the latter metabolite was excreted when compound (41) was given.[73] The phenolic acids formed from coumarin are excreted partly free, partly as glucuronide conjugates and, in the case of coumaric (43) and melilotic (34) acids, also as glycine conjugates.[73,76,78]

Coumarin contains an αβ-unsaturated group in common with many compounds of this type which are known to be conjugated with glutathione. However, Boyland and Chasseaud[90] found that this glutathione *S*-transferase activity from rat liver did not utilize coumarin as a substrate. On the other hand, Lake et al.[91] reported that single doses (125 mg/kg, i.p.) of coumarin reduced the levels of hepatic glutathione in rats. They suggested that a reactive metabolite of coumarin, possibly coumarin-3,4-epoxide, and not coumarin itself forms a glutathione conjugate.

Considerable species variations in the metabolism of coumarin have been observed. Quantitative data by Kaighen and Williams[75] and by Feuer[74] show that the excretion of hydroxycoumarins in rats is relatively low (~5 to 10% of the dose) compared with 55 to 60% in rabbits,[75] ~60 to 65% in the baboon,[92,93] and 80% in man.[81] In the last two species, ring hydroxylation takes place largely at the 7-position. Ritschel et al.[94] found that although the absorption of coumarin in man was rapid, little was available systemically due to an extensive first-pass effect. Thus, the bulk of the material present in the systemic circulation was 7-hydroxycoumarin and, mainly, its glucuronide conjugate. Moran et al.[95] found that an oral dose of coumarin was rapidly converted by human subjects to 7-hydroxycoumarin and its glucuronide. About 63% of the dose was recovered in the urine in these forms (mainly as glucuronide) in 24 h. Ritschel and Hardt[96] reported that the formation of 7-hydroxycoumarin in the gerbil is similar to that seen in humans. The 7-hydroxylation of coumarin in mice is of interest because of the pronounced variations observed in different strains. This has been demonstrated both using hepatic microsomes[97,98] and *in vivo*.[99] The amount of umbelliferone (7-hydroxycoumarin) excreted in the urine of the 19 strains of mice tested varied by a factor of ~8 between the low and high metabolizers. Wood and Taylor[100] investigated the genetic regulation of this hepatic coumarin 7-hydroxylase activity in mice.

The pathway via 3-hydroxycoumarin leading to acidic metabolites is of minor importance in man and accounts for only ~6% of the dose.[81] The same is true in the baboon which excreted in the urine less than 1% of the dose as *o*-hydroxyphenylacetic acid.[93] In rats, differing results have been reported for acidic metabolites, ranging from ~20%[75] to nearly 70%.[74] Feuer[101] showed that the coumarin 3-hydroxylase activity in rat liver microsomes was greater than that of coumarin 7-hydroxylase by about a factor of 300. In rabbits, the corresponding value for the acidic metabolites ((41) and (39)) is 20 to 25%.[75] Another species difference shown by coumarin is the high degree of fecal excretion of metabolites seen in the rat only.[75] Similarly, Piller[102] found that about a third of the radioactivity was excreted in the feces of rats given [3-^{14}C]coumarin by injection. Also, loss of radioactivity from the body was relatively slow, with 20 and 7% of the dose remaining after 24 h and 100 h, respectively. The results of Kaighen and Williams[75] indicated that the fecal metabolites included compound (39) and unidentified compounds but little unchanged compound. Their presence in the feces is no doubt explained by the greater facility of the rat in excreting compounds in the bile.

COUMARINS CONTAINING HYDROXY AND/OR METHOXY GROUPS

As noted above, the true parent of most of the naturally occurring coumarins is 7-hydroxycoumarin (umbelliferone) (46). This section will therefore deal mainly with coumarin derivatives containing an oxygen function in the 7-position although, for comparative purposes, the metabolism of the other isomeric monohydroxylated compounds will be briefly reviewed.

The most comprehensive study of the metabolism of the hydroxycoumarins was carried out by Mead et al.[103] who administered the compounds orally (200 mg/kg) to rabbits. After giving the 3-, 4-, 5-, 6-, and 8-hydroxycoumarins the 24 h urines were found to contain 40 to 75% of the dose as the corresponding conjugates, although in several cases the values obtained may have been low due to incomplete hydrolysis of the glucuronide conjugates. The compounds administered were excreted mainly as their glucuronides but ~10 to 30% of the dose was accounted for as ethereal sulfates, except with 4-hydroxycoumarin which did not form a sulfate. This was considered likely to be due to the higher acidity of the 4-hydroxy group compared with that of the other isomers. An early study of the metabolism of 3-hydroxycoumarin in man[104] showed that it was excreted nearly entirely as the corresponding glucuronide. More recently, Kaighen and Williams[75] found that 3-hydroxycoumarin (~250 mg/kg, p.o.) was excreted in rabbits not only as the corresponding glucuronide and sulfate but also as metabolites formed by the scission of the heterocyclic ring. These products were shown to be o-hydroxyphenylpyruvic (40), o-hydroxyphenyllactic (41), and o-hydroxyphenylacetic (39) acids (see Figure 5).

4-Hydroxycoumarin (84 mg/kg) given i.v. to dogs was excreted in the urine within 24 h as unchanged compound (~50%) and as the corresponding glucuronide (25%).[105] The fate of the remaining 25% was not determined although it was shown that the urine did not contain increased amounts of steam-distillable phenols or ethereal sulfates and that no salicylic acid was formed. Hydroxylation of the aforementioned monohydroxy coumarins does not appear to take place readily and this reaction has been reported only with 6-hydroxycoumarin which is partly converted to 6,7-dihydroxycoumarin (esculetin) (38) (see Figure 5) in rabbits.[103]

The metabolism of **umbelliferone** (46) has been the subject of many metabolic investigations. Sieburg,[106] using rabbits, found that umbelliferone (300 to 500 mg/kg, i.p.) was excreted in the urine as conjugated material which formed the original compound upon hydrolysis. Similar experiments by Mead et al.,[107] employing a dose of 200 mg/kg, confirmed this and showed that the conjugates consisted of ethereal sulfate (~20%) and twice as much or more of umbelliferone glucuronide. These findings were further substantiated by Fujita and Furuya[108] in similar experiments. They found that excretion was complete within 48 h and also that ~20% of the dose was excreted unchanged. The formation of the glucuronide and sulfate conjugates of umbelliferone has also been shown to occur in isolated epithelial cells from the rat small intestine.[109] Dawson and Bridges[110] similarly demonstrated these conjugation reactions using intact intestinal preparations from rats and guinea pigs. They found that the highest activities were located in the upper part of the small intestine and that the conjugates formed in the epithelium could be released into the intestinal lumen. Hepatic and extrahepatic (kidney, small intestine, and lung) conjugation of umbelliferone with sulfate was studied using isolated cells.[111]

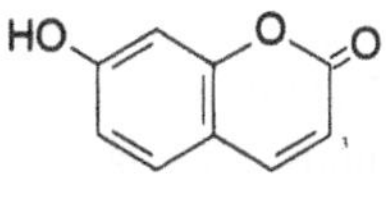

(46)
Umbelliferone

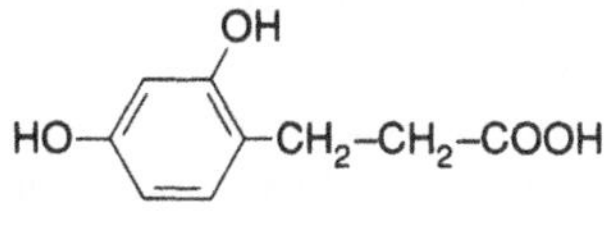

(47)

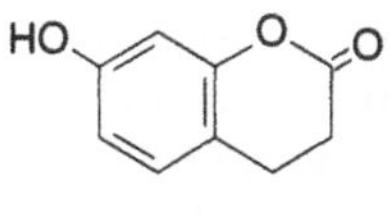

(48)

None of the aforementioned investigations using rabbits demonstrated the occurrence of biological hydroxylation or ring cleavage of umbelliferone. However, Indahl and Scheline[112] found that both of these reactions took place when the compound (100 mg/kg) was administered to rats. In addition to confirming the excretion of both free and conjugated umbelliferone, this study showed that ring scission to give 2,4-dihydroxyphenylpropionic acid (47) occurred and that, after i.p. but not oral dosage, a small amount of 3,7-dihydroxycoumarin was formed. Legrum and Netter[113] found that this dihydroxycoumarin, together with the 6,7- and 7,8-isomers, were formed when umbelliferone was incubated with hepatic microsomes from mice which had been pretreated with cobalt chloride in order to increase the activity of the mixed-function oxidases. The formation of 6,7-dihydroxycoumarin (esculetin) from umbelliferone during incubation with liver microsomes was also reported by Kerékjártó.[87] Further evidence of hydroxylation of the umbelliferone structure in rats was obtained with its homologue, 4-methylumbelliferone.[114] In this study, the major urinary metabolites, both as free compounds and glucuronides, were found to be the 3,5,7-trihydroxy-, 6-hydroxy-7-methoxy-, and 7-hydroxy-6-methoxy-derivatives of 4-methylcoumarin.

Ring cleavage of umbelliferone to the phenylpropionic acid derivative (47) was found to be carried out by the intestinal microflora,[112] a result which supported the earlier conclusion noted above with coumarin itself.[77] Again, formation of the C_6–C_3-acid appears to be due to reduction of the coumarin to the corresponding 3,4-dihydro compound (48) which is easily hydrolyzed to compound (47). However, there was no indication of ring cleavage of umbelliferone leading to C_6–C_2-acids as described above with coumarin. Another investigation of umbelliferone metabolism in the rat[82] showed that it was excreted in the urine mainly free and in conjugated form. These metabolites accounted for about half of the dose (~45 mg/rat, i.p.) and were excreted within 24 h. The only other urinary metabolite detected was 2,4-dihydroxybenzoic acid (β-resorcylic acid) which may arise via β-oxidation of 2,4-dihydroxyphenylpropionic acid formed as a result of the opening of the lactone ring. The formation of a C_6–C_1-phenolic acid has not otherwise been reported in studies of the metabolism of coumarin and its derivatives.

The metabolism of the methyl ether of umbelliferone, **herniarin** (49) (see Figure 6), was investigated in rabbits[115] and rats.[112] In addition, Legrum and Frahseck[116] specifically studied its oxidative demethylation in mice. Using material labeled with ^{14}C in the methoxy group, they found that 16% of the dose (~3 mg/kg, i.p.) was lost in 60 min as respiratory $^{14}CO_2$. In the study using rabbits only ~30% of the dose (200 mg/kg) could be accounted for, chiefly as umbelliferone (46) which was excreted free and conjugated in a ratio of ~2:1. A small amount of herniarin itself was excreted. As with umbelliferone, no hydroxylation or ring cleavage of herniarin was noted in the rabbit. In contrast, the pathways of its metabolism in the rat are more extensive and are shown in Figure 6. The major metabolite of herniarin was 2-hydroxy-4-methoxyphenylacetic acid (50) which, in a manner analogous to that described above for the formation of o-hydroxyphenylacetic acid from coumarin, most likely arises via 3-hydroxylation of herniarin as shown in Figure 6. A lesser amount of hydroxylation occurred at the 6-position to give metabolite (51) and the results also suggested that trace amounts of the 8-hydroxy derivative (not shown) were also excreted. As with umbelliferone, herniarin was metabolized by the intestinal bacteria to C_6–C_3-phenolic acids (compounds (52) and (47)). These metabolites were not excreted in the urine of germ-free rats given herniarin. Both (52) and (47) were readily formed when herniarin was incubated with rat cecal microorganisms *in vitro*. Matsubara et al.[117,118] carried out extensive studies on the demethylation of herniarin to umbelliferone (46) using liver microsomes and homogenates. Using rat liver, these systems also formed additional metabolites including umbelliferone glucuronide and three unidentified compounds. The latter were probably two hydroxylated derivatives of herniarin and one hydroxylated derivative of umbelliferone. Comparative studies of herniarin demethylation using mouse, guinea pig, rabbit, dog, and Cynomolgus monkey preparations were also carried out. Ullrich and Kremers[119] noted the influence of various inducers (phenobarbital and polycyclic aromatic hydrocarbons) of hepatic

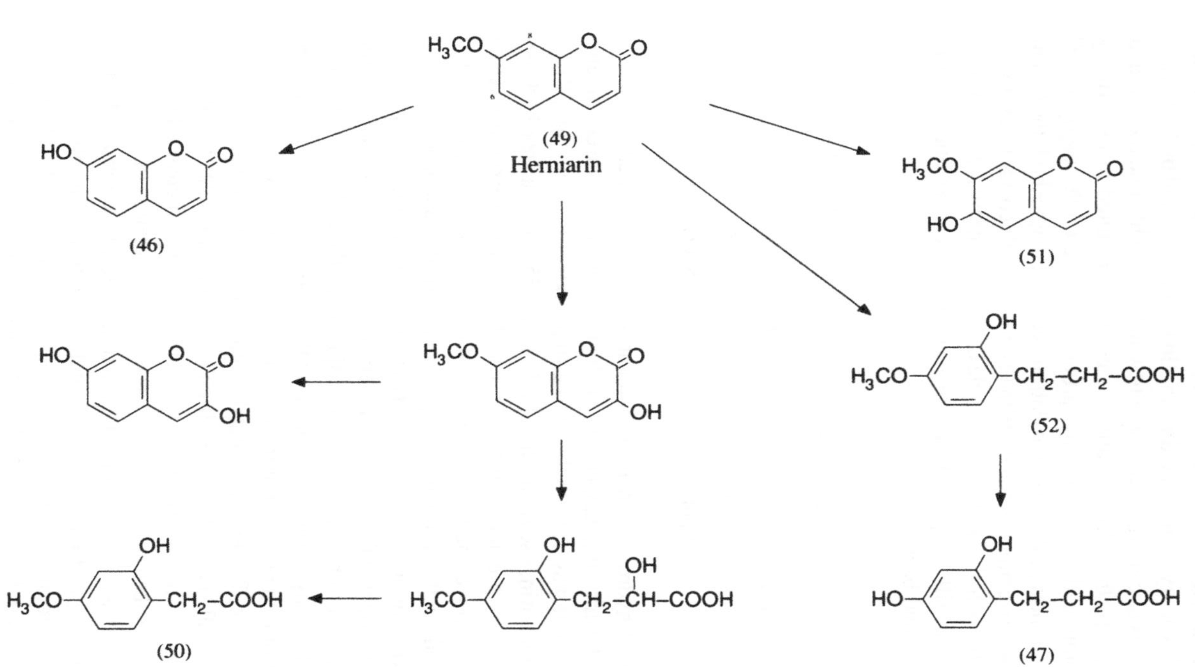

FIGURE 6.   Metabolic pathways of herniarin in rats.

microsomal activity on the relative extents of *O*-demethylation or 6-hydroxylation of herniarin.

Armillarisin A is 3-acetyl-5-hydroxymethylumbelliferone and Shao et al.[120] studied the biological disposition of the ^{14}C-labeled compound in rats. Absorption and excretion were rapid and 89 and 8% of the radioactivity was found in the urine and feces, respectively, in 48 h. Four metabolites were detected, however their identities were not reported.

The remaining plant coumarins in this group that have been studied metabolically are dihydroxy compounds and their derivatives. **Esculetin** (6,7-dihydroxycoumarin) (38) (see Figure 5) and the isomeric 7,8-dihydroxy compound **daphnetin** (53) were found by Sieburg[106] to be excreted in the urine of rabbits as conjugation products which would liberate the original compounds upon hydrolysis. In rats, the administration of esculetin resulted in the urinary excretion of some unchanged compound[121] as well as its sulfate and glucuronide conjugates.[122] In addition, the latter investigation showed that *O*-methylation occurred to give both the 6- and 7-methyl ethers. Similar results were obtained when **esculin** (esculetin-6-glucoside) was given, a finding undoubtedly related to the hydrolysis of the glycoside in the intestine by the microflora. Hawksworth et al.[123] and Drasar and Hill[124] reported that this reaction was carried out by nearly all of the common groups of intestinal bacteria.

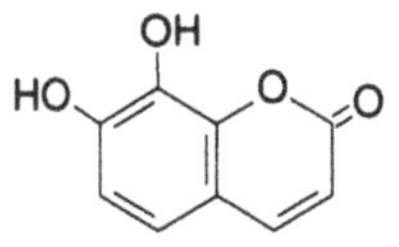

(53)
Daphnetin

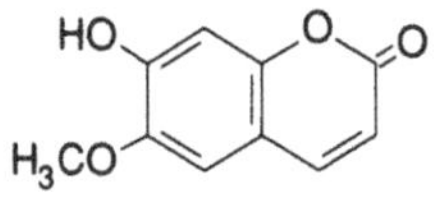

(54)
Scopoletin

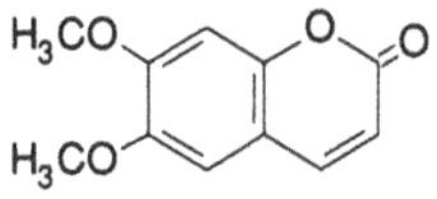

(55)
Scoparone

The metabolism of **scopoletin** (7-hydroxy-6-methoxycoumarin) (54) was studied in rats[122,125] and rabbits.[126] It was excreted in the urine unchanged and as glucuronide and sulfate conjugates in both animal species. In addition, it was demethylated to a small extent to esculetin (38).

Scoparone (6,7-dimethoxycoumarin, esculetin dimethyl ether) (55) lacks a free hydroxy group and is of necessity subjected to a greater degree of demethylation prior to excretion. Both monodemethylated isomers were formed and then excreted in the urine of rabbits given an oral dose of 200 mg/kg.[126] However, demethylation at the 6-position (forming isoscopoletin) was found to be about five times as extensive as that occurring at the 7-position (forming scopoletin). Most of the recovered dose was accounted for as the monodemethylated metabolites, both free and conjugated, but ~3% was shown to be excreted as a conjugate of the fully demethylated esculetin (38). The preferential demethylation of scoparone at the 6-position was also reported by Müller-Enoch et al.[127] who employed rat liver microsomes. Using a substrate labeled with ^{14}C in the methoxy group at either C-6 or C-7, they found that the ratio of isoscopoletin to scopoletin formed was 1.8:1. Legrum et al.[128] also used these specifically labeled substrates to study the demethylation of scoparone in mice. An i.p. dose of only 0.24 mg/kg, which ensured that the enzyme system was not in a state of substrate saturation, was used and excretion of [14]CO$_2$ in the expired air was essentially finished within 60 min. The maximal rate of excretion of [14]CO$_2$ occurred after 4 min and was found to be four times higher with [6-*methyl*-^{14}C]scoparone than with the 7-methyl radioisomer. The total excretion (60 min) of [14]CO$_2$ was ~27% for the former compound and only 11% for the latter. Contrariwise, the amounts of radioactivity in the urine or in the bile and gastrointestinal tract after 60 min were significantly higher with [7-*methyl*-^{14}C]scoparone. Scoparone metabolism by rat hepatocytes similarly showed preferential 6-*O*-demethylation; however, the formation of isoscopoletin and scopoletin were nearly equal when the hepatocytes were obtained from rats pretreated with phenobarbital.[129]

COUMARINS DERIVATIVES CONTAINING MORE COMPLEX SUBSTITUENTS

Compounds in this group may contain substituents attached to either the carbocyclic or heterocyclic rings. An important example of the former type is the furocoumarins (psoralens). Little is known of the metabolism of **psoralen** (56) itself, however Pathak et al.[130,131] reported that it was rapidly excreted in the urine following oral or i.p. administration to mice. Using ^{3}H- or ^{14}C-labeled material, they found that >90% of the radioactivity was excreted in 12 h. Similar extensive and rapid urinary excretion of psoralen was found in man. Some information on the metabolism of psoralen was obtained which suggested that it underwent hydroxylation, glucuronidation, and scission of the lactone ring.

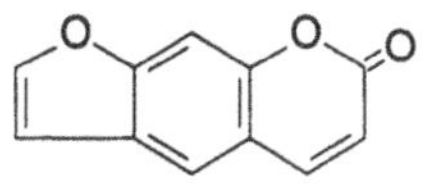

(56)

Psoralen

The most widely studied psoralen derivative is **8-methoxypsoralen** (57) (see Figure 7). Investigations have been carried out using mice, rats, and dogs, however most of the data on its biological disposition derive from studies in man. 8-Methoxypsoralen is rapidly absorbed from the gastrointestinal tract after oral dosage and its metabolism and excretion are also rapid.[132-135] The first of these reports indicated that ~80% of the radioactivity (^{3}H-labeled material) was excreted in the urine and the remainder in the feces. However, different values may be obtained depending on the physical form of the preparation administered. Using 8-*methoxy*-^{14}C-labeled 8-methoxypsoralen, Busch et al.[133] found that ~75% of the radioactivity was excreted in the urine when subjects were given an oral dose (40 mg) in solution, whereas only slightly more than 30% was excreted in the urine when coarse crystalline material was given. This oral dose in solution resulted in an additional excretion of ~14% in the feces.[135] These studies also showed that virtually no unchanged compound was excreted. The latter finding was confirmed by Ehrsson et al.[136] who found <0.1% of the dose (45 mg) as unchanged compound in the 24 h urine. However, ~2% of the dose was liberated as 8-methoxypsoralen following hydrolysis of the urine with a β-glucuronidase + sulfatase preparation. The latter finding is probably due to the hydrolysis of conjugates of the ring-opened form of 8-methoxypsoralen. The liberated hydroxy acid may then lactonize to give the original compound. This phenomenon was also reported by Busch et al.[133] with samples of rat bile which showed a fivefold increase in the amount of 8-methoxypsoralen present following hydrolysis. Negligible (~1%) excretion of unchanged 8-methoxypsoralen in the urine and feces was also recorded in mice given the compound orally (250 mg/kg).[137] Muni et al.[138] gave ^{14}C-labeled 8-methoxypsoralen (6 mg/kg) orally to hairless mice and found that 55% of the radioactivity was excreted in the urine in 24 h. A further 30% was found in the feces which was believed to be due to biliary excretion of material. The latter phenomenon has been shown to occur in rats.[133-140] The last two studies reported values of 30 to 35% of the dose (0.5 or 6 mg/kg of ^{3}H-labeled compound) for the biliary excretion. Nozu et al.[139] stated that the absorption and excretion patterns of 8-methoxypsoralen were similar in mice and rats. They found that 65 and 22% of the radioactivity were excreted by rats in the urine and feces, respectively, and that most of this excretion occurred within 24 h. Busch et al.[133] noted that dogs excreted more of the ^{14}C in the feces (40%) than in the urine (30%) following an oral dose (0.5 mg/kg) of 8-methoxypsoralen. Extensive fecal excretion of metabolites in dogs was also reported by Kolis et al.[141] following i.v. dosage (5 mg/kg) of ^{14}C-labeled compound. The values were 45% (urine) and 40% (feces) in 3 d and point to an extensive biliary excretion of metabolites also in this species.

The identities of the metabolites of 8-methoxypsoralen have been determined in rats,[142,143] dogs,[141] and man.[135] These investigations indicated that the pathways of metabolism in the three species are qualitatively similar. These pathways are shown in Figure 7 which indicates that 8-methoxypsoralen undergoes *O*-demethylation, hydroxylation, hydrolysis of the lactone ring,

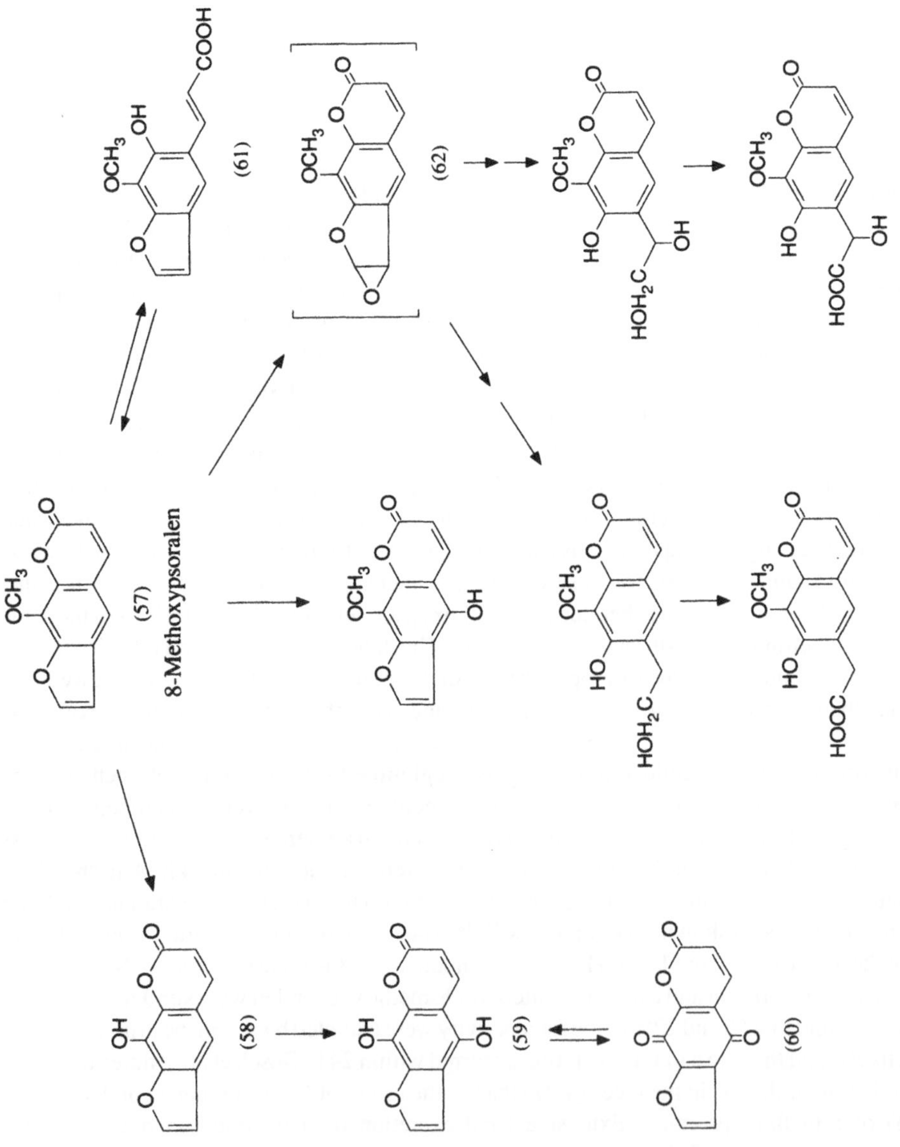

FIGURE 7.   Metabolic pathways of 8-methoxypsoralen. Postulated intermediate shown in brackets.

and oxidative cleavage of the furan ring. In addition, many of the metabolites shown are excreted as conjugates including glucuronides, sulfates, and other types.[141] The excretion of the demethylated metabolite (58) is not extensive in any of the species studied and accounted for only ~1 to 2% of the dose. A value of 1% was recorded in man by Ehrsson et al.[136] Nevertheless, the reaction of O-demethylation is probably much more extensive than this. Busch et al.[133] reported that ~20% of the radioactivity was lost in the respiratory air of rats when 8-*methoxy*-^{14}C-labeled compound was administered. This corresponds closely to the value of 21% reported by Mays et al.[143] for the urinary excretion in rats of the demethylated metabolites (58), (59), and (60). Metabolite (59) was the most abundant of these compounds and it was excreted mainly as a sulfate conjugate. Scission of the lactone ring to give metabolite (61) was fairly extensive in the dog but not in man, in which case oxidation of the furan ring to give metabolites which arise via the postulated intermediate (62) was most important.

Relatively little information is available on the *in vitro* metabolism of 8-methoxypsoralen. Mandula and Pathak[144] found minimal metabolism by mouse liver fractions or by the 10,000-g supernatant from guinea pig liver. Only in the case of incubates using mouse liver microsomes were small amounts of two metabolites formed which retained the fluorescent properties of the starting material. However, Mays et al.[143] reported that metabolites were formed when ^{14}C-labeled 8-methoxypsoralen was incubated with the 9000-g supernatant or microsomes from rat liver. They noted that some of the radioactivity became unextractable from the incubates and they concluded that this was caused by covalent binding of metabolite to microsomal protein.

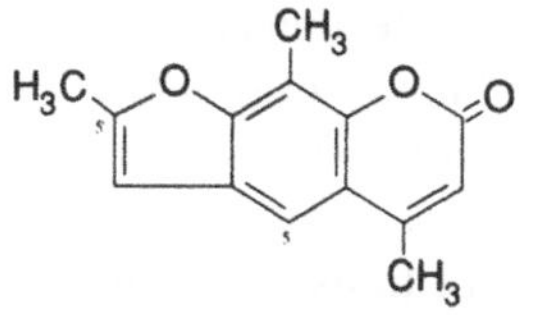

(63)   (64)

4,5',8-Trimethylpsoralen

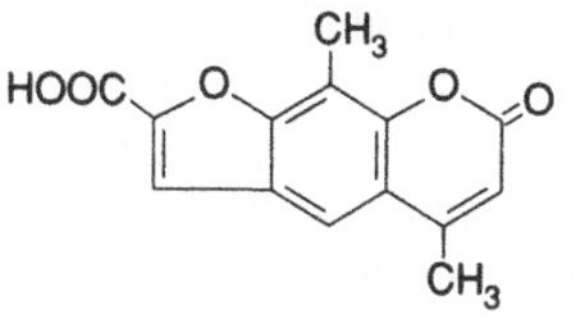

Another psoralen derivative is **4,5',8-trimethylpsoralen** (63) which Pathak et al.[130,131] found to be rapidly excreted in the urine of mice following oral or i.p. administration. Rapid urinary excretion was also noted when human subjects were given the compound orally. As with psoralen itself, they found that the trimethyl derivative was metabolized by hydroxylation and glucuronidation. Additionally, they detected a carboxy derivative which they suggested might have arisen by opening of the lactone ring. Further study of the nature of this acidic metabolite revealed, however, that it was not formed via scission of the heterocyclic ring, but by oxidation of the methyl group at C-5' to give compound (64).[145] Using ^{3}H-labeled compound, they determined that about half of the urinary radioactivity was in the form of metabolite (64). The same metabolite was found in the urine of human subjects given 4,5',8-trimethylpsoralen. An additional minor metabolite present in mouse urine was a hydroxylated derivative of (64). Mandula and Pathak[144] reported on further studies of the nature of these metabolites and found that an intermediate, the 5'-hydroxymethyl derivative formed in the sequence –CH$_3$ $\rightarrow$ –CH$_2$OH $\rightarrow$ –COOH, was present in incubates using mouse liver homogenates. This intermediate could not be detected in urine samples. In addition, a third metabolite was detected which appeared to be a hydroxylated derivative of the hydroxymethyl intermediate. It was proposed that this compound could be the precursor of the hydroxylated derivative of (64) found in the earlier study to be present in mouse urine. The alternative pathway of hydroxylation of (64) itself seems unlikely in view of its lack of further metabolism when incubated with mouse liver homogenates or 10,000-g supernatant fractions. The position of the hydroxy group was not determined, however it is tempting to speculate that it may be in the 5-position in accord with the corresponding hydroxylation of 8-methoxypsoralen noted above. Additional experiments showed that the 10,000-g fraction from guinea pig liver similarly metabolized 4,5',8-trimethylpsoralen, whereas no detectable metabolism occurred when fractions from mouse kidney or guinea pig epidermis were used.

As noted above, some coumarin derivatives contain more complex substituents attached to the heterocyclic (pyrone) ring. 3-Phenylcoumarins are uncommon natural products and metabolic data are available only on some of the simpler synthetic analogues.[66] 3-Phenyl-7-hydroxycoumarin (3-phenylumbelliferone) (65) was completely excreted within 48 h following oral dosage (200 mg/kg) to rabbits. About 20% was excreted unchanged while most of the dose (~70%) was found as the corresponding glucuronide and sulfate conjugates. With 3-phenylherniarin, the methyl ether of (65), only ~30% of the dose was accounted for, nearly entirely as 3-phenylumbelliferone. In this case the ratio of free to conjugated compound was 2:1. Furthermore, a trace (0.2%) of the administered 3-phenylherniarin was detected in the urine.

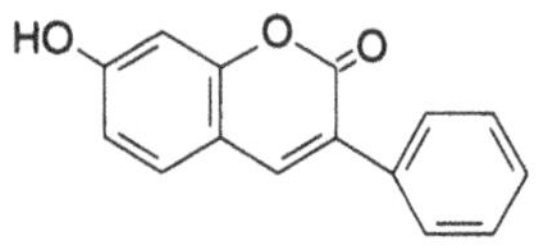

(65) 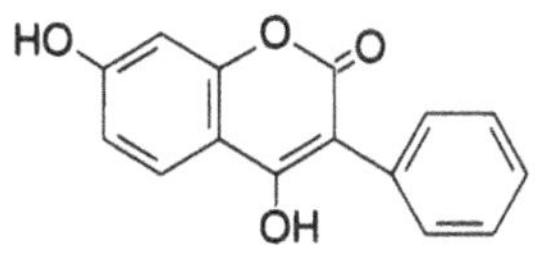   (66)

The metabolism of 4,7-dihydroxy-3-phenylcoumarin (66) in rabbits was found to differ from that observed with the monohydroxy compound (65).[66] The main urinary metabolite detected was the unchanged compound (22%) and no conjugates were found in the 48 h urines. A small amount (0.2%) of the 4-methyl ether of (66) was found in two of five 24-h urine samples.

Another derivative of 3-phenylcoumarin is **coumestrol** (67), however little information is available on its metabolism. Kelly[146] found that sheep fed on a coumestan-rich diet (containing coumestrol and its 4'-methyl ether) developed resistance to its estrogenic activity. This was believed to be due to development of an inactivation mechanism. Cox and Braden[147] noted that a greater proportion of the absorbed coumestrol was found in the blood of ruminants in unconjugated form compared with that normally found with phytoestrogens from forages.

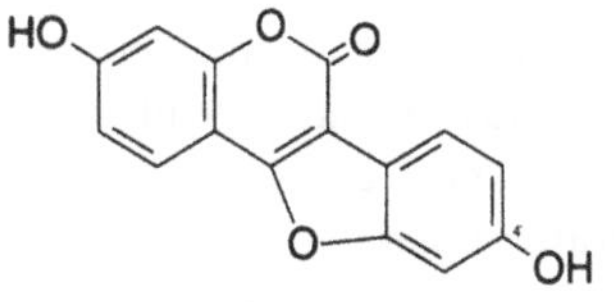

(67)   (68)   (69)

Coumestrol   Calophyllolide

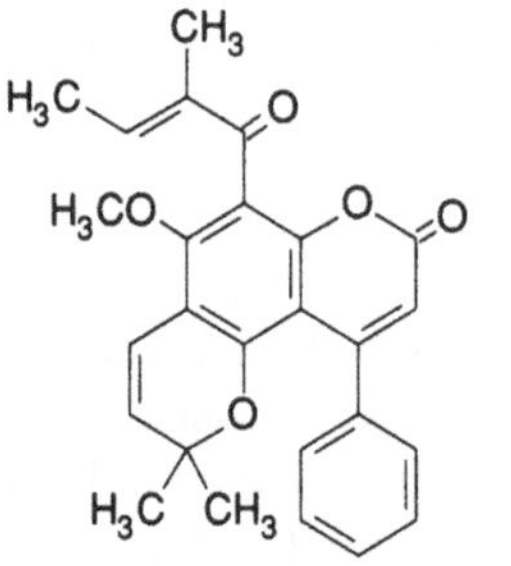

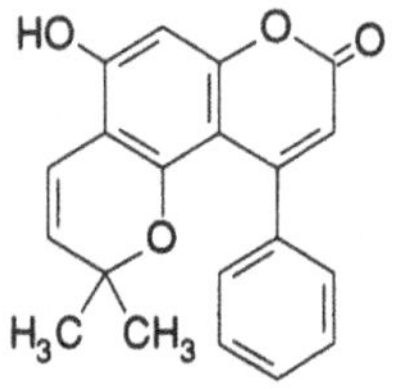

4-Phenylcoumarins are also uncommon plant compounds and the metabolism of only a single representative, calophyllolide (68), has been reported together with data on two simpler synthetic analogues, 4-phenylumbelliferone and its methyl ether, 4-phenylherniarin.[66,148] 4-Phenylumbelliferone (200 mg/kg, p.o.) was metabolized in rabbits similarly to that described above for both umbelliferone and its 3-phenyl derivative. Thus 75% of the dose was accounted for in the 48 h urine, partly free (25%) and partly as glucuronide and sulfate conjugates (50%). When the methoxy compound was given, only ~30% of the dose was recovered in the urine, nearly exclusively as the demethylated compound. The ratio of free to conjugated metabolite was 2:1, similar to that noted above with the isomeric 3-phenylherniarin. No information on the fate of the remainder of these compounds was obtained, but it was stated that oxidation or cleavage of the lactone ring was not observed.

When **calophyllolide** (68) (50 mg/kg) was given orally to rabbits only ~2% of the dose could be accounted for.[66] This was divided fairly evenly between a free metabolite and its conjugate. The metabolite (69) is a simpler derivative in which demethylation has occurred at C-7 and the acyl group lost at C-8. The latter reaction is unusual as the 8-acyl moiety forms, in fact, an aromatic ketone. This provides a metabolically active site and it is possible that part of the great majority of the calophyllolide still unaccounted for may be the carbinol derivative. Additionally, allylic hydroxylation at the two available sites in the 8-acyl group may occur.

FLAVONOIDS

The flavonoids, which are of widespread occurrence in plants, are commonly classified into several major and a few minor groups. Information on their metabolism in mammals is available to varying extents for all the major and for several of the minor types. The following summaries are based on this classification into groups, however general surveys of the metabolism of flavonoids are available in the reviews of DeEds[149] and Griffiths.[150]

FLAVONES

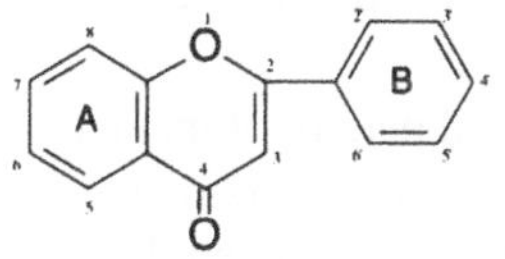

(70)

Flavone

Flavone (70) occurs naturally, in contrast to that found with the unsubstituted parent compound of most of the other flavonoid groups. Demole[151] found that it was not excreted in the urine unchanged following its administration to mice, rats, guinea pigs, and rabbits. However, two unidentified urinary metabolites were detected when flavone (200 mg/kg) was given to rabbits. Das and Griffiths[152] reported that 4'-hydroxyflavone was excreted in the urine of guinea pigs given flavone orally or intraperitoneally. However, only ~3% of the dose (100 mg/kg) was accounted for as the 4'-hydroxylated derivative. Smaller amounts of 3',4'-dihydroxyflavone were detected in the urine after oral administration of flavone and, in contrast to the results of Demole,[151] a small amount of unchanged compound was also found. With the possible exception of salicylic acid which was detected in trace quantities, no phenolic acids arising from ring fission of the flavone molecule were found in the urine. The biological disposition of [2-¹⁴C]flavone in rats was studied by Svardal et al.[153] Following an oral dose (220 mg/kg) they found complete excretion of the radioactivity in the urine (21%) and feces (78%) in 5 d. Most of the excretion occurred within 48 h and no radioactivity was lost as respiratory CO_2. Experiments with bile duct cannulated rats showed that nearly a fourth of the dose was excreted in the bile. The urinary radioactivity consisted of neutral (i.e., hydroxylated flavones) and acidic metabolites, both free and conjugated. In addition, some polar material which was not hydrolyzed by β-glucuronidase + sulfatase was present. The nature of the metabolites present in the acidic and polar fractions was not ascertained, however no evidence was found for the presence of phenolic acids which might have arisen via cleavage of the flavonoid. The neutral urinary metabolites, which corresponded to ~8% of the dose, contained mainly 3'-hydroxyflavone and somewhat greater amounts of 4'-hydroxyflavone. Traces of the 2'-hydroxy- and 3',4'-dihydroxy-derivatives were sometimes detected. The preferential hydroxylation of flavone in the B-ring contrasts to that seen with flavanone in which hydroxylation of the A-ring is dominant (see below). The fecal metabolites of flavone were mainly unchanged compound and its 3'-hydroxy-derivative.

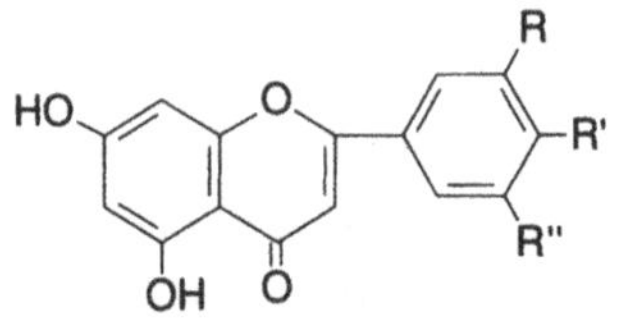

(71)

(a) Chrysin,   R = R' = R" = H

(b) Apigenin,   R = R" = H , R' = OH

(c) Acacetin,   R = R" = H , R' = OCH$_3$

(d) Diosmetin,   R = OH , R' = OCH$_3$, R" = H

(e) Tricetin,   R = R' = R" = OH

(f) Tricin,   R = R" = OCH$_3$, R' = OH

(g)   R = R' = R" = OCH$_3$

(h) Luteolin,   R = R' = OH , R" = H

The first study to reveal the general pattern of metabolism of substituted flavones was that of Booth et al.[154] who found that the major urinary metabolite in rats of **diosmetin** (71d) and its 7-rhamnoglucoside (**diosmin**) was *m*-hydroxyphenylpropionic acid. Also detected were traces of the corresponding cinnamic acid derivative, *m*-coumaric acid, and of diosmetin. When the latter compound was administered, some diosmetin glucuronide was also excreted. Thus, the metabolism of this flavone derivative proceeds, in effect, by way of fission at the 1,2-bond and at the carbonyl group in the 4,5-region leading to a C_6–C_3-phenolic acid. Using ³H-labeled diosmin, Oustrin et al.[155] determined the organ distribution and urinary and fecal excretion of radioactivity following oral (30 mg/kg) or i.v. (15 mg/kg) doses in rats. In the former case nearly 90% of the dose was excreted in 48 h in about equal amounts in the urine and feces. Urinary excretion predominated when the diosmin was given by injection, however about a fourth of the dose was nonetheless lost in the feces due to biliary excretion of conjugated material. The excretion of radioactivity in the urine and feces of human subjects given oral doses (500 mg) of [2-¹⁴C]diosmin was studied by Winternitz et al.[156] Fecal excretion of radioactivity accounted for 80% of the dose, however excretion was delayed and reached maximum values after 24 to 72 h. About 14% of the radioactivity was recovered in the urine. Urine samples from three subjects were investigated for the presence of phenolic metabolites and radioactive 3,4-dihydroxyphenylacetic acid (homoprotocatechuic acid) was identified in all cases. It is noteworthy that formation of this C_6–C_2-phenolic acid differs from the general pathway of flavone metabolism reported in most studies which instead results in C_6–C_3-derivatives.

The metabolism of numerous flavones was studied by Griffiths and Smith[157,158] and the results obtained greatly clarified the structural features required for ring fission. The simplest flavone investigated was **chrysin** (71a), which is unsubstituted in the B-ring. After its oral administration to rats, some unchanged chrysin was excreted in the urine together with its 4'-hydroxy derivative, apigenin (71b). No urinary phenolic acids resulting from ring fission were observed and no detectable metabolites were found following the anaerobic incubation of chrysin with rat cecal microorganisms. **Tectochrysin** is the 7-*O*-methyl ether of chrysin (71a) and similar experiments produced the same general results. Thus, no phenolic acids were excreted and no detectable metabolism occurred in the incubates. The main urinary metabolite appeared to be apigenin (71b) but small amounts of its 7-*O*-methyl ether (genkwanin) and unchanged tectochrysin were also excreted. Lack of ring fission to acidic metabolites and lack of microbial degradation were also noted with 7,4'-dihydroxyflavone. Oral administration in rats led to the urinary excretion of unchanged compound and an unidentified metabolite. DeEds[149] reported that **nobiletin** (5,6,7,8,3',4'-hexamethoxyflavone) was resistant to metabolic degradation.

In contrast to the general metabolic picture summarized above showing that the dihydroxyflavones are resistant to fission of the heterocyclic ring system, administration of the commonly occurring trihydroxyflavone **apigenin** (71b) orally to rats led to quite different results.[157] The major urinary metabolites were *p*-hydroxyphenylpropionic acid, *p*-coumaric acid, and *p*-hydroxybenzoic acid together with unchanged apigenin and two conjugates of apigenin, one of which was a glucuronide. Incubation experiments with apigenin showed that it was degraded to *p*-hydroxyphenylpropionic acid by the intestinal microorganisms. Apigenin is not converted to ring-fission products in germ-free rats.[159] When **acacetin** (71c), the 4-*O*-methyl ether of apigenin, was given to normal rats, the extent of ring fission was greatly reduced and only traces of *p*-methoxyphenylpropionic acid were excreted.[157] This degradation was also observed in incubates. The results summarized above suggest that a free hydroxy group at the 4'-position favors ring fission in the flavones. This point was investigated further in studies using the pentahydroxyflavone **tricetin** (71e) and its methyl ethers **tricin** (71f) and compound (71g).[158] The 3',4',5'-trihydroxy compound tricetin was excreted in the urine partly unchanged after its oral administration to rats but some 3,5-dihydroxyphenylpropionic acid was also detected. This phenolic acid and some *m*-hydroxyphenylpropionic acid were formed *in vitro* in incubates using rat cecal microorganisms. In experiments using the 4'-hydroxy-3',5'-dimethoxy compound tricin, similar results were obtained. Both unchanged tricin and 3,5-dihydroxyphenylpropionic acid were excreted in the urine and small amounts of the latter compound were formed in the incubates. This report is in variance with those of Bickoff et al.[160] and Stelzig and Ribeiro[161] who were unable to detect degradation products in the urine of tricin-treated rats. With the trimethoxy compound (71g), Griffiths and Smith[158] found that fission of the heterocyclic ring system was less extensive. No phenylpropionic acid derivative was formed in incubation experiments and only traces of 3,5-dihydroxyphenylpropionic acid were detected in the urine of rats given compound (71g). Interestingly, large amounts of both unchanged compound and its *O*-demethylated derivative (71f) were excreted in the feces. The extensive fecal excretion of tricin (73f) by rats was also reported by Stelzig and Ribeiro.[161]

Limited information of the biological disposition of some rare flavones is also available. **Baicalin** is the 7-*O*-glucuronide of **baicalein** (5,6,7-trihydroxyflavone) and it was reported by Mao et al.[162] to be absorbed to the extent of 20 to 40% from the rat small intestine. **Scutellarin** differs from baicalin only by the presence of an additional hydroxy group at C-4'. Cai[163] administered ³H-labeled material (10 mg/kg, i.v.) to mice and recovered 19% (urine) and 24% (feces) of the radioactivity in 24 h. Han et al.[164,165] gave **nevadensin** (5,7-dihydroxy-6,8,4'-trimethoxyflavone) to rats and noted poor absorption after intragastric administration. Only 1% of the administered compound was found in the urine in 24 h, however its glucuronide and sulfate conjugates as well as a more polar metabolite containing the flavonoid structure were also detected in the urine. Nevadensin was also excreted in both the bile and feces, also after i.v. dosage.

The flavone glycosides noted above are *O*-glycosides, however some derivatives with a *C*-glycosyl linkage are known. **Orientin** and **isoorientin** (homoorientin) are such examples. These glycosides are derivatives of luteolin (73h) in which the β-D-glucopyranoside moiety is attached to the ring at C-8 or C-6, respectively. Laparra et al.[166] gave both of these compounds intravenously to mice. They employed ¹⁴C-labeled compounds and noted a marked uptake of radioactivity in the bladder, bile, and intestinal tract in 1 h. Elimination of radioactivity was rapid and removal of the sugar group did not appear to occur. The metabolism of isoorientin by human intestinal microorganisms was studied by Hattori et al.[167] They found that incubation under anaerobic conditions resulted in the reduction of the double bond at C-2,3 to give the corresponding flavanone derivative, 6-*C*-glucosyleriodictyol. Further metabolism via loss of the sugar groups formed the aglycons luteolin and eriodictyol. The flavone luteolin was not metabolized further under these conditions, however eriodictyol was converted to 3,4-dihydroxyphenylpropionic acid and phloroglucinol. These findings indicate that *C*-glycosylflavonoids may undergo loss of the sugar residues in a manner similar to that found with the more common *O*-glycosides and that further cleavage to phenols and phenolic acid may also occur.

The above findings clearly indicate that the presence of a 4′-hydroxy group in the flavones is important with regard to the degree of ring fission and production of phenolic acids that can be expected. The results mentioned above with diosmetin (71d), a 4′-methoxy compound, need not contradict this because its extensive metabolism to *m*-hydroxyphenylpropionic acid most likely occurs following initial *O*-demethylation of the flavone. Based upon the finding noted above that 7,4′-dihydroxyflavone is not degraded, it appears that the susceptibility to ring fission may be related to the presence of a free hydroxy group at position 5, as is observed with the flavonols (see below). Nothing definite is presently known of the intermediates formed in the degradation of flavones to phenolic acids. However, the general similarities in end-products of metabolism between this group of flavonoids and the flavanones and dihydrochalcones noted below suggest that these three groups may be degraded in a similar fashion. It is of interest to note that intermediates have been identified only in those flavonoid groups containing a 3-hydroxy group (flavonols and catechins). Although the evidence for the key role of the intestinal bacteria in the degradation of flavones is clear, Takács and Gábor[168] reported that the isolated, perfused rat liver was able to convert apigenin (71b) to the *p*-hydroxylated derivatives of benzoic, phenylacetic, cinnamic, and phenylpropionic acids. No explanation of these results is at hand, however it seems unlikely that this site is a significant factor in the degradative metabolism of flavone derivatives.

FLAVONOLS

Many derivatives of flavonol (3-hydroxyflavone) (72) are found in higher plants, however only kaempferol (73a), quercetin (73b), and myricetin (73c) are common. While metabolic information is available on all three of these compounds, most investigations have dealt with quercetin or its glycoside rutin (quercetin 3-rutinoside). Flavonols generally occur in plants as glycosides and rutin is the most common of these.

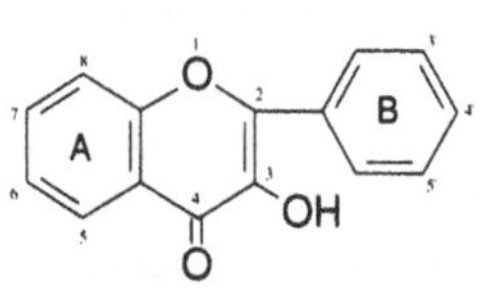

(72)

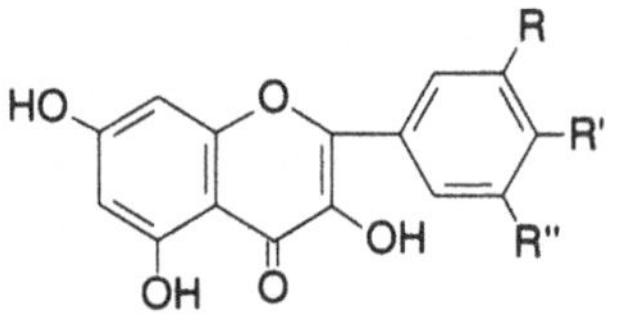

(73)

(a) Kaempferol, R = R″ = H , R′ = OH

(b) Quercetin, R = R′ = OH , R″ = H

(c) Myricetin, R = R′ = R″ = OH

Flavonol itself does not occur naturally and nothing is known of its metabolism in animals. The simplest flavonol studied metabolically is 3′,4′-dihydroxyflavonol which is also not known to occur naturally. An early report by Ozawa[169] indicated that only about a tenth of the administered dose was excreted unchanged in the urine and feces of rabbits. However, three unidentified metabolites were excreted in the urine.

As noted above, most of the data on flavonol metabolism deals with **quercetin** (73b) and **rutin** (quercetin 3-rutinoside). The often conflicting results of a number of early studies in this field were reviewed by DeEds.[149] In brief, the findings pointed to little or no urinary excretion of rutin taking place following its oral administration. On the other hand, injection of rutin led to the appearance of some of the compound in the urine. Clark and MacKay[170] found that large oral doses of rutin (50 mg/kg) in humans resulted neither in its urinary excretion nor its recovery in the feces. A similar experiment by Baba et al.[171] confirmed that no unchanged rutin was excreted in the urine following oral dosage (10 or 50 mg/kg) of the glycoside. They also showed that no quercetin was excreted by these subjects. The same negative results were found in experiments in which rutin was given orally to rats[172] or to rats and mice.[173] The latter investigation, which employed ^{3}H-labeled rutin (50 mg/kg), showed that rats excreted ~65% of the radioactivity in the urine and ~25% in the feces. Contrariwise, mice excreted ~75% in the feces and 18% in the urine. Similar investigations using the aglycon quercetin have been carried out in man and rats. Gugler et al.[174] studied the disposition of quercetin in man and found that an oral dose (4 g) was absorbed to an extent of 1% or less. No unchanged material was excreted in the urine, however about half the dose was recovered in the feces. On the other hand, Peter et al.[175] claimed that unchanged quercetin could be detected in the urine of rats after the oral administration of an extract of *Ginkgo biloba* which contains a mixture of several flavonols including quercetin. Ueno et al.[176] obtained evidence for the urinary excretion of both the sulfate and glucuronide conjugates of quercetin when rats were given large oral doses (630 mg/kg) of the flavonol.

The first investigation to clearly identify metabolites of quercetin and rutin was carried out by DeEds' group.[177-179] Oral administration of quercetin to rats, guinea pigs, rabbits, or humans resulted in the urinary excretion of *m*-hydroxyphenylacetic acid, 3,4-dihydroxyphenylacetic acid (homoprotocatechuic acid), and 4-hydroxy-3-methoxyphenylacetic acid (homovanillic acid). These metabolites were believed to arise from fission of the heterocyclic ring at the 1,2- and 3,4-bonds. Thus, these metabolites are derived from the B-ring and not from the phloroglucinol part of the molecule (A-ring). The formation of these C_6–C_2-phenolic acids from quercetin in rats has been confirmed in studies using randomly labeled [^{14}C]quercetin[180,181] and by others using nonlabeled quercetin.[122,182] Baba et al.[171,172] used deuterium-labeled rutin and showed that these three C_6–C_2-phenolic acids were excreted in the urine of rats and humans. They found that the total urinary excretion of these compounds in two subjects given of rutin orally was 7 and 13.5% of the dose (10 or 3 mg/kg).[183] However, the overall metabolic picture is more complex as derivatives of benzoic, phenylpropionic, and cinnamic acids as well as a neutral compound characterized as an *o*-dihydroxyphenyl lactone[122] have been detected in the urine of rats given quercetin orally. These metabolites include *m*-hydroxybenzoic acid, 4-hydroxy-3-methoxybenzoic acid (vanillic acid), *m*-hydroxyphenylpropionic acid, and *m*-coumaric acid.[122,180,182] Baba et al.[172] confirmed the excretion of *m*-hydroxyphenylpropionic acid and also showed that 4-methylcatechol (3,4-dihydroxytoluene) was a urinary metabolite of rutin in rats. Similar studies in humans[171] showed that 4-methylcatechol and β-*m*-hydroxyphenylhydracrylic acid (see structure (92)) were excreted in the urine in addition to the three C_6–C_2-phenolic acids noted above. The serum concentrations of these phenolic acids and of 4-methylcatechol in humans given rutin (75 mg) orally were measured by Sawai et al.[184] They also determined the extent of urinary excretion of these metabolites in 48 h. Homovanillic acid was the most abundant metabolite (19% of the dose) and was excreted mainly free. This was also found with *m*-hydroxyphenylacetic acid (13%), however homoprotocatechuic acid (11%) was excreted mainly conjugated. Nearly all the 4-methylcatechol detected (8%) was conjugated. A total of ~50% of the dose was accounted for by these metabolites. An interesting urinary metabolite of [4-^{14}C]quercetin reported by Ueno et al.[176] was phloroglucinol carboxylic acid. This metabolite was radioactive and clearly was formed from the A-ring of the flavonol.

Most of the reports on the metabolites of quercetin have described degradation products; however, Peter et al.[175] reported that isorhamnetin (3'-*O*-methylquercetin) was excreted in the urine of rats given quercetin orally. Interestingly, Brown and Griffiths[185] reported that isorhamnetin, in conjugated form, was a biliary metabolite of quercetin in rats. A conjugate of quercetin was also detected. These biliary conjugates were most abundant when the flavonol was given i.p., however they were also found following oral dosage (~30 mg/kg). Ueno et al.[176] similarly reported that glucuronide conjugates of quercetin and *O*-methylquercetin were excreted in the bile of rats given [4-^{14}C]quercetin orally. However, they found that the methylated product consisted of a mixture of the 3'- and 4'-*O*-methyl ethers, i.e., isorhamnetin and tamarixetin, respectively. These two *O*-methylated metabolites, in conjugated form, were also excreted in the urine of rats fed on a 10%-quercetin diet.

The finding of Petrakis et al.[180] that ~15% of the radioactivity was recovered in the respiratory CO_2 following oral administration of randomly labeled [^{14}C]quercetin indicates that at least a portion of the molecule undergoes extensive metabolism. Ueno et al.[176] administered 4-^{14}C-labeled quercetin and found that about a third of the oral dose (630 mg/kg) was lost as respiratory $^{14}CO_2$ in rats. They also showed that anaerobic incubation of the preparation with bacteria from the cecum and colon resulted in the formation of large amounts of $^{14}CO_2$.

Following the initial reports describing some of the metabolic products of quercetin and rutin, interest was directed towards the site or sites of their formation. Lang and Weyland[186] described a mitochondrial enzyme system that utilized quercetin and rutin anaerobically. Highest activities were found in liver and kidney with lower amounts present in heart, brain, and muscle. Takács and Gábor[168] claimed that the isolated, perfused rat liver metabolized rutin to a number of the C_6–C_1-, C_6–C_2-, and C_6–C_3-phenolic acids previously identified as metabolites of rutin or quercetin. Douglass and Hogan[187] reported that rat kidney homogenates were able to metabolize quercetin to 3,4-dihydroxybenzoic acid (protocatechuic acid) under aerobic conditions. However, the phenylacetic acid derivatives mentioned above were not formed by these preparations and it was suggested that they may arise from bacterial or digestive action in the gastrointestinal tract. The possibility of gastrointestinal formation was partially tested by Braymer[122] who found that quercetin was not degraded *in vivo* in the ligated rat stomach. Westlake et al.[188] reported that various molds, streptomycetes, and bacteria were able to degrade rutin, and Booth and Williams[189] made the significant finding that anaerobic incubates of rat cecal or fecal microorganisms converted rutin to *m*-hydroxyphenylpropionic acid. As noted above, this C_6–C_3-phenolic acid is a urinary metabolite of quercetin in rats. Further studies of the anaerobic degradation of rutin and also quercetin by rat cecal microorganisms were carried out by Scheline[190] who found that both compounds were degraded to *m*-hydroxyphenylacetic acid and *m*-hydroxyphenylpropionic acid. Baba et al.[172] confirmed that these two *m*-hydroxy compounds, in addition to homoprotocatechuic acid, were formed when rutin was incubated anaerobically with rat intestinal microorganisms. From this it is clear that several types of the main urinary metabolites of these flavonols can be formed entirely by the intestinal microflora. Variations in ring substitution due to dehydroxylation also arise in this way whereas subsequent tissue reactions including dehydrogenation or β-oxidation of the side chain and *O*-methylation explain the formation of other urinary metabolites. Further evidence for the key role played by the intestinal microflora in the degradation of quercetin or rutin was obtained by Nakagawa et al.[182] who reported that the C_6–C_2-phenolic acids normally appearing in the urine of quercetin-treated rats were absent when the animals were also given neomycin orally. This was confirmed by Baba et al.[172] who reported that the phenolic metabolites normally excreted in the urine of quercetin-treated rats were absent when the flavonol was given 2 d after treating the animals for 4 d with neomycin. Significantly, metabolite production returned to normal when dosing was repeated 12 d later. Also, phenolic acid metabolites of rutin were not excreted by germ-free rats.[159]

Interest in the degradation of flavonols by gastrointestinal bacteria has also been directed towards the activity of rumen microorganisms. Incubation of rutin with bovine rumen fluid resulted in its degradation to phenolic compounds which appeared to be similar to those formed in the rat intestinal tract.[191] Numerous strains of *Butyrivibrio* sp. were subsequently isolated from bovine rumen contents and shown to degrade rutin anaerobically.[192] The metabolic products formed included phloroglucinol, 3,4-dihydroxybenzaldehyde, and 3,4-dihydroxyphenylacetic acid.[193] These latter results suggest that different patterns of degradation may be produced by bovine rumen and rat intestinal microorganisms. Krumholz and Bryant[194] isolated an anaerobic bacterium, *Eubacterium oxidoreducens*, from the rumen of a steer which metabolized quercetin to acetic, butyric, and homoprotocatechuic acids. The finding noted above that 4-methylcatechol was a urinary metabolite of rutin in rats and humans[171,172] has also been reported in sheep.[195] When a dose of 7.3 g/d was infused into the rumen, 18% of the administered rutin was excreted as 4-methylcatechol.

The finding that rumen microorganisms degrade rutin to phloroglucinol[191,193] is of interest in regard to the fate of the A-ring. While Kallianos et al.[196] reported that quercetin was converted to phloroglucinol, phloroglucinol carboxylic acid, and 3,4-dihydroxybenzoic acid in the stomach of the rat, there is good reason to believe that these compounds may also be formed as a result of chemical degradation during sample preparation.[181] Nonetheless, the latter group found subsequently[149] that small amounts of phloroglucinol may be formed from quercetin when incubated for short times with rat fecal microorganisms. The transient nature of this metabolite has been noted in experiments with bovine rumen microorganisms.[191] The report by Ueno et al.[176] that phloroglucinol carboxylic acid was a urinary metabolite of [4-^{14}C]quercetin is noteworthy in this regard. This metabolite was radioactive and was most likely the immediate precursor of the large amounts (>30% of the dose) of $^{14}CO_2$ which were expired.

The detailed pathways involved in the degradation of quercetin remain unclear although it seems reasonable to assume that multiple pathways exist. One of these allows the A-ring to remain intact and give rise to phloroglucinol carboxylic acid and/or phloroglucinol. Another pathway which has been prominent in most studies of flavonol metabolism to date involves destruction of the A-ring leading to CO_2 and excretion of the remainder of the quercetin molecule as C_6–C_2- and sometimes C_6–C_3-phenolic acids. It is also possible that phenolic lactone intermediates[122] similar to those encountered in catechin degradation (see below) may be involved in this pathway. It is likely that the relative significance of these routes may be a reflection of the types and relative numbers of microorganisms in the particular intestinal microflora.

Experiments with normal and antibiotic-treated animals and with cultures of intestinal microorganisms have shown that the pathways of metabolism described above for quercetin and rutin are similar to those which are involved with other flavonols. **Kaempferol** (73a) and its 7-glycoside **robinin** were degraded to the expected *p*-hydroxyphenylacetic acid when incubated with rat cecal microorganisms and this phenolic acid was excreted in the urine of rats given these flavonols.[157] **Myricetin** (73c) and its 3-rhamnoside **myricitrin** were metabolized in rats and by rat cecal microorganisms to phenylacetic acid derivatives, mainly 3,5-dihydroxyphenylacetic acid.[158] The conversion in rats was prevented when the animals were treated with neomycin before and during administration of the flavonols. The degradation of myricetin was also absent in germ-free rats.[159]

Some information is available on the structural requirements for flavonol degradation. The conclusions are similar to those noted above on flavone metabolism. The presence of a 5-hydroxy group appears to be essential as **robinetin** (3,7,3′,4′,5′-pentahydroxyflavone), which differs from myricetin only by the absence of this group, was not degraded when given orally to rats or incubated with cecal microorganisms.[158] DeEds[149] reported that **rhamnetin** (quercetin-7-methyl ether), **azaleatin** (quercetin-5-methyl ether), and quercetin-3-methyl ether were not degraded in the rat. Likewise, no urinary metabolites of **tangeretin** (3,5,6,7,4′-penta-methoxyflavone) were detected. Lack of degradation to phenolic metabolites was also noted by Braymer[122] with the 5,7,3′,4′-tetramethyl ether of quercetin in the rat.

FLAVANONES

Flavanone (74) is chemically closely related to flavone (70), differing only in the lack of a double bond between C-2 and C-3. Furthermore, flavanones are isomeric with chalcones which are formed by ring opening at the 1,2-position. The structural relationship of 3-hydroxyflavanones (flavanonols, dihydroflavonols) (75) to flavanones is similar to that found between flavonols and flavones. However, in contrast to the relatively large number of investigations dealing with the metabolism of flavonols, the literature on the 3-hydroxyflavanones is restricted to a few reports. Therefore, both the flavanones and their 3-hydroxy derivatives are treated together in this section.

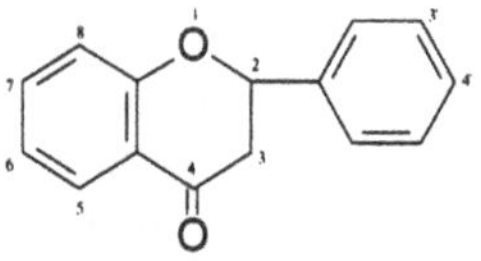

(74) 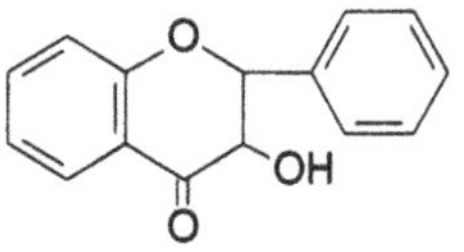   (75)

Flavanone (74) itself does not occur naturally but its metabolism in animals is described here in order to allow a comparison with that observed with flavone (70) described above. Das et al.[197] administered flavanone to rats and found that the urine contained unchanged compound, flavone (70) and flav-3-ene (76). These three compounds accounted for ~25% of the dose and, although several unidentified metabolites were detected, the fate of the remainder was unknown. However, no evidence was obtained for the formation of aromatic acids arising from ring fission. Interestingly, formation of a 4'-hydroxylated metabolite, as seen with flavone, was not noted. The biological disposition of flavanone in rats was subsequently studied by Buset and Scheline[198] using 2-^{14}C-labeled compound. Following an oral dose (100 mg/kg) the radioactivity was excreted in the urine (27%) and feces (67%) in 48 h. Excretion of metabolites was extensive in the bile, however no radioactivity was detected in the expired CO_2. About half of the radioactivity in the 24 h urine was due to neutral metabolites (free and conjugated) and these compounds were also prominent in the bile, which also contained other polar metabolites. Neutral compounds made up the bulk of the fecal radioactivity and both unchanged flavanone and its 6-hydroxy derivative were identified. The nature of the urinary metabolites of flavanone in rats was investigated by Buset and Scheline[199] who detected more than 40 compounds. The most common metabolic reactions were reduction of the keto group and hydroxylation at the 3- or 6-positions. The major metabolites were flavan-4α-ol (77), *trans*-3-hydroxyflavan-4β-ol (78), 6-hydroxyflavanone, and 6-hydroxyflavan-4β-ol. Hydroxylation occurred in ring B, however this was not extensive. The product formed, 4'-hydroxyflavanone, was accompanied by the corresponding 4-α-hydroxy and 4-β-hydroxy epimers formed by reduction of the keto group. Interestingly, this study revealed that gas chromatographic analysis of underivatized flavanone metabolites could in some cases result in the formation of dehydrated products including flavone, flav-3-ene, and flavanone itself. Thus, it seems evident that these three compounds are artifacts and not urinary metabolites of flavanone.

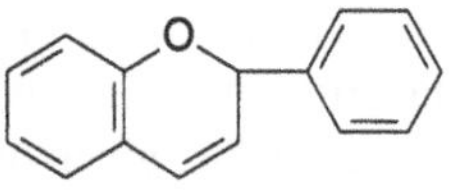

(76) 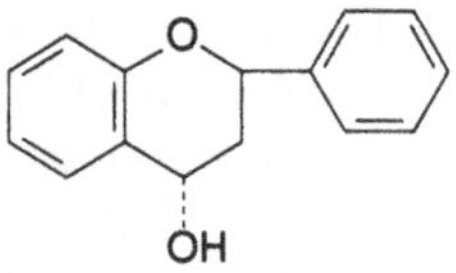   (77) 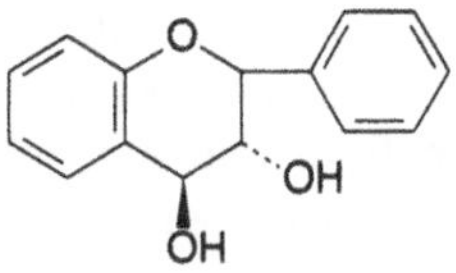   (78)

Metabolic studies of flavanones have dealt mainly with four closely related compounds: naringenin (79a), eriodictyol (79b), hesperetin (79c), and homoeriodictyol (79d) as well as the 7-rhamnoglucosides of (79a) and (79c). The latter compounds are known as naringin and hesperidin, respectively. In addition, some information is available on the metabolism of farrerol, a *C*-methyl flavanone.

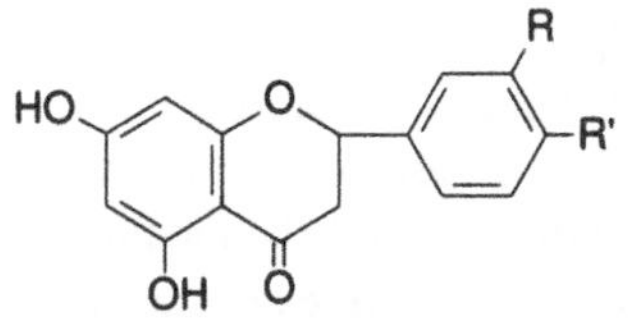

(79)

(a) Naringenin, R = H , R' = OH

(b) Eriodictyol, R = R' = OH

(c) Hesperetin, R = OH , R' = OCH$_3$

(d) Homoeriodictyol, R = OCH$_3$, R' = OH

Naringenin (79a) was first shown by Booth et al.[200] to be degraded to *p*-hydroxyphenylpro-pionic acid, which was excreted in the urine of rabbits given the flavanone orally. In a more extensive report, Booth et al.[201] showed that, in rats, smaller amounts of *p*-coumaric acid (*p*-hydroxycinnamic acid) and the ethereal sulfate of *p*-hydroxybenzoic acid were also excreted. Both the unchanged flavanone and its glucuronide conjugate were also detected in the urine. The same pattern of urinary metabolites was seen when **naringin** (naringenin-7-rhamnoglucoside) was given. The latter results were largely confirmed by Griffiths and Smith[157] who, with a large oral dose (~600 mg/kg) in rats, detected *p*-hydroxyphenylpropionic, *p*-coumaric, and *p*-hydroxybenzoic acids and the aglycon naringenin in the urines. Hackett et al.[202] found that the urinary excretion of intact flavanone material in rats was dependent on the dose. None was detected when oral doses of ~30 mg/kg of naringin or its glycoside were given. However, values of 2 to 3% were recorded when the dose was increased to ~150 mg/kg. A species difference in metabolism has been noted with naringin which, when given orally to a human volunteer, was excreted only as naringenin and its glucuronide in the urine.[201]

The excretion by rats of naringin or naringenin via the biliary route was studied by Hackett et al.[202] who found extensive excretion in some cases. Both the route of administration and the dose were found to influence the extent of biliary excretion. Using small i.p. doses (~15 or 30 mg/kg), 90 to 100% of the dose was recovered in the bile in 24 h. At a higher dose (~150 mg/kg) the percentage excreted was lower, however this was due to retention of the poorly soluble flavanones in the tissues. Biliary excretion of flavanone material was considerably reduced when the compounds were given orally. No flavanones were detected in the bile when doses of ~30 mg/kg were given and this was believed to be due to complete degradation of the compound in the gastrointestinal tract prior to absorption. At the highest dose level, ~8 to 11% was recovered in the bile in 48 h. Investigation of the nature of the biliary metabolites showed that naringin was mainly excreted as the unchanged glycoside at the lowest dosage level. However, large amounts of a glucuronide conjugate, probably the 4'-*O*-glucuronide, were formed at higher doses. The main biliary metabolite of naringenin was a glucuronide conjugate. Small amounts of two unidentified conjugates were also detected and, at the highest dosage level, small amounts of eriodictyol were formed from naringenin via hydroxylation at C-3'.

Booth et al.[154,200] carried out *in vivo* metabolic studies with **hesperetin** (79c) and its glycoside **hesperidin, eriodictyol** (79b), and **homoeriodictyol** (79d). The main finding in rats was that the major urinary metabolite in all cases was *m*-hydroxyphenylpropionic acid. Thus, all of the flavanones studied are metabolized to phenylpropionic acid derivatives, i.e., C$_6$–C$_3$-phenolic acids. It is noteworthy that the flavone derivatives, which differ only in having a double bond at the 2,3-position, are also degraded in this fashion. In addition to the major metabolite noted above, other urinary metabolites in rats were: from hesperidin and hesperetin, *m*-hydroxycinnamic acid and a conjugate of hesperetin; from eriodictyol, *m*-hydroxycinnamic acid, eriodictyol glucuronide, and homoeriodictyol; from homoeriodictyol, *m*-hydroxycinnamic acid, 3-methoxy-4-hydroxyphenylpropionic acid (dihydroferulic acid), unchanged compound, and homoeriodictyol glucuronide. Species variations in metabolism have been noted with hesperidin in rabbits and man.[154] In the case of rabbits, the nature and extent of the urinary metabolites varied with the type of diet given. When maintained on a purified diet and given hesperidin orally, the following compounds were detected in the urine: hesperetin, hesperetin glucuronide, 3,4-dihydroxyphenylpropionic acid, 3-methoxy-4-hydroxyphenylpropionic acid, *m*-hydroxycinnamic acid, *m*-hydroxyphenylpropionic acid, *m*-hydroxyhippuric acid, *m*-hydroxybenzoic acid, and vanillic acid. In man, the major urinary metabolite of hesperidin and hesperetin was shown to be 3-hydroxy-4-methoxyphenylhydracrylic acid (80).

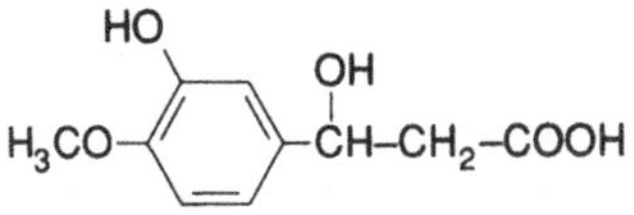

(80)

Honohan et al.[203] studied the metabolism of [3-^{14}C]hesperetin in rats. The doses given (~1.6 mg/kg, p.o. or 0.8 mg/kg, i.p.) were 1% or less of those employed in the earlier studies. Excretion of radioactivity in the urine, feces, bile, and respiratory air was maximal during the first 24 h and essentially complete after 48 h. A third of the radioactivity was excreted in the urine following oral dosage and most of this was shown to be due to the hydroxymethoxy-, dihydroxy-, and *m*-hydroxy-derivatives of phenylpropionic acid. A further 40% of the dose was lost as $^{14}CO_2$ and this indicates that these C_6–C_3-metabolites also underwent extensive β-oxidation to benzoic acid derivatives. Biliary excretion of radioactivity was studied after both oral and i.p. administration. Recoveries were 57 and 100%, respectively, in cannulated rats. The latter value is noteworthy as it indicates conclusively that if any tissue metabolism of hesperetin had occurred, no C_6–C_1-metabolites could have been formed as this would have entailed loss of radioactivity as CO_2. Therefore, no conversion of hesperetin to C_6–C_3-derivatives could have occurred in this experiment either. The *in vivo* experiments of Honohan et al. indicated that the metabolism of hesperetin is mediated by the intestinal microflora. This conclusion was further substantiated by *in vitro* experiments using cecal microorganisms as noted below. Further information on the biliary excretion of hesperidin and hesperetin in rats was given by Hackett et al.[202] In experiments similar to those described above with naringin and naringenin, they found that extensive biliary excretion of flavanone material occurred following i.p. injection of the compounds. With a dose of ~30 mg/kg the values were 97% for hesperidin and 84% for hesperetin in 12 h. Most of the biliary material following hesperidin dosage was a glucuronide conjugate, probably the 3′-*O*-glucuronide of hesperidin. In addition, both unchanged hesperidin and hesperetin glucuronide were excreted. The major biliary metabolite of hesperetin was its glucuronide. No evidence for the 4′-*O*-demethylation of hesperetin to eriodictyol was obtained.

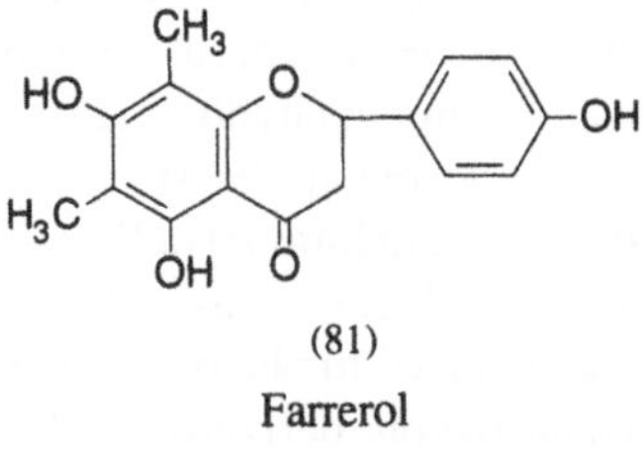

(81)

Farrerol

The metabolism in rats of the *C*-methyl flavanone **farrerol** (81) was studied by Feng and Zhu.[204] Using an assay method specific for unchanged compound, they found that 70 to 80% of the orally administered farrerol disappeared from the gastrointestinal tract within 6 to 12 h. About 30% and <2% of the dose was recovered unchanged in the feces and urine, respectively, in 4 to 5 d. Four urinary metabolites of farrerol were detected and one of these was shown to be a glucuronide conjugate.

The investigations summarized above in which flavanones were administered to animals have been supplemented by studies which show that these compounds and their glycosides can be fully degraded to phenylpropionic acid derivatives by intestinal microorganisms. Thus, the anaerobic incubation of hesperidin or hesperetin with rat cecal microorganisms resulted in their degradation to m-hydroxyphenylpropionic acid.[190] Similar experiments with naringin resulted in the formation of p-hydroxyphenylpropionic acid.[157] Anaerobic incubation of eriodictyol with human intestinal bacteria converted it to 3,4-dihydroxyphenylpropionic acid and phloroglu-cinol[167]. Honohan et al.[203] incubated [3-^{14}C]hesperetin anaerobically with mixed cultures of rat cecal microorganisms and found extensive degradation to varying proportions of the three phenylpropionic acid metabolites noted above. Very little radioactivity was lost as $^{14}CO_2$, indicating that further bacterial metabolism to C_6–C_1-metabolites was insignificant. Both hesperidin and naringin were degraded to water-soluble products by bovine rumen microorgan-isms.[191] Cheng et al.[205] showed that the hydrolysis and further degradation of naringin was carried out by *Butyrivibrio* sp. C_3 isolated from the rumen. The metabolites formed were phloroglucinol and p-hydroxyphenylpropionic acid. These products are the same as those formed from the corresponding dihydrochalcone phloretin (see below). This finding suggests that the metabolic pathway may involve an initial reduction followed by a final hydrolytic stage. However, the postulated intermediate phloretin was not detected in the incubates.

When hesperidin and naringin were given orally to germ-free rats, none of the phenolic acid metabolites were detected in the urine.[159] Also, Hackett et al.[202] showed that no phenolic acids were excreted in the urine when the flavanones were administered intraperitoneally to bile duct-cannulated rats. It is therefore evident that the intestinal reactions are able to fully account for the major urinary metabolites of the flavanones. Subsequent dehydrogenation and β-oxidation of the phenylpropionic acids in the tissues give rise to the cinnamic and benzoic acid derivatives that are also usually detected.

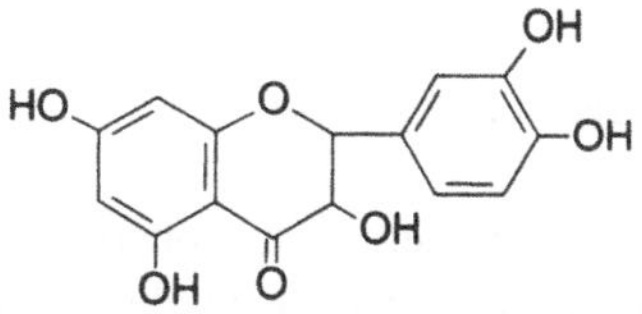

(82)

Taxifolin

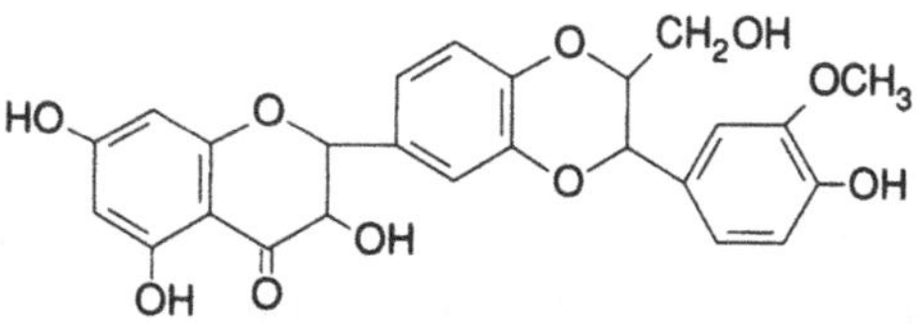

(83)

Silybin

Limited information is available on the metabolism of flavanonols (75). Booth and DeEds[206] and DeEds[149] reported that **taxifolin** (dihydroquercetin) (82) was metabolized in rats and humans to m-hydroxyphenylacetic acid, 3,4-dihydroxyphenylacetic acid (homoprotocatechuic acid), and 4-hydroxy-3-methoxyphenylacetic acid (homovanillic acid). In addition to these C_6–C_2-phenolic acids, a small amount of m-hydroxyphenylpropionic acid was also detected in rat urine. Thus, the metabolism of this 3-hydroxyflavanone resembles that observed with the flavonols rather than that of the flavanones. The excretion of taxifolin metabolites in the bile and urine of rats was studied by Brown and Griffiths.[185] Following an i.p. dose (~30 mg/kg) the bile contained several metabolites which, when hydrolyzed by β-glucuronidase + sulfatase, liber-ated three aglycons. These were found to be unchanged taxifolin, an O-methylated derivative (probably 3'-O-methyltaxifolin) and a compound which appeared to be a hydroxylation product of taxifolin. The flavanonol conjugates were also present in the urine, however at much lower levels. This finding is in agreement with the usual observation that flavonoid conjugates are preferentially excreted via the biliary route. The same biliary metabolites of taxifolin were detected following oral dosage, however the amounts present were lower.

Silybin (83) is a more complex flavanonol which, after oral or i.v. administration to rats, was excreted partly in the urine but mainly in the bile.[207] The urinary material was mainly unchanged silybin whereas glucuronide and sulfate conjugates of silybin and, perhaps, 2,3-dehydrosilybin were excreted in the bile. Ognyanova et al.[208] also administered a preparation containing silybin to rats. They similarly found low urinary excretion of metabolites and extensive excretion of glucuronide and sulfate conjugates in the bile. Mennicke et al.[209] stated that a fourth to a third of the biliary material underwent reabsorption. They also noted that two structural isomers of silybin, **silydianin** and **silychristin**, similarly underwent partial absorption from the gastrointestinal tract followed by extensive excretion in the bile. Mennicke et al.[209] and Flory et al.[210] found that silybin and silychristin underwent extensive biliary excretion in humans, again as glucuronide and sulfate conjugates. The excretion of the flavanonols in the urine was negligible.

ANTHOCYANINS

The anthocyanins are intensely colored, water-soluble pigments which are to a large extent responsible for the attractive scarlet, pink, red, mauve, violet, and blue colors in flowers, leaves, and fruits of higher plants. Anthocyanins are glycosides of anthocyanidins and all of the latter compounds of natural occurrence are derivatives of the parent flavylium cation structure (84).

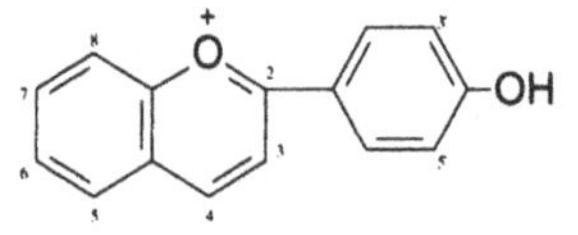

(84)

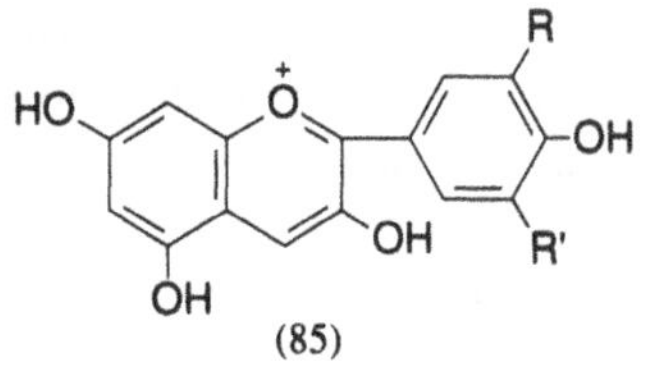

(85)

(a) Cyanidin, R = OH , R' = H

(b) Delphinidin, R = R' = OH

(c) Petunidin, R = OCH_3 , R' = OH

(d) Malvidin, R = R' = OCH_3

(e) Pelargonin, R = R' = H

An early investigation of the metabolism of the anthocyanin pigment from Concord grapes was reported by Horwitt.[211] The precise nature of these anthocyanins was then unknown but a subsequent phytochemical study showed that the fruit of the Concord grape contains several glycosides of cyanidin (85a), delphinidin (85b), petunidin (85c), and malvidin (85d).[212] Horwitt found that s.c. injection of the pigment (~100 mg/kg) in rats resulted in considerable excretion in the urine, apparently in unchanged form. However, oral administration at about twice this dose did not lead to detectable urinary excretion of the pigment. This was also the case in dogs, however a few percent of the dose was excreted in the urine of rabbits fed the pigment. Some pigment was excreted in the feces in both rats and rabbits following oral administration of the anthocyanins. In the dog, biliary excretion was found to be an alternative route of excretion of injected pigment. The possibility that the pigment was metabolized by intestinal bacteria was studied using incubates containing extracts of human feces. However, no loss of color was observed.

The degradation of **cyanidin** chloride (85a) by rat cecal microorganisms was studied by Scheline[190] and Griffiths and Smith[157] who found that it was not metabolized to phenolic compounds. The aforementioned results suggest that the flavylium cation is metabolically stable, not an unreasonable situation in view of its ionic and hydrophilic nature. Nonetheless, studies by Griffiths and Smith[157,158] showed that some of these compounds can be metabolized when given orally to rats or incubated with rat intestinal microorganisms. Thus **pelargonin**, the 3,5-diglucoside of pelargonidin (85e), was converted in incubation experiments to a phenolic compound tentatively identified as *p*-hydroxyphenyllactic acid (86). Similar experiments with **delphinidin** (85b) resulted in the formation of two unidentified metabolites having, respectively, neutral and acidic properties. The neutral metabolite was also detected in the urine of a rat given delphinidin (100 mg) orally. **Malvin,** the 3,5-diglucoside of malvidin (85d), gave rise *in vivo* to three unidentified neutral urinary metabolites but none of these was detected in the incubation experiments.

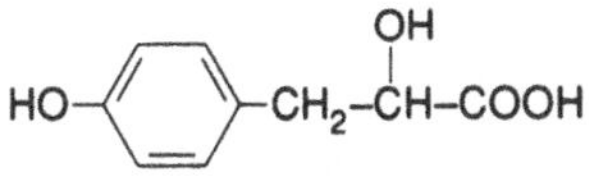

$$HO-\underset{}{\bigcirc}-CH_2-\underset{\overset{OH}{|}}{CH}-COOH$$

(86)

The limited results presently available indicate that flavylium compounds may undergo metabolic alteration but that this occurs to a much more limited extent than is the case with related flavonoids, e.g., catechins, which lack the cationic group.

CATECHINS

Catechins are derivatives of flavan-3-ol (87) and differ structurally from many other types of flavonoids by their lack of a 4-keto group. The flavan-3,4-diols (leucoanthocyanidins) (88) are close relatives of the catechins but are also closely related to the anthocyanidins by virtue of the fact that they are readily converted to the latter in the presence of acid. However, knowledge of the metabolism of leucoanthocyanidins in animals is limited to a single compound, leucocyanidin (98), which will be dealt with at the end of this section.

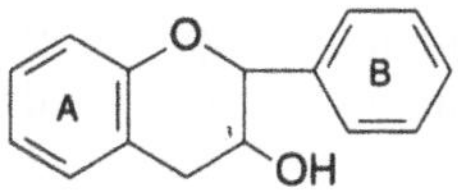

(87) 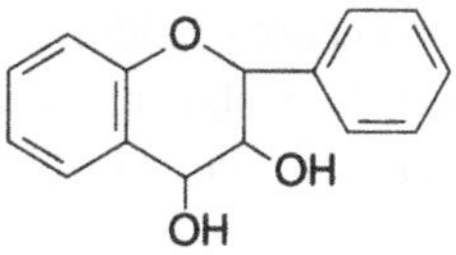   (88)

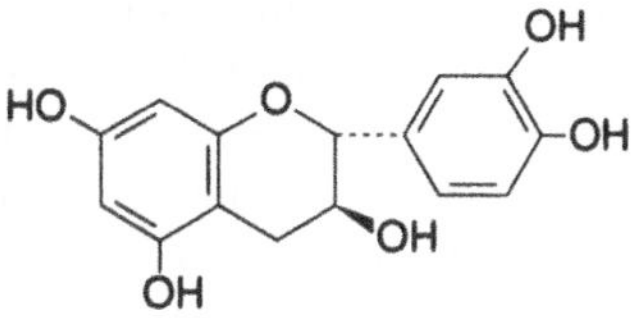

(89)   (90)

(+)-Catechin   (−)-Epicatechin

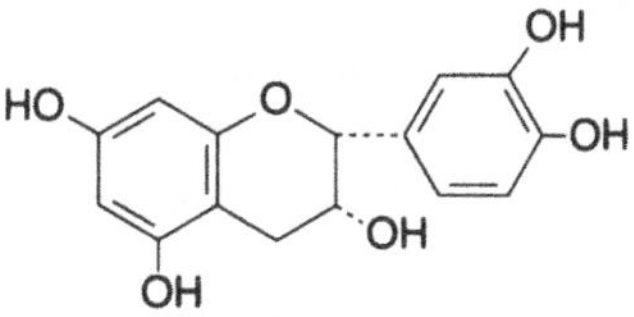

The most common flavan-3-ols are the diastereoisomeric pair (+)-catechin (89) and (−)-epicatechin (90). The metabolism of both has been studied but most of the investigations have dealt with the former compound. In fact, our knowledge of the metabolism of (+)-catechin is more extensive than that of any other single flavonoid compound. The metabolism of (+)-**catechin** (cyanidol, (+)-cyanidanol-3) (89) was first studied in rabbits by Oshima et al.[213] and Oshima and Watanabe[214] who found that it was degraded to several simple phenolic acids and to neutral compounds. The major acidic metabolites were *m*-hydroxybenzoic acid, 3,4-dihydroxybenzoic acid (protocatechuic acid), and 4-hydroxy-3-methoxybenzoic acid (vanillic acid) and the neutral metabolites were postulated to be phenolic derivatives of phenyl-γ-valerolactone. A detailed investigation of the latter compounds showed that they were δ-(3-

hydroxyphenyl-γ-valerolactone (91a), δ-(3,4-dihydroxyphenyl)-γ-valerolactone (91b), and δ-(4-hydroxy-3-methoxyphenyl)-γ-valerolactone (91c).[215-218] All three of these valerolactones have also been identified as urinary metabolites when (+)-catechin was administered to rats,[219] rats and guinea pigs,[220] monkeys (*Macaca iris*),[221] and man.[222]

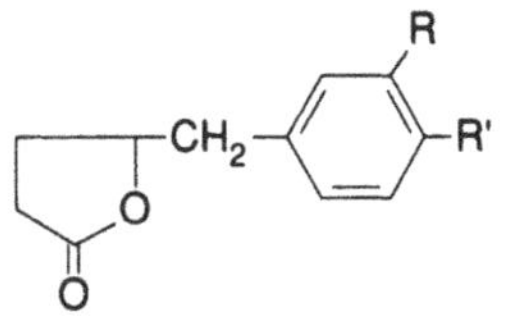

(91)

(a) R = OH , R' = H

(b) R = R' = OH

(c) R = OCH$_3$, R' = OH

While the initial investigation of (+)-catechin metabolism indicated that C_6–C_1-phenolic acids were excreted in the urine of rabbits, a subsequent study showed that a C_6–C_3-acid, *m*-hydroxyphenylpropionic acid, was a major urinary metabolite of (+)-catechin in rats.[223] As the metabolite showed a somewhat delayed excretion profile, it was suggested that it may be formed by the metabolic action of intestinal microorganisms. This prediction was confirmed by Booth and Williams[189] who detected *m*-hydroxyphenylpropionic acid following the anaerobic incubation of (+)-catechin with rat fecal or cecal microorganisms. This reaction sequence was confirmed using rat intestinal microorganisms by Scheline[190] and by Das.[224] Interestingly, two of the phenylvalerolactones (compounds (91a) and (91b)) were also detected in the incubates in the latter investigation. These compounds were shown to be intermediates in the degradation of (+)-catechin to phenolic acids as the oral administration of compound (91a) to guinea pigs resulted in its partial metabolism to *m*-hydroxyphenylpropionic acid, *m*-hydroxybenzoic acid, and *m*-hydroxyhippuric acid which were excreted in the urine.[220] Similar experiments in rats resulted in the excretion of the first two of these phenolic acids.[225]

From the above results it appears that a similar pattern of degradation of (+)-catechin exists in the different animal species studied. Thus, the flavonoid is metabolized to phenylvalerolactones and partly further to C_6–C_3-phenolic acids by the intestinal microflora. Following the absorption of these metabolites, some of the catecholic compounds may be *O*-methylated and the acids may be metabolized further by β-oxidation to benzoic acid derivatives which may then be conjugated with glycine giving hippuric acid derivatives. Not unexpectedly, phenolic acid metabolites were not detected in the urine of germ-free rats given (+)-catechin.[159] An extensive study of the effects of antibiotic pretreatment on the metabolism and excretion of (+)-catechin in rats was carried out by Gott and Griffiths.[226] Excretion of C_6–C_1-phenolic acids and/or their conjugates is seen in rabbits[213] and also predominates in guinea pigs.[220] On the other hand, *m*-hydroxyphenylpropionic acid is excreted in increased amounts in rats[225,227] and is the major urinary phenolic acid in man following the oral administration of (+)-catechin.[222] Nonetheless, Hackett et al.[228] found that only a very small percentage of the oral dose (2 g) of (+)-catechin was excreted in the urine in the form of *m*-hydroxylated phenolic acids in man. The compounds detected were *m*-hydroxybenzoic, *m*-hydroxyhippuric, and *m*-hydroxyphenylpropionic acids. *m*-Hydroxyphenylhydracrylic acid (92), the β-oxidation product of *m*-hydroxyphenylpropionic acid, is also excreted by man and this compound is the major urinary phenolic acid in the monkey (*Macaca iris*).[221] These findings may indicate a species difference in flavonoid metabolism in higher animals and man, as a phenylhydracrylic acid derivative has also been identified as a urinary metabolite of flavonol and flavanone derivatives in man but not in laboratory animals.

Much of the initial work on elucidating the metabolic fate of (+)-catechin was directed towards products of degradation. However, it is now clear that conjugates of catechin play an important part in the metabolic picture. Griffiths[227] reported that conjugates were found in the urine of (+)-catechin-treated rats and similar findings were noted with guinea pigs,[220] monkeys,[221] and man.[222] Shaw and Griffiths[229] reported that more than 40% of the oral dose (40 mg/kg) was excreted in the urine of rats as conjugates of 3'-O-methyl-(+)-catechin. In similar experiments, Hackett et al.[230] showed that the major urinary metabolites were the glucuronide of (+)-catechin and the glucuronide and sulfate of 3'-O-methyl-(+)-catechin. Noteworthy in this study, which employed uniformly ^{14}C-labeled material, was the finding that the well-known ring fission products accounted for only one-sixth of the 50 to 63% of the radioactivity which appeared in the urine. This reduced formation of fission products compared with that found in earlier studies was suggested to be a result of the use of a moderate dose (40 mg/kg) which would allow for more complete absorption from the intestine. Additionally, the use of specific pathogen-free rats in this study might be expected to reduce the significance of the bacterial reactions due to a more restricted microflora. Another interesting finding in this study was that the glucuronide of 3'-O-methyl-(+)-catechin was not identical with the corresponding conjugate shown to be excreted in rat bile (see below). Hackett et al. also showed that the marmoset, under similar conditions, excreted only ~22% of the radioactivity in the urine. The main conjugate formed in this species was the sulfate of 3'-O-methyl-(+)-catechin. Wermeille et al.[231] and Hackett et al.[228] reported that glucuronide conjugates of both (+)-catechin and its 3'-O-methyl ether as well as the sulfate conjugate of the latter were urinary metabolites of (+)-catechin in man. About 20 to 40% of an oral dose (1 or 2 g) was excreted in the urine as these conjugates. On the other hand, very small amounts (usually ~0.1 to 0.3%) of the nonconjugated forms of these metabolites were detected. This finding agrees closely with the value of ~0.5% for unchanged urinary (+)-catechin reported by Balant et al.[232] in subjects given doses of 0.5 to 2 g.

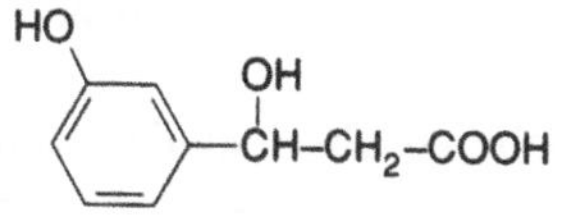

(92)

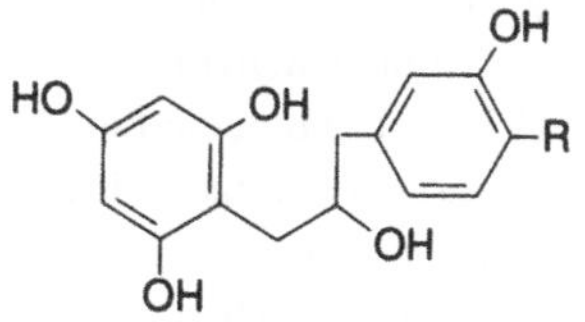

(93)

(a) R = OH

(b) R = H

Although the details of the degradation of (+)-catechin to phenylvalerolactones and of the latter compounds to phenolic acids are not entirely clear, some data on these points are available. Watanabe[218] proposed that the initial step in the degradation of (+)-catechin is the scission of the O–C-bond in the heterocyclic ring to form the diphenylpropanol intermediate (93a). However, this compound remained undetected in all subsequent investigations until Groenewoud and Hundt[233] isolated it from the anaerobic incubates of (+)-catechin with rat cecal microorganisms. They also isolated the corresponding dehydroxylated compound (93b). Watanabe suggested that the following step was fission of the phloroglucinol moiety of the diphenylpropanol derivative to give an acidic intermediate which lactonized to the phenylvalerolactones which have been commonly detected as (+)-catechin metabolites. This scheme would result in the loss of part of the phloroglucinol moiety as CO_2 and is thus in accord with the finding that $^{14}CO_2$ was found in the expired air of rats or guinea pigs given ^{14}C-labeled (+)-catechin.[225] Also, the production of $^{14}CO_2$ was greater with A-ring-labeled compound than with randomly-labeled compound. Using a product of the latter type, Hackett et al.[230] recovered 7% of the oral dose (40 mg/kg) in rats as respiratory $^{14}CO_2$ in 4 d. The corresponding values in marmosets was 3% in 3 d. Awata et al.[234] gave ^{14}C-labeled (+)-catechin orally to rats and recovered 8% of the radioactivity in the expired air in 3 d. These results show that part of the A-ring can be completely oxidized. While the degradation of (+)-catechin and the phenylvalerolactones to phenolic phenylpropionic acid derivatives is well established, it is also known that the flavonoid, when incubated anaerobically with rat or especially rabbit intestinal microorganisms, may be converted to C_6–C_5 phenolic acids.[235] These compounds were shown to be 5-(3-hydroxyphenyl)valeric acid (94a) and 5-(3,4-dihydroxyphenyl)valeric acid (94b). However, it is not known if these metabolites are intermediates leading to the C_6–C_3-phenolic acids or if they represent terminal metabolites formed under special conditions. They have not been reported to be excreted in the urine of animals given (+)-catechin.

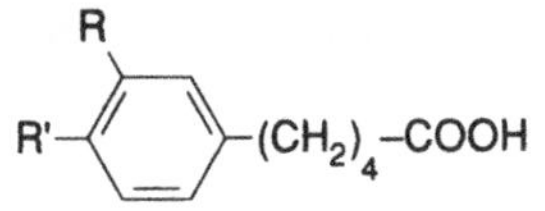

R
R'—⟨benzene ring⟩—(CH$_2$)$_4$—COOH

(94)

(a)   R = OH , R' = H

(b)   R = R' = OH

While the results summarized above deal mainly with the urinary excretion of (+)-catechin and its metabolites, excretion in both the feces and bile has also been studied. Das and Griffiths[225] found that only ~1% of the radioactivity was excreted in the 48-h feces following the oral administration of [^{14}C]-(+)-catechin to rats or guinea pigs. Similar experiments in monkeys indicated ~2% excretion in 5 d.[221] (+)-Catechin and *m*-hydroxyphenylpropionic acid were identified as fecal metabolites in monkeys, a finding similar to that reported earlier in man.[222] However, nearly 20% of the orally administered (+)-catechin (83 mg/kg) was recovered unchanged in the feces in man. In rats, *m*-hydroxyphenylpropionic acid and the phenylvalerolactones (91a) and (91b) were detected in the feces.[225,227]

The biliary excretion of (+)-catechin and its metabolites has been studied in rats. Griffiths and Barrow[236] found that unchanged compound, two conjugates of (+)-catechin, and an unknown compound were excreted in the bile following oral or i.p. administration. Two further conjugates were also detected in the latter case. Das and Sothy[237] reported that 33 to 44% of the radioactivity was excreted in the bile of rats within 24 h after i.v. injection of [^{14}C]-(+)-catechin. The biliary metabolites were of the same types as those reported by Griffiths and Barrrow.[236] Further study of the nature of the material in rat bile was made by Shaw and Griffiths[229] who found that the major metabolite was the glucuronide of 3'-*O*-methyl-(+)-catechin. Although several compounds were detected in the bile, the major metabolite accounted for nearly half of the total. These findings were confirmed by Miura et al.[219] As noted above, this glucuronide was not identical with the urinary glucuronide of 3'-*O*-methyl-(+)-catechin.[230] The findings of van der Merwe and Hundt[238] may be of significance in this regard. Using the isolated perfused pig liver they found that (+)-catechin was metabolized to two types of diglucuronide conjugates. These were the 3,4'-diglucuronide and either the 3,5- or 3,7-diglucuronide, both of (+)-catechin itself, and also of 3'-*O*-methyl-(+)-catechin. Thus, four distinct conjugates were identified. Additionally, they showed that the *O*-methylation of the B-ring also occurred to a slight extent at C-4'.

The metabolism of (–)-**epicatechin** (90), a stereoisomer of (+)-catechin, was studied in rabbits.[239] Following its oral administration, the same acidic metabolites (*m*-hydroxybenzoic acid, protocatechuic acid, and vanillic acid) and the same neutral lactones (compounds (91a), (91b), and (91c)) were excreted in the urine as were found earlier with (+)-catechin.[213,214] (+)-Catechin and (–)-epicatechin have different optical properties and this difference was also confirmed in metabolite (91b). When (–)-epicatechin was given, this metabolite was levorotatory whereas the dextrorotatory form was produced from (+)-catechin.

On the basis of incubation experiments using rat intestinal microorganisms, it appears that the degradation of (–)-**epiafzelechin** (95) follows the same general pattern as that described above with the isomers of catechin. Thus, Griffiths and Smith[157] detected p-hydroxyphenylpropionic acid and two neutral metabolites in the incubates. Subsequently, Griffiths and Barrow[236] identified a neutral metabolite of (–)-epiafzelechin as δ-(p-hydroxyphenyl)-γ-valerolactone, i.e., isomeric with compound (91a).

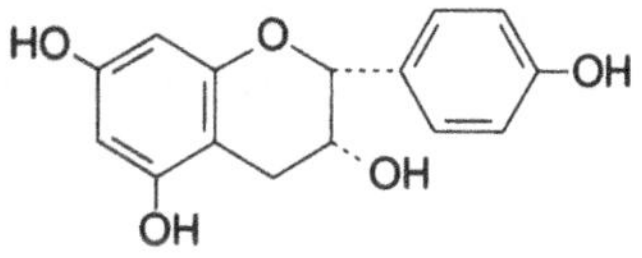

(95)	(96)
(–)-Epiafzelechin	Proanthocyanidin B-3

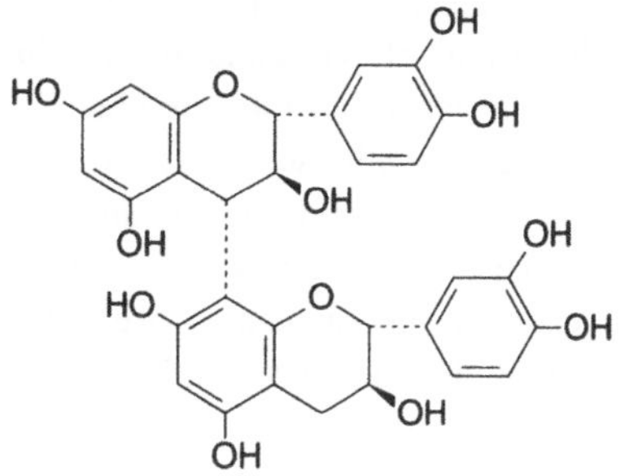

In addition to the many studies on the metabolism of the monomeric catechins summarized above, some data are available on the fate of the more complex condensed catechins. Groenewoud and Hundt[240] investigated the metabolism of **proanthocyanidin B-3** (96) by the rat cecal microflora. Prolonged anaerobic incubation of this dimeric compound resulted in the formation of numerous metabolites, of which seven were identified. In general, the metabolic picture was similar to that seen with (+)-catechin. The metabolites detected included derivatives of phenylvalerolactone and diphenylpropanol which as noted above are formed from (+)-catechin. However, several noteworthy differences were also found. Firstly, the phenolic phenylpropionic acid derivative detected was the p-hydroxy, rather than the m-hydroxy, compound. The fully dehydroxylated derivative, phenylpropionic acid, was also detected. Secondly, phenylacetic acid and its p-hydroxy derivative were found. These C_6–C_2-compounds are not typical catechin metabolites. Finally, phloroglucinol was detected. It is possible that some of these atypical results were due to the abnormally long incubation times employed (~120 h) or to significant differences in the nature of the cecal microflora of the rats used compared with that present in other studies.

In addition to the above investigations using pure catechins, the metabolism of mixtures of catechins in animals has also been reported. Harmand and Blanquet[241] employed ^{14}C-labeled products from grapes which contained mainly (+)-catechin, (–)-epicatechin, and the proanthocyanidin dimers B-1, B-2, B-3, and B-4. According to the labeling procedure used, products which were uniformly labeled, or selectively labeled in the A- or B-ring were obtained. Significantly, administration to rats of the uniformly and A-ring-labeled products gave rise to respiratory ^{14}CO$_2$ whereas this metabolite was not produced from the product labeled in the B-ring. Not unexpectedly, the degradation of these products was shown to be dependent on the activity of the intestinal microflora. Anaerobic incubation with cecal contents resulted in the formation of numerous metabolites including benzoic acid, 4-vinylcatechol, 4-ethylcatechol, and the m-hydroxy derivatives of phenylacetic and phenylpropionic acids. Because of the complex nature of the products employed, the identities of the precursors of these metabolites are a matter of speculation. Many of them, including ethylcatechol which was the major metabolite in the incubates, are not typical catechin metabolites. In fact, several of them are identical with those formed from caffeic acid when the latter is given to rats or incubated with rat cecal microorganisms (see the chapter entitled Metabolism of Acids, Lactones, and Esters). Numerous urinary metabolites of the catechin products were identified which included large amounts of conjugates of the administered substances. Among the many degradation products detected in the urine were the m-hydroxy derivatives of benzoic, hippuric, phenylacetic, phenylpropionic, and cinnamic acids. Also found were vanillic, homovanillic, and hippuric acids, ethylcatechol, and δ-(3,4-dihydroxyphenyl)-γ-valerolactone (91b).

The fate of tea catechins was studied in rabbits[242] and in guinea pigs[243] and some metabolic results of feeding lucerne tannins to mice have been reported.[244] However, these products contained a mixture of catechins as well as other related compounds and it is difficult to determine the origin of particular metabolites. Both products contained epicatechins and gallocatechins, the general formula of which is shown in structure (97). In addition, the lucerne preparation contained gallic acid.[245] It is therefore not surprising that a wide variety of C_6-C_1- and C_6-C_3-phenolic acids and catechin derivatives were detected in the feces of mice given the lucerne tannin. The urinary metabolites of the tea catechin preparation in both rabbits and guinea pigs were generally the same as those derived from gallic acid (see the chapter entitled Metabolism of Acids, Lactones, and Esters).

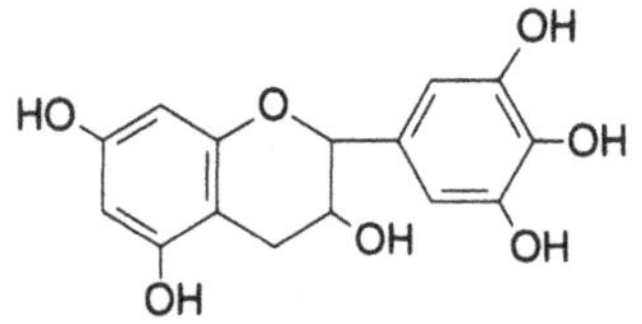

(97) 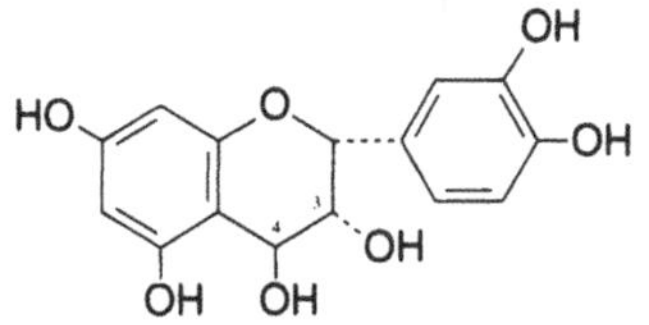 (98)

Leucocyanidin

Information on the metabolism of leucoanthocyanidins is limited to **leucocyanidin** (98). Masquelier et al.[246] and Claveau and Masquelier[247] administered it orally or i.p. to rats and detected homoprotocatechuic acid, homovanillic acid, *m*-hydroxyphenylacetic acid, and phloroglucinol glucuronide in the urine. The formation of these phenolic acids indicates that the 3,4-bond has been broken and that the metabolic degradation leading to phenolic acids is similar to that summarized above with the flavonols. The excretion of phloroglucinol is also similar to that reported in some studies of flavonol metabolism.

DIHYDROCHALCONES AND CHALCONES

Dihydrochalcone (99) is the parent of a small group of plant compounds which are structurally closely related to the flavanones. This similarity is more easily seen when the structural formula is drawn as in (100). Although dihydrochalcone itself occurs naturally, most of these compounds contain hydroxy groups or their derivatives in one or usually both of the aromatic rings.

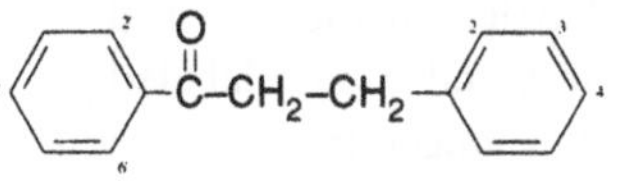

(99) (100)

Dihydrochalcone

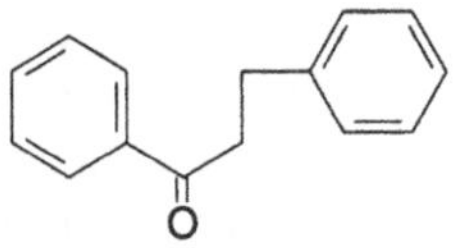

Information on the metabolism of dihydrochalcones and chalcones is not extensive and deals mainly with phloretin (101) and its 2′-*O*-glucoside, **phloridzin**. The metabolism of the latter compound in rabbits was the subject of an early investigation by Schüller[248] who reported that it was excreted in the urine mainly as phloridzin glucuronide. This reaction also occurs in the dog,[249] although somewhat less of the glucuronide than of phloridzin itself was excreted. The location of attachment of the glucuronic acid moiety is not known although Schüller believed that this was at the 4-position rather than at one of the hydroxy groups of the phloroglucinol ring. Booth et al.[201] reported that phloridzin-treated rats excreted the same urinary metabolites as those found with phloretin described below. These results were confirmed by Griffiths and Smith.[157] Griffiths and Barrow[236,250] showed that biliary excretion was also involved in phloridzin elimination in rats. Following its i.p. administration, unchanged compound as well as three acid-labile conjugates of increased polarity were detected in the bile. Oral administration of phloridzin led to two different biliary metabolites, of which one was an acid-labile conjugate. The metabolism of phloridzin by intestinal bacteria is noted below.

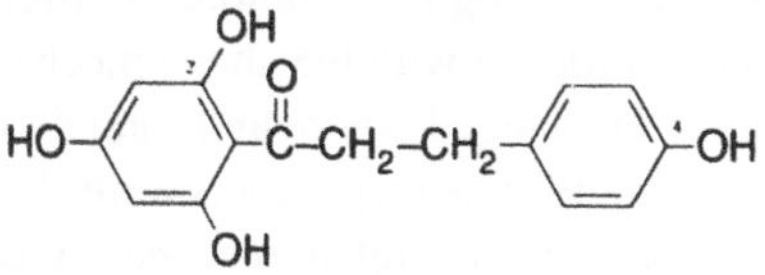

(101)

Phloretin

Phloretin (101) is structurally similar to the flavanone naringenin (79a) and it is therefore interesting to note that the same urinary metabolites are produced from both compounds.[201] This investigation showed that rats given large doses of phloretin (100 mg/rat, p.o., or s.c.) excreted *p*-hydroxyphenylpropionic acid (phloretic acid) as the major urinary metabolite together with some *p*-coumaric acid and the ethereal sulfate of *p*-hydroxybenzoic acid. In addition to these degradation products, some phloretin and its glucuronide were excreted in the urine following phloretin administration. A subsequent study of the metabolism of phloretin in rats by Monge et al.[251] dealt with the quantitative aspects of metabolite excretion. They confirmed the excretion of phloretic, *p*-coumaric, and *p*-hydroxybenzoic acids and of phloretin itself. Following an intragastric dose (~200 mg/kg), the urinary excretion values of these metabolites were 33, 6, 5, and 4%, respectively, in 4 d. Phloretin itself was detected only during the first 24 h, however the phenolic acids showed prolonged excretion patterns. A small amount of the phloretic acid excreted was shown to be conjugated with glycine. Additionally, 5% of the dose was found in the form of phloroglucinol. These urinary metabolites accounted for about half of the administered dose. An undetermined amount of excretion of phloretic acid and phloretin in the feces was also noted.

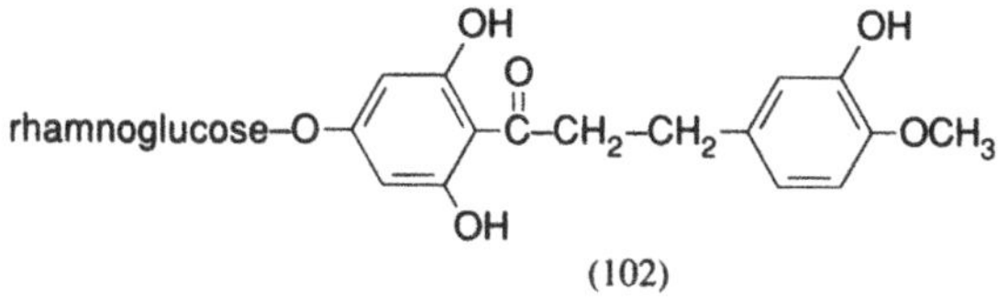

(102)

Preceding sections have clearly shown that the intestinal microflora play an essential part in the degradative metabolism of flavonoids. This is also true with the dihydrochalcone derivatives. DeEds[149] described experiments in which two compounds not occurring naturally, the dihydrochalcones of naringin and neohesperidin, were incubated anaerobically with cecal microorganisms. The former compound, which is phloretin-4-rhamnoglucoside and therefore closely related to phloridzin, was degraded to phloretic acid. This phenolic acid was a major urinary metabolite of the dihydrochalcone in rats. Incubation of the dihydrochalcone of neohesperidin (102) led to the formation of 3-hydroxy-4-methoxyphenylpropionic and 3-hydroxyphenylpropionic acids. A subsequent report by Griffiths and Smith[157] described the metabolism of phloridzin in anaerobic cultures of rat cecal microorganisms. These incubates contained the aglycon phloretin (101), the expected phenolic acid, phloretic acid, and also a third metabolite which was identified as phloroglucinol. Similar experiments by Monge et al.[251] showed that both phloretin and phloridzin were metabolized to phloretic acid and phloroglucinol. It was also shown that the latter metabolite was susceptible to further bacterial metabolism. This fact may explain why much less phloroglucinol than phloretic acid was found in the urine of rats given phloretin. The finding that both phloretic acid and phloroglucinol are formed from phloretin is of considerable interest because it suggests that the degradation of dihydrochalcones may depend upon the action of a hydrolase which splits the molecule between the ketone group and the phloroglucinol ring. This view was strengthened by the findings of Skjevrak et al.[252] who studied the metabolism of 2′,4′,4-trihydroxydihydrochalcone and 2′,6′,4-trihydroxydihydrochalcone in rats. These analogues of phloretin differ only in their lack of a ring hydroxy group and cleavage in accordance with the above mechanism will produce phloretic acid and resorcinol. Although the unchanged compounds and their conjugates were the most abundant urinary metabolites, both of the fission products were also detected. Interestingly, the amounts of phloretic acid and resorcinol excreted were nearly equal and it was found that resorcinol, unlike phloroglucinol, did not undergo further metabolism by the intestinal microflora.

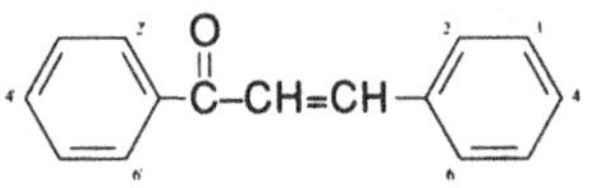

(103)

The naturally occurring derivatives of chalcone (103) contain multiple hydroxy or methoxy groups. These substituents may be present in one or both rings, however a hydroxy group is commonly found at the 2′-position. Formanek and Höller[253] administered several nonplant chalcones (4′,2,4-trihydroxy-, 2′,2,4-trihydroxy-, 2′,4-dihydroxy-, and 4-hydroxychalcone) orally or i.v. to rats. They reported that a series of unidentified compounds was rapidly excreted in the urine and bile. Chanal et al.[254] studied the disposition in rats of the semisynthetic derivative hesperidin methyl chalcone which is formed by methylation of the flavanone glycoside hesperidin. They found that biliary excretion of the *methoxy*-^{14}C-labeled compound amounted to 70% of the dose (10 mg/kg) in 24 h. When this dose was given orally to noncannulated rats, ~30% of the radioactivity was detected in the urine and 26% in the feces in 48 h. The nature of the metabolites was not determined.

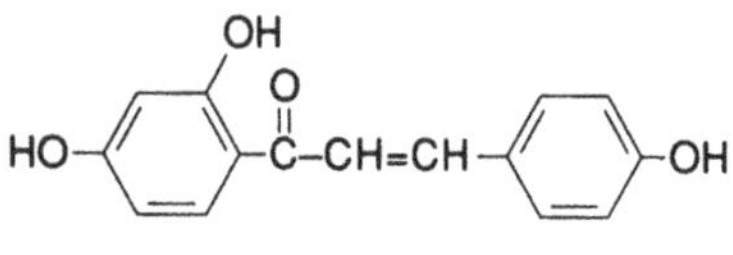

(104)   (105)

Isoliquiritigenin   Butein

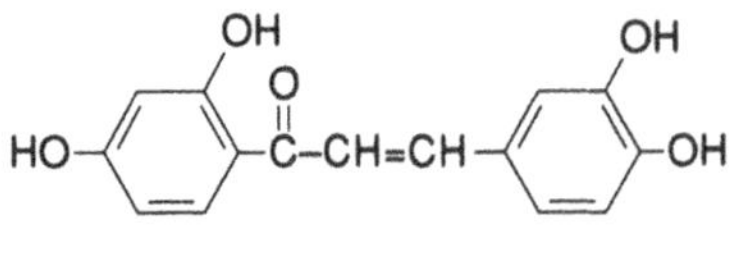

The metabolism of **isoliquiritigenin** (2′,4′,4-trihydroxychalcone) (104) and **butein** (2′,4′,3,4-tetrahydroxychalcone) (105) in the rat was studied by Brown and Griffiths[255] using compounds labeled with ^{14}C in the carbonyl group. This investigation also included 2′,3,4-trihydroxychalcone and the 3-*O*-methyl- and 4-*O*-methyl ethers of butein. Following an i.p. dose (~60 mg/kg) of isoliquiritigenin, nearly 100% of the radioactivity was excreted in the bile in 24 h. This material consisted of sulfate conjugates, mainly of the administered chalcone but also of the corresponding flavanone liquiritigenin (isomeric with and formed by ring-closure of the chalcone) and of trace amounts of the hydroxylated product butein. When the dose of isoliquiritigenin was given orally to noncannulated rats, 56% of the radioactivity was recovered in the urine and 40% in the feces. Unchanged isoliquiritigenin was a prominent fecal metabolite, however two additional compounds were detected which were resistant to acid or enzymatic hydrolysis and which did not correspond to the flavanone or dihydrochalcone derivatives of isoliquiritigenin. Similar results were obtained when butein was given. Biliary excretion of radioactivity was 53% following an i.p. dose and this material consisted largely of sulfate conjugates. The aglycons of these conjugates were mainly butein and the 3-*O*-methyl ether of butein, however small amounts of the 4-*O*-methyl ether were also formed. Small amounts of the corresponding flavanones formed by ring-closure of the chalcones were also identified. When the dose (~60 mg/kg) of butein was given orally to noncannulated rats, the excretion of radioactivity in the urine (49%) and feces (51%) was similar to that noted above with isoliquiritigenin. The major urinary metabolite following hydrolysis of the samples was butein, however small amounts of methyl ether derivatives were also present. The latter were also found in the feces, however the major fecal metabolites were nonhydrolyzable sulfate conjugates of unknown structure. Neither butein nor isoliquiritigenin were metabolized by the rat cecal microflora and no respiratory $^{14}CO_2$ was detected following oral dosage of the chalcones. This negative finding is to be expected in view of the lack of a 6'-hydroxy group in these two compounds. The 6'-hydroxy group is equivalent to the 5-hydroxy group in most other groups of flavonoids. As noted in preceding sections, this substituent is required if ring scission by the microflora is to take place.

ISOFLAVONOIDS

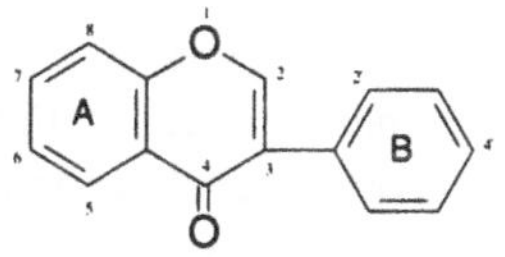

(106)

Isoflavonoids differ structurally from the flavonoids in having the phenyl ring (B-ring) attached at the 3- rather than at the 2-position of the heterocyclic ring. This is illustrated in the structure of isoflavone (106), derivatives of which form the most common group of isoflavonoids. Our knowledge of isoflavonoid metabolism is largely limited to these isoflavones and owes its existence mainly to the fact that several of these compounds exert effects on the reproductive system of animals. As these isoflavones occur naturally in several species of forage legumes, knowledge of their metabolism as well as other biological properties is of practical interest.[256]

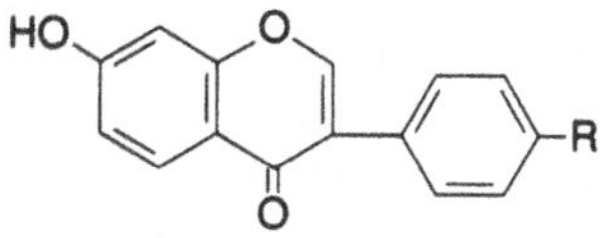

(107)

(a) Daidzein, R = OH

(b) Formononetin, R = OCH₃

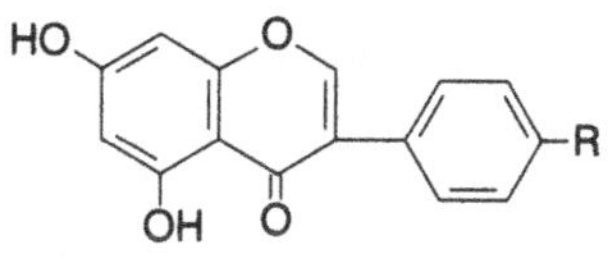

(108)

(a) Genistein, R = OH

(b) Biochanin A, R = OCH₃

Most of the data on isoflavone metabolism deals with the compounds **daidzein** (107a), **formononetin** (107b), **genistein** (108a), and **biochanin A** (108b). The structural formulas show that formononetin and biochanin A are the 4'-*O*-methyl ethers of daidzein and genistein, respectively. The first published report on isoflavone metabolism indicated that *O*-demethylation could occur.[257] When biochanin A (108b) was incubated *in vitro* with rumen fluid from cattle or sheep, genistein (108a) was formed along with variable amounts of two unidentified metabolites. Subsequent reports confirmed the *O*-demethylation of biochanin A when incubated with rumen fluid.[258-260] Also, the corresponding reaction with formononetin (107b) to give daidzein (107a) has been demonstrated.[258-261] In contrast, the two methoxy compounds have been shown to be largely resistant to demethylation when incubated with microorganisms from rat intestine.[157,262] However, administration of biochanin A (25 mg/kg, i.p.) to rats resulted in the excretion of unchanged compound and its demethylation product in the urine and feces.[262] This reaction was shown to be carried out by the liver. Further studies in rats using ^{3}H-labeled biochanin A given i.p. showed that ~30% of the radioactivity appeared in the urine and 60% in the feces.[263] Most of the urinary material was in the form of conjugates, but ~8% free biochanin A and 1% free genistein were detected. Fecal radioactivity was entirely unconjugated and consisted of ~40% unchanged compound and 14% genistein. These values are explainable in terms of the biliary excretion of biochanin A and its metabolites. The isoflavone conjugates appear to be glucuronides and sulfates.[263] The conjugates, chiefly glucuronides, comprise 98 to 99% of the total plasma isoflavones in sheep fed on diets of estrogenic clover.[264] Labow and Layne[265] demonstrated that biochanin A, genistein, formononetin, and daidzein as well as the related compound equol (109), form monoglucuronides when incubated with rabbit liver microsomal fractions and UDP-glucuronic acid. The structures of the glucuronides were not determined but with formononetin (107b) the attachment must be at the 7-position which is the only available site. Interestingly, Labow and Layne also showed that these isoflavones were converted in low yield to monoglucosides when UDP-glucose was substituted for UDP-glucuronic acid in the incubates.

A comprehensive study of the liver microsomal system catalyzing the O-demethylation of biochanin A, formononetin, and other methoxylated isoflavones was carried out by Nilsson.[266] Appreciable demethylating activity was present in liver preparations from rabbits, rats, and mice with lower activities found in swine, cattle, and sheep. The lower activities in the preparations from the larger animals may, however, be due to technical factors rather than intrinsic species differences. The following isoflavones underwent O-demethylation when incubated with a cofactor-fortified supernatant fraction (14,000 g) from rabbit liver: **prunetin** (5,4′-dihydroxy-7-methoxyisoflavone), **prunusetin** (7,4′-dihydroxy-5-methoxyisoflavone), **muningin** (6,4′-dihydroxy-5,7-dimethoxyisoflavone), 7-methoxyisoflavone, and 5,7,2′-trimethoxyisoflavone. These results indicate that O-demethylation may take place at sites in both the A-ring and B-ring. In addition, **ferreirin** (5,7,2′-trihydroxy-4′-methoxyisoflavanone) was demethylated by the rabbit liver preparation. Both biochanin A and formononetin underwent demethylation in sheep following their intraruminal or i.m. administration.[267]

Braden et al.,[268] in an attempt to explain why cattle are less susceptible than sheep to the infertility syndrome caused by estrogenic isoflavonoids, studied the plasma concentrations of these compounds and their known metabolites. Their results suggested that the isoflavonoids and their metabolites were more efficiently conjugated in cattle than in sheep and also that the metabolism of formononetin was faster in cattle. However, Lundh et al.[269] found only small differences in the capacity of sheep and cattle liver microsomes to demethylate formononetin or to conjugate formononetin or daidzein with glucuronide.

From the foregoing it can be seen that methoxylated isoflavones are subject to demethylation both by tissue enzymes and by some intestinal bacteria. However, several other and far more profound metabolic changes also occur with isoflavones. Batterham et al.[258] and Braden et al.[259] reported that biochanin A and genistein, when given intraruminally to sheep, were degraded to p-ethylphenol which was excreted in the urine. This pathway appeared to account for ~80% of the ingested isoflavone. Interestingly, i.m. injection of these isoflavones did not lead to an increased excretion of p-ethylphenol. Nekby[270] reported that ~58% of the daily intake of biochanin A from a diet of clover (containing ~5 g of total isoflavones) was excreted by sheep as p-ethylphenol, mainly in conjugated form. Similar experiments in sheep employing [4-^{14}C]biochanin A[271] showed that the p-ethylphenol excreted was not radioactive, indicating that this metabolite is derived from the isoflavone B-ring. Lindsay and Francis[272] concluded that the rate of genistein inactivation in sheep increased during a period of continuous ingestion of the isoflavone.

The steps leading to p-ethylphenol are not known although it seems possible that scission of the isoflavone to a phenyl-α-methylbenzyl ketone intermediate followed by hydrolysis of the latter may be involved. This would also give rise to phenolic acids and, in fact, evidence is available indicating the formation of several such compounds, although the small amounts detected were insufficient to permit identification.[271] Evidence concerning the site of *p*-ethylphenol formation is conflicting. Braden et al.[259] looked for but did not find degradation of biochanin A and genistein to simple phenols in incubation experiments using sheep rumen fluid. However, Griffiths and Smith[157] showed that genistein was extensively converted to *p*-ethylphenol in incubates of rat intestinal bacteria. Also, Nekby[270] showed that sheep rumen fluid converted biochanin A to *p*-ethylphenol and also that the protozoal fraction was much more active than the bacterial fraction in this regard. In contrast, the rumen bacteria were mainly capable of demethylating biochanin A.

Formononetin and daidzein differ from biochanin A and genistein only by their lack of a 5-hydroxy group. Nonetheless, this deficiency is sufficient to give rise to profound differences in their overall metabolic fates. Su and Zhu[273] studied the excretion of radioactivity in the urine and feces of rats given [^{14}C]daidzein. About two-thirds of the radioactivity was absorbed from the gastrointestinal tract after oral administration and 39% of the dose was excreted in the bile (24 h). This resulted in about equal amounts of radioactivity being excreted in the urine and feces. The importance of biliary excretion was also seen in experiments which employed i.v. doses. Although over 70% of the radioactivity was excreted in the urine, the excretion (24 h) in the feces was 17%. The radioactive material in the gastrointestinal tract, urine, and bile consisted mainly of metabolites, however the nature of these was not determined. The absorption and excretion of daidzein in rats was also studied by Yueh and Chu,[274] however their method specifically measured unchanged daidzein. Only 0.7% of a large oral dose (500 mg/kg) was excreted in the urine in 24 h and a third of the dose was found in the feces at this time. When the oral dose was reduced to 50 mg/kg, nearly 4% was excreted in the urine as unchanged daidzein with a further 12% present in a conjugated form which liberated daidzein upon acid hydrolysis. Biliary excretion of unchanged daidzein was low (~0.9% in 6 h) after i.v. administration (20 mg/kg), however it was noted that a larger amount of conjugated daidzein was present in the bile. A similar picture of excretion of unchanged compound was seen in two human subjects who received 300 mg daidzein orally. Only 1.8% of the dose was excreted unchanged in the urine in 24 h and 12.5% was recovered in the feces (60 h). The amount of conjugated daidzein in these samples was not determined.

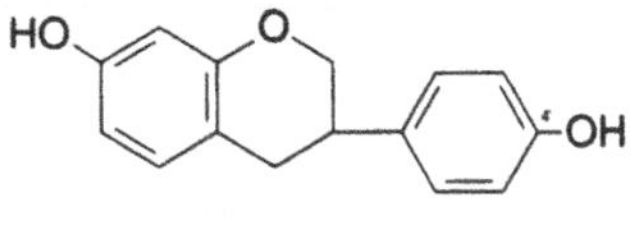

(109) 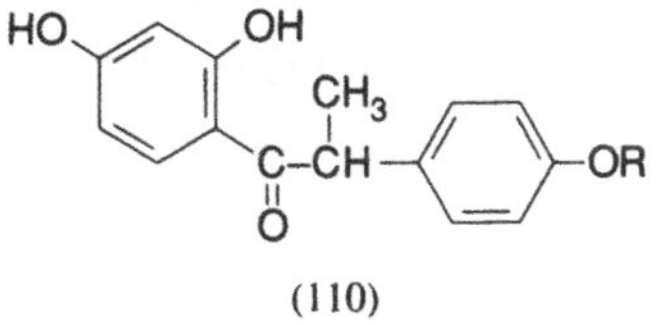 (110)

(a) R = H

(b) R = CH₃

In the case of formononetin, several metabolites have been identified. Following its intraruminal administration to sheep, equol (7,4′-dihydroxyisoflavan) (109) was excreted in the urine.[258,259] This did not occur when the isoflavone was given by i.m. injection. The metabolism of formononetin to equol in sheep was confirmed by Shutt and Braden,[275] Batterham et al.,[271] and Nekby[270] who reported that ~9% of the daily intake of formononetin from a diet of clover (containing ~5 g of isoflavones) was excreted by sheep as free urinary equol. Slightly more was excreted in conjugated form. The 4′-*O*-methyl ether of equol has also been identified as a urinary metabolite of formononetin.[147] Batterham et al.[271] also identified 2,4-dihydroxyphenyl-α-(4′-hydroxyphenyl)ethyl ketone (*O*-desmethyl angolensin) (110a) as a urinary metabolite of formononetin in sheep. Cox and Braden[147] stated that small amounts of the corresponding *O*-

methyl compound, angolensin (110b), were also excreted. These reactions are not confined to sheep as guinea pigs, fed a diet containing formononetin, were found to have both of these metabolites in the blood.[275] Equol and *O*-desmethyl angolensin are evidently produced by different metabolic sequences, but none of these investigations has found simpler end-products such as *p*-ethylphenol which, as noted above, is formed from biochanin A and genistein. Incubation of formononetin or daidzein with sheep rumen fluid[260] or rat intestinal microorganisms[157] resulted in the formation of equol. The results of Nekby[270] indicated that the protozoal fraction from sheep rumen fluid was much more active than the bacterial fraction in converting formononetin to equol. As with biochanin A, the metabolism of formononetin by rumen bacteria was mainly limited to demethylation.

XANTHONES

Derivatives of xanthone (111) are not widespread in nature and relatively little is known of their metabolism in animals. For this reason the metabolism of xanthone itself will be included although all of the naturally occurring compounds have a hydroxy group at C-1.

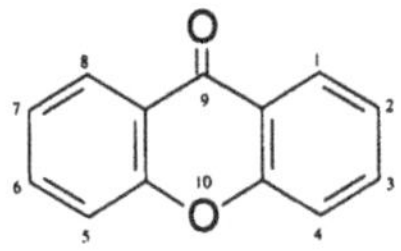

(111)

The metabolism of xanthone (111) in rats was studied by Griffiths[276] who found that the major unconjugated urinary metabolite was 4-hydroxyxanthone, accounting for about one-fourth of the 300 mg/kg dose. The other monohydroxy derivatives formed were 2-hydroxyxanthone (13%) and 3-hydroxyxanthone (6%). Trace amounts of five additional urinary metabolites were present, one of which was an acid-labile conjugate and another an *o*-dihydroxy derivative. Unchanged compound was not excreted. Conjugates of both 2- and 4-hydroxyxanthone were excreted in the urine and bile, the latter occurrence no doubt responsible for the finding that metabolite excretion was prolonged. Rupture of the heterocyclic ring, a reaction known to occur with coumarin and its derivatives and many hydroxylated flavonoids, was not detected. This finding is similar to that seen with the structurally related compound flavone, which undergoes ring hydroxylation but not ring fission.

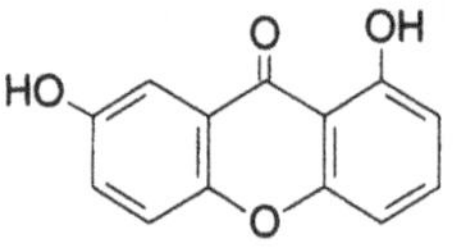

(112)

Euxanthone

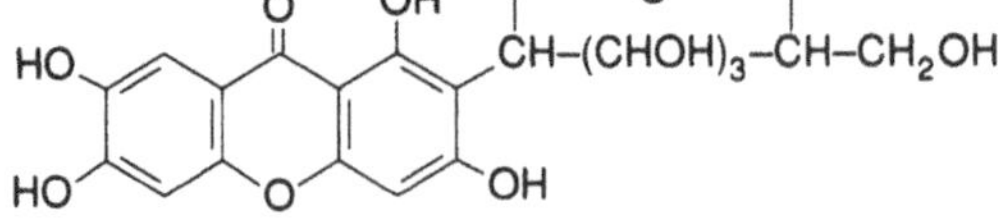

(113)

Mangiferin

Of the naturally occurring xanthones, only **euxanthone** (1,7-dihydroxyxanthone) (112) and mangiferin (113) have been studied metabolically. Very early experiments[277,278] showed that oral administration of euxanthone to rabbits and dogs resulted in the urinary excretion of the 7-glucuronide derivative, euxanthic acid. It appears that the 1-glucuronide (isoeuxanthic acid) was not formed. The same metabolite, euxanthic acid, was shown by Wiechowski[279] to be excreted in the urine of rabbits fed the crude yellow coloring matter from mango leaves. Subsequently, pure samples of **mangiferin** (113), the yellow principle from the leaves of *Mangifera indica*, were administered to animals. Iseda[280] showed that euxanthic acid was excreted in the urine of rabbits given mangiferin orally. Similar experiments were reported by Krishnaswamy et al.[66] No free mangiferin or 1,3,6,7-tetrahydroxyxanthone (norathyriol) was detected in the urine but both euxanthone (10%) and its glucuronide, euxanthic acid (12%), were found. The elimination of mangiferin from the organism was slow and a relatively large amount of benzoic acid was excreted by the treated but not the control animals. The unusual sequence in the metabolism of mangiferin involves loss of the *C*-glucoside moiety and removal of two phenolic hydroxy groups. Both types of reactions are characteristic of the intestinal microflora and Hattori et al.[281] showed that anaerobic incubation of mangiferin with human fecal bacteria resulted in the formation of the aglycon, 1,3,6,7-tetrahydroxyxanthone.

OTHER OXYGEN HETEROCYCLIC COMPOUNDS

The metabolism of **kadsurenone** (114) and its semisynthetic derivative 9,10-dihydrokadsurenone was studied *in vivo* in rhesus monkeys and *in vitro* with rat liver microsomes.[282] Evidence for an epoxide-diol pathway was obtained in both situations and the 9,10-dihydroxy derivative was identified as a urinary metabolite and a product of the microsomal incubations. In the former case the diol was excreted as its monoglucuronide conjugate and accounted for 4% of the dose (10 mg/kg, i.v.) in 24 h. The diol was shown to exist as a mixture of C-9-epimers. Oxidation of the alkyl group at C-5 was also noted in similar experiments with the reduced derivative labeled with ^{3}H. Hydroxylation at C-8, C-9, or C-10 was found *in vitro* and the 9-hydroxy and 10-hydroxy derivatives were also isolated (as glucuronide conjugates) from the urine. Numerous minor products of 9,10-dihydrokadsurenone were detected both in the urine and in the microsomal incubates. Most of the radioactivity following p.o. or i.v. doses (10 mg/kg) was slowly excreted in the urine, however nearly 25% of the dose was recovered in the feces following both routes of administration.

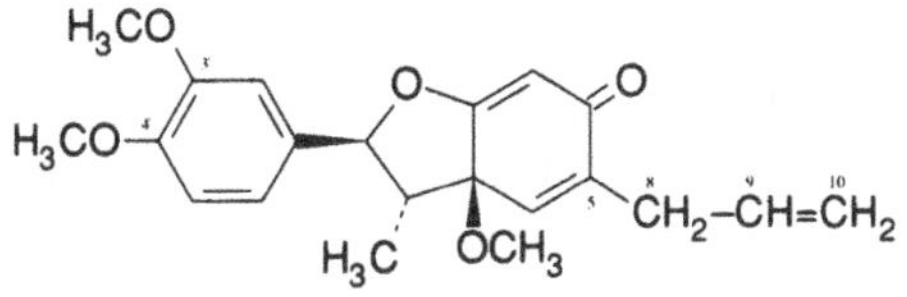

(114)

Kadsurenone

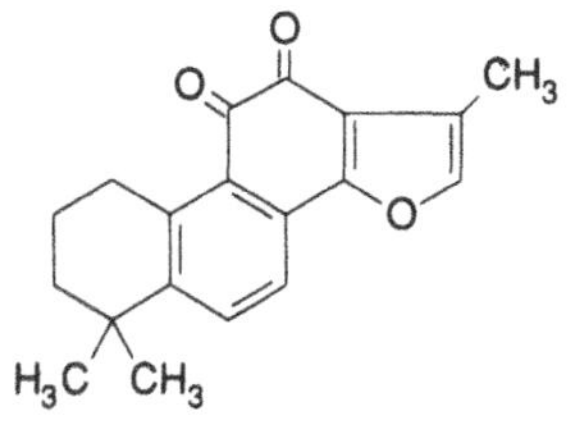

(115)	(116)
Tanshinone IIA	Cryptotanshinone

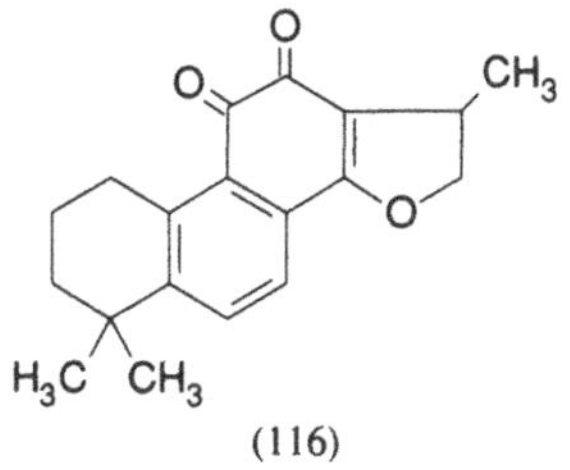

Luo et al.[283] administered **tanshinone I, tanshinone IIA** (115), and **cryptotanshinone** (116) intraduodenally to rats. It was reported that tanshinone IIA was excreted in the bile and that cryptotanshinone was dehydrogenated in the liver to tanshinone IIA.

The metabolism of several iridoid derivatives, mainly by intestinal microorganisms, has been studied. Miyagoshi et al.[284] found that **geniposide** (117) was hydrolyzed to its aglycon (genipin) when incubated with large intestine contents from rats. No structural changes occurred when it was incubated with gastric juice (pH 1.8) at 37° C, however similar treatment of the closely-related **gardenoside** (118) resulted in the formation of an epimeric pair having structures (119) and (120). The latter compounds were hydrolyzed to their corresponding aglycons when incubated with intestinal contents. Kohashi et al.[285] noted the hydrolysis of geniposide by a β-glucosidase from an intestinal *Eubacterium* and the hydrolysis of the aglycon genipin by bacterial and microsomal esterases.

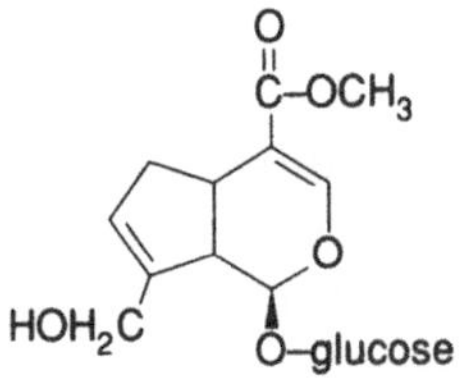

(117)

Geniposide

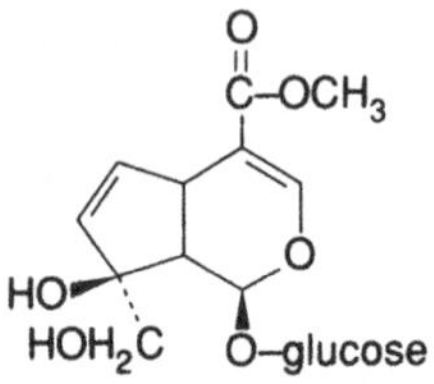

(118)

Gardenoside

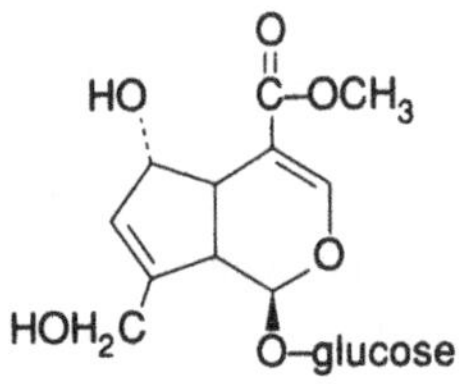

(119)

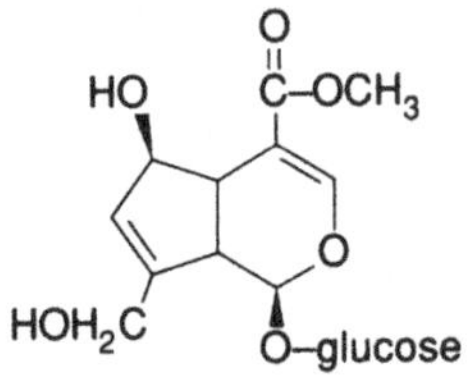

(120)

Similar studies were carried out with the closely related seco-iridoid glucosides **gentiopicroside** (gentiopicrin) (121) and **swertiamarin** (122), however with the use of isolated strains of human intestinal bacteria.[286,287] Anaerobic incubation of the glycosides with numerous bacterial strains resulted in the formation of at least five metabolites from gentiopicroside and three from swertiamarin. Two of these were formed from both compounds and were identified as erythrocentaurin (123) and 5-hydroxymethylisochroman-1-one (124). Additional metabolites from gentiopicroside were a compound identical with (124) except for the presence of a double bond at C-3,4 and two metabolites which retained the general heterocyclic ring structure of gentiopicroside. An unexpected metabolite of swertiamarin was shown to contain nitrogen in one of the rings.

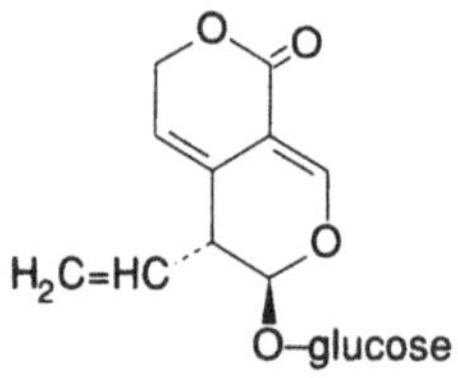

(121)

Gentiopicroside

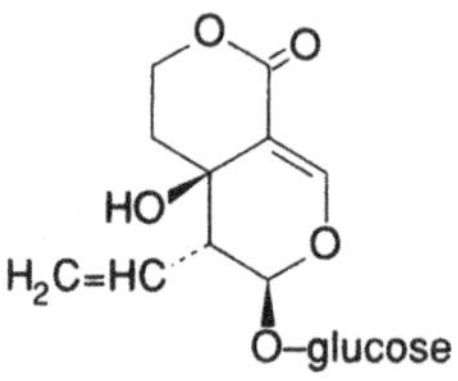

(122)

Swertiamarin

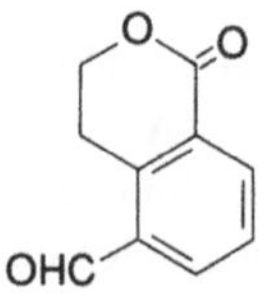

(123)

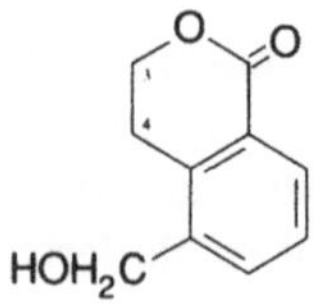

(124)

Peoniflorin (paeoniflorin) and albiflorin are two closely related glucosides found in peony roots. Their metabolism in mammals has not been studied, however several investigations have been carried out using human intestinal microorganisms. Hattori et al.[288] incubated **peoniflorin** (125) anaerobically with a human fecal suspension and detected three metabolites which they named paeonimetabolines I, II, and III. The same metabolites were formed from the closely related peony constituents oxypeoniflorin (hydroxybenzoyl group attached at C-8) or benzoylpeoniflorin (additional benzoyl group attached to terminal hydroxy group of the glucose moiety). Numerous intestinal bacteria were tested for their ability to metabolize peoniflorin. In many cases considerable decomposition of the substrate was found, however only *Peptostreptococcus anaerobius* formed large amounts of paeonimetaboline I. They also showed that the major metabolite (paeonimetaboline I) has structure (126). A more comprehensive investigation[289] determined that this metabolite has the *S*-configuration at C-7. It was also shown that the 7*R*-epimer of this compound (structure (127)) was a major metabolite. Additionally, the minor metabolite paeonimetaboline II was found to consist of a mixture of the 7*R*- and 7*S*-epimers of compound (128).

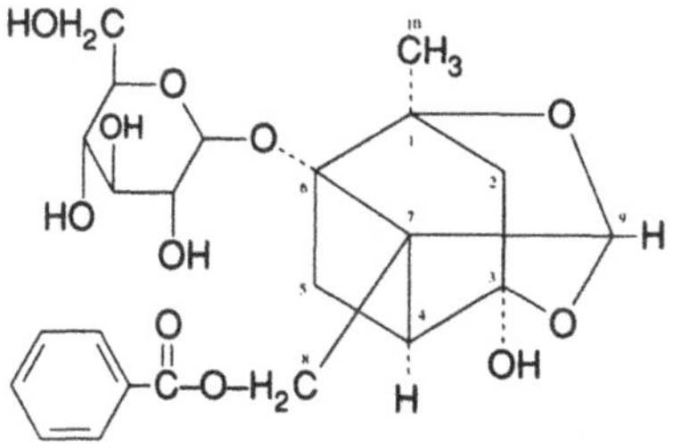

(125)

Peoniflorin

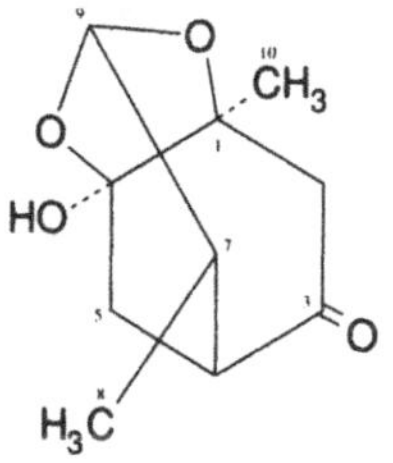

(126)

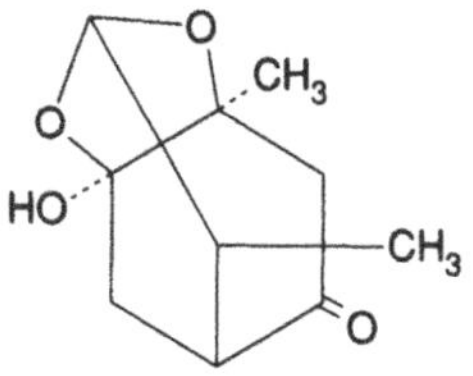

(127)

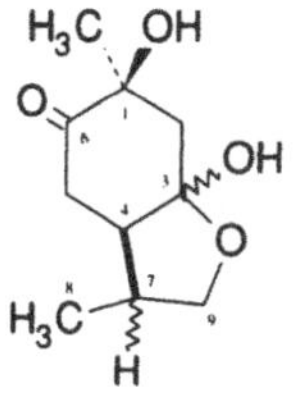

(128)

Study of the metabolism of peoniflorin by individual strains of intestinal bacteria was extended by Shu et al.[290] who measured the formation of the 7*R*- and 7*S*-epimers of paeonimetaboline I. Appreciable activity was detected among species belonging to the genera *Bacteroides*, *Bifidobacterium*, *Clostridium*, *Lactobacillus*, and *Streptococcus*. Both epimers were usually formed in nearly equal amounts, however, preferential formation of either the 7*R* or 7*S* epimers was noted in a few cases. Incubation studies with the fecal microflora from 15 individuals showed the formation of both epimers, however production of the 7*S*-epimer was usually most abundant. This investigation suggested that the age of the individual influenced the composition of the bacterial metabolites formed.

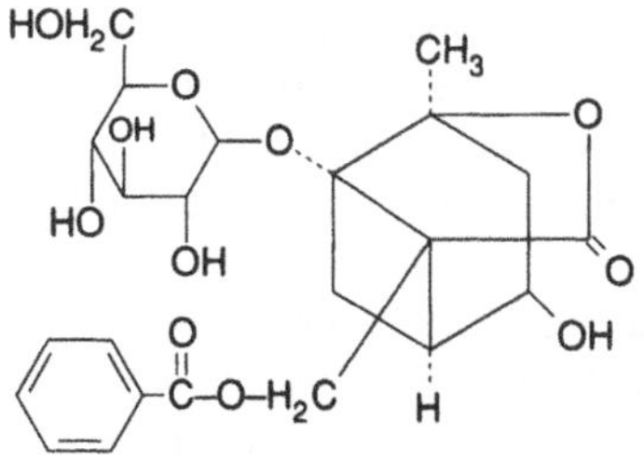

(129)

Albiflorin

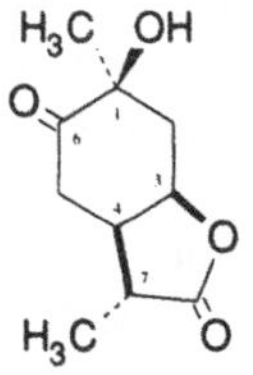

(130)

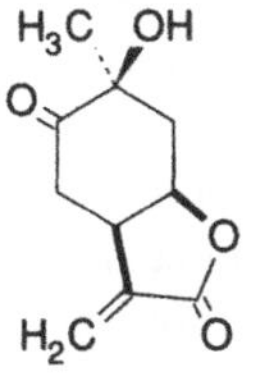

(131)

Investigations similar to those described above for peoniflorin were carried out by Hattori et al.[291] with **albiflorin** (129). They found that anaerobic intestinal bacteria converted albiflorin to two lactone derivatives. These were shown by spectroscopic methods to be identical with paeonilactone-A (130) and paeonilactone-B (131), both of which are constituents of fresh peony root. The results indicated that the former was formed via reduction of the latter.

Phyllanthoside is a complex oxygen heterocyclic compound containing two ester groups which link a cinnamic acid residue and a disaccharide moiety. It was rapidly degraded to an unidentified less polar metabolite when incubated with mouse or rat plasma.[292] This reaction occurred more slowly when plasma from dog or monkey (cynomologous macaque) was employed and several less polar metabolites were formed when human plasma was used.

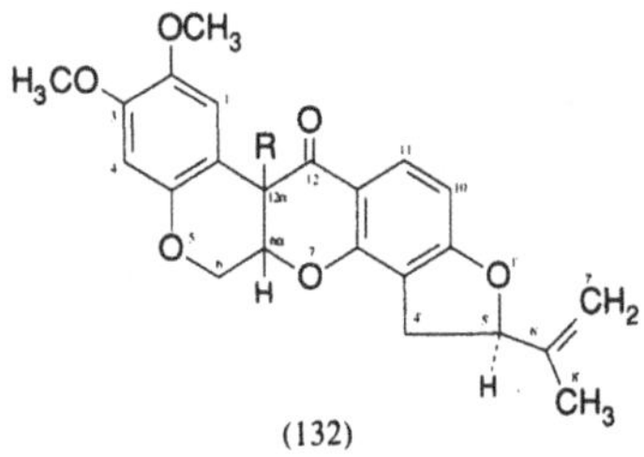

(132)

(a) Rotenone, R = H

(b) R = —OH

(c) R = ····OH

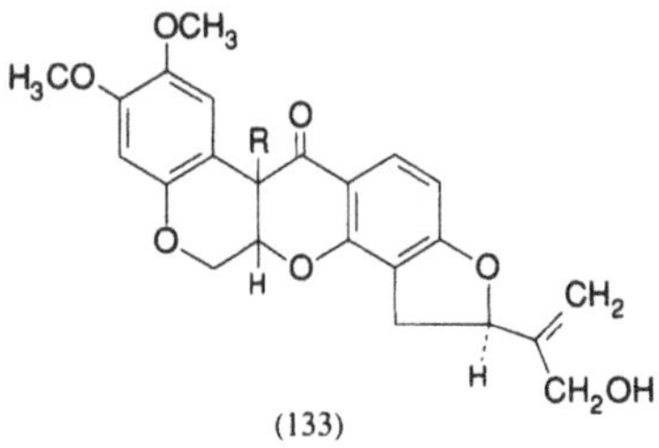

(133)

(a) R = H

(b) R = —OH

(c) R = ····OH

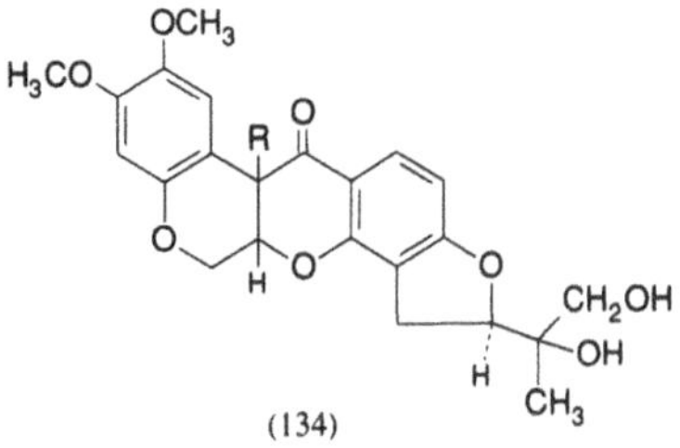

(134)

(a) R = H

(b) R = —OH

(c) R = ····OH

The metabolism of **rotenone** (132a) has been studied using material labeled with ^{14}C, either in the 6α or the 3-methoxy position. With the former compound, Fukami et al.[293] and Yamamoto[294] showed that compounds (134a) and (134b) were urinary metabolites in mice. Excretion of radioactivity occurred predominantly in the feces following the intragastric administration of rotenone. After 48 h, 20% of the ^{14}C was found in the urine, 0.3% in the expired air, 5% in the body, and the remainder in the feces. The nature of the fecal metabolites was not determined. Similar findings were made by Wessel et al.[295] who reported that more than 80% (48 h) and 95% (144 h) of the radioactivity was excreted in the feces of rats after oral or i.p. administration of rotenone labeled with ^{14}C in the 6α-position. Urinary excretion accounted for only 2 to 4% of the dose. More than 80% of the fecal radioactivity was present in the form of 6–8 metabolites.

Fukami et al.[293] and Yamamoto[294] also showed that eight metabolites were formed *in vitro* during incubation with the microsome-NADPH system from rat or mouse liver. Although all of the metabolites are more polar than rotenone, the toxicity of several of these to mice was similar to that of rotenone. The major metabolites *in vitro* were rotenolone I (132b), (133a), (133b),

(134a), and (134b) and, of these, metabolite (134a) was the most important. The corresponding epimers (132c) (rotenolone II), (133c), and (134c) were also detected. Fukami et al.[296] discussed differences in metabolic rate or route in relation to the selective toxicity of rotenone in various animal species.

Further information on the metabolism of natural rotenone (5′-β-rotenone) (132a) was obtained using material labeled with ^{14}C in the 3-methoxy group.[297] These experiments showed that extensive demethylation occurred. In mice, 27% of the radioactivity was found in the expired air within 50 h, after either oral or i.p. administration. The corresponding value in rats was 12.5%. The finding that the extent of formation of $^{14}CO_2$ was similar following both routes of administration suggests that rotenone absorption from the gut is extensive and that the high fecal excretion of radioactivity noted above may be due to biliary excretion of rotenone and its metabolites.

REFERENCES

1. **Burka, L. T. and Boyd, M. R.,** Furans, in *Bioactivation of Foreign Compounds*, Anders, M. W., Ed., Academic Press, Orlando, 1985, 243.
2. **Strolin Benedetti, M. and Dostert, P.,** Metabolic activation of furans as a basis for their toxicity, *Actual. Chim. Thér.*, 14, 237, 1987.
3. **Ravindranath, V., Burka, L. T., and Boyd, M. R.,** Reactive metabolites from the bioactivation of toxic methylfurans, *Science*, 224, 884, 1984.
4. **Ravindranath, V. and Boyd, M. R.,** Metabolic activation of 2-methylfuran by rat microsomal systems, *Toxicol. Appl. Pharmacol.*, 78, 370, 1985.
5. **Ravindranath, V., McMenamin, M. G., Dees, J. H., and Boyd, M. R.,** 2-Methylfuran toxicity in rats — role of metabolic activation *in vivo*, *Toxicol. Appl. Pharmacol.*, 85, 78, 1986.
6. **Hirata, T., Sakano, S., and Suga, T.,** Biotransformation of aloenin, a bitter glucoside constituent of *Aloe arborescens*, by rats, *Experientia*, 37, 1252, 1981.
7. **Rasmussen, A. K., Scheline, R. R., Solheim, E., and Hänsel, R.,** Metabolism of some kava pyrones in the rat, *Xenobiotica*, 9, 1, 1979.
8. **Duffield, A. M., Jamieson, D. D., Lidgard, R. O., Duffield, P. H., and Bourne, D. J.,** Identification of some human urinary metabolites of the intoxication beverage kava, *J. Chromatogr.*, 475, 273, 1989.
9. **Rennhard, H. N.,** The metabolism of ethyl maltol and maltol in the dog, *J. Agric. Food Chem.*, 19, 152, 1971.
10. **Gautrelet, J. and Gravellat, H.,** De l'action physiologique de quelques couleurs d'origine végétale, *C. R. Séanc. Soc. Biol.*, 61, 134, 1906.
11. **Simon, E. J., Gross, C. S., and Milhorat, A. T.,** The metabolism of vitamin E. I. The absorption and excretion of d-α-tocopheryl-5-methyl-C^{14}-succinate, *J. Biol. Chem.*, 221, 797, 1956.
12. **Simon, E. J., Eisengart, A., Sundheim, L. and Milhorat, A. T.,** The metabolism of vitamin. E II. Purification and characterization of urinary metabolites of α-tocopherol, *J. Biol. Chem.*, 221, 807, 1956.
13. **Csallany, A. S., Draper, H. H., and Shah, S. N.,** Conversion of d-α-tocopherol-C^{14} to tocopherol-*p*-quinone *in vivo*, *Arch. Biochem. Biophys.*, 98, 142, 1962.
14. **Wiss, O. and Gloor, H.,** Absorption, distribution, storage, and metabolites of vitamins K and related quinones, *Vitam. Hormone*, 24, 575, 1966.
15. **Draper, H. H. and Csallany, A. S.,** Metabolism and function of vitamin E, *Fed. Proc. Fed. Am. Soc. Exp. Biol.*, 28, 1690, 1969.
16. **Watanabe, M., Toyoda, M., Imada, I., and Morimoto, H.,** Ubiquinone and related compounds. XXVI. The urinary metabolites of phylloquinone and α-tocopherol, *Chem. Pharm. Bull.*, 22, 176, 1974.
17. **Mechoulam, R.,** Chemistry of cannabis, in *Psychotropic Agents Part III: Alcohol and Psychomimetics, Psychotropic Effects of Central Acting Drugs*, Hoffmeister, F. and Stille, G., Eds., Springer Verlag, Berlin, 1982, 119.
18. **Turner, C. E., Elsohly, M. A., and Boeren, E. G.,** Constituents of *Cannabis sativa* L. XVII. A review of the natural constituents, *J. Natl. Prod.*, 43, 169, 1980.
19. **Harvey, D. J.,** Chemistry, metabolism, and pharmacokinetics of the cannabinoids, in *Marihuana in Science and Medicine*, Nahas, G. G., Ed., Raven Press, New York, 1984, 37.

20. **Mechoulam, R.,** Cannabinoid chemistry, in *Marihuana. Chemistry, Pharmacology, Metabolism, and Clinical Effects*, Mechoulam, R., Ed., Academic Press, New York, 1973, 1.

21. **Mechoulam, R., McCallum, N. K., and Burstein, S.,** Recent advances in the chemistry and biochemistry of cannabis, *Chem. Rev.*, 76, 75, 1976.

22. **Agurell, S., Haldin, M., Lindgren, J.-E., Ohlsson, A., Widman, M., Gillespie, H., and Hollister, L.,** Pharmacokinetics and metabolism of Δ^1-tetrahydrocannabinol and other cannabinoids with emphasis on man, *Pharmacol. Rev.*, 38, 21, 1986.

23. **Harvey, D. J. and Paton, W. D. M.,** Metabolism of the cannabinoids, *Rev. Biochem. Toxicol.*, 6, 221, 1984.

24. **Lemberger, L. and Rubin, A.,** The cannabinoids, in *Physiologic Disposition of Drugs of Abuse*, Spectrum Publications, New York, 1976, 269.

25. **Nordqvist, M., Agurell, S., Rydberg, M., Falk, L., and Ryman, T.,** More acidic metabolites of Δ^1-tetrahydrocannabinol isolated form rabbit urine, *J. Pharm. Pharmacol.*, 31, 238, 1979.

26. **Halldin, M. M. and Widman, M.,** Glucuronic acid conjugate of Δ^1-tetrahydrocannabinol identified in the urine of man, *Arzneim. Forsch.*, 33(I), 177, 1983.

27. **Harvey, D. J. and Leuschner, J. T. A.,** Studies on the β-oxidative metabolism of Δ^1- and Δ^9-tetrahydrocannabinol in the mouse. The *in vivo* biotransformation of metabolites oxidized in the side chain, *Drug Metab. Dispos.*, 13, 215, 1985.

28. **Harvey, D. J.,** Further studies on the oxidative cleavage of the pentyl side-chain of cannabinoids: identification of new biotransformation pathways in the metabolism of 3'-hydroxy-delta-9-tetrahydrocannabinol by the mouse, *Xenobiotica*, 19, 1437, 1989.

29. **Nordqvist, M., Lindgren, J.-E., and Agurell, S.,** Acidic metabolites of Δ^1-tetrahydrocannabinol isolated from rabbit urine, *J. Pharm. Pharmacol.*, 31, 231, 1979.

30. **Halldin, M. M., Widman, M., Agurell, S., Hollister, L. E., and Kanter, S. L.,** Acidic metabolites of delta-1-tetrahydrocannabinol excreted in the urine of man, in *The Cannabinoids: Chemical, Pharmacologic, and Therapeutic Aspects*, Agurell, S., Dewey, W. L., and Willette, R. E., Eds., Academic Press, New York, 1984, 211.

31. **Ben-Zvi, Z.,** Epoxide-diol pathway of Δ^1-THC in the rat *in vitro*, *Xenobiotica*, 10, 805, 1980.

32. **Narimatsu, S., Yamamoto, I., Watanabe, K., and Yoshimura, H.,** $9\alpha,10\alpha$-Epoxyhexahydrocannabinol formation from Δ^9-tetrahydrocannabinol by liver microsomes of phenobarbital-treated mice and its pharmacological activities in mice, *J. Pharmacobiol. Dyn.*, 6, 558, 1983.

33. **Yamamoto, I., Narimatsu, S., Shimonishi, T., Watanabe, K., and Yoshimura, H.,** Difference in epoxides formation and their further metabolism between Δ^9- and Δ^8-tetrahydrocannabinols by human liver microsomes, *J. Pharmacobiol. Dyn.*, 7, 254, 1984.

34. **Leighty, E. G.,** Metabolism and distribution of cannabinoids in rats after different methods of administration, *Biochem. Pharmacol.*, 22, 1613, 1973.

35. **Leighty, E. G., Fentiman, A. F., and Foltz, R. L.,** Long-retaining metabolites of Δ^9- and Δ^8-tetrahydrocannabinols identified as novel fatty acid conjugates, *Res. Commun. Chem. Pathol. Pharmacol.*, 14, 13, 1976.

36. **Leighty, E. G.,** In vitro rat liver microsomal conjugation of several cannabinoids to fatty acids, *Res. Commun. Subst. Abuse*, 1, 169, 1980.

37. **Leighty, E. G.,** An *in vitro* rat liver microsomal system for conjugating fatty acids to 11-hydroxy-Δ^9-tetrahydrocannabinol, *Res. Commun. Chem. Pathol. Pharmacol.*, 23, 483, 1979.

38. **Leighty, E. G.,** Hydrolysis of 11-palmitoyloxy-Δ^9-tetrahydrocannabinol to 11-hydroxy-Δ^9-tetrahydrocannabinol by cholesterol esterase and a lipase, *Res. Commun. Chem. Pathol. Pharmacol.*, 24, 393, 1979.

39. **Jones, G., Pertwee, R. G., Gill, E. W., Paton, W. D. M., Nilsson, I. M., Widman, M., and Agurell, S.,** Relative pharmacological potency in mice of optical isomers of Δ^1-tetrahydrocannabinol, *Biochem. Pharmacol.*, 23, 439, 1974.

40. **Harvey, D. J.,** *In vivo* metabolism of (+)-*trans*-delta-9-tetrahydrocannabinol in the mouse, *Biomed. Environ. Mass Spectrom.*, 15, 117, 1988.

41. **Halldin, M. M., Isaac, H., Widman, M., Nilsson, E., and Ryrfeldt, Å.,** A comparison between the metabolism of Δ^1-tetrahydrocannabinol by perfused lung and liver of rat and guinea pig, *Xenobiotica*, 14, 277, 1984.

42. **Watanabe, K., Tanaka, T., Yamamoto, I., and Yoshimura, H.,** Brain microsomal oxidation of Δ^8- and Δ^9-tetrahycrocannabinol, *Biochem. Biophys. Res. Commun.*, 157, 75, 1988.

43. **Marriage, H. J. and Harvey, D. J.,** Metabolism of delta-9-tetrahydrocannabinol by fractionated isozymes of mouse hepatic cytochrome P-450, *Res. Commun. Subst. Abuse*, 7, 89, 1986.

44. **Wall, M. E., Sadler, B. M., Brine, D., Taylor, H., and Perez-Reyes, M.,** Metabolism, disposition, and kinetics of delta-9-tetrahydrocannabinol in men and women, *Clin. Pharmacol. Ther.*, 34, 352, 1983.

45. **Wall, M. E., Sadler, B. M., Brine, D., Taylor, H., and Perez-Reyes, M.,** Metabolism, disposition and pharmacokinetics of delta-9-tetrahydrocannabinol in male and female subjects, in *The Cannabinoids: Chemical, Pharmacologic, and Therapeutic Aspects*, Agurell, S., Dewey, W. L., and Willette, R. E., Eds., Academic Press, New York, 1984, 185.

46. **Narimatsu, S., Yamamoto, I., and Yoshimura, H.,** Stereospecific hydrolysis of $8\alpha,9\alpha$- and $8\beta,9\beta$-epoxyhexahydrocannabinols in the mouse *in vivo* and *in vitro*, *Xenobiotica*, 15, 227, 1985.

47. **Yagen, B., Levy, S., Mechoulam, R., and Ben-Zvi, Z.,** Synthesis and enzymatic formation of a C-glucuronide of Δ^6-tetrahydrocannabinol, *J. Am. Chem. Soc.*, 99, 6444, 1977.

48. **Levy, S., Yagen, B., and Mechoulam, R.,** Identification of a C-glucuronide of Δ^6-tetrahydrocannabinol as a mouse liver conjugate in vivo, *Science*, 200, 1391, 1978.

49. **Harvey, D. J. and Marriage, H. J.,** Metabolism of $(+)$-*trans*-Δ^8-tetrahydrocannabinol in the mouse *in vitro* and *in vivo*, *Drug Metab. Dispos.*, 15, 914, 1987.

50. **Harvey, D. J. and Mechoulam, R.,** Metabolites of cannabidiol identified in human urine, *Xenobiotica*, 20, 303, 1990.

51. **Yamamoto, I., Gohda, H., Narimatsu, S., and Yoshimura, H.,** Identification of cannabielsoin, a new metabolite of cannabidiol formed by guinea-pig hepatic microsomal enzymes, and its pharmacological activity in mice, *J. Pharmacobiol. Dyn.*, 11, 833, 1988.

52. **Yamamoto, I., Gohda, H., Narimatsu, S., and Yoshimura, H.,** Mechanism of biological formation of cannabielsoin from cannabidiol in the guinea-pig, mouse, rat and rabbit, *J. Pharmacobiol. Dyn.*, 12, 488, 1989.

53. **Nilsson, I., Agurell, S., Nilsson, J. L. G., Widman, M., and Leander, K.,** Two cannabidiol metabolites formed by rat liver, *J. Pharm. Pharmacol.*, 25, 486, 1973.

54. **Harvey, D. J., Martin, B. R., and Paton, W. D. M.,** Comparative *in vivo* metabolism of Δ^1-tetrahydrocannabinol (Δ^1-THC), cannabidiol (CBD) and cannabinol (CBN) by several species, in *Recent Developments in Mass Spectrometry in Biochemistry and Medicine*, Vol. 1, Frigerio, A., Ed., Plenum Press, New York, 1978, 161.

55. **Samara, E., Bialer, M., and Harvey, D. J.,** Identification of glucose conjugates as major urinary metabolites of cannabidiol in the dog, *Xenobiotica*, 20, 177, 1990.

56. **Yisak, W.-A., Widman, M., Lindgren, J.-E., and Agurell, S.,** Neutral *in vivo* metabolites of cannabinol isolated from rat faeces, *J. Pharm. Pharmacol.*, 29, 487, 1977.

57. **Wall, M. E., Brine, D. R., and Perez-Reyes, M.,** Metabolism of cannabinoids in man, in *Pharmacology of Marihuana*, Braude, M. C. and Szara, S., Eds., Raven Press, New York, 1976, 93.

58. **Yisak, W.-A., Agurell, S., Lindgren, J.-E., and Widman, M.,** *In vivo* metabolites of cannabinol identified as fatty acid conjugates, *J. Pharm. Pharmacol.*, 30, 462, 1978.

59. **Harvey, D. J., Martin, B. R., and Paton, W. D. M.,** Identification and measurement of cannabinoids and their *in vivo* metabolites in liver by gas chromatography-mass spectrometry, in *Marihuana: Biological Effects*, Nahas, G. G. and Paton, W. D. M., Eds., Pergamon Press, Oxford, 1979, 45.

60. **Brown, N. K. and Harvey, D. J.,** *In vivo* metabolism of the methyl homologues of delta-8-tetrahydrocannabinol, delta-9-tetrahydrocannabinol and *abn*-delta-8-tetrahydrocannabinol in the mouse, *Biomed. Environ. Mass Spectrom.*, 15, 389, 1988.

61. **Brown, N. K. and Harvey, D. J.,** *In vivo* metabolism of the *n*-propyl homologues of delta-8- and delta-9-tetrahydrocannabinol in the mouse, *Biomed. Environ. Mass Spectrom.*, 15, 403, 1988.

62. **Brown, N. K. and Harvey, D. J.,** *In vivo* metabolism of the *n*-butyl homologues of Δ^9-tetrahydrocannabinol and Δ^8-tetrahydrocannabinol by the mouse, *Xenobiotica*, 18, 417, 1988.

63. **Brown, N. K. and Harvey, D. J.,** Metabolism of n-hexyl-homologues of delta-8-tetrahydrocannabinol and delta-9-tetrahydrocannabinol in the mouse, *Eur. J. Drug Metab. Pharmacokinet.*, 13, 165, 1988.

64. **Harvey, D. J.,** Identification of hepatic metabolites of *n*-heptyl-delta-1-tetrahydrocannabinol in the mouse, *Xenobiotica*, 15, 187, 1985.

65. **Harvey, D. J., Martin, B. R., and Paton, W. D. M.,** Identification of glucuronides as *in vivo* liver conjugates of seven cannabinoids and some of their hydroxy and acid metabolites, *Res. Commun. Chem. Pathol. Pharmacol.*, 16, 265, 1977.

66. **Krishnaswamy, N. R., Seshadri, T. R., and Tahir, P. J.,** Metabolism of some polyphenolic compounds related to calophyllolide in rabbits, *Indian J. Exp. Biol.*, 9, 458, 1971.

67. **Crew, M. C., Szpiech, J. M., and DiCarlo, F. J.,** Metabolism of 3-(hydroxymethyl)-8-methoxychromone in the rat. I. Absorption, tissue distribution and excretion, *Xenobiotica*, 6, 83, 1976.

68. **Crew, M. C., Melgar, M. D., George, S., Greenough, R. C., Szpiech, J. M., and DiCarlo, F. J.,** Metabolism of 3-(hydroxymethyl)-8-methoxychromone in the rat. II. Classification and identification of urinary drug metabolites, *Xenobiotica*, 6, 89, 1976.

69. **DiCarlo, F. J., Herzig, D. J., Kusner, E. J., Schumann, P. R., Melgar, M. D., George, S., and Crew, M. C.,** 3-(Hydroxymethyl)-8-methoxychromone and unconjugated metabolites in rat plasma. Identification of the biologically active species, *Drug Metab. Dispos.*, 4, 368, 1976.

70. **Solheim, E. and Scheline, R. R.,** Metabolism of alkenebenzene derivatives in the rat. I. *p*-Methoxyallylbenzene (estragole) and *p*-methoxypropenylbenzene (anethole), *Xenobiotica*, 3, 493, 1973.

71. **Solheim, E. and Scheline, R. R.,** Metabolism of alkenebenzene derivatives in the rat. II. Eugenol and isoeugenol methyl ethers, *Xenobiotica*, 6, 137, 1976.

72. **Cohen, A. J.,** Critical review of the toxicology of coumarin with special reference to interspecies differences in metabolism and hepatotoxic response and their significance to man, *Food Cosmet. Toxicol.*, 17, 277, 1979.

73. **Booth, A. N., Masri, M. S., Robbins, D. J., Emerson, O. H., Jones, F. T., and DeEds, F.,** Urinary metabolites of coumarin and *o*-coumaric acid, *J. Biol. Chem.*, 234, 946, 1959.

74. **Feuer, G.,** The metabolism and biological actions of coumarins, in *Progress in Medicinal Chemistry*, Vol. 10, Ellis, G. P. and West, G. B., Eds., North-Holland, Amsterdam, 1974, 85.

75. **Kaighen, M. and Williams, R. T.,** The metabolism of [3-¹⁴C]coumarin, *J. Med. Chem.*, 3, 25, 1961.

76. **Mead, J. A. R., Smith, J. N., and Williams, R. T.,** Studies in detoxication. LXXII. The metabolism of coumarin and of *o*-coumaric acid, *Biochem. J.*, 68, 67, 1958.

77. **Scheline, R. R.,** Studies on the role of the intestinal microflora in the metabolism of coumarin in rats, *Acta Pharmacol. Toxicol.*, 26, 325, 1968.

78. **Furuya, T.,** Studies on the metabolism of naturally occurring coumarins. V. Urinary metabolites of coumarin and dihydrocoumarin, *Chem. Pharm. Bull.*, 6, 701, 1958.

79. **Feuer, G., Golberg, L., and Gibson, K. I.,** Liver response tests. VII. Coumarin metabolism in relation to the inhibition of rat-liver glucose 6-phosphatase, *Food Cosmet. Toxicol.*, 4, 157, 1966.

80. **Pekker, I. and Schäfer, E.-A.,** Vergleich von enteraler und perkutaner Resorption von Cumarin, *Arzneim. Forsch.*, 19, 1744, 1969.

81. **Shilling, W. H., Crampton, R. F., and Longland, R. C.,** Metabolism of coumarin in man, *Nature (London)*, 221, 664, 1969.

82. **Van Sumere, C. F. and Teuchy, H.,** The metabolism of [2-¹⁴C]coumarin and [2-¹⁴C]-7-hydroxycoumarin in the rat, *Arch. Int. Physiol. Biochim.*, 79, 665, 1971.

83. **Norman, R. L. and Wood, A. W.,** *o*-Hydroxyphenylethanol, a novel lactone ring-opened metabolite of coumarin, *Drug Metab. Dispos.*, 12, 543, 1984.

84. **Creaven, P. J., Parke, D. V., and Williams, R. T.,** A spectrophotofluorimetric study of the 7-hydroxylation of coumarin by liver microsomes, *Biochem. J.*, 96, 390, 1965.

85. **Kratz, F. and Staudinger, H.,** Kinetische Untersuchungen zur Hydroxylierung von Cumarin mit Lebermikrosomen von Kaninchen, *Hoppe-Seyler's Z. Physiol. Chem.*, 343, 27, 1965.

86. **Fink, P.-C. and Kerékjártó, B. v.,** Zur enzymatischen Hydroxylierung von Cumarin. II. Die mikrosomalen Reaktionsprodukte bei verschiedenen Tierarten, *Hoppe-Seyler's Z. Physiol. Chem.*, 345, 272, 1966.

87. **Kerékjártó, B. v.,** Zur enzymatischen Hydroxylierung von Cumarin. I. Nachweis der Reaktionsprodukte und Reaktionswege, *Hoppe-Seyler's Z. Physiol. Chem.*, 345, 264, 1966.

88. **Gibbs, P. A., Janakidevi, K., and Feuer, G.,** Metabolism of coumarin and 4-methylcoumarin by rat-liver microsomes, *Can. J. Biochem.*, 49, 177, 1971.

89. **Walters, D. G., Lake, B. G., and Cottrell, R. C.,** High-performance liquid chromatography of coumarin and its metabolites, *J. Chromatogr.*, 196, 501, 1980.

90. **Boyland, E. and Chasseaud, L. F.,** Enzyme-catalysed conjugations of glutathione with unsaturated compounds, *Biochem. J.*, 104, 95, 1967.

91. **Lake, B. G., Gray, T. J. B., Evans, J. G., Lewis, D. F. V., Beamand, J. A., and Hue, K. L.,** Studies on the mechanism of coumarin-induced toxicity in rat hepatocytes: comparison with dihydrocoumarin and other coumarin metabolites, *Toxicol. Appl. Pharmacol.*, 97, 311, 1989.

92. **Gangolli, S. D., Shilling, W. N., Grasso, P., and Gaunt, I. F.,** Studies on the metabolism and hepatotoxicity of coumarin in the baboon, *Biochem. Soc. Trans.*, 2, 310, 1974.

93. **Waller, A. R. and Chasseaud, L. F.,** The metabolic fate of [¹⁴C]coumarin in baboons, *Food Cosmet. Toxicol.*, 19, 1, 1981.

94. **Ritschel, W. A., Brady, M. E., Tan, H. S. I., Hoffman, K. A., Yiu, I. M., and Grummich, K. W.,** Pharmacokinetics of coumarin and its 7-hydroxy-metabolites upon intravenous and peroral administration of coumarin in man, *Eur. J. Clin. Pharmacol.*, 12, 457, 1977.

95. **Moran, E., O'Kennedy, R., and Thornes, R. D.,** Analysis of coumarin and its urinary metabolites by high-performance liquid chromatography, *J. Chromatogr.*, 416, 165, 1987.

96. **Ritschel, W. A. and Hardt, T. J.,** Pharmacokinetics of coumarin, 7-hydroxycoumarin and 7-hydroxycoumarin glucuronide in the blood and brain of gerbils following intraperitoneal administration of coumarin, *Arzneim. Forsch.*, 33(II), 1254, 1983.

97. **Wood, A. W. and Conney, A. H.,** Genetic variation in coumarin hydroxylase activity in the mouse (*Mus musculus*), *Science*, 185, 612, 1974.

98. **Wudl, L. R., Taylor, B. A., and Wood, A. W.,** Metabolism of coumarin in C57BL/6J (B6) and B6.L-*Coh^h* (congenic) mice, *Fed. Proc. Fed. Am. Soc. Exp. Biol.*, 39, 523, 1980.

99. **Lush, I. E. and Andrews, K. M.,** Genetic variation between mice in their metabolism of coumarin and its derivatives, *Genet. Res.*, 31, 177, 1978.

100. **Wood, A. W. and Taylor, B. A.,** Genetic regulation of coumarin hydroxylase activity in mice, *J. Biol. Chem.*, 254, 5647, 1979.

101. **Feuer, G.,** 3-Hydroxylation of coumarin or 4-methylcoumarin by rat-liver microsomes and its induction by 4-methylcoumarin given orally, *Chem. Biol. Interact.*, 2, 203, 1970.

102. **Piller, N. B.,** Tissue levels of (3-¹⁴C) coumarin in the rat: distribution and excretion, *Br. J. Exp. Pathol.*, 58, 28, 1977.

103. **Mead, J. A. R., Smith, J. N., and Williams, R. T.,** Studies in detoxication. LXXI. The metabolism of hydroxycoumarins, *Biochem. J.*, 68, 61, 1958.

104. **Flatow, L.,** Über den Abbau von Aminosäuren im Organismus, *Hoppe-Seyler's Z. Physiol. Chem.*, 64, 367, 1910.

105. **Roseman, S., Huebner, C. F., Pankratz, R., and Link, K. P.,** Studies on 4-hydroxycoumarins. XVI. The metabolism of 4-hydroxycoumarin in the dog, *J. Am. Chem. Soc.*, 76, 1650, 1954.

106. **Sieburg, E.,** Über die physiologische Wirkung einiger natürlich vorkommender Oxycumarine (Umbelliferone, Daphnetin, Äsculetin, Chrysatropasäure und Herniarin), *Biochem. Z.*, 113, 176, 1921.

107. **Mead, J. A. R., Smith, J. N., and Williams, R. T.,** Studies in detoxication. LXVII. The biosynthesis of the glucuronides of umbelliferone and 4-methylumbelliferone and their use in fluorimetric determination of β-glucuronidase, *Biochem. J.*, 61, 569, 1955.

108. **Fujita, M. and Furuya, T.,** Studies on the metabolism of naturally occurring coumarins. II. Urinary metabolites of umbelliferone, *Chem. Pharm. Bull.*, 6, 517, 1958.

109. **Shirkey, R. J., Kao, J., Fry, J. R., and Bridges, J. W.,** A comparison of xenobiotic metabolism in cells isolated from rat liver and small intestinal mucosa, *Biochem. Pharmacol.*, 28, 1461, 1979.

110. **Dawson, J. R. and Bridges, J. W.,** Conjugation and excretion of metabolites of 7-hydroxycoumarin in the small intestine of rats and guinea-pigs, *Biochem. Pharmacol.*, 28, 3291, 1979.

111. **Dawson, J., Berggren, M., and Moldeus, P.,** Isolated cells as a model for studying sulphate conjugation and its regulation, in *Sulfate Metabolism and Sulfate Conjugation*, Mulder, G. J., Caldwell, J., Van Kempen, G. M. J., and Vonk, R. J., Eds., Taylor and Francis, London, 1982, 135.

112. **Indahl, S. R. and Scheline, R. R.,** The metabolism of umbelliferone and herniarin in rats and by the rat intestinal microflora, *Xenobiotica*, 1, 13, 1971.

113. **Legrum, W. and Netter, K. J.,** Characteristics of coumarin metabolism by liver microsomes from cobalt-pretreated mice, *Xenobiotica*, 10, 271, 1980.

114. **Tatematsu, A., Nadai, T., Yoshizumi, H., Sakurai, H., Furukawa, H., and Hayashi, M.,** Study of drug metabolites by mass spectrometry. III. Metabolites of 7-hydroxy-4-methylcoumarin in rats, *Shitsuryo Bunseki*, 20, 339, 1972 (Chem. Abstr. 78, 119062x, 1973).

115. **Fujita, M. and Furuya, T.,** Studies on the metabolism of naturally occurring coumarins. III. Urinary metabolites of herniarin, *Chem. Pharm. Bull.*, 6, 520, 1958.

116. **Legrum, W. and Frahseck, J.,** Acceleration of 7-[*methoxy*-^{14}C]coumarin-derived carbon dioxide exhalation by cobalt pretreatment in mice, *J. Pharmacol. Exp. Ther.*, 221, 790, 1982.

117. **Matsubara, T., Yoshihara, E., Iwata, T., Tochino, Y., and Hachino, T.,** Biotransformation of coumarin derivatives (1) 7-Alkoxycoumarin O-dealkylase in liver microsomes, *Jpn.. J. Pharmacol.*, 32, 9, 1982.

118. **Matsubara, T., Otsubo, S., Yoshihara, E., and Touchi, A.,** Biotransformation of coumarin derivatives. II. Oxidative metabolism of 7-alkoxycoumarin by microsomal enzymes and a simple assay procedure for 7-alkoxycoumarin O-dealkylase, *Jpn.. J. Pharmacol.*, 33, 41, 1983.

119. **Ullrich, V. and Kremers, P.,** Multiple forms of cytochrome P450 in the microsomal monooxygenase system, *Arch. Toxicol.*, 39, 41, 1977.

120. **Shao, H.-S., Zhang, S.-R., Pan, S.-R., Jing, X.-N., Ying, L.-Q., Zhu, H.-Y., and Gu, C.-H.,** Absorption, distirbution, excretion and metabolites of [^{14}C]-armillarisin A in rats, *Chung-kuo Yao Li Hsueh Pao*, 1, 120, 1980 (Chem. Abstr. 94, 76540n, 1981).

121. **Yang, C.-N., Braymer, H. D., Petrakis, P. L., Shetlar, M. R., and Wender, S. H.,** Formation of scopeletin from esculin and esculetin in the rat, *Arch. Biochem. Biophys.*, 75, 538, 1958.

122. **Braymer, H. D.,** The metabolic fate of some phenolic compounds in the rat, Ph.D. Thesis, University of Oklahoma, Norman, OK, 1960.

123. **Hawksworth, G., Drasar, B. S., and Hill, M. J.,** Intestinal bacteria and the hydrolysis of glycosidic bonds, *J. Med. Microbiol.*, 4, 451, 1971.

124. **Drasar, B. S. and Hill, M. J.,** *Human Intestinal Flora*, Academic Press, London, 1974, 63.

125. **Braymer, H. D., Shetlar, M. R., and Wender, S. H.,** Urinary products of scopoletin in laboratory rats, *Biochim. Biophys. Acta*, 44, 606, 1960.

126. **Furuya, T.,** Studies on the metabolism of naturally occurring coumarins. IV. Urinary metabolites of dimethylesculetin and scopoletin, *Chem. Pharm. Bull.*, 6, 696, 1958.

127. **Müller-Enoch, D., Sato, N., and Thomas, H.,** O-Demethylation of scoparone and studies on the scoparone-induced spectral change of cytochrome P-450 in rat liver microsomes, *Hoppe-Seyler's Z. Physiol. Chem.*, 362, 1091, 1981.

128. **Legrum, W., Kling, L., and Funke, E.,** Radioisomers of scoparone (6,7-dimethoxycoumarin) as a tool for *in vivo* differentiation of various hepatic monooxygenase inducers in mice using the breath test technique, *J. Pharmacol. Exp. Ther.*, 228, 769, 1984.

129. **Van Pelt, F. N. A. M., Mennes, W. C., Hassing, I. G. A. M., and Blaauboer, B. J.,** Application of high-performance liquid chromatography analysis of scoparone and its metabolites in the study of cytochrome P450 differentiation in vitro, *J. Chromatogr.*, 487, 489, 1989.

130. **Pathak, M. A., Dall'Acqua, F., Rodighiero, G., and Parrish, J. A.,** Metabolism of psoralens, *J. Invest. Dermatol.*, 62, 347, 1974.

131. **Pathak, M. A., Mandula, B., Nakayama, Y., Parrish, J. A., and Fitzpatrick, T. B.,** Cutaneous photosensitization and in vivo metabolism of psoralens, *J. Invest. Dermatol.*, 64, 279, 1975.

132. **Schalla, W., Schaeffer, H., Kammerau, B., and Zesch, A.,** Pharmacokinetics of 8-methoxypsoralen (8-MOP) after oral and local application, *J. Invest. Dermatol.*, 66, 258, 1976.

133. **Busch, U., Schmid, J., Koss, F. W., Zipp, H., and Zimmer, A.,** Pharmacokinetics and metabolite-pattern of 8-methoxypsoralen in man following oral administration as compared to the pharmacokinetics in rat and dog, *Arch. Dermatol. Res.*, 262, 255, 1978.

134. **Gazith, J., Schalla, W., and Schaeffer, H.,** 8-Methoxypsoralen-gas chromatographic determination and serum kinetics, *Arch. Dermatol. Res.*, 263, 215, 1978.

135. **Schmid, J., Prox, A., Reuter, A., Zipp, H., and Koss, F. W.,** The metabolism of 8-methoxypsoralen in man, *Eur. J. Drug Metab. Pharmacokinet.*, 5, 81, 1980.

136. **Ehrsson, H., Eksborg, S., and Wallin, I.,** Metabolism of 8-methoxypsoralen in man: identification and quantification of 8-hydroxypsoralen, *Eur. J. Drug Metab. Pharmacokinet.*, 3, 125, 1978.

137. **Mei, X.-R., Yeh, Y.-H., Lui, H.-H., Chang, C.-J., Cheng, H.-F., and Fung, S.-C.,** Study on the metabolism of 8-methoxypsoralen in animals, *Chung Ts'ao Yao*, 11, 29, 1980 (Chem. Abstr. 93, 160929h, 1980).

138. **Muni, I. A., Schneider, F. H., Olsson, T. A., and King, M.,** Absorption, distribution, and excretion of 8-methoxypsoralen in HRA/Skh mice, *Natl. Cancer Inst. Monogr.*, 66, 85, 1984.

139. **Nozu, T., Setoyama, K., Suwa, T., and Tanaka, I.,** Absorption, distribution and excretion of ^{3}H-labeled 8-methoxypsoralen, *Oyo Yakuri*, 18, 489, 1979 (Chem. Abstr. 92, 157484h, 1980).

140. **Smyth, R. D., Van Harken, D. R., Pfeffer, M., Nardella, P. A., Vasiljev, M., Pinto, J. S., and Hottendorf, G. H.,** Biological disposition of 8-methoxsalen in rat and man, *Arzneim. Forsch.*, 30(II), 1725, 1980.

141. **Kolis, S. J., Williams, T. H., Postma, E. J., Sasso, G. J., Confalone, P. N., and Schwartz, M. A.,** The metabolism of ^{14}C-methoxsalen by the dog, *Drug Metab. Dispos.*, 7, 220, 1979.

142. **Mays, D. C., Rogers, S. L., Guiler, R. C., Sharp, D. E., Hecht, S. G., Staubus, A. E., and Gerber, N.,** Disposition of 8-methoxypsoralen in the rat: methodology for measurement, dose-dependent pharmacokinetics, tissue distribution and identification of metabolites, *J. Pharmacol. Exp. Ther.*, 236, 364, 1986.

143. **Mays, D. C., Hecht, S. G., Unger, S. E., Pacula, C. M., Climie, J. M., Sharp, D. E., and Gerber, N.,** Disposition of 8-methoxypsoralen in the rat. Induction of metabolism *in vivo* and *in vitro* and identification of urinary metabolites by thermospray mass spectrometry, *Drug Metab. Dispos.*, 15, 318, 1987.

144. **Mandula, B. B. and Pathak, M. A.,** Metabolic reactions *in vitro* of psoralens with liver and epidermis, *Biochem. Pharmacol.*, 28, 127, 1979.

145. **Mandula, B. B., Pathak, M. A., and Dudek, G.,** Photochemotherapy: identification of a metabolite of 4,5′,8-trimethylpsoralen, *Science*, 193, 1131, 1976.

146. **Kelly, R. W.,** The oestrogenic activity of coumestans in ovariectomized ewes, *J. Reprod. Fertil.*, 28, 159, 1972.

147. **Cox, R. I. and Braden, A. W.,** The metabolism and physiological effects of phyto-oestrogens, *Proc. Aust. Soc. Anim. Prod.*, 10, 122, 1974.

148. **Arora, R. B., Tahir, P. J., Krishnaswamy, N. R., and Seshadri, T. R.,** Metabolism of coumarin derivatives: a note on the urinary metabolites of calophyllolide, 4-phenyllumbelliferone and 4-phenyl-7-methylumbelliferone, *Indian J. Biochem.*, 3, 58, 1966.

149. **DeEds, F.,** Flavonoid metabolism, in *Comprehensive Biochemistry*, Vol. 20, Florkin, M. and Stotz, E. H., Eds., Elsevier, Amsterdam, 1968, 127.

150. **Griffiths, L. A.,** Mammalian metabolism of flavonoids, in *The Flavonoids: Advances in Research*, Harborne, J. B. and Mabry, T. J., Eds., Chapman and Hall, London, 1982, 681.

151. **Demole, V.,** Toxicité, résorption, élimination de la Flavone synthétique, *Helv. Physiol. Acta*, 20, 93, 1962.

152. **Das, N. P. and Griffiths, L. A.,** Studies on flavonoid metabolism. Metabolism of flavone in the guinea pig, *Biochem. J.*, 98, 488, 1966.

153. **Svardal, A., Buset, H., and Scheline, R. R.,** Disposition of [2-^{14}C]flavone in the rat, *Acta Pharm. Suec.*, 18, 55, 1981.

154. **Booth, A. N., Jones, F. T., and DeEds, F.,** Metabolic fate of hesperidin, eriodictyol, homoeriodictyol, and diosmin, *J. Biol. Chem.*, 230, 661, 1958.

155. **Oustrin, J., Fauran, M. J., and Commanay, L.,** A pharmacokinetic study of ^{3}H-diosmine, *Arzneim. Forsch.*, 27(II), 1688, 1977.

156. **Winternitz, P. F., Dulcire, B., and Crawley, Y.,** Pharmacokinetics and metabolism of diosmin, *Eur. Congr. Biopharm. Pharmacokinet., 3rd,* Clermont-Ferrand, France, 1987.

157. **Griffiths, L. A. and Smith, G. E.,** Metabolism of apigenin and related compounds in the rat, *Biochem. J.*, 128, 901, 1972.

158. **Griffiths, L. A. and Smith, G. E.,** Metabolism of myricetin and related compounds in the rat. Metabolite formation *in vivo* and by the intestinal microflora *in vitro*, *Biochem. J.*, 130, 141, 1972.

159. **Griffiths, L. A. and Barrow, A.,** Metabolism of flavonoid compounds in germ-free rats, *Biochem. J.*, 130, 1161, 1972.

160. **Bickoff, E. M., Livingston, A. L., and Booth, A. N.,** Tricin from alfalfa—isolation and physiological activity, *J. Pharm. Sci.*, 53, 1411, 1964.

161. **Stelzig, D. A. and Ribiero, S.,** Metabolism of quercetin and tricin in the male rat, *Proc. Soc. Exp. Biol. Med.,* 141, 346, 1972.

162. **Mao, F., Tu, X., Zhu, J., Zhao, L., and Hu, Z.,** Absorption of baicalin in the small intestine of rats, *Nanjing Yaoxueyuan Xuebao,* 15, 61, 1984 (Chem. Abstr. 101, 103530f, 1984).

163. **Cai, X.-L.,** Absorption, distribution and excretion of ³H-scutellarin in vivo, *Chung Ts'ao Yao,* 12, 26, 1981 (Chem. Abstr. 95, 90742g, 1981).

164. **Han, G., Su, C., and Zhang, Y.,** Absorption, distribution and elimination of nevadensin in rats, *Yaoxue Tongbao,* 17, 306, 1982 (Chem. Abstr. 97, 192664s, 1982).

165. **Han, G., Su, C., and Zhang, Y.,** Absorption, distribution and elimination of nevadensin in the rat, and the relation between plasma concentration of the drug and its hypotensive effect, *Yaoxue Xuebao,* 17, 572, 1982 (Chem. Abstr. 97, 207696e, 1982).

166. **Laparra, J., Michaud, J., Masquelier, J., Linard, A., and Paris, R.,** Study on the distribution and tissue uptake of C-flavonoids from *Lespedeza capitata, Bull. Liaison, Groupe Polyphenols,* 8, 404, 1978 (Chem. Abstr. 91, 33504z, 1979).

167. **Hattori, M., Shu, Y.-Z., El-Sedawy, A. I., Namba, T., Kobashi, K., and Tomimori, T.,** Metabolism of homoorientin by human intestinal bacteria, *J. Nat. Prod.,* 51, 874, 1988.

168. **Takács, Ö. and Gábor, M.,** New data on the metabolism of flavonoids, in *Topics in Flavonoid Chemistry and Biochemistry,* Farkas, L., Gábor, M., and Kállay, F., Eds., Elsevier, Amsterdam, 1975, 227.

169. **Ozawa, H.,** Pharmacological and chemical studies on rutin-like compounds. VI. Observation of fate, excretion and toxicity of 3',4'-dihydroxyflavonol, *J. Pharm. Soc. Jpn..,* 71, 1191, 1951.

170. **Clark, W. G. and MacKay, E. M.,** The absorption and excretion of rutin and related flavonoid substances, *J. Am. Med. Assoc.,* 143, 1411, 1950.

171. **Baba, S., Furuta, T., Horie, M., and Nakagawa, H.,** Studies on drug metabolism by use of isotopes. XXVI. Determination of urinary metabolites of rutin in humans, *J. Pharm. Sci.,* 70, 780, 1981.

172. **Baba, S., Furuta, T., Fujioka, M., and Goromaru, T.,** Studies on drug metabolism by use of isotopes. XXVII. Urinary metabolites of rutin in rats and the role of intestinal microflora in the metabolism of rutin, *J. Pharm. Sci.,* 72, 1155, 1983.

173. **Tamemasa, O., Goto, R., and Ogura, S.,** Metabolic fate of ³H-rutin, *Pharmacometrics,* 12, 193, 1976.

174. **Gugler, R., Leschik, M., and Dengler, H. J.,** Disposition of quercetin in man after single oral and intravenous doses, *Eur. J. Clin. Pharmacol.,* 9, 229, 1975.

175. **Peter, H., Fisel, J., and Weisser, W.,** Zur Pharmakologie der Wirkstoffe aus *Ginko biloba, Arzneim. Forsch.,* 16, 719, 1966.

176. **Ueno, I., Nakano, N., and Hirono, I.,** Metabolic fate of [¹⁴C]quercetin in the ACI rat, *Jpn.. J. Exp. Med.,* 53, 41, 1983.

177. **Murray, C. W., Booth, A. N., DeEds, F., and Jones, F. T.,** Absorption and metabolism of rutin and quercetin in the rabbit, *J. Am. Pharm. Assoc. Sci. Ed.,* 43, 361, 1954.

178. **Booth, A. N., Murray, C. W., DeEds, F., and Jones, F. T.,** Metabolic fate of rutin and quercetin, *Fed. Proc. Fed. Am. Soc. Exp. Biol.,* 14, 321, 1955.

179. **Booth, A. N., Murray, C. W., Jones, F. T., and DeEds, F.,** The metabolic fate of rutin and quercetin in the animal body, *J. Biol. Chem.,* 223, 251, 1956.

180. **Petrakis, P. L., Kallianos, A. G., Wender, S. H., and Shetlar, M. R.,** Metabolic studies of quercetin labelled with C¹⁴, *Arch. Biochem. Biophys.,* 85, 264, 1959.

181. **Masri, M. S., Booth, A. N., and DeEds, F.,** The metabolism and acid degradation of quercetin, *Arch. Biochem. Biophys.,* 85, 284, 1959.

182. **Nakagawa, Y., Shetlar, M. R., and Wender, S. H.,** Urinary products from quercetin in neomycin-treated rats, *Biochim. Biophys. Acta,* 97, 233, 1965.

183. **Baba, S., Furuta, T., Fujioka, M., Goromaru, T., and Horie, M.,** Studies on metabolism of rutin in man and rats by stable isotope tracer technique, *J. Pharmacobiol. Dyn.,* 4, 28, 1981.

184. **Sawai, Y., Kohsaka, K., Nishiyama, Y., and Ando, K.,** Serum concentrations of rutoside metabolites after oral administration of a rutoside formulation to humans, *Arzneim. Forsch.,* 37(I), 729, 1987.

185. **Brown, S. and Griffiths, L. A.,** New metabolites of the naturally-occurring mutagen, quercetin, the pro-mutagen, rutin and of taxifolin, *Experientia,* 39, 198, 1983.

186. **Lang, K. and Weyland, H.,** Über die Stoffwechsel des Rutins und Quercetins, *Biochem. Z.,* 327, 109, 1955.

187. **Douglass, C. D. and Hogan, R.,** Formation of protocatechuic acid from quercetin by rat kidney in vitro, *J. Biol. Chem.,* 230, 625, 1958.

188. **Westlake, D. W. S., Talbot, G., Blakley, E. R., and Simpson, F. J.,** Microbial decomposition of rutin, *Can. J. Microbiol.,* 5, 621, 1959.

189. **Booth, A. N. and Williams, R. T.,** Dehydroxylation of catechol acids by intestinal contents, *Biochem. J.,* 88, 66, 1963.

190. **Scheline, R. R.,** The metabolism of drugs and other organic compounds by the intestinal microflora, *Acta Pharmacol. Toxicol.,* 26, 332, 1968.

191. **Simpson, F. J., Jones, G. A., and Wolin, E. A.,** Anaerobic degradation of some bioflavonoids by microflora of the rumen, *Can. J. Microbiol.,* 15, 972, 1969.

192. **Cheng, K.-J., Jones, G. A., Simpson, F. J., and Bryant, M. P.,** Isolation and identification of rumen bacteria capable of anaerobic rutin degradation, *Can. J. Microbiol.,* 15, 1365, 1969.

193. **Krishnamurty, H. G., Cheng, K.-J., Jones, G. A., Simpson, F. J., and Watkin, J. E.,** Identification of products produced by the anaerobic degradation of rutin and related flavonoids in *Butyrivibrio* sp. C_3, *Can. J. Microbiol.,* 16, 759, 1970.

194. **Krumholz, L. R. and Bryant, M. P.,** *Eubacterium oxidoreducens* sp. nov. requiring H_2 or formate to degrade gallate, pyrogallol, phloroglucinol and quercetin, *Arch. Microbiol.,* 144, 8, 1986.

195. **Martin, A. K.,** The origin of urinary aromatic compounds excreted by ruminants. III. The metabolism of phenolic compounds to simple phenols, *Br. J. Nutr.,* 48, 497, 1982.

196. **Kallianos, A. G., Petrakis, P. L., Shetlar, M. R., and Wender, S. H.,** Preliminary studies on degradation products of quercetin in the rat's gastrointestinal tract, *Arch. Biochem. Biophys.,* 81, 430, 1959.

197. **Das, N. P., Scott, K. N., and Duncan, J. H.,** Identification of flavanone metabolites in rat urine by combined gas-liquid chromatography and mass spectrometry, *Biochem. J.,* 136, 903, 1973.

198. **Buset, H. and Scheline, R. R.,** Disposition of [2-^{14}C]flavanone in the rat, *Acta Pharm. Suec.,* 17, 157, 1980.

199. **Buset, H. and Scheline, R. R.,** Identification of urinary metabolites of flavanone in the rat, *Biomed. Mass Spectrom.,* 6, 212, 1979.

200. **Booth, A. N., Jones, F. T., and DeEds, F.,** Metabolic fate of some flavanones, *Fed. Proc. Fed. Am. Soc. Exp. Biol.,* 15, 223, 1956.

201. **Booth, A. N., Jones, F. T., and DeEds, F.,** Metabolic and glucosuria studies on naringin and phloridzin, *J. Biol. Chem.,* 233, 280, 1958.

202. **Hackett, A. M., Marsh, I., Barrow, A., and Griffiths, L. A.,** The biliary excretion of flavanones in the rat, *Xenobiotica,* 9, 491, 1979.

203. **Honohan, T., Hale, R. L., Brown, J. P., and Wingard, R. E.,** Synthesis and metabolic fate of hesperetin-3-^{14}C, *J. Agric. Food Chem.,* 24, 906, 1976.

204. **Feng, Y.-S. and Zhu, D.-Z.,** Metabolism of farrerol, *Acta Pharm. Sin.,* 14, 149, 1979 (Chem. Abstr. 92, 33726u, 1980).

205. **Cheng, K.-J., Krishnamurty, H. G., Jones, G. A., and Simpson, F. J.,** Identification of products produced by the anaerobic degradation of naringin by *Butyrivibrio* sp C_3, *Can. J. Microbiol.,* 17, 129, 1971.

206. **Booth, A. N. and DeEds, F.,** The toxicity and metabolism of dihydroquercetin, *J. Am. Pharm. Assoc. Sci. Ed.,* 47, 183, 1958.

207. **Bülles, H., Bülles, J., Krumbiegel, G., Mennicke, W. H., and Nitz, D.,** Untersuchungen zur Verstoffwechselung und zur Ausscheidung von Silybin bei der Ratte, *Arzneim. Forsch.,* 25, 902, 1975.

208. **Ognyanova, V., Drenska, A., Mikhailova, D., and Nachev, I.,** Karsil (silymarin) pharmacokinetics in experimental animals, *Eksp. Med. Morfol.,* 20, 214, 1981 (Chem. Abstr. 96, 135308j, 1982).

209. **Mennicke, W. H., Lang, W., and Lorenz, D.,** Studies on the pharmacokinetics of silymarin, *Hepatology,* 79, 411, 1979.

210. **Flory, P. J., Krug, G., Lorenz, D., and Mennicke, W. H.,** Studies on elimination of silymarin in cholecystectomized patients. I. Biliary and renal elimination after a single dose, *Planta Med.,* 38, 227, 1980 (Chem. Abstr. 93, 18929q, 1980).

211. **Horwitt, M. K.,** Observations on behavior of the anthocyan pigment from Concord grapes in the animal body, *Proc. Soc. Exp. Biol. Med.,* 30, 949, 1933.

212. **Timberlake, C. F. and Bridle, P.,** The anthocyanins, in *The Flavonoids,* Harborne, J. B., Mabry, T. J., and Mabry, H., Eds., Chapman and Hall, London, 1975, 214.

213. **Oshima, Y., Watanabe, H., and Isakari, S.,** The mechanisms of catechins metabolism. I. Acidic substances in the urine of rabbits administered (+)-catechin, *J. Biochem. (Tokyo),* 45, 861, 1958.

214. **Oshima, Y. and Watanabe, H.,** The mechanisms of catechins metabolism. II. Neutral substances in the urine of rabbits administered (+)-catechin, *J. Biochem. (Tokyo),* 45, 973, 1958.

215. **Watanabe, H.,** The chemical structure of the intermediate metabolites of catechin. I. Chemical properties of the intermediate metabolites (G and H) and their derivatives, *Bull. Agric. Chem. Soc. Jpn.,* 23, 257, 1959.

216. **Watanabe, H.,** The chemical structure of the intermediate metabolites of catechin. II. Oxidative decomposition of the intermediate metabolites (G and H), *Bull. Agric. Chem. Soc. Jpn.,* 23, 260, 1959.

217. **Watanabe, H.,** The chemical structure of the intermediate metabolites of catechin. III. Synthesis of the intermediate metabolites (G and H), *Bull. Agric. Chem. Soc. Jpn.,* 23, 263, 1959.

218. **Watanabe, H.,** The chemical structure of the intermediate metabolites of catechin. III. Structure of the intermediate metabolite (F), *Bull. Agric. Chem. Soc. Jpn.,* 23, 268, 1959.

219. **Miura, S., Hamada, T., Satomi, O., Midorikawa, T., and Awata, N.,** Metabolic fate of cianidanol (KB-53). III. In vivo and in vitro metabolism in rats, *Oyo Yakuri,* 25, 1015, 1983 (Chem. Abstr. 99, 151663p, 1983).

220. **Das, N. P. and Griffiths, L. A.,** Studies on flavonoid metabolism. Metabolism of (+)-catechin in the guinea pig, *Biochem. J.,* 110, 449, 1968.

221. **Das, N. P.,** Studies on flavonoid metabolism. Excretion of *m*-hydroxyphenylhydracrylic acid from (+)-catechin in the monkey (*Macaca iris* sp.), *Drug Metab. Dispos.*, 2, 209, 1974.

222. **Das, N. P.,** Studies on flavonoid metabolism. Absorption and metabolism of (+)-catechin in man, *Biochem. Pharmacol.*, 20, 3435, 1971.

223. **Griffiths, L. A.,** *m*-Hydroxyphenylpropionic acid, a major urinary metabolite of (+)-catechin in the rat, *Nature (London)*, 194, 869, 1962.

224. **Das, N. P.,** Studies on flavonoid metabolism. Degradation of (+)-catechin by rat intestinal contents, *Biochim. Biophys. Acta*, 177, 668, 1969.

225. **Das, N. P. and Griffiths, L. A.,** Studies on flavonoid metabolism. Metabolism of (+)-[^{14}C]catechin in the rat and guinea pig, *Biochem. J.*, 115, 831, 1969.

226. **Gott, D. M. and Griffiths, L. A.,** Effects of antibiotic pretreatments on the metabolism and excretion of [U^{14}C](+)-catechin ([U^{14}C](+)-cyanidanol-3) and its metabolite, 3′-*O*-methyl-(+)-catechin, *Xenobiotica*, 17, 423, 1987.

227. **Griffiths, L. A.,** Studies on flavonoid metabolism. Identification of the metabolites of (+)-catechin in rat urine, *Biochem. J.*, 92, 173, 1964.

228. **Hackett, A. M., Griffiths, L. A., Broillet, A., and Wermeille, M.,** The metabolism and excretion of (+)-[^{14}C]cyanidanol-3 in man following oral administration, *Xenobiotica*, 13, 279, 1983.

229. **Shaw, I. C. and Griffiths, L. A.,** Identification of the major biliary metabolite of (+)-catechin in the rat, *Xenobiotica*, 10, 905, 1980.

230. **Hackett, A. M., Shaw, I. C., and Griffiths, L. A.,** 3′-*O*-Methyl-(+)-catechin glucuronide and 3′-*O*-methyl-(+)-catechin sulphate: new urinary metabolites of (+)-catechin in the rat and the marmoset, *Experientia*, 38, 538, 1982.

231. **Wermeille, M., Turin, E. and Griffiths, L. A.,** Identification of the major urinary metabolites of (+)-catechin and 3-*O*-methyl-(+)-catechin in man, *Eur. J. Drug Metab. Pharmacokinet.*, 8, 77, 1983.

232. **Balant, L., Burki, B., Wermeille, M., and Golden, G.,** Comparison of some pharmacokinetic parameters of (+)-cyanidanol-3 obtained with specific and non-specific analytical methods, *Arzneim. Forsch.*, 29(II), 1758, 1979.

233. **Groenewoud, G. and Hundt, H. K. L.,** The microbial metabolism of (+)-catechin to two novel diarylpropan-2-ol metabolites *in vitro*, *Xenobiotica*, 14, 711, 1984.

234. **Awata, N., Miura, S., Hamada, T., Satomi, O., and Midorikawa, T.,** Metabolic fate of cianidanol (KB-53). I. Absorption, distribution and excretion in rats, *Oyo Yakuri*, 25, 993, 1983 (Chem. Abstr. 99, 151661m, 1983).

235. **Scheline, R. R.,** The metabolism of (+)-catechin to hydroxyphenylvaleric acids by the intestinal microflora, *Biochim. Biophys. Acta*, 222, 228, 1970.

236. **Griffiths, L. A. and Barrow, A.,** The fate of orally and parenterally administered flavonoids in the mammal, *Angiologica*, 9, 162, 1972.

237. **Das, N. P. and Sothy, S. P.,** Studies on flavonoid metabolism. Biliary and urinary excretion of metabolites of (+)-[U-^{14}C]catechin, *Biochem. J.*, 125, 417, 1971.

238. **van der Merwe, P. J. and Hundt, H. K. L.,** Metabolism of (+)-catechin and some of its C-6 and C-8 substituted derivatives in the isolated perfused pig liver, *Xenobiotica*, 14, 795, 1984.

239. **Ohshima, Y., Watanabe, H., and Kuwazuka, S.,** On the mechanism of metabolism of (−)-epicatechin, *Bull. Agric. Chem. Soc. Jpn.*, 24, 497, 1960.

240. **Groenewoud, G. and Hundt, H. K. L.,** The microbial metabolism of condensed (+)-catechins by rat-caecal microflora, *Xenobiotica*, 16, 99, 1986.

241. **Harmand, M. F. and Blanquet, P.,** The fate of total flavanolic oligomers (OFT) extracted from "Vitis vinifera L." in the rat, *Eur. J. Drug Metab. Pharmacokinet.*, 3, 15, 1978.

242. **Watanabe, A. and Oshima, Y.,** Metabolism of gallic acid and tea catechin by rabbit, *Agric. Biol. Chem.*, 29, 90, 1965.

243. **Fedurov, V. V.,** Metabolism of bioflavonoids in animals, *Fenol' nye Soedin. Ikh Biol. Funkts. Mater Vses Simp., 1st (Publ. 1968)*, 371, 1966 (Chem. Abstr. 71, 11414e, 1969).

244. **Milić, B. L. and Stojanović, S.,** Lucerne tannins. III. Metabolic fate of lucerne tannins in mice, *J. Sci. Food Agric.*, 23, 1163, 1972.

245. **Milić, B. L.,** Lucerne tannins. I. Content and composition during growth, *J. Sci. Food Agric.*, 23, 1151, 1972.

246. **Masquelier, J., Claveau, P., and Colse, J.,** Metabolism of leucocyanidol in the rat, *Bull. Soc. Pharm. Bordeaux*, 104, 193, 1965 (Chem. Abstr. 65, 9531g, 1966).

247. **Claveau, P. and Masquelier, J.,** Le metabolisme du leucocyanidol chez le rat, *Can. J. Pharm. Sci.*, 1, 74, 1966.

248. **Schüller, J.,** Über Phlorhizin- und Phloretin-Glukuronsäure, *Z. Biol.*, 56, 274, 1911.

249. **Braun, W., Whittaker, V. P., and Lotspeich, W. G.,** Renal excretion of phlorizin and phlorizin glucuronide, *Am. J. Physiol.*, 190, 563, 1957.

250. **Barrow, A. and Griffiths, L. A.,** The biliary excretion of hydroxyethylrutosides and other flavonoids in the rat, *Biochem. J.*, 125, 24, 1971.

251. **Monge, P., Solheim, E., and Scheline, R. R.,** Dihydrochalcone metabolism in the rat: phloretin, *Xenobiotica*, 14, 917, 1984.

252. **Skjevrak, I., Solheim, E., and Scheline, R. R.,** Dihydrochalcone metabolism in the rat: trihydroxylated derivatives related to phloretin, *Xenobiotica,* 16, 35, 1986.

253. **Formanek, K. and Höller, H.,** Adsorption and excretion of chalcones, *Sci. Pharm.,* 29, 102, 1961 (Chem. Abstr. 55, 22619h, 1961).

254. **Chanal, J. L., Cousse, H., Sicart, M. T., Bonnaud, B., and Marignan, R.,** Absorption and elimination of (^{14}C) hesperidin methyl chalcone in the rat, *Eur. J. Drug Metab. Dispos.,* 6, 171, 1981.

255. **Brown, S. and Griffiths, L. A.,** Metabolism and excretion of butein, 2',3,4-trihydroxychalcone, 3-O-methylbutein, 4-O-methylbutein and 2',4',4-trihydroxychalcone in the rat, *Xenobiotica,* 13, 669, 1983.

256. **Shutt, D. A.,** The effects of plant oestrogens on animal reproduction, *Endeavour,* 35, 110, 1976.

257. **Nilsson, A.,** On the *in vitro* metabolism of the plant estrogen biochanin A in rumen fluid, *Ark. Kemi,* 17, 305, 1961.

258. **Batterham, T. J., Hart, N. K., Lamberton, J. A., and Braden, A. W. H.,** Metabolism of oestrogenic isoflavones in sheep, *Nature (London),* 206, 509, 1965.

259. **Braden, A. W. H., Hart, N. K., and Lamberton, J. A.,** The oestrogenic activity and metabolism of certain isoflavones in sheep, *Aust. J. Agric. Res.,* 18, 335, 1967.

260. **Nilsson, A., Hill, J. L., and Lloyd Davies, H.,** An *in vitro* study of formononetin and biochanin A metabolism in rumen fluid from sheep, *Biochim. Biophys. Acta,* 148, 92, 1967.

261. **Nilsson, A.,** Demethylation of the plant estrogen formononetin to daidzein in rumen fluid, *Ark. Kemi,* 19, 549, 1962.

262. **Nilsson, A.,** Demethylation of the plant œstrogen biochanin A in the rat, *Nature (London),* 192, 358, 1961.

263. **Nilsson, A.,** Fractionation of some plant estrogens and their animal excretion metabolites on dextran gels, *Acta Chem. Scand.,* 16, 31, 1962.

264. **Shutt, D. A., Axelsen, A., and Lindner, H. R.,** Free and conjugated isoflavones in the plasma of sheep following ingestion of oestrogenic clover, *Aust. J. Agric. Res.,* 18, 647, 1967.

265. **Labow, R. S. and Layne, D. S.,** The formation of glucosides of isoflavones and of some other phenols by rabbit liver microsomal fractions, *Biochem. J.,* 128, 491, 1972.

266. **Nilsson, A.,** O-Demethylaion of biochanin A and some other isoflavones and methylated estrogens by micorsomal liver enzymes, *Ark. Kemi,* 21, 97, 1963.

267. **Lindner, H. R.,** Study on the fate of phyto-oestrogens in the sheep by determination of isoflavones and coumestrol in the plasma and adipose tissue, *Aust. J. Agric. Res.,* 18, 305, 1967.

268. **Braden, A. W. H., Thain, R. I., and Shutt, D. A.,** Comparison of plasma phyto-oestrogen levels in sheep and cattle after feeding on fresh clover, *Aust. J. Agric. Res.,* 22, 663, 1971.

269. **Lundh, T. J.-O., Pettersson, H., and Kiessling, K.-H.,** Demethylation and conjugation of formononetin and daidzein in sheep and cow liver microsomes, *J. Agric. Food Chem.,* 36, 22, 1988.

270. **Nekby, C.,** Plant estrogens — their metabolism and physiological actions in ruminants. A literature review and metabolic studies in sheep, Master of Science, Swedish University of Agricultural Sciences, 1985.

271. **Batterham, T. J., Shutt, D. A., Hart, N. K., Braden, A. W. H., and Tweeddale, H. J.,** Metabolism of intraruminally administered [4-^{14}C]formononetin and [4-^{14}C]biochanin A in sheep, *Aust. J. Agric. Res.,* 22, 131, 1971.

272. **Lindsay, D. R. and Francis, C. M.,** Effect of progesterone and duration of pasture intake on cervical mucus response to phyto-oestrogens in sheep, *Aust. J. Agric. Res.,* 20, 719, 1969.

273. **Su, C.-Y. and Zhu, X.-Y.,** Metabolic fate of the effective components of *Radix puerariae.* II. Absorption, distribution and elimination of ^{14}C-daidzein, *Acta Pharm. Sin.,* 14, 129, 1979 (Chem. Abstr. 92, 33725t, 1980).

274. **Yueh, T.-L. and Chu, H.-Y.,** The metabolic fate of daidzein, *Sci. Sin.,* 20, 513, 1977.

275. **Shutt, D. A. and Braden, A. W. H.,** The significance of equol in relating to the oestrogenic responses in sheep ingesting clover with a high formononetin content, *Aust. J. Agric. Res.,* 19, 545, 1968.

276. **Griffiths, L. A.,** Metabolism of xanthone in the rat, *Xenobiotica,* 4, 375, 1974.

277. **Kostanecki, S. V.,** Ueber die Bildung von Euxanthinsäure aus dem Euxanthon mit Hülfe des thierischen Organismus, *Ber. Dtsch. Chem. Ges.,* 19, 2918, 1886.

278. **Külz, E.,** Zur Kenntnis des Indischgelb und der Glykuronsäure, *Z. Biol.,* 23, 475, 1887.

279. **Wiechowski, W.,** Über die Muttersubstanz des Indischgelb, *Arch. Exp. Path. Pharmak.,* 97, 462, 1923.

280. **Iseda, S.,** The structure of mangiferin, *Nippon Kagaku Zasshi,* 77, 1629, 1956 (Chem. Abstr. 53, 21916b, 1959).

281. **Hattori, M., Shu, Y.-Z., Tomimori, T., Kobashi, K., and Namba, T.,** A bacterial cleavage of the C-glucosyl bond of mangiferin and bergenin, *Phytochemistry,* 28, 1289, 1989.

282. **Thompson, K. L., Chang, M. N., Chabala, J. C., Chiu, S.-H. L., Eline, D., Hucker, H. B., Sweeney, B. M., White, S. D., Arison, B. H., and Smith, J. L.,** Metabolism of kadsurenone and 9,10-dihydrokadsurenone in rhesus monkeys and rat liver microsomes, *Drug Metab. Dispos.,* 16, 737, 1988.

283. **Luo, H.-W., Sheng, L.-S., Zhang, S.-Q., Xu, L.-F., and Wei, P.,** Tanshinones: antimycobacterial agents — bile excretion and biotransformation in rat liver, *Yaoxue Xuebao,* 18, 1, 1983 (Chem. Abstr. 99, 16073f, 1983).

284. **Miyagoshi, M., Amagaya, S., and Ogihara, Y.,** Structural transformation of geniposide, gardenoside, and related compounds in rat gastrointestinal contents, *Planta Med.,* 54, 556, 1988.

285. **Kohashi, K., Akao, M., and Yuda, M.,** Metabolism of geniposide by β-glucuronidase of intestinal microorganisms, *Wakan Iyaku Gakkaishi*, 5, 400, 1988 (Chem. Abstr. 111, 49872k, 1989).

286. **El-Sedawy, A. I., Hattori, M., Kobashi, K., and Namba, T.,** Metabolism of gentiopicroside (gentiopicrin) by human intestinal bacteria, *Chem. Pharm. Bull.*, 37, 2435, 1989.

287. **El-Sedawy, A. I., Shu, Y.-Z., Hattori, M., Kobashi, K., and Namba, T.,** Metabolism of swertiamarin from *Swertia japonica* by human intestinal bacteria, *Planta Med.*, 55, 147, 1989.

288. **Hattori, M., Shu, Y.-Z., Shimizu, M., Hayashi, T., Morita, N., Kobashi, K., Xu, G.-J., and Namba, T.,** Metabolism of paeoniflorin and related compounds by human intestinal bacteria, *Chem. Pharm. Bull.*, 33, 3838, 1985.

289. **Shu, Y.-Z., Hattori, M., Akao, T., Kobashi, K., Kagei, K., Fukayama, K., Tsukihara, T., and Namba, T.,** Metabolism of paeoniflorin and related compounds by human intestinal bacteria. II. Structures of 7S- and 7R-paeonimetabolines I and II formed by *Bacteroides fragilis* and *Lactobacillus brevis*, *Chem. Pharm. Bull.*, 35, 3726, 1987.

290. **Shu, Y.-Z., Hattori, M., Akao, T., Kobashi, K., and Namba, T.,** Metabolism of paeoniflorin and related compounds by human intestinal bacteria. III. Metabolic activity of intestinal bacterial strains and fecal flora from different individuals, *J. Med. Pharm. Soc. Wakan Yaku*, 4, 82, 1987.

291. **Hattori, M., Shu, Y.-Z., Kobashi, K., and Namba, T.,** Metabolism of albiflorin by human intestinal bacteria, *J. Med. Pharm. Soc. Wakan Yaku*, 2, 398, 1985.

292. **Powis, G. and Moore, D. J.,** High-performance liquid chromatographic assay for the antitumor glycoside phyllanthoside and its stability in plasma of several species, *J. Chromatogr.*, 342, 129, 1985.

293. **Fukami, J.-I., Yamamato, I., and Casida, J. E.,** Metabolism of rotenone in vitro by tissue homogenates from mammals and insects, *Science*, 155, 713, 1967.

294. **Yamamoto, I.,** Mode of action of natural insecticides, in *Residue Reviews*, Vol. 25, Gunther, F. A., Ed., Springer Verlag, Berlin, 1969, 161.

295. **Wessel, R., Thakur, A., Lutz, M., and Eiseman, J.,** The pharmacokinetic evaluation of ^{14}C-rotenone in the rat, *Pharmacologist*, 25, 116, 1983.

296. **Fukami, J.-I., Shishido, T., Fukunaga, K., and Casida, J. E.,** Oxidative metabolism of rotenone in mammals, fish, and insects and its relation to selective toxicity, *J. Agric. Food Chem.*, 17, 1217, 1969.

297. **Unai, T., Cheng, H.-M., Yamamoto, I., and Casida, J. E.,** Chemical and biological O-demethylation of rotenone derivatives, *Agric. Biol. Chem.*, 37, 1937, 1973.

METABOLISM OF AMINO COMPOUNDS

The major group of plant amines consists of aliphatic amines with aromatic substituents, however some aliphatic monoamines and aliphatic polyamines are also known. Most of these compounds are primary amines, having the general formula $R-NH_2$, although some secondary and tertiary amines are also encountered. The pathway of metabolism for these compounds is oxidative deamination which involves cleavage of the C–N bond. This may be carried out by both mitochondrial and microsomal enzymes. The former include monoamine oxidase and diamine oxidase which are involved in the metabolism of both endogenous substrates and xenobiotic compounds. Tipton[1] reviewed the literature on monoamine oxidase including the subjects of substrate specificity and multiple forms of the enzyme. Aliphatic amines in which the amino group is attached to an unsubstituted methylene group are subject to metabolism by monoamine oxidase. The best substrates are primary amines ($R-CH_2-NH_2$), although mono- and dimethylated amino compounds are also deaminated. The microsomal oxidation of primary and secondary aliphatic amines produces hydroxylamines and other N-oxygenated products. This subject was reviewed by Coutts and Beckett[2] and by Damani.[3] Additional information on the cleavage of compounds containing the C–N bond was summarized by Lindeke and Cho[4] who discussed the close metabolic relationship between deamination and N-dealkylation.

The aliphatic monoamines are volatile compounds ranging from the simplest homologue, methylamine, to n-hexylamine and also include several examples of branched-chain compounds. Among the latter, isoamylamine ($(H_3C)_2-CH-CH_2-CH_2-NH_2$) has widespread occurrence. The metabolic data on these simple amines derives mainly from older investigations, several having been carried out over a century ago. This information on aliphatic monoamines, as well as that dealing with the polyamines, was conveniently summarized by Williams,[5] from which much of the present summary is taken. Alles and Heegaard[6] showed that optimal rates of oxidation are achieved with higher homologues of alkylamines, however other data have shown that **methylamine** appears to be largely metabolized in rabbits, dogs, and man. Only a very small percentage of the dose can be recovered unchanged in the urine in these species. In accordance with the general reaction for the oxidative deamination of primary alkylamines by monoamine oxidase and the subsequent metabolism of the products (Figure 1), both the carboxylic acid (formic acid) and urea have been identified as metabolites of methylamine. Another prominent metabolite of methylamine is respiratory CO_2. Dar et al.[7] found that about half of the administered dose (~35 µg/kg, i.p.) was lost in this form when rats were given ^{14}C-labeled material. Interestingly, their use of various inhibitors of monoamine oxidase indicated that the oxidation of methylamine is mediated by enzyme systems other than monoamine oxidase. Their results also showed that the intestinal microorganisms play a minor role in methylamine metabolism in rats. Methylurea was reported by Dar and Bowman[8] to be a minor urinary metabolite of methylamine. They recorded values of ~1.8 and 5.6% of the dose (0.10 to 0.24 mg/kg, i.p.) in rats and rabbits, respectively. They found that the carbonyl group in methylurea was derived from the same metabolic CO_2 pool that is involved in the formation of urinary urea and respiratory CO_2.

In contrast to that seen with methylamine, **dimethylamine** is poorly metabolized and instead excreted unchanged nearly quantitatively in the urine by man. Ishiwata et al.[9] also recorded rapid and extensive excretion of dimethylamine in the urine of rats. **Ethylamine** appears to be partly metabolized and partly excreted unchanged. With higher homologues the latter pathway is reduced and both ***n*-propylamine** and ***n*-butylamine** are largely or entirely metabolized. Alles and Heegaard[6] found n-butylamine to be a good substrate for monoamine oxidase *in vitro*, the rate of oxidation being about half of that observed with ***n*-amylamine** and ***n*-hexylamine**. Both of the latter compounds gave near maximal rates among the aliphatic amines tested. A branched side chain may reduce metabolism slightly and **isobutylamine** ($(H_3C)_2-CH-CH_2-NH_2$) underwent some excretion unchanged in humans. With **isoamylamine** ($(H_3C)_2-CH-CH_2-CH_2-NH_2$),

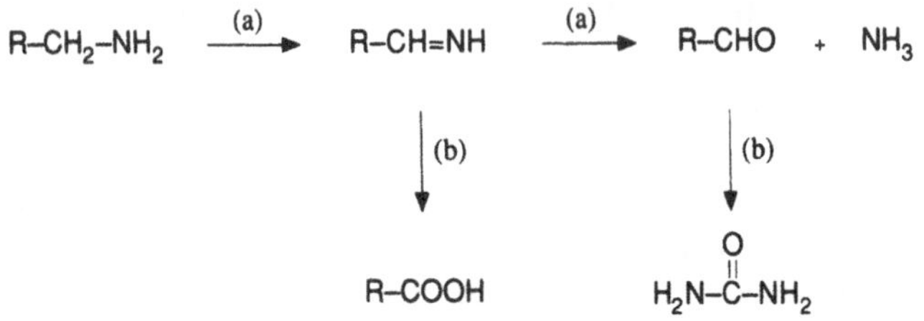

$$R-CH_2-NH_2 \xrightarrow{\text{(a)}} R-CH=NH \xrightarrow{\text{(a)}} R-CHO + NH_3$$

$$\downarrow \text{(b)} \qquad\qquad \downarrow \text{(b)}$$

$$R-COOH \qquad\qquad H_2N-\overset{\overset{\displaystyle O}{\|}}{C}-NH_2$$

FIGURE 1.   Oxidative deamination (a) and further metabolism (b) of aliphatic amines.

however, metabolism is complete as Richter[10] found no unchanged amine in the urine of humans given 100 mg of the compound orally. Earlier work demonstrated the conversion of isoamylamine to the expected isovaleric acid and urea and Richter[11] isolated isovaleraldehyde as its 2,4-dinitrophenylhydrazone from incubates of the amine with a monoamine oxidase preparation from guinea pig liver.

Aliphatic polyamines are often diamines with the general formula $H_2N-(CH_2)_n-NH_2$. **Putrescine** (1,4-butanediamine) ($n = 4$) is of fairly widespread occurrence in plants and **cadaverine** (1,4-pentanediamine) ($n = 5$) is also sometimes found. However, most of the interest in their physiological and metabolic properties relates to animal organisms. Putrescine, in fact, appears to be ubiquitously present in animal tissues. Decarboxylation of the amino acids ornithine and lysine are responsible for the formation of putrescine and cadaverine, respectively. Diamine oxidase (histaminase) oxidatively deaminates these compounds and, analogous to that described above with monoamines, the reaction products are an aldehyde and ammonia. Putrescine is an excellent substrate for this enzyme and quite large oral doses are required before some unchanged compound will escape metabolism and be excreted in the urine. Monoamine oxidase shows little activity towards putrescine and cadaverine. Further information on the metabolism of these diamines is available in the reviews by Tabor and Tabor[12] and Raina and Jänne.[13] Subsequent investigations have dealt with putrescine metabolites formed by rat liver slices[14,15] and those found in rat urine.[16]

As noted above, the main group of plant amines consists of aliphatic amines containing aromatic substituents. It is in this area that a great number of metabolic investigations have been carried out, primarily because of the important physiological and pharmacological properties of many of these compounds. A point of interest with this group is the fact that several of these compounds occur both in plants and animals. This is true of histamine, noradrenaline (norepinephrine), and serotonin (5-hydroxytryptamine), and by far the greatest number of investigations with these compounds have dealt with them in the context of mammalian physiology. Accordingly, and because many other sources of information on their metabolism are available, the present summary will deal with compounds which are primarily considered to be of plant origin. A further noteworthy point is that the compounds to be covered are nearly exclusively aromatic (phenyl or indolyl) derivatives of ethylamine or occasionally isopropylamine. **Benzylamine** provides an exception to this general rule. However, this compound which was formerly known as moringine seems to be of limited occurrence in plants. Its mammalian metabolism is straightforward according to earlier reports.[17] These showed that it is converted to benzaldehyde which, in the dog, is further oxidized and then excreted in the urine as hippuric acid. All three metabolites, benzaldehyde, benzoic acid, and hippuric acid, were identified in incubation experiments using rat hepatocytes[18] The capacity of the body to metabolize benzylamine is large and an oral dose of 160 mg to humans did not result in the urinary excretion of unchanged compound. Its metabolism proceeds rapidly and Wood et al.,[19] using material labeled with ^{14}C in the methylene group, found that more than 90% of an oral dose (75 mg) was excreted in man as urinary hippuric acid. This value increased to over 98% in 24 h and no other radioactive metabolites were detected. In the *in vitro* liver system used by Alles and Heegard,[6] benzylamine was oxidized at about a third of the rate observed with the higher homologues of ω-phenylalkylamines. Benzylamine metabolism by type A and B monoamine oxidase was studied by Parkinson et al.[20] who found that the type B enzyme was the most active in mitochondrial preparations from rats. Lewinsohn et al.[21] described an additional enzyme, present in many rat and human tissues, which they named benzylamine oxidase and which differed from monoamine oxidase B in both tissue distribution and inhibitor sensitivity. Additional studies on the metabolism of benzylamine by this enzyme, termed semicarbazide-sensitive amine oxidase, were reported by Elliott et al.[22]

The best represented group of plant alkylamines containing aromatic substituents is that of the β-phenylethylamines. Their metabolic fate is largely determined by the reactions of the amino group, however ring substituents may also be directly involved. In the first case, oxidative deamination leading via the aldehyde to a phenylacetic acid derivative is likely, possibly followed by conjugation of the acid with an amino acid. The amino group can also undergo acetylation or methylation. It is also possible that these pathways may not proceed readily, in which case the compound is excreted unchanged. Ring substituents include hydroxy or methoxy groups which may undergo conjugation or *O*-demethylation reactions, respectively. A further possibility involves the β-hydroxylase pathway found in the synthesis of catecholamines and which gives ethanolamine derivatives.

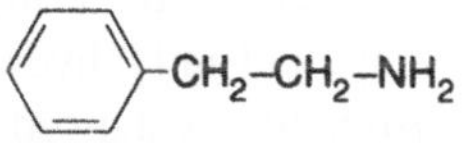

(1)

β-Phenylethylamine

The simplest member of this group is **β-phenylethylamine** (phenethylamine) (1) itself. An early study by Guggenheim and Löffler[23] showed that when administered to rabbits or dogs it was converted to both phenylacetic acid and phenylethanol. Richter[11] found that it was metabolized by guinea pig liver and intestine preparations to phenylacetic acid and ammonia. Similarly, Alles and Heegaard[6] reported that β-phenylethylamine was rapidly oxidized by liver amine oxidase. Yang and Neff,[24] using type A and B monoamine oxidase from rat brain, showed that the deamination of β-phenylethylamine was carried out by type B. While this behavior is found in most tissues, differences have been noted and Oguchi et al.[25] found that human placenta also contains a type A monoamine oxidase which has low-affinity sites for β-phenylethylamine.

β-Phenylethylamine undergoes rapid deamination *in vivo* and Richter[10] found that no unchanged compound was excreted in the urine in man following an oral dose of 300 mg (~4 to 5 mg/kg). Citing earlier work which indicated that the amine was converted to phenylacetic acid in rabbits, Richter found that 62% of the dose was excreted in 4.5 h as this acid, a portion of which may have been conjugated with glutamine. Block,[26] using β-phenylethylamine labeled with ^{14}C in the α-position, showed that radioactivity was rapidly excreted in the urine (64% in 2 h) by mice given an i.p. dose of ~80 mg/kg. Total urinary excretion of radioactivity reached 80% and most of this material consisted of phenylacetic acid. Only traces of $^{14}CO_2$ were detected in the respiratory air, indicating that decarboxylation of the acid or other metabolic pathways leading to C_6-C_1-compounds were not of importance. Wu and Boulton[27] carried out an extensive study of the distribution and metabolism of injected β-phenylethylamine in rats. The major metabolite in tissues and urine was phenylacetic acid and significant amounts of an unidentified metabolite were also detected. This compound was not the intermediate aldehyde, however, and no aldehydes were detected. A very small amount (0.1 to 0.3%) of phenylethanolamine was detected and further indication that a β-hydroxylase enzyme is involved in the metabolism of β-phenylethylamine was the evidence for the formation of octopamine. This involves aliphatic hydroxylation at the β-position and aromatic hydroxylation at the *p*-position. Aromatic hydroxylation was also indicated by the finding of radioactivity associated with *m*- and *p*-

tyramine. Similarly, Davis and Boulton[28] found these two tyramine isomers in the urine of humans fed ^{2}H-labeled phenylethylamine. Creveling et al.[29] described a β-oxidase capable of forming β-hydroxylated products with numerous phenylethylamines including β-phenylethylamine. This pathway has also been observed in humans.[30] Following an oral dose (150 mg) of β-phenylethylamine, the increase in urinary excretion of mandelic acid accounted for 4.5% of the administered amine. They suggested that phenylethanolamine was the intermediate in this metabolic pathway. Axelrod[31] reported that enzyme preparations from several tissues, among which lung was most active, were able to transfer the methyl group of S-adenosylmethionine to numerous amino compounds. This N-methylating enzyme showed moderate activity towards β-phenylethylamine, however the product thus formed has not been reported as an actual metabolite of β-phenylethylamine and it is likely that this pathway has little significance *in vivo*.

Tyramine (2) (see Figure 2) is β-p-hydroxyphenylethylamine and its metabolism is similar to that noted above with β-phenylethylamine. Ewins and Laidlaw[32] carried out experiments *in vivo* in dogs and *in vitro* using perfused rabbit or cat livers which showed that p-hydroxyphenylacetic acid (3) was formed. When dogs were given tyramine (~60 mg/kg, p.o.) ~25% of the dose was isolated in the urine as the acid, however other data indicated that the actual extent of oxidative deamination was about twice this amount. Other early studies showed that p-hydroxyphenylacetic acid and p-hydroxyphenylethanol (4) were tyramine metabolites in rabbits and dogs.[23] A study using tyramine labeled with ^{14}C in the α-position indicated that mice excreted in the urine only unchanged compound and p-hydroxyphenylacetic acid.[33]

Investigations carried out using *in vitro* methods showed that guinea pig liver preparations metabolized tyramine to p-hydroxyphenylacetaldehyde (5) and ammonia,[11] and that active deaminating systems are present in the liver and kidneys of rats, guinea pigs, rabbits, cats, and dogs.[34] Similar results were reported by Blaschko and Philpot.[35] The monoamine oxidase activity towards tyramine which is present in rat intestine was shown by Strolin Benedetti et al.[36] to be due to both forms, type A and B. Contrariwise, Elliott et al.[22] found that the latter type made little contribution to the metabolism of tyramine by rat mesentery arteries. Instead, type A monoamine oxidase and an additional system, termed semicarbazide-sensitive amine oxidase, were responsible for its metabolism.

A subject of continuing interest with tyramine metabolism has been its conversion to sulfate derivatives. Grimes[37] showed that sulfate conjugation occurred when the amine was incubated with a rat liver preparation, however the conjugate did not appear to be the O-sulfate. On the other hand, Spencer[38] found that active sulfate was transferred to tyramine by a rat liver sulfotransferase preparation to give three detectable metabolites. The identities of these compounds were not determined. The properties of a purified form of this enzyme were studied by Mattock and Jones.[39] Their results indicated that it was not identical with phenol sulfotransferase. Additional information on the properties of this enzyme was given by Wong[40] who carried out a comparative investigation using liver preparations from mice, rats, guinea pigs, monkeys, and man. Experiments using small intestine preparations from all species except man showed low activities in all species except the monkey (*Macaca sascicularis*) which gave values which were nearly three times those from the liver preparations. It was suggested that this situation, provided it is also found in humans, may explain the large differences in the urinary excretion of conjugated tyramine when oral or i.v. doses are given (see Table 1). Mullen and Smith[41] recorded a value of ~15% of the dose for the urinary excretion of tyramine O-sulfate following an unspecified oral dose of tyramine in man. Similarly, values of 6 to 13% in six subjects given oral doses of 100 mg tyramine were reported by Smith and Mitchell.[42] Additional data on the influence of the route of administration on the relative proportions of free and conjugated tyramine excreted in the urine of humans were presented by Boulton and Marjerrison.[43] Their values were lower than those summarized in Table 1 and totaled only ~1 to 2% of the dose. However, the conjugated form predominated by a factor of 13 when a tracer dose (0.076 mg) was given orally or by a factor of eight when a 100 mg dose was employed. Conversely, the

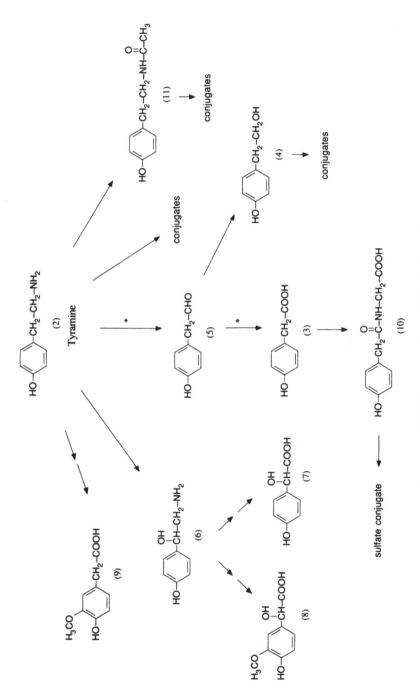

FIGURE 2.   Metabolic pathways of tyramine. (*) Major pathway.

TABLE 1
Urinary Metabolites of Tyramine

Compound	Structure (see Figure 2)	Rat[44] (1.2 mg/kg, i.p.)	Rabbit[45] (5 mg/kg, i.p.)	Man[46] (1.3 mg, i.v.)	Man[47] (100 mg, p.o.)
Tyramine	(2)	1.4[a]		3.5	0.4
Conjugates of (2)		0.6		0	8.7
p-Hydroxyphenylacetaldehyde	(5)	0.2		<0.3	
p-Hydroxyphenylacetic acid	(3)	74	71	49	57
O-Conjugates of (3)		4			6
p-Hydroxyphenylacetylglycine + sulfate	(10)	10.8		0	
p-Hydroxyphenylethanol + O-conjugates	(4)	0.7	3.4	0	
N-Acetyltyramine	(11)	0.2		<0.3	
Conjugates of (11)		2.9		0	
p-Hydroxymandelic acid	(7)	0	6	0	~0.1
3-Hydroxy-4-methoxymandelic acid	(8)	0	0.9	0	
Homovanillic acid	(9)	0	0.5	0	
Unidentified metabolites		1[b]	3[b]	~6[c]	

[a] All values equal percent of dose.
[b] Two compounds.
[c] Seven compounds.

i.v. administration of a tracer dose of tyramine resulted in the excretion of more than 80% of this material in the free form.

Several studies using tracer methods have aimed to give a comprehensive assessment of the urinary metabolites of tyramine. These investigations were carried out in rats,[44] rabbits,[45] and man[46,47] and the findings are summarized in Table 1. These studies also showed that the material was rapidly excreted in the urine, essentially none being lost by fecal or respiratory routes. The metabolic pathways involved are shown in Figure 2. In accordance with the earlier findings noted above, the most striking result is that which shows the extensive excretion of p-hydroxyphenylacetic acid (3), mainly in the free form. The results from rats and man showed that pathways involving β-hydroxylation and/or aromatic hydroxylation giving rise initially to octopamine (6), dopamine, and noradrenaline and then to the acidic metabolites (7), (8), and (9), were not of quantitative importance. Similar findings were reported by Wiseman-Distler et al.[48] who showed that neither metabolite (9) nor its 3-O-demethylated precursor (3,4-dihydroxyphenylacetic acid) was excreted in the urine of rats given tyramine (9.4 mg/kg, i.p.). However, Creveling et al.[49] obtained results which suggested that several percent of the tyramine present in the tissues may be metabolized via noradrenaline. Furthermore, Jones et al.[50] found that small amounts of p-hydroxymandelic acid (7) were excreted in the urine when human subjects were given [²H]tyramine intravenously. Trace amounts of 3,4-dihydroxyphenylacetic acid were also detected in these experiments. Octopamine (6), which is readily metabolized to p-hydroxymandelic acid (7) by oxidative deamination,[51] has been detected as a urinary metabolite of tyramine in rats pretreated with a monoamine oxidase inhibitor.[52] As shown in Table 1, these pathways involving hydroxylation reactions are responsible for the metabolism of at least 7 to 8% of the dose in rabbits. The β-oxidase system studied by Creveling et al.[29] was very active in β-hydroxylating p-tyramine. Dehydroxylation, a reaction not shown in Figure 2, appears to be a minor reaction of tyramine. Mosnaim et al.[53] reported that this conversion to β-phenylethylamine (1) was carried out by the 10,000-g fraction from rabbit brain preparations.

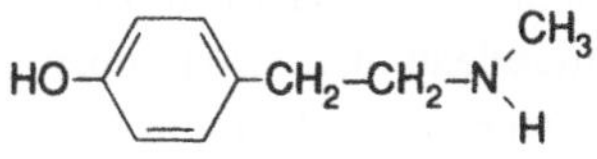

$$HO\!-\!\langle\!\text{ring}\!\rangle\!-\!CH_2\!-\!CH_2\!-\!N\!\begin{smallmatrix}CH_3\\H\end{smallmatrix}$$

(12)

N-Methyltyramine

The metabolic disposition of *N*-**methyltyramine** (12) in mice and rabbits was studied by Hai et al.[54] Using material labeled with ³H in the ring and i.v. doses of 2 mg/kg (rabbits) or 5 mg/kg (mice), they found extremely rapid metabolism and urinary excretion of the radioactivity. Mice excreted 79% of the label in the urine within 1 h and nearly complete excretion took place in 6 h. Chromatographic analysis of the plasma and urine of treated animals revealed the presence of at least three metabolites, however their identities were not determined. A less polar metabolite appeared to be formed first, followed by conversion to two metabolites of higher polarity which were excreted in the urine.

$$HO\!-\!\langle\!\text{ring}\!\rangle\!-\!CH_2\!-\!CH_2\!-\!N\!\begin{smallmatrix}CH_3\\CH_3\end{smallmatrix}$$

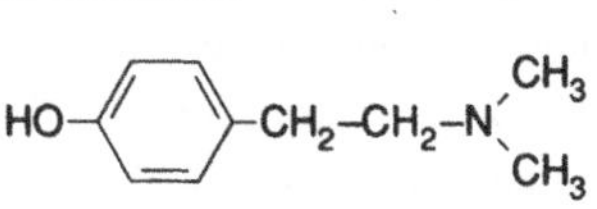

(13)

Hordenine

Hordenine (13) is *N,N*-dimethyltyramine and was the subject of an early study by Ewins and Laidlaw[32] who both administered it orally to dogs (~125 mg/kg) and perfused it through cat liver. In both cases a relatively small amount of unchanged compound was recovered, however little p-hydroxyphenylacetic acid was found. The low recoveries of metabolites indicated that metabolism had taken place although the presence of the tertiary amino group clearly depressed the pathway of oxidative deamination. That metabolism along this latter route can occur, however, was shown by Richter[11] who identified dimethylamine in the incubates of hordenine using an amine oxidase preparation from guinea pig liver. An unidentified aldehyde was also detected in the incubates. The inhibiting influence of *N*-methyl groups on amine oxidase activity was reported by Alles and Heegaard[6] in an extensive study of substrate specificity. Barwell et al.[55] found that hordenine was a highly selective substrate for the B form of monoamine oxidase found in rat liver but that it was not deaminated by the A form present in the intestinal epithelium.

A minor metabolic pathway in the metabolism of hordenine may involve ring hydroxylation. Wiseman-Distler et al.[48] reported that nearly 2% of the dose (23 mg/kg, i.p.) was excreted in the urine as 4-hydroxy-3-methoxyphenylacetic acid (9) (see Figure 2). Also, Daly et al.[56] described a microsomal enzyme system which readily formed a catechol derivative with hordenine.

The metabolism of **mescaline** (14) (see Figure 3) has received considerable attention because of the possibility that this factor may be closely associated with pharmacological effects, perhaps through the formation of one or more active metabolites. Lemberger and Rubin[57] reviewed this and related aspects of the biological disposition of mescaline. Metabolic studies with mescaline date from the investigation of Slotta and Müller[58] who administered it orally to rabbits and dogs and intravenously to humans. Interestingly, the 24-h urine samples from the first two species contained approximately a third of the dose as 3,4,5-trimethoxyphenylacetic acid (15), whereas this metabolite was not found in human urine. In the latter case a monomethoxy derivative was detected. This early indication of species differences in the metabolism of mescaline has spurred the subsequent investigation, both *in vitro* and *in vivo*, of its fate in many animal species. In order to summarize this abundant and widespread material in a way that more clearly underlines the metabolic features, the results from numerous *in vivo* studies are tabulated in Table 2, which lists the metabolites found, their amounts, and other relevant points. The pathways in the metabolism of mescaline are illustrated in Figure 3.

The data in Table 2, in spite of numerous differences in experimental details, allow several general conclusions to be drawn about the metabolism of mescaline. It is evident that this compound is more resistant to oxidative deamination than are, for example, β-phenylethylamine and tyramine. This fact is reflected in the generally high values for the excretion of unchanged compound, although some species differences do appear to exist, notably with the rather low values from rats compared with those from mice. However, the urinary pH values of the experimental urines in these studies were not reported and possible variations in these values, influencing as they do the urinary excretion of the unchanged compound, make interpretation difficult. Another point of interest which emerges from the data shown in Table 2 is that 3,4,5-trimethoxyphenylacetic acid (15) has been reported as a urinary metabolite in all of the species studied (mouse, rat, rabbit, cat, dog, and man). This metabolite and unchanged compound generally furnish the bulk of the excreted material with little contribution apparently being made by *N*-acetylation. However, the latter reaction has been reported in the mouse, rat, and man (Table 2) and as a liver metabolite of mescaline in monkeys.[59,60] Musacchio and Goldstein[61] reported that *N*-acetyl derivatives accounted for ~30% of the urinary metabolites in rats.

The study of Musacchio and Goldstein[61] also showed that β-hydroxymescaline was not an excretory product of mescaline in rats. This metabolite or other β-hydroxylated derivatives were not reported in any of the other investigations summarized in Table 2 and it seems likely that β-hydroxylation is not of importance in the metabolism of mescaline. Goldstein and Contrera[62] reported that mescaline is a weak substrate of phenylamine-β-hydroxylase *in vitro* and that *O*-methylation in the aromatic ring is a factor which decreases substrate affinity for the hydroxylating enzyme.

The role of *O*-demethylation in mescaline metabolism is rather difficult to assess, however several such metabolites have been identified (Table 2) and an unidentified metabolite of this type was reported in the early work of Slotta and Müller.[58] However, the amounts of these metabolites formed are usually small. This subject was studied *in vitro* by Axelrod[63] and Daly

315

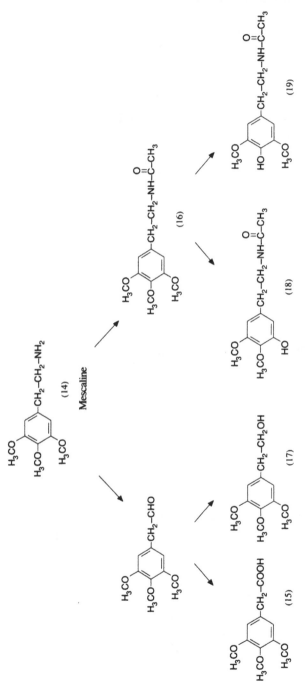

FIGURE 3. Metabolic pathways of mescaline. See text for description of minor additional metabolites.

TABLE 2
Urinary Metabolites of Mescaline

Species	Dose	Collection period (h)	Compound (see Figure 3 for structures): Mescaline (14)	(15)	(16)	(17)	(18)	(19)	Other	Unknown	Ref.
Mouse	80 mg/kg, i.p.	40	79[a]	16						4	152
	50 mg/kg, i.p.	3	68[a]	20 (31 in 24 h)	8					5 (neutral cpds)	153
Rat	24 mg/kg, i.p.	40	18[a]	72						9	152
	0.4 mg/kg	24	+[b]	+	+						154
	0.2 or 40 mg/kg, i.p.	24	20[a]	42 (incl. (17))	1.7		14	15		6	61
Rabbit	340 mg, p.o.	24		30—50[c]							58
Cat	25 mg/kg, i.v.	0.5—6	+	+					None detected		155
Dog	340 mg, p.o.	24		30—50[c]							58
	20 mg/kg p.o., i.v., i.m.	24	28—46[c]	trace							156
	20 mg/kg, i.p.	13	~ 40[a]	~ 60							157
Man	~ 3 mg/kg i.v., p.o.	24	52, 58[c]								10
	400 mg	24	~ 35[c]	—					[d]		158
	400 mg, p.o.	5	+ (large amount)						[e]		159
	5 mg/kg, i.v.	6	31[c] (12—67)	7							160
	350 mg, p.o.	12		26[c]							161
	6 mg/kg	24	23[c]	18							162
	500 mg, p.o.	24	55—60[a]	27—30	0.1		5		A hydroxymethoxy-phenylacetic acid	~ 10 (5 compounds)	163
	~325 mg, p.o.	24		+					3,4,5-Trimethoxy-benzoic acid		66

[a] % of radioactivity in urine.
[b] (+) Present, (—) absent.
[c] Percent of dose.
[d] 1 to 2% of glutamine conjugate of 3,4-dihydroxy-5-methoxy phenylacetic acid.
[e] Small amount of 3,4-dimethoxy-5-hydroxyphenylethylamine.

et al.[64] Demethylation was effected by liver microsomes from many species and experiments using rabbit liver microsomes showed that the reaction took place at both the 4- and 5-positions. The demethylated metabolites, of which the 5-hydroxy derivative predominated, were formed in relatively small amounts.

An additional minor metabolic pathway of mescaline involves side-chain degradation giving small amounts of 3,4,5-trimethoxybenzoic acid. Demisch and Seiler[65] reported that this metabolite was formed *in vitro* in mouse tissue preparations, especially those from brain. The system, which is localized in the nuclear and microsomal fractions, also formed the corresponding phenylacetic acid derivative (15). Demisch et al.[66] subsequently isolated this benzoic acid derivative from the urine of human subjects given mescaline (~325 mg) orally. However, the amount excreted was only ~0.01% of the quantity of 3,4,5-trimethoxyphenylacetic acid found. The *N*-methylation of mescaline has not been reported in *in vivo* studies, however rabbit lung preparations are capable of transferring the methyl group of *S*-adenosylmethionine to numerous amines including mescaline.[31]

Attention has also been given to the presence of mescaline metabolites in various tissues, especially the brain. Ho et al.[67] administered [14C]mescaline (3.5 to 25 mg/kg, i.v.) to rats and found that *N*-acetylmescaline (16) accounted for as much as a third of the radioactivity in the brain after 30 min. Unchanged mescaline was the major component, however small amounts (2 to 6% each) of the acid (15) and alcohol (17) were also present. Similar results were reported by Taska and Schoolar[60] in monkeys (*Saimiri sciureus*) given mescaline (~5 mg/kg, i.v.). The radioactivity in the liver during the first 6 h consisted mainly of unchanged compound but 10 to 20% was due to the *N*-acetyl derivative (16). Smaller amounts of (15) and (17) were also present. *N*-Acetylmescaline has also been identified in the brains of mice treated with mescaline (120 mg/kg, i.p.) labeled with 14C in the α-position.[68] Interestingly, this investigation showed that 0.05% of the administered radioactivity appeared in the respiratory air. As noted above, the expected metabolite resulting from side-chain decarboxylation, 3,4,5-trimethoxybenzoic acid, was identified in brain and liver. The small amounts of this benzoic acid derivative which were found confirm that this route is a very minor pathway in the metabolism of mescaline. Three other metabolites, shown to be nondemethylated amines but otherwise only partially characterized, were also identified in small amounts in the brain.

A development which coincided with the *in vivo* studies of mescaline metabolism was the elucidation of the enzyme systems responsible for its metabolism and their tissue and species distribution. Several early studies indicated that the systems which readily metabolized β-phenylethylamine or tyramine were often without significant effects towards mescaline. Pugh and Quastel[69] showed that this was the case with the tyramine oxidase in rat brain slices and guinea pig kidney slices. Bernheim and Bernheim[34] reported that liver and kidneys from most animals contain an active tyramine oxidase, however only the preparation from rabbit liver showed appreciable activity towards mescaline. Later, Roth et al.[70] found that mescaline-oxidizing activity was several times higher in rabbit lung homogenates than in similar preparations from liver and kidney. Blaschko[71] concluded that the activity of rabbit liver was not due to amine oxidase. The liver amine oxidase studied by Alles and Heegaard,[6] and which oxidized many phenylethylamine derivatives, was without activity towards mescaline. Steensholt[72] investigated the oxidation of mescaline by rat and rabbit liver homogenates. Little activity was found in the former case whereas appreciable oxidation occurred with the latter. However, it was concluded that mescaline oxidase was distinct from monoamine oxidase, although many similarities exist between the two. This problem was re-examined in detail by Zeller et al.[73] who studied the mescaline oxidase activity of various tissues in seven mammalian species. This enzyme in rabbit liver was found to be present in the mitochondrial and microsomal fractions and inhibitor studies indicated that it is a typical diamine oxidase rather than a monoamine oxidase. This was confirmed by Huszti and Borsy[74] using rabbit liver mitochondria. Zeller et al. showed that mescaline was also oxidized by the diamine oxidases of pig kidney cortex and sheep

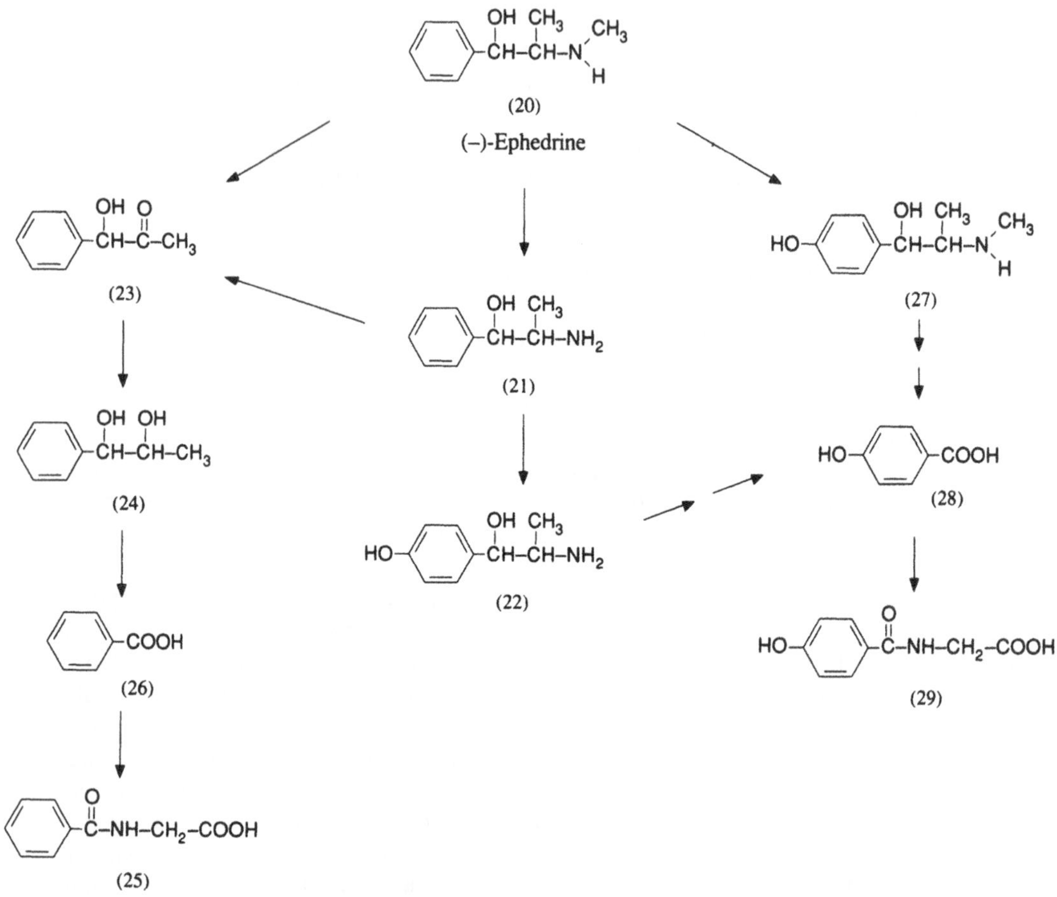

FIGURE 4. Metabolic pathways of (–)-ephedrine.

plasma. Nonetheless, examination of other tissue preparations from these species (e.g., mouse or pig liver mitochondria) indicated that the enzyme system involved in mescaline deamination had an inhibitor pattern typical for monoamine oxidase. Interestingly, Riceberg et al.[75] reported that the enzymes oxidizing mescaline in rabbits are related to the copper-containing plasma amine oxidases and are not the flavin-containing mitochondrial monoamine oxidases. A monoamine oxidase was shown to be responsible for mescaline oxidation in mouse brain mitochondria.[76-78] Guha and Mitra[79] reported the presence of a dehydrogenase system localized in rat and guinea pig brain mitochondria which was capable of dehydrogenating mescaline.

Ephedrine contains two asymmetric carbon atoms and can therefore exist in four isomeric forms. These are (–)- and (+)-ephedrine and (–)- and (+)-pseudoephedrine, of which the first and the last are naturally occurring. Interest in the metabolism of (**–)ephedrine** (20) (see Figure 4) has paralleled that seen with other phenylisopropylamine drugs. Most members of this group, of which amphetamine is the simplest example, are synthetic compounds and a discussion of their metabolism therefore falls outside the scope of this book. However, the reviews by Caldwell[80,81] and Lemberger and Rubin[57] cover numerous aspects of the metabolism of these compounds and serve as supplements to the present summary of ephedrine metabolism.

The major pathways in the metabolism of (–)-ephedrine are illustrated in Figure 4 which shows that the main reactions are oxidative deamination which may lead to degradation of the aliphatic side chain, *N*-demethylation, and aromatic hydroxylation. The quantitative aspects of urinary metabolite excretion in seven mammalian species are summarized in Table 3. These data show that although all three pathways are often operative, their relative importance varies

TABLE 3
Urinary Metabolites of (−)-Ephedrine

Species	Dose	Collection period (h)	%a of dose recovered in urine	(−)-Ephedrine (20)	(21)	(22)	(22+27)	(27)	(23)	(24)	(26)	(26+25)	(25)	(28)	(29)	Other	Ref.
				Compound (see Figure 4 for structures)													
Mouse	5 mg/kg, s.c.	24	93	86[a]	2.9			2.3		0	0.4		1.1	0	0		164
Rat	41 mg/kg, i.p.	24		32	7.5		13										108
	i.p.	24	64	43	0.5			2			0.3		1.8				84
	2 mg/kg, i.p.[b]	40	80	53			20					4					165
	5 mg/kg, s.c.	24	71	56	4.6			31		0.1	0.4		2.7				164
	8 mg/kg, s.c.	24	70	33—37				11—14									166
Guinea pig	41 mg/kg, i.p.	24		2	39		0.9									1	08
	5 mg/kg, s.c.	24	83	1.6	51.9			0		8.6	23		2.8	0.5			164
	1 mg/kg, i.v.	40	85	13	51							17				2.6 as p-hydroxy cpds.	167
Rabbit	41 mg/kg, i.p.	24		0.1	1.8		1.9										108
	5 mg/kg, p.o.	24	95	0.7	1.3	0		0.7	0.8	29.1	19.1		32.9	1.3	8.8		164,168
	8 mg/kg, s.c.	24	94							13			31				166
	8 mg/kg, s.c.	24								14—17			25				169
	3 mg/kg, s.c.	24	71	0.5	1.1				0.2	3.3	14		21				86
Dog	i.p.	24			~80	1		0.3									170
	41 mg/kg, i.p.	24		6.5	58		1.5										108
Horse	~0.4 mg/kg, s.c.	3[c]		6													171
	~1 mg/kg, p.o.,i.v.	24		0	7[d](15[e])		3									mandelic acid not present	172
Man	~1 mg/kg, p.o.	24	87														173
	i.v.	24		80—100[f]		18[f]			0	19							174
	0.4—1.5 mg/kg, p.o.	48		79	4.3												104
	27 mg/kg, p.o.	24		27[d](88[e])	18[d](7[e])						tr[g]					10	82
	20—25 mg/kg, p.o.	24		77	4.0								6.3				175

TABLE 3 (continued)
Urinary Metabolites of (−)-Ephedrine

Species	Dose	Collection period (h)	% of dose recovered in urine	(−)-Ephedrine (20)	(21)	(22)	(22+27)	(27)	(23)	(24)	(26)	(26+25)	(25)	(28)	(29)	Other	Ref.
Man	30 mg, p.o.	24	88	61[h]	13	0		0	0	1—10	+[i]		4.6				176
	0.82 mg/kg, p.o.	24		75	3.6								3.5				177

Compound (see Figure 4 for structures)

[a] Figures equal percent of dose unless otherwise indicated.
[b] Used (±)-ephedrine.
[c] No amine excreted after 3 h.
[d] Alkaline urine.
[e] Acidic urine.
[f] Includes compound (21).
[g] (tr) Trace.
[h] Urine pH 6.5 to 6.8.
[i] (+) Present.

greatly from species to species. In fact, the large species variations in (–)-ephedrine metabolism have provided perhaps the most interesting aspect of this subject. The data in Table 3 indicate that the mouse and man excrete most of the dose unchanged in the urine. It should be noted that urinary pH influences the extent of excretion of the unchanged compound and that this may be reduced appreciably when the urine is alkaline.[82] The rat is intermediate (~30 to 50% of the dose) in the amount of compound which is excreted unchanged whereas guinea pigs, rabbits, dogs, and horses all metabolize the compound extensively.

The summarized data show that N-demethylation occurs to an appreciable extent in rats, guinea pigs, and dogs with the demethylated metabolites (21) and (22) accounting for perhaps ~20% of the dose in rats, 40 to 50% in guinea pigs, and 60 to 80% in dogs. The actual extent of N-demethylation in rabbits is masked by the extensive metabolism to phenylpropane derivatives and to benzoic and hippuric acids. However, it is not unreasonable to expect that the extent of N-demethylation may be considerably higher than that suggested by the small amounts of urinary N-demethylated compounds since it is known that norephedrine (21) undergoes extensive further metabolism in rabbits to compounds (23), (24), and (25).[83]

As seen in Table 3, aromatic hydroxylation of (–)-ephedrine is a prominent pathway only in the rat. On the other hand, that of oxidative deamination and side-chain shortening is a minor route in this species. The latter pathway is the major feature of (–)-ephedrine metabolism in rabbits and an important route in both guinea pigs and horses. It appears to account for ~10% of the dose in man.

Several investigations have been carried out to assess possible differences in the metabolism of (–)- and (+)-ephedrine. Nagase et al.[84] studied the excretion and metabolism of [^{14}C]-(+)-ephedrine in rats and found that the excretion rate was similar to that of the (–)-isomer but that the recovery of urinary radioactivity was ~20% less with the former. The N-demethylation and aromatic hydroxylation which occurred with (–)-ephedrine were not detected with the (+)-isomer. A subsequent study using deuterium-labeled isomers[85] confirmed that the (–)-isomer was more easily hydroxylated in rats than was the (+)-isomer. The formation of glucuronide conjugates was also greater with the (–)-isomer. On the other hand, N-demethylation was more extensive with the (+)-isomer. Feller and Malspeis[86] compared the *in vitro* and *in vivo* metabolism of (–)- and (+)-ephedrine in rabbits. The (–)-isomer was found to be metabolized at a faster rate by liver microsomes and the rate of benzoic acid formation was about three times as large. The relative amounts of demethylated compound (21) and diol intermediate (24) formed from the two isomers were nearly identical. The *in vivo* data showed that both isomers were mainly metabolized by N-demethylation and oxidative deamination of the side chain. However, some quantitative differences were noted in urinary metabolite excretion. About 91% of the administered dose of (+)-ephedrine was excreted in the urine in 24 h compared with the 71% found with the (–)-isomer (Table 3). Other differences with (+)-ephedrine were a higher excretion of benzoic acid (26) and lower excretion of hippuric acid (25), although the total amounts of these metabolites excreted in the two sets of experiments were not very different. About five times as much unchanged compound and diol (24) were excreted when the (+)-isomer was given.

Information is available on the biological disposition of **pseudoephedrine** in rats and humans. Also, Kuntzman et al.[87] studied the influence of urinary pH on the plasma half-life of (+)-pseudoephedrine in dogs and humans. As expected, the half-lives decreased in both species when the urinary pH-values were lowered by administration of NH_4Cl. Jacquot et al.[88] administered racemic pseudoephedrine (5 mg/kg, i.p.) labeled with ^{14}C to rats. About 80% of the radioactivity was eliminated in the urine in 18 h and, of this, ~42% was unchanged compound, 44% was *p*-hydroxypseudoephedrine and 12% was benzoic and hippuric acids. Several studies have shown that 100 mg oral doses of (+)-pseudoephedrine in humans were mainly excreted in the urine as unchanged compound. This was found by Baaske et al.[89] who recorded values of 91%. Most of this material was excreted during the first 24 h and minute amounts of N-

demethylated compound were also detected. Corresponding values of 72 to 74% (24 h) were reported by Nieder and Jaeger,[90] however Lo et al.[91] found 43% of the dose as unchanged compound and ~1% as norpseudoephedrine in 24 h.

The cellular and enzymatic aspects of ephedrine metabolism have been studied using a number of *in vitro* preparations. An early study[6] showed that the amine oxidase preparation from rabbit liver which effectively oxidized aliphatic amines and β-phenylethylamine and its derivatives, did not oxidize the racemic mixtures of ephedrine, pseudoephedrine, or their demethylated derivatives. It was later shown that liver microsomes are able to carry out numerous metabolic reactions with ephedrine including *N*-demethylation and oxidative deamination. Axelrod[92] showed that rabbit liver microsomes which oxidized amphetamine to phenylacetone and ammonia actively metabolized (–)-ephedrine. Baba et al.[93,94] used NADPH-fortified rabbit liver preparations or rabbit liver slices and showed conversion of (–)-ephedrine to the ketone (23) and its reduction product (24). These two metabolites were also reported by Beckett and Al-Sarraj[95] using fortified 10,000-g fractions from rabbit liver. A subsequent extensive study of this reaction, using norephedrine as the substrate, was also carried out.[96] The dehydrogenase system in brain mitochondria noted above in the section on mescaline was also found to dehydrogenate ephedrine.[79]

Microsomal *N*-demethylation of (–)-ephedrine and its isomers has been reported in several studies. In an important early investigation Axelrod[97] showed that rabbit liver microsomes in the presence of NADPH catalyzed the conversion of (–)-ephedrine to norephedrine and formaldehyde. This reaction was also carried out by dog, guinea pig, and rat liver preparations, although with lower relative activities. The high *N*-demethylase activity of rabbit liver microsomes was confirmed by Gaudette and Brodie[98] in a study relating oxidation to lipid solubility. McMahon[99] reported the *N*-demethylation of (–)-ephedrine by rat, guinea pig, and rabbit liver microsomes and studied the effect of inhibitors on this reaction. Dann et al.[100] and Feller et al.[101] studied the rabbit microsomal *N*-demethylation of the four stereoisomers of ephedrine and noted small differences in the relative reaction rates. An investigation of the mechanism of *N*-dealkylation of several secondary amines including (–)-ephedrine and pseudoephedrine by a purified microsomal mixed function oxidase from pig liver was carried out by Ziegler et al.[102] The reaction involves an initial *N*-oxidation followed by the breakdown of this unstable intermediate to the primary amine and formaldehyde. Beckett and Al-Sarraj[95] and Beckett et al.[96] studied the *N*-oxidation of ephedrine and norephedrine using 10,000-g rabbit liver preparations. Their results showed that in addition to the α-*C*-oxidation noted above which leads to the formation of metabolites (23) and (24), *N*-oxidation giving *N*-hydroxy compounds (hydroxylamines) also occurred.

The oxygenation of (+)-norpseudoephedrine by dopamine-β-hydroxylase and its further conversion to the ketone analogue cathinone is noted below. However, the secondary amines (+)-ephedrine and (+)-pseudoephedrine were shown to be very weak substrates for this enzyme.[103]

A few *in vitro* studies have shown the formation of most of the known metabolites of ephedrine in the species studied. Baba et al.[93] incubated (–)-ephedrine with an NADPH-fortified rabbit liver preparation and detected the demethylated compound (21), the α-*C*-oxidized products (23) and (24), and benzoic acid (26). With rabbit liver slices Baba et al.[94] noted the formation of hippuric acid (25) in addition to the above four metabolites.

The oral administration of **methylephedrine** (30) (see Figure 5) to man results in appreciable excretion of unchanged compound in the urine.[104] Like that observed with (–)-ephedrine as noted above, the extent its urinary excretion is influenced by urinary pH. Wilkinson and Beckett[82] recorded values for unchanged compound of 2 to 6% of the dose (~27 mg, p.o.) in three subjects when the urine was alkaline. These values increased to 64 to 80% when the urine was acidic. They also noted that methylephedrine was extensively metabolized and that *N*-demethylation to both ephedrine and norephedrine occurred. The metabolism of (±)-methylephedrine in rats and man was recently studied more fully by Inoue and Suzuki.[105] Slightly more than 50% of the

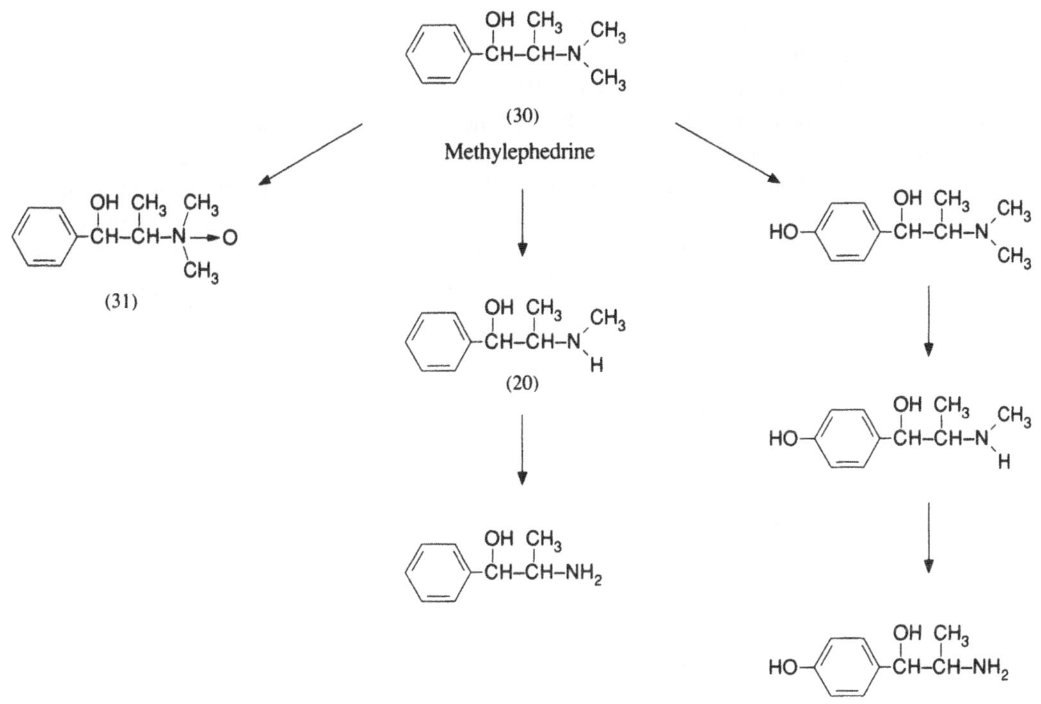

FIGURE 5.   Metabolic pathways of methylephedrine in rats.

dose (100 mg/kg, p.o. of methylephedrine HCl) was recovered as unchanged methylephedrine and its metabolites in rats. The corresponding value in humans given an oral dose (20 mg) was ~70%. The compounds identified in rat urine are shown in Figure 5, however formation of hydroxylated metabolites was a minor pathway and accounted for ~2% of the dose. Formation of these p-hydroxylated derivatives was even less pronounced in man and their presence could only be demonstrated by the use of a sensitive technique employing gas chromatography and mass spectrometry. The major metabolite in rat urine was methylephedrine-N-oxide (31) which accounted for a third of the dose. Unchanged compound and ephedrine (20) furnished ~8% each and the remaining metabolites each accounted for <1% of the dose. The results in man showed that unchanged compound was the major urinary metabolite. The values in two subjects were 33 and 40% and the urinary pH values were in the range of 5.5 to 6.5. Additional major metabolites were ephedrine (21 and 14%) and methylephedrine-N-oxide (15%). Experiments using rat liver microsomes showed that the N-oxidation of methylephedrine was carried out by the flavin-containing monooxygenase system and that its demethylation to ephedrine was mediated by the cytochrome P-450 system.[106]

Metabolic data are also available on the N-demethylated derivatives of ephedrine. (+)-**Norpseudoephedrine** (cathine) (32) is a constituent of khat, a stimulant drug obtained from the leaves of an East African plant. This aminoalcohol is structurally and metabolically closely related to cathinone, the corresponding ketone analogue noted below. Maitai and Mugera[107] gave (+)-norpseudoephedrine (~30 mg, p.o.) to human subjects and found that ~40% of the dose was excreted unchanged in the urine in 6 h. This material was not excreted in conjugated form. (+)-Norpseudoephedrine was shown to be oxygenated by dopamine-β-hydroxylase.[103] The product formed is presumably a *gem*-diol which, via loss of water, would exist predominately as the ketone cathinone (33).

As noted above, norephedrine (21) is formed in widely varying amounts from ephedrine in various species and several investigations have dealt with its excretion and metabolism. Axelrod[108] found that (−)-norephedrine was mainly excreted unchanged in the dog. The same result was found in man.[82,104] Sinsheimer et al.[83] administered (±)-norephedrine to rats, rabbits, and man and confirmed that most (86%) of the dose (25 mg) was excreted unchanged in the urine by man. They identified 4-hydroxynorephedrine (22) and hippuric acid (25) (see Figure 4) as minor metabolites. Rabbits excreted less than 10% of the dose (12 mg/kg) unchanged. Oxidative deamination was the main pathway and ~80% of the dose was converted to metabolites (23), (24), and (25). At the same dose level, rats excreted about half unchanged and 28% as the *p*-hydroxylated metabolite (22). Thiercelin et al.[109] also administered racemic [^{14}C]norephedrine to rats (0.535 mg/kg, i.v.) and found that ~75% of the radioactivity was excreted in the urine in 18 h. This material consisted of unchanged compound and metabolite (22) (54 and 14% of the dose, respectively). This investigation also dealt with the kinetics of norephedrine distribution and elimination.

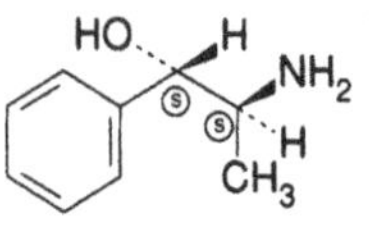

(32)

(+)-Norpseudoephedrine

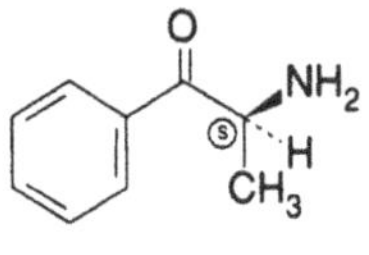

(33)

(–)-Cathinone

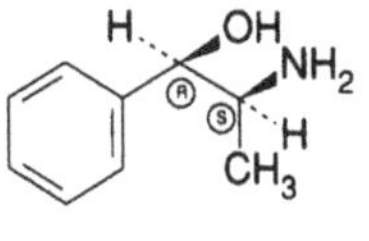

(34)

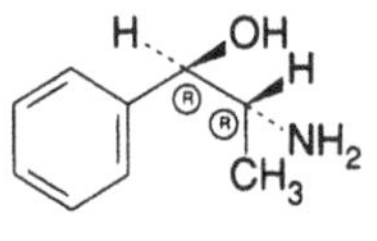

(35)

It is noted above that (+)-norpseudoephedrine (cathine) is an alkaloid from khat, however this plant also contains the ketone analogue **(–)-cathinone** (33). An interconversion between the aminoalcohol and aminoketone forms occurs which may alter their relative proportions at different stages of plant development or during storage.[110] Also noted above was the finding that cathine undergoes ketonization to cathinone by dopamine-β-hydroxylase. Two investigations have dealt with the metabolism of cathinone in humans, however the results are conflicting. Guantai and Maitai[111] used an oral dose (16 mg) of cathinone isolated from *Catha edulis* as well as synthetic material. In the former case the material must have been mainly or entirely *S*-(–)-cathinone which is the isomer found in the plant. They recovered 11% of the dose in the urine as unchanged compound in 24 h and a further 82% as (+)-norpseudoephedrine (32). Two additional, unidentified metabolites were also noted. Similar experiments were carried out by Brenneisen et al.[112] who employed 24 mg doses of the *S*-(–)-, *R*-(+)-, or (±)- forms of cathinone. In all cases the 24-h urines contained ~1 to 3% of the dose as unchanged compound and 21 to 50% as aminoalcohols formed by the reduction of the ketone group. This reductive step was stereospecific and the *R*-forms of the aminoalcohols were formed from both (–)- and (+)-cathinone. The main metabolite formed from (–)-cathinone was *R/S*-(–)-norephedrine (34), however smaller amounts of *R/R*-(–)-norpseudoephedrine (35) were also detected. These metabolites occurred in a ratio of ~8:1 and the presence of the latter compound was explained by the fact that the cathinone isomers employed were not completely pure and also by the ease of racemization of cathinone in aqueous solution.

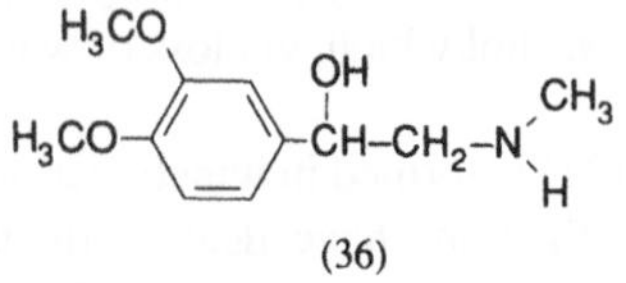

(36)

Normacromerine

The biological disposition of **normacromerine** (36) in rats was studied by Ferguson et al.[113] using material labeled with ^{14}C in the β position. Following an oral dose of 85 mg/kg about half of the radioactivity was excreted in the urine in 24 h. An additional 10% was found in the feces, however no respiratory $^{14}CO_2$ was detected. About a third of the metabolites present in the urine samples obtained 8 h after dosing consisted of unchanged normacromerine, however these values decreased to 3 and 0% after 16 and 24 h, respectively. The identities of the metabolites were not determined.

Another major group of amines containing aromatic substituents is that of the indoleamines. These compounds are usually derivatives of ethylamine (tryptamines), although simpler homologues are sometimes found. Gramine (3-(dimethylaminomethyl)indole) is such an example, however its metabolism in mammals has not been investigated. The tryptamine derivatives which occur in higher plants include tryptamine (37) itself as well as *N*-methylated and/or 5-hydroxylated or 5-methoxylated compounds.

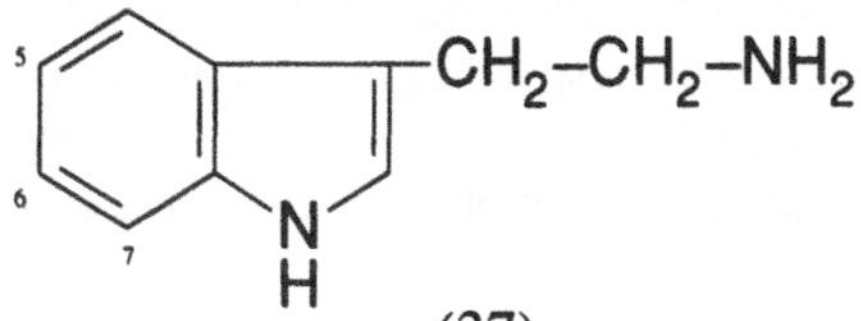

(37)

Tryptamine

Tryptamine (37) was the subject of early metabolic studies by Ewins and Laidlaw[114] and Guggenheim and Löffler.[23] The latter investigation showed that the amine was converted to indole-3-acetic acid (38) following its administration (p.o. or i.v.) to rabbits and dogs. This metabolite was also detected by Ewins and Laidlaw in perfusion experiments with rabbit or cat liver. However, they found that 20 to 30% of the dose (50 to 100 mg/kg) was excreted in the urine of dogs as indoleaceturic acid (39), the glycine conjugate of (38). These results were confirmed by Erspamer[115] in rats. Following a dose of ~16 mg/kg (s.c.), the urine (11 h) contained 25 and 59% as metabolites (38) and (39), respectively. These results agree well with those of Blaschko and Philpot[35] who demonstrated the *in vitro* oxidation of tryptamine by liver and kidney preparations from several mammalian species.

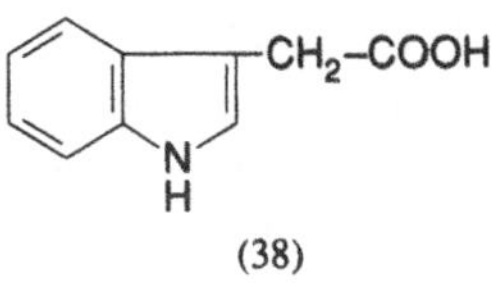

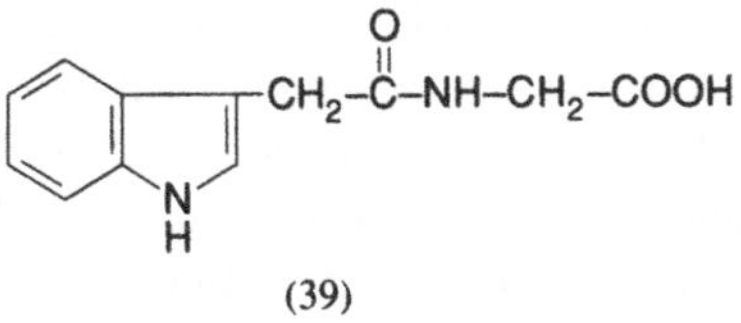

Another metabolic pathway for tryptamine involves *N*-methylation. Axelrod[31] reported the presence of an enzyme system in the soluble supernatant fraction of several tissues from rabbits, but mainly in the lung, which utilized *S*-adenosylmethionine in forming mono- and dimethylated derivatives from several tryptamines including tryptamine itself. However, this enzyme is very species dependent, probably being present in human lung but not in that from mouse, rat, guinea pig, cat, and monkey. Another enzyme system with a narrower range of substrate specificity was found in rat brain[116] and rat and human brain[117] which formed the mono- and dimethylated derivatives of tryptamine. An *N*-methyltransferase system from rabbit liver was believed to methylate the ring nitrogen of tryptamine.[118] However, reinvestigation of this reaction using both crude rabbit liver and lung homogenates as well as purified *N*-methyltransferases showed that the products formed were methylated on the aliphatic amino group.[119] The major metabolite was the monomethyl derivative, however smaller amounts of dimethyltryptamine were also detected. Additional information on the *N*-methylation of tryptamine derivatives is found below in the summary of *N*-methyltryptamine metabolism.

A third metabolic pathway of tryptamine involves aromatic hydroxylation. Jepson et al.[120] reported that rabbit liver microsomes supplemented with NADPH and O_2 hydroxylated tryptamine in the 6-position. The 5- and 7-hydroxytryptamines were not formed. However, this pathway appears to have little *in vivo* significance.

A subject of some interest has been the formation of 1,2,3,4-tetrahydro-β-carboline (tetrahydronorharmane) derivatives from the tryptamines. This synthetic pathway, which involves 5-methyltetrahydrofolate as the source of formaldehyde needed in the condensation reaction, converts tryptamine to 1,2,3,4-tetrahydro-β-carboline (40) (see Figure 6). The reaction has been shown to occur in brain tissue fractions,[121] with most of the activity found in the cytosol. This metabolite was also shown to be formed *in vivo* in the rat brain following injection of tryptamine i.p.[122] or into the lateral brain ventricle.[123] Further metabolism of 1,2,3,4-tetrahydro-β-carboline may lead to hydroxylated and fully aromatic metabolites.[124,125] The formation of tetrahydro-β-carbolines from *N*-methylated tryptamines is noted below.

The pathways of metabolism of the *N*-methyl and *N,N*-dimethyl derivatives of tryptamine are similar to those noted above with the corresponding primary amine. In experiments similar to that described above with tryptamine, Erspamer[115] found that rats excreted metabolites (38) and (39) (see Figure 6) in the urine after injection of *N*-**methyltryptamine** (41) or *N,N*-**dimethyltryptamine** (42). As with tryptamine, most of these metabolites consisted of the glycine conjugate (39) but the total amount of oxidized metabolites ((38) + (39)) excreted decreased from ~85% with tryptamine to 36 and 18% for the secondary and tertiary amines, respectively. Compound (38) was reported to be the major excretory product of *N,N*-dimethyltryptamine in rats[126] and in man.[127] Experiments in mice treated with inhibitors of monoamine oxidase or the hepatic microsomal enzymes indicated that *N,N*-dimethyltryptamine is metabolized chiefly by the former system.[128]

Conversion of the two methylated tryptamines to indole-3-acetic acid (38) in *in vitro* experiments has also been demonstrated. Fish et al.[129] found that the conversion rate was higher with the secondary amine when using a mitochondrial preparation from mouse liver. When fortified mouse liver homogenates were employed, three additional metabolites including the *N*-oxide (43) of the added *N,N*-dimethyltryptamine were formed. Experiments with metabolite (43) indicated that it is not an intermediate in the oxidation of the parent amine by the mitochondrial monoamine oxidase. This finding was confirmed by Smith et al.[130] who employed a monoamine oxidase preparation from guinea pig liver mitochondria. Szara and Axelrod[131] reported the formation of metabolite (43) by rabbit liver microsomes. Rat liver homogenates converted *N,N*-dimethyltryptamine to its *N*-oxide derivative and to indole-3-acetic acid (38).[132] Additionally, the precursor to compound (38), indoleacetaldehyde (44), was also formed in these incubates. The *N*-oxide derivative was found to be both a prominent tissue and urinary metabolite of *N,N*-dimethyltryptamine in rats.[133,134] Most of the metabolites shown in Figure 6 were found to be formed when *N,N*-dimethyltryptamine was incubated with whole homogenates from rat brain.[135]

Szara and Axelrod,[131] using rabbit liver microsomes, found that *N,N*-dimethyltryptamine was demethylated to the monomethyl derivative but not further to the primary amine. However, rats given the dimethyl compound (10 mg, i.p.) excreted in the urine unchanged compound and both of the demethylated derivatives. Other experiments, with rat liver homogenates[132] or by determination of the metabolites in the tissues[133] or urine[134] of rats, indicated that *N,N*-dimethyltryptamine underwent *N*-demethylation to the monomethyl derivative (41) to a small extent. In fact, only 0.02% of the dose (10 mg/kg, i.p.) was found in the urine as metabolite (41) in 24 h.

As noted above in the summary of tryptamine metabolism, hydroxylation of the indole nucleus may occur at the 6-position. This pathway appears to be of somewhat greater importance with the methylated tryptamines, which are less susceptible to oxidative deamination, than with the primary amine. Szara and Axelrod[131] obtained evidence for the 6-hydroxylation of *N,N*-

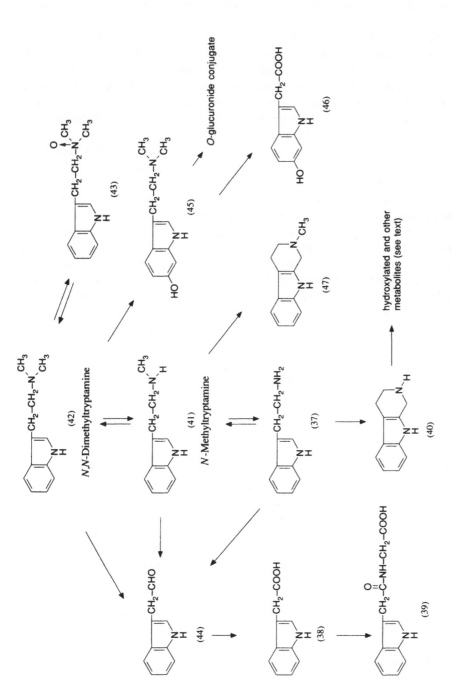

FIGURE 6. Metabolic pathways of *N*-methyltryptamine and *N,N*-dimethyltryptamine in rats.

dimethyltryptamine *in vitro* using rabbit liver microsome preparations and *in vivo* in rats. In the latter experiments, the urines collected for 48 h after dosage (10 mg, i.p.) contained 6-hydroxy-*N,N*-dimethyltryptamine (45) and 6-hydroxyindole-3-acetic acid (46), as well as the *N*-demethylated metabolites noted above. Further information on the 6-hydroxylation pathway is available,[136] however it appears to be less developed in man than in rodents.[137]

The formation of 1,2,3,4-tetrahydro-β-carboline derivatives via tryptamine or *N*-methyltryptamine is shown in Figure 6. The reaction giving (47) was shown by Hsu and Mandell[121] to be carried out by fractions of rat brain and Barker et al.[138] found that it and the demethylated derivative (40) were present in the brains of rats given i.p. injections of *N,N*-dimethyltryptamine. However, the amounts of these condensation products detected were very small. Sitaram et al.[132] failed to detect these two 1,2,3,4-tetrahydro-β-carboline derivatives when *N,N*-dimethyltryptamine was incubated with rat liver preparations.

It was pointed out above in the summary of tryptamine metabolism that some tissue preparations from a few species effect the *N*-methylation of tryptamine to the mono- and dimethyl derivatives. An enzyme obtained from human lung catalyzed the methylation of *N*-methyltryptamine when incubated in the presence of *S*-adenosylmethionine.[139,140] Formation of dimethyltryptamine by human red blood cells, blood plasma, and platelets has also been reported.[141] Ahn et al.[142] reported that this reaction occurred *in vivo* when *N*-methyltryptamine was administered intravenously to rabbits pretreated with a monoamine oxidase inhibitor. The dimethylated product was found in several tissues and especially in the lungs; however, this reaction was not detected when rats were similarly treated. This difference in rabbits and rats was confirmed by Mandel et al.[143] who also found that the formation of dimethyltryptamine did not occur in the rhesus monkey. These species differences in metabolism must be kept in mind when viewing the metabolic pathways of *N,N*-dimethyltryptamine shown in Figure 6.

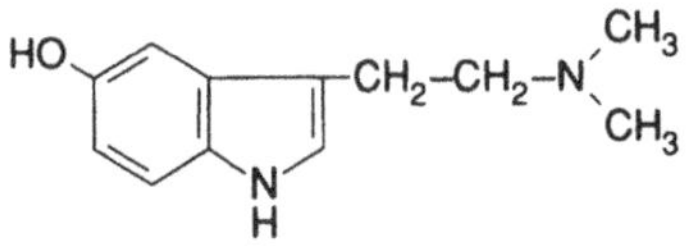

(48)
Bufotenine

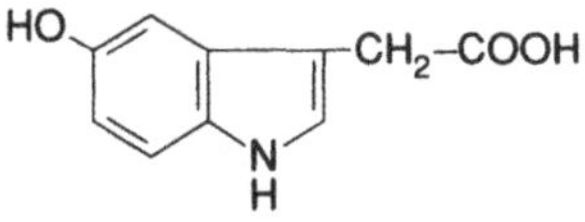

(49)

Several studies on the metabolism of **bufotenine** (5-hydroxy-*N,N*-dimethyltryptamine) (48) have been carried out in rats, with the result that its metabolism in this species is partially understood. Erspamer[115] detected unchanged compound and 5-hydroxyindole-3-acetic acid (49) as urinary metabolites after small doses (1.2 to 1.5 mg/kg, s.c.) of bufotenine. Compound (49) was found to be a major urinary metabolite by Ahlborg et al.[126] Quantitative data on its excretion were obtained by Gessner et al.[144] and by Sanders and Bush.[145] In the former study, ~7% of a large dose (125 mg/kg, given s.c. in portions over 6 h) was excreted as the acid in 48 h. The latter study employed a much smaller dose (~1 mg/kg, i.v.) and obtained a value of 14% of the dose for urinary 5-hydroxyindole-3-acetic acid. These low values for the excretion of the deaminated product are not unexpected in view of the reports by Blaschko and Philpot[35] and Govier et al.[146] showing that bufotenine is only slowly oxidized by monoamine oxidase. Oxidative deamination was shown to be an important metabolic pathway of bufotenine in rat brain.[147] Sanders and Bush found that bufotenine glucuronide is the major metabolite, accounting for 35% of the dose. The earlier study by Gessner et al. indicated that the increase in urinary glucuronide output following dosing could account for up to 25% of the dose. These studies also confirmed the excretion of unchanged compound. With the higher dose this amounted to ~22% while 6% was excreted unchanged with the lower dose. This difference is hardly surprising in view of the difference in dosage of over two orders of magnitude. The data of Sanders and Bush indicated that bufotenine metabolism in the rat is both rapid and extensive. No evidence was obtained for the O-methylation of bufotenine, however a small amount of a metabolite tentatively identified as 5-methoxyindole-3-acetic acid was detected. It is conceivable that the latter metabolite may arise from methylation of 5-hydroxyindole-3-acetic acid.

These studies of the metabolism of bufotenine in rats have not accounted for the entire dose and the fate of the remainder is not known. However, unidentified metabolites have been reported by Erspamer[115] and by Gessner et al.[144] It was noted above that N,N-dimethyltryptamine undergoes aromatic hydroxylation and this route may also be involved in the metabolism of bufotenine. While this reaction has not been demonstrated *in vivo*, Daly et al.[56] showed that a rabbit liver preparation containing microsomes, soluble supernatant, and added S-adenosylmethionine had low but detectable activity to first hydroxylate bufotenine and then methylate the catechol formed. By analogy with results obtained using other indoles, the hydroxylation most likely occurred at the 6-position.

The above picture of bufotenine metabolism in rats has been extended to humans. Sanders-Bush et al.[148] administered ^{14}C-labeled material (0.2 and 1 mg, i.v.) to two subjects and found that essentially quantitative excretion of the radioactivity in the urine occurred within the first 24 h. Most (68 and 74%) of this material was 5-hydroxyindole-3-acetic acid (49) and only 1 and 6% were found to be due to unchanged bufotenine. The remainder consisted of very polar metabolites (22 and 20%) and neutral metabolites (9 and <1%, respectively). It was believed that the alcohol analogue of compound (49), 5-hydroxytryptophol, occurred in the latter fraction.

When **5-methoxy-N-methyltryptamine** labeled with ^{14}C was administered to rats using a dose of 3 mg, ~90% of the radioactivity was excreted in the urine in 24 h.[149] No unchanged compound was present and ~95% of the urinary material was due to 5-methoxyindole-3-acetic acid. In some cases, a few percent of the dose corresponded to a compound which was probably the O-glucuronide of the 6-hydroxylated derivative. An alternative metabolic pathway, however probably not in the rat, may be the N-methylation of 5-methoxy-N-methyltryptamine. Mandel and Walker[150] reported on an indoleamine-N-methyltransferase from rabbit and human lung which carried out this reaction.

The results summarized above indicate that nonmethylated tryptamines undergo rapid and extensive oxidative deamination. Dimethylated derivatives are more resistant in this regard with the result that the rate of metabolism is lower and extent of conversion to the indoleacetic acid derivative is less. This has been shown with **5-methoxy-N,N-dimethyltryptamine**. Ahlborg et al.[126] found that ~50% of the radioactivity appeared in the urine in 12 h when rats were given the ^{14}C-labeled compound (10 mg/kg, i.p.). Most of this material was 5-methoxyindole-3-acetic acid. Agurell et al.,[151] in similar experiments using half the above dose level, found that 59 to 65% of the radioactivity was excreted during the first 24 h. The urinary material consisted of four compounds: 5-methoxyindole-3-acetic acid (33% of the dose), bufotenine (5-hydroxy-N,N-dimethyltryptamine) (5%), bufotenine glucuronide (14%), and 5-hydroxyindole-3-acetic acid (8%). Several possible urinary metabolites were shown to be absent. These included the unchanged compound and its N-demethylated products, the mono- and didemethylated products of bufotenine, 5-methoxyindoleaceturic acid, and the 6-hydroxy derivative of the administered compound. The data indicated that 5-hydroxyindole-3-acetic acid was formed via bufotenine as 5-methoxyindole-3-acetic acid did not undergo demethylation. Additional information on the urinary metabolites of 5-methoxy-N,N-dimethyltryptamine in rats was reported by Sitaram et al.[134] Following an i.p. dose of 10 mg/kg the excretion of bufotenine in the 24-h urines was confirmed, however it accounted for a mere 0.1% of the dose. Also confirmed was the absence of the demethylated product, 5-methoxy-N-methyltryptamine. A very small amount of unchanged compound (<0.2%) was detected, however the most abundant metabolite among those reported in this study was 5-methoxy-N,N-dimethyltryptamine-N-oxide (~7% of the dose). Pretreatment of the rats with monoamine oxidase inhibitors greatly increased the amount of N-oxide derivative excreted. Similar experiments which assessed the tissue levels of these metabolites[133] confirmed the importance of the route of N-oxidation. Additional *in vitro* experiments were carried out using rat liver homogenates.[132] These incubates showed that 5-methoxy-*N*,*N*-dimethyltryptamine was metabolized via oxidative deamination, *N*-demethylation, *O*-demethylation, and *N*-oxidation. The first of these reactions led to the formation of both 5-methoxyindoleacetic acid and the corresponding aldehyde intermediate. *N*-oxidation occurred extensively under these conditions, however only small amounts of the two types of demethylated products were found. The metabolism of 5-methoxy-*N*,*N*-dimethyltryptamine by rat liver tissue was very similar to that seen with *N*,*N*-dimethyltryptamine.

REFERENCES

1. **Tipton, K. F.,** Monoamine oxidase, in *Enzymatic Basis of Detoxication*, Vol. 1, Jakoby, W. B., Ed., Academic Press, New York, 1980, 355.
2. **Coutts, R. T. and Beckett, A. H.,** Metabolic N-oxidation of primary and secondary aliphatic medicinal amines, *Drug Metab. Rev.*, 6, 51, 1977.
3. **Damani, L. A.,** Oxidation at nitrogen centers, in *Metabolic Basis of Detoxication*, Jakoby, W. B., Bend, J. R., and Caldwell, J., Eds., Academic Press, New York, 1982, 127.
4. **Lindeke, B. and Cho, A. K.,** *N*-Dealkylation and deamination, in *Metabolic Basis of Detoxication*, Jakoby, W. B., Bend, J. R., and Caldwell, J., Eds., Academic Press, New York, 1982, 105.
5. **Williams, R. T.,** *Detoxication Mechanisms*, Chapman and Hall, London, 1959, 127.
6. **Alles, G. A. and Heegaard, E. V.,** Substrate specificity of amine oxidase, *J. Biol. Chem.*, 147, 487, 1943.
7. **Dar, M. S., Morselli, P. L., and Bowman, E. R.,** The enzymatic systems involved in the mammalian metabolism of methylamine, *Gen. Pharmacol.*, 16, 557, 1985.
8. **Dar, M. S. and Bowman, E. R.,** *In vivo* mammalian metabolism of methylamine and methylurea and their metabolic interrelationship, *Drug Metab. Dispos.*, 13, 682, 1985.
9. **Ishiwata, H., Iwata, R., and Tanimura, A.,** Intestinal distribution, absorption and secretion of dimethylamine and its biliary and urinary excretion in rats, *Food Chem. Toxicol.*, 22, 649, 1984.
10. **Richter, D.,** Elimination of amines in man, *Biochem. J.*, 32, 1763, 1938.
11. **Richter, D.,** Adrenaline and amine oxidase, *Biochem. J.*, 31, 2022, 1937.
12. **Tabor, H. and Tabor, C. W.,** Spermidine, spermine, and related amines, *Pharmacol. Rev.*, 16, 245, 1964.
13. **Raina, A. and Jänne, J.,** Physiology of the natural polyamines putrescine, spermidine and spermine, *Med. Biol.*, 53, 121, 1975.
14. **Lundgren, D. W. and Hankins, J.,** Metabolism of putrescine to 2-pyrrolidine by rat liver slices, *J. Biol. Chem.*, 253, 7130, 1978.
15. **Lundgren, D. W. and Fales, H. M.,** Metabolism of putrescine to 5-hydroxy-2-pyrrolidone via 2-pyrrolidone, *J. Biol. Chem.*, 255, 4481, 1980.
16. **Noto, T., Tanaka, T., and Nakajima, T.,** Urinary metaboites of polyamines in rats, *J. Biochem. (Tokyo)*, 83, 543, 1978.
17. **Williams, R. T.,** *Detoxication Mechanisms*, Chapman and Hall, London, 1959, 130.
18. **Jones, D. P.,** Simple high-performance liquid chromatographic assay for benzylamine oxidation products in cell suspensions, *J. Chromatogr.*, 305, 256, 1984.
19. **Wood, S. G., Al-Ani, M. R., and Lawson, A.,** Hippuric acid excretion after benzylamine ingestion in man, *Br. J. Ind. Med.*, 35, 230, 1978.
20. **Parkinson, D., Lyles, G. A., Browne, B. J., and Callingham, B. A.,** Some factors influencing the metabolism of benzylamine by type A and B monoamine oxidase in rat heart and liver, *J. Pharm. Pharmacol.*, 32, 844, 1980.
21. **Lewinsohn, R., Böhm, K.-H., Glover, V., and Sandler, M.,** A benzylamine oxidase distinct from monoamine oxidase B—widespread distribution in man and rat, *Biochem. Pharmacol.*, 27, 1857, 1978.
22. **Elliott, J., Callingham, B. A., and Sharman, D. F.,** Metabolism of amines in the isolated perfused mesenteric arterial bed of the rat, *Br. J. Pharmacol.*, 98, 507, 1989.
23. **Guggenheim, M. and Löffler, W.,** Das Schicksal proteinogener Amine im Tierkörper, *Biochem. Z.*, 72, 325, 1915.
24. **Yang, H.-Y. T. and Neff, N. H.,** β-Phenylethylamine: a specific substrate for type B monoamine oxidase of brain, *J. Pharmacol. Exp. Ther.*, 187, 365, 1973.
25. **Oguchi, K., Kobayashi, S., Uesato, T., and Kamijo, K.,** Studies on β-phenylethylamine deamination by human placental monoamine oxidase, *Jpn. J. Pharmacol.*, 31, 7, 1981.
26. **Block, W.,** Zur Physiologie des ^{14}C-radioaktiven Mescalins im Tierversuch. IV. Mitt. Vergleichende Untersuchungen mit ^{14}C-Mescalin und ^{14}C-β-Phenyl-äthylamin, *Z. Naturforsch.*, 8b, 440, 1953.

27. **Wu, P. H. and Boulton, A. A.,** Metabolism, distribution and disappearance of injected β-phenylethylamine in the rat, *Can. J. Biochem.*, 53, 42, 1975.
28. **Davis, B. A. and Boulton, A. A.,** Longitudinal urinary excretion of some "trace" acids in a human male, *J. Chromatogr.*, 222, 161, 1981.
29. **Creveling, C. R., Daly, J. W., Witkop, B., and Udenfriend, S.,** Substrates and inhibitors of dopamine-β-oxidase, *Biochim. Biophys. Acta*, 64, 125, 1962.
30. **Hoag, G. N., Hill, A., and Zaleski, W.,** Urinary mandelic acid: identified in normal individuals following a single oral load of phenylethylamine, *Clin. Biochem.*, 10, 181, 1977.
31. **Axelrod, J.,** The enzymatic *N*-methylation of serotonin and other amines, *J. Pharmacol. Exp. Ther.*, 138, 28, 1962.
32. **Ewins, A. J. and Laidlaw, P. P.,** The fate of parahydroxyphenylethylamine in the organism, *J. Physiol. (London)*, 41, 78, 1910.
33. **Schayer, R. W.,** *In vivo* inhibition of monoamine oxidase studied with radioactive tyramine, *Proc. Soc. Exp. Biol. Med.*, 84, 60, 1953.
34. **Bernheim, F. and Bernheim, M. L. C.,** The oxidation of mescaline and certain other amines, *J. Biol. Chem.*, 123, 317, 1938.
35. **Blaschko, H. and Philpot, F. J.,** Enzymic oxidation of tryptamine derivatives, *J. Physiol. (London)*, 122, 403, 1953.
36. **Strolin Benedetti, M., Boucher, T., Carlsson, A., and Fowler, C. J.,** Intestinal metabolism of tyramine by both forms of monoamine oxidase in the rat, *Biochem. Pharmacol.*, 32, 47, 1983.
37. **Grimes, A. J.,** Synthesis of ^{35}S-labelled arylsulphates by intact animals and by tissue preparations, with particular reference to L-tyrosine *O*-sulphate, *Biochem. J.*, 73, 723, 1959.
38. **Spencer, B.,** Endogenous sulphate acceptors in rat liver, *Biochem. J.*, 77, 294, 1960.
39. **Mattock, P. and Jones, J. G.,** Partial purification and properties of an enzyme from rat liver that catalyses the sulphation of L-tyrosyl derivatives, *Biochem. J.*, 116, 797, 1970.
40. **Wong, K. P.,** The conjugation of tyramine with sulphate by liver and intestine of different animals, *Biochem. J.*, 160, 491, 1976.
41. **Mullen, P. E. and Smith, I.,** Tyramine metabolism and migraine: a metabolic defect, *Br. J. Pharmacol.*, 41, 413, 1971.
42. **Smith, I. and Mitchell, P. D.,** The effect of oral inorganic sulphate on the metabolism of 4-hydroxyphenethylamine (tyramine) in man: tyramine *O*-sulphate measurement in human urine, *Biochem. J.*, 142, 189, 1974.
43. **Boulton, A. A. and Marjerrison, G. L.,** Effect of L-dopa therapy on urinary *p*-tyramine excretion and EEG changes in Parkinson's disease, *Nature (London)*, 236, 76, 1972.
44. **Tacker, M., McIsaac, W. M., and Creaven, P. J.,** Metabolism of tyramine-1-^{14}C by the rat, *Biochem. Pharmacol.*, 19, 2763, 1970.
45. **Lemberger, L., Klutch, A., and Kuntzman, R.,** The metabolism of tyramine in rabbits, *J. Pharmacol. Exp. Ther.*, 153, 183, 1966.
46. **Tacker, M., Creaven, P. J., and McIsaac, W. M.,** Preliminary observations on the metabolism of [1-^{14}C]tyramine in man, *J. Pharm. Pharmacol.*, 24, 247, 1972.
47. **Boulton, A. A. and Davis, B. A.,** The metabolism of ingested deuterium-labelled *p*-tyramine in normal subjects, *Biomed. Environ. Mass Spectrom.*, 14, 207, 1987.
48. **Wiseman-Distler, M. H., Sourkes, T. L., and Carabin, S.,** Precursors of 3,4-dihydroxyphenylacetic acid and 4-hydroxy-3-methoxyphenylacetic acid in the rat, *Clin. Chim. Acta*, 12, 335, 1965.
49. **Creveling, C. R., Levitt, M., and Udenfriend, S.,** An alternative route for biosynthesis of norepinephrine, *Life Sci.*, 1, 523, 1962.
50. **Jones, R. A. D., Lee, C. R., and Pollitt, R. J.,** The metabolism of deuterium-labelled *p*-tyramine in man, in *Stable Isotopes*, Baillie, T. A., Ed., Macmillan, London, 1978, 247.
51. **Hengstmann, J. H., Konen, W., Konen, C., Eichelbaum, M., and Dengler, H. J.,** The physiological disposition of *p*-octopamine in man, *Naunyn-Schmiedeberg's Arch. Pharmacol.*, 283, 93, 1974.
52. **Tacker, M., McIsaac, W. M., and Creaven, P. J.,** Effect of tranylcypromine sulphate on the metabolism of [^{14}C]tyramine *in vivo* in the rat, *J. Pharm. Pharmacol.*, 24, 245, 1972.
53. **Mosnaim, A. D., Edstrand, D. L., Wolf, M. E., and Silkaitis, R. P.,** Evidence for an enzymatic *p*-tyramine-dehydroxylating system in rabbit brain preparations *in vitro*, *Biochem. Pharmacol.*, 26, 1725, 1977.
54. **Hai, H., Guo, Z.-G., and Wang, J.-M.,** Disposition of *N*-methyl-[*ring*-3,5-^{3}H]tyramine in rabbits and mice, *Acta Pharmacol. Sin.*, 10, 41, 1989.
55. **Barwell, C. J., Basma, A. N., Lafi, M. A. K., and Leake, L. D.,** Deamination of hordenine by monoamine oxidase and its action on vasa deferentia of the rat, *J. Pharm. Pharmacol.*, 41, 421, 1989.
56. **Daly, J., Inscoe, J. K., and Axelrod, J.,** The formation of *O*-methylated catechols by microsomal hydroxylation of phenols and subsequent enzymatic catechol *O*-methylation. Substrate specificity, *J. Med. Chem.*, 8, 153, 1965.
57. **Lemberger, L. and Rubin, A.,** Amphetamine, in *Physiologic Disposition of Drugs of Abuse*, Spectrum Publications, New York, 1976, 31.

58. **Slotta, K. H. and Müller, J.,** Über den Abbau des Mescalins und mescalinähnlicher Stoffe im Organismus, *Hoppe-Seyler's Z. Physiol. Chem.*, 238, 14, 1936.
59. **Taska, R. J. and Schoolar, J. C.,** Placental transfer and fetal distribution of mescaline-^{14}C in monkeys, *J. Pharmacol. Exp. Ther.*, 183, 427, 1972.
60. **Taska, R. J. and Schoolar, J. C.,** Peripheral tissue distribution, brain distribution and metabolism of mescaline-^{14}C in monkeys, *Arch. Int. Pharmacodyn. Ther.*, 202, 66, 1973.
61. **Musacchio, J. M. and Goldstein, M.,** The metabolism of mescaline-^{14}C in rats, *Biochem. Pharmacol.*, 16, 963, 1967.
62. **Goldstein, M. and Contrera, J. F.,** The substrate specificity of phenylamine-β-hydroxylase, *J. Biol. Chem.*, 237, 1898, 1962.
63. **Axelrod, J.,** The enzymic cleavage of aromatic ethers, *Biochem. J.*, 63, 634, 1956.
64. **Daly, J., Axelrod, J., and Witkop, B.,** Methylation and demethylation in relation to the *in vitro* metabolism of mescaline, *Ann. N.Y. Acad. Sci.*, 96, 37, 1962.
65. **Demisch, L. and Seiler, N.,** Oxidative metabolism of mescaline in the central nervous system. V. *In vitro* deamination of mescaline to 3,4,5-trimethoxy-benzoic acid, *Biochem. Pharmacol.*, 24, 575, 1975.
66. **Demisch, L., Kaczmarczyk, P., and Seiler, N.,** 3,4,5-Trimethyoxybenzoic acid, a new mescaline metabolite in humans, *Drug Metab. Dispos.*, 6, 507, 1978.
67. **Ho, B. T., Pong, S. F., Browne, R. G., and Walker, K. E.,** Acetylation of mescaline in rat brains, *Experientia*, 29, 275, 1973.
68. **Seiler, N. and Demisch, L.,** Oxidative metabolism of mescaline in the central nervous system. III. Side chain degradation of mescaline and formation of 3,4,5-trimethoxy-benzoic acid *in vivo*, *Biochem. Pharmacol.*, 23, 259, 1974.
69. **Pugh, C. E. M. and Quastel, J. H.,** Oxidation of amines by animal tissues, *Biochem. J.*, 31, 2306, 1937.
70. **Roth, R. A., Roth, J. A., and Gillis, C. N.,** Disposition of ^{14}C-mescaline by rabbit lung, *J. Pharmacol. Exp. Ther.*, 200, 394, 1977.
71. **Blaschko, H.,** Enzymic oxidation of mescaline in the rabbit's liver, *J. Physiol. (London)*, 103, 13, 1944.
72. **Steensholt, G.,** On an amine oxidase in rabbit's liver, *Acta Physiol. Scand.*, 14, 356, 1947.
73. **Zeller, E. A., Borsy, J., Berman, E. R., Cherkas, M. S., and Fouts, J. R.,** Degradation of mescaline by amine oxidases, *J. Pharmacol. Exp. Ther.*, 124, 282, 1958.
74. **Huszti, Z. and Borsy, J.,** Differences between amine oxidases deaminating mescaline and the structurally related 3,4-dimethoxyphenylethyl amine, *Biochem. Pharmacol.*, 15, 475, 1966.
75. **Riceberg, L. J., Simon, M., Van Vunakis, H., and Abeles, R. H.,** Effects of aminoacetonitrile, an amine oxidase inhibitor, on mescaline metabolism in the rabbit, *Biochem. Pharmacol.*, 24, 119, 1975.
76. **Seiler, N.,** Der oxydative Abbau des Mezcalins im Zentralnervensystem. I. Charakterisierung des Mezcalin abbauenden Fermentes, *Hoppe-Seyler's Z. Physiol. Chem.*, 341, 105, 1965.
77. **Seiler, N. and Demisch, L.,** Oxidative metabolism of mescaline in the central nervous system. II. Oxidative deamination of mescaline and 2,3,4-trimethoxy-β-phenylethylamine by different mouse brain area *in vitro*, *Biochem. Pharmacol.*, 20, 2485, 1971.
78. **Seiler, N. and Demisch, L.,** Oxidative metabolism of mescaline in the central nervous system. IV. *In vivo* metabolism of mescaline and 2,3,4-trimethoxy-β-phenylethylamine, *Biochem. Pharmacol.*, 23, 273, 1974.
79. **Guha, S. R. and Mitra, C.,** Amphetamine-tetrazolium reductase activity in brain, *Biochem. Pharmacol.*, 20, 3539, 1971.
80. **Caldwell, J.,** The metabolism of amphetamines in mammals, *Drug. Metab. Rev.*, 5, 219, 1976.
81. **Caldwell, J.,** The metabolism of amphetamines and related stimulants in animals and man, in *Amphetamine and Related Stimulants: Chemical, Biological, Clinical, and Sociological Aspects*, Caldwell, J., Ed., CRC Press, Boca Raton, FL, 1980, 29.
82. **Wilkinson, G. R. and Beckett, A. H.,** Absorption, metabolism and excretion of the ephedrines in man. I. The influence of urinary pH and urine volume output, *J. Pharmacol. Exp. Ther.*, 162, 139, 1968.
83. **Sinsheimer, J. E., Dring, L. G., and Williams, R. T.,** Species differences in the metabolism of norephedrine in man, rabbit and rat, *Biochem. J.*, 136, 763, 1973.
84. **Nagase, Y., Baba, S., and Matsuda, A.,** Studies on the analysis of drugs by use of isotopes. IV. Reverse dilution analysis of the metabolites of ^{14}C-ephedrine, *J. Pharm. Soc. Jpn.*, 87, 123, 1967.
85. **Baba, S., Kuroda, Y., and Horie, M.,** Studies on drug metabolism by use of isotopes. XXIX. Studies of the differences in biological fates of ephedrine isomers by use of a pseudo-racemic mixture, *Biomed. Environ. Mass Spectrom.*, 13, 141, 1986.
86. **Feller, D. R. and Malspeis, L.,** Biotransformation of D(−)-ephedrine and L(+)-ephedrine in the rabbit, in vivo and in vitro, *Drug Metab. Dispos.*, 5, 37, 1977.
87. **Kuntzman, R. G., Tsai, I., Brand, L., and Mark, L. C.,** The influence of urinary pH on the plasma half-life of pseudoephedrine in man and dog and a senstive assay for its determination in human plasma, *Clin. Pharmacol. Ther.*, 12, 62, 1971.
88. **Jacquot, C., Bralet, J., and Cohen, Y.,** Métabolisme, distribution tissulaire et mécanisme de fixation cardiaque de la *dl* pseudoéphédrine ^{14}C chez le Rat, *C.R. Séanc. Soc. Biol.*, 167, 1789, 1973.

89. Baaske, D. M., Lai, C.-M., Klein, L., Look, Z. M., and Yacobi, A., Comparison of GLC and high-pressure liquid chromatographic methods for analysis of urinary pseudoephedrine, *J. Pharm. Sci.*, 68, 1472, 1979.

90. Nieder, M. and Jaeger, H., Sensitive quantification of pseudoephedrine in human plasma and urine by high-performance liquid chromatography, *J. Chromatogr.*, 424, 73, 1988.

91. Lo, L. Y., Land, G., and Bye, A., Sensitive assay for pseudoephedrine and its metabolite, norpseudoephedrine in plasma and urine using gas-liquid chromatography with electron-capture detection, *J. Chromatogr.*, 222, 297, 1981.

92. Axelrod, J., The enzymatic deamination of amphetamine (benzedrine), *J. Biol. Chem.*, 214, 753, 1955.

93. Baba, S., Matsuda, A., and Nagase, Y., Analysis of drugs by use of radioisotopes. V. Identification of metabolites of ephedrine in the rabbit liver, *J. Pharm. Soc. Jpn.*, 89, 833, 1969 (Chem. Abstr. 71, 100055f, 1969).

94. Baba, S., Matsuda, A., Nagase, Y., and Kawai, K., Studies on the analysis of drugs by use of radioisotopes. VI. Identification of metabolites of ephedrine in the rabbit liver slices, *J. Pharm. Soc. Jpn.*, 91, 584, 1971.

95. Beckett, A. H. and Al-Sarraj, S., Metabolism of ephedrine on and near its basic centre, *J. Pharm. Pharmacol.*, 25, 169, 1973.

96. Beckett, A. H., Jones, G. R., and Al-Sarraj, S., Metabolic *N*- and α-*C*-oxidation of norephedrine by rabbit liver microsomal fractions and synthesis of the metabolic products, *J. Pharm. Pharmacol.*, 26, 945, 1974.

97. Axelrod, J., The enzymatic demethylation of ephedrine, *J. Pharmacol. Exp. Ther.*, 114, 430, 1955.

98. Gaudette, L. E. and Brodie, B. B., Relationship between the lipid solubility of drugs and their oxidation by liver microsomes, *Biochem. Pharmacol.*, 2, 89, 1959.

99. McMahon, R. E., Some observations on the *in vitro* demethylation of secondary *N*-methyl amines by liver microsomes, *Life Sci.*, 3, 235, 1964.

100. Dann, R. E., Feller, D. R., and Snell, J. F., The microsomal *N*-demethylation of the stereoisomers of ephedrine, *Eur. J. Pharmacol.*, 16, 233, 1971.

101. Feller, D. R., Basu, P., Mellon, W., Curott, J., and Malspeis, L., Metabolism of ephedrine isomers in rabbit liver, *Arch. Int. Pharmacodyn. Ther.*, 203, 187, 1973.

102. Ziegler, D. M., Mitchell, C. H., and Jollow, D., The properties of a purified hepatic microsomal mixed function amine oxidase, in *Microsomes and Drug Oxidations*, Gillette, J. R., Conney, A. H., Cosmides, G. J., Estabrook, R. W., Fouts, J. R., and Mannering, G. J., Eds., Academic Press, New York, 1969, 173.

103. May, S. W., Phillips, R. S., Herman, H. H., and Mueller, P. W., Bioactivation of *Catha edulis* alkaloids: enzymatic ketonization of norpseudoephedrine, *Biochem. Biophys. Res. Commun.*, 104, 38, 1982.

104. Beckett, A. H. and Wilkinson, G. R., Urinary excretion of (–)-methylephedrine, (–)-ephedrine and (–)-norephedrine in man, *J. Pharm. Pharmacol.*, 17, 107, 1965.

105. Inoue, T. and Suzuki, S., The metabolism of (±)-methylephedrine in rat and man, *Xenobiotica*, 20, 99, 1990.

106. Inoue, T. and Tanaka, K., *N*-Oxidation and *N*-demethylation of methylephedrine by rat-liver microsomes, *Xenobiotica*, 20, 265, 1990.

107. Maitai, C. K. and Mugera, G. M., Excretion of the active principle of *Catha edulis* (Miraa) in human urine, *J. Pharm. Sci.*, 64, 702, 1975.

108. Axelrod, J., Studies on sympathomimetic amines. I. The biotransformation and physiological disposition of *l*-ephedrine and *l*-norephedrine, *J. Pharmacol. Exp. Ther.*, 109, 62, 1953.

109. Thiercelin, J. F., Jacquot, C., Rapin, J. R., and Cohen, Y., Pharmacocinétique de la DL-noréphedrine ^{14}C chez le rat, *Arch. Int. Pharmacodyn. Ther.*, 220, 153, 1976.

110. Kalix, P. and Braenden, O., Pharmacological aspects of the chewing of khat leaves, *Pharmacol. Rev.*, 37, 149, 1985.

111. Guantai, A. N. and Maitai, C. K., Metabolism of cathinone to *d*-norpseudoephedrine in humans, *J. Pharm. Sci.*, 72, 1217, 1983.

112. Brenneisen, R., Geisshüsler, S., and Schorno, X., Metabolism of cathinone to (–)-norephedrine and (–)-norpseudoephedrine, *J. Pharm. Pharmacol.*, 38, 298, 1986.

113. Ferguson, P. W., Keller, W. J., and Risinger, F. O., ^{14}C]Normacromerine fate in the rat, *J. Pharm. Sci.*, 73, 692, 1984.

114. Ewins, A. J. and Laidlaw, P. P., The fate of indolethylamine in the organism, *Biochem. J.*, 7, 18, 1913.

115. Erspamer, V., Observations on the fate of indolalkylamines in the organism, *J. Physiol. (London)*, 127, 118, 1955.

116. Morgan, M. and Mandell, A. J., Indole(ethyl)amine *N*-methyltransferase in the brain, *Science*, 165, 492, 1969.

117. Saavedra, J. M. and Axelrod, J., Psychotomimetic *N*-methylated tryptamines: formation in brain in vivo and in vitro, *Science*, 175, 1365, 1972.

118. Lyon, E. S. and Jakoby, W. B., Arylamine *N*-methyltransferase. Methylation of the indole ring, *J. Biol. Chem.*, 257, 7531, 1982.

119. Crooks, P. A., Godin, C. S., Nwosu, C. G., Ansher, S. S., and Jakoby, W. B., Reevaluation of the products of tryptamine catalyzed by rabbit liver *N*-methyltransferase, *Biochem. Pharmacol.*, 35, 1600, 1986.

120. Jepson, J. B., Zaltzman, P., and Udenfriend, S., Microsomal hydroxylation of tryptamine, indoleacetic acid and related compounds, to 6-hydroxy derivatives, *Biochim. Biophys. Acta*, 62, 91, 1962.

121. **Hsu, L. L. and Mandell, A. J.,** Enzymatic formation of tetrahydro-β-carboline from tryptamine and 5-methyltetrahydrofolic acid in rat brain fractions: regional and subcellular distribution, *J. Neurochem.,* 24, 631, 1975.

122. **Honecker, H. and Rommelspacher, H.,** Tetrahydronorharmane (tetrahydro-β-carboline), a physiologically occurring compound of indole metabolism, *Naunyn Schmiedeberg's Arch. Pharmacol.,* 305, 135, 1978.

123. **Hsu, L. L.,** *In vivo* formation of 1,2,3,4-tetrahydro-β-carboline from [^{14}C]-tryptamine in the brain, *IRCS Med. Sci.,* 13, 1054, 1985.

124. **Greiner, B. and Rommelspacher, H.,** Urinary metabolites of tetrahydronorharmane in the rat, *Prog. Clin. Biol. Res.,* 90, 201, 1982.

125. **Greiner, B. and Rommelspacher, H.,** Two metabolic pathways of tetrahydronorharmane (tetrahydro-β-carboline) in rats, *Naunyn Schmiedeberg's Arch. Pharmacol.,* 325, 349, 1984.

126. **Ahlborg, U., Holmstedt, B., and Lindgren, J.-E.,** Fate and metabolism of some hallucinogenic indolealkylamines, in *Advances in Pharmacology,* Vol. 6, Part B, Garattini, S. and Shore, P. A., Eds., Academic Press, New York, 1968, 213.

127. **Szára, S.,** Dimethyltryptamin: its metabolism in man; the relation of its psychotic effect to the serotonin metabolism, *Experientia,* 12, 441, 1956.

128. **Shah, N. S. and Hedden, M. P.,** Behavioral effects and metabolic fate of N,N-dimethyltryptamine in mice pretreated with β-diethylaminoethyl-diphenylpropylacetate (SKF 525-A), isoniazid and chlorpromazine, *Pharmacol. Biochem. Behav.,* 8, 351, 1978.

129. **Fish, M. S., Johnson, N. M., Lawrence, E. P., and Horning, E. C.,** Oxidative N-dealkylation, *Biochim. Biophys. Acta,* 18, 564, 1955.

130. **Smith, T. E., Weissbach, H., and Udenfriend, S.,** Studies on the mechanism of action of monoamine oxidase: metabolism of N,N-dimethyltryptamine and N,N-dimethyltryptamine-N-oxide, *Biochemistry,* 1, 137, 1962.

131. **Szara, S. and Axelrod, J.,** Hydroxylation and N-demethylation of N,N-dimethyltryptamine, *Experientia,* 15, 216, 1959.

132. **Sitaram, B. R., Talomsin, R., Blackman, G. L., and McLeod, W. R.,** Study of metabolism of psychotomimetic indolealkylamines by rat tissue extracts using liquid chromatography, *Biochem. Pharmacol.,* 36, 1503, 1987.

133. **Sitaram, B. R., Lockett, L., Talomsin, R., Blackman, G. L., and McLeod, W. R.,** *In vivo* metabolism of 5-methoxy-N,N-dimethyltryptamine and N,N-dimethyltryptamine in the rat, *Biochem. Pharmacol.,* 36, 1509, 1987.

134. **Sitaram, B. R., Lockett, L., Blackman, G. L., and McLeod, W. R.,** Urinary excretion of 5-methoxy-N,N-dimethyltryptamine, N,N-dimethyltryptamine and their N-oxides in the rat, *Biochem. Pharmacol.,* 36, 2235, 1987.

135. **Barker, S. A., Monti, J. A., and Christian, S. T.,** Metabolism of the hallucinogen N,N-dimethyltryptamine in rat brain homogenates, *Biochem. Pharmacol.,* 29, 1049, 1980.

136. **Szara, S.,** Hallucinogenic effects and metabolism of tryptamine derivatives in man, *Fed. Proc. Fed. Am. Soc. Exp. Biol.,* 20, 885, 1961.

137. **Szara, S.,** Discussion of the fate and metabolism of some hallucinogenic indolealkylamines, in *Advances in Pharmacology,* Vol. 6, Part B, Garattini, S. and Shore, P. A., Eds., Academic Press, New York, 1968, 230.

138. **Barker, S. A., Beaton, J. M., Christian, S. T., Monti, J. A., and Morris, P. E.,** *In vivo* metabolism of α,α,β,β-tetradeutero-N,N-dimethyltryptamine in rodent brain, *Biochem. Pharmacol.,* 33, 1395, 1984.

139. **Mandel, L. R., Ahn, H. S., VandenHeuvel, W. J. A., and Walker, R. W.,** Indoleamine-N-methyl transferase in human lung, *Biochem. Pharmacol.,* 21, 1197, 1972.

140. **Walker, R. W., Ahn, H. S., Mandel, L. R., and VandenHeuvel, W. J. A.,** Identification of N, N-dimethyltryptamine as the product of an *in vitro* enzymic methylation, *Anal. Biochem.,* 47, 228, 1972.

141. **Wyatt, R. J., Saavedra, J. M., and Axelrod, J.,** A dimethyltryptamine-forming enzyme in human blood, *Am. J. Psychiatry,* 130, 754, 1973.

142. **Ahn, H. S., Walker, R. W., VandenHeuvel, W. J. A., Rosegay, A., and Mandel, L. R.,** Studies on the *in vivo* biosynthesis of N,N-dimethyltryptamine (DMT) in the rabbit and rat, *Fed. Proc. Fed. Am. Soc. Exp. Biol.,* 32, 511, 1973.

143. **Mandel, L. R., Prasad, R., Lopez-Ramos, B., and Walker, R. W.,** The biosynthesis of dimethyltryptamine *in vivo, Res. Commun. Chem. Pathol. Pharmacol.,* 16, 47, 1977.

144. **Gessner, P. K., Khairallah, P. A., McIsaac, M. W., and Page, I. H.,** The relationship between the metabolic fate and pharmacological actions of serotonin, bufotenine and psilocybin, *J. Pharmacol. Exp. Ther.,* 130, 126, 1960.

145. **Sanders, E. and Bush, M. T.,** Distribution, metabolism and excretion of bufotenine in the rat with preliminary studies of its O-methyl derivative, *J. Pharmacol. Exp. Ther.,* 158, 340, 1967.

146. **Govier, W. M., Howes, B. G., and Gibbons, A. J.,** The oxidative deamination of serotonin and other 3-(beta-aminoethyl)-indoles by monamine oxidase and the effect of these compounds on the deamination of tyramine, *Science,* 118, 596, 1953.

147. **Gomes, U. C. R., McCarthy, B. W., and Shanley, B. C.,** Effect of neuroleptics on indoleamine-N-methyltransferase activity and brain metabolism of bufotenin, *Biochem. Pharmacol.*, 30, 571, 1981.

148. **Sanders-Bush, E., Oates, J. A., and Bush, M. T.,** Metabolism of bufotenine-2'-^{14}C in human volunteers, *Life Sci.*, 19, 1407, 1976.

149. **Taborsky, R. G. and McIsaac, W. M.,** The relationship between the metabolic fate and pharmacological action of 5-methoxy-N-methyltryptamine, *Biochem. Pharmacol.*, 13, 531, 1964.

150. **Mandel, L. R. and Walker, R. W.,** The biosynthesis of 5-methoxy-N,N-dimethyltryptamine *in vitro*, *Life Sci.*, 15, 1457, 1974.

151. **Agurell, S., Holmstedt, B., and Lindgren, J. E.,** Metabolism of 5-methoxy-N,N-dimethyltryptamine-^{14}C in the rat, *Biochem. Pharmacol.*, 18, 2271, 1969.

152. **Block, W., Block, K., and Patzig, B.,** Zur Physiologie des ^{14}C-radioaktiven Mescalins im Tierversuch. I. Mitteilung. Fermentversuche und Ausscheidungsprodukte, *Hoppe-Seyler's Z. Physiol. Chem.*, 290, 160, 1952.

153. **Shah, N. S. and Himwich, H. E.,** Study with mescaline-8-C^{14} in mice: effect of amine oxidase inhibitors on metabolism, *Neuropharmacology*, 10, 547, 1971.

154. **Goldstein, M., Friedhoff, A. J., Pomeranz, S., Simmons, C., and Contrera, J. F.,** Formation of 3:4:5-trimethoxyphenylethanol from mescaline, *J. Neurochem.*, 6, 253, 1961.

155. **Neff, N., Rossi, G. V., Chase, G. D., and Rabinowitz, J. L.,** Distribution and metabolism of mescaline-C^{14} in the cat brain, *J. Pharmacol. Exp. Ther.*, 144, 1, 1964.

156. **Cochin, J., Woods, L. A., and Seevers, M. H.,** The absorption, distribution and urinary excretion of mescaline in the dog, *J. Pharmacol. Exp. Ther.*, 101, 205, 1951.

157. **Spector, E.,** Identification of 3,4,5-trimethoxyphenylacetic acid as the major metabolite of mescaline in the dog, *Nature (London)*, 189, 751, 1961.

158. **Harley-Mason, J., Laird, A. H., and Smythies, J. R.,** I. The metabolism of mescalin in the human. II. Delayed clinical reactions to mescalin, *Confin. Neurol.*, 18, 152, 1958.

159. **Ratcliffe, J. and Smith, P.,** Metabolism of mescaline, *Chem. Ind.*, 925, 1959.

160. **Mokrasch, L. C. and Stevenson, I.,** The metabolism of mescaline with a note on correlations between metabolism and psychological effects, *J. Nerv. Ment. Dis.*, 129, 177, 1959.

161. **Charalampous, K. D., Orengo, A., Walker, K. E., and Kinross-Wright, J.,** Metabolic fate of β-(3,4,5-trimethoxyphenyl)-ethylamine (mescaline) in humans: isolation and identification of 3,4,5-trimethoxyphenylacetic acid, *J. Pharmacol. Exp. Ther.*, 145, 242, 1964.

162. **Friedhoff, A. J. and Hollister, L. E.,** Comparison of the metabolism of 3,4-dimethoxyphenylethylamine and mescaline in humans, *Biochem. Pharmacol.*, 15, 269, 1966.

163. **Charalampous, K. D., Walker, K. E., and Kinross-Wright, J.,** Metabolic fate of mescaline in man, *Psychopharmacologia*, 9, 48, 1966.

164. **Baba, S., Enogaki, K., Matsuda, A., and Nagase, Y.,** Studies on the analysis of drugs by use of radioisotopes. VIII. Species differences in *l*-ephedrine metabolism, *J. Pharm. Soc. Jpn.*, 92, 1270, 1972.

165. **Bralet, J., Cohen, Y., and Valette, G.,** Metabolisme et distribution tissulaire de la *dL* ephedrine ^{14}C, *Biochem. Pharmacol.*, 17, 2319, 1968.

166. **Kawai, K. and Baba, S.,** Studies on drug metabolism by use of isotopes. XIII. Isotope effect on metabolism of deuterated *l*-ephedrine, *Chem. Pharm. Bull.*, 22, 2372, 1974.

167. **Jacquot, C., Rapin, J. R., Wepierre, J., and Cohen, Y.,** Métabolisme et distribution tissulaire de l'éphédrine ^{14}C chez le Cobaye, *Arch. Int. Pharmacodyn. Ther.*, 207, 298, 1974.

168. **Matsuda, A., Baba, S., and Nagase, Y.,** Studies on the analysis of drugs by use of radioisotopes. VII. Identification of metabolites of ephedrine in rabbit, *J. Pharm. Soc. Jpn.*, 91, 542, 1971.

169. **Kawai, K. and Baba, S.,** Studies on drug metabolism by use of isotopes. XIV. Mass spectrometric quantification of urinary metabolites of deuterated *l*-ephedrine in rabbits, *Chem. Pharm. Bull.*, 23, 289, 1975.

170. **Axelrod, J.,** The biotransformation of ephedrine, *J. Pharmacol. Exp. Ther.*, 106, 372, 1952.

171. **Karawya, M. S., El-Keiy, M. A., Wahba, S. K., and Kozman, A. R.,** A note on a simple estimation of amphetamine, methylamphetamine and ephedrine in horse urine, *J. Pharm. Pharmacol.*, 20, 650, 1968.

172. **Nicholson, J. D.,** The metabolism of *l*-ephedrine in ponies, *Arch. Int. Pharmacodyn. Ther.*, 188, 375, 1970.

173. **Nicholson, J. D.,** 1,2-Dihydroxy-1-phenylpropane: a metabolite of ephedrine in ponies, *Arch. Int. Pharmacodyn. Ther.*, 192, 291, 1971.

174. **Chapman, D. I. and Marcroft, J.,** Unpublished data cited by M. S. Moss, in *Drug Metabolism—from Microbe to Man*, Parke, D. V. and Smith, R. L., Eds., Taylor and Francis, London, 1977, 263.

175. **Baba, S.,** Analysis of drug metabolites in human bodies using deuterium and carbon-13 labeled drugs, *Nippon Aisotopu Kaigi Hobunshu*, 11, 171, 1973 (Chem. Abstr. 84, 83892a, 1976).

176. **Sever, P. S., Dring, L. G., and Williams, R. T.,** The metabolism of (–)-ephedrine in man, *Eur. J. Clin. Pharmacol.*, 9, 193, 1975.

177. **Kawai, K. and Baba, S.,** Studies on drug metabolism by use of isotopes. XVII. Mass spectrometric quantification of urinary metabolites of deuterated *l*-ephedrine in man, *Chem. Pharm. Bull.*, 24, 2728, 1976.

METABOLISM OF NITRILES, NONPROTEIN AMINO ACIDS, NITRO COMPOUNDS, AMIDES, AND OTHER ALIPHATIC NITROGEN COMPOUNDS

NITRILES

Many nitriles (organic cyanides) found in plants are cyanogenetic glycosides.[1] These compounds, which have the general formula shown in Figure 1, are widely distributed and occur in several hundred genera belonging to dozens of plant families. The best-known cyanogenetic glycoside is amygdalin (1), the toxic principle of bitter almonds which is also found in the kernels of fruits including cherries, apricots, and plums. Other common examples include linamarin (2) and lotaustralin (3), which are often found together, and prunasin (4). The latter compound contains the same aglycon (mandelonitrile) (5) as amygdalin, however the sugar moiety consists of a single glucose unit.

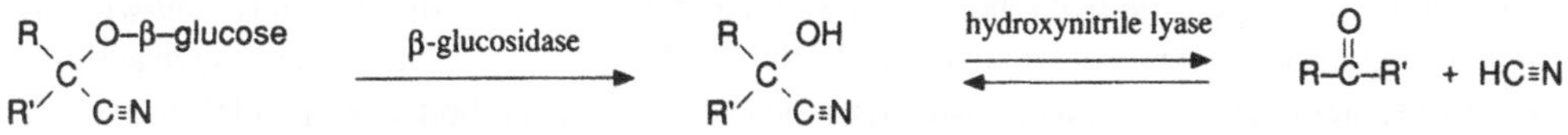

FIGURE 1.   Hydrolysis of cyanogenetic glycosides. (R′) = H in many cases.

All of these compounds are β-glycosides and will undergo hydrolysis when subjected to the actions of β-glycosidases also present in the plants. Normally, glycoside and enzyme are separated in the plant tissue but damage by cutting, crushing, or soaking in water permits their contact and results in the liberation of HCN as illustrated in Figure 1. With amygdalin this occurs via the sequential loss of two glucose units to give the aglycon mandelonitrile (5) (see Figure 2) followed by the decomposition of this unstable cyanohydrin (α-hydroxynitrile) to benzaldehyde and HCN.[2] The further metabolism in animals of the HCN formed leads mainly to thiocyanate, however some CO_2 is also produced. Johnson and Isom[3] reported that a few percent of administered cyanide may be lost as expired HCN and as CO_2. Further details on the mammalian metabolism of hydrogen cyanide are found in the articles of Conn,[1] Montgomery[4,5] and Way.[6]

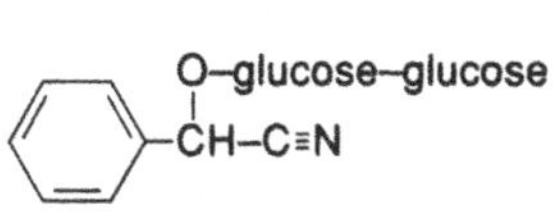

(1)

Amygdalin

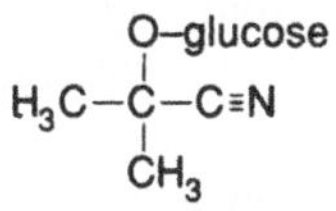

(2)

Linamarin

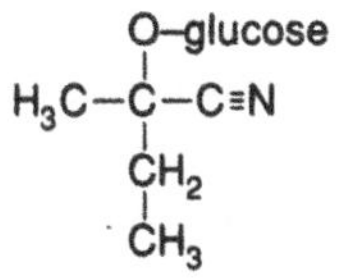

(3)

Lotaustralin

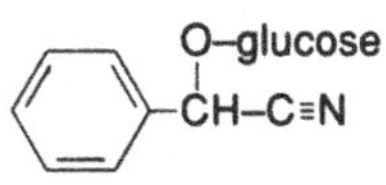

(4)

Prunasin

The metabolism of **mandelonitrile** (5) in rats was studied by Singh et al.[7] They found that an oral dose (30 mg/kg) was metabolized via a main pathway (see Figure 2) which ultimately formed hippuric acid (6) and thiocyanate ion (7) and along a minor pathway which gave mainly mandelic acid (8). More than 85% of the dose was recovered as urinary cyanide and thiocyanate, in about equal amounts. Most of the cyanide excretion occurred during the first 24 h, however thiocyanate excretion was more prolonged. The benzaldehyde formed concomitantly was metabolized to hippuric acid and 71% of the dose was recovered as such. The minor pathway leading to mandelic acid accounted for 13% of the dose. Additional unidentified trace metabolites were detected which were possibly phenylglyoxylic acid and *p*-hydroxymandelic acid, known metabolites of mandelic acid (see the chapter entitled Metabolism of Acids, Lactones, and Esters).

Considerable attention has been given to the subject of cyanide poisoning by cyanogenetic glycosides, however there appears to have been a good deal of confusion about the site and nature of the hydrolysis process. The belief lingered that hydrolysis following ingestion of the glycosides by animals was dependent upon the concomitant ingestion of the β-glucosidase. While this is not the case, it is known that the intake of β-glucosidase will increase the toxicity of these glycosides. This was demonstrated in rats,[8] dogs,[9] and humans[10] and also discussed by Stavric and Klassen.[11] Some have maintained that hydrolysis probably takes place after the absorption of the glycoside from the intestine.[1] Indeed, several investigations have shown that β-glycosidase activity is present in mammalian tissues. Freese et al.[12] demonstrated such activity in homogenates of cat kidney, with lower activities found in rat or rabbit kidney and little or none in mouse, guinea pig, sheep, cow, or human kidney. This enzyme catalyzed the removal of the terminal glucose moiety of **amygdalin** (1) to yield **prunasin** (4). Newmark et al.[13] found that small intestine mucosa from rats and humans contained glucosidases that removed glucose from both amygdalin and prunasin. Interestingly, this study failed to find such glucosidase activity in a wide variety of human tumors, a finding similar to that reported by Hill et al.[14] using a mouse tumor. Strugala et al.[15] also noted the ability of rat small intestine to remove the terminal glucose unit of amygdalin to form prunasin. The hydrolysis of amygdalin by rat small and large intestine was shown by Adewusi and Oke[16] to vary with age. The activities from both sites were highest at an age of 15 d and later decreased to zero.

In spite of the findings noted above, the results from numerous investigations indicate that intestinal microorganisms are responsible for the release of cyanide from cyanogenetic glycosides and the toxicity which follows. An early investigation which pointed to this conclusion was that of Coop and Blakley[17] who, using a sheep with a rumen fistula, measured the formation of HCN from amygdalin, linamarin, or lotaustralin administered intraruminally. They found that the rumen microflora rapidly hydrolyzed the glycosides. These findings are also useful in explaining why the cyanogenetic glycosides are somewhat less toxic than HCN or an equivalent amount of cyanide ion. In the latter case, symptoms of poisoning appear very rapidly whereas Coop and Blakley reported that periods from 5 min to several hours elapsed following glycoside administration. In this case the HCN will be liberated over a certain period of time and thus allow the normal thiocyanate pathway of detoxification to more effectively remove the absorbed HCN. Coop and Blakley[18] also showed that relatively large amounts of HCN or cyanogenetic glycoside were tolerated when they were given repeatedly in small divided doses. It is therefore important that the absorption rate of HCN, which is rapid, does not exceed the rate of conversion of cyanide to thiocyanate. It was also found that feeding, which promotes the bacterial population of the rumen, increased the toxicity of the glycosides whereas starvation reduced it. The metabolism of cyanogenetic glycosides by rumen microorganisms was also studied by Majak and Cheng[19,20] who showed that amygdalin and prunasin were hydrolyzed when incubated with bovine rumen fluid or with about half of the nearly 70 strains of rumen bacteria tested. They noted that hydrolysis of the glycoside appears to be the rate-limiting step in cyanide formation in the rumen because degradation of the glycosides occurred more slowly than that of the corresponding cyanohydrins.

Further evidence of the key role of the gastrointestinal microflora in the metabolism of cyanogenetic glycosides was obtained by Smith[21] who administered amygdalin to mice. No

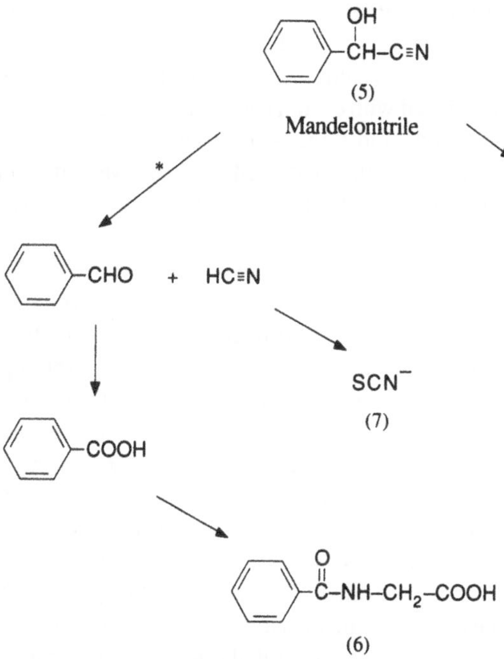

FIGURE 2.   Metabolic pathways of mandelonitrile. (*) Major pathway.

toxicity was observed following i.p. injection of doses as large as 5 g/kg whereas the oral LD_{50} was 350 mg/kg. While the latter value was considerably higher than that given by potassium cyanide, the symptoms of toxicity were similar to those produced by inorganic cyanide. It was found that pretreatment of the mice with lactose, another β-glycoside, conferred protection against a lethal dose of amygdalin but not against cyanide. This protection was not achieved when an α-glycoside, maltose, was used. Thus, competition for the available β-glycosidase present in the intestine will reduce the rate of cyanide formation. That the source of the enzyme is the microflora was indicated by the fact that its suppression by pretreatment with the antibiotic kanamycin reduced amygdalin toxicity, especially in fasted mice. Finally, ten strains of enterobacteria and enterococci were isolated from mouse intestine and were shown to degrade amygdalin to cyanide *in vitro*. Similar results using isolated strains of intestinal bacteria were reported by Drasar and Hill[22] who also found that amygdalin hydrolysis could be carried out by some nonsporing anaerobes (e.g., bifidobacteria). The release of cyanide from amygdalin is not effected by β-glucuronidase.[23]

Conclusive evidence of the importance of the intestinal bacteria in cyanogenetic glycoside metabolism was furnished by Carter et al.[24] who employed normal and germ-free rats. Oral doses of amygdalin which were fatal in the former group were not toxic in germ-free animals. The normal rats had high blood concentrations of cyanide whereas the germ-free animals had levels equal to those found in normal rats not given amygdalin. The intestinal contents of conventional rats, but not germ-free rats, catalyzed the release of benzaldehyde from amygdalin. Strugala et al.[15] found that, whereas the isolated perfused rat liver showed no metabolic activity towards amygdalin or prunasin, hydrolysis of these glucosides occurred when incubated with rat cecal contents. They concluded that the toxicity of amygdalin due to cyanide release is dependent upon metabolism by the intestinal bacteria. Stavric and Klassen[11] found large differences in the relative potency of fecal samples from different individuals of the same species to hydrolyze amygdalin. Even greater differences were noted between different species (mouse, rat, hamster, guinea pig, monkey, and man). Large interindividual differences in the ability to effect the release of cyanide from amygdalin were confirmed in mice.[14] Little activity was found in the contents of the stomach and upper intestine, whereas incubation with the contents of the lower intestine released large amounts of cyanide. Cyanide was liberated from both amygdalin and prunasin when the glucosides were incubated with cecum contents from hamsters.[25] This study also included measurements of the cyanide and thiocyanate concentrations in the blood of hamsters given an oral dose (~200 mg/kg) of amygdalin. Maximum levels of cyanide were reached 1 to 3 h after dosing and declined rapidly thereafter. Thiocyanate levels increased steadily during the 4-h test period.

The excretion of unchanged amygdalin in the urine is greatly influenced by the route of administration. Flora et al.[26] reported that mice excreted ~70% of an i.v. dose (100 mg/kg) unchanged whereas a value of ~20% was found after oral dosage. Most of the excretion occurred during the first 24 h. Rauws et al.[27] gave amygdalin to rats (50 mg) and dogs (50 mg/kg) both orally and i.v. and found that only ~1% of the dose was excreted unchanged in the urine of both species by the former route. A value of ~70% was noted when the glucoside was injected. Excretion of the monoglucoside prunasin was also measured in these experiments. Virtually none was detected in the urine of dogs following i.v. dosage of amygdalin, however 21% of the dose was excreted in this form following oral administration. Prunasin excretion in rats was more extensive and gave values of ~7% (i.v.) and 39% (oral). Additional experiments with prunasin itself[28] in dogs showed that its oral bioavailability was ~50%. Zhang and Jin[29] recorded a value of 62% of the dose for the urinary excretion of unchanged amygdalin by rabbits following an i.v. dose of 500 mg/kg. Studies in humans receiving daily doses of amygdalin[10] showed similar results. Values of 62 to 96% were recorded after i.v. dosage compared with 8 to 32% after oral administration.

Linamarin (2), the glucoside of acetone cyanohydrin, is found in several foodstuffs including cassava and lima beans. Most investigations on its metabolism have emphasized the role of the intestinal microorganisms. Winkler[30] reported that cooked lima beans released significant amounts of HCN when incubated with extracts of feces or with *E. coli*. Whereas Jansz et al.[31] reported that seven types of coliform bacteria failed to liberate cyanide from linamarin, this reaction was seen when a β-glucosidase preparation from hamster cecum was used.[25] Linamarin was found to be rapidly hydrolyzed when incubated with sheep rumen fluid,[17] a result also reported by Majak and Cheng[20] using bovine rumen fluid. The latter investigation showed that cyanide production occurred with half of the 50 strains of rumen bacteria tested. Barrett et al.[32] administered linamarin (~300 mg/kg) orally to rats and recovered nearly 20% of the dose unchanged in the urine within 24 h. No unchanged glycoside was identified in the blood or feces, however an additional 10% of the dose was accounted for as urinary thiocyanate ion. A subsequent report[33] dealt with the urinary and plasma thiocyanate levels in rats given linamarin in the diet. Both levels were higher in these animals than in animals which did not receive the glycoside. Other studies of linamarin metabolism in rats[34] showed that oral doses (10 to 350 mg/ kg) resulted in the urinary excretion (24 h) of only ~2% of the dose as unchanged compound. Small but increasing amounts of free cyanide and its detoxication product thiocyanate were found in the urine following these doses. It was also shown that linamarin could be degraded *in vitro* by rat liver microsomes via a pathway which did not lead to the formation of HCN. The possibility of tissue metabolism of linamarin was also suggested by the findings of Maduagwu and Umoh[35] who detected unchanged compound and several nonglucosidic cyanide metabolites in the bile of rats given linamarin (300 mg/kg, i.p.). The cyanide and thiocyanate concentrations in the blood of hamsters given an oral dose (~110 mg/kg) of linamarin were determined by Frakes et al.[25] Maximum levels of cyanide were reached 1 to 3 h after dosing and declined rapidly thereafter. Thiocyanate levels increased steadily during the 4-h test period.

It is noted above that **lotaustralin** (3) underwent hydrolysis when given intraruminally to sheep.[17] They also isolated two rumen microorganisms, a Gram-positive diplococcus and a Gram-negative bacillus, which readily hydrolyzed lotaustralin *in vitro*.

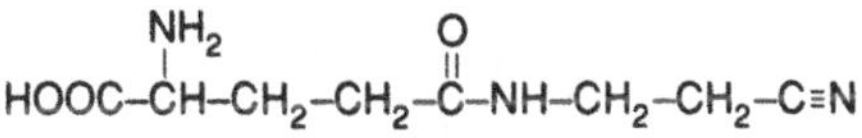

$$\underset{(9)}{\text{HOOC–}\overset{\overset{\displaystyle NH_2}{|}}{\text{CH}}\text{–CH}_2\text{–CH}_2\text{–}\overset{\overset{\displaystyle O}{\|}}{\text{C}}\text{–NH–CH}_2\text{–CH}_2\text{–C}\equiv\text{N}}$$

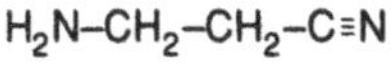

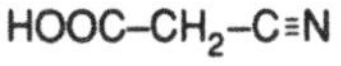

$$\underset{(10)}{\text{H}_2\text{N–CH}_2\text{–CH}_2\text{–C}\equiv\text{N}} \qquad\qquad \underset{(11)}{\text{HOOC–CH}_2\text{–C}\equiv\text{N}}$$

Another type of nitrile is represented by the osteolathyrogen found in the seeds of *Lathyrus* species (vetches). The causative agent, which produces changes in bone and connective tissue structure, is β-(γ-L-glutamyl)aminopropionitrile (9). Its metabolism has not been studied, however several investigations have dealt with β-aminopropionitrile (10) which also produces the symptoms of osteolathyrism (odoratism). The glutamyl residue is therefore not necessary for the biological activity of the compound[36] and it seems reasonable to assume that the amino acid (9) undergoes hydrolysis of the amide bond in the body to the simple aminonitrile. An analogous reaction has been reported with the closely related amino acid theanine (19) (see below).

Lipton et al.[37,38] reported that β-aminopropionitrile was metabolized in rats and rabbits to cyanoacetic acid (11), a metabolite which does not share the toxic properties of the aminonitrile. This study indicated that about equal amounts of the administered compound and the metabolite were excreted in the urine of rats following an i.p. dose of ~135 mg/kg. Keiser et al.,[39] using a small dose (7 mg/rat), reported that approximately six times as much cyanoacetic acid as β-aminopropionitrile was excreted in the urine. No other urinary metabolites were detected. Disappearance of the aminonitrile from the body was rather slow and appears to involve the participation of monoamine oxidase. This was shown in an experiment in mice in which administration of a monoamine oxidase inhibitor greatly reduced the rate of disappearance of the compound from the body. Fleisher et al.[40] similarly showed that the use of a monoamine oxidase inhibitor reduced the extent of formation of cyanoacetic acid from β-aminopropionitrile in the perfused rabbit liver and also resulted in the urinary excretion of more nitrile and less cyanoacetic acid in rabbits. The oxidative deamination of β-aminopropionitrile by rat liver homogenates was shown to occur at a low rate.[41] Also, Sievert et al.[42] reported that rat liver homogenates carried out this conversion more slowly than did preparations from several other species (mouse, cotton rat, guinea pig, rabbit, cow, and horse). Human liver homogenates showed the lowest activity of all while rabbit preparations were most active. Lipton et al.[38] reported that rabbits are very efficient in converting β-aminopropionitrile to cyanoacetic acid *in vivo*. Unlike that noted above with rats, no unchanged compound was detected in the urine of rabbits given the nitrile. The metabolism of β-aminopropionitrile in rats was also studied by Fleisher et al.[41] Following a large dose (400 mg/kg, i.p.), excretion of unchanged compound in the urine accounted for 30% of the dose in 12 h and 32% in 24 h, after which no further excretion occurred. Urinary excretion of cyanoacetic acid was 12 and 22% in 24 h and 48 h, respectively. When cyanoacetic acid itself was administered it was rapidly excreted. These findings, as well as that showing the presence of the metabolite but not the administered nitrile in liver and brain after 24 h, suggest that the metabolically formed acid is only slowly released from tissues. Fleisher et al.[43] reported a similar prolonged excretion of cyanoacetic acid in humans given the nitrile. They found that 16% of the 1 g of nitrile given daily in divided doses was excreted in the urine unchanged whereas about three times this amount appeared as cyanoacetic acid.

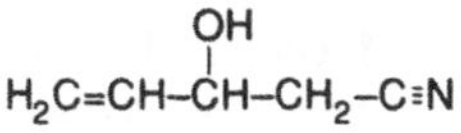

$$\underset{(12)}{\text{H}_2\text{C=CH–}\overset{\overset{\displaystyle OH}{|}}{\text{CH}}\text{–CH}_2\text{–C}\equiv\text{N}}$$

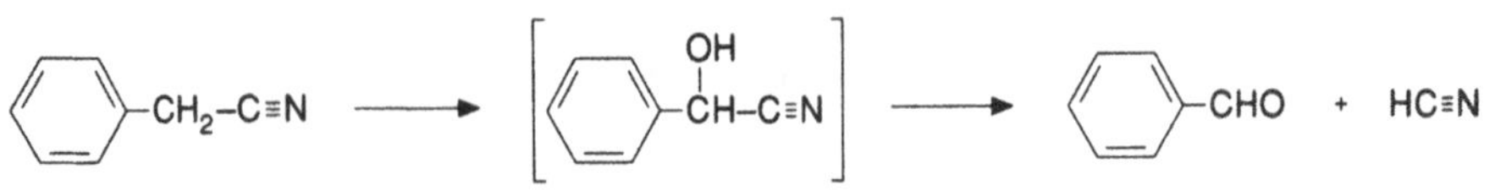

(13)

Benzyl cyanide

FIGURE 3.   Metabolism of benzyl cyanide by mouse liver microsomes.

As noted in the chapter entitled Metabolism of Sulfur Compounds, the breakdown of glucosinolates can lead to the formation of organic nitriles. Examples of these include aliphatic nitriles (e.g., **1-cyano-2-hydroxy-3-butene** (12)) and aromatic nitriles (e.g., **benzyl cyanide** (13)). Little is known about the metabolism of compound (12), however Wallig et al.[44] reported that its oral administration to rats resulted in a moderate increase in the excretion of urinary thiocyanate. This indicates that some release of cyanide from the nitriles had taken place. Ohkawa et al.[45] studied the metabolism of several nitriles using mouse liver microsomes. It appears that this system hydroxylates the methylene position to produce an unstable cyanohydrin which breaks down to the aldehyde and HCN. This reaction sequence with benzyl cyanide (13) is illustrated in Figure 3. Thus, the intermediate formed (mandelonitrile) (5) is identical to that produced from the hydrolysis of the cyanogenetic glycoside amygdalin. The hypothesis that nitriles are metabolized to cyanide via a cyanohydrin intermediate was discussed by Kaplita and Smith.[46]

NONPROTEIN AMINO ACIDS

Of the more than 200 amino acids isolated from plants, only ~10% consist of those which are incorporated into the proteins of living matter. Information describing the mammalian metabolism of nonprotein amino acids is limited, both in regard to the number of compounds studied and the amount of information available on specific compounds.

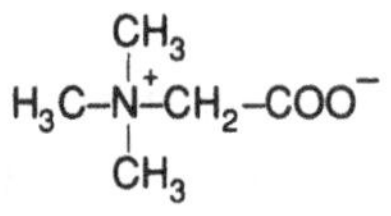

(14)

Betaine

An early study on the metabolism of **betaine** (14) suggested that it was excreted mainly unchanged in the urine following its administration (p.o. or s.c.) to rabbits, cats, and dogs.[47] It was also stated that some conversion to trimethylamine occurred in rabbits. However, subsequent studies using isotopically labeled betaine have shown that it undergoes extensive metabolism in nonruminant and ruminant animals. Stetten,[48] using ^{15}N-labeled compound, showed that its oral administration to rats resulted in its conversion to urea and ammonia which were found in the urine. The results also showed that betaine was rapidly demethylated to glycine and further metabolized to ethanolamine, the former compound being found in organ proteins and the latter in phospholipids. Ericson et al.[49] studied the enzyme system from rat and pig liver which demethylated betaine. This system, a betaine-homocysteine transmethylase, thus effects the formation of methionine from homocysteine. The activity of this transmethylase system in calf liver is much less than in rat liver.[50] Mitchell et al.[51] studied the metabolism of betaine labeled either with ^{14}C or ^{15}N in a steer or in sheep. *In vitro* studies using rumen fluid showed that [*carboxy*-^{14}C]betaine was converted to labeled acetate whereas labeled trimethylamine, methane, and CO_2 were formed from [*methyl*-^{14}C]betaine. Labeled ammonia was formed when the compound labeled with ^{15}N was used. The ^{14}C-labeled compounds were also given intraruminally to sheep. With [*carboxy*-^{14}C]betaine, labeled acetate was detected in both the rumen and blood and nearly half of the radioactivity was lost as expired CO_2. When the label was in the methyl group, nearly a fourth of the radioactivity was excreted in the urine, mainly in the form of trimethylamine and trimethylamine-*N*-oxide.

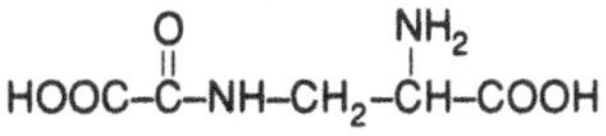

$$\underset{\text{(15)}}{\text{HOOC}-\overset{\overset{\textstyle O}{\|}}{\text{C}}-\text{NH}-\text{CH}_2-\overset{\overset{\textstyle NH_2}{|}}{\text{CH}}-\text{COOH}}$$

β-*N*-Oxalyl-L-α,β-diaminopropionic acid (15) is an acidic amino acid believed to be responsible for causing neurolathyrism. Cheema et al.,[52] using ^{14}C-labeled compound, found that 50 to 70% of the amino acid was excreted in the urine within 24 h after i.p. injection. In 1-h experiments, most of the radioactivity was located in the kidneys and liver with ~70% found in unchanged form. About 5 to 10% of the radioactivity in the kidneys was in a keto form and this transamination product was shown to be *N*-oxalyl-β-aminopyruvic acid. This indicates that the α-amino group underwent transamination while the *N*-oxalyl moiety remained intact.

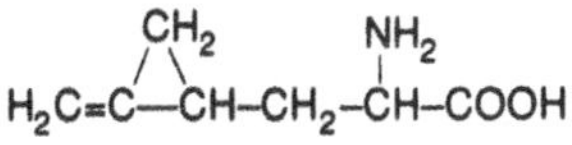

$$\underset{\text{(16)}}{\text{H}_2\text{C}{=}\text{C}\overset{\overset{\displaystyle CH_2}{\diagup\diagdown}}{}\text{CH}-\text{CH}_2-\overset{\overset{\textstyle NH_2}{|}}{\text{CH}}-\text{COOH}} \qquad \underset{\text{(17)}}{\text{H}_2\text{C}{=}\text{C}\overset{\overset{\displaystyle CH_2}{\diagup\diagdown}}{}\text{CH}-\text{CH}_2-\overset{\overset{\textstyle O}{\|}}{\text{C}}-\text{COOH}} \qquad \underset{\text{(18)}}{\text{H}_2\text{C}{=}\text{C}\overset{\overset{\displaystyle CH_2}{\diagup\diagdown}}{}\text{CH}-\text{CH}_2-\text{COOH}}$$

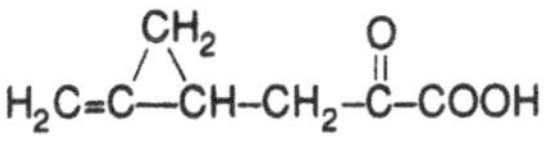

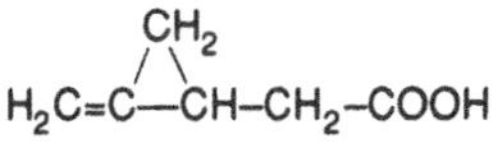

Hypoglycin

The metabolism of **hypoglycin** (β-(methylenecyclopropyl)alanine) (16) was studied in rats and in rat liver homogenates.[53,54] The results indicated that the first step in the metabolism of hypoglycin involved transamination to the keto acid (17). When rats were given [^{14}C]hypoglycin by injection, ~60% of the radioactivity was lost within 2 h as respiratory $^{14}CO_2$. The data indicated that the decarboxylating enzyme was associated with the mitochondria. The decarboxylation product, which was detected in the homogenates and in the livers of rats treated with hypoglycin, was identified as methylenecyclopropaneacetic acid (18).

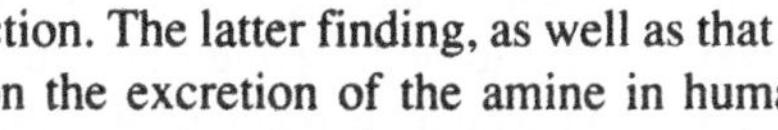

$$\underset{\text{(19)}}{\text{H}_3\text{C}-\text{CH}_2-\text{NH}-\overset{\overset{\textstyle O}{\|}}{\text{C}}-\text{CH}_2-\text{CH}_2-\overset{\overset{\textstyle NH_2}{|}}{\text{CH}}-\text{COOH}}$$

Theanine

Theanine (*N*-ethyl-γ-glutamine) (19), a tea constituent, was found by Asatoor[55] to be a source of urinary ethylamine in humans. Ethylamine was also excreted in the urine of rats given tea extract orally or by s.c. injection. The latter finding, as well as that showing that neomycin did not have a marked effect on the excretion of the amine in humans, suggests that theanine hydrolysis occurred in the tissues rather than in the gut as a result of bacterial metabolism.

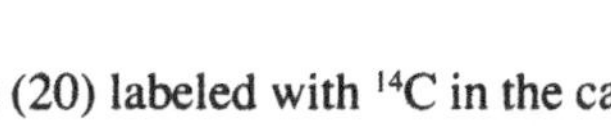

$$\underset{\text{(20)}}{\text{H}_2\text{N}-\overset{\overset{\textstyle O}{\|}}{\text{C}}-\text{NH}-(\text{CH}_2)_3-\overset{\overset{\textstyle NH_2}{|}}{\text{CH}}-\text{COOH}}$$

Citrulline

The metabolism of **citrulline** (20) labeled with ^{14}C in the carbamoyl moiety was studied in the perfused sheep and goat udder.[56] The results indicated that it was metabolized in the mammary tissue via arginine to urea, ornithine, and proline.

The metabolism of **L-canavanine** (21) (see Figure 4) has been the subject of several

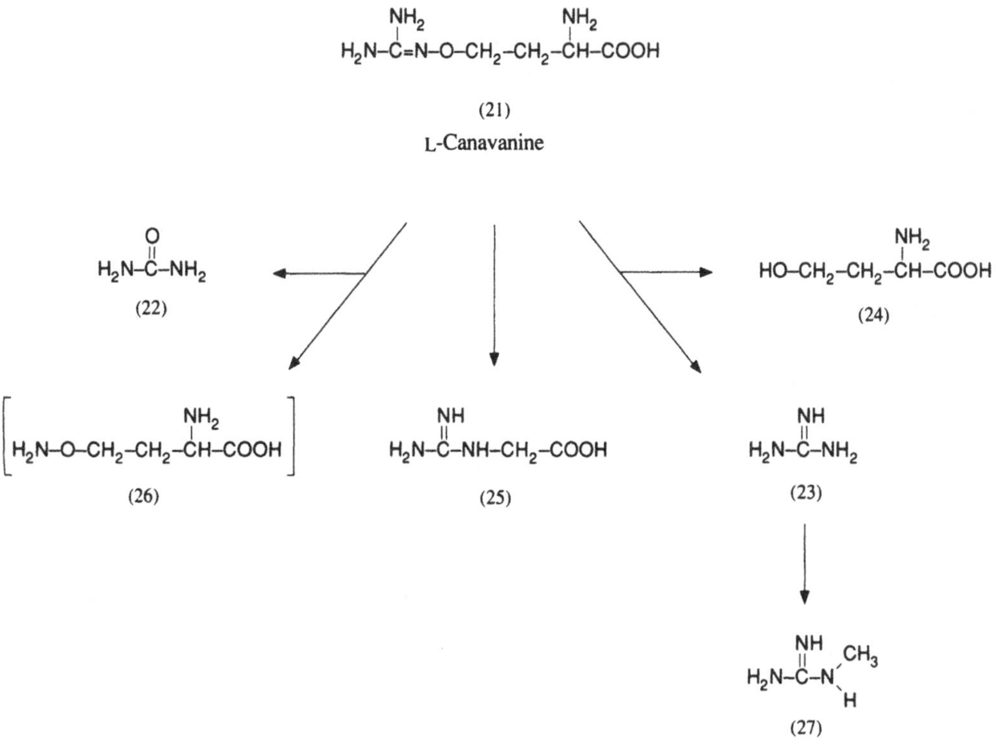

$$\underset{(21)}{\underset{\text{L-Canavanine}}{\text{H}_2\text{N}-\overset{\overset{\displaystyle \text{NH}_2}{|}}{\text{C}}=\text{N}-\text{O}-\text{CH}_2-\text{CH}_2-\overset{\overset{\displaystyle \text{NH}_2}{|}}{\text{CH}}-\text{COOH}}}$$

$$\underset{(22)}{\text{H}_2\text{N}-\overset{\overset{\displaystyle \text{O}}{||}}{\text{C}}-\text{NH}_2} \qquad \underset{(24)}{\text{HO}-\text{CH}_2-\text{CH}_2-\overset{\overset{\displaystyle \text{NH}_2}{|}}{\text{CH}}-\text{COOH}}$$

$$\underset{(26)}{\left[\text{H}_2\text{N}-\text{O}-\text{CH}_2-\text{CH}_2-\overset{\overset{\displaystyle \text{NH}_2}{|}}{\text{CH}}-\text{COOH}\right]} \qquad \underset{(25)}{\text{H}_2\text{N}-\overset{\overset{\displaystyle \text{NH}}{||}}{\text{C}}-\text{NH}-\text{CH}_2-\text{COOH}} \qquad \underset{(23)}{\text{H}_2\text{N}-\overset{\overset{\displaystyle \text{NH}}{||}}{\text{C}}-\text{NH}_2}$$

$$\underset{(27)}{\text{H}_2\text{N}-\overset{\overset{\displaystyle \text{NH}}{||}}{\text{C}}-\text{N}\underset{\text{H}}{\overset{\text{CH}_3}{\diagdown}}}$$

FIGURE 4.   Metabolic pathways of L-canavanine in rats.

investigations. Reiter and Horner[57] employed very small s.c. doses of material labeled with ^{14}C in the guanidino moiety and found that radioactive urea (22) and guanidine (23) were excreted in the urine. Most of the urinary radioactivity was associated with urea. Natelson[58] used larger doses (~90 mg/kg, i.p.) and detected guanidine, homoserine (24), and guanidinoacetic acid (25) as urinary metabolites. A subsequent study by Thomas and Rosenthal[59] using L-[*guanidinooxy*-^{14}C]canavanine (2 g/kg) showed that the radioactivity was mainly eliminated in the urine regardless of the route of administration (p.o., s.c., or i.v.). Fecal excretion was very small (2% or less) and 5 to 8% of the radioactivity was recovered as ^{14}CO$_2$. The extent of excretion of unchanged canavanine in the urine (24 h) varied markedly. Values of 1% of the urinary radioactivity (p.o. dosage), 12% (s.c.), and 33% (i.v.) were recorded. The most prominent urinary metabolite was urea (22) which furnished 50 to 88% of the urinary radioactivity. Urea and L-canaline (26) were presumably formed via the hydrolytic cleavage of canavanine by arginase. However, canaline itself was not detected in the urine and its metabolic fate is not known. The reductive pathway leading to homoserine (24), guanidine (23), and its methyl derivative (27) was a minor reaction which accounted for only ~6 to 13% or the urinary radioactivity. Likewise, formation of guanidinoacetic acid (25), which probably arose via a transamidation reaction involving glycine, was a minor pathway (1 to 2%).

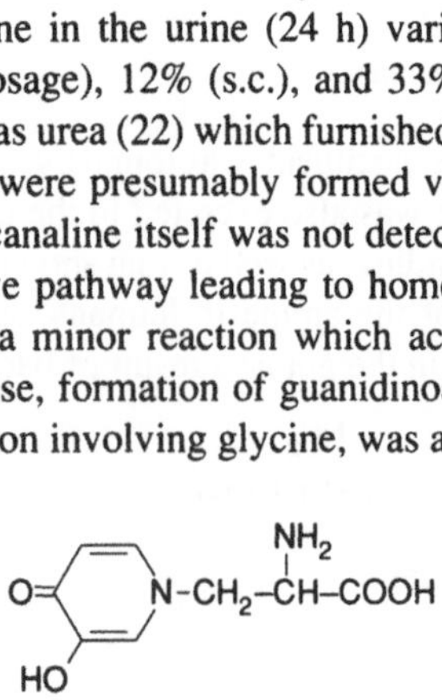

$$\underset{(28)}{\underset{\text{Mimosine}}{\text{(3-hydroxy-4-oxo-1-pyridyl structure)} \quad \text{N}-\text{CH}_2-\overset{\overset{\displaystyle \text{NH}_2}{|}}{\text{CH}}-\text{COOH}}} \qquad \underset{(29)}{\text{(3,4-dihydroxypyridine structure)}}$$

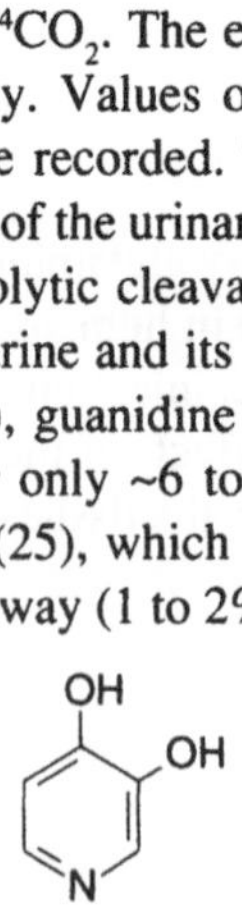

The metabolism of **mimosine** (β-(3-hydroxy-4-oxo-1-pyridyl)alanine) (28) has mainly been studied in ruminants because of the toxic effects of the forage plant in which it occurs. Hegarty et al.[60] showed that little degradation of the amino acid occurred in sheep following i.v. or intra-abomasal administration and that most of the compound was excreted unchanged in the urine. Following intraruminal administration, however, only 5 to 10% of the dose was recovered in the urine as unchanged compound and a metabolite, 3,4-dihydroxypyridine (29), was excreted which accounted for over 40% of the dose. The results indicated that metabolite (29) is the end product of mimosine metabolism in the rumen and that this degradation is probably carried out by the microflora. Reis et al.,[61] also using sheep, found that the main urinary metabolite during the first day after daily dosing was unchanged mimosine, however 3,4-dihydroxypyridine made a significant contribution from the second day. They also detected small amounts of urinary mimosinamine, formed by decarboxylation of mimosine. This metabolite was probably formed in the tissues because it was also excreted after i.v. administration of mimosine. Sharry et al.,[62] who administered [^{3}H]mimosine to sheep by an unspecified route, found that the main urinary metabolite (~40 to 80% of the radioactivity) was mimosine. Small amounts of 3,4-dihydroxypyridine (3 to 6%) and mimosinamine (2 to 10%) were found. A conjugate of mimosine was also detected which accounted for nearly half of the radioactivity in some experiments. The ruminal metabolism of mimosine and 3,4-dihydroxypyridine was reviewed by Jones[63] who noted that the latter compound may also be metabolized to 2,3-dihydroxypyridine and that all three of these compounds may undergo further degradation to unidentified metabolites.

The metabolism of mimosine in nonruminants has also been studied. Tsai and Ling[64] administered it orally to rats and found that it was excreted unchanged in both the urine and feces. They did not find evidence for the excretion of 3,4-dihydroxypyridine. However, Tang and Ling[65] reported that mice excreted trace amounts of this metabolite in the urine following oral administration of mimosine. They also reported that the urine contained mimosinamine, mimosinic acid, and unchanged mimosine.

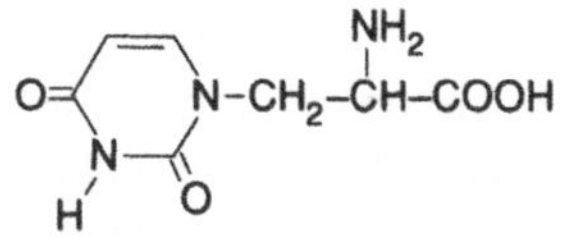

(30)

Willardiine

Al-Baldawi[66] administered **willardiine** (α-amino-β-uracil-1-ylpropionic acid) (30) and **isowillardiine** to rabbits by i.v. injection and found that they were excreted intact in the urine.

Some information is available on the metabolism of a few sulfur-containing amino acids and derivatives. These compounds are included among the sulfides in the chapter entitled Metabolism of Sulfur Compounds.

NITRO COMPOUNDS

Nitro compounds are not widespread among higher plants, however a few aliphatic and aromatic examples are known. Simple examples of the former type are **3-nitropropanol** (31) and **3-nitropropionic acid** (32). These nitro compounds may occur in plants as such, however the alcohol is usually found as the β-D-glucoside **miserotoxin** and the acid esterified with glucose. Most of the metabolic studies in this field have been carried out in cattle and sheep, a fact related to the occurrence of these toxic nitro derivatives in loco weeds and poison vetches which may be present in forages. A major point of interest is the influence of the gastrointestinal microflora on the metabolism and toxicity of these compounds. Williams et al.[67] reported that miserotoxin was metabolized to 3-nitropropanol by rumen fluid from cattle and sheep. No other nitro compounds were detected. The rapid hydrolysis of the glycoside when administered intraruminally to cattle was reported by Majak et al.[68] In rats, however, miserotoxin was readily absorbed from the upper regions of the gastrointestinal tract and the degree of its gastrointestinal hydrolysis was small.[69] Its toxicity following oral administration was therefore found be be much lower than that of the aglycon, 3-nitropropanol. Furthermore, the toxic symptoms produced by miserotoxin showed a delayed onset, suggesting hydrolysis of the glycoside to the toxic nitroalcohol in the lower regions of the intestine. There was no indication that additional nitro compounds, including 3-nitropropionic acid, were formed in the intestine. Gustine et al.[70] reported that glucose esters of 3-nitropropionic acid were hydrolyzed to the free acid when incubated with bovine rumen fluid.

Metabolism by the rumen microorganisms extends to the free nitro compounds. Majak and Cheng[71,72] showed that both 3-nitropropanol and 3-nitropropionic acid underwent degradation when incubated anaerobically with rumen fluid or several pure strains of rumen bacteria. The nitro group was metabolized to inorganic nitrite which was subsequently reduced to ammonia. The *in vitro* or *in vivo* metabolism of 3-nitropropanol by rumen microorganisms was enhanced when nitroethane was added to the incubates or administered intraruminally.[73]

$$HOH_2C-CH_2-CH_2-NO_2 \qquad HOOC-CH_2-CH_2-NO_2$$

(31)   (32)

3-Nitropropanol   3-Nitropropionic acid

The administration of 3-nitropropanol into the gastrointestinal tract of sheep resulted in its rapid absorption and conversion to 3-nitropropionic acid.[74] The same results were obtained when the alcohol was formed by intraruminal hydrolysis of miserotoxin.[68] Similarly, the i.v. administration of 3-nitropropanol in sheep and cattle resulted in its rapid conversion to 3-nitropropionic acid.[75] The disappearance of the acid from plasma was slower than that of the alcohol. Furthermore, the plasma levels of nitrite were more closely associated with those of the acid and it was believed that nitrite is derived from 3-nitropropionic acid. Alston et al.[76] suggested that the other product in this decomposition reaction is acrolein, a cytotoxic aldehyde. Studies on the oxidation of 3-nitropropanol by alcohol dehydrogenase showed that the aldehyde intermediate formed, 3-nitropropanal, underwent both spontaneous decomposition to nitrite and acrolein and further metabolism to 3-nitropropionic acid.[77] However, it was believed that the favored metabolic pathway *in vivo* involves formation of the latter metabolite. 3-Nitropropionic acid is itself toxic because it acts as an irreversible inhibitor of succinate dehydrogenase, thus blocking the citric acid cycle.[78] Majak et al.[79] reviewed the subjects of absorption and metabolism of these nitrotoxins in ruminants.

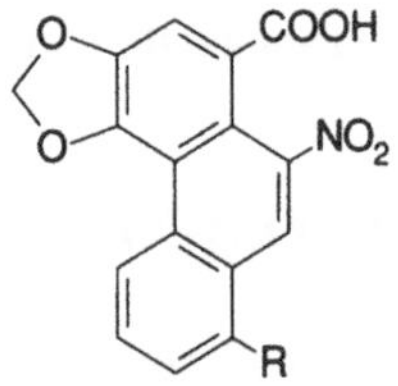

(33)

(a) Aristolochic acid I,   R = OCH₃

(b) Aristolochic acid II,   R = H

(c) R = OH

The metabolism of **aristolochic acid I** (33a) and **aristolochic acid II** (33b) has been studied both *in vitro* using rat liver preparations and *in vivo* in several animal species. Interestingly, these nitrophenanthrene derivatives gave two distinct patterns of metabolism following anaerobic or aerobic incubation with a 9000-g supernatant fraction from liver.[80] The major metabolites of both compounds were the corresponding aristolactam derivatives (34a) and (34b) when incubated under anaerobic conditions. In contrast to this similarity in reductive metabolism, aerobic incubation of aristolochic acid I resulted in the formation of the corresponding O-demethylated derivative aristolochic acid Ia (33c) whereas aristolochic acid II was not converted to detectable metabolites.

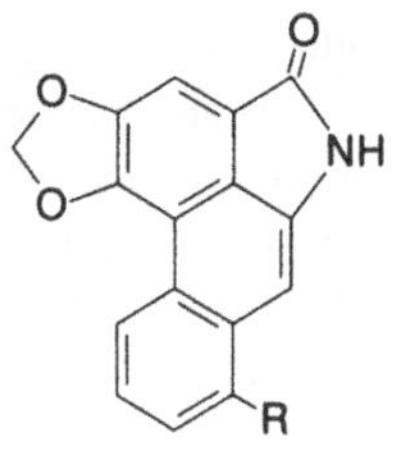

(34)   (35)

(a) R = OCH$_3$ (a) R = OCH$_3$

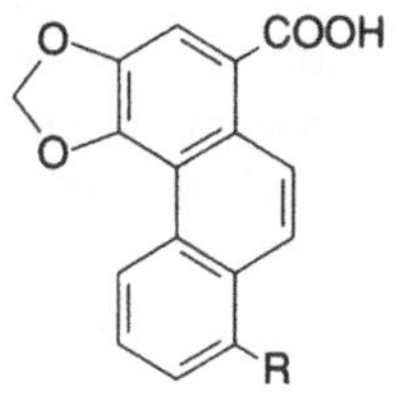

(b) R = H   (b) R = H

(c) R = OH   (c) R = OH

Quantitative studies of the metabolism of aristolochic acids I and II in rats and qualitative studies in mice, guinea pigs, rabbits, dogs, and man were reported by Krumbiegel et al.[81] The recoveries of urinary and fecal metabolites of the two compounds in rats were markedly different. Aristolochic acid I was largely excreted as aristolactam Ia (34c) in the urine (46% of the dose) and feces (37%). The urinary aristolactam Ia appeared to exist in about equal amounts as the N- and O-glucuronide conjugates. The urinary metabolites, which were only excreted during the first 24 h, also included aristolochic acid Ia (33c) (~3%) and small amounts or traces of aristolactam I (34a), aristolic acid I (35a), and 3,4-methylenedioxy-8-hydroxy-1-phenanthrenecarboxylic acid (35c). The bulk of the metabolites are thus reduction products, however no evidence was obtained for the presence of metabolites which contained a free amino group. Nitroso or hydroxyamine derivatives were also not detected. Another reaction which was not observed was demethylenation of the 3,4-methylenedioxy moiety. However, aristolochic acid I is a substituted piperonylic acid and, as such, would not be expected to undergo this reaction.[82] The unusual reaction of reductive replacement of the nitro group to form metabolites (35a) and (35c) was found to be carried out by the intestinal microflora. In contrast to this picture of abundant metabolism of aristolochic acid I, very small amounts of only a few metabolites of aristolochic acid II were detected. Aristolactam II (34b) was found in both the urine (~4%) and feces (~9%), as were small quantities of 3,4-methylenedioxy-1-phenanthrenecarboxylic acid (35b) and traces of aristolactam Ia (34c) in the urine only. An additional metabolite, either a positional isomer or hydroxylated derivative of aristolactam Ia, was also detected as a metabolite of aristolochic acid II. Studies in the other animal species showed that mice metabolize the two aristolochic acids to the same metabolites noted above in rats. Fewer metabolites were detected in the experiments with guinea pigs and rabbits, and fewer still in dogs and man. The metabolites lacking the nitro group were not formed in the last two species.

Additional *in vitro* experiments using the 9000-g supernatant fraction from rat liver showed that both aristolochic acids I and II were metabolized to reactive intermediates which formed adducts with DNA.[83] Adduct formation did not occur when the corresponding lactam derivatives (34a) and (34b) were used. When the liver supernatant was substituted with xanthine oxidase, which is known to reduce aromatic nitro groups, nearly identical patterns of adducts were obtained with the two aristolochic acids. Also, *in vivo* experiments showed that similar adduct patterns were seen in the DNA from many organs when rats were given oral doses of these two compounds.

AMIDES

Plant compounds which are characterized primarily by the presence of an amide group appear to be quite restricted in their occurrence, except, of course, for peptides which fall outside the scope of this book. A few amides consisting of various amines linked to the γ-carboxy group of L-glutamic acid have been reported. These include β-(γ-L-glutamyl)aminopropionitrile (9) (see above) and the tea constituent theanine (19). As noted above, theanine undergoes hydrolysis in the body and serves as a source of the ethylamine excreted in the urine by tea drinkers. Other examples of this type include amides derived from other simple aliphatic amines or from aromatic amines (e.g., *p*-hydroxyphenylamine, *o*-, and *p*-hydroxybenzylamine), however the mammalian metabolism of these compounds has not been reported.

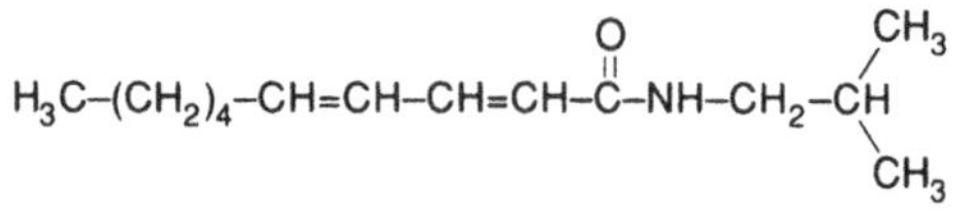

$$H_3C-(CH_2)_4-CH=CH-CH=CH-\overset{\overset{\displaystyle O}{\|}}{C}-NH-CH_2-\overset{\overset{\displaystyle CH_3}{\diagup}}{\underset{\underset{\displaystyle CH_3}{\diagdown}}{CH}}$$

(36)

Another group of amides deserving mention consists of derivatives of isobutylamine and various unsaturated fatty acids. These products have insecticidal properties and include pellitorine, a mixture of several isobutylamides which consists mainly of *N*-isobutyldeca-*trans*-2-*trans*-4-dienamide (36). The metabolism of this group has not been studied, however Kuhn et al.[84] reported on the oxidation of amides of several related unsaturated fatty acids in rabbits. This investigation determined the extent of ω-oxidation of the compounds to the semiamides of the corresponding dicarboxylic acid following large repeated doses ranging from 7 to 20 g. Based on the isolation of the corresponding ω-carboxy derivatives from urine, the yields of the metabolites were (% of dose): sorbamide, $H_3C-(CH=CH)_2-CO-NH_2$ (32%), octatrienoic amide, $H_3C-(CH=CH)_3-CO-NH_2$ (42%), and decatetraenoic amide, $H_3C-(CH=CH)_4-CO-NH_2$ (20%). The only *N*-substituted amide, sorbic acid *N*-methyl amide, underwent ω-oxidation to an extent of 44% of the dose. Nonetheless, all of these examples involve allylic oxidation, a situation which will not be encountered in the ω-oxidation of the naturally occurring isobutyl-amides. Also, the extent of hydrolysis of the amide bond in the above compounds was undetermined and this is a possible route of metabolism with the isobutylamides. An amide showing complex structural features is physostigmine. However, it also contains heterocyclic nitrogen atoms and is therefore covered in the chapter entitled Metabolism of Nitrogen Heterocyclic Compounds.

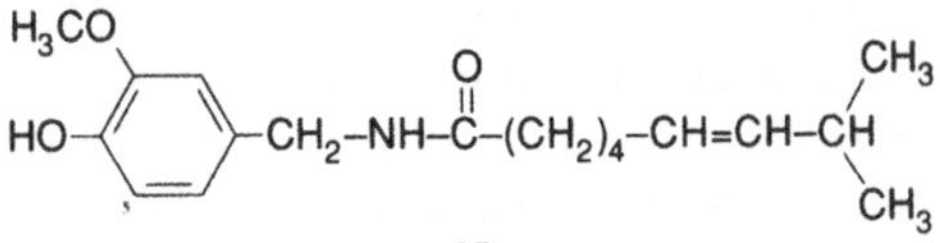

$$\underset{HO}{\overset{H_3CO}{\diagdown}}\underset{\diagup}{\diagup}-CH_2-NH-\overset{\overset{\displaystyle O}{\|}}{C}-(CH_2)_4-CH=CH-\overset{\overset{\displaystyle CH_3}{\diagup}}{\underset{\underset{\displaystyle CH_3}{\diagdown}}{CH}}$$

(37)

Capsaicin

Capsaicin (8-methyl-*N*-vanillyl-6-nonenamide) (37) and **dihydrocapsaicin** (38) (see Figure 5) are the main pungent principles from hot peppers. Sambaiah et al.[85] reported that no unchanged capsaicin was detected in the urine or feces of rabbits following an oral dose (100 mg). An increase in urinary glucuronides and conjugated phenols was noted, however hydrolysis of this conjugated material did not yield capsaicin or vanillylamine (39). They found that capsaicin was metabolized to several unidentified metabolites, both more and less polar than capsaicin, when incubated with rat liver homogenates or with mitochondrial or microsomal fractions from rat liver. Hydrolysis of the amide bond and formation of 8-methyl-6-nonenoic acid was shown following incubation of capsaicin with rat intestine homogenates. Similar results were reported by Kawada et al.[86] who incubated dihydrocapsaicin (38) with rat jejunal

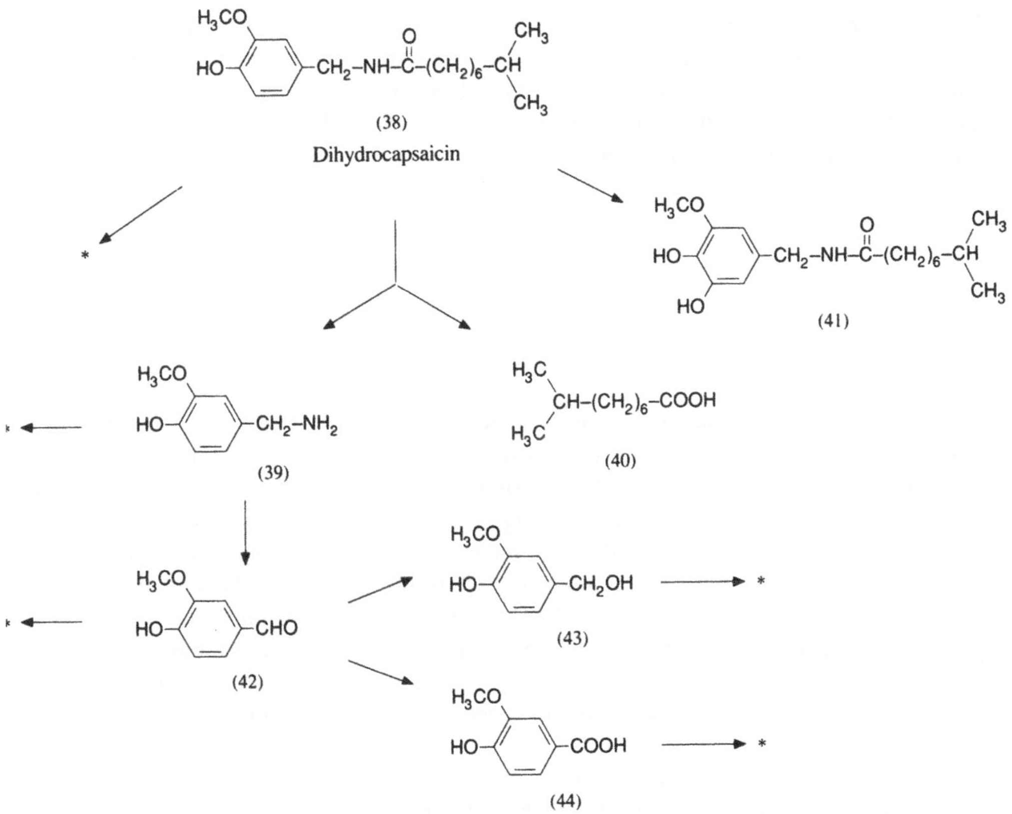

FIGURE 5.   Metabolic pathways of dihydrocapsaicin. (*) Glucuronide conjugates.

tissue. They detected the expected hydrolysis products, vanillylamine (39) and 8-methylnonanoic acid (40) in the incubates. They suggested that capsaicin and its analogues are partly metabolized during absorption in the intestine. Kawada et al. also found that capsaicin and dihydrocapsaicin were rapidly and extensively absorbed from the gastrointestinal tract. This finding contradicts the suggestion of Kim and Park[87] that capsaicin is poorly absorbed from the intestine. Their opinion was based on the finding that the oral LD_{50} value of capsaicin in mice was more than 100-times as large as the value found with i.v. administration. However, the reason for this difference is probably explained partly by the metabolism of capsaicin during the absorption process and, as discussed by Saria et al.,[88] partly by the higher concentrations reached in the central nervous system following i.v. injection.

Additional investigations have emphasized the nature of the metabolic products of capsaicin and dihydrocapsaicin. Lee and Kumar[89] showed that capsaicin and certain analogues were metabolized by a mixed function oxidase system in rat liver. The metabolite was formed by monohydroxylation and their results indicated that this occurred at the 5-position of the ring to give N-(3,4-dihydroxy-5-methoxybenzyl)acylamides (see structure (41) in Figure 5). The findings of Miller et al.[90] support this metabolic pathway for capsaicin and dihydrocapsaicin. They showed that a reactive metabolite was formed which bound irreversibly to hepatic microsomal protein. This intermediate is most likely an epoxide which may either bind with protein or form the hydroxylated metabolite noted above. On the other hand, Kawada and Iwai[91] did not detect a ring-hydroxylated metabolite when dihydrocapsaicin was incubated with homogenates or cell-free extracts of rat liver. The metabolites formed using cell-free extracts were 8-methylnonanoic acid (40), vanillylamine (39), and vanillin (42). Experiments using various organs indicated that the liver is the main site of metabolism of dihydrocapsaicin. This investigation also dealt with the *in vivo* metabolism of dihydrocapsaicin in rats. About 75% of an oral dose (20 mg/kg) was found in the urine in 48 h, mainly in the form of glucuronide conjugates. An additional 10% of the dose was recovered in the feces as unchanged compound.

The urinary metabolites included dihydrocapsaicin (~9% of the dose), vanillylamine (~5%), vanillin (~5%), vanillyl alcohol (43) (~38%, nearly entirely in conjugated form), and vanillic acid (44) (~19%). The pathways of metabolism of dihydrocapsaicin, including the hydroxylation step noted above, are shown in Figure 5. Kawada and Iwai[91] believed that this is a general pathway for capsaicin and its analogues.

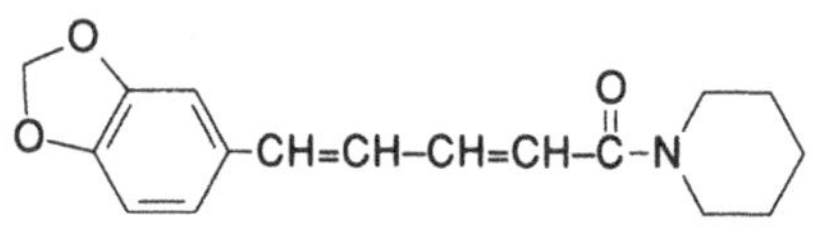

(45)

Piperine

The biological disposition of **piperine** (1-piperoylpiperidine) (45) in rats was studied by Bhat and Chandrasekhara.[92,93] They showed that it was extensively absorbed and metabolized following oral or i.p. administration. No unchanged compound was detected in the urine and ~3% of the dose (170 mg/kg, p.o. or 85 mg/kg, i.p.) was found in the feces as piperine after either route of administration. About a third of the oral dose was excreted in the urine in the form of conjugated phenols and a further 60% as methylenedioxyphenyl compounds. A relatively small portion of the latter group (~11% of the dose) was found as free metabolites which included piperonylic acid, the corresponding aldehyde piperonal, and the alcohol piperonyl alcohol. Also found among the free metabolites was vanillic acid which accounted for an additional 4% of the dose. Formation of this metabolite requires demethylenation of the methylenedioxyphenyl group, however it seems likely that this occurred prior to hydrolysis of the amide bond. Hydrolysis of this bond gives carboxylic acid derivatives which would not be expected to undergo demethylenation due to their hydrophilicity.[82] Biliary excretion of metabolites was not extensive and only piperic acid (~1% of the dose) was found in the bile in 6 h.

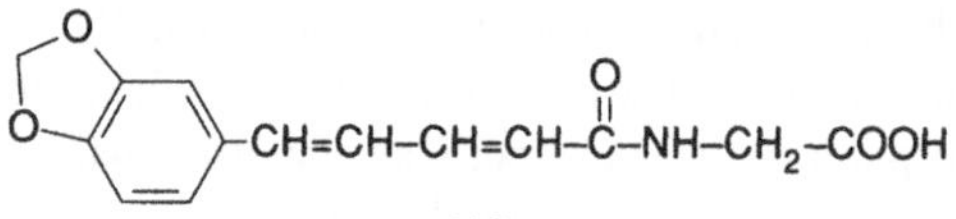

(46)

A compound showing structural similarity to piperine is the amide of piperic acid and glycine, piperoylglycine (46). Acheson and Atkins[94] investigated the metabolism in rats of this nonplant amide. Following a total oral dose of 20 mg (~200 mg/kg), the 48-h urines contained some unchanged compound as well as the glycine conjugates of the homologous acids, 3,4-methylenedioxybenzoic acid and 3,4-methylenedioxycinnamic acid. Although free C_6–C_1-, C_6–C_3-, or C_6–C_5-acids were not detected, the presence of the glycine conjugates of the first two compounds indicates that amide hydrolysis of the administered compound must have taken place.

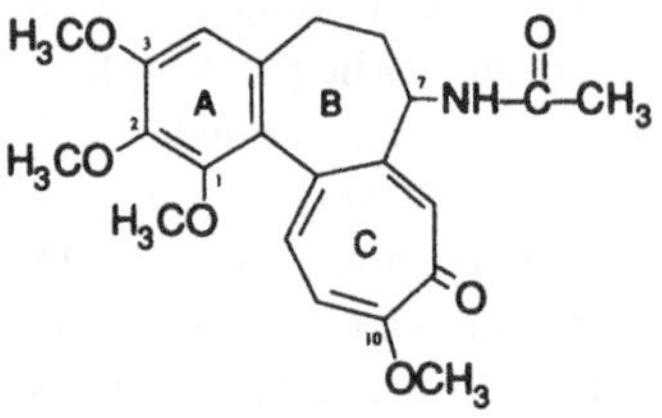

(47)

Colchicine

The final compound to be included among this small group of amides is **colchicine** (47). This complex polymethoxylated tropolone alkaloid has a long history of use, especially in the treatment of acute attacks of gouty arthritis. Nonetheless, its metabolic fate is far from clear. As sporadic studies of its metabolism have been carried out over a relatively long period of time, it is natural that several different analytical methods have been employed. Earlier methods were based on chemical or biological assays and, using the former, Boyland and Mawson[95] concluded that hydrolysis of colchicine to the C-10-demethylated derivative colchiceine in the acidic environment of the stomach is not a significant factor in its metabolism. Most of the present knowledge concerning colchicine metabolism has been obtained subsequent to the availability of the ³H- or ¹⁴C-labeled compounds and the present summary of results deals largely with these more recent findings. A short review of earlier data is available in the article by Wallace et al.[96]

The first study of the metabolism of radioactive colchicine was carried out by Walaszek et al.[97] using a product either randomly labeled by biosynthetic means or labeled specifically in the acetyl moiety at C-7. With the former compound given at dose levels of 0.1 to 10 mg/kg, they found that ~3 to 9% was excreted unchanged in the urine in 48 h in mice, rats, hamsters, and guinea pigs. The corresponding fecal values were lower (1 to 2%) except for the rat which showed urinary and fecal excretion values for colchicine of ~3 and 6%, respectively. In addition, smaller amounts of chloroform-soluble metabolites were found in the excreta in most cases. Loss of radioactivity as respiratory $^{14}CO_2$ was also noted. This was only 2 and 9% in mice and hamsters, respectively, whereas guinea pigs excreted 28% and rats 32% of the dose by this route. Administration of the preparation labeled in the N-acetyl moiety to a patient with gout resulted in excretion of radioactivity as $^{14}CO_2$, indicating hydrolysis of the amide bond. Urinary excretion of unchanged colchicine in three humans averaged more than a fourth of the dose (2 to 3 mg); however, Wallace and Ertel[98] reported that less than 10% of the administered compound or its metabolites was excreted in the urine in 24 h following i.v. injection. Recently, Achtert et al.[99] reported that healthy male subjects excreted 17% of an oral dose (1 mg) of colchicine in the urine as unchanged compound.

An important factor in the metabolic disposition of colchicine appears to be its biliary excretion. This excretory route was recognized in rats in an early report by Brues.[100] Later, Fleischmann et al.[101,102] reported that appreciable biliary excretion of colchicine occurred in the golden hamster as well as in the related rodent, the Mongolian gerbil.[103] An extensive study of the biliary excretion of colchicine labeled with ³H or ¹⁴C in the methoxy group at C-10 was carried out by Hunter and Klaassen.[104] Initial experiments with rats receiving a dose of 0.2 mg/kg (i.v.) indicated appreciable biliary excretion. The fecal:urinary ratio of radioactivity was 5:1 for the initial period of 3 d. More than 80% of the administered radioactivity was excreted during this time. At a dose level of 2 mg/kg (i.v.), the biliary excretion of radioactivity in 2 h in rats, hamsters, rabbits, and dogs was 50, 32, 16, and 20%, respectively. The percentages of this material excreted as unchanged colchicine were 53 in rats, 45 in hamsters, 72 in rabbits, and 34 in dogs. In addition to unchanged colchicine, the bile of all four animal species contained metabolites including polar compounds, some of which was a glucuronide conjugate of a colchicine derivative formed by demethylation in the A-ring. In all species except the rabbit, some free demethylated metabolite was also detected. This amounted to 15, 10, and 35% of the biliary material in rats, hamsters, and dogs, respectively. There was no evidence indicating that colchiceine, formed by demethylation at C-10, was a biliary metabolite or that N-deacetylated metabolites were excreted by this route. The lower capacity of newborn rats to conjugate and excrete colchicine in the bile was studied by Hunter and Klaassen.[105] This deficiency was related to the higher toxicity of colchicine in very young rats than in older rats.

An early report by Axelrod[106] on the mechanism of cleavage of aromatic ethers indicated that colchicine is a substrate for the liver microsomal system dependent on NADPH and O_2. This reaction was subsequently studied in detail by Schönharting et al.[107] who found that mouse, rat, and hamster liver microsomes oxidatively demethylated colchicine at the C-2 and C-3 positions. Additionally, demethylation at C-10 to colchiceine was carried out by mouse and rat preparations. The monodemethylated metabolites were not metabolized further by the microsomes except for glucuronidation which took place at the 2-position with rat microsomes and the 3-position with hamster microsomes. The three demethylated metabolites and a fourth derivative formed by rearrangement of the tropolone ring (C-ring) were formed by a modified Udenfriend system.[108]

Demecolcine is a tropolone derivative closely related to colchicine (47), differing only in the group located at the 7-position. In demecolcine the *N*-acetylamino group of colchicine is replaced by an *N*-methylamino group. Incubation of demecolcine with rat liver microsomes resulted in the formation of three mono-*O*-demethylated metabolites (at C-2, C-3, or C-10).[109]

OTHER ALIPHATIC NITROGEN COMPOUNDS

Cycad plants contain toxic glycosides of methylazoxymethanol (48), the most important being **cycasin** (methylazoxymethanol-β-D-glucoside) (49). The toxic properties of cycasin were reviewed by Laqueur[110] and Wogan and Busby.[111] Kobayashi and Matsumoto[112] reported that the i.p. injection of cycasin did not lead to toxicity and that the glucoside was excreted nearly entirely unchanged in the urine. In contrast, oral administration resulted in toxic effects and a much lower urinary excretion of unchanged compound. Because injection of the aglycon, methylazoxymethanol, also proved to be toxic, it was clear that this property is associated with the latter compound rather than with the glucoside itself.

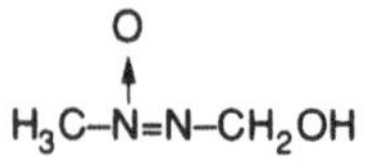

$$\underset{(48)}{H_3C-\overset{\overset{\displaystyle O}{\uparrow}}{N}=N-CH_2OH} \qquad \underset{(49)}{H_3C-\overset{\overset{\displaystyle O}{\uparrow}}{N}=N-CH_2-O-glucose}$$

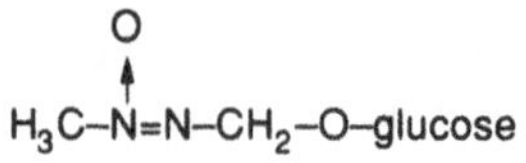

Cycasin

Laqueur[113] and Spatz et al.[114] studied cycasin toxicity and excretion in conventional and germ-free rats. They found that the latter group lacked the β-glucosidase needed to hydrolyze cycasin and that these animals excreted it unchanged without showing signs of toxicity. These findings clearly implicated the intestinal microflora and Spatz et al.[115] reported that the toxic properties of cycasin reappeared when germ-free rats were monocontaminated with various strains of bacteria possessing β-glucosidase activity. Furthermore, the degree of toxicity observed and the amounts of unchanged cycasin excreted in the urine were positively correlated with the *in vivo* levels of β-glucosidase activity measured in various bacteria. Thus, *Streptococcus faecalis* contained high enzyme levels and its use as a monocontaminant resulted in both relatively low excretion of unchanged cycasin and severe toxicity.

Feinberg and Zedeck[116] described findings which indicated that the toxic actions of methylazoxymethanol are associated with its metabolism via NAD- or NADP-dependent dehydrogenase reactions. Metabolism by liver alcohol dehydrogenase formed the probable initial metabolite, methylazoxymethanal, and this highly unstable aldehyde was found to rapidly release carbonium ions. This sequence was accompanied by interactions with cellular macromolecules. The data indicated that the observed variations in organ toxicity were related to the ability of the organs to carry out the dehydrogenation reaction.

REFERENCES

1. **Conn, E. E.,** Cyanogenic glycosides, in *International Review of Biochemistry. Biochemistry of Nutrition IA,* Vol. 27, Neuberger, A. and Jukes, T. H., Eds., University Park Press, Baltimore, MD, 1979, 21.

2. **Haisman, D. R. and Knight, D. J.,** The enzymic hydrolysis of amygdalin, *Biochem. J.,* 103, 528, 1967.

3. **Johnson, J. D. and Isom, G. E.,** Quantification of expired metabolites following potassium cyanide administration: a new method, *J. Anal. Toxicol.,* 9, 112, 1985.

4. **Montgomery, R. D.,** Cyanogenetic glucosides, *Handb. Clin. Neurol.,* 36, 515, 1979.

5. **Montgomery, R. D.,** Cyanogens, in *Toxic Constituents of Plant Foodstuffs,* 2nd ed., Liener, I. E., Ed., Academic Press, New York, 1980, 143.

6. **Way, J. L.,** Cyanide intoxication and its mechanism of antagonism, *Annu. Rev. Pharmacol. Toxicol.,* 24, 451, 1984.

7. **Singh, P. D. A., Jackson, J. R., and James, S. P.,** Metabolism of mandelonitrile in the rat, *Biochem. Pharmacol.,* 34, 2207, 1985.

8. **Adewusi, S. R. A. and Oke, O. L.,** On the metabolism of amygdalin. I. The LD_{50} and biochemical changes in rats, *Can. J. Physiol. Pharmacol.,* 63, 1080, 1985.

9. **Schmidt, E. C., Newton, G. W., Sanders, S. M., Lewis, J. P., and Conn, E. E.,** Laetrile toxicity studies in dogs, *J. Am. Med. Assoc.,* 239, 943, 1978.

10. **Moertel, C. G., Ames, M. M., Kovach, J. S., Moyer, T. P., Rubin, J. R., and Tinker, J. H.,** A pharmacologic and toxicological study of amygdalin, *J. Am. Med. Assoc.,* 245, 591, 1981.

11. **Stavric, B. and Klassen, R.,** Enzymatic hydrolysis of amygdalin by fecal samples and some foods, in *Nat. Toxins, Proc. Int. Symp. Anim. Plant Microb. Toxins, 6th, 1979,* Eaker, D., Ed., Pergamon Press, Oxford, 1980, 655.

12. **Freese, A., Brady, R. O., and Gal, A. E.,** A β-glucosidase in feline kidney that hydrolyzes amygdalin (laetrile), *Arch. Biochem. Biophys.,* 201, 363, 1980.

13. **Newmark, J., Brady, R. O., Grimley, P. M., Gal, A. E., Waller, S. G., and Thistlewaite, J. R.,** Amygdalin (laetrile) and prunasin β-glucosidases: distribution in germ-free rat and in human tumor tissue, *Proc. Natl. Acad. Sci. USA,* 78, 6513, 1981.

14. **Hill, H. Z., Backer, R., and Hill, G. J.,** Blood cyanide levels in mice after administration of amygdalin, *Biopharm. Drug Dispos.,* 1, 211, 1980.

15. **Strugala, G. J., Rauws, A. G., and Elbers, R.,** Intestinal first pass metabolism of amygdalin in the rat *in vitro,* *Biochem. Pharmacol.,* 35, 2123, 1986.

16. **Adewusi, S. R. A. and Oke, O. L.,** On the metabolism of amygdalin. II. The distribution of β-glucosidase activity and orally administered amygdalin in rats, *Can. J. Physiol. Pharmacol.,* 63, 1084, 1985.

17. **Coop, I. E. and Blakley, R. L.,** The metabolism and toxicity of cyanides and cyanogenetic glucosides in sheep I. Activity in the rumen, *N. Z. J. Sci. Technol. Sect. A,* 30, 277, 1949.

18. **Coop, I. E. and Blakley, R. L.,** The metabolism and toxicity of cyanides and cyanogenetic glucosides in sheep. III. The toxicity of cyanides and cyanogenetic glucosides, *N. Z. J. Sci. Technol. Sect. A,* 31, 44, 1950.

19. **Majak, W. and Cheng, K.-J.,** Cyanogenesis in bovine rumen fluid and pure cultures of rumen bacteria, *J. Anim. Sci.,* 59, 784, 1984.

20. **Majak, W. and Cheng, K.-J.,** Hydrolysis of the cyanogenic glycosides amygdalin, prunasin and linamarin by ruminal microorganisms, *Can. J. Anim. Sci.,* 67, 1133, 1987.

21. **Smith, R. L.,** The role of the gut flora in the conversion of inactive compounds to active metabolites, in *A Symposium on Mechanisms of Toxicity,* Aldridge, W. N., Ed., Macmillan, London, 1971, 229.

22. **Drasar, B. S. and Hill, M. J.,** *Human Intestinal Flora,* Academic Press, London, 1974, 63.

23. **Winstead, M. B., Ciccarelli, C. A., and Winchell, H. S.,** Liberation of cyanide from α-aminonitriles relative to amygdalin, *J. Pharmacol. Exp. Ther.,* 205, 751, 1978.

24. **Carter, J. H., McLafferty, M. A., and Goldman, P.,** Role of the gastrointestinal microflora in amygdalin (laetrile)-induced cyanide toxicity, *Biochem. Pharmacol.,* 29, 301, 1980.

25. **Frakes, R. A., Sharma, R. P., and Willhite, C. C.,** Comparative metabolism of linamarin and amygdalin in hamsters, *Food Chem. Toxicol.,* 24, 417, 1986.

26. **Flora, K. P., Cradock, J. C., and Ames, M. M.,** A simple method for the estimation of amygdalin in urine, *Res. Commun. Chem. Pathol. Pharmacol.,* 20, 367, 1978.

27. **Rauws, A. G., Olling, M., and Timmerman, A.,** The pharmacokinetics of amygdalin, *Arch. Toxicol.,* 49, 311, 1982.

28. **Rauws, A. G., Olling, M., and Timmerman, A.,** The pharmacokinetics of prunasin, a metabolite of amygdalin, *J. Toxicol. Clin. Toxicol.,* 19, 851, 1983.

29. **Zhang, G.-M. and Jin, B.-Q.,** Pharmacokinetics of amygdalin in rabbits, *Acta Pharmacol. Sin.,* 7, 460, 1986.

30. **Winkler, W. O.,** Report on methods for glucosidal HCN in lima beans, *J. Assoc. Off. Agric. Chem.,* 41, 282, 1958.

31. **Jansz, E. R., JeyaRaj, E. E., Pieris, N., and Abeyratne, D. J.,** Cyanide liberation from linamarin, *J. Natl. Sci. Council Sri Lanka*, 2, 57, 1974.

32. **Barrett, M. D., Hill, D. C., Alexander, J. C., and Zitnak, A.,** Fate of orally dosed linamarin in the rat, *Can. J. Physiol. Pharmacol.*, 55, 134, 1977.

33. **Barrett, M. D. P., Alexander, J. C., and Hill, D. C.,** Effect of linamarin on thiocyanate production and thyroid activity in rats, *J. Toxicol. Environ. Health*, 4, 735, 1978.

34. **Maduagwu, E. N.,** Metabolism of linamarin in rats, *Food Chem. Toxicol.*, 27, 451, 1989.

35. **Maduagwu, E. N. and Umoh, I. B.,** Biliary excretion of linamarin in the Wistar rat after a single dose, *Biochem. Pharmacol.*, 35, 3003, 1986.

36. **Padmanaban, G.,** Lathyrogens, in *Toxic Constituents of Plant Foodstuffs*, 2nd ed., Liener, I. E., Ed., Academic Press, New York, 1980, 239.

37. **Lipton, S. N., Lalich, J. J., and Strong, F. M.,** Identification of cyanoacetic acid as a metabolite of β-aminopropionitrile (BAPN) and other nitriles, *J. Am. Chem. Soc.*, 80, 2022, 1958.

38. **Lipton, S. H., Lalich, J. J., Garbutt, J. T., and Strong, F. M.,** Identification of cyanoacetic acid as a urinary metabolite of β-aminopropionitrile, *J. Am. Chem. Soc.*, 80, 6594, 1958.

39. **Keiser, H. R., Harris, E. D., and Sjoerdsma, A.,** Studies on beta-aminopropionitrile in animals, *Clin. Pharmacol. Ther.*, 8, 587, 1967.

40. **Fleisher, J. H., Speer, D., Brendel, K., and Chvapil, M.,** Effect of pargyline on the metabolism of β-aminopropionitrile (BAPN) by rabbits, *Toxicol. Appl. Pharmacol.*, 47, 61, 1979.

41. **Fleisher, J. H., Arem, A. J., Chvapil, M., and Peacock, E. E.,** Metabolic disposition of β-aminopropionitrile in the rat, *Proc. Soc. Exp. Biol. Med.*, 152, 469, 1976.

42. **Sievert, H. W., Lipton, S. H., and Strong, F. M.,** Quantitative determination of cyanoacetic acid as an enzymic product of β-aminopropionitrile, *Arch. Biochem. Biophys.*, 86, 311, 1960.

43. **Fleisher, J. H., Peacock, E. E., and Chvapil, M.,** Urinary excretion of β-aminopropionitrile and cyanoacetic acid, *Clin. Pharmacol. Ther.*, 23, 520, 1978.

44. **Wallig, M. A., Gould, D. H., and Fettman, M. J.,** Selective pancreatotoxicity in the rat induced by the naturally occurring plant nitrile 1-cyano-2-hydroxy-3-butene, *Food Chem. Toxicol.*, 26, 137, 1988.

45. **Ohkawa, H., Ohkawa, R., Yamamoto, I., and Casida, J. E.,** Enzymatic mechanisms and toxicological significance of hydrogen cyanide liberation from various organothiocyanates and organonitriles in mice and houseflies, *Pestic. Biochem. Physiol.*, 2, 95, 1972.

46. **Kaplita, P. V. and Smith, R. P.,** Pathways for the bioactivation of aliphatic nitriles to free cyanide in mice, *Toxicol. Appl. Pharmacol.*, 84, 533, 1986.

47. **Kohlrausch, A.,** Untersuchungen über das Verhalten von Betain, Trigonellin und Methylpyridylammoniumhydroxyd im tierischen Organismus, *Z. Biol.*, 57, 273, 1912.

48. **Stetten, D.,** Biological relationships of choline, ethanolamine, and related compounds, *J. Biol. Chem.*, 138, 437, 1941.

49. **Ericson, L.-E., Williams, J. N., and Elvehjem, C. A.,** Studies on partially purified betaine-homocysteine transmethylase of liver, *J. Biol. Chem.*, 212, 537, 1955.

50. **Hopper, J. H. and Johnson, B. C.,** Transmethylation reactions *in vivo* and *in vitro* in the young calf, *Proc. Soc. Exp. Biol. Med.*, 91, 497, 1956.

51. **Mitchell, A. D., Chappell, A., and Knox, K. L.,** Metabolism of betaine in the ruminant, *J. Anim. Sci.*, 49, 764, 1979.

52. **Cheema, P. S., Padmanaban, G., and Sarma, P. S.,** Transamination of β-N-oxalyl-L-α,β-diaminopropionic acid, the *Lathyrus sativus* neurotoxin, in tisssues of the rat, *Indian J. Biochem. Biophys.*, 8, 16, 1971.

53. **Von Holt, C., Chang, J., Von Holt, M., and Böhm, H.,** Metabolism and metabolic effects of hypoglycin, *Biochim. Biophys. Acta*, 90, 611, 1964.

54. **Von Holt, C.,** Methylenecyclopropaneacetic acid, a metabolite of hypoglycin, *Biochim. Biophys. Acta*, 125, 1, 1966.

55. **Asatoor, A. M.,** Tea as a source of urinary ethylamine, *Nature (London)*, 210, 1358, 1966.

56. **Roets, E., Verbeke, R., Massart-Leën, A.-M., and Peeters, G.,** Metabolism of [^{14}C]citrulline in the perfused sheep and goat udder, *Biochem. J.*, 144, 435, 1974.

57. **Reiter, A. J. and Horner, W. H.,** Studies on the metabolism of guanidine compounds in mammals. Formation of guanidine and hydroxyguanidine in the rat, *Arch. Biochem. Biophys.*, 197, 126, 1979.

58. **Natelson, S.,** Metabolic relationship between urea and guanidino compounds as studied by automated fluorimetry of guanidino compounds in urine, *Clin. Chem.*, 30, 252, 1984.

59. **Thomas, D. A. and Rosenthal, G. A.,** Metabolism of L-[*guanidinooxy*-^{14}C]canavanine in the rat, *Toxicol. Appl. Pharmacol.*, 91, 406, 1987.

60. **Hegarty, M. P., Schinckel, P. G., and Court, R. D.,** Reaction of sheep to the consumption of *Leucaena glauca* Benth. and to its toxic principle mimosine, *Aust. J. Agric. Res.*, 15, 153, 1964.

61. **Reis, P. J., Tunks, D. A., and Hegarty, M. P.,** Fate of mimosine administered orally to sheep and its effectiveness as a defleecing agent, *Aust. J. Biol. Sci.*, 28, 495, 1975.

62. **Sharry, L. F., O'Keefe, J. H., and Downes, A. M.,** The fate of L-[³H]mimosine in the sheep, *Proc. Aust. Biochem. Soc.,* 14, 45, 1981.

63. **Jones, R. J.,** Leucaena toxicity and the ruminal degradation of mimosine, in *Plant Toxicology Proceedings of the Australia – U.S.A. Poisonous Plants Symposium Brisbane, Australia, May 14-18, 1984,* Seawright, A. A., Hegarty, M. P., James, L. F., and Keeler, R. F., Eds., Queensland Poisonous Plants Committee, Yeerongpilly, Australia, 1985, 111.

64. **Tsai, W.-C. and Ling, K.-H.,** Effect of metals on the absorption and excretion of mimosine and 3,4-dihydroxypyridine in rat in vivo, *J. Formosan Med. Assoc.,* 73, 543, 1974.

65. **Tang, S.-Y. and Ling, K.-H.,** Studies on the metabolism of mimosine, *T'ai-wan I Hseuh Hui Tsa Chih,* 76, 587, 1977 (Chem. Abstr. 88, 99829d, 1978).

66. **Al-Baldawi, N. F.,** Aspects of willardiine and isowillardiine metabolism in living systems, *Iraqi J. Agric. Sci.,* 13, 24, 1978 (Chem. Abstr. 93, 180253m, 1980).

67. **Williams, M. C., Norris, F. A., and Van Kampen, K. R.,** Metabolism of miserotoxin to 3-nitro-1-propanol in bovine and ovine ruminal fluid, *Am. J. Vet. Res.,* 31, 259, 1970.

68. **Majak, W., Pass, M. A., Muir, A. D., and Rode, L. M.,** Absorption of 3-nitropropanol (miserotoxin aglycone) from the compound stomach of cattle, *Toxicol. Lett.,* 23, 9, 1984.

69. **Majak, W., Pass, M. A., and Madryga, F. J.,** Toxicity of misertoxin and its aglycone (3-nitropropanol) to rats, *Toxicol. Lett.,* 19, 171, 1983.

70. **Gustine, D. L., Moyer, B. G., Wangsness, P. J., and Shenk, J. S.,** Ruminal metabolism of 3-nitropropanoyl-D-glucopyranoses from crownvetch, *J. Anim. Sci.,* 44, 1107, 1977.

71. **Majak, W. and Clark, L. J.,** Metabolism of aliphatic nitro compounds in bovine rumen fluid, *Can. J. Anim. Sci.,* 60, 319, 1980.

72. **Majak, W. and Cheng, K.-J.,** Identification of rumen bacteria that anaerobically degrade aliphatic nitrotoxins, *Can. J. Microbiol.,* 27, 646, 1981.

73. **Majak, W., Cheng, K.-J., and Hall, J. W.,** Enhanced degradation of nitropropanol by ruminal microorganisms, *J. Anim. Sci.,* 62, 1072, 1986.

74. **Pass, M. A., Majak, W., Muir, A. D., and Yost, G. S.,** Absorption of 3-nitropropanol and 3-nitropropionic acid from the digestive system of sheep, *Toxicol. Lett.,* 23, 1, 1984.

75. **Muir, A. D., Majak, W., Pass, M. A., and Yost, G. S.,** Conversion of 3-nitropropanol (miserotoxin aglycone) to 3-nitropropionic acid in cattle and sheep, *Toxicol. Lett.,* 20, 137, 1984.

76. **Alston, T. A., Seitz, S. P., and Bright, H. J.,** Conversion of 3-nitro-1-propanol (miserotoxin aglycone) to cytotoxic acrolein by alcohol dehydrogenase, *Biochem. Pharmacol.,* 30, 2719, 1981.

77. **Benn, M. H., McDiarmid, R. E., and Majak, W.,** In-vitro biotransformation of 3-nitropropanol (miserotoxin aglycone) by horse liver alcohol dehydrogenase, *Toxicol. Lett.,* 47, 165, 1989.

78. **Alston, T. A., Mela, L., and Bright, H. J.,** 3-Nitropropionate, the toxic substance of *Indigofera,* is a suicide inactivator of succinate dehydrogenase, *Proc. Natl. Acad. Sci. USA,* 74, 3767, 1977.

79. **Majak, W., Cheng, K.-J., Muir, A. D., and Pass, M. A.,** Analysis and metabolism of nitrotoxins in cattle and sheep, in *Plant Toxicology Proceedings of the Australia – U.S.A. Poisonous Plants Symposium Brisbane, Australia, May 14-18, 1984,* Seawright, A. A., Hegarty, M. P., James, L. F., and Keeler, R. F., Eds., Queensland Poisonous Plants Committee, Yeerongpilly, Australia, 1985, 446.

80. **Schmeiser, H. H., Pool, B. L., and Wiessler, M.,** Identification and mutagenicity of metabolites of aristolochic acid formed by rat liver, *Carcinogenesis,* 7, 59, 1986.

81. **Krumbiegel, G., Hallensleben, J., Mennicke, W. H., Rittman, N., and Roth, H. J.,** Studies on the metabolism of aristolochic acids I and II, *Xenobiotica,* 17, 981, 1987.

82. **Klungsøyr, J. and Scheline, R. R.,** Metabolism in rats of several carboxylic acid derivatives containing the 3,4-methylenedioxyphenyl group, *Acta Pharmacol. Toxicol.,* 49, 305, 1981.

83. **Schmeiser, H. H., Schoepe, K.-B., and Wiessler, M.,** DNA adduct formation of aristolochic acid I and II *in vitro* and *in vivo, Carcinogenesis,* 9, 297, 1988.

84. **Kuhn, R., Köhler, F., and Köhler, L.,** Über die biologische Oxydation hochungsättiger Fettsäuren. Ein neuer Weg zur Darstellung von Polyen-dicarbonsäuren, *Hoppe-Seyler's Z. Physiol. Chem.,* 247, 197, 1937.

85. **Sambaiah, K., Srinivasan, M. R., Vijayalakshmi, R., Rao, M. V. L., Chandrasekhara, N., Satyanarayanan, M. N., and Raghavendra Rao, M. R.,** Biochemistry and metabolism of spices with special reference to tumeric (*Curcuma longa*) and red chillies (*Capsicum* species), Proceedings of the 2nd Symposium of the Federation of Asian and Oceanian Biochemists, Kuala Lumpur, 1979, 50.

86. **Kawada, T., Suzuki, T., Takahashi, M., and Iwai, K.,** Gastrointestinal absorption and metabolism of capsaicin and dihydrocapsaicin in rats, *Toxicol. Appl. Pharmacol.,* 72, 449, 1984.

87. **Kim, N. D. and Park, C. Y.,** Study on the absorption and excretion of capsaicin in rabbits, *J. Pharm. Soc. Korea,* 25, 101, 1981 (Chem. Abstr. 96, 135245m, 1982).

88. **Saria, A., Skofitsch, G., and Lembeck, F.,** Distribution of capsaicin in rat tissues after systemic administration, *J. Pharm. Pharmacol.,* 34, 273, 1982.

89. **Lee, S. S. and Kumar, S.,** Metabolism *in vitro* of capsaicin, a pungent principle of red pepper, with rat liver homogenates, in *Microsomes, Drug Oxidations, and Chemical Carcinogenesis,* Vol. II, Coon, M. J., Conney, A. H., Estabrook, R. W., Gelboin, H. V., Gillette, J. R., and O'Brien, P. J., Eds., Academic Press, New York, 1980, 1009.

90. **Miller, M. S., Brendel, K., Burks, T. F., and Sipes, I. G.,** Interaction of capsaicinoids with drug-metabolizing systems. Relationship to toxicity, *Biochem. Pharmacol.,* 32, 547, 1983.

91. **Kawada, T. and Iwai, K.,** *In vivo* and *in vitro* metabolism of dihydrocapsaicin, a pungent principle of hot pepper, in rats, *Agric. Biol. Chem.,* 49, 441, 1985.

92. **Bhat, B. G. and Chandrasekhara, N.,** Studies on the metabolism of piperine: absorption, tissue distribution and excretion of urinary conjugates in rats, *Toxicology,* 40, 83, 1986.

93. **Bhat, B. G. and Chandrasekhara, N.,** Metabolic disposition of piperine in the rat, *Toxicology,* 44, 99, 1987.

94. **Acheson, R. M. and Atkins, G. L.,** The metabolites of piperic acid and some related compounds in the rat, *Biochem. J.,* 79, 268, 1961.

95. **Boyland, E. and Mawson, E. H.,** The conversion of colchicine into colchiceine, *Biochem. J.,* 32, 1204, 1938.

96. **Wallace, S. L., Omokoku, B., and Ertel, N. H.,** Colchicine plasma levels, *Am. J. Med.,* 48, 443, 1970.

97. **Walaszek, E. J., Kocsis, J. J., Leroy, G. V., and Geiling, E. M. K.,** Studies on the excretion of radioactivie colchicine, *Arch. Int. Pharmacodyn. Ther.,* 125, 371, 1960.

98. **Wallace, S. L. and Ertel, N. H.,** Colchicine: current problems, *Bull. Rheum. Dis.,* 20, 582, 1969.

99. **Achtert, G., Scherrmann, J. M., and Christen, M. O.,** Pharmacokinetics/bioavailability of colchicine in healthy male volunteers, *Eur. J. Drug Metab. Pharmacokinet.,* 14, 317, 1989.

100. **Brues, A. M.,** The fate of colchicine in the body, *J. Clin. Invest.,* 21, 646, 1942.

101. **Fleischmann, W., Price, H. G., and Fleischmann, S. K.,** Fate of intraperitoneally administered colchicine in the golden hamster, *Med. Pharmacol. Exp.,* 12, 172, 1965.

102. **Fleischmann, W., Price, H. G., and Fleischmann, S. K.,** Pathways of excretion of colchicine in the golden hamster, *Pharmacology,* 1, 48, 1968.

103. **Fleischmann, W., Price, H. G., and Fleischmann, S. K.,** Fate of intraperitoneally administered colchicine in the Mongolian gerbil, *Med. Pharmacol. Exp.,* 17, 323, 1967.

104. **Hunter, A. L. and Klaassen, C. D.,** Biliary excretion of colchicine, *J. Pharmacol. Exp. Ther.,* 192, 605, 1975.

105. **Hunter, A. L. and Klaassen, C. D.,** Biliary excretion of colchicine in newborn rats, *Drug Metab. Dispos.,* 3, 530, 1975.

106. **Axelrod, J.,** The enzymic cleavage of aromatic ethers, *Biochem. J.,* 63, 634, 1956.

107. **Schönharting, M., Mende, G., and Siebert, G.,** Metabolic transformation of colchicine. II. The metabolism of colchicine by mammalian liver microsomes, *Hoppe-Seyler's Z. Physiol. Chem.,* 355, 1391, 1974.

108. **Schönharting, M., Pfaender, P., Rieker, A., and Siebert, G.,** Metabolic transformation of colchicine. I. The oxidative formation of products from colchicine in the Udenfriend system, *Hoppe-Seyler's Z. Physiol. Chem.,* 354, 421, 1973.

109. **Lukič, V., Walterová, D., Husek, A., Gašič, O., and Simánek, V.,** Biotransformation of demecolcine by rat liver microsomes, *Acta Univ. Palacki. Olomuc., Fac. Med.,* 120, 429, 1988.

110. **Laqueur, G. L.,** Oncogenicity of cycads and its implications, *Adv. Mod. Toxicol.,* 3, 231, 1977.

111. **Wogan, G. N. and Busby, W. F.,** Naturally occurring carcinogens, in *Toxic Constituents of Plant Foodstuffs,* 2nd ed., Liener, I. E., Ed., Academic Press, New York, 1980, 329.

112. **Kobayashi, A. and Matsumoto, H.,** Studies on methylazoxymethanol, the aglycone of cycasin. Isolation, biological, and chemical properties, *Arch. Biochem. Biophys.,* 110, 373, 1965.

113. **Laqueur, G. L.,** Carcinogenic effects of cycad meal and cycasin, methylaxoxymethanol glycoside, in rats and effects of cycasin in germfree rats, *Fed. Proc. Fed. Am. Soc. Exp. Biol.,* 23, 1386, 1964.

114. **Spatz, M., McDaniel, E. G., and Laqueur, G. L.,** Cycasin excretion in conventional and germfree rats, *Proc. Soc. Exp. Biol. Med.,* 121, 417, 1966.

115. **Spatz, M., Smith, D. W. E., McDaniel, E. G., and Laqueur, G. L.,** Role of intestinal microorganisms in determining cycasin toxicity, *Proc. Soc. Exp. Biol. Med.,* 124, 691, 1967.

116. **Feinberg, A. and Zedeck, M. S.,** Production of a highly reactive alkylating agent from the organospecific carcinogen methylazoxymethanol by alcohol dehydrogenase, *Cancer Res.,* 40, 4446, 1980.

METABOLISM OF NITROGEN HETEROCYCLIC COMPOUNDS

PYRROLIDINE DERIVATIVES

Alkaloids based on pyrrolidine are usually subdivided into simple derivatives of pyrrolidine (tetrahydropyrrole) (1) and more complex types containing the pyrrolidine or pyrrole ring systems. Of the first type, only a few compounds are known (e.g., hygrine, cuscohygrine, and stachydrine) but no reports on their metabolism are available. Nicotine and several related tobacco alkaloids contain the pyrrolidine ring. These compounds are included in this section because it is their pyrrolidine moiety which is the site of greatest metabolic activity. Additional tobacco alkaloids contain the piperidine ring rather than the pyrrolidine ring and are therefore included in the following section. The more complex types of pyrrolidine derivatives include the tropane, indole, and pyrrolizidine alkaloids which are covered below in the appropriate sections.

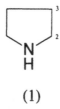

(1)

The tobacco alkaloids typically contain a pyridine ring attached at the β-position to the α-position of a pyrrolidine or piperidine ring. The metabolism of a few examples of the latter type are summarized in the following section. The most important example of the former type is S-(–)-**nicotine** (2) (see Figure 1), the chief alkaloid in tobacco. Many studies of its biological disposition have been carried out during the past five decades with the result that the literature on this subject is extensive. Exhaustive coverage of this wealth of data is obviously beyond the scope of the present summary, however more detailed treatments of the earlier findings are available in the works of Larson et al.,[1] Larson and Silvette,[2-4] and McKennis.[5] More recent reviews of nicotine metabolism include those of Gorrod and Jenner,[6] Schievelbein,[7] and Nakayama.[8] The last of these emphasized the enzyme systems responsible for the metabolism of nicotine. An additional review of nicotine metabolism included a tabular compilation of the metabolites of nicotine, together with the animal species and experimental conditions under which they were found.[9]

It is well established that only a minor part of the elimination of nicotine from the body can be ascribed to the urinary excretion of unchanged compound.[10,11] However, the extent of this excretion is dependent upon the urinary pH. Haag and Larson[12] found that nicotine excretion was about four times as great when the urine was acidic than when it was alkaline. Beckett et al.[13] showed that an increase in the urinary pH to a range of 7.5 to 8.5 caused the excretion of nicotine in man to fall to zero. In contrast, reducing the pH to 4.8 resulted in excretion values of 9 to 28% of the dose (2 mg, p.o.). Similar results were found in a group of cigarette smokers.[14] Whereas no significant amount of nicotine was excreted in the urine when the pH was above 7.5, substantial amounts were excreted when the pH was <7.4. Another variable which may affect the urinary excretion of unchanged nicotine is the dose. Finnegan et al.[11] reported a linear increase with increasing dosage when they administered relatively large doses (15 to 48 mg/kg) to dogs. On the other hand, Kyerematen et al.[15] found no dose-related changes in the urinary excretion of nicotine in rats. The latter experiment employed a range of doses (1 to 10 mg/kg) which was closer to that normally used in metabolism studies of nicotine. In general, values of 5 to 15% of the dose may be considered as reasonably normal for the urinary excretion of unchanged nicotine, however a range of 1 to 35% has been noted in some cases.[16]

The pathways in the mammalian metabolism of (–)-nicotine (2) are shown in Figure 1, which

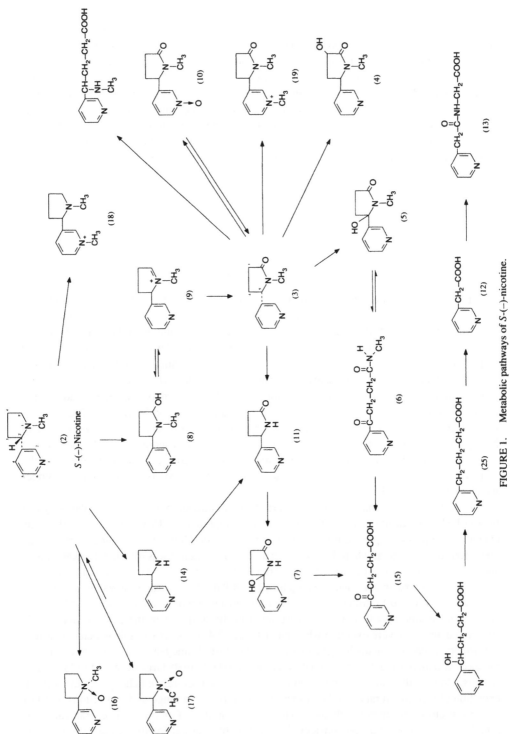

FIGURE 1.   Metabolic pathways of *S*-(−)-nicotine.

outlines the most important features of nicotine metabolism. The compounds illustrated are metabolites identified in a large number of *in vivo* and *in vitro* studies in several animal species in which not only nicotine, but also cotinine (3) and several other metabolites were employed. Thus, the diverse spectrum of metabolites shown represents the possibilities available rather than the actual findings in any one study. A large number of nicotine metabolites may nonetheless be detected when sensitive and selective methods are used. Kyerematen et al.[15] identified nicotine and 12 metabolites in the urine of rats given nicotine. These products were formed via one or more of all of the important pathways shown in Figure 1 including ring *C*-oxidation followed by degradation to pyridyl acids, *N*-demethylation, *N*-oxidation, and *N*-methylation to quaternary derivatives. These four main routes of metabolism will serve as the basis for the following summary of nicotine metabolism.

A major pathway in the metabolism of nicotine is its *C*-oxidation to cotinine (3). This reaction has been repeatedly demonstrated in many *in vitro* and *in vivo* studies using a large number of animal species. In contrast to that found with nicotine, the urinary excretion of cotinine is not influenced by the urinary pH.[17] Representative studies using *S*-(–)-nicotine indicated urinary cotinine excretion values (% of dose) of ~7 to 9% in rats,[15,18] 0% in hamsters,[18] 2% in guinea pigs,[18] 8 to 10% in rabbits,[18,19] and 5 to 20% in man.[13,20] However, the importance of this metabolic pathway cannot be judged by the extent of cotinine formation alone. As shown in Figure 1, further extensive metabolism of cotinine occurs and some of these products may contribute significantly to the overall recovery of metabolites. This is especially true with 3′-hydroxycotinine (4) which is a major metabolite of nicotine in several animal species. It should be noted that the numbering of the pyrrolidine ring in cotinine differs from that used with nicotine, the 3′-position of cotinine corresponding to the 4′-position in nicotine. Nwosu and Crooks[18] found that the urinary excretion of 3′-hydroxycotinine following nicotine administration was higher than that of cotinine in rats, hamsters, guinea pigs, and rabbits. The most striking examples were seen in hamsters and guinea pigs which excreted little or no cotinine but 37 and 24% of the dose, respectively, as the 3′-hydroxy derivative. Also, 3′-hydroxycotinine was the major urinary metabolite when *S*-(–)-cotinine itself was given to guinea pigs.[21] 3′-Hydroxycotinine has also been shown to be a major metabolite of nicotine in smokers.[22-24] Neurath and Pein[25] calculated that ~20% of the nicotine absorbed by smokers was excreted as 3′-hydroxycotinine. This hydroxycotinine was shown to have the *trans* configuration.[23,26] An additional hydroxylated derivative of cotinine is 5′-hydroxycotinine (5) (allohydroxycotinine). This hydroxy lactam is tautomeric with the open-chain keto amide (6) (γ-(3-pyridyl)-γ-oxo-*N*-methylbutyramide) which has been detected as a urinary metabolite of nicotine or cotinine in mice,[27] rats[28,29] and monkeys.[30] The last study showed that the equilibrium mixture consisted mainly of the cyclic tautomer (5). The corresponding *N*-demethylated metabolite, allohydroxydemethylcotinine (7) was recently detected as a minor metabolite of nicotine in rats.[15]

As shown in Figure 1, cotinine is formed by two sequential reactions. This pathway, first proposed by Hucker et al.,[31] involves an initial oxidation at the α-position of the pyrrolidine ring to form 5′-hydroxynicotine (8). This reaction is dependent on the cytochrome P-450 system and is followed by the action of an aldehyde oxidase which converts the intermediate to cotinine. Murphy[32] showed that the oxidized intermediate is nicotine $\Delta^{1'(5')}$ iminium ion (9). The subsequent availability of this intermediate compound has shown that it is the predominating intermediate species[33] and that it is readily transformed by aldehyde oxidase to cotinine.[34] This was confirmed by Gorrod and Hibberd[35] who also showed that the iminium ion (9) was converted to cotinine when administered to rats or guinea pigs. A study of the enzymology of this conversion added further evidence for the role of this hepatic soluble enzyme, aldehyde oxidase.[36] However, Obach and Van Vunakis[37] reported that a second pathway exists for the formation of cotinine from the iminium ion (9). This reaction was catalyzed by rabbit liver microsomes and required NAD. Biochemical studies showed that it differed in several respects from aldehyde oxidase. Although this system had a lower capacity to oxidize the iminium ion than that shown by the soluble system, it was believed that it may an important role *in vivo* because of its location in the endoplasmatic reticulum where the iminium ion is formed. Liver fractions or microsomes have been used to study the stereochemistry of the microsomal oxidation of S-(–)-nicotine[38] and the effect of inhibitors on the principal microsomal reactions (i.e., 5′-hydroxylation and N-oxidation).[39] McCoy et al.[40] recently reported on the ability of six purified cytochrome P-450 isoenzymes from rabbit liver to carry out these reactions.

Figure 1 also shows that the cotinine formed from nicotine may be metabolized along several additional pathways. These include the formation of the N-oxide (10) of cotinine which, unlike the corresponding derivative of nicotine, is oxidized at the pyridine nitrogen. It is also evident that cotinine, and its demethylated derivative (11), are subject to scission of the pyrrolidine ring to give pyridyl acids. The ultimate metabolite of nicotine along this route is 3-pyridylacetic acid (12), however its glycine conjugate (13) has also been described. [41]

The N-demethylation of nicotine to nornicotine (14) has been demonstrated in many *in vitro* and *in vivo* studies, however some investigations have failed to show its formation.[6] Several early studies using [*methyl*-^{14}C]nicotine showed that $^{14}CO_2$ was produced *in vitro* with mouse tissue slices[42] and *in vivo* in mice and rats.[28,43] However, nornicotine was not specifically detected in these investigations and $^{14}CO_2$ formation is also possible at other stages in the metabolism of nicotine, e.g., the demethylation of cotinine (3) to demethylcotinine (11) or in the further metabolism of methylamine formed when the amide (6) is hydrolyzed to the acid (15). Several recent studies have shown that nornicotine is a minor urinary metabolite (~2 to 3% of the dose) of nicotine in guinea pigs[44,45] and accounts for 7 to 9% of the dose in rats.[15] On the other hand, microsomes from hamster liver which showed both N'-oxidase and 5′-hydroxylase activity did not demethylate nicotine to nornicotine.[39] Studies with rabbit liver post-mitochondrial supernatant fractions indicated that nicotine demethylation occurred via an intermediate carbinol (not shown in Figure 1).[46,47] This reaction may proceed via an iminium ion in an manner analogous to that seen in the conversion of nicotine to cotinine (3) in which the 5′-hydroxynicotine intermediate (8) is converted to the isomeric iminium ion (9).

The formation of N-oxides of nicotine has been demonstrated *in vitro* in numerous organs and *in vivo* in many mammalian species, however the extent of this reaction is highly variable. Typical values for the urinary excretion of nicotine-1′-N-oxide are ~2% of the dose in mice[48] and 3 to 4% in man.[13] A comparative study in four mammalian species[18] showed that, whereas rats and guinea pigs excreted ~20 to 25% of the injected dose of S-(–)-nicotine as the N-oxide, the metabolite was absent from the urine of hamsters and rabbits. Both the *cis* (16) and *trans* (17) diastereoisomers of nicotine-1′-N-oxide are formed.[13,49-51] Their formation is stereospecific and the ratios of isomers produced varies in different tissues and animal species. Marked species differences in the *in vitro* metabolism of both (–)-nicotine and (+)-nicotine to their diastereoisomeric N-oxides have been encountered.[52] The stereochemistry of the diastereoisomers of nicotine-1′-oxide formed from the (–)- and (+)-isomers of nicotine was investigated by Beckett et al.[53] and a reappraisal of some of the data was given by Testa et al.[54] Another point of interest with nicotine-1′-N-oxide is that it can be reduced to nicotine. The reducing conditions essential for this reaction are found in the lower intestine[55,56] and *in vitro* studies using rat intestinal contents demonstrated considerable reduction of the N-oxide to nicotine.[57] The latter investigation, which also utilized germ-free rats, pointed additionally to the involvement of the liver and/or other tissues in the reduction of nicotine-1′-N-oxide. Other reports have shown that many tissues, and especially the liver, can reduce the N-oxide to nicotine under anaerobic conditions.[50,58,59] Cotinine-N-oxide (10), a minor metabolite of nicotine in humans,[60] similarly undergoes reduction. Yi et al.,[61] found that it was excreted in the urine of rabbits and dogs partly unchanged but also as cotinine and several further metabolites of cotinine.

The N-oxidation of nicotine, which Gorrod et al.[62] showed was carried out by a enzyme system different than that responsible for C-oxidation, was studied using a flavin-containing monooxygenase from pig liver microsomes.[63] No stereoselectivity was found in the formation of the *cis* and *trans* N'-oxides when S-(−)-nicotine was used, however only the *trans* diastereoisomer was formed from R-(+)-nicotine.

The N-methylation of the pyridine ring of nicotine was first demonstrated by McKennis et al.[64] They detected isomethylnicotinium ion (N-methylnicotinium ion) (18) in the urine of dogs and also found that cotinine was similarly methylated to cotinine methonium ion (19). A series of recent investigations showed that the methylation of nicotine is species dependent and that regiospecific and stereospecific factors are involved. Most of these studies were carried out in guinea pigs and it was found that the enzyme responsible was a cytosolic, S-adenosylmethionine-dependent N-methyltransferase present in lung, liver, and other tissues.[65,66] Interestingly, this enzyme from guinea pigs showed absolute stereospecificity for R-(+)-nicotine and no methylation was found with the natural S-(−)-isomer. In fact, it was found that the S-(−)-enantiomer acted as a competitive inhibitor of the methylation of R-(+)-nicotine.[67] Nwosu and Crooks[18] reported that no methylated metabolite of S-(−)-nicotine could be detected in the urine from the rat, hamster, guinea pig, or rabbit. The N-methylation of R-(+)-nicotine was confirmed *in vivo* using tritiated material and the N-methylated metabolite accounted for ~10% of the urinary radioactivity.[68] Pool and Crooks[69] showed that N-demethylation of this metabolite did not occur *in vivo* in the guinea pig. They used ^{14}C-N-CH$_3$-labeled material and detected no ^{14}CO$_2$ in the expired air. However, N-methylnicotinium ion was transformed in the guinea pig by N'-demethylation to N-methylnornicotinium ion and by oxidation to N-methylcotininium ion.[70,71] The total urinary excretion of these quaternary ammonium metabolites accounted for 15 to 20% of the dose of R-(+)-nicotine.[45] R-(+)-nicotine and R-(+)-N-methylnicotinium ion were also shown to be oxidized at the N'-position to give the *cis* and *trans* diastereoisomers of N-methyl-N'-oxonicotinium ion.[72] This reaction was not seen with S-(−)-nicotine. Neither cytochrome P-450 nor the flavin-containing monooxygenase, which normally effects N-oxidation of tertiary amines, was able to catalyze the formation of these oxo compounds.[63]

Additional studies on the N-methylation of nicotine enantiomers have been carried out using liver cytosol from rats and humans[73] or purified N-methyltransferases from rabbit liver.[74] Whereas human liver cytosol methylated both the R-(+)- and S-(−)-isomers, the cytosol from rats showed no activity towards either enantiomer. Both of the purified N-methyltransferases from rabbit liver methylated R-(+)-nicotine at the pyridyl nitrogen, however when used together some methylation of the pyrrolidine nitrogen was also noted. The mixture of N-methyltransferases also methylated S-(−)-nicotine at the pyridyl nitrogen, a reaction which was not detected when the enzymes were used separately.

In addition to the metabolic pathways of nicotine shown in Figure 1, occasional mention has been made of other metabolites. This topic was discussed by McKennis.[5] Total recovery of the administered dose of nicotine is seldom attained and additional pathways may exist. For example, Shen and Van Vunakis[75,76] showed that nicotine, in the presence of NADase, can be exchanged for the nicotinamide moiety of NAD and NADP. These analogues are inactive as coenzymes but can act as competitive inhibitors of them with certain dehydrogenases. Similar analogues were formed from cotinine[77] and it was found that cotinine nucleotides were present in the tissues of rabbits injected with nicotine or cotinine.[78] An additional pathway appears to involve the covalent binding of nicotine metabolites to macromolecules.[79] Binding to microsomal macromolecules was detected using preparations from rabbit liver and lung and human liver. This binding, which was not of a high magnitude, appeared to be associated with cytochrome P-450 activity and not with flavin-containing monooxygenase activity which is responsible for the N-oxidation of nicotine. A correlation was noted between the extent of covalent binding and the extent of formation of S-nicotine-$\Delta^{1',5'}$-iminium ion (9), however it was not determined if this metabolite is the actual reactive intermediate involved. Interestingly, Obach and Van Vunakis[80] showed that this iminium ion can bind nonenzymatically to sulfhydryl groups of polycysteine.

The subject of sex differences in nicotine metabolism was discussed by Kyerematen et al.[81]

who also investigated these differences *in vitro* and *in vivo* in rats. They found that urinary recoveries of unchanged nicotine were higher in female rats and that the urinary output of nicotine metabolites was greater in male rats. The latter finding was consistent with the enhanced *C*- and *N*-oxidation which was recorded *in vitro* in male rats. The effect of age on nicotine metabolism was studied in mice.[82] No such differences were found when the liver microsomal enzymes from animals 2, 12, or 18 months of age were used.

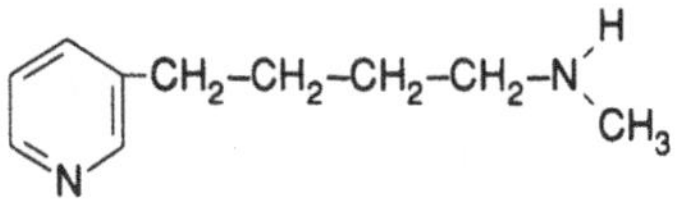

(20)

Dihydrometanicotine

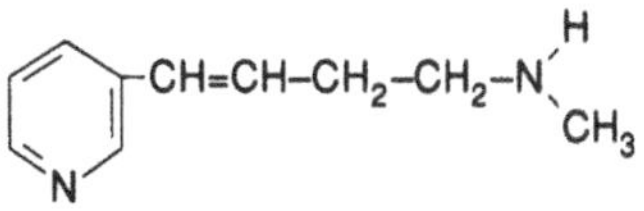

(21)

Metanicotine

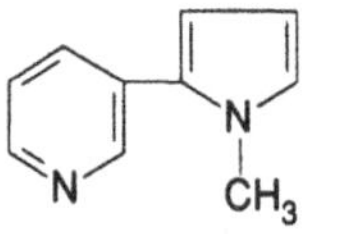

(22)

β-Nicotyrine

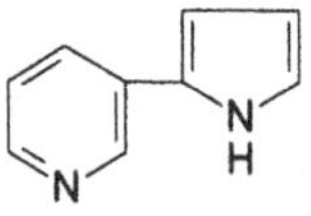

(23)

β-Nornicotyrine

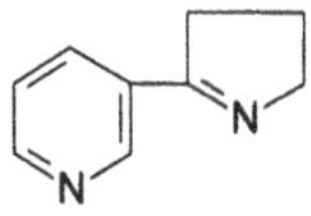

(24)

Myosmine

In addition to nicotine, tobacco contains small quantities of closely related alkaloids which are based on the pyridine and pyrrolidine moieties. Those for which some metabolic information is available include (–)-nornicotine (14), dihydrometanicotine (20), metanicotine (21), β-nicotyrine (22), β-nornicotyrine (23), and myosmine (24).

The urinary excretion of unchanged **nornicotine** (14) (see Figure 1) by dogs was found to be several times greater than that seen with nicotine.[83] The extent of excretion was dose-dependent and reached values of ~60% at high levels (~20 mg/kg). The more extensive urinary excretion of unchanged nornicotine compared with unchanged nicotine was also found in humans.[84] Also, excretion was enhanced when the urinary pH was reduced. The metabolism of nornicotine in dogs was shown to give γ-(3-pyridyl)-γ-aminobutyric acid (not shown in Figure 1) and norcotinine (11) as urinary products.[85] Axelrod[86] showed that the *N*-methylation of nornicotine to nicotine was carried out by an enzyme system from rabbit lung which required *S*-adenosylmethionine. Lower activities were found in other tissues from rabbits, however none was detected in the lungs of several animal species including man.

The minor tobacco alkaloid **dihydrometanicotine** (20) was shown by Meacham et al.[87] to be metabolized to the pyridyl acids (25) and (12) (see Figure 1) which were excreted in the urine by rats and dogs. **Metanicotine** (21), the dehydro derivative of compound (20), underwent similar metabolism. When the *trans* isomer (15 mg/kg) was administered orally to dogs, the urine was found to contain pyridyl acids including 3-pyridylacetic acid (12).[88]

In contrast to that observed with many tobacco alkaloids, neither β-**nicotyrine** (22) or β-**nornicotyrine** (23) was found to be excreted in unchanged form in the urine following oral administration in man.[84] The *in vitro* metabolism of β-nicotyrine by hepatic preparations from several animal species was studied by Jenner and Gorrod.[51] Extensive, and in some cases complete, metabolism of this compound occurred in all species. However, the nature of the metabolic products was not ascertained.

Myosmine (24) is another minor tobacco alkaloid which was not found to be excreted unchanged when administered orally to man.[84]

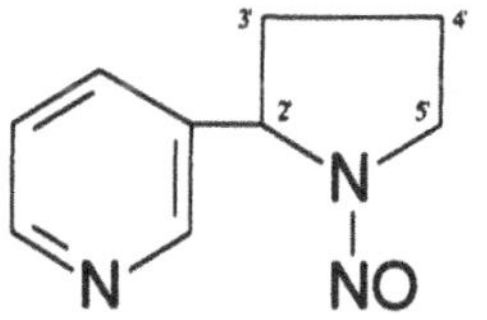

(26) 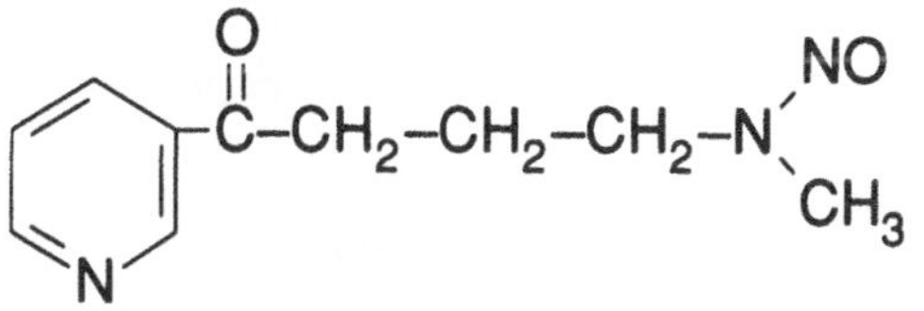   (27)

Nicotine and other tobacco alkaloids undergo nitrosation reactions during curing and processing. Nicotine may thus be converted to N'-nitrosonornicotine (26) and 4-(methylnitrosoamino)-1-(3-pyridyl)-1-butanone (27). The metabolism in rats of (26) leads to α-hydroxylation of the pyrrolidine ring and further to ring-opened products.[89] Also, it undergoes β-hydroxylation at the 3'- and 4'-positions as well as N'oxidation of the pyridyl moiety.[90] The urinary metabolites of N'-nitrosonornicotine were studied in mice and golden hamsters and also in rats given a wide range of doses.[91] Its metabolism was subsequently studied in the miniature pig.[92] Similar studies of the metabolism of (methylnitrosoamino)-1-(3-pyridyl)-1-butanone (27) have been carried out in rats[93] and golden hamsters.[94] The reactions of α-hydroxylation, N-oxidation, and carbonyl reduction of compound (27) were demonstrated in *in vitro* studies with respiratory tissues from golden hamsters.[95] The effects of various inducers or inhibitors of monooxygenase activity on these metabolic pathways were studied.[96] Carbonyl reduction of compound (27) was a major metabolic pathway in fetal lung tissue of hamsters and it was found that guanine residues in DNA were methylated.[97] Both of these nitroso compounds were found to form adducts with hemoglobin in rats and it is likely that this reaction results from the formation of α-hydroxy intermediates.[98] Hoffmann and Hecht[99] reviewed the subject of nicotine-derived N-nitrosamines including their metabolic activation.

PIPERIDINE DERIVATIVES

Pipecolic acid (28) is of metabolic interest because it is an intermediate in the pathway from lysine to α-aminoadipic acid (29) and therefore a normal constituent found in blood and urine.[100,101] In addition, pipecolic acid is present in many plants, often in the fruits and seeds. Intraperitoneal administration of the DL-compound to rats resulted in its excretion unchanged in the urine, no metabolites being detected.[102] However, small amounts of unidentified compounds were formed when it was incubated with rat liver homogenates. Other *in vitro* experiments employing guinea pig brain and kidney homogenates have indicated that pipecolic acid can undergo decarboxylation to piperidine.[103]

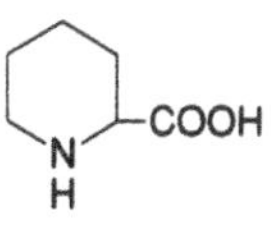

(28)   (29) 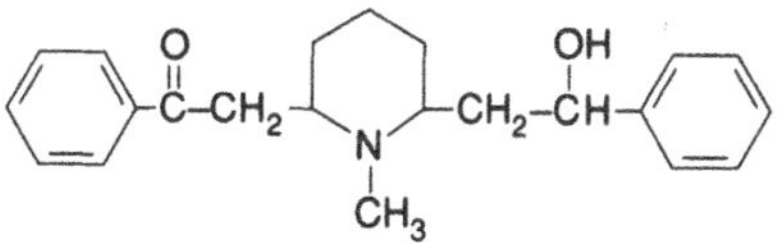   (30)

Pipecolic acid   Lobeline

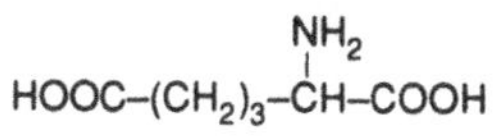

Uehleke[104] reported that **lobeline** (30) disappeared rapidly from the blood of rats following i.v. administration and that some unchanged compound was excreted in the urine. In addition, two unidentified metabolites including one having phenolic properties were excreted.

The metabolism of **arecoline** (31) and **arecaidine** (32), two of the betel or *Areca* alkaloids, was investigated in rats by Boyland and Nery[105] and Nery.[106] These findings are summarized in Figure 2 which shows that arecoline is metabolized along three pathways. These include ester

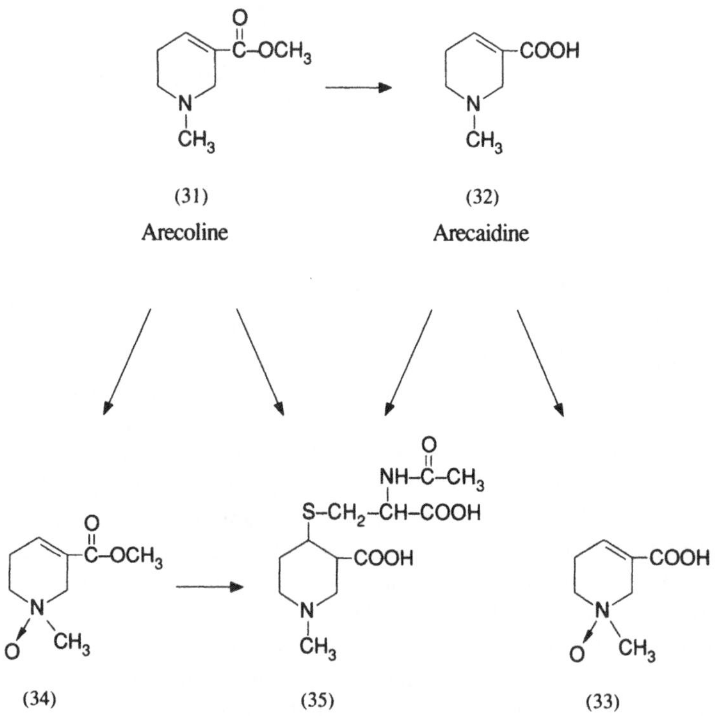

FIGURE 2.   Metabolic pathways of arecoline and arecaidine.

hydrolysis to give arecaidine (32), N-oxide formation with or without ester hydrolysis to give metabolites (33) and (34), and conjugation of the α,β-unsaturated moiety with glutathione followed by subsequent conversion to the mercapturic acid (35). Conjugation with glutathione occurs spontaneously. Arecoline was shown to be rapidly hydrolyzed to arecaidine by rat liver homogenates.[107] Rats given ³H-labeled arecoline orally excreted little radioactivity in the feces and ~20% of the dose in the urine within 18 h.[106] This material was mainly unchanged compound (5 to 6%) and metabolite (33) (~4%). The remaining metabolites and also an unidentified metabolite each accounted for 1 to 2% of the dose. However, it was felt that the actual extent of production of N-oxides was greater due to the ease of the *in vivo* reduction of these compounds to the tertiary amines.

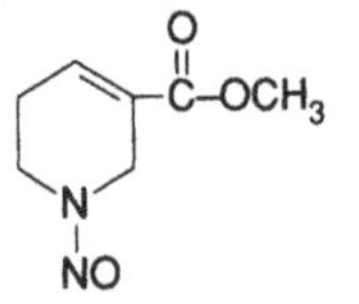

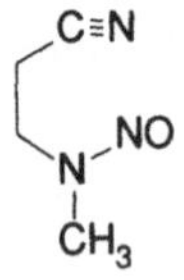

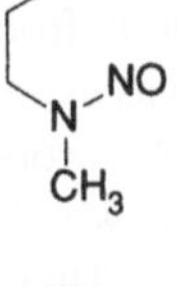

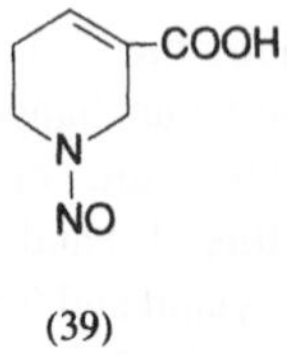

The possible formation of carcinogenic nitrosamines from arecoline has received attention because of the finding that the saliva of users of betel quid contained nitrite.[108] Wenke and Hoffmann[109] showed that nitrosation of arecoline *in vitro* resulted in the formation of *N*-nitrosoguvacoline (36), 3-(methylnitrosamino)propionitrile (37) and 3-(methylnitrosamino)propionaldehyde (38). Subsequent studies[110,111] showed that the saliva from individuals who chewed betel quid contained *N*-nitrosoguvacoline (36), its hydrolysis product *N*-nitrosoguvacine (39) and 3-(methylnitrosamino)propionitrile (37).

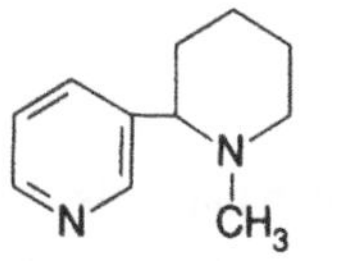

(40)

Methylanabasine

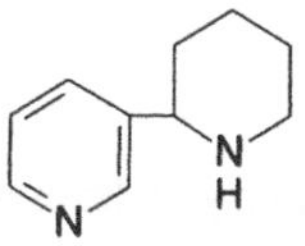

(41)

Anabasine

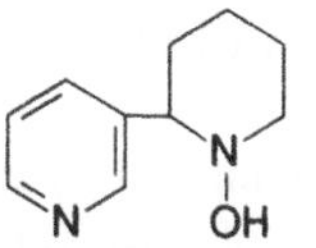

(42)

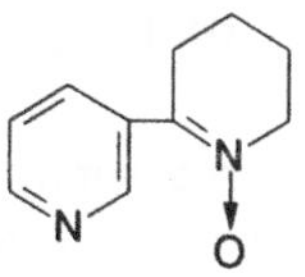

(43)

Methylanabasine (40) was reported by Gvishiani[112] to be metabolized by the liver following its i.v. administration to cats, however the metabolic products were not investigated. Beckett and Sheikh[113] studied the metabolism of both (–)-**anabasine** (41) and (–)-methylanabasine in liver and lung homogenates from rats, rabbits, and guinea pigs. The main metabolites of (–)-methylanabasine in both tissue preparations were its two diastereoisomeric *N*-oxides. Also formed, especially with the liver preparations, were some *N*-demethylated compound, anabasine, and its *N*-oxidation products 1′-hydroxyanabasine (42) and anabasine-1′Δ-nitrone (43). Metabolites (42) and (43) were also formed when (–)-anabasine was used as the substrate in the incubations. Similar experiments with the *N*-hydroxy compound (42) showed that it was converted to the nitrone (43) whereas the latter compound was not metabolized. The metabolism of (–)-methylanabasine (40) by liver preparations from mice, rats, hamster, guinea pigs, and rabbits was also studied by Jenner and Gorrod.[51] The formation of *N*-oxides was small in all species except the guinea pig. Furthermore, no evidence was obtained for the *N*-demethylation of methylanabasine to anabasine or for the α-*C*-oxidation of the substrate to a cotinine-like metabolite. The former finding thus differs from that reported by Beckett and Sheikh.[113] The extent of urinary excretion of unchanged (–)-anabasine and (–)-methylanabasine following oral administration to man was shown to be dependent on urinary pH.[84] In a typical experiment, 28% of the dose (2 mg) of anabasine was recovered unchanged when the urinary pH was allowed to fluctuate between 5.2 and 7.6. This value increased to 70% when the pH was maintained at 4.8. The corresponding values for methylanabasine, a weaker base, were ~4% and 17%. No *N*-oxides of these alkaloids were detected in the urine and methylanabasine was not converted to *N*-demethylated metabolites.

PYRIDINE DERIVATIVES

Tobacco alkaloids furnish the most important group of pyridine derivatives. Because they also contain pyrrolidine or piperidine moieties, which provide the most active sites of metabolism, these compounds are treated above.

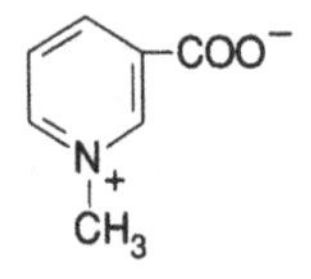

(44)

Trigonelline

Trigonelline (44), of fairly widespread occurrence in plants, is also a urinary metabolite of nicotinic acid. An early report[114] indicated that injected trigonelline was excreted unchanged in the urine of rabbits and cats. This was confirmed in rabbits by Chattopadhyay et al.[115] who also showed that trigonelline was not excreted in the feces and that it was not demethylated to nicotinic acid. However, only about half of the injected trigonelline was recovered in the urine and it was assumed that some was metabolized.

PYRAZINE DERIVATIVES

Pyrazine compounds are noteworthy because of their high odor potency. The only report on the metabolic fate of this group of compounds is that of Hawksworth and Scheline[116] who investigated the metabolism of four alkyl- and two alkoxy-derivatives of pyrazine (45) in rats. When **2-methyl-**, **2,3-dimethyl-**, **2,5-dimethyl-,** or **2,6-dimethylpyrazine** were administered orally at a dose level of 100 mg/kg it was found that they were oxidized to monocarboxylic acids which were then excreted in the urine as such or sometimes partly as their glycine conjugates. This oxidative pathway accounted for nearly all of the dose with all of these pyrazines except the 2,3-dimethyl derivative. In the latter case 10 to 15% of the dose was oxidized to 2-methylpyrazine-3-carboxylic acid. Ring hydroxylation, a reaction which was not detected with the three other alkyl pyrazines, resulted in nearly 40% of the dose being excreted as conjugates of 2,3-dimethyl-5-hydroxypyrazine.

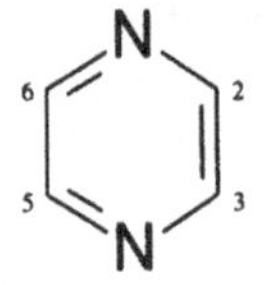

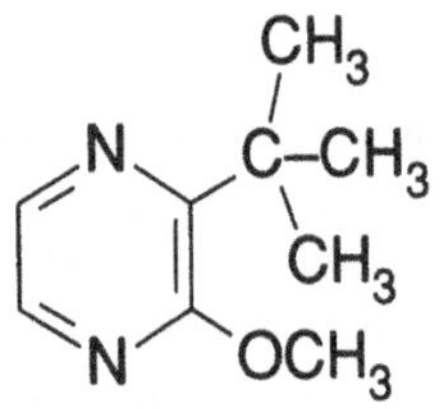

(45)   (46)

2-Isobutyl-3-methoxypyrazine

The two methoxypyrazines investigated were 2-methoxypyrazine and **2-isobutyl-3-methoxypyrazine** (46), the latter compound being the major characteristic aroma component of bell peppers. While less precise quantitative information was obtained with these compounds, *O*-demethylation was noted in both cases. With the 2-methoxy compound, ~20% of the identified urinary metabolites consisted of 2-hydroxypyrazine while the remainder was a single monohydroxy derivative of 2-methoxypyrazine. Both metabolites were excreted as glucuronide and/or sulfate conjugates. With the bell pepper compound, the demethylated product was a major urinary metabolite. Aliphatic side-chain oxidation to 2-methoxy-3-(2-

carboxypropyl)pyrazine was also noted, but this reaction was not extensive. In this regard the compound resembled 2,3- rather than 2,5- or 2,6-dimethylpyrazine. Three unidentified urinary metabolites were detected but these accounted for only a minor part of the dose. These appeared to be compounds in which the isobutyl group had undergone modification but in which the methoxy group was intact and ring hydroxylation was lacking.

TROPANE DERIVATIVES

Plant compounds based on tropane (47) are usually divided into two main groups, the solanaceous alkaloids and the coca alkaloids. The most common compounds in the solanaceous group are (–)-hyoscyamine (48) and its 6,7-epoxide derivative, (–)-scopolamine (hyoscine). These compounds are esters of tropic acid (49) and the amino alcohols tropine (50) or scopine (51), respectively. Racemization of the esters occurs fairly readily and dl-hyoscyamine, known as atropine, is therefore generally used rather than the (–)-isomer.

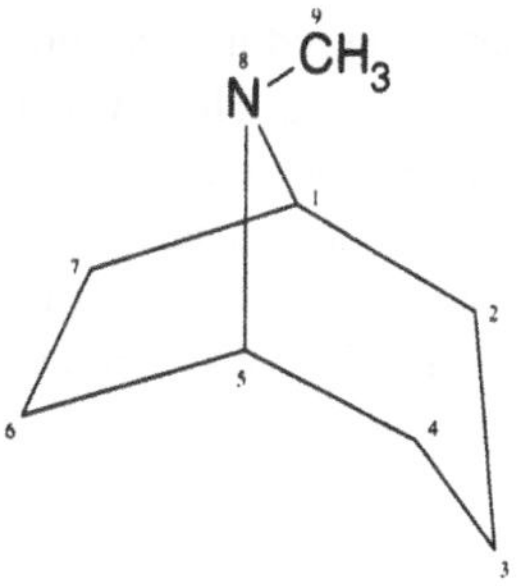

(47)

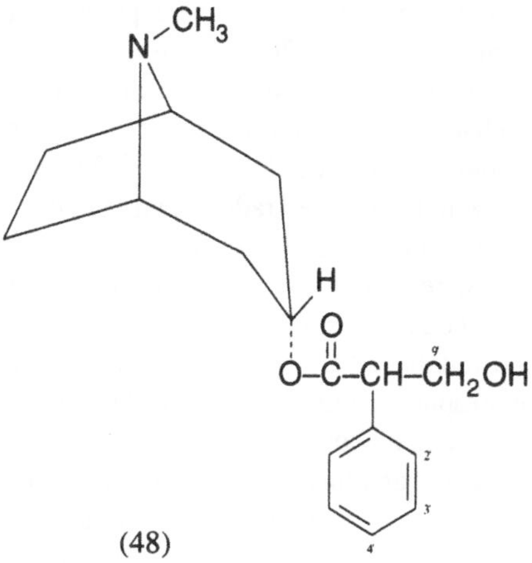

$O-\overset{O}{\overset{\|}{C}}-CH-CH_2OH$

(48)

(–)-Hyoscyamine

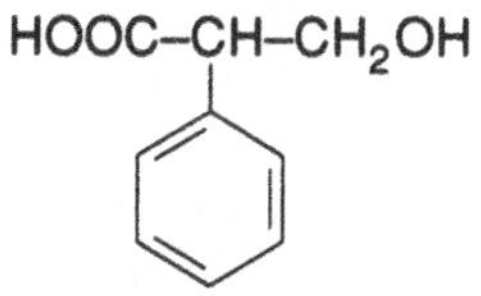

$HOOC-CH-CH_2OH$

(49)

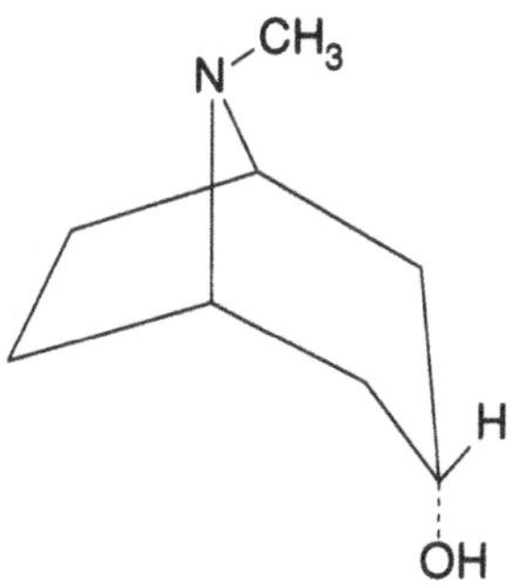

(50)

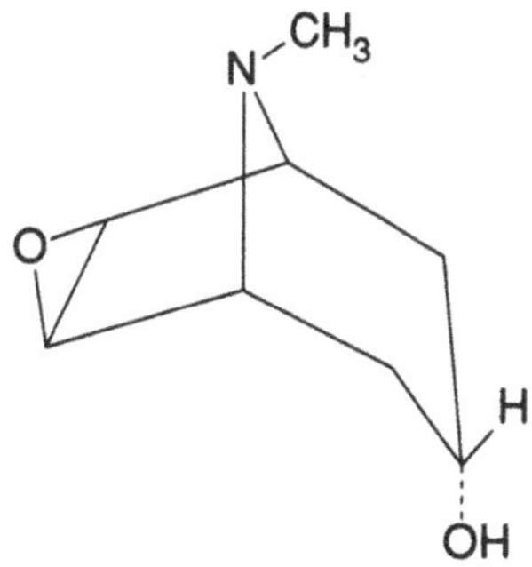

(51)

Information on the metabolic fate of **atropine** derives from a fairly large number of studies carried out over a lengthy period of time. Several general points stand out from this body of information. Firstly, a relatively high proportion of the administered compound is excreted in the urine in unchanged form. Secondly, the patterns of metabolism vary considerably among different animal species and it is therefore not possible to summarize the metabolism of atropine in a scheme which has general applicability. Thirdly, knowledge of the precise nature of the metabolites of atropine is often scanty.

The appreciable urinary excretion of unchanged atropine was registered at an early date as Wiechowski[117] found that its administration to dogs (~30 mg/kg, i.p.) resulted in the excretion of 33% of the dose unchanged. Subsequent investigations with a much lower dose (0.5 mg/kg, s.c.) indicated that the urinary excretion of unchanged compound in dogs was ~23% within 2 h[118] or 20 to 35% within 4 h.[119] Appreciable urinary excretion of unchanged atropine is commonly seen in other animal species. In the mouse, the reported values for the amount of atropine excreted unchanged in the urine are ~25% in 48 h[120,121] and ~18% in 2 h.[122] Similar values (23 to 28%) were obtained in rabbits.[123] The latter investigation also showed that atropine excretion in rats varied from 10% (s.c.) to 33% (i.v.) and that the excretion values were reduced to ~0.5–1% of the dose in rats and rabbits following oral administration of the same amount of atropine. Kalser et al.[124] reported that the urinary excretion of what appeared to be unchanged atropine accounted for 39% of the dose (i.v.) in rats and 25% in guinea pigs. They also found a considerable but undetermined amount of unchanged atropine in the urine of cats. The situation in man does not appear to be appreciably different from that observed in laboratory animals. Tønnesen[123] found 20% of the dose (s.c.) to be excreted unchanged. However, the data of Gosselin et al.[125] indicate that about half of the atropine given by i.m. injection appeared unchanged in the urine. A similar value for the urinary excretion of unchanged atropine was also reported by Van der Meer et al.[126] They used [^{3}H]atropine sulfate (2 mg, i.v.) and recovered ~80% of the injected radioactivity in 24 h. Most of this material (57%) was atropine, however this was present in the form of (+)-hyoscyamine. This finding suggests that (–)-hyoscyamine was stereoselectively metabolized and the (+)-enantiomer excreted unchanged. Recognition must be given to the influence of urinary pH on the excretion of atropine and Albanus et al.[118] showed that alkalinization of the urine to pH 8 completely abolished the net tubular transport of atropine and more than halved the amount excreted by dogs.

As noted above, appreciable species differences in the metabolism of atropine have been registered and this, together with the fact that our knowledge of the identity of metabolites is often fairly sketchy, makes it difficult to summarize schematically the pathways of atropine metabolism generally or, in fact, in most animal species. Much of our knowledge of the nature of atropine metabolites has been obtained in studies using mice. In this species the following reactions have been established: N-demethylation to noratropine, ester hydrolysis to tropic acid (49) and tropine (50), conjugation with glucuronic acid, and aromatic hydroxylation followed by conjugation with glucuronic acid. The use of [N-*methyl*-^{14}C]atropine showed that 7 to 10% of the dose was excreted as respiratory CO_2.[127,128] This extent of N-demethylation was confirmed in the former investigation by showing that the nortropine formed by hydrolysis of the urinary noratropine and closely related metabolites also accounted for ~10% of the dose. In the mouse, ester hydrolysis to tropic acid and tropine is not extensive. It is known that the administration of tropic acid leads to the excretion of essentially all of the compound unchanged in the urine.[120] Therefore, the extent of hydrolysis of atropine to these compounds should be reflected in the amount of tropic acid excreted in the urine. While the evidence regarding tropic acid formation in atropine-treated mice is conflicting, it does appear that this pathway is of minor importance. Gosselin et al.[120] reported its absence as a urinary metabolite while Gabourel and Gosselin[121] found that it comprised ~1% of the dose. Its presence as a urinary metabolite was also noted by Eling[127] and Werner and Schmidt.[128] In contrast to the limited formation of the metabolites noted above, it appears that conjugation of atropine or its hydroxylation products with glucuronic acid is a major metabolic pathway in the mouse.[121,122,128] A major hydroxylated product is 4'-hydroxyatropine but Gabourel and Gosselin[121] obtained evidence for the excretion of large amounts (40 to 50% of the dose) of glucuronides of 3',4'-dihydroxyatropine.

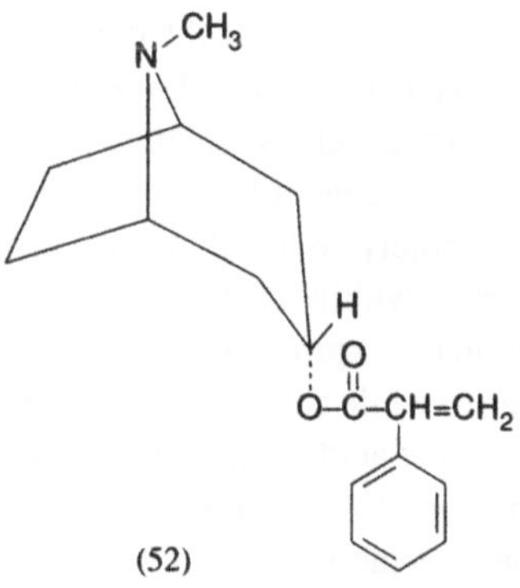

(52)

With the above outline of atropine metabolism in mice as a guideline, the main similarities and differences found in other animal species can be summarized. Gabourel and Gosselin[121] found that rats and mice metabolized atropine similarly and Werner[129] also found a fairly close similarity between these two species. Truhaut and Yonger[130] reported that atropine was converted by rat liver *in vitro* to noratropine, apoatropine (52), and hydroxylated derivatives, especially 2'-hydroxyatropine. Franklin[131] reported that atropine was demethylated when incubated with a rat liver microsomal preparation. Activity was seen with several *N*-methyl compounds of plant or synthetic origin but not with endogenous mammalian substrates. In the rat, as in the mouse,[132] biliary excretion of atropine metabolites was shown to be especially extensive.[133,134] As much as 50% of an injected dose appeared in the bile within 4 h, entirely as metabolites more polar than atropine. Bernheim and Bernheim[135] found that rat liver was not very active in hydrolyzing atropine. On the other hand, this activity was pronounced in guinea pig liver. Several other investigations have noted the ease of atropine hydrolysis in this species.[124,129,136] Liver preparations from guinea pigs hydrolyzed atropine, a reaction which was not observed in similar experiments using rat liver.[130] Otherwise, the oxidation products noted above for the rat (noratropine, apoatropine, and hydroxylated derivatives) were also formed *in vitro* by guinea pig liver. Phillipson et al.[137] incubated atropine with microsomal preparations from guinea pig liver and found that noratropine was formed. Additionally, these incubates contained the two isomeric *N*-oxides of atropine. When noratropine was used as a substrate, this secondary amine underwent *N*-oxidation to give the hydroxylamine derivative.

The metabolism of atropine in rabbits is of interest because of the presence of a genetically determined serum esterase in some animals. Bernheim and Bernheim[135] found the enzyme to be lacking in about a third of the rabbits they investigated. Other studies gave values of ~70%,[138] slightly over 50%,[139] and ~40%[140] for the proportions of animals which did not hydrolyze atropine. It is known that this esterase is not identical with that responsible for the hydrolysis of cocaine in rabbits.[141] This conclusion was confirmed by Stormont and Suzuki[142] and by Cauthen et al.[143] who purified and characterized the atropinesterase and cocainesterase present in rabbit serum. A clear difference in the patterns of urinary atropine metabolites between the two different types of rabbits is shown in the data of Werner.[129] Information on the metabolism of atropine in cats is both scanty and apparently contradictory. While Bernheim and Bernhein[135] found that its hydrolysis by cat liver occurred slowly, Langecker and Lewit[136] concluded that cats could readily hydrolyze the alkaloid. The latter finding was confirmed by Godeaux and Tønnesen.[139] However, the data given by Werner[129] on the urinary metabolites of [¹⁴C]atropine in six different animal species indicate that the amount of urinary products is lowest in the cat. Knowledge of the metabolic fate of atropine in dogs is very limited. Albanus et al.[118] concluded that the pattern in dogs is qualitatively similar to that seen in mice. They detected four urinary metabolites of atropine, two of which were probably tropic acid and tropine.

Considerable uncertainty also exists regarding the metabolism of atropine in man. In order to overcome the technical problems associated with the detection and identification of small quantities of metabolites arising from the relatively small dose of atropine tolerated by man, the use of radioactive alkaloid seems to be essential. Gosselin et al.[125] employed a preparation labeled with ^{14}C in the tropic acid moiety. When 2 mg was given by i.m. injection, most of the radioactivity appeared in the urine within 24 h, a trace was found in the feces and, not unexpectedly, none was detected in the expired air. As mentioned above, about half of the dose was excreted in the urine unchanged while about one-third was found to consist of unidentified metabolites and less than 2% as tropic acid (49). The unidentified metabolites appeared to be esters of tropic acid but neither aromatic hydroxylation nor glucuronide formation could be demonstrated, indicating that these metabolites are not the same as those found in rat or mouse urine. A subsequent study was carried out by Kalser and McLain[144] using two different preparations labeled with ^{14}C either in the ring or the *N*-methyl group of the tropine moiety. With the latter compound, *N*-demethylation was detected and ~3% of the dose was recovered as $^{14}CO_2$, although it was considered possible that as much as 5 to 10% may be lost by this route. In contrast to that mentioned above, this study indicated that considerable amounts of a glucuronide conjugate were excreted. In addition, evidence was obtained for the excretion of four other metabolites. Of these, one was not characterized, one was tentatively identified as tropine, and the others were converted upon alkaline hydrolysis to a compound resembling tropine. The metabolic pattern of atropine in man was clarified by Van der Meer et al.[126] who administered [3H]atropine sulfate (2 mg, i.v.). As noted above, about half of the dose was excreted unchanged, presumably in the form of (+)-hyoscyamine. The (–)-enantiomer was metabolized by *N*-demethylation, *N*-oxidation, and ester hydrolysis. They confirmed that hydrolysis to tropic acid (49) was a minor reaction. Only 2 to 3% of the dose was found in this form and as tropine (50). The most abundant metabolic product was noratropine which accounted for 24% of the dose. Atropine-*N*-oxide (equatorial isomer) contributed a further 15%. None of the urinary metabolites appeared to be excreted in the form of glucuronide or sulfate conjugates.

The investigations noted above utilized atropine, the racemic form of hyoscyamine, but a few reports have appeared in which the optically active **(–)-hyoscyamine** was studied. In regard to the amount of alkaloid excreted unchanged in the urine, Tønnesen[123] reported values similar to those for atropine in rabbits and man but not in rats. Similar values were obtained for the hydrolysis of (–)-hyoscyamine and atropine in the perfused rabbit liver[139] and rabbit serum.[141] However, Bernheim and Bernheim[135] reported the preferential hydrolysis of the (–)-isomer by guinea pig liver *in vitro* and the data of Glick and Glaubach[138] suggest a slightly greater hydrolysis of the (–)-isomer using rabbit serum. An esterase found in rabbit serum and liver which hydrolyzes (–)-hyoscyamine has been studied by Werner[129] and Werner and Brehmer.[145,146]

The biological disposition of **(–)-scopolamine**, the 6,7-epoxide of (–)-hyoscyamine, is similar to that seen with atropine. Tønnesen[123] determined the extent of urinary excretion of unchanged scopolamine in rats and rabbits and found that the values following various routes of administration were generally similar to those found with atropine or (–)-hyoscyamine. Werner and Schmidt[147] gave i.p. doses of 10, 100, and 1000 μg to mice and found that ~9 to 16% was excreted unchanged in the urine in 20 h. Excretion of unchanged scopolamine in man was reported to be much lower than that of atropine.[123] Shaw and Urquhart[148] reported a value of 10% of a parenteral dose (~0.2 mg) and Putcha et al.,[149] using i.v. or oral doses (0.4 mg) of scopolamine, found 6 and 1%, respectively, of the dose in the urine as unchanged compound.

Werner and Schmidt[147] carried out a comprehensive investigation of scopolamine metabolism in mice, rats, guinea pigs, and marmosets. Using [9-^{14}C](–)-scopolamine, they found that 62 to 68% of the doses noted above were excreted in the urine of mice in 20 h. These values are similar to those reported by Yue et al.[150] who recovered 62% (urine) and 25% (feces) of the radioactivity in 48 h following the i.v. administration of [3H]scopolamine to rats. Scopolamine metabolism was most thoroughly studied by Werner and Schmidt[147] in the mouse and they identified six urinary metabolites in addition to unchanged compound. These included norscopolamine, the 9'-glucuronides of both scopolamine and norscopolamine, aposcopolamine (analogous to compound (52)), 6-hydroxyhyoscyamine, and scopine (51). The dominant urinary metabolite was scopolamine-9'-glucuronide which accounted for ~40 to 50% of the dose. Norscopolamine-9'-glucuronide was also an abundant metabolite and experiments using [N-methyl-^{14}C]scopolamine showed that 23% of the radioactivity was expired as $^{14}CO_2$ in 24 h. The third most abundant metabolite was aposcopolamine which accounted for ~6 to 10% of the dose. The other compounds were minor metabolites (each <1% of the dose). No racemization of the (–)-scopolamine or its optically active metabolites was noted. Interestingly, injection of scopolamine-9'-glucuronide led to the formation of norscopolamine. This undoubtedly occurred as a result of biliary excretion of the conjugate followed by release of the aglycon due to the action of bacterial β-glucuronidase in the intestine. This would allow for absorption of the liberated scopolamine and subsequent demethylation in the tissues. Appreciable quantities of scopolamine-9'-glucuronide were found in the feces and intestinal contents following i.p. injection of scopolamine. The results in rats, guinea pigs, and marmosets were also obtained using scopolamine labeled with ^{14}C in the 9- and 9'-positions. With the former compound, N-demethylation was found to be lower in these species than in the mouse. In contrast to that noted above (23% of the radioactivity expired as $^{14}CO_2$ in 24 h), the corresponding values in rats, guinea pigs, and marmosets were ~6, 13, and 4%, respectively. Also, both qualitative and quantitative differences were found in the excretion patterns of the remaining metabolites. 6-Hydroxyhyoscyamine was a prominent urinary metabolite in rats (20% of the urinary radioactivity) but was absent in the urines of guinea pigs and marmosets. Marmosets excreted mainly scopolamine-9'-glucuronide and formed only small amounts of scopine and aposcopolamine. In contrast, >70% of the urinary metabolites from guinea pigs was scopine. An additional 20% was scopolamine-9'-glucuronide, and aposcopolamine furnished only a few percent. In general, rats showed a more balanced excretion of the metabolites and also had the highest percentage of unchanged scopolamine in the urine.

The possibility of formation of N-oxides from scopolamine and its metabolites was not investigated in the study described above, however Ziegler et al.[151] isolated scopolamine-N-oxide from an incubation mixture of the alkaloid and a purified amine oxidase preparation. Interestingly, one isomeric N-oxide of (–)-scopolamine (N-oxide moiety in the equatorial position) was formed when (–)-scopolamine was incubated with microsomal preparations from guinea pig liver.[137] These preparations also formed norscopolamine and the latter compound, when used as a substrate, underwent N-oxidation to give the hydroxylamine derivative.

It is not known if scopolamine undergoes aromatic hydroxylation, however Sano and Hakusui[152] showed that the quaternary methyl derivative of scopolamine underwent appreciable hydroxylation in rats to the 4'-hydroxy derivative. Another possible reaction of scopolamine is conversion of the epoxide moiety to a diol by the action of epoxide hydrolase. In vitro studies with this enzyme indicated, however, that this does not occur.[153]

Conflicting results have been published on the metabolism of scopolamine by esterases. Glick and Glaubach[138] found a slightly lower rate of hydrolysis with scopolamine than with (–)-hyoscyamine in rabbit serum and Godeaux and Tønnesen[139] reported similar rates in the perfused rabbit liver. However, Bernheim and Bernheim[135] claimed that scopolamine was usually not hydrolyzed by liver preparations, even those from guinea pigs which otherwise showed high esterase activity. The atropinesterase from rabbit serum, which also hydrolyzes scopolamine, was purified and shown to be distinct from the cocainesterase also present in rabbit serum.[143]

Anisodamine (6-hydroxyatropine) was found to be excreted unchanged in the urine of rats to a greater extent than was atropine following i.v. dosage.[154] Values of 39% (rats) and 32 to 48% (humans) for the unchanged compound were recorded in 24 h. Yue et al.[150] used ^{3}H-labeled anisodamine and found that 78% of the radioactivity was excreted in the urine and ~13% in the feces of rats in 48 h following i.v. dosage. These results suggest that a considerable proportion of the dose was excreted in the form of metabolites.

The second major group of tropane derivatives is that of the coca alkaloids. These consist of several types including those based on ecgonine, tropine, and hygrine. Ecgonine derivatives furnish the principle type and, among these, **cocaine** (53) (see Figure 3) is the most important and most widely studied. Early investigations were largely devoted to the question of the extent of urinary excretion of cocaine. Wiechowski[117] administered cocaine i.p. to rabbits and dogs and found that ~5% of the dose could be recovered in the 48-h urine in the latter species. However, no cocaine was detected in rabbit urine. Oelkers and Vincke[155] studied the urinary excretion of cocaine following its injection in mice, rats, guinea pigs, rabbits, cats, and dogs. In general, excretion values of 1 to 5% were obtained but as much as 16% of the dose was found excreted unchanged in rabbits when the urine was acidic. A low level of cocaine excretion in the rat was confirmed by Nayak et al.[156] who found 0.75% of the dose (20 mg/kg, s.c.) in the 96-h urine. Fecal excretion during this period was a mere 0.25%. Likewise, Woods et al.[157] detected 1 to 12% of the dose (10 to 15 mg/kg, s.c. or i.v.) unchanged in the 24-h urine in dogs. It was also noted that much less cocaine was absorbed following oral administration. They found that no or at most traces of cocaine were excreted in the urine of rabbits. This publication is also of interest for its summary of the early work on cocaine metabolism and excretion. Misra et al.,[158] using ring-labeled [³H]cocaine (5 mg/kg, i.v.), recorded a value of 2.9% of the dose in dogs. The limited extent of urinary excretion of unchanged cocaine in monkeys is similar to that found in other species.[159] Information on the urinary excretion of unchanged cocaine in man has been obtained in many studies, in both addicts and nonaddicts. Woods et al.[157] referred to unpublished results which gave values of 1 to 21% in nonaddicts. The urinary excretion (6 h) following an i.v. dose (100 mg) of cocaine hydrochloride was found to be ~1 to 9%,[160] however a mean value of only 0.7% in 9 h was recorded in similar experiments which employed doses of 4 to 24 mg.[161] Jindal et al.[162] reported that ~1% of the dose (100 mg of cocaine hydrochloride, i.v.) was recovered unchanged in the urine in 6 h. An average value of 1.7% of the dose (1.5 mg/kg, intranasally) in the urine (24 h) was recorded in five subjects.[163] In addicts, the excretion has been reported to be 1 to 9%,[164] 4 to 9%,[165] and 7 to 21%.[166] Fish and Wilson[164] obtained the highest excretion values when the urinary pH was reduced.

From the foregoing it is apparent that cocaine is largely metabolized in animals. This alkaloid contains two ester linkages and much of the information on its metabolism deals with its hydrolysis at these metabolically active sites. In addition, *N*-demethylated metabolites may be formed as well as ring-hydroxylated derivatives. The general pathways of cocaine metabolism are summarized in Figure 3. Of course, the quantitative picture may vary greatly from species to species and, in some cases, certain pathways may be inconsequential or unexplored. Several early reports revealed certain aspects of the mammalian metabolism of cocaine. Langecker and Lewit[136] were the first to propose that the probable metabolic sequence involves the initial hydrolysis of the methyl ester group to give benzoylecgonine (54) followed by further hydrolysis to ecgonine (55) and benzoic acid (56). Ortiz[167] detected both benzoylecgonine and ecgonine as urinary metabolites in rats. Interestingly, it was noted that only a small proportion of the dose (45 mg/kg, i.p.) was excreted as these metabolites in rats. This was also noted by Sánchez[165] who found only 1 to 3% of the 90 to 220 mg daily dose taken orally by addicts to be excreted in the urine as ecgonine. Valanju et al.[168] noted that whenever ecgonine was present in the urine of addicts, benzoylecgonine was also found. However, the latter metabolite was often present in the absence of ecgonine. Several comprehensive studies of the metabolism of cocaine have more fully clarified this subject. These investigations have been carried out in rats,[169,170] dogs,[158,170] monkeys,[159,170] and man.[171-173]

Nayak et al.[169] found benzoylecgonine (54), ecgonine (55), ecgonine methyl ester (57), and benzoylnorecgonine (58) as urinary metabolites of [³H]cocaine (20 mg/kg, s.c.) in rats. Neither the likely precursor of benzoylnorecgonine, norcocaine (59), nor its possible hydrolysis product norecgonine (60) was detected. It is possible that the nor compound (58) might be formed via an alternative pathway from benzoylecgonine (54), however Misra et al.[174] showed that the

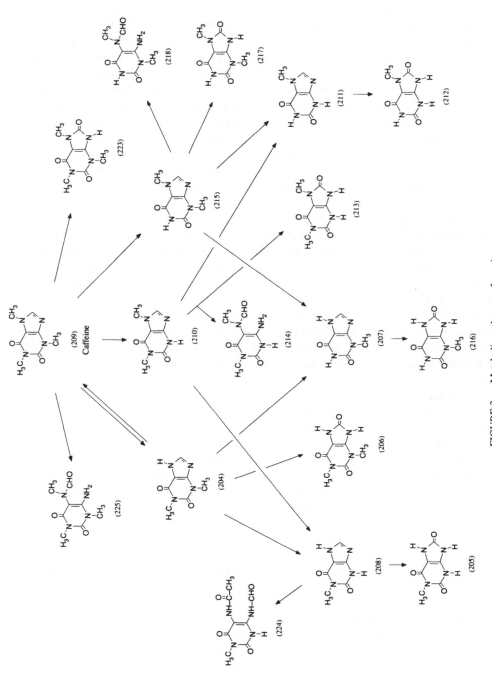

FIGURE 3. Metabolic pathways of cocaine.

major urinary products of administered benzoylecgonine in rats were the unchanged compound and ecgonine, its hydrolysis product, together with minor amounts of a phenolic metabolite and its glucuronide. Nevertheless, Mulé et al.[175] reported that benzoylnorecgonine was present in the brains of rats given benzoylecgonine. When the nor compound was given, however, no further metabolism was detected. This confirms the earlier result that benzoylnorecgonine is not metabolized further to norecgonine in rats. Misra et al.[176] administered benzoylnorecgonine to rats and found that it was rapidly and extensively excreted unchanged in the urine. A small amount of an unidentified compound was also detected, however this was not norecgonine. On the other hand, both benzoylnorecgonine and norecgonine were detected in brain when rats were given norcocaine (59).[177] These differences may be related to the variations in lipophilicity of the different compounds employed. Formation of norecgonine appears not to occur via *N*-demethylation of ecgonine (55). Misra et al.[178] found that the latter compound was rapidly and extensively excreted in the urine of rats without undergoing metabolism. The routes of excretion of cocaine and its metabolites in rats were also studied by Nayak et al.[169]. At the dose level noted above, ~22% of the radioactivity was lost by the fecal route. This finding is explained by the considerable excretion of cocaine metabolites in the bile. Following an i.v. dose of 5 mg/kg, ~36% was excreted by this route as metabolites in 3.5 h. The amount of unchanged cocaine excreted in the bile was very small.

The biological disposition of cocaine in dogs was studied by Misra et al.[158] Using ring-labeled [³H]cocaine (5 mg/kg, i.v.), 58% of the radioactivity was excreted in the urine in 48 h. In addition to the small amount of unchanged cocaine noted above, small amounts (<1% of the dose) of ecgonine methyl ester (57), norcocaine (59), and norecgonine (60) were found. The major urinary metabolite was ecgonine (55) (37%) followed by benzoylnorecgonine (58) (9%) and benzoylecgonine (54) (2.5%). These findings indicate that the metabolism of cocaine in dogs is qualitatively similar to that seen in rats. However, some additional metabolites were also detected. These include the demethylated intermediate norcocaine (59) and the fully hydrolyzed and demethylated norecgonine (60). Nevertheless, the amount of the latter metabolite found was quantitatively small compared with that of ecgonine (55). Rapid metabolism of cocaine was also indicated in dogs which had high levels of radioactivity in the bile shortly after administration of the compound.

The excretion of cocaine and its metabolites in monkeys occurs rapidly.[159] Following an i.v. injection of [³H]cocaine (1 mg/kg), most of the radioactivity recovered was excreted within 24 h. Both norcocaine (59) and ecgonine methyl ester (57) were found in small amounts, however the finding that urinary benzoylnorecgonine (58) accounted for >40% of the dose reveals that the overall extent of *N*-demethylation was high. Appreciable amounts of benzoylecgonine were also excreted. These results indicate that the methyl ester group was preferentially hydrolyzed. A small amount of highly polar metabolites was also detected, however these were not identified.

A noteworthy feature of the metabolism of cocaine in humans is the extensive urinary excretion of ecgonine methyl ester (57). Whereas this metabolite is excreted in small amounts in the animal species noted above (rats, dogs, and monkeys), Inaba et al.[171] found that it accounted for a third to half of the urinary metabolites in man. Values of 26 to 60% of the dose (16 to 96 mg, i.v. or intranasally) were recorded by Ambre et al.[172] for the excretion of ecgonine methyl ester. In this study, benzoylecgonine excretion accounted for a further 33 to 55% of the dose. A subsequent study employing infusions of larger doses of cocaine,[173] again showed that ecgonine methyl ester (12 to 21% of the dose) and benzoylecgonine (14 to 17%) were major urinary metabolites. These values for urinary benzoylecgonine excretion agree well with those of Hamilton et al.[163] who recovered 26% of the dose (1.5 mg/kg, intranasally) in the urine (24 h) in five subjects. The hydroxylation and *N*-demethylation of cocaine are discussed below.

The formation of hydroxylated metabolites of cocaine has been conclusively demonstrated only in rats and man. When rats were given cocaine (20 mg/kg, s.c.), a phenolic and two nonphenolic hydroxylated metabolites were detected as urinary metabolites.[169] These are the *p*-hydroxylated metabolite (61) and compounds which were presumably hydroxylated in the 6- and/or 7-positions of the pyrrolidine ring. A hydroxycocaine, thought to be the 3'-hydroxy derivative (62), was detected in the bile of a cocaine user.[179] Subsequent studies using urine samples from cocaine users showed that 3'-hydroxycocaine was present, however the major hydroxylated metabolite of cocaine was the 4'-hydroxy derivative (61).[180] Additionally, three hydroxymethoxy derivatives of cocaine were detected in similar urine samples.[181] Two of these were identified as 3'-hydroxy-4'-methoxycocaine (63) and 4'-hydroxy-3'-methoxycocaine (64). Interestingly, the ethyl ester homologues of these hydroxymethoxy derivatives were found in the urine of cocaine users who had taken ethanol concurrently.[182] Jindal and Lutz[183] also reported the excretion of ring-hydroxylated metabolites of cocaine in the urine of rats and humans. They identified compounds (62) and (63) and also a metabolite which contained a hydroxydimethoxyphenyl moiety.

As noted above, cocaine undergoes *N*-demethylation to norcocaine (59) and other demethylated metabolites. This metabolic pathway, which has been studied both *in vitro* and *in vivo*, is of added interest because it leads to the formation of biologically active metabolites. The *N*-demethylation of cocaine was carried out by rat liver microsomes[184] and by similar preparations from mouse, rat, guinea pig, and dog liver.[185] Estevez et al.[186] reported a reduction in the amounts of demethylated metabolites (norcocaine and benzoylnorecgonine) in the livers of rats given the inhibitor SKF-525A prior to cocaine administration. Leighty and Fentiman[187] also demonstrated norcocaine formation by rat liver microsomes, however this compound and benzoylecgonine, the only metabolites identified, made up together less than 20% of the cocaine metabolized. A similar value was found for the *N*-demethylation of cocaine by isolated rat hepatocytes.[188] Additional *in vitro* studies using mouse hepatic microsomes revealed that the oxidative demethylation of cocaine to norcocaine may take place via two alternative pathways.[189] One of these involves a direct *N*-demethylation and is dependent on cytochrome P-450, whereas the second is a two-step reaction in which a FAD-containing monooxygenase converts cocaine to its *N*-oxide (65) and this intermediate is then metabolized via a cytochrome P-450-catalyzed demethylation to norcocaine. Norcocaine has been shown to undergo further metabolism by microsomes from mouse liver[190] or rat brain[191] to *N*-hydroxynorcocaine (66). The formation of this hydroxylamine metabolite is significant because it may be further oxidized to a nitroxy free radical (norcocaine nitroxide) (67).[191,192] Formation of the nitroxide derivative is species dependent, occurring in mice but not in rats, hamsters, guinea pigs, or rabbits.[193,194] This metabolite binds covalently to hepatic protein and appears to be responsible for the hepatic toxicity of cocaine in mice.[195,196] The *in vivo* metabolism of *N*-hydroxynorcocaine and cocaine-*N*-oxide in rats was investigated by Misra et al.[197] Ester hydrolysis was the major metabolic pathway with both compounds (see structures (68), (69), and (70)).

The *in vivo* formation of norcocaine from cocaine has been demonstrated in several animal species. Using [*N-methyl*-^{14}C]cocaine, Werner[129] found that 45% of the injected radioactivity was recovered as respiratory $^{14}CO_2$ in mice. This suggests that *N*-demethylation may be a prominent metabolic pathway in this species.[198] Also, norcocaine was identified as a metabolite of cocaine in the brains of rats[199] and dogs[169] and in the brains and other organs of monkeys.[198,200] The rate of *N*-demethylation of cocaine in chronically treated rats was not significantly different than that found after a single dose of cocaine.[201] Only small amounts of norcocaine were found in the urine of human subjects given cocaine hydrochloride (100 mg) by i.v. dosage.[162] The minor importance of this pathway in humans was confirmed in a study which employed [*N-methyl*-^{14}C]cocaine. Following an oral dose (10 mg) of this material, only ~2 and 6% of the dose was recovered from two subjects as expired $^{14}CO_2$.[171] Misra et al.[197] found that the further metabolism of norcocaine, at least in the rat, leads to the formation of the nor derivatives of benzoylecgonine (58), ecgonine methyl ester (71), and ecgonine (60) as urinary metabolites.

Numerous studies have been carried out on the enzymatic hydrolysis of cocaine, both in serum and tissues. In an early report, Glick and Glaubach[138] showed that hydrolysis took place with rabbit serum. On the other hand, Glick et al.[202] noted that horse and human serum did not possess this activity. Blaschko et al.[203] confirmed the lack of cocainesterase activity in horse serum and human plasma. The belief that cocainesterase and atropinesterase (tropinesterase) in rabbit serum were separate entities[138] was confirmed by Blaschko et al.,[204] Stormont and Suzuki[142] and Cauthen et al.[143] who purified and characterized these two enzymes. The presence of two enzymes was also reported by Ammon and Savelsberg[141] who found that about a third of the rabbits tested were able to hydrolyze cocaine. This activity was also seen when ecgonine methyl ester (57) was used as the substrate. The usual hydrolytic pathway was via benzoylecgonine (54) to ecgonine (55), however Werner and Brehmer[145] showed the presence of a cocainesterase in rabbit serum which, alternatively, produced ecgonine via ecgonine methyl ester (57). In contrast to the earlier negative results in humans, Taylor et al.[205] reported that the *in vitro* hydrolysis of cocaine to ecgonine by human serum occurred via both hemiesters, benzoylecgonine (54), and ecgonine methyl ester (57). The initial reaction in one of these pathways, the formation of ecgonine methyl ester, was shown to be mediated by human plasma cholinesterase.[206,207] The alternative intermediate, benzoylecgonine, was not formed enzymatically. In fact, the rate of spontaneous formation of the latter compound at physiological pH suggested that it may arise nonenzymatically in the body.

The ability of other tissues to hydrolyze cocaine has also been studied. Heim and Haas[208] investigated its hydrolysis to benzoylecgonine in guinea pig tissues and found that the liver had greater activity than kidney, brain, or muscle tissue. Mikami[209] studied the formation of benzoic acid from cocaine in the liver *in vitro* and found the activity to decrease in the following order: rabbit, guinea pig, cat, mouse, dog, and rat. Cocaine hydrolyzing activity was also studied by Severi et al.[210] who also found a decreasing order of activity in rabbits, guinea pigs, and rats. Iwatsubo[211] studied the hydrolysis of various esters by liver microsomes solubilized with sodium deoxycholate. Preparations from rats, guinea pigs, and rabbits showed good activity with several simple, synthetic esters, however cocaine hydrolysis was seen only with rabbit liver microsomes which had weak activity. Leighty and Fentiman[187] reported that the observed hydrolysis of cocaine to benzoylecgonine by rat liver microsomes accounted for only a small part of the cocaine metabolized in these incubates. On the other hand, Estevez et al.[186] found that the *in vitro* and *in vivo* hydrolysis of cocaine by rat liver was significantly reduced by the microsomal enzyme inhibitor SKF-525A. Stewart et al.[207] reported that the rate of enzymatic hydrolysis of cocaine to ecgonine methyl ester was lower in human liver than serum.

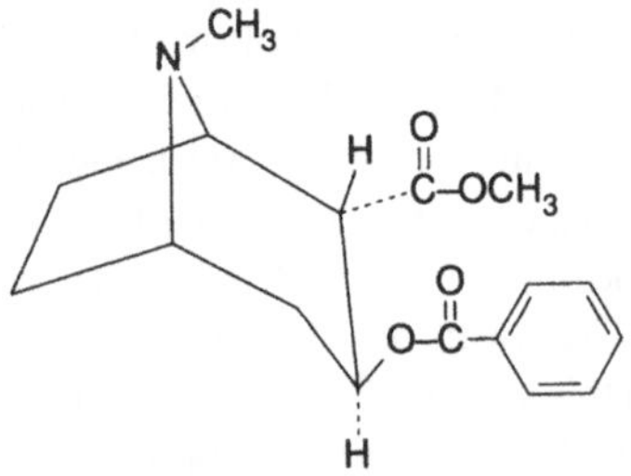

(72)

Little information is available on the metabolism of other coca alkaloids. Based on the excretion of the unchanged compounds in cats, Gruhn[212] suggested that a larger proportion of (+)-cocaine (pseudococaine) (72) than (−)-cocaine (53), the natural isomer, was metabolized. The greater metabolism of the (+)-isomer than of the (−)-isomer was also noted by Misra and Pontani[213] who found that it was excreted unchanged in very small amounts in the urine and feces of rats following i.v. dosage. The urinary metabolites detected were the pseudo isomers of norcocaine (59), ecgonine methyl ester (57), benzoylecgonine (54), benzoylnorecgonine (58), and ecgonine (55). Gatley et al.[214] reported that (+)-cocaine was hydrolyzed by baboon plasma more than 1000 times faster than the natural (–)-enantiomer. Serum butyrylcholinesterase appeared to be the enzyme responsible.

Cinnamoylcocaine (73), a constituent of coca leaves and some illegal cocaine preparations, was shown to be present in the postmortem fluids (urine and bile) of a cocaine user.[215] Some hydrolysis of the methyl ester group also occurred as cinnamoylecgonine was detected.

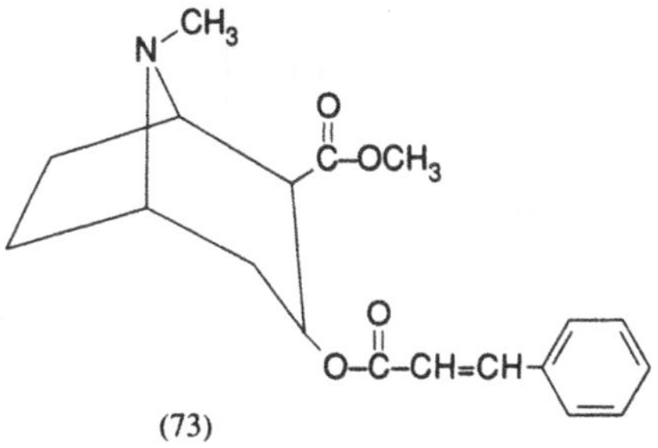

(73)

Cinnamoylcocaine

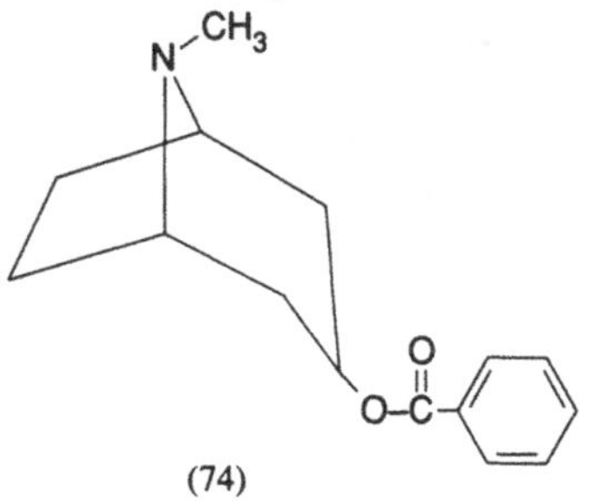

(74)

Tropacocaine

Glick and Glaubach[138] found that **tropacocaine** (74) was hydrolyzed by rabbit and horse serum and believed that this activity may be different from that hydrolyzing atropine, cocaine, and choline esters. Blaschko et al.[204] confirmed the presence of this enzyme in horse serum and showed that the hydrolysis of tropacocaine was not inhibited by several other esters including atropine and cocaine. They concluded that the tropacocainesterase of horse serum is not identical with pseudocholinesterase. Other studies of the enzymatic hydrolysis of tropacocaine[216] and nortropacocaine[217] indicated that the esterase activity involved is not identical with that responsible for the hydrolysis of benzoylcholine.

INDOLE DERIVATIVES

Indole (75) (see Figure 4) is the parent of an extensive and sometimes complex group of nitrogen heterocyclic compounds. In fact, perhaps as many as one fourth of all known alkaloids contain the indole moiety. One important type of indole compound is the indolethylamine group (tryptamine derivatives). However, their metabolism is governed more by the alkylamine than the indole moiety and these compounds are therefore covered in the chapter entitled Metabolism of Amino Compounds.

In contrast to that seen with the parent compounds of most of the other classes included in this chapter, indole itself occurs in a number of essential oils. Although attempts to determine its metabolic fate were made in the nineteenth century, it is only during the last few decades that an adequate understanding of indole metabolism has been obtained. Earlier investigations, the results of which were summarized by Williams,[218] showed mainly that indole is hydroxylated to 3-hydroxyindole (indoxyl) (76) which is then excreted in the urine of several animal species, partly as the glucuronide (77b) but mainly as the sulfate conjugate (indican) (77a). The enzymology of two purified phenol sulfotransferases from rat liver which carry out the latter reaction was studied by Sekura and Jakoby.[219]

The metabolic pathways of indole in rats are shown in Figure 4. The major route involves hydroxylation at the 3-position giving indoxyl (76), its conjugates, and further oxidation products leading to N-formylanthranilic acid (78) and anthranilic acid (79). King et al.[220] carried out a detailed investigation of the metabolism of [2-¹⁴C]indole in rats. They employed oral doses (60 to 70 mg/kg) and found that ~80% of the radioactivity was excreted in the urine in 2 d. Only traces of unchanged compound were detected in the urine. Similar experiments in rats using unlabeled indole also showed that some unchanged compound was excreted in the urine.[221] The

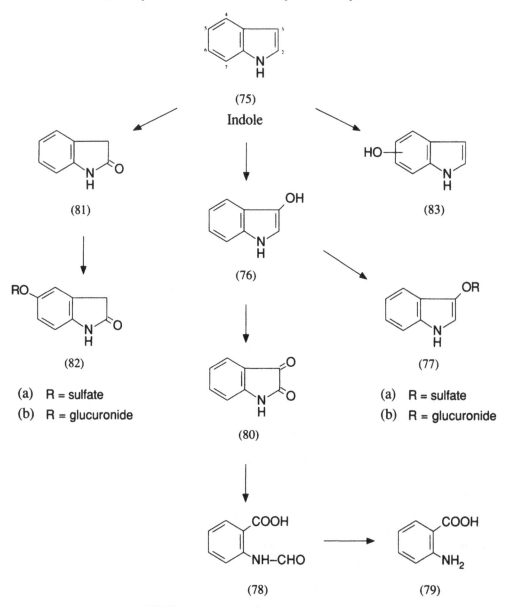

FIGURE 4.   Metabolic pathways of indole in rats.

major urinary metabolite in the former study was indoxyl sulfate (77a) which accounted for 50% of the dose. About 6% of the dose was excreted as isatin (80). This compound was further metabolized via scission of the pyrrole ring to give metabolites (78) and (79). Formation of the latter metabolite led to loss of the label and no value for the urinary excretion of anthranilic acid (79) was therefore obtained. However, a value of ~2% appears reasonable as this corresponds to the amount of $^{14}CO_2$ lost in the expired air. *In vitro* experiments indicated that 3-hydroxylation occurred under aerobic conditions, suggesting that this reaction is carried out by the microsomal hydroxylating enzymes. This conversion of indole to indoxyl (76) was also demonstrated by Posner et al.[222] and Beckett and Morton[223] using the microsomal hydroxylating system from rabbit liver. All of these reports noted the rapid further oxidation of indoxyl to indigo blue (indigotin), although Beckett and Morton found this reaction to take place with rabbit liver preparations but not with those from rat liver which formed oxindole (81) instead. Frydman et al.[224] described an oxygenase system from rat liver, designated pyrrolooxygenases, which oxidized the pyrrole ring of various indole derivatives. This system, which readily cleaves the pyrrole ring to form 2-formamidophenylacyl derivatives, is described more fully in the discussion of the metabolism of 3-methylindole given below. Indole itself was found to be a weak substrate of this system.

Another metabolic pathway of indole involves the initial oxidation at the 2-position to form oxindole (81). As noted above, Beckett and Morton[223] showed that this reaction was carried out by rat liver microsomes and King et al.[220] demonstrated it in aerobic and anaerobic incubates of rat liver preparations. The available data suggest that this pathway is of minor quantitative importance. Beckett and Morton did not detect oxindole in the urine of rats given indole intraperitoneally, however they did find small amounts of conjugates of 5-hydroxyoxindole. King et al. showed that the main urinary metabolites resulting from this reaction sequence were the sulfate (82a) and glucuronide (82b) conjugates of 5-hydroxyoxindole. Beckett and Morton found that microsomal preparations from rat, guinea pig, and rabbit liver carried out the 5-hydroxylation of oxindole. Most of the investigations of indole metabolism indicated that indole does not undergo direct hydroxylation at the 5- or 6-positions. Posner et al.[222] specifically failed to find evidence for the formation of 5-hydroxyindole from indole when the microsomal hydroxylating system from rabbit liver was used. On the other hand, Acheson and Nwankwo[221] found that both the 5- and 6-hydroxy derivatives (83) were excreted in the urine of rats given an oral dose (80 mg/kg) of indole. This study generally confirmed the pattern of metabolism as revealed in the earlier investigation of King et al.[220]

The results of King et al.[220] indicated that biliary excretion of indole and its metabolites was not extensive in the rat (~5% of the dose) and that about one-tenth of the dose was excreted in the feces, mainly as unidentified metabolites.

Whereas metabolic indican is indoxyl sulfate (77a), the indican found in plants (plant indican) is **indoxyl-β-D-glucoside**. Li et al.[225] gave oral doses of plant indican to rabbits, dogs, and humans and found that the 24-h urines contained unchanged compound, indoxyl sulfate, and indoxyl glucuronide as well as some highly polar compounds. In the urine of rabbits only, some anthranilic acid (79) and its glycine conjugate were also detected.

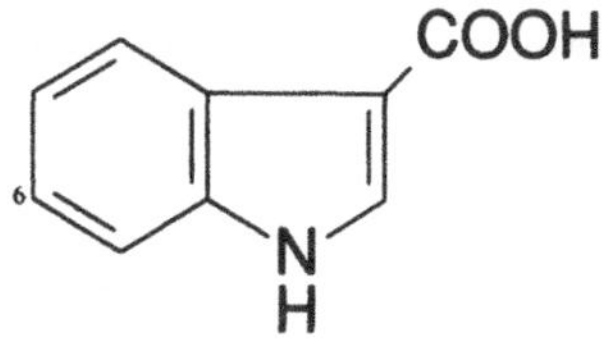

(84)

Indole-3-carboxylic acid

The metabolism of **indole-3-carboxylic acid** (84) in rats was studied by Acheson and King.[226] The pattern of urinary metabolites was similar after oral or i.p. administration. Following doses of ~150 mg/kg, ~20% was excreted as unchanged compound with 90% of this appearing within 24 h. The major urinary metabolite was the glucuronide conjugate of indole-3-carboxylic acid. Small amounts of 6-hydroxyindole-3-carboxylic acid were excreted, however the glycine conjugate of indole-3-carboxylic acid was not detected. In a subsequent study in which rats were given indole-3-carboxylic acid orally at the dose level noted above,[221] the glucuronide conjugate was again shown to be the major urinary metabolite. The 6-hydroxy derivative was not detected, however three unidentified metabolites were observed on the chromatograms.

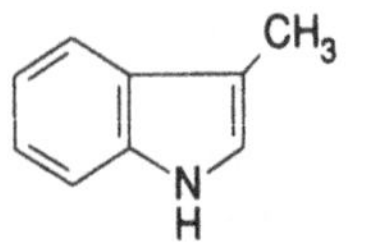

(85) 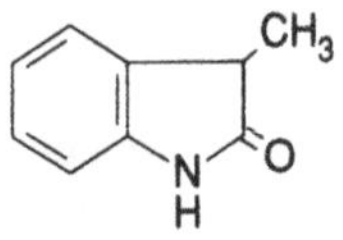   (86)

3-Methylindole

3-Methylindole (skatole) (85) has been identified as a plant constituent, however it is perhaps more commonly associated with other sources. It is evident that hydroxylation at C-3, which is the major metabolic pathway with its lower homologue indole, is no longer possible. The resultant shift to the oxindolic pathway was clearly seen in the results of Hammond et al.[227] who administered [*methyl*-^{14}C]3-methylindole to goats. About 90% of the radioactivity was excreted in the urine in 72 h following i.v. doses (30 to 40 mg/kg). Chromatography of this material revealed that a fraction corresponding to 80% of the urinary radioactivity could be separated into ten discrete peaks. The most prominent of these was shown to be 3-methyloxindole (86). Indole-3-carboxylic acid (84) was also detected and several of the unidentified metabolites were shown to be glucuronide conjugates. This investigation did not find evidence for the urinary excretion of acetophenone derivatives which could have arisen via scission of the heterocyclic ring, however their formation could not be ruled out. Earlier studies by Frydman et al.[224,228] showed that 3-methylindole was metabolized to such compounds by microsomal mixed-function oxidases which they designated pyrrolooxygenases. The main product formed was 2-formamidoacetophenone (87), however some oxindole derivative (86) was also produced. These enzyme preparations from rat liver also contained amidase (formylase) activity which effectively converted compound (87) to 2-aminoacetophenone (88). Frydman et al.[228] suggested that the initial intermediate in the reaction sequence may be a 2,3-indoleepoxide which, upon hydration to the 2,3-diol, could either undergo dehydration to the oxindole derivative (86) or oxidation to the acetophenone derivative (87). Skiles et al.[229] reported an interesting species difference in the metabolism of 3-methylindole. 3-Methyloxindole (86), the major urinary metabolite in goats, was not detected in the urine from mice. Instead, the major metabolite was shown to be 3-hydroxy-3-methyloxindole which accounted for ~10% of the dose (400 mg/kg, i.p.).

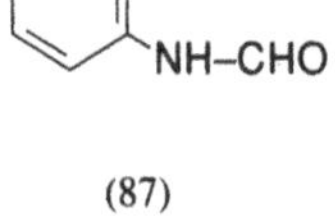

(87) 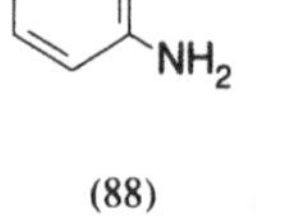   (88)

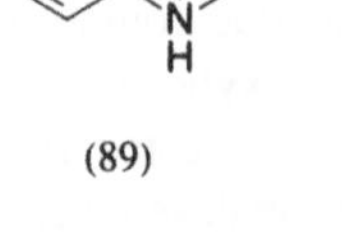

(89) 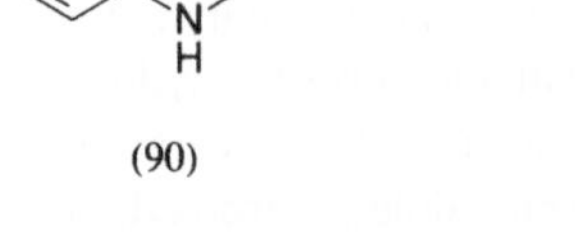   (90)

Considerable interest has recently been shown in the metabolism of 3-methylindole, which is also a ruminal metabolite of tryptophan, because of the lung toxicity which it may cause in ruminants.[230,231] Toxicity was found to result from 3-methylindole administration in goats, but not from dosage with its metabolite 3-methyloxindole (86) or with indole-3-carbinol, the probable precursor of indole-3-carboxylic acid.[232] These findings indicate that the lung toxicity is associated with a metabolic step prior to the formation of these metabolites. A study by Nocerini et al.[233] indicated that the toxic intermediate was an electrophilic metabolite which also formed a glutathione adduct. Studies of the formation of this glutathione conjugate by lung and liver microsomes from several species (mouse, rat, monkey, goat, and horse) suggested that species and organ-selective toxicity is related to differences in the rates of oxidative metabolism of 3-methylindole to the electrophilic metabolite.[234] Additional studies[235] indicated that this metabolite is probably not an epoxide but an imine methide intermediate (89). This belief is supported by the finding that the glutathione adduct formed was shown to have structure (90), i.e., glutathione addition at the methyl group without incorporation of oxygen into the product. They postulated that the imine methide could undergo hydration to give 3-methyloxindole (86) or indole-3-carbinol. Huijzer et al.[236] investigated the different forms of cytochrome P-450 in goat lung which participate in the bioactivation of 3-methylindole. Other studies[237,238] have shown that an additional system present in lung and liver microsomes, prostaglandin H synthase, contributes to the bioactivation of 3-methylindole.

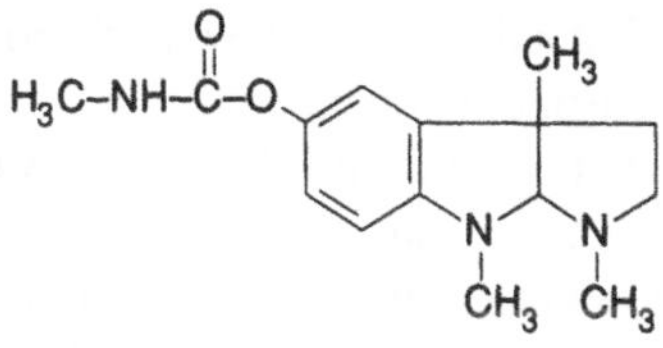

(91)   (92)

Physostigmine

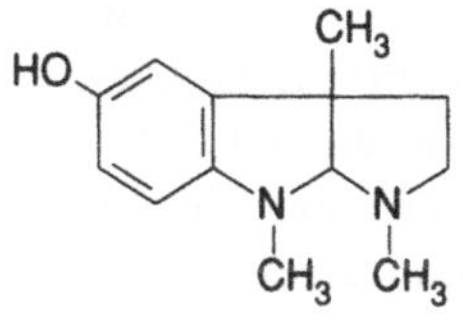

Although **physostigmine** (eserine) (91) has a history of use which extends over 125 years, relatively little is known about its metabolic fate. However, some information is available from investigations of the pharmacokinetics of physostigmine in rats following oral,[239] i.m.,[240,241] or i.v. dosage,[242] in dogs following i.v. dosage,[243] and in humans following several routes of administration.[244] These studies have demonstrated that physostigmine is rapidly metabolized. The urinary excretion of unchanged compound is not extensive. Somani and Boyer[240] recorded a value of only 2.5 to 4% of the dose (0.65 mg/kg, i.m.) in rats. This study, which employed [³H]physostigmine, also showed that the excretion of radioactivity was 44% in 24 h and 53% in 7 d. Small amounts of the hydrolysis product, eseroline (92), were detected, especially during the first few hours following dosage. Most of the urinary radioactivity consisted of an unidentified metabolite which showed a prolonged excretion profile. This persistence may be related to the extensive excretion of physostigmine and its metabolites in the bile of rats. Boyer and Somani[245] found that ~60% of the dose (0.1 mg/kg, i.v.) of [³H]physostigmine was excreted by this route in 3 h. The biliary material consisted of unchanged compound and metabolites in about a 2:3 ratio. Studies of the plasma metabolites of physostigmine indicated that an additional unidentified metabolite was formed in small amounts.[241,242] The unidentified metabolites did not appear to be degradation products of eseroline, e.g., its oxidation product rubreserine or colored products.[239] This pattern of metabolites was also found in the dog,[243] however the extent of metabolism may be less than in the rat. In addition to these results from *in vivo* investigations, the *in vitro* metabolism of physostigmine by mouse liver microsomes has been studied.[246] The results demonstrated that cytochrome P-450 was involved and no indication was found that the metabolites were conjugated with glucuronic acid or glutathione. Two major and six minor metabolites were detected and, in contrast to the results noted above, none of these appeared to be identical with eseroline (92). Also, none of these compounds corresponded to rubreserine, the oxidation product of eseroline. In fact, the evidence suggested that no hydrolysis of the urethane moiety had occurred with these metabolites. This finding suggested that other reactions including *N*-dealkylation, *N*-hydroxylation, or aromatic hydroxylation may be involved.

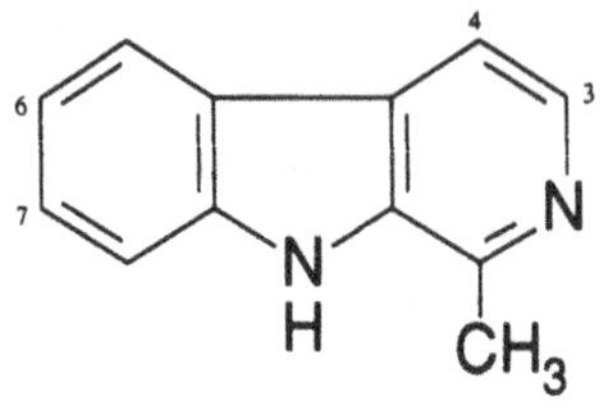

(93)

Harman

Several groups of indole alkaloids are derivatives of β-carboline. The more simple types include **harman** (1-methyl-β-carboline) (93) and the closely related harmala alkaloids, harmine (94), harmaline (95), and harmalol (96) (see Figure 5). The metabolism of harman by mouse liver microsomes was studied by Tweedie et al.[247] They detected three metabolites and found that the major product formed was 6-hydroxyharman. Induction of mice with phenobarbital or 3-methylcholanthrene increased the rate of metabolism of harman.

The metabolic fate of the harmala alkaloids is not complex, involving mainly *O*-demethylation and/or conjugation. These pathways are shown in Figure 5, however hydroxylation may provide an additional route in the case of harmine and harmaline. Very little **harmine** (94) was excreted unchanged in the urine following its injection in rats or man.[248-252] In the rat, nearly 75% of the dose (5 mg/kg, i.v.) was excreted in the bile and the remainder in the urine.[251] This study showed that the same metabolites were found in both urine and bile and that their relative proportions were fairly similar. Urinary excretion of metabolites in noncannulated rats was ~30%,[252] which suggests that reabsorption of the biliary metabolites is not extensive. In man, the urinary excretion of harmine metabolites amounted to about a third of the dose. These investigations also showed that the major metabolic product is harmol (97). A kinetic study of the liver microsomal *O*-demethylation of harmine to harmol in six mammalian species was reported by Burke and Upshall.[253] The *in vivo* studies revealed that only small amounts of free harmol were excreted and that most of the urinary material consisted of its glucuronide and sulfate conjugates. In man, somewhat more glucuronide than sulfate was excreted in the urine whereas, in rats given the same or a larger dose (0.5 or 5 mg/kg), excretion of the sulfate conjugate was 2.5 to 3 times greater than excretion of the glucuronide.[252] Additional information on the conjugation of harmol is summarized below in the discussion of harmalol metabolism.

Hydroxylation may provide an additional metabolic pathway for harmine, however this has only been demonstrated *in vitro* using liver microsomes from mice induced with 3-methylcholanthrene.[254] The major metabolite produced in these experiments was 6-hydroxyharmine (6-hydroxy-7-methoxyharman). An additional metabolite, either 3- or 4-hydroxy-7-methoxyharman, was also detected and both of these compounds, as well as the harmol which was also formed, were further metabolized to unidentified compounds.

The metabolic pathways shown in Figure 5 indicate that **harmaline** (95) and harmine are subject to similar pathways of metabolism. In addition, dehydrogenation at the 3,4-position in harmalol occurs to form the fully aromatic derivative harmol (97). This reaction was suggested by the results of Villeneuve and Sourkes.[249] Ho et al.[255] obtained evidence which indicated that the dehydrogenation reaction occurred in rats with harmalol but not with harmaline. This contradicted an early report[256] which stated that harmaline was metabolized to harmine in rabbits and dogs. The latter investigation also reported the degradation of harmaline to harminic acid (6-azaindole-2,3-dicarboxylic acid), but this metabolite was not detected in rat urine.[255] Villeneuve and Sourkes[249] found that harmaline was metabolized more slowly in rats than was harmine and that little unchanged compound was excreted, a finding in agreement with that of Flury[256] in rabbits. However, Ho et al.[255] found that nearly 10% of the dose (40 mg/kg, s.c.) of

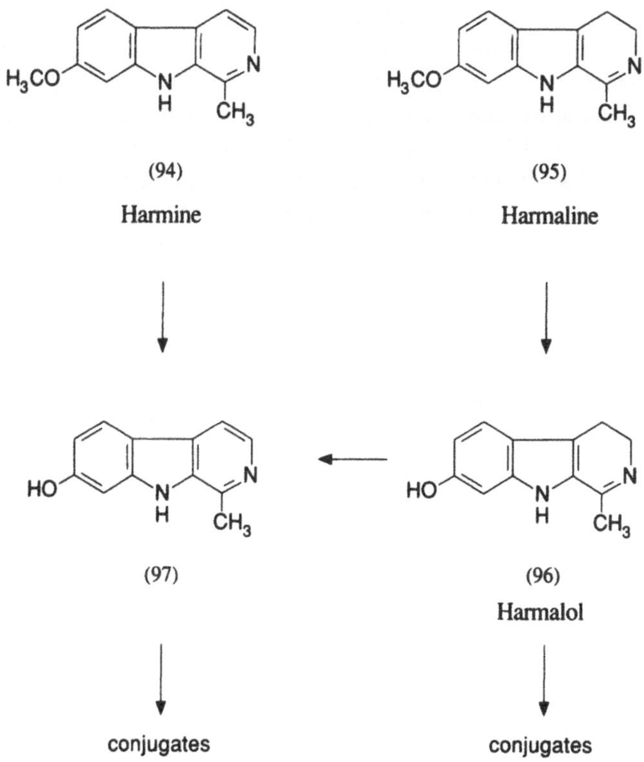

FIGURE 5. Metabolic pathways of harmine, harmaline, and harmalol.

harmaline was excreted unchanged in the urine of rats within 8 h. Very little free harmalol or free harmol was found in the urine of the harmaline-treated rats. The bulk of the urinary metabolites, which amounted to slightly more than 60% of the dose in 4 d, consisted of conjugated material. Interestingly, this was mainly the glucuronide conjugate of harmalol rather than the sulfate which is the case with harmol. Daly et al.[257] reported that harmalol is an excellent substrate for the hydroxylating system in rabbit liver microsomes. The hydroxylated product was not identified, however it was readily converted by the catechol-O-methyltransferase also present in the preparation to a hydroxymethoxy derivative. Oxidation has therefore taken place at either the 6- or 8-position and from the knowledge that indoles are commonly hydroxylated at the 6-position,[258] it seems likely that the product is the 6,7-dihydroxy derivative which is then methylated, probably to a mixture of the two possible monomethyl ethers. This reaction, which is analogous to that noted above with harmine which is hydroxylated at the 6-position by mouse liver microsomes, does not appear to have been demonstrated with harmalol or related compounds *in vivo*.

It is noted above that conjugated harmalol is predominantly the glucuronide whereas the sulfate is the major conjugate of harmol. This subject was studied in more detail by Mulder and Hagedoorn.[259] They found that i.v. administration to rats resulted in extensive excretion of conjugates of both compounds in the urine and bile. In agreement with the earlier results, harmol was found to be excreted as the sulfate (70%) and the glucuronide (30%) whereas harmalol was excreted mainly as the glucuronide, only a trace of the sulfate being found. In addition, *in vitro* studies showed that both substrates were readily glucuronidated by UDP-glucuronosyltransferase and that harmol was a good substrate of phenol sulfotransferase. However, harmalol was only poorly converted to its sulfate conjugate by this enzyme. The glucuronidation of harmol and harmalol using UDP-glucuronic acid and the glucuronosyltransferase enzyme from guinea pig liver microsomes was studied by Wong and Sourkes.[260,261] Studies on the conjugation and excretion of harmol have subsequently become a subject of widespread interest. Several investigations dealt with the kinetics of sulfation and glucuronidation of harmol in the isolated rat liver[262,263] and in the rat *in vivo*.[264] Additional studies examined these reactions in isolated rat hepatocytes[265-267] and at extrahepatic sites.[268] These conjugation reactions in relation to the biliary excretion of harmol *in vivo* and in the perfused liver were studied by Mulder et al.[269] Mulder and Bleeker[270] studied the differences in the biliary and urinary excretion of the sulfate and glucuronide conjugates of harmol in the mouse, rat, guinea pig, rabbit, and cat. They found that large species variations occurred with respect to the proportions of the two conjugates which were excreted.

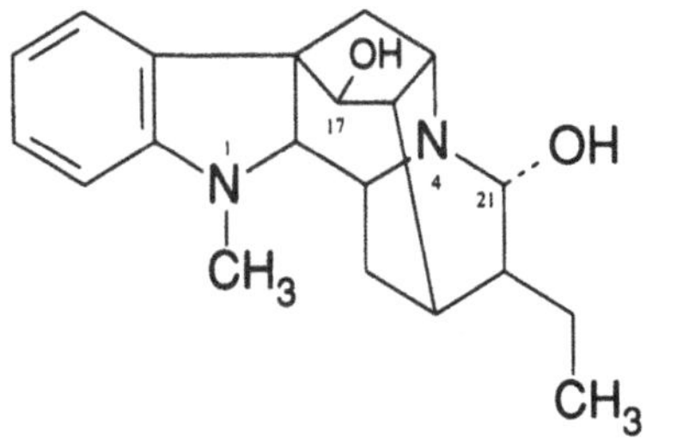

(98)

Ajmaline

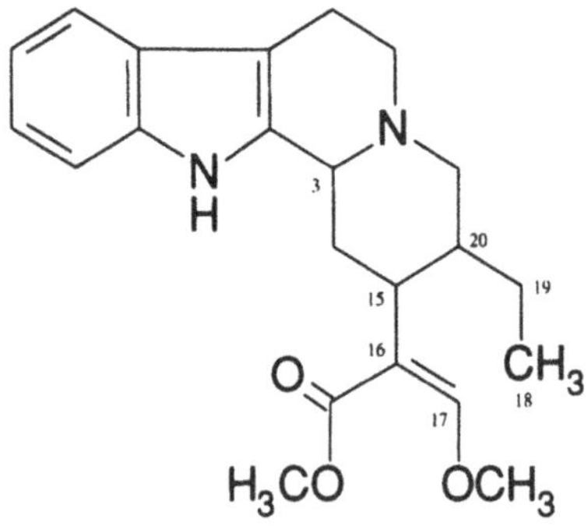

(99)

Corynantheidine

Iven[271] found that **ajmaline** (98) underwent extensive metabolism in mice and that only 5% of the dose (10 mg/kg, i.v.) was excreted unchanged in the urine. The metabolism of ajmaline in patients receiving the drug by continuous infusion (480 mg/d) was studied by Köppel et al.[272] They identified ajmaline and numerous metabolites in the urine and found that these were excreted mainly in conjugated form. The major metabolic pathway was hydroxylation of the aromatic ring which formed the main metabolite, hydroxyajmaline, and which additionally gave a dihydroxylated derivative including an *O*-methylated metabolite formed from the latter. Another main pathway was oxidation at N-4 to give the *N*-oxide derivative. Additional metabolic steps included conversion of the hydroxyl groups at C-17 and C-21 to keto derivatives, *N*-demethylation at N-1, and reduction at C-21 to give a ring-opened derivative resulting from fission of the C–N bond. Several of the metabolites identified were formed by a combination of two or three of these reactions.

Corynantheidine (99) differs from **corynantheine** by its lack of a double bond at C-18,19. Three other diastereoisomeric forms of (99), **dihydrocorynantheine, isocorynantheidine,** and **hirsutine,** have different configurations at the C-3, C-15, and C-20 positions. The metabolism of these four mitragyna alkaloids by rabbit liver microsome preparations was studied by Beckett and Morton.[273] They found that the major metabolic reaction *in vitro* with corynantheidine, dihydrocorynantheine, and isocorynantheidine was *O*-demethylation at C-17. This metabolic route accounted for most of the metabolism observed, especially with the first two compounds, and no other metabolic products were detected. With hirsutine, however, *O*-demethylation accounted for only ~5% of the total metabolism and another route is therefore mainly involved in the metabolism of this compound. These variations in the extent of *O*-demethylation were explained in terms of the preferred conformations of the various diastereoisomers, *O*-demethylation being greater in those cases where the indole and piperidine ring systems lie in the same plane.

The metabolism and excretion of **echitamine** (100) was studied in rats.[274] When the alkaloid was injected (~28 mg/kg, i.m.), blood levels increased for 2 h and then decreased to undetectable levels after 6 h. Unchanged compound equal to ~80% of the dose was present in urine samples taken during the first 48 h. Two unidentified metabolites were detected in urine samples obtained at 6 h and later. Chromatographic evidence indicated that these compounds were more polar than echitamine.

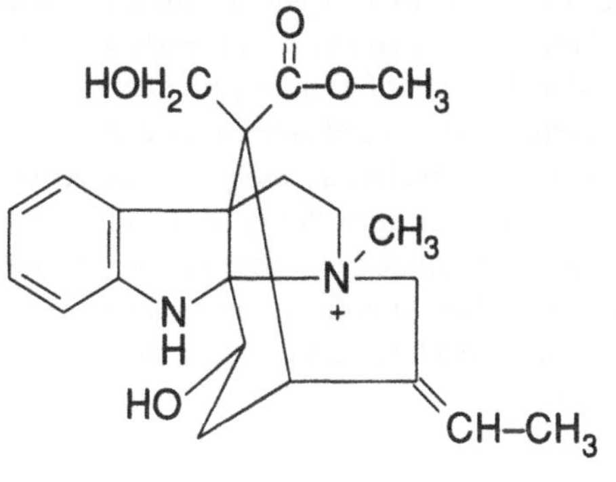

(100)

Echitamine

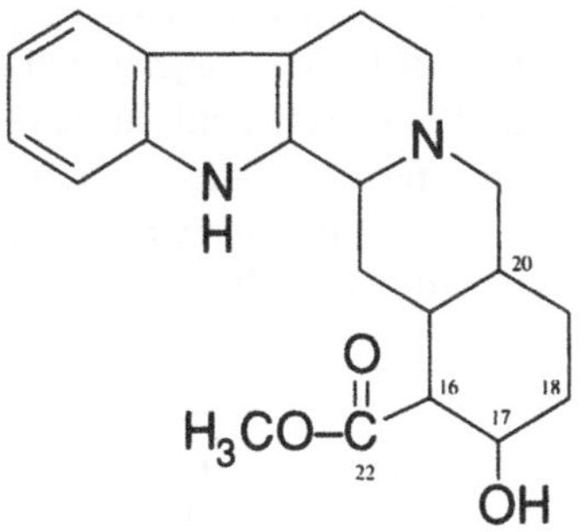

(101)

Yohimbine

The metabolism of **yohimbine** (101) in mice was investigated by Ho et al.[275] using ^{3}H-labeled material. The compound (10 mg/kg, i.p.) was found to be metabolized rapidly although some unchanged yohimbine was excreted in the urine. Of the 22% of the radioactivity excreted in the urine in 24 h, ~35 to 40% (i.e., 8% of the dose) was unchanged compound. Several unidentified metabolites were detected in liver and spleen extracts and two metabolites in addition to yohimbine were found in the urine at 3 and 24 h. Interestingly, distribution studies of the radioactivity showed that large amounts of activity were found in the intestine. Shortly after injection this consisted of metabolites, including yohimbic acid which results from the hydrolysis of the ester group at C-22. However, at 0.75 and 1.5 h, but not from 3 h onwards, most of the intestinal radioactivity was due to yohimbine itself. While fecal excretion of radioactivity was not studied, the above results and the finding that activity in the tissues after 24 h was low suggest that this is an important excretory route for yohimbine in mice. The disposition of an oral dose (10 mg) of yohimbine HCl in humans was studied by Owen et al.[276] Their results showed that absorption was rapid and that only ~0.35% of the dose was excreted unchanged in the urine in 24 h. These findings indicate that yohimbine underwent extensive metabolism, however the nature of the metabolites is not known. However, Owen et al. believed that hydrolysis of the ester moiety at C-22 is not a major metabolic route in man.

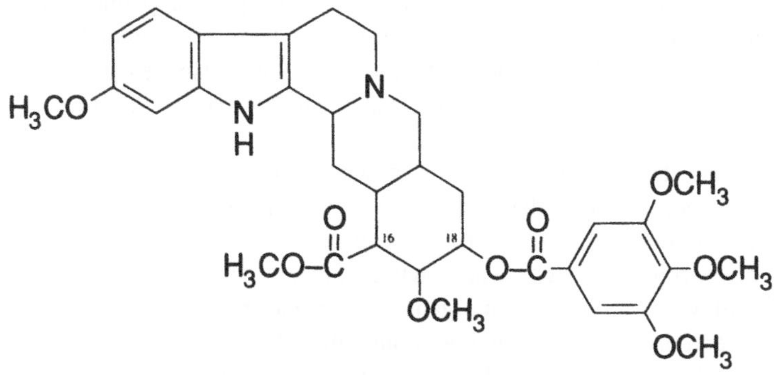

(102)

Reserpine

Reserpine (102) shows structural similarity to yohimbine, however it contains an additional ester group at C-18 which furnishes an important site of metabolism. Nonetheless, some elimination from the body in unchanged form also occurs. Sheppard et al.[277] found as much as 2.5% of the reserpine given i.v. to rats in the feces within 24 h as reserpine-like material. Similar values were found in guinea pigs[278,279] and the latter report also noted that nearly 10% of the administered reserpine was recovered unchanged in the feces of a rat only 6 h after dosing. In mice, as much as 35% of the dose (i.v.) may be excreted unchanged in the feces within 24 h.[280] Maass et al.[281] reported that an average of 62% of the dose, mainly as reserpine, was recovered in the feces of human subjects within 4 d after administering 0.25 mg of reserpine orally. However, they detected little unchanged compound in the urine from these subjects. Maronde et al.[282] also reported little or no urinary excretion of free reserpine in man. Similar results showing amounts of urinary reserpine ranging from nil to a few tenths of a percent of the dose were reported in mice,[280] rats,[277,283] and guinea pigs.[278,279]

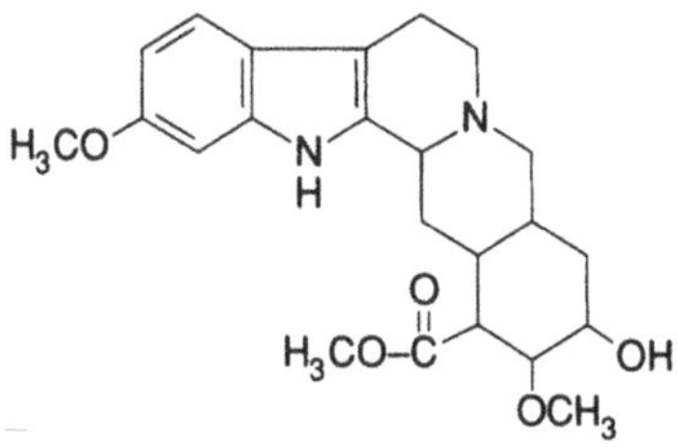

(103) 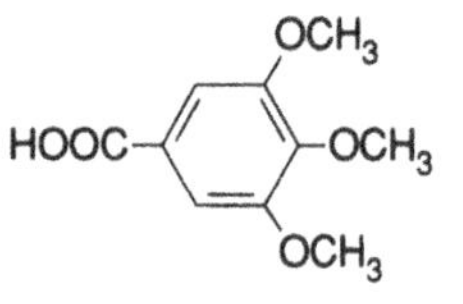   (104)

Although our present understanding of the metabolism of reserpine is incomplete, it appears that hydrolysis of the ester linkage at C-18 is the dominating metabolic step. This leads to the formation of methyl reserpate (103) and 3,4,5-trimethoxybenzoic acid (104), both of which have been detected in various organs or excreta in numerous investigations. Dhar et al.[284,285] reported detecting both metabolites in various organs, urine, and feces from rats given reserpine by both the oral and parenteral routes. Sheppard et al.,[277] using reserpine labeled with ^{14}C in the 4-methoxy group of the trimethoxybenzoic acid moiety, found that 7 to 14% of the radioactivity was excreted in the urine of rats as trimethoxybenzoic acid-like material within 24 h of injection. The corresponding values in guinea pigs were 5 to 24%, of which the bulk of the radioactivity was metabolite (104) or similar material.[278] Similar results were reported when reserpine labeled with ^{14}C in the carbonyl group of the trimethoxybenzoic acid moiety was employed.[279] Using the latter labeled compound in mice, Numerof et al.[280] found that essentially all of the radioactivity excreted in the urine in 24 h (40 to 70% of the dose) consisted of trimethoxybenzoic acid. Maggiolo and Haley[286] detected the latter metabolite and methyl reserpate (103) in the brains of mice given reserpine and they also showed that the concentration of metabolites increased during the 5-d experimental period whereas that of the unchanged compound decreased. In man given radioactive reserpine orally, trimethoxybenzoic acid was detected in the blood, where it appeared to be a major metabolite shortly after dosing.[287] It was also the main metabolite found in the urine 24 h following dosing.[281] Maronde et al.[282] calculated that ~6% of the daily oral dose was converted to urinary methyl reserpate.

In contrast to the abundant documentation showing that reserpine is hydrolyzed at the C-18 ester linkage, no reports are available which show that the monoester resulting from hydrolysis at C-16 is formed. However, this reaction may occur as reserpic acid (105) has been reported to be formed from reserpine, although the data on this point are conflicting. Dhar et al.[284,285] claimed that reserpic acid was found in several organs of reserpine-treated rats as well as in intestinal contents, feces, and urine. In addition, both it and methyl reserpate (103) were formed when reserpine was incubated with rat liver slices. However, Glazko et al.[283] found no trace of reserpic acid in the urine of reserpine-treated rats.

(105)

(106)   (107)

An important consideration in the metabolism of reserpine involves the enzyme systems responsible for its hydrolysis. Interestingly, Glazko et al.[283] found that much higher levels of urinary methyl reserpate (103) were found in rats after oral than after parenteral administration. This hydrolysis was carried out *in vitro* by rat intestinal mucosa but not by similar preparations from dogs and monkeys. The two latter species also formed relatively small amounts of methyl reserpate *in vivo*. It is also known that the liver is active in hydrolyzing reserpine in several animal species.[278] Stitzel et al.[288] and Stawarz and Stitzel[289] carried out a detailed investigation of the *in vitro* metabolism of reserpine by liver preparations. They found that most of the esterase activity was associated with the microsomal fraction and also that the mechanism is dependent upon the monooxygenase system, as indicated by the requirement for NADPH and O_2 as well as by inhibition of the reaction by inhibitors of this system. These results suggest that the hydrolysis of reserpine is preceded by an oxidative step, most likely *O*-demethylation of the methyl ether group in the 4-position of the trimethoxybenzoyl moiety. This reaction sequence will form syringoyl methyl reserpate (106) and then syringic acid (107), although the reports do not make it clear if these metabolites were actually formed. The study of Sheppard et al.[279] is pertinent in this regard as they found that rat liver slices converted significant amounts of reserpine to syringoyl methyl reserpate. This ester was also excreted in the feces of reserpine-treated rats. Further information supporting the existence of this metabolic pathway has been obtained using reserpine labeled with ^{14}C in the 4-methoxy group of the trimethoxybenzoyl moiety.[277,278] They showed that as much as 20% of the dose was converted to [14]CO_2 *in vitro* using rat liver slices, with lesser amounts in preparations from dogs, mice, and guinea pigs. Also, nearly a fourth of the radioactivity was recovered in the respiratory CO_2 within 6 h after giving rats the labeled compound by i.v. injection. Nonetheless, Sheppard et al.[279] found that the urinary trimethoxybenzoic acid-like material in guinea pigs and rats was trimethoxybenzoic acid with only a small contribution by syringic acid being noted in the case of rats.

Interest in the metabolism of reserpine has declined in recent years and no new data of importance has emerged. It therefore appears that several uncertainties about the exact nature of these reserpine metabolites will remain. A comprehensive summary of the biological fate of reserpine was given by Stitzel.[290]

Raubasine (ajmalicine) (108) is a *Rauwolfia* alkaloid which shows structural similarity to reserpine. The metabolic products of raubasine have not been identified and the available data deal with its excretion in several animal species. Löhr and Bartsch[291] found that less than 0.1%

of the compound was excreted unchanged in the urine following oral dosage to rats. Both raubasine and some unidentified metabolites were excreted in the bile. Marzo et al.[292] studied the absorption and excretion of [³H]raubasine in dogs and humans. Using a dose of 0.8 mg/kg, the cumulative urinary excretion of radioactivity in dogs was 22% (i.v.) and 13% (p.o.) in 3 d. The corresponding fecal excretion was 51 and 57%, respectively. The extensive excretion of radioactivity in the feces following i.v. dosage was shown to be due to substantial biliary excretion of the compound. Following oral dosage (0.24 mg/kg) of labeled raubasine to humans, the urinary and fecal excretion of radioactivity was 29 and 24%, respectively, in 3 d.

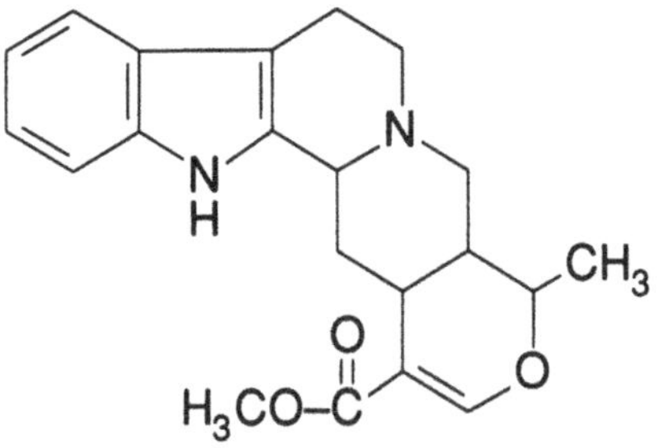

(108)

Raubasine

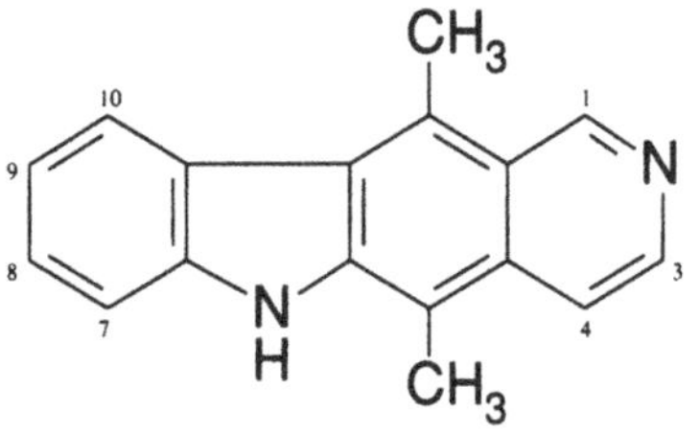

(109)

Ellipticine

The biological disposition of **ellipticine** (109) has been studied in several animal species. An early investigation in mice[293] showed that fecal excretion was the main route of elimination and that only trace amounts were present in the urine. The dose used was large (250 mg/kg, p.o., s.c., or i.p.) and it was found that the rate and extent of excretion was dependent on the formulation of the preparation given. Chadwick et al.[294] carried out an extensive investigation of [1-¹⁴C]ellipticine disposition in mice, rats, and dogs (6 mg/kg, i.v.) and monkeys (3 mg/kg, i.v.). They found that all species excreted 80% of the dose in the feces and 10% in the urine in 24 h. No radioactivity was detected in the expired CO_2 of rats. Unchanged ellipticine was not a major excretory product and, in the rat, evidence pointed to the formation of 9-hydroxyellipticine and its subsequent conjugation and excretion in the bile as a dominant pathway. At least nine metabolites from rat liver or bile were detected, but the two major ones appeared to be the glucuronide and sulfate conjugates of 9-hydroxyellipticine. Characterization of these two conjugates was made by Branfman et al.[295] and it was demonstrated that hydroxylation occurred without involving an arene oxide/NIH shift mechanism.[296] The formation of 9-hydroxyellipticine was shown by Lesca et al.[297,298] to be carried out by rat liver microsomes. Additionally, formation of minor amounts of 7-hydroxyellipticine in rats has also been reported.[299,300] Administration of 9-hydroxyellipticine itself to mice and rats showed that it underwent extensive metabolism, mainly to its glucuronide conjugate.[301]

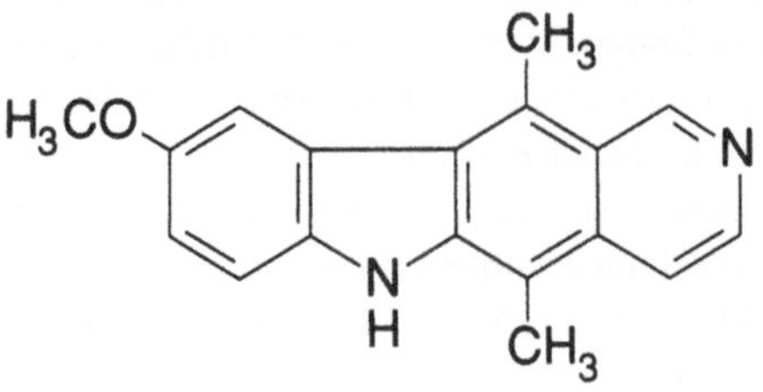

(110)

9-Methoxyellipticine

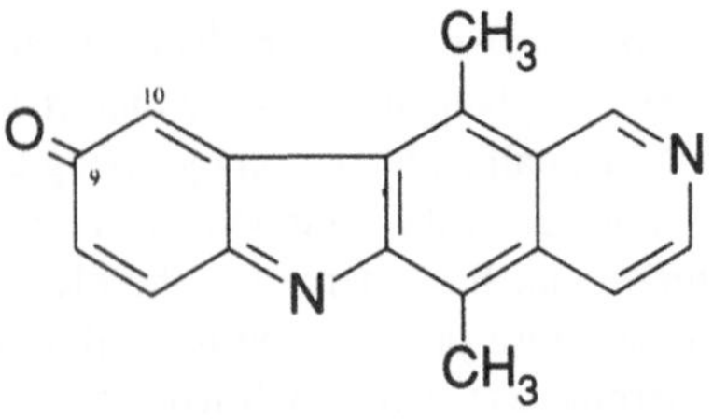

(111)

Administration of **9-methoxyellipticine** (110) to rats resulted in the urinary and biliary excretion of its demethylated derivative, 9-hydroxyellipticine.[302] Following i.v. administration (~10 mg/kg), about one-fourth of the dose was excreted in the urine in 48 h. Most of this material was in conjugated form. In bile-duct cannulated rats, the value for biliary excretion (15 h) was over 70% and this material was entirely conjugated. The O-demethylation of 9-methoxyellipticine by liver microsomes from mice and rats was reported by Roy et al.,[303] however a finding of greater potential interest was that the demethylation reaction may be catalyzed by a peroxidase in the presence of hydrogen peroxide.[304-306] This reaction involves cleavage of the oxygen-aromatic carbon bond and is thus a demethoxylation. The products formed are the corresponding quinone-imine (111) and methanol. Meunier et al.[307] discussed this reaction and the ability of the electrophilic quinone-imine to form adducts with nucleophiles. The results of Braham et al.[308] showed that such an adduct, a glutathione derivative, was excreted in the bile of rats given 9-methoxyellipticine and that it was formed when 9-methoxyellipticine or 9-hydroxyellipticine was incubated with rat liver microsomes fortified with glutathione. They suggested that the adduct was formed from the quinone-imine to give a 9-hydroxyellipticine derivative containing a glutathione residue at the 10-position.

Numerous closely related investigations have been carried out on the metabolism of some semisynthetic N^2-methyl quaternary derivatives of ellipticine. The 9-hydroxy derivative has been prominent among these and the results have confirmed and extended the findings summarized above on the naturally occurring ellipticine alkaloids. The topics covered with this group of compounds include the formation of the electrophilic quinone-imine derivative[309,310] and the formation of O-conjugates (glucuronide and sulfate) and glutathione conjugates including the resultant cysteine derivatives.[311-314]

Ibogaine (112) was excreted in unchanged form in the urine of rabbits and man following oral doses of 20 mg/kg and 5 mg, respectively.[315] An unidentified urinary metabolite of ibogaine was also detected in the unhydrolyzed fraction. It seems reasonable to expect that most of the urinary metabolites will be found as conjugates, however this fraction was not investigated.

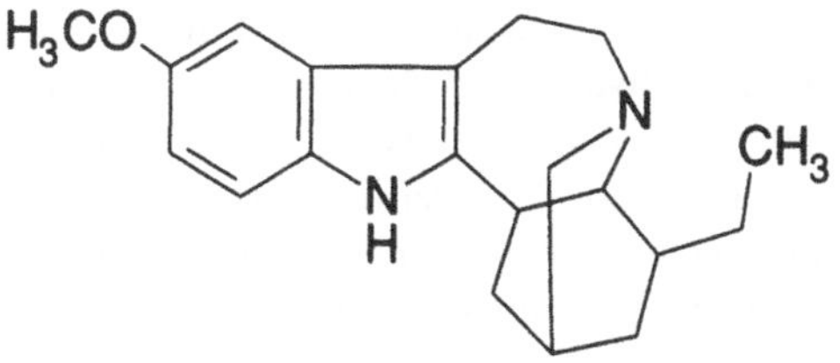

(112)

Ibogaine

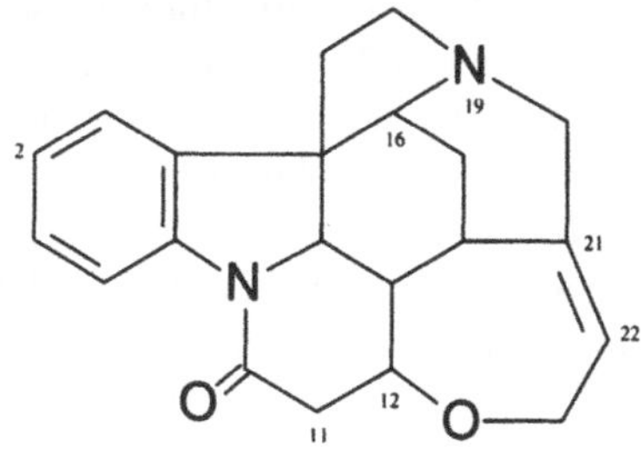

(113)

Strychnine

Although the metabolism of **strychnine** (113) has been the subject of sporadic investigations over a long period of time, relatively little was known about the identity of the metabolites until recently. An important early development in this area was the investigation of Hatcher and Eggleston,[316] whose report is also a useful guide to the earlier extensive and often conflicting literature. They found that only small amounts of unchanged compound were excreted in the urine 24 to 48 h after administering strychnine to guinea pigs, cats, and dogs and that none was excreted in the feces. Their evidence pointed to a rapid destruction of the compound in the body, especially in the case of guinea pigs. Perfusion experiments using guinea pig or dog liver showed rapid metabolism of strychnine by this organ. No unchanged strychnine was excreted in the bile of dogs given the compound i.v. The low urinary excretion of unchanged strychnine was also observed in experiments in a young human.[317] Excretion values of ~1 to 5% in 24 h were recorded with oral doses of ~1 to 3 mg taken at the ages of 6 to 8 months. Later, at ages of 36 or 42 months, values of 3% and nearly 14% were found with a dose of ~11 mg.

The *in vitro* metabolism of strychnine was studied by Adamson and Fouts[318] using liver preparations from several species. With homogenates, increasing activity was found in dog, rat, mouse, rabbit, and guinea pig liver, the activity in rabbits being about ten times greater than that in dogs. This activity was associated with the monooxygenase system. It was found that strychnine metabolism led to a loss in its biological activity but the nature of the metabolic products formed was not studied. Kato et al.[319] showed that the prior administration of inducing substances, especially phenobarbital, resulted in increased strychnine metabolism in rats together with increased tolerance to its toxic effects. The metabolism of strychnine was also shown to increase in rabbits following treatment with phenobarbital.[320] The latter investigation also dealt with the metabolic products formed when the alkaloid was incubated with rabbit liver slices or 9000-g supernatant fractions. Four metabolites were detected, a minor one being 2-hydroxystrychnine which was found to possess only ~1% of the toxicity shown by strychnine itself. The identities of the two major *in vitro* metabolites of strychnine were not determined.

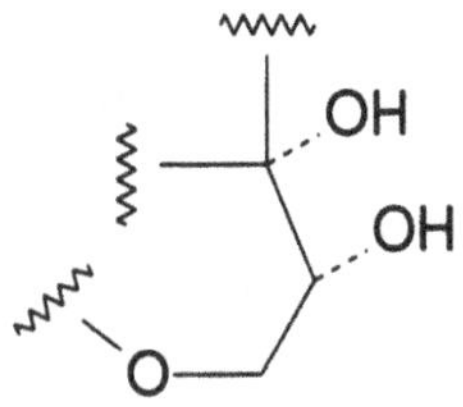

(114) 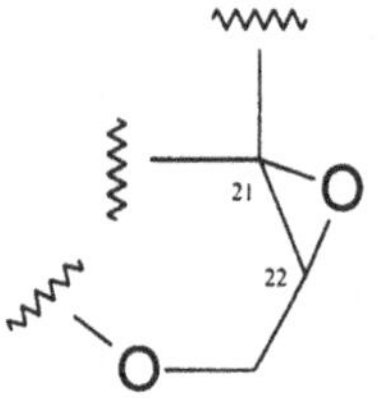   (115) 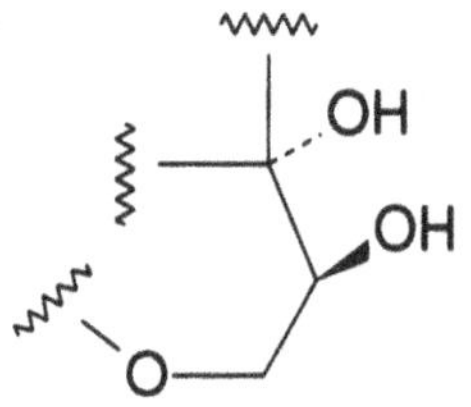   (116)

Two recent studies have confirmed the rapid metabolism of strychnine *in vitro*[321] and *in vivo*.[322] Also, the structures of several strychnine metabolites were determined in these investigations. The *in vitro* studies used the 9000-g supernatant fraction from rat and rabbit liver and five strychnine metabolites were isolated from the latter incubates. Three of these were identified as strychnine-*N*-oxide (oxidation at N-19), 2-hydroxystrychnine, and 21α,22α-dihydroxy-22-hydrostrychnine (114). The *N*-oxide was the major metabolite. The remaining metabolites were believed to be the 21,22-epoxide (115) and 11,12-dehydrostrychnine. The disappearance of strychnine from the incubates was much greater than the formation of these five metabolites. This suggests that relatively large amounts of unidentified compounds may also be formed. The *in vivo* experiments were carried out in rats given [³H]strychnine (0.5 mg/kg, s.c.). Nearly 90% of the radioactivity was excreted in 24 h. The ratio between fecal and urinary excretion was ~2:1 and the amounts of unchanged strychnine in the feces (~3%) and urine (~6%) were small. Fractionation of the urinary and fecal samples revealed that more than half of the radioactivity was due to free metabolites, however additional acid-hydrolyzable and residual polar compounds were also present. In agreement with the *in vitro* results described above, strychnine-*N*-oxide, 2-hydroxystrychnine, strychnine-21,22-epoxide (115), and 21α,22α-dihydroxy-22-hydrostrychnine (114) were detected in the urine and feces. The 11,12-dehydro derivative was not found in the urine or feces, however an additional metabolite, 16-hydroxystrychnine, was identified. The major identified metabolite was the epoxide (115) and about equal amounts of two stereoisomeric diols, 21α,22α-dihydroxy-22-hydrostrychnine (114) and 21α,22β-dihydroxy-22-hydrostrychnine (116), were formed.

Brucine (117) is 2,3-dimethoxystrychnine and its metabolism was studied in regard to its demethylation. Tsukamoto et al.[323] reported that brucine underwent selective *O*-demethylation in the rabbit, giving mainly 3-hydroxy-2-methoxystrychnine which was excreted in the urine largely as its glucuronide conjugate. In addition, small amounts of the isomeric 2-hydroxy-3-

methoxy compound, a nonphenolic base, and unchanged brucine were also detected in the urine. Fecal excretion of all of these metabolites occurred, but to a very minor extent. A similar metabolic pattern showing preferential demethylation *meta* to the nitrogen atom was found when brucine was incubated with the 9000-g supernatant fraction from rabbit liver.[324]

The vinca alkaloids include both simpler types such as vincamine and dimeric compounds of greater structural complexity such as vinblastine and vincristine. When ^{14}C-labeled **vincamine** (118) was given to rats (50 mg/kg, i.v.), most of the material was eliminated within 24 h, partly unchanged and partly as unidentified metabolites.[325] Other studies using [3H]vincamine indicated that the urinary excretion of unchanged compound in rats ranged from 3 to 7% following doses of 10 or 20 mg/kg (p.o., i.v., or i.p.).[326,327] Vigano et al.[328] also reported little excretion of unchanged vincamine in the urine of rats. Ventouras et al.[329] found that humans excreted ~70% of the dose (15 mg) in the urine in 48 h when [3H]vincamine was given orally. Little of this material appears to be unchanged compound, however, and Siegers et al.[330] reported a value (24 h) of only 6 to 7% of the dose (30 or 60 mg, p.o.) in a group of five subjects. Iven and Siegers[326] found at least four metabolites in rat urine, none of which was vincaminic acid which could have arisen by hydrolysis of the methyl ester group. The main urinary metabolites were sulfate and glucuronide conjugates. Two of the aglycons were shown to be the 6-hydroxy and 6-keto derivatives of vincamine.[328] These metabolites were also formed by dogs and humans given vincamine.[331] Vereczkey et al.[327,332] confirmed the identities of these metabolites following the administration of vincamine (10 mg/kg) to rats or its incubation with rat liver homogenates and also showed that the 6-hydroxylated compound consisted of both the α- and β-isomers. About 70% of the urinary material was due to these metabolites. The hydrolysis product, vincaminic acid, was not detected as a metabolite, however a derivative containing a keto group at position 14 was found. This compound, (+)-eburnamonine (vincamone), was considered to be a degradation product of vincaminic acid. Vereczkey[333] subsequently reported that vincaminic acid was detected in the plasma and urine of vincamine-treated rats when a sufficiently sensitive method of detection was employed. Vereczkey et al.[327] found that biliary excretion of vincamine metabolites was extensive in the rat. Using ^{3}H-labeled compound (10 mg/kg, i.p.), ~40% of the radioactivity was recovered in the bile in 6 h. About 30% of this material was due to the metabolites noted above. Other results indicate that only a few percent of the dose was excreted unchanged in the bile of rats.[326]

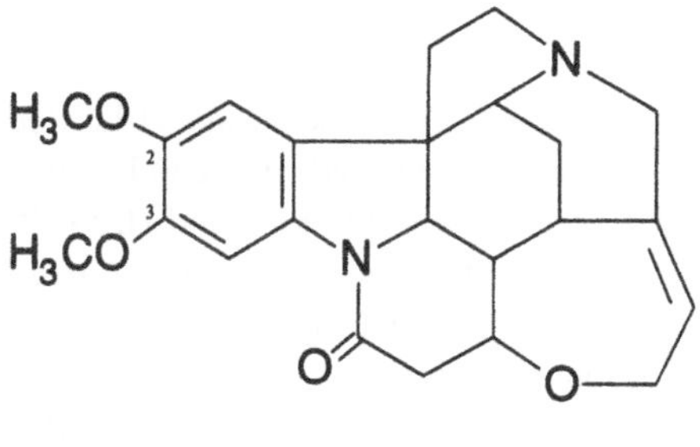

(117)

Brucine

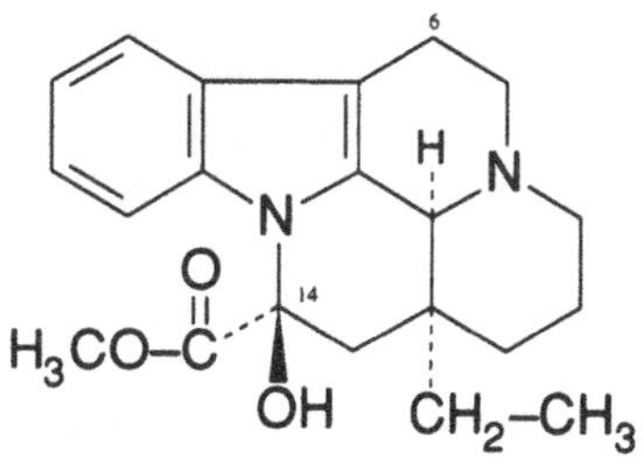

(118)

Vincamine

Most of the studies on **vinblastine** (119a) have dealt with its biological disposition rather than with the metabolic products formed. These investigations thus provide information on the routes and extent of excretion of material from the body. The fate of injected [3H]vinblastine was investigated in rats.[334-337] Urinary excretion of radioactivity accounted for ~5% of the dose in 12 h and little was subsequently lost by this route. Only ~1% of the dose was found in the urine as unchanged vinblastine. On the other hand, biliary excretion of ~20% of the radioactivity occurred within 24 h, about a tenth of which was unchanged compound. Creasey et al.[338] carried out similar experiments in dogs. Following a dose of 0.15 mg/kg (i.v.), the urinary excretion of radioactivity levelled off at 12 to 17% of the dose after 3 d. Fecal excretion was more prolonged, amounting to 30 to 36% after 9 d and still increasing. Three unidentified metabolites were detected chromatographically in urine, bile, and feces. The bile was a major excretory route of the injected vinblastine and most of the material was due to unchanged compound. The urinary and fecal excretion of unchanged vinblastine in rhesus monkeys amounted to ~11 and 7%, respectively, in 4 d following an i.v. dose (0.2 mg/kg).[339] The findings suggested that vinblastine underwent extensive binding in the tissues.

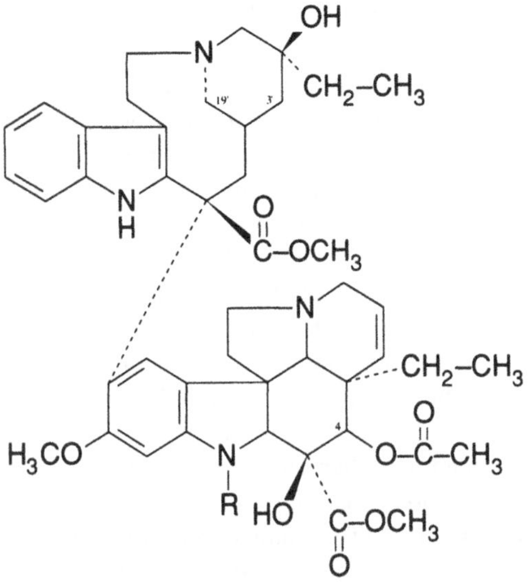

(119)

(a) Vinblastine, R = CH$_3$

(b) Vincristine, R = CHO

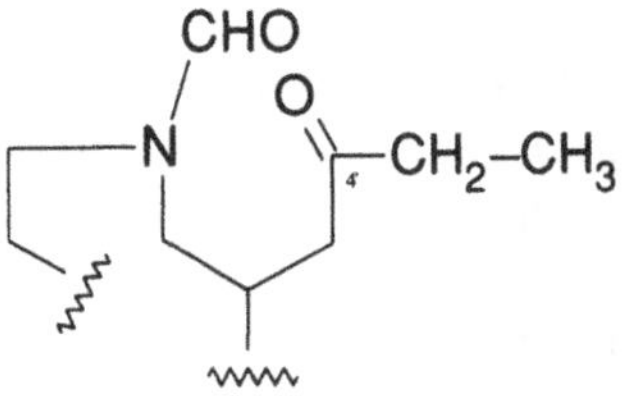

Pharmacokinetic studies of vinblastine in humans have employed nonlabeled and ^{3}H-labeled material, thus allowing for the determination of the excretion of unchanged compound as well as of total metabolites. Using material labeled in the 4-acetyl group, Owellen and Hartke[340] showed that most of the radioactivity was excreted in the feces. The two individuals excreted 25 and 41% of the dose (0.2 to 0.25 mg/kg, i.v.) by this route in 72 h. The corresponding values for urinary excretion were 19 and 23%. Appreciable amounts of unchanged compound were found in the urine, however little was detected in the feces. A similar study using vinblastine labeled in the aromatic ring showed that >70% of the radiolabel was retained in the body after 6 d.[341] About 5% of the dose was excreted as unchanged vinblastine in the urine in 48 h. This value is similar to that reported later (3.5% in 36 h) by Vendrig et al.[342] for unchanged compound following the injection of unlabeled vinblastine. Lu et al.[343] reported that <10% of the generally labeled vinblastine given as a single i.v. injection was excreted in the urine in 24 h as unchanged compound. Additionally, metabolites accounted for a further 5%. Several studies have shown that small amounts of the deacetylated metabolite, 4-deacetylvinblastine, were excreted in the urine of patients.[341,342] This metabolite was also detected in tissues of vinblastine-treated mice and in tumor xenografts in mice.[344] Interestingly, 4-deacetylvinblastine and four other derivatives have been shown to be degradation products of vinblastine.[345] They were formed when the alkaloid was incubated in glycine buffer at pH 7.4 or 8.8. The four compounds were tentatively identified as an isomer of vinblastine, 19'-hydroxy-3',4'-dehydrovinblastine, 19'-oxovinblastine, and 19'-oxo-3',4'-dehydrovinblastine. The *in vitro* metabolism of vinblastine by peroxidase[346] or by the human serum copper oxidase ceruloplasmin[347] resulted in the formation of catharinine and, in the latter instance, of additional metabolites. Catharinine results from scission of the piperidine ring of vinblastine as shown in structure (120).

The metabolic disposition of tritiated **vincristine** (119b) in rats and dogs was studied by Castle et al.[348] The bile was found to be the major excretory route. Biliary excretion was dose-dependent, relatively more being excreted via this route when a higher dose (1 mg/kg, i.v.) than when a lower dose (0.1 mg/kg) was given. When ^{3}H-labeled vincristine was administered at the lower dosage to rats the cumulative urinary and fecal excretion of radioactivity in 72 h was 15 to 17 and 70%, respectively. Relatively little (<10%) of the radioactivity in the bile and urine was due to metabolites. These were not identified but the urinary metabolites appeared to be more polar than those found in the bile. It was concluded that vincristine is not metabolized extensively in the rat. A study in mice, rats, dogs, and rhesus monkeys showed that [^{3}H]vincristine was excreted unchanged and as metabolites in the urine of all of these species.[349] The urinary radioactivity generally consisted of about equal amounts of unchanged vincristine and metabolites. The total values were 47% in mice (48 h), 14% in rats (6 h), and 10 to 25% in dogs and monkeys (24 h). Another study in the rhesus monkey gave a value of 9% for urinary excretion (4 d) and 28% for fecal excretion.[339] These findings indicate that vincristine undergoes considerable tissue binding. Pharmacokinetic studies in humans indicated that [^{3}H]vincristine and its metabolites were excreted mainly in the feces.[350] Nearly 70% of the radioactivity was recovered in the feces in 72 h compared with only 12% in the urine. A similar value (~10% in 24 h) was recorded in adults[351] and ~25% of the dose was recovered in the urine in 48 h in two pediatric patients.[352]

As noted above, the bile was shown to be an important route of vincristine excretion in rats and dogs.[348] This has been confirmed in studies using [^{3}H]vincristine which showed that dogs excreted ~10% of the radioactivity in 24 h and monkeys ~30% in 4 h, both as unchanged compound and as metabolites.[349] Rats excreted 25 to 35% of the radioactivity in the bile in 4 h, and about a half to two-thirds of this appeared to be unchanged vincristine.[353] Castle and Mead[354] also found rapid and extensive excretion of radioactivity in the bile of rats following injection of vincristine. They noted that the bile initially contained mainly unchanged compound, however the proportion consisting of metabolites increased with time. An experiment in a human patient showed that 22 and 50% of the injected radioactivity was excreted in the bile in 24 h and 72 h, respectively.[355] Much of the excreted material was in the form of metabolites which rapidly appeared in the bile.

The presence of numerous metabolic or breakdown products of vincristine has been reported in plasma, bile, and urine of rats[354] and in plasma and urine[350] and bile[355] of humans. The formation of several vincristine metabolites by rat hepatocytes was also reported.[356] Interestingly, rapid decomposition of vincristine occurred when the alkaloid was merely incubated in buffer or biological fluids.[355] Sethi and Thimmaiah[357] studied the degradation products of vincristine formed by incubation in glycine buffer (pH 7.4 or 8.8). 4-Deacetylvincristine was definitely identified and other products were believed to be an isomer of vincristine, an isomer of 4-deacetylvincristine, 4-deacetyl-3-deoxyvincristine, and *N*-formylleurosine (identical with vincristine except that the oxygen atom of the 4'-hydroxy group is connected via an epoxide bond to C-3'). 4-Deacetylvincristine and several other compounds were detected in tissues of vinblastine-treated mice.[344] In addition, *N*-deformylvincristine was identified as a urinary metabolite of vincristine in humans.[358] No detectable metabolism of vincristine was found in rat liver homogenates;[354] however, incubation with rat intestinal microorganisms was shown to result in its conversion to two metabolites more polar that vincristine itself.[359]

QUINOLINE DERIVATIVES

The only examples in this group are provided by the cinchona alkaloids, the most important of which are the stereoisomeric pairs cinchonine and cinchonidine with structure (121) and their 6′-methoxy derivatives quinidine and quinine.

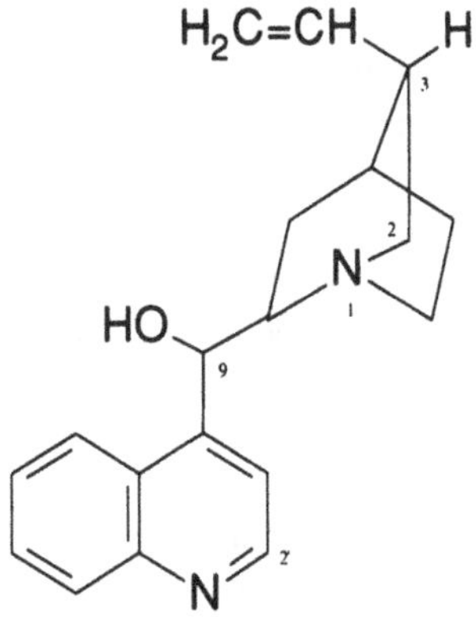

(121)

Cinchonine and cinchonidine

Information on the metabolism of **cinchonine** (121) has been obtained in studies in man. Urinary metabolites were shown to account for ~85% of the daily dose.[360,361] This material consisted of unchanged compound (4 to 5%), 2′-hydroxycinchonine (cinchonidine carbostyril) (55%) and a further oxidation product in which an additional oxygen atom was found in the quinuclidine moiety. Renal clearance of the unchanged compound but not the 2′-hydroxy derivative was depressed upon alkalinization of the urine. It was noted that cinchonine was metabolized much more rapidly in man than were the other cinchona alkaloids, a finding also discussed by Taggart et al.[362] The results clearly pointed to the initial step in cinchonine metabolism being oxidation at the 2′-position followed by hydroxylation in the quinuclidine moiety. This metabolic sequence was also demonstrated by Brodie et al.[363] While the site of the second hydroxy group was not determined, it was believed to be in the 2-position. In view of the abundant findings which indicate that the analogous reaction with quinine and quinidine (see below) gives rise via allylic hydroxylation to the 3-hydroxy derivative, it seems reasonable to assume that the latter pathway is also operative in the case of cinchonine.

The pattern of urinary metabolites of **cinchonidine** (121) in man is similar to that described above for its stereoisomer. No quantitative data are available but Brodie et al.[363] showed that the 2′-hydroxy product (carbostyril) was the major metabolite, being about five times as abundant as unchanged compound and the dihydroxy metabolite which were excreted in approximately equal amounts. In addition to these *in vivo* experiments in man, *in vitro* studies employing the quinoline oxidizing enzyme from rabbit liver (see below) were carried out by Knox.[364] Cinchonidine was oxidized much more rapidly than the other three cinchona alkaloids, the rates for the (−)-isomers being 15 to 25 times greater than those for the corresponding (+)-isomers (cinchonine and quinidine).

Studies of the metabolism of quinine and quinidine are far more numerous than those with cinchonine and cinchonidine. The older literature on **quinine** (122) metabolism, sometimes conflicting, was summarized by Williams.[365] In short, the most noteworthy features of these early investigations were the claims that quinetine and hemoquinic acid were urinary metabolites. The former compound differs from quinine in having the vinyl group at C-3 replaced with a carboxy group while the latter compound was believed to be 6-methoxyquinoline-4-ketocarboxylic acid. Williams considered that their being true metabolites of quinine was an open question and Watabe and Kiyonaga[366] suggested that quinetine is not a urinary metabolite of quinine in rabbits.

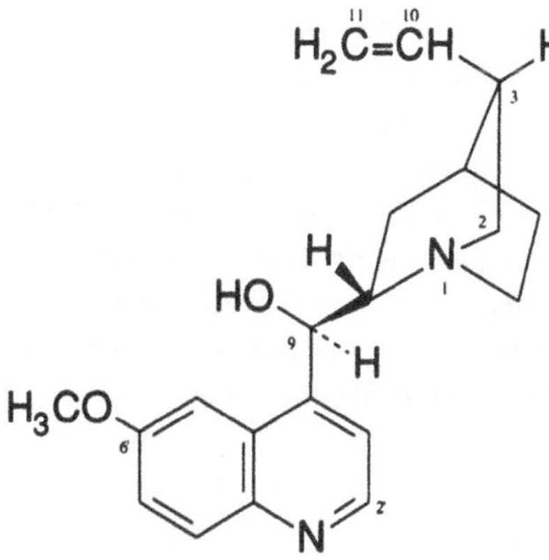

(122)

Quinine

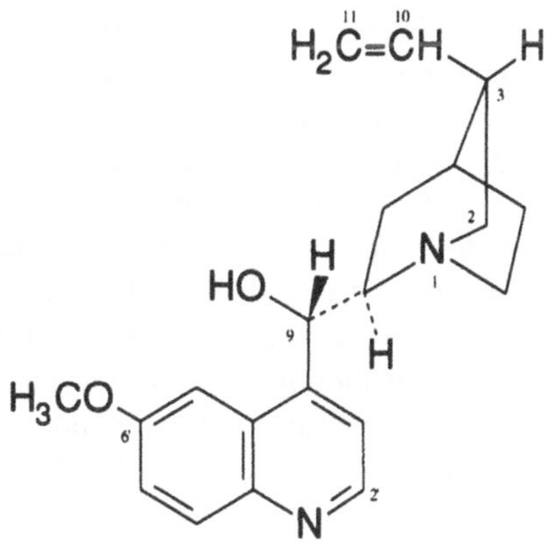

(123)

Quinidine

A subject of interest in many of the early studies of quinine metabolism was quinine oxidase, an enzyme system which hydroxylates the quinoline moiety and which is especially active in rabbit liver. While quinine oxidase activity is found in many tissues from numerous animal species, Kelsey and Oldham[367] whose article is also a useful guide to the early literature on this subject, found that rabbits contained the greatest amount of enzyme of all the species studied (mouse, rat, guinea pig, rabbit, cat, dog, sheep, pig, steer, monkey, and man). The liver was the most abundant source of the enzyme. Using the rabbit liver enzyme, Kelsey et al.[368] isolated the metabolite which was shown by Mead and Koepfli[369] to be 2'-hydroxyquinine. Kelsey and Oldham[367] reported that little or no enzyme activity was present in rabbit liver before birth, after which it increased to adult levels sometime after the age of 6 weeks. Also, quinine oxidase activity was reduced in rabbits during late pregnancy and the early postpartum period. Several subsequent investigations have dealt with the properties and substrate specificity of rabbit liver quinine oxidase. Knox[364] showed that oxidation is limited to the α-position of quinoline and derivatives and that the rate of oxidation correlated with the activity of the α-hydrogen. The enzyme has properties similar to, and is associated with, the flavoprotein liver aldehyde oxidase. Lang and Keuer[370] found that quinine oxidase activity was localized mainly in the microsome plus soluble fraction with little activity being found in the mitochondrial fraction of rabbit liver. Führ and Kaczmarczyk[371] demonstrated considerable quinine oxidase activity *in vitro* using human liver and also showed that quinine metabolism was inhibited by the simultaneous administration of N-methylnicotinamide, another excellent substrate of the enzyme. Pertinent related data on quinoline oxidation *in vitro* by rabbit liver preparations were given by Sax and Lynch.[372]

The urinary excretion of unchanged quinine was demonstrated in rats[373] after i.p. dosage and in humans,[363,374-376] but not in rabbits,[366] after oral dosage. The value found by Haag et al.[374] was ~17% of the dose (0.5 g) when the urine was acidic, decreasing to approximately half of this following alkalinization of the urine.

The metabolic pathways of quinine in man were studied by Brodie et al.[363] who found that the major urinary metabolite was hydroxylated in the quinuclidine moiety and not at the 2'-position to give a carbostyril derivative as is the case with cinchonine and cinchonidine. About three times as much of this compound was excreted as was the carbostyril derivative and some of it appeared to undergo an additional hydroxylation in the quinuclidine group. As noted above with the two cinchona alkaloids lacking the 6'-methoxy group, the location of the quinuclidine hydroxy group was not determined, although it also was believed to be at the 2-position. A subsequent study[366] of quinine metabolism in rabbits suggested that the site of this hydroxylation is more likely the 3-position as a result of allylic hydroxylation. Six urinary metabolites of quinine (200 mg/kg, p.o.) were detected in this study. These were excreted mainly in conjugated form and all were hydroxylated in the 2'-position. The major urinary metabolite was 2'-hydroxyquinine and spectral data indicated that it exists as 2'-quininone, i.e., in the 2-quinolone form. The two other main metabolites were identified as 2',3-dihydroxyquinine and 2',6'-dihydroxycinchonidine. Quantitative measurements indicated that conversion to these metabolites accounted for 35 and 6% of the dose, respectively. The identification of the latter metabolite shows that O-demethylation occurred. The data did not indicate if this reaction occurred before or after oxidation at the 2'-position, but no direct evidence was obtained which indicated that the O-demethylated product of quinine, 6'-hydroxycinchonidine, was excreted in rabbit urine. However, it was not excluded that this metabolite could be one of the three unidentified minor metabolites. Axelrod[377] reported the O-demethylation of quinine by a rabbit liver preparation containing microsomes and soluble fraction, but the nature of the product or products was not determined.

Subsequent studies[373,376,378] of quinine metabolism in rats and man confirmed the formation of 3-hydroxyquinine, 2'-quininone, and the O-demethylated metabolite (6'-hydroxycinchonidine). Additionally, these investigations demonstrated that quinine is also metabolized via the epoxide-diol pathway. Both quinine-10,11-epoxide and its hydrolysis product, quinine-10,11-dihydrodiol, were identified as urinary metabolites. The latter compound was present in the form of two diastereoisomers.[373] Horning et al.[378] characterized several additional metabolites of quinine in the urine of rats and humans. These were a hydroxyquinine (not the 3'-hydroxy derivative), a dihydroxyquinine, two hydroxy-O-desmethylquinines (hydroxy-6'-hydroxycinchonidines), and a trihydroxy-10,11-dihydroquinine. An additional metabolic pathway of quinine in man is oxidation to an N-oxide derivative,[379] however its quantitative importance is not known.

Commercial samples of quinine are known to contain dihydroquinine, usually in amounts ranging from 3 to 6% of the total alkaloidal content.[380] Studies of quinine metabolism will therefore reflect this fact. Liddle et al.[376] detected, in addition to the metabolites noted above, both 3-hydroxydihydroquinine and 6'-hydroxydihydrocinchonidine in the urine of a patient receiving quinine sulfate.

The metabolism of **quinidine** (123) closely follows the pathways described above for its stereoisomer quinine. A similar qualitative pattern of metabolism of the two isomers in man was noted in an early study by Brodie et al.,[363] however they found that excretion of unchanged compound relative to that of its metabolites was greater with quinidine than with quinine. The urinary excretion of unchanged quinidine has also been reported in mice,[271] rats,[373,378] and rabbits.[381]

Brodie et al.[363] demonstrated that the major metabolite in man of quinidine, like quinine, is a compound hydroxylated in the quinuclidine moiety. The 2'-hydroxy (carbostyril) derivative was also excreted, but in smaller amounts than with quinine. This preponderance of excretion of the former metabolite in human urine has been noted in numerous subsequent investigations.[382-384] Palmer et al.[385] reported that the urinary metabolites were excreted mainly unconjugated, however some metabolites were also found as glucuronides. The latter did not include a conjugate of quinidine itself, indicating that no conjugation took place at the 9-position. Two metabolites were isolated which were found to be polyoxygenated compounds having lower molecular weights than quinidine, however the two major characterized metabolites were derivatives containing oxygen in the 2'-position or in the quinuclidine moiety. The former metabolite was shown to be 2'-quinidinone, i.e., the keto form of 2'-hydroxyquinidine. Carroll et al.[386] confirmed this structure and also showed that the second metabolite was 3-hydroxyquinidine. Beermann et al.[387] also reported that the latter compound was the 3-hydroxy derivative and its stereochemistry was established as (3S)-3-hydroxyquinidine.[388] In contrast to that seen in man, rats given quinidine did not excrete 2'-quinidinone in the urine.[373]

Quinidine also undergoes dealkylation to *O*-desmethylquinidine (6-hydroxycinchonine). This metabolite, both in the free and conjugated form, was detected in human urine[389] and accounted for ~1 to 2% of the dose.[390] Both free and conjugated *O*-desmethylquinidine were excreted in the urine of rabbits dosed with quinidine.[381] Horning et al.[378] and Barrow et al.[373] detected, in addition to quinidine, 2'-quinidinone, 3-hydroxyquinidine, and this demethylated derivative, several new metabolites of quinidine in the urine of rats and man. Three of these were formed via the epoxide-diol pathway and both quinidine-10,11-epoxide and a diastereoisomeric pair of quinidine-10,11-dihydrodiols formed by hydrolysis of the epoxide were identified as urinary metabolites. Additional metabolites which were partially characterized were a hydroxyquinidine (not the 3'-hydroxy derivative), a dihydroxyquinidine, two hydroxy-*O*-desmethylquinidines (hydroxy-6'-hydroxycinchonines), and a trihydroxy-10,11-dihydroquinidine. The dihydrodiol derivatives, which were formed in significant amounts, were subsequently shown to be major quinidine metabolites in rabbit plasma and urine[381] and also the sole biliary metabolite in dogs given quinidine by infusion.[383,391] An additional point of interest with the formation of quinidine epoxide is the finding that this intermediate appears to be responsible for the ability of quinidine to bind with liver microsomal proteins following metabolic activation.[392] Dihydroquinidine, which lacks the double bond at C-10,11, did not appear to bind to proteins.

An additional quinidine metabolite which was detected in relatively small amounts in human plasma[384] or urine[382,383] is the *N*-oxide derivative which arises from oxidation of the aliphatic nitrogen. Guentert et al.[393] confirmed its chemical structure. The *N*-oxide was also found as a minor quinidine metabolite in rabbit plasma and urine.[381] Guengerich et al.[394] studied the formation of this metabolite and of 3-hydroxyquinidine by various forms of liver microsomal P-450 enzymes. The major oxidative reactions of quinidine are not catalyzed by the monooxygenase isoenzyme which is responsible for the oxidation of sparteine or debrisoquine in humans.[395,396]

Two novel metabolites of quinidine and its 3-hydroxy derivative were identified as esters of lactic acid, probably attached at the 9-hydroxy group.[397] These interesting conjugates were identified in the urine and in several tissues of a patient following acute quinidine intoxication. It remains to be seen if they are formed following normal quinidine dosage.

As with quinine which contains some dihydro compound, commercial samples of quinidine typically contain 5 to 9% of dihydroquinidine.[380] The biological disposition of the dihydro derivative is not significantly different from that of quinidine[398] and many of the metabolic pathways are identical for both compounds. Guentert et al.[393] found that dihydroquinidine was metabolized to its *N*-oxide derivative and Resplandy et al.[399] detected this metabolite and 3-hydroxydihydroquinidine in human plasma and urine. A third metabolite, formed by ω-oxidation, was identified as 11-hydroxydihydroquinidine.

ISOQUINOLINE DERIVATIVES

(±)-**Reticuline** (124) was shown by Kametani et al.[400,401] to undergo a cyclization reaction to give a tetrahydroprotoberberine derivative when administered (200 mg/kg, i.p.) to rats. The urine, following hydrolysis with β-glucuronidase, contained unchanged reticuline and a small amount of coreximine (125a). This metabolite and its positional isomer scoulerine (125b) were formed when reticuline was incubated with rat liver homogenate. When the alkaloid was incubated with the 9000-g supernatant fraction from rat liver, formation of coreximine and *N*-demethylation to norreticuline was observed.[401,402] Two additional metabolites formed under the latter incubation conditions were shown to be the morphinandienone alkaloid pallidine (126) and the aporphine alkaloid isoboldine (127). An experiment with deuterium-labeled reticuline indicated that the *N*-methyl group was not retained in the tetrahydroprotoberberines (125a) and (125b). Pallidine, isoboldine and scoulerine were not detected as urinary metabolites when rats were given reticuline by i.p. injection.[400]

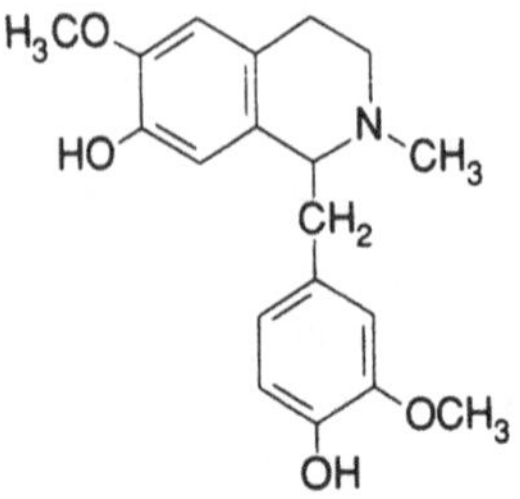

(124)

Reticuline

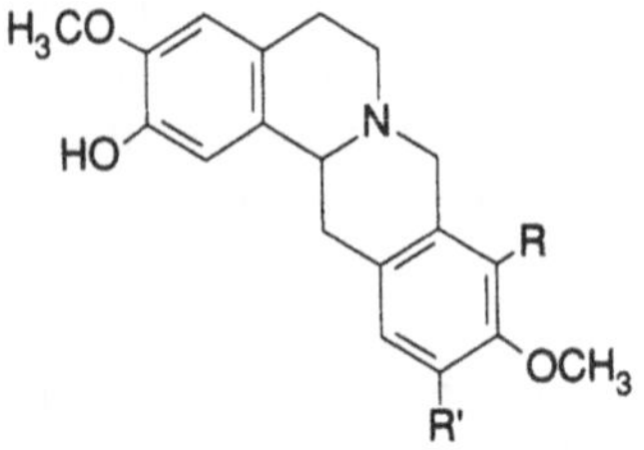

(125)

(a) R = H , R' = OH

(b) R = OH , R' = H

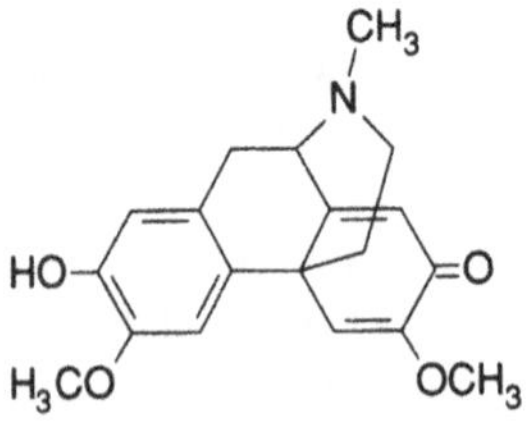

(126)

Pallidine

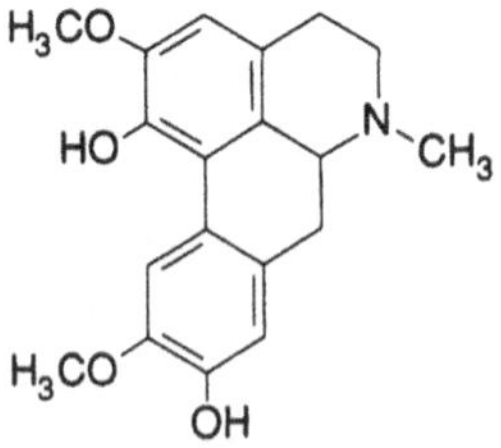

(127)

The above example of the mammalian synthesis of tetrahydroberberine derivatives from benzyltetrahydroisoquinolines is analogous to that known to occur with alkaloids derived from neuroamines[403,404] and closely related compounds.[405] Meyerson and Davis[406] described an *S*-adenosylmethionine-dependent methyltransferase system from rat liver which was capable of inserting the methylene bridge required in this transformation.

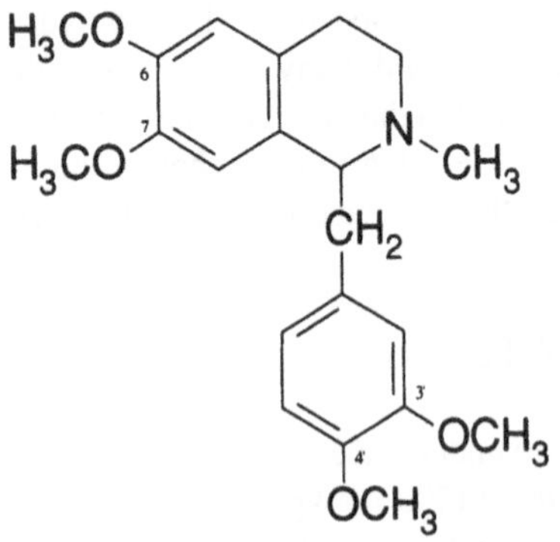

(128)

Laudanosine

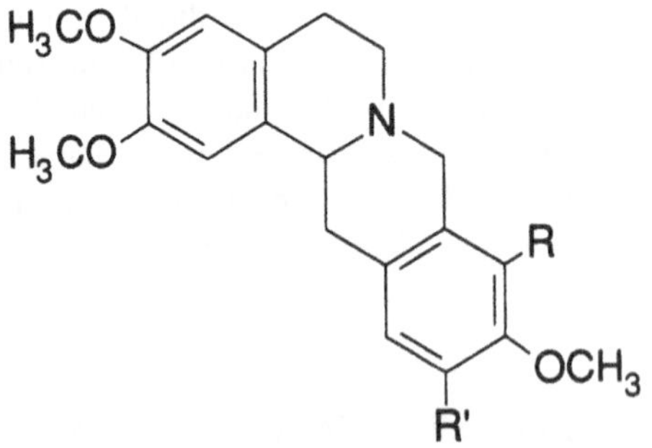

(129)

(a) R = H , R' = OCH$_3$

(b) R = OCH$_3$, R' = H

The metabolism of **laudanosine** (128) was studied in rabbits, dogs, and man.[407] Urine and bile were collected from rabbits and dogs given the alkaloid by i.v. dosage (1 to 3 mg/kg). In human patients, urine was collected following the administration of atracurium, a neuromuscular blocking agent which spontaneously decomposes to laudanosine. The results indicated that laudanosine was extensively metabolized in all species. The 4-h urine and bile samples from dogs contained <3 and <0.1%, respectively, of the dose as unchanged compound. The patterns of metabolites excreted were similar in all three species and demonstrated extensive demethylation. In fact, all five of the possible monodemethylated metabolites were detected. These included the major metabolite in all species, pseudocodamine (4′-desmethyllaudanosine), the trace metabolite codamine (7-desmethyllaudanosine), laudanine (3′-desmethyllaudanosine), pseudolaudanine (6-desmethyllaudanosine), and norlaudanosine. In addition, two didemethylated derivatives, norpseudolaudanine and norpseudocodamine, were also identified. The demethylation of laudanosine to norlaudanosine and a didemethylated derivative by rabbit liver homogenates and microsomal preparations was reported by Sharma et al.[408] The *in vitro N*-demethylation of laudanosine to norlaudanosine was shown to be carried out by a 9000-g supernatant fraction from rat liver.[401] Very small amounts of two protoberberine derivatives were also identified in these experiments. These were the isomeric tetramethoxy derivatives xylopinine (129a) and tetrahydropalmatine (129b). This pathway is similar to that noted above with reticuline.

Studies on the metabolism of **papaverine** (130) extend over a period of about 75 years, although useful data on the transformations which occur are of fairly recent date. In an early study, Zahn[409] was not able to detect unchanged compound or metabolites in the excreta of rabbits, cats, and dogs given papaverine i.p. Only traces (<1% of the dose) of unchanged papaverine were found in the urine of rabbits 24 h after i.v. administration (20 mg/kg).[410] Axelrod et al.[411] similarly reported values of <1% for urinary papaverine in rats, guinea pigs, and dogs given the alkaloid by injection. Belpaire and Bogaert[412] found only trace amounts of unchanged compound in the urine of rats, guinea pigs, rabbits, cats, and dogs following i.v. dosage of the alkaloid. In man, Elek and Bergman[413] failed to find any unchanged papaverine in the urine following oral or parenteral dosage (~200 to 600 mg), although one or more probable derivatives were noted. Axelrod et al.[411] noted that <1% of the dose was recovered unchanged in the 48-h urine of subjects receiving papaverine orally (10 mg/kg). This value for the urinary excretion of unchanged papaverine in man was confirmed by Guttman et al.[414] and Wilén and Ylitalo.[415]

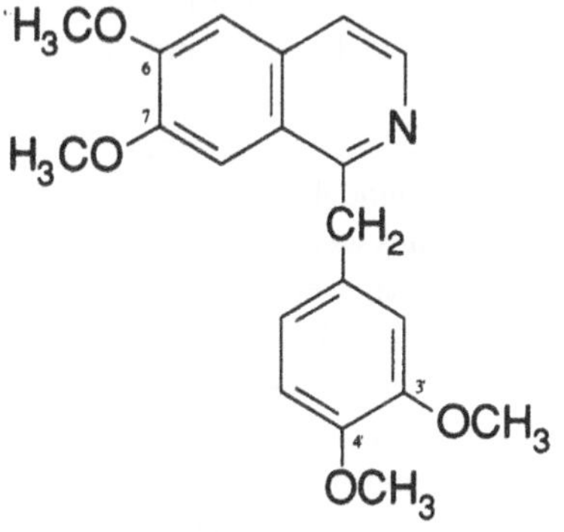

(130)

Papaverine

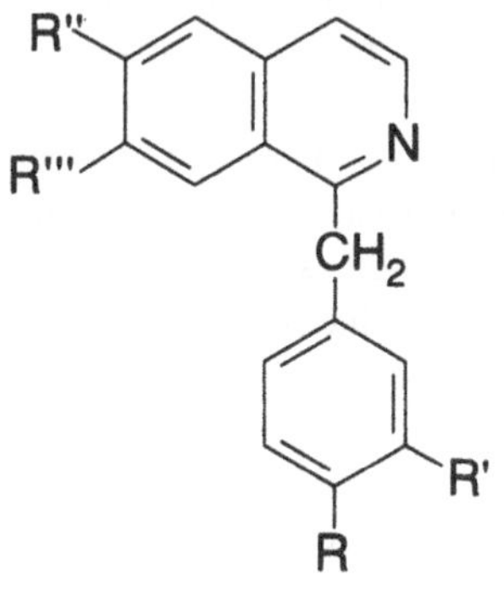

(131)

(a) R = OH , R' = R" = R''' = OCH$_3$

(b) R" = OH , R = R' = R''' = OCH$_3$

(c) R''' = OH , R = R' = R" = OCH$_3$

(d) R' = OH , R = R" = R''' = OCH$_3$

(e) R = R" = OH , R' = R''' = OCH$_3$

Axelrod[377] reported that papaverine was a good substrate for the *O*-demethylating enzyme found in a rabbit liver preparation containing the microsomal and soluble fractions. Using tissue preparations from guinea pigs and rabbits, Axelrod et al.[411] subsequently showed that the liver is the most active site of papaverine metabolism. Formation of the monodemethylated metabolites by the 9000-g fraction of rat liver was stimulated by the inducing agent phenobarbital and inhibited by the compound SKF 525 A.[416] Rosazza et al.[417] found that the three desmethyl metabolites (131a), (131b), and (131c) were formed when papaverine was incubated with phenobarbital-induced rat liver microsomes. The 7-desmethyl derivative (131c) was the major metabolite in these studies, however about equal amounts of metabolites (131a) and (131b) were formed when liver microsomes from guinea pigs were used.

Axelrod[377] also found that urine from guinea pigs given papaverine (100 mg/kg, i.p.) contained three phenolic metabolites. The amounts of these metabolites increased following acid hydrolysis of the samples. One compound was identified as the 4'-*O*-demethylated derivative (131a), which accounted for ~70% of the phenolic material. Additional experiments in rats, guinea pigs, dogs, and man indicated that all species except the dog were fairly similar with respect to the amounts of conjugated phenolic material excreted. Typical values of ~45 to 65% of the dose for the amount of conjugated metabolites excreted were obtained. It was also noted that guinea pigs excreted much more unconjugated phenolic metabolites than was seen in the other species. Dogs excreted relatively little free and conjugated phenolic metabolites in the urine. As with the guinea pig, the major urinary phenolic metabolite in man was the 4'-*O*-demethylated compound.

Additional investigations, mainly using ³H- or ¹⁴C-labeled compound, have expanded our understanding of the biological disposition of papaverine and especially of the nature of the metabolites formed. Belpaire and Bogaert[418,419] showed that most of the radioactivity was eliminated within 24 h in the feces of rats given doses of 5 mg/kg, p.o., i.m., or i.v. After 4 d, the total fecal excretion of radioactivity was 68 to 76% whereas 11 to 18% was found in the urine, mainly in the first day. About 70% of an i.v. or i.m. dose of papaverine was excreted in the bile in 6 h. This result, together with the finding that absorption of radioactivity did not occur when the biliary papaverine metabolites were administered duodenally, indicates that enterohepatic circulation is not an important feature in the biological disposition of papaverine in rats. Hydrolysis of the biliary metabolites with a preparation containing β-glucuronidase and sulfatase gave rise to four metabolites. Belpaire et al.[420] showed that these were the three monodemethylated derivatives 4'-desmethylpapaverine (131a), 6-desmethylpapaverine (131b), and 7-desmethylpapaverine (131c), and the didemethylated compound 4',6-didesmethylpapaverine (131e). Studies on the excretion of the papaverine metabolites were extended to other species including guinea pigs, rabbits, cats, and dogs.[412,421] ³H-labeled papaverine (5 mg/kg) was given i.v. to these animals and rats and the biliary and urinary excretion of radioactivity measured for 5 to 6 h. Biliary excretion of radioactivity, which accounted for ~80% of the dose in rats in these experiments, averaged ~50% in the other species. The average values for urinary excretion of radioactivity were 8% (rat), 29% (guinea pig), 45% (rabbit), 7% (cat), and 10% (dog). These values, perhaps due to the shorter collection period, are mostly lower than those noted above for the urinary excretion of phenolic metabolites.[411] The bulk of the excreted material in both the bile and urine consisted of conjugates of the four phenolic metabolites noted above. However, quantitative differences were found, metabolite (131b) being most important in cats and metabolite (131a) in the other species. An unidentified fifth metabolite, which was not a monodemethylated derivative, was detected in the bile of dogs and, especially, cats.

Subsequent experiments were carried out in humans[422,423] and it was found that oral administration of papaverine led to the urinary excretion of four monodemethylated and one didemethylated derivatives. These included 3'-desmethylpapaverine (palaudine) (131d) in addition to the compounds noted above (131a, 131b, 131c, and 131e). The major metabolite was 6-desmethylpapaverine (131b), followed by 4'-desmethylpapaverine (131a) and lesser amounts of the other derivatives. About 50% of the dose was excreted in the urine in 48 h as these metabolites. These results show that excretion of papaverine metabolites by the urinary route is

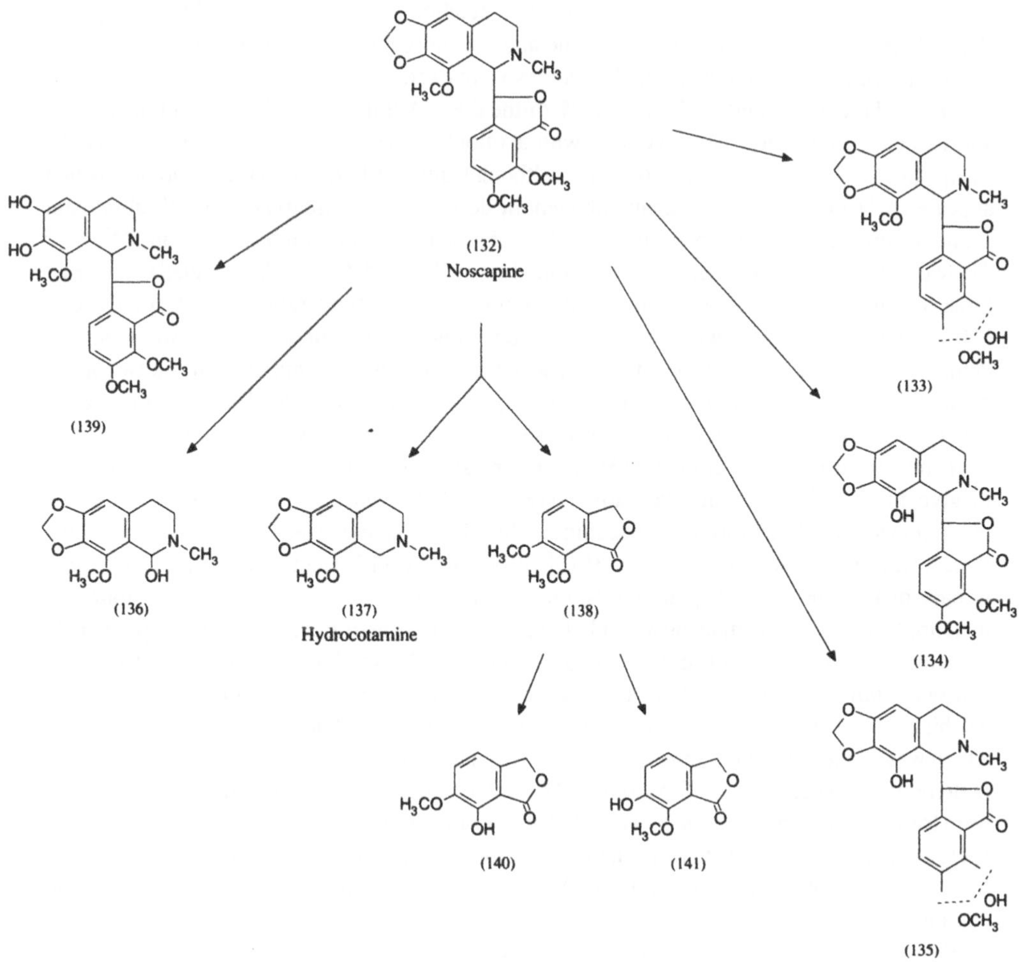

FIGURE 6. Metabolic pathways of noscapine.

more important in man than in experimental animals. Wilén and Ylitalo[415] gave [^{14}C]papaverine (200 mg, p.o.) to humans and found that 80% of the radioactivity was excreted in the urine in 48 h. They confirmed that the urinary metabolites were excreted in conjugated form (glucuronides and lesser but variable amounts of sulfates). They detected at least six metabolites and showed that two of these were 4'-desmethylpapaverine (131a) and 3'-desmethylpapaverine (131d).

Noscapine (narcotine) (132) (see Figure 6) is another opium alkaloid related chemically to papaverine. In an early study in cats, Cooper and Hatcher[424] found that only 0.3 to 0.5% of a total i.m. dose (200 to 300 mg/kg) was excreted in the urine in 4 to 5 d as unchanged compound. Nayak et al.[425] reported that essentially no unchanged noscapine was excreted in the feces and <1% of the dose (15 mg/kg, i.v.) was found in the urine of rabbits in 24 h. A value of ~5% was found for the urinary excretion of unchanged noscapine (0.5 mg/kg, i.v.) in rats.[426] In man, the urinary excretion of free and conjugated noscapine did not exceed 1% over a 6-h period following a 500 mg oral dose.[427] These results showing very low values for the urinary excretion of unchanged noscapine were confirmed by Tsunoda and Yoshimura.[428,429] Following oral doses in rats (120 or 150 mg/kg), rabbits (150 mg/kg), and man (10 mg), most of the noscapine excretion occurred within 24 h and accounted for <1% of the dose in the first two species and usually <0.1% in man. The bile was shown to be an important route of excretion of noscapine and/or its metabolites in mice[430] and rats.[426] This factor appears to explain the abundant excretion of radioactivity in the feces of rats when [^{3}H]noscapine was given by either oral or i.v. dosage.[426] Little excretion of unchanged noscapine was noted in breast milk of human mothers.[431]

Several reports have shown that noscapine is metabolized at a rapid rate.[425,426,432] Nayak[432] showed that rabbit liver and kidney slices metabolized noscapine and also that two glucuronide conjugates of unidentified phenolic derivatives were excreted in the urine in rats and rabbits. However, these accounted for less than 5% of the dose. A third highly fluorescent metabolite was detected which gave negative tests with alkaloidal, phenolic, and glucuronide reagents. Idänpään-Heikkilä[426] reported two major, although unidentified, biliary metabolites which appeared to be glucuronide conjugates of phenolic derivatives of noscapine. Several subsequent studies have characterized numerous metabolites of noscapine in rats, rabbits, and man.[428,429,433,434] These compounds are shown in Figure 6. Tsunoda et al.[433] found that rabbits excreted in the urine three free and three conjugated compounds following oral administration of noscapine (80 to 150 mg/kg). The former were shown to be unchanged compound and the two mono-*O*-demethylated metabolites (133) and (134) while the latter were conjugates of the two mon-odemethylated compounds and of a di-*O*-demethylated metabolite (135). The major metabolite was (133), however the total amount of these urinary metabolites was small. Göber et al.[434] confirmed the urinary excretion of (135) in rats and also showed that C–C-cleavage took place to give the two isoquinoline derivatives cotarnine (136) and hydrocotarnine (137) and a monodemethylated derivative of meconine (138). Tsunoda and Yoshimura[428] found that the three cleavage products cotarnine (136), hydrocotarnine (137), and meconine (138) were urinary metabolites of noscapine in rats, rabbits, and man. Meconine was the most abundant metabolite in rabbits and man, however both meconine and cotarnine were found in appreciable amounts (~4% of the oral dose of 120 mg/kg) in rats. Only small amounts (usually <0.1%) of hydrocotarnine were detected in all species. In a subsequent investigation, Tsunoda and Yoshimura[429] characterized three additional urinary metabolites of noscapine. These were (139) which was formed in rabbits and man by demethylenation of the methylenedioxy group and two *O*-demethylated derivatives of meconine. These two derivatives, compounds (140) and (141), were formed in all three species, however (140) was only a trace metabolite. Its isomer (141) was excreted mainly in conjugated form and was a major urinary metabolite of noscapine in humans. More than half of the dose (10 mg, p.o.) in one female subject was recovered as free and conjugated (141).

Hydrocotarnine (137) (see Figure 6) is itself an opium alkaloid and was found to be metabolized in rats to several *N*-demethylated, ring dehydrogenated, and ring oxygenated metabolites.[435] It has not been determined if these metabolites of hydrocotarnine are also formed when noscapine is administered.

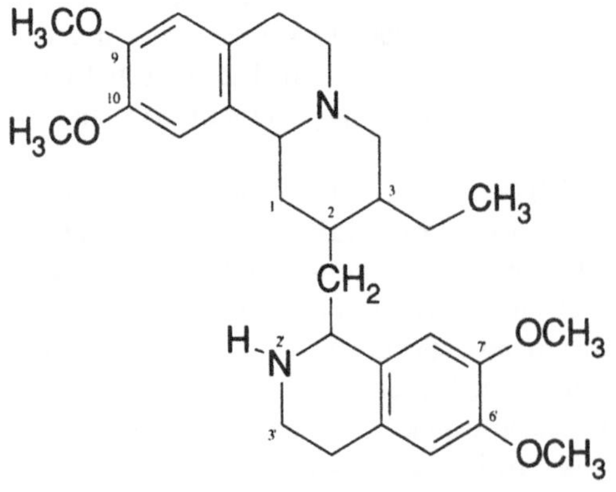

(142)

Emetine

Emetine (142), the major ipecac alkaloid, appears to be slowly eliminated from the body. Gimble et al.[436] found that little, if any, unchanged compound was present in the urine of rats and dogs receiving the alkaloid (1 to 2 mg/kg) by injection or in the urine of man after ten daily oral doses totaling 600 mg. Gehlert et al.[437] gave [¹⁴C]emetine to rats (1 or 5 mg/kg, i.v.) and found that only 47% of the dose was excreted in the urine and feces in 10 d. Slow excretion of radioactivity was also found in guinea pigs given [¹⁴C]emetine.[438] Following a series of six daily injections (1.87 mg base/kg, i.p.), 13% of the radioactivity remained in the body 8 d after the last injection. Of the remainder, 83% of the dose was found in the feces and only ~6% in the urine. No information was obtained concerning the nature of the possible metabolites, but most of the fecal material appeared to be present as metabolites as only 40% of the radioactivity was extracted when methods used for the determination of emetine were employed. Peeples and Dalvi[439] noted that little or no demethylation occurred when emetine was incubated with the demethylation system from rat liver microsomes.

Some relevant information on the metabolism of emetine may be obtained from studies which employed the semisynthetic analogue, 2,3-dehydroemetine.[440,441] Material labeled with ¹⁴C in the 3'-position was given to bile-duct cannulated rats. About 35% of the dose (5 mg/kg, i.v.) was found in approximately equal amounts in the bile and urine and ~1% in the feces in 72 h. Interestingly, 30 to 40% of the urinary radioactivity appeared to consist of unchanged 2,3-dehydroemetine while the remainder included several polar metabolites not susceptible to hydrolysis with β-glucuronidase and sulfatase. In the bile, however, only a trace of the excreted material was unchanged compound and most of these metabolites appeared to be glucuronide conjugates. One of these released 2,3-dehydroemetine upon hydrolysis, suggesting that it may be a 2'-N-glucuronide derivative. Other aglycons more polar than the original compound were detected but it is not clear if these were derivatives formed by O-demethylation or ring hydroxylation. The former possibility appears reasonable by analogy with the metabolism of other methoxyisoquinolines, however incubation of 2,3-dehydroemetine with the rat liver microsomal drug metabolizing system failed to show appreciable O-demethylation.

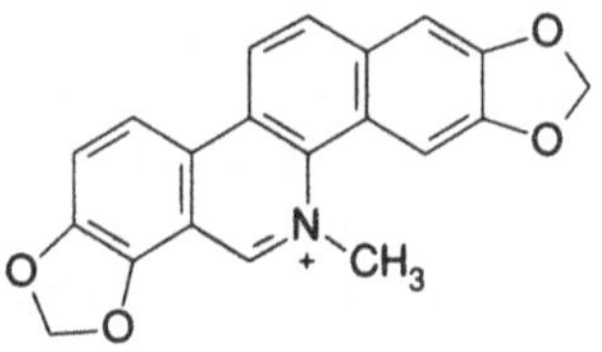

(143)

Sanguinarine

A single report on the metabolism of the argemone alkaloid **sanguinarine** (143) described its conversion to 3,4-benzacridine.[442] The metabolite was detected in the blood in isolated rat liver experiments, in the milk of lactating rabbits, and in rabbit urine after s.c. injection of the alkaloid (~2 mg/kg). No further reports have appeared to confirm this unusual reaction or to otherwise clarify the metabolic pathways of sanguinarine.

The urinary excretion of unchanged d-**glaucine** (144) following an oral dose (60 mg) in man is very low.[443] The values from three subjects ranged from 0.1 to 0.3% of the dose and most of the excretion occurred within 8 h.

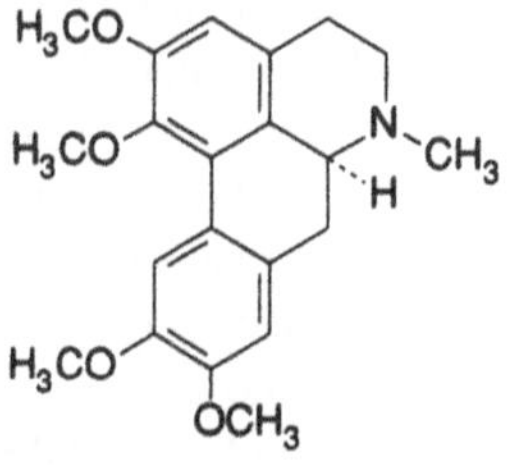

(144)

d-Glaucine

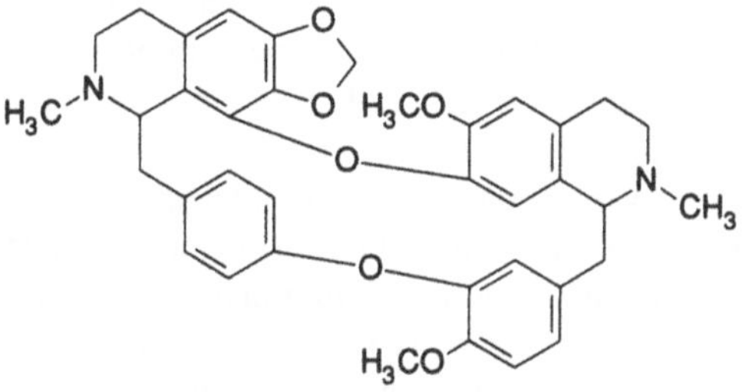

(145)

Cepharanthine

Cepharanthine (145), labeled with ^{14}C in the methylenedioxy moiety, was administered (5 mg/kg, p.o. and i.v.) to rats.[444] Following oral dosage, nearly 70% of the dose was absorbed, however ~85% of the radioactivity was excreted in the feces due to biliary excretion of the absorbed material. The same extent of fecal excretion was found after i.v. dosage. The metabolism of cepharanthine appeared to be rapid and polar metabolites were excreted in the feces.

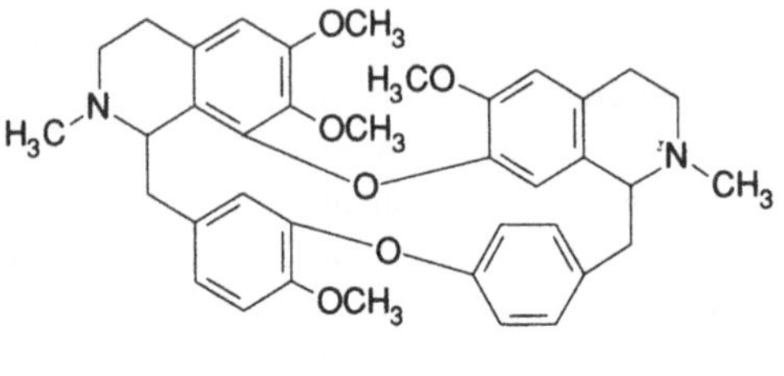

(146)

Tetrandrine

The metabolism of **tetrandrine** (146) in rats and man was studied by Lin et al.[445] Unchanged compound was found in rat liver and lung and in the urine of both species. In addition, smaller amounts of two 2'-*N*-oxide isomers and the *N*-demethylated 2'-nortetrandrine were detected in these samples.

d-Tubocurarine and related curarines are relatively large molecules which possess two quaternary ammonium groups. The resultant ionized nature of these compounds is a major determinant with regard to their biological disposition, as well as the feature which dictates the use of the i.v. route when administering the compounds to effect muscular relaxation. Waser et al.[446] reported that the i.v., s.c., and p.o. doses of *C*-curarine required to produce the same degree of paralysis in cats increased in the ratio 1:10:100. Most of the studies on the biological disposition of these compounds have directed attention to their distribution and plasma levels. Information on these topics, which falls outside the scope of this book, is available in the articles of Mahfouz,[447] Marsh,[448] Foldes,[449] Kalow,[450] Chagas,[451] Cohen et al.,[452,453] and Crankshaw and Cohen.[454]

The results from several early studies of the biological disposition of *d*-**tubocurarine**, which indicated that only a fraction of the injected dose was excreted in the urine, led to the suggestion that it is metabolized.[447,448,455] Matteo et al.[456] found that <50% of the dose (0.3 mg/kg, i.v.) was excreted in the urine of man in 24 h. However, results obtained with radioactive compounds suggest that this is not the case. Cohen et al.[457] administered [^{3}H]tubocurarine (0.3 mg/kg, i.v.) to dogs and recovered ~75% of the radioactivity in the urine and 11% in the bile in 24 h. No metabolites were detected in the urine, even after hydrolysis with β-glucuronidase, although a demethylated derivative, accounting for 0.1 to 1% of the dose, was found in the bile together with the unchanged compound. Biliary excretion of tubocurarine was also demonstrated by Robelet et al.[458] using the perfused, isolated rat liver. The material excreted was probably unchanged compound as it did not show any loss in biological activity. Meijer et al.[459] carried out similar experiments and found extensive excretion of tubocurarine in the bile. Neither *d*-tubocurarine nor its methylated derivative, trimethyltubocurarine, appeared to undergo metabolism. The latter compound, formerly called dimethyltubocurarine, was found by Chagas[451] to be excreted to the extent of 50 to 70% in the urine and 2% in the bile in 3 to 6 h when radioactive material was given by injection to dogs. Chromatography failed to show the presence of any metabolized compounds. Waser and Lüthi,[460] using material labeled with ^{14}C in the *N*-methyl group, found that both *C*-**curarine** and *C*-**toxiferine** were mainly excreted in the urine and that no *N*-demethylation occurred in cats following i.v. doses (100 to 200 mg/kg). About 10% of the dose of the former compound was excreted in the bile.

Thalicarpine was administered i.v. to cats, dogs, and monkeys[461] and to humans.[462] Little urinary or biliary excretion was found in the experimental animals and only ~0.3 to 0.6% of the dose was found as unchanged thalicarpine in the urine of patients. These findings suggest that thalicarpine undergoes extensive metabolism, however no metabolites were identified.

MORPHINAN DERIVATIVES

Morphinan (147) is the parent substance of several of the principal opium alkaloids. The most important of these are morphine, codeine, and, to a lesser extent, thebaine. As morphine was the first alkaloid to be isolated in pure form as well as one of the few reliable and effective drugs available prior to the present century, it is understandable that interest in its biological properties spans a relatively long period of time. In fact, studies of its metabolic fate in animals were carried out over a century ago. Many of these studies had their origin in the desire to clarify the enigma of the addiction and tolerance which result from repeated use of morphine and several of its relatives. It is therefore hardly surprising that there exists a vast literature on the biological disposition of these opium alkaloids. In the following summary, their metabolic pathways will be outlined with emphasis being placed on investigations of a more recent date. It is these which furnish reliable information on the identity of the metabolites and their pathways of formation. Additional information on the biological disposition of the opium alkaloids is available in the reviews by Way and Adler,[463,464] Mulé,[465] Scrafani and Clouet,[466] Misra,[467] Boerner et al.,[468] and Lemberger and Rubin.[469]

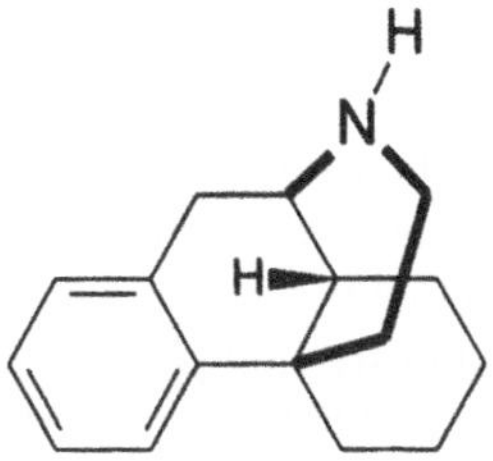

(147) 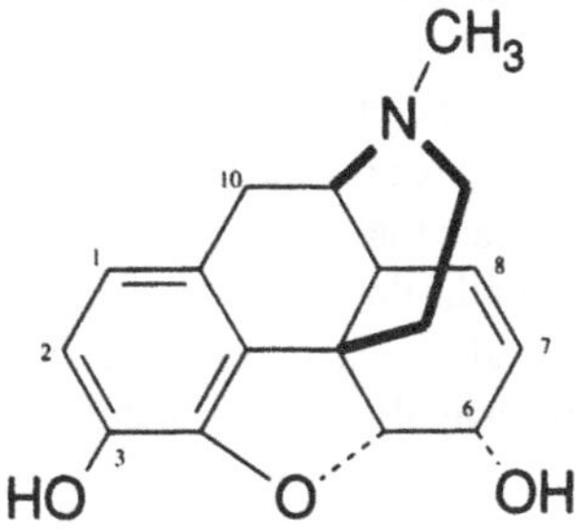   (148)

Morphine

It is noteworthy that injections are nearly always employed when studying the metabolism of **morphine** (148). This practice is related to the well-known unpredictability of effects when morphine is given orally. This had previously been considered to be due to erratic absorption from the intestine, however an increasing body of evidence indicates that its absorption from this site is nearly complete and that the low and variable effects observed are more likely due to a first-pass effect. Iwamoto and Klaassen[470] studied the phenomenon in rats and reviewed the relevant literature. They recorded an overall first-pass effect of more than 80% of the dose (5 mg/kg). Of this, the ratio of the extent of extraction and/or metabolism by the intestine compared with the liver was found to be ~2:1. Brunk and Delle[471] showed that morphine was well absorbed from the intestine in man, however it was rapidly conjugated with glucuronic acid in the cells of the intestinal mucosa.

Although most of the administered morphine undergoes metabolic transformation prior to its excretion from the body, numerous investigations using many animal species have indicated that some excretion of unchanged compound also occurs, mainly in the urine. The reported values for free urinary morphine generally fall in the range of 5 to 30% of the dose. Species differences are not pronounced but it appears that free morphine excretion in rats and cats is slightly higher than that found in other species. Zauder[472] reported a value of 28% in the 24-h urine of rats given morphine (12 mg/kg, s.c.). Values of 24% in 48 h following a similar dose,[473] 17% daily for Wistar rats and 24% daily for Gunn rats following repeated daily injections of 25 to 35 mg/kg[474] and 20% after a dose of 10 mg/kg[475] have also been found. On the other hand, Klutch[476] reported that 7 to 11% of repeated i.p. doses (10 mg/kg) were excreted unchanged each day. Similarly, 10 to 11% of a single s.c. dose of morphine (20 mg/kg) was excreted in the urine of rats as such, mainly in the first 24 h.[477] Yeh et al.[478] recorded a value of 9% in similar experiments. Yeh et al.[478] also studied the extent of excretion of free morphine in several other species including the mouse which gave values of 6.5 to 10% in 144 h following a s.c. dose of 75 mg. Guinea pigs (25 mg/ kg, s.c.) excreted slightly more than 3% in 24 h. Axelrod and Inscoe[479] found 7% of the dose (90 mg/kg, i.p.) excreted unchanged in the 24-h urine in guinea pigs. Values of 3 to 12% in 48 h were obtained in rabbits given large (100 to 200 mg/kg) injections of morphine[480] and Yeh et al.[478] recorded values of ~4 to 5% in 24 h following a s.c. dose of 25 mg/kg. In cats, Yosikawa[481] reported values of ~11 to 27% and Yeh et al.[482] found that 19% of a 20 mg/kg injection was excreted as free morphine in 48 h. More recently, Yeh et al.[478] reported 24-h values of 20 to 27% in similar experiments. Dogs have often been employed in studies of the biological disposition of morphine and abundant data are therefore available on its urinary excretion in this species. A comprehensive summary of this work was made by Way and Adler[463] who noted that ~15% of the dose is excreted as free urinary morphine within 24 h. Misra et al.[483] subsequently reported that ~14% of a 20 mg/kg dose injected s.c. was excreted in the urine of dogs in 12 h and Yeh et al.[478] recorded a 24-h value of 20% in similar experiments. In the monkey (*Macaca mulatta*) only 3 to 7% of the dose of morphine (30 mg/kg, s.c.) was excreted unchanged in the urine.[484] A subsequent study using a smaller dose (2 mg/kg) gave slightly higher values (6 to 14%).[485] The data on morphine excretion is most extensive in the case of man. These results, both from addicts and nonaddicts, were tabulated by Way and Adler[463] and more recently by Boerner et al.[468] Their summaries indicate that there is fairly good agreement among these results which show that an average of 7% of the dose is excreted in the urine as free morphine. The range of values reported is 1 to 14%. An additional investigation in morphine-dependent subjects gave a value of 10%.[486] These studies have also shown that the percentage excretion of free morphine does not vary greatly with varying doses and also that the differences between addicts and nonaddicts are not significant.

In addition to the urinary excretion of morphine and its metabolites, some loss from the body occurs in the feces. The data summarized by Way and Adler[463] indicate that most species including man excrete from a few to ~10% of the dose by this route. Subsequent data using cats showed that 2 to 20% of the dose was found as total morphine in 48 h in the feces.[482] The nature of the fecal metabolites has not been investigated in detail but it appears that it consists mainly of free and conjugated morphine. It seems clear that the fecal excretion of morphine is dependent upon its excretion in the bile, mainly as morphine-3-glucuronide.[487,488] Interestingly, species variations in the biliary excretion of morphine and its metabolites are large, in contrast to that found with the fecal excretion. Approximately 60% of the dose (5 mg/kg, s.c.) was found in the bile of rats in 6 h.[489] Similar results were obtained in this species by Smith et al.[490] More than 35% of the injected morphine was excreted in conjugated form in the bile of dogs.[491] Biliary excretion is less extensive in the cat, the reported values being not more than 10%,[492] 5 to 23% in 3 h,[482] and 14% in 3 h.[490] In monkeys (*Macaca mulatta*), 10 to 20% of the dose was recovered within a few hours in the bile, mainly in conjugated form.[484] The biliary route is of less importance in man, the results of Elliott et al.[493] suggesting a value of ~7%. These findings showing that large amounts of morphine, mainly in conjugated form, are sometimes excreted in the bile whereas much less, usually in the free form, is found in the feces indicates that an enterohepatic circulation of the compound takes place. Walsh and Levine[494] and Parker et al.[495] studied the enterohepatic circulation of morphine in rats. The latter study showed that ~40% of the material excreted in the bile was reabsorbed from the intestine.

As noted above, most of the morphine administered is metabolized. The two major general metabolic routes consist of synthetic reactions, of which glucuronide formation is usually the most important, and oxidative reactions, of which *N*-demethylation is the most important. Suggestions that morphine is excreted in the urine in what was called a bound form were made in some of the earliest reports on morphine metabolism. However, these beliefs were first placed on a firm foundation by the investigations of Gross and Thompson[496] in dogs and Oberst[497] in man. The former study showed that the substance released upon acid hydrolysis of the urine was actually morphine. The results in man indicated that acidic hydrolysis of the urine increased the amount of morphine present by factors of three to 36. Oberst[498] found that the glucuronic acid concentration in urine increased proportionally with increasing doses of morphine and concluded therefore that the bound form is a glucuronide. This was later confirmed when several groups demonstrated conclusively that the bound form, or at least the major bound form, is morphine-3-glucuronide. Woods[491] isolated this conjugate from the urine and bile of morphine-treated dogs and further confirmation was provided by the results of Seibert et al.[499] and Fujimoto and Way.[500,501] The main pathway in the glucuronidation of morphine in mammals leads to the 3-glucuronide, although this is a minor reaction in cats which form instead the ethereal sulfate (see below). Capel et al.[502] did not detect morphine-3-glucuronide in the urine 24 h following the i.p. administration of morphine to cats; however, Yeh et al.[482] reported that 1 to 2% of the dose was excreted in this form in similar experiments. In addition to the 3-glucuronide, it is now clear that the 6-glucuronide is also formed in small amounts. This minor metabolite was detected in the urine of morphine-treated rabbits[503] and dogs[483] and subsequently in the urine of mice, rats, guinea pigs, rabbits, and man.[475] The last study showed that the ratio of 3-glucuronide to 6-glucuronide found in guinea pigs, rabbits, and man is ~100:1. The same result was found by Yeh et al.[504] who administered large daily doses of morphine to morphine-dependent humans. Additionally, they identified morphine-3,6-diglucuronide which also was present in the urine to the extent of ~1% of that of the major conjugate. Hand et al.[505] reported that the plasma concentrations of morphine-3-glucuronide were ~10 times greater than those of morphine-6-glucuronide in patients who had received a single dose (10 mg, i.v.) of morphine. This relationship between the plasma concentrations appears to be constant in individual patients and independent of the dose or duration of treatment.[506] The plasma concentrations of morphine-3-glucuronide in the rhesus monkey were found to be at least 25 times greater than those of the 6-isomer.[507] The metabolism of these glucuronides in rats[508] and in mice, guinea pigs, and rabbits[509] has been studied. In general, the patterns of urinary metabolites varied according to the route of administration of the glucuronides in the latter two species, whereas this was seen to a much lesser degree in rats and mice due to their greater ability to excrete the conjugates in the bile and thus expose them to the action of bacterial β-glucuronidase in the lower intestine.

The extent of urinary excretion of morphine glucuronide has been determined in many investigations. In their summary of earlier data, Way and Adler[463] reported values of 11 to 45% of the dose in man. In a study using addicts, Yeh[486] reported a value of ~65% for the excretion of urinary morphine glucuronide. Depending on the dosage route, extrahepatic conjugation may contribute to the formation of the glucuronide. Thus, Brunk and Delle[471] obtained evidence which indicated that morphine glucuronide is rapidly formed in the cells of the intestinal mucosa following oral administration of the alkaloid to man. Jacqz et al.[510] reported that morphine glucuronidation in dogs occurred about equally at hepatic and extrahepatic sites. Additional information on extrahepatic sites of morphine glucuronidation is given below in the summary of the *in vitro* formation of these conjugates. The value for the urinary excretion of morphine glucuronide in the monkey (*Macaca mulatta*) is 70 to 80%.[484] Typical values in dogs fall in the range of 38 to 75%.[463] Morphine-3-glucuronide is a major urinary metabolite of morphine in rabbits,[511] however quantitative data are lacking. Guinea pigs, following a relatively large i.p. dose (90 mg/kg), excreted 38% of the dose as glucuronide conjugated material.[479] In rats, values between 20 and 50% were typical in earlier investigations[463] although Misra et al.[473] reported that 14% of the dose was excreted as conjugated morphine in Sprague-Dawley rats. Abrams and

Elliott[474] used Gunn and Wistar rats and found no differences in the amounts of urinary morphine glucuronide, both strains excreting ~9% of the repeated daily doses in this form.

Several studies have dealt with the *in vitro* formation of the glucuronide conjugates of morphine, both in the liver and in extrahepatic tissues. Early investigations[479,512] showed that morphine served as an acceptor in a glucuronide-synthesizing system consisting of calf or guinea pig liver microsomes and UDPGA. Sanchez and Tephly[513] obtained results which indicated that different glucuronosyltransferases may catalyze the conjugation of morphine and *p*-nitrophenol in rat liver microsomes. At least two different forms of UDP-glucuronosyltransferase from human liver appear to be involved in morphine glucuronidation.[514] This reaction has been shown to take place in microsomes from human fetal and adult liver,[515] human fetal and adult kidney,[516,517] and from some samples of human brain.[518] Koster et al.[519] investigated the glucuronidation of morphine by isolated mucosal cells from various regions of the rat intestine. They found that the glucuronidation capacity was higher in the duodenum than in the colon and rectum and that the UDP-glucuronosyltransferase activities in these regions followed the same pattern. Rane et al.[507] reported that the relative rates of formation of morphine-3-glucuronide and morphine-6-glucuronide by liver microsomes from the rhesus monkey were similar to the ratio of these metabolites observed in the blood of animals given morphine. Abrams and Elliott[474] found that glucuronidation of morphine in Gunn rats, which show varying degrees of deficiency towards various substrates, was not significantly different than that shown by preparations from Wistar rats. Morphine glucuronidation in *Octodon degus*, a South American rodent which shows natural resistance to the central nervous system effects of morphine, was found to be higher than that measured in Wistar rats due to higher concentrations of liver microsomal protein and cytochrome P-450 in the former species.[520] A similar finding of enhanced formation of morphine-3-glucuronide was reported in the Afghan pika (*Ochotona rufescens*), a naturally morphine-tolerant lagomorph mammal related to the rabbit.[521]

In the mammalian species studied, ethereal sulfate conjugation appears to be a major synthetic reaction only in cats. Morphine ethereal sulfate was first isolated from the urine of morphine-treated cats by Woods and Chernov[522] and this conjugate was shown by Fujimoto and Haarstad[511] to be morphine-3-*O*-sulfate. This metabolite was also shown to be present in the urine of cats given injections of morphine by Mori et al.[523] who found no evidence for the excretion of the 6-*O*-sulfate conjugate. Quantitative data on the excretion of morphine-3-*O*-sulfate in cats was reported by Yeh et al.[482] and Capel et al.[502] In the former study, ~48% of the dose (20 mg/kg, s.c.) was excreted in the urine in 48 h as conjugated material, mainly the 3-*O*-sulfate. Values of ~70% for this metabolite were recorded in the latter investigation, which employed a dose of 5 mg/kg and a 24-h collection period. Also, no morphine glucuronide was found but another sulfate conjugate, possibly morphine-6-*O*-sulfate which accounted for ~1% of the dose, was detected. Other species in which ethereal sulfate conjugation of morphine has been demonstrated include the dog[483] and man.[524] The latter study indicated that the ratio of urinary morphine ethereal sulfate to morphine glucuronide was ~1:4, however other results indicated a ratio of ~1:100 when large daily doses of morphine were given to morphine-dependent subjects.[504]

Relatively little is known regarding the possibility that morphine may be metabolized to other conjugated derivatives. Misra et al.[483] detected two minor metabolites in the urine of morphine-treated dogs which gave purple colors when sprayed with ninhydrin reagent. Misra and Woods[525] obtained evidence which suggested that morphine and codeine can form a covalently-linked conjugate with glutathione under nonenzymatic conditions, particularly in the presence of ferrous ions. Interestingly, morphine administration to mice[526] or rats[527] was shown to result in the depletion of hepatic glutathione. When morphine and glutathione were incubated with rat liver microsomes an adduct was formed which was identified as 10α-*S*-glutathionylmorphine (149).[528,529] This metabolite was formed via benzylic oxidation of morphine at C-10 to form an electrophilic species which reacted with the thiol glutathione. An additional glutathione conjugate of morphine was shown by Ishida et al.[530] to be formed from morphinone (150) as an intermediate. This adduct was found to be 8-*S*-(glutathion-*S*-yl)dihydromorphinone (151) and was present in the bile of guinea pigs given morphine. About 9% of the dose (25 mg/kg, s.c.) was found in the bile as the glutathione conjugate. The morphine 6-dehydrogenase from guinea pig liver required for the conversion of morphine to morphinone was purified and characterized.[531] This reaction sequence has also been demonstrated in isolated rat hepatocytes[532] and *in vivo* in mice[533] and it appeared that the morphinone formed could bind covalently with the thiol group of cysteine residues in protein.

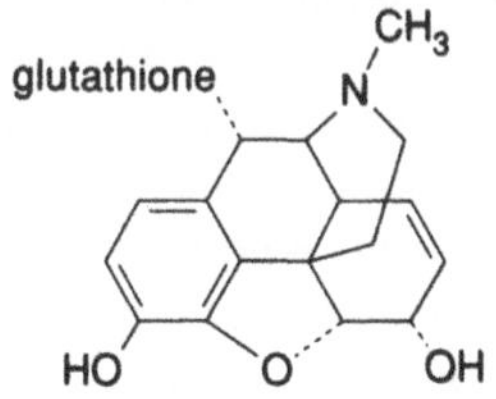

(149) 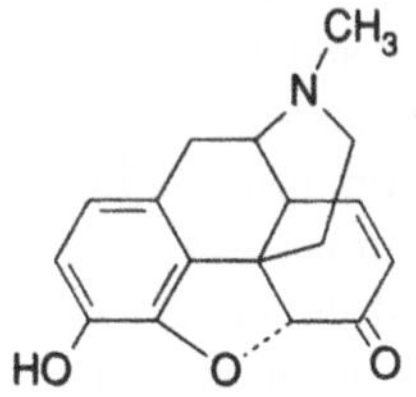   (150) 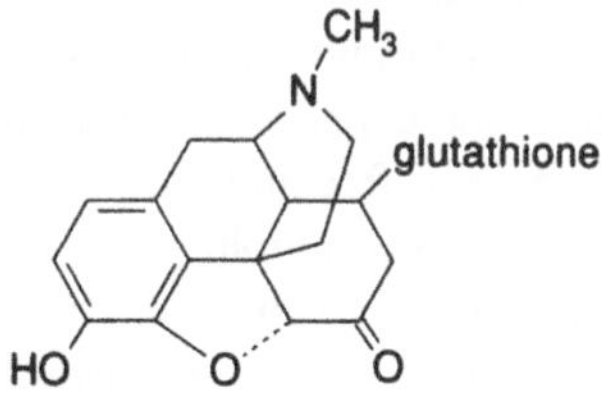   (151)

The most widely studied oxidative reaction occurring with morphine is *N*-demethylation. The initial evidence for its oxidative demethylation to normorphine was obtained using [*N*-methyl-^{14}C]morphine. March and Elliott[489,534] found that the injection of labeled morphine (5 mg/kg, s.c.) in rats resulted in loss of some of the radioactivity as respiratory $^{14}CO_2$. When male rats were used, ~5% of the radioactivity was excreted in this manner whereas, with female rats, the values were only ~10% as large. Also, rat liver slices metabolized morphine to $^{14}CO_2$ and the preparations showed these sex-related differences. Similar *in vivo* studies in man confirmed the *N*-demethylation reaction, with 3.5 to 6% of the radioactivity being lost as respiratory $^{14}CO_2$ in a 24-h period following the injection of morphine (15 mg/kg, i.m.). Another early but different approach to this problem was taken by Axelrod[535] who found that enzyme systems were present in the livers of several mammalian species which were capable of demethylating morphine to normorphine and formaldehyde. These enzyme systems are located in the liver microsomal fraction and require NADPH, O_2, and various cofactors. Gutierrez and Flaine[536] detected normorphine when morphine was incubated with rat liver and brain homogenates. Using morphine labeled with ^{14}C in the *N*-methyl group, Elison and Elliott[537] found that aerobic incubation with rat brain slices resulted in the formation of $^{14}CO_2$, however relatively low activity was present compared with that present in the liver. Guinea pig liver microsomes convert morphine to normorphine.[538,539] The increased levels of liver microsomal protein and cytochrome P-450 in *Octodon degus*, a South American rodent which shows natural resistance to the central nervous system effects of morphine, was noted above regarding the increased formation of morphine glucuronide in this species compared with Wistar rats. Letelier et al.[520] found that the *N*-demethylation of morphine was similarly greater in *Octodon degus*.

Numerous *in vivo* investigations have indicated that normorphine is a metabolite of morphine in many mammalian species. This was demonstrated in mice[475,540] and Adler[541] recovered 1 to 2% of the dose as respiratory $^{14}CO_2$ in 4 h following injection of [*N-methyl*-^{14}C]morphine (5 mg/kg, s.c.). In rats, many investigations have demonstrated the *N*-demethylation of morphine, although Abrams and Elliott[474] did not find evidence for normorphine excretion in the urine of Gunn and Wistar rats given repeated daily injections of morphine (20 to 35 mg/kg). However, several other studies indicate that values of a few to ~10% can be expected with single or repeated doses of this amount.[476,477,542,543] Further evidence for the conversion of morphine to normorphine in rats was obtained by Misra et al.[473] and Oguri et al.[475] Normorphine was also shown to be formed from morphine in the brain of rats *in vivo*.[544] Fishman et al.[545] carried out *in vivo* studies in rats which indicated that the *N*-demethylation of morphine in brain occurs at sites which have a high content of opiate receptors. This enzyme in brain is different from the liver enzyme which N-demethylates morphine.[546] Oguri et al.[475] reported the N-demethylation of morphine in rabbits and normorphine was detected both in the liver[547] and urine[482] of cats dosed with morphine. Misra et al.[483] obtained evidence in morphine-treated dogs which suggested the urinary excretion of normorphine glucuronide. It appears that the N-demethylation of morphine in this species may be rather limited as Mellett and Woods[485] found that only ~0.2% of the dose (2 mg/kg, s.c.) of [N-methyl-¹⁴C]morphine was excreted as respiratory $^{14}CO_2$. They obtained a value of ~1% in similar experiments in monkeys. Oguri et al.[475] detected normorphine in the urine of a human given three 10-mg injections of morphine and several subsequent investigations have confirmed and extended this finding. Yeh[486,524] reported that morphine-dependent subjects receiving four 60-mg injections daily excreted from 0.5 to 1% of the dose as free and ~3% as conjugated normorphine. Boerner et al.[548] estimated that ~5% of large doses of morphine given chronically to a patient was excreted in the urine as normorphine. Additionally, Yeh et al.[478,549] investigated the urinary excretion of normorphine in several mammalian species following the acute or chronic administration of morphine. Free and conjugated normorphine were found in all species (mouse, rat, guinea pig, rabbit, cat, dog, monkey, and man), however the total excretion values were not large (approximate range of 0.5 to 7%).

The normorphine formed from morphine as noted above may be metabolized in other ways, including N-methylation which regenerates morphine. This capability is pronounced in rabbit lung, as shown by Axelrod[86,550] using an *in vitro* system incorporating S-adenosylmethionine. Clouet[551,552] reported that normorphine injected into the rat brain was N-methylated and also that this reaction was carried out *in vitro* using rat liver and brain preparations containing S-adenosylmethionine. The N-methyltransferase involved was studied by Clouet et al.[553] On the other hand, Elison and Elliott[554] did not find evidence for the N-methylation of normorphine *in vitro* in rat liver preparations or for the urinary excretion of morphine in rats dosed with normorphine. The same negative finding was reported by Yeh et al.[555] who administered normorphine (20 mg/kg, s.c.) to dogs. The major urinary metabolites were free and conjugated normorphine which each accounted for about a third of the dose.

Morphine is converted to codeine (154) as a result of O-methylation at the 3-position. This reaction was first demonstrated by Elison and Elliott[554] who found small amounts of the metabolite in the urine of rats given morphine. Increased formation of codeine was found when rats deficient in liver glucuronide conjugating ability were used. Codeine formation was also noted *in vitro* using rat liver preparations, however the nature of the methyl donor was unclear as [methyl-¹⁴C]-S-adenosylmethionine added to the incubates did not result in the formation of radioactive product. An investigation of the metabolism of morphine in both Gunn and Wistar rats failed to detect codeine in the urine, thereby raising doubts as to the significance of this metabolic pathway.[474] Disagreement has also been registered on the formation of codeine from morphine in man. Börner and Abbott[556] reported that 0.7 to 0.9% of the dose of morphine given p.o. or i.v. to nontolerant subjects was recovered as urinary codeine. The codeine excretion was significantly increased in tolerant subjects. Subsequently, Boerner et al.[548] identified norcodeine in the urine of nontolerant and tolerant subjects. Yeh[557] detected only very small amounts of codeine in the urine of subjects receiving morphine chronically and concluded that these were more likely due to the small amounts of codeine present in the morphine used than to O-methylation of the latter. Also, Yeh[558] questioned the interpretation that norcodeine is a metabolite of morphine, suggesting instead that it arises from the metabolism of the small amounts of codeine present in morphine samples. This viewpoint was not, however, accepted by Boerner and co-workers.[468,559] Mari and Bertol[560] studied the possible excretion of codeine by a large number of human subjects who took daily doses (60 to 120 mg) of morphine. None of the urine samples from these individuals contained detectable amounts of codeine. On the other hand, Phillipson et al.,[538,539] using morphine which was shown to be free of codeine, reported that codeine was formed in incubates using microsomes from guinea pig liver and also excreted in the urine of rats given morphine.

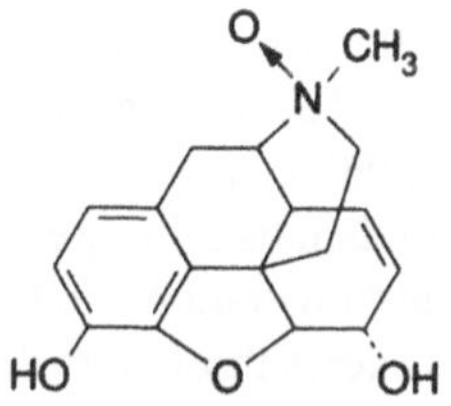

(152)

Another metabolite of morphine formed by an oxidative reaction is morphine-N-oxide (152), although its appearance seems to usually require special circumstances. Woo et al.[561] identified morphine-N-oxide in the urine from patients receiving morphine together with amiphenazole (2,4-diamino-5-phenylthiazole) or tacrine (1,2,3,4-tetrahydro-9-aminoacridine). The metabolite was not detected when these compounds were given alone. It is possible that the metabolite is formed chemically since it was produced when morphine and amiphenazole were mixed in ammoniacal methanol, although the authors believed that its excretion may be due to inhibition of its further metabolism or to inhibition of an alternative pathway in the metabolism of morphine. Heimans et al.[542] found that 46% of the total urinary opiates consisted of morphine-N-oxide when rats were given morphine and tacrine. Also, the excretion of normorphine was greatly reduced compared with that seen in rats receiving morphine alone. These results are in sharp contrast to those of Phillipson et al.[538,539] who found that rats excreted 0.2% of the dose in the urine as morphine-N-oxide, with or without the concomitant administration of tacrine. Misra and Mitchell[562] found no evidence indicating that morphine is converted to its N-oxide in the central nervous system of rats. Yeh et al.[549] investigated the possible formation of morphine-N-oxide in seven species (rat, guinea pig, rabbit, cat, dog, monkey, and human) and found that only the guinea pig excreted this metabolite in the urine. Morphine-N-oxide has been detected as a metabolite of morphine under *in vitro* conditions. Misra et al.[543] found that small amounts of the N-oxide were formed when morphine was incubated aerobically with rat liver homogenates fortified with NADPH. Also, Ziegler et al.[151] showed that a purified microsomal amine oxidase from pig liver converted morphine to its N-oxide. When morphine-N-oxide was administered to rats, it was excreted in the urine to a large extent as unchanged compound but mainly as morphine and morphine-3-glucuronide together with some normorphine.[542,562]

Daly et al.[257] investigated a microsomal hydroxylating system from rabbit liver which metabolized phenols to catechols. This activity was assayed by converting the latter compounds to O-methyl derivatives using [*methyl*-^{14}C]-S-adenosylmethionine and catechol-O-methyltransferase. They reported that morphine was a substrate for this system, although the activity was low compared with that seen with many compounds. Misra et al.[543] found that rat brain and liver homogenates were able to hydroxylate morphine to a derivative having the properties of a 2,3-catechol structure (2-hydroxymorphine). In the urine of morphine-treated rats, however, a different metabolite was detected and this was tentatively identified as the 2,3-dihydrodiol derivative. Yeh et al.[549] provided further evidence for the formation of mono- and dihydroxylated derivatives of morphine in rats, guinea pigs, rabbits, and cats but not in dogs, monkeys, or humans. The identities of these metabolites remain uncertain, however they appear to include the 2-hydroxy-, 3-hydroxy-, and 8-hydroxy-derivatives of morphine.

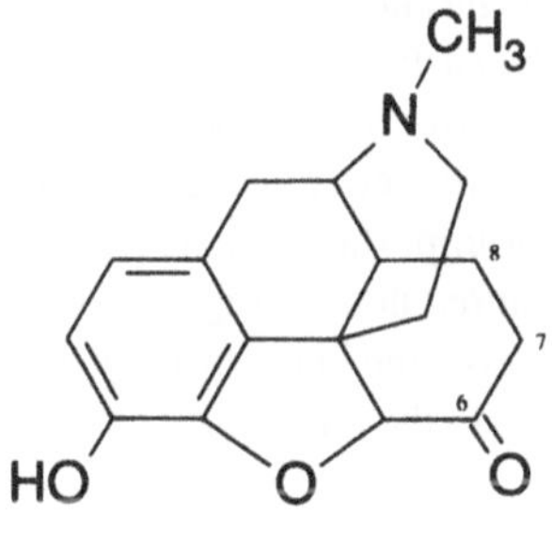

(153)

A few investigations have shown that the 7,8-double bond of morphine may be reduced. Klutch[476] isolated dihydromorphinone (153) from the urine of rats injected with repeated daily doses (25 mg/kg) of morphine. Quantitative data indicated that ~4% of the dose was excreted as dihydromorphinone. Yeh et al.[478] studied its urinary excretion in several mammalian species given morphine by acute or chronic administration. No free dihydromorphinone was detected, however all acid-hydrolyzed samples except those from dogs and morphine-dependent humans contained this metabolite. The amounts found were generally ~1% or less of the administered dose, however a value of 6 to 7% was obtained using chronically treated monkeys. The excretion of dihydromorphinone in morphine-treated guinea pigs was not increased following chronic administration. Reduction of dihydromorphinone (153) to the corresponding 6-hydroxy derivative (dihydromorphine) was observed in rabbits[563] and in guinea pigs.[564] Yeh et al.[549] found that both 6α- and 6β-isomers of dihydromorphine (hydromorphol) were formed in the guinea pig. Cone et al.[565] confirmed the excretion of small amounts of these two isomers in the urine of guinea pigs given morphine (8 mg/kg, s.c.).

Pseudomorphine (2,2′-bimorphine) is an oxidation product of morphine which has long been suggested as a possible morphine metabolite. However, little evidence is available to support this view.[463] Misra and Mulé[566] found no evidence for the formation of pseudomorphine *in vivo* in the brain of rats injected with [³H]morphine (10 mg/kg). Pseudomorphine itself was excreted mainly unchanged in the urine.

The metabolism of (+)-morphine, the unnatural isomer of morphine, was studied in rat liver microsomes.[567] It showed preferential glucuronidation at the 6-position compared with the 3-position which was found with the natural (−)-isomer of morphine. This was believed to be due to different isoenzymes responsible for the two glucuronidation reactions. Also, the rate of *N*-demethylation of (+)-morphine was only about half of that observed with its isomer. Additional studies on the glucuronidation of the (−)- and (+)-enantiomers of morphine clearly indicated the presence of two UDP-glucuronosyltransferase isoenzymes in rat liver and also showed that human liver microsomes carried out this reaction at both the 3- and 6-positions.[568] The major products with human liver microsomes were the 3-*O*-glucuronide with (−)-morphine and the 6-*O*-glucuronide with (+)-morphine.

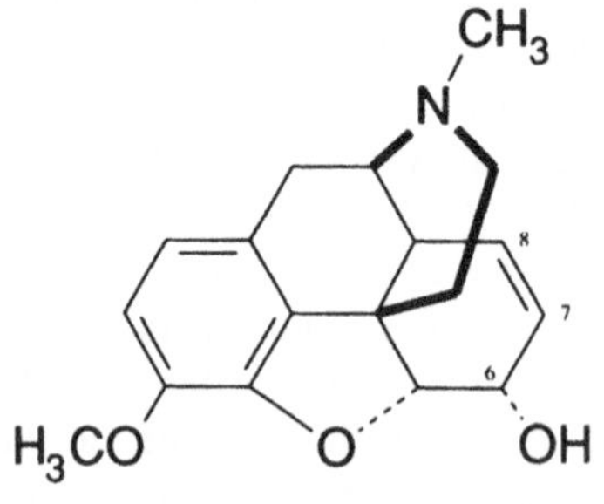

(154)

Codeine

The metabolism of **codeine** (154) has also been the subject of a large number of investigations. A comprehensive review of the earlier studies of its biological disposition was published by Way and Adler[464] and other summaries have appeared.[466,467] The general picture of codeine metabolism is one of the predominant excretion of the alkaloid and its metabolites in the urine, other excretory routes being of minor importance in most species. However, Yeh and Woods[569] found that 13 to 14% of the administered dose (2 mg/kg, s.c.) was excreted in the feces in rats. With respect to the excretion of free codeine in the urine of animals, the results summarized by Way and Adler[464] indicate that values in the range of a few to 10 to 15% of the dose are typical.

Thus, in rats ~3% of the dose (33 mg/kg, s.c.) was excreted free, a finding which agrees reasonably well with the 6 to 7% reported by Yoshimura et al.[570] and Yeh and Woods,[571] using doses of 10 and 20 mg/kg, respectively. Cone et al.[572] measured the urinary excretion (24 h) of free codeine in rats and guinea pigs given s.c. doses of 5 mg. The value in rats was 6.5% and that in guinea pigs was ~4%. Additional values of 8 to 9% in guinea pigs were reported by Axelrod and Inscoe[479] and Yoshimura et al.[570] with doses of 90 and 10 mg/kg, respectively. Values of 0.1% of the dose (5 mg, s.c.),[572] 3.5% of the dose (10 mg/kg), and 5 to 7% of the dose (20 mg/kg)[570,571] have been found for free urinary codeine in rabbits. A value of ~30% was recorded in cats.[571] Excretion of free urinary codeine in dogs has been given as 4 to 11% of the dose (20 mg/kg, s.c.)[573] and ~3% of the dose (10 mg, s.c.).[572] Corresponding values in rhesus monkeys (*Macaca mulatta*) were 3 to 10%.[573] A range of 3 to 17% was reported for the urinary excretion of unchanged codeine in man in the earlier work summarized by Way and Adler.[464] Additional studies using oral doses have given values of 14% following a dose of 100 mg,[574] 5 to 6% after 30 mg,[572] and ~5% after 1 mg/kg.[575] Interethnic differences were reported in the urinary excretion of unchanged codeine.[576] Following an oral dose of 25 mg, the 8-h excretion values were 7.2% in a group of Chinese and 4.3% in a group of Caucasians.

The above results indicate that codeine, like morphine, is largely metabolized prior to its excretion from the body. Not unexpectedly, the main pathways of metabolism seen with morphine are also seen with codeine. Thus, glucuronide conjugation and *N*-demethylation are important metabolic reactions but, additionally, codeine can be *O*-demethylated. The bound or conjugated material excreted in the urine of animals receiving codeine may consist of derivatives of either codeine itself or its *O*-demethylated metabolite morphine (see below). Yeh and Woods[577] isolated codeine-6-glucuronide from the urine of codeine-treated dogs, thus identifying one of these conjugated metabolites. This article also provides useful information on the earlier investigations of the urinary excretion of bound codeine in various animal species. These workers subsequently showed that male dogs excreted 36% and female dogs 78% of the dose (20 mg/kg, s.c.) as codeine-6-glucuronide.[571] Similar previous experiments indicated the urinary excretion of 42 to 58% of the dose as conjugated codeine[573] and it is evident that this route is the principle metabolic pathway for codeine in dogs. The data summarized by Way and Adler[464] indicate that large amounts of conjugated codeine (20 to 50% of the dose) are excreted by man. Newer data of Chen et al.,[578] obtained with improved methodology, indicated that the values may be even higher. They recovered 67% of the dose as urinary codeine-6-glucuronide in a subject given codeine phosphate (30 mg, p.o.). More than 90% of the total urinary codeine in man was found to be conjugated.[575] Interethnic differences have been reported and Chinese subjects were less able to metabolize codeine by glucuronidation than were Caucasians.[576] Woods et al.[573] reported values of ~30 to 40% for codeine glucuronidation in the rhesus monkey. The values found in the other animal species studied are usually much lower, especially in cats which excrete very little codeine-6-glucuronide (1 to 2% of the dose)[571] and in rats (1% of the dose).[570] Yoshimura[570] also reported values of 11% in rabbits and 13% in guinea pigs, the latter result agreeing well with the value of 10% reported earlier by Axelrod and Inscoe.[479] Of course, those species showing a minor urinary excretion of codeine-6-glucuronide may have a correspondingly higher excretion of conjugated morphine. This appears to be the case in rabbits[571] and rats.[569] The main glucuronide excreted in the urine and bile of codeine-treated rats was shown to be morphine-3-glucuronide.[579]

The first investigations which demonstrated that codeine undergoes oxidative *N*-demethylation utilized [*N-methyl*-^{14}C]codeine. Adler[580] found that ~13% of the radioactivity appeared as respiratory $^{14}CO_2$ in rats within 30 h of dosing (40 mg/kg). In experiments in man, the *N*-demethylated product, norcodeine, was detected in the urine both free and conjugated. The amount of norcodeine excreted in the urine was later found to be ~14% of the dose, although the values for respiratory $^{14}CO_2$ were calculated to be only about half of this.[581] The results did not differ appreciably when the dose (20 to 30 mg/kg) was given by different routes of administration (p.o., s.c., i.m.). Also, urinary excretion of metabolites predominated with only ~1% of the dose being found in the feces within 24 h.

In addition to these initial experiments in rats and man, numerous *in vitro* and *in vivo* investigations using various animal species have shown that the *N*-demethylation of codeine to norcodeine is a general metabolic reaction. In an early report, Axelrod[535] described the enzyme system in rat liver microsomes which converted codeine to norcodeine and formaldehyde in the presence of NADPH, O_2, and other cofactors. Bodd et al.[582] demonstrated that the major metabolite of norcodeine in isolated rat liver hepatocytes was the fully demethylated metabolite, normorphine. Kuhn and Friebel[583] found that the urine of codeine-treated rats contained norcodeine and also small amounts of normorphine. Similar results were obtained by Yeh and Woods[579] using both rat urine and bile. However, Phillipson et al.[538,539] reported that appreciable amounts of norcodeine but no normorphine were present in the urine of rats given codeine by i.p. injection. The finding by Cone et al.[572] that only ~0.2% of the dose (5 mg, s.c.) was excreted in the urine of rats as free norcodeine, suggests that most of the norcodeine is excreted in conjugated form. The results of Elison and Elliott[554] pointed to the existence of different enzymes in rat liver being responsible for the *N*-demethylation and *O*-demethylation (see below) of codeine.

The *N*-demethylation of codeine in mice appears to be fairly pronounced. Adler[541] found that 27 to 31% of the dose (33 mg/kg) was recovered in 24 h as respiratory $^{14}CO_2$. Cone et al.[572] reported that guinea pigs, like rats, excreted only ~0.2% of the dose (5 mg, s.c.) in the urine as free norcodeine. The urinary excretion of norcodeine by guinea pigs was also reported by Kuhn and Friebel.[583] Phillipson et al.[538,539] showed that microsomal preparations from guinea pig liver carried out the *N*-demethylation of codeine. Yeh and Woods[571] investigated the urinary excretion of norcodeine in rabbits, cats, and dogs following s.c. administration of codeine (20 mg/kg). The ability to *N*-demethylate codeine varied appreciably among these three species. It was most pronounced in cats which excreted 48% of the dose as norcodeine, entirely in the free form, in the 48-h urine. The values for rabbits were 1 to 3% (free) and 7 to 14% (conjugated). The least *N*-demethylation was found in dogs which excreted ~4% of the dose in each fraction. Also, 9 to 14% of the dose in rabbits was excreted as conjugated normorphine. This fully demethylated metabolite was not detected in the experiments with cats. The urinary excretion of free norcodeine in rabbits and dogs following s.c. administration of codeine (5 or 10 mg/kg, respectively) was also determined by Cone et al.[572] The values (24 h) were 0.1% of the dose (rabbits) and 9.5% (dogs). This investigation also studied norcodeine formation in two human subjects who received codeine (30 mg, p.o.). The urinary excretion (24 h) of free norcodeine was ~1 and ~5%. In another study in man it was estimated that ~9% of the dose (10 to 20 mg) was excreted as total urinary norcodeine.[584] A further related point of interest is that norcodeine can be *N*-methylated by preparations of rat liver and brain[551] and of rabbit lung.[550]

In addition to the *N*-demethylation described above, codeine can alternatively undergo oxidative demethylation of the *O*-methyl group to form morphine. The first direct evidence showing that the latter type of demethylation takes place was obtained using codeine labeled with ^{14}C in the 3-methoxy group. Adler and Latham[585] and Latham and Elliott[586] found that 50% of the radioactivity was lost as respiratory $^{14}CO_2$ in 31 h in rats given this material (40 mg/kg, s.c.). *In vitro* experiments in the former study showed that this conversion took place mainly in the liver. While the identity of the alkaloidal metabolite remaining after demethylation was not determined in these investigations, Adler and Shaw[587] soon showed that it was morphine. They isolated and characterized the metabolite following its formation *in vitro* using rat liver slices, which were able to convert a large proportion of the codeine to morphine in these experiments. Axelrod[377,588] studied the *in vitro* *O*-demethylation of codeine using various liver preparations from several animal species. The reaction, which produced morphine and formaldehyde, was catalyzed by an enzyme system located in the liver microsomes and required NADPH, O_2, and other cofactors. Greatest activity was found in rabbit liver preparations, intermediate activity in rat liver preparations, low activity in those from guinea pigs and dogs, and none in those from mouse liver. However, Takemori and Mannering[589] reinvestigated this reaction in microsomal preparations from the livers of several species including the mouse and found high activity in this species. It was noted that this apparent discrepancy was due to marked differences in demethylating ability of the livers of various strains of mice. These results also suggested that the N- and O-demethylating systems may be different. This view is also supported by the results of Elison and Elliott[554] and Duquette et al.[590] who carried out kinetic studies of the N- and O-demethylation of codeine by rat liver preparations and rat liver microsomes, respectively.

The findings noted above which indicate that the O-demethylation of codeine is a prominent pathway in rats and rabbits have subsequently been confirmed in several investigations. Findlay et al.[591] found that codeine was rapidly converted to morphine in rats. Yeh and Woods[569] found that rats given s.c. doses of codeine (2 mg/kg) excreted 3 to 8% of the dose as free morphine and 19 to 44% as conjugated morphine in the urine in 24 h. Yoshimura et al.[570] reported values of 7 and 19% for free morphine and morphine-3-glucuronide, respectively, in a similar experiment employing a dose of 10 mg/kg. Yeh and Woods[579] isolated morphine-3-glucuronide from the urine and bile of codeine-treated rats. The 48-h urinary metabolites of codeine (20 mg/kg, s.c.) in rabbits included free morphine (1 to 2%) and conjugated morphine (47 to 51%).[571] In another investigation with rabbits using half this dose, Yoshimura et al.[570] found that the 24-h excretion values of morphine and morphine-3-glucuronide were 5 and 29%, respectively.

Extensive O-demethylation of codeine also occurs in mice. Adler[541] reported that ~25% of the dose (33 mg/kg) of [3-O-methyl-^{14}C]codeine appeared as respiratory ^{14}CO$_2$ in 24 h. Yoshimura et al.[570] showed that this pathway was less important in guinea pigs, which gave 24-h excretion values of only 2% (morphine) and 4.5% (morphine-3-glucuronide). The conversion of codeine to morphine in guinea pigs was also reported by Kuhn and Friebel,[583] and Phillipson et al.[538,539] showed that both morphine and normorphine were formed upon incubation of codeine with guinea pig liver microsomes. Both cats and dogs appear to be particularly deficient in their ability to O-demethylate codeine. Using a s.c. dose of 20 mg/kg, Yeh and Woods[571] found only ~1% free morphine and no conjugated morphine in the urine in 48 h in the former species while, with dogs, the corresponding values were ~1 and <0.1%. In an earlier study, Woods et al.[573] found neither free nor conjugated morphine in the plasma, urine, feces, or bile from dogs dosed with codeine (20 mg/kg, s.c.). Pærregaard[592] recorded a value of ~2% for free urinary morphine in dogs following the administration of codeine in doses of 2 to 40 mg/kg.

The remaining species studied, monkey and man, seem to occupy an intermediate position with regard to their ability to O-demethylate codeine. Woods et al.[573] found that the rhesus monkey excreted in the urine ~0.5 to 2% of the dose as free and 6 to 10% as conjugated morphine following an injection of codeine (20 to 40 mg/kg, s.c.). The initial report on the O-demethylation of codeine in man stated that, following the s.c. injection of 30 mg of [3-O-methyl-^{14}C]codeine, 15% of the radioactivity was recovered in 24 to 30 h as respiratory ^{14}CO$_2$.[593] Mannering et al.[594] isolated and identified morphine in the urine following the oral administration of 130 mg of codeine sulfate to three subjects. They estimated that the amount of total morphine present corresponded, at most, to ~3% of the dose. However, Adler et al.[581] reported a value of ~10% of the dose (20 mg, i.m. or p.o.) being excreted as total urinary morphine in 24 h. Redmond and Parker[595] found this excretion to account for 14% of the dose (45 mg, p.o.) in 32 h. Pærregaard[592] reported a value of 4% (24-h urine) in a similar experiment, but noted that this probably represented the actual conversion of nearly 10% of the dose into morphine. Ebbighausen et al.[584] confirmed this general level of urinary excretion of morphine in codeine-treated subjects. Following administration of an oral dose of 20 mg, they found that urinary total morphine accounted for ~10% of the dose. In addition, the fully demethylated metabolite, normorphine, was detected and calculated to account for <4% of the dose. Further information on the conversion of codeine to morphine in man is available from studies in which their plasma concentrations were determined. These results indicated that morphine amounted to 2 to 3%[596] or ~10%[597] of the codeine present in the plasma.

The *O*-demethylation of codeine by human liver microsomes was shown to be carried out by the polymorphic monooxygenase which catalyzes debrisoquine hydroxylation and which is lacking in some subjects.[598] This result was confirmed in an *in vivo* study in which a large group of Caucasian subjects was given codeine orally.[599] A similar correlation and bimodal distribution were not found with the glucuronidation and *N*-demethylation of codeine in these subjects. Interestingly, the occurrence of genetically poor *O*-demethylators of codeine was rare in a group of Chinese subjects.[576]

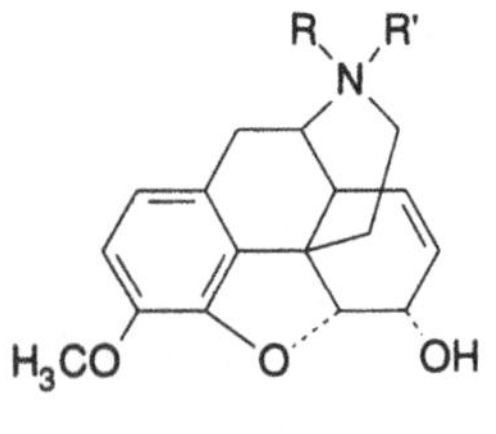

(155)

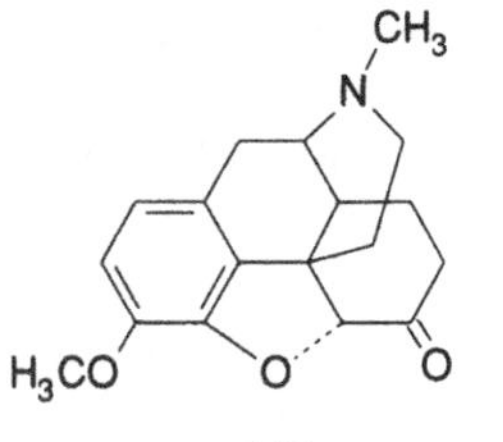

(a) R = CH₃ , R' = →O

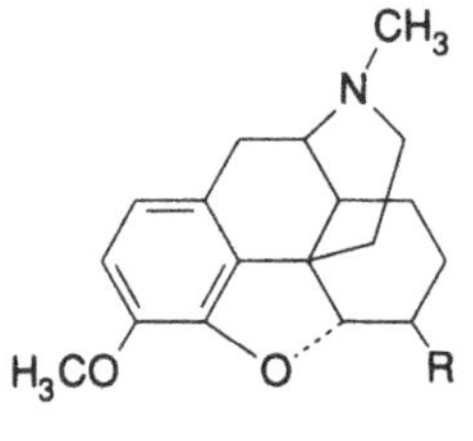

(b) R = →O , R' = CH₃

(156)

(157)

(a) R = ····OH

(b) R = —OH

In addition to the major pathways of codeine metabolism summarized above, several supplementary routes have been described which may produce minor metabolites. Uba et al.[600] showed that the olefinic bond at C-7,8 was subject to epoxidation. Using hepatic microsomes from rats induced with phenobarbital and incubates containing an inhibitor of epoxide hydrolase, they demonstrated the formation of codeine-7,8-epoxide. Small amounts of this metabolite were also found in the urine of rats given codeine by i.p. injection.[601] Phillipson et al.[538,539] showed that codeine was oxidized to two isomeric *N*-oxides when incubated with liver microsomes from guinea pigs. Only trace amounts of the minor isomer (155b) were detected compared with the major isomer (155a) which accounted for ~20% of the total alkaloidal content in the incubates. The latter compound was also found to be a minor urinary metabolite of codeine in rats. Cone et al.[572] identified four novel metabolites of codeine in which the double bond at C-7,8 was reduced. These were hydrocodone (156), its *N*-demethylated derivative, norhydrocodone, and the two reduced compounds 6α-hydrocodol (157a) and 6β-hydrocodol (157b). Their urinary excretion in rats, guinea pigs, rabbits, dogs, and man was studied, however in no instance was any of these metabolites excreted in amounts exceeding 0.1% of the dose (5 mg s.c. in the first three species, 10 mg s.c. in dogs, and 30 mg p.o. in man). All four of these metabolites were detected in guinea pig and human urine, however only hydrocodone and norhydrocodone were excreted by dogs. None of these metabolites were detected in the urine of rats or rabbits after codeine administration.

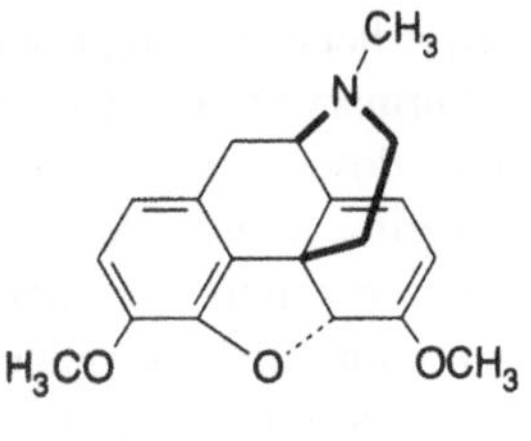

(158)

Thebaine

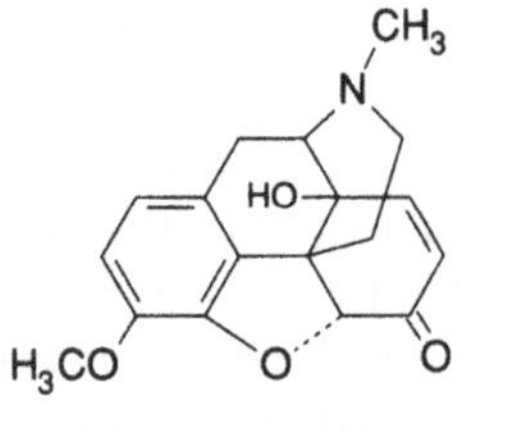

(159)

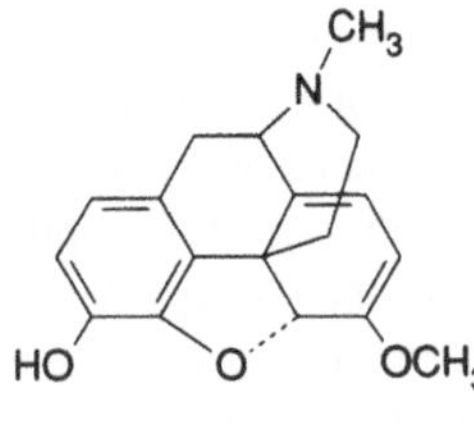

(160)

Studies on the metabolism of **thebaine** (158) are limited. Axelrod[588] reported that it was demethylated to an unidentified phenolic metabolite when incubated with a preparation containing rabbit liver microsomes. Misra et al.[602] carried out a detailed investigation of the metabolism of [³H]thebaine in rats. They found that it was excreted unchanged in the urine and feces in 96 h to an extent of 17 and 4% of the dose (5 mg/kg, s.c.), respectively. The metabolism of thebaine in the rat was extensive, with the glucuronides of normorphine and norcodeine being identified as the major urinary metabolites. Codeine, norcodeine, morphine, normorphine, and 14-hydroxycodeinone (159) were identified as minor free metabolites. Three additional metabolites were detected, one of which appeared to be the 3-*O*-demethylated compound, oripavine (160). Oripavine was also found to be a major urinary metabolite of thebaine (8 mg/kg, s.c.) in rhesus monkeys, accounting for ~10% of the dose.[603] Also identified was its *N*-demethylated product, nororipavine. These metabolites did not appear to be excreted in the conjugated form. The results suggested that several of the previously identified metabolites (codeine, morphine, and the nor derivatives of codeine or morphine) were also excreted. Unchanged thebaine, northebaine, thebaine-*N*-oxide, and dihydrothebaine were not detected in the urine. Experiments with monkey liver microsomes showed the formation of both of the monodemethylated metabolites, oripavine and northebaine, and the didemethylated compound nororipavine.

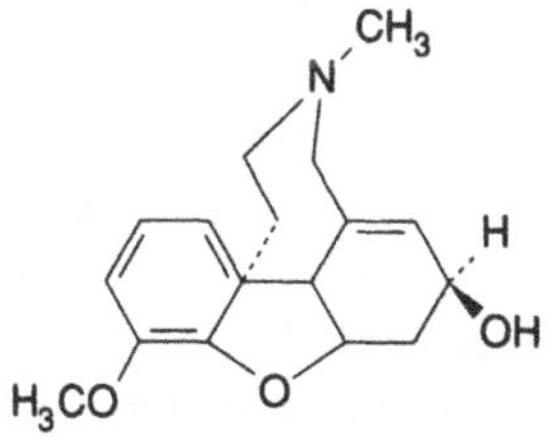

(161)
Galanthamine

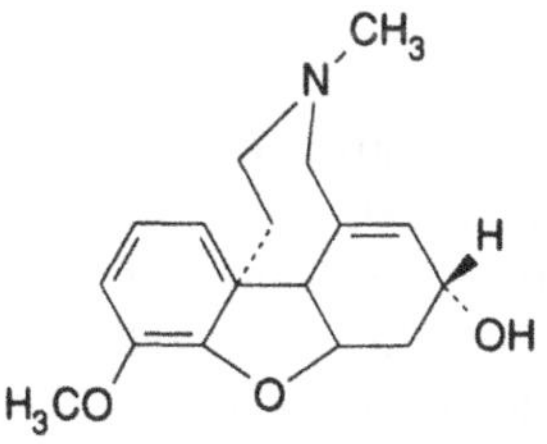

(162)

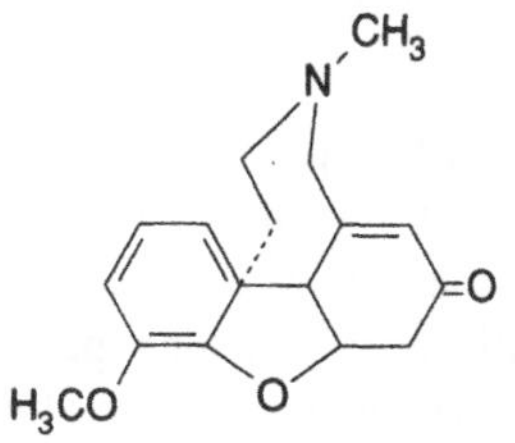

(163)

Galanthamine (161) is not a true morphinan derivative, however it shows considerable structural likeness to the opium alkaloids covered above. Its metabolism was studied *in vivo* in humans[604] and *in vitro* using rat and rabbit liver homogenates.[605] Little, if any, *O*- or *N*-demethylation of galanthamine occurred in these experiments. When the alkaloid was given orally (10 mg) to humans, the plasma obtained 3 h after dosing contained unchanged compound and epigalanthamine. The relevant change in the stereochemistry of the epimer is seen in structure (162). When the same dose was given s.c., the urine (12 h) contained the two aforementioned compounds and the corresponding keto derivative galanthaminone (163). Negligible metabolism was found in the rat liver homogenates, however those from rabbit liver readily formed epigalanthamine and the oxidized product galanthaminone. Incubation of the keto derivative resulted in the formation of both of the reduced products, with epigalanthamine production being favored by a ratio of 8:1.

PYRROLIZIDINE DERIVATIVES

Pyrrolizidine alkaloids are sometimes referred to as *Senecio* alkaloids, however these compounds are found in many genera of several plant families which show widespread geographical distribution. The pyrrolizidine alkaloids have long been the subject of interest, mainly because of the poisonous nature of many of these compounds when ingested by domestic animals or man. Their toxicological properties were reviewed in the monographs of Bull et al.[606] and Mattocks[607] and the reviews of McLean,[608] Huxtable[609] and Peterson and Culvenor.[610]

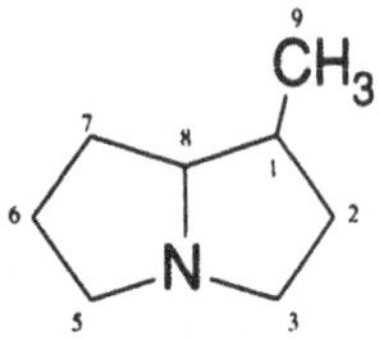

(164)

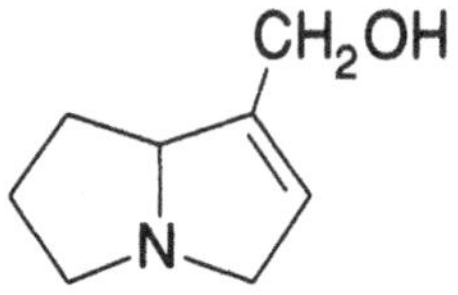

(165)

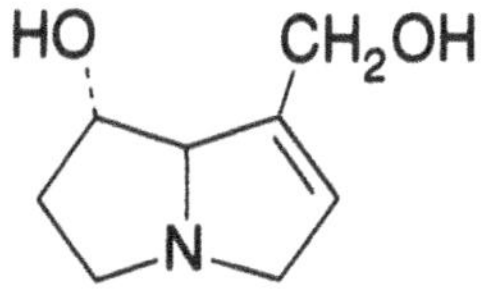

(166)

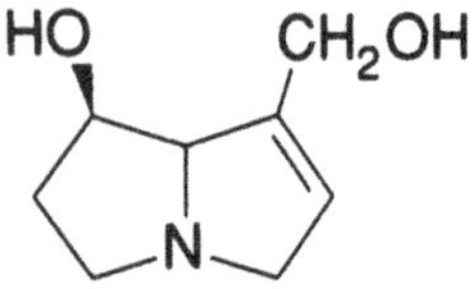

(167)

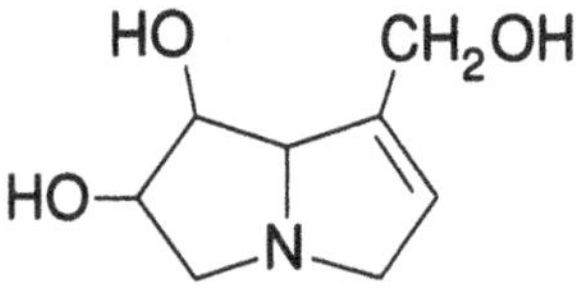

(168)

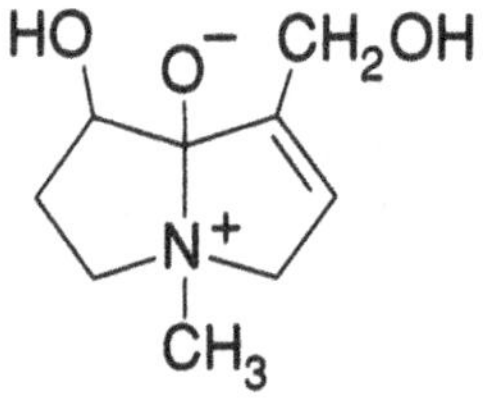

(169)

Most pyrrolizidine alkaloids are relatives of 1-methylpyrrolizidine (164). Those which exhibit toxicity are derivatives of unsaturated amino alcohols known as necines. These unsaturated derivatives are also termed pyrrolizines. Some common examples of necines which illustrate the main types are supinidine (165), heliotridine (166), retronecine (167), crotanecine (168), and otonecine (169). An additional requirement for toxicity is the presence of an acid moiety connected via an ester linkage to the necine. These are branched-chain hydroxylated acids which are called necic acids. The products formed may be monoesters of monocarboxylic acids (e.g., supinine (170a), heleurine (170b), or heliotrine (171)), diesters of monocarboxylic acids (e.g., lasiocarpine (172)), or macrocyclic diesters of dicarboxylic acids (e.g., monocrotaline (173), retrorsine (174), or senecionine (175)). Also, many of the pyrrolizidine alkaloids are present in the form of their *N*-oxides. The chemical properties of the pyrrolizidine alkaloids are described in detail in the books of Bull et al.[606] and Mattocks.[607]

Studies of the biological disposition of pyrrolizidine alkaloids, mainly using ^{3}H- or ^{14}C-labeled compounds, have shown that both the unchanged compound and its metabolites are rapidly and extensively excreted. However, small amounts of material are bound to tissue components and elicit toxic reactions via mechanisms described below. A study of the fate of nonlabeled and ^{3}H-labeled **monocrotaline** (173) in rats showed that 50 to 70% of the dose (s.c.) was excreted in the urine as unchanged compound.[611] A fairly extensive excretion of this compound in the unchanged form may also be inferred from the data presented by Bull et al.[612] who showed that monocrotaline has a relatively low partition coefficient (lipid phase:aqueous phase) and that, with **heliotrine** (171) and several related compounds, increasing proportions of the dose were excreted unchanged as the hydrophilicity of the alkaloids increased.[613] Thus, heliotrine which has a slightly greater partition coefficient than monocrotaline was found to be excreted unchanged in the urine to the extent of 30% of the dose in 16 h following i.p. injection to rats. The corresponding value for **lasiocarpine** (172), which has a partition coefficient more than 20-fold higher, was only 1 to 1.5%.

Several studies have shown that the bile is an important route of excretion of pyrrolizidine metabolites in some species. Using [^{3}H]monocrotaline, Hayashi[611] found that ~30% of the radioactivity was excreted as an unknown metabolite in the bile of rats. Only a trace of the unchanged alkaloid was detected. About 25% of the dose of **retrorsine** (174) was excreted in the bile of rats as pyrrolic metabolites, however only a small amount was found as unchanged compound.[614] Lafranconi et al.[615] detected metabolites of monocrotaline in rat bile using the isolated perfused liver. Extensive biliary excretion of dehydroretronecine (183), a major pyrrole metabolite of monocrotaline, was observed in rhesus monkeys.[616] In contrast to these data, Jago et al.[617] found that injections of heliotrine (171) in sheep resulted in the biliary excretion of only a few tenths of a percent of the dose.

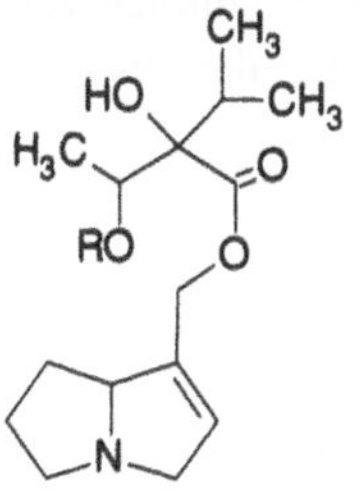

(170)

(a) Supinine, R = H

(b) Heleurine, R = CH$_3$

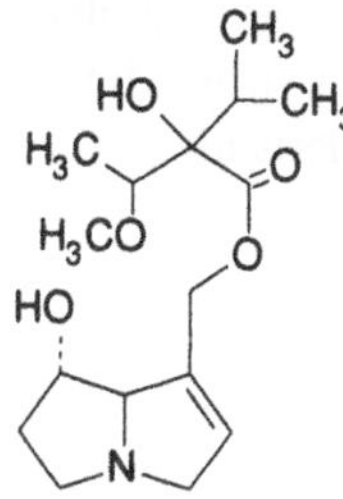

(171)

Heliotrine

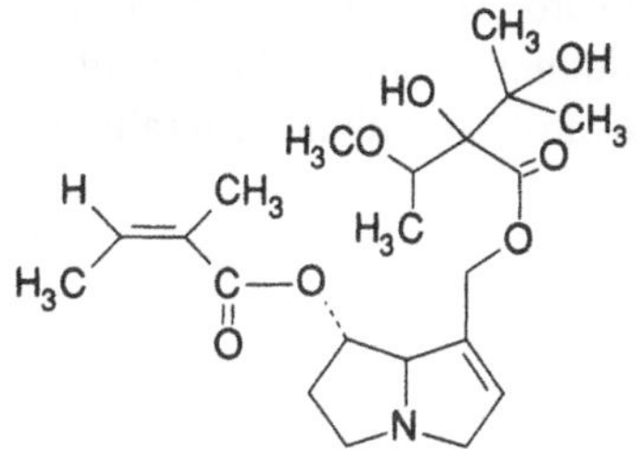

(172)

Lasiocarpine

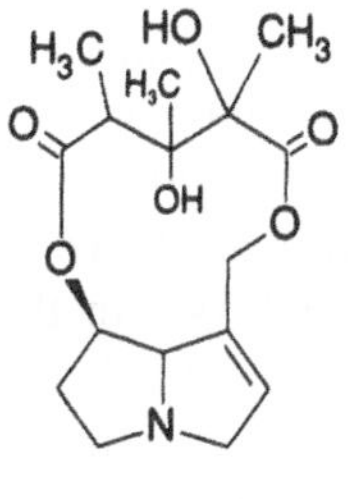

(173)

Monocrotaline

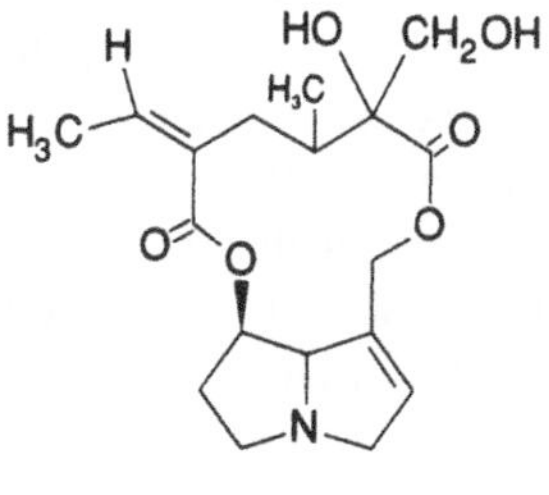

(174)

Retrorsine

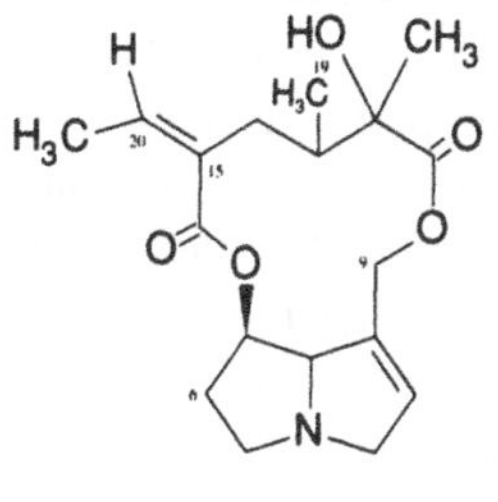

(175)

Senecionine

Extensive metabolism of the pyrrolizidine alkaloids, resulting in degradation to CO_2, has been shown to occur. Bull et al.[618] studied the fate of randomly-labeled [^{14}C]lasiocarpine (172) injected i.p. in rats. They recovered ~9% of the radioactivity in the expired CO_2 in slightly more than 4 h. On the other hand, an experiment in mice with a similarly labeled preparation of **senecionine** (175) indicated that only ~0.2% of the dose was lost as expired $^{14}CO_2$.[619]

Studies of the metabolism of pyrrolizidine alkaloids have shown that three major pathways are involved. These are cleavage of the ester linkages, either by hydrolysis or by reductive fission, nitrogen oxidation to give *N*-oxides, and oxidation of the ring to form pyrrole derivatives. In addition, several minor reactions including hydroxylation, demethylation, and, apparently, scission reactions leading to opening of the heterocyclic ring have been described. Reviews of the metabolism of the pyrrolizidine alkaloids are found in the references cited above.

A conspicuous chemical feature of the pyrrolizidine alkaloids is the presence of one or two ester linkages. Mattocks[620] reported that these compounds were generally fairly resistant to hydrolysis by esterases from rat liver or serum *in vitro*. Mouse liver microsomes were shown to hydrolyze senecionine (175)[621,622] and monocrotaline (173).[622] The respective necic acids formed, senecic acid (176) and monocrotalic acid (177), were identified and quantified in these incubates. Small interspecies differences were found in the *in vitro* hydrolysis of senecionine to senecic acid in liver microsomes from several animal species (rat, guinea pig, sheep, cow, and horse).[623] A few reports have appeared which demonstrated the hydrolysis of the pyrrolizidine alkaloids *in vivo*. Bull et al.[613] found that rats hydrolyzed heliotrine (171) and lasiocarpine (172) to a small extent. Approximately 3% of the dose (i.p.) was found as heliotridine (166) in the urine in 16 h. The latter metabolite was also identified in the urine of sheep given heliotrine i.v. or duodenally,[617] although the metabolite in this species likewise appeared to account for only a few percent of the administered dose. As the necine and necic acid components produced by hydrolysis are not cytotoxic, this reaction is one of detoxication. Mattocks[624] summarized the findings on the hydrolysis of numerous pyrrolizidine derivatives and noted the low rates of hydrolysis with the naturally occurring derivatives.

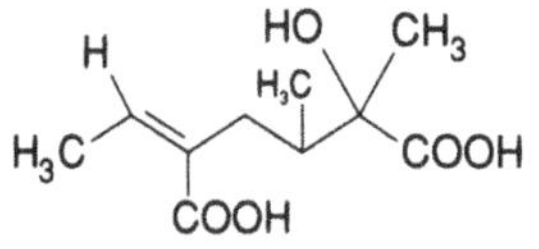

(176) 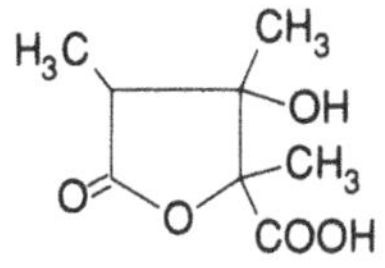   (177)

An interesting alternative pathway resulting in cleavage of the ester linkage involves reductive fission. This reaction was found to be carried out *in vivo* and *in vitro* by sheep rumen microorganisms. Dick et al.[625] postulated that sheep, which may ingest appreciable quantities of heliotrine and yet show relatively slight liver damage, are likely able to destroy much of the alkaloid in the rumen. They found that the rumen of these animals contained, in addition to some unchanged compound and its hydrolysis product heliotridine (166), 7α-hydroxy-1-methylene-8α-pyrrolizidine (178) as the major metabolite. This compound was devoid of acute hepatotoxic properties. Heliotrine was also converted to compound (178) *in vitro* by rumen microorganisms and the conversion rate was increased when the vitamin B_{12} concentration in the medium was increased. Under the latter conditions several other pyrrolizidine alkaloids were metabolized to 1-methylene derivatives. Thus, lasiocarpine (172) was converted to 7α-angelyloxy-1-methylene-8α-pyrrolizidine and supinine (170a), and heleurine (170b) gave 1-methylenepyrrolizidine. Russell and Smith[626] isolated a small Gram-negative coccus from sheep rumen contents which was able to carry out this reaction. They found that one mole of heliotrine was reduced by one mole of hydrogen gas to give one mole each of compound (178) and heliotric acid (179). Formate could also act as the hydrogen donor, in which case stoichiometric amounts of CO_2 were formed. Further investigation of this detoxication reaction by Lanigan and Smith[627] indicated that the 1-methylene derivatives are not the ultimate metabolic products but are further reduced to 1-methyl compounds. Thus, *in vitro* incubation of heliotrine with sheep rumen microorganisms resulted in the formation of 7α-hydroxy-1α-methyl-8α-pyrrolizidine (180) and *in vivo* experiments showed that this metabolite was formed from compound (178). In addition, the fully reduced metabolite (180) was also the final metabolic product of lasiocarpine (172). Lanigan[628] discovered that the ability of sheep rumen fluid to effect reductive fission of pyrrolizidine alkaloids varied according to the diet employed. Adaption of the rumen microflora resulting in greater rates of alkaloid metabolism occurred when the animals were fed plants containing the alkaloids and this property was rapidly lost when these plants were removed from the diet. It is worthy of note that this reductive pathway which is seen with open mono- and diesters is not of significance with the macrocyclic diesters. Shull et al.[629] and Swick et al.[630] found that *Senecio* alkaloids, which are closed esters, were not detoxified when incubated with sheep rumen fluid.

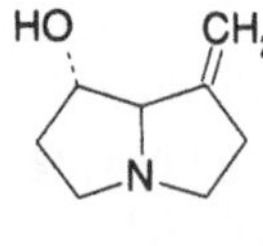

(178) 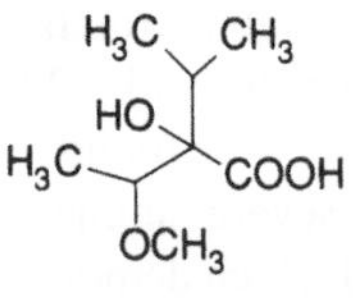   (179) 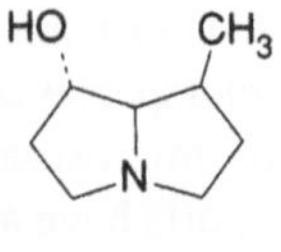   (180)

The oxidative pathways of metabolism with the pyrrolizidine alkaloids include the formation of N-oxides and the production of pyrrole derivatives via dehydrogenation reactions. Bull et al.[613] reported that the urine of rats given i.p. doses of heliotrine (171), its O-demethylated derivative (heliotridine trachelanthate), or lasiocarpine (172) contained very small amounts of the N-oxides of the first two of these compounds. However, from 6 to 20% of the dose was recovered as the N-oxide of heliotridine (N-oxide of (166)), the necine derived from these alkaloids. Mattocks[631] reported that ~14% of the dose (60 mg/kg, i.p.) of retrorsine (174) was excreted in the urine as the N-oxide derivative, mainly within 3 to 4 h. The rapid urinary excretion of these oxidation products was noted to occur generally with several pyrrolizidine alkaloids, not an unexpected finding in view of the higher water solubility of N-oxides. Chesney and Allen[632] reported that both rats and guinea pigs converted monocrotaline to its N-oxide. The oxidation of pyrrolizidine alkaloids to N-oxides was also noted in sheep, although in this species heliotrine-N-oxide was the sole metabolite of this type detected following heliotrine administration.[617]

The N-oxidation of pyrrolizidine alkaloids has been most thoroughly studied using hepatic microsomes. The results of Mattocks and White[633] showed that the mixed-function oxidases in rat microsomes were involved; however, recent findings indicate that both the cytochrome P-450 and flavin-containing monooxygenase systems may participate in N-oxidation.[634] The relative contributions of these two systems may be species and tissue dependent, however cytochrome P-450 appears to be the major system in rats[635]. Williams et al.[636] presented evidence which indicated that two cytochrome P-450 isoenzymes are involved in the N-oxidation of senecionine in rats and that the marked sex differences in metabolism which are observed result from differences in activities of these isoenzymes. Several investigations have indicated that the extent of N-oxidation increases with increasing partition coefficients of the alkaloids.[622,637,638] These studies also revealed that the two oxidative reactions, N-oxidation and pyrrole formation, are mediated by different enzymes. Minor interspecies differences were found in the N-oxidation of senecionine by liver microsomes from several animal species (rat, guinea pig, sheep, cow, and horse).[623]

A major advance in the understanding of the events involved in production of toxicity by pyrrolizidine alkaloids was the demonstration by Mattocks[631] of a new type of metabolite containing a pyrrole structure. Much of the subsequent work in this field has concerned itself with the formation and properties of these metabolites. Mattocks[631] found that numerous pyrrolizidine alkaloids were converted to compounds termed metabolic pyrroles. This reaction was observed *in vitro* using rat liver slices and the metabolites were found in the liver and urine of rats injected with the alkaloids. Values for the urinary excretion (24 h) of metabolic pyrroles in rats fell into the range of ~5 to 15% of the dose.

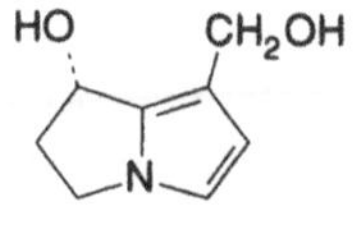

(181) 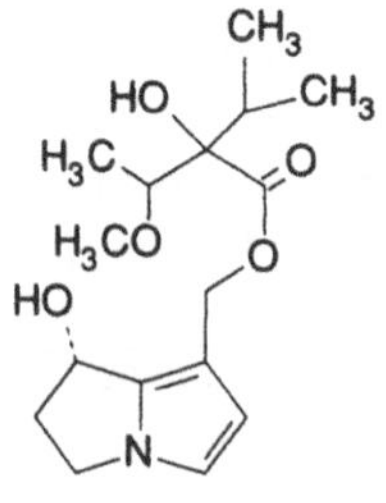   (182) 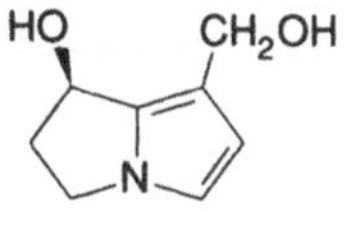   (183)

Numerous subsequent investigations have confirmed the formation of these pyrrole derivatives. Culvenor et al.[639] and Jago et al.[640] described the formation of pyrroles from heliotrine (171) and lasiocarpine (172) using rat liver homogenates and microsomal preparations. They identified the main water-soluble pyrrolic derivative, which was also one of the main urinary metabolites, as dehydroheliotridine (6,7-dihydro-7α-hydroxy-1-hydroxymethyl-5H-pyrrolizine)

(181). The formation of pyrroles was stimulated several fold when livers from rats pretreated with phenobarbital were used. Pyrrole derivatives containing the original ester groups were not detected, however this is not surprising in view of their lability.[631] It was therefore suggested that heliotrine is first converted by liver microsomal enzymes to dehydroheliotrine (182), the active alkylating species. Some of the active metabolite may undergo hydrolysis to dehydroheliotridine (181) which is stable and thus detected. A similar sequence was suggested by the results obtained with lasiocarpine and supinine. Likewise, Hsu et al.[641] demonstrated that dehydroretronecine (6,7-dihydro-7β-hydroxy-1-hydroxymethyl-5*H*-pyrrolizine) (183) was the major pyrrole metabolite in liver, blood, and urine of rats given monocrotaline (173). The toxic pyrrole, 6,7-dihydro-7-hydroxy-1-hydroxymethyl-5*H*-pyrrolizine, was likewise shown to be a microsomal metabolite of the macrocyclic alkaloids monocrotaline and senecionine (175) in preparations from mice.[622] This metabolite was also formed from the latter alkaloid by hepatic microsomes from several other mammalian species (rat, guinea pig, sheep, cow, and horse)[623] and by rat liver microsomes from several pyrrolizidine alkaloids which show close structural similarity to senecionine.[638] These were **jacobine** (senecionine-15,20-epoxide), **jacoline** (senecionine-15,20-diol), and **seneciphylline** (CH$_2$ at C-19) which differ slightly in the necic acid portion of the molecule but have a common necine moiety. Segall et al.[642] found that the metabolism of senecionine by mouse liver microsomes may lead to the formation of other dehydronecine derivatives. They identified both compound (184), the 7-methyl ether of dehydroretronicine, and hydroxydanaidal (185). Kedzierski and Buhler[643] showed that breakdown of the intermediate dehydroalkaloid formed by microsomes from senecionine to give the main pyrrolic metabolite, 6,7-dihydro-7-hydroxy-1-hydroxymethyl-5*H*-pyrrolizine (186), involved fission of the C–O bonds at C-7 and C-9. Thus, both oxygen atoms in this compound are derived from the solvent and not from the original alkaloid. This suggests the involvement of a C-7 carbonium ion in the activation sequence. This belief is supported by the finding that lasiocarpine and jacobine, which differ in configuration at C-7, did not form the corresponding pyrrolic enantiomers, dehydroheliotridine (181) and dehydroretronecine (183).[644] Instead, the product isolated from these microsomal incubations was racemic (186).

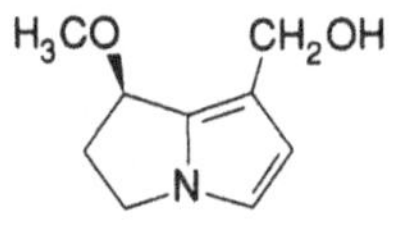

(184) 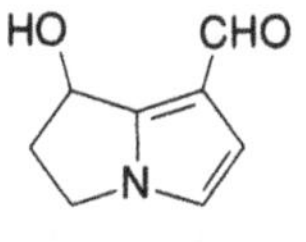   (185) 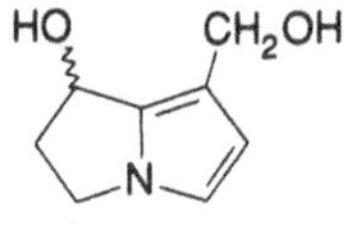   (186)

The possibility that pyrrolizidine alkaloid *N*-oxides may be intermediates in the formation of metabolic pyrroles was investigated by Mattocks and White[633] using rat liver microsomes and numerous pyrrolizidine alkaloids. They concluded that *N*-oxides are not intermediates in the formation of pyrroles. Similar conclusions were reached with monocrotaline and its *N*-oxide in a study involving the stimulation and inhibition of metabolism by rat liver microsomes[645] and in an *in vivo* and *in vitro* study with rats and guinea pigs.[632]

Although attention has mainly been centered on the toxicological role of the metabolic pyrroles, recent findings indicate that further metabolism of the pyrroles may occur. Segall et al.[646] reported that senecionine (175) was metabolized by a hepatic microsomal system to *trans*-4-hydroxy-2-hexenal (187). It was determined that this alkenal was derived from the necine portion of the alkaloid, apparently by loss of the atoms at positions 2, 3, and 4. Significantly, 4-hydroxyalkenals are highly reactive compounds which react with thiol groups. Griffin and Segall[647] found similarities in the mechanism of action of the toxic pyrrolizidine alkaloids and *trans*-4-hydroxy-2-hexenal and suggested that this alkenal may be an important toxic metabolite of these alkaloids. Studies in which ³H-labeled *trans*-4-hydroxy-2-hexenal was injected i.v. in rats showed that most of the dose was rapidly excreted in the urine. However, a few percent of the recovered radioactivity remained in the liver after 24 h. Most of the urinary radioactivity consisted of acidic metabolites and a mercapturic acid derivative, having the hemiacetal structure (188), was identified in this fraction.[648]

$$H_3C\text{-}CH_2\text{-}\underset{\underset{OH}{|}}{CH}\text{-}CH{=}CH\text{-}CHO$$

(187)

$$H_3C\text{-}CH_2\text{-}\underset{\underset{O}{|}}{CH}\text{-}CH\text{-}S\text{-}CH_2\text{-}\underset{\underset{NH\text{-}\overset{\overset{O}{\|}}{C}\text{-}CH_3}{|}}{CH}\text{-}COOH$$

(188)

Mattocks[631] found that the metabolic pyrroles were strongly bound in the liver, especially to the microsomal fraction and the fraction containing nuclei. Robertson et al.[649] suggested that the mechanism of interaction with cellular components involves alkylation of sulfhydryl groups. They demonstrated that the monocrotaline metabolite dehydroretronecine (186) reacted *in vitro* and nonenzymatically with cysteine and glutathione to form 7-thio derivatives of dehydroretronecine. White[650] showed that glutathione levels in the liver influenced the levels of pyrrolic metabolites in this organ following the administration of retrorsine to rats. Whereas an increase in glutathione decreased the levels of pyrroles, depletion of liver glutathione prior to alkaloid administration resulted in large increases in the levels of pyrroles in the liver. Several additional investigations have demonstrated that the pyrrolizidine alkaloids are metabolized to compounds which form covalent bonds with tissue macromolecules. Hsu et al.[651] reported that the pyrrolic metabolite dehydroretronecine (186) binds to proteins and nucleic acids *in vitro* and *in vivo*. Using senecionine (175) and seneciphylline, this was similarly demonstrated in mouse liver DNA, RNA, and protein and in calf thymus DNA[619] as well as in DNA from rat liver, lung, and kidney.[652] The former alkaloid was also shown to form covalent derivatives with cellular macromolecules when incubated with isolated rat hepatocytes.[653] Robertson[654] showed that the covalent adducts formed by the pyrrole metabolites may also involve formation of C–N bonds. When the monocrotaline metabolite dehydroretronecine (186) reacted nonenzymatically with deoxyguanosine, two major adducts were formed. These were shown to be an enantiomeric pair of derivatives in which the necine moiety was connected at C-7 to a nitrogen atom at N^2 of the nucleoside.

Mattocks[655] investigated the relationships between acute hepatotoxicity and pyrrolic metabolite formation in rats given pyrrolizidine alkaloids. Liver pyrrole levels were found to be proportional to the dose of alkaloid given 2 h earlier and in many cases the acute toxicity correlated with the amounts of pyrroles formed. However, it was pointed out that not all of the toxic metabolite formed may be involved in the production of toxicity as alternative pathways (e.g., hydrolysis, reaction with glutathione or amino acids) are available. Mattocks[656] showed earlier that the acid moieties of the pyrrolizidine alkaloids influence their toxicity. In alkaloids having a common base moiety, esterification with different acids results in alkaloids showing both quantitative and qualitative differences in toxicity. These findings suggested that these differences are due to differences in the amounts and stabilities of pyrrolic metabolites formed. In a further study on the influence of structural features on pyrrole formation, Mattocks[657] concluded that pyrrole formation was promoted by steric hindrance or chemical properties which provide resistance to ester hydrolysis, by lipophilic character which allows access to the microsomal enzymes and by a conformation which favors ring oxidation over *N*-oxidation.

In view of the fact that liver microsomal enzyme activity is low in newborn animals, Culvenor et al.[658] and Mattocks and White[659] studied the ability of newborn and young rats to convert pyrrolizidine alkaloids to pyrrolic derivatives *in vitro*. This capacity was very low in newborn rats but increased rapidly. The former study indicated that the rate of conversion of lasiocarpine in day-old rats was ~60% of that seen in adults and the latter study showed that the conversion of retrorsine reached very high levels within 5 d. In the case of human embryo tissue, liver but not lung was found to metabolize pyrrolizidine alkaloids including lasiocarpine and retrorsine to pyrroles *in vitro*.[660]

The possible oxidation of pyrrolizidine alkaloids by the cytochrome P-450 enzymes to give metabolites other than the *N*-oxides and pyrrole derivatives has received little attention. Eastman and Segall[621] found that senecionine (175) was metabolized to 19-hydroxysenecionine by mouse liver microsomes.

The heliotric acid moiety of heliotrine contains a methyl ether. This group has been shown to undergo cleavage in rats[613] and sheep.[617] In the former species 10% of the dose (i.p.) was excreted in the urine in 16 h as the corresponding alcohol, heliotridine trachelanthate. A further 5% was recovered as the *N*-oxide of the latter metabolite and, together, these demethylated products accounted for about one-fourth of the recovered dose. In sheep, the total urinary recovery was only 24% after i.v. injection and the quantity of heliotridine trachelanthate excreted was not determined. McLean[608] reported unpublished observations by Jago which showed that cleavage of the methyl ether can be carried out by rat liver microsomes *in vitro*. This reaction is one of detoxication as the alcoholic derivatives are less toxic than heliotrine.

Because of the presence of a double bond in the necine moiety of pyrrolizidine alkaloids, the possibility was raised that metabolically formed 1,2-epoxide derivatives might be implicated in the toxicity.[639,661] This view was developed further by Schoental[662] who believed that the epoxides are the proximal active forms of the alkaloids. However, these metabolites do not appear to have been detected as metabolic products. Furthermore, Culvenor et al.[639,658] prepared the α- and β-epoxides of monocrotaline and found them to be devoid of relevant biological activity.

Species differences in the toxicity and metabolism of pyrrolizidine alkaloids was reviewed by Cheeke.[663] Large variations in the susceptibility to these compounds are found among animal species. Some large herbivores, including horses and cattle, are highly susceptible whereas others, including goats and sheep, show a high degree of resistance. Resistance is also found among several small herbivores (hamsters, gerbils, guinea pigs, and rabbits) but not among mice and rats which are highly susceptible. A major determinant of these variations appears to be differences in the hepatic metabolism of the alkaloids, in particular the rates of formation of the pyrrole metabolites.[664] Chesney and Allen[632] reported that guinea pigs were much less able than rats to form pyrrolic derivatives. This was shown with monocrotaline both *in vivo* and *in vitro*. Both animal species were highly susceptible to the effects of the pyrrolic metabolites when these were given by i.v. administration. Shull et al.[629] studied the ability of liver microsomes from several species to produce pyrroles from monocrotaline or from the alkaloids of *Senecio jacobaea*. In most cases, this ability appeared to be directly related to the species differences in susceptibility to pyrrolizidine alkaloids. Nonetheless, some species deviate from this pattern (e.g., rabbits) and Cheeke[663] noted that additional metabolic pathways, which may increase detoxication processes, may explain exceptions of this type. Also, Winter et al.[623] studied the metabolism of senecionine in hepatic microsomes from guinea pigs and found that extensive formation of metabolic pyrroles occurred. This difference from that noted in the earlier findings may be explained by metabolic differences between the individual pyrrolizidine alkaloids used in the various studies or, perhaps, by a higher degree of resistance by guinea pigs to pyrrole toxicity.

As noted above, the toxic effects of pyrrolizidine alkaloids are associated with their metabolic dehydrogenation to pyrrole esters. These metabolites are chemically reactive and function as alkylating agents. Major sites of toxicity are the liver and lung and the stability of the reactive metabolites influences the patterns of toxicity seen. This was found with **anacrotine**, the 6-hydroxy derivative of senecionine, which produced less liver toxicity but more lung toxicity than that normally observed with pyrrolizidine alkaloids.[665] This was ascribed to a greater degree of stability of the dehydroanacrotine formed by the dehydrogenation reaction. Some pyrrolizidine alkaloids show different patterns of saturation in the necine moiety. **Senaetnine** (189), which shows close structural similarity to senecionine, contains the pyrrolic structure. Mattocks and Driver[666] studied its toxicity and metabolism in rats. It was less toxic than senecionine and appeared to have only mild alkylating activity. It was rapidly excreted in the urine, either as such or as metabolites. The metabolism of the nontoxic saturated alkaloids **platyphylline** (190a) and **rosmarinine** (190b) also results in the formation of pyrroles, however these are of a different type.[631,639] Mattocks and White,[667] employing rat liver slices or microsomes, obtained evidence which indicated that these alkaloids were dehydrogenated in the other ring. Thus, the metabolites formed were suggested to be compound (191a) from platyphylline and compound (191b) from rosmarinine. These pyrroles are not reactive. A few of the pyrrolizidine alkaloids are *N*-methyl derivatives and the metabolism of one of these, **otosenine** (192), was studied by Culvenor et al.[658] using rat liver microsomes. A major pyrrolic metabolite was identified as dehydroretronecine (186). This finding indicates that the *N*-methyl derivatives are subject to demethylation.

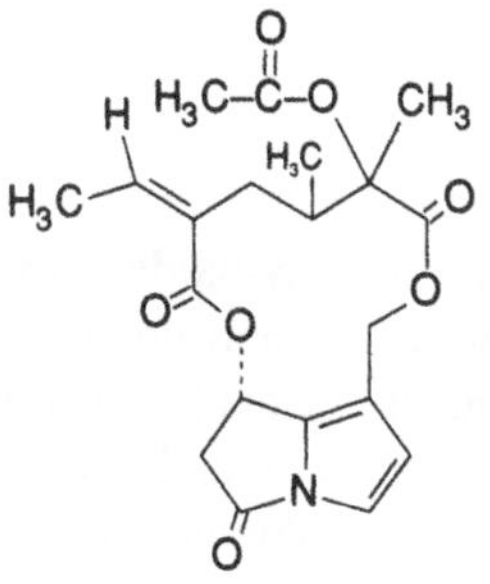

(189)

Senaetine

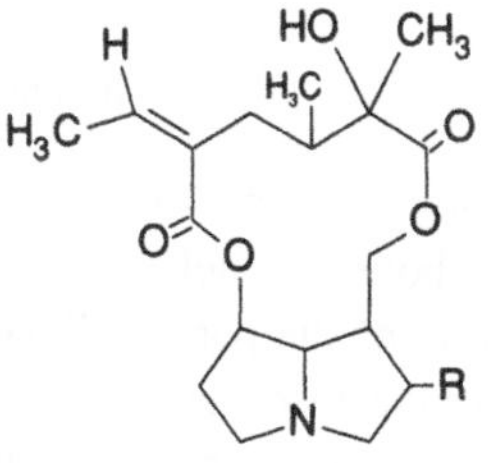

(190)

(a) Platyphylline, R = H

(b) Rosmarinine, R = OH

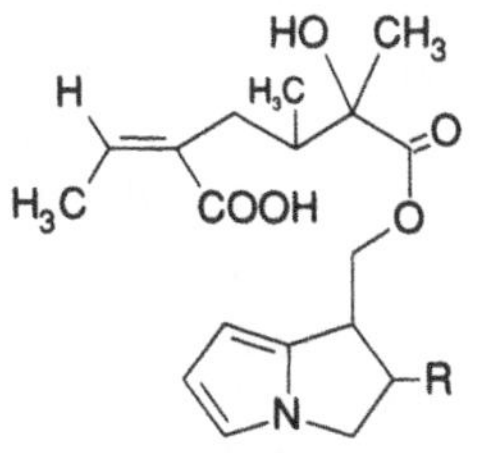

(191)

(a) R = H

(b) R = OH

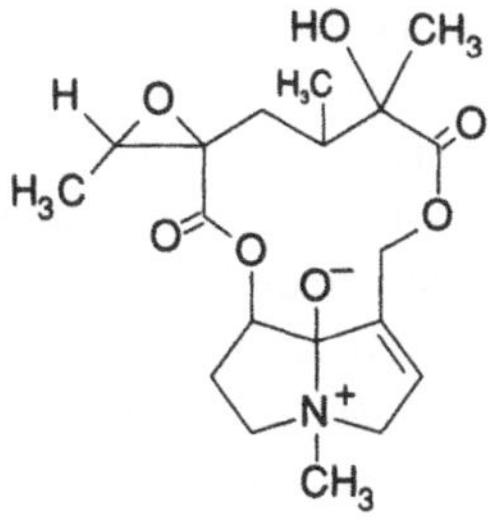

(192)

Otosenine

It was noted above that many of the pyrrolizidine alkaloids occur naturally in the form of their *N*-oxides. Additionally, an important oxidative pathway is formation of the *N*-oxides from the tertiary bases. As discussed by Mattocks,[668] the *N*-oxide derivatives are usually less toxic than the parent alkaloids, however the effects produced may be greatly influenced by the route of administration. Thus, oral administration may expose these compounds to the reducing effects of the intestinal microflora which generate the alkaloidal base, thus reducing or eliminating any difference in toxicity seen between the latter compound and its *N*-oxide. This was demonstrated by Mattocks[655,669] who also reported that the compound employed, **retrorsine-N-oxide**, was rapidly reduced to the base when incubated *in vitro* with the contents from rat intestine.

Lanigan[628] found that pyrrolizidine *N*-oxides were rapidly reduced to the tertiary bases by sheep rumen fluid. This reaction was noted earlier with **heliotrine-N-oxide** by Dick et al.[625] When heliotrine-*N*-oxide was administered orally and by i.p. injection to mice, the urinary excretion (24 h) of free and conjugated parent alkaloid was ~19 and 3% of the dose, respectively.[670]

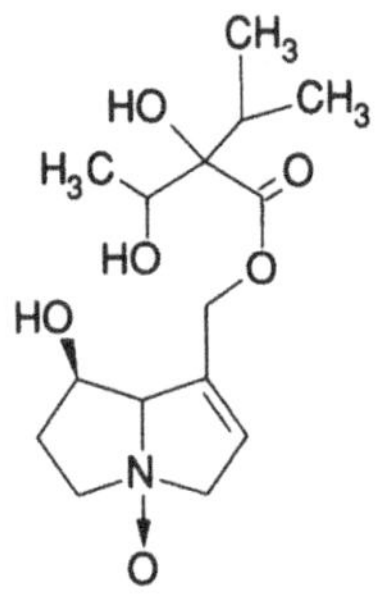

(193)

Indicine- *N*-oxide

The most thoroughly studied *N*-oxide derivative is **indicine-*N*-oxide** (193). Experiments similar to those noted above with heliotrine-*N*-oxide showed even more pronounced differences in the urinary excretion of the parent alkaloid when indicine-*N*-oxide was given to mice.[670] Approximately 98% of the dose was excreted unchanged in the urine and, again, 3% of the dose (468 mg/kg) was found in the urine as parent alkaloid after i.p. dosage. In contrast, 50% was recovered as parent alkaloid (free and conjugated) following oral dosage. El Dareer et al.[671] also studied the excretion of [³H]indicine-*N*-oxide in mice. They found that >80% of the dose (100 or 500 mg/kg, i.v.) was excreted in the urine as unchanged compound in 24 h. Similar experiments in rhesus monkeys given lower doses (0.24, 2.4, or 24 mg/kg) gave corresponding values of 88 to 96%. Extensive urinary excretion of unchanged indicine-*N*-oxide in rabbits following i.v. dosage (25 mg/kg) was reported by Ames and Powis.[672] They also detected excretion of some indicine, however the amounts of the reduced metabolite were small (~1% of the dose). A considerable proportion of the dose was also found as conjugated indicine-*N*-oxide and indicine.[673] These experiments in rabbits clearly revealed the major role of the intestinal microflora in the reduction of indicine-*N*-oxide. It was found that the formation of indicine also after i.v. dosage of the *N*-oxide was not due to extensive excretion of the latter in the bile. Instead, it appeared that it was able to enter the intestine by passive diffusion. Reduction in the numbers of anaerobic intestinal bacteria by the use of antibiotics resulted in a large decrease in the plasma indicine level and in the urinary excretion of indicine after dosage with indicine-*N*-oxide. The excretion of indicine-*N*-oxide and its parent alkaloid following i.v. administration of the *N*-oxide in humans shows a similar pattern, however the excretion values (24 h) were only 40 and 2%[674] or 53 and 2%[673] for the *N*-oxide and parent compound, respectively. A small additional amount was excreted in the form of conjugates. Another study showed that ~96% of the dose (i.v.) was excreted as unchanged indicine-*N*-oxide in 20 h.[675]

Although reduction of the pyrrolizidine *N*-oxides to the tertiary amines occurs largely via bacterial action, some reduction by liver enzymes may also take place. Mattocks[631] noted the formation of small amounts of pyrrole derivatives when retrorsine-*N*-oxide was incubated with rat liver slices and Jago et al.[640] reported similar results when the *N*-oxides of lasiocarpine or heliotrine were incubated with rat liver microsomes. Jago et al. suggested that formation of the pyrroles occurred after the *N*-oxides were reduced to the corresponding tertiary amines. The reduction of indicine-*N*-oxide to indicine was carried out by crude rat liver homogenate.[676] Indicine-*N*-oxide was reduced under anaerobic conditions by hepatic microsomes from rats[677] and rabbits.[673] Cytochrome P-450 was shown to be involved in the reaction. However, in contrast to that seen with cytochrome P-450-catalyzed oxidative reactions which largely depend on

NADPH, NADH was nearly as effective as NADPH in supporting the reduction of indicine-*N*-oxide.

IMIDAZOLE DERIVATIVES

The sole imidazole alkaloid for which metabolic data are available is **pilocarpine** (194).

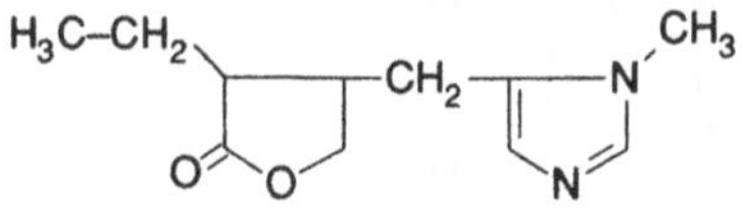

(194)

Pilocarpine

A likely site of metabolism of pilocarpine is the ester linkage in the lactone ring and most of the available data deals with the hydrolysis of this group. Lavallee and Rosenkrantz[678] found that the alkaloid was rapidly metabolized by rat serum *in vitro* to one or more metabolites which no longer gave positive reactions for the lactone or imidazole ring systems. Serum from rabbits, dogs, monkeys, and humans also metabolized pilocarpine *in vitro*. Preliminary *in vivo* data using rats (10 mg/kg, i.v.) and dogs (3 mg/kg, i.v.) suggested that pilocarpine rapidly disappeared from the blood and that small amounts (3 to 4% in 2 h) were excreted unchanged in the urine. The enzymatic hydrolysis of pilocarpine by rabbit and human serum, ocular tissues, and liver was studied by Schonberg and Ellis[679] and Ellis et al.[680] Pilocarpine inactivation in the samples was not due to binding with serum proteins but to an esterase which is not cholinesterase. This enzyme shows high stability and its activity can be prevented by the addition of a number of chelating agents. The enzyme levels in rabbit serum are considerably higher than those in human serum. Contrariwise, Newsome and Stern[681] reported that pilocarpine was not destroyed enzymatically by rabbit serum or several ocular tissues but was instead bound to serum and tissue components. Release of active material was effected by heating.

In view of the ophthalmologic use of pilocarpine, it is understandable that several investigations have dealt specifically with its disposition and metabolism in ocular tissue. Lazare and Horlington[682] found that [³H]pilocarpine applied topically accumulated to a greater extent in the pigmented rabbit iris than in the iris from albino rabbits. Their chromatographic results indicated that most of the radioactivity was due to unchanged compound and that little metabolism had occurred. Other studies have indicated that hydrolysis of pilocarpine to pilocarpic acid does occur in ocular tissue. This was reported by Schmitz et al.[683] and by Lee et al.[684] who found that metabolism was more extensive in pigmented rabbits. On the other hand, Salminen and Urtti[685] reported that pilocarpine disposition, but not metabolism, was influenced by ocular pigmentation. The conversion of the alkaloid to pilocarpic acid was equally efficient in the albino and pigmented rabbit eye. The sites of pilocarpine metabolism in rabbit ocular tissues was studied by Wood and Robinson.[686]

QUINOLIZIDINE DERIVATIVES

The metabolism of the quinolizidine alkaloids **matrine** (sophocarpidine) (195) and **matrine-*N*-oxide** was studied in rats.[687] When the *N*-oxide was given by i.m. injection, high concentrations of unchanged compound were detected in tissues, bile, and urine. However, following its oral administration the most abundant compound in the tissues was matrine. This finding suggests that the *N*-oxide was reduced to matrine in the gastrointestinal tract, following which the latter compound was absorbed and excreted in the urine. Song et al.[688] similarly concluded that the reduction to matrine occurred in the gastrointestinal tract, however they reported that this reaction could also take place in the liver. They administered the *N*-oxide labeled with ^{3}H by i.m. injection to mice and found that ~83 and 6% of the radioactivity were excreted in the urine and feces, respectively, in 48 h.

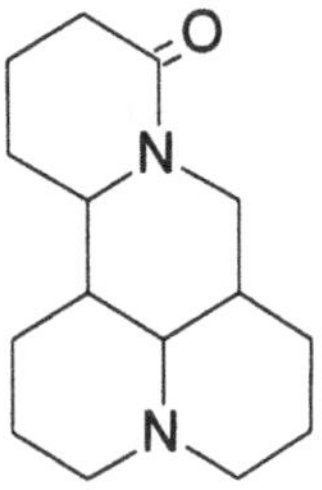

(195)	(196)	(197)
Matrine	(–)-Sparteine	(+)-Sparteine

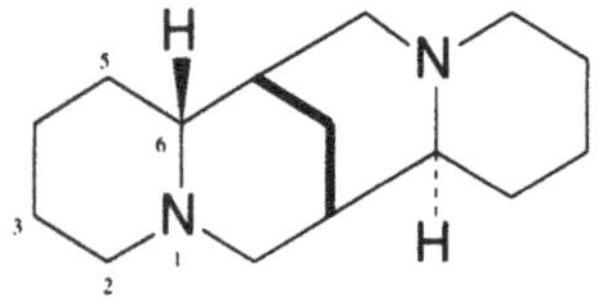

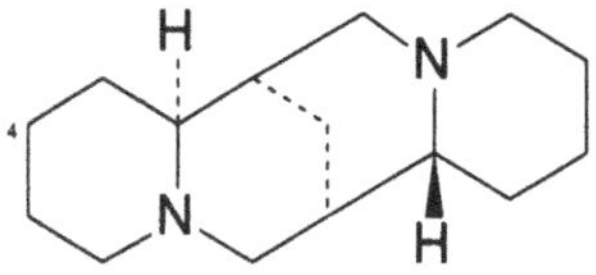

Dengler et al.[689] carried out a pharmacokinetic study in man of **sparteine** (196) which indicated that the plasma and urine of most subjects contained two metabolites in addition to unchanged sparteine. Although the metabolites were not identified, it was shown that they were amines which resembled sparteine and that they were probably not hydroxysparteines or conjugates. A mass spectrometric study[690] of the metabolites indicated that they are 2,3-dehydrosparteine (Δ^2-sparteine) and 5,6-dehydrosparteine (Δ^5-sparteine). It was suggested that these compounds may arise via dehydration of a single primary oxidation product, sparteine-1-*N*-oxide. However, Guengerich[691] presented evidence against the participation of the *N*-oxide. It was shown that rat liver microsomes did not convert the *N*-oxides of sparteine to dehydro derivatives. The results were also not consistent with an initial *C*-hydroxylation in the overall reaction. Eichelbaum et al.[692] cited findings which similarly rule out the possibility that an *N*-oxide is the primary metabolite. However, they believed that α-hydroxysparteines, which readily undergo dehydration, may nonetheless be involved in the reaction sequence. This publication[692] conveniently summarized the chemical and structural properties of sparteine and reviewed the findings on its metabolism. The formation of the dehydrosparteines is a reaction dependent on cytochrome P-450 and has been demonstrated in both rat and human liver microsomes. The urinary excretion of the 2-dehydro compound in man is usually three or four times greater than that of the 5-dehydro compound. The ratio is even more disproportionate in rats which excrete only traces of the latter derivative. Studies in humans have shown that 20 to 40% of the dose is usually unaccounted for by urinary sparteine and its dehydro derivatives. Eichelbaum et al.[692] summarized some recent findings which suggest that further metabolites, possibly including dimeric products, are formed from the dehydro compounds.

As described by Eichelbaum et al.,[693] the finding that some individuals did not excrete sparteine metabolites led to the discovery that they were unable to metabolize the alkaloid. This report indicated that ~5% of the subjects lacked this ability and that ~99% of the administered sparteine could be recovered unchanged in the urine of these individuals. These initial reports have triggered an extensive number of investigations designed to elucidate the nature of the phenomenon involved. The article by Eichelbaum et al.[692] provides a convenient summary of the genetic polymorphism of sparteine metabolism. Two phenotypes, termed extensive metabolizers and poor metabolizers, have been recognized in Caucasian and Japanese populations. The poor metabolizers comprised from 2.3 to 9% of these populations, a range of values which was also recorded in studies of Danish Caucasians,[694] Canadian Caucasians,[695,696] British Caucasians,[697] and Greenlanders living in Denmark.[698] Newer studies in Japanese subjects[699,700] confirm the existence of a poor metabolizer group, however the incidence of this phenotype (~2%) seems to be slightly lower than that found in Caucasian populations. Interestingly, studies of Ghanaians[701] and Cuna Amerindians from Panama[702] showed that these groups contained no deficient metabolizers of sparteine. The familial relationships in the polymorphic metabolism of sparteine was studied in 20 families with poor metabolizers.[703] Inaba et al.[704] showed that the genetic defect in the metabolism of sparteine is not a generalized deficiency of drug oxidation or of the cytochrome P-450 system. Similarly, the *N*-demethylation and *C*-oxidation of theophylline is not under the same genetic control as the oxidation of sparteine.[705] Interestingly, drug administration may influence the metabolism of sparteine. Brinn et al.[706] found that patients initially classified as extensive metabolizers exhibited the phenotype of poor metabolizers during quinidine treatment.

The metabolism of **(+)-sparteine** (pachycarpine) (197), the stereoisomer of (–)-sparteine, was studied in rats.[707] Little urinary excretion of the dehydro metabolites was found and the major urinary metabolite was identified as (+)-4-*S*-hydroxysparteine.

The excretion of racemic **tetrahydropalmatine** (198) labeled with ^{3}H was studied in rats.[708] No metabolites were identified, however 70 and 15% of the radioactivity were excreted in the urine and feces, respectively, in 136 h after i.v. dosage.

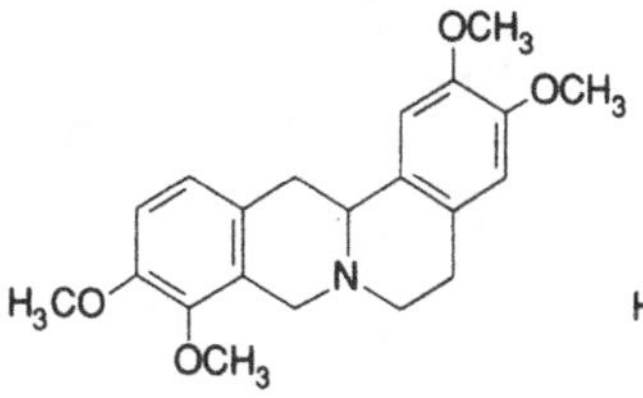

(198)
Tetrahydropalmatine

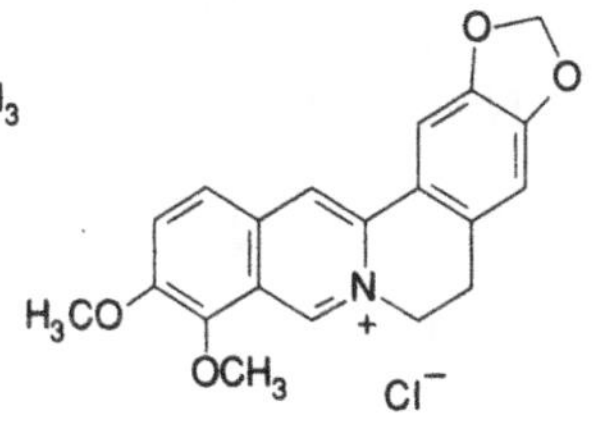

(199)
Berberine

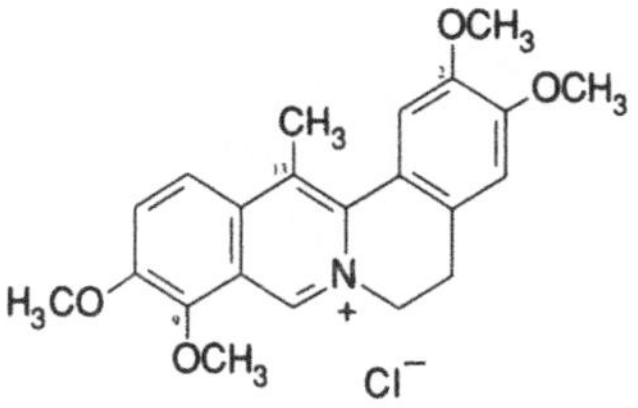

(200)
Dehydrocorydaline

Berberine chloride (199), a pyridinium derivative, was found to be slowly absorbed following oral administration to rats.[709] This was also noted in similar experiments using [^{3}H]berberine chloride.[710] In 48 h, ~3% of the radioactivity was excreted in the urine and 86% in the feces. To distinguish between unchanged berberine and its metabolites, Miyazaki et al.[711,712] developed methods which allowed its specific quantitation in urine. Following oral dosage of berberine·chloride (100 mg) to human subjects, <0.1% of the dose was recovered unchanged in the urine in 24 h. Additional results indicate that berberine excretion in the urine is low also following its injection. Only ~1% of the dose (50 mg/kg, s.c.) appeared in the urine of rats in 24 h.[713] Chin et al.[714] recovered ~15% of the dose (60 mg/kg, i.p.) in the urine of rabbits in 48 h. They believed that this material was unchanged berberine, however related compounds may have contributed to the value as their assay method was not specific for berberine. They also found that the alkaloid was metabolized by liver slices. The ability of the liver to metabolize berberine has also been reported by Furuya[715] in perfusion experiments. The metabolite was an unidentified oxidation product which was also isolated from the urines of rabbits and dogs given berberine chloride by s.c. injection.

The biological disposition of **dehydrocorydaline** (200) labeled with ^{14}C in the methyl group at C-13 was studied in mice and rats.[716] Blood levels were low after oral dosage (50 mg/kg), probably due to poor absorption, and most of the radioactivity was restricted to the gastrointestinal tract and liver before being excreted in the feces. Urinary excretion of radioactivity was <2% of the dose in mice and <0.2% in rats. Unchanged dehydrocorydaline was the most prominent compound in urine, feces, and bile, however smaller amounts of compounds formed via *O*-demethylation at the 2- and/or 9-positions were also detected.

PURINE DERIVATIVES

Derivatives of purine (201) are constituents of nucleotides and nucleic acids and therefore ubiquitous in plants. However, a discussion of the metabolism of these types falls outside the scope of this book, which limits itself to compounds classed as alkaloidal purines. These alkaloids consist of methylated derivatives of xanthine (202) and uric acid (203). The compounds which have been studied metabolically are theophylline (204), paraxanthine (210), theobromine (215), and caffeine (209). Because of their important pharmacological properties as well as their widespread occurrence in plants, interest in their metabolic fate developed from an early date. It was reported in the middle of the nineteenth century that no unchanged caffeine could be detected in the urine following oral administration of the alkaloid. This finding was generally borne out in later experiments which also indicated that demethylated products were formed. However, the analytical methods available earlier were often unsatisfactory by modern standards and the data obtained, often conflicting, are therefore not treated in detail in the present summary. The publications of Myers and Wardell[717] and Buchanan et al.[718] are useful sources for references to and discussion of this early work. Several recent reviews of the metabolism of methylxanthines are cited in the following discussions of the individual alkaloids.

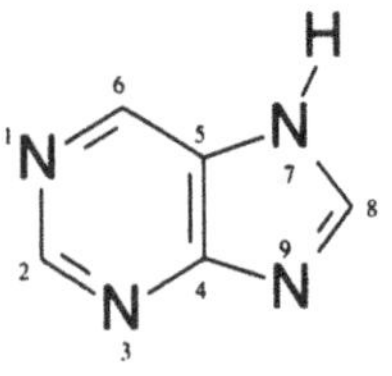

(201) 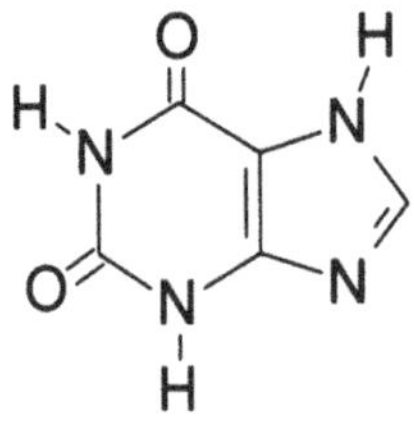   (202) 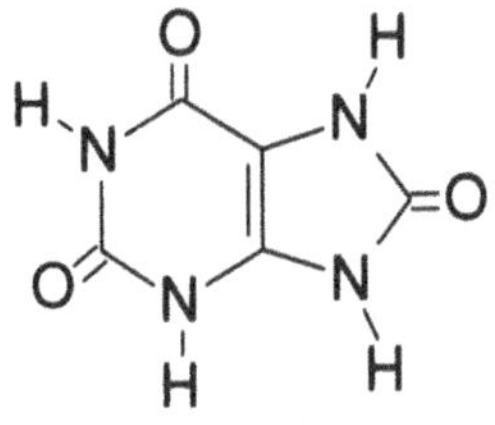   (203)

The metabolism of **theophylline** (1,3-dimethylxanthine) (204) (see Figure 7) has been studied in detail, especially in humans due to the therapeutic usefulness of this drug. The following summary will therefore emphasize the findings in man, however results from investigations using laboratory and domestic animals are also covered.

Early studies by Myers and Wardell[717] and Buchanan et al.[718] indicated that oxidation at C-8 to give uric acid derivatives was an important pathway of theophylline metabolism in man. The latter workers felt that about half of the dose (3 g) was converted to methyluric acids, of which about half might be 1-methyluric acid (205). Myers and Hanzal[719] also left open the possibility that some of the methylated uric acid excreted could be this monomethyl derivative. Weinfeld and Christman[720] subsequently confirmed that 1-methyluric acid (205) and 1,3-dimethyluric acid (206) are urinary metabolites of theophylline in man and Brodie et al.[721] showed that as much as 50% of an oral dose (750 mg) was excreted in the urine as 1,3-dimethyluric acid. The excretion value for unchanged compound was ~10% but little or no monomethyluric acid was detected.

The classic investigation of Cornish and Christman[722] first placed the subject of theophylline metabolism in man on a firm footing. They administered an oral dose (1 g) and found that the urine (48 h) contained unchanged compound (10% of the dose), 1,3-dimethyluric acid (206) (35%), 1-methyluric acid (205) (19%), and 3-methylxanthine (207) (13%). These results have stood the test of time and are in close agreement with values obtained from a long series of subsequent investigations which have employed i.v. doses[723] or single or steady-state oral therapeutic doses.[724-732] Thus, the average values from all of these studies are: unchanged theophylline, 11%; 1,3-dimethyluric acid, 38%; 1-methyluric acid, 19%; and 3-methylxanthine, 16%. In addition to this total of ~85% of the dose, small amounts of 1-methylxanthine (208) or unidentified compounds have sometimes been detected. The main pathways of theophylline metabolism are shown in Figure 7. 3-Methylxanthine (207) is an end-product of theophylline metabolism and was shown to be excreted unchanged in the urine following its administration.[733] The formation of 1-methyluric acid (205) proceeds via 1-methylxanthine (208) and not by demethylation of 1,3-dimethyluric acid (206).[732] This finding was confirmed by Birkett et al.[734]

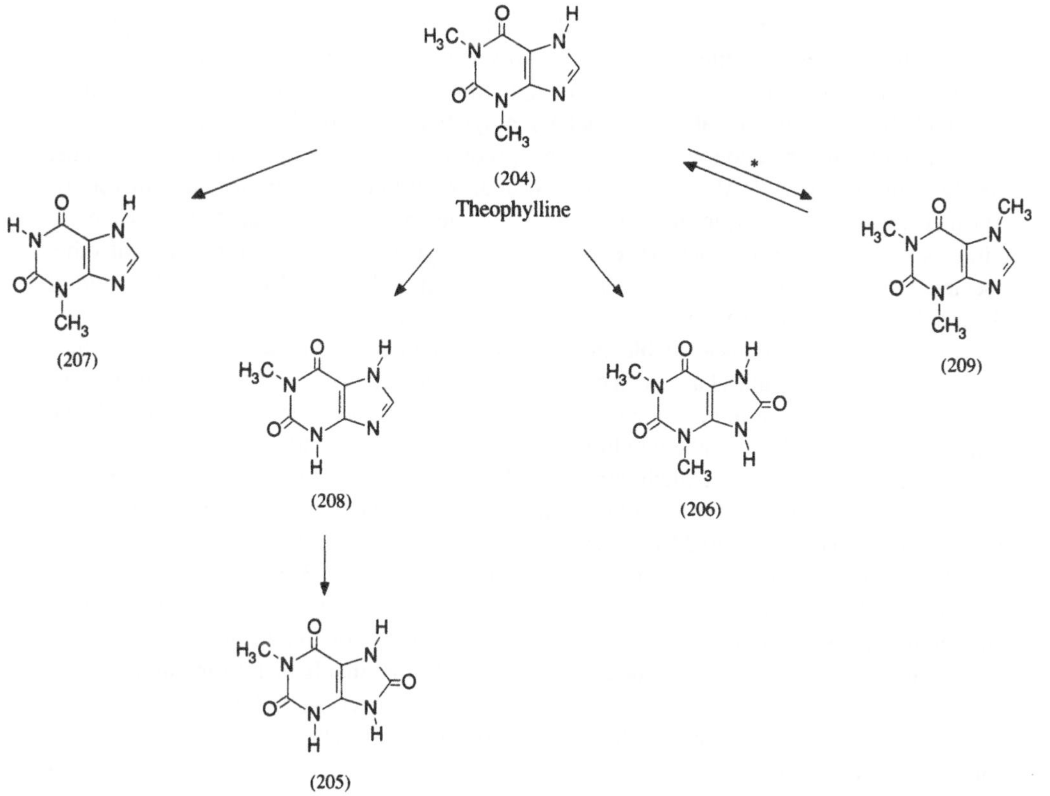

FIGURE 7. Metabolic pathways of theophylline. (*) See text.

who administered compounds (208) and (206). The former compound was largely recovered in the urine as 1-methyluric acid (205) whereas the latter was excreted unchanged. Also, inhibition of xanthine oxidase during steady-state administration of theophylline resulted in a sharp decrease in the excretion of 1-methyluric acid and a corresponding increase in the excretion of the precursor, 1-methylxanthine.[735] No change in the excretion of 1,3-dimethyluric acid was noted in this experiment. Excretion values similar to the averages noted above were obtained when the alkaloid was given by i.v. injection.[724,736] Subjects placed on diets lacking the methylxanthines normally present in foods and beverages showed increased rates of theophylline metabolism and greater amounts of urinary 3-methylxanthine and 1,3-dimethyluric acid.[724,725] However, diets containing greater than normal amounts of methylxanthines did not influence the disposition of theophylline.[737]

As described above, theophylline metabolism in adult humans involves ring oxidation and *N*-demethylation. In premature newborn infants, however, *N*-methylation to caffeine (209) has been demonstrated.[738-740] This finding has been confirmed in numerous subsequent studies, including one which employed theophylline labeled with stable isotopes (^{13}C and ^{15}N).[741] This investigation showed that caffeine, and also small amounts of its demethylated metabolites paraxanthine and theobromine, were present in both the plasma and urine. The metabolic profiles in plasma and urine were studied in premature infants under steady-state conditions.[742] Caffeine amounted to ~10 and ~20% of the metabolites found in urine and plasma, respectively. A major urinary metabolite in these infants was 1,3-dimethyluric acid (206) and lesser amounts of 1-methyluric acid (205) were also found. This indicates that ring oxidation at C-8 readily occurred in these subjects. This finding was confirmed in a study of theophylline metabolism in premature newborns, a young infant, children, and adults.[743] It appeared that the capacity to methylate theophylline was lost within the first few months of life. On the other hand, another study of premature infants with similar gestational ages showed that unchanged theophylline (98% of the dose) and caffeine (2%) were the only urinary metabolites.[744] Also, children (2 to 12 years) excreted the same metabolites of theophylline in the same amounts as found in adults. Although these studies generally agree that *N*-methylation of theophylline to caffeine does not occur in adults, this reaction was reported in a group of individuals given the alkaloid under steady-state conditions.[745] About 6% of the dose underwent conversion to caffeine in this study. The metabolism of theophylline in the newborn was reviewed by Aranda et al.[746] and by Haley.[747] Little is known of the possible existence of this methylation pathway in neonates of other species, however it was shown to be absent in neonatal piglets.[748] Also, the methylation of theophylline to caffeine was shown to occur in rat liver slices.[749]

Much less information is available on the metabolism of theophylline in species other than man, however a few studies have been carried out in mice, rats, rabbits, dogs, and horses. The pathways of metabolism in mice are similar to those summarized above in man. Betlach and Tozer[750] administered [8-¹⁴C]theophylline by i.p. injection and found unchanged compound, 1,3-dimethyluric acid (206), 1-methyluric acid (205), and 3-methylxanthine (207) in the urine. They employed eight inbred strains of mice and found that the urinary excretion of unchanged theophylline ranged from 6 to 20%. However, the variations among the strains were attributed mainly to variations in the formation of the major metabolite, 1,3-dimethyluric acid. This metabolic pathway was also shown to be responsible for the variations in induction observed among the strains in response to polycyclic aromatic hydrocarbon treatment.[751]

An early study of theophylline metabolism in rats showed that both 1,3-dimethyluric acid (206) and 1-methyluric acid (205) were urinary metabolites.[720] These findings were confirmed by Lohmann and Miech[752] and by Christensen and Whitsett.[753] The latter investigations also showed that little or no 3-methylxanthine (207) was excreted. Similar results were reported by Van Gennip et al.[732] who found that 1,3-dimethyluric acid was the most abundant urinary metabolite. Using [8-¹⁴C]theophylline, Arnaud et al.[754] obtained quantitative values for the urinary excretion of theophylline and its metabolites in rats. The values (48 h) were: theophylline, 25% of the dose; 1,3-dimethyluric acid, 24%; 1-methyluric acid, 13%; and 3-methylxanthine, 2%. Additionally, very small amounts of 1-methylxanthine, 3-methyluric acid, and two unidentified polar metabolites were detected. Interestingly, similar experiments in pregnant rats showed marked reductions in the amounts of oxidized or demethylated metabolites excreted and a doubling of the amount of unchanged theophylline in the urine. Gorodischer et al.[749] studied the metabolism of theophylline in liver slices of young and adult rats.

Weinfield and Christman[720] identified 1-methyluric acid (205) and 1,3-dimethyluric acid (206) as urinary metabolites of theophylline in rabbits. These results were confirmed by Celardo et al.,[755] however the most prominent metabolite by far was compound (206). They recovered 88% of the dose in the urine in 48 h and found that 1,3-dimethyluric acid (206) accounted for 75% of this (66% of the dose). Unchanged theophylline furnished 15% and 1-methyluric acid (205), 1-methylxanthine (208), and 3-methylxanthine (207) a further 2, 1, and 1%, respectively. Interestingly, several additional metabolites accounting for traces to <1% of the dose were detected in this investigation. These included a polar metabolite and the three possible *N*-methylated aminouracil compounds derived from theophylline (see structures (214), (218), and (225) for examples of related aminouracil derivatives). Myers and Hanzal[719] reported that the Dalmatian dog, a species which resembles man in its excretion of uric acid, metabolized theophylline mainly to 1,3-dimethyluric acid (206). Recently, Kuze et al.[756] found that dogs excreted ~85% of the dose in the urine in 24 h. The urinary material included unchanged compound (18%), 1,3-dimethyluric acid (26%), and 3-methylxanthine (40%). Moss[757] reported that 80% of the radioactivity was excreted in the urine following the oral administration of [¹⁴C]theophylline to horses. Excretion was complete within 4 d and ~21% of the dose was excreted as unchanged compound and 52% presumably as 1,3-dimethyluric acid (206).

A number of investigations have been carried out using *in vitro* systems in order to elucidate the enzyme systems involved in theophylline metabolism. An early report by Brodie et al.[721]

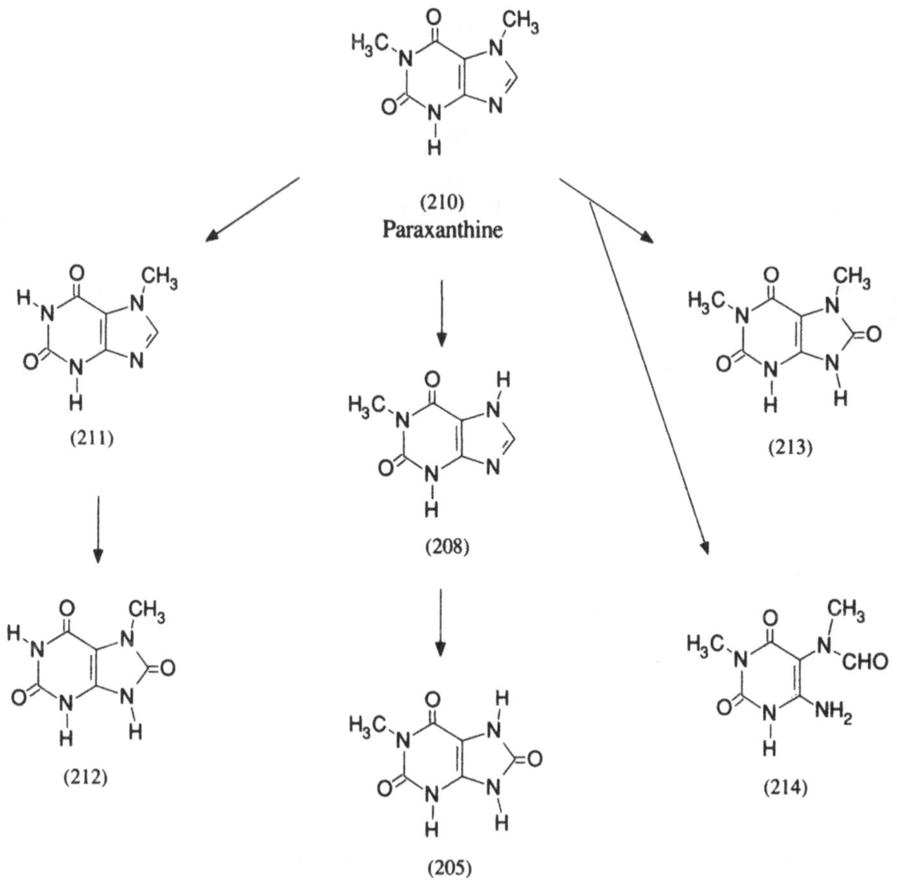

FIGURE 8.   Metabolic pathways of paraxanthine.

showed that oxidation at C-8 is not catalyzed *in vitro* by milk xanthine oxidase. No *N*-demethylation of the alkaloid was observed in an *in vitro* system using rabbit liver microsomes;[758] however, a study using mouse liver microsomes showed slight demethylation activity.[759] Subsequently, an extensive investigation of this subject in rats by Lohmann and Miech[752] clearly indicated that theophylline is metabolized by the liver microsomal system. On the other hand, it is not a substrate for liver xanthine oxidase or aldehyde oxidase, the latter result confirming an earlier report by Krenitsky et al.[760] Among the many types of tissue slices used, only those from liver were able to metabolize theophylline. Fractionation studies localized this activity in the microsomal fraction and the results from stimulation and inhibition experiments were in accordance with the other data implicating the liver microsomal system. Furthermore, they found that this system converted theophylline to 1,3-dimethyluric acid (206), a major urinary metabolite of the alkaloid, and apparently to 1-methylxanthine (208), which is usually not found in the urine. The latter metabolite is normally metabolized further by xanthine oxidase to 1-methyluric acid (205). More recent investigations have shown that at least two distinct P-450 isoenzymes from human liver are involved in theophylline metabolism.[761-763] A similar situation is found in liver microsomes from rats and rabbits.[764] Distinct forms of cytochrome P-450 in human liver mediate *N*-demethylation or oxidation at C-8.[729,744,762] The genetic control of theophylline metabolism is not the same as that responsible for sparteine or debrisoquine oxidations.[705] Theophylline metabolism by cultured hepatocytes from rats and humans was shown to follow the same pathways as those observed *in vivo*.[765] Thus, 3-methylxanthine and 1,3-dimethyluric acid were preferentially formed.

Paraxanthine (210) (see Figure 8) is the dimethylxanthine which shows the most restricted occurrence and for which the least metabolic information is available. The metabolic scheme

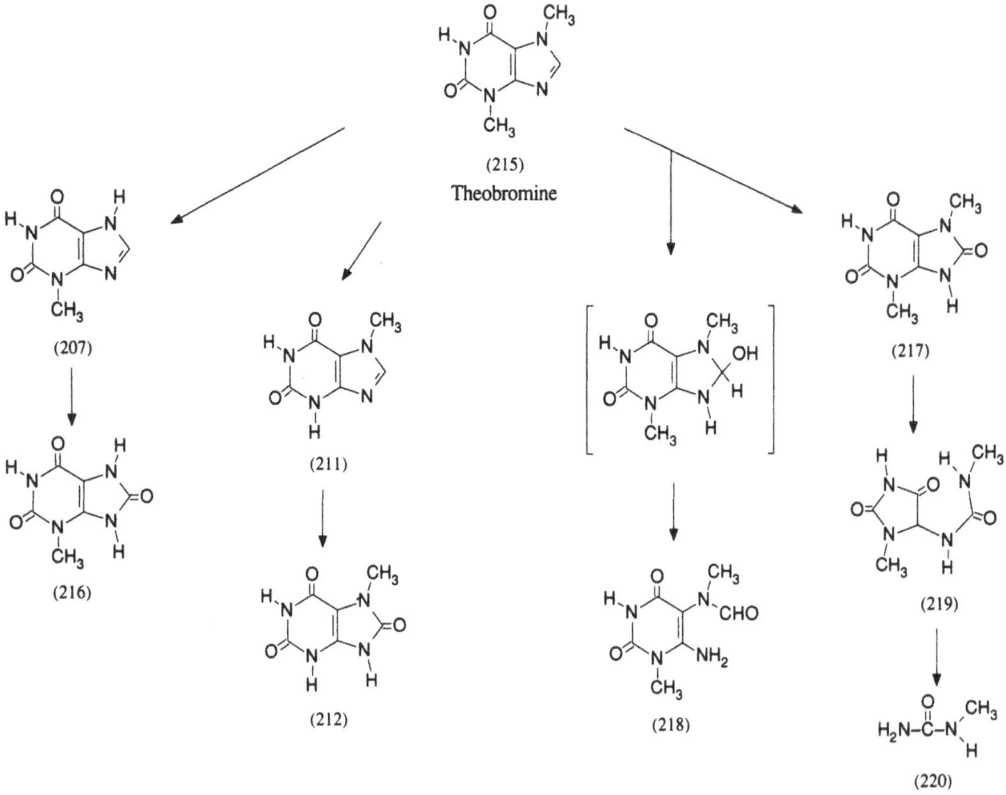

FIGURE 9.   Metabolic pathways of theobromine. Postulated intermediate in brackets.

shown in Figure 8 is based on the findings of Arnaud and Welsch in rats[766] and man.[767] The former study employed [1-*methyl*-^{14}C]paraxanthine and showed that little loss of the 1-methyl group occurred. Only ~2% of the dose was recovered as expired $^{14}CO_2$ and small amounts of the resultant xanthine (211) and uric acid (212) derivatives were excreted in the urine. The urine contained 85% of the administered radioactivity and the major urinary metabolite (52%) was unchanged compound. The remainder consisted of 1-methylxanthine (208) (11%), 1-methyluric acid (205) (21%), 1,7-dimethyluric acid (213) (15%), and the ring-opened compound, 6-amino-5(N-formylmethylamino)-3-methyluracil (214) (15%). The results in man indicated that N-demethylation was more extensive than in the rat and that excretion of 1,7-dimethyluric acid and the uracil derivative was reduced. A third report[768] quoted values of 7% (rats) and 4% (man) for the urinary excretion of the uracil metabolite (214).

Theobromine (3,7-dimethylxanthine) (215) (see Figure 9) is readily and extensively absorbed following oral administration.[726,769] As noted below, the subsequent excretion of unchanged theobromine and its metabolites occurs mainly in the urine. The pathways of metabolism of theobromine (Figure 9) involve N-demethylation, ring oxidation, and ring scission and are thus similar to those described above for theophylline and paraxanthine. However, oxidation of theobromine at C-8 to give uric acid derivatives is less extensive than that found with theophylline. It seems likely that this reaction is hindered by the methyl group at C-7.[722] Additionally, the urinary excretion of unchanged theobromine is fairly extensive in some species.

Because of the reduced importance of ring oxidation at C-8, metabolites formed by N-demethylation are prominent excretory products of theobromine. This was reported in an early study in man by Krüger and Schmidt[770] who found that 7-methylxanthine and 3-methylxanthine were excreted in a ratio of ~2:1. Both of these monomethylxanthines were excreted in the urine of rabbits and dogs following large doses of theobromine.[770,771] In rabbits the ratio of urinary 7-methylxanthine:3-methylxanthine was ~15:1 whereas it was ~1:5 in dogs, which also converted less of the dose to methylxanthines than did rabbits or humans. Interestingly, this early indication of a species difference in theobromine metabolism has been confirmed by newer studies using [^{14}C]theobromine.[772,773] These and other reports have clearly shown that the pathways of theobromine metabolism are qualitatively similar in several animal species including man. However, significant quantitative differences among these pathways have been noted. This was clearly demonstrated by Miller et al.[772] who studied the metabolism of theobromine in mice, rats, hamsters, rabbits, and dogs. The values for excretion of unchanged compound varied from ~14 to 16% of the dose in male mice and rabbits to as much as a third of the dose in rats and dogs. Both monomethylxanthines (207) and (211) were formed in all species as were all of the uric acid derivatives (216), (212), and (217). However, 3-methyluric acid (216) was a minor (<1% of the dose) metabolite of theobromine in all species. 6-Amino-5-(N-formylmethylamino)-1-methyluracil (218) was an additional ubiquitous metabolite which accounted for 10 to 20% of the dose in most species except the dog (~8%) and male mice (~28%). The observed variations between species were attributed to large differences in N-demethylation which, however, did not proceed beyond the monomethyl stage. The pathway of ring oxidation was restricted in extent and showed little species variation.

Several additional investigations in rats and rabbits largely confirmed these findings. Arnaud and Welsch[774] administered [7-methyl-^{14}C]theobromine orally to rats and recovered 6% of the dose as ^{14}CO$_2$. This value corresponded to the amount of 3-methylxanthine (207) and 3-methyluric acid (216) formed. The corresponding 7-methyl derivatives (211) and (212) furnished a further 5 and 3%, respectively, however the main urinary metabolites were unchanged compound (~40%) and the uracil derivative (218) which gave 30%. About 2% of the dose was recovered as 3,7-dimethyluric acid (217) and only trace amounts of dimethylallantoin (219) and N-methylurea (220) were excreted. Similar values were recorded in subsequent experiments using female[775] or male[769,776] rats. These studies employed oral doses of a few milligrams to 100 mg and noted only small differences in the amounts of the individual urinary metabolites. Also, the metabolic profiles did not differ significantly between male and female rats or between pregnant and nonpregnant rats. Newer studies of theobromine metabolism in rabbits confirmed the predominance of N-demethylation to 7-methylxanthine over the competing reaction to 3-methylxanthine.[773,777] The values for the excretion of most of the metabolites were similar to those found in rats, however less unchanged compound was recovered.

Appreciable excretion of unchanged theobromine occurs in horses. Moss[757] reported that when ^{14}C-labeled compound was given orally, 75% of the radioactivity was recovered in the urine in 4 d. This material contained theobromine (42% of the dose) and a metabolite (21% of the dose) which was presumed to be 3,7-dimethyluric acid (217).

A significant advance in the understanding of theobromine metabolism in man was made by Cornish and Christman[722] who identified 3-methylxanthine (207), 7-methylxanthine (211), and unchanged compound as major urinary metabolites. A small amount of 7-methyluric acid (212) was also detected. Subsequently, Arnaud and Welsch[774] identified 3,7-dimethyluric acid (217) and the uracil derivative (218) in human urine following theobromine administration. These findings were confirmed in several later studies[726,778,779] which show fairly consistent values for the amounts of urinary metabolites excreted. Typical ranges of values for unchanged theobromine, 3-methylxanthine, and 7-methylxanthine are 11 to 18, 14 to 22, and 30 to 50%, respectively. Lesser quantities of 7-methyluric acid (212) (7 to 12%) and 6-amino-5-(N-formylmethylamino)-1-methyluracil (218) (6 to 13%) were found and very little (usually <1%) 3-methyluric acid (216) or 3,7-dimethyluric acid (217) were excreted.

Although early studies on the in vitro metabolism of theobromine showed no demethylation with rabbit liver microsomes[758] or very slight activity with mouse liver microsomes,[759] it is now evident that liver microsomes effect both its ring oxidation and N-demethylation. Campbell et al.[763] demonstrated that theobromine, like theophylline and caffeine, is demethylated by one or more isoenzymes of cytochrome P-450 which are inducible by polycyclic aromatic hydrocarbons. Shivley and Vesell[776] found that incubation of theobromine with rat liver microsomes resulted in the formation of 3,7-dimethyluric acid (217) as the major metabolite. When the cytosol fraction or thiols were added to these incubates, the ring-opened uracil derivative (218) was also formed. This suggests that these two metabolites may derive from a common precursor and that the proportions formed may be regulated by thiol status. Inhibition of xanthine oxidase was shown to have no effect on the plasma clearance of theobromine in man.[780] The excretion of 3-methylxanthine, the uracil derivative (218) and unchanged compound was not affected, however the excretion of 7-methyluric acid was abolished. This reduction was compensated for by a concomitant increase in the excretion of 7-methylxanthine. Additional information on the *in vitro* metabolism of methylxanthines is found in the summaries of theophylline and caffeine metabolism.

Studies on the metabolism of **caffeine** (1,3,7-trimethylxanthine) (209) (see Figure 10), the most widely used of the methylxanthines, have a long history. This interest has resulted in an extensive literature, especially that from the past two decades which has largely clarified the qualitative and quantitative aspects of caffeine metabolism in man and several animal species. This subject has been reviewed in varying detail many times in recent years, however the articles on the biological disposition of caffeine by Burg,[781] Yesair et al.,[782] and Bonati and Garattini[783] merit special note. Additionally, the outstanding review by Arnaud[784] presents the most thorough summary of caffeine metabolism and is especially useful in tracing the historical developments in this field. In order to avoid an extensive duplication of these surveys of caffeine metabolism, the present summary instead summarizes the metabolic data in tabular form. This compilation lists the urinary metabolites of caffeine and, in most cases, the amounts formed in numerous animal species. In addition, the pathways of metabolism are shown (Figure 10), however this composite scheme includes metabolites identified in many studies using many animal species and may not therefore have universal applicability.

Caffeine is readily and extensively absorbed from the gastrointestinal tract of most animal species including the mouse,[785,786] rat,[787] pig,[788] rhesus and squirrel monkey,[789] and man.[790] However, poor absorption was noted after oral administration to horses.[791] The values for unchanged caffeine excreted in the urine (Table 1) clearly illustrate that it undergoes extensive metabolism in the species studied. Table 1 lists the main urinary metabolites of caffeine together with information on species and amounts excreted. The metabolic pathways are shown in Figure 10. Several additional minor or occasional metabolites have been reported, especially in the rat. A notable example of these minor metabolites is the sulfoxide derivative (221) which was identified as a urinary metabolite of caffeine in the mouse, rat, rabbit, and horse.[792] The corresponding sulfide (R–S–CH$_3$) and sulfone (R–SO$_2$–CH$_3$) derivatives were also present in mouse urine. The formation of compound (221) in rats given caffeine was confirmed[793] and it was found that its urinary excretion was greatly reduced in germ-free animals. It was proposed that the formation of compound (221) involves conjugation of caffeine with glutathione and excretion of the conjugate in the bile. The conjugate may then undergo scission by C-S-lyase from the intestinal microflora and the resultant thiol methylated and oxidized in the tissues. Interestingly, Arnaud[784] was not able to confirm the formation of the sulfur-containing metabolites. Another metabolic pathway of caffeine, which has been identified in rats, results in the formation of trimethylallantoin (222), probably via 1,3,7-trimethyluric acid (223).[784,794,795] Further degradation of the ring-opened metabolite probably occurs as trace amounts of both *N*-methylurea and *N,N'*-dimethylurea were excreted in the urine. In a similar manner, small amounts of the dimethylallantoin derivatives which arise from the corresponding uric acid derivatives may be formed, as shown in Figure 9.

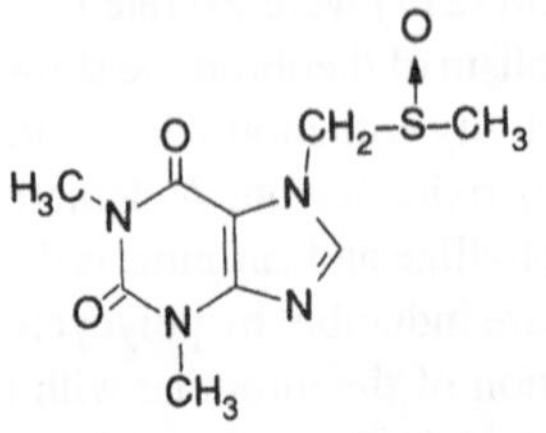

(221) 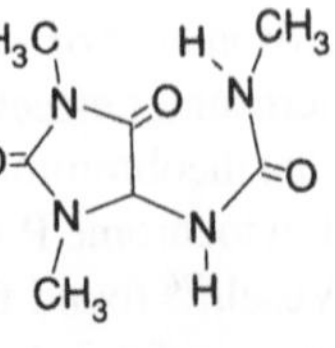   (222)

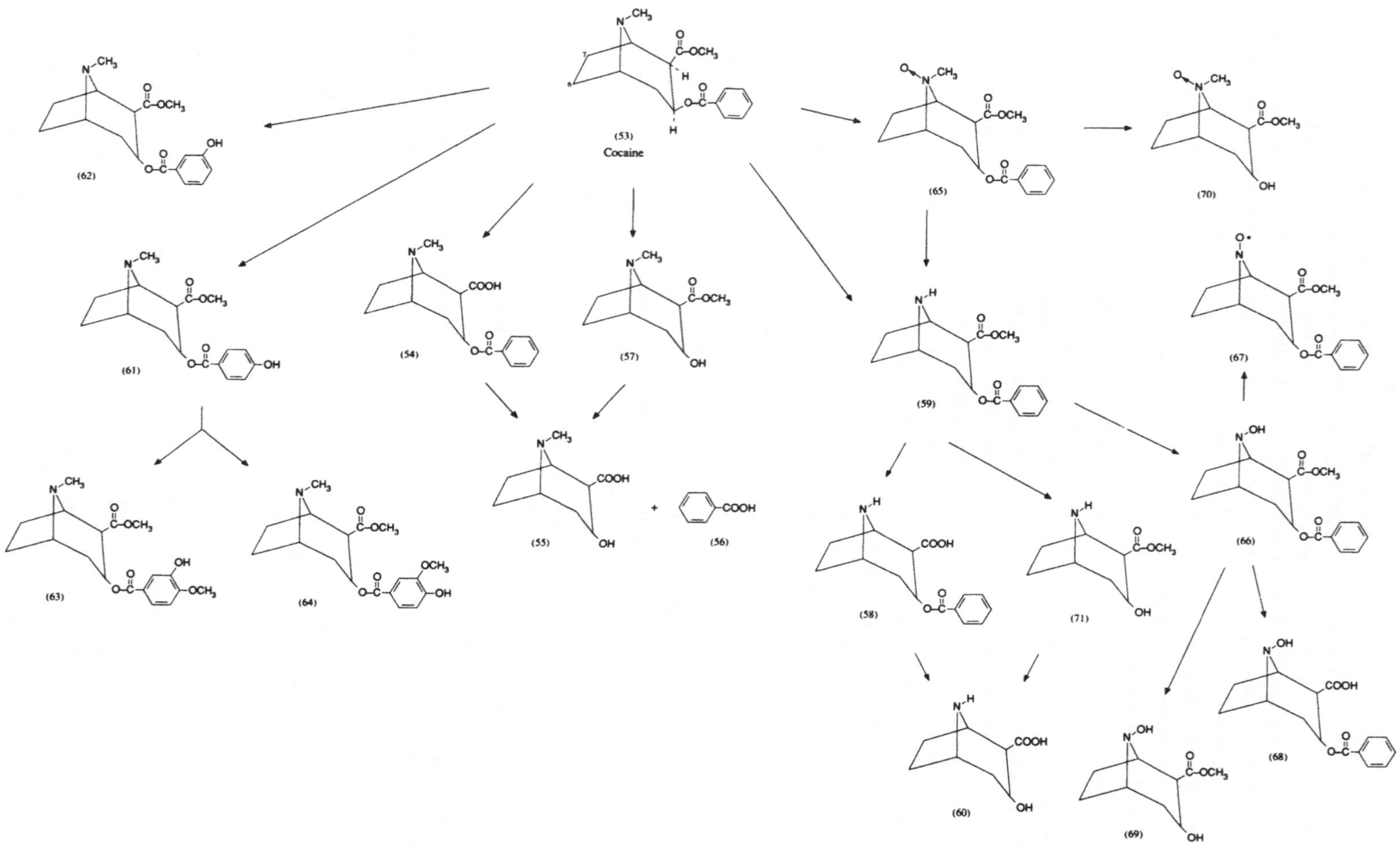

FIGURE 10. Major metabolic pathways of caffeine. See text for discussion of additional minor metabolites.

The urinary excretion values listed in Table 1 indicate that, although the patterns of metabolism show qualitative similarity in man and two nonhuman primates (cynomolgus monkey and baboon), large quantitative differences exist. Most notable is the extensive demethylation at the 7-position in the latter species compared with man. Caffeine metabolism in three additional nonhuman primates, the chimpanzee, the rhesus monkey, and the galago (bushbaby), a prosimian, was studied using 8-^{14}C-labeled material.[796,797] Metabolism was extensive in all cases and the excretion of unchanged compound accounted for <1% of the urinary radioactivity. The quantities of the types of metabolites excreted (monomethylxanthines, dimethylxanthines, and dimethyluric acids) differed considerably among the three species. The chimpanzee, which is taxonomically closest to man, showed the most divergent pattern of metabolism.

The data summarized in Table 1 reveal that all of the possible mono- and dimethylxanthines have been detected as caffeine metabolites in several animal species. Also, uric acid (203) is not a metabolite of caffeine. This point was the subject of much disagreement in the earlier literature until clarified by Buchanan et al.[718] The ring-opened metabolite 5-acetylamino-6-formylamino-3-methyluracil (224) is notable because of its lability. It is easily deformylated to 5-acetylamino-6-amino-3-methyluracil.[798] The latter compound was found to be a major urinary metabolite of caffeine in humans[799-801] and it seems evident that this pathway may be of greater quantitative importance in man than indicated by the value of metabolite (224) in Table 1. It should also be recognized that the uracil derivative (225), which is a major caffeine metabolite in several animal species, was earlier designated 1,3,7-trimethyldihydrouric acid.

The formation of N-demethylated metabolites of caffeine has also been studied by measuring the amount of labeled CO_2 expired following administration of the alkaloid appropriately labeled with ^{13}C or [14]C. Arnaud et al.[802] gave [1,3,7-*methyl*-^{13}C]caffeine to humans and recovered 21 to 26% of the ^{13}C in the expired CO_2 in 24 h. This result reflects the mean rate of N-demethylation of the methyl groups. A subsequent investigation in rats using material labeled with ^{14}C at each of the methyl groups showed that loss at the 1- or 7-positions was equal (17% of the dose) whereas that at the 3-position was significantly higher (22.5%).[803] Interestingly, induction of the animals with a polycyclic aromatic hydrocarbon but not with phenobarbital resulted in increased amounts of expired [14]CO_2. A similar experiment in mice using [1-*methyl*-^{14}C]caffeine showed that 12% of the dose was lost as [14]CO_2.[804] Arnaud[784] reviewed the subject of the formation of labeled CO_2 following the administration of methyl-labeled caffeine.

Numerous factors have been shown to influence the metabolism of caffeine. As noted above, species differences are important in determining the patterns of metabolism. However, intraspecies differences may also be significant. A study in man showed >10-fold variations in the amounts of several urinary metabolites of caffeine.[805] Also, interethnic (Caucasian and Oriental) differences were found in the metabolic pathways dependent on the cytochrome P-450 enzymes but not on xanthine oxidase activity. Also, both an interethnic variability and a bimodal distribution in the acetylation reaction leading to 5-acetylamino-6-formylamino-3-methyluracil (224) has been demonstrated.[806] This subject was discussed by Kalow.[807]

Several investigations have shown that age, at least in young animals, greatly influences the metabolism of caffeine. In human infants, the plasma clearance rate reaches adult levels at an age of 3 to 4.5 months.[808] This development is reflected in the overall pattern of metabolism observed in infants which reaches the adult stage by 7 to 9 months.[809] Unchanged caffeine is initially the major urinary metabolite in infants. Carrier et al.[810] found large differences in infants in the maturation of the individual metabolic pathways (N-demethylation, ring oxidation, and acetylation). The urinary profiles of caffeine metabolites did not vary substantially in young and elderly males, however greater amounts of 1-methyluric acid, 7-methyluric acid, and 1,7-dimethyluric acid were excreted in the latter group.[811] Limited capacity to metabolize caffeine was also reported in young beagle dogs.[812] The changes which occur during development are due primarily to an increase in the 7-demethylase activity.[813] These age-related changes were also demonstrated *in vitro* using rat liver slices.[814]

TABLE 1
Urinary Metabolites of Caffeine[a]

Metabolite	Structure (see Figure 10)	Species	% of dose	Ref.
1,3,7-Trimethylxanthine (Caffeine)	(209)	Mouse	5	785,792,851,852
		Rat	2—9	781,794,851,853,854
		Guinea pig	15	855
		Rabbit	~1	851,856
		Dog	3—7	812,820,857,858
		Horse	1—3	757,791,857
		Monkey[b]	2—5	851,859
		Baboon	7	860
		Man	~1	722,767,790,800,811,851,861-863
1,3-Dimethylxanthine (Theophylline)	(204)	Mouse	5	851,852
		Rat	1—5	794,851,854
		Guinea pig	7	855
		Rabbit	1	856
		Dog	4—7	812,858
		Horse	~6	757
		Monkey	21—28	851,859,864
		Baboon	25	860
		Man	1—2	767,800,805,811,851,863
1,7-Dimethylxanthine (Paraxanthine)	(210)	Mouse	10	785,792,851,852
		Rat	9	794,851,854
		Guinea pig	14	855
		Rabbit	14	851,856
		Dog	1—5	812,858
		Monkey	0—3	851,859
		Baboon	2	860
		Man	3—7	722,767,800,805,811,851,861-863
3,7-Dimethylxanthine (Theobromine)	(215)	Mouse	+[c]	792,852
		Rat	5—7	794,851,854
		Guinea pig	12	855
		Rabbit	4—5	851,856
		Dog	2—5	820,858,865
		Horse	~6	757
		Monkey	1—3	851,859
		Man	1—3	767,800,805,811,851,861-863
1-Methylxanthine	(208)	Mouse	4	851
		Rat	3—4	794,851
		Guinea pig	2	855
		Rabbit	12	851,856
		Dog	1	820
		Baboon	2.5	860
		Man	8—20	722,767,800,805,811,851,861-863
3-Methylxanthine	(207)	Mouse	2	785,792,851,852
		Rat	0.5—1.5	794,851
		Guinea pig	3	855
		Rabbit	2	856
		Dog	21	820
		Monkey	2—8	851,859
		Man	1—4	767,800,805,811,851,861-863
7-Methylxanthine	(211)	Mouse	4	785,792,851,852
		Rat	tr[d]	794
		Rabbit	9	856
		Man	2—9	722,767,800,805,811,851,861-863
1,3,7-Trimethyluric acid	(223)	Mouse	15	851,852
		Rat	4—6	794,851
		Rabbit	1—2	851,856

TABLE 1 (continued)
Urinary Metabolites of Caffeine[a]

Metabolite	Structure (see Figure 10)	Species	% of dose	Ref.
		Dog	2.5	812,820
		Monkey	1	851
		Baboon	2	860
		Man	1—3	767,800,805,811,851,863
1,3-Dimethyluric acid	(206)	Mouse	15	785,851
		Rat	1—2	794,851
		Rabbit	1.5	856
		Dog	13,28	812,820
		Monkey	9—44	851,859,864
		Baboon	10.5	860
		Man	1—14	722,767,800,805,811,851,861,863
1,7-Dimethyluric acid	(213)	Mouse	6	851
		Rat	2	794,851
		Rabbit	2—3	851,856
		Dog	tr—2	812,820
		Baboon	1	860
		Man	5—9	767,800,805,811,851,861,863
3,7-Dimethyluric acid	(217)	Rat	tr	794
		Rabbit	6.5	851
		Man	0.8	767,851
1-Methyluric acid	(205)	Mouse	10	785,851
		Rat	4	851
		Rabbit	10	856
		Dog	8—11	812,820
		Monkey	2	851
		Baboon	13	860
		Man	12—44	722,767,800,805,811,851,861,863
3-Methyluric acid	(216)	Mouse	8	851
		Rat	tr	854
		Rabbit	~2	856
		Man	0.1	767,851
7-Methyluric acid	(212)	Mouse	2	851
		Dog	2	820
		Man	0—7	811,863
6-Amino-5-(N-formyl-methylamino)-1,3-di-methyluracil	(225)	Rat	~18	851,859
		Guinea pig	8	855
		Rabbit	2	851
		Dog	15—20	812
		Monkey	0.4—0.6	851,859
		Baboon	2.8	860
		Man	1	767,851,859,861
6-Amino-5-(N-formyl-methylamino)-3-methyluracil	(214)	Dog	tr	812
		Man	2.4	851
6-Amino-5-(N-formyl-methylamino)-1-methyluracil	(218)	Rat	3	851
		Man	1.9	851
5-Acetylamino-6-formyl-amino-3-methyluracil	(224)	Man	4	805

[a]   See text for discussion of additional minor metabolites in some species.
[b]   Cynomolgus monkey. See text for data in other nonhuman primates.
[c]   (+) Present.
[d]   (tr) Trace.

The effects of pregnancy, cigarette smoking, or inducing agents on the metabolism of caffeine have also been studied. Scott et al.[815] found that pregnant and nonpregnant women excreted caffeine and its metabolites to an equal extent in the urine, however the proportions of metabolites were different in the two groups. The amounts of 1-methylxanthine and 1-methyluric acid were reduced in the pregnant individuals, whereas those of 3-methylxanthine and 3,7-dimethylxanthine were increased. The urinary metabolites and their quantities in pregnant rats given caffeine were generally similar to the values shown in Table 1, however the excretion of the uracil derivative (225) was lower (3 to 4% of the dose).[816] The plasma clearance of caffeine was lower among women taking oral contraceptives than among males or ovulating females, however the metabolic pathways were not significantly altered.[801] The body clearance of caffeine among smokers is greater than that found in nonsmokers, probably due to induction of the cytochrome P-450 monooxygenase system.[817] Inducing agents have been shown to influence the clearance and metabolism of caffeine in both rats[818,819] and dogs.[820]

In comparison with the extensive literature on the *in vivo* metabolism of caffeine, relatively few studies have been carried out using *in vitro* systems. Several early reports on the *N*-demethylation of caffeine by liver microsomes reported low activity by mouse liver microsomes[759] and moderate rates of activity by microsomes from rats[131] or rabbits.[758] Burg and Stein[785] reported that mouse liver microsomes failed to demethylate caffeine and confirmed an earlier report[821] that caffeine was not oxidized to its uric acid derivative by bovine xanthine oxidase. Nonetheless, subsequent studies have clearly indicated that the primary metabolism of caffeine (demethylation and oxidation at C-8) is catalyzed by the microsomal mixed-function oxidases. This was demonstrated with hepatic microsomes from mice,[822] rats,[823,824] and man.[825,826] These studies demonstrated that an isoenzyme of cytochrome P-450 which is inducible by polycyclic aromatic hydrocarbons plays a key role in this process. Caffeine metabolism by cultured hepatocytes from rats and humans showed the same pathways as those found *in vivo*.[765] The *in vitro* metabolism of caffeine to the uracil derivative (6-amino-5-(*N*-formylmethylamino)-1,3-dimethyluracil (225) has been demonstrated both in mouse liver microsomes[822] and in the isolated, perfused rat liver.[827] The former study showed that the yield of the uracil derivative was increased at the expense of the other main metabolite, 1,3,7-trimethyluric acid, when the cytosol fraction or glutathione or other thiols were added to the incubation mixture. Kalow and Campbell[828] recently reviewed the microsomal metabolism of caffeine.

OTHER ALKALOIDS

Divicine and **isouramil** are closely related pyrimidine derivatives which are present in broad beans as their respective β-glucosides, **vicine** and **convicine**. The reactions of these unstable aglycons and the resultant effects on red blood cells was studied by Chevion et al.[829] Studies on the nature of the intermediate formed from divicine indicated that a free radical is involved.[830]

Betanin (226), the glucoside of betanidin, is the red pigment of beetroot. Watson[831] reported that excretion of the pigment in the urine was not observed in the majority of normal individuals, even after the consumption of large amounts of beetroot. However, the urinary excretion of betanin was very rapid in normal individuals after its injection (5 to 40 mg, i.v.). The maximum rate was reached within minutes and diminishing excretion was seen for 3 to 10 h. Recovery ranged from 45 to 75% but it is not known if the material unaccounted for was due to metabolites. Rats given the pigment (5 to 10 mg/kg, i.v.) also rapidly excreted it unchanged in the urine. Similar results were reported by Krantz et al.[832] who administered betanin (~2.5 mg) to rats. Nearly complete recovery in the urine was found following i.v. dosage whereas only 3% of the dose was recovered in the urine and 3% in the feces in 24 h after oral dosage. Liver perfusion studies indicated that little excretion of betanin occurred in the bile of rats. Incubation experiments using gastrointestinal tract tissues or contents suggested that betanin was metabolized by the tissues. However, the incubations using intestinal contents were not carried out under conditions favorable to the anaerobic bacteria which dominate the normal intestinal flora. Accordingly, these results do not provide unequivocal evidence about the possibility of betanin metabolism in the gastrointestinal tract.

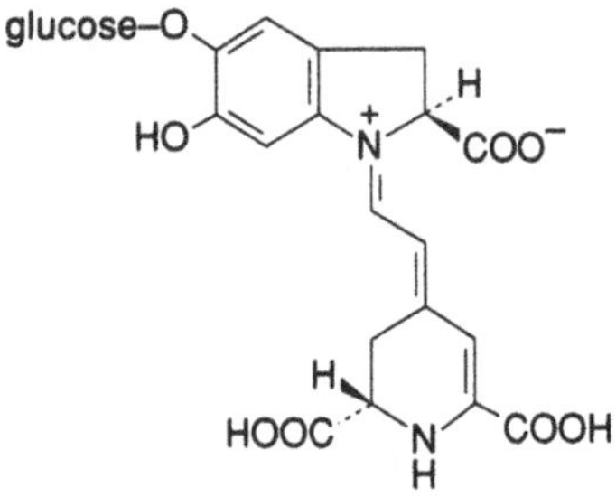

(226)

Betanin

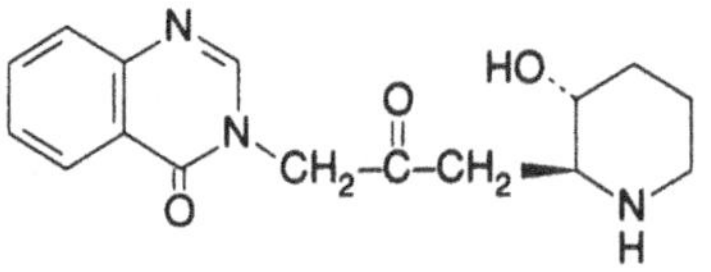

(227)

Febrifugine

The disposition of **febrifugine** (β-dichroine) (227) was studied in rats.[833] Following oral dosage (10 mg/kg), 58 and 30% of the alkaloid remained in the gastrointestinal tract after 1 or 4 h, respectively. Intravenous dosage resulted in only moderate excretion of febrifugine in the urine (16% of the dose in 24 h). No unchanged compound was excreted in the bile and little was detected in the feces.

The metabolism of **acronycine** (acronine) (228a) labeled with ^{14}C in the *N*-methyl group was investigated in rats and humans by Sullivan et al.[834] They found that radioactive metabolites were excreted mainly in the bile in rats. Five hydroxylated biliary metabolites were shown to be compounds (228b), (228c), (228d), (228e), and (228f). All of these metabolites except (228e) were excreted in the urine of patients given acronine. Liu and Ji[835] administered acronine (200 mg) intragastrically to rats and found that 59 and 5% of the dose were excreted in the urine and feces, respectively, in 72 h.

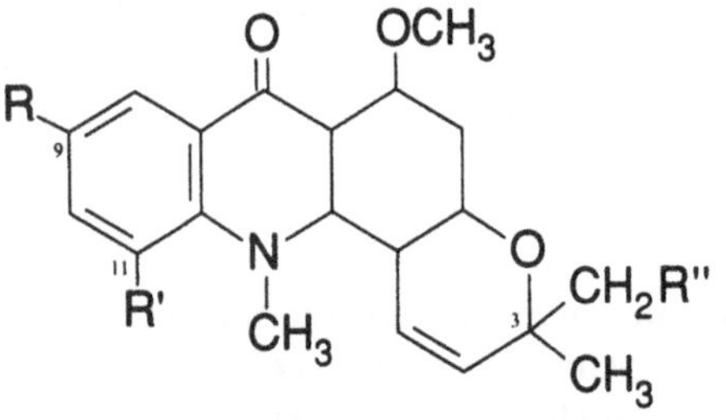

(228)

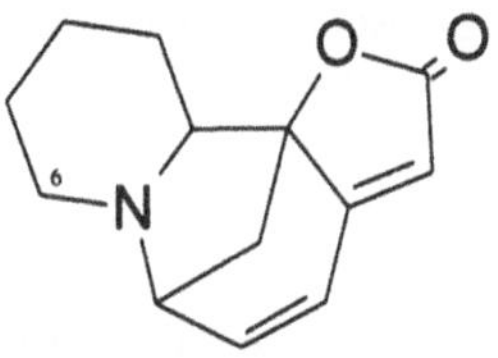

(229)

Securinine

(a) Acronycine, R = R' = R" = H

(b) R = OH , R' = R" = H

(c) R' = OH , R = R" = H

(d) R" = OH , R = R' = H

(e) R = R' = OH , R" = H

(f) R' = R" = OH , R = H

The biological disposition of **securinine** (229) in several animal species was studied by Yao and Sung.[836] Disappearance of the alkaloid from the tissues was rapid following oral or parenteral administration. Very little unchanged compound was excreted in the urine and none was detected in the feces. Securinine was found to be rapidly metabolized by rat liver slices or homogenates and also by the erythrocytes in this and several other species. The metabolic reaction involved was shown not to be simple cleavage of the lactone ring. Paper chromatography of the urine from rats treated with securinine indicated the presence of four metabolites in addition to some unchanged compound. Two of these metabolites were identified as the 6-hydroxy and 6-keto derivatives of securinine. These two metabolites were also detected in the urine of monkeys, however both were absent from human urine.

Plowman et al.[837] studied the disposition in mice and rats of the sulfoacetate salt of the quaternary nitrogen alkaloid **coralyne** (230). Following administration (i.p. or i.v.) of the ^{14}C-labeled compound, the major route of excretion of radioactivity was the bile. Rats excreted 18 to 42% of the ^{14}C in the bile and only 1 to 2% in the urine in 24 h. In noncannulated animals the cumulative excretion of urinary radioactivity after 96 h was 4 to 7% (rats) and 9 to 12% (mice), whereas the corresponding values for fecal radioactivity were ~40 to 50 and 70%, respectively. Unchanged coralyne was identified as the major component excreted in the bile, however a minor metabolite less polar than coralyne was also detected.

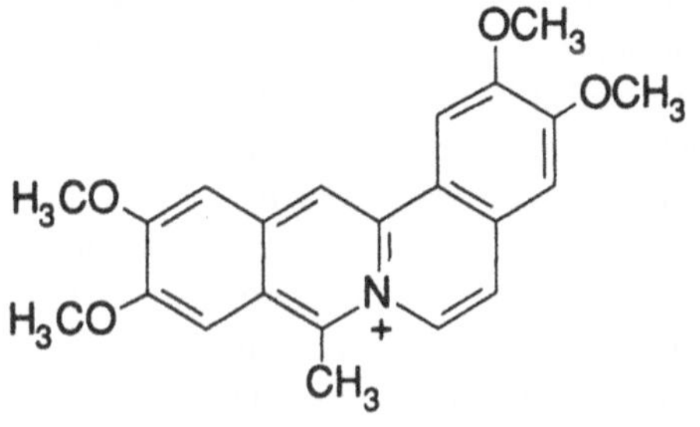

(230)

Coralyne

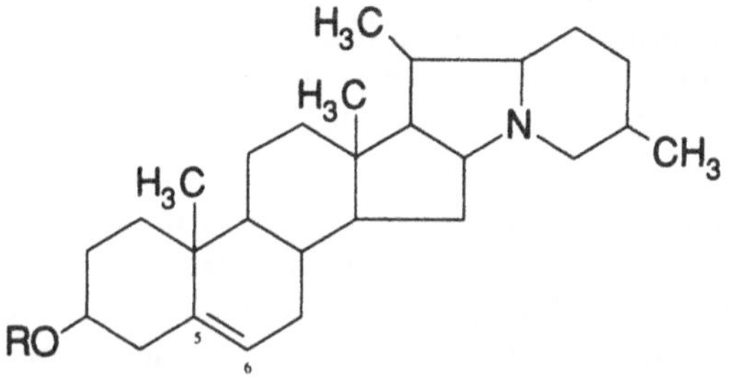

(231)

(a) Solanidine, R = H

(b) Solanine, R = –galactose
$\begin{cases} \text{glucose} \\ \text{rhamnose} \end{cases}$

(c) α-Chaconine, R = –glucose
$\begin{cases} \text{rhamnose} \\ \text{rhamnose} \end{cases}$

Metabolic data are available on two closely related glycoalkaloids, α-solanine (231b) and α-chaconine (231c). These two compounds consist of a common aglycon (solanidine) (231a) and differ only in one of the three sugar moieties. Also, information is available on the metabolism of solasodine (232), a steroid alkaloid aglycon which is structurally similar to solanidine.

Nishie et al.[838] administered ^{3}H-labeled α-**solanine** orally and by i.p. injection to rats. After oral dosage (5 mg/kg), nearly 80% of the radioactivity was excreted in the urine and feces within 24 h and, of this, over 90% was lost by the latter route. About 65% of the fecal radioactivity was shown to be the aglycon solanidine. This metabolite appeared to be poorly absorbed from the intestine and less than 1% of the dose was excreted in this form in the urine. The urinary radioactivity (~10% of the dose) consisted mainly of two basic metabolites with polarity characteristics intermediate between those of solanine and solanidine. These were believed to be intermediate products of hydrolysis containing one or two hexose units. After i.p. injection (5 to 15 mg/kg), approximately equal amounts (15 to 20%) of the radioactivity were found in the urine and feces in 24 h. Slightly less than half of the fecal radioactivity was due to solanidine (231a) and in this case, the two metabolites with polarity intermediate between unchanged compound and aglycon were also present. The urine from these animals also contained solanidine and small amounts of the two tentatively identified basic metabolites in addition to some unchanged compound. Other experiments[839] in which [³H]solanidine was given to human subjects by i.v. injection showed low rates of excretion of radioactivity in the urine and feces. Values of only 1 to 4% (urine) and 1 to 3% (feces) in 24 h were recorded. The additional daily loss of radioactivity during the following week averaged only ~2% of the dose. These results indicate that continued ingestion of solanidine may promote its concentration and retention in the body.

Metabolic experiments with tritiated α-**chaconine** (231c) have been carried out in rats[840] and golden hamsters.[841] Although the doses used (5 to 10 mg/kg) and routes of administration (oral or injection) were similar in the two studies, the results obtained showed major differences. The most striking of these was the finding that 80% of the radioactivity following oral dosage was lost in the feces of rats in 48 h whereas the value in hamsters was only ~0.2%. On the other hand, fairly similar values for the radioactivity ultimately excreted in the urine were noted. These were ~10% in rats and ~20% in hamsters, although the value in rats reached a plateau in only 12 to 24 h compared with 7 d in hamsters. When the α-chaconine was given by i.p. injection radioactivity was excreted primarily in the urine of rats (24 h), however high levels in the intestinal tract indicate that biliary excretion of material occurred. The major urinary and fecal metabolite of α-chaconine in rats following both oral and i.p. administration (5 mg/kg) was the aglycon solanidine (231a). Some unchanged compound appeared to be excreted in the feces following oral dosage of the glycoalkaloid. Two minor metabolites, representing 1 to 5% of the administered radioactivity, were detected in the urine and feces. These were less polar than α-chaconine but more polar than its aglycon.

The intestinal metabolism of the glycoalkaloids α-solanine and α-chaconine was studied using bovine rumen microorganisms.[842] Hydrolysis of the glycosides to the aglycon solanidine occurred rapidly in the incubation mixtures. An additional product, which appeared subsequent to the formation of solanidine, was identified as the 5,6-dehydro derivative which arose via reduction of the double bond at the 5,6-position. More than 80% of the glycoalkaloid mixture was converted to these two compounds, a finding which rules out appreciable conversion to other metabolites.

The metabolism of **solasodine** (232) was studied in man and the golden hamster.[843] The results showed that ³H-labeled compound was rapidly taken up and retained by extravascular sites. Only 7% of the radioactivity was excreted in the urine of humans given the compound i.v. and this value was reached in ~24 h. Excretion in the feces accounted for ~20% of the dose in 4 to 5 d. The general picture of urinary excretion in hamsters was similar, however the amount of radioactivity excreted (~26%) and the time required for this (~200 h) were greater. Although some of the urinary material was due to unchanged compound, most consisted of glucuronide conjugates and other compounds of unknown structure. The radioactivity present in the livers of treated hamsters was shown to contain unchanged solasodine and about an equal amount of a related compound which was tentatively identified as the 5,6-dihydro derivative, soladulcidine. It would be of interest to determine if the prolonged retention of these alkaloidal steroid aglycons in the tissues may be due to their conversion to more lipophilic compounds, e.g., metabolites formed from the esterification of the 3β-hydroxy group with endogenous acids.

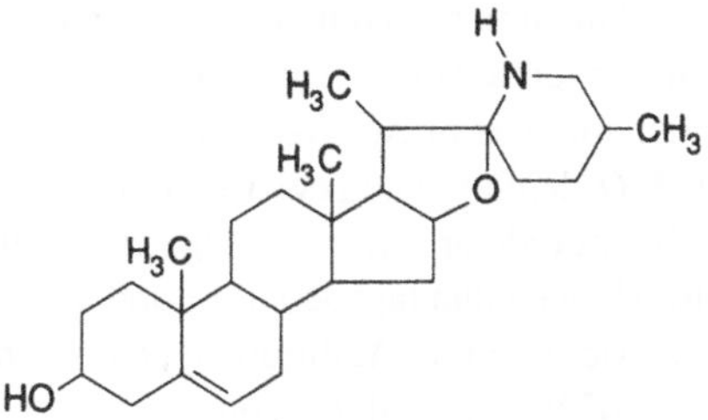

(232)

Solasodine

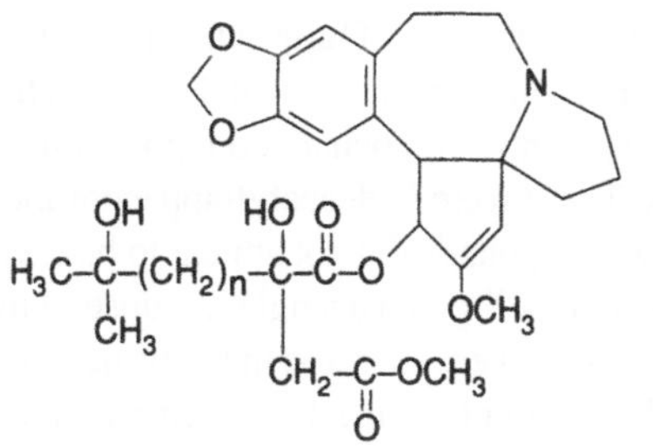

(233)

(a) Harringtonine, $n = 2$

(b) Homoharringtonine, $n = 3$

The biological disposition of **harringtonine** (233a) and **homoharringtonine** (233b) has been studied in several animal species. Following the i.v. injection of [³H]harringtonine in rats, ~30% of the radioactivity was recovered in the urine and 17% in the feces in 24 h.[844] Nearly 15% of the dose was excreted as unchanged compound during this period. Appreciable biliary excretion of radioactivity, mainly as unchanged harringtonine, was observed. Similar experiments with homoharringtonine gave results which showed comparable values for the excretion of radioactivity.[845] About 42 and 6% were recovered in the urine and feces, respectively, in 24 h and unchanged compound accounted for ~16% of the dose. Again, biliary excretion of radioactivity was extensive (~58% in 48 h), however only about a third of this material consisted of unchanged homoharringtonine. Additional investigations on the disposition of [³H]homoharringtonine in dogs[846] and humans[847] have been carried out. Both studies showed that unchanged alkaloid was not excreted extensively in the urine. The values (72 h) were ~18% in dogs and 11% in humans. A greater part of the total urinary radioactivity (40% in dogs and 28% in humans in 72 h) was due to metabolites. The nature of these was not determined, however one major and two minor compounds were detected. A major metabolite was found in human plasma which showed concentration peaks at 2 to 6 and 24 to 48 h. These results may indicate an enterohepatic circulation of the material. Experiments in dogs indicated that ~14% of the radioactivity was excreted in the bile in 5 h. These results pointed to a rapid and extensive metabolism of homoharringtonine in dogs and man. Studies with rat hepatic microsomal enzymes indicated that these may play a major role in the *in vivo* metabolism of homoharringtonine.[848]

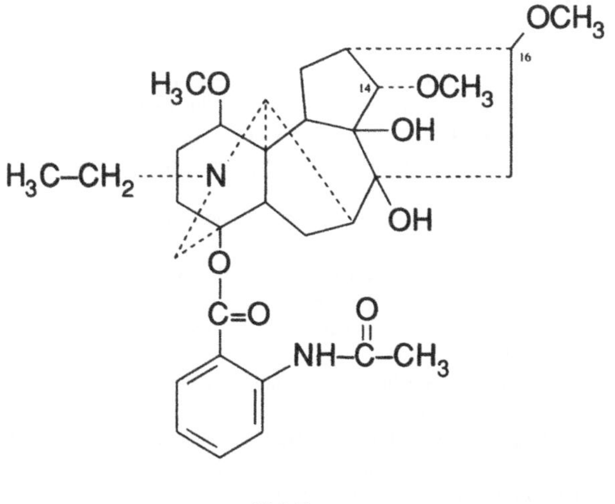

(234)

Lappaconitine

The metabolism of **lappaconitine** (234) was studied in humans given the compound by i.m. injection (three daily doses of 4 mg).[849] The compounds excreted in the urine during this period were shown to be unchanged lappaconitine and small amounts of two *N*-deacetylated derivatives. These were *N*-deacetyllappaconitine and the 16-*O*-demethylated derivative of the latter. These compounds did not appear to be excreted as glucuronide or sulfate conjugates. A similar study in rats[850] given a single i.v. dose (4 mg/kg) also showed that lappaconitine and these two metabolites were excreted in the urine, however in a different ratio. Additionally, rats excreted smaller amounts of the 14-*O*-demethylated derivative of *N*-deacetyllappaconitine.

REFERENCES

1. **Larson, P. S., Haag, H. B., and Silvette, K.,** *Tobacco: Experimental and Clinical Studies*, Williams and Wilkins, Baltimore, MD, 1961.
2. **Larson, P. S. and Silvette, H.,** *Tobacco: Experimental and Clinical Studies. Supplement I*, Williams and Wilkins, Baltimore, MD, 1968.
3. **Larson, P. S. and Silvette, H.,** *Tobacco: Experimental and Clinical Studies. Supplement II*, Williams and Wilkins, Baltimore, MD, 1971.
4. **Larson, P. S. and Silvette, H.,** *Tobacco: Experimental and Clinical Studies. Supplement III*, Williams and Wilkins, Baltimore, MD, 1975.
5. **McKennis, H.,** Disposition and fate of nicotine in animals, in *Tobacco Alkaloids and Related Compounds*, von Euler, U. S., Ed., Pergamon Press, Oxford, 1965, 53.
6. **Gorrod, J. W. and Jenner, P.,** The metabolism of tobacco alkaloids, in *Essays in Toxicology*, Vol. 6, Hayes, W. J., Ed., Academic Press, New York, 1975, 35.
7. **Schievelbein, H.,** Nicotine, resorption and fate, *Pharmacol. Ther.*, 18, 233, 1982.
8. **Nakayama, H.,** Nicotine metabolism in mammals, *Drug Metab. Drug Interact.*, 6, 95, 1988.
9. **Scheline, R. R.,** *Mammalian Metabolism of Plant Xenobiotics*, Academic Press, London, 1978, 372.
10. **Larson, P. S. and Haag, H. B.,** Studies on the fate of nicotine in the body. II. On the fate of nicotine in the dog, *J. Pharmacol. Exp. Ther.*, 76, 240, 1942.
11. **Finnegan, J. K., Larson, P. S., and Haag, H. B.,** Studies on the fate of nicotine in the body. V. Observations on relation of nicotine dosage to per cent excreted in urine, rate of excretion and rate of detoxication, *J. Pharmacol. Exp. Ther.*, 91, 357, 1947.
12. **Haag, H. B. and Larson, P. S.,** Studies on the fate of nicotine in the body. I. The effect of pH on the urinary excretion of nicotine by tobacco smokers, *J. Pharmacol. Exp. Ther.*, 76, 235, 1942.
13. **Beckett, A. H., Gorrod, J. W., and Jenner, P.,** The analysis of nicotine-1'-*N*-oxide in urine, in the presence of nicotine and cotinine, and its application to the study of *in vivo* nicotine metabolism in man, *J. Pharm. Pharmacol.*, 23, 55, 1971.
14. **Matsukura, S., Sakamoto, N., Takahashi, K., Matsuyama, H., and Muranaka, H.,** Effect of pH and urine flow on urinary nictotine excretion after smoking cigarettes, *Clin. Pharmacol. Ther.*, 25, 549, 1979.
15. **Kyerematen, G. A., Taylor, L. H., deBethizy, J. D., and Vesell, E. S.,** Pharmacokinetics of nicotine and 12 metabolites in the rat. Application of a new radiometric high performance liquid chromatography assay, *Drug Metab. Dispos.*, 16, 125, 1988.
16. **Benowitz, N. L. and Jacob, P.,** Metabolism, pharmacokinetics, and pharmacodynamics of nicotine in man, *Adv. Behav. Biol.*, 31, 357, 1987.
17. **Matsukura, S., Sakamoto, N., Seino, Y., Tamada, T., Matsuyama, H., and Muranaka, H.,** Cotinine excretion and daily cigarette smoking in habituated smokers, *Clin. Pharmacol. Ther.*, 25, 555, 1979.
18. **Nwosu, C. G. and Crooks, P. A.,** Species variation and stereoselectivity in the metabolism of nicotine enantiomers, *Xenobiotica*, 18, 1361, 1988.
19. **Hucker, H. B., Gillette, J. R., and Brodie, B. B.,** Cotinine: an oxidaion product of nicotine formed by rabbit liver, *Nature (London)*, 183, 47, 1959.
20. **Bowman, E. R., Turnbull, L. B., and McKennis, H.,** Metabolism of nicotine in the human and excretion of pyridine compounds by smokers, *J. Pharmacol. Exp. Ther.*, 127, 92, 1959.
21. **Cundy, K. C. and Crooks, P. A.,** Biotransformation of primary nicotine metabolites. II. Metabolism of [^{3}H]-S-(–)-cotinine in the guinea pig: determination of *in vivo* urinary metabolites by high-performance liquid-radiochromatography, *Xenobiotica*, 17, 785, 1987.

22. **Neurath, G. B., Dünger, M., Orth, D., and Pein, F. G.,** Trans-3′-hydroxycotinine as a main metabolite of urine of smokers, *Int. Arch. Occup. Environ. Health*, 59, 199, 1987.
23. **O'Doherty, S., Revans, A., Smith, C. L., McBride, M., and Cooke, M.,** Determination of *cis*- and *trans*-3-hydroxycotinine by high performance liquid chromatography, *J. High Resolut. Chromatogr. Chromatogr. Commun.*, 11, 723, 1988.
24. **Jacob, P., Benowitz, N. L., and Shulgin, A. T.,** Recent studies of nicotine metabolism in humans, *Pharmacol. Biochem. Behav.*, 30, 249, 1988.
25. **Neurath, G. B. and Pein, F. G.,** Gas chromatographic determination of *trans*-3′-hydroxycotinine, a major metabolite of nicotine in smokers, *J. Chromatogr.*, 415, 400, 1987.
26. **Dagne, E. and Castagnoli, N.,** Structure of hydroxycotinine, a nicotine metabolite, *J. Med. Chem.*, 15, 356, 1972.
27. **Stålhandske, T.,** The metabolism of nicotine and cotinine by a mouse liver preparation, *Acta Physiol. Scand.*, 78, 236, 1970.
28. **McKennis, H., Turnbull, L. B., Schwartz, S. L., Takami, E., and Bowman, E. R.,** Demethylation in the metabolism of (–)-nicotine, *J. Biol. Chem.*, 237, 541, 1962.
29. **Morselli, P. L., Ong, N. H., Bowman, E. R., and McKennis, H.,** Metabolism of (±)-cotinine-2-^{14}C in the rat, *J. Med. Chem.*, 10, 1033, 1967.
30. **Nguyen, T.-L., Dagne, E., Gruenke, L., Bhargava, H., and Castagnoli, N.,** The tautomeric structures of 5-hydroxycotinine, a secondary mammalian metabolite of nicotine, *J. Org. Chem.*, 46, 758, 1981.
31. **Hucker, H. B., Gillette, J. R., and Brodie, B. B.,** Enzymatic pathway for the formation of cotinine, a major metabolite of nicotine in rabbit liver, *J. Pharmacol. Exp. Ther.*, 129, 94, 1960.
32. **Murphy, P. J.,** Enzymatic oxidation of nicotine to nicotine $\Delta^{1'(5')}$ iminium ion. A newly discovered intermediate in the metabolism of nicotine, *J. Biol. Chem.*, 248, 2796, 1973.
33. **Brandänge, S. and Lindblom, L.,** Synthesis, structure and stability of nicotine $\Delta^{1'(5')}$ iminium ion, an intermediary metabolite of nicotine, *Acta Chem. Scand.*, B33, 187, 1979.
34. **Brandänge, S. and Lindblom, L.,** The enzyme "aldehyde oxidase" is an iminium oxidase. Reaction with nicotine $\Delta^{1'(5')}$ iminium ion, *Biochem. Biophys. Res. Commun.*, 91, 991, 1979.
35. **Gorrod, J. W. and Hibberd, A. R.,** The metabolism of nicotine-$\Delta^{1'(5')}$-iminium ion, *in vivo* and *in vitro*, *Eur. J. Drug Metab. Pharmacokinet.*, 7, 293, 1982.
36. **Hibberd, A. R. and Gorrod, J. W.,** Enzymology of the metabolic pathway from nicotine to cotinine, *in vitro*, *Eur. J. Drug Metab. Pharmacokinet.*, 8, 151, 1983.
37. **Obach, R. S. and Van Vunakis, H.,** Nicotinamide adenine dinucleotide (NAD)-dependent oxidation of nicotine-$\Delta^{1'(5')}$-iminium ion to cotinine by rabbit liver microsomes, *Biochem. Pharmacol.*, 39, R1, 1990.
38. **Peterson, L. A., Trevor, A., and Castagnoli, N.,** Stereochemical studies on the cytochrome P-450 catalyzed oxidation of (*S*)-nicotine to the (*S*)-nicotine $\Delta^{1'(5')}$-iminium species, *J. Med. Chem.*, 30, 249, 1987.
39. **McCoy, G. D., Howard, P. C., and DeMarco, G. J.,** Characterization of hamster liver nicotine metabolism. I. Relative rates of microsomal C and N oxidation, *Biochem. Pharmacol.*, 35, 2767, 1986.
40. **McCoy, G. D., DeMarco, G. J., and Koop, D. R.,** Microsomal nicotine metabolism: a comparison of relative activities of six purified rabbit cytochrome P-450 isozymes, *Biochem. Pharmacol.*, 38, 1185, 1989.
41. **McKennis, H., Bowman, E. R., Yi, J. M., and Sprouse, C. T.,** Participation of pyridino-N-oxides in the metabolism of nicotine in vivo — a preliminary study, in *Biological Oxidation of Nitrogen*, Gorrod, J. W., Ed., Elsevier/North-Holland Biomedical Press, Amsterdam, 1978, 163.
42. **Hansson, E., Hoffman, P. C., and Schmitterlöw, C. G.,** Metabolism of nicotine in mouse tissue slices, *Acta Physiol. Scand.*, 61, 380, 1964.
43. **Hansson, E. and Schmitterlöw, C. G.,** Physiological disposition and fate of C^{14}-labelled nicotine in mice and rats, *J. Pharmacol. Exp. Ther.*, 137, 91, 1962.
44. **Cundy, K. C. and Crooks, P. A.,** High-performance liquid chromatographic method for the determination of N-methylated metabolites of nicotine, *J. Chromatogr.*, 306, 291, 1984.
45. **Nwosu, C. G., Godin, C. S., Houdi, A. A., Damani, L. A., and Crooks, P. A.,** Enantioselective metabolism during continuous administration of *S*-(–)- and *R*-(+)-nicotine isomers to guinea-pigs, *J. Pharm. Pharmacol.*, 40, 862, 1988.
46. **Nguyen, T.-L., Gruenke, L. D., and Castagnoli, N.,** Metabolic N-demethylation of nicotine. Trapping of a reactive iminium species with cyanide ion, *J. Med. Chem.*, 19, 1168, 1976.
47. **Nguyen, T.-L., Gruenke, L. D., and Castagnoli, N.,** Metabolic oxidation of nicotine to chemically reactive intermediates, *J. Med. Chem.*, 22, 259, 1979.
48. **Thompson, J. A., Norris, K. J., and Petersen, D. R.,** Isolation and analysis of N-oxide metabolites of tertiary amines: quantiation of nicotine-1′-N-oxide formation in mice, *J. Chromatogr.*, 341, 349, 1985.
49. **Booth, J. and Boyland, E.,** The metabolism of nicotine into two optically-active stereoisomers of nicotine-1′-oxide by animal tissues *in vitro* and by cigarette smokers, *Biochem. Pharmacol.*, 19, 733, 1970.
50. **Booth, J. and Boyland, E.,** Enzymic oxidation of (–)-nicotine by guinea-pig tissues *in vitro*, *Biochem. Pharmacol.*, 20, 407, 1971.

51. **Jenner, P. and Gorrod, J. W.,** Comparative in vitro hepatic metabolism of some tertiary *N*-methyl tobacco alkaloids in various species, *Res. Commun. Chem. Pathol. Pharmacol.*, 6, 829, 1973.

52. **Jenner, P., Gorrod, J. W., and Beckett, A. H.,** Species variation in the metabolism of R-(+)- and S-(–)-nicotine by α-*C*- and *N*-oxidation *in vitro*, *Xenobiotica*, 3, 573, 1973.

53. **Beckett, A. H., Jenner, P., and Gorrod, J. W.,** Characterization of the diastereoisomers of nicotine-1'-*N*-oxide, a metabolite of nicotine, and other possible oxidation products by nuclear magnetic resonance spectroscopy, *Xenobiotica*, 3, 557, 1973.

54. **Testa, B., Jenner, P., Beckett, A. H., and Gorrod, J. W.,** A reappraisal of the stereoselective metabolism of nicotine to nicotine-1'-*N*-oxide, *Xenobiotica*, 6, 553, 1976.

55. **Beckett, A. H., Gorrod, J. W., and Jenner, P.,** Absorption of (–)-nicotine-1'-*N*-oxide in man and its reduction in the gastrointestinal tract, *J. Pharm. Pharmacol.*, 22, 722, 1970.

56. **Jenner, P., Gorrod, J. W., and Beckett, A. W.,** The absorption of nicotine-1'-*N*-oxide and its reduction in the gastro-intestinal tract in man, *Xenobiotica*, 3, 341, 1973.

57. **Dajani, R. M., Gorrod, J. W., and Beckett, A. H.,** Reduction *in vivo* of (–)-nicotine-1'-*N*-oxide by germ-free and conventional rats, *Biochem. Pharmacol.*, 24, 648, 1975.

58. **Dajani, R. M., Gorrod, J. W., and Beckett, A. H.,** Hepatic and extrahepatic reduction of nicotine 1'-*N*-oxide in rats, *Biochem. J.*, 130, 88, 1972.

59. **Dajani, R. N., Gorrod, J. W., and Beckett, A. H.,** *In vitro* hepatic and extra-hepatic reduction of (–)-nicotine-1-*N*-oxide in rats, *Biochem. Pharmacol.*, 24, 109, 1975.

60. **Shulgin, A. T., Jacob, P., Benowitz, N. L., and Lau, D.,** Identification and quantitative analysis of cotinine-*N*-oxide in human urine, *J. Chromatogr.*, 423, 365, 1987.

61. **Yi, J. M., Sprouse, C. T., Bowman, E. R., and McKennis, H.,** The interrelationship between the metabolism of (S)-cotinine-*N*-oxide and (S)-cotinine, *Drug Metab. Dispos.*, 5, 355, 1977.

62. **Gorrod, J. W., Jenner, P., Keysell, G., and Beckett, A. H.,** Selective inhibition of alternative oxidative pathways of nicotine metabolism *in vitro*, *Chem. Biol. Interact.*, 3, 269, 1971.

63. **Damani, L. A., Pool, W. F., Crooks, P. A., Kaderlik, R. K., and Ziegler, D. M.,** Stereoselectivity in the *N'*-oxidation of nicotine isomers by flavin-containing monooxygenase, *Mol. Pharmacol.*, 33, 702, 1988.

64. **McKennis, H., Turnbull, L. B., and Bowman, E. R.,** *N*-Methylation of nicotine and cotinine *in vivo*, *J. Biol. Chem.*, 238, 719, 1963.

65. **Cundy, K. C., Godin, C. S., and Crooks, P. A.,** Stereospecific *in vitro* *N*-methylation of nicotine in guinea pig tissues by an *S*-adenosylmethionine-dependent *N*-methyltransferase, *Biochem. Pharmacol.*, 34, 281, 1985.

66. **Cundy, K. C. and Crooks, P. A.,** *In vitro* characteristics of guinea pig lung aromatic azaheterocycle *N*-methyltransferase, *Drug Metab. Dispos.*, 13, 658, 1985.

67. **Cundy, K. C., Crooks, P. A., and Godin, C. S.,** Remarkable substrate-inhibitor properties of nicotine enantiomers towards a guinea pig lung aromatic azaheterocycle *N*-methyltransferase, *Biochem. Biophys. Res. Commun.*, 128, 312, 1985.

68. **Cundy, K. C., Sato, M., and Crooks, P. A.,** Stereospecific *in vivo* *N*-methylation of nicotine in the guinea pig, *Drug Metab. Dispos.*, 13, 175, 1985.

69. **Pool, W. F. and Crooks, P. A.,** Biotransformation of primary nicotine metabolites: metabolism of R-(+)-[³H-*N'*–CH₃; ¹⁴C-*N*-CH₃]*N*-methylnicotinium acetate—the use of double isotope studies to determine the in-vivo stability of the *N*-methyl groups of *N*-methylnicotinium ion, *J. Pharm. Pharmacol.*, 40, 758, 1988.

70. **Sato, M. and Crooks, P. A.,** *N*-Methylnornicotinium ion, a new *in vivo* metabolite of R-(+)-nicotine, *Drug Metab. Dispos.*, 13, 348, 1985.

71. **Pool, W. F. and Crooks, P. A.,** Biotransformation of primary nicotine metabolites. I. *In vivo* metabolism of R-(+)-[¹⁴C-NCH3]*N*-methylnicotinium ion in the guinea pig, *Drug Metab. Dispos.*, 13, 578, 1985.

72. **Pool, W. F., Houdi, A. A., Damani, L. A., Layton, W. J., and Crooks, P. A.,** Isolation and characterization of *N*-methyl-*N'*-oxonicotinium ion, a new urinary metabolite of R-(+)-nicotine in the guinea pig, *Drug Metab. Dispos.*, 14, 574, 1986.

73. **Crooks, P. A. and Godin, C. S.,** *N*-Methylation of nicotine enantiomers by human liver cytosol, *J. Pharm. Pharmacol.*, 40, 153, 1987.

74. **Damani, L. A., Shaker, M. S., Godin, C. S., Crooks, P. A., Ansher, S. S., and Jakoby, W. B.,** The ability of amine *N*-methyltransferases from rabbit liver to *N*-methylate azaheterocycles, *J. Pharm. Pharmacol.*, 38, 547, 1986.

75. **Shen, W.-C. and Van Vunakis, N.,** Nicotine and its metabolites. IV. Formation of the nicotine analogue of DPN by pig brain DPNase, *Res. Commun. Chem. Pathol. Pharmacol.*, 9, 405, 1974.

76. **Shen, W.-C. and Van Vunakis, H.,** The formation and characterization of the nicotine analog of triphosphopyridine nucleotide, *Biochemistry*, 13, 5362, 1974.

77. **Shen, W.-C., Franke, J., and Van Vunakis, N.,** Nicotinamide nucleotide analogues of nicotine and cotinine—enzymic studies, *Biochem. Pharmacol.*, 26, 1835, 1977.

78. **Shen, W.-C., Greene, K. M., and Van Vunakis, N.,** Detection by radioimmunoassay of nicotinamide nucleotide analogues in tissues of rabbits injected with nicotine and cotinine, *Biochem. Pharmacol.*, 26, 1841, 1977.

79. **Shigenaga, M. K., Trevor, A. J., and Castagnoli, N.,** Metabolism-dependent covalent binding of (S)-[5-³H]nicotine to liver and lung microsomal macromolecules, *Drug Metab. Dispos.*, 16, 397, 1988.

80. **Obach, R. S. and Van Vunakis, H.,** Non-metabolic covalent binding of nicotine-Δ¹′⁽⁵′⁾-iminium ion to liver microsomes and sulfhydryl-containing polyamino acids, *Biochem. Pharmacol.*, 37, 4601, 1988.

81. **Kyerematen, G. A., Owens, G. F., Chattopadhyay, B., deBethizy, J. D., and Vesell, E. S.,** Sexual dimorphism of nicotine metabolism and distribution in the rat. Studies *in vivo* and *in vitro*, *Drug Metab. Dispos.*, 16, 823, 1988.

82. **Slanina, P. and Stålhandske, T.,** In vitro metabolism of nicotine in liver of ageing mice, *Arch. Int. Pharmacodyn. Ther.*, 226, 258, 1977.

83. **Hucker, H. B. and Larson, P. S.,** Studies on the metabolism of nornicotine in the dog, *J. Pharmacol. Exp. Ther.*, 123, 259, 1958.

84. **Beckett, A. H., Gorrod, J. W., and Jenner, P.,** A possible relation between pKa, and lipid solubility and the amounts excreted in urine of some tobacco alkaloids given to man, *J. Pharm. Pharmacol.*, 24, 115, 1972.

85. **Wada, E., Bowman, E. R., Turnbull, L. B., and McKennis, H.,** Nornicotine (desmethylcotinine) as a urinary metabolite of nornicotine, *J. Med. Pharm. Chem.*, 4, 21, 1961.

86. **Axelrod, J.,** The enzymatic N-methylation of serotonin and other amines, *J. Pharmacol. Exp. Ther.*, 138, 28, 1962.

87. **Meacham, R. H., Bowman, E. R., and McKennis, H.,** Additional routes in the metabolism of nicotine to 3-pyridylacetate The metabolism of dihydrometanicotine, *J. Biol. Chem.*, 247, 902, 1972.

88. **Meacham, R. H., Sprouse, C. T., Bowman, E. R., and McKennis, H.,** Pyridyl acids from the mammalian metabolism of *trans*-meta-nicotine, *Fed. Proc. Fed. Am. Soc. Exp. Biol.*, 32, 511, 1973.

89. **Chen, C.-H. B., Hecht, S. S., and Hoffmann, D.,** Metabolic α-hydroxylation of the tobacco-specific carcinogen, N′-nitrosonornicotine, *Cancer Res.*, 38, 3639, 1978.

90. **Hecht, S. S., Chen, C.-H. B., and Hoffmann, D.,** Metabolic β-hydroxylation and N-oxidation of N′-nitrosonornicotine, *J. Med. Chem.*, 23, 1175, 1980.

91. **Hecht, S. S., Lin, D., and Chen, C.-H. B.,** Comprehensive analysis of urinary metabolites of N′-nitrosonornicotine, *Carcinogenesis*, 2, 833, 1981.

92. **Domellöf, L., Andersson, M., Tjälve, H., Veals, S., Trushin, N., and Hecht, S. S.,** Distribution and metabolism of N′-nitrosonornicotine in the miniature pig, *Carcinogenesis*, 8, 1741, 1987.

93. **Hecht, S. S., Young, R., and Chen, C. B.,** Metabolism in the F344 rat of 4-(N-methyl-N-nitrosamino)-1-(3-pyridyl)-1-butanone, a tobacco-specific carcinogen, *Cancer Res.*, 40, 4144, 1980.

94. **Hoffmann, D., Castonguay, A., Rivenson, A., and Hecht, S. S.,** Comparative carcinogenicity and metabolism of 4-(methylnitrosamino)-1-(3-pyridyl)-1-butanone and N′-nitrosonornicotine in Syrian golden hamsters, *Cancer Res.*, 41, 2386, 1981.

95. **Castonguay, A., Allaire, L., Charest, M., Rossignol, G., and Boutet, M.,** Metabolism of 4-(methylnitrosoamino)-1-(3-pyridyl)-1-butanone by hamster respiratory tissues cultured with ellagic acid, *Cancer Lett.*, 46, 93, 1989.

96. **Charest, M., Rossignol, G., and Castonguay, A.,** In vitro and in vivo modulation of the bioactivation of 4-(methylnitrosamino)-1-(3-pyridyl)-1-butanone in hamster lung tissues, *Chem. Biol. Interact.*, 71, 265, 1989.

97. **Rossignol, G., Alaoui-Jamali, M. A., Castonguay, A., and Schuller, H. M.,** Metabolism and DNA damage induced by 4-(methylnitrosamino)-1-(3-pyridyl)-1-butanone in fetal tissues of the Syrian golden hamster, *Cancer Res.*, 49, 5671, 1989.

98. **Carmella, S. G. and Hecht, S. S.,** Formation of hemoglobin adducts upon treatment of F344 rats with the tobacco-specific nitrosamines 4-(methylnitroso)-1-(3-pyridyl)-1-butanone and N′-nitrosonornicotine, *Cancer Res.*, 47, 2626, 1987.

99. **Hoffmann, D. and Hecht, S. S.,** Nicotine-derived N-nitrosamines and tobacco-related cancer: current status and future directions, *Cancer Res.*, 45, 935, 1985.

100. **Rothstein, M. and Miller, L. L.,** The conversion of lysine to pipecolic acid in the rat, *J. Biol. Chem.*, 211, 851, 1954.

101. **Rothstein, M. and Greenberg, D. M.,** Metabolism of DL-pipecolic acid-2-C¹⁴, *J. Am. Chem. Soc.*, 81, 4756, 1959.

102. **Boulanger, P. and Osteux, R.,** Stoffwechsel der DL-Pipecolinsäure bei den Vertebraten, *Hoppe-Seyler's Z. Physiol. Chem.*, 321, 79, 1960.

103. **Kasé, Y., Kataoka, M., and Miyata, T.,** *In vitro* production of piperidine from pipecolic acid in the presence of brain tissue, *Life Sci.*, 6, 2427, 1967.

104. **Uehleke, H.,** Lobelin: Aufnahme, Verteilung und Ausscheidung, *Naunyn Schmiedebergs Arch. Exp. Pathol. Pharmakol.*, 246, 34, 1963.

105. **Boyland, E. and Nery, R.,** Mercapturic acid formation during the metabolism of arecoline and arecaidine in the rat, *Biochem. J.*, 113, 123, 1969.

106. **Nery, R.,** The metabolic interconversion of arecoline and arecoline 1-oxide in the rat, *Biochem. J.*, 122, 503, 1971.

107. **Nieschulz, O. and Schmersahl, P.,** Zur Pharmakologie der Wirkstoff des Betels. II. Mitteilung: Ummandlung des Arecolin in Arecaidin, *Arzneim. Forsch.*, 18, 222, 1968.

108. **Shivapurkar, N. M., D'Souza, A. V., and Bhide, S. V.,** Effect of betel-quid chewing on nitrite levels in saliva, *Food Cosmet. Toxicol.*, 18, 277, 1980.

109. **Wenke, G. and Hoffmann, D.,** A study of betel quid carcinogenesis. I. On the *in vitro* N-nitrosation of arecoline, *Carcinogenesis*, 4, 169, 1983.

110. **Nair, J., Ohshima, H., Friesen, M., Croisy, A., Bhide, S. V., and Bartsch, H.,** Tobacco-specific and betel nut-specific N-nitroso compounds: occurrence in saliva and urine of betel quid chewers and formation *in vitro* by nitrosation of betel quid, *Carcinogenesis*, 6, 295, 1985.

111. **Prokopczyk, B., Rivenson, A., Bertinato, P., Brunnemann, K. D., and Hoffmann, D.,** 3-(Methylnitrosamino)propionitrile: occurrence in saliva of betel quid chewers, carcinogenicity, and DNA methylation in F344 rats, *Cancer Res.*, 47, 467, 1987.

112. **Gvishiani, G. S.,** The inactivation of certain pharmacological substances in the liver, *Soobshcheniya Akad. Nauk Gruzinskoi SSR*, 24, 225, 1960 (Chem. Abstr. 55, 1912c, 1961).

113. **Beckett, A. H. and Sheikh, A. H.,** *In vitro* metabolic N-oxidation of the minor tobacco alkaloids, (−)-methylanabasine and (−)-anabasine to yield a hydroxylamine and a nitrone in lung and liver homogenates, *J. Pharm. Pharmacol.*, 25, 171, 1973.

114. **Kohlrausch, A.,** Untersuchungen über das Verhalten von Betain, Trigonellin und Methylpyridylammoniumhydroxyd im tierischen Organismus, *Z. Biol.*, 57, 273, 1912.

115. **Chattopadhyay, D., Ghosh, N. C., Chattopadhyay, H., and Banerjee, S.,** Studies on the metabolism of nicotinic acid in rabbits, *J. Biol. Chem.*, 201, 529, 1953.

116. **Hawksworth, G. and Scheline, R. R.,** Metabolism in the rat of some pyrazine derivatives having flavour importance in foods, *Xenobiotica*, 5, 389, 1975.

117. **Wiechowski, W.,** Ueber das Schiksal des Cocains und Atropins im Thierkörper, *Arch. Exp. Pathol. Pharmak.*, 45, 155, 1901.

118. **Albanus, L., Sundwall, A., Vangbo, B., and Winbladh, B.,** The fate of atropine in the dog, *Acta Pharmacol. Toxicol.*, 26, 571, 1968.

119. **Winbladh, B.,** The fate of atropine in the puppy, *Acta Pharmacol. Toxicol.*, 32, 46, 1973.

120. **Gosselin, R. E., Gabourel, J. D., Kalser, S. C., and Wills, J. H.,** The metabolism of C¹⁴-labeled atropine and tropic acid in mice, *J. Pharmacol. Exp. Ther.*, 115, 217, 1955.

121. **Gabourel, J. D. and Gosselin, R. E.,** The mechanism of atropine detoxication in mice and rats, *Arch. Int. Pharmacodyn. Ther.*, 115, 416, 1958.

122. **Albanus, L., Hammarström, L., Sundwall, A., Ullberg, S., and Vangbo, B.,** Distribution and metabolism of H³-atropine in mice, *Acta Physiol. Scand.*, 73, 447, 1968.

123. **Tønnesen, M.,** The excretion of atropine and allied alkaloids in urine, *Acta Pharmacol. Toxicol.*, 6, 147, 1950.

124. **Kalser, S. C., Wills, J. H., Gabourel, J. D., Gosselin, R. E., and Epes, C. F.,** Further studies on the excretion of atropine-*alpha*-C¹⁴, *J. Pharmacol. Exp. Ther.*, 121, 449, 1957.

125. **Gosselin, R. E., Gabourel, J. D., and Wills, J. H.,** The fate of atropine in man, *Clin. Pharmacol. Ther.*, 1, 597, 1960.

126. **Van der Meer, M. J., Hundt, H. K. L., and Müller, F. O.,** The metabolism of atropine in man, *J. Pharm. Pharmacol.*, 38, 781, 1986.

127. **Eling, T. E.,** Synthesis and metabolism of tropine-labeled atropine, *Diss. Abstr. B*, 29, 1778, 1968.

128. **Werner, G. and Schmidt, H.-L.,** Chemische Analyse des Stoffwechsels von Atropin bei der Maus, *Hoppe-Seyler's Z. Physiol. Chem.*, 349, 677, 1968.

129. **Werner, G.,** Untersuchungen zum Stoffwechsel von Tropan-Alkaloiden bei einigen Säugetieren, *Planta Med.*, 9, 293, 1961.

130. **Truhaut, R. and Yonger, J.,** Catabolism of atropine in rat and guinea pig liver. Identification of metabolites formed by oxidation, *C. R. Hebd. Séanc. Acad. Sci. Paris Ser. D.*, 264, 2526, 1967 (Chem. Abstr. 67, 52399q, 1967).

131. **Franklin, M.,** Studies on the N-demethylation of morphine and other compounds, *Can. J. Biochem.*, 43, 1053, 1965.

132. **Evertsbusch, V. and Geiling, E. M. K.,** Studies with radioactive atropine. I. Distribution and excretion patterns in the mouse, *Arch. Int. Pharmacodyn. Ther.*, 105, 175, 1956.

133. **Kalser, S. C., Kelvington, E. J., Randolph, M. M., and Santomenna, D. M.,** Drug metabolism in hypothermia. I. Biliary excretion of C¹⁴-atropine metabolites in the intact and nephrectomized rat, *J. Pharmacol. Exp. Ther.*, 147, 252, 1965.

134. **Kalser, S. C., Kelvington, E. J., Randolph, M. M., and Santomenna, D. M.,** Drug metabolism in hypothermia. II. C¹⁴-Atropine uptake, metabolism and excretion by the isolated, perfused rat liver, *J. Pharmacol. Exp. Ther.*, 147, 260, 1965.

135. **Bernheim, F. and Bernheim, M. L. C.,** The hydrolysis of homatropine and atropine by various tissues, *J. Pharmacol. Exp. Ther.*, 64, 209, 1938.

136. **Langecker, H. and Lewit, K.,** Über die Entgiftung von Cocain und Percain, sowie von Atropin im Organismus, *Naunyn Schmiedebergs Arch. Exp. Pathol. Pharmakol.*, 190, 492, 1938.

137. **Phillipson, J. D., Handa, S. S., and Gorrod, J. W.,** Metabolic *N*-oxidation of atropine, hyoscine and the corresponding nor-alkaloids by guinea-pig liver microsomal preparations, *J. Pharm. Pharmacol.*, 28, 687, 1976.

138. **Glick, D. and Glaubach, S.,** The occurrence and distribution of atropinesterase, and the specificity of tropinesterases, *J. Gen. Physiol.*, 25, 197, 1941.

139. **Godeaux, J. and Tønnesen, M.,** Investigations into atropine metabolism in the animal organism, *Acta Pharmacol. Toxicol.*, 5, 95, 1949.

140. **Ellis, S.,** Enzymic hydrolysis of morphine esters, *J. Pharmacol. Exp. Ther.*, 94, 130, 1948.

141. **Ammon, R. and Savelsberg, W.,** Die enzymatische Spaltung von Atropin, Cocain und Chemisch verwandten Estern, *Hoppe-Seyler's Z. Physiol. Chem.*, 284, 135, 1949.

142. **Stormont, C. and Suzuki, Y.,** Atropinesterase and cocainesterase of rabbit serum: localization of the enzyme activity in isozymes, *Science*, 167, 200, 1970.

143. **Cauthen, S. E., Ellis, R. D., Larrison, S. B., and Kidd, M. R.,** Resolution, purification and characterization of rabbit serum atropinesterase and cocainesterase, *Biochem. Pharmacol.*, 25, 181, 1976.

144. **Kalser, S. C. and McLain, P. L.,** Atropine metabolism in man, *Clin. Pharmacol. Ther.*, 11, 214, 1970.

145. **Werner, G. and Brehmer, G.,** Degradation of tropane alkaloids by esterases and its significance for some pharmacological effects, *Abh. Dtsch. Akad. Wiss. Berlin*, 4, 217, 1963 (Chem. Abstr. 61, 6213b, 1964).

146. **Werner, G. and Brehmer, G.,** Zur Stereospezifität der (–)-Hyoscyamin-Acylhydrolase des Kaninchins, *Hoppe-Seyler's Z. Physiol. Chem.*, 348, 1640, 1967.

147. **Werner, G. and Schmidt, K.-H.,** Untersuchungen zum Stoffwechsel von Tropan-Alkaloiden. VIII. Chemische Analyse des Stoffwechsels von (–)-Scopolamin bei einigen Säugetieren, *Hoppe-Seyler's Z. Physiol. Chem.*, 349, 741, 1968.

148. **Shaw, J. and Urquhart, J.,** Programmed, systemic drug delivery by the transdermal route, *Trends Pharmacol. Sci.*, 1, 208, 1980.

149. **Putcha, L., Cintrón, N. M., Tsui, J., Vanderploeg, J. M., and Kramer, W. G.,** Pharmacokinetics and oral bioavailability of scopolamine in normal subjects, *Pharm. Res.*, 6, 481, 1989.

150. **Yue, T.-L., Wang, G.-F., and Song, Z.-Y.,** Metabolism of ^{3}H-scopolamine and ^{3}H-anisodamine, *Yao Hsueh Hsueh Pao*, 14, 208, 1979 (Chem. Abstr. 92, 51661y, 1980).

151. **Ziegler, D. M., Mitchell, C. H., and Jollow, D.,** The properties of a purified hepatic microsomal mixed function amine oxidase, in *Microsomes and Drug Oxidations*, Gillette, J. R., Conney, A. H., Cosmides, G. J., Estabrook, R. W., Fouts, J. R., and Mannering, G. J., Eds., Academic Press, New York, 1969, 173.

152. **Sano, M. and Hakusui, H.,** Metabolism of methylscopolammonium methylsufate (DD-234) in rats, *Chem. Pharm. Bull.*, 22, 696, 1974.

153. **Oesch, F.,** Personal communication, 1977.

154. **Anon.,** Absorption, distribution and excretion of anisodamine, *China Med. J.*, 5, 274, 1973.

155. **Oelkers, H. A. and Vincke, E.,** Untersuchungen über die Ausscheidung von Cocain, *Naunyn-Schmiedebergs Arch. Exp. Pathol. Pharmakol.*, 179, 341, 1935.

156. **Nayak, P. K., Misra, A. L., and Mulé, S. J.,** Physiologic disposition and metabolism of [^{3}H]-cocaine in the rat, *Fed. Proc. Fed. Am. Soc. Exp. Biol.*, 33, 527, 1974.

157. **Woods, L. A., McMahon, F. G., and Seevers, M. H.,** Distribution and metabolism of cocaine in the dog and rabbit, *J. Pharmacol. Exp. Ther.*, 101, 200, 1951.

158. **Misra, A. L., Patel, M. N., Alluri, V. R., Mulé, S. J., and Nayak, P. K.,** Disposition and metabolism of [^{3}H]cocaine in acutely and chronically treated dogs, *Xenobiotica*, 6, 537, 1976.

159. **Misra, A. L., Giri, V. V., Patel, M. N., Alluri, V. R., and Mulé, S. J.,** Disposition and metabolism of [^{3}H] cocaine in acutely and chronically treated monkeys, *Drug Alcohol Depend.*, 2, 261, 1977.

160. **Kogan, M. J., Verebey, K. G., DePace, A. C., Resnick, R. B., and Mulé, S. J.,** Quantitative determination of benzoylecgonine and cocaine in human biofluids by gas-liquid chromatography, *Anal. Chem.*, 49, 1965, 1977.

161. **Javaid, J. I., Dekirmenjian, H., Davis, J. M., and Schuster, C. R.,** Determination of cocaine in human urine, plasma and red blood cells by gas-liquid chromatography, *J. Chromatogr.*, 152, 105, 1978.

162. **Jindal, S. P., Lutz, T., and Vestergaard, P.,** Mass spectrometric determination of cocaine and its biologically active metabolite, norcocaine, in human urine, *Biomed. Mass Spectrom.*, 5, 658, 1978.

163. **Hamilton, H. E., Wallace, J. E., Shimek, E. L., Land, P., Harris, S. C., and Christenson, J. G.,** Cocaine and benzoylecgonine excretion in humans, *J. Forensic Sci.*, 22, 697, 1977.

164. **Fish, F. and Wilson, W. D. C.,** Excretion of cocaine and its metabolites in man, *J. Pharm. Pharmacol.*, 21, 135, 1969.

165. **Sánchez, C. A.,** Chromatographic analysis of ecgonine in the urine of subjects habitually chewing coca leaves, *An. Fac. Farm. Bioquim. Univ. Nacl. Mayor San Marcos*, 8, 82, 1957 (Chem. Abstr. 53, 22466h, 1959).

166. **Ortiz, V. C.,** The problem of the chewing of the coca leaf in Peru, *Bull. Narcot.*, 4, 26, 1952.

167. **Ortiz, R. V.,** Distribution and metabolism of cocaine in the rat, *An. Fac. Quim. Farm. Univ. Chile*, 18, 15, 1966 (Chem. Abstr. 68, 48055c, 1968).

168. **Valanju, N. N., Baden, M. M., Valanju, S. N., Mulligan, D., and Verma, S. K.,** Detection of biotransformed cocaine in urine from drug abusers, *J. Chromatogr.*, 81, 170, 1973.

169. **Nayak, P. K., Misra, A. L., and Mulé, S. J.,** Physiological disposition and biotransformation of [³H]cocaine in acutely and chronically treated rats, *J. Pharmacol. Exp. Ther.*, 196, 556, 1976.

170. **Misra, A. L.,** Disposition and biotransformation of cocaine, in *Cocaine: Chemical, Biological, Clinical, Social and Treatment Aspects*, Mulé, S. J., Ed., CRC Press, Boca Raton, FL, 1976, 73.

171. **Inaba, T., Stewart, D. J., and Kalow, W.,** Metabolism of cocaine in man, *Clin. Pharmacol. Ther.*, 23, 547, 1978.

172. **Ambre, J., Fischman, M., and Ruo, T.-I.,** Urinary excretion of ecgonine methyl ester, a major metabolite of cocaine in humans, *J. Anal. Toxicol.*, 8, 23, 1984.

173. **Ambre, J., Ruo, T. I., Nelson, J., and Belknap, S.,** Urinary excretion of cocaine, benzoylecgonine, and ecgonine methyl ester in humans, *J. Anal. Toxicol.*, 12, 301, 1988.

174. **Misra, A. L., Nayak, P. K., Bloch, R., and Mulé, S. J.,** Estimation and disposition of [³H]benzoylecgonine and pharmacological activity of some cocaine metabolites, *J. Pharm. Pharmacol.*, 27, 784, 1975.

175. **Mulé, S. J., Casella, G. A., and Misra, A. L.,** Intracellular disposition of [³H]-cocaine, [³H]-norcocaine, [³H]-benzoylecgonine and [³H]-benzoylnorecgonine in the brain of rats, *Life Sci.*, 19, 1585, 1976.

176. **Misra, A. L., Giri, V. V., Patel, M. N., Alluri, V. R., Pontani, R. B., and Mulé, S. J.,** Disposition of [³H] benzoylecgonine (cocaine metabolite) in the rat, *Res. Commun. Chem. Pathol. Pharmacol.*, 13, 579, 1976.

177. **Misra, A. L., Pontani, R. B., and Mulé, S. J.,** [³H]-Norcocaine and [³H]-pseudococaine: effect of N-demethylation and C_2-epimerization of cocaine on its pharmacokinetics in the rat, *Experientia*, 32, 895, 1976.

178. **Misra, A. L., Vadlamani, N. L., Bloch, R., Nayak, P. K., and Mulé, S. J.,** Physiologic disposition and metabolism of [³H] ecgonine (cocaine metabolite) in the rat, *Res. Commun. Chem. Pathol. Pharmacol.*, 8, 55, 1974.

179. **Lowry, W. T., Lomonte, J. N., Hatchett, D., and Garriott, J. C.,** Identification of two novel cocaine metabolites in bile by gas chromatography and gas chromatography/mass spectrometry in a case of acute intravenous cocaine overdose, *J. Anal. Toxicol.*, 3, 91, 1979.

180. **Smith, R. M.,** Arylhydroxy metabolites of cocaine in the urine of cocaine users, *J. Anal. Toxicol.*, 8, 35, 1984.

181. **Smith, R. M., Poquette, M. A., and Smith, P. J.,** Hydroxymethoxybenzoylmethylecgonines: new metabolites of cocaine from human urine, *J. Anal. Toxicol.*, 8, 29, 1984.

182. **Smith, R. M.,** Ethyl esters of arylhydroxy- and arylhydroxymethoxycocaines in the urine of simultaneous cocaine and ethanol users, *J. Anal. Toxicol.*, 8, 38, 1984.

183. **Jindal, S. P. and Lutz, T.,** Mass spectrometric studies of cocaine disposition in animals and humans using stable isotope-labeled analogues, *J. Pharm. Sci.*, 78, 1009, 1989.

184. **Axelrod, J. and Cochin, J.,** The inhibitory action of nalophine on the enzymatic N-demethylation of narcotic drugs, *J. Pharmacol. Exp. Ther.*, 121, 107, 1957.

185. **Ramos-Aliaga, R. and Chiriboga, J.,** Enzymic N-demethylation of cocaine and nutritional status, *Arch. Latinoam. Nutr.*, 20, 415, 1970 (Chem. Abstr. 74, 108861y, 1971).

186. **Estevez, V. C., Ho, B. T., and Englert, L. F.,** Inhibition of the metabolism of cocaine by SKF-525A, *Res. Commun. Chem. Pathol. Pharmacol.*, 17, 179, 1977.

187. **Leighty, E. G. and Fentiman, A. F.,** Metabolism of cocaine to norcocaine and benzoyl ecgonine by an *in vitro* microsomal enzyme system, *Res. Commun. Chem. Pathol. Pharmacol.*, 8, 65, 1974.

188. **Stewart, D. J., Inaba, T., and Kalow, W.,** N-Demethylation of cocaine in the rat and in isolated rat hepatocytes: comparison with aminopyrine demethylation, *J. Pharmacol. Exp. Ther.*, 207, 171, 1978.

189. **Kloss, M. W., Rosen, G. M., and Rauckman, E. J.,** N-Demethylation of cocaine to norcocaine. Evidence for participation by cytochrome P-450 and FAD-containing monooxygenase, *Mol. Pharmacol.*, 23, 482, 1983.

190. **Shuster, L., Casey, E., and Welankiwar, S. S.,** Metabolism of cocaine and norcocaine to N-hydroxynorcocaine, *Biochem. Pharmacol.*, 32, 3045, 1983.

191. **Kloss, M. W., Rosen, G. M., and Rauckman, E. J.,** Biotransformation of norcocaine to norcocaine nitroxide by rat brain microsomes, *Psychopharmacology*, 84, 221, 1984.

192. **Rauckman, E. J., Rosen, G. M., and Cavagnaro, J.,** Norcocaine nitroxide. A potential hepatotoxic metabolite of cocaine, *Mol. Pharmacol.*, 21, 458, 1982.

193. **Evans, M. A. and Johnson, M. E.,** The role of a reactive nitroxide radical in cocaine-induced hepatic necrosis, *Fed. Proc. Fed. Am. Soc. Exp. Biol.*, 40, 638, 1981.

194. **Evans, M. A.,** Microsomal activation of N-hydroxy norcocaine to a reactive nitroxide, *Toxicologist*, 1, 1, 1981.

195. **Evans, M. A. and Harbison, R. D.,** Cocaine-induced hepatotoxicity in mice, *Toxicol. Appl. Pharmacol.*, 45, 739, 1978.

196. **Thompson, M. L., Shuster, L., and Shaw, K.,** Cocaine-induced hepatic necrosis in mice—the role of cocaine metabolism, *Biochem. Pharmacol.*, 28, 2389, 1979.

197. **Misra, A. L., Pontani, R. B., and Vadlamami, N. L.,** Metabolism of norcocaine, N-hydroxy norcocaine and cocaine-N-oxide in the rat, *Xenobiotica*, 9, 189, 1979.

453

198. **Just, W. W., Grafenburg, L., Thel, S., and Werner, G.,** Comparative metabolic, autoradiographic and pharmacologic studies of cocaine and its metabolite norcocaine, *Arch. Pharmacol.*, 293, R56, 1976.

199. **Misra, A. L., Nayak, P. K., Patel, M. N., Vadlamami, N. L., and Mulé, S. J.,** Identification of norcocaine as a metabolite of [³H]-cocaine in rat brain, *Experientia*, 30, 1312, 1974.

200. **Hawks, R. L., Kopin, I. J., Colburn, R. W., and Thoa, N. B.,** Norcocaine: a pharmacologically acitive metabolite of cocaine found in brain, *Life Sci.*, 15, 2189, 1974.

201. **Englert, L. F., Estevez, V. S., and Ho, B. T.,** Demethylation in rats chronically treated with cocaine, *Res. Commun. Chem. Pathol. Pharmacol.*, 13, 555, 1976.

202. **Glick, D., Glaubach, S., and Moore, D. H.,** Azolesterase activities of electrophoretically separated proteins of serum, *J. Biol. Chem.*, 144, 525, 1942.

203. **Blaschko, H., Himms, J. M., and Strömblad, B. C. R.,** The enzymic hydrolysis of cocaine and *alpha*-cocaine, *Br. J. Pharmacol.*, 10, 442, 1955.

204. **Blaschko, H., Chou, T. C., and Wajda, I.,** The affinity of atropine-like esters for esterases, *Br. J. Pharmacol.*, 2, 108, 1947.

205. **Taylor, D., Estevez, V. S., Englert, L. F., and Ho, B. T.,** Hydrolysis of carbon-labeled cocaine in human serum, *Res. Commun. Chem. Pathol. Pharmacol.*, 14, 249, 1976.

206. **Stewart, D. J., Inaba, T., Tang, B. K., and Kalow, W.,** Hydrolysis of cocaine in human plasma by cholinesterase, *Life Sci.*, 20, 1557, 1977.

207. **Stewart, D. J., Inaba, T., Lucassen, M., and Kalow, W.,** Cocaine metabolism: cocaine and norcocaine hydrolysis by liver and serum esterases, *Clin. Pharmacol. Ther.*, 25, 464, 1979.

208. **Heim, F. and Haas, A.,** Über den fermentativen Abbau von Pantokain, Novokain und Kokain durch Extrakte asu Meerschweinchenleber, -niere, -gehirn und -muskulatur, *Naunyn Schmiedebergs Arch. Exp. Pathol. Pharmakol.*, 211, 458, 1950.

209. **Mikami, K.,** The fate of cocaine in the body, *Folia Pharmacol. Jpn.*, 47, 51, 1951 (Chem. Abstr. 46, 7656f, 1952).

210. **Severi, A., Longhi, A., and Pomarelli, P.,** Detoxication of cocaine, *Atti Accad. Med. Lomb.*, 22, 485, 1967 (Chem. Abstr. 71, 29032p, 1969).

211. **Iwatsubo, K.,** Studies on the classification of the enzymes hydrolyzing ester-form drugs in liver microsomes, *Jpn. J. Pharmacol.*, 15, 244, 1965.

212. **Gruhn, E.,** Über die Ausscheidung der sterio-isomeren Kokains im Harn und ihre Beziehung zur Toxizität, *Naunyn Schmiedebergs Arch. Exp. Pathol. Pharmakol.*, 106, 115, 1925.

213. **Misra, A. L. and Pontani, R. B.,** Disposition and metabolism of ψ-cocaine (*dextro*-cocaine) in the rat, *Drug Metab. Dispos.*, 5, 556, 1977.

214. **Gatley, S. J., MacGregor, R. R., Fowler, J. S., Wolf, A. P., Dewey, S. L., and Schlyer, D. J.,** Rapid stereoselective hydrolysis of (+)-cocaine in baboon plasma prevents its uptake in the brain: implications for behavioral studies, *J. Neurochem.*, 54, 720, 1990.

215. **Valentour, J. C., Aggarwal, V., McGee, M. P., and Goza, S. W.,** Cocaine and benzoylecgonine determinations in postmortem samples by gas chromatography, *J. Anal. Toxicol.*, 2, 134, 1978.

216. **Seiler, N., Kameniková, L., and Werner, G.,** Untersuchungen zum Stoffwechsel von Tropan-Alkaloiden. VII. Versuche zur Charakterisierung der Tropacocainesterase, *Hoppe Seyler's Z. Physiol. Chem.*, 349, 692, 1968.

217. **Seiler, N., Kameniková, L., and Werner, G.,** Examination of the metabolism of tropane alkaloids. X. Activity of enzymes splitting nortropacocaine in mouse organs during development, *Colln. Czech. Chem. Commun.*, 34, 719, 1968.

218. **Williams, R. T.,** *Detoxication Mechanisms*, Chapman and Hall, London, 1959, 668.

219. **Sekura, R. D. and Jakoby, W. B.,** Phenol sulfotransferases, *J. Biol. Chem.*, 254, 5658, 1979.

220. **King, L. J., Parke, D. V., and Williams, R. T.,** The metabolism of [2-¹⁴C]indole in the rat, *Biochem. J.*, 98, 266, 1966.

221. **Acheson, R. M. and Nwankwo, J. O.,** The metabolism of some 1-hydroxylated indoles in the rat, *Xenobiotica*, 14, 877, 1984.

222. **Posner, H. S., Mitoma, C., and Udenfriend, S.,** Enzymic hydroxylation of aromatic compounds. II. Further studies of the properties of the microsomal hydroxylating system, *Arch. Biochem. Biophys.*, 94, 269, 1961.

223. **Beckett, A. H. and Morton, D. M.,** The metabolism of oxindole and related compounds, *Biochem. Pharmacol.*, 15, 937, 1966.

224. **Frydman, B., Frydman, R. B., and Tomaro, M. L.,** Pyrrolooxygenases: a new type of oxidases, *Mol. Cell. Biochem.*, 2, 121, 1973.

225. **Li, C.-H., Ho, S.-H., and Wang, S.-H.,** The metabolism of plant indican, *Yao Hsueh Hsueh Pao*, 10, 581, 1963 (Chem. Abstr. 60, 7309f, 1964).

226. **Acheson, R. M. and King, L. J.,** The metabolism of indole-3-carboxylic acid by the rat, *Biochim. Biophys. Acta*, 71, 643, 1963.

227. **Hammond, A. C., Carlson, J. R., and Willett, J. D.,** The metabolism and disposition of 3-methylindole in goats, *Life Sci.*, 25, 1301, 1979.

228. **Frydman, R. B., Tomaro, M. L., and Frydman, B.,** Pyrrolooxygenases: isolation, properties, and products formed, *Biochim. Biophys. Acta*, 284, 63, 1972.

229. **Skiles, G. L., Adams, J. D., and Yost, G. S.,** Isolation and identification of 3-hydroxy-3-methyloxindole, the major murine metabolite of 3-methylindole, *Chem. Res. Toxicol.*, 2, 254, 1989.

230. **Yost, G. S.,** Mechanisms of 3-methylindole pneumotoxicity, *Chem. Res. Toxicol.*, 2, 273, 1989.

231. **Bray, T. M. and Kirkland, J. B.,** The metabolic basis of 3-methylindole-induced pneumotoxicity, *Pharmacol. Ther.*, 46, 105, 1990.

232. **Potchoiba, M. J., Carlson, J. R., and Breeze, R. G.,** Metabolism and pneumotoxicity of 3-methyloxindole, indole-3-carbinol, and 3-methylindole in goats, *Am. J. Vet. Res.*, 43, 1418, 1982.

233. **Nocerini, M. R., Carlson, J. R., and Yost, G. S.,** Electrophilic metabolites of 3-methylindole as toxic intermediates in pulmonary oedema, *Xenobiotica*, 14, 561, 1984.

234. **Nocerini, M. R., Carlson, J. R., and Yost, G. S.,** Adducts of 3-methylindole and glutathione: species differences in organ-selective bioactivation, *Toxicol. Lett.*, 28, 79, 1985.

235. **Nocerini, M. R., Yost, G. S., Carlson, J. R., Liberato, J., and Breeze, R. G.,** Structure of the glutathione adduct of activated 3-methylindole indicates that an imine methide is the electrophilic intermediate, *Drug Metab. Dispos.*, 13, 690, 1985.

236. **Huijzer, J. C., Adams, J. D., Jaw, J.-Y., and Yost, G. S.,** Inhibition of 3-methylindole bioactivation by the cytochrome P-450 suicide substrates 1-aminobenzotriazole and α-methylaminobenzotriazole, *Drug Metab. Dispos.*, 17, 37, 1989.

237. **Formosa, P. J. and Bray, T. M.,** Evidence for metabolism of 3-methylindole by prostaglandin H synthase and mixed-function oxidases in goat lung and liver microsomes, *Biochem. Pharmacol.*, 37, 4359, 1988.

238. **Formosa, P. J., Bray, T. M., and Kubow, S.,** Metabolism of 3-methylindole by prostaglandin H synthase in ram seminal vesicles, *Can. J. Physiol. Pharmacol.*, 66, 1524, 1988.

239. **Somani, S. M.,** Pharmacokinetics and pharmacodynamics of physostigmine in the rat after oral administration, *Biopharm. Drug Dispos.*, 10, 187, 1989.

240. **Somani, S. M. and Boyer, A.,** Urinary pharmacokinetics of physostigmine in the rat, *Eur. J. Drug Metab. Pharmacokinet.*, 10, 343, 1985.

241. **Somani, S. M. and Khalique, A.,** Distribution and pharmacokinetics of physostigmine in rat after intramuscular administration, *Fundam. Appl. Toxicol.*, 6, 327, 1986.

242. **Somani, S. M. and Khalique, A.,** Pharmacokinetics and pharmacodynamics of physostigmine in the rat after intravenous administration, *Drug Metab. Dispos.*, 15, 627, 1987.

243. **Giacobini, E., Somani, S., McIlhany, M., Downen, M., and Hallak, M.,** Pharmacokinetics and pharmacodynamics of physostigmine after intravenous administration in beagle dogs, *Neuropharmacology*, 26, 831, 1987.

244. **Hartvig, P., Wiklund, L., and Lindström, B.,** Pharmacokinetics of physostigmine after intravenous, intramuscular and subcutaneous administration in surgical patients, *Acta Anaesthesiol. Scand.*, 30, 177, 1986.

245. **Boyer, A. W. and Somani, S. M.,** Pharmacokinetics of biliary excretion of physostigmine in rat, *Arch. Int. Pharmacodyn. Ther.*, 278, 180, 1985.

246. **Isaksson, K. and Kissinger, P. T.,** Metabolism of physostigmine in mouse liver microsomal incubations studied by liquid chromatography with dual-electrode amperometric detection, *J. Chromatogr.*, 419, 165, 1987.

247. **Tweedie, D. J., Prough, R. A., and Burke, M. D.,** Effects of induction on the metabolism and cytochrome P-450 binding of harman and other β-carbolines, *Xenobiotica*, 18, 785, 1988.

248. **Zetler, G., Back, G., and Iven, H.,** Pharmacokinetics in the rat of the hallucinogenic alkaloids harmine and harmaline, *Naunyn Schmiedeberg's Arch. Pharmacol.*, 285, 273, 1974.

249. **Villeneuve, A. and Sourkes, T. L.,** Metabolism of harmaline and harmine in the rat, *Rev. Can. Biol.*, 25, 231, 1966.

250. **Slotkin, T. and DiStefano, V.,** Urinary metabolites of harmine in the rat and their inhibition of monoamine oxidase, *Biochem. Pharmacol.*, 19, 125, 1970.

251. **Slotkin, T. A. and DiStefano, V.,** A model of harmine metabolism in the rat, *J. Pharmacol. Exp. Ther.*, 174, 456, 1970.

252. **Slotkin, T. A., DiStefano, V., and Au, W. Y. W.,** Blood levels and urinary excretion of harmine and its metabolites in man and rats, *J. Pharmacol. Exp. Ther.*, 173, 26, 1970.

253. **Burke, M. D. and Upshall, D. G.,** Species and phenobarbital-induced differences in the kinetic constants of liver microsomal harmine O-demethylation, *Xenobiotica*, 6, 321, 1976.

254. **Tweedie, D. J. and Burke, M. D.,** Metabolism of the β-carbolines, harmine and harmol, by liver microsomes from phenobarbitone- or 3-methylcholanthrene-treated mice Identification and quantitation of two novel harmine metabolites, *Drug Metab. Dispos.*, 15, 74, 1987.

255. **Ho, B. T., Estevez, V., Fritchie, G. E., Tansey, L.-W., Idänpään-Heikkilä, J., and McIsaac, W. M.,** Metabolism of harmaline in rats, *Biochem. Pharmacol.*, 20, 1313, 1971.

256. **Flury, F.,** Beiträge zur Pharmakologie der Steppenraute (Peganum harmala), *Arch. Exp. Pathol. Pharmak.*, 64, 105, 1911.

257. **Daly, J., Inscoe, J. K., and Axelrod, J.,** The formation of O-methylated catechols by microsomal hydroxylation of phenols and subsequent enzymatic catechol O-methylation. Substrate specificity, *J. Med. Chem.*, 8, 153, 1965.

258. **Daly, J. and Witkop, B.,** Recent studies on the centrally active endogenous amines, *Angew. Chem. Int. Ed. Engl.*, 2, 421, 1963.

259. **Mulder, G. J. and Hagedoorn, A. H.,** UDP glucuronyltransferase and phenolsulfotransferase *in vivo* and *in vitro*. Conjugation of harmol and harmalol, *Biochem. Pharmacol.*, 23, 2101, 1974.

260. **Wong, K. P. and Sourkes, T. L.,** Determination of UDPG and UDPGA in tissues, *Anal. Biochem.*, 21, 444, 1967.

261. **Wong, K. P. and Sourkes, T. L.,** Glucuronidation of 3-O-methylnoradrenaline, harmalol and some related compounds, *Biochem. J.*, 110, 99, 1968.

262. **Pang, K. S., Koster, H., Halsema, I. C. M., Scholtens, E., and Mulder, G. J.,** Aberrant pharmacokinetics of harmol in the perfused rat liver preparation: sulfate and glucuronide conjugations, *J. Pharmacol. Exp. Ther.*, 219, 134, 1981.

263. **Koster, H., Halsema, I., Scholtens, E., Pang, K. S., and Mulder, G. J.,** Kinetics of sulfation and glucuronidation of harmol in the perfused rat liver preparation, *Biochem. Pharmacol.*, 31, 3023, 1982.

264. **Mulder, G. J., Dawson, J. R., and Pang, K. S.,** Competition between sulphation and glucuronidation in the rat *in vivo*: enzyme kinetics and pharmacokinetics of conjugation, *Biochem. Soc. Trans.*, 12, 17, 1984.

265. **Dawson, J., Berggren, M., and Moldeus, P.,** Isolated cells as a model for studying sulphate conjugation and its regulation, in *Sulfate Metabolism and Sulfate Conjugation*, Mulder, G. J., Caldwell, J., Van Kempen, G. M. J., and Vonk, R. J., Eds., Taylor and Francis, London, 1982, 135.

266. **Sundheimer, D. W. and Brendel, K.,** Metabolism of harmol and transport of harmol conjugates in isolated rat hepatocytes, *Drug Metab. Dispos.*, 11, 433, 1983.

267. **Araya, H., Mizuma, T., Horie, T., Hayashi, M., and Awazu, S.,** Heterogeneous distribution of conjugation activites of harmol in isolated rat liver cells, *J. Pharmacobiol. Dyn.*, 7, 624, 1984.

268. **Mulder, G. J., Weitering, J. G., Scholtens, E., Dawson, J. R., and Pang, K. S.,** Extrahepatic sulfation and glucuronidation in the rat *in vivo*. Determination of the hepatic extraction ratio of harmol and the extrahepatic contribution to harmol conjugation, *Biochem. Pharmacol.*, 33, 3081, 1984.

269. **Mulder, G. J., Hayen-Keulemans, K., and Sluiter, N. E.,** UDP glucuronyltransferase and phenolsulfotransferase from rat liver *in vivo* and *in vitro*. Characterization of conjugation and biliary excretion of harmol *in vivo* and in the perfused liver, *Biochem. Pharmacol.*, 24, 103, 1975.

270. **Mulder, G. J. and Bleeker, B.,** UDP glucuronyltransferase and phenolsulfotransferase from rat liver *in vivo* and *in vitro*. IV. Species differences in harmol conjugation and elimination in bile and urine *in vivo*, *Biochem. Pharmacol.*, 24, 1481, 1975.

271. **Iven, H.,** The pharmacokinetics and organ distribution of ajmaline and quinidine in the mouse, *Naunyn Schmiedeberg's Arch. Pharmacol.*, 298, 43, 1977.

272. **Köppel, C., Tenczer, J., and Arndt, I.,** Metabolic disposition of ajmaline, *Eur. J. Drug Metab. Pharmacokinet.*, 14, 309, 1989.

273. **Beckett, A. N. and Morton, D. M.,** The *in vitro* metabolism of mitragyna alkaloids of corynantheidine structure, *J. Pharm. Pharmacol.*, 18, 88, 1966.

274. **Chandrasekaran, B. and Nagarajan, B.,** Metabolism of echitamine and plumbagin in rats, *J. Biosci.*, 3, 395, 1981.

275. **Ho, A. K. S., Hoffman, D. B., Gershon, S., and Loh, H. H.,** Distribution and metabolism of tritiated yohimbine in mice, *Arch. Int. Pharmacodyn. Ther.*, 194, 304, 1971.

276. **Owen, J. A., Nakatsu, S. L., Fenemore, J., Condra, M., Surridge, D. H. C., and Morales, A.,** The pharmacokinetics of yohimbine in man, *Eur. J. Clin. Pharmacol.*, 32, 577, 1987.

277. **Sheppard, H., Lucas, R. C., and Tsien, W. H.,** The metabolism of reserpine-C[14], *Arch. Int. Pharmacodyn. Ther.*, 103, 256, 1955.

278. **Sheppard, H. and Tsien, W. H.,** Metabolism of reserpine-C[14]. II. Species differences as studied *in vitro*, *Proc. Soc. Exp. Biol. Med.*, 90, 437, 1955.

279. **Sheppard, H., Tsien, W. H., Sigg, E. B., Lucas, R. A., and Plummer, A. J.,** The metabolism of reserpine-C[14]. III. C[14]-Concentration vs. time in the brains and other tissues of rats and guinea pigs, *Arch. Int. Pharmacodyn. Ther.*, 113, 160, 1957.

280. **Numerof, P., Gordon, M., and Kelly, J. M.,** The metabolism of reserpine. I. Studies in the mouse with C-14 labeled reserpine, *J. Pharmacol. Exp. Ther.*, 115, 427, 1955.

281. **Maass, A. R., Jenkins, B., Shen, Y., and Tannenbaum, P.,** Studies on absorption, excretion, and metabolism of ^{3}H-reserpine in man, *Clin. Pharmacol. Ther.*, 10, 366, 1969.

282. **Maronde, R. F., Haywood, L. J., Feinstein, D., and Sobel, C.,** The monoamine oxidase inhibitor, pargyline hydrochloride, and reserpine, *J. Am. Med. Assoc.*, 184, 7, 1963.

283. **Glazko, A. J., Dill, W. A., Wolf, L. M., and Kazenko, A.,** Studies on the metabolism of reserpine, *J. Pharmacol. Exp. Ther.*, 118, 377, 1956.

284. **Dhar, M. M., Kohli, J. D., and Srivastava, S. K.,** Studies on the metabolism of reserpine. I. Identification of metabolites, *J. Sci. Ind. Res.*, 14C, 179, 1955.

285. **Dhar, M. M., Kohli, J. D., and Srivastava, S. K.,** Reserpine metabolism, *Indian J. Pharm.*, 18, 293, 1956.

286. **Maggiolo, C. and Haley, T. J.,** Brain concentration of reserpine-H³ and its metabolites in the mouse, *Proc. Soc. Exp. Biol. Med.*, 115, 149, 1964.

287. **Numerof, P., Virgona, A. J., Cranswick, E. N., Cunningham, R. N., and Kline, N. S.,** The metabolism of reserpine. II. Studies in schizophrenic patients, *Psychiat. Res. Rep.*, 9, 139, 1958.

288. **Stitzel, R. E., Wagner, L. A., and Stawarz, R. J.,** Studies on the microsomal metabolism of ³H-reserpine, *J. Pharmacol. Exp. Ther.*, 182, 500, 1972.

289. **Stawarz, R. J. and Stitzel, R. E.,** Factors influencing the microsomal metabolism of ³H-reserpine, *Pharmacology*, 11, 178, 1974.

290. **Stitzel, R. E.,** The biological fate of reserpine, *Pharmacol. Rev.*, 28, 179, 1977.

291. **Löhr, J. P. and Bartsch, G. G.,** Untersuchungen zur Resorption und Ausscheidung des Raubasins in Ratten unter Verwendung einer direkten Nachweismethode des Alkaloids, *Arzneim. Forsch.*, 25, 870, 1975.

292. **Marzo, A., Ghirardi, P., Alessio, R., and Villa, A.,** Absorption and excretion of ³H-raubasine in human subjects and dogs, *Arzneim. Forsch.*, 27(II), 2343, 1977.

293. **Hardesty, C. T., Chaney, N. A., and Mead, J. A. R.,** The effect of route of administration on the distribution of ellipticine in mice, *Cancer Res.*, 32, 1884, 1972.

294. **Chadwick, M., Silveira, D. M., Platz, B. B., and Hayes, D.,** Comparative physiological disposition of ellipticine in several animal species after intravenous administration, *Drug Metab. Dispos.*, 6, 528, 1978.

295. **Branfman, A. R., Bruni, R. J., Reinhold, V. N., Silveira, D. M., Chadwick, M., and Yesair, D. W.,** Characterization of the metabolites of ellipticine in rat bile, *Drug Metab. Dispos.*, 6, 542, 1978.

296. **Reinhold, V. N. and Bruni, R. J.,** Aromatic hydroxylation of ellipticine in rats: lack of an NIH shift, *Biomed. Mass Spectrom.*, 3, 335, 1976.

297. **Lesca, P., Lecointe, P., Paoletti, C., and Mansuy, D.,** Induction des mono-oxygénases hépatiques par l'ellipticine chez le rat: formation de cytochrome P_{448}. Activité hydroxylante, *C.R. Hebd. Séanc. Acad. Sci. Paris*, 282D, 1457, 1976.

298. **Lesca, P., Lecointe, P., Paoletti, C., and Mansuy, D.,** The hydroxylation of the antitumor agent, ellipticine, by liver microsomes from differently pretreated rats, *Biochem. Pharmacol.*, 26, 2169, 1977.

299. **Lallemand, J.-Y., Lemaitre, P., Beeley, L., Lesca, P., and Mansuy, D.,** Hydroxylation de l'ellipticine chez le rat: structure des deux principaux métabolites. Synthése de l'hydroxy-7 ellipticine, *Tetrahedron Lett.*, 1261, 1978.

300. **Paoletti, C., Lecointe, P., Lesca, P., Cros, S., Mansuy, D., and Dat Xuong, N.,** Metabolism of ellipticine and derivatives and its involvement in the antitumor action of these drugs, *Biochimie*, 60, 1003, 1978.

301. **Van-Bac, N., Moisand, C., Gouyette, A., Muzard, G., Dat-Xuong, N., Le Pecq, J.-B., and Paoletti, C.,** Metabolism and disposition studies of 9-hydroxyellipticine and 2-methyl-9-hydroxyellipticinium acetate in animals, *Cancer Treat. Rep.*, 64, 879, 1980.

302. **Lecointe, P. and Puget, A.,** Métabolisme et pharmacocinétique de la méthoxy-9-ellipticine et son métabolite principal, l'hydroxy-9-ellipticine, *C. R. Acad. Sci. Paris*, 296, 279, 1983.

303. **Roy, M., Monsarrat, B., Cros, S., Lecointe, P., Rivalle, C., and Bisagni, E.,** Cytochrome P-450-mediated *O*-demethylation of two ellipticine derivatives. Differential effect of the murine *Ah* locus phenotype, *Drug Metab. Dispos.*, 13, 497, 1985.

304. **Meunier, G., Paoletti, C., and Meunier, B.,** Mise en évidence de la réaction de *O*-déméthylation catalysée par une peroxydase: application au cas de la méthoxy-9-ellipticine, *C. R. Acad. Sci. Ser. 3*, 299, 629, 1984.

305. **Meunier, G. and Meunier, B.,** Evidences for an efficient demethylation of methoxyellipticine derivatives catalyzed by a peroxidase, *J. Am. Chem. Soc.*, 107, 2558, 1985.

306. **Meunier, G. and Meunier, B.,** Peroxidase-catalyzed *O*-demethylation reactions. Quinone-imine formation from 9-methoxyellipticine derivatives, *J. Biol Chem.*, 260, 10576, 1985.

307. **Meunier, G., De Montauzon, D., Bernadou, J., Grassy, G., Bonnafous, M., Cros, S., and Meunier, B.,** The biooxidation of cytotoxic ellipicine derivatives: a key to structure-activity relationship studies?, *Mol. Pharmacol.*, 33, 93, 1988.

308. **Braham, Y., Meunier, G., and Meunier, B.,** Mise en évidence d'une biotransformation oxydante de la méthoxy-9-ellipticine. Comparaison avec le cas de l'hydroxy-9-ellipticine, *C. R. Acad. Sci. Paris*, 304, 301, 1987.

309. **Bernadou, J., Meunier, G., Paoletti, C., and Meunier, B.,** *o*-Quinone formation in the biochemical oxidation of the antitumor drug N^2-methyl-9-hydroxyellipticinium acetate, *J. Med. Chem.*, 26, 574, 1983.

310. **Meunier, G., Bernadou, J., and Meunier, B.,** Peroxidase-catalysed oxidation of N^2,N^6-dimethyl-9-hydroxyellipticinium acetate. Evidence for the formation of an electrophilic quinone-iminium derivative, *Biochem. Pharmacol.*, 36, 2599, 1987.

311. **Maftouh, M., Monsarrat, B., Rao, R. C., Meunier, B., and Paoletti, C.,** Identification of the glucuronide and glutathione conjugates of the antitumor drug N^2-methyl-9-hydroxyellipticinium acetate (celiptium) Comparative disposition of this drug with its olivacinium isomer in rat bile, *Drug Metab. Dispos.*, 12, 111, 1984.

312. **Monsarrat, B., Maftouh, M., Meunier, G., Bernadou, J., Armand, J. P., Paoletti, C., and Meunier, B.,** Oxidative biotransformation of the antitumour agent elliptinium acetate: structural characterization of its human and rat urinary metabolites, *J. Pharm. Biomed. Anal.*, 5, 341, 1987.

313. **Gouyette, A.,** Synthesis of deuterium-labelled elliptinium and its use in metabolic studies, *Biomed. Environ. Mass Spectrom.*, 15, 243, 1988.

314. **Braham, Y., Meunier, G., and Meunier, B.,** The rat biliary and urinary metabolism of the N^6-methylated derivative of elliptinium acetate, an antitumor agent, *Drug Metab. Dispos.*, 16, 316, 1988.

315. **Cartoni, G. P. and Giarusso, A.,** Gas chromatographic determination of ibogaine in biological fluids, *J. Chromatogr.*, 71, 154, 1972.

316. **Hatcher, R. A. and Eggleston, C.,** The fate of strychnin in the body, *J. Pharmacol. Exp. Ther.*, 10, 281, 1917.

317. **Egloff, T., Niederwieser, A., Pfister, K., Otten, A., Steinmann, B., Steiner, W., and Gitzelmann, R.,** A new high performance liquid chromatography (HPLC) method for the quantitation of strychnine in urine and tissue extracts, *J. Clin. Chem. Clin. Biochem.*, 20, 203, 1982.

318. **Adamson, R. H. and Fouts, J. R.,** Enzymatic metabolism of strychnine, *J. Pharmacol. Exp. Ther.*, 127, 87, 1959.

319. **Kato, R., Chiesara, E., and Vassanelli, P.,** Increased activity of microsomal strychnine-metabolizing enzyme induced by phenobarbital and other drugs, *Biochem. Pharmacol.*, 11, 913, 1962.

320. **Tsukamoto, H., Oguri, K., Watabe, T., and Yoshimura, H.,** Metabolism of drugs. XLI. The metabolic fate of strychnine in rabbit liver, *J. Biochem. (Tokyo)*, 55, 394, 1964.

321. **Mishima, M., Tanimoto, Y., Oguri, K., and Yoshimura, H.,** Metabolism of strychnine *in vitro*, *Drug Metab. Dispos.*, 13, 716, 1985.

322. **Oguri, K., Tanimoto, Y., Mishima, M., and Yoshimura, H.,** Metabolic fate of strychnine in rats, *Xenobiotica*, 19, 171, 1989.

323. **Tsukamoto, H., Yoshimura, H., Watabe, T., and Oguri, K.,** Metabolism of drugs. XLVIII. The study of selective demethylation of brucine *in vivo*, *Biochem. Pharmacol.*, 13, 1577, 1964.

324. **Watabe, T., Yoshimura, H., and Tsukamoto, H.,** Metabolism of drugs. L. The *in vitro* study on metabolism of brucine and 4-substituted veratroles, *Chem. Pharm. Bull.*, 12, 1151, 1964.

325. **Ezer, E. and Szporny, L.,** Uptake, elimination and distribution of vincamine-^{14}C in animals, *Kiserl. Orvostud.*, 19, 67, 1967 (Chem. Abstr. 67, 52404n, 1967).

326. **Iven, H. and Siegers, C.-P.,** Fluorometrischer Nachweis und Pharmakokinetik von Vincamin bei Ratten, *Arzneim. Forsch.*, 27(I), 1248, 1977.

327. **Vereczkey, L., Tamás, J., Czira, G., and Szporny, L.,** Metabolism of vincamine in the rat in vivo and in vitro, *Arzneim. Forsch.*, 30(II), 1860, 1980.

328. **Vigano, V., Paracchini, S., Piacenza, G., and Pesce, E.,** Metabolismo della vincamina nel ratto, *Il Farmaco Ed. Sci.*, 33, 583, 1978.

329. **Ventouras, K., Schulz, P., Doelker, E., Boucherat, J., and Buri, P.,** Etude pharmacocinétique de la vincamine administrée par voie orale, *Pharm. Acta Helv.*, 51, 334, 1976.

330. **Siegers, C.-P., Iven, H., and Strubelt, O.,** Plasmaspiegel und renale Ausscheidung von Vincamin nach oraler Applikation bei Probanden, *Arzneim. Forsch.*, 27(I), 1271, 1977.

331. **Vigano, V., Paracchini, S., Pesce, E., and Piacenza, G.,** MS identification of the vincamine metabolites, *Quant. Mass Spectrom. Life Sci.*, 2, 357, 1978 (Chem. Abstr. 91, 13415t, 1979).

332. **Vereczkey, L., Tamás, J., Czira, G., Rosdy, B., and Szporny, L.,** Metabolism of vincamine and some of related compounds, in *Recent Developments in Mass Spectrometry in Biochemistry and Medicine*, Vol. 6, Frigerio, A. and McCamish, M., Eds., Elsevier, Amsterdam, 1980, 267.

333. **Vereczkey, L.,** Pharmacokinetics and metabolism of vincamine and related compounds, *Eur. J. Drug Metab. Pharmacokinet.*, 10, 89, 1985.

334. **Beer, C. T. and Richards, J. F.,** The metabolism of vinca alkaloids. II. The fate of tritiated vinblastine in rats, *Lloydia*, 27, 352, 1964.

335. **Beer, C. T., Wilson, M. L., and Bell, J.,** A preliminary investigation of the fate of tritiated vinblastine in rats, *Can. J. Physiol. Pharmacol.*, 42, 368, 1964.

336. **Beer, C. T., Wilson, M. L., and Bell, J.,** The metabolism of vinca alkaloids. I. Preparation of tritiated vinblastine; the rate of urinary excretion of radioactivity by rats receiving the compound, *Can. J. Physiol. Pharmacol.*, 42, 1, 1964.

337. **Greenius, H. F., McIntyre, R. W., and Beer, C. T.,** The preparation of vinblastine-4-acetyl-*t* and its distribution in the blood of rats, *J. Med. Chem.*, 11, 254, 1968.

338. **Creasey, W. A., Scott, A. I., Wei, C.-C., Kutcher, J., Schwartz, A., and Marsh, J. C.,** Pharmacological studies with vinblastine in the dog, *Cancer Res.*, 35, 1116, 1975.

339. **Sethi, V. S., Surratt, P., and Spurr, C. L.,** Pharmacokinetics of vincristine, vinblastine, and vindesine in rhesus monkeys, *Cancer Chemother. Pharmacol.*, 12, 31, 1984.

340. **Owellen, R. J. and Hartke, C. A.,** The pharmacokinetics of 4-acetyl tritium vinblastine in two patients, *Cancer Res.*, 35, 975, 1975.

341. **Owellen, R. J., Hartke, C. A., and Hains, F. O.,** Pharmacokinetics and metabolism of vinblastine in humans, *Cancer Res.*, 37, 2597, 1977.

342. **Vendrig, D. E. M. M., Teeuwsen, J., Holthuis, J. J. M., de Vries, E. G. E., and Mulder, N. H.,** Pharmacokinetics of vinblastine, *Proc. Am. Assoc. Cancer Res.*, 28, 190, 1987.

343. **Lu, K., Yap, H.-Y., and Loo, T. L.,** Clinical pharmacokinetics of vinblastine by continuous intravenous infusion, *Cancer Res.*, 43, 1405, 1983.

344. **Houghton, J. A., Williams, L. G., Torrance, P. M., and Houghton, P. J.,** Determinants of intrinsic sensitivity to *Vinca* alkaloids in xenographs of pediatric rhabdomyosarcomas, *Cancer Res.*, 44, 582, 1984.

345. **Thimmaiah, K. N. and Sethi, V. S.,** Chemical characterization of the degradation products of vincristine dihydrogen sulfate, *Cancer Res.*, 45, 5382, 1985.

346. **Elmarakby, S. A., Duffel, M. W., Goswami, A., Sariaslani, F. S., and Rosazza, J. P. N.,** In vitro metabolic transformations of vinblastine: oxidations catalyzed by peroxidase, *J. Med. Chem.*, 32, 674, 1989.

347. **Elmarakby, S. A., Duffel, M. W., and Rozazza, J. P. N.,** In vitro metabolic transformations of vinblastine: oxidations catalyzed by human ceruloplasmin, *J. Med. Chem.*, 32, 2158, 1989.

348. **Castle, M. C., Margileth, D. A., and Oliverio, V. T.,** Distribution and excretion of [³H]vincristine in the rat and the dog, *Cancer Res.*, 36, 3684, 1976.

349. **El Dareer, S. M., White, V. M., Chen, F. P., Mellett, L. B., and Hill, D. L.,** Distribution and metabolism of vincristine in mice, rats, dogs, and monkeys, *Cancer Treat. Rep.*, 61, 1269, 1977.

350. **Bender, R. A., Castle, M. C., Margileth, D. A., and Oliverio, V. T.,** The pharmacokinetics of [³H]-vincristine in man, *Clin. Pharmacol. Ther.*, 22, 430, 1977.

351. **Owellen, R. J., Root, M. A., and Hains, F. O.,** Pharmacokinetics of vindesine and vincristine in humans, *Cancer Res.*, 37, 2603, 1977.

352. **Sethi, V. S. and Kimball, J. C.,** Pharmacokinetics of vincristine sulfate in children, *Cancer Chemother. Pharmacol.*, 6, 111, 1981.

353. **Owellen, R. J. and Donigan, D. W.,** [³H]Vincristine. Preparation and preliminary pharmacology, *J. Med. Chem.*, 15, 894, 1972.

354. **Castle, M. C. and Mead, J. A. R.,** Investigations of the metabolic fate of tritiated vincristine in the rat by high-pressure liquid chromatography, *Biochem. Pharmacol.*, 27, 37, 1978.

355. **Jackson, D. V., Castle, M. C., and Bender, R. A.,** Biliary excretion of vincristine, *Clin. Pharmacol. Ther.*, 24, 101, 1978.

356. **Rahmani, R., Somadossi, J. P., Martin, M., Cano, J. P., and Barbet, J.,** Evidence for vincristine (VCR) metabolism in freshly isolated rat hepatocytes, *Proc. Am. Assoc. Cancer Res.*, 26, 237, 1985.

357. **Sethi, V. S. and Thimmaiah, K. N.,** Structural studies on the degradation products of vincristine dihydrogen sulfate, *Cancer Res.*, 45, 5386, 1985.

358. **Sethi, V. S., Castle, M. C., Surratt, P., Jackson, D. V., and Spurr, C. L.,** Isolation and partial characterization of human urinary metabolites of vincristine sulphate, *Proc. Am. Assoc. Cancer Res.*, 22, 173, 1981.

359. **Kennedy, D. G., Armstrong, P., Karim, A., Van Den Berg, H. W., and Murphy, R. F.,** The metabolism of vincristine *in vitro, Biochem. Soc. Trans.*, 14, 450, 1986.

360. **Earle, D. P.,** Cinchona alkaloids. V. Physiological disposition of cinchonine metabolic products in man, *Fed. Proc. Fed. Am. Soc. Exp. Biol.*, 5, 175, 1946.

361. **Earle, D. P., Welch, W. J., and Shannon, J. A.,** Studies on the chemotherapy of the human malarias. IV. The metabolism of cinchonine in relation to its antimalarial activity, *J. Clin. Invest.*, 27, 87, 1948.

362. **Taggart, J. V., Earle, D. P., Berliner, R. W., Zubrod, C. G., Welch, W. J., Wise, N. B., Schroeder, E. F., London, I. M., and Shannon, J. A.,** Studies on the chemotherapy of the human malarias. III. The physiological disposition and antimalarial activity of the cinchona alkaloids, *J. Clin. Invest.*, 27, 80, 1948.

363. **Brodie, B. B., Baer, J. E., and Craig, L. C.,** Metabolic products of the cinchona alkaloids in human urine, *J. Biol. Chem.*, 188, 567, 1951.

364. **Knox, W. E.,** The quinine-oxidizing enzyme and liver aldehyde oxidase, *J. Biol. Chem.*, 163, 699, 1946.

365. **Williams, R. T.,** *Detoxication Mechanisms*, Chapman and Hall, London, 1959, 657.

366. **Watabe, T. and Kiyonaga, K.,** The metabolic fate of quinine in rabbits, *J. Pharm. Pharmacol.*, 24, 625, 1972.

367. **Kelsey, F. E. and Oldham, F. K.,** Studies on antimalarial drugs. The distribution of quinine oxidase in animal tissues, *J. Pharmacol. Exp. Ther.*, 79, 77, 1943.

368. **Kelsey, F. E., Geiling, E. M. K., Oldham, F. K., and Dearborn, E. N.,** Studies on antimalarial drugs. The preparation and properties of a metabolic derivative of quinine, *J. Pharmacol. Exp. Ther.*, 80, 391, 1944.

369. **Mead, J. and Koepfli, J. B.,** The structure of a new metabolic derivative of quinine, *J. Biol. Chem.*, 154, 507, 1944.

370. **Lang, K. and Keuer, N.,** Über die Chininoxydase. I, *Biochem. Z.*, 329, 277, 1957.

371. **Führ, J. and Kaczmarczyk, J.,** Das Schicksal injizierten Chinins im menschlichen Organismus, *Arzneim. Forsch.*, 5, 705, 1955.

372. **Sax, S. M. and Lynch, H. J.,** Oxidation of quinoline by rabbit liver, *J. Pharmacol. Exp. Ther.*, 145, 113, 1964.

373. **Barrow, S. E., Taylor, A. A., Horning, E. C., and Horning, M. G.,** High-performance liquid chromatographic separation and isolation of quinidine and quinine metabolites in rat urine, *J. Chromatogr.*, 181, 219, 1980.

459

374. **Haag, H. B., Larson, P. S., and Schwartz, J. J.,** The effect of urinary pH on the elimination of quinine in man, *J. Pharmacol. Exp. Ther.*, 79, 136, 1943.

375. **Schütz, H. and Hempel, J.,** Renale Ausscheidungsprofile nach der Einnahme chininhaltiger Tonic-Water, *Arch. Toxicol.*, 32, 143, 1974.

376. **Liddle, C., Graham, G. G., Christopher, R. K., Bhuwapathanapun, S., and Duffield, A. M.,** Identification of new urinary metabolites in man of quinine using methane chemical ionization gas chromatography-mass spectrometry, *Xenobiotica*, 11, 81, 1981.

377. **Axelrod, J.,** The enzymic cleavage of aromatic ethers, *Biochem. J.*, 63, 634, 1956.

378. **Horning, M. G., Taylor, A. A., Horning, E. C., and Barrow, S. E.,** Isolation and identification of quinidine metabolites by HPLC, GC and GC-MS procedures, *Fed. Proc. Fed. Am. Soc. Exp. Biol.*, 39, 307, 1980.

379. **Jovanović, J., Remberg, G., Ende, M., and Spitelier, G.,** Quinine-N-oxide—a urinary component after the consumption of quinine beverages, *Arch. Toxicol.*, 35, 137, 1976.

380. **Smith, E., Barkan, S., Ross, B., Maienthal, M., and Levine, J.,** Examination of quinidine and quinine and their pharmaceutical preparations, *J. Pharm. Sci.*, 62, 1151, 1973.

381. **Guentert, T. W., Huang, J.-D., and Øye, S.,** Disposition of quinidine in the rabbit, *J. Pharm. Sci.*, 71, 812, 1982.

382. **Bonora, M. R., Guentert, T. W., Upton, R. A., and Riegelman, S.,** Determination of quinidine and metabolites in urine by reverse-phase high-pressure liquid chromatography, *Clin. Chim. Acta*, 91, 277, 1979.

383. **Rakhit, A., Kunitani, M., Holford, N. H. G., and Riegelman, S.,** Improved liquid-chromatographic assay of quinidine and its metabolites in biological fluids, *Clin. Chem.*, 28, 1505, 1982.

384. **Thompson, K. A., Murray, J. J., Blair, I. A., Woosley, R. L., and Roden, D. M.,** Plasma concentrations of quinidine, its major metabolites, and dihydroquinidine in patients with torsades de pointes, *Clin. Pharmacol. Ther.*, 43, 636, 1988.

385. **Palmer, K. H., Martin, B., Baggett, B., and Wall, M. E.,** The metabolic fate of orally administered quinidine gluconate in humans, *Biochem. Pharmacol.*, 18, 1845, 1969.

386. **Carroll, F. I., Smith, D., Wall, M. E., and Moreland, C. G.,** Carbon-13 magnetic resonance study. Structure of the metabolites of orally administered quinidine in humans, *J. Med. Chem.*, 17, 985, 1974.

387. **Beermann, B., Leander, K., and Lindström, B.,** The metabolism of quinidine in man: structure of a main metabolite, *Acta Chem. Scand.*, B30, 465, 1976.

388. **Carroll, F. I., Philip, A., and Coleman, M. C.,** Synthesis and stereochemistry of a metabolite resulting from the biotransformation of quinidine in man, *Tetrahedron Lett.*, 1757, 1976.

389. **Drayer, D. E., Cook, C. E., and Reidenberg, M. M.,** Active quinidine metabolites, *Clin. Res.*, 24, 623, 1976.

390. **Drayer, D. E., Lowenthal, D. T., Restivo, K. M., Schwartz, A., Cook, C. E., and Reidenberg, M. M.,** Steady-state serum levels of quinidine and active metabolites in cardiac patients with varying degrees of renal function, *Clin. Pharmacol. Ther.*, 24, 31, 1978.

391. **Rakhit, A., Holford, N. H. G., Effeney, D. J., and Riegelman, S.,** Induction of quinidine metabolism and plasma protein binding by phenobarbital in dogs, *J. Pharmacokinet. Biopharm.*, 12, 495, 1984.

392. **Maloney, K., Sadee, W., and Castagnoli, N.,** Covalent binding of quinidine, in *Pharmacokinetics, Proc. Sidney Riegelman Memorial Symposium, 1982*, Benet, L. Z., Levy, G. and Ferraiolo, B. L., Eds., Plenum Press, New York, 1984, 471 (Chem. Abstr. 103, 101d, 1985).

393. **Guentert, T. W., Daly, J. J., and Riegelman, S.,** Isolation, characterisation and synthesis of a new quinidine metabolite, *Eur. J. Drug Metab. Pharmacokinet.*, 7, 31, 1982.

394. **Guengerich, F. P., Müller-Enoch, D., and Blair, I. A.,** Oxidation of quinidine by human liver cytochrome P-450, *Mol. Pharmacol.*, 30, 287, 1986.

395. **Mikus, G., Ha, H. R., Vožeh, S., Zekorn, C., Follath, F., and Eichelbaum, M.,** Pharmacokinetics and metabolism of quinidine in extensive and poor metabolisers of sparteine, *Eur. J. Clin. Pharmacol.*, 31, 69, 1986.

396. **Otton, S. V., Brinn, R. U., and Gram, L. F.,** *In vitro* evidence against the oxidation of quinidine by the sparteine/debrisoquine monooxygenase of human liver, *Drug Metab. Dispos.*, 16, 15, 1988.

397. **Leferink, J. G., Maes, R. A. A., Sunshine, I., and Forney, R. B.,** A novel quinidine metabolism in a suicide case with quinidine sulphate detected by gas chromatography-mass spectrometry, *J. Anal. Toxicol.*, 1, 62, 1977.

398. **Ueda, C. T., Williamson, B. J., and Dzindzio, B. S.,** Disposition kinetics of dihydroquinidine following quinidine administration, *Res. Commun. Chem. Pathol. Pharmacol.*, 14, 215, 1976.

399. **Resplandy, G., Roux, A., Dupas, I., Viel, C., Plat, M., and Flouvat, B.,** Specific determination of hydroquinidine and its major metabolites in biological fluids by high performance liquid chromatography, *J. Liq. Chromatogr.*, 11, 1495, 1988.

400. **Kametani, T., Ihara, M., and Takahashi, K.,** Biotransformation of (+)-reticuline in rat, *Chem. Pharm. Bull.*, 20, 1587, 1972.

401. **Kametani, T., Takemura, M., Ihara, M., Takahashi, K., and Fukumoto, K.,** Biotransformation of 1-benzyl-1,2,3,4-tetrahydro-2-methylisoquinolines into tetrahydroprotoberberines with rat liver enzymes, *J. Am. Chem. Soc.*, 98, 1956, 1976.

402. **Kametani, T., Ohta, T., Takemura, M., Ihara, M., and Fukumoto, K.,** Biotransformation of reticuline into morphinandienone, aporphine and protoberberine alkaloids with rat liver enzyme, *Heterocycles*, 6, 415, 1977.

403. **Cashaw, J. L., McMurtrey, K. D., Brown, H., and Davis, V. E.,** Identification of catecholamine-derived alkaloids in mammals by gas chromatography and mass spectrometry, *J. Chromatogr.*, 99, 567, 1974.

404. **Davis, V. E., Cashaw, J. L., and McMurtrey, K. D.,** Disposition of catecholamine-derived alkaloids in mammalian systems, in *Alcohol Intoxication and Withdrawal. Experimental Studies II*, Vol. 59, Gross, M. M., Ed., Plenum Press, New York, 1975, 65.

405. **Williams, D. A., Maher, T. J., and Zaveri, D. C.,** Identification of a tetrahydroprotoberberine as a metabolite of trimetoquinol in the rat, *Biochem. Pharmacol.*, 32, 1447, 1983.

406. **Meyerson, L. R. and Davis, V. E.,** Purification and characterization of a benzyltetrahydroisoquinoline methyltransferase from rat liver, *Fed. Proc. Fed. Am. Soc. Exp. Biol.*, 34, 508, 1975.

407. **Canfell, P. C., Castagnoli, N., Fahey, M. R., Hennis, P. J., and Miller, R. D.,** The metabolic disposition of laudanosine in dog, rabbit, and man, *Drug Metab. Dispos.*, 14, 703, 1986.

408. **Sharma, M., Fahey, M. R., Castagnoli, K., Shi, W.-Z., and Miller, R. D.,** In vitro metabolic studies of atracurium with rabbit liver preparations, *Anesthesiology*, 61, A304, 1984.

409. **Zahn, K.,** Über das Schicksal des Papaverins im tierischen Organismus, *Biochem. Z.*, 68, 444, 1915.

410. **Lévy, J.,** Sur la destinée de la papavérine dans l'organisme, *Bull. Soc. Chim. Biol.*, 27, 578, 1945.

411. **Axelrod, J., Shofer, R., Inscoe, J. K., King, W. M., and Sjoerdsma, A.,** The fate of papaverine in man and other mammals, *J. Pharmacol. Exp. Ther.*, 124, 9, 1958.

412. **Belpaire, F. M. and Bogaert, M. G.,** Metabolism of papaverine. II. Species differences, *Xenobiotica*, 5, 421, 1975.

413. **Elek, S. R. and Bergman, H. C.,** Rate of papaverine metabolism in man, *J. Appl. Physiol.*, 6, 168, 1953.

414. **Guttman, D. E., Kostenbauder, H. B., Wilkinson, G. R., and Dubé, P. H.,** GLC determination of papaverine in biological fluids, *J. Pharm. Sci.*, 63, 1625, 1974.

415. **Wilén, G. and Ylitalo, P.,** Metabolism of [14C]papaverine in man, *J. Pharm. Pharmacol.*, 34, 264, 1981.

416. **Belpaire, F. M. and Bogaert, M. G.,** Metabolism of papaverine. III. Effect of phenobarbital, 3-methylcholanthrene and SKF 525. A pre-treatment *in vivo* and *in vitro*, *Xenobiotica*, 5, 431, 1975.

417. **Rosazza, J. P., Kammer, M., Youel, L., Smith, R. V., Erhardt, P. W., Truong, D. H., and Leslie, S. W.,** Microbial models of mammalian metabolism. *O*-Demethylation of papaverine, *Xenobiotica*, 7, 133, 1977.

418. **Belpaire, F. M. and Bogaert, M. G.,** Excretion of H3-papaverine in the rat, *Arch. Int. Pharmacodyn. Ther.*, 199, 191, 1972.

419. **Belpaire, F. M. and Bogaert, M. G.,** The excretion of 3H-papaverine in the rat, *Biochem. Pharmacol.*, 22, 59, 1973.

420. **Belpaire, F. M., Bogaert, M. G., Rosseel, M. T., and Anteunis, M.,** Metabolism of papaverine. I. Identification of metabolites in rat bile, *Xenobiotica*, 5, 413, 1975.

421. **Belpaire, F. M. and Bogaert, M. G.,** Species differences in the excretion of papaverine metabolites, *Arch. Int. Pharmacodyn. Ther.*, 208, 362, 1974.

422. **Rosseel, M. T. and Belpaire, F. M.,** Identification of papaverine metabolites in human urine by mass spectrometry, in *Quantitative Mass Spectrometry in Life Sciences*, De Leenheer, A. P. and Roncucci, R. R., Eds., Elsevier, Amsterdam, 1977, 133.

423. **Belpaire, F. M., Rosseel, M. T., and Bogaert, M. G.,** Metabolism of papaverine. IV. Urinary elimination of papaverine metabolites in man, *Xenobiotica*, 8, 297, 1978.

424. **Cooper, N. and Hatcher, R. A.,** A contribution to the pharmacology of narcotine, *J. Pharmacol. Exp. Ther.*, 51, 411, 1934.

425. **Nayak, K. P., Brochmann-Hanssen, E., and Way, E. L.,** Biological disposition of noscapine. I. Kinetics of metabolism, urinary excretion, and organ distribution, *J. Pharm. Sci.*, 54, 191, 1965.

426. **Idänpään-Heikkilä, J. E.,** Studies on the fate of 3H-noscapine in mice and rats, *Ann. Med. Exp. Biol. Fenn.*, 46, 201, 1968.

427. **Vedsø, S.,** Absorption and excretion of noscapine, *Acta Pharmacol. Toxicol.*, 18, 157, 1961.

428. **Tsunoda, N. and Yoshimura, H.,** Metabolic fate of noscapine. II. Isolation and identification of novel metabolites produced by C–C bond cleavage, *Xenobiotica*, 9, 181, 1979.

429. **Tsunoda, N. and Yoshimura, H.,** Metabolic fate of noscapine. III. Further studies on identification and determination of the metabolites, *Xenobiotica*, 11, 23, 1981.

430. **Idänpään-Heikkilä, J. E.,** An autoradiographic study on the distribution of 3H-noscapine in mice, *Eur. J. Pharmacol.*, 2, 26, 1967.

431. **Olsson, B., Bolme, P., Dahlström, B., and Marcus, C.,** Excretion of noscapine in human breast milk, *Eur. J. Clin. Pharmacol.*, 30, 213, 1986.

432. **Nayak, P. K.,** Studies on the metabolism of noscapine, *Diss. Abstr.*, 26, 5735, 1966.

433. **Tsunoda, N., Yoshimura, H., and Kozuka, H.,** Metabolic fate of noscapine. I. Detection of the metabolites in rabbit urine, *J. Hyg. Chem.*, 22, 280, 1976.

434. **Göber, B., Brandt, K.-P., Pfeifer, S., and Otto, A.,** Zur Biotransformation von Narcotin, *Pharmazie*, 32, 543, 1977.

435. **Göber, B., Pfeiffer, S., Pankow, K., and Kraft, R.,** Biotransformation und Stabilität von Hydrocotarnin und Diazocarbonylhydrocotarnin-bzw.-hydrohydrastininverbindungen, *Pharmazie*, 34, 830, 1979.

436. **Gimble, A. I., Davison, C., and Smith, P. K.,** Studies on the toxicity, distribution and excretion of emetine, *J. Pharmacol. Exp. Ther.*, 94, 431, 1952.

437. **Gehlert, D. R., Ferrell, K. G., and Rollins, D. E.,** Disposition of emetine in the rat, *Pharmacologist*, 25, 112, 1983.

438. **Schwartz, D. E. and Herrero, J.,** Comparative pharmacokinetic studies of dehydroemetine and emetine in guinea pigs using spectrofluorometric and radioactive methods, *Am. J. Trop. Med.*, 14, 78, 1965.

439. **Peeples, A. and Dalvi, R. R.,** Biotransformation of some alkaloids by liver microsomes, *Toxicol. Appl. Pharmacol.*, 41, 167, 1977.

440. **Johnson, R. K., Wynn, W. T., and Jondorf, W. R.,** Some aspects of the metabolism of ^{14}C-labelled (±)-2,3-dehydroemetine in the rat, *Biochem. J.*, 125, 26, 1971.

441. **Johnson, R. K. and Jondorf, W. R.,** Studies on the metabolic fate of (^{14}C)-labelled (±)-2,3-dehydroemetine in the rat, *Xenobiotica*, 3, 85, 1973.

442. **Hakim, S. A. E., Mijović, V., and Walker, J.,** Experimental transmission of sanguinarine in milk: detection of a metabolic product, *Nature (London)*, 189, 201, 1961.

443. **Fels, J.-P., Lechat, P., Rispe, R., and Cautreels, W.,** Determination of glaucine in plasma and urine by high-performance liquid chromatography, *J. Chromatogr.*, 308, 273, 1984.

444. **Yokoshima, T., Tsutsumi, S., Ohtsuki, T., Takaichi, M., Nakajima, T., and Akasu, M.,** Studies on the metabolic fate of cepharanthine. Absorption, distribution, metabolism and excretion in rats, *Iyakuhin Kenkyu*, 17, 458, 1986 (Chem. Abstr. 105, 90761n, 1986).

445. **Lin, M.-B., Zhang, W., Zhao, X.-W., Lu, J.-X., Wang, M., and Chen, L.-G.,** Biotransformation of tetrandrine in rats and in men, *Acta Pharm. Sin.*, 17, 728, 1982.

446. **Waser, P., Schmid, H., and Schmid, K.,** Resorption, Verteilung und Ausscheidung von radio-Calebassen-Curarin bei Katzen, *Naunyn Schmiedebergs Arch. Exp. Pathol. Pharmakol.*, 96, 368, 1954.

447. **Mahfouz, M.,** The fate of tubocurarine in the body, *Br. J. Pharmacol.*, 4, 295, 1949.

448. **Marsh, D. F.,** The distribution, metabolism, and excretion of D-tubocurarine chloride and related compounds in man and other animals, *J. Pharmacol. Exp. Ther.*, 105, 299, 1952.

449. **Foldes, F. F.,** The fate of muscle relaxants in man, *Acta Anaesth. Scand.*, 1, 63, 1957.

450. **Kalow, W.,** The distribution, destruction and elimination of muscle relaxants, *Anesthesiology*, 20, 505, 1959.

451. **Chagas, C.,** The fate of curare during curarization, in *Curare and Curare-like Agents*, De Reuck, A. V. S., Ed., J. & A. Churchill, London, 1962, 2.

452. **Cohen, E. N., Corbascio, A., and Fleischli, G.,** The distribution and fate of *d*-tubocurarine, *J. Pharmacol. Exp. Ther.*, 147, 120, 1965.

453. **Cohen, E. N., Hood, N., and Golling, R.,** Use of whole-body autoradiography for determination of uptake and distribution of labeled muscle relaxants in the rat, *Anesthesiology*, 29, 987, 1968.

454. **Crankshaw, D. P. and Cohen, E. N.,** Uptake, distribution and elimination of skeletal muscle relaxants, in *Monographs in Anesthesiology, Muscle Relaxants*, Vol. 3, Katz, R. L., Ed., Excerpta Medica, Amsterdam, 1975, 125.

455. **Kalow, W.,** Urinary excretion of *d*-tubocurarine in man, *J. Pharmacol. Exp. Ther.*, 109, 74, 1953.

456. **Matteo, R. S., Nishitateno, K., Pua, E. K., and Spector, S.,** Pharmacokinetics of *d*-tubocurarine in man: effect of an osmotic diuretic on urinary excretion, *Anesthesiology*, 52, 335, 1980.

457. **Cohen, E. N., Brewer, H. W., and Smith, D.,** The metabolism and elimination of *d*-tubocurarine-H^3, *Anesthesiology*, 28, 309, 1967.

458. **Robelet, A., Bizard-Gregoire, N., and Bizard, J.,** Fate of *d*-tubocurarine, gallamine, and succinylcholine during a perfusion of isolated rat liver, *C. R. Séanc. Soc. Biol.*, 158, 1100, 1964.

459. **Meijer, D. K. F., Weitering, J. G., and Vonk, R. J.,** Hepatic uptake and biliary excretion of *d*-tubocurarine and trimethyltubocurarine in the rat *in vivo* and in isolated perfused rat livers, *J. Pharmacol. Exp. Ther.*, 198, 229, 1976.

460. **Waser, P. G. and Lüthi, U.,** Distribution and metabolism of curarizing and depolarizing drugs in cats, *J. Nucl. Biol. Med.*, 12, 4, 1968.

461. **Palm, P. E., Nick, M. S., Arnold, E. P., Yesair, D. W., and Callahan, M. M.,** Single- and repeated-dose toxicity of thalicarpine (NSC 68075) in monkeys. (A comparison with dogs), including physiological disposition studies, in *U. S. Natl. Tech. Inform. Serv.*, PB Rep. 1971, No. 201914, 1971, 164 (Chem. Abstr. 76, 68093g, 1972).

462. **Smellie, M., Corder, M., and Rosazza, J. P.,** High-performance liquid chromatography of thalicarpine, hernandaline, hernandalinol and dehydrothalicarpine, *J. Chromatogr.*, 155, 439, 1978.

463. **Way, E. L. and Adler, T. K.,** The biological disposition of morphine and its surrogates. I, *Bull. WHO*, 25, 227, 1961.

464. **Way, E. L. and Adler, T. K.,** The biological disposition of morphine and its surrogates. II, *Bull. WHO*, 26, 51, 1962.

465. **Mulé, S. J.,** Physiological disposition of narcotic agonists and antagonists, in *Narcotic Drugs. Biochemical Pharmacology*, Clouet, D. H., Ed., Plenum Press, New York, 1971, 99.

466. **Scrafani, J. T. and Clouet, D. H.,** Biotransformations, in *Narcotic Drugs. Biochemical Pharmacology,* Clouet, D. H., Ed., Plenum Press, New York, 1971, 137.

467. **Misra, A. L.,** Disposition and metabolism of drugs of dependence, in *Chemical and Biological Aspects of Drug Dependence,* Mulé, S. J. and Brill, H., Eds., CRC Press, Boca Raton, FL, 1972, 219.

468. **Boerner, U., Abbott, S., and Roe, R. L.,** The metabolism of morphine and heroin in man, *Drug Metab. Rev.,* 4, 39, 1975.

469. **Lemberger, L. and Rubin, A.,** Morphine and morphine substitutes, in *Physiologic Disposition of Drugs of Abuse,* Spectrum Publications, New York, 1976, 125.

470. **Iwamoto, K. and Klaassen, C. D.,** First-pass effect of morphine in rats, *J. Pharmacol. Exp. Ther.,* 200, 236, 1976.

471. **Brunk, S. F. and Delle, M.,** Morphine metabolism in man, *Clin. Pharmacol. Ther.,* 16, 51, 1974.

472. **Zauder, H. L.,** The effect of prolonged morphine administration on the *in vivo* and *in vitro* conjugation of morphine by rats, *J. Pharmacol. Exp. Ther.,* 104, 11, 1952.

473. **Misra, A. L., Mulé, S. J., and Woods, L. A.,** The preparation of tritium nuclear-labeled morphine and evidence for its *in vivo* biotransformation to normorphine in the rat, *J. Pharmacol. Exp. Ther.,* 132, 317, 1961.

474. **Abrams, L. S. and Elliott, H. W.,** Morphine metabolism *in vivo* and *in vitro* by homozygous Gunn rats, *J. Pharmacol. Exp. Ther.,* 189, 285, 1974.

475. **Oguri, K., Ida, S., Yoshimura, H., and Tsukamoto, H.,** Metabolism of drugs. LXIX. Studies on the urinary metabolites of morphine in several mammalian species, *Chem. Pharm. Bull.,* 18, 2414, 1970.

476. **Klutch, A.,** A chromatographic investigation of morphine metabolism in rats. Confirmation of *N*-demethylation of morphine and isolation of a new metabolite, *Drug Metab. Dispos.,* 2, 23, 1974.

477. **Milthers, K.,** The *N*-demethylation of morphine in rats. Quantitative determination of normorphine and morphine in the urine and faeces of rats given subcutaneous morphine, *Acta Pharmacol. Toxicol.,* 19, 149, 1962.

478. **Yeh, S. Y., McQuinn, R. L., and Gorodetzky, C. W.,** Biotransformation of morphine to dihydromorphinone and normorphine in the mouse, rat, rabbit, guinea pig, cat, dog, and monkey, *Drug Metab. Dispos.,* 5, 335, 1977.

479. **Axelrod, J. and Inscoe, J. K.,** Glucuronide formation of narcotic drugs *in vitro* and *in vivo, Proc. Soc. Exp. Biol. Med.,* 103, 675, 1960.

480. **Keeser, E., Oelkers, H. A., and Raetz, W.,** Über das Schicksal des Morphins im Tierkörper, *Naunyn Schmiedebergs Arch. Exp. Pathol. Pharmakol.,* 173, 622, 1933.

481. **Yosikawa, N.,** Morphine elimination in the urine and feces by various animal species, *Jpn. J. Med. Sci.,* 12, 74, 1940 (Chem. Abstr. 34, 74294, 1940).

482. **Yeh, S. Y., Chernov, H. I., and Woods, L. A.,** Metabolism of morphine by cats, *J. Pharm. Sci.,* 60, 469, 1971.

483. **Misra, A. L., Yeh, S. Y., and Woods, L. A.,** Morphine conjugates in the dog, *Biochem. Pharmacol.,* 19, 1536, 1970.

484. **Mellett, L. B. and Woods, L. A.,** The distribution and fate of morphine in the non-tolerant and tolerant monkey, *J. Pharmacol. Exp. Ther.,* 116, 77, 1956.

485. **Mellett, L. B. and Woods, L. A.,** Excretion of urinary *N*-C[14]-methyl morphine and pulmonary C[14]O$_2$ in the monkey and dog after subcutaneous injection of the labelled drug, *Proc. Soc. Exp. Biol. Med.,* 106, 221, 1961.

486. **Yeh, S. Y.,** Urinary excretion of morphine and its metabolites in morphine-dependent subjects, *J. Pharmacol. Exp. Ther.,* 192, 201, 1975.

487. **Smith, R. L.,** *The Excretory Function of Bile. The Elimination of Drugs and Toxic Substances in Bile,* Chapman and Hall, London, 1973.

488. **Peterson, R. E. and Fujimoto, J. M.,** Biliary excretion of morphine-3-glucuronide and morphine-3-ethereal sulfate by different pathways in the rat, *J. Pharmacol. Exp. Ther.,* 184, 409, 1973.

489. **March, C. N. and Elliott, H. W.,** Distribution and excretion of radioactivity after administration of morphine-*N*-methyl C[14] to rats, *Proc. Soc. Exp. Biol. Med.,* 86, 494, 1954.

490. **Smith, D. S., Peterson, R. E. and Fujimoto, J. M.,** Species differences in the biliary excretion of morphine, morphine-3-glucuronide and morphine-3-ethereal sulfate in the cat and rat, *Biochem. Pharmacol.,* 22, 485, 1973.

491. **Woods, L. A.,** Distribution and fate of morphine in non-tolerant and tolerant dogs and rats, *J. Pharmacol. Exp. Ther.,* 112, 158, 1954.

492. **Chernov, H. I. and Woods, L. A.,** Central nervous system distribution and metabolism of C[14]-morphine during morphine-induced feline mania, *J. Pharmacol. Exp. Ther.,* 149, 146, 1965.

493. **Elliott, H. W., Tolbert, B. M., Adler, T. K., and Anderson, H. H.,** Excretion of carbon-14 by man after administration of morphine-*N*-methyl-C[14], *Proc. Soc. Exp. Biol. Med.,* 85, 77, 1954.

494. **Walsh, C. T. and Levine, R. R.,** Studies of the enterohepatic circulation of morphine in the rat, *J. Pharmacol. Exp. Ther.,* 195, 303, 1975.

495. **Parker, R. J., Hirom, P. C., and Millburn, P.,** Enterohepatic recycling of phenolphthalein, morphine, lysergic acid diethylamide (LSD) and diphenylacetic acid in the rat. Hydrolysis of glucuronic acid conjugates in the gut lumen, *Xenobiotica,* 10, 689, 1980.

496. **Gross, E. G. and Thompson, V.,** The excretion of a combined form of morphine in tolerant and non-tolerant dogs, *J. Pharmacol. Exp. Ther.,* 68, 413, 1940.

497. **Oberst, F. W.,** Free and bound morphine in the urine of morphine addicts, *J. Pharmacol. Exp. Ther.*, 69, 240, 1940.

498. **Oberst, F. W.,** Relationship of the chemical structure of morphine derivatives to their urinary excretion in free and bound forms, *J. Pharmacol. Exp. Ther.*, 73, 401, 1941.

499. **Seibert, R. A., Williams, C. E., and Huggins, R. A.,** The isolation and identification of "bound" morphine, *Science*, 120, 222, 1954.

500. **Fujimoto, J. M. and Way, E. L.,** Characterization of a bound form of morphine in human urine, *Fed. Proc. Fed. Am. Soc. Exp. Biol.*, 13, 356, 1954.

501. **Fujimoto, J. M. and Way, E. L.,** Isolation and crystallization of "bound" morphine from urine of human addicts, *J. Pharmacol. Exp. Ther.*, 121, 340, 1957.

502. **Capel, I. D., Millburn, P., and Williams, R. T.,** The conjugation of 1- and 2-naphthols and other phenols in the cat and pig, *Xenobiotica*, 4, 601, 1974.

503. **Yoshimura, H., Oguri, K., and Tsukamoto, H.,** Metabolism of drugs. LXII. Isolation and identification of morphine glucuronides in urine and bile of rabbits, *Biochem. Pharmacol.*, 18, 279, 1969.

504. **Yeh, S. Y., Gorodetzky, C. W., and Krebs, H. A.,** Isolation and identification of morphine 3- and 6-glucuronides, morphine 3,6-diglucuronide, morphine 3-ethereal sulfate, normorphine, and normorphine 6-glucuronide as morphine metabolites in humans, *J. Pharm. Sci.*, 66, 1288, 1977.

505. **Hand, C. W., Moore, R. A., McQuay, H. J., Allen, M. C., and Sear, J. W.,** Analysis of morphine and its major metabolites by differential radioimmunoassay, *Ann. Clin. Biochem.*, 24, 153, 1987.

506. **Säwe, J., Svensson, J. O., and Rane, A.,** Morphine metabolism in cancer patients on increasing oral doses—no evidence for autoinduction or dose-dependence, *Br. J. Clin. Pharmacol.*, 16, 85, 1983.

507. **Rane, A., Säwe, J., Lindberg, B., Svensson, J.-O., Garle, M., Erwald, R., and Jorulf, H.,** Morphine glucuronidation in the rhesus monkey: a comparative *in vivo* and *in vitro* study, *J. Pharmacol. Exp. Ther.*, 229, 571, 1984.

508. **Ida, S., Oguri, K., and Yoshimura, H.,** Metabolism of drugs. LXXXIII. Metabolism of morphine glucuronides in rats, *J. Pharm. Soc. Jpn.*, 95, 564, 1975.

509. **Ida, S., Oguri, K., and Yoshimura, H.,** Metabolism of drugs. LXXXIV. Metabolism of morphine glucuronides in rabbits, guinea pigs and mice, *J. Pharm. Soc. Jpn.*, 95, 570, 1975.

510. **Jacqz, E., Ward, S., Johnson, R., Schenker, S., Gerkens, J., and Branch, R. A.,** Extrahepatic glucuronidation of morphine in the dog, *Drug Metab. Dispos.*, 14, 627, 1986.

511. **Fujimoto, J. M. and Haarstad, V. B.,** The isolation of morphine ethereal sulfate from urine of the chicken and cat, *J. Pharmacol. Exp. Ther.*, 165, 45, 1969.

512. **Strominger, J. L., Kalckar, H. M., Axelrod, J., and Maxwell, E. S.,** Enzymatic oxidation of uridine diphosphate glucose to uridine diphosphate glucuronic acid, *J. Am. Chem. Soc.*, 76, 6411, 1954.

513. **Sanchez, E. and Tephly, T. R.,** Studies on morphine glucuronide synthesis in rat liver microsomes, *Fed. Proc. Fed. Am. Soc. Exp. Biol.*, 32, 763, 1973.

514. **Miners, J. O., Lillywhite, K. J., and Birkett, D. J.,** *In vitro* evidence for the involvement of at least two forms of human liver UDP-glucuronosyltransferase in morphine 3-glucuronidation, *Biochem. Pharmacol.*, 37, 2839, 1988.

515. **Pacifici, G. M., Säwe, J., Kager, L., and Rane, A.,** Morphine glucuronidation in human fetal and adult liver, *Eur. J. Clin. Pharmacol.*, 22, 553, 1982.

516. **Pacifici, G. M. and Rane, A.,** Renal glucuronidation of morphine in the human foetus, *Acta Pharmacol. Toxicol.*, 50, 155, 1982.

517. **Yue, Q., Odar-Cederlöf, I., Svensson, J.-O., and Säve, J.,** Glucuronidation of morphine in human kidney microsomes, *Pharmacol. Toxicol.*, 63, 337, 1988.

518. **Wahlström, A., Winblad, B., Bixo, M., and Rane, A.,** Human brain metabolism of morphine and naloxone, *Pain*, 35, 121, 1988.

519. **Koster, A. S., Frankhuijzen-Sierevogel, A. C., and Noordhoek, J.,** Distribution of glucuronidation capacity (1-naphthol and morphine) along the rat intestine, *Biochem. Pharmacol.*, 34, 3527, 1985.

520. **Letelier, M. E., Sánchez, E., and del Villar, E.,** Enhanced metabolism of morphine in *Octodon degus* compared to Wistar rats, *Gen. Pharmacol.*, 15, 403, 1984.

521. **Coimbra-Farges, R., Puget, A., Monsarrat, B., Moisand, C., and Meunier, J.-C.,** Morphine metabolism in the naturally morphine-tolerant Afghan pika: a preliminary study, *Life. Sci.*, 46, 663, 1990.

522. **Woods, L. A. and Chernov, H. I.,** A new morphine metabolite—from cat urine, *Pharmacologist*, 8, 206, 1966.

523. **Mori, M.-A., Oguri, K., Yoshimura, H., Shimomura, K., Kamata, O., and Ueki, S.,** Chemical synthesis and analgesic effect of morphine ethereal sulfates, *Life Sci.*, 11(I), 525, 1972.

524. **Yeh, S. Y.,** Isolation and identification of morphine ethereal sulfate (MES) normorphine (NM) and normorphine conjugate (NMC) as morphine (M) metabolites in man, *Fed. Proc. Fed. Am. Soc. Exp. Biol.*, 32, 763, 1973.

525. **Misra, A. L. and Woods, L. A.,** Evidence for interaction *in vitro* of morphine with glutathione, *Nature (London)*, 228, 1226, 1970.

526. **James, R. C., Goodman, D. R., and Harbison, R. D.,** Hepatic glutathione and hepatotoxicity: changes induced by selected narcotics, *J. Pharmacol. Exp. Ther.*, 221, 708, 1982.

527. **Correia, M. A., Wong, J. S., and Soliven, E.,** Morphine metabolism revisted. I. Metabolic activation of morphine to a reactive species in rats, *Chem. Biol. Interact.*, 49, 255, 1984.

528. **Correira, M. A., Krowech, G., Caldera-Munoz, P., Yee, S. L., Straub, K., and Castagnoli, N.,** Morphine metabolism revisited. II. Isolation and chemical characterization of a glutathionylmorphine adduct from rat liver microsomal preparations, *Chem. Biol. Interact.*, 51, 13, 1984.

529. **Krowech, G., Caldera-Munoz, P. S., Straub, K., Castagnoli, N., and Correia, M. A.,** Morphine metabolism revisited. III. Confirmation of a novel metabolic pathway, *Chem. Biol. Interact.*, 58, 29, 1986.

530. **Ishida, T., Kumagai, Y., Ikeda, Y., Ito, K., Yano, M., Toki, S., Mihashi, K., Fujioka, T., Iwase, Y., and Hachiyama, S.,** (8S)-(Glutathion-S-yl)dihydromorphininone, a novel metabolite of morphine from guinea pig bile, *Drug Metab. Dispos.*, 17, 77, 1989.

531. **Yamano, S., Kageura, E., Ishida, T., and Toki, S.,** Purification and characterization of guinea pig liver morphine 6-dehydrogenase, *J. Biol Chem.*, 260, 5259, 1985.

532. **Nagamatsu, K., Ohno, Y., Ikebuchi, H., Takahashi, A., Terao, T., and Takanaka, A.,** Morphine metabolism in isolated rat hepatocytes and its implications for hepatotoxicity, *Biochem. Pharmacol.*, 35, 3543, 1986.

533. **Nagamatsu, K., Kido, Y., Terao, T., Ishida, T., and Toki, S.,** Studies on the mechanism of covalent binding of morphine metabolites to proteins in mouse, *Drug Metab. Dispos.*, 11, 190, 1983.

534. **March, C. H. and Elliott, H. W.,** Biological studies with radioactive morphine, *Fed. Proc. Fed. Am. Soc. Exp. Biol.*, 11, 373, 1952.

535. **Axelrod, J.,** The enzymatic N-demethylation of narcotic drugs, *J. Pharmacol. Exp. Ther.*, 117, 322, 1956.

536. **Gutierrez, I. F. and Flaine, L. A.,** Metabolism of morphine: normorphine and derivatives as metabolites of morphine, *An. Fac. Quim. Farm. Univ. Chile*, 10, 191, 1958 (Chem. Abstr. 54, 14454i, 1960).

537. **Elison, C. and Elliott, H. W.,** N- and O-demethylation of some narcotic analgesics by brain slices from male and female Long-Evans rats, *Biochem. Pharmacol.*, 12, 1363, 1963.

538. **Phillipson, J. D., El-Dabbas, S. W., and Gorrod, J. W.,** *In vivo* and *in vitro* N-oxidation of morphine and codeine, in *Biological Oxidation of Nitrogen*, Gorrod, J. W., Ed., Elsevier/North-Holland Biomedical Press, Amsterdam, 1978, 125.

539. **Phillipson, J. D., El-Dabbas, S. W., and Gorrod, J. W.,** Metabolism of morphine and codeine by guinea-pig liver microsomal preparations and by rats, *Eur. J. Drug Metab. Pharmacokinet.*, 3, 119, 1978.

540. **Maggiolo, C. and Huidobro, F.,** Studies on morphine metabolism in mice, *Arch. Biol. Med. Exp. (Santiago)*, 4, 99, 1967.

541. **Adler, T. K.,** Studies on morphine tolerance in mice. I. *In vivo* N-demethylation of morphine and N- and O-demethylation of codeine, *J. Pharmacol. Exp. Ther.*, 156, 585, 1967.

542. **Heimans, R. L. H., Fennessy, M. R., and Gaff, G. A.,** Some aspects of the metabolism of morphine-N-oxide, *J. Pharm. Pharmacol.*, 23, 831, 1971.

543. **Misra, A. L., Vadlamani, N. L., Pontani, R. B., and Mulé, S. J.,** Evidence for a new metabolite of morphine-N-methyl-^{14}C in the rat, *Biochem. Pharmacol.*, 22, 2129, 1973.

544. **Milthers, K.,** The in vivo transformation of morphine and nalorphine into normorphine in the brain of rats, *Acta Pharmacol. Toxicol.*, 19, 235, 1962.

545. **Fishman, J., Hahn, E. F., and Norton, B. I.,** N-Demethylation of morphine in rat brain is localized in sites with high opiate receptor content, *Nature (London)*, 261, 64, 1976.

546. **Hahn, E. F., Norton, B. I., and Fishman, J.,** Opiate target site N-demethylase enzymes: differences from the liver N-demethylase, *Biochem. Biophys. Res. Commun.*, 89, 233, 1979.

547. **Tampier, L. and Penna-Herreros, A.,** Normorphine, a metabolite of morphine and nalorphine in the cat, *Arch. Biol. Med. Exp. (Chile)*, 3, 146, 1966.

548. **Boerner, U., Roe, R. L., and Becker, C. E.,** Detection, isolation and chacterization of normorphine and norcodeine as morphine metabolites in man, *J. Pharm. Pharmacol.*, 26, 393, 1974.

549. **Yeh, S. Y., Krebs, H. A., and Gorodetzky, C. W.,** Isolation and identification of morphine N-oxide α- and β-dihydromorphines, β- or γ-isomorphine, and hydroxylated morphine as morphine metabolites in several mammalian species, *J. Pharm. Sci.*, 68, 133, 1979.

550. **Axelrod, J.,** Enzymatic formation of morphine and nicotine in a mammal, *Life Sci.*, 1, 29, 1962.

551. **Clouet, D. H.,** The methylation of normorphine in rat brain and liver, *Life Sci.*, 1, 31, 1962.

552. **Clouet, D. H.,** The methylation of normorphine in rat brain *in vivo*, *Biochem. Pharmacol.*, 12, 967, 1963.

553. **Clouet, D. H., Ratner, M., and Kurzman, M.,** N-Methylation of normorphine by rat tissues *in vitro*, *Biochem. Pharmacol.*, 12, 957, 1963.

554. **Elison, C. and Elliott, H. W.,** Studies on the enzymatic N- and O-demethylation of narcotic analgesics and evidence for the formation of codeine from morphine in rats and dogs, *J. Pharmacol. Exp. Ther.*, 144, 265, 1964.

555. **Yeh, S. Y., McQuinn, R. L., Krebs, H. A., and Gorodetzky, C. W.,** Metabolism and excretion of normorphine in dogs, *J. Pharm. Sci.*, 67, 878, 1978.

556. **Börner, U. and Abbott, S.,** New observations in the metabolism of morphine. The formation of codeine from morphine in man, *Experientia*, 29, 180, 1973.

557. **Yeh, S. Y.,** Absence of evidence of biotransformation of morphine to codeine in man, *Experientia*, 30, 264, 1974.

558. **Yeh, S. Y.,** Question about the formation of norcodeine from morphine in man, *J. Pharm. Pharmacol.*, 27, 214, 1975.

559. **Boerner, U. and Roe, R. L.,** The formation of norcodeine from morphine in man, *J. Pharm. Pharmacol.*, 27, 215, 1975.

560. **Mari, F. and Bertol, E.,** Observations on urinary excretion of codeine in illicit heroin addicts, *J. Pharm. Pharmacol.*, 33, 814, 1981.

561. **Woo, J. T. C., Gaff, G. A., and Fennessy, M. R.,** A note on the effects of 2,4-diamino-5-phenylthiazole and 1,2,3,4-tetrahydro-9-aminoacridine on morphine metabolism, *J. Pharm. Pharmacol.*, 20, 763, 1968.

562. **Misra, A. L. and Mitchell, C. L.,** Determination of morphine-N-methyl-^{14}C oxide in biological materials, its excretion and metabolites in the rat, *Biochem. Med.*, 5, 379, 1971.

563. **Roerig, S., Fujimoto, J. M., and Wang, R. I. H.,** Isolation of hydromorphone and dihydromorphine glucuronides from urine of the rabbit after hydromorphone administration, *Proc. Soc. Exp. Biol. Med.*, 143, 230, 1973.

564. **Yeh, S. Y.,** Isolation of morphine-N-oxide and dihydromorphine as morphine metabolites in the urine of guinea pig, *Pharmacologist*, 18, 178, 1976.

565. **Cone, E. J., Darwin, W. D., and Buchwald, W. F.,** Assay for codeine, morphine and ten potential urinary metabolites by gas chromatography-mass fragmentography, *J. Chromatogr.*, 275, 307, 1983.

566. **Misra, A. L. and Mulé, S. J.,** Disposition and metabolism of [^{3}H]pseudomorphine in the rat, *Biochem. Pharmacol.*, 21, 103, 1972.

567. **Rane, A., Gawronska-Szklarz, B., and Svensson, J.-O.,** Natural (−)- and unnatural (+)-enantiomers of morphine: comparative metabolism and effect of morphine and phenobarbital treatment, *J. Pharmacol. Exp. Ther.*, 234, 761, 1985.

568. **Coughtrie, M. W. H., Ask, B., Rane, A., Burchell, B., and Hume, R.,** The enantioselective glucuronidation of morphine in rats and humans. Evidence for the involvement of more than one UDP-glucuronosyltransferase isoenzyme, *Biochem. Pharmacol.*, 38, 3273, 1989.

569. **Yeh, S. Y. and Woods, L. A.,** Physiologic disposition of N-C^{14}-methyl-codeine in the rat, *J. Pharmacol. Exp. Ther.*, 166, 86, 1969.

570. **Yoshimura, H., Mori, M.-A., Oguri, K., and Tsuksmoto, H.,** Metabolism of drugs. LXV. Studies on the urinary conjugated metabolites of codeine, *Biochem. Pharmacol.*, 19, 2353, 1970.

571. **Yeh, S. Y. and Woods, L. A.,** Excretion of codeine and its metabolites by dogs, rabbits and cats, *Arch. Int. Pharmacodyn. Ther.*, 191, 231, 1971.

572. **Cone, E. J., Darwin, W. D., and Gorodetzky, C. W.,** Comparative metabolism of codeine in man, rat, dog, guinea pig and rabbit: identification of four new metabolites, *J. Pharm. Pharmacol.*, 31, 314, 1979.

573. **Woods, L. A., Muehlenbeck, H. E., and Mellett, L. B.,** Plasma levels and excretion of codeine and metabolites in the dog and monkey, *J. Pharmacol. Exp. Ther.*, 117, 117, 1956.

574. **Nariyuki, H. and Asaki, K.,** Excretion of codeine in human urine, *Kagaku to Sosa*, 12, 175, 1959 (Chem. Abstr. 55, 1921i, 1961).

575. **Bodd, E., Beylich, K. M., Christopherson, A. S., and Mørland, J.,** Oral administration of codeine in the presence of ethanol: a pharmacokinetic study in man, *Pharmacol. Toxicol.*, 61, 297, 1987.

576. **Yue, Q. Y., Svensson, J.-O., Alm, C., Sjöqvist, F., and Säwe, J.,** Interindividual and interethnic differences in the demethylation and glucuronidation of codeine, *Br. J. Clin. Pharmacol.*, 28, 629, 1989.

577. **Yeh, S. Y. and Woods, L. A.,** Isolation and characterization of codeine-6-glucuronide from dog urine, *J. Pharmacol. Exp. Ther.*, 173, 21, 1970.

578. **Chen, Z. R., Reynolds, G., Bochner, F., and Somogyi, A.,** Direct determination of codeine-6-glucuronide in plasma and urine using solid-phase extraction and high-performance liquid chromatography with fluorescence detection, *J. Chromatogr.*, 493, 313, 1989.

579. **Yeh, S. Y. and Woods, L. A.,** Isolation of morphine-3-glucuronide from urine and bile of rats injected with codeine, *J. Pharmacol. Exp. Ther.*, 175, 69, 1970.

580. **Adler, T. K.,** A newly identified metabolic product of codeine: N-demethylated codeine, *J. Pharmacol. Exp. Ther.*, 106, 371, 1952.

581. **Adler, T. K., Fujimoto, J. M., Way, E. L., and Baker, E. M.,** The metabolic fate of codeine in man, *J. Pharmacol. Exp. Ther.*, 114, 251, 1955.

582. **Bodd, E., Christopherson, A. S., and Fongen, U.,** Normorphine is the major metabolite of norcodeine in isolated rat hepatocytes, *Acta Pharmacol. Toxicol.*, 59, 252, 1986.

583. **Kuhn, H. F. and Friebel, H.,** Excretion of codeine and codeine metabolites in urine in codeine-adapted rats and guinea pigs, *Med. Exptl.*, 7, 255, 1962 (Chem. Abstr. 59, 2072f, 1963).

584. **Ebbighausen, W. O. R., Mowat, J., and Vestergaard, P.,** Mass fragmentographic detection of normorphine in urine of man after codeine intake, *J. Pharm. Sci.*, 62, 146, 1973.

585. **Adler, T. K. and Latham, M. E.,** Demethylation of C^{14}-labelled codeine in the rat, *Proc. Soc. Exp. Biol. Med.*, 73, 401, 1950.

586. **Latham, M. E. and Elliott, H. W.,** Some aspects of the metabolism of codeine in the rat, *J. Pharmacol. Exp. Ther.,* 101, 259, 1951.

587. **Adler, T. K. and Shaw, F. H.,** The biological liberation of morphine from codeine in the rat, *J. Pharmacol. Exp. Ther.,* 104, 1, 1952.

588. **Axelrod, J.,** The enzymatic conversion of codeine to morphine, *J. Pharmacol. Exp. Ther.,* 115, 259, 1955.

589. **Takemori, A. E. and Mannering, G. J.,** Metabolic *N*- and *O*-demethylation of morphine and morphinan-type analgesics, *J. Pharmacol. Exp. Ther.,* 123, 171, 1958.

590. **Duquette, P. H., Peterson, F. J., Crankshaw, D. L., Lindemann, N. J., and Holtzman, J. L.,** Studies on the metabolic *N*- and *O*-demethylation of [6-^{3}H]codeine, *Drug Metab. Dispos.,* 11, 477, 1983.

591. **Findlay, J. W. A., Butz, R. F., and Welch, R. M.,** Specific radioimmunoassays for codeine and morphine. Metabolism of codeine to morphine in the rat, *Res. Commun. Chem. Pathol. Pharmacol.,* 17, 595, 1977.

592. **Pærregaard, P.,** The liberation of morphine from codeine in man and dog, *Acta Pharmacol. Toxicol.,* 14, 394, 1958.

593. **Adler, T. K.,** Studies on radiocodeine metabolism in man and in the rat, *J. Pharmacol. Exp. Ther.,* 110, 1, 1954.

594. **Mannering, G. J., Dixon, A. C., Baker, E. M., and Asami, T.,** The in *vivo* liberation of morphine from codeine in man, *J. Pharmacol. Exp. Ther.,* 111, 142, 1954.

595. **Redmond, N. and Parker, J. M.,** Metabolism of codeine: urinary excretion rate of codeine and morphine, *Can. J. Biochem. Physiol.,* 41, 243, 1963.

596. **Quiding, H., Anderson, P., Bondesson, U., Boréus, L. O., and Hynning, P.-Å.,** Plasma concentrations of codeine and its metabolite, morphine, after single and repeated oral administration, *Eur. J. Clin. Pharmacol.,* 30, 673, 1986.

597. **Findlay, J. W. A., Jones, E. C., Butz, R. F., and Welch, R. M.,** Plasma codeine and morphine concentrations after therapeutic oral doses of codeine-containing analgesics, *Clin. Pharmacol. Ther.,* 24, 60, 1978.

598. **Dayer, P., Desmeules, J., Leeman, T., and Striberni, R.,** Bioactivation of the narcotic drug codeine in human liver is mediated by the polymorphic monooxygenase catalyzing debrisoquine 4-hydroxylation (cytochrome P-450 db1/buf1), *Biochem. Biophys. Res. Commun.,* 152, 411, 1988.

599. **Yue, Q. Y., Svensson, J.-O., Alm, C., Sjöqvist, F., and Säwe, J.,** Codeine *O*-demethylation co-segregates with polymorphic debrisoquine hydroxylation, *Br. J. Clin. Pharmacol.,* 28, 639, 1989.

600. **Uba, K., Miyata, N., Watanabe, K., and Hirobe, M.,** Codeine-7,8-oxide (4,5α-epoxy-7β,8β-epoxy-3-methoxy-17-methylmorphinan-6α-ol): identification as a metabolite of codeine, *Chem. Pharm. Bull.,* 28, 356, 1980.

601. **Miyata, N., Uba, K., Watanabe, K., and Hirobe, M.,** Studies on the metabolism of morphine alkaloids: synthesis and analgesic activity of 7,8-epoxide as a new metabolite, *J. Pharmacobio-Dyn.,* 4, 7, 1981.

602. **Misra, A. L., Pontani, R. B., and Mulé, S. J.,** Pharmacokinetics and metabolism of [^{3}H]thebaine, *Xenobiotica,* 4, 17, 1974.

603. **Yamazoe, Y., Numata, H., and Yanagita, T.,** Thebaine metabolites in the urine of rhesus monkeys, *Jpn. J. Pharmacol.,* 31, 433, 1981.

604. **Tencheva, J., Yamboliev, I., and Zhivkova, Z.,** Reversed-phase liquid chromatography for the determination of galanthamine and its metabolites in human plasma and urine, *J. Chromatogr.,* 421, 396, 1987.

605. **Mihailova, D., Velkov, M., and Zhivkova, Z.,** In vitro metabolism of galanthamine hydrobromide (Nivalin) by rat and rabbit liver homogenates, *Eur. J. Drug Metab. Pharmacokinet.,* 12, 25, 1987.

606. **Bull, L. B., Culvenor, C. C. J., and Dick, A. T.,** *The Pyrrolizidine Alkaloids,* North-Holland, Amsterdam, 1968.

607. **Mattocks, A. R.,** *Chemistry and Toxicology of Pyrrolizidine Alkaloids,* Academic Press, London, 1986.

608. **McLean, E. K.,** The toxic action of pyrrolizidine (Senecio) alkaloids, *Pharmacol. Rev.,* 22, 429, 1970.

609. **Huxtable, R. J.,** New aspects of the toxicology and pharmacology of pyrrolizidine alkaloids, *Gen. Pharmacol.,* 10, 159, 1979.

610. **Peterson, J. E. and Culvenor, C. C. J.,** Hepatotoxic pyrrolizidine alkaloids, in *Plant and Fungal Toxins,* Keeler, R. F. and Tu, A. T., Eds., Marcel Dekker, New York, 1983, 637.

611. **Hayashi, Y.,** Excretion and alteration of monocrotaline in rats after a subcutaneous injection, *Fed. Proc. Fed. Am. Soc. Exp. Biol.,* 25, 688, 1966.

612. **Bull, L. B., Culvenor, C. C. J., and Dick, A. T.,** *The Pyrrolizidine Alkaloids,* North-Holland, Amsterdam, 1968, 34.

613. **Bull, L. B., Culvenor, C. C. J., and Dick, A. T.,** *The Pyrrolizidine Alkaloids,* North-Holland, Amsterdam, 1968, 215.

614. **White, I. N. H.,** Excretion of pyrrolic metabolites in the bile of rats given the pyrrolizidine alkaloid retrorsine or the bis-*N*-ethylcarbamate of synthanecine A, *Chem. Biol. Interact.,* 16, 169, 1977.

615. **Lafranconi, W. M., Ohkuma, S., and Huxtable, R. J.,** Biliary excretion of novel pneumotoxic metabolites of the pyrrolizidine alkaloid, monocrotaline, *Toxicon,* 23, 983, 1985.

616. **Hsu, I. C., Robertson, K. A., and Allen, J. R.,** Tissue distribution, binding properties and lesions produced by dehydroretronecine in the nonhuman primate, *Chem. Biol. Interact.,* 12, 19, 1976.

617. **Jago, M. V., Lanigan, G. W., Bingley, J. B., Piercy, D. W. T., Whittem, J. H., and Titchen, D. A.,** Excretion of the pyrrolizidine alkaloid heliotrine in the urine and bile of sheep, *J. Pathol.*, 98, 115, 1969.

618. **Bull, L. B., Culvenor, C. C. J., and Dick, A. T.,** *The Pyrrolizidine Alkaloids*, North-Holland, Amsterdam, 1968, 213.

619. **Eastman, D. F., Dimenna, G. P., and Segall, H. J.,** Covalent binding of two pyrrolizidine alkaloids, senecionine and seneciphylline, to hepatic macromolecules and their distribution, excretion, and transfer into milk of lactating mice, *Drug Metab. Dispos.*, 10, 236, 1982.

620. **Mattocks, A. R.,** Toxicity and metabolism of *Senecio* alkaloids, in *Phytochemical Ecology*, Harborne, J. B., Ed., Academic Press, London, 1972, 179.

621. **Eastman, D. F. and Segall, H. J.,** A new pyrrolizidine alkaloid metabolite, 19-hydroxysenecionine isolated from mouse hepatic microsomes *in vitro*, *Drug Metab. Dispos.*, 10, 696, 1982.

622. **Winter, C. K., Segall, H. J., and Jones, A. D.,** Determination of pyrrolizidine alkaloid metabolites from mouse liver microsomes using tandem mass spectrometry and gas chromatography/mass spectrometry, *Biomed. Environ. Mass Spectrom.*, 15, 265, 1988.

623. **Winter, C. K., Segall, H. J., and Jones, A. D.,** Species differences in the hepatic microsomal metabolism of the pyrrolizidine alkaloid senecionine, *Comp. Biochem. Physiol.*, 90C, 429, 1988.

624. **Mattocks, A. R.,** *Chemistry and Toxicology of Pyrrolizidine Alkaloids*, Academic Press, London, 1986, 159.

625. **Dick, A. T., Dann, A. T., Bull, L. B., and Culvenor, C. C. J.,** Vitamin B_{12} and the detoxification of hepatotoxic pyrrolizidine alkaloids in rumen liquor, *Nature (London)*, 197, 207, 1963.

626. **Russell, G. R. and Smith, R. M.,** Reduction of heliotrine by a rumen microorganism, *Aust. J. Biol. Sci.*, 21, 1277, 1968.

627. **Lanigan, G. W. and Smith, L. W.,** Metabolism of pyrrolizidine alkaloids in the ovine rumen. I. Formation of 7α-hydroxy-1α-methyl-8α-pyrrolizidine from heliotrine and lasiocarpine, *Aust. J. Agric. Res.*, 21, 493, 1970.

628. **Lanigan, G. W.,** Metabolism of pyrrolizidine alkaloids in the ovine rumen. II. Some factors affecting rate of alkaloid breakdown by rumen fluid *in vitro*, *Aust. J. Agric. Res.*, 21, 633, 1970.

629. **Shull, L. R., Buckmaster, G. W., and Cheeke, P. R.,** Factors influencing pyrrolizidine (Senecio) alkaloid metabolism: species, liver sulfhydryls and rumen fermentation, *J. Anim. Sci.*, 43, 1247, 1976.

630. **Swick, R. A., Cheeke, P. R., Ramsdell, H. S., and Buhler, D. R.,** Effect of sheep rumen fermentation and methane inhibition on the toxicity of *Senecio jacobaea*, *J. Anim. Sci.*, 56, 645, 1983.

631. **Mattocks, A. R.,** Toxicity of pyrrolizidine alkaloids, *Nature (London)*, 217, 723, 1968.

632. **Chesney, C. F. and Allen, J. R.,** Resistance of the guinea pig to pyrrolizidine alkaloid intoxication, *Toxicol. Appl. Pharmacol.*, 26, 385, 1973.

633. **Mattocks, A. R. and White, I. N. H.,** The conversion of pyrrolizidine alkaloids to *N*-oxides and to dihydropyrrolizidine derivatives by rat-liver microsomes *in vitro*, *Chem. Biol. Interact.*, 3, 383, 1971.

634. **Williams, D. E., Reed, R. L., Kedzierski, B., Ziegler, D. M., and Buhler, D. R.,** The role of flavin-containing monooxygenase in the *N*-oxidation of the pyrrolizidine alkaloid senecionine, *Drug Metab. Dispos.*, 17, 380, 1989.

635. **Ramsdell, H. S. and Buhler, D. R.,** Microsomal metabolism of pyrrolizidine alkaloids: *N*-oxidation of seneciphylline and senecionine, *Toxicol. Lett.*, 37, 241, 1987.

636. **Williams, D. E., Reed, R. L., Kedzierski, B., Dannan, G. A., Guengerich, F. P., and Buhler, D. R.,** Bioactivation and detoxication of the pyrrolizidine alkaloid senecionine by cytochrome P-450 enzymes in rat liver, *Drug Metab. Dispos.*, 17, 387, 1989.

637. **Mattocks, A. R. and Bird, I.,** Pyrrolic and *N*-oxide metabolites formed from pyrrolizidine alkaloids by hepatic microsomes in vitro: relevance to in vivo hepatotoxicity, *Chem. Biol. Interact.*, 43, 209, 1983.

638. **Ramsdell, H. S., Kedzierski, B., and Buhler, D. R.,** Microsomal metabolism of pyrrolizidine alkaloids from *Senecio jacobaea*. Isolation and quantification of 6,7-dihydro-7-hydroxy-1-hydroxymethyl-5H-pyrrolizidine and *N*-oxides by high performance liquid chromatography, *Drug Metab. Dispos.*, 15, 32, 1987.

639. **Culvenor, C. C. J., Downing, D. T., Edgar, J. A., and Jago, M. V.,** Pyrrolizidine alkaloids as alkylating and antimitotic agents, *Ann. N.Y. Acad. Sci.*, 163, 837, 1969.

640. **Jago, M. V., Edgar, J. A., Smith, L. V., and Culvenor, C. C. J.,** Metabolic conversion of heliotridine-based pyrrolizidine alkaloids to dehydroheliotridine, *Mol. Pharmacol.*, 6, 402, 1970.

641. **Hsu, I. C., Allen, J. R., and Chesney, C. F.,** Identification and toxicological effects of dehydroretronecine, a metabolite of monocrotaline, *Proc. Soc. Exp. Biol. Med.*, 144, 834, 1973.

642. **Segall, H. J., Dallas, J. L., and Haddon, W. F.,** Two dihydropyrrolizine alkaloid metabolites isolated from mouse hepatic microsomes *in vitro*, *Drug Metab. Dispos.*, 12, 68, 1984.

643. **Kedzierski, B. and Buhler, D. R.,** The formation of 6,7-dihydro-7-hydroxymethyl-5H-pyrrolizine, a metabolite of pyrrolizidine alkaloids, *Chem. Biol. Interact.*, 57, 217, 1986.

644. **Kedzierski, B. and Buhler, D. R.,** Configuration of necine pyrroles — toxic metabolites of pyrrolizidine alkaloids, *Toxicol. Lett.*, 25, 115, 1985.

645. **Chesney, C. F., Hsu, I. C., and Allen, J. R.,** Modifications of the *in vitro* metabolism of the hepatotoxic pyrrolizidine alkaloid, monocrotaline, *Res. Commun. Chem. Pathol. Pharmacol.*, 8, 567, 1974.

646. **Segall, H. J., Wilson, D. W., Dallas, J. L., and Haddon, W. F.,** *Trans*-4-hydroxy-2-hexenal: a reactive metabolite from the macrocyclic pyrrolizidine alkaloid senecionine, *Science*, 229, 472, 1985.

647. **Griffin, D. S. and Segall, H. J.,** Genotoxicity and cytotoxicity of selected pyrrolizidine alkaloids, a possible alkenal metabolite of the alkaloids, and related alkenals, *Toxicol. Appl. Pharmacol.*, 86, 227, 1986.

648. **Winter, C. K., Segall, H. J., and Jones, A. D.,** Distribution of *trans*-4-hydroxy-2-hexenal and tandem mass spectrometric detection of its urinary mercapturic acid in the rat, *Drug Metab. Dispos.*, 15, 608, 1987.

649. **Robertson, K. A., Seymour, J. L., Hsia, M.-T., and Allen, J. R.,** Covalent interaction of dehydroretronicine, a carcinogenic metabolite of the pyrrolizidine alkaloid monocrotaline, with cysteine and glutathione, *Cancer Res.*, 37, 3141, 1977.

650. **White, I. N. H.,** The role of liver glutathione in the acute toxicity of retrorsine to rats, *Chem. Biol. Interact.*, 13, 333, 1976.

651. **Hsu, I. C., Robertson, K. A., Shumaker, R. C., and Allen, J. R.,** Binding of tritiated dehydroretronecine to macromolecules, *Res. Commun. Chem. Pathol. Pharmacol.*, 11, 99, 1975.

652. **Candrian, U., Lüthy, J., and Schlatter, C.,** In vivo covalent binding of retronecine-labelled [³H]seneciphylline and [³H]senecionine to DNA of rat liver, lung and kidney, *Chem. Biol. Interact.*, 54, 57, 1985.

653. **Green, C. E., Segall, H. J., and Byard, J. L.,** Metabolism, cytotoxicity, and genotoxicity of the pyrrolizidine alkaloid senecionine in primary cultures of rat hepatocytes, *Toxicol. Appl. Pharmacol.*, 60, 176, 1981.

654. **Robertson, K. A.,** Alkylation of N^2 in deoxyguanosine by dehydroretronecine, a carcinogenic metabolite of the pyrrolizidine alkaloid monocrotaline, *Cancer Res.*, 42, 8, 1982.

655. **Mattocks, A. R.,** Acute hepatotoxicity and pyrrolic metabolites in rats dosed with pyrrolizidine alkaloids, *Chem. Biol. Interact.*, 5, 227, 1972.

656. **Mattocks, A. R.,** Role of the acid moieties in the toxic actions of pyrrolizidine alkaloids on liver and lung, *Nature (London)*, 228, 174, 1970.

657. **Mattocks, A. R.,** Relation of structural features to pyrrolic metabolites in livers of rats given pyrrolizidine alkaloids and derivatives, *Chem. Biol. Interact.*, 35, 301, 1981.

658. **Culvenor, C. C. J., Edgar, J. A., Smith, L. W., Jago, M. W., and Peterson, J. E.,** Active metabolites in the chronic hepatotoxicity of pyrrolizidine alkaloids, including otonecine esters, *Nature (London) New Biol.*, 229, 255, 1971.

659. **Mattocks, A. R. and White, I. N. H.,** Toxic effects and pyrrolic metabolites in the liver of young rats given the pyrrolizidine alkaloid retrorsine, *Chem. Biol. Interact.*, 6, 297, 1973.

660. **Armstrong, S. J. and Zuckerman, A. J.,** Production of pyrroles from pyrrolizidine alkaloids by human embryo tissue, *Nature (London)*, 228, 569, 1970.

661. **Bull, L. B., Culvenor, C. C. J., and Dick, A. T.,** *The Pyrrolizidine Alkaloids*, North-Holland, Amsterdam, 1968, 214.

662. **Schoental, R.,** Hepatotoxic activity of retrorsine, senkirkine and hydroxysenkirkine in newborn rats, and the role of epoxides in carcinogenesis by pyrrolizidine alkaloids and aflatoxins, *Nature (London)*, 227, 401, 1970.

663. **Cheeke, P. R.,** Toxicity and metabolism of pyrrolizidine alkaloids, *J. Anim. Sci.*, 66, 2343, 1988.

664. **White, I. N. H., Mattocks, A. R., and Butler, W. H.,** The conversion of the pyrrolizidine alkaloid retrorsine to pyrrolic derivatives *in vivo* and *in vitro* and its acute toxicity to various animal species, *Chem. Biol. Interact.*, 6, 207, 1973.

665. **Mattocks, A. R. and Driver, H. E.,** Metabolism and toxicity of anacrotine, a pyrrolizidine alkaloid, in rats, *Chem. Biol. Interact.*, 63, 91, 1987.

666. **Mattocks, A. R. and Driver, E.,** Toxic actions of senaetnine, a new pyrrolizidine-type alkaloid, in rats, *Toxicol. Lett.*, 38, 315, 1987.

667. **Mattocks, A. R. and White, I. N. H.,** Pyrrolic metabolites from non-toxic pyrrolizidine alkaloids, *Nature (London) New Biol.*, 231, 114, 1971.

668. **Mattocks, A. R.,** *Chemistry and Toxicology of Pyrrolizidine Alkaloids*, Academic Press, London, 1986, 212.

669. **Mattocks, A. R.,** Hepatotoxic effects due to pyrrolizidine alkaloid N-oxides, *Xenobiotica*, 1, 563, 1971.

670. **Powis, G., Ames, M. M., and Kovach, J. S.,** Relationship of the reductive metabolism of indicine N-oxide to its antitumor activity, *Res. Commun. Chem. Pathol. Pharmacol.*, 24, 559, 1979.

671. **El Dareer, S. M., Tillery, K. F., Lloyd, H. H., and Hill, D. L.,** Disposition of indicine N-oxide in mice and monkeys, *Cancer Treat. Rep.*, 66, 183, 1982.

672. **Ames, M. M. and Powis, G.,** Determination of indicine N-oxide and indicine in plasma and urine by electron-capture gas-liquid chromatography, *J. Chromatogr.*, 166, 519, 1978.

673. **Powis, G., Ames, M. M., and Kovach, J. S.,** Metabolic conversion of indicine N-oxide to indicine in rabbits and humans, *Cancer Res.*, 39, 3564, 1979.

674. **Kovach, J. S., Ames, M. M., Powis, G., Moertel, C. G., Hahn, R. G., and Creagan, E. T.,** Toxicity and pharmacokinetics of pyrrolizidine alkaloid, indicine N-oxide, in humans, *Cancer Res.*, 39, 4540, 1979.

675. **Evans, J. V., Daley, S. K., McClusky, G. A., and Nielsen, C. J.,** Direct quantitative analysis of indicine-N-oxide in cancer patient samples by gas chromatography using the internal standard heliotrine-N-oxide including a mass spectral comparison of the trimethylsilyl derivatives, *Biomed. Mass Spectrom.*, 7, 65, 1980.

676. **Nielsen, C. J., Daley, S. K., and Emmert, S. E.,** Comparative studies on indicine-*N*-oxide in cultured L1210 cells and rat liver homogenate, *Proc. Am. Assoc. Cancer Res.*, 21, 268, 1980.

677. **Powis, G. and Wincentsen, L.,** Pyridine nucleotide cofactor requirements of indicine *N*-oxide reduction by hepatic microsomal cytochrome P-450, *Biochem. Pharmacol.*, 29, 347, 1980.

678. **Lavallee, W. F. and Rosenkrantz, H.,** Evidence for pilocarpine transformation by serum, *Biochem. Pharmacol.*, 15, 206, 1966.

679. **Schonberg, S. S. and Ellis, P. P.,** Pilocarpine inactivation, *Arch. Ophthalmol.*, 82, 351, 1969.

680. **Ellis, P. P., Littlejohn, K., and Deitrich, R. A.,** Enzymatic hydrolysis of pilocarpine, *Invest. Ophthal.*, 11, 747, 1972.

681. **Newsome, D. A. and Stern, R.,** Pilocarpine adsorption by serum and ocular tissues, *Am. J. Ophthal.*, 77, 918, 1974.

682. **Lazare, R. and Horlington, M.,** Pilocarpine levels in the eyes of rabbits following topical application, *Exp. Eye Res.*, 21, 281, 1975.

683. **Schmitz, J., Mueller, R., Ferber, C., and Mosebach, K. O.,** The metabolism of pilocarpine in the rabbit eye following local application, *Fortschr. Ophthalmol.*, 82, 613, 1986 (Chem. Abstr. 104, 101929f, 1986).

684. **Lee, V. H.-L., Hui, H.-W., and Robinson, J. R.,** Corneal metabolism of pilocarpine in pigmented rabbits, *Invest. Ophthalmol. Vis. Sci.*, 19, 210, 1980.

685. **Salminen, L. and Urtti, A.,** Effect of ocular pigmentation on ocular disposition, metabolism and biophasic availablilty of pilocarpine, *Doc. Ophthalmol. Proc. Ser.*, 43, 91, 1985 (Chem. Abstr. 104, 14449f, 1986).

686. **Wood, R. W. and Robinson, J. R.,** Ocular disposition of pilocarpine in the pigmented rabbit: site(s) of metabolism, *Int. J. Pharm.*, 29, 127, 1986.

687. **Xie, M., Zhou, W., and Zhang, Y.,** The metabolic fate of oxymatrine, *Acta Pharm. Sin.*, 16, 481, 1981.

688. **Song, Y., Dong, Q., Chen, H., and Jin, Y.,** The metabolism of ^{3}H-oxymatrine in mice and rats, *Zhongguo Yixue Kexueyuan Xuebao*, 8, 261, 1986 (Chem. Abstr. 107, 190310z, 1987).

689. **Dengler, H. J., Eichelbaum, M., Hengstmann, J., and Wieber, J.,** Pharmacokinetic studies in man with sparteine, *Pharmacol. Clin.*, 2, 189, 1970.

690. **Spiteller, M. and Spiteller, G.,** Structural elucidation of sparteine metabolites, in *Biological Oxidation of Nitrogen*, Gorrod, J. W., Ed., Elsevier/North-Holland Biomedical Press, Amsterdam, 1978, 109.

691. **Guengerich, F. P.,** Oxidation of sparteines by cytochrome P-450: evidence against the formation of *N*-oxides, *J. Med. Chem.*, 27, 1101, 1984.

692. **Eichelbaum, M., Reetz, K.-P., Schmidt, E. K., and Zekorn, C.,** The genetic polymorphism of sparteine metabolism, *Xenobiotica*, 16, 465, 1986.

693. **Eichelbaum, M., Spannbrucker, N., and Dengler, H. J.,** A probably genetic defect of the metabolism of sparteine, in *Biological Oxidation of Nitrogen*, Gorrod, J. W., Ed., Elsevier/North-Holland Biomedical Press, Amsterdam, 1978, 113.

694. **Brøsen, K., Otton, S. V., and Gram, L. F.,** Sparteine oxidation polymorphism in Denmark, *Acta Pharmacol. Toxicol.*, 57, 357, 1985.

695. **Vinks, A., Inaba, T., Otton, S. V., and Kalow, W.,** Sparteine metabolism in Canadian Caucasians, *Clin. Pharmacol. Ther.*, 31, 23, 1982.

696. **Inaba, T., Vinks, A., Otton, S. V., and Kalow, W.,** Comparative pharmacogenetics of sparteine and debrisoquine, *Clin. Pharmacol. Ther.*, 33, 394, 1983.

697. **Evans, D. A. P., Harmer, D., Downham, D. Y., Whibley, E. J., Idle, J. R., Ritchie, J., and Smith, R. L.,** The genetic control of sparteine and debrisoquine metabolism in man with new methods of analysing bimodal distributions, *J. Med. Genet.*, 20, 321, 1983.

698. **Brøsen, K.,** Sparteine oxidation polymorphism in Greenlanders living in Denmark, *Br. J. Clin. Pharmacol.*, 22, 415, 1986.

699. **Ishizaki, T., Eichelbaum, M., Horai, Y., Hashimoto, K., Chiba, K., and Dengler, H. J.,** Evidence for polymorphic oxidation of sparteine in Japanese subjects, *Br. J. Clin. Pharmacol.*, 23, 482, 1987.

700. **Horai, Y., Ishizaki, T., Eichelbaum, M., Hashimoto, K., Chiba, K., and Dengler, H. J.,** Further analysis of sparteine oxidation in a Japanese population and comparison with data observed in different ethnic populations, *Xenobiotica*, 18, 1077, 1988.

701. **Eichelbaum, M. and Woolhouse, N. M.,** Inter-ethnic difference in sparteine oxidation among Ghanaians and Germans, *Eur. J. Clin. Pharmacol.*, 28, 79, 1985.

702. **Arias, T. D., Jorge, L. F., Lee, D., Barrantes, R., and Inaba, T.,** The oxidative metabolism of sparteine in the Cuna Amerindians of Panama: absence of evidence for deficient metabolizers, *Clin. Pharmacol. Ther.*, 43, 456, 1988.

703. **Brøsen, K., Otton, S. V., and Gram, L. F.,** Sparteine oxidation polymorphism: a family study, *Br. J. Clin. Pharmacol.*, 21, 661, 1986.

704. **Inaba, T., Otton, S. V., and Kalow, W.,** Deficient metabolism of debrisoquine and sparteine, *Clin. Pharmacol. Ther.*, 27, 547, 1980.

705. **Dahlqvist, R., Bertilsson, L., Birkett, D. J., Eichelbaum, M., Säwe, J., and Sjöqvist, F.,** Theophylline metabolism in relation to antipyrine, debrisoquine, and sparteine metabolism, *Clin. Pharmacol. Ther.*, 35, 815, 1984.

706. **Brinn, R., Brøsen, K., Gram, L. F., Haghfelt, T., and Otton, S. V.,** Sparteine oxidation is practically abolished in quinidine-treated patients, *Br. J. Clin. Pharmacol.,* 22, 194, 1986.

707. **Ebner, T., Eichelbaum, M., Fischer, P., and Meese, C. O.,** Über die stereospezifische Hydroxylierung von (+)-Spartein (Pachycarpin) bei der Ratte, *Arch. Pharm. (Weinheim, Ger.),* 322, 399, 1989.

708. **Shen, Z., Li, K., Ma, Z., Yi, M., and Liu, G.,** Distribution and excretion of ³H-DL-tetrahydropalmatine in rats, *Nanjing Yaoxueyuan Xuebao,* 17, 24, 1986 (Chem. Abstr. 105, 54065e, 1986).

709. **Mrozikiewcz, A., Kowalewski, Z., and Drost-Karwowska, K.,** Pharmacokinetics of berberine chloride, *Herba Pol.,* 26, 123, 1980 (Chem. Abstr. 94, 185343w, 1981).

710. **Sakurai, S., Tezuka, M., and Tamemasa, O.,** Studies on the absorption, distribution, and excretion of ³H-berberine chloride, *Oyo Yakuri,* 11, 351, 1976 (Chem. Abstr. 88, 145908d, 1978).

711. **Miyazaki, H., Shirai, E., Ishibashi, M., Niizima, K., and Kamura, Y.,** Quantitative analysis of berberine in urine samples by chemical ionization mass fragmentography, *J. Chromatogr.,* 152, 79, 1978.

712. **Miyazaki, H., Shirai, E., Ishibashi, M., Hosoi, K., Shibata, S., and Iwanaga, M.,** Quantitation of berberine chloride in human urine by use of selected ion monitoring in the field desorption mode, *Biomed. Mass Spectrom,* 5, 559, 1978.

713. **Schein, F. T. and Hanna, C.,** The absorption, distribution and excretion of berberine, *Arch. Int. Pharmacodyn. Ther.,* 124, 317, 1960.

714. **Chin, K.-C., Wang, Y.-N., Pao, T.-Y., and Hsu, P.,** Physiological disposition of tetrahydroberberine, *Sheng Li Hsueh Pao,* 28, 72, 1965 (Chem. Abstr. 63, 12151e, 1965).

715. **Furuya, T.,** Metabolic conversion of berberine hydrochloride, *Bull. Osaka Med. Sch.,* 2, 18, 1956 (Chem. Abstr. 52, 2274e, 1958).

716. **Fujii, T., Miyazaki, H., Nambu, K., Kagemoto, A., and Hashimoto, M.,** Disposition and metabolism of ¹⁴C-dehydrocorydaline in mice and rats, *Radioisotopes,* 33, 519, 1984.

717. **Myers, V. C. and Wardell, E. L.,** The influence of the ingestion of methyl xanthines on the excretion of uric acid, *J. Biol. Chem.,* 77, 697, 1928.

718. **Buchanan, O. H., Christman, A. A., and Block, W. D.,** The metabolism of the methylated purines. II. Uric acid excretion following the ingestion of caffeine, theophylline, and theobromine, *J. Biol. Chem.,* 157, 189, 1945.

719. **Myers, V. C. and Hanzal, R. F.,** The metabolism of methylxanthines and their related methyluric acids, *J. Biol. Chem.,* 162, 309, 1946.

720. **Weinfeld, H. and Christman, A. A.,** The metabolism of caffeine and theophylline, *J. Biol. Chem.,* 200, 345, 1953.

721. **Brodie, B. B., Axelrod, J., and Reichenthal, J.,** Metabolism of theophylline (1,3-dimethylxanthine) in man, *J. Biol. Chem.,* 194, 215, 1952.

722. **Cornish, H. H. and Christman, A. A.,** A study of the metabolism of theobromine, theophylline, and caffeine in man, *J. Biol. Chem.,* 228, 315, 1957.

723. **Choi, H. R., Shin, S. G., and Lee, K. S.,** Theophylline pharmacokinetics and metabolism in young normal adults, *Chungang Uidaechi,* 14, 175, 1989 (Chem. Abstr. 111, 166741c, 1989).

724. **Monks, T. J., Caldwell, J., and Smith, R. L.,** Influence of methylxanthine-containing foods on theophylline metabolism and kinetics, *Clin. Pharmacol. Ther.,* 26, 513, 1979.

725. **Caldwell, J., Cotgreave, I. A., Lawrie, C. A., and Monks, T. J.,** Origins of interindividual variations in theophylline metabolism in man, in *Theophylline and other Methylxanthines,* Rietbrock, N., Woodstock, B. G., and Staib, A. H., Eds., F. Vieweg & Sohn, Braunschweig, W. Germany, 1982, 159.

726. **Birkett, D. J., Dahlqvist, R., Miners, J. O., Lelo, A., and Billing, B.,** Comparison of theophylline and theobromine metabolism in man, *Drug Metab. Dispos.,* 13, 725, 1985.

727. **Miller, M., Opheim, K. E., Raisys, V. A., and Motulsky, A. G.,** Theophylline metabolism: variation and genetics, *Clin. Pharmacol. Ther.,* 35, 170, 1984.

728. **Jonkman, J. H. G., Tang, D., Upton, R. A., and Riegelman, S.,** Measurement of excretion characteristics of theophylline and it major metabolites, *Eur. J. Clin. Pharmacol.,* 20, 435, 1981.

729. **Grygiel, J. J. and Birkett, D. J.,** Cigarette smoking and theophylline clearance and metabolism, *Clin. Pharmacol. Ther.,* 30, 491, 1981.

730. **Rovei, V., Chanoine, F., and Strolin Benedetti, M.,** Pharmacokinetics of theophylline: a dose-range study, *Br. J. Clin. Pharmacol.,* 14, 769, 1982.

731. **Jenne, J. W., Nagasawa, H. T., and Thompson, R. D.,** Relationship of urinary metabolites of theophylline to serum theophylline levels, *Clin. Pharmacol. Ther.,* 19, 375, 1976.

732. **Van Gennip, A. H., Grift, J., and Van Bree-Blom, E. J.,** Urinary excretion of methylated purines in man and in the rat after the administration of theophylline, *J. Chromatogr.,* 163, 351, 1979.

733. **Bonati, M. and Latini, R.,** Theophylline (T) and 3-methylxanthine (3-MX) kinetics in man, First World Conference Clinical Pharmacology and Therapeutics, London, 1980.

734. **Birkett, D. J., Miners, J. O., and Attwood, J.,** Secondary metabolism of theophylline. Biotransformation products in man—route of formation of 1-methyluric acid, *Br. J. Clin. Pharmacol.,* 15, 117, 1983.

735. **Grygiel, J. J., Wing, L. M. H., Farkas, J., and Birkett, D. J.**, Effects of allopurinol on theophylline metabolism and clearance, *Clin. Pharmacol. Ther.*, 26, 660, 1979.

736. **Monks, T. J., Smith, R. L., and Caldwell, J.**, A metabolic and pharmacokinetic comparison of theophylline and aminophylline (theophylline ethylenediamine), *J. Pharm. Pharmacol.*, 33, 93, 1980.

737. **Monks, T. J., Lawrie, C. A., and Caldwell, J.**, The effect of increased caffeine intake on the metabolism and pharmacokinetics of theopylline in man, *Biopharm. Drug Dispos.*, 2, 31, 1981.

738. **Bory, C., Baltassat, P., Porthault, M., Bethenod, M., Frederich, A., and Aranda, J. V.**, Metabolism of theophylline to caffeine in premature newborn infants, *J. Pediatr.*, 94, 988, 1979.

739. **Bada, H. S., Khanna, N. N., Somani, S. M., and Tin, A. A.**, Interconversion of theophylline and caffeine in newborn infants, *J. Pediatr.*, 94, 993, 1979.

740. **Boutroy, M. J., Vert, P., Royer, R. J., Monin, P., and Royer-Morrot, M. J.**, Caffeine, a metabolite of theophylline during the treatment of apnea in the premature infant, *J. Pediatr.*, 94, 996, 1979.

741. **Brazier, J. L., Ribon, B., and Desage, M.**, Study of theophylline metabolism in premature human newborns using stable isotope labelling, *Biomed. Mass Spectrom.*, 7, 189, 1980.

742. **Tserng, K.-Y., King, K. C., and Takieddine, F. N.**, Theopylline metabolism in premature infants, *Clin. Pharmacol. Ther.*, 29, 594, 1981.

743. **Bonati, M., Latini, R., Marra, G., Assael, B. M., and Parini, R.**, Theophylline metabolism during the first month of life and development, *Pediatr. Res.*, 15, 304, 1981.

744. **Grygiel, J. J. and Birkett, D. J.**, Effect of age on patterns of theophylline metabolism, *Clin. Pharmacol. Ther.*, 28, 456, 1980.

745. **Tang-Liu, D. D.-S. and Riegelman, S.**, Metabolism of theophylline to caffeine in adults, *Res. Commun. Chem. Pathol. Pharmacol.*, 34, 371, 1981.

746. **Aranda, J. V., Brazier, J. L., Louridas, A. T., and Sasyniuk, B. I.**, Methylxanthine metabolism in the newborn infant, in *Drug Metabolism in the Immature Human*, Soyka, L. F. and Redmond, G. P., Eds., Raven Press, New York, 1981, 183.

747. **Haley, T. J.**, Metabolism and pharmacokinetics of theophylline in human neonate, children, and adults, *Drug Metab. Rev.*, 14, 295, 1983.

748. **Kearns, G. L., Hill, D. E., and Tumbelson, M. E.**, Theophylline and caffeine disposition in the neonatal piglet, *Dev. Pharmacol. Ther.*, 9, 389, 1986.

749. **Gorodischer, R., Yaari, A., Margalith, M., Warszawski, D., and Ben-Zvi, Z.**, Changes in theophylline metabolism during postnatal development in rat liver slices, *Biochem. Pharmacol.*, 35, 3077, 1986.

750. **Betlach, C. J. and Tozer, T. N.**, Biodisposition of theophylline. I. Genetic variation in inbred mice, *Drug Metab. Dispos.*, 8, 268, 1980.

751. **Betlach, C. J. and Tozer, T. N.**, Biodisposition of theophylline. II. Effect of aromatic hydrocarbon treatment in mice, *Drug Metab. Dispos.*, 8, 271, 1980.

752. **Lohmann, S. M. and Miech, R. P.**, Theophylline metabolism by the rat liver microsomal system, *J. Pharmacol. Exp. Ther.*, 196, 213, 1976.

753. **Christensen, H. D. and Whitsett, T. L.**, Measurements of xanthines and their metabolites by means of high pressure liquid chromatography, in *Biological/Biomedical Applications of Liquid Chromatography*, Hawk, G. L., Ed., Marcel Dekker, New York, 1979, 507.

754. **Arnaud, M. J., Bracco, I., and Welsch, C.**, Metabolism and distribution of labeled theophylline in the pregnant rat. Impairment of theophylline metabolism by pregnancy and absence of a blood-brain barrier in the fetus, *Pediatr. Res.*, 16, 167, 1982.

755. **Celardo, A., Taina, G. L., Jankowski, A., and Bonati, M.**, Pharmacokinetics of theophylline and its metabolites in rabbits, *Eur. J. Drug Metab. Pharmacokinet.*, 10, 279, 1985.

756. **Kuze, T., Miyazaki, H., and Taneike, T.**, Theophylline: pharmacokinetics, metabolism and urinary excretion in dogs, *Folia Pharmacol. Jpn.*, 91, 325, 1988.

757. **Moss, M. S.**, The metabolism and urinary and salivary excretion of drugs in the horse and their relevance to detection of dope, in *Drug Metabolism — from Microbe to Man*, Parke, D. V. and Smith, R. L., Eds., Taylor and Francis, London, 1977, 263.

758. **Gaudette, L. E. and Brodie, B. B.**, Relationship between the lipid solubility of drugs and their oxidation by liver microsomes, *Biochem. Pharmacol.*, 2, 89, 1959.

759. **Mazel, P. and Henderson, J. F.**, On the relationship between lipid solubility and microsomal metabolism of drugs, *Biochem. Pharmacol.*, 14, 92, 1965.

760. **Krenitsky, T. A., Neil, S. M., Elion, G. B., and Hitchings, G. H.**, A comparison of the specificities of xanthine oxidase and aldehyde oxidase, *Arch. Biochem. Biophys.*, 150, 585, 1972.

761. **Robson, R. A., Matthews, A. P., Miners, J. O., McManus, M. E., Meyer, U. A., Hall, P. D. L. M. and Birkett, D. J.**, Characterisation of theophylline metabolism in human liver microsomes, *Br. J. Clin. Pharmacol.*, 24, 293, 1987.

762. **Robson, R. A., Miners, J. O., Matthews, A. P., Stupans, I., Meller, D., McManus, M. E., and Birkett, D. J.**, Characterisation of theopylline metabolism by human liver microsomes, *Biochem. Pharmacol.*, 37, 1651, 1988.

763. **Campbell, M. E., Grant, D. M., Inaba, T., and Kalow, W.,** Biotransformation of caffeine, paraxanthine, theophylline, and theobromine by polycyclic aromatic hydrocarbon-inducible cytochrome(s) P-450 in human liver microsomes, *Drug Metab. Dispos.*, 15, 237, 1987.

764. **McManus, M. E., Miners, J. O., Gregor, D., Stupans, I., and Birkett, D. J.,** Theophylline metabolism by human, rabbit and rat liver microsomes and by purified forms of cytochrome P-450, *J. Pharm. Pharmacol.*, 40, 388, 1987.

765. **Berthou, F., Ratanasavanh, D., Alix, D., Carlhant, D., Riche, C., and Guillouzo, A.,** Caffeine and theophylline metabolism in newborn and adult human hepatocytes; comparison with adult rat hepatocytes, *Biochem. Pharmacol.*, 37, 3691, 1988.

766. **Arnaud, M. J. and Welsch, C.,** Metabolism of [1-Me-^{14}C]paraxanthine in the rat: identification of a new metabolite, *Experientia*, 35, 946, 1979.

767. **Arnaud, M. J. and Welsch, C.,** Caffeine metabolism in human subjects, Proc. 9th Int. Coll. Sci. Technol. of Coffee, London, 1980, 385.

768. **Arnaud, M. J., Welsch, C., and Sauvageat, J. L.,** Quantification in urine of uracil metabolites after oral human administration of caffeine, theobromine, thephylline and paraxanthine, 8th Int. Congr. Pharmacology, Tokyo, 1981, 560.

769. **Bonati, M., Latini, R., Sadurska, B., Riva, E., Galletti, F., Borzelleca, J. F., Tarka, S. M., Arnaud, M. J., and Garattini, S.,** I. Kinetics and metabolism of theobromine in male rats, *Toxicology*, 30, 327, 1984.

770. **Krüger, M. and Schmidt, P.,** Das Verhalten von Theobromin im Organismus des Menschen, *Arch. Exp. Pathol. Pharmak.*, 45, 259, 1901.

771. **Krüger, M. and Schmidt, P.,** Ueber das Verhalten von Theobromin, Paraxanthin und 3-Methylxanthin im Organismus, *Ber. Dtsch. Chem. Ges.*, 32, 2677, 1899.

772. **Miller, G. E., Radulovic, L. L., Dewit, R. H., Brabec, M. J., Tarka, S. M., and Cornish, H. H.,** Comparative theobromine metabolism in five mammalian species, *Drug Metab. Dispos.*, 12, 154, 1984.

773. **Latini, R., Bonati, M., Gaspari, F., Traina, G. L., Jiritano, L., Bortolotti, A., Borzelleca, J. F., Tarka, S. M., Arnaud, M. J., and Garattini, S.,** II. Kinetics and metabolism of theobromine in male and female nonpregnant and pregnant rabbits, *Toxicology*, 30, 343, 1984.

774. **Arnaud, M. J. and Welsch, C.,** Metabolic pathway of theobromine in the rat and identification of two new metabolites in human urine, *J. Agric. Food Chem.*, 27, 524, 1979.

775. **Shively, C. A. and Tarka, S. M.,** Theobromine metabolism and pharmacokinetics in pregnant and nonpregnant Sprague-Dawley rats, *Toxicol. Appl. Pharmacol.*, 67, 376, 1983.

776. **Shively, C. A. and Vesell, E. S.,** In vivo and in vitro biotransformation of theobromine by phenobarbital- and 3-methylcholanthrene-inducible cytochrome P-450 monooxygenases in rat liver. Role of thiol compounds, *Drug Metab. Dispos.*, 15, 217, 1987.

777. **Traina, G. L. and Bonati, M.,** Pharmacokinetics of theobromine and its metabolites in rabbits, *J. Pharmacokinet. Biopharm.*, 13, 41, 1985.

778. **Tarka, S. M., Arnaud, M. J., Dvorchik, B. H., and Vesell, E. S.,** Theobromine kinetics and metabolic disposition, *Clin. Pharmacol. Ther.*, 34, 546, 1983.

779. **Birkett, D. J., Miners, J. O., Wing, L. M. H., Lelo, A., and Robson, R. A.,** Methylxanthine metabolism in man, in *Anti-Asthma Xanthines and Adenosine*, Andersson, K.-E. and Persson, C. G. A., Eds., Excerpta Medica, Amsterdam, 1985, 230.

780. **Miners, J. O., Attwood, J., and Birkett, D. J.,** Theobromine metabolism in man, *Drug Metab. Dispos.*, 10, 672, 1982.

781. **Burg, A. W.,** Physiological disposition of caffeine, *Drug Metab. Rev.*, 4, 199, 1975.

782. **Yesair, D. W., Branfman, A. R., and Callahan, M. M.,** Human disposition and some biochemical aspects of methylxanthines, in *The Methylxanthine Beverages and Food: Chemistry, Consumption, and Health Effects*, Spiller, G. A., Ed., Alan R. Liss, New York, 1984, 215.

783. **Bonati, M. and Garattini, S.,** Pharmacokinetics of caffeine, *ISI Atlas Sci. Pharmacol.*, 2, 33, 1988.

784. **Arnaud, M. J.,** Products of metabolism of caffeine, in *Caffeine: Perspectives from Recent Research*, Dews, P. B., Ed., Springer Verlag, Berlin, 1984, 3.

785. **Burg, A. W. and Stein, M. E.,** Urinary excretion of caffeine and its metabolites in the mouse, *Biochem. Pharmacol.*, 21, 909, 1972.

786. **Burg, A. W. and Werner, E.,** Tissue distribution of caffeine and its metabolites in the mouse, *Biochem. Pharmacol.*, 21, 923, 1972.

787. **Czok, G., Schmidt, B., and Lang, K.,** Verteilung von 8-^{14}C-Coffein im Organismus der Rotte, *Z. Ernährwiss.*, 9, 109, 1969.

788. **Cunningham, H. M.,** Biological half-life of caffeine in pigs, *Can. J. Anim. Sci.*, 50, 49, 1970.

789. **Burg, A. W., Burrows, R., and Kensler, C. J.,** Unusual metabolism of caffeine in the squirrel monkey, *Toxicol. Appl. Pharmacol.*, 28, 162, 1974.

790. **Axelrod, J. and Reichenthal, J.,** The fate of caffeine in man and a method for its estimation in biological material, *J. Pharmacol. Exp. Ther.*, 107, 519, 1953.

791. **Greene, E. W., Woods, W. E., and Tobin, T.,** Pharmacology, pharmacokinetics, and behavioral effects of caffeine in horses, *Am. J. Vet. Res.*, 44, 57, 1983.

792. **Kamei, K., Matsuda, M., and Momose, A.,** New sulfur-containing metabolites of caffeine, *Chem. Pharm. Bull.*, 23, 683, 1975.

793. **Rafter, J. J. and Nilsson, L.,** Involvement of the intestinal microflora in the formation of sulphur-containing metabolites of caffeine, *Xenobiotica*, 11, 771, 1981.

794. **Arnaud, M. J.,** Identification, kinetic and quantitative study of [2-^{14}C] and [1-Me-^{14}C]caffeine metabolites in rat's urine by chromatographic separations, *Biochem. Med.*, 16, 67, 1976.

795. **Arnaud, M. J., Ben Zvi, Z., Yaari, A., and Gorodischer, R.,** 1,3,8-Trimethylallantoin: a major caffeine metabolite formed by rat liver, *Res. Commun. Chem. Pathol. Pharmacol.*, 52, 407, 1986.

796. **Caldwell, J., O'Gorman, J., and Adamson, R.,** The fate of caffeine in three non-human primates, *Biochem. Soc. Trans.*, 9, 215, 1981.

797. **Caldwell, J., O'Gorman, J., and Adamson, R. H.,** The metabolism of caffeine in three non-human primate species, in *Theophylline and Other Methylxanthines*, Rietbrock, N., Woodstock, B. G., and Staib, A. H., Eds., F. Vieweg & Sohn, Braunschweig, W. Germany, 1982, 181.

798. **Tang, B. K., Grant, D. M., and Kalow, W.,** Isolation and identification of 5-acetylamino-6-formylamino-3-methyluracil as a major metabolite of caffeine in man, *Drug Metab. Dispos.*, 11, 218, 1983.

799. **Branfman, A. R., McComish, M. F., Bruni, R. J., Callahan, M. M., Robertson, R., and Yesair, D. W.,** Characterization of diaminouracil metabolites of caffeine in human urine, *Drug Metab. Dispos.*, 11, 206, 1983.

800. **Callahan, M. M., Robertson, R. S., Arnaud, M. J., Branfman, A. R., McComish, M. F., and Yesair, D. W.,** Human metabolism of [1-*methyl*-^{14}C]- and [2-^{14}C]caffeine after oral administration, *Drug Metab. Dispos.*, 10, 417, 1982.

801. **Callahan, M. M., Robertson, R. S., Branfman, A. R., McComish, M. F., and Yesair, D. W.,** Comparison of caffeine metabolism in three nonsmoking populations after oral administration of radiolabeled caffeine, *Drug Metab. Dispos.*, 11, 211, 1983.

802. **Arnaud, M. J., Thelin-Doerner, A., Ravussin, E., and Acheson, K. J.,** Study of demethylation of [1,3,7-Me-^{13}C]caffeine in man using respiratory exchange measurements, *Biomed. Mass Spectrom.*, 7, 521, 1980.

803. **Wietholtz, H., Voegelin, M., Arnaud, M. J., Bircher, J., and Preisig, R.,** Assessment of the cytochrome P-448 dependent liver enzyme system by a caffeine breath test, *Eur. J. Clin. Pharmacol.*, 21, 53, 1981.

804. **Gescher, A. and Raymont, C.,** Studies of the metabolism of N-methyl containing anti-tumour agents ^{14}CO$_2$ breath analysis after administration of ^{14}C-labelled N-methyl drugs, formaldehyde and formate in mice, *Biochem. Pharmacol.*, 30, 1245, 1981.

805. **Grant, D. M., Tang, B. K., and Kalow, W.,** Variability in caffeine metabolism, *Clin. Pharmacol. Ther.*, 33, 591, 1983.

806. **Grant, D. M., Tang, B. K., and Kalow, W.,** Polymorphic N-acetylation of a caffeine metabolite, *Clin. Pharmacol. Ther.*, 33, 355, 1983.

807. **Kalow, W.,** Variability of caffeine metabolism in humans, *Arzneim. Forsch.*, 35(I), 319, 1985.

808. **Aranda, J. V., Collinge, J. M., Zinman, R., and Watters, G.,** Maturation of caffeine elimination in infancy, *Arch. Dis. Child.*, 54, 946, 1979.

809. **Aldridge, A., Aranda, J. V., and Neims, A. H.,** Caffeine metabolism in the newborn, *Clin. Pharmacol. Ther.*, 25, 447, 1979.

810. **Carrier, O., Pons, G., Rey, E., Richard, M.-O., Moran, C., Badoual, J. and Olive, G.,** Maturation of caffeine metabolic pathways in infancy, *Clin. Pharm. Ther.*, 44, 145, 1988.

811. **Blanchard, J., Sawers, S. J. A., Jonkman, J. H. G., and Tang-Liu, D. D.-S.,** Comparison of the urinary metabolite profile of caffeine in young and elderly males, *Br. J. Clin. Pharmacol.*, 19, 225, 1985.

812. **Warszawski, D., Ben-Zvi, Z., Gorodischer, R., Arnaud, M. J., and Bracco, I.,** Urinary metabolites of caffeine in young dogs, *Drug Metab. Disp.*, 10, 424, 1982.

813. **Aldridge, A. and Niems, A. H.,** Relationship between the clearance of caffeine and its 7-N-demethylation in developing beagle puppies, *Biochem. Pharmacol.*, 29, 1909, 1980.

814. **Warszawski, D., Ben-Zvi, Z., and Gorodischer, R.,** Caffeine metabolism in liver slices during postnatal development in the rat, *Biochem. Pharmacol.*, 30, 3145, 1981.

815. **Scott, N. R., Chakraborty, J., and Marks, V.,** Urinary metabolites of caffeine in pregnant women, *Br. J. Clin. Pharmacol.*, 22, 475, 1986.

816. **Jiritano, L., Bortolotti, A., Gaspari, F., and Bonati, M.,** Caffeine disposition after oral administration to pregnant rats, *Xenobiotica*, 15, 1045, 1985.

817. **Parsons, W. D. and Neims, A. H.,** Effect of smoking on caffeine clearance, *Clin. Pharmacol. Ther.*, 24, 40, 1978.

818. **Aldridge, A., Parsons, W. D., and Neims, A. H.,** Stimulation of caffeine metabolism in the rat by 3-methylcholanthrene, *Life Sci.*, 21, 967, 1977.

819. **Welch, R. M., Hsu, S.-Y. Y., and DeAngelis, R. L.,** Effect of Aroclor 1254, phenobarbital, and polycyclic aromatic hydrocarbons on the plasma clearance of caffeine in the rat, *Clin. Pharmacol. Ther.*, 22, 791, 1977.

820. **Aldridge, A. and Niems, A. H.,** The effects of phenobarbital and β-naphthoflavone on the elimination kinetics and metabolite pattern of caffeine in the beagle dog, *Drug Metab. Dispos.*, 7, 378, 1979.

821. **Bergmann, F. and Dikstein, S.,** Studies on uric acid and related compounds. III. Observations on the specificity of mammalian xanthine oxidases, *J. Biol. Chem.*, 223, 765, 1956.

822. **Ferrero, J. L. and Neims, A. H.,** Metabolism of caffeine by mouse liver microsomes: GSH or cytosol causes a shift in products from 1,3,7-trimethylurate to a substituted diaminouracil, *Life Sci.*, 33, 1173, 1983.

823. **Arnaud, M. J. and Welsch, C.,** Comparison of caffeine metabolism by perfused rat liver and isolated microsomes, in *Microsomes, Drug Oxidations, and Chemical Carcinogenesis*, Vol. 2, Coon, M. J., Conney, A. H., Estabrook, R. W., Gelboin, H. V., Gillette, J. R., and O'Brien, P. J., Eds., Academic Press, New York, 1980, 813.

824. **Bonati, M., Latini, R., Marzi, E., Cantoni, R., and Belvedere, G.,** [2-¹⁴C]Caffeine metabolism in control and 3-methylcholanthrene induced rat liver microsomes by high pressure liquid chromatography, *Toxicol. Lett.*, 7, 1, 1980.

825. **Grant, D. M., Campbell, M. E., Tang, B. K., and Kalow, W.,** Biotransformation of caffeine by microsomes from human liver, *Biochem. Pharmacol.*, 36, 1251, 1987.

826. **Berthou, F., Ratanasavanh, D., Riche, C., Picart, D., Voirin, T., and Guillouzo, A.,** Comparison of caffeine metabolism by slices, microsomes and hepatocyte cultures from adult human liver, *Xenobiotica*, 19, 401, 1989.

827. **Guaitani, A., Abbruzzi, R., Bastone, A., Bianchi, M., Bonati, M., Catalani, P., Latini, R., Pantarotto, C., and Szczawinska, K.,** Metabolism of caffeine to 6-amino-5-[*N*-methylformylamino]-1,3-dimethyluracil in the isolated, perfused liver from control or phenobarbital-, β-naphthoflavone- and 3-methylcholanthrene-pretreated rats, *Toxicol. Lett.*, 38, 55, 1987.

828. **Kalow, W. and Campbell, M.,** Biotransformation of caffeine by microsomes, *ISI Atlas Sci. Pharmacol.*, 2, 381, 1988.

829. **Chevion, M., Navok, T., Glaser, G., and Mager, J.,** The chemistry of favism-inducing compounds. The properties of isouramil and divicine and their reaction with glutathione, *Eur. J. Biochem.*, 127, 405, 1982.

830. **Albano, E., Tomasi, A., Mannuzzu, L., and Arese, O.,** Detection of a free radical intermediate from divicine of *Vicia faba*, *Biochem. Pharmacol.*, 33, 1701, 1984.

831. **Watson, W. C.,** The metabolism of betanin, *Biochem. J.*, 90, 3, 1964.

832. **Krantz, C., Monier, M., and Wahlström, B.,** Absorption, excretion, metabolism and cardiovascular effects of beetroot extract in the rat, *Food Cosmet. Toxicol.*, 18, 363, 1980.

833. **Sung, C.-Y. and Ho, C.-F.,** The physiological disposition of β-dichroine, an alkaloid of *Dichroa febrifuga* Lour., *Acta Pharm. Sin.*, 11, 437, 1964.

834. **Sullivan, H. R., Billings, R. E., Jansen, C. J., Occolowitz, J. L., Boaz, H. E., Marshall, F. J., and McMahon, R. E.,** In vivo hydroxylation of acronine, *Pharmacologist*, 11, 241, 1969.

835. **Liu, C. and Ji, X.,** Absorption, distribution and excretion of acronycine in rats, *Zhongguo Yaoli Xuebao*, 5, 137, 1984 (Chem. Abstr. 101, 48108x, 1984).

836. **Yao, P.-P. and Sung, C.-Y.,** The metabolic fate of securinine, *Chin. Med. J.*, 1, 205, 1975.

837. **Plowman, J., Cysyk, R. L., and Adamson, R. H.,** The disposition of coralyne sulphoacetate in rodents, *Xenobiotica*, 6, 281, 1976.

838. **Nishie, K., Gumbmann, M. R., and Keyl, A. C.,** Pharmacology of solanine, *Toxicol. Appl. Pharmacol.*, 19, 81, 1971.

839. **Claringbold, W. D. B., Few, J. D., and Renwick, J. H.,** Kinetics and retention of solanidine in man, *Xenobiotica*, 12, 293, 1982.

840. **Norred, W. P., Nishie, K. and Osman, S. F.,** Excretion, distribution and metabolic fate of ³H-α-chaconine, *Res. Commun. Chem. Pathol. Pharmacol.*, 13, 161, 1976.

841. **Alozie, S. O., Sharma, R. P., and Salunkhe, D. K.,** Excretion of α-chaconine-³H, a steroidal glycoalkaloid from *Solanum tuberosum* L. and its metabolites in hamsters, *Pharmacol. Res. Commun.*, 11, 483, 1979.

842. **King, R. R. and McQueen, R. E.,** Transformations of potato glycoalkaloids by rumen microorganisms, *J. Agric. Food Chem.*, 29, 1101, 1981.

843. **Claringbold, W. D. B., Few, J. D., Brace, C. J., and Renwick, J. H.,** The disposition and metabolism of [³H]-solasodine in man and syrian hamster, *J. Steroid Biochem.*, 13, 889, 1980.

844. **Ji, X., Liu, J., and Liu, Z.,** Metabolism of harringtonine, *Acta Sin. Pharm.*, 14, 234, 1979.

845. **Ji, X., Liu, Y., Lin, H., and Liu, Z.,** Metabolism of homoharringtonine in rats and mice, *Yaoxue Xuebao*, 17, 881, 1982 (Chem. Abstr. 98, 100715d, 1983).

846. **Lu, K., Savaraj, N., Feun, L. G., Zhengang, G., Umsawasdi, T., and Loo, T. L.,** Pharmacokinetics of homoharringtonine in dogs, *Cancer Chemother. Pharmacol.*, 21, 139, 1988.

847. **Savaraj, N., Lu, K., Dimery, I., Feun, L. G., Burgess, M., Keating, M., and Loo, T. L.,** Clinical pharmacology of homoharringtonine, *Cancer Treat. Rep.*, 70, 1403, 1986.

848. **Loo, T. L., Savaraj, N., and Lu, K.,** Metabolism of triciribine phosphate (TCN-P), homoharringtonine (HHT), and nafidimide (BIDA) by rat hepatic microsomal enzymes, *Proc. Am. Assoc. Cancer Res.*, 27, 422, 1986.

849. **Xie, F., Wang, H.-C., Li, J.-H., Shu, H.-L., Jiang, J.-R., Chang, J.-P., and Hsieh, Y.-Y.,** Studies on the metabolism of lappaconitine in humans. Identification of the metabolites of lappaconitine in human urine by high performance liquid chromatography, *Biomed. Chromatogr.*, 4, 43, 1990.

850. **Xie, F., Wang, H., Shu, H., Li, J., Jiang, J., Chang, J., and Hsieh, Y.,** Separation and characterization of the metabolic products of lappaconitine in rat urine by high-performance liquid chromatography, *J. Chromatogr.*, 526, 109, 1990.

851. **Bonati, M., Latini, R., Tognoni, G., Young, J. F., and Garattini, S.,** Interspecies comparison of in vivo caffeine pharmacokinetics in man, monkey, rabbit, rat, and mouse, *Drug Metab. Rev.*, 15, 1355, 1985.

852. **Beach, C. A., Bianchine, J. R., and Gerber, N.,** Metabolism of caffeine in the mouse, *Pharmacologist*, 25, 120, 1983.

853. **Usanova, M. I. and Snol, S. E.,** The distribution of radioactive caffeine in the animal organism and its penetration from the mother into the fetus, *Byull. Exp. Biol. Med.*, 40, 41, 1955 (Chem. Abstr. 50, 488c, 1956).

854. **Khanna, K. L., Rao, G. S., and Cornish, H. H.,** Metabolism of caffeine-^{3}H in the rat, *Toxicol. Appl. Pharmacol.*, 23, 720, 1972.

855. **Bonati, M. and Garattini, S.,** Interspecies comparison of caffeine disposition, in *Caffeine: Perspectives from Recent Research*, Dews, P. B., Ed., Springer Verlag, Berlin, 1984, 48.

856. **Beach, C. A., Mays, D. C., Sterman, B. M., and Gerber, N.,** Metabolism, distribution, seminal excretion and pharmacokinetics of caffeine in the rabbit, *J. Pharmacol. Exp. Ther.*, 233, 18, 1985.

857. **Fisher, R. S., Algeri, E. J., and Walker, J. T.,** The determination and the urinary excretion of caffeine in animals, *J. Biol. Chem.*, 179, 71, 1949.

858. **Krüger, M.,** Ueber den Abbau des Caffeïns im Organismus des Hundes, *Ber. Dtsch. Chem. Ges.*, 32, 2818, 1899.

859. **Latini, R., Bonati, M., Marzi, E., and Garattini, S.,** Urinary excretion of an uracilic metabolite from caffeine by rat, monkey and man, *Toxicol. Lett.*, 7, 267, 1981.

860. **Christensen, H. D., Manion, C. V., and Kling, O. R.,** Caffeine kinetics during late pregnancy, in *Drug Metabolism in the Immature Human*, Souka, L. F. and Redmond, G. P., Eds., Raven Press, New York, 1981, 163.

861. **Bonati, M., Latini, R., Galletti, F., Young, J. F., Tognoni, G., and Garattini, S.,** Caffeine disposition after oral doses, *Clin. Pharmacol Ther.*, 32, 98, 1982.

862. **Schmidt, G. and Schoyerer, R.,** Zum nachweis von Coffein und seinen Metaboliten im Harn, *Dtsch. Z. Gerichtl. Med.*, 57, 402, 1966.

863. **Tang-Liu, D. D.-S., Williams, R. L., and Riegelman, S.,** Disposition of caffeine and its metabolites in man, *J. Pharmacol. Exp. Ther.*, 224, 180, 1983.

864. **Gilbert, S. G., Stavric, B., Klassen, R. D., and Rice, D. C.,** The fate of chronically consumed caffeine in the monkey (*Macaca fascicularis*), *Fundam. Appl. Toxicol.*, 5, 578, 1985.

865. **Tse, F. L. S. and Valia, K. H.,** Plasma theobromine after oral administration of caffeine to dogs, *J. Pharm. Sci.*, 70, 579, 1981.

METABOLISM OF SULFUR COMPOUNDS

Sulfur is ubiquitous in plants as it is a component of vitamins, coenzymes, and proteins which are essential in intermediary metabolism. In addition, some plants contain secondary organic sulfur compounds, usually in small amounts, which often are of interest because of their powerful and characteristic odors. Many of these have been shown to be simple aliphatic sulfur compounds including thiols (mercaptans),[1] sulfides, and disulfides. Other aliphatic sulfur compounds from plants include sulfonium compounds, sulfoxides, and sulfones. The general structures of these compounds are illustrated in Figure 1. Derivatives in which sulfur is found in a higher oxidation state include sulfonic acids and sulfates, however, these types appear to be found mainly in lower plants. A noteworthy class consists of thiophene derivatives containing substituent groups in the 2- and 5-positions (Figure 1) which include highly unsaturated moieties (e.g., 1-propynyl, R = –C≡C–CH$_3$). One of the most important types of sulfur compound is that of glucosinolates. Their general structure is shown in Figure 1 and R may include simple alkyl groups, various aromatic substituents, or more complex groups. These anionic compounds are normally found as potassium salts, however salts of organic bases also occur, e.g., the basic compound sinapine found in sinalbin. Recently, many thiazole (Figure 1) derivatives have been identified as volatile aromatic constituents of foods.[2] These are generally fairly simple derivatives containing one, two, or three alkyl groups attached to the ring carbons. Benzothiazole, the 2-phenyl derivative, also occurs naturally. Various other sulfur compounds have also been reported, including thioketones and sulfur-containing alkaloids.

The above brief comments indicate that organic sulfur compounds from plants constitute a diverse group of substances. However, knowledge of their metabolism in mammals is scanty.

THIOLS

The simple thiols are volatile compounds having offensive odors. The metabolism of the simplest member of the series, **methanethiol** (methyl mercaptan), was studied *in vivo* in rats by Canellakis and Tarver[3] using ^{35}S- or ^{14}C-labeled compound. The respiratory air and urine were monitored for 6 h following the administration of the latter compound (~1.7 mg/kg, i.p.). Only ~2% of the radioactivity appeared in the urine during this period whereas ~40% was lost as respiratory CO$_2$, mainly during the first hour (29%). About 6% of the dose was lost as volatile sulfur, presumably unchanged compound, during the first hour only. The ^{35}S-compound was administered similarly in four divided doses totaling 10 mg/kg. The urine collected over an 8-h period ending 2 h after the last dose contained 32% of the dose as total sulfur. The values for inorganic sulfate and total sulfate were 29 and 31% of the dose, respectively. These results show that both the carbon and the sulfur of methanethiol are rapidly oxidized to CO$_2$ and sulfate, respectively. A similar study by Derr and Draves[4] in which rats were given [^{35}S]methanethiol (1.8 mg/kg, i.p.) showed that 94% of the radioactivity was excreted in the urine in 21 h. This material appeared to consist entirely of ^{35}S-sulfate. The metabolism of methanethiol to sulfate has also been shown to occur *in vitro* by erythrocytes.[5] Information on this cleavage of the C–S bond was obtained by Mazel et al.[6] in a study of the S-demethylation of several S-methyl compounds including methanethiol by a microsomal system from rat liver. This system, requiring NADPH and O$_2$, converted the thiol to formaldehyde and H$_2$S. However, the S-demethylating system differs from the well-known O- and N-demethylating systems in some respects (e.g., inhibitor effects and inducibility). It is likely that more than one S-demethylating enzyme exists.

Derr and Draves[4] found that an additional 0.5% of the administered [^{35}S]methanethiol was recovered as unchanged compound in the expired air. The loss of some unchanged methanethiol in the expired air following i.p. injection in a mouse was reported by Susman et al.[7] Additionally, they found that some of the thiol was metabolized to dimethyl sulfide which was also lost by this

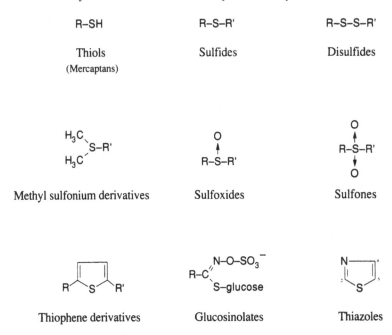

FIGURE 1.   General structures of some sulfur-containing plant compounds.

route. The methylation of methanethiol by a transmethylating enzyme system employing *S*-adenosylmethionine was described by Bremer and Greenberg.[8] This system was found in the microsomal fraction of several rat tissues and in the livers of mice, guinea pigs, rabbits, sheep, and cattle. It methylates xenobiotic but not physiologic thiols. The purification and characterization of this thiol methyltransferase enzyme from rat liver was reported by Borchardt and Cheng[9] and Weisiger and Jakoby.[10] The results of Weisiger et al.[11] on the tissue distribution of this system suggest that its function is to metabolize toxic xenobiotic mercaptans absorbed from the intestine.

The study of Canellakis and Tarver[3] showed that considerable radioactivity from ^{14}C-labeled methanethiol was retained in the tissues and it was found that the methyl carbon was converted to the β-carbon of serine and the methyl groups of methionine, choline, and creatinine. On the other hand, the experiments using [^{35}S]methanethiol indicated that the sulfur did not appear to any significant extent in liver methionine or cysteine. This study indicates that the fate of the methyl group is similar to that of methyl alcohol, entering into the 1-carbon metabolic pool. A metabolic possibility not studied in this investigation is the *S*-methylation of the thiol followed by oxidation to dimethyl sulfone, a reaction which occurs with the closely related ethanethiol which is formed from diethyl disulfide (see below). It seems likely that the method employed for the determination of total sulfur, based on oxidation to sulfate, would not detect the presence of the highly stable sulfone. The nonplant aromatic thiol, thiophenol, undergoes *S*-methylation and oxidation in rats with the result that methylphenylsulfone is excreted in the urine.[12]

Little information is available on the metabolism of **ethanethiol** and **1-propanethiol**. Relevant information on the metabolism of the former compound may be inferred from the investigation by Snow[13] which dealt with the metabolism of diethyldisulfide (see below). Sandmeyer[14] claimed that 1-propanethiol does not enter into the normal pathways of mammalian metabolism. Instead, it appears to undergo sulfur oxidation prior to excretion.

A dithiol, dihydroasparagusic acid, is noted below in the discussion of the metabolism of the disulfide asparagusic acid.

SULFIDES

Some simple sulfides including the dimethyl, diallyl, and dibutyl derivatives occur in cabbage and garlic, however metabolic data on this group are limited. Maw[15] reported that **dimethyl sulfide**, when given orally or by injection to rats, did not result in an increase in the urinary sulfate excretion. Some unchanged compound was detected in the expired air 1 h following injection. Following an i.p. injection of dimethyl sulfide in a mouse, unchanged compound but no dimethyl disulfide or methanethiol was detected in the expired air.[7] Expiration of unchanged dimethyl sulfide was similarly reported in rabbits by Williams et al.[16] who also found that some oxidation to the corresponding sulfoxide and sulfone occurred. In this experiment a total dose (~1.4 g/kg) of dimethyl sulfide was injected subcutaneously in four daily portions and the urine collected for 6 d. About 20% of the dose in the pooled urine was recovered as dimethyl sulfoxide and a further 10% was present as dimethyl sulfone. Lee et al.[17] described a microsomal system from rat liver which oxidized sulfides to the corresponding sulfoxides. Several sulfides including dimethylsulfide were studied by Mazel et al.[6] using the microsomal S-demethylating system from rat liver. As noted in the previous section, this system was active in demethylating the corresponding thiol, however no demethylation was observed when dimethylsulfide was used as the substrate.

Diethyl sulfide has also been reported to be metabolized by sulfur oxidation.[18] When rats were given the sulfide (50 mg/kg), ~4 and 20% of the dose were excreted in the urine as the sulfoxide and sulfone derivatives, respectively.

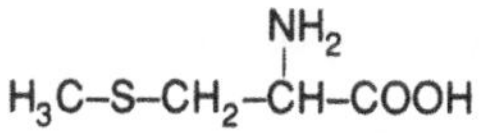

$$\overset{\displaystyle NH_2}{H_3C\text{–}S\text{–}CH_2\text{–}\overset{|}{C}H\text{–}COOH}$$

(1)

S-Methylcysteine

The aforementioned liver microsomal system[6] also demethylated S-**methylcysteine** (1), however other metabolic pathways are available. Binkley[19] described its cleavage by rat liver preparations to form methanethiol and Canellakis and Tarver[20] reported the formation of the same metabolite by a different system from liver mitochondria. Horner and Kuchinskas[21] administered S-methylcysteine labeled with ^{14}C in the methyl group to rats but were unable to detect methanethiol in the expired air. However, it was believed that this metabolite would be rapidly oxidized to CO_2 and, in fact, nearly 40% of the administered radioactivity was lost as $^{14}CO_2$ in 24 h. Smith et al.[22] studied the ability of goat rumen contents to metabolize S-methylcysteine *in vitro*. Analysis of the head-space gases from these incubations showed an initial production of about equal amounts of dimethyl disulfide and methanethiol, with smaller amounts of dimethyl sulfide. The formation of methanethiol from S-methylcysteine by bovine rumen microorganisms was reported by Zikakis and Salsbury.[23]

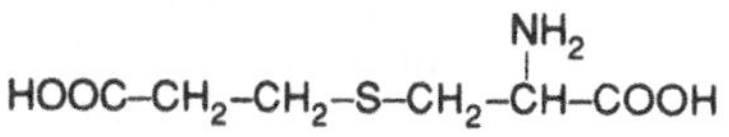

$$\overset{\displaystyle NH_2}{HOOC\text{–}CH_2\text{–}CH_2\text{–}S\text{–}CH_2\text{–}\overset{|}{C}H\text{–}COOH}$$

(2)

$$\overset{\displaystyle NH_2}{HOOC\text{–}\overset{|}{C}H\text{–}CH_2\text{–}S\text{–}CH_2\text{–}S\text{–}CH_2\text{–}\overset{|}{C}H\text{–}COOH}$$

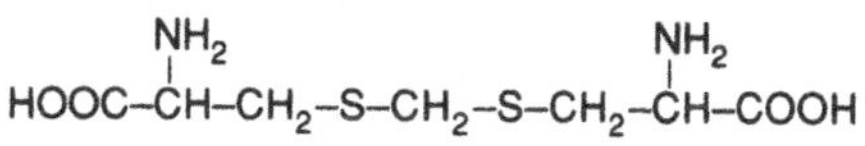

(3)

Djenkolic acid

The metabolism of two sulfur-containing amino acids, S-**2-carboxyethyl-L-cysteine** (2) and **djenkolic acid** (3) was studied by Binkley[19] using an enzyme preparation from rat liver capable of cleaving thioethers. With the former compound, very little cleavage of the carbon-sulfur bond of the cysteine moiety was observed whereas the symmetrical molecule of djenkolic acid was readily split to form cysteine.

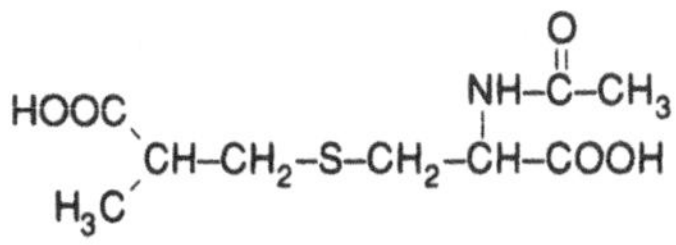

$$\begin{array}{c} O \\ \parallel \\ NH-C-CH_3 \end{array}$$

HOOC
  \
   CH-CH₂-S-CH₂-CH-COOH
  /
H₃C

(4)

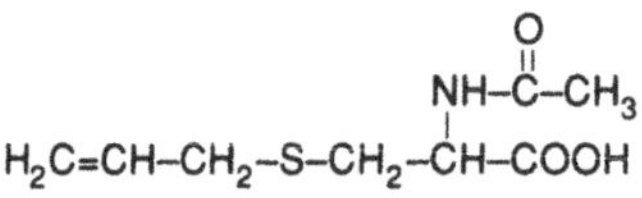

$$\begin{array}{c} O \\ \parallel \\ NH-C-CH_3 \end{array}$$

H₂C=CH-CH₂-S-CH₂-CH-COOH

(5)

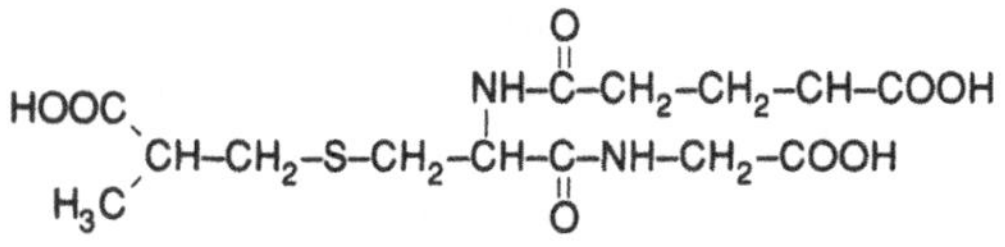

$$\begin{array}{c} O \\ \parallel \\ NH-C-CH_2-CH_2-CH-COOH \end{array}$$

HOOC
  \
   CH-CH₂-S-CH₂-CH-C-NH-CH₂-COOH
  /   ‖
H₃C   O

(6)

$$\begin{array}{c} O \qquad NH_2 \\ \parallel \qquad | \\ NH-C-CH_2-CH_2-CH-COOH \end{array}$$

H₂C=CH-CH₂-S-CH₂-CH-COOH

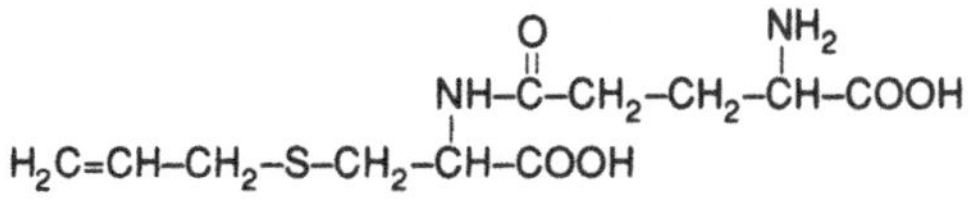

(7)

Jandke and Spiteller[24] identified two mercapturic acid derivatives, *N*-acetyl-*S*-(2-carboxypropyl)cysteine (4) and *N*-acetyl-*S*-allylcysteine (5), in human urine following the consumption of onions or garlic. In view of the fact that these plants are known to contain the glutathione derivative **γ-glutamyl-*S*-(2-carboxypropyl)cysteinylglycine** (6) and the dipeptide **γ-glutamyl-*S*-allylcysteinylglycine** (7) and the knowledge that derivatives of glutathione are normally metabolized to mercapturic acids, they proposed that compounds (4) and (5) are metabolites of the peptides (6) and (7), respectively.

DISULFIDES

Plant disulfides, which are mainly found in cabbage, onion, and garlic, are simple symmetrical or unsymmetrical derivatives containing small groups (e.g., methyl, ethyl, propyl, allyl, and butyl). When **dimethyl disulfide** was given by i.p. administration (~35 to 40 mg/kg) to mice ~6% of the dose was excreted via the lungs as unchanged compound.[7] Excretion was greatest 3 to 6 min after dosing and was complete in less than 30 min. Additionally, both dimethyl sulfide and methanethiol were excreted in the expired air. Each of these accounted for ~0.5% of the dose and excretion was greatest ~6 min after injection of the disulfide.

The metabolism of **diethyl disulfide** (8) (see Figure 2) was studied using ³⁵S-labeled compound.[13] Following two s.c. injections in mice using a total dose of 800 mg/kg, excretion of radioactivity occurred mainly in the urine. Little radioactivity was detected in the feces, however as much as 14% of the dose was found in the respiratory air. This was believed to be due to the reduction product, ethanethiol. In both mice and guinea pigs, the 24-h urines contained 40 to 65% of the dose as relatively nonvolatile radioactive compounds as well as an undetermined amount of ethanethiol. Sulfate accounted for 80 to 90% of the urinary radioactivity and two organic metabolites were also detected chromatographically. The major metabolite was shown to be ethyl methyl sulfone and the other remained unidentified although it was not the corresponding sulfoxide. The sulfone was also found in mouse tissues. The radioactive components separated on chromatograms of rabbit urine were similar except that an additional substance with radioactivity about equal to that in the sulfone peak and having slightly greater polarity was observed. In view of these findings it seems likely that the major metabolic pathways of diethyl disulfide are those shown in Figure 2.

Pushpendran et al.[25] administered [³⁵S]**diallyl disulfide** to mice by i.p. injection and studied the distribution and nature of the radioactivity in the liver. Subcellular fractionation of liver homogenates after 2 h showed that >70% of the radioactivity was found in the cytosol fraction and, of this, ~90% remained in the supernatant after the proteins were precipitated. More than 80% of the radioactivity present in the cytosol was due to sulfate and only 8% was unchanged compound.

481

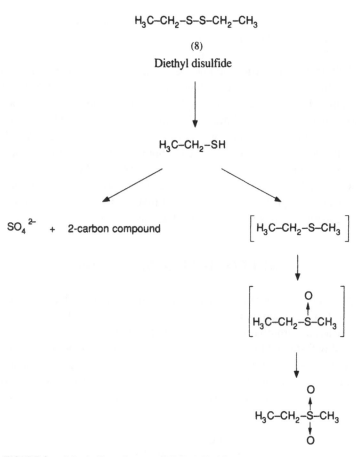

FIGURE 2. Metabolic pathways of diethyl disulfide. Postulated intermediates shown in brackets.

(9)

Asparagusic acid

(10)

Dihydroasparagusic acid

(11)

(12)

Waring et al.[26] studied the nature of the urinary odor produced by man after eating asparagus. This capacity, which is genetically determined and occurred in ~40% of the subjects in a British study,[27] results in the formation of up to six sulfur-containing compounds (methanethiol, dimethyl sulfide, dimethyl disulfide, bis-(methylthio)methane, dimethyl sulfoxide, and dimethyl sulfone). Interestingly, a similar study employing French subjects showed that all of these individuals produced the characteristic odorous urines.[28] The former study indicated that these compounds were also produced when susceptible subjects were given the asparagus constituent **asparagusic acid** (9) orally. No urinary odor was produced when asparagusic acid was administered to an individual who did not possess the odor-producing ability. It was considered likely that asparagusic acid may be metabolized by reduction to **dihydroasparagusic acid** (2,2′-dithiolisobutyric acid) (10) which undergoes further metabolism to the urinary sulfur compounds noted above. Interestingly, Jansen[29] found no urinary odor when two individuals were given oral doses (10 mg) of dihydroasparagusic acid. This negative finding was taken as evidence that compound (10) was not the odor-producing precursor. The newer data demonstrating the polymorphic nature of the phenomenon shows that this conclusion may be incorrect because it is not known to which group the two subjects belonged. Results of an earlier study[30] indicated that the agents responsible for the urinary odor following asparagus ingestion were the thioesters methyl thioacrylate (11) and methyl 3-(methylthio)thiopropionate (12). However, the asparagus constituents from which these compounds are derived were not identified.

SULFONIUM COMPOUNDS

This group is represented by the thetins, the general formula of which is shown in structure (13). Past interest in these compounds derived partly from the fact that they share some of the properties of quaternary ammonium compounds and also that they can enter into transmethylation reactions. Maw[15] studied the metabolism of four simple alkyl thetins including **dimethyl-β-propiothetin** (14) (see Figure 3) which has been reported to occur in some lower plants. In thetins containing one or two methyl groups, oxidation to sulfate was observed. With dimethyl-β-propiothetin two-thirds of the sulfur was excreted in the urine as inorganic sulfate in 2 d after an oral dose (~13 mg S/rat).

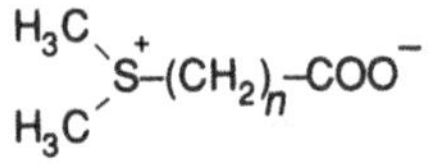

$$H_3C \diagdown \overset{+}{S} - (CH_2)_n - COO^-$$
$$H_3C \diagup$$

(13)

Maw[15] also directed attention to the possible intermediates in sulfate formation from the thetins, noting earlier work which showed that one methyl group may undergo transmethylation and subsequent oxidation to CO_2. The products thus formed are S-alkyl derivatives of acetic or propionic acid. In the former case the metabolites (e.g., S-methyl- or S-ethylthioglycollate), when themselves administered to rats, gave rise to appreciable excretion of urinary sulfate. This result of high inorganic sulfate excretion was also obtained with the corresponding thiol, thioglycollic acid, a finding which was subsequently confirmed by Freeman et al.[31] Interestingly, the latter investigation showed that species differences in thioglycollate metabolism occur. Monkeys excreted the sulfur mainly in the neutral sulfur fraction whereas rabbits excreted it in the neutral sulfur and organic sulfate fractions. In view of the results obtained in rats, the probable metabolic pathway for dimethyl-β-propiothetin is that shown in Figure 3.

The participation of the intestinal microflora in the metabolism of thetins must also be considered and Zikakis and Salsbury[23] found that dimethyl thetin (structure (13), $n = 1$) was converted to dimethyl sulfide when incubated with bovine rumen microorganisms.

SULFOXIDES

Little information is available on the metabolism of plant sulfoxides, however both oxidation to sulfones and reduction to sulfides may be expected. This was shown to be the case with the simplest sulfoxide, dimethyl sulfoxide, which was metabolized to the corresponding sulfone and

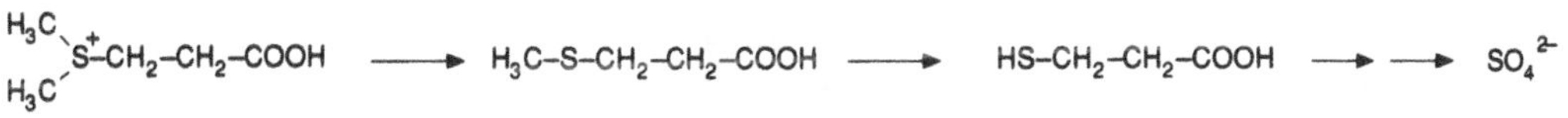

$$H_3C\overset{+}{\underset{H_3C}{\diagup}}S-CH_2-CH_2-COOH \longrightarrow H_3C-S-CH_2-CH_2-COOH \longrightarrow HS-CH_2-CH_2-COOH \longrightarrow \longrightarrow SO_4^{2-}$$

(14)

Dimethyl- β-propiothetin

FIGURE 3.   Probable metabolic pathways of dimethyl-b-propiothetin in rats.

sulfide when given by s.c. injection to rabbits.[16] Dimethyl sulfide was rapidly excreted in the expired air from cats given this sulfoxide by i.p dosage.[32] Reduction of the sulfoxide moiety may be carried out by tissue enzymes or by the intestinal microflora.[33] Aymard et al.[34] described a system in rat liver and kidney which reduced methionine sulfoxide.

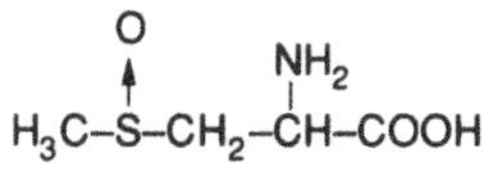

$$H_3C-\overset{O}{\overset{\uparrow}{S}}-CH_2-\overset{NH_2}{\underset{|}{C}H}-COOH$$

(15)

S-Methylcysteine sulfoxide

The best known sulfoxides from plants are derivatives of cysteine. The lachrymatory factor in onion results from the breakdown of S-(prop-1-enyl)cysteine sulfoxide. Its isomer, the corresponding S-allyl derivative (alliin), is the precursor of the active principle of garlic. The only metabolic data from animals available deal with the simpler analogue, **S-methylcysteine sulfoxide** (15), which occurs in onions and several crucifers including turnips and kale. Smith et al.[22] studied the ability of goat rumen contents to metabolize S-methylcysteine sulfoxide *in vitro*. Analysis of the head-space gases from these incubations showed the presence of dimethyl disulfide, methanethiol, and dimethyl sulfide. The latter compound, a minor component, showed a proportionate increase in later samples, as did methanethiol. The disulfide was the dominant component during the earlier stages.

GLUCOSINOLATES

The glucosinolates (thioglucosides, mustard oil glycosides) are important flavor constituents in many plants, especially crucifers (e.g., mustard, horseradish, and water cress). Also present in the plant is an enzyme system, known as myrosinase or thioglucosidase, which is released upon injury to the plant and effects the hydrolysis of the glycoside. As shown in Figure 4 this reaction leads to the formation of isothiocyanates or mustard oils. These secondary products are responsible for the notable physiological properties of this group of compounds. In addition to the major route leading to isothiocyanates, breakdown of the glucosinolates can also lead to thiocyanates (R–S–C≡N), nitriles (R–C≡N), and, in some cases, heterocyclic ring compounds (e.g., the oxazolidine-2-thione derivative goitrin and epithionitriles). The chemical and biological properties of the glucosinolates was reviewed by Tookey et al.[35]

The metabolism of **sinalbin** (16) given to rats (~400 mg/kg, p.o.) was studied by Griffiths,[36] however interest was shown mainly in the sinapine (17) moiety of sinalbin. It was found that it underwent ester hydrolysis with the result that sinapic acid (18) and dihydrosinapic acid (19) were excreted in the 24-h urines. On the second day a further metabolite, 3-hydroxy-5-methoxyphenylpropionic acid (20) was excreted. A discussion of the metabolism of these and related methoxylated C_6–C_3-acids, including the role played by the intestinal microflora, is found in the chapter entitled Metabolism of Acids, Lactones, and Esters. A fourth metabolite, p-hydroxybenzoic acid, was detected in the 24-h urines and it seems evident that this compound arises from the glucosinalbate (4-hydroxybenzylglucosinolate) moiety.

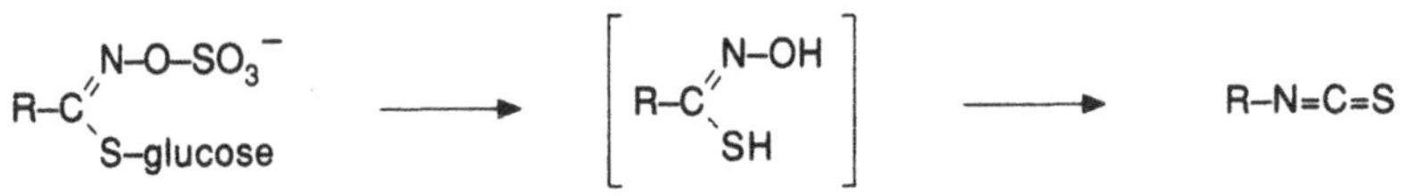

FIGURE 4.   Hydrolysis of glucosinolates to isothiocyanates.

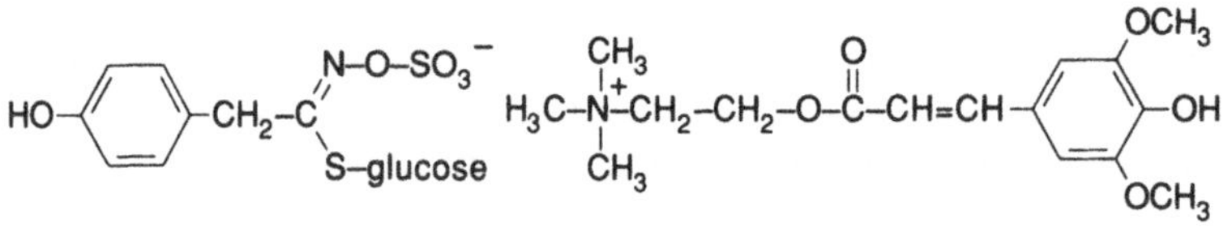

(16)

Sinalbin

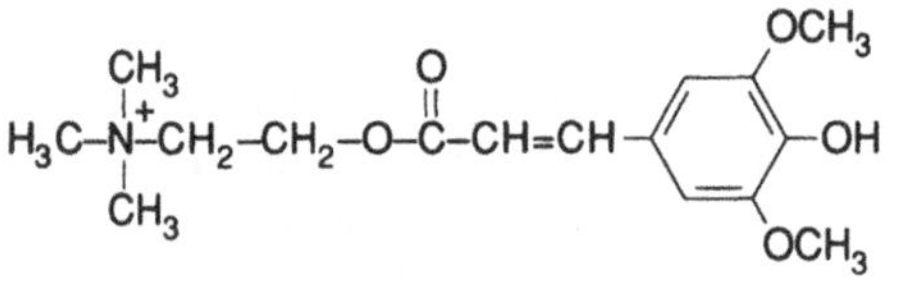

(17)

Sinapine

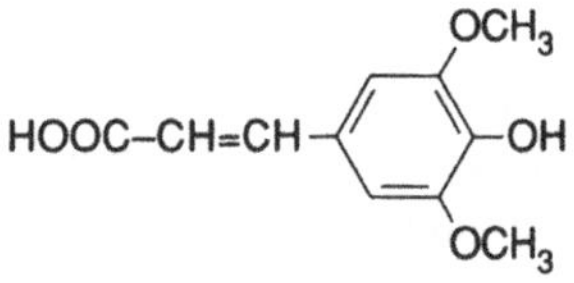

(18)

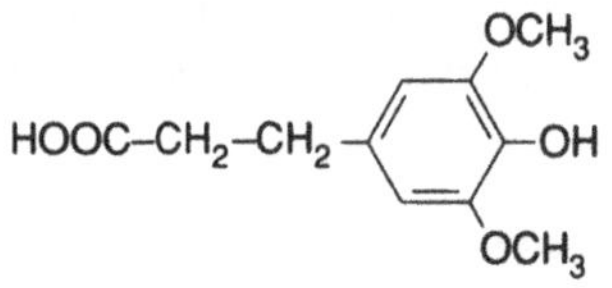

(19)

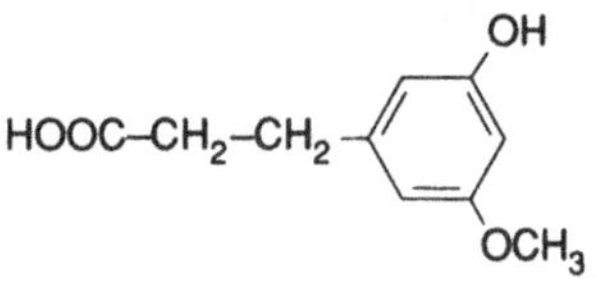

(20)

The site of glucosinolate hydrolysis is probably the intestinal lumen, the reaction being carried out by the intestinal bacteria. The ionized nature of the glucosinolate would be expected to delay its absorption from the intestine, thus allowing it to reach the lower intestine where it could be metabolized by bacterial enzymes. This conclusion was reached by Greer[37] who studied the conversion of **progoitrin** (21) to the active goitrogenic compound **goitrin** (22) in rats and humans. Oginsky et al.[38] isolated representatives from several genera of human fecal bacteria which were able to carry out this reaction *in vitro*. Thioglucosidase activity appears to be widely distributed among intestinal bacteria and, in the strains tested, was higher in *Paracolobactrum*, *Proteus vulgaris*, and *Bacillus subtilis*. When oral doses (50 to 100 mg/kg) of progoitrin were given to germ-free and conventional rats, significantly more unchanged compound was excreted in the urine and feces in the former group.[39] The conversion of progoitrin to goitrin was also shown to be carried out by sheep rumen bacteria.[40] They found that the thioglucoside was also converted to 5-vinyl-3-thiazolidine-2-one (23), an isomer of goitrin, under these conditions. However, goitrin was not metabolized to compound (23) when incubated with rumen fluid.

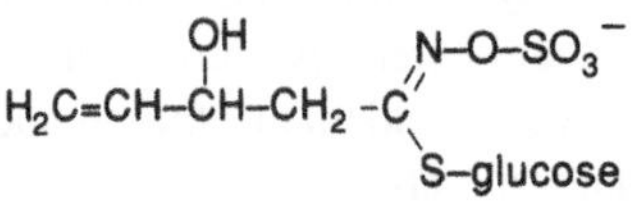

(21)

Progoitrin

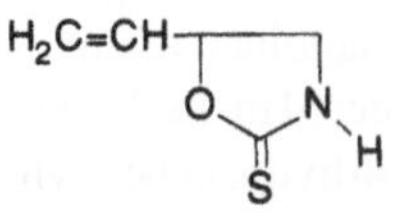

(22)

Goitrin

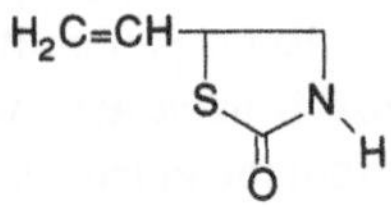

(23)

An important metabolic pathway for the isothiocyanates liberated from glucosinolates involves conjugation with glutathione. The tripeptide derivatives formed will undergo further metabolism to mercapturic acids which are excreted in the urine. Brüsewitz et al.[41] investigated this pathway using **benzyl isothiocyanate** (24). It reacted rapidly with glutathione both spontaneously and enzymatically to form the tripeptide conjugate. The latter was degraded to the cysteine conjugate and then acetylated to give the corresponding mercapturic acid (*N*-acetyl-*S*-(*N*-benzylthiocarbamoyl)-L-cysteine) (25). This sequence of reactions was observed *in vitro* when rat liver or kidney homogenates were used. Also, metabolite (25) was excreted in the urine of rats given benzyl isothiocyanate orally. Mennicke et al.[42] reported that rats given benzyl isothiocyanate (~30 mg/kg, p.o.) excreted 40% of the dose in the urine in 24 h in the form of metabolite (25). This compound is also the major metabolite of benzyl isothiocyanate in man. Mennicke et al.[43] found that ~54% of the dose (~0.2 mg/kg, p.o.) was excreted in the urine in 24 h as the cysteine derivative (25). The metabolite was excreted rapidly with maximal amounts appearing in 2 to 6 h. No other metabolites, including the cyclic derivative noted below, were detected.

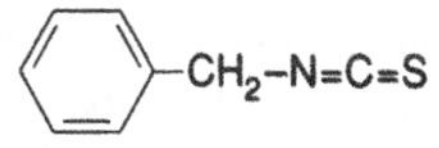

(24)

Benzyl isothiocyanate

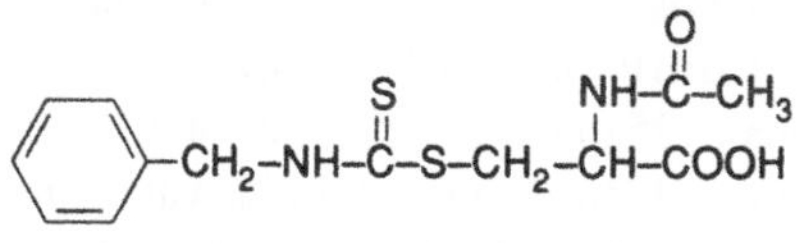

(25)

In contrast to the results summarized above, the metabolism of benzyl isothiocyanate in dogs[41] and guinea pigs and rabbits[44] leads to the formation of other major metabolites. In fact, when dogs were given the cysteine conjugate of benzyl isothiocyanate orally they failed to form the *N*-acetylated metabolite (25). The major urinary metabolite was found to be hippuric acid. This suggests that the cysteine conjugate may undergo cleavage in dogs to benzylamine, which is known to be oxidized via benzaldehyde to benzoic acid and then conjugated with glycine to give hippuric acid (see the chapter entitled Metabolism of Amino Compounds).

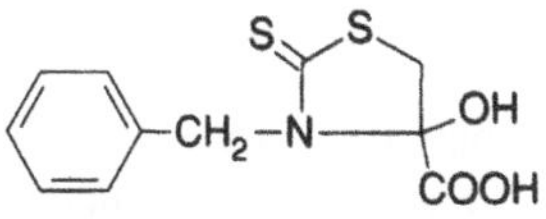

(26)

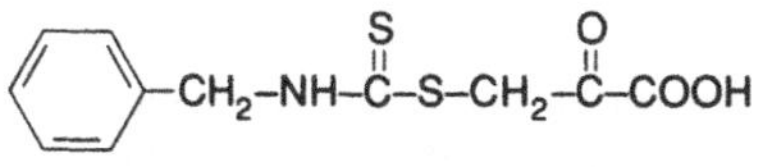

(27)

The metabolism of benzyl isothiocyanate in guinea pigs and rabbits, two species known to be poor producers of mercapturic acids, also reveals a difference in the final step leading to the formation of metabolite (25).[44] Although small amounts of this metabolite were excreted in the urine of guinea pigs or rabbits given benzyl isothiocyanate or its cysteine conjugate, the major metabolite was a cyclic compound 4-hydroxy-4-carboxy-3-benzylthiazolidin-2-thione (26). This metabolite is formed by the transamination of the cysteine conjugate to the *S*-substituted mercaptopyruvate derivative (27) which, via the enol form, spontaneously cyclizes to compound (26).

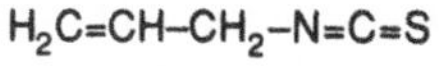

(28)

Allyl isothiocyanate

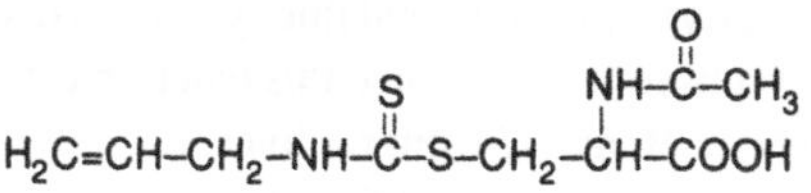

(29)

Mercapturic acid derivatives similar to compound (25) are also excreted by rats given **methyl isothiocyanate, ethyl isothiocyanate, allyl isothiocyanate** (28), and **butyl isothiocyanate.**[45] Following oral doses (~30 to 40 mg/kg) the urinary excretion values (24 h) for the mercapturic acids were 47, 30, 37, and 24%, respectively.[42] More extensive studies on the metabolism of allyl isothiocyanate in mice and rats were reported by Ioannou et al.[46] The compound they employed was uniformly labeled with ^{14}C and was given to male and female animals at dose levels of 2.5 and 25 mg/kg (p.o. and i.p.). In all cases ~75% of the radioactivity was excreted in the urine in 24 h. The finding that 13 to 15% was recovered in the expired CO_2 from rats given the larger dose (p.o.) indicates that the allyl group may also undergo extensive metabolism. Little (1 to 5%) of the radioactivity was found in the feces, however the values for biliary excretion were three- to fivefold higher. This suggests that some of the material underwent enterohepatic circulation. The major urinary metabolite in rats was the mercapturic acid derivative N-acetyl-S-(N-allylthiocarbamoyl)-L-cysteine (29), however four additional metabolites of undetermined nature were also detected. Mice excreted four major (including compound (29)) and two minor metabolites in the urine. Some sex-related differences were observed in the relative amounts of these compounds. A single unidentified metabolite, which was also the major metabolite in rat bile, was found in the feces of both species. Additional experiments using uniformly labeled allyl isothiocyanate (25 mg/kg, p.o.) in rats of different ages (3, 16, and 27 months) were carried out by Borghoff and Birnbaum.[47] Similar values for the urinary excretion of radioactivity were obtained (65 to 80% of the dose in 72 h) in these groups. The main urinary metabolite was the mercapturic acid derivative (29) and the extent of its excretion did not change with age. The relative amounts of the minor urinary metabolites noted above did not change significantly with age, however an additional compound was excreted in the two older age groups. Some age-related changes were noted in the excretion of radioactivity in the feces and in the expired air in the form of CO_2 or as volatile compounds.

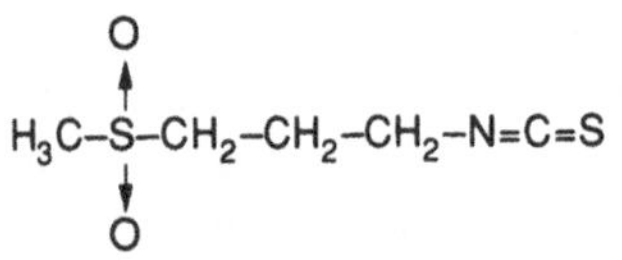

(30)

Cheiroline

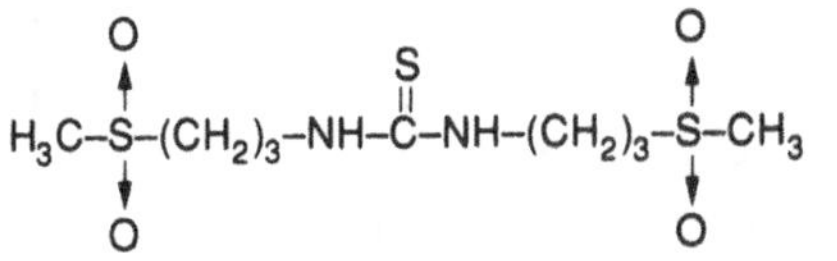

(31)

Evidence for another route of metabolism for isothiocyanate derivatives was obtained by Bachelard and Trikojus[48] in a study of the metabolism of the goitrogenic compound **cheiroline** (30). They found that incubation of cheiroline with rumen liquor resulted in the formation of dicheiroline thiourea (31). It is not known if this intestinal reaction of cheiroline is a general reaction of isothiocyanates.

As noted above, glucosinolates may undergo degradation to thiocyanates (R–S≡C=N) and it is possible that their mammalian metabolism also involves these intermediates. Ohkawa and Casida[49] and Ohkawa et al.[50] studied the ability of mouse liver homogenates and subcellular fractions to liberate HCN from numerous synthetic organic thiocyanates. These included several lower alkyl thiocyanates and **benzyl thiocyanate**. These studies showed that the formation of HCN from the thiocyanates was catalyzed by glutathione S-transferases which bring about the attack of glutathione at the thiocyanate sulfur. This activity was found in the soluble and not the microsomal fraction, and the reaction products detected were HCN, oxidized glutathione, and the thiol derivative of the original thiocyanate. This reaction sequence is illustrated in Figure 5. Habig et al.[51] subsequently studied this reaction using homogeneous preparations of glutathione S-transferases A, B, and C from rat liver. Employing ethyl thiocyanate, octyl thiocyanate, and benzyl thiocyanate, they found that the reaction products were HCN and the mixed disulfide formed from the substrate and glutathione. Thus, these purified preparations carry out the first step shown in Figure 5.

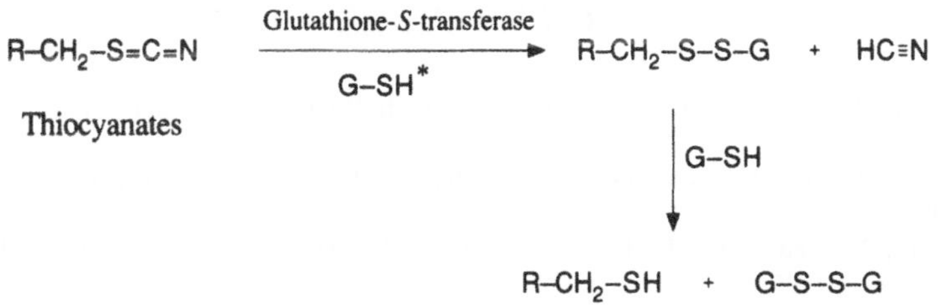

$$R-CH_2-S=C=N \xrightarrow[\text{G-SH}^*]{\text{Glutathione-}S\text{-transferase}} R-CH_2-S-S-G \quad + \quad HC\equiv N$$

Thiocyanates

$$\downarrow G\text{-SH}$$

$$R-CH_2-SH \quad + \quad G-S-S-G$$

FIGURE 5.   Metabolism of organic thiocyanates. (*G-SH) Glutathione.

The sequence of metabolic events occurring with organic thiocyanates *in vivo* is not known, however Ohkawa et al.[50] also administered the compounds noted above to mice. After 15 min they measured the concentrations of HCN and its metabolite thiocyanate (SCN⁻) present in the liver and brain. High levels of these compounds were found, especially in the liver, with most of the thiocyanates tested and these results indicated that HCN liberation occurred *in vivo*. The fate of the disulfides and thiols formed is not known, however it is possible that these compounds may be metabolized along the pathways shown above in Figure 2 for diethyl disulfide and its thiol derivative.

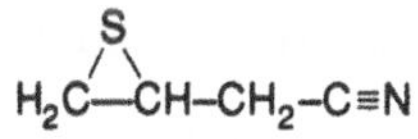

$$H_2C-\overset{\displaystyle S}{\underset{\displaystyle }{\diagup\!\!\!\diagdown}}CH-CH_2-C\equiv N$$

(32)

3,4-Epithiobutanenitrile

$$N\equiv C-CH_2-\underset{\underset{\displaystyle SH}{|}}{CH}-CH_2-S-CH_2-\underset{\underset{\displaystyle NH-\overset{\overset{\displaystyle O}{||}}{C}-CH_3}{|}}{CH}-COOH$$

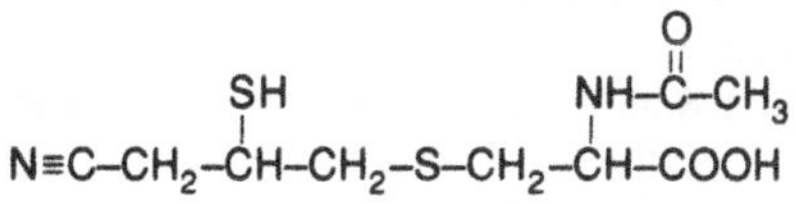

(33)

The nitriles (R–C≡N) formed from the breakdown of glucosinolates may include sulfur compounds of the thiirane type. Thus, **3,4-epithiobutanenitrile** (32) is formed from allylglucosinolate. Brocker et al.[52] administered ¹⁴C- and ³⁵S-labeled 3,4-epithiobutanenitrile to rats and found that nearly half of the dose (~11 mg/kg, p.o. or i.p.) was excreted in the urine as a mercapturic acid derivative having the probable structure (33). Using the ¹⁴C-labeled compound, ~95% of the radioactivity was excreted in the urine in 72 h. Six additional urinary metabolites of undetermined structure were detected. Wallig et al.[53] found that large oral doses (125 mg/kg) of compound (32) given daily for 3 d produced only a slight increase in the excretion of urinary thiocyanate in rats. No significant increases were noted following one or two doses.

THIAZOLES

As noted above, numerous thiazole derivatives are known to be volatile aroma constituents in foods. While no published reports on the mammalian metabolism of these compounds are available, Scheline[54] found mass spectrometric evidence for the presence of thiazole-4-carboxylic acid in the urine of rats treated orally with 4-methylthiazole (34) at a dose level of 200 mg/kg.

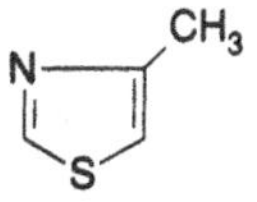

(34)

4-Methylthiazole

REFERENCES

1. **Maga, J. A.,** The role of sulfur compounds in food flavor. III. Thiols, *CRC Critical Rev. Food Sci. Nutr.*, 7, 147, 1976.
2. **Maga, J. A.,** The role of sulfur compounds in food flavor. I. Thiazoles, *CRC Critical Rev. Food Sci. Nutr.*, 6, 153, 1975.
3. **Canellakis, E. S. and Tarver, H.,** The metabolism of methyl mercaptan in the intact animal, *Arch. Biochem. Biophys.*, 42, 446, 1953.
4. **Derr, R. F. and Draves, K.,** Methanethiol metabolism in the rat, *Res. Commun. Chem. Pathol. Pharmacol.*, 39, 503, 1983.
5. **Blom, H. J. and Tangerman, A.,** Methanethiol metabolism in whole blood, *J. Lab. Clin. Med.*, 111, 606, 1988.
6. **Mazel, P., Henderson, J. F., and Axelrod, J.,** S-Demethylation by microsomal enzymes, *J. Pharmacol. Exp. Ther.*, 143, 1, 1964.
7. **Susman, J. L., Hornig, J. F., Thomae, S. C., and Smith, R. P.,** Pulmonary excretion of hydrogen sulfide, methanethiol, dimethyl sulfide and dimethyl disulfide in mice, *Drug Chem. Toxicol.*, 1, 327, 1978.
8. **Bremer, J. and Greenberg, D. M.,** Enzymic methylation of foreign sulfhydryl compounds, *Biochim. Biophys. Acta*, 46, 217, 1961.
9. **Borchardt, R. T. and Cheng, C. F.,** Purification and characterization of rat liver microsomal thiol methyltransferase, *Biochim. Biophys. Acta*, 522, 340, 1978.
10. **Weisiger, R. A. and Jakoby, W. B.,** Thiol S-methyltransferase from rat liver, *Arch. Biochem. Biophys.*, 196, 631, 1979.
11. **Weisiger, R. A., Pinkus, L. M., and Jakoby, W. B.,** Thiol S-methyltransferase: suggested role in detoxication of intestinal hydrogen sulfide, *Biochem. Pharmacol.*, 29, 2885, 1980.
12. **McBain, J. B. and Menn, J. J.,** S-Methylation, oxidation, hydroxylation and conjugation of thiophenol in the rat, *Biochem. Pharmacol.*, 18, 2282, 1969.
13. **Snow, G. A.,** The metabolism of compounds related to ethanethiol, *Biochem. J.*, 65, 77, 1957.
14. **Sandmeyer, E. E.,** Organic sulfur compounds, in *Patty's Industrial Hygiene and Toxicology*, Vol. 2A, 3rd ed., Clayton, G. D. and Clayton, F. E., Eds., Wiley-Interscience, New York, 1981, 2061.
15. **Maw, G. A.,** The oxidation of dimethylthetin and related compounds to sulphate in the rat, *Biochem. J.*, 55, 42, 1953.
16. **Williams, K. I. H., Burstein, S. H., and Layne, D. S.,** Metabolism of dimethyl sulfide, dimethyl sulfoxide, and dimethyl sulfone in the rabbit, *Arch. Biochem. Biophys.*, 117, 84, 1966.
17. **Lee, Y. C., Joiner Hayes, M. G., and McCormick, D. B.,** Microsomal oxidation of α-thiocarboxylic acids to sulfoxides, *Biochem. Pharmacol.*, 19, 2825, 1970.
18. **Hoodi, A. A. and Damani, L. A.,** Sulphoxidation of alicyclic and aliphatic sulphides, First Internat. Symp. Foreign Comp. Metab., Oct. 30 to Nov. 4, 1983, West Palm Beach, FL, 1983.
19. **Binkley, F.,** Enzymatic cleavage of thioethers, *J. Biol. Chem.*, 186, 287, 1950.
20. **Canellakis, E. S. and Tarver, H.,** Studies on protein synthesis *in vitro*. IV. Concerning the apparent uptake of methionine by particulate preparation from liver, *Arch. Biochem. Biophys.*, 42, 387, 1953.
21. **Horner, W. H. and Kuchinskas, E. J.,** Metabolism of methyl-labeled S-methylcysteine in the rat, *J. Biol. Chem.*, 234, 2935, 1959.
22. **Smith, R. H., Earl, C. R., and Matheson, N. A.,** The probable role of S-methylcysteine sulphoxide in kale poisoning in ruminants, *Biochem. Soc. Trans.*, 2, 101, 1974.
23. **Zikakis, J. P. and Salsbury, R. L.,** Metabolism of sulfur amino acids by rumen microorganisms, *J. Dairy Sci.*, 52, 2014, 1969.
24. **Jandke, J. and Spiteller, G.,** Unusual conjugates in biological profiles originating from consumption of onions and garlic, *J. Chromatogr.*, 421, 1, 1987.
25. **Pushpendran, C. K., Devasagayam, T. P. A., Chintalwar, G. J., Banerji, A., and Eapen, J.,** The metabolic fate of [^{35}S]-diallyl disulphide in mice, *Experientia*, 36, 1000, 1980.
26. **Waring, R. H., Mitchell, S. C., and Fenwick, G. R.,** The chemical nature of the urinary odour produced by man after asparagus ingestion, *Xenobiotica*, 17, 1363, 1987.
27. **Mitchell, S. C., Waring, R. H., Land, D., and Thorpe, W. V.,** Odorous urine following asparagus ingestion in man, *Experientia*, 43, 382, 1987.
28. **Richer, C., Decker, N., Belin, J., Imbs, J. L., Montastruc, J. L., and Giudicelli, J. F.,** Odorous urine in man after asparagus, *Br. J. Clin. Pharmacol.*, 27, 640, 1989.
29. **Jansen, E. F.,** The isolation and identification of 2,2′-dithiolisobutyric acid from asparagus, *J. Biol. Chem.*, 176, 657, 1948.
30. **White, R. H.,** Occurrence of S-methyl thioesters in urines of humans after they have eaten asparagus, *Science*, 189, 810, 1975.

31. **Freeman, M. V., Draize, J. H., and Smith, P. K.,** Some aspects of the absorption, distribution and excretion of sodium thioglycolate, *J. Pharmacol. Exp. Ther.,* 118, 304, 1956.

32. **Distefano, V. and Borgstedt, H. H.,** Reduction of dimethylsulfoxide to dimethylsulfide in the cat, *Science,* 144, 1137, 1964.

33. **Renwick, A. G., Strong, H. A., and George, C. F.,** The role of the gut flora in the reduction of sulphoxide containing drugs, *Biochem. Pharmacol.,* 35, 64, 1986.

34. **Aymard, C., Seyer, L., and Cheftel, J.-C.,** Enzymatic reduction of methionine sulfoxide. *In vitro* experiments with rat liver and kidney, *Agric. Biol. Chem.,* 43, 1869, 1979.

35. **Tookey, H. L., VanEtten, C. H., and Daxenbichler, M. E.,** Glucosinolates, in *Toxic Constituents of Plant Foodstuffs,* 2nd ed., Liener, I. E., Ed., Academic Press, New York, 1980, 103.

36. **Griffiths, L. A.,** Metabolism of sinapic acid and related compounds in the rat, *Biochem. J.,* 113, 603, 1969.

37. **Greer, M. A.,** The natural occurrence of goitrogenic agents, *Recent Prog. Horm. Res.,* 18, 187, 1962.

38. **Oginsky, E. L., Stein, A. E., and Greer, M. A.,** Myrosinase activity in bacteria as demonstrated by the conversion of progoitrin to goitrin, *Proc. Soc. Exp. Biol. Med.,* 119, 360, 1965.

39. **Macholz, R., Kujawa, M., Schulze, J., Lewerenz, H. J., and Schnaak, W.,** The metabolism of some xenobiotics in germ-free and conventional rats, *Arch. Toxicol.,* Suppl. 8, 373, 1985.

40. **Lanzani, A. and Piana, G.,** Changes in *Brassica napus* progoitrin induced by sheep rumen fluid, *J. Am. Oil Chem. Soc.,* 51, 517, 1974.

41. **Brüsewitz, G., Cameron, B. D., Chasseaud, L. F., Görler, K., Hawkins, D. R., Koch, H., and Mennicke, W. H.,** The metabolism of benzyl isothiocyanate and its cysteine conjugate, *Biochem. J.,* 162, 99, 1977.

42. **Mennicke, W. H., Kral, T., Krumbiegel, G., and Rittmann, N.,** Determination of *N*-acetyl-*S*-(*N*-alkylthiocarbamoyl)-L-aysteine, a principal metabolite of alkyl isothiocyanates, in rat urine, *J. Chromatogr.,* 414, 19, 1987.

43. **Mennicke, W. H., Görler, K., Krumbiegel, G., Lorenz, D., and Rittmann, N.,** Studies on the metabolism and excretion of benzyl isothiocyanate in man, *Xenobiotica,* 18, 441, 1988.

44. **Görler, K., Krumbiegel, G., Mennicke, W. H., and Siehl, H.-U.,** The metabolism of benzyl isothiocyanate and its cysteine conjugate in guinea pigs and rabbits, *Xenobiotica,* 12, 535, 1982.

45. **Mennicke, W. H., Görler, K., and Krumbiegel, G.,** Metabolism of some naturally occurring isothiocyanates in the rat, *Xenobiotica,* 13, 203, 1983.

46. **Ioannou, Y. M., Burka, L. T., and Matthews, H. B.,** Allyl isothiocyanate: comparative disposition in rats and mice, *Toxicol. Appl. Pharmacol.,* 75, 173, 1984.

47. **Borghoff, S. J. and Birnbaum, L. S.,** Age-related changes in the metabolism and excretion of allyl isothiocyanate. A model compound for glutathione conjugation, *Drug Metab. Dispos.,* 14, 417, 1986.

48. **Bachelard, H. S. and Trikojus, V. M.,** Plant thioglycosides and the problem of endemic goitre in Australia, *Nature (London),* 185, 80, 1960.

49. **Ohkawa, H. and Casida, J. E.,** Glutathione *S*-transferases liberate hydrogen cyanide from organic thiocyanates, *Biochem. Pharmacol.,* 20, 1708, 1971.

50. **Ohkawa, H., Ohkawa, R., Yamamoto, I., and Casida, J. E.,** Enzymatic mechanisms and toxicological significance of hydrogen cyanide liberation from various organothiocyanates and organonitriles in mice and houseflies, *Pestic. Biochem. Physiol.,* 2, 95, 1972.

51. **Habig, W. H., Keen, J. H., and Jakoby, W. B.,** Glutathione *S*-transferase in the formation of cyanide from organic thiocyanates and as an organic nitrite reductase, *Biochem. Biophys. Res. Commun.,* 64, 501, 1975.

52. **Brocker, E. R., Benn, M. H., Lüthy, J., and von Däniken, A.,** Metabolism and distribution of 3,4-epithiobutanenitrile in the rat, *Food Chem. Toxicol.,* 22, 227, 1984.

53. **Wallig, M. A., Gould, D. H., Fettman, M. J., and Willhite, C. C.,** Comparative toxicities of the naturally occurring nitrile 1-cyano-3,4-epithiobutane and the synthetic nitrile *n*-valeronitrile in rats: differences in target organs, metabolism and toxic mechanisms, *Food Chem. Toxicol.,* 26, 149, 1988.

54. **Scheline, R. R.,** Unpublished data, 1973.

INDEX

A

Abietane derivatives, 206
(–)-Abietic acid, 206
Acacetin, 269
Acetaldehyde, 85
Acetic acid, quercetin and, 273
Acetoin, 102
Acetoisovanillone, 116, 117
Acetone, 99, 100
Acetone cyanohydrin, 340
Acetophenone, 114
  cinnamic acid and, 150
Acetophenone derivatives, 75, 114—118
Acetovanillone, 116, 117
  γ-oryzanol and, 216
Acetoveratrone, 116—117
Acetylacrolein, 243
5-Acetylamino-6-amino-3-methyluracil, 438
5-Acetylamino-6-formylamino-3-methyluracil, 438
Acetylenes, 4
3-Acetyl-5-hydroxymethylumbelliferone, see
  Armillarisin A
N-Acetylisojuripidine, 231
N-Acetylmescaline, 317
Acetylsalicylic acid, 158, see also Salicylic acid
N-Acetyl-S-allylcysteine, 480
N-Acetyl-S-(2-carboxypropyl)cysteine, 480
N-Acetyl-S-(N-allylthiocarbamoyl)-L-cysteine, 486
N-Acetyl-S-(N-benzylthiocarbamoyl)-L-cysteine, 485
Acids, 139—175, see also specific acids
  alicyclic, 142—145
  aliphatic, 139—142
  aromatic, 146—175
Acorn sugar, see D-Quercitol
Acrolein, nitro compounds and, 346
Acronine, see Acronycine
Acronycine, 442
Acyclic terpene alcohols, 33—35
Acyclic terpene aldehydes, 86—87
Acyclic terpene hydrocarbons, 4—5
S-Adenosylmethionine, 162
  morphine and, 410
Adipic acid, 141
Adonitoxin, 229
Aerobacter aerogenes, 160, 162, 168, 170
Ajmalicine, see Raubasine
Ajmaline, 384
Albiflorin, 292, 294
Alcohol dehydrogenase, 27, 30
Alcohols, 27—44, see also specific alcohols
  aliphatic, 28, 30—33
  allyl, isomeric, 72
  aromatic, 41—44
  direct conjugation of, 28
  isomeric, n-alkanes and, 4
  monoterpenoid, 33—41
    acyclic, 33—35

bicyclic, 39—41
monocyclic, 35—39
oxidation of, 27
primary vs. secondary, 27
triterpenoid, 208
Aldehydes, 27, 85—99
  aliphatic, 85—86
  aromatic, 89—99
  monoterpenoid, 86—89
    acyclic, 86—87
    bicyclic, 88—89
    monocyclic, 87—88
Alicyclic acids, 142—145
Alicyclic esters, 181—182
Alicyclic ketones, 102—103
Aliphatic acids, 139—142
Aliphatic alcohols, 28, 30—33
Aliphatic aldehydes, 85—86
Aliphatic hydrocarbons, 1—4
  polyacetylenic, 21
Aliphatic ketones, 99—103
Aliphatic monoamines, 307, see also specific
  monoamines
Aliphatic nitriles, 342
Aliphatic nitrogen compounds, 352
Aliphatic polyamines, 308, see also specific
  polyamines
Alizarin, 123—124
Alizarin-1-methyl ether, 124
Alkaloidal purines, 429—441, see also specific
  purines
Alkaloids, 441—446, see also specific alkaloid or
  type of alkaloid
n-Alkanes, 1—4
Alkenebenzene structure, ethers based on, 62—79,
  see also specific ethers
5-Alkyl resosrcinols, 54
Allethrin I, 181
Allethrin II, 181
Alliin, 483
Allohydroxycotinine, 359
Allohydroxydemethylcotinine, 359
Allyl alcohol, isomeric, 72
4-Allylanisole, see Estagole
Allylbenzene derivatives, 62, see also specific
  derivatives
4-Allylcatechol epoxide, 68
Allylglucosinolate, 487
Allyl isothiocyanate, 486
4-Allyl-2-methoxyphenol, see Eugenol
4-Allylphenol, see Chavicol
4-Allylveratrole, see Eugenol methyl ether
Aloe-emodin, 125, 129
Aloe-emodin anthrone, 130
Aloenin, 243—244
Aloin, see Barbaloin
Ambrettolide, 175
Amides, 348—352

Printed and bound by CPI Group (UK) Ltd, Croydon, CR0 4YY

22/10/2024

01777638-0012